Artificial Intelligence in Medical Imaging and Image Processing

Artificial Intelligence in Medical Imaging and Image Processing

Editors

Rafał Obuchowicz
Michał Strzelecki
Adam Piórkowski

MDPI

Basel • Beijing • Wuhan • Barcelona • Belgrade • Novi Sad • Cluj • Manchester

Editors
Rafał Obuchowicz
Jagiellonian University
Medical College
Cracow
Poland

Michał Strzelecki
Lodz University of
Technology
Łódź
Poland

Adam Piórkowski
AGH University of Science
and Technology
Cracow
Poland

Editorial Office
MDPI
St. Alban-Anlage 66
4052 Basel, Switzerland

This is a reprint of articles from the Topic published online in the open access journals *Cancers* (ISSN 2072-6694), *Tomography* (ISSN 2379-139X), *Journal of Clinical Medicine* (ISSN 2077-0383), and *Diagnostics* (ISSN 2075-4418) (available at: https://www.mdpi.com/topics/UQER5BX2R5).

For citation purposes, cite each article independently as indicated on the article page online and as indicated below:

Lastname, A.A.; Lastname, B.B. Article Title. *Journal Name* **Year**, *Volume Number*, Page Range.

ISBN 978-3-7258-1259-2 (Hbk)
ISBN 978-3-7258-1260-8 (PDF)
doi.org/10.3390/books978-3-7258-1260-8

Contents

About the Editors ix

Preface xi

Rafał Obuchowicz, Michał Strzelecki and Adam Piórkowski
Clinical Applications of Artificial Intelligence in Medical Imaging and Image Processing—A Review
Reprinted from: *Cancers* **2024**, *16*, 1870, doi:10.3390/cancers16101870 1

Burak B. Ozkara, Melissa M. Chen, Christian Federau, Mert Karabacak, Tina M. Briere, Jing Li and Max Wintermark
Deep Learning for Detecting Brain Metastases on MRI: A Systematic Review and Meta-Analysis
Reprinted from: *Cancers* **2023**, *15*, 334, doi:10.3390/cancers15020334 17

Syed Rahmanuddin, Asma Jamil, Ammar Chaudhry, Tyler Seto, Jordyn Brase, Pejman Motarjem, et al.
COVID and Cancer: A Complete 3D Advanced Radiological CT-Based Analysis to Predict the Outcome
Reprinted from: *Cancers* **2023**, *15*, 651, doi:10.3390/cancers15030651 37

Hee Seung Nam, Sang Hyun Park, Jade Pei Yuik Ho, Seong Yun Park, Joon Hee Cho and Yong Seuk Lee
Key-Point Detection Algorithm of Deep Learning Can Predict Lower Limb Alignment with Simple Knee Radiographs
Reprinted from: *J. Clin. Med.* **2023**, *12*, 1455, doi:10.3390/jcm12041455 48

João Mendes, Nuno Matela and Nuno Garcia
Avoiding Tissue Overlap in 2D Images: Single-Slice DBT Classification Using Convolutional Neural Networks
Reprinted from: *Tomography* **2023**, *9*, 32, doi:10.3390/tomography9010032 59

Nuno M. Rodrigues, Sara Silva, Leonardo Vanneschi and Nickolas Papanikolaou
A Comparative Study of Automated Deep Learning Segmentation Models for Prostate MRI
Reprinted from: *Cancers* **2023**, *15*, 1467, doi:10.3390/cancers15051467 74

Haichao Ma, Chao Xu, Chao Nie, Jubao Han, Yingjie Li and Chuanxu Liu
DBE-Net: Dual Boundary-Guided Attention Exploration Network for Polyp Segmentation
Reprinted from: *Diagnostics* **2023**, *13*, 896, doi:10.3390/diagnostics13050896 95

Yajiao Zhang, Chao Wu, Zhibo Xiao, Furong Lv and Yanbing Liu
A Deep Learning Radiomics Nomogram to Predict Response to Neoadjuvant Chemotherapy for Locally Advanced Cervical Cancer: A Two-Center Study
Reprinted from: *Diagnostics* **2023**, *13*, 1073, doi:10.3390/diagnostics13061073 114

Wei-Chih Lien, Chung-Hsing Yeh, Chun-Yang Chang, Chien-Hsiang Chang, Wei-Ming Wang, Chien-Hsu Chen and Yang-Cheng Lin
Convolutional Neural Networks to Classify Alzheimer's Disease Severity Based on SPECT Images: A Comparative Study
Reprinted from: *J. Clin. Med.* **2023**, *12*, 2218, doi:10.3390/jcm12062218 133

Chun-Chao Huang, Hsin-Fan Chiang, Cheng-Chih Hsieh, Chao-Liang Chou, Zong-Yi Jhou, Ting-Yi Hou and Jin-Siang Shaw
Using Deep-Learning-Based Artificial Intelligence Technique to Automatically Evaluate the Collateral Status of Multiphase CTA in Acute Ischemic Stroke
Reprinted from: *Tomography* **2023**, *9*, 52, doi:10.3390/tomography9020052 . . . . **147**

Myrthe A. D. Buser, Alida F. W. van der Steeg, Marc H. W. A. Wijnen, Matthijs Fitski, Harm van Tinteren, Marry M. van den Heuvel-Eibrink, et al.
Radiologic versus Segmentation Measurements to Quantify Wilms Tumor Volume on MRI in Pediatric Patients
Reprinted from: *Cancers* **2023**, *15*, 2115, doi:10.3390/cancers15072115 . . . . **157**

Sebastian Gibala, Rafal Obuchowicz, Julia Lasek, Zofia Schneider, Adam Piorkowski, Elżbieta Pociask and Karolina Nurzynska
Textural Features of MR Images Correlate with an Increased Risk of Clinically Significant Cancer in Patients with High PSA Levels
Reprinted from: *J. Clin. Med.* **2023**, *12*, 2836, doi:10.3390/jcm12082836 . . . . **168**

Zineb Farahat, Nabila Zrira, Nissrine Souissi, Safia Benamar, Mohammed Belmekki, Mohamed Nabil Ngote and Kawtar Megdiche
Application of Deep Learning Methods in a Moroccan Ophthalmic Center: Analysis and Discussion
Reprinted from: *Diagnostics* **2023**, *13*, 1694, doi:10.3390/diagnostics13101694 . . . . **185**

Paulina Kukla, Karolina Maciejewska, Iga Strojna, Małgorzata Zapał, Grzegorz Zwierzchowski and Bartosz Bąk
Extended Reality in Diagnostic Imaging—A Literature Review
Reprinted from: *Tomography* **2023**, *9*, 88, doi:10.3390/tomography9030088 . . . . **203**

Hepseeba Kode and Buket D. Barkana
Deep Learning- and Expert Knowledge-Based Feature Extraction and Performance Evaluation in Breast Histopathology Images
Reprinted from: *Cancers* **2023**, *15*, 3075, doi:10.3390/cancers15123075 . . . . **215**

Debbie Anaby, David Shavin, Gali Zimmerman-Moreno, Noam Nissan, Eitan Friedman and Miri Sklair-Levy
'Earlier than Early' Detection of Breast Cancer in Israeli *BRCA* Mutation Carriers Applying AI-Based Analysis to Consecutive MRI Scans
Reprinted from: *Cancers* **2023**, *15*, 3120, doi:10.3390/cancers15123120 . . . . **236**

Maciej Skwirczyński, Zbisław Tabor, Julia Lasek, Zofia Schneider, Sebastian Gibała, Iwona Kucybała, et al.
Deep Learning Algorithm for Differentiating Patients with a Healthy Liver from Patients with Liver Lesions Based on MR Images
Reprinted from: *Diagnostics* **2023**, *15*, 3142, doi:10.3390/cancers15123142 . . . . **251**

Emilien Valat, Katayoun Farrahi and Thomas Blumensath
Sinogram Inpainting with Generative Adversarial Networks and Shape Priors
Reprinted from: *Tomography* **2023**, *9*, 94, doi:10.3390/tomography9030094 . . . . **266**

Dongkil Jeong, Wonjoon Jeong, Ji Han Lee and Sin-Youl Park
Use of Automated Machine Learning for Classifying Hemoperitoneum on Ultrasonographic Images of Morrison's Pouch: A Multicenter Retrospective Study
Reprinted from: *J. Clin. Med.* **2023**, *12*, 4043, doi:10.3390/jcm12124043 . . . . **282**

Rohan Nadkarni, Darin P. Clark, Alex J. Allphin and Cristian T. Badea
A Deep Learning Approach for Rapid and Generalizable Denoising of Photon-Counting Micro-CT Images
Reprinted from: *Tomography* **2023**, *9*, 102, doi:10.3390/tomography9040102 . . . . **294**

Genshen Song, Ziyue Xie, Haoran Wang, Shiman Li, Demin Yao, Shiyao Chen and Yonghong Shi
Segmentation of Portal Vein in Multiphase CTA Image Based on Unsupervised Domain Transfer and Pseudo Label
Reprinted from: *Diagnostics* **2023**, *13*, 2250, doi:10.3390/diagnostics13132250 . . . . **311**

Joanna Ożga, Michał Wyka, Agata Raczko, Zbisław Tabor, Zuzanna Oleniacz, Michał Korman and Wadim Wojciechowski
Performance of Fully Automated Algorithm Detecting Bone Marrow Edema in Sacroiliac Joints
Reprinted from: *J. Clin. Med.* **2023**, *12*, 4852, doi:10.3390/jcm12144852 . . . . **327**

Se-On Kim and Yoon-Chul Kim
Effects of Path-Finding Algorithms on the Labeling of the Centerlines of Circle of Willis Arteries
Reprinted from: *Tomography* **2023**, *9*, 113, doi:10.3390/tomography9040113 . . . . **346**

Michal Pula, Emilia Kucharczyk, Agata Zdanowicz and Maciej Guzinski
Image Quality Improvement in Deep Learning Image Reconstruction of Head Computed Tomography Examination
Reprinted from: *Tomography* **2023**, *9*, 118, doi:10.3390/tomography9040118 . . . . **357**

Robert Poel, Amith J. Kamath, Jonas Willmann, Nicolaus Andratschke, Ekin Ermiş, Daniel M. Aebersold, et al.
Deep-Learning-Based Dose Predictor for Glioblastoma–Assessing the Sensitivity and Robustness for Dose Awareness in Contouring
Reprinted from: *Cancers* **2023**, *15*, 4226, doi:10.3390/cancers15174226 . . . . **366**

Reabal Najjar
Redefining Radiology: A Review of Artificial Intelligence Integration in Medical Imaging
Reprinted from: *Diagnostics* **2023**, *13*, 2760, doi:10.3390/diagnostics13172760 . . . . **379**

Andrea Cozzi, Maurizio Cè, Giuseppe De Padova, Dario Libri, Nazarena Caldarelli, Fabio Zucconi, et al.
Deep Learning-Based Versus Iterative Image Reconstruction for Unenhanced Brain CT: A Quantitative Comparison of Image Quality
Reprinted from: *Tomography* **2023**, *9*, 130, doi:10.3390/tomography9050130 . . . . **404**

Jakub Kufel, Katarzyna Bargieł-Łączek, Maciej Koźlik, Łukasz Czogalik, Piotr Dudek, Mikołaj Magiera, et al.
Chest X-ray Foreign Objects Detection Using Artificial Intelligence
Reprinted from: *J. Clin. Med.* **2023**, *12*, 5841, doi:10.3390/jcm12185841 . . . . **413**

Luca Peretti, Graziella Donatelli, Matteo Cencini, Paolo Cecchi, Guido Buonincontri, Mirco Cosottini, et al.
Generating Synthetic Radiological Images with PySynthMRI: An Open-Source Cross-Platform Tool
Reprinted from: *Tomography* **2023**, *9*, 137, doi:10.3390/tomography9050137 . . . . **424**

Johannes Raspe, Felix Harder, Selina Rupp, Sean McTavish, Johannes M. Peeters, Kilian Weiss, et al.
Retrospective Motion Artifact Reduction by Spatial Scaling of Liver Diffusion-Weighted Images
Reprinted from: *Tomography* **2023**, *9*, 146, doi:10.3390/tomography9050146 . . . . **435**

Karthikeyan Shanmugam and Harikumar Rajaguru
Exploration and Enhancement of Classifiers in the Detection of Lung Cancer from Histopathological Images
Reprinted from: *Diagnostics* **2023**, *13*, 3289, doi:10.3390/diagnostics13203289 . . . . **453**

Max C. Lindemann, Lukas Glänzer, Anjali A. Roeth, Thomas Schmitz-Rode and Ioana Slabu
Towards Realistic 3D Models of Tumor Vascular Networks
Reprinted from: *Cancers* **2023**, *15*, 5352, doi:10.3390/cancers15225352 . . . . **491**

Rafael Bayareh-Mancilla, Luis Alberto Medina-Ramos, Alfonso Toriz-Vázquez, Yazmín Mariela Hernández-Rodríguez and Oscar Eduardo Cigarroa-Mayorga
Automated Computer-Assisted Medical Decision-Making System Based on Morphological Shape and Skin Thickness Analysis for Asymmetry Detection in Mammographic Images
Reprinted from: *Diagnostics* **2023**, *13*, 3440, doi:10.3390/diagnostics13223440 . . . . **510**

Dongrong Yang, Yuhua Huang, Bing Li, Jing Cai and Ge Ren
Dynamic Chest Radiograph Simulation Technique with Deep Convolutional Neural Networks: A Proof-of-Concept Study
Reprinted from: *Cancers* **2023**, *15*, 5768, doi:10.3390/cancers15245768 . . . . **536**

Md. Mamun Hossain, Md. Moazzem Hossain, Most. Binoee Arefin, Fahima Akhtar and John Blake
Combining State-of-the-Art Pre-Trained Deep Learning Models: A Noble Approach for Skin Cancer Detection Using Max Voting Ensemble
Reprinted from: *Diagnostics* **2023**, *14*, 89, doi:10.3390/diagnostics14010089 . . . . **548**

About the Editors

Rafał Obuchowicz

Rafał Obuchowicz currently works at the Department of Radiology of the University Hospital, Krakow, and heads the Imaging Diagnostics Department at Lux Med, Krakow. His professional interests include the problems of focal change detection and shape detection in various radiological techniques using classical methods and deep learning machine algorithms. The result of the work that he has undertaken in this direction is the NCBiR project "XR-AI A diagnostic browser for radiology with computer-aided use of artificial intelligence," of which he is the originator, as well as head of the medical team. He also has a professional interest in analyzing image quality using MR and X-ray techniques, resulting in a completed habilitation procedure entitled "Analysis and parametrization of information in selected techniques of radiological diagnostics." His hobbies include environmental psychology, history, ecology, and urbanization, and he is the husband of Barbra and father of Isabella and Wiktor.

Michał Strzelecki

Prof. Michał Strzelecki has been employed at the Institute of Electronics, Lodz University of Technology (TUL), since 1988, and since 2007, he has worked as a full professor. In the years 2004–2019, he was Deputy Director of the Institute for Teaching and Research. He also served as Vice Dean for Research at the Faculty of Electrical, Electronic, Computer, and Control Engineering, TUL, until 2020. His scientific interests include the processing and analysis of biomedical signals and images, data analysis methods, and artificial intelligence. In particular, he is interested in issues related to image texture analysis and the applications and hardware implementations of artificial neural networks. He also works on projects related to the development of software aimed at supporting medical imaging diagnostics. Within these interests, Prof. Strzelecki conducts extensive cooperation with domestic and foreign medical universities for the construction and development of systems supporting the diagnostic process. He has participated in numerous international research programs, including COST B21, BM1103, TD1007 (a member of the Management Committee), DGF, and 7FP and projects funded by the National Science Centre (Poland) and the National Centre for Research and Development (Poland). He is the author and co-author of more than 200 scientific publications. In 2006–2008, he was employed at the Jeonbuk National University, Jeonju, Republic of Korea, as a visiting professor.

Adam Piórkowski

Adam Piórkowski obtained a PhD in computer science in 2005. In 2015, he obtained a post-doctoral degree in biomedical engineering, and in 2019, he became a professor at AGH in Krakow. He has been involved in medical image processing since 2003. He is particularly interested in the segmentation, normalization, and textural analysis of this type of data. He conducts scientific projects generally related to medical informatics. He is also an active programmer.

Preface

The following Reprint focuses on the increasingly critical role of computer-aided diagnosis in modern healthcare, emphasizing the integration of artificial intelligence (AI) for automating tasks traditionally performed by medical professionals. This includes the application of AI in enhancing the accuracy and efficiency of medical imaging analysis through advanced pattern recognition and analysis.

Scope: This discussion extends to the use of sophisticated neural networks capable of performing multilayered analyses, essential for minimizing error bias in diagnostic suggestions. AI's contribution is highlighted across three key phases of the medical imaging workflow: detection (image segmentation), recognition (classification), and description (natural language interpretation of results). The scope also covers the impact of AI in terms of streamlining diagnostic processes and reducing the time required for pathology detection and result interpretation.

The aim of this Reprint is to present the latest advancements in and knowledge regarding AI methodologies as applied to medical systems, particularly in diagnostic imaging. The presented papers are intended to showcase a variety of AI applications across different diagnostic fields, highlighting the versatility and adaptability of AI technologies in healthcare. The purpose of this Reprint is to foster a comprehensive exchange of knowledge regarding the technical aspects and practical applications of AI in radiology. By compiling a collection of works on this topic, the goal is to contribute to a broader understanding and appreciation of AI's potential to revolutionize modern radiology, ultimately leading to enhanced healthcare quality and efficiency.

Rafał Obuchowicz, Michał Strzelecki, and Adam Piórkowski
Editors

cancers

MDPI

Editorial

Clinical Applications of Artificial Intelligence in Medical Imaging and Image Processing—A Review

Rafał Obuchowicz [1], Michał Strzelecki [2,*] and Adam Piórkowski [3]

1 Department of Diagnostic Imaging, Jagiellonian University Medical College, 31-008 Krakow, Poland; rafalobuchowicz@su.krakow.pl
2 Institute of Electronics, Lodz University of Technology, 93-590 Lodz, Poland
3 Department of Biocybernetics and Biomedical Engineering, AGH University of Krakow, 30-059 Krakow, Poland
* Correspondence: michal.strzelecki@p.lodz.pl

Citation: Obuchowicz, R.; Strzelecki, M.; Piórkowski, A. Clinical Applications of Artificial Intelligence in Medical Imaging and Image Processing—A Review. *Cancers* **2024**, *16*, 1870. https://doi.org/10.3390/cancers16101870

Received: 19 March 2024
Accepted: 28 April 2024
Published: 14 May 2024

1. Introduction

Artificial intelligence (AI) is currently becoming a leading field in data processing. While the results it produces are spectacular when it comes to simple statistical relationships, in the case of medical imaging, the search for optimal methods is still ongoing. This collection of articles [1–34] focuses on the Topic "Artificial Intelligence in Medical Imaging and Image Processing" and discusses the latest achievements in this area.

Artificial intelligence refers to the simulation of human intelligence in machines that are programmed to think and learn like humans. It involves the development of algorithms and computer systems that can perform tasks which typically require human reasoning, adaptability, and simulation of cognitive processes. AI encompasses a broad range of techniques and technologies aimed at creating systems capable of reasoning, problem-solving, perception, and learning. AI has found various applications in medical imaging [35–37], revolutionizing the field by enhancing diagnostic accuracy, efficiency, and patient care. Some important applications of AI in medical imaging are as follows:

- Image segmentation—identification and delineation of specific structures or regions of interest within medical images;
- Disease detection and diagnosis—identification of abnormalities by highlighting potential areas of concern; early detection and diagnosis of various medical conditions by analyzing medical images regardless of imaging modality;
- Image preprocessing—enhancing the quality of medical images and reconstructing from incomplete or noisy data to improve their overall clarity and diagnostic value;
- Personalized treatment planning—tailoring treatment plans based on individual patient characteristics and response to therapy;
- Predictive analytics—analysis of medical imaging data along with other clinical information to predict disease progression, treatment response, and potential complications, enabling better-informed decision-making;
- Quality control—maintaining the quality of medical images by detecting artifacts, ensuring that images used for diagnosis are of the highest quality;
- Monitoring and follow-Up—assisting in the continuous monitoring of disease progression and treatment response over time, enabling timely adjustments to the treatment plan.

Currently, there are two main groups of AI-based approaches which are widely used in medical imaging. The first group encompasses machine learning (ML), i.e., algorithms that enable a system to learn independently based on data. The other approach, which in fact is a subset of ML, involves using complex artificial neural networks to model and solve problems. The main difference between these groups lies in the fact that ML models require the definition of a set of features that characterize the objects under analysis. In the case of analyzing medical images, radiomics is a significant source of such features. Radiomics is a field of study within medical imaging that involves the extraction and analysis of quantitative features from medical images. These features are derived from the analysis of intensity, shape, texture, and other characteristics present in medical data [38]. In particular, the analysis of image texture provides a wealth of information about the morphology and physiology of organs and tissues. Texture analysis is an effective tool that provides machine learning models with diagnostically relevant parameters obtaied from images of various modalities [39]. In contrast to classical machine learning models, deep networks are capable of determining significant parameters of the examined organs on their own within their internal convolutional layers. The input for such algorithms are images, and the output is the response to a given research problem (e.g., a segmented image or the classification of a pathological change). Based on an analysis of the literature, it can be stated that, currently, deep learning algorithms are more prevalent than classical ML methods. Usually, deep learning provides more accurate and reliable outcomes. There are, however, exceptions. For example, in [40], it was demonstrated that the texture analysis of selected vertebrae in CT images provided a more accurate prediction of a patient's age than deep learning methods.

This collection includes eight articles on detection, five on classification, four on segmentation, seven on prediction, five on quality improvement, and three on simulation. The following organs were of interest: the lungs (4), breasts (4), liver (3), brain (9), prostate (3), and others (8). Two studies were conducted on phantoms. The most popular imaging modalities are MRI (11) and CT (8), but there are also works on radiography, ultrasound, and microscopic images. The achieved efficiencies cannot be clearly compared due to different research methodologies, but, as an example, in **detection tasks, efficiency ranged from approx. 65% to even 100%**. The exact results can be found in the relevant table (Appendix A).

2. Review of AI Applications in Medical Imaging

2.1. *Deep Learning for Detecting Brain Metastases on MRI: A Systematic Review and Meta-Analysis*

In this investigation, a systematic review and meta-analysis were undertaken to evaluate the diagnostic efficacy of deep learning algorithms utilizing magnetic resonance imaging (MRI) for the identification of cerebral metastases in oncological patients. The methodology encompassed a comprehensive literature search across MEDLINE, EMBASE, and Web of Science, culminating on 30 September 2022 and adhering to pre-specified inclusion and exclusion criteria to ensure the selection of relevant original research articles. The inclusion criteria mandated the investigation of patients with brain metastases using deep learning models applied to MRI data, with a clear articulation of diagnostic performance metrics. Conversely, the exclusion criteria filtered out non-original articles, small case series, studies with overlapping patient cohorts, and papers with insufficient details on diagnostic performance.

This qualitative synthesis of 24 studies, following rigorous quality assessment via the Quality Assessment of Diagnostic Accuracy Studies-2 (QUADAS-2) and the Checklist for Artificial Intelligence in Medical Imaging, yielded a pooled detectability rate of 89% at both the patient and lesion levels. This synthesis underscores the potent utility of deep learning paradigms in enhancing the precision of cerebral metastases detection, thereby facilitating timely and informed clinical decision-making.

The findings advocate for stringent adherence to established reporting standards within the domain of artificial intelligence in medical imaging in order to augment the reproducibility and interpretability of diagnostic accuracy studies. The evident heterogeneity in reported false-positive rates precluded a unified analysis, highlighting an area for future standardization. Collectively, this study elucidates the significant potential of deep learning models in advancing our diagnostic capabilities in the field of brain metastases, thereby potentially ameliorating patient management and prognostic outcomes in the oncological setting.

2.2. Extended Reality in Diagnostic Imaging—A Literature Review

The past decade has witnessed an unprecedented surge in the adoption of Extended Reality (ER) in healthcare, with a particular focus on its applications in diagnostic imaging, patient positioning, and medical education. A comprehensive analysis of scientific publications was conducted to explore ER's potential benefits and utilization in various medical contexts, including ultrasound, interventional radiology, and computed tomography. Additionally, this study investigated ER's role as a potential alternative to traditional anesthesia and sedation during medical examinations.

The integration of ER into medical education has garnered significant attention due to its ability to create immersive learning environments, particularly in anatomy and patient positioning. However, questions have arisen regarding the economic feasibility of investing in ER technology and its ongoing maintenance costs.

Overall, the analysis suggests that implementing ER in clinical practice holds promise for expanding diagnostic capabilities, improving medical education outcomes, and enhancing patient experiences through increased visualization and comprehension of medical conditions. While further research is needed to fully realize ER's potential in healthcare and address associated challenges, its adoption represents an innovative approach to advancing diagnostic imaging, medical education, and patient care.

2.3. Redefining Radiology: A Review of Artificial Intelligence Integration in Medical Imaging

This review delves into the integration of artificial intelligence (AI) into radiology, highlighting its transformative impact on healthcare. It covers the evolution from traditional X-ray discovery to the adoption of machine learning and deep learning for medical image analysis. The key AI applications it discusses include image segmentation, computer-aided diagnosis, predictive analytics, and workflow optimization, all contributing to improved diagnostic accuracy, personalized treatment, and increased clinical efficiency. Despite these benefits, challenges such as data integrity, algorithm transparency, and ethical concerns are acknowledged. The review concludes optimistically, advocating for ongoing research, technological advancement, and collaboration between radiologists and AI developers to harness AI's full potential in radiology, underscoring the importance of innovation and ethical responsibility.

3. Modality: X-ray

3.1. Key-Point Detection Algorithm of Deep Learning Can Predict Lower Limb Alignment with Simple Knee Radiographs

This study leveraged a convolutional neural network (CNN) to predict the weight-bearing line (WBL) ratio from knee anteroposterior (AP) radiographs, enhancing the diagnosis of conditions like osteoarthritis. Employing stratified random sampling, this study analyzed 4790 knee AP radiographs from 2410 patients, using annotated and cropped images to focus the CNN's analysis. The CNN model's key-point detection capability to identify crucial tibial plateau points facilitated accurate WBL ratio calculations. This method demonstrated an accuracy comparable to that of direct measurements from full-leg radiographs and promises significant benefits for the medical community and patients by offering a more precise, efficient, and accessible tool for assessing knee alignment, potentially improving the early detection and treatment of related conditions.

3.2. Dynamic Chest Radiograph Simulation Technique with Deep Convolutional Neural Networks: A Proof-of-Concept Study

This research study introduces an innovative Radiograph Motion Simulation (RMS) network which integrates U-Net with LSTM networks to simulate and predict respiratory lung motion from single-phase chest X-rays. A Spatial Transformer Network is also applied for precise image deformation, reflecting true respiratory motion. The network's performance is evaluated through qualitative and quantitative methods, including Dice score calculations for motion prediction accuracy. This approach enhances diagnostic capabilities by providing insights into lung dynamics from static X-rays, offers a non-invasive alternative for lung function assessment, and increases diagnostic efficiency by extracting detailed information from routine chest X-rays. Such advancements hold significant potential for improving pulmonary diagnostics, patient comfort, and healthcare efficiency.

3.3. *Chest X-ray Foreign Objects Detection Using Artificial Intelligence*

This study leveraged a deep convolutional neural network to identify objects in chest X-ray images using data from the NIH Chest X-ray Dataset. The employed techniques included manual image review for object identification, computer-assisted annotations, and preprocessing steps like resizing and normalization. The object detection model, based on the YOLOv8 architecture and Ultralytics framework, demonstrated an average precision of 0.815 in detecting foreign objects. This AI model presents several benefits for patients, including improved diagnostic accuracy, increased efficiency in processing X-ray images, support for radiologists by automating routine detections, and the potential for earlier disease detection, thereby enhancing patient care and treatment outcomes.

4. Modality: MRI

4.1. *A Comparative Study of Automated Deep Learning Segmentation Models for Prostate MRI*

This study focuses on enhancing prostate cancer diagnoses through advanced imaging techniques. It involves a comparative analysis of various segmentation models for the accurate delineation of the prostate gland and its zones, coupled with an innovative approach incorporating an object detection algorithm to aid the segmentation process. Conducted on two public datasets, this study finds that while most models yield similar results, the nnU-Net framework stands out for its superior performance. Additionally, models augmented with object detection preprocessing show enhanced generalizability. This research study offers significant benefits for medical practitioners, including improved diagnostic accuracy, better treatment planning, early detection, and increased efficiency, ultimately leading to better patient outcomes and quality of life.

4.2. *A Deep Learning Radiomics Nomogram to Predict Response to Neoadjuvant Chemotherapy for Locally Advanced Cervical Cancer: A Two-Center Study*

This study introduced a deep learning radiomics nomogram (DLRN) leveraging multiparametric MR imaging to predict responses to neoadjuvant chemotherapy (NACT) in locally advanced cervical cancer (LACC) patients. By extracting and analyzing both hand-selected and deep learning-based radiomics features from MR sequences, this study developed DLRN-integrated predictive radiomics signatures along with clinical features. Validated through AUC, calibration curves, and Decision Curve Analysis, the DLRN demonstrated superior predictive accuracy over traditional clinical models and radiomics signatures, highlighting its potential for personalizing treatment plans and improving prognostication. This approach supports doctors by enabling more informed treatment decisions and optimizing resource use, while patients benefit from tailored therapies potentially leading to better outcomes and reduced unnecessary treatments.

4.3. *Radiologic vs. Segmentation Measurements to Quantify Wilms Tumor Volume on MRI in Pediatric Patients*

This study explored the accuracy of measuring tumor volume in Wilms tumor patients using manual segmentation and traditional radiological methods and investigated the potential of automating the segmentation process with deep learning (nnU-Net). Manual segmentation typically resulted in larger tumor volume estimates compared to radiological measurements, highlighting a potential underestimation of tumor size when using traditional methods. nnU-Net-based automated segmentation achieved high accuracy, with a median Dice coefficient of 0.90 and a median 95th percentile Hausdorff distance of 7.2 mm. This suggests that deep learning can offer precise volume measurements efficiently while avoiding the time and observer variabilities associated with manual methods. Such advancements could lead to more accurate and personalized treatment planning, benefiting both the medical community and patients by enabling better treatment outcomes and potentially less invasive treatment options.

4.4. Textural Features of MR Images Correlate with an Increased Risk of Clinically Significant Cancer in Patients with High PSA Levels

This study employs a comprehensive approach to improve prostate cancer detection by combining multiparametric MRI guided by the PIRADS protocol with advanced image analysis techniques. Key methods include the manual delineation of prostate zones and tumor foci, extraction of over 7000 textural features using MaZda software (http://www.eletel.p.lodz.pl/programy/mazda/, accessed on 18 March 2024), and statistical analysis to correlate these features with PSA levels. The use of MIL-SVM machine learning further enhances diagnostic accuracy. This integrative approach yields a 92% accuracy rate in identifying prostate cancer, offering significant benefits such as improved diagnostic precision, early detection capabilities, personalized treatment options, a reduced need for invasive procedures, and optimized healthcare resources. This advancement in diagnostic technology holds promise for both the medical community and patients, potentially transforming prostate cancer diagnosis and treatment.

4.5. 'Earlier than Early' Detection of Breast Cancer in Israeli BRCA Mutation Carriers Applying AI-Based Analysis to Consecutive MRI Scans

This study developed an artificial intelligence (AI)-based approach, specifically a convolutional neural network (CNN), to improve the classification of enhancing foci in MRI scans of BRCA pathogenic variant carriers, aiming to reduce false-negative interpretations in breast cancer (BC) detection. The method involved manually segmenting retrospectively identified enhancing foci from previous MRIs, which were then used to train the CNN for accurate differentiation between malignant and benign/normal findings. This AI model successfully identified 65% of cancerous foci, particularly excelling in the detection of triple-negative BC. The application of this technology promises to enhance early BC detection in high-risk individuals, reduce diagnostic errors, and pave the way for more personalized and effective surveillance strategies, thereby offering significant benefits to the medical community in terms of improved patient care and outcomes.

4.6. Deep Learning Algorithm for Differentiating Patients with a Healthy Liver from Patients with Liver Lesions Based on MR Images

In this study, advanced diagnostic methodologies were employed to address the complexities of liver disease diagnosis using multiparametric magnetic resonance (MR) imaging. The core of this approach involved a sophisticated multiclass segmentation algorithm designed to differentiate between various types of liver lesions and healthy tissue. This comprehensive algorithmic pipeline, from data preprocessing to final classification, was rigorously evaluated using statistical metrics, notably the area under the receiver operating characteristic (AUC ROC) curve, demonstrating high accuracy in distinguishing between healthy and diseased liver states. This study's findings offer significant benefits to patients, including the potential for early disease detection, improved diagnostic precision, and the standardization of liver disease assessments. These advancements promise to enhance patient outcomes through more timely and accurate treatment interventions in hepatology.

4.7. Performance of Fully Automated Algorithm Detecting Bone Marrow Edema in Sacroiliac Joints

This study utilized advanced MRI techniques and an automated algorithm to detect bone marrow edema (BME) in axial spondyloarthritis (axSpA) patients, focusing on the sacral and iliac bones. Key methods included MRI examination, deviation angle measurement for image accuracy, manual and automated bone segmentations, and BME lesion assessment using the SPARCC scale. The algorithm's performance was evaluated through sensitivity and specificity analysis, showing high accuracy regardless of image acquisition angles.

The benefits of the methodology developed in this study are significant from a patient care perspective. Such methodology enhances diagnostic accuracy and efficiency, provides consistent results, enables early detection of axSpA, supports large-scale patient screening, and serves as an educational tool for clinicians. This approach promises to streamline the diagnostic process, improve patient outcomes, and foster a more efficient workflow in clinical settings.

4.8. Effects of Path-Finding Algorithms on the Labeling of the Centerlines of Circle of Willis Arteries

This study focused on the automated identification of centerlines in intracranial vessel segments using path-finding algorithms applied to 3D time-of-flight MRA images. Three algorithms were compared: depth-first search, Dijkstra's, and A*. Among 840 vessel segments, Dijkstra's algorithm exhibited the highest accuracy (97.1%), closely followed by the A* algorithm (96.1%), with the depth-first search algorithm showing lower accuracy (83.5%). This study highlighted Dijkstra's and A* algorithms as effective and comparable in both accuracy and speed for delineating pathways in the circle of Willis arteries, offering significant benefits for medical imaging analysis in terms of diagnostic accuracy, treatment planning efficiency, and standardization of vascular imaging assessments.

4.9. Retrospective Motion Artifact Reduction by Spatial Scaling of Liver Diffusion-Weighted Images

This study introduced a data-driven algorithm to counteract motion-induced signal loss in diffusion-weighted MRI (DWI) of the liver. This algorithm enhances DWI by excluding heavily distorted images and applying spatially variable image scaling, based on a signal-loss model, to improve signal uniformity and accuracy, especially in the left lobe, which is most commonly affected by this phenomenon. Advantages for practitioners include improved diagnostic accuracy of liver DWI, more homogeneous liver imaging, accurate apparent diffusion coefficient (ADC) measurements, and expanded clinical utility of DWI for liver disease assessment.

4.10. Generating Synthetic Radiological Images with PySynthMRI: An Open-Source Cross-Platform Tool

PySynthMRI is an open-source, user-friendly software designed to create synthetic MR images with variable radiological contrasts by adjusting parameters like echo time, repetition time, and inversion time(s). It calculates pixelwise signal intensity from input images such as T1 and T2 maps and is compatible with Linux, Windows, and MacOS. The tool supports customization, allowing users to add new features and contrasts, and can export images in DICOM or NiFTI formats. In terms of its benefits for the medical community and for patients, PySynthMRI streamlines imaging by reducing scan times, enhances flexibility for both research and clinical applications, and creates the potential for enhanced diagnostic accuracy, making it a valuable resource for improving patient care and advancing medical research.

5. Modality: SPECT

Convolutional Neural Networks to Classify Alzheimer's Disease Severity Based on SPECT Images: A Comparative Study

This study employed convolutional neural networks (CNNs) and single-photon emission computed tomography (SPECT) imaging to analyze the progression of Alzheimer's disease (AD). Utilizing a range of CNN models, from lightweight (MobileNet V2, NASNet-Mobile) to heavier models (VGG16, Inception V3, ResNet), this research study aimed to enhance the accuracy of AD diagnosis and early detection by analyzing complex patterns in neuroimaging data. The effectiveness of transfer learning was also demonstrated in this context, highlighting its potential to efficiently leverage limited datasets in medical imaging. This methodology offers a number of benefits for medical practitioners and for patients, including improved diagnostic accuracy, early detection, personalized treatment plans, optimized resources, and reduced cognitive assessment burden. This approach is a promising advancement in neurodegenerative disease diagnostics and patient care.

6. Modality: CT

6.1. COVID-19 and Cancer: A Complete 3D Advanced Radiological CT-Based Analysis to Predict the Outcome

Utilizing a retrospective cohort design, this study scrutinized computed tomography (CT)-derived pulmonary morphological alterations in 22 oncological subjects post SARS-CoV-2 infection. Employing quantitative metrics, such as Hounsfield Units for tissue density and volumetric analyses for pulmonary capacity, alongside three-dimensional radiometric evaluations, this study delineated significant post-infectious pulmonary sequelae characterized by fibrotic densification and volumetric reduction. These findings elucidate the exacerbation of pulmonary vulnerability in the oncological demographic when confronted with SARS-CoV-2, thereby underscoring the imperative for bespoke therapeutic stratagems. This research study contributes to the corpus of evidence-based medicine, facilitating optimized clinical decision-making and prognostication for oncological patients amidst the COVID-19 pandemic.

6.2. Using Deep-Learning-Based Artificial Intelligence Technique to Automatically Evaluate the Collateral Status of Multiphase CTA in Acute Ischemic Stroke

This study utilized multiphase computed tomography angiography (mCTA) and convolutional neural network (CNN) techniques to develop an AI model for the automatic prediction of collateral status in acute ischemic stroke patients. The AI approach aimed to enhance the efficiency and accuracy of collateral evaluation, traditionally performed through time-consuming manual assessments. This study demonstrated that CNN-based AI models could provide rapid and consistent evaluations of mCTA images, offering significant benefits for both clinicians and patients. For doctors, this method reduces the workload and enables quicker decision-making, which is crucial in stroke management. Patients stand to benefit from timely and personalized treatment interventions, potentially improving recovery outcomes and reducing the risk of long-term disability.

6.3. Sinogram Inpainting with Generative Adversarial Networks and Shape Priors

This study introduces a novel method utilizing generative adversarial networks (GANs) to enhance X-ray computed tomography (CT) image reconstruction, particularly in cases with limited tomographic measurements due to incomplete scanning coverage. This method integrates shape priors to accurately infer missing data, thereby reducing artifacts and improving image quality. Unlike traditional approaches that address the interpolation of evenly spaced missing angles, this method is adept at reconstructing images with substantial gaps in consecutive scanning angles. The application of this technique has demonstrated significant improvements in image fidelity, evidenced by a 7 dB increase in Peak Signal-to-Noise Ratio (PSNR) compared to existing sinogram-inpainting methods. For medical professionals and patients, this advancement offers multiple benefits, including higher-quality images for more reliable diagnoses, increased efficiency of imaging procedures, and reduced patient exposure to ionizing radiation. This approach is a significant step forward in medical imaging, potentially transforming CT imaging practices by optimizing the use of limited or suboptimal datasets.

6.4. A Deep Learning Approach for Rapid and Generalizable Denoising of Photon-Counting Micro-CT Images

This study introduces UnetU, a deep learning model based on a 2D U-net convolutional neural network designed to denoise photon-counting CT (PCCT) images by approximating iterative reconstruction from weighted filtered backprojection (wFBP) images. UnetU employs a custom loss function and a specific transformation of wFBP to improve material decomposition accuracy across various energy thresholds. This method offers significant benefits to the medical community, including faster reconstruction times, improved image quality, enhanced material decomposition, broad applicability, and increased clinical workflow efficiency. UnetU's potential to accelerate and refine PCCT imaging promises substantial advancements in both clinical diagnostics and preclinical research.

6.5. Segmentation of Portal Vein in Multiphase CTA Image Based on Unsupervised Domain Transfer and Pseudo Label

This study introduces an innovative approach for segmenting the portal vein in multiphase CT angiography (CTA) images, leveraging unsupervised domain transfer and pseudo labeling techniques. The process begins by stylistically aligning hepatic arterial phase (H-phase) and equilibrium phase (E-phase) images with portal vein phase (P-phase) images to mitigate contrast media discrepancies. Pseudo labels generated from P-phase annotations guide the segmentation of H-phase and E-phase images, resulting in accurate portal vein delineation across all phases.

This methodology presents several benefits for patient care: it provides a comprehensive diagnostic view by enabling portal vein analysis across multiple imaging phases, enhances efficiency by reducing the need for manual annotation, and improves segmentation accuracy through style normalization. Furthermore, it optimizes resource use by extending the utility of existing annotations and ultimately supports improved patient care through more precise diagnostic insights.

6.6. Deep Learning-Based vs. Iterative Image Reconstruction for Unenhanced Brain CT: A Quantitative Comparison of Image Quality

This study investigated image quality in brain CT scans using two reconstruction algorithms: the iterative AIDR-3D and the deep learning-based AiCE. Through a preliminary phantom study and a retrospective analysis of 100 emergency brain CTs, it assessed image noise, artifact presence, and contrast-to-noise ratios. The findings revealed AiCE's superiority in reducing image noise and enhancing contrast-to-noise ratios, while AIDR-3D showed lower artifact indices. These insights into the differential benefits of each algorithm can aid patient care, optimizing CT imaging for more accurate diagnoses by offering enhanced image quality, better visualization of brain structures, and informed choices in algorithm selection for clinical practice.

6.7. Image Quality Improvement in Deep Learning Image Reconstruction of Head Computed Tomography Examination

This study compared the image quality of cranial CT scans reconstructed using deep learning image reconstruction (DLIR) with those using adaptive statistical iterative reconstruction (ASIR-V). Through both objective measurements (SNR and CNR in the brain's grey and white matter) and subjective evaluations by experienced radiologists, DLIR demonstrated superior image quality, showing significant improvements in SNR (up to 54% for grey matter and 60% for white matter) and CNR (58% in BGA and 50% in PCF) compared to ASIR-V. Subjectively, DLIR also received higher rating scores from radiologists. This suggests that DLIR offers notable advantages in enhancing cranial CT scan quality, potentially improving diagnostic accuracy and patient care in medical imaging.

7. Modality: Ultrasonography

Use of Automated Machine Learning for Classifying Hemoperitoneum on Ultrasonographic Images of Morrison's Pouch: A Multicenter Retrospective Study

This study explored the use of automated machine learning (AutoML) for identifying the presence of hemoperitoneum in ultrasonography (USG) images of Morrison's pouch in trauma patients. Utilizing a dataset of 2200 USG images from 864 patients across multiple South Korean trauma centers, this research study employed Google's open-source AutoML for model training and validation. The process involved training the model with 1800 images, internal validation with 200 images, and external validation with an additional 200 images from an independent center. The AutoML model demonstrated high accuracy, with sensitivity and specificity rates exceeding 94% in both internal and external validations and an area under the receiver operating characteristic (AUROC) curve of 0.97. These results highlight AutoML's potential to efficiently and accurately classify medical images, offering significant benefits for healthcare professionals, such as improved diagnostic reliability, efficiency in emergency care, scalability, and accessibility, thereby promising to enhance patient care in trauma and emergency settings.

8. Modality: Mammography

8.1. Avoiding Tissue Overlap in 2D Images: Single-Slice DBT Classification Using Convolutional Neural Networks

This study leverages digital breast tomosynthesis (DBT) and deep learning to enhance breast cancer screening, addressing the limitations of mammography, like tissue overlap. By modifying a deep learning framework with adjustments to the fully connected layers and regularization, as well as by employing data augmentation techniques to increase dataset variability, the model effectively classifies DBT slices as benign or malignant. Utilizing 2772 augmented images for training, the model achieved a 93.2% accuracy rate on the test set, with high sensitivity, specificity, precision, F1-score, and Cohen's kappa values. These results underscore the potential of DBT coupled with deep learning to improve diagnostic accuracy, reduce false positives/negatives, and potentially surpass traditional mammography in terms of screening efficacy, offering substantial benefits to the medical community and patients.

8.2. Automated Computer-Assisted Medical Decision-Making System Based on Morphological Shape and Skin Thickness Analysis for Asymmetry Detection in Mammographic Images

This study employed Dynamic Time Warping (DTW) for morphological analysis and the Growing Seed Region (GSR) method for skin segmentation in mammographic images to detect asymmetries, which are indicative of potential breast cancer. DTW assesses the similarity between a patient's two breasts by analyzing the distances from the breast image centroid to its perimeter, taking into account possible geometric distortions. GSR identifies skin-related asymmetries by expanding a seed set on skin pixels based on intensity and depth similarity. These methods enhance early cancer detection, increase diagnostic efficiency by speeding up the analysis process, improve accuracy in identifying asymmetries, and offer patient-specific insights, thereby supporting more personalized diagnostic and treatment approaches. Overall, the DTW and GSR methods significantly contribute to the early detection and diagnosis of breast cancer, offering substantial benefits to healthcare professionals in delivering effective patient care.

9. Other Imaging Techniques (Histological Analyses, Comparative Studies)

9.1. DBE-Net: Dual Boundary-Guided Attention Exploration Network for Polyp Segmentation

The Dual Boundary-guided Attention Exploration Network (DBE-Net) introduces innovative methods for polyp segmentation in colonoscopy, addressing key challenges such as indistinct boundaries, size variability, and resemblance to adjacent tissues. Incorporating a dual boundary-guided module, multi-scale enhancement, and low-level detail enhancement, DBE-Net significantly improves the precision of polyp detection. This advancement offers substantial benefits, including enhanced diagnostic accuracy, improved treatment outcomes, increased procedural efficiency, and reduced patient burden. Demonstrating superior performance on benchmark datasets, DBE-Net holds promise for advancing colorectal care by aiding in early and accurate polyp identification, contributing positively to both medical practice and the patient experience.

9.2. Application of Deep Learning Methods in a Moroccan Ophthalmic Center: Analysis and Discussion

This study introduces the application of two artificial intelligence (AI) methodologies—the U-Net algorithm for segmentation and the You Look Only Once Version 5 (YOLOv5) for detection—to diagnose diabetic retinopathy (DR) in color fundus images. U-Net is utilized to distinguish and color-code hemorrhages and exudates, enhancing visual assessment, while YOLOv5 detects these DR indicators, assigning a confidence score to each finding. These AI-driven approaches significantly benefit clinicians by improving diagnostic accuracy, increasing efficiency, enabling early detection of DR, reducing workload, and facilitating greater access to quality eye care. This study demonstrates that these AI tools can outperform traditional diagnostic methods, with the detection algorithm successfully identifying 100% of DR signs compared to lower detection rates by expert and resident doctors.

9.3. Deep Learning- and Expert Knowledge-Based Feature Extraction and Performance Evaluation in Breast Histopathology Images

This study explored three feature extraction methods for breast cancer diagnosis using histopathology images: a convolutional neural network (CNN), transfer learning with the VGG16 architecture, and a knowledge-based system. These methods were evaluated using seven classifiers on the BreakHis 400× image dataset. The CNN and VGG16 approaches showed promising results, with accuracies up to 85% and 86%, respectively, while the knowledge-based system outperformed them both, reaching an accuracy of up to 98%. These advancements have a significant impact on patient care, including improved diagnostic accuracy, increased efficiency in processing histopathology images, and the potential for standardizing diagnostic criteria. Such innovations hold the promise of enhancing breast cancer detection and treatment, thereby improving patient outcomes.

9.4. Deep-Learning-Based Dose Predictor for Glioblastoma–Assessing the Sensitivity and Robustness for Dose Awareness in Contouring

This study focused on developing a 3D VMAT dose prediction model for glioblastoma treatment using deep learning techniques, specifically cascaded 3D U-Nets. This model was trained and tested on a dataset of 125 glioblastoma patients, with the aim of improving the efficiency and quality of radiation therapy planning. The model demonstrated good sensitivity to realistic contour variations and was further refined for robustness against out-of-distribution cases. The successful implementation of this deep learning model in radiation therapy planning presents significant benefits for practitioners and for patients, including enhanced treatment planning efficiency, increased accuracy in dose predictions, and improved quality assurance in automated contouring processes, leading to more personalized and effective treatment strategies for patients.

9.5. Exploration and Enhancement of Classifiers in the Detection of Lung Cancer from Histopathological Images

This study utilized advanced computational methods to enhance lung cancer detection from histopathological images, employing techniques like enhanced Kernel Fuzzy C-Means segmentation, Particle Swarm, and Grey Wolf Optimization for dimensional reduction, and feature selection algorithms such as Kullback–Leibler Divergence and Invasive Weed Optimization. Seven classifiers, including SVM, KNN, and decision trees, were used for categorizing images, with hyperparameter tuning to improve accuracy. This approach offers significant benefits for medical professionals, including early cancer detection, higher diagnostic accuracy, efficiency in handling large image sets, objective assessments, personalized treatment planning, and educational opportunities in digital pathology.

9.6. Towards Realistic 3D Models of Tumor Vascular Networks

This study employed sophisticated computational techniques for the reconstruction and examination of neoplastic vascular architectures derived from histological sections. The methodological framework encompassed image registration algorithms, which were instrumental in rectifying spatial discrepancies across sequential histological slices, ensuring fidelity in the reconstitution of the three-dimensional vascular continuum. The procedure entailed a preliminary feature- and area-based alignment, succeeded by an exhaustive parallel registration for holistic dataset alignment. This study harnessed both intensity- and color-thresholding methods, supplemented by heuristic analyses, to precisely demarcate vascular entities within histological specimens. The capability to generate intricate three-dimensional models of tumor vasculature augments our comprehension of tumor biology, particularly of the spatial and morphological characteristics of vascular networks, which are pivotal in understanding tumor growth, metastasis, and angiogenesis. The detailed visualization of the vascular architecture within tumors aids in the strategic planning of diagnostic and therapeutic interventions, enabling clinicians to identify viable routes for targeted therapies or surgical excisions, thereby optimizing patient-specific treatment protocols.

9.7. Combining State-of-the-Art Pre-Trained Deep Learning Models: A Noble Approach for Skin Cancer Detection Using Max Voting Ensemble

This study implements a max voting ensemble technique, combining predictions from various pre-trained deep learning models, such as MobileNetV2, AlexNet, VGG16, ResNet50, DenseNet201, DenseNet121, InceptionV3, ResNet50V2, InceptionResNetV2, and Xception, for skin cancer classification. These models, pre-trained on skin cancer datasets, contribute individual predictions that are aggregated through max voting to achieve a final classification. This ensemble approach enhances diagnostic accuracy by leveraging the diverse strengths of each model, offering significant benefits to healthcare professionals. It ensures more reliable skin cancer classification, boosts confidence in diagnostic decisions, saves time in diagnosis, and provides robust support for complex cases, thereby improving patient care and treatment efficacy.

Author Contributions: Conceptualization, R.O., M.S. and A.P.; methodology, R.O., M.S. and A.P.; software, R.O., M.S. and A.P.; validation, R.O., M.S. and A.P.; formal analysis, R.O., M.S. and A.P.; investigation, R.O., M.S. and A.P.; resources, R.O., M.S. and A.P.; data curation, R.O., M.S. and A.P.; writing—original draft preparation, R.O., M.S. and A.P.; writing—review and editing, R.O., M.S. and A.P.; visualization, R.O., M.S. and A.P.; supervision, R.O., M.S. and A.P.; project administration, R.O., M.S. and A.P.; funding acquisition, R.O., M.S. and A.P. All authors have read and agreed to the published version of the manuscript.

Funding: This research received no external funding.

Conflicts of Interest: The authors declare no conflicts of interest.

Appendix A

Table A1. Summary of research objectives, analysed organs (locations), imaging modalities, methods, and results described in papers published in this collection.

No.	Objective	Localization	Imaging	Methods	Results
1	A systematic review of the performance of DL models that use MRI to detect BMs in cancer patients	Brain	MRI	25 DL algorithms	Deep learning algorithms effectively detect BMs with a pooled sensitivity of 89%
2	Outcome prediction for cancer patients infected with COVID-19	Lung	CT	3D radiometric analysis	COVID-19 infection may have further detrimental effects on the lungs of cancer patients
3	Prediction of the weight-bearing line (WBL) ratio using simple knee radiographs	Knee	X-ray	CNN	Proposed method for predicting lower limb alignment demonstrated comparable accuracy to that of direct measurement using whole-leg radiographs

Table A1. *Cont.*

No.	Objective	Localization	Imaging	Methods	Results
4	Breast cancer detection using digital breast tomosynthesis	Breast	Digital breast tomosynthesis	CNN	Accuracy of 93.2%; sensitivity, specificity, precision, and F1-score of 92%, 94%, 94%, and 94%, respectively
5	Comparative study of different models for prostate segmentation	Prostate	MRI	Various segmentation deep networks	nnUNet gives most accurate segmentation results
6	New approach for polyp segmentation	Polyp	Endoscopic images	Dual boundary-guided attention exploration network (DBE-Net) for polyp segmentation	mDice of the proposed model reaches 82.4% and 80.6% for 2 datasets
7	Construction of a nomogram based on multiparametric MR images for predicting the response to neoadjuvant chemotherapy in patients with locally advanced cervical cancer	Cervix	MRI	Extraction of hand-selected (including texture) and DL-based radiomics features.	DL features slightly overlap with hand-selected ones in response prediction
8	Testing various CNN architectures to assess Alzheimer's disease severity	Brain	SPECT	MobileNet V2, NASNetMobile, VGG16, Inception V3, and ResNet	Good results for all networks; the best was Resnet, with an average accuracy 65%
9	Using an artificial intelligence (AI) technique to develop an automatic AI prediction model for the collateral status of mCTA	Brain	Multiphase CTA	CNN	Prediction model reached an accuracy of 0.746 ± 0.008
10	Application of deep learning to evaluate Wilms tumor volume	Brain	MRI	nnUnet for tumor segmentation compared to manual segmentation	nnUnet: median Dice of 0.90, median HD95 of 7.2 mm. Deep learning shows potential to replace manual segmentation
11	Estimation of PSA level based on MR prostate image texture	Prostate	MRI	Classical texture analysis	Multiparametric classification using MIL-SVM: 92% accuracy
12	Detection of both exudates and hemorrhages in color fundus images	Eye	Non-mydriatic retinal camera	Unet for exudate and hemorrhage segmentation, YOLO5 for detection	Segmentation: Dice 85%; detection: 100%
13	Assessment of ER applications in the field of diagnostic imaging	All	USG, X-ray, CT	ER technologies	ER has significant potential to improve accuracy and efficiency in diagnostic imaging procedures and enhance the patient experience
14	Evaluation of three feature extraction methods and their performance in breast cancer diagnosis	Breast	Histopathology images	Three feature extraction methods: basic CNN transfer learning, VGG16, and a knowledge-based approach. All tested on standard classifiers	Accuracy: CNN: 85%, VGG16: 86%, knowledge-based (geometrical, directional, intensity features): 98%
15	More accurate classification of enhancing foci in MRIs of BRCA PV carriers for early breast cancer detection	Breast	MRI	CNN for tumor and non-tumor ROI detection	Correct classification of ~65% of tumors at an early time point

Table A1. *Cont.*

No.	Objective	Localization	Imaging	Methods	Results
16	Multiclass segmentation and classification of liver images into lesions (6?) and normal tissue	Liver	Multiparametric MRI	nnUnet for segmentation, image registration, final UNet segmentation	Healthy liver/lesion classification: AUC ROC: 0.85, sensitivity and specificity: 0.79
17	Reducing image artefacts by inferring missing measurements using shape priors	Phantoms	CT	Deep convolutional GAN (DCGAN) architecture combining limited acquisition data and shape information	Improvement of reconstructed image quality by 7 dB Peak Signal-to-Noise Ratio compared to other methods
18	Classification of presence or absence of hemoperitoneum in USG images of Morrison's pouch	Peritoneum	USG	Open-source DL provided by Google (https: //teachablemachine. withgoogle.co, accessed on 1 March 2023)	External validation: sensitivity, specificity, and AUROC were 94%, 99%, and 0.97, respectively.
19	Denoising weighted filtered backprojection (wFBP) reconstructions	Phantom images, mouse scans	Photon-counting CT	Deep learning (UnetU) model for iterative reconstruction estimation from weighted filtered backprojection	UnetU provides higher SSIM PSNR when compared to classical ME NLM approach
20	Segmentation of liver portal veins from unlabeled H-phase and E-phase images by using the label of P-phase	Liver portal veins	CTA	Portal vein segmentation network (PVSegNet) applied to multiphase images	Portal vein segmented from H-phase and E-phase images achieved DSC 0.76 and 0.86 and Jaccard 0.61 and 0.76, respectively
21	Detection of active inflammation in the form of bone marrow edema (BME) in iliac and sacral bones	Sacroiliac joints	MRI	Own segmentation algorithm based on joint morphology	The Dice coefficient for automated bone segmentations with respect to reference manual segmentations was 0.9820
22	Comparison of the performance of path-finding algorithms for vessel labeling	Brain	3D TOF MRA	Three path-finding methods: depth-first search, Dijkstra's, and A* algorithm	The best accuracy was observed using Dijkstra's method
23	Quality comparison of two reconstruction algorithms of brain CT images	Brain	CT	DLIR (deep learning image reconstruction) and iterative reconstruction ASIR-V algorithms	DLIR shows superiority in both subjective and objective (SNR, CNR) assessments of image quality improvement
24	Building a 3D dose prediction model for glioblastoma VMAT treatment and testing its robustness and sensitivity for the purpose of quality assurance of automatic contouring	Brain	CT	Two-level cascaded 3D U-net trained for dose prediction	Improvement in dose and DVH score values, improvement in the spatial distribution of the predicted dose in the updated models at exactly the locations of concern
25	Overview of AI in radiology	All	All	Classical and deep ML	AI in radiology improves the quality of healthcare; however, several limitations exist
26	Quality comparison of two reconstruction algorithms of brain CT images	Brain	CT	Iterative (AIDR-3D) and deep learning-based (AiCE) reconstruction algorithms	AIDR-3D: lower artifact index, AiCE: higher CNR, lower median image noise, dependence on brain area
27	Object detection in chest X-ray images	Chest	Chest X-ray	YOLO v3 deep network	Average precision for 4 object classes: 0.815, APR: 0.987

Table A1. *Cont.*

No.	Objective	Localization	Imaging	Methods	Results
28	Development of software tool for generation of MR synthetic images (various sequences)	Brain	MRI T1, T2, T2*, PD	Algorithm that modifies sequence parameters of input images	Qualitative visual assessment
29	Mitigation of motion-induced signal loss in liver DWI images	Liver	MRI DWI	Algorithm based on spatial scaling of average diffusion-weighted images	Reduced ADC bias, improved homogeneity of liver DWIs
30	Detection of lung cancer using histopathological images (LC25000 Dataset: Benign Lung tissue and Lung Adenocarcinomas)	Lungs	Optical histopathological images of lung and colon cancer cases	Kernel Fuzzy C-Means segmentation, Particle Swarm Optimization, and Grey Wolf Optimization for feature extraction/selection; classification based on classical ML models	Overall accuracy of 91.57% (in classifying benign and adenocarcinoma classes)
31	Reconstruction of a 3D tumor vascular network from histologic slices	Pancreatic ductal adenocarcinoma	Microscopic images of histological samples	Image registration (Fiji, Improved CWR), own vessel segmentation algorithm, 3D reconstruction by interpolation	Visual quality assessment of the obtained vascular network
32	Asymmetry detection in mammographic images for skin cancer detection support	Breast	Mammography	Dynamic Time Warping (DTW) for shape analysis; Growing Seed Region (GSR) method for breast skin segmentation	Accuracy of asymmetry detection: 83%, accuracy of skin segmentation: 66.7–90.5%
33	Simulation of respiratory lung motion and extraction of information for early diagnosis of lung cancer	Lungs	Chest X-ray	Combination of U-Net and a long short-term memory (LSTM) network for image generation and sequential prediction	prediction of respiratory motion: average Dice 0.96
34	Skin cancer classification (malignant vs. benign) in available datasets	Skin	Dermoscopic images	Ensemble of deep networks with different architectures	Classification accuracy: 93.18%, AUC: 0.932

References

1. Ozkara, B.; Chen, M.; Federau, C.; Karabacak, M.; Briere, T.; Li, J.; Wintermark, M. Deep Learning for Detecting Brain Metastases on MRI: A Systematic Review and Meta-Analysis. *Cancers* **2023**, *15*, 334. [CrossRef]
2. Rahmanuddin, S.; Jamil, A.; Chaudhry, A.; Seto, T.; Brase, J.; Motarjem, P.; Khan, M.; Tomasetti, C.; Farwa, U.; Boswell, W.; et al. COVID and Cancer: A Complete 3D Advanced Radiological CT-Based Analysis to Predict the Outcome. *Cancers* **2023**, *15*, 651. [CrossRef]
3. Nam, H.; Park, S.; Ho, J.; Park, S.; Cho, J.; Lee, Y. Key-Point Detection Algorithm of Deep Learning Can Predict Lower Limb Alignment with Simple Knee Radiographs. *J. Clin. Med.* **2023**, *12*, 1455. [CrossRef] [PubMed]
4. Mendes, J.; Matela, N.; Garcia, N. Avoiding Tissue Overlap in 2D Images: Single-Slice DBT Classification Using Convolutional Neural Networks. *Tomography* **2023**, *9*, 398–412. [CrossRef] [PubMed]
5. Rodrigues, N.; Silva, S.; Vanneschi, L.; Papanikolaou, N. A Comparative Study of Automated Deep Learning Segmentation Models for Prostate MRI. *Cancers* **2023**, *15*, 1467. [CrossRef]
6. Ma, H.; Xu, C.; Nie, C.; Han, J.; Li, Y.; Liu, C. DBE-Net: Dual Boundary-Guided Attention Exploration Network for Polyp Segmentation. *Diagnostics* **2023**, *13*, 896. [CrossRef] [PubMed]
7. Zhang, Y.; Wu, C.; Xiao, Z.; Lv, F.; Liu, Y. A Deep Learning Radiomics Nomogram to Predict Response to Neoadjuvant Chemotherapy for Locally Advanced Cervical Cancer: A Two-Center Study. *Diagnostics* **2023**, *13*, 1073. [CrossRef]
8. Lien, W.; Yeh, C.; Chang, C.; Chang, C.; Wang, W.; Chen, C.; Lin, Y. Convolutional Neural Networks to Classify Alzheimer' Disease Severity Based on SPECT Images: A Comparative Study. *J. Clin. Med.* **2023**, *12*, 2218. [CrossRef]

9. Huang, C.; Chiang, H.; Hsieh, C.; Chou, C.; Jhou, Z.; Hou, T.; Shaw, J. Using Deep-Learning-Based Artificial Intelligence Technique to Automatically Evaluate the Collateral Status of Multiphase CTA in Acute Ischemic Stroke. *Tomography* **2023**, *9*, 647–656. [CrossRef]
10. Buser, M.; van der Steeg, A.; Wijnen, M.; Fitski, M.; van Tinteren, H.; van den Heuvel-Eibrink, M.; Littooij, A.; van der Velden, B. Radiologic versus Segmentation Measurements to Quantify Wilms Tumor Volume on MRI in Pediatric Patients. *Cancers* **2023**, *15*, 2115. [CrossRef]
11. Gibala, S.; Obuchowicz, R.; Lasek, J.; Schneider, Z.; Piorkowski, A.; Pociask, E.; Nurzynska, K. Textural Features of MR Images Correlate with an Increased Risk of Clinically Significant Cancer in Patients with High PSA Levels. *J. Clin. Med.* **2023**, *12*, 2836. [CrossRef] [PubMed]
12. Farahat, Z.; Zrira, N.; Souissi, N.; Benamar, S.; Belmekki, M.; Ngote, M.; Megdiche, K. Application of Deep Learning Methods in a Moroccan Ophthalmic Center: Analysis and Discussion. *Diagnostics* **2023**, *13*, 1694. [CrossRef] [PubMed]
13. Kukla, P.; Maciejewska, K.; Strojna, I.; Zapał, M.; Zwierzchowski, G.; Bąk, B. Extended Reality in Diagnostic Imaging-A Literature Review. *Tomography* **2023**, *9*, 1071–1082. [CrossRef] [PubMed]
14. Kode, H.; Barkana, B. Deep Learning- and Expert Knowledge-Based Feature Extraction and Performance Evaluation in Breast Histopathology Images. *Cancers* **2023**, *15*, 3075. [CrossRef]
15. Anaby, D.; Shavin, D.; Zimmerman-Moreno, G.; Nissan, N.; Friedman, E.; Sklair-Levy, M. 'Earlier than Early' Detection of Breast Cancer in Israeli BRCA Mutation Carriers Applying AI-Based Analysis to Consecutive MRI Scans. *Cancers* **2023**, *15*, 3120. [CrossRef] [PubMed]
16. Skwirczyński, M.; Tabor, Z.; Lasek, J.; Schneider, Z.; Gibała, S.; Kucybała, I.; Urbanik, A.; Obuchowicz, R. Deep Learning Algorithm for Differentiating Patients with a Healthy Liver from Patients with Liver Lesions Based on MR Images. *Cancers* **2023**, *15*, 3142. [CrossRef] [PubMed]
17. Valat, E.; Farrahi, K.; Blumensath, T. Sinogram Inpainting with Generative Adversarial Networks and Shape Priors. *Tomography* **2023**, *9*, 1137–1152. [CrossRef] [PubMed]
18. Jeong, D.; Jeong, W.; Lee, J.; Park, S. Use of Automated Machine Learning for Classifying Hemoperitoneum on Ultrasonographic Images of Morrison's Pouch: A Multicenter Retrospective Study. *J. Clin. Med.* **2023**, *12*, 4043. [CrossRef] [PubMed]
19. Nadkarni, R.; Clark, D.; Allphin, A.; Badea, C. A Deep Learning Approach for Rapid and Generalizable Denoising of Photon-Counting Micro-CT Images. *Tomography* **2023**, *9*, 1286–1302. [CrossRef]
20. Song, G.; Xie, Z.; Wang, H.; Li, S.; Yao, D.; Chen, S.; Shi, Y. Segmentation of Portal Vein in Multiphase CTA Image Based on Unsupervised Domain Transfer and Pseudo Label. *Diagnostics* **2023**, *13*, 2250. [CrossRef]
21. Ożga, J.; Wyka, M.; Raczko, A.; Tabor, Z.; Oleniacz, Z.; Korman, M.; Wojciechowski, W. Performance of Fully Automated Algorithm Detecting Bone Marrow Edema in Sacroiliac Joints. *J. Clin. Med.* **2023**, *12*, 4852. [CrossRef] [PubMed]
22. Kim, S.; Kim, Y. Effects of Path-Finding Algorithms on the Labeling of the Centerlines of Circle of Willis Arteries. *Tomography* **2023**, *9*, 1423–1433. [CrossRef] [PubMed]
23. Pula, M.; Kucharczyk, E.; Zdanowicz, A.; Guzinski, M. Image Quality Improvement in Deep Learning Image Reconstruction of Head Computed Tomography Examination. *Tomography* **2023**, *9*, 1485–1493. [CrossRef] [PubMed]
24. Poel, R.; Kamath, A.; Willmann, J.; Andratschke, N.; Ermiş, E.; Aebersold, D.; Manser, P.; Reyes, M. Deep-Learning-Based Dose Predictor for Glioblastoma-Assessing the Sensitivity and Robustness for Dose Awareness in Contouring. *Cancers* **2023**, *15*, 4226. [CrossRef]
25. Najjar, R. Redefining Radiology: A Review of Artificial Intelligence Integration in Medical Imaging. *Diagnostics* **2023**, *13*, 2760. [CrossRef] [PubMed]
26. Cozzi, A.; Cè, M.; De Padova, G.; Libri, D.; Caldarelli, N.; Zucconi, F.; Oliva, G.; Cellina, M. Deep Learning-Based Versus Iterative Image Reconstruction for Unenhanced Brain CT: A Quantitative Comparison of Image Quality. *Tomography* **2023**, *9*, 1629–1637. [CrossRef] [PubMed]
27. Kufel, J.; Bargieł-Łączek, K.; Koźlik, M.; Czogalik, Ł.; Dudek, P.; Magiera, M.; Bartnikowska, W.; Lis, A.; Paszkiewicz, I.; Kocot, S.; et al. Chest X-ray Foreign Objects Detection Using Artificial Intelligence. *J. Clin. Med.* **2023**, *12*, 5841. [CrossRef] [PubMed]
28. Peretti, L.; Donatelli, G.; Cencini, M.; Cecchi, P.; Buonincontri, G.; Cosottini, M.; Tosetti, M.; Costagli, M. Generating Synthetic Radiological Images with PySynthMRI: An Open-Source Cross-Platform Tool. *Tomography* **2023**, *9*, 1723–1733. [CrossRef]
29. Raspe, J.; Harder, F.; Rupp, S.; McTavish, S.; Peeters, J.; Weiss, K.; Makowski, M.; Braren, R.; Karampinos, D.; Van, A. Retrospective Motion Artifact Reduction by Spatial Scaling of Liver Diffusion-Weighted Images. *Tomography* **2023**, *9*, 1839–1856. [CrossRef]
30. Shanmugam, K.; Rajaguru, H. Exploration and Enhancement of Classifiers in the Detection of Lung Cancer from Histopathological Images. *Diagnostics* **2023**, *13*, 3289. [CrossRef]
31. Lindemann, M.; Glänzer, L.; Roeth, A.; Schmitz-Rode, T.; Slabu, I. Towards Realistic 3D Models of Tumor Vascular Networks. *Cancers* **2023**, *15*, 5352. [CrossRef] [PubMed]
32. Bayareh-Mancilla, R.; Medina-Ramos, L.; Toriz-Vázquez, A.; Hernández-Rodríguez, Y.; Cigarroa-Mayorga, O. Automated Computer-Assisted Medical Decision-Making System Based on Morphological Shape and Skin Thickness Analysis for Asymmetry Detection in Mammographic Images. *Diagnostics* **2023**, *13*, 3440. [CrossRef] [PubMed]
33. Yang, D.; Huang, Y.; Li, B.; Cai, J.; Ren, G. Dynamic Chest Radiograph Simulation Technique with Deep Convolutional Neural Networks: A Proof-of-Concept Study. *Cancers* **2023**, *15*, 5768. [CrossRef] [PubMed]

34. Hossain, M.; Hossain, M.; Arefin, M.; Akhtar, F.; Blake, J. Combining State-of-the-Art Pre-Trained Deep Learning Models: A Noble Approach for Skin Cancer Detection Using Max Voting Ensemble. *Diagnostics* **2024**, *14*, 89. [CrossRef]
35. Strzelecki, M.; Badura, P. Machine Learning for Biomedical Application. *Appl. Sci.* **2022**, *12*, 2022. [CrossRef]
36. Piórkowski, A.; Obuchowicz, R.; Urbanik, A.; Strzelecki, M. Advances in Musculoskeletal Imaging and Their Applications. *J. Clin. Med.* **2023**, *12*, 6585. [CrossRef] [PubMed]
37. Strzelecki, M.; Kociołek, M.; Strąkowska, M.; Kozłowski, M.; Grzybowski, A.; Szczypiński, P.M. Artificial intelligence in the detection of skin cancer: State of the art. In *Clinics in Dermatology*; Elsevier: Amsterdam, The Netherlands, 2024. [CrossRef]
38. Mayerhoefer, M.E.; Materka, A.; Langs, G.; Häggström, I.; Szczypiński, P.; Gibbs, P.; Cook, G. Introduction to Radiomics. *J. Nucl. Med.* **2020**, *61*, 488–495. [CrossRef] [PubMed]
39. Obuchowicz, R.; Kruszyńska, J.; Strzelecki, M. Classifying median nerves in carpal tunnel syndrome: Ultrasound image analysis. *Biocybern. Biomed. Eng.* **2021**, *41*, 335–351. [CrossRef]
40. Nurzynska, K.; Piórkowski, A.; Strzelecki, M.; Kociołek, M.; Banyś, R.P.; Obuchowicz, R. Differentiating age and sex in vertebral body CT scans–Texture analysis versus deep learning approach. *Biocybern. Biomed. Eng.* **2024**, *44*, 20–30. [CrossRef]

cancers

MDPI

Systematic Review

Deep Learning for Detecting Brain Metastases on MRI: A Systematic Review and Meta-Analysis

Burak B. Ozkara [1], Melissa M. Chen [1], Christian Federau [2], Mert Karabacak [3], Tina M. Briere [4], Jing Li [5] and Max Wintermark [1,*]

1 Department of Neuroradiology, MD Anderson Cancer Center, 1400 Pressler Street, Houston, TX 77030, USA
2 Faculty of Medicine, University of Zurich, Pestalozzistrasse 3, CH-8032 Zurich, Switzerland
3 Department of Neurosurgery, Mount Sinai Health System, 1468 Madison Avenue, New York, NY 10029, USA
4 Department of Radiation Physics, MD Anderson Cancer Center, 1515 Holcombe Boulevard, Houston, TX 77030, USA
5 Department of Radiation Oncology, MD Anderson Cancer Center, 1515 Holcombe Boulevard, Houston, TX 77030, USA
* Correspondence: mwintermark@mdanderson.org

Simple Summary: Manual detection and delineation of brain metastases are time consuming and variable. Studies have therefore been conducted to automate this process using imaging studies and artificial intelligence. To the best of our knowledge, no study has conducted a systematic review and meta-analysis on brain metastasis detection using only deep learning and MRI. As a result, a systematic review of this topic is required, as well as an assessment of the quality of the studies and a meta-analysis to determine the strength of the current evidence. The purpose of this study was to perform a systematic review and meta-analysis of the performance of deep learning models that use MRI to detect brain metastases in cancer patients.

Abstract: Since manual detection of brain metastases (BMs) is time consuming, studies have been conducted to automate this process using deep learning. The purpose of this study was to conduct a systematic review and meta-analysis of the performance of deep learning models that use magnetic resonance imaging (MRI) to detect BMs in cancer patients. A systematic search of MEDLINE, EMBASE, and Web of Science was conducted until 30 September 2022. Inclusion criteria were: patients with BMs; deep learning using MRI images was applied to detect the BMs; sufficient data were present in terms of detective performance; original research articles. Exclusion criteria were: reviews, letters, guidelines, editorials, or errata; case reports or series with less than 20 patients; studies with overlapping cohorts; insufficient data in terms of detective performance; machine learning was used to detect BMs; articles not written in English. Quality Assessment of Diagnostic Accuracy Studies-2 and Checklist for Artificial Intelligence in Medical Imaging was used to assess the quality. Finally, 24 eligible studies were identified for the quantitative analysis. The pooled proportion of patient-wise and lesion-wise detectability was 89%. Articles should adhere to the checklists more strictly. Deep learning algorithms effectively detect BMs. Pooled analysis of false positive rates could not be estimated due to reporting differences.

Keywords: artificial intelligence; deep learning; brain metastasis; magnetic resonance imaging; pooled analysis

Citation: Ozkara, B.B.; Chen, M.M.; Federau, C.; Karabacak, M.; Briere, T.M.; Li, J.; Wintermark, M. Deep Learning for Detecting Brain Metastases on MRI: A Systematic Review and Meta-Analysis. *Cancers* **2023**, *15*, 334. https://doi.org/10.3390/cancers15020334

Academic Editors: Rafał Obuchowicz, Michał Strzelecki and Adam Piorkowski

Received: 11 December 2022
Revised: 31 December 2022
Accepted: 31 December 2022
Published: 4 January 2023

1. Introduction

Brain metastases (BMs) are observed in nearly 20% of adult cancer patients [1]. They are the most common type of intracranial neoplasm in adults [2]. Although the brain parenchyma is the most common intracranial metastatic site, BMs frequently occur in conjunction with metastases to other sites such as the cranium, dura, or leptomeninges [3]. Therefore, making accurate diagnosis of BM is critical.

Contrast-enhanced magnetic resonance imaging (MRI) is the preferred imaging examination for diagnosing BMs. It is more sensitive than either nonenhanced MRI or computed tomography (CT) in detecting lesions [4–7]. Although whole-brain radiotherapy (WBRT), which can cause cognitive decline, has historically been the mainstay of radiotherapy for the treatment of BMs, stereotactic radiosurgery (SRS) has now become the standard of care in many clinical situations. In 2014, a multi-institutional prospective study demonstrated that SRS without WBRT in patients with five to ten brain metastases is non-inferior to that in patients with two to four brain metastases [8]. Therefore, SRS plays a more prominent role today and new treatment methods, such as hippocampal avoidance-WBRT and newer systemic agents, are being developed [9]. Before SRS, in the current standard practice, each BM must be correctly detected and manually delineated for treatment planning, which is time consuming and subject to considerable inter- and intra-observer variability [10].

Because manual detection and delineation of BMs is time consuming and variable, research has been done to make this process automated using imaging studies with the help of artificial intelligence [11]. Deep learning, a type of representation learning method, has emerged as the cutting-edge machine learning method [12]. Because it does not require an intermediary feature extraction or an engineering phase to learn the relationship between the input and the output, deep learning has accelerated the development of computer-aided detection in various fields [13]. Furthermore, with advances in network architectures and increased imaging data quantity and quality in recent years, the performance of deep learning-based approaches has been greatly enhanced [14]. A meta-analysis of 12 studies examined computer-aided detection in BMs and showed a pooled proportion of detectability of 0.90 (95% confidence interval [CI], 85–93%) [11]. However, they included articles that used machine learning or deep learning algorithms, totaling twelve studies in their meta-analysis, with only seven implementing deep learning.

To the best of our knowledge, no study in the literature has performed a systematic review and meta-analysis on BM detection with exclusively deep learning using MRI. As a result, it is necessary to conduct a systematic review of this topic, assess the quality of the studies, and run a meta-analysis to evaluate the strength of the current evidence. This study aimed to conduct a systematic review and meta-analysis of the performance of deep learning models that implement MRI in detecting BMs in cancer patients.

2. Materials and Methods

This study followed the Preferred Reporting Items for Systematic Reviews and Meta-Analyses extension for Diagnostic Test Accuracy Studies (PRISMA-DTA) [15]. Checklists for PRISMA-DTA for Abstracts and PRISMA-DTA are available online in Supplemental Tables S1 and S2, respectively. The study has not been registered. This study is exempt from ethical approval of the Institutional Review Board since the analysis only involves anonymized data, and all the included studies have received local ethics approval.

2.1. Literature Search

A systematic search of articles was performed by two independent readers (B.B.O. and M.K., with more than 2 years and 4 years of experience in conducting systematic reviews and meta-analyses, respectively) from inception until September 30, 2022. The following search terms were used to search MEDLINE: ("deep learning" OR "deep architecture" OR "artificial neural network" OR "convolutional neural network" OR "convolutional network" OR CNN OR "recurrent neural network" OR RNN OR Auto-Encoder OR Autoencoder OR "Deep belief network" OR DBN OR "Restricted Boltzmann machine" RBM OR "Long Short Term Memory" OR "Long Short-Term Memory" OR LSTM OR "Gated Recurrent Units" OR GRU) AND (("brain metastasis") OR ("brain metastases") OR ("metastatic brain tumor") OR ("intra-axial metastatic tumor") OR ("cerebral metastasis") OR ("cerebral metastases")). The search strategy was then adapted for EMBASE and Web of Science before searching these databases.

2.2. Study Selection

All search results were exported to the Rayyan online platform [16]. After removing duplicates, two authors (B.B.O. and M.K.) independently screened titles and abstracts using the Rayyan platform and reviewed the full text of potentially relevant articles. A senior author compared and examined the results of each search and analysis step (M.W., 25 years of experience in neuroradiology). Any disagreement was resolved via discussion with the senior author (M.W.).

Articles were included based on the satisfaction of all the following criteria: (I) inclusion of patients with BMs; (II) deep learning using MRI images was applied to detect the BMs; (III) sufficient data were present in terms of detective performance of the deep learning algorithms; (IV) original research articles.

Articles were excluded if they fulfilled any of the following criteria: (I) reviews, letters, guidelines, editorials, or errata; (II) case reports or series with less than 20 patients; (III) studies with overlapping cohorts; (IV) insufficient data in terms of detective performance of the deep learning algorithms; (V) machine learning, not deep learning, was used to detect BMs; (VI) articles not written in English. In the case of overlapping cohorts, the article with the largest sample size was included. If articles have the same number of patients, the most recent one was preferred.

2.3. Data Extraction

The following variables were collected by the two authors (B.B.O. and M.K.): (I) study characteristics (first author, publication year, study design, number of patients in each dataset, sex of patients in each dataset, number of metastatic lesions in each dataset, mean or median size of lesions, reference standard for metastasis detection, validation method, and primary tumors); (II) deep learning details and statistics (detectability, false positive rate, deep learning algorithm, and data augmentation); (III) MRI and scanner characteristics; and (IV) inclusion and exclusion criteria of each study.

2.4. Quality Assessment

Two authors (B.B.O. and M.K.) independently performed the quality assessments using the Checklist for Artificial Intelligence in Medical Imaging (CLAIM) and Quality Assessment of Diagnostic Accuracy Studies-2 (QUADAS-2) [17,18]. The CLAIM is a new 42-item checklist for evaluating artificial intelligence studies in medical imaging. For each item, studies were given a score of 0 or 1 on a 2-point scale. The CLAIM score was calculated by adding the scores from each study. The items were all equally weighted. QUADAS-2 was used to assess four domains: (I) patient selection, (II) index test, (III) reference standard, and (IV) flow and timing. During the quality assessment, any disagreements were resolved with the assistance of the senior author.

2.5. Meta-Analysis

The primary goal was to assess the detectability (sensitivity) of deep learning algorithms in detecting BMs. For the pooled proportion analysis of detectability of all included studies in the meta-analysis, the reported sensitivity per study was multiplied by the total number of patients in each study to estimate the number of detected events. In other words, reported sensitivity was converted to patient-wise sensitivity. Another pooled proportion analysis of detectability was performed for studies reporting their sensitivity lesion-wise. The detectability of each study in this group was calculated by dividing the number of correctly identified metastases in the test set (using given true positive values or calculated with reported sensitivities) by the total number of metastases in the test set. Following statistical analyses were performed in both pooled analyses.

The pooled proportion analysis of detectability estimates with 95% CIs was performed with the random-effects model. The random-effects model, using the inverse variance method, was chosen since it captures uncertainty due to heterogeneity among studies [19,20]. The Freeman–Tukey double arcsine transformation was performed to stabilize the variances before pooling. Forest plots were created to provide a visual representation of the results. Meta-regression analyses were performed to determine if training size was associated with the sensitivity. Heterogeneity among all included studies was evaluated using Q-test with $p < 0.05$ suggesting the presence of study heterogeneity and I2 statistics. I2 values were defined as follows: heterogeneity that might not be important (0–25%), low heterogeneity (26–50%), moderate heterogeneity (51–75%), and high heterogeneity (76–100%) [21]. Since heterogeneity might indicate subgroup effects, we also explored heterogeneity in the pooled results using subgroup analysis [22]. Subgroup analyses were conducted in terms of the validation method used in the study (split training-test sets versus cross-validation), the plurality of MRI sequences utilized (single sequence versus multi-sequence), the calculation method of the reported sensitivity rates (lesion-wise versus other [patient-wise, voxel-wise]), the plurality of the primary tumor type (single versus multiple), dimension of the included images (2D versus 3D versus both), study design (single-center versus multi-center), and deep learning algorithms (DeepMedic versus U-Net versus others)—Figure 1. Meta-regression analyses were also performed to determine if training size was associated with heterogeneity. Publication bias occurs when the findings of a study influence the study's likelihood of publication. The Egger method was applied to test the funnel plot asymmetry for publication bias [23]. R version 4.2.1 was used for all statistical analysis, and the function metaprop from package meta was utilized to perform meta-analysis and generate pooled estimates [24]. An alpha level of 0.05 was considered statistically significant.

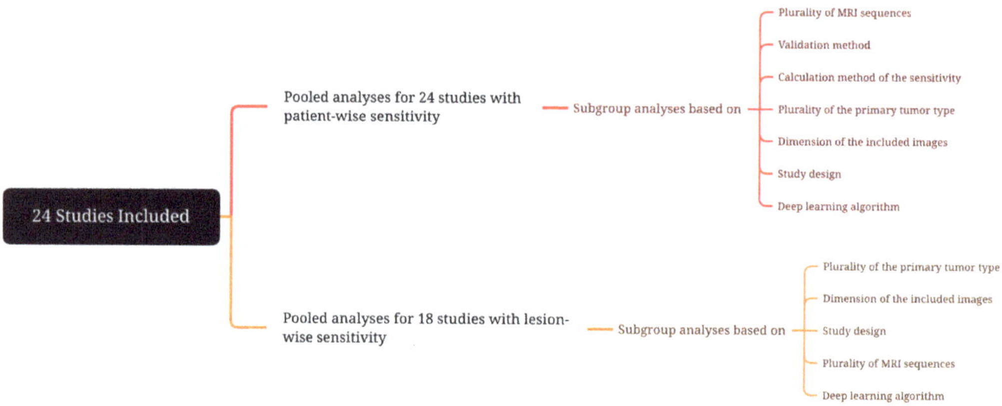

Figure 1. Meta-analysis design.

3. Results

3.1. Literature Search

The study selection process is shown in the PRISMA flowchart (Figure 2). The literature search yielded 401 studies: 90 from MEDLINE, 189 from EMBASE, and 122 from Web of Science. After removing 191 duplicate articles, the remaining 210 were screened based on their title and abstract on the Rayyan platform, and 111 were excluded. Full texts of the remaining 99 articles were acquired and reviewed.

Seventy-four articles were excluded because: irrelevant studies or studies lacking significant amount of data ($n = 15$); overlapping patient cohort ($n = 4$); reviews, letters, guidelines, editorials, conference abstracts, and poster presentations ($n = 30$); focused on glioma and metastasis differentiation ($n = 12$); focused on segmentation performance ($n = 5$); another imaging modality was used rather than MRI ($n = 6$); machine learning was used, not deep learning ($n = 2$).

Finally, 25 eligible studies were identified [14,25–48]. Due to a lack of sensitivity reporting, one study was included in the review but not in the quantitative analysis.

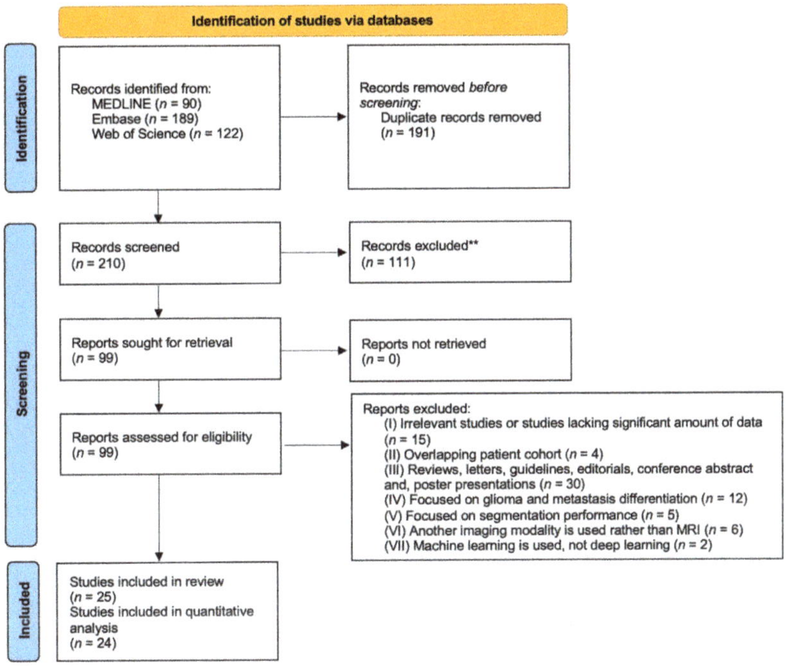

Figure 2. Study selection process.

3.2. Quality Assessment

Table 1 shows a quality assessment summary of the included studies using the CLAIM. The mean CLAIM score of the four studies was 26.56 with a standard deviation (SD) of 3.19 (range, 18.00–31.00). The mean scores of the subsections of the CLAIM were 1.52 (SD = 0.71) for the title/abstract section, 2.00 (SD = 0.00) for the introduction section, 18.16 (SD = 2.25) for the methods section, 2.00 (SD = 0.82) for the results section, 1.96 (SD = 0.20) for the discussion section, and 0.92 (SD = 0.28) for the other information section.

Figure 3 illustrates a quality assessment summary of the included studies using the QUADAS-2 tool. In terms of patient selection, eight studies showed an unclear risk of bias because they did not specify the inclusion criteria for patient enrollment, or they did not report the excluded patients or exclusion criteria [26,30,33–35,41,45,47]. Six studies were found to have a high risk of bias due to patient exclusion based on lesion size or the number of lesions [25,29,31,32,37,38]. The inclusion and exclusion criteria of the studies can be found online in Supplementary Table S3. All included studies in the index test section were determined to have a low bias risk since the algorithm was blinded to the reference standard. Since the lesions were annotated on contrast-enhanced MRI manually or semi-manually with manual correction, which is the method of choice for assessing and delineating brain metastases, all studies in the reference standard section were considered to have a low risk of bias [5]. No risk of bias was found concerning the flow and timing. There were no concerns regarding the applicability of patient selection, and index tests. Two studies were found to have high concerns about the applicability of the reference test because they used 1.0 Tesla (T) scanners [36,40]. In current practice, scanners with magnets less than 1.5T are not recommended for BM detection [49].

Table 1. The Checklist for Artificial Intelligence in Medical Imaging scores.

Study	Title/Abstract Score (*n*/2)	Introduction Score (*n*/2)	Methods Score (*n*/28)	Results Score (*n*/5)	Discussion Score (*n*/2)	Other Information Score (*n*/3)	Total Score (*n*/42)
Amemiya et al.—2022 [25]	2	2	21	3	2	1	31
Bousabarah et al.—2020 [26]	2	2	18	2	2	1	27
Charron et al.—2018 [27]	1	2	17	0	2	1	23
Chartrand et al.—2022 [28]	2	2	18	1	2	1	26
Cho et al.—2021 [29]	2	2	19	2	2	1	28
Deike-Hofmann et al.—2021 [30]	1	2	17	2	2	1	25
Dikici et al.—2022 [31]	0	2	17	2	2	1	24
Grøvik et al.—2021 [32]	1	2	22	3	2	1	31
Han et al.—2019 [33]	0	2	13	1	1	1	18
Hsu et al.—2021 [34]	0	2	19	2	2	1	26
Huang et al.—2022 [35]	2	2	16	2	2	0	24
Jünger et al.—2021 [36]	2	2	19	2	2	1	28
Kikuchi et al.—2022 [37]	2	2	14	3	2	1	24
Kottlors et al.—2021 [38]	1	2	14	2	2	1	22
Liang et al.—2022 [14]	2	2	19	2	2	1	28
Park et al.—2021 [39]	2	2	18	3	2	1	28
Pennig et al.—2021 [40]	2	2	18	1	2	1	26
Pflüger et al.—2022 [41]	2	2	21	3	2	1	31
Rudie et al.—2021 [42]	2	2	21	2	2	1	30
Xue et al.—2020 [43]	2	2	20	3	2	1	30
Yin et al.—2022 [44]	2	2	20	3	2	1	30
Yoo et al.—2022 [45]	1	2	18	1	2	1	25
Yoo et al.—2021 [46]	2	2	17	1	2	0	24
Zhang et al.—2020 [47]	2	2	19	2	2	1	28
Zhou et al.—2020 [48]	1	2	19	2	2	1	27

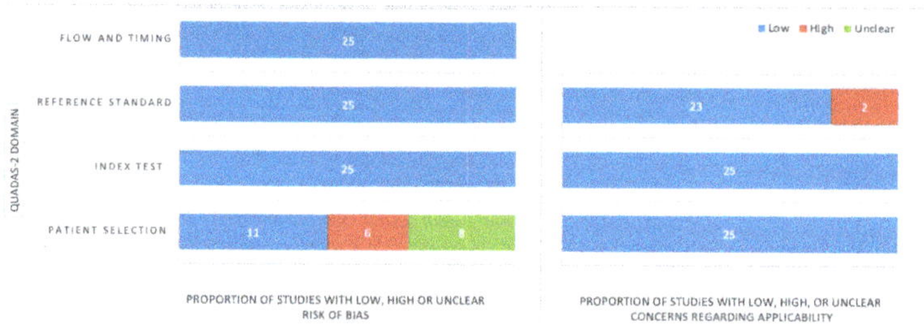

Figure 3. A summary of the Quality Assessment of Diagnostic Accuracy Studies-2 results.

3.3. Characteristics of Included Studies

The patient and study characteristics are shown in Table 2. All studies were conducted retrospectively. Five studies were multi-center studies [14,32,36,43,44], and the others were single-center. Four studies used cross-validation [30,31,38,43], and others used split training-test sets to evaluate their models. Delineation of brain metastases was done semi-automatically in one study to serve as a reference standard [25], and it was done manually in others. There were 6840, 1419, and 643 patients in the training and validation sets combined, test sets, and other sets, respectively. Ultimately, 25 studies included a total of 8902 patients. The number of metastatic lesions in the training sets was reported in 20 studies, totaling 31,530. Furthermore, the number of metastatic lesions in the test sets was documented in 18 studies, making a total of 5565. The total reported number of metastatic lesions was 40,654. Twenty-two studies included multiple primary tumor types, and three studies included only one primary tumor type. Two articles included only malignant melanoma patients, and one included non-small cell lung cancer (NSCLC) patients [30,36,40]. Table 2 also displays the means or medians of the volumes of the lesions or the longest diameter of the lesions.

Table 2. The patient and study characteristics.

Study	Year	Design	No of Patients in the Training Set (M:F)	No of Patients in the Validation Set (M:F)	No of Patients in the Test Set (M:F)	No of Patients in the Other Sets (M:F)	No of Metastatic Lesions (Training/Validation/Test/ Other Sets)	Mean or Median—Whole Volume or Longest Diameter of Lesions (Training/Validation/Test/ Other Sets)	Reference Standard	Validation Method	Primary Tumor
Amemiya et al. [25]	2022	Single-center	178 (96:82)	NA	56 (30:26)	NA	1249/NA/228/NA	4.1/NA/10.4/NA mm—mean	Semi-automatic	Split training-test	Multiple
Bousabarah et al. [26]	2020	Single-center	469 (244:225)	NA	40 (26:14)	NA	524/NA/47/NA	1.29/NA/1.92/NA cm^3—mean	Manual	Split training-test	Multiple
Charron et al. [27]	2018	Single-center	164 (NR)	NA	18 (NR)	NA	374/NA/38/NA	8.1 mm (whole sample)—mean	Manual	Split training-test	Multiple
Chartrand et al. [28]	2022	Single-center	383 (NR)	50 (NR)	97 (NR) [1]	NA	1460/NR/296/NA [2]	NR/NR/NR/NA	Manual	Split training-test	Multiple
Cho et al. [29]	2021	Single-center	127 (61:66)	NA	20 (12:8)	47 (25:22)	1298/NA/412/NR	6.5/NA/6/NR mm—median	Manual	Split training-test	Multiple
Deike-Hofmann et al. [30]	2021	Single-center	43 (35:8)	NA	NA	NA	494/NA/NA/NA	4.2/NA/NA/NA mm—median	Manual	Cross-validation	Single/Malignant melanoma
Dikici et al. [31]	2022	Single-center	158 (M:F = 0.89) [3]	NA	NA	NA	932/NA/NA/NA	5.45/NA/NA/NA mm—mean	Manual	Cross-validation	Multiple
Grøvik et al. [32]	2021	Multi-center	100 (29:71)	10 (NR) [4]	55 (NR)	NA	NR/NR/NR/NA	NR/NR/NR/NA	Manual	Split training-test	Multiple
Han et al. [33]	2019	Single-center	126 (NR) [5]	18 (NR)	36 (NR)	NA	NR/NR/NR/NA	NR/NR/NR/NA	Manual	Split training-test	Multiple
Hsu et al. [34]	2021	Single-center	409 (NR)	NA	102 (NR)	NA	1345/NA/367/NA	NR/NA/NR/NA	Manual	Split training-test	Multiple
Huang et al. [35]	2022	Single-center	135 (NR)	9 (NR)	32 (NR)	NA	1503/NR/278/NA	NR/NR/NR/NA	Manual	Split training-test	Multiple
Jünger et al. [36]	2021	Multi-center [6]	66 (24:42)	NA	17 (6:11)	15 (5:10)	248/NA/67/0	0.99/NA/0.96/NA cm^3—mean	Manual	Split training-test	Single/NSCLC
Kikuchi et al. [37]	2022	Single-center	50 (30:20)	NA	34 (16:18) [7]	NA	165/NA/48/NA	4/NA/2.9/NA mm—median	Manual	Split training-test	Multiple
Kottlors et al. [8] [38]	2021	Single-center	85 (52.3%:47.7%) [9]	NA	NA	NA	47/NA/NA/NA	NR/NA/NA/NA	Manual	Cross-validation	Multiple
Liang et al. [14]	2022	Multi-center	326 (127:150) [10]	NA	81 (31:50)	NA	1284/NA/327/NA	15.9/NA/17.6/NA mm—median	Manual	Split training-test	Multiple
Park et al. [39]	2021	Single-center	188 (98:90)	NA	94 (55:39) [11]	NA	917/NA/203/NA	1.6/NA/1.9/NA cm^3—mean	Manual	Split training-test	Multiple
Pennig et al. [40]	2021	Single-center	55 (?) [12]	NA	14 (NR)	NA	103/NA/32/NA	2.6/NA/1/NA cm^3—mean	Manual	Split training-test	Single/Malignant melanoma
Pflüger et al. [41]	2022	Single-center	246 (134:112)	NA	62 (29:33)	30 (15:15)	1682/NA/384/155	1.23/NA/1.24/1.03 cm^3—mean	Manual	Split training-test	Multiple
Rudie et al. [42]	2021	Single-center	313 (127:186)	NA	100 (48:52)	NA	4494/NA/708/NA	0.57/NA/0.50/NA cm^3—mean	Manual	Split training-test	Multiple
Xue et al. [43]	2020	Multi-center	1201 (684:517)	NA	NA	251 (236:215)	NR/NA/NA/NR	4.01/NA/NA/NR cm^3—mean	Manual	Cross-validation	Multiple
Yin et al. [13] [44]	2022	Multi-center	680 (374:306)	NA	270 (144:2126)	300 (161:139)	9630/NA/818/1066	5.5/NA/7.5/5.8 mm—mean	Manual	Split training-test	Multiple
Yoo et al. [45]	2022	Single-center	53 (29:24)	NA	12 (6:6)	NA	545/NA/58/NA	0.592/NA/0.158/NA cm^3—mean	Manual	Split training-test	Multiple

Table 2. *Cont.*

Study	Year	Design	No of Patients in the Training Set (M:F)	No of Patients in the Validation Set (M:F)	No of Patients in the Test Set (M:F)	No of Patients in the Other Sets (M:F)	No of Metastatic Lesions (Training/Validation/Test/Other Sets)	Mean or Median—Whole Volume or Longest Diameter of Lesions (Training/Validation/Test/Other Sets)	Reference Standard	Validation Method	Primary Tumor
Yoo et al. [46]	2021	Single-center	341 (NR)	36 (NR)	45 (NR)	NA	NR/NR/NR/NA [14]	NR/NR/NA/4.17 cm^3—mean	Manual	Split training-test	Multiple
Zhang et al. [47]	2020	Single-center	73 (NR) [15]	NA	48 (NR)	NA	1565/NA/488/NA	NR/NA/NR/NA	Manual	Split training-test	Multiple
Zhou et al. [48]	2020	Single-center	748 (NR) [16]	NA	186 (NR)	NA	3131/NA/766/NA	NR/NA/NR/NA [17]	Manual	Split training-test	Multiple

Abbreviations: No, number; M, male; F, female; NA, not applicable; mm, millimeter; cm^3, cubic centimeter; NR, not reported. [1] The total dataset contains 291 females and 239 males. [2] The combined training and validation sets contain 1460 lesions. [3] Unlabeled data contained 867 patients. [4] Validation and testing sets included 35 female and 30 male patients combined. [5] Synthetic images are not included in our study. [6] The authors stated, "this study included imaging data from referring institutions, it does not approximate to a true multicenter approach" in the limitations. [7] The test group included 17 patients with brain metastases and 17 without enhancing lesions. [8] This study is not included in the quantitative analysis since they did not report the sensitivity. [9] Fifty-nine patients do not have brain metastases. [10] Due to image anonymization, gender information was missing in 49 patients. [11] The test set included 45 patients with brain metastases and 49 patients without. [12] The total dataset contains 30 females and 39 males. [13] The sets included 269, 108, and 529 patients without brain metastases, respectively. [14] The whole dataset included 878 metastatic lesions. [15] The total dataset contains 46 males and 75 females. [16] The total dataset contains 460 males and 474 females. [17] The mean size of the metastases in the entire dataset was 9 mm.

A summary of the scanning details of the studies is shown in Table 3. Eleven studies included both 3.0T and 1.5T scanners [25,26,28,29,32,34,41,42,44,47,48]; two studies included only 1.5T scanners [27,30]; five studies included only 3.0T scanners [37–39,43,46]; two studies included 1.0T, 1.5T, and 3.0T scanners [36,40]. No information was found in five studies regarding their scanners' magnets [14,31,33,35,45]. Ten studies used multiple MRI sequences in their models [14,26,27,32,36,38–42], whereas the remaining fifteen only used one. Two studies used 2D images [30,33], and seven used 2D and 3D images [14,26,27,36,38,40,41]. The remaining sixteen studies only used 3D images. All the studies included at least one sequence with contrast enhancement. With twenty studies, the most common included sequence was 3D contrast-enhanced T1 weighted imaging (WI). The slice thickness of images ranged from 0.43 mm to 7.22 mm (Table 3).

Table 3. Scanning characteristics.

Study	Slice Thickness	Scanning Sequences	Scanner	Tesla
Amemiya et al.—2022 [25]	1 mm	3D CE T1WI	SIEMENS MAGNETOM Skyra; SIEMENS MAGNETOM Avanto; GE Signa EXCITE HDxt x2; GE Premier; GE Signa EXCITE HDxt; GE Signa EXCITE HD; Toshiba Excelart Vantage; Phillips Ingenia CX	3T; 1.5T; 3T; 3T; 1.5T; 1.5T; 1.5T; 3T
Bousabarah et al.—2020 [26]	NR	2D/3D CE T1WI, 2D/3D T2WI, 2D/3D FLAIR	Philips Ingenia; Philips Ingenia; Philips Archieva; Philips Intera	3T; 1.5T; 3T; 1.5T
Charron et al.—2018 [27]	1.02 mm	3D CE T1WI, 2D FLAIR	NR	1.5T
Chartrand et al.—2022 [28]	1, 1.50 or 2 mm	3D CE T1WI	Philips Ingenia Elition; Philips Achieva; Siemens MAGNETOM® Aera; Siemens MAGNETOM® Avanto fit; GE SIGNA™ Explorer; GE Optima™ MR450w GEM; GE Discovery™ MR750; Philips Intera; Siemens MAGNETOM Skyra	3T; 3T; 1.5T; 1.5T; 1.5T; 1.5T; 3T; 1.5T; 3T
Cho et al.—2021 [29]	1 mm	3D CE T1WI	Philips Intera; Philips Achieva; Philips Ingenia; SIEMENS Verio	1.5T; 3T; 3T; 3T
Deike-Hofmann et al.—2021 [30]	4 mm	2D CE T1WI	SIEMENS MAGNETOM Symphony	1.5T
Dikici et al.—2022 [31]	NR	3D CE T1WI	NR	NR
Grøvik et al.—2021 [32]	0.90 to 1.60 mm	3D T1WI, 3D CE T1WI, 3D Black Blood, 3D CE FLAIR	GE TwinSpeed; GE SIGNA Explorer; GE SIGNA Architect; GE Discovery 750 and 750w; SIEMENS Skyra	1.5T; 1.5T; 3T; 3T; 3T
Han et al.—2019 [33]	NR	2D CE T1WI	NR	NR
Hsu et al.—2021 [34]	1 to 1.98 mm	3D CE T1WI	GE Discovery MR750w; GE Optima MR450w; GE Signa PET/MR; GE Signa HDxt; GE Signa Architect; GE Signa Artist; GE Signa Excite; Philips Ingenia; SIEMENS Aera	3T; 1.5T; 3T; 1.5T; 3T; 1.5T; 1.5T; 3T; 1.5T
Huang et al.—2022 [35]	1 mm	3D CE T1WI	NR	NR
Jünger et al.—2021 [36]	2 to 6 mm	2D/3D T1WI, 2D/3D CE T1WI, 2D/3D T2WI, 2D/3D FLAIR	Philips Achieva; GE Optima; Philips Ingenia; Philips Intera; Philips Panorama; Siemens Aera; Siemens Amira; Siemens Avanto; Siemens Espree; Siemens Skyra; Siemens Symphony; Siemens Prisma	1T or 1.5T or 3T
Kikuchi et al.—2022 [37]	2 mm	3D Black Blood	Philips Achieva; Philips Ingenia	3T; 3T
Kottlors et al.—2021 [38]	1 mm, 5 mm	2D CE T1WI; 3D Black Blood	Philips Ingenia	3T
Liang et al.—2022 [14]	0.43 to 7.22 mm	2D/3D CE T1WI; 2D/3D FLAIR	NR (The MR images were acquired on 14 types of scanners from 4 major vendors—Siemens, GE, Philips, and Toshiba).	NR
Park et al.—2021 [39]	1 mm	3D Black Blood, 3D GRE	Philips Achieva; Philips Ingenia; Philips Ingenia CX; Philips Ingenia Elition X	3T; 3T; 3T; 3T
Pennig et al.—2021 [40]	2.30 to 5.20 mm	2D T1WI, 2D/3D CE T1WI, 2D T2WI, 2D FLAIR	Philips Achieva; Philips Gyroscan; Philips Ingenia; Philips Intera; Philips Panorama; SIEMENS Avanto; SIEMENS Biograph; GE Optima; GE Genesis Signa	1T or 1.5T or 3T
Pflüger et al.—2022 [41]	1 to 5 mm	2D T1WI, 3D CE T1WI, 2D FLAIR	SIEMENS Magnetom Verio; SIEMENS Skyra; SIEMENS Trio TIM; SIEMENS Magnetom Avanto	3T; 3T; 3T; 1.5T
Rudie et al.—2021 [42]	1.50 mm	3D T1WI, 3D CE T1WI	GE Signa HDxt; Philips Achieva; GE Discovery MR750	1.5T; 1.5T; 3T
Xue et al.—2020 [43]	1.50 mm	3D CE T1WI	SIEMENS MAGNETOM Skyra	3T
Yin et al.—2022 [44]	1 mm	3D CE T1WI	MAGNETOM Aera; Discovery MR750; Discovery MR750W; SIGNA Pioneer; SIGNA Premier; SIGNA Architect; Ingenia CX; MAGNETOM Trio Tim; MAGNETOM Prisma; uMR560; uMR780; uMR790; Optima MR360; MAGNETOM Skyra; MAGNETOM Verio	1.5T; 3T; 3T; 3T; 3T; 3T; 3T; 3T; 3T; 3T; 3T; 3T; 1.5T; 3T; 3T
Yoo et al.—2022 [45]	1 mm	3D CE T1WI	NR	NR

Table 3. *Cont.*

Study	Slice Thickness	Scanning Sequences	Scanner	Tesla
Yoo et al.—2021 [46]	0.90 mm	3D CE T1WI	SIEMENS MAGNETOM Skyra	3T
Zhang et al.—2020 [47]	0.89 to 3.84 mm	3D CE T1WI	NR	1.5T or 3T
Zhou et al.—2020 [48]	1 mm	3D CE T1WI	GE Signa HDxt; GE Discovery MR750w	1.5T; 3T

Abbreviations: CE, contrast-enhanced; WI, weighted imaging; GE, General Electric; T, Tesla; FLAIR, fluid attenuated inversion recovery; GRE, gradient echo.

3.4. Deep Learning Algorithms

Table 4 provides a summary of detectability statistics, false-positive rates, and deep learning algorithm details. U-Net was used in ten studies [14,26,28–30,39,41,42,45,46], and DeepMedic was used in five [27,35–37,40]. Two studies implemented a single-shot detector [25,48]. Each of the following algorithms was used in a single study: You Only Look Once v3, fully convolution network, convolutional neural network (CNN), feature pyramid network, faster region-based CNN, input-level dropout model, Modified V-Net 3D, CropNet and noisy student. Data augmentation was not implemented in 6 studies [32,37,41,43,46,48], and in 19 studies, it was. False positive rates were reported in studies either per scan, per patient, per slice, per case, or per lesion. In two studies, false positive rates were not reported. Detailed information can be found in Table 4.

Table 4. Algorithms and statistics.

Study	Detectability/Test Level	False Positive Rate	DL Algorithm	Data Augmentation
Studies with lesion-wise sensitivity reporting				
Amemiya et al.—2022 [25]	0.86/lesion-wise	4.3 per scan	Single-shot detector	Yes
Bousabarah et al.—2020 [26]	0.82/lesion-wise	0.35 per lesion	Conventional U-Net and modified U-Net	Yes
Charron et al.—2018 [27]	0.98/lesion-wise	14.2 per patient	DeepMedic	Yes
Chartrand et al.—2022 [28]	0.909/lesion-wise	0.66 per scan	3D U-Net	Yes
Cho et al.—2021 [29]	0.58/lesion-wise	2.5 per scan	3D U-Net	Yes
Han et al.—2019 [33]	0.83/lesion-wise	3.59 per slice	You Only Look Once v3	Yes
Hsu et al.—2021 [34]	0.90/lesion-wise	3.4 per patient	Modified V-Net 3D	Yes
Huang et al.—2022 [35]	0.975/lesion-wise	6.97 per patient *	DeepMedic	Yes
Jünger et al.—2021 [36]	0.851/lesion-wise	1.5 per scan	DeepMedic	Yes
Kikuchi et al.—2022 [37]	0.917/lesion-wise	1.5 per case	DeepMedic	No
Liang et al.—2022 [14]	0.91/lesion-wise	1.7 per scan	U-Net	Yes
Park et al.—2021 [39]	0.931/lesion-wise	0.59 per patient	3D U-Net	Yes
Pennig et al.—2021 [40]	0.88/lesion-wise	0.71 per scan	DeepMedic	Yes
Pflüger et al.—2022 [41]	0.81/lesion-wise	0.87 per scan	nnU-Net	No
Rudie et al.—2021 [42]	0.70/lesion-wise	0.46 per scan	3D U-Net	Yes
Yin et al.—2022 [43]	0.958/lesion-wise	0.39 per scan	FPN	Yes
Yoo et al.—2022 [45]	0.966/lesion-wise	1.25 per patient	2D U-Net	Yes
Yoo et al.—2021 [46]	0.91/lesion-wise	7.67 per patient	3D U-Net	No
Zhang et al.—2020 [47]	0.956/lesion-wise	19.9 per scan	Faster R-CNN *	Yes
Zhou et al.—2020 [48]	0.85/lesion-wise	3 per patient	Single-shot detector	No
Studies with patient-wise or voxel-wise sensitivity reporting				
Deike-Hofmann et al.—2021 [30]	0.727/patient-wise	6.6 per case	U-Net	Yes
Dikici et al.—2022 [31]	0.9/patient-wise	8.44 per patient	CropNet and noisy student	Yes
Grovik et al.—2021 [32]	0.671/voxel-wise	12.3 per lesion	Input-level dropout model	No
Xue et al.—2019 [43]	0.96/voxel-wise	Not reported	3D FCN	No
Studies not included in the quantitative analysis				
Kottlors et al.—2021 [38]	NR	NR	CNN	Yes

* Calculated using the study's presented data.

3.5. Assessment of Detectability Performance

In our study, we included the internal test set results or the cross-validation results in the pooled detectability analysis. If more than one algorithm was used in a study, we utilized the best algorithm in our pooled analysis. Furthermore, if more than one level of input was present, the one that yielded the most successful outcome was included in our study.

3.5.1. Patient-Wise

Twenty studies evaluated and reported their detectability lesion-wise, whereas the remaining four assessed and reported their detectability voxel-wise or patient-wise. The detectability of the 24 included studies ranged from 58% to 98%. The pooled proportion of detectability (patient-wise) of deep learning algorithms in all 24 included studies was 89% (95% CI, 84–92%) (Figure 4). The meta-regression did not find a significant impact of the training sample size on the sensitivity ($p = 0.12$). The Q-test indicated that heterogeneity was present across the studies ($Q = 111.13$, $p < 0.01$), and the Higgins I^2 statistic showed the presence of high heterogeneity in detectability ($I^2 = 79\%$).

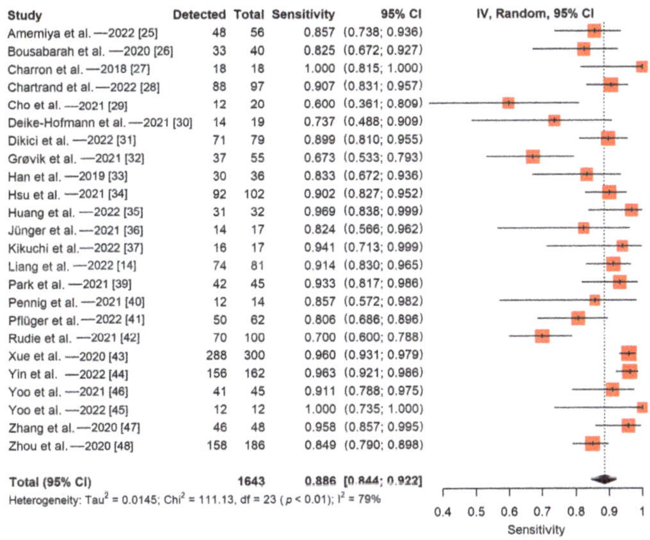

Figure 4. Forest plot of deep learning algorithms' patient-wise detectability.

There was no statistically significant difference in performance between the groups in any subgroup analyses (p-values ranged from 0.08 to 0.69). Heterogeneity was moderate or high in subgroups that are separated based on the plurality of MRI sequences utilized (single MRI sequence; $Q = 50.86$, $p < 0.01$; $I^2 = 72\%$/multiple MRI sequences; $Q = 33.09$, $p < 0.01$; $I^2 = 76\%$), validation method (split training-test; $Q = 83.18$, $p < 0.01$; $I^2 = 76\%$/cross-validation; $Q = 11.63$, $p < 0.01$; $I^2 = 83\%$), study design (single-center; $Q = 52.65$, $p < 0.01$; $I^2 = 66\%$/multi-center; $Q = 36.83$, $p < 0.01$; $I^2 = 89\%$), and the calculation method of the reported sensitivity (lesion-wise; $Q = 68.78$, $p < 0.01$; $I^2 = 72\%$/others; $Q = 37.60$, $p < 0.01$; $I^2 = 92\%$). Subgroup analyses based on the plurality of the primary tumor type revealed a high heterogeneity in studies with multiple primary tumor types ($Q = 105.92$, $p < 0.01$; $I^2 = 81\%$). There was no evidence of heterogeneity in studies with a single primary tumor type ($Q = 0.66$, $p = 0.72$; $I^2 = 0\%$). Furthermore, subgroup analyses based on the dimension of the included images revealed high heterogeneity in studies with 3D images ($Q = 94.2$, $p < 0.01$; $I^2 = 84\%$). There was no evidence of heterogeneity in studies with a mixture of 2D and 3D images ($Q = 9.64$, $p = 0.09$; $I^2 = 48\%$) and studies with 2D images ($Q = 0.72$, $p = 0.39$; $I^2 = 0\%$). The meta-regression revealed that the heterogeneity not explained by the training sample size was significant, indicating that the training sample size did not influence heterogeneity significantly ($p < 0.0001$).

The funnel plot was asymmetrical, indicating publication bias among the included studies (Figure 5). Furthermore, not all studies were plotted within the area under the curve of the pseudo-95% CI, showing a possible publication bias [50]. However, the Egger test did not indicate obvious publication bias (regression intercept = 1.32, $p = 0.15$).

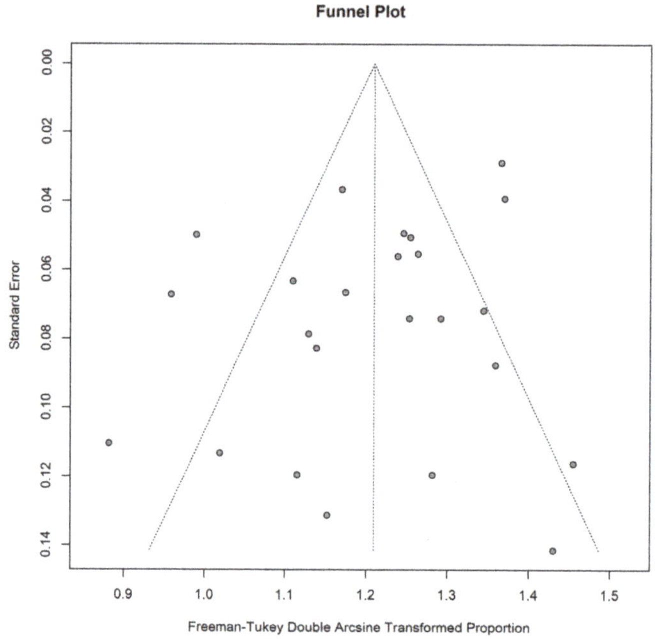

Figure 5. Funnel plot for studies included in the pooled analysis for patient-wise detectability.

3.5.2. Lesion-Wise

Among 20 studies evaluated and reported their detectability lesion-wise, 18 of them reported the number of metastases in the test sets. All studies in this group used split training-test sets to evaluate their models. The detectability of the 18 included studies ranged from 58% to 98%. The pooled proportion of detectability of deep learning algorithms (lesion-wise) was 89% (95% CI, 83–93%) (Figure 6). The meta-regression did not find a significant impact of the training sample size on the sensitivity ($p = 0.86$). The Q-test indicated that heterogeneity was present across the studies ($Q = 551.71$, $p < 0.01$), and the Higgins I^2 statistic showed the presence of high heterogeneity in detectability ($I^2 = 97\%$).

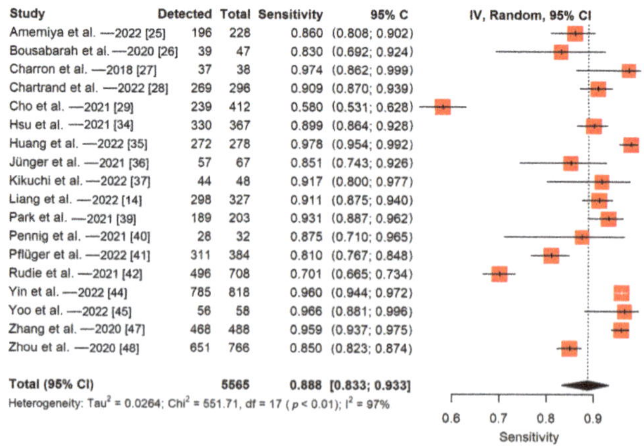

Figure 6. Forest plot of deep learning algorithms' lesion-wise detectability.

There was no statistically significant difference in performance between the groups in any subgroup analyses (p-values ranged from 0.26 to 0.68). Heterogeneity was high in subgroups that were separated based on the plurality of MRI sequences (single MRI sequence; Q = 370.72, $p < 0.01$; $I^2 = 98\%$/multiple MRI sequences; Q = 112.75, $p < 0.01$; $I^2 = 94\%$), study design (single-center; Q = 439.53, $p < 0.01$; $I^2 = 97\%$/multi-center; Q = 17.01, $p < 0.01$; $I^2 = 88\%$), and the dimension of the included images (3D images; Q = 529.39, $p < 0.01$; $I^2 = 98\%$/both 2D and 3D images; Q = 21.47, $p < 0.01$; $I^2 = 77\%$). Subgroup analyses based on the plurality of the primary tumor type revealed a high heterogeneity in studies with multiple primary tumor types (Q = 551.3, $p < 0.01$; $I^2 = 97\%$). There was no evidence of heterogeneity in studies with a single primary tumor type (Q = 0.06, $p = 0.81$; $I^2 = 0\%$). In this group, two subgroup analyses to investigate heterogeneity could not be carried out: based on validation method (all studies used split training-test method) and detection level (all studies reported lesion-wise sensitivity). The meta-regression showed that the heterogeneity not explained by the training sample size was significant, indicating that the training sample size had no significant effect on heterogeneity ($p < 0.0001$).

The asymmetrical funnel plot indicated publication bias among the included studies in this group (Figure 7). In addition, only some studies were plotted within the area under the curve of the pseudo-95% CI, indicating possible publication bias [50]. On the other hand, the Egger test revealed no obvious publication bias (regression intercept = 1.17, $p = 0.66$).

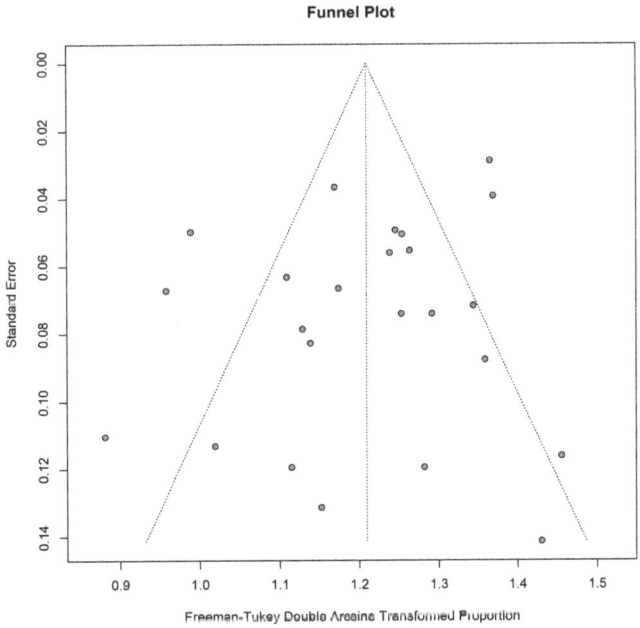

Figure 7. Funnel plot for studies included in the pooled analysis for lesion-wise detectability.

4. Discussion

BMs are ten times more common than primary malignant brain tumors, and patients with a history of BM should be followed up by imaging every three months and whenever clinically indicated [51,52]. Therefore, there is a massive demand for radiologists to detect and follow-up on these lesions. However, radiologists face several challenges in detecting BMs accurately. Among the challenges are a massive workload, difficulties in differentiation of BMs from noises and blood vessels, BMs with small sizes, significant variations in lesion shape and size among patients, weak signal intensities, and multiple locations and lesions in a patient [11]. As a result of these challenges, the use of deep learning methods in detecting BMs has recently increased in research. Because the recent meta-analysis by

Cho et al. included just 12 studies, only 7 of which were deep learning studies, we decided to conduct a systematic review and assess the strength of the present evidence [11]. Their study also found a statistically significant difference in false positives between the deep learning and machine learning groups, necessitating the focused study on each group. Our analysis showed that studies with deep learning models using MR images perform well in detecting BMs, with a pooled sensitivity of 89% (95% CI, 84–92%). The detectability ranged from 58% to 98%.

The size of the lesions is an important determinant of the success of detecting BMs with deep learning algorithms [53]. For instance, in a study by Rudie et al., detection sensitivity for BMs smaller than 3 mm was only 14.6%, and sensitivity for BMs greater than 3 mm was 84.3% [42]. In another study, the detection sensitivity was reduced by 71% for BMs < 3 mm compared to all BMs [48]. Various approaches to solving this problem have been proposed in the literature. Amemiya et al. reinforced their SSD model for small lesion detections with feature fusion and managed to increase its sensitivity from 35.4% to 45.8% for lesions smaller than 3 mm [25]. In another study, Bousabarah et al. showed that the sensitivity of their models for small BMs ranged from 0.40 to 0.51 [26]. They trained another model on a subsample containing only small BMs (smaller than 0.4 mL), showing an increased sensitivity of 0.62–0.68 for small BMs. It is worth noting that this model's sensitivity was 0.43–0.53 for all sizes of BMs. As stated, Dikici et al. used a sample of smaller BMs and acquired results similar to theirs [54]. Although training deep learning models with small BMs would help to improve sensitivity in detecting BMs, there are concerns when applying this method to a heterogeneous group of BMs. Furthermore, since a meta-analysis demonstrated that black blood images successfully detect small lesions [55], Park et al. added black blood images to their model with gradient echo sequence images and increased their sensitivity by 23.5% for detecting BMs < 3 mm [39]. Kottlors et al. also showed that smaller BMs (<5 mm) could be detected better with deep learning models using black blood images [38]. Yin et al. detected BMs < 3 mm with an impressive sensitivity of 89% by only using contrast-enhanced T1WI, and having a sensitivity of 95.8% in all sizes of BMs [44]. We believe, in part, this success was due to a large number of metastases in the cohort (11514 BMs) and a high proportion of small BMs (<5 mm, 58.6%) in their sample. Therefore, more models that are trained with real-world data representation and successful on both small and large lesions are required. Moreover, surgery or SRS planning scan with a higher spatial resolution can be more sensitive for detecting metastases than a conventional scan, revealing lesions not previously detected [56]. Therefore, utilizing MRI scanners with a higher spatial resolution might improve the efficacy of deep learning models, particularly for smaller lesions. Furthermore, we also analyzed the impact of training sample size on detectability, but no statistically significant effect was observed. Studies in the literature demonstrate that the success of deep learning models typically increases with sample size [57,58]. Although our results contradict this, one possible explanation could be that even though some papers have a small number of patients, the number of lesions in such a small number of patients can be quite high. For instance, Charron et al. included 164 patients with 374 metastatic lesions, whereas Deike-Hofmann et al. included 43 patients with 494 metastatic lesions. However, training sample size (number of lesions) on detectability in the group reporting sensitivity lesions-wise was not statistically significant either, indicating further research is needed.

Various medical imaging modalities have unique characteristics and different reactions to different tissues in the human body. Our review showed that ten studies used multiple MRI sequences in their models whereas the remaining fifteen only used one. Our meta-analysis showed that there was no statistically significant difference between the pooled proportion of detectability of studies with a single MRI sequence and the pooled proportion of detectability of studies with multiple MRI sequences. Park et al. combined and separately used two MRI sequences in their model [39]. Gradient echo sequence and black blood images showed 76.8% and 92.6% sensitivity, respectively. Their combination showed 93.1% sensitivity. There was a statistically significant difference between the models using only gradient echo sequence and combined sequences, but no difference between the black blood model and the combined model. This was also evident in the models' sensitivity in lesions smaller than 3 mm, with the gradient echo sequence model, black blood model, and the combined model showing a sensitivity of 23.5%, 82.4%, and 82.4%, respectively. Charron et al. showed that the model combining different MRI modalities surpassed the model using single modalities in their study [27]. Contrary to these findings, our subgroup analyses showed no statistically significant difference between models using single or multiple MRI sequences. Therefore, additional research would be needed to determine which MRI scanners and scanner combinations would be the optimal strategy for automatically detecting brain metastases with deep learning. Furthermore, there was no statistically significant difference in the performance of models that included only 2D, 3D, or a combination of them; however, this comparison is not ideal due to the small number of studies that included only 2D images and studies that included both 2D and 3D images.

Most BMs are caused by lung cancer, breast cancer, melanoma, and renal cell carcinoma [2]. Our review showed that two articles included only malignant melanoma patients, and one included NSCLC patients. Others included BMs patients with any primary tumor. The appearance of BMs on MRI may vary between primary tumors, particularly in NSCLC [36,59]. Due to that, Jünger et al. stated that different deep learning models for different BMs originating from different primary tumors should be designed to obtain a satisfactory detection performance [36]. Although this may be helpful to increase the performance of the model detecting BMs originating from the particular primary tumor, the primary tumor is unknown in 5–40% of patients demonstrating the symptoms of BMs [60]. Furthermore, the primary tumor in up to 15% of BM patients cannot be identified [61]. Therefore, generalized deep learning models might be required to detect BMs with unknown primary tumors. Furthermore, we performed subgroup analyses to compare the performance of studies with a single primary tumor type and studies with multiple primary tumor types. There was no statistically significant difference in their performance; however, this comparison was not ideal due to the small number of studies with a single primary tumor type compared to studies with multiple primary tumor types. This is still an open research area, and further studies would be needed.

Besides sensitivity, false positives were the other important measure reported by the included studies. An important cause of false positives was the similarity between BMs and blood vessels, as both may present with a small focus of hyperintensity on a contrast-enhanced T1WI [46]. Several studies showed that most or at least some of their false positives were in and near vascular structures [25,32,37,39,44,46,47]. There have been several proposed solutions to this problem in the literature. Grøvik et al. hypothesized that adding other MRI sequences, such as diffusion-weighted MRI, may help deal with false positives [32]. Furthermore, black blood imaging can be helpful since it can suppress blood vessel signals, which is also increasingly applied in clinical practice [25,38,39,44]. It is worth noting that Park et al. demonstrated that when only black blood imaging is used, false positives are significantly higher than when black blood and gradient echo sequences images are used together [39]. Furthermore, Deike-Hofmann et al. demonstrated that including a pre-diagnosis scan in their model greatly reduced false positives, but including additional sequences contrarily decreased the specificity [30]. They hypothesized that this result was due to models interpreting that lesions that change over time are more likely to be BMs, whereas stable structures such as blood vessels are less likely, as humans do. Finally, skull stripping was another method that significantly reduced false positives, particularly extra-axial ones [14,25,42]. We were unable to conduct a pooled analysis of false positive rates in our study since included studies reported false positive rates differently.

Various deep learning algorithms were applied to MR images to detect BMs. Our review revealed that the most commonly used deep learning algorithm in BMs detection was U-Net, with different versions [62]. U-Net is an algorithm for semantic segmentation, also known as pixel-based classification. A contracting encoder and an expanding decoder comprise the U-Net. The expanding decoder creates the label map after the contracting encoder extracts low-level features. The second most commonly used deep learning algorithm was DeepMedic (Biomedical Image Analysis Group, Department of Computing Imperial College London, London, UK) [63]. DeepMedic is built around a 3D deep CNN and a 3D conditional random field. Unsurprisingly, both algorithms are the most widely used since they are easily accessible. They are likely to be used more in the future, and there is room for research into combining these models in BMs detection [64]. The performance of the deep learning models was also compared. However, there was no statistically significant difference between the patient-wise sensitivity group and the lesion-wise sensitivity group ($p = 0.25$ and $p = 0.26$, respectively).

The performance of the deep learning models should be compared to the performance of radiologists in detecting BMs. Kikuchi et al. compared their model with twelve radiologists; seven were board-certified, and five were residents [37]. Their model exhibited higher sensitivity (91.7%) than the radiologists (88.7 $\pm$ 3.7%), however the article did not compare how the algorithms performed compared to faculty versus trainees, and the lumping of the trainees with the faculty may explain the overall lower sensitivity of the radiologists. Rudie et al. compared the performance of two initial manual annotations with the deep learning model [42]. Contrary to the findings of the aforementioned study, the detection sensitivity was higher and statistically significant for radiologists than for the deep learning model for metastases smaller than 3 mm (63.2% versus 14.7%) and metastases between 3 and 6 mm (90.8% versus 76.7%). Liang et al. discovered that their deep learning model detected 2% of BMs that were overlooked during the manual annotation, even though the manual annotations were checked by two investigators and reevaluated by a senior radiation oncologist [14]. In addition, Yin et al. showed that with the assistance of the deep learning model, readers' mean sensitivity increased by 21% [44]. Another meta-analysis that compared deep learning models' performance with healthcare professionals in detecting diseases from medical images found that the diagnostic performance of deep learning models and healthcare professionals was equivalent [65]. We believe these results are auspicious, and although completely automated models cannot be fully implemented into clinical practice today, they may serve as a robust assistant for radiologists. We could not conduct a pooled analysis to compare deep learning models and radiologists in our study due to the small number of studies reporting radiologists' sensitivity. The literature is lacking in this regard, and more papers comparing the two groups are needed, especially for small, difficult to detect lesions.

Our study was not without limitations. The main limitation was the lack of pooled false positive rate analysis since there was no uniform unit of reporting them in the included studies. Studies reported false positives in per-patient or per-scan or per-lesion. Second, subgroup analyses based on different MRI scanners and slice thicknesses were not possible because the results for different scanners and slice thicknesses were not included in any of the included studies. Furthermore, study heterogeneity was high in our analyses; but it was commonly observed in meta-analyses on imaging-based deep learning studies [13,66–69]. In addition, in our study, the subgroups with no evidence of heterogeneity included very few studies. However, it is known that Q test has inadequate power to detect true heterogeneity when the meta-analysis includes a small number of studies [70]. Therefore, it is likely that heterogeneity is not due to subgroups. It is worth noting that, even though it is not a limitation of our study, our quality assessment showed that reporting was poor in some studies.

5. Conclusions

Our study revealed that deep learning algorithms effectively detect BMs with a pooled sensitivity of 89%. In the future, deep learning studies should adhere to CLAIM and QUADAS-2 checklists more strictly. Uniform reporting standards that clearly explain the results of deep learning models in BMs detection are needed. Since all the included studies were conducted retrospectively, there is a need for additional large-scale prospective studies.

Supplementary Materials: The following supporting information can be downloaded at: https://www.mdpi.com/article/10.3390/cancers15020334/s1, Table S1: PRISMA-DTA Abstract Checklist; Table S2: PRISMA-DTA Checklist; Table S3: Inclusion and exclusion criteria.

Author Contributions: Conceptualization, B.B.O., M.M.C., M.K. and M.W.; methodology, B.B.O. and M.K.; software, B.B.O. and M.K.; validation, M.M.C., C.F. and M.W.; formal analysis, B.B.O., C.F. and M.K.; investigation, B.B.O., C.F. and M.K.; resources, T.M.B., J.L. and M.W.; data curation, B.B.O.; writing—original draft preparation, B.B.O., M.M.C. and M.K.; writing—review and editing, C.F., T.M.B., J.L. and M.W.; visualization, B.B.O. and M.K.; supervision, M.M.C., T.M.B., J.L. and M.W.; project administration, M.W. All authors have read and agreed to the published version of the manuscript.

Funding: This research received no external funding.

Conflicts of Interest: The authors declare no conflict of interest.

References

1. Lin, X.; DeAngelis, L.M. Treatment of Brain Metastases. *J. Clin. Oncol.* **2015**, *33*, 3475–3484. [CrossRef]
2. Sacks, P.; Rahman, M. Epidemiology of Brain Metastases. *Neurosurg. Clin. N. Am.* **2020**, *31*, 481–488. [CrossRef]
3. Nayak, L.; Lee, E.Q.; Wen, P.Y. Epidemiology of Brain Metastases. *Curr. Oncol. Rep.* **2012**, *14*, 48–54. [CrossRef]
4. Suh, J.H.; Kotecha, R.; Chao, S.T.; Ahluwalia, M.S.; Sahgal, A.; Chang, E.L. Current Approaches to the Management of Brain Metastases. *Nat. Rev. Clin. Oncol.* **2020**, *17*, 279–299. [CrossRef]
5. Soffietti, R.; Abacioglu, U.; Baumert, B.; Combs, S.E.; Kinhult, S.; Kros, J.M.; Marosi, C.; Metellus, P.; Radbruch, A.; Villa Freixa, S.S.; et al. Diagnosis and Treatment of Brain Metastases from Solid Tumors: Guidelines from the European Association of Neuro-Oncology (EANO). *Neuro-Oncol.* **2017**, *19*, 162–174. [CrossRef]
6. Sze, G.; Milano, E.; Johnson, C.; Heier, L. Detection of Brain Metastases: Comparison of Contrast-Enhanced MR with Unenhanced MR and Enhanced CT. *AJNR Am. J. Neuroradiol.* **1990**, *11*, 785–791.
7. Davis, P.C.; Hudgins, P.A.; Peterman, S.B.; Hoffman, J.C. Diagnosis of Cerebral Metastases: Double-Dose Delayed CT vs. Contrast-Enhanced MR Imaging. *AJNR Am. J. Neuroradiol.* **1991**, *12*, 293–300.
8. Yamamoto, M.; Serizawa, T.; Shuto, T.; Akabane, A.; Higuchi, Y.; Kawagishi, J.; Yamanaka, K.; Sato, Y.; Jokura, H.; Yomo, S.; et al. Stereotactic Radiosurgery for Patients with Multiple Brain Metastases (JLGK0901): A Multi-Institutional Prospective Observational Study. *Lancet Oncol.* **2014**, *15*, 387–395. [CrossRef]
9. Gondi, V.; Bauman, G.; Bradfield, L.; Burri, S.H.; Cabrera, A.R.; Cunningham, D.A.; Eaton, B.R.; Hattangadi-Gluth, J.A.; Kim, M.M.; Kotecha, R.; et al. Radiation Therapy for Brain Metastases: An ASTRO Clinical Practice Guideline. *Pract. Radiat. Oncol.* **2022**, *12*, 265–282. [CrossRef]
10. Growcott, S.; Dembrey, T.; Patel, R.; Eaton, D.; Cameron, A. Inter-Observer Variability in Target Volume Delineations of Benign and Metastatic Brain Tumours for Stereotactic Radiosurgery: Results of a National Quality Assurance Programme. *Clin. Oncol.* **2020**, *32*, 13–25. [CrossRef]
11. Cho, S.J.; Sunwoo, L.; Baik, S.H.; Bae, Y.J.; Choi, B.S.; Kim, J.H. Brain Metastasis Detection Using Machine Learning: A Systematic Review and Meta-Analysis. *Neuro-Oncol.* **2021**, *23*, 214–225. [CrossRef]
12. Chan, H.-P.; Samala, R.K.; Hadjiiski, L.M.; Zhou, C. Deep Learning in Medical Image Analysis. *Adv. Exp. Med. Biol.* **2020**, *1213*, 3–21. [CrossRef]
13. Karabacak, M.; Ozkara, B.B.; Mordag, S.; Bisdas, S. Deep Learning for Prediction of Isocitrate Dehydrogenase Mutation in Gliomas: A Critical Approach, Systematic Review and Meta-Analysis of the Diagnostic Test Performance Using a Bayesian Approach. *Quant. Imaging Med. Surg.* **2022**, *12*, 4033–4046. [CrossRef]
14. Liang, Y.; Lee, K.; Bovi, J.A.; Palmer, J.D.; Brown, P.D.; Gondi, V.; Tome, W.A.; Benzinger, T.L.S.; Mehta, M.P.; Li, X.A. Deep Learning-Based Automatic Detection of Brain Metastases in Heterogenous Multi-Institutional Magnetic Resonance Imaging Sets: An Exploratory Analysis of NRG-CC001. *Int. J. Radiat. Oncol. Biol. Phys.* **2022**, *114*, 529–536. [CrossRef] [PubMed]
15. McInnes, M.D.F.; Moher, D.; Thombs, B.D.; McGrath, T.A.; Bossuyt, P.M.; the PRISMA-DTA Group; Clifford, T.; Cohen, J.F.; Deeks, J.J.; Gatsonis, C.; et al. Preferred Reporting Items for a Systematic Review and Meta-Analysis of Diagnostic Test Accuracy Studies: The PRISMA-DTA Statement. *JAMA* **2018**, *319*, 388–396. [CrossRef]
16. Ouzzani, M.; Hammady, H.; Fedorowicz, Z.; Elmagarmid, A. Rayyan—A Web and Mobile App for Systematic Reviews. *Syst. Rev.* **2016**, *5*, 210. [CrossRef] [PubMed]
17. Mongan, J.; Moy, L.; Kahn, C.E. Checklist for Artificial Intelligence in Medical Imaging (CLAIM): A Guide for Authors and Reviewers. *Radiol. Artif. Intell.* **2020**, *2*, e200029. [CrossRef] [PubMed]
18. Whiting, P.F.; Rutjes, A.W.S.; Westwood, M.E.; Mallett, S.; Deeks, J.J.; Reitsma, J.B.; Leeflang, M.M.G.; Sterne, J.A.C.; Bossuyt, P.M.M. QUADAS-2 Group QUADAS-2: A Revised Tool for the Quality Assessment of Diagnostic Accuracy Studies. *Ann. Intern. Med.* **2011**, *155*, 529–536. [CrossRef] [PubMed]
19. Dettori, J.R.; Norvell, D.C.; Chapman, J.R. Fixed-Effect vs Random-Effects Models for Meta-Analysis: 3 Points to Consider. *Global Spine J.* **2022**, *12*, 1624–1626. [CrossRef]

20. Borenstein, M.; Hedges, L.V.; Higgins, J.P.T.; Rothstein, H.R. A Basic Introduction to Fixed-Effect and Random-Effects Models for Meta-Analysis. *Res. Synth. Methods* **2010**, *1*, 97–111. [CrossRef]
21. Higgins, J.P.T.; Thompson, S.G.; Deeks, J.J.; Altman, D.G. Measuring Inconsistency in Meta-Analyses. *BMJ* **2003**, *327*, 557–560. [CrossRef] [PubMed]
22. Groenwold, R.H.H.; Rovers, M.M.; Lubsen, J.; van der Heijden, G.J. Subgroup Effects despite Homogeneous Heterogeneity Test Results. *BMC Med. Res. Methodol.* **2010**, *10*, 43. [CrossRef] [PubMed]
23. Egger, M.; Davey Smith, G.; Schneider, M.; Minder, C. Bias in Meta-Analysis Detected by a Simple, Graphical Test. *BMJ* **1997**, *315*, 629–634. [CrossRef] [PubMed]
24. Balduzzi, S.; Rücker, G.; Schwarzer, G. How to Perform a Meta-Analysis with R: A Practical Tutorial. *Evid. Based Ment. Health* **2019**, *22*, 153–160. [CrossRef]
25. Amemiya, S.; Takao, H.; Kato, S.; Yamashita, H.; Sakamoto, N.; Abe, O. Feature-Fusion Improves MRI Single-Shot Deep Learning Detection of Small Brain Metastases. *J. Neuroimaging* **2022**, *32*, 111–119. [CrossRef]
26. Bousabarah, K.; Ruge, M.; Brand, J.-S.; Hoevels, M.; Rues, D.; Borggrefe, J.; Grose Hokamp, N.; Visser-Vandewalle, V.; Maintz, D.; Treuer, H.; et al. Deep Convolutional Neural Networks for Automated Segmentation of Brain Metastases Trained on Clinical Data. *Radiat. Oncol.* **2020**, *15*, 87. [CrossRef]
27. Charron, O.; Lallement, A.; Jarnet, D.; Noblet, V.; Clavier, J.B.; Meyer, P. Automatic Detection and Segmentation of Brain Metastases on Multimodal MR Images with a Deep Convolutional Neural Network. *Comput. Biol. Med.* **2018**, *95*, 43–54. [CrossRef]
28. Chartrand, G.; Emiliani, R.D.; Pawlowski, S.A.; Markel, D.A.; Bahig, H.; Cengarle-Samak, A.; Rajakesari, S.; Lavoie, J.; Ducharme, S.; Roberge, D. Automated Detection of Brain Metastases on T1-Weighted MRI Using a Convolutional Neural Network: Impact of Volume Aware Loss and Sampling Strategy. *J. Magn. Reson. Imaging* **2022**, *56*, 1885–1898. [CrossRef]
29. Cho, J.; Kim, Y.J.; Sunwoo, L.; Lee, G.P.; Nguyen, T.Q.; Cho, S.J.; Baik, S.H.; Bae, Y.J.; Choi, B.S.; Jung, C.; et al. Deep Learning-Based Computer-Aided Detection System for Automated Treatment Response Assessment of Brain Metastases on 3D MRI. *Front. Oncol.* **2021**, *11*, 739639. [CrossRef]
30. Deike-Hofmann, K.; Dancs, D.; Paech, D.; Schlemmer, H.P.; Maier-Hein, K.; Baumer, P.; Radbruch, A.; Gotz, M. Pre-Examinations Improve Automated Metastases Detection on Cranial MRI. *Investig. Radiol.* **2021**, *56*, 320–327. [CrossRef]
31. Dikici, E.; Nguyen, X.V.; Bigelow, M.; Ryu, J.L.; Prevedello, L.M. Advancing Brain Metastases Detection in T1-Weighted Contrast-Enhanced 3D MRI Using Noisy Student-Based Training. *Diagnostics* **2022**, *12*, 2023. [CrossRef] [PubMed]
32. Grovik, E.; Yi, D.; Iv, M.; Tong, E.; Nilsen, L.B.; Latysheva, A.; Saxhaug, C.; Jacobsen, K.D.; Hell, A.; Emblem, K.E.; et al. Handling Missing MRI Sequences in Deep Learning Segmentation of Brain Metastases: A Multicenter Study. *NPJ Digit. Med.* **2021**, *4*, 33. [CrossRef] [PubMed]
33. Han, C.; Murao, K.; Noguchi, T.; Kawata, Y.; Uchiyama, F.; Rundo, L.; Nakayarna, H.; Satoh, S. Learning More with Less: Conditional PGGAN-Based Data Augmentation for Brain Metastases Detection Using Highly-Rough Annotation on MR Images. In Proceedings of the 28th ACM International Conference on Information and Knowledge Management (CIKM), Beijing, China, 3–7 November 2019; pp. 119–127. [CrossRef]
34. Hsu, D.G.; Ballangrud, A.; Shamseddine, A.; Deasy, J.O.; Veeraraghavan, H.; Cervino, L.; Beal, K.; Aristophanous, M. Automatic Segmentation of Brain Metastases Using T1 Magnetic Resonance and Computed Tomography Images. *Phys. Med. Biol.* **2021**, *66*, 175014. [CrossRef] [PubMed]
35. Huang, Y.X.; Bert, C.; Sommer, P.; Frey, B.; Gaipl, U.; Distel, L.V.; Weissmann, T.; Uder, M.; Schmidt, M.A.; Dorfler, A.; et al. Deep Learning for Brain Metastasis Detection and Segmentation in Longitudinal MRI Data. *Med. Phys.* **2022**, *49*, 5773–5786. [CrossRef] [PubMed]
36. Jünger, S.T.; Hoyer, U.C.I.; Schaufler, D.; Laukamp, K.R.; Goertz, L.; Thiele, F.; Grunz, J.-P.; Schlamann, M.; Perkuhn, M.; Kabbasch, C.; et al. Fully Automated MR Detection and Segmentation of Brain Metastases in Non-Small Cell Lung Cancer Using Deep Learning. *J. Magn. Reson. Imaging* **2021**, *54*, 1608–1622. [CrossRef] [PubMed]
37. Kikuchi, Y.; Togao, O.; Kikuchi, K.; Momosaka, D.; Obara, M.; Van Cauteren, M.; Fischer, A.; Ishigami, K.; Hiwatashi, A. A Deep Convolutional Neural Network-Based Automatic Detection of Brain Metastases with and without Blood Vessel Suppression. *Eur. Radiol.* **2022**, *32*, 2998–3005. [CrossRef] [PubMed]
38. Kottlors, J.; Geissen, S.; Jendreizik, H.; Grose Hokamp, N.; Fervers, P.; Pennig, L.; Laukamp, K.; Kabbasch, C.; Maintz, D.; Schlamann, M.; et al. Contrast-Enhanced Black Blood MRI Sequence Is Superior to Conventional T1 Sequence in Automated Detection of Brain Metastases by Convolutional Neural Networks. *Diagnostics* **2021**, *11*, 1016. [CrossRef]
39. Park, Y.W.; Jun, Y.; Lee, Y.; Han, K.; An, C.; Ahn, S.S.; Hwang, D.; Lee, S.K. Robust Performance of Deep Learning for Automatic Detection and Segmentation of Brain Metastases Using Three-Dimensional Black-Blood and Three-Dimensional Gradient Echo Imaging. *Eur. Radiol.* **2021**, *31*, 6686–6695. [CrossRef]
40. Pennig, L.; Shahzad, R.; Caldeira, L.; Lennartz, S.; Thiele, F.; Goertz, L.; Zopfs, D.; Meisner, A.K.; Furtjes, G.; Perkuhn, M.; et al. Automated Detection and Segmentation of Brain Metastases in Malignant Melanoma: Evaluation of a Dedicated Deep Learning Model. *AJNR Am. J. Neuroradiol.* **2021**, *42*, 655–662. [CrossRef]
41. Pfluger, I.; Wald, T.; Isensee, F.; Schell, M.; Meredig, H.; Schlamp, K.; Bernhardt, D.; Brugnara, G.; Heusel, C.P.; Debus, J.; et al. Automated Detection and Quantification of Brain Metastases on Clinical MRI Data Using Artificial Neural Networks. *Neuro-Oncol. Adv.* **2022**, *4*, vdac138. [CrossRef]

42. Rudie, J.D.; Weiss, D.A.; Colby, J.B.; Rauschecker, A.M.; Laguna, B.; Braunstein, S.; Sugrue, L.P.; Hess, C.P.; Villanueva-Meyer, J.E. Three-Dimensional U-Net Convolutional Neural Network for Detection and Segmentation of Intracranial Metastases. *Radiol.-Artif. Intell.* **2021**, *3*, e200204. [CrossRef] [PubMed]
43. Xue, J.; Wang, B.; Ming, Y.; Jiang, Z.; Wang, C.; Liu, X.; Chen, L.; Qu, J.; Xu, S.; Tang, X.; et al. Deep Learning-Based Detection and Segmentation-Assisted Management of Brain Metastases. *Neuro-Oncol.* **2020**, *22*, 505–514. [CrossRef] [PubMed]
44. Yin, S.H.; Luo, X.; Yang, Y.D.; Shao, Y.; Ma, L.D.; Lin, C.P.; Yang, Q.X.; Wang, D.L.; Luo, Y.W.; Mai, Z.J.; et al. Development and Validation of a Deep-Learning Model for Detecting Brain Metastases on 3D Post-Contrast MRI: A Multi-Center Multi-Reader Evaluation Study. *Neuro-Oncol.* **2022**, *24*, 1559–1570. [CrossRef] [PubMed]
45. Yoo, S.K.; Kim, T.H.; Kim, H.J.; Yoon, H.I.; Kim, J.S. Deep Learning-Based Automatic Detection and Segmentation of Brain Metastases for Stereotactic Ablative Radiotherapy Using Black-Blood Magnetic Resonance Imaging. *Int. J. Radiat. Oncol. Biol. Phys.* **2022**, *114*, e558. [CrossRef]
46. Yoo, Y.J.; Ceccaldi, P.; Liu, S.Q.; Re, T.J.; Cao, Y.; Balter, J.M.; Gibson, E. Evaluating Deep Learning Methods in Detecting and Segmenting Different Sizes of Brain Metastases on 3D Post-Contrast T1-Weighted Images. *J. Med. Imaging* **2021**, *8*, 037001. [CrossRef]
47. Zhang, M.; Young, G.S.; Chen, H.; Li, J.; Qin, L.; McFaline-Figueroa, J.R.; Reardon, D.A.; Cao, X.H.; Wu, X.; Xu, X.Y. Deep-Learning Detection of Cancer Metastases to the Brain on MRI. *J. Magn. Reson. Imaging* **2020**, *52*, 1227–1236. [CrossRef]
48. Zhou, Z.J.; Saners, J.W.; Johnson, J.M.; Gule-Monroe, M.; Chen, M.; Briere, T.M.; Wang, Y.; Son, J.B.; Pagel, M.D.; Ma, J.; et al. MetNet: Computer-Aided Segmentation of Brain Metastases in Post-Contrast T1-Weighted Magnetic Resonance Imaging. *Radiother. Oncol.* **2020**, *153*, 189–196. [CrossRef]
49. Ellingson, B.M.; Bendszus, M.; Boxerman, J.; Barboriak, D.; Erickson, B.J.; Smits, M.; Nelson, S.J.; Gerstner, E.; Alexander, B.; Goldmacher, G.; et al. Consensus Recommendations for a Standardized Brain Tumor Imaging Protocol in Clinical Trials. *Neuro-Oncol.* **2015**, *17*, 1188–1198. [CrossRef]
50. Cochrane Handbook for Systematic Reviews of Interventions. Available online: https://training.cochrane.org/handbook (accessed on 17 October 2022).
51. Ostrom, Q.T.; Wright, C.H.; Barnholtz-Sloan, J.S. Brain Metastases: Epidemiology. *Handb. Clin. Neurol.* **2018**, *149*, 27–42. [CrossRef]
52. Le Rhun, E.; Guckenberger, M.; Smits, M.; Dummer, R.; Bachelot, T.; Sahm, F.; Galldiks, N.; de Azambuja, E.; Berghoff, A.S.; Metellus, P.; et al. EANO-ESMO Clinical Practice Guidelines for Diagnosis, Treatment and Follow-up of Patients with Brain Metastasis from Solid Tumours. *Ann. Oncol.* **2021**, *32*, 1332–1347. [CrossRef]
53. Zhou, Z.; Sanders, J.W.; Johnson, J.M.; Gule-Monroe, M.K.; Chen, M.M.; Briere, T.M.; Wang, Y.; Son, J.B.; Pagel, M.D.; Li, J.; et al. Computer-Aided Detection of Brain Metastases in T1-Weighted MRI for Stereotactic Radiosurgery Using Deep Learning Single-Shot Detectors. *Radiology* **2020**, *295*, 407–415. [CrossRef] [PubMed]
54. Dikici, E.; Ryu, J.L.; Demirer, M.; Bigelow, M.; White, R.D.; Slone, W.; Erdal, B.S.; Prevedello, L.M. Automated Brain Metastases Detection Framework for T1-Weighted Contrast-Enhanced 3D MRI. *IEEE J. Biomed. Health Inform.* **2020**, *24*, 2883–2893. [CrossRef]
55. Suh, C.H.; Jung, S.C.; Kim, K.W.; Pyo, J. The Detectability of Brain Metastases Using Contrast-Enhanced Spin-Echo or Gradient-Echo Images: A Systematic Review and Meta-Analysis. *J. Neurooncol.* **2016**, *129*, 363–371. [CrossRef] [PubMed]
56. Zakaria, R.; Das, K.; Bhojak, M.; Radon, M.; Walker, C.; Jenkinson, M.D. The Role of Magnetic Resonance Imaging in the Management of Brain Metastases: Diagnosis to Prognosis. *Cancer Imaging* **2014**, *14*, 8. [CrossRef] [PubMed]
57. Cho, J.; Lee, K.; Shin, E.; Choy, G.; Do, S. How Much Data Is Needed to Train a Medical Image Deep Learning System to Achieve Necessary High Accuracy? *arXiv* **2015**, arXiv:1511.06348. [CrossRef]
58. Fang, Y.; Wang, J.; Ou, X.; Ying, H.; Hu, C.; Zhang, Z.; Hu, W. The Impact of Training Sample Size on Deep Learning-Based Organ Auto-Segmentation for Head-and-Neck Patients. *Phys. Med. Biol.* **2021**, *66*, 185012. [CrossRef]
59. Jena, A.; Taneja, S.; Talwar, V.; Sharma, J.B. Magnetic Resonance (MR) Patterns of Brain Metastasis in Lung Cancer Patients: Correlation of Imaging Findings with Symptom. *J. Thorac. Oncol.* **2008**, *3*, 140–144. [CrossRef]
60. Vuong, D.A.; Rades, D.; Vo, S.Q.; Busse, R. Extracranial Metastatic Patterns on Occurrence of Brain Metastases. *J. Neurooncol.* **2011**, *105*, 83–90. [CrossRef]
61. Balestrino, R.; Rudà, R.; Soffietti, R. Brain Metastasis from Unknown Primary Tumour: Moving from Old Retrospective Studies to Clinical Trials on Targeted Agents. *Cancers* **2020**, *12*, 3350. [CrossRef]
62. Ronneberger, O.; Fischer, P.; Brox, T. U-Net: Convolutional Networks for Biomedical Image Segmentation. In *Proceedings of the Medical Image Computing and Computer-Assisted Intervention—MICCAI 2015*; Navab, N., Hornegger, J., Wells, W.M., Frangi, A.F., Eds.; Springer International Publishing: Cham, Switzerland, 2015; pp. 234–241.
63. Kamnitsas, K.; Ledig, C.; Newcombe, V.F.J.; Simpson, J.P.; Kane, A.D.; Menon, D.K.; Rueckert, D.; Glocker, B. Efficient Multi-Scale 3D CNN with Fully Connected CRF for Accurate Brain Lesion Segmentation. *Med. Image Anal.* **2017**, *36*, 61–78. [CrossRef]
64. Nie, D.; Wang, L.; Adeli, E.; Lao, C.; Lin, W.; Shen, D. 3-D Fully Convolutional Networks for Multimodal Isointense Infant Brain Image Segmentation. *IEEE Trans. Cybern.* **2019**, *49*, 1123–1136. [CrossRef]
65. Liu, X.; Faes, L.; Kale, A.U.; Wagner, S.K.; Fu, D.J.; Bruynseels, A.; Mahendiran, T.; Moraes, G.; Shamdas, M.; Kern, C.; et al. A Comparison of Deep Learning Performance against Health-Care Professionals in Detecting Diseases from Medical Imaging: A Systematic Review and Meta-Analysis. *Lancet Digit. Health* **2019**, *1*, e271–e297. [CrossRef] [PubMed]

66. Aggarwal, R.; Sounderajah, V.; Martin, G.; Ting, D.S.W.; Karthikesalingam, A.; King, D.; Ashrafian, H.; Darzi, A. Diagnosti Accuracy of Deep Learning in Medical Imaging: A Systematic Review and Meta-Analysis. *NPJ Digit. Med.* **2021**, *4*, 65. [CrossRef [PubMed]
67. Bedrikovetski, S.; Dudi-Venkata, N.N.; Kroon, H.M.; Seow, W.; Vather, R.; Carneiro, G.; Moore, J.W.; Sammour, T. Artificia Intelligence for Pre-Operative Lymph Node Staging in Colorectal Cancer: A Systematic Review and Meta-Analysis. *BMC Cance* **2021**, *21*, 1058. [CrossRef]
68. Decharatanachart, P.; Chaiteerakij, R.; Tiyarattanachai, T.; Treeprasertsuk, S. Application of Artificial Intelligence in Chronic Live Diseases: A Systematic Review and Meta-Analysis. *BMC Gastroenterol.* **2021**, *21*, 10. [CrossRef] [PubMed]
69. Kim, H.Y.; Cho, S.J.; Sunwoo, L.; Baik, S.H.; Bae, Y.J.; Choi, B.S.; Jung, C.; Kim, J.H. Classification of True Progression afte Radiotherapy of Brain Metastasis on MRI Using Artificial Intelligence: A Systematic Review and Meta-Analysis. *Neuro-Onco Adv.* **2021**, *3*, vdab080. [CrossRef]
70. Huedo-Medina, T.B.; Sánchez-Meca, J.; Marín-Martínez, F.; Botella, J. Assessing Heterogeneity in Meta-Analysis: Q Statistic or I^2 Index? *Psychol. Methods* **2006**, *11*, 193–206. [CrossRef]

cancers MDPI

Article

COVID and Cancer: A Complete 3D Advanced Radiological CT-Based Analysis to Predict the Outcome

Syed Rahmanuddin [1,*], Asma Jamil [1], Ammar Chaudhry [1], Tyler Seto [2], Jordyn Brase [1], Pejman Motarjem [1], Marjaan Khan [1], Cristian Tomasetti [3], Umme Farwa [1], William Boswell [1], Haris Ali [4], Danielle Guidaben [1], Rafay Haseeb [1], Guibo Luo [5], Guido Marcucci [4], Steven T. Rosen [4] and Wenli Cai [5]

1 Department of Diagnostic Radiology, City of Hope Comprehensive Cancer Center National Medical Center, Duarte, CA 91010, USA
2 Department of Quality, Risk and Regulatory Management, City of Hope National Medical Center, Duarte, CA 91010, USA
3 Center for Cancer Prevention and Early Detection, City of Hope National Medical Center, Duarte, CA 91010, USA
4 Department of Hematology-Medical Oncology, City of Hope National Medical Center, Duarte, CA 91010, USA
5 Department of Radiology, Massachusetts General Hospital, Harvard University, Boston, MA 02114, USA
* Correspondence: srahmanuddin@coh.org

Simple Summary: Coronavirus-19 is caused by SARS-CoV-2 which is a fatal disease if untreated. The chance of mortality is higher in cancer patients because, in immunocompromised cancer patients, lung damage occurs more rapidly than in normal non-cancer patients which needs more precision based diagnostic criteria. Our study highlights the significance of CT-based radiomics findings to detect morphological damage in the lungs due to COVID-19 in cancer patients. This methodology will help clinicians to investigate the disease more precisely for more accurate precision-based diagnosis.

Abstract: Background: Cancer patients infected with COVID-19 were shown in a multitude of studies to have poor outcomes on the basis of older age and weak immune systems from cancer as well as chemotherapy. In this study, the CT examinations of 22 confirmed COVID-19 cancer patients were analyzed. Methodology: A retrospective analysis was conducted on 28 cancer patients, of which 22 patients were COVID positive. The CT scan changes before and after treatment and the extent of structural damage to the lungs after COVID-19 infection was analyzed. Structural damage to a lung was indicated by a change in density measured in Hounsfield units (HUs) and by lung volume reduction. A 3D radiometric analysis was also performed and lung and lesion histograms were compared. Results: A total of 22 cancer patients were diagnosed with COVID-19 infection. A repeat CT scan were performed in 15 patients after they recovered from infection. Most of the study patients were diagnosed with leukemia. A secondary clinical analysis was performed to show the associations of COVID treatment on the study subjects, lab data, and outcome on mortality. It was found that post COVID there was a decrease of >50% in lung volume and a higher density in the form of HUs due to scar tissue formation post infection. Conclusion: It was concluded that COVID-19 infection may have further detrimental effects on the lungs of cancer patients, thereby, decreasing their lung volume and increasing their lung density due to scar formation.

Keywords: cancer; COVID-19; CT scan; corona virus; treatment

Citation: Rahmanuddin, S.; Jamil, A.; Chaudhry, A.; Seto, T.; Brase, J.; Motarjem, P.; Khan, M.; Tomasetti, C.; Farwa, U.; Boswell, W.; et al. COVID and Cancer: A Complete 3D Advanced Radiological CT-Based Analysis to Predict the Outcome. *Cancers* **2023**, *15*, 651. https://doi.org/10.3390/cancers15030651

Academic Editors: Rafał Obuchowicz, Michał Strzelecki and Adam Piorkowski

Received: 23 December 2022
Revised: 13 January 2023
Accepted: 16 January 2023
Published: 20 January 2023

1. Introduction

Severe acute respiratory syndrome coronavirus-2 (SARS-CoV-2) is a novel coronavirus that emerged from Wuhan, China, which can cause severe respiratory illness and a myriad of other complications now referred to as COVID-19. COVID-19 became a pandemic with more than three million confirmed cases and more than 220,000 deaths around the globe [1].

It can lead to severe symptoms ending in hospitalization, often due to acute respiratory distress syndrome (ARDS) [2–4].

In addition, cancer affects a severely vast proportion of the population around the world contributing to more than 18 million cases recorded every year. Individuals with cancer are more prone to infections because of their poor immune and health status due to the prevailing cancer and its anticancer treatment [5]. To date, cancer patients have been considered to be one of the most vulnerable groups in the current COVID-19 pandemic. Although cancer patients infected with SARS-CoV-2 have been shown, in a multitude of studies, to have poor outcomes on the basis of older age and weakened immune systems from cancer as well as chemotherapy [3–12], nevertheless, the clinical characteristics of SARS-CoV-2-infected cancer patients remain largely unknown [8].

In addition, cancer patients have increased contact with healthcare systems as they visit hospitals for monitoring, anticancer therapy, as well as supportive and preventive care [13].

Recently, a study was conducted among cancer patients to determine the severity of COVID-19 as compared with patients without cancer [8]; however, due to the limited sample size of this study, the results were not applicable to the overall population, and thereby, no insightful analyses were performed [14,15]. Another study concluded that COVID-19 infection in cancer patients posed increased risk for more deteriorating conditions and severe health outcomes [8]. Another study, conducted in China, revealed that as compared with non-cancer patients, COVID-19 cancer patients had less favorable outcomes and a higher risk of disease course with increased severity [6–10].

In another study, the authors concluded that, as per the results of epidemiological studies in which the majority of patients included in the COVID-19 cohort also had cancer, cancer patients were more likely to develop COVID-19 infection [14]. In contrast, another study concluded that COVID-19 was a very contagious infection to which almost every individual was susceptible, once they were exposed to an infection source [16].

Additionally, it has been concluded by researchers that, in this COVID-19 outbreak, the major risk for the morbidity and mortality of the infected individuals was access to medical care services in terms of access to a healthcare facility and provision of best possible treatment [17]. Moreover, increased severity and ultimate death from COVID-19 in cancer patients were principally driven by gender; age; need for ICU support; elevated levels of D-dimer, lactate dehydrogenase, and lactate; and comorbidities such as diabetes mellitus, liver disease, cardiovascular disease, chronic kidney disease, respiratory disease, and immunosuppression in multivariate analyses [11,18–20].

Furthermore, hematologic and lung cancer patients are more susceptible. In addition, patients in the metastatic stages of their cancer have a higher frequency of severity as compared with patients without cancer. Further, patients who have undergone cancer surgery have increased mortality rates and increased chances of devastating symptoms [9].

Among COVID-19 cancer patients, high all-cause mortality associated with both general and unique risk factors related to cancer have been reported. In this study, the CT scans of 13 confirmed COVID-19 cancer patients were analyzed to determine the CT scan changes before and after treatment and the extent of structural damage to the lungs after COVID-19 infection in terms of changes in density and lung volume reduction. The main purpose of this study is to see the COVID-related morphological changes in the lungs of cancer patients who already have immune deficiencies. Cancer patients have less immune response and receive multiple chemotherapies which places them in the category of high-risk individuals for COVID-related diseases. Our research aim is to analyze the CT-based precision imaging findings to accurately estimate the damage in the lungs which could be helpful in clinical diagnosis & management.

2. Materials and Methods

2.1. Patients

This retrospective study was approved by the City of Hope National Medical Center Comprehensive Cancer Center Institutional Review Board. A total of 28 cancer patients were recruited for the study, including 18 males (65.5%) and 10 females (34.5%) with ages ranging from 18 to 83 years and an overall mean age of 55.6 ± 18.54 years. The data was collected from City of Hope patients during the period from January 2020 to June 2021. All adult patients who were identified with COVID-19 infection were included in this study; all others were excluded.

2.2. Study Procedure

A retrospective analysis was conducted on the 28 cancer patients. The CT scan changes before and after treatment and the extent of structural damage to the lungs after COVID-19 infection were analyzed. Structural damage to the lung was indicated by changes in density measured in Hounsfield units (HUs) and lung volume reduction. A 3D radiometric analysis of data was also performed, and density histograms were constructed. Further, secondary clinical analysis was also completed to show the association of COVID-19 treatment on the study subjects, lab data, and outcome on mortality.

2.3. Methods

The data was collected from City of Hope patients during the period of 01 January 2020 till 30 June 2021. Volumetric quantification of lung and pneumonia lesions was carried out using the deep-learning segmentation model of lung and lesions on the 3DQI platform (https://3dqi.mgh.harvard.edu, accessed on 30 June 2021) [18]. After automated segmentation of lung and lesions, an image analyst, who was blinded to clinical data, reviewed and confirmed the segmentation results. A 3D radiometric analysis of the data was performed on the 3DQI Radiometrics tool based on extracted image features and clinical data.

2.4. Statistics and Data Analysis

Demographic, pathological, and clinical characteristics of the patients were summarized using frequencies and percentages for categorical variables; however, mean and standard deviation (mean ± SD) were obtained for continuous variables. The potential changes in the patients were analyzed using means, standard deviations, and kurtosis values. Statistical analyses were conducted using IBM SPSS, version 20 (Statistical Package for the Social Science, IBM Corp, Armonk, NY, USA).

3. Results

3.1. Patients' Characteristics

The patients' demographic characteristics are summarized in Table 1. The patients' ages ranged from 18 to 83 years and the overall mean age was 55.6 ± 18.54 years.

Table 1. Patients' demographic characteristics (n = 28).

Demographic Characteristics		Frequency	Percentage
Age	Mean ± SD	55.6 ± 18.54	
Gender	Male	18	65.5
	Female	10	34.5

SD, standard deviation.

3.2. Metastatic Profile of Cancer Patients

A CT image analysis was performed to analyze structural damage to the lungs of the cancer patients. Structural damage to the lung was indicated by changes in density measured in Hounsfield units (HUs) and lung volume reduction (Table 2).

Table 2. Metastatic profile of cancer patients measured in Hounsfield units (HUs) and lung volume reduction (n = 28).

	Age (Years)	Gender	Lung (CC)	Lesion (CC)	Lesion (%)	<0.25 (HU)	>0.75 (HU)	Kurtosis
Patient 1	38	Male	3417.75	251.32	7.35	−668.40	−232.80	−1.06
Patient 2	64	Female	4842.12	28.34	0.59	−801.50	−577.70	0.89
Patient 3	79	Male	3983.27	1214.42	30.49	−677.90	−375.40	−0.23
Patient 4	54	Female	3822.75	63.32	1.66	−705.60	−288.80	−0.76
Patient 5	39	Female	4759.91	3030.76	63.67	−604.70	−198.80	−0.96
Patient 6	37	Male	3646.28	535.72	14.69	−679.20	−372.30	−0.18
Patient 7	73	Male	2998.78	2395.55	79.88	−590.30	−207.60	−1.00
Patient 8	59	Male	2772.49	239.01	8.62	−685.80	−241.20	−1.07
Patient 9	19	Male	4331.45	603.58	13.93	−628.40	−355.90	−0.59
Patient 10	34	Female	1625.70	725.48	44.63	−603.40	−209.60	−1.07
Patient 11	68	Female	4928.02	956.12	19.40	−760.10	−348.30	−0.66
Patient 12	66	Male	5173.26	10.36	0.20	−713.00	−447.40	−0.30
Patient 13	31	Female	2550.71	16.47	0.65	−763.63	−766.04	0.26
Patient 14	18	Male	3538.70	142.05	4.01	−704.75	−687.97	0.38
Patient 15	62	Male	2707.00	0.00	0.00	−710.00	−721.00	−1.06
Patient 16	51	Female	2387.56	313.39	13.13	−699.78	−722.27	0.75
Patient 17	83	Female	1984.52	1487.44	74.95	−413.59	−307.48	0.27
Patient 18	23	Male	1782.86	1031.47	57.85	−398.49	−450.33	1.24
Patient 19	83	Male	3749.23	73.12	1.95	−812.62	−810.27	0.34
Patient 20	65	Male	3474.40	238.29	6.86	−685.86	−693.05	0.51
Patient 21	62	Male	4233.50	146.46	3.46	−776.89	−784.54	0.36
Patient 22	73	Male	2699.41	185.86	6.89	−653.43	−689.60	0.27
Patient 23	64	Female	4507.96	51.79	1.15	−795.79	−813.74	0.27
Patient 24	34	Female	3341.41	0.00	0.00	−797.91	−810.44	0.26
Patient 25	74	Male	4202.10	230.08	5.48	−719.93	−700.45	0.45
Patient 26	63	Male	1842.53	70.84	3.84	−750.79	−787.42	0.24
Patient 27	63	Male	4486.89	351.70	7.84	−664.92	−691.91	0.57
Patient 28	70	Male	3090.97	1716.68	55.54	−493.92	−466.78	0.92

HU, Hounsfield unit.

3.3. Metastatic Profile of Cancer Patients Post COVID-19 Infection

A total of 22 cancer patients diagnosed with COVID-19 infection were included; most of the study patients were also diagnosed with leukemia. A repeat CT scan was performed in the patients after they recovered from infection. It was found that post COVID there was a decrease of >50% in lung volume and higher density in the form of HUs due to scar tissue formation post infection (Table 3).

Table 3. Correlation of metastatic profile post COVID infection (n = 15).

		Lung (CC)	Lesion (CC)	Lesion (%)
Patient 2	Pre	4842.12	28.34	0.59
	Post	4609.06	164.81	3.58
Patient 3	Pre	3983.27	1214.42	30.49
	Post	5821.05	457.45	7.86
Patient 6	Pre	3646.28	535.72	14.69
	Post	4399.97	242.51	5.51
Patient 7	Pre	2998.78	2395.55	79.88
	Post	3390.13	2970.58	87.62
Patient 15	Pre	2707.00	0.00	0.00
	Post	1455.00	0.00	0.00
Patient 16	Pre	2387.56	313.39	13.13
	Post	2598.47	243.58	9.37
Patient 18	Pre	1782.86	1031.47	57.85
	Post	2733.87	0.00	0.00
Patient 19	Pre	3749.23	73.12	1.95
	Post	3578.28	524.87	14.67
Patient 20	Pre	3474.40	238.29	6.86
	Post	2942.11	133.54	4.54
Patient 22	Pre	2699.41	185.86	6.89
	Post	3964.46	48.37	1.22
Patient 23	Pre	4507.96	51.79	1.15
	Post	4634.40	247.88	5.35
Patient 24	Pre	3341.41	0.00	0.00
	Post	1749.26	856.68	48.97
Patient 26	Pre	1842.53	70.84	3.84
	Post	3025.89	1271.02	42.00
Patient 27	Pre	4486.89	351.70	7.84
	Post	5492.09	7.94	0.14
Patient 28	Pre	3090.97	1716.68	55.54
	Post	4482.71	18.68	0.42

Statistical analysis (Table 4) and 3D radiometric analysis of data were also performed, and density histograms were also constructed for the patients (Figures 1–5). To analyze the changes in lung volume and lesion volumes between the pre- and post-treatment, we calculated the change in the percentage of lung volume, the change in the lesion volume, and the change in lesion percentage of each patient. We have reported the means and standard errors.

Table 4. Statistical analysis of pre and post COVID infection (n = 15).

n = 15	ΔLung Vol % (Post-Pre)/Pre	ΔLesion Vol (cc) (Post-Pre)/Pre	ΔLesion % (Post-Pre)
Mean	13.65%	−67.95	−3.92
Standard Error	34.23%	729.77	28.18

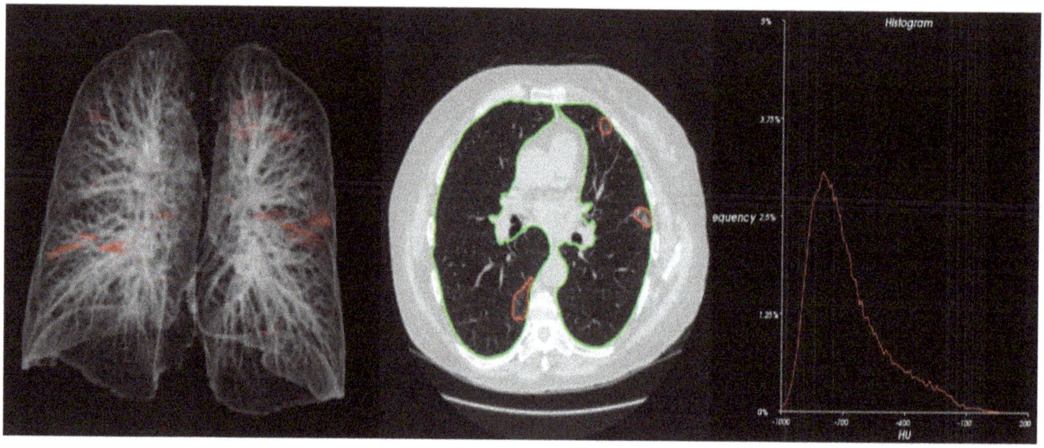

Figure 1. Radiometric 3D analysis (red circle part indicate COVID related damage in the lungs) and time density histograms constructed for Patient 2.

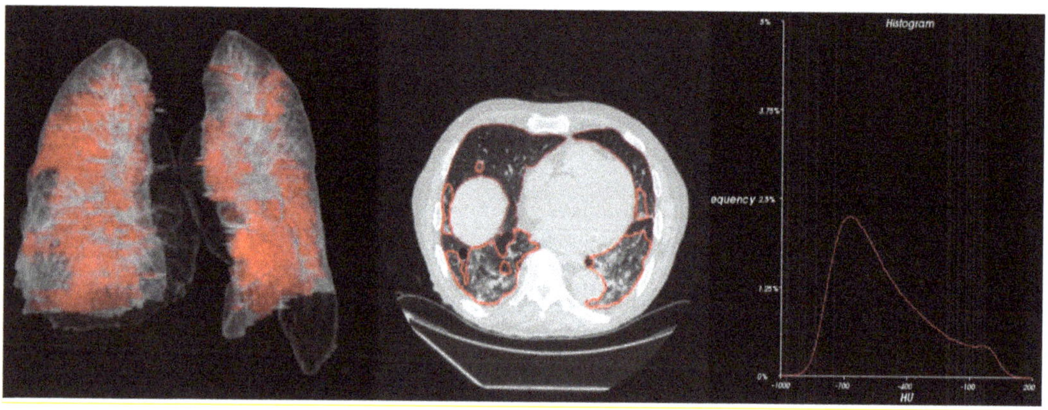

Figure 2. Radiometric 3D analysis (red circle part indicate COVID related damage in the lungs) and time density histograms constructed for Patient 3.

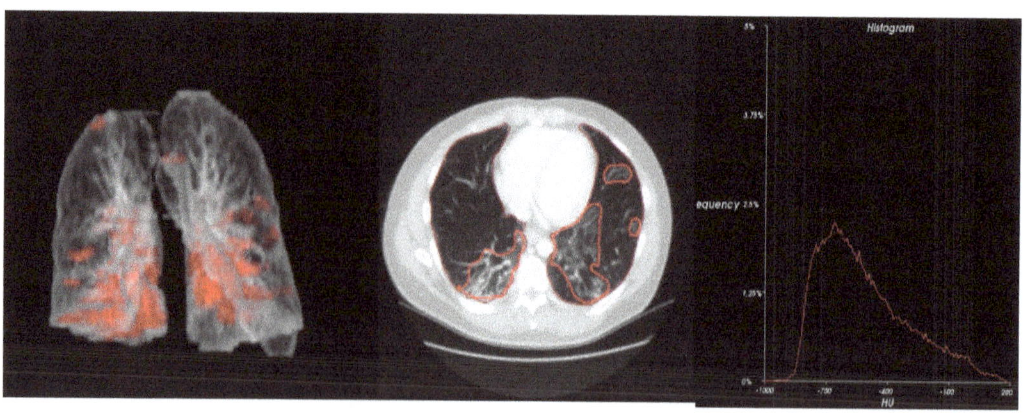

Figure 3. Radiometric 3D analysis (red circle part indicate COVID related damage in the lungs) and time density histograms constructed for Patient 6.

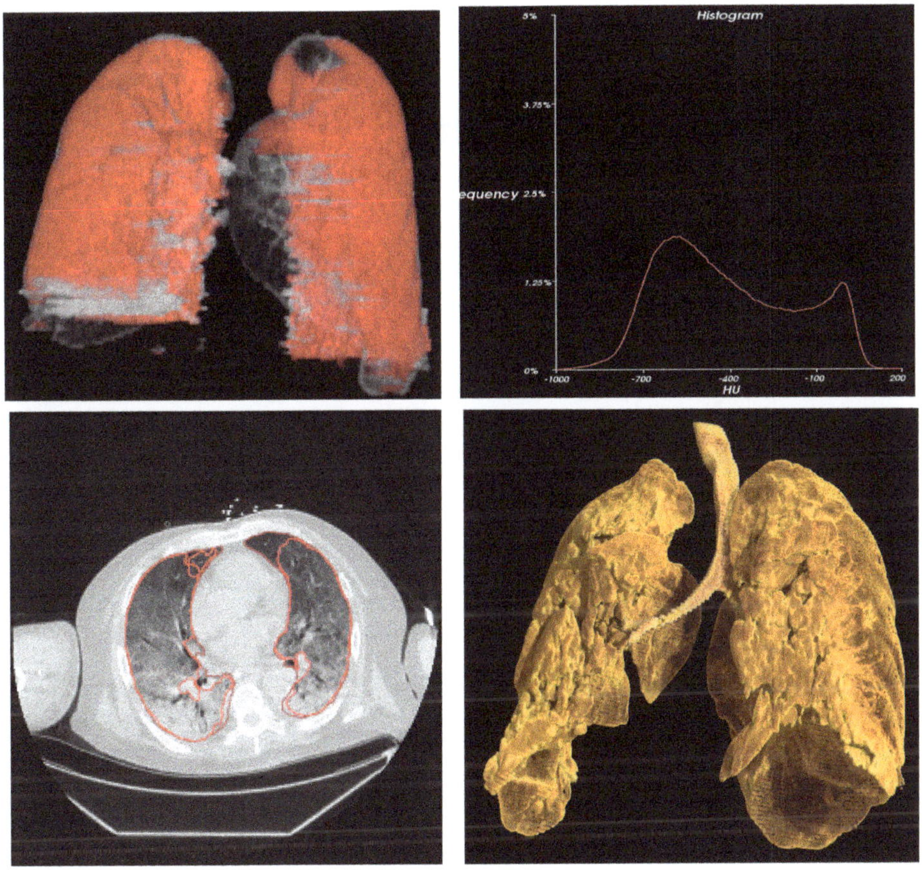

Figure 4. Radiometric 3D analysis (red circle part indicate COVID related damage in the lungs) and time density histograms constructed for Patient 7.

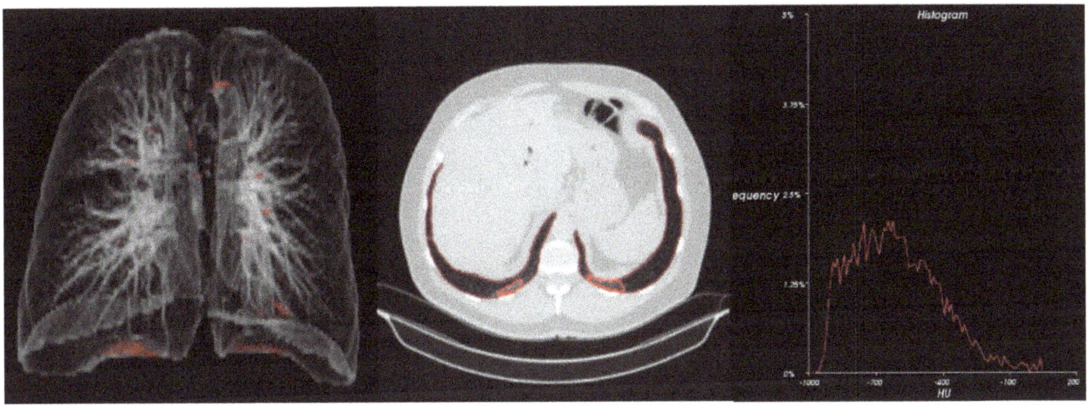

Figure 5. Radiometric 3D analysis(red circle part indicate COVID related damage in the lungs) and time density histograms constructed for Patient 15.

4. Discussion

Due to the rapid spread of the COVID-19 infection around the world, many countries have not been prepared to handle the large number of people affected due to a lack of awareness about the general pathophysiology of the COVID-19 infection and its impact on individuals' lives, in addition to weak health systems and poor access of people to healthcare facilities [9]. Cancer patients are not only more likely to be exposed to COVID-19 due to their need to access the healthcare system, but they are also more susceptible to the disease itself. For instance, COVID-19 cancer patients experience more severe outcomes as compared with the general population both in terms of morbidity and mortality rates. In the general population, the mortality rate for the COVID-19 infection is only 2–3%; however, it increases up to three fold in COVID-19 cancer patients [7,21].

Factors associated with increased severity and ultimately death from COVID-19 in cancer patients include gender; age; need for ICU support; elevated levels of D-dimer, lactate dehydrogenase, and lactate; as well as comorbidities [11].

Previous studies have analyzed the severity and case fatality of cancer patients if co-infected with COVID-19 and have also found that male sex was the prominent risk factor for the infection [20,22]. Other researchers have also concluded that there is a higher incident and mortality rate of COVID-19 infections in males in general. Although the pathogenesis behind this is not established yet, several hypotheses have been proposed [23,24]. For instance, a hypothesis was established that the androgenic hormones have a potential role in the pathogenesis of COVID-19 infection in cancer patients, thereby, posing a higher risk in males for severe outcomes [25]. Another hypothesis proposed that the sex differences in the multistep immune-pathogenic pathway among both males and females leads to sex biases causing the male gender to be more prone to the COVID-19 infection [26,27]. Because none of these theories have been accepted or established in the medical literature, scientists are still searching for the most appropriate reason to explain why male sex appears to be a risk factor for COVID-19 cancer patients [28].

Similar to our study, a study conducted in Wuhan China among COVID-19-infected lung cancer patients found reduced lung volume due to the presence of tumor and also the features of pneumonia through CT scan images [8]. This study concluded that COVID-19-infected lung cancer patients had worse endurance and lung function at baseline and were prone to develop anoxia which progressed more rapidly due to COVID-19, thereby, creating an urgent need for the treatment of COVID-19-infected lung cancer patients [8]. Further, it was concluded that decreased lung function and severe infection in lung cancer patients contributed to the more severe and devastating outcomes in COVID-19 cancer patients [29,30]. Moreover, patients with lung cancer are more likely to deteriorate in a very short span of time when infected with COVID-19 infection due to [9] pulmonary edema, focal reactive hyperplasia of pneumocytes, proteinaceous exudate, hyaline membrane formation, and multinucleated giant cells with patchy inflammatory cellular infiltration.

In this retrospective analysis, the CT scans of 22 confirmed COVID-19 patients with leukemia were analyzed to determine the changes before and after treatment and the extent of structural damage to the lungs after COVID-19. Structural damage to the lung was indicated by changes in density measured in Hounsfield units (HUs) and lung volume reduction. Repeat CT scans were performed in 15 of the patients after they recovered from infection. It was found that, post COVID, there was a decrease of >50% in lung volume and a higher density in the form of HUs due to scar tissue formation post infection.

Similar to other studies of COVID-19, the current study also showed increased cases of males (65.5%) as compared with females (34.5%). Similar to previous findings [19,20], the current results also indicated that the increased severity of COVID-19 cancer patients was related to older age as most of our patients were above 50 years of age. This might be due to the fact that in older age, the host response is diminished and the immune system is compromised and suffering from multi-organ complications due to the presence of different comorbidities. Thus, multiple factors are involved that lead to the increased severity of COVID-19 risk factor in advanced age cancer patients [11].

The current study is one of the novel studies whose findings have suggested that cancer patients, particularly leukemia patients, may be more vulnerable to COVID-19 and have more pronounced lung structural damage due to the co-infection. Based on the results of this study related to leukemia patients who subsequently developed COVID-19, it is suggestive that these patients may be at a higher risk of long-term effects on the lungs. Other studies have shown increased COVID-19 mortality in hematologic malignancies; however, no studies, to the best of our knowledge, have looked at lung damage specifically as in this study. Studies have shown likely links between high rates of immune dysfunction in hematologic malignancies and increased mortality from COVID-19. Leukemia patients, in particular, are at a high risk for immune dysfunction, whether solely related to the primary disease process or therapy-related immunosuppression [31]. However, additional studies are required to understand the link between leukemia and COVID-19 as well as links between other types of cancers and COVID-19.

However, there are some limitations of the current study including that the study was a non-randomized, retrospective analysis conducted on a small sample size. Further, some confounders such as tumor stage and COVID-19 infection severity were not included in the study, as they might affect the clinical outcomes of the cancer patients co-infected with COVID-19. Furthermore, due to retrospective study design, only fatal cases of COVID-19 cancer patients were reported in this study and a comparison between the COVID-19 infected cancer and non-cancer patients was not conducted.

Thus, it is recommended to conduct future studies with a larger sample size and prospective study design to better explore the comparison between cancer and non-cancer patients infected with COVID-19 and their severe events.

Further, it is also suggested that cancer patients should receive online medical counselling in addition to vigorous screening for COVID-19 to help identify critical cases for proper treatment. In cancer patients who do develop COVID-19, it may be necessary to avoid further immunosuppressive treatments or at least consider decreased dosages [8].

Thus, COVID-19 cancer patients reinforce several important considerations for clinical care, and thereby, there should be more emphasis on the urgent need for more research in this area.

5. Conclusions

This was a retrospective analysis study conducted on COVID-19 cancer patients to analyze the CT imaging changes before and after treatment and the extent of structural damage to the lungs post COVID-19 infection. It was found that, post COVID, there was a decrease of >50% in lung volume and a higher density in the form of HUs. Thus, it was concluded that COVID-19 infection had further detrimental effects on the lungs of cancer patients, thereby, decreasing their lung volume and increasing their lung density due to scar formation post COVID-19 infection. The study also supports the notion that patients with hematologic malignancies such as leukemia may be at particularly high risk for adverse outcomes from COVID-19 infection. Proactive strategies are required to reduce the likelihood of infection, and therefore, improved early identification of COVID-19 positivity in cancer patients is clearly warranted. Hence, increased follow-up is required to better understand the effects of COVID-19 on outcomes in cancer patients, considering the ability to follow specific cancer treatments. In general, during active treatment, cancer patients receive frequent follow-up which should be based on the extent of damage. However, we suggest that patients be followed weekly with labs and low-dose CT scans as needed until resolution of symptoms. If there is extensive fibrosis, monthly, then every 3 months, and then every 6 months followup is needed, but so far, no such protocol has been established.

Author Contributions: Conceptualization, (S.R.; A.J.; A.C.; T.S.; J.B.; P.M.; M.K.; C.T.; U.F.; W.B.; H.A.; D.G.; S.T.R.; G.M.; W.C.).; methodology, (S.R.; A.J.; A.C.; T.S.; J.B.; P.M.; M.K.; C.T.; U.F.; W.B.; H.A.; D.G.; R.H.; S.T.R.; W.C.); software, (S.R.; A.J.; A.C.; T.S.; J.B.; P.M.; M.K.; U.F.; W.B.; H.A.; G.L.; S.T.R.; W.C.).; validation, S.R.; A.J.; A.C.; T.S.; J.B.; P.M.; M.K.; C.T.; U.F.; W.B.; H.A.; G.L.; S.T.R.; G.M.; W.C.; formal analysis, S.R.; A.J.; A.C.; T.S.; J.B.; P.M.; M.K.; C.T.; U.F.; W.B.; H.A.; D.G.; G.L.; S.T.R.; G.M.; W.C.; investigation, S.R.; A.J.; A.C.; T.S.; P.M.; M.K.; C.T.; U.F.;H.A.; D.G.; S.T.R.; W.C.; resources, S.R.; A.J.; A.C.; T.S.; J.B.; P.M.; M.K.; U.F.; W.B.; H.A.; D.G.; R.H.; G.L.; S.T.R.; W.C. data curation, S.R.; A.J.; A.C.; T.S.; J.B.; P.M.; M.K.; U.F.; W.B.; H.A.; R.H.; G.L.; S.T.R.; W.C. writing—original draft preparation, S.R.; A.J.; A.C.; T.S.; J.B.; P.M.; M.K.; C.T.; U.F.; W.B.; H.A.; D.G.; R.H.; G.L.; S.T.R.; G.M.; W.C. writing—review and editing, S.R.; A.J.; A.C.; T.S.; J.B.; P.M.; M.K.; C.T.; U.F.; W.B.; H.A.; D.G.; R.H.; G.L.; S.T.R.; W.C. visualization, S.R.; A.J.; A.C.; T.S.; J.B.; P.M.; M.K.; C.T.; U.F.; W.B.; S.T.R.; W.C. supervision, S.R.; A.J.; A.C.; T.S.; W.B.; S.T.R.; W.C.; project administration, S.R.; A.J.; A.C.; T.S.; W.B.; S.T.R.; W.C.; funding acqui-sition, none. All authors have read and agreed to the published version of the manuscript.

Funding: This research received no external funding.

Institutional Review Board Statement: The study was conducted in accordance with the Declaration of Helsinki, and approved by the Institutional Review Board (or Ethics Committee) of City of Hope (COH Protocol #/Ref #: 20703 / 201272- Date- 12/09/2020).

Informed Consent Statement: Patient consent was waived due to nature of retrospective imaging study.

Data Availability Statement: The data presented in this study are available on request from the corresponding author.

Conflicts of Interest: The authors declare no conflict of interest.

References

1. Engelhardt, M.; Shoumariyeh, K.; Rösner, A.; Ihorst, G.; Biavasco, F.; Meckel, K.; von Metzler, I.; Theurich, S.; Hebart, H.; Grube, M.; et al. Clinical characteristics and outcome of multiple myeloma patients with concomitant COVID-19 at Comprehensive Cancer Centers in Germany. *Haematologica* **2020**, *105*, 2872. [CrossRef] [PubMed]
2. Wu, Z.; McGoogan, J.M. Characteristics of and important lessons from the coronavirus disease 2019 (COVID-19) outbreak in China: Summary of a report of 72,314 cases from the Chinese Center for Disease Control and Prevention. *Jama* **2020**, *323*, 1239–1242. [CrossRef] [PubMed]
3. Terpos, E.; Engelhardt, M.; Cook, G.; Gay, F.; Mateos, M.-V.; Ntanasis-Stathopoulos, I.; van de Donk, N.W.; Avet-Loiseau, H.; Hajek, R.; Vangsted, A.J.; et al. Management of patients with multiple myeloma in the era of COVID-19 pandemic: A consensus paper from the European Myeloma Network (EMN). *Leukemia* **2020**, *34*, 2000–2011. [CrossRef] [PubMed]
4. Mian, H.; Grant, S.J.; Engelhardt, M.; Pawlyn, C.; Bringhen, S.; Zweegman, S.; Stege, C.A.; Rosko, A.E.; von Lilienfeld-Toal, M.; Wildes, T.M. Caring for older adults with multiple myeloma during the COVID-19 Pandemic: Perspective from the International Forum for Optimizing Care of Older Adults with Myeloma. *J. Geriatr. Oncol.* **2020**, *11*, 764–768. [CrossRef]
5. Chen, W.; Zheng, R.; Baade, P.D.; Zhang, S.; Zeng, H.; Bray, F.; Jemal, A.; Yu, X.Q.; He, J. Cancer statistics in China, 2015. *CA A Cancer J. Clin.* **2016**, *66*, 115–132. [CrossRef]
6. Guan, W.J.; Ni, Z.Y.; Hu, Y.; Liang, W.H.; Ou, C.Q.; He, J.X.; Liu, L.; Shan, H.; Lei, C.L.; Hui, D.S.; et al. Clinical characteristics of coronavirus disease 2019 in China. *N. Engl. J. Med.* **2020**, *382*, 1708–1720. [CrossRef]
7. Liang, W.; Guan, W.; Chen, R.; Wang, W.; Li, J.; Xu, K.; Li, C.; Ai, Q.; Lu, W.; Liang, H.; et al. Cancer patients in SARS-CoV-2 infection: A nationwide analysis in China. *Lancet Oncol.* **2020**, *21*, 335–337. [CrossRef]
8. Zhang, L.; Zhu, F.; Xie, L.; Wang, C.; Wang, J.; Chen, R.; Jia, P.; Guan, H.Q.; Peng, L.; Chen, Y.; et al. Clinical characteristics of COVID-19-infected cancer patients: A retrospective case study in three hospitals within Wuhan, China. *Ann. Oncol.* **2020**, *31*, 894–901. [CrossRef]
9. Dai, M.; Liu, D.; Liu, M.; Zhou, F.; Li, G.; Chen, Z.; Zhang, Z.; You, H.; Wu, M.; Zheng, Q.; et al. Patients with cancer appear more vulnerable to SARS-CoV-2: A multicenter study during the COVID-19 outbreakpatients with cancer in SARS-COV-2 infection. *Cancer Discov.* **2020**, *10*, 783–791. [CrossRef]
10. Omarini, C.; Maur, M.; Luppi, G.; Narni, F.; Luppi, M.; Dominici, M.; Longo, G.; Piacentini, F. Cancer treatment during the coronavirus disease 2019 pandemic: Do not postpone, do it! *Eur. J. Cancer* **2020**, *133*, 29–32. [CrossRef]
11. Lee, L.Y.; Cazier, J.B.; Starkey, T.; Turnbull, C.D.; The UK Coronavirus Monitoring Project Team; Kerr, R.; Middleton, G. COVID-19 mortality in patients with cancer on chemotherapy or other anticancer treatments: A prospective cohort study. *Lancet* **2020**, *395*, 1919–1926. [CrossRef] [PubMed]
12. Hultcrantz, M.; Richter, J.; Rosenbaum, C.A.; Patel, D.; Smith, E.L.; Korde, N.; Lu, S.X.; Mailankody, S.; Shah, U.A.; Lesokhin, A.M.; et al. COVID-19 infections and clinical outcomes in patients with multiple myeloma in New York City: A cohort study from five academic centers. *Blood Cancer Discov.* **2020**, *1*, 234–243. [CrossRef] [PubMed]

.3. Kuderer, N.M.; Choueiri, T.K.; Shah, D.P.; Shyr, Y.; Rubinstein, S.M.; Rivera, D.R.; Shete, S.; Hsu, C.Y.; Desai, A.; de Lima Lopes, G., Jr.; et al. Clinical impact of COVID-19 on patients with cancer (CCC19): A cohort study. *Lancet* **2020**, *395*, 1907–1918. [CrossRef]
.4. Wang, H.; Zhang, L. Risk of COVID-19 for patients with cancer. *Lancet Oncol.* **2020**, *21*, e181. [CrossRef] [PubMed]
.5. Xia, Y.; Jin, R.; Zhao, J.; Li, W.; Shen, H. Risk of COVID-19 for patients with cancer. *Lancet Oncol.* **2020**, *21*, e180. [CrossRef] [PubMed]
.6. Chen, N.; Zhou, M.; Dong, X.; Qu, J.; Gong, F.; Han, Y.; Qiu, Y.; Wang, J.; Liu, Y.; Wei, Y.; et al. Epidemiological and clinical characteristics of 99 cases of 2019 novel coronavirus pneumonia in Wuhan, China: A descriptive study. *Lancet* **2020**, *395*, 507–513. [CrossRef] [PubMed]
17. Wang, Y.; Zhou, S.; Yang, F.; Qi, X.; Wang, X.; Guan, X.; Shen, C.; Duma, N.; Aguilera, J.V.; Chintakuntlawar, A.; et al. Treatment-related adverse events of PD-1 and PD-L1 inhibitors in clinical trials: A systematic review and meta-analysis. *JAMA Oncol.* **2019**, *5*, 1008–1019. [CrossRef]
18. Cai, W.; Liu, T.; Xue, X.; Luo, G.; Wang, X.; Shen, Y.; Fang, Q.; Sheng, J.; Chen, F.; Liang, T. CT Quantification and Machine-learning Models for Assessment of Disease Severity and Prognosis of COVID-19 Patients. *Acad. Radiol.* **2020**, *27*, 1665–1678. [CrossRef]
19. von Lilienfeld-Toal, M.; Vehreschild, J.J.; Cornely, O.; Pagano, L.; Compagno, F.; Hirsch, H.H. Frequently asked questions regarding SARS-CoV-2 in cancer patients—Recommendations for clinicians caring for patients with malignant diseases. *Leukemia* **2020**, *34*, 1487–1494. [CrossRef]
20. Wang, B.; Van Oekelen, O.; Mouhieddine, T. A tertiary center experience of multiple myeloma patients with COVID-19: Lessons learned and the path forward. *J. Hematol. Oncol.* **2020**, *13*, 94. [CrossRef]
21. Zheng, R.S.; Sun, K.X.; Zhang, S.W.; Zeng, H.M.; Zou, X.N.; Chen, R.; Gu, X.Y.; Wei, W.W.; He, J. Report of cancer epidemiology in China, 2015. *Zhonghua Zhong Liu Za Zhi [Chin. J. Oncol.]* **2019**, *41*, 19–28. [PubMed]
22. Mehta, V.; Goel, S.; Kabarriti, R.; Cole, D.; Goldfinger, M.; Acuna-Villaorduna, A.; Pradhan, K.; Thota, R.; Reissman, S.; Sparano, J.A.; et al. Case fatality rate of cancer patients with COVID-19 in a New York hospital system. *Cancer Discov.* **2020**, *10*, 935–941. [CrossRef]
23. Devlin, H. Men are much more likely to die from coronavirus—But why. *Guardian* **2020**, *26*.
24. Peckham, H.; de Gruijter, N.M.; Raine, C.; Radziszewska, A.; Ciurtin, C.; Wedderburn, L.R.; Rosser, E.C.; Deakin, C.T.; Webb, K. Sex-bias in COVID-19: A Meta-Analysis and Review of Sex Differences in Disease and Immunity. Available online: https://ssrn.com/abstract=3572881 (accessed on 20 December 2022). [CrossRef]
25. Wambier, C.G.; Vaño-Galván, S.; McCoy, J.; Gomez-Zubiaur, A.; Herrera, S.; Hermosa-Gelbard, Á.; Moreno-Arrones, O.M.; Jiménez-Gómez, N.; González-Cantero, A.; Fonda-Pascual, P.; et al. Androgenetic alopecia present in the majority of patients hospitalized with COVID-19: The "Gabrin sign". *J. Am. Acad. Dermatol.* **2020**, *83*, 680–682. [CrossRef] [PubMed]
26. Scully, E.P.; Haverfield, J.; Ursin, R.L.; Tannenbaum, C.; Klein, S.L. Considering how biological sex impacts immune responses and COVID-19 outcomes. *Nat. Rev. Immunol.* **2020**, *20*, 442–447. [CrossRef]
27. Klein, S.L.; Dhakal, S.; Ursin, R.L.; Deshpande, S.; Sandberg, K.; Mauvais-Jarvis, F. Biological sex impacts COVID-19 outcomes. *PLoS Pathog.* **2020**, *16*, e1008570. [CrossRef]
28. Park, R.; Chidharla, A.; Mehta, K.; Sun, W.; Wulff-Burchfield, E.; Kasi, A. Sex-bias in COVID-19-associated illness severity and mortality in cancer patients: A systematic review and meta-analysis. *EClinicalMedicine* **2020**, *26*, 100519. [CrossRef] [PubMed]
29. Nicholls, J.M.; Poon, L.L.; Lee, K.C.; Ng, W.F.; Lai, S.T.; Leung, C.Y.; Chu, C.M.; Hui, P.K.; Mak, K.L.; Lim, W.; et al. Lung pathology of fatal severe acute respiratory syndrome. *Lancet* **2003**, *361*, 1773–1778. [CrossRef]
30. Russell, C.D.; Millar, J.E.; Baillie, J.K. Clinical evidence does not support corticosteroid treatment for 2019-nCoV lung injury. *Lancet* **2020**, *395*, 473–475. [CrossRef]
31. Passamonti, F.; Romano, A.; Salvini, M.; Merli, F.; Della Porta, M.G.; Bruna, R.; Coviello, E.; Romano, I.; Cairoli, R.; Lemoli, R.; et al. COVID-19 elicits an impaired antibody response against SARS-CoV-2 in patients with haematological malignancies. *Br. J. Haematol.* **2021**, *195*, 371–377. [CrossRef]

Journal of
Clinical Medicine

MDPI

Article

Key-Point Detection Algorithm of Deep Learning Can Predict Lower Limb Alignment with Simple Knee Radiographs

Hee Seung Nam †, Sang Hyun Park †, Jade Pei Yuik Ho, Seong Yun Park, Joon Hee Cho and Yong Seuk Lee *

Department of Orthopaedic Surgery, Seoul National University College of Medicine, Seoul National University Bundang Hospital, Seongnam-si 13620 82, Republic of Korea
* Correspondence: smcos1@hanmail.net; Tel.: +82-31-787-7199
† These authors contributed equally to this work.

Abstract: (1) Background: There have been many attempts to predict the weight-bearing line (WBL) ratio using simple knee radiographs. Using a convolutional neural network (CNN), we focused on predicting the WBL ratio quantitatively. (2) Methods: From March 2003 to December 2021 2410 patients with 4790 knee AP radiographs were randomly selected using stratified random sampling. Our dataset was cropped by four points annotated by a specialist with a 10-pixel margin The model predicted our interest points, which were both plateau points, i.e., starting WBL point and exit WBL point. The resulting value of the model was analyzed in two ways: pixel units and WBL error values. (3) Results: The mean accuracy (MA) was increased from around 0.5 using a 2-pixel unit to around 0.8 using 6 pixels in both the validation and the test sets. When the tibial plateau length was taken as 100%, the MA was increased from approximately 0.1, using 1%, to approximately 0.5 using 5% in both the validation and the test sets. (4) Conclusions: The DL-based key-point detection algorithm for predicting lower limb alignment through labeling using simple knee AP radiographs demonstrated comparable accuracy to that of the direct measurement using whole leg radiographs Using this algorithm, the WBL ratio prediction with simple knee AP radiographs could be useful to diagnose lower limb alignment in osteoarthritis patients in primary care.

Keywords: knee; weight-bearing line; machine learning; convolutional neural network; prediction

Citation: Nam, H.S.; Park, S.H.; Ho, J.P.Y.; Park, S.Y.; Cho, J.H.; Lee, Y.S. Key-Point Detection Algorithm of Deep Learning Can Predict Lower Limb Alignment with Simple Knee Radiographs. *J. Clin. Med.* **2023**, *12*, 1455. https://doi.org/10.3390/jcm12041455

Academic Editors: Rafał Obuchowicz, Michał Strzelecki and Adam Piorkowski

Received: 16 December 2022
Revised: 7 February 2023
Accepted: 10 February 2023
Published: 11 February 2023

1. Introduction

Osteoarthritis (OA) is the most common form of arthritis, affecting millions of people worldwide [1]. It occurs when the protective cartilage that cushions the ends of the bones wears down [2]. Therefore, weight-bearing joints, such as the knee joint, are more vulnerable. The medial compartment of the knee joint is the most commonly affected site in OA, and the medial joint space narrows as OA progresses. This induces varus deformity of the lower limb. Consequently, the adduction moment, which is the magnitude of the ground reaction force, moves medially from the center of the knee joint during ambulation, and the moment arm of the ground reaction force increases [3]. In this manner, the varus deformity enters a vicious cycle. Therefore, it is important to intervene through treatment before such vicious cycle develops [4,5].

The evaluation of the weight-bearing axis of the lower limb is a fundamental step in the identification, classification, and treatment of lower limb deformities, which may result from degeneration, trauma, inflammation, or congenital diseases. When deciding between treatment options, such as conservative treatment, osteotomy, and arthroplasty, the weight-bearing axis should be considered in addition to the patient's basic demographics [6]. Several methods assess the weight-bearing axis of the lower limbs. Techniques to determine the hip–knee–ankle angle, mechanical axis, and weight-bearing line (WBL) ratio are commonly used methods [7,8]. These parameters are usually measured on whole leg radiographs (WLR) and in institutions, such as large hospitals or community clinics [9]. Therefore, they may not be readily available because of the high costs involved.

Moreover, primary physicians may face difficulties in identifying such deformities, and several attempts have been made to predict the WBL ratio using simple standing knee radiographs [10].

The application of artificial intelligence in medicine has gained popularity in recent years because of its ability to improve the efficiency of healthcare delivery and patient diagnosis. Karnuta et al. used machine learning to identify knee arthroplasty implants from X-rays [11]. In addition, Joseph et al. employed machine learning to forecast the development of osteoarthritis over 8 years using combined MR imaging features, demographics, and clinical factors as input [12]. Convolutional neural networks (CNNs) are a subtype of deep learning (DL) that have shown impressive outcomes in image classification and recognition [13,14]. A CNN was used in a previous study that predicted the WBL ratio as a parameter for lower limb alignment in simple knee radiographs. However, a limitation of this study was that the prediction was only possible within intervals. Therefore, a quantitative assessment that can predict the WBL ratio and translate it to an accurate point on the tibial plateau may be more intuitive for clinical use [15].

This study aimed to develop a DL algorithm to predict the point at which the weight-bearing axis of the lower limb crosses the tibial plateau. The hypothesis of this study was that the WBL ratio obtained from weight-bearing WLR could be predicted by a specially designed DL model using standing simple knee anteroposterior (AP) radiographs with high predictive value.

2. Materials and Methods

With the assumption that standing simple knee radiographs are a part of WLR and may be related to the lower limb alignment, the most appropriate and accurate model to predict the WBL ratio after learning the WBL ratio was designed using simple knee radiographs. This was based on the key-point detection model [16,17]. The key-point detection model involves locating the key object parts that represent the underlying object in a feature-rich manner. After directly labeling the WLR picture according to how we draw the WBL in a real clinical situation, we cut it into a simple knee radiograph picture and trained the DL model. Subsequently, the accuracy of the learned DL algorithm for measuring the WBL ratio was investigated. For the analysis of the accuracy of the DL model, its mean absolute error (MAE) and intra-class correlation coefficients (ICC) were evaluated. In addition, the accuracy values of the DL model were compared with the ICCs of the rater, using real measurements of the WBL ratio in WLR to check whether the DL algorithm was reliable as the measurements of the rater. All procedures involving human participants were performed following the ethical standards of the institutional review board (IRB No. B-2210-784-101) and the Helsinki Declaration (1964) and its later amendments. Consent was not sought because this study retrospectively reviewed the medical record image data of patients who underwent X-ray examination at our hospital, and personal identification information was not included in the data analysis process. Therefore, it was difficult to evaluate if the risk for the patients included in this study was increased compared to that of other patients. There is no reason to presume the refusal of consent.

2.1. Data Set

From March 2003 to December 2021, 89,709 patients with knee pain and standing knee AP radiographs were obtained from the clinical database of our hospital. Among them, 3515 patients (3.9% of standing knee AP acquisition) who underwent weight-bearing WLR were included. The exclusion criteria were as follows: (1) previous ipsilateral surgery (hip, knee, or ankle joint); (2) children with remaining growth plates; and (3) patients with deformity due to previous trauma or congenital diseases. After excluding patients who met the exclusion criteria, finally, 2410 patients with 4790 knee AP radiographs were randomly selected using stratified random sampling. To avoid the cluster effect between multiple radiographs in a single patient, only the initial knee AP radiograph was used.

2.2. WBL Ratio Measurement and Labeling

Our dataset was created by one specialist, and the data were cropped by four points annotated by a specialist (tibial plateau at both ends, WBL starting and exit points) with a 10-pixel margin. Although four points varied on the cropped data, and the data size was variable, the data were resized uniformly in the training phase. Therefore, our model was invariant in relation to the image size to some degree. The model's robustness of the data size was evaluated with various random margins of 5 to 10 pixels. This experiment showed that the WBL ratio did not affect the data size, and our model predicted the line's tendency. This revealed that even if the users crop the image data abnormally, the model's prediction result will not decline.

The WBL ratio was measured using the weight-bearing WLR of all 2410 patients with 4790 knees for labeling the training set and analysis of prediction accuracy in the validation and test sets. The WBL was drawn from the center of the femoral head to the center of the superior articular surface of the talus. The WBL ratio was calculated as the ratio of the crossing point of the mechanical axis, from the medial edge to the entire width of the tibial plateau.

2.3. Image Preprocessing

A standing knee AP radiograph was chosen as the research object. The PyDicom library (version 1.3.0) was used for the preprocessing of DICOM images. The right or left knee was cropped in a knee radiograph that included both knees. Strong augmentation, such as sheer, distortion, and high rate of random brightness, contrast, equalization, and hue saturation for robustness, was used to improve the performance of the algorithm.

2.4. DL Algorithm

The DL algorithm consists of two stages. In the first stage, key-points heat maps were predicted using the WBL prediction model. Our WBL prediction model is similar to conventional pose estimation, which predicts some key-points through heat maps and is composed of a feature extractor and a simple convolutional decoder. The model predicts our points of interest, which are both plateau points, i.e., the starting WBL point and the exit WBL point. Using the logits calculated from the model, we trained our model in an end-to-end manner with the Adam W optimizer and binary cross entropy loss. The other setting was similar to that of the conventional DL. In the second stage, the WBL ratio was calculated using the four predicted points. Two lines from four points were drawn to determine the intersection of the two lines and calculate the WBL ratio (Figure 1).

2.5. Experiment

In the experiment, our model for 10,000 iterations using Adam W with $\beta 1 = 0.9$, $\beta 2 = 0.999$ was trained; learning rate = 1×10^{-3}, weight decay = 1×10^{-2}, cosine decaying scheduler, and binary cross entropy loss. In the training phase, the cropped image data were transformed using random brightness, contrast, equalization, and hue saturation for robustness. Our model was on four 2080 ti GPU with an 8-batch size per GPU. The resulting value of the model was analyzed in two ways: WBL error values and pixel units. When analyzing the results with the WBL error value, a value of 1 was assigned if the DL prediction value was within each error value of the tibial plateau, and a value of 0 was assigned otherwise. When analyzing the results in pixel units, the accuracy was calculated by assigning a value of 1 if the DL prediction was received in each pixel unit, and a value of 0 otherwise. The algorithm was first trained by making as many landmarks as possible, i.e., 27 dots, around the knee joint. According to the traditional WBL calculation method, radiological landmarks that can symbolize the bony anatomy of both femur and tibia were used, so there were 27 of them, as follows. The starting point and exit point of the WBL line, both endpoints of the femur cortex in X-ray images, both endpoints of the tibia cortex in X-ray images, both endpoints of femur and tibia at the tibiofemoral joint—the most distant points medial and lateral from femur and tibia and transepicondyle points at the tibiofemoral joint—trochlear notch center—midpoint of both tibial spines and tibial spines, intersection of tibia and fibula, inflection points at femur and tibia. The marking was reduced by identifying the most appropriate dots that showed the best performance (Figure 1).

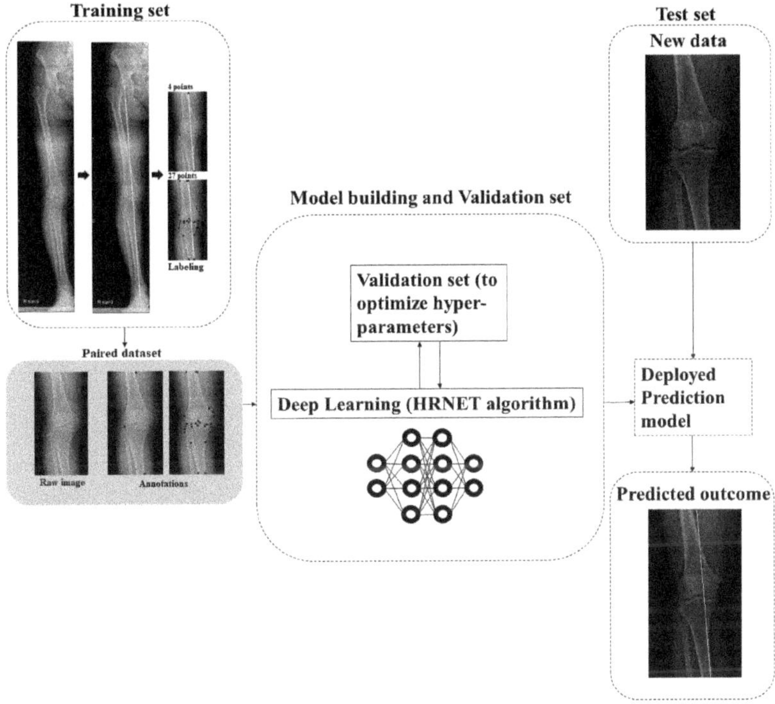

Figure 1. Pipeline of the study. Annotations: labeled image. Training set: the data sample used to fit the model. Validation set: the data sample used to provide an unbiased evaluation of a model fit on the training dataset while tuning the model hyperparameters. Test set: the data sample used to provide an unbiased evaluation of the final model fit on the training dataset.

2.6. Statistical Analysis

Data are presented as means and standard deviations for continuous variables. One-way analysis of variance was performed to compare the quantitative variables (i.e., age, body mass index (BMI), and WBL ratio). Pearson's chi-squared test or Fisher's exact test was used to compare the qualitative variables (i.e., gender). Statistical significance was set at $p < 0.05$. The data were analyzed using SPSS 25.0 (IBM, Armonk, NY, USA). To examine the reproducibility of the calculation of the WBL ratio using WLR, two observers were chosen: A, an orthopedic surgeon with 5 years of experience; B. an orthopedic surgeon with 20 years of experience. Independent measurements obtained by each of the two raters for each data set (raters A, B) and two independent measurements (A1 and A2) obtained by a single rater for each data set were compared. The mean difference between the independent measurements obtained by the raters was evaluated. The inter- and intra-observer reliabilities of the measurements were analyzed using ICC, with ICC < 0.40 indicating poor agreement, in the range 0.40–0.75 indicating fair to good (moderate) agreement, and in the range 0.76–1.00 indicating excellent agreement. MAE was used as a measure to determine how well the CNN fit the WBL ratio [13–15]. MAE is a measure that indicates the difference between the actual labeled WBL ratio by A (AL) using WLR and the WBL ratio predicted by the CNN using simple knee radiographs, $MAE = \frac{1}{N}\sum_{i=1}^{N}|\hat{y}_i - y_i|$, with $\hat{y}_i$ being the estimated WBL ratio of the ith data, and y_i being the ground-truth WBL ratio of the ith data [18].

3. Results

The baseline characteristics of the patients and the distribution of the labels in the training, validation, and test sets are summarized in Table 1. Age, sex, BMI, and WBL ratio were not significantly different between the datasets. The performance was improved through simple labeling that marked both ends of the tibia and the starting and exit points of the WBL line. Four points indicated the best performance among the trials. A comparison of the WBL mean accuracy at 4 and 27 points is shown in Figure 2. As the threshold of the WBL error percentage value increased, the accuracy of taking four points approached 0.6, whereas when 27 points were taken, a steady state was reached with an accuracy not exceeding 0.1. These results were also similarly obtained in the pixel unit, and as the threshold increased, the accuracy when taking four points was close to 0.8 in 6 pixels, while that achieved when taking 27 points was approximately 0.2 in 6 pixels. A representative example of a patient's difference between the predicted and the correct points on an X-ray image is shown in Figure 3. The differences between the actual measured WBL ratio and the WBL ratio predicted by the CNN in this patient were 0.03 in the right knee and 0.05 in the left knee.

Table 1. Baseline characteristics of the dataset.

	Training Set	Validation Set	Test Set	Total	p-Value
Age (year)	65.1 ± 9.31	65.0 ± 9.42	65.4 ± 9.27	65.2 ± 12.2	0.521
Gender (M/F)	391/1928	46/241	49/241	486/2410	0.215
BMI (kg/m^2)	25.7 ± 3.22	25.8 ± 2.51	25.3 ± 2.37	25.7 ± 3.21	0.333
WBL ratio	0.33 ± 0.17	0.31 ± 0.15	0.32 ± 0.12	0.32 ± 0.15	0.087

Values are presented as number or mean ± standard deviation. M: men, F: female, BMI: body mass index, WBL: weight-bearing line. Training set: the sample of data used to fit the model. Validation set: the sample of data used to provide an unbiased evaluation of a model fit on the training dataset while tuning the model hyper-parameters. Test set: the sample of data used to provide an unbiased evaluation of a final model fit on the training dataset.

The prediction of the algorithm implemented by learning is shown in Table 2 using units of pixels and WBL error values. The mean accuracy (MA) was increased from around 0.5 using a 2-pixel unit to around 0.8 using 6 pixels in both the validation and the test sets. The probability of the prediction and target values entering within 6 pixels was close to 0.8. When the tibial plateau length was taken as 100%, the MA was increased from approximately 0.1, using 1%, to approximately 0.5, using 5%, in both the validation and the test sets. The probability of obtaining a value within 5% exceeded 0.5.

The mean difference, ICC, and MAE value are shown in Table 3. The mean difference of the validation and test sets between intra-observer and inter-observer measurements of the WBL ratio using a WLR ranged from 0.023 to 0.036. The MAE of the validation and test sets, with the DL model measuring the WLB ratio using simple knee radiographs were 0.064 (95% CI, 0.057–0.071) and 0.051 (95% CI, 0.044–0.058), respectively. The ICCs of the validation and test sets were all over 0.8, which indicated excellent agreement. The distributions of the WBL predictions in the validation and test sets are shown in Figure 4. The distribution of the DL model showed less difference in the high incidence area, where the WBL ratio was between 0.25 and 0.50. However, higher differences in the section that showed a lower incidence of the WBL ratio were noted. These values were similar for both the validation and the test sets. The MA of the WBL ratio for the validation and test sets is shown in Figure 5. As the thresholds of the WBL error percentage value and pixel units increased, the accuracy increased in both the validation and the test sets. The accuracy increased from around 0.1 to 0.5 with the increase in the WBL error percentage value thresholds, and from around 0.5 to 0.8 with the increase in the pixel threshold.

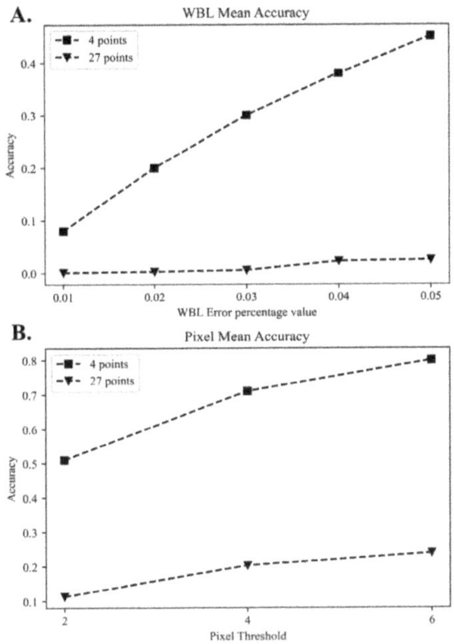

Figure 2. Comparison of WBL mean accuracy in 4 and 27 points. (**A**) Comparison graph of 4 and 27 points in WBL error percentage. (**B**) Comparison graph of 4 and 27 points in pixel threshold. Validation set: the data sample used to provide an unbiased evaluation of a model fit on the training dataset while tuning the model hyperparameters. Test set: the data sample used to provide an unbiased evaluation of a final model fit on the training dataset.

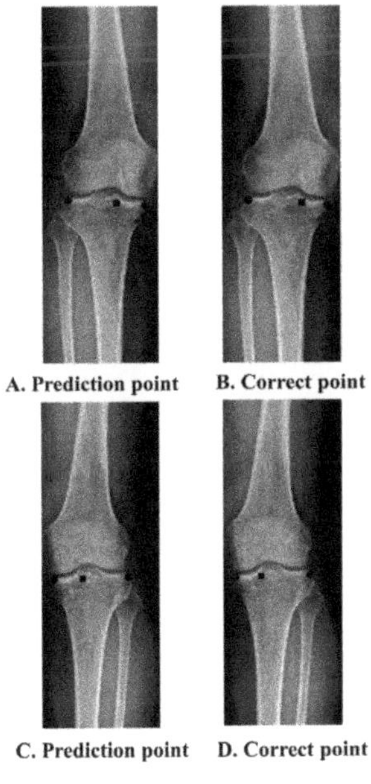

Figure 3. Prediction and correct point. (**A–D**) Prediction point vs. correct point. (**A**) Prediction point in Rt knee, (**B**) correct point in Rt knee, (**C**) prediction point in Lt knee, (**D**) correct point in Lt knee. Prediction point: point predicted by the deep learning algorithm. Correct point: labeled point.

Table 2. Results regarding the prediction of the deep learning algorithm.

	Validation Set					Test Set				
Pixel	2 pixels		4 pixels		6 pixels	2 pixels		4 pixels		6 pixels
MA	0.51 ± 0.07		0.71 ± 0.06		0.80 ± 0.08	0.52 ± 0.06		0.73 ± 0.06		0.84 ± 0.04
WEV	1%	2%	3%	4%	5 %	1%	2%	3%	4%	5%
MA	0.08 ± 0.04	0.20 ± 0.05	0.30 ± 0.06	0.38 ± 0.07	0.45 ± 0.07	0.10 ± 0.05	0.22 ± 0.07	0.37 ± 0.09	0.46 ± 0.11	0.53 ± 0.11

Values are presented as number or mean ± standard deviation. MA: Mean accuracy, WEV: WBL error value. Validation set: the sample of data used to provide an unbiased evaluation of a model fit on the training dataset while tuning model hyper-parameters. Test set: the sample of data used to provide an unbiased evaluation of a final model fit on the training dataset.

Table 3. Mean difference, mean absolute error, and ICC in the validation and test sets.

Validation Set	Mean Difference	ICC
A1 and A2	0.023 ± 0.005	0.96 ± 0.02
A1 and B	0.034 ± 0.003	0.94 ± 0.05
Test set	Mean difference	ICC
A1 and A2	0.024 ± 0.006	0.97 ± 0.03
A1 and B	0.036 ± 0.004	0.95 ± 0.05
Validation set	Mean absolute error	ICC
AL and DL	0.064 ± 0.007	0.89 ± 0.09
Test set	Mean absolute error	ICC
AL and DL	0.051 ± 0.007	0.88 ± 0.08

Values are presented as number or mean ± standard deviation. ICC, intra-class correlation coefficients. Validation set, the sample of data used to provide an unbiased evaluation of a model fit on the training dataset while tuning the model hyper-parameters. Test set, the sample of data used to provide an unbiased evaluation of a final model fit on the training dataset; A1, rater 1, A2: rater 1 at different times; B, rater 2; AL, rater 1 with labeling on the WLR; DL, deep learning.

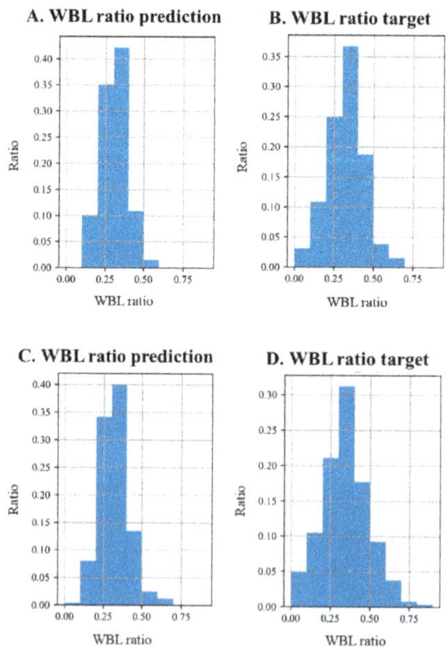

Figure 4. Histogram of the WBL ratio distribution in the validation and test sets. (**A–D**) Prediction distribution vs. target distribution. (**A**) WBL ratio prediction distribution in the validation set, (**B**) WBL ratio target distribution in the validation set, (**C**) WBL ratio prediction distribution in the test set, (**D**) WBL ratio target distribution in the test set. WBL ratio prediction: predicted value of the WBL, WBL ratio target: calculated WBL target.

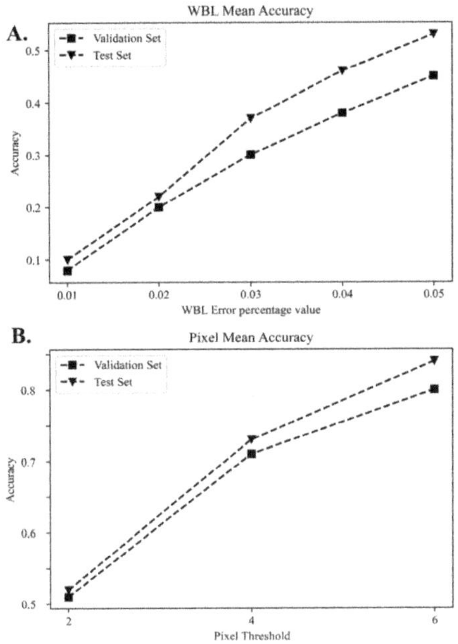

Figure 5. Graph of the WBL mean accuracy. (**A**) Accuracy graph of the WBL error percentage value, (**B**) accuracy graph of the pixel threshold. Validation set: the data sample used to provide an unbiased evaluation of a model fit on the training dataset while tuning the model hyperparameters. Test set: the sample of data used to provide an unbiased evaluation of a final model fit on the training dataset.

4. Discussion

The principal finding of this study is that the novel approach using the four-point marked key-point detection algorithm could predict the alignment of the lower limb using standing knee AP radiographs with high accuracy, comparable to that achieved with the direct measurement of the WLR. As one pixel was 0.265 mm, approximately 80% of the test and validation set prediction values were entered into 6 pixels. Therefore, it was assumed that approximately 80% of patients could be correctly evaluated within approximately 1 mm intervals.

The mean difference, MAE, and ICC values were used to test the accuracy and reliability of this study [19]. The accuracy of DL was indirectly estimated by comparing the mean difference between the values measured by the raters using WLR and the MAE of the values of DL predicting the WBL ratio using simple knee radiographs. The WBL ratio showed a slightly increase in the MAE when compared with the mean difference, because the WLR was not provided. In addition, the ICC of DL was also lower than the ICC of the values measured by the raters using WLR, but we observed that it provided a relatively good prediction at 0.8 or more.

Several trials to predict lower limb alignment using simple knee radiographs by linear regression analysis exist [10,20]. However, the results are not satisfactory, and to obtain satisfactory results, an X-ray image approximately 20 cm in length above and below the knee joint was required [21]. To solve these problems, a new prediction method based on DL using a key-point detection algorithm is proposed in our study. The key-point detection algorithm was considered the most suitable model because of the ability of the algorithm to find a specific point where the WBL passes through the tibia plateau [22,23]. Key-point detection algorithms are often used for pose estimation, face detection, and object detection [17,23]. Interestingly, the prediction accuracy decreased when the marking increased. This seems to be due to the characteristics of the DL model. DL is so automatic and high dimensional that the process of calculating the output by extracting features from the input is represented as a black box [13,24,25]. Therefore, it can be understood that if more labeling is performed, the labeling error increases, and the automatic feature extraction process is hindered; thus, the prediction accuracy could be lowered, as occurred in our study.

From the patient's point of view, visiting a tertiary hospital for WLR is time-consuming, expensive, and leads to high radiation exposure [26]. If the WBL ratio can be predicted through simple knee AP radiographs using this algorithm in primary care, it will be possible to easily determine the treatment process, as well as the degree of arthritis in more detail. This will also be useful for follow-up evaluation of patients who underwent re-alignment procedures such as osteotomy [19,27]. Expecting lower limb alignment using only simple knee radiographs has a lot of pros in the decision of patient-specific treatment protocols in various kinds of institutions. A DL model for predicting the WBL ratio through a simple knee radiograph was not attempted in the past, but it will become an essential medical technique in the future society characterized by the use of artificial intelligence [28].

The strength of this study is that a more accurate prediction of lower limb alignment can be obtained using the DL key-point detection model. Our study is meaningful in that it not only uses a key-point detection model, but also takes a significant point and trains ML to make a prediction using this model. In addition, this study has significance as it allows predicting the WBL ratio using a simple knee AP radiograph in a situation where WLR imaging is limited. Primary care physicians can properly diagnose patients with knee OA using the DL model. In addition, it is expected that planning the realignment procedure would be possible with high accuracy.

This study has several limitations. First, it was difficult to interpret the developed CNN model itself; therefore, it was hard to determine whether the CNN model focused on the WBL prediction. Second, the number of test sets was small, although the ratio between the validation and the test sets was adequate. Third, the WBL predictions of our model were relatively distributed around the center compared with the target WBL value. Thus, our model was trained on general WBL values and has limitations in predicting infrequent WBL values. This is because of the small amount of WBL data corresponding to outliers, as shown in Figure 4. This phenomenon is expected to decrease as the amount of data increases. Fourth, the prediction would be inaccurate if there is a deformity in the proximal femur or distal tibia, because this cannot be checked on standing simple knee radiographs. Fifth, since the study was conducted on patients who presented to tertiary medical institutions, there may be a selection bias in the patient group. This is because these patients were referred from primary and secondary medical institutions. The severity of disease in patients visiting tertiary care may be higher than that of patients visiting primary or secondary health care institutions.

5. Conclusions

The DL-based key-point detection algorithm to predict lower limb alignment through labeling using simple knee AP radiographs demonstrated comparable accuracy to that of the direct measurement using whole leg radiographs. Using this algorithm, the WBL ratio prediction with simple knee AP radiographs could be useful to diagnose lower limb alignment in osteoarthritis patients in primary care.

Author Contributions: All authors were involved in drafting the article or revising it critically for important intellectual content. Detailed contribution: study conception and design (S.H.P., H.S.N. and Y.S.L.), acquisition and analysis of data (S.H.P., H.S.N., J.P.Y.H. and Y.S.L.), discussions and interpretation of data (H.S.N., S.Y.P., J.H.C. and Y.S.L.), drafting of the article (S.H.P., H.S.N. and Y.S.L.). All authors have read and agreed to the published version of the manuscript.

Funding: This work was supported by a National Research Foundation of Korea (NRF) grant funded by the Korea government (MSIT) No.2021R1A2C1092657.

Institutional Review Board Statement: Institutional Review Board approval was obtained before performing the study (IRB NO: B-2210-784-101, Approval date: 15 September 2022).

Informed Consent Statement: Not applicable.

Data Availability Statement: Y.S.L. (smcos1@hanmail.net) had full access to all of the data in the study and takes responsibility for the integrity of the data and the accuracy of the data analysis. All authors have read and agreed to the published version of the manuscript.

Conflicts of Interest: The authors declare no conflict of interest.

References

1. Neogi, T. The epidemiology and impact of pain in osteoarthritis. *Osteoarthr. Cartil.* **2013**, *21*, 1145–1153. [CrossRef] [PubMed]
2. Lee, D.H.; Kim, S.J.; Kim, S.A.; Ju, G.I. Past, present, and future of cartilage restoration: From localized defect to arthritis. *Knee Surg. Relat. Res.* **2022**, *34*, 1. [CrossRef] [PubMed]
3. Costello, K.E.; Felson, D.T.; Neogi, T.; Segal, N.A.; Lewis, C.E.; Gross, K.D.; Nevitt, M.C.; Lewis, C.L.; Kumar, D. Ground reaction force patterns in knees with and without radiographic osteoarthritis and pain: Descriptive analyses of a large cohort (the Multicenter Osteoarthritis Study). *Osteoarthr. Cartil.* **2021**, *29*, 1138–1146. [CrossRef] [PubMed]
4. Fürmetz, J.; Patzler, S.; Wolf, F.; Degen, N.; Prall, W.C.; Soo, C.; Böcker, W.; Thaller, P.H. Tibial and femoral osteotomies in varus deformities—Radiological and clinical outcome. *BMC Musculoskelet. Disord.* **2020**, *21*, 201. [CrossRef]
5. Matsumoto, T.; Hashimura, M.; Takayama, K.; Ishida, K.; Kawakami, Y.; Matsuzaki, T.; Nakano, N.; Matsushita, T.; Kuroda, R.; Kurosaka, M. A radiographic analysis of alignment of the lower extremities–initiation and progression of varus-type knee osteoarthritis. *Osteoarthr. Cartil.* **2015**, *23*, 217–223. [CrossRef]
6. Wang, J.H.; Shin, J.M.; Kim, H.H.; Kang, S.H.; Lee, B.H. Discrepancy of alignment in different weight bearing conditions before and after high tibial osteotomy. *Int. Orthop.* **2017**, *41*, 85–92. [CrossRef]
7. Lin, Y.H.; Chang, F.S.; Chen, K.H.; Huang, K.C.; Su, K.C. Mismatch between femur and tibia coronal alignment in the knee joint: Classification of five lower limb types according to femoral and tibial mechanical alignment. *BMC Musculoskelet. Disord.* **2018**, *19*, 411. [CrossRef]
8. Zampogna, B.; Vasta, S.; Amendola, A.; Uribe-Echevarria Marbach, B.; Gao, Y.; Papalia, R.; Denaro, V. Assessing Lower Limb Alignment: Comparison of Standard Knee Xray vs Long Leg View. *Iowa Orthop. J.* **2015**, *35*, 49–54.
9. Choi, H.U.; Kim, D.H.; Lee, S.W.; Choi, B.C.; Bae, K.C. Comparison of Lower-Limb Alignment in Patients with Advanced Knee Osteoarthritis: EOS Biplanar Stereoradiography versus Conventional Scanography. *Clin. Orthop. Surg.* **2022**, *14*, 370–376. [CrossRef]
10. Pan, S.; Huang, C.; Zhang, X.; Ruan, R.; Yan, Z.; Li, Z.; Pang, Y.; Guo, K.; Zheng, X. Non-weight-bearing short knee radiographs to evaluate coronal alignment before total knee arthroplasty. *Quant. Imaging Med. Surg.* **2022**, *12*, 1214–1222. [CrossRef]
11. Karnuta, J.M.; Luu, B.C.; Roth, A.L.; Haeberle, H.S.; Chen, A.F.; Iorio, R.; Schaffer, J.L.; Mont, M.A.; Patterson, B.M.; Krebs, V.E.; et al. Artificial Intelligence to Identify Arthroplasty Implants from Radiographs of the Knee. *J. Arthroplast.* **2021**, *36*, 935–940. [CrossRef]
12. Joseph, G.B.; McCulloch, C.E.; Nevitt, M.C.; Link, T.M.; Sohn, J.H. Machine learning to predict incident radiographic knee osteoarthritis over 8 Years using combined MR imaging features, demographics, and clinical factors: Data from the Osteoarthritis Initiative. *Osteoarthr. Cartil.* **2022**, *30*, 270–279. [CrossRef] [PubMed]
13. Ko, S.; Pareek, A.; Ro, D.H.; Lu, Y.; Camp, C.L.; Martin, R.K.; Krych, A.J. Artificial intelligence in orthopedics: Three strategies for deep learning with orthopedic specific imaging. *Knee Surg. Sport. Traumatol. Arthrosc.* **2022**, *30*, 758–761. [CrossRef] [PubMed]
14. Ye, Q.; Shen, Q.; Yang, W.; Huang, S.; Jiang, Z.; He, L.; Gong, X. Development of automatic measurement for patellar height based on deep learning and knee radiographs. *Eur. Radiol.* **2020**, *30*, 4974–4984. [CrossRef]
15. Moon, H.D.; Choi, H.G.; Lee, K.J.; Choi, D.J.; Yoo, H.J.; Lee, Y.S. Can Deep Learning Using Weight Bearing Knee Anterio-Posterior Radiograph Alone Replace a Whole-Leg Radiograph in the Interpretation of Weight Bearing Line Ratio? *J. Clin. Med.* **2021**, *10*, 1772. [CrossRef] [PubMed]
16. Qorchi, S.; Vray, D.; Orkisz, M. Estimating Arterial Wall Deformations from Automatic Key-Point Detection and Matching. *Ultrasound Med. Biol.* **2021**, *47*, 1367–1376. [CrossRef] [PubMed]
17. Gao, J.; Yang, T. Research on Real-Time Face Key Point Detection Algorithm Based on Attention Mechanism. *Comput. Intell. Neurosci.* **2022**, *2022*, 6205108. [CrossRef] [PubMed]
18. Qi, J.; Du, J.; Siniscalchi, S.M.; Ma, X.; Lee, C.-H. On mean absolute error for deep neural network based vector-to-vector regression. *IEEE Signal Process. Lett.* **2020**, *27*, 1485–1489. [CrossRef]
19. Lee, Y.S.; Lee, B.K.; Kwon, J.H.; Kim, J.I.; Reyes, F.J.; Suh, D.W.; Nha, K.W. Serial assessment of weight-bearing lower extremity alignment radiographs after open-wedge high tibial osteotomy. *Arthroscopy* **2014**, *30*, 319–325. [CrossRef] [PubMed]
20. Oh, S.M.; Bin, S.I.; Kim, J.Y.; Lee, B.S.; Kim, J.M. Short knee radiographs can be inadequate for estimating TKA alignment in knees with bowing. *Knee Surg. Relat. Res.* **2020**, *32*, 9. [CrossRef]
21. Sgroi, M.; Faschingbauer, M.; Reichel, H.; Kappe, T. Can the frontal tibiofemoral alignment be assessed on anteroposterior knee radiographs? *J. Orthop. Traumatol.* **2016**, *17*, 339–343. [CrossRef]
22. Wang, R.; Zeng, L.; Wu, S.; Cao, W.; Wong, K. Illumination-Invariant Feature Point Detection Based on Neighborhood Information. *Sensors* **2020**, *20*, 6630. [CrossRef] [PubMed]

23. Sun, Y.; Xing, Y.; Zhao, Z.; Meng, X.; Xu, G.; Hai, Y. Comparison of manual versus automated measurement of Cobb angle in idiopathic scoliosis based on a deep learning keypoint detection technology. *Eur. Spine J.* **2022**, *31*, 1969–1978. [CrossRef] [PubMed]
24. Hsu, W.; Elmore, J.G. Shining Light Into the Black Box of Machine Learning. *J. Natl. Cancer Inst.* **2019**, *111*, 877–879. [CrossRef]
25. Yi, P.H.; Wei, J.; Kim, T.K.; Sair, H.I.; Hui, F.K.; Hager, G.D.; Fritz, J.; Oni, J.K. Automated detection & classification of knee arthroplasty using deep learning. *Knee* **2020**, *27*, 535–542. [CrossRef]
26. Perka, N.; Kopf, S.; Hommel, H. A whole leg radiograph is not necessary for postoperative determination of the mechanical leg axis after total knee arthroplasty. *Arch. Orthop. Trauma Surg.* **2019**, *139*, 1455–1460. [CrossRef]
27. Yoon, T.H.; Choi, C.H.; Kim, S.J.; Kim, S.H.; Kim, N.H.; Jung, M. Effect of Medial Open-Wedge High Tibial Osteotomy on the Patellofemoral Joint According to Postoperative Realignment. *Am. J. Sports Med.* **2019**, *47*, 1863–1873. [CrossRef]
28. Richardson, J.P.; Smith, C.; Curtis, S.; Watson, S.; Zhu, X.; Barry, B.; Sharp, R.R. Patient apprehensions about the use of artificial intelligence in healthcare. *NPJ Digit. Med.* **2021**, *4*, 140. [CrossRef]

tomography

MDPI

Article

Avoiding Tissue Overlap in 2D Images: Single-Slice DBT Classification Using Convolutional Neural Networks

João Mendes [1,2], Nuno Matela [1,*] and Nuno Garcia [2]

[1] Faculdade de Ciências, Instituto de Biofísica e Engenharia Biomédica, Universidade de Lisboa, 1749-016 Lisboa, Portugal
[2] Faculdade de Ciências, LASIGE, Universidade de Lisboa, 1749-016 Lisboa, Portugal
* Correspondence: nmatela@fc.ul.pt

Abstract: Breast cancer was the most diagnosed cancer around the world in 2020. Screening programs, based on mammography, aim to achieve early diagnosis which is of extreme importance when it comes to cancer. There are several flaws associated with mammography, with one of the most important being tissue overlapping that can result in both lesion masking and fake-lesion appearance. To overcome this, digital breast tomosynthesis takes images (slices) at different angles that are later reconstructed into a 3D image. Having in mind that the slices are planar images where tissue overlapping does not occur, the goal of the work done here was to develop a deep learning model that could, based on the said slices, classify lesions as benign or malignant. The developed model was based on the work done by Muduli et. al, with a slight change in the fully connected layers and in the regularization done. In total, 77 DBT volumes—39 benign and 38 malignant—were available. From each volume, nine slices were taken, one where the lesion was most visible and four above/below. To increase the quantity and the variability of the data, common data augmentation techniques (rotation, translation, mirroring) were applied to the original images three times. Therefore, 2772 images were used for training. Data augmentation techniques were then applied two more times—one set used for validation and one set used for testing. Our model achieved, on the testing set, an accuracy of 93.2% while the values of sensitivity, specificity, precision, F1-score, and Cohen's kappa were 92%, 94%, 94%, 94%, and 0.86, respectively. Given these results, the work done here suggests that the use of single-slice DBT can compare to state-of-the-art studies and gives a hint that with more data, better augmentation techniques and the use of transfer learning might overcome the use of mammograms in this type of studies.

Keywords: carcinoma of the breast; CNN; DBT; deep learning

Citation: Mendes, J.; Matela, N.; Garcia, N. Avoiding Tissue Overlap in 2D Images: Single-Slice DBT Classification Using Convolutional Neural Networks. *Tomography* **2023**, *9*, 398–412. https://doi.org/10.3390/tomography9010032

Academic Editors: Rafał Obuchowicz, Michał Strzelecki and Adam Piorkowski

Received: 23 December 2022
Revised: 8 February 2023
Accepted: 13 February 2023
Published: 14 February 2023

1. Introduction

The prevalence of breast cancer (BC) has been growing for several years and, in 2020, it became the most commonly diagnosed type of cancer [1]. The nefarious effects of this disease also represent a significant weight in the cancer paradigm: one in six deaths by cancer in women is caused by BC [2].

The emergence of breast cancer is related to both genetic predisposition (BRCA1 and BRCA 2 gene mutations put women at higher risk [3]) and environmental risk factors. Age and increased breast density are two of the most studied factors that contribute to the risk of developing this disease [4]. Several studies related dense breast patterns with an increased risk of BC [5].

One of the best weapons to fight the potential effects of breast cancer is to have an early diagnosis, which is currently aimed through generalized screening programs. Most of the worldwide screening programs use mammography as the standard imaging technique; however, its benefits and harms have been a topic of discussion in the scientific community [6].

The probability, in Europe, of a woman of age 50–69 with a biennial screening having a false positive result is 20%. On the other hand, in the United States of America, 50% of women will suffer the consequences of a false positive. Nonetheless, getting a false negative is also a possible outcome of screening with mammography: 28–33% of the cancers detected in women that undergo screening are interval cancer [6].

Since mammography is a two-dimensional image of the breast—which is a 3D volume—there will be tissue overlapping [7]. This fact can, consequently, lead to both tumour masking or false lesions appearance, hence contributing to false positive/negative rates. One way to surpass the barriers imposed by overlapping tissue is to use digital breast tomosynthesis (DBT). The fact that the DBT slices are so thin is what guarantees that the problem with overlapping tissue does not occur in DBT in the same way that it appears in mammography. Despite the fact that a DBT exam consists of several slice acquisitions, the dose of radiation is similar to that used in common mammography routines [8].

Nonetheless, it is important to verify how DBT compares with full-field digital mammography (FFDMG) in several other parameters. The use of DBT alongside FFDMG compared to the use of FFDMG alone increases the cancer detection rate [9]. These results should be looked upon carefully; even though detection of more cancers, and earlier, might positively impact women's lives (less aggressive treatment and better outcomes), it is important to assure that this increase in detection rate is not related to overdiagnosis [10] The use of DBT+FFDMG also has positive impacts in recall rates, which results in a higher specificity compared to the use of FFDMG alone [10].

Therefore, given that DBT alone can overcome the problem of tissue overlapping and that its use in clinical practice helps to improve the detection rate while decreasing recall rates, this study focuses on the use of DBT.

Artificial intelligence (AI) has made its way into medical diagnosis and more specifically into the field of breast cancer imaging. A review of several applications of AI to breast imaging, done by our team, can be found elsewhere [11]. Given that, the classification of DBT images into healthy/diseased classes or into benign/malignant lesions can also be done through AI.

A group of researchers [12] aimed to classify DBT images and whole mammograms using convolutional neural networks (CNN). In order to do that, they used both well-established algorithms—AlexNet [13] and ResNet [14]—and self-developed models. Different variations of the established models were used depending on the type of classification being made: 3D models if they wanted to classify DBT, and 2D models if the goal was to classify mammograms. Moreover, the authors also aimed to compare model performance for the AlexNet and ResNet with and without transfer learning [15].

The idea of combining the information of 2D mammograms and 3D DBT was proposed by Liang et al. [16]. There, 2D models were used both for mammogram and DBT classification. In order to do that, the authors did not use the entire volume to make a classification but rather extracted a "fixed slice" from the volume, which aimed to represent the observed changes across the different slices. Their model could be divided into a "backbone" architecture—a fully convolutional network (AlexNet, ResNet, SqueezeNet, or DenseNet) that served as a feature extractor—and an ensemble of classifiers, each one composed of a convolutional layer succeeded by two fully connected layers. There were three classifiers: one that classified the features extracted from mammography images, another that classified the features driven from DBT, and a third classifier that was focused on classifying the concatenation of the features extracted from each imaging modality. To decide the final output, a majority voting algorithm was applied.

The concept of retrieved fixed slices or dynamic images from a DBT volume has been used in other studies with the rationale of diminishing the computational burden that is associated with big 3D volumes. Zhang et al. [17] aimed to implement that and compare it with an alternative methodology where features were extracted from each slice and a final feature map was obtained by a pooling procedure across the different feature maps. Besides comparing these methods between themselves, the authors also compared both their methodologies with classic AlexNet architectures for 3D volumes. The authors found that any of the variations of their proposed methodology outperformed the classical architectures, which achieved 0.63 as the maximum AUC value. With their approach, they found out that the extraction of features from each slice and further pooling them into a final feature map produced better results than using the dynamic image; the maximum AUC values for each approach were 0.854 and 0.792, respectively, both using AlexNet architectures.

Furthermore, there are several different researches that aim to differentiate benign and malignant lesions. The work by Muduli et al. [18] that served as motivation for this research, aimed to classify lesions present in mammograms and ultrasound. To do that, the authors applied classic data augmentation techniques (translation, rotation, etc.) and used a CNN model with five learnable layers. Another group of researchers [19], to avoid the inflexible receptive field of 3D convolutions for DBT classification, developed a hybrid model that was able to extract hierarchical feature representations using 2D convolution in a slice-by-slice approach. Moreover, a self-attention module was used to learn structural representations of benign and malignant lesions. The idea of combining 2D and 3D convolution was also pursued by Xiao et al. [20] for slice feature extraction and lesion structural information, respectively. While many works define a region of interest to analyse the lesions, some authors [21] aim to use entire DBT slices for identifying the presence of lesions through the use of self-built CNN.

As it can be seen, there are several studies that aim to use the advantages given by DBT in relation to mammography, while maintaining the 2D convolution approach. These approaches allow the authors to overcome the problem of overlapping tissue, leading to lesion masking or fake lesion creation, while maintaining the low computational burden that is associated with 2D CNN architectures.

Given that, the aim of this work was to develop a CNN model to differentiate malignant lesions from benign lesions, using single DBT slices.

2. Materials and Methods

2.1. Dataset

The data used in this study came from the Breast Cancer Screening-Digital Breast Tomosynthesis (BCS-DBT) dataset, publicly available in the Cancer Imaging Archive [22]. This dataset contains DBT images from healthy, actionable, benign, and malignant cases. The cases were collected at Duke Health [23] System from 2014 to 2018 and were made available in anonymized DICOM format. The number of patients present in this dataset goes up to almost 14,000. Each case was annotated by expert radiologists, as was the DBT slice where the lesion is most visible. The patient IDs with the associated class, the available views, and the lesion slice information are also available at the referred website. The available data consist of three sets of images: training, validation, and testing. However, only the labels for the training set are available, so all of the data used in this study were derived from the training set. This set is composed of nearly 22,000 scans but not all of them were used. The rationale for the dataset constructed and used here was a balance amongst the different classes used, meaning that it was aimed to have approximately the same number of benign and malignant cases. Given that, the assembly of the dataset was limited by the class with the fewest observations. There were 39 available volumes of malignant images and, for that reason, 39 volumes from benign cases were also retrieved, making a total of 78 volumes considered for this work. As already mentioned, the goal was to make a single-slice classification and, for that reason, the single slice chosen was the one where the lesion was most visible. Thus, each volume contributed with a 2D slice that served to either train or test the developed model.

2.2. Image Preparation

Each of the 2D slices used in this study had to undergo a preprocessing methodology before being fed to the model. First, and following the information present in the dataset, all the images were segmented in order to retrieve small regions of interest (ROIs) that contained the lesions. For each DBT volume, experts annotated the slice where the lesion was most visible. Moreover, a small region that encapsulated the lesion was also defined. This information was used to manually extract the ROIs from each slice. After that, in order to improve image quality to ensure that the model was not misled, image adjustment techniques were applied to all ROIs. This technique aimed to improve the contrast of the image by saturating both the bottom and superior 1% of intensity values. Finally, the extracted regions of interest were normalized in terms of pixel intensity, so that the model did not learn by mistake random pixel intensity patterns that could be present in the images. A "min-max" normalization was applied (Equation (1)), which resulted in an image with pixel values ranging from 0 to 1. In the equation, I is the original image, $min(I)$ is the overall minimum pixel intensity of image I, $max(I)$ is the overall maximum pixel intensity of image I, and $I_{normalized}$ is the normalized image.

$$I_{normalized} = \frac{I - min(I)}{max(I) - min(I)} \quad (1)$$

After that, each pixel value was multiplied by a scaling factor of 255. Given that each volume contributed with an image, 78 slices were available, which was not a substantial number of images for the proposed task. Actually, one of the malignant volumes was compromised so only 77 slices could be used. In order to overcome this problem, common data augmentation techniques were designed to increase both the variability and the number of available images. First, for each volume, instead of taking just the lesion slice, four slices above and four slices below the reference one were also extracted. The fact that the slices were retrieved from a relatively small neighbourhood of the lesion slice allowed us to guarantee that the lesion was still clearly visible, while still showing a slightly different breast configuration. On the other hand, these slices shared a lot of characteristics with the original lesion slice so the simple addition of them to the data pool could bias classification. With the purpose of surpassing this obstacle, an augmentation routine was applied three times to each of the slices. This routine consisted of randomly applying a transformation or a combination of transformations to each of the slices. The possible transformations defined were: a rotation ranging from $-50°$ to $50°$; a vertical translation ranging from 1 pixel to 5% of the vertical size of the image; a horizontal translation ranging from 1 pixel to 5% of the horizontal size of the image; a horizontal reflection; and a vertical reflection. The limits of the rotation and translation transformations were imposed with the rationale of guaranteeing that while the variation of the images was increased, the lesion remained within the image. Given the wide range of possibilities for the rotation transform and for both translations, there was great confidence that besides augmenting the number of available images, there was also considerable variability in the data used. This data augmentation procedure was done four times to each image, hence increasing four times the number of available 2D images. The original images and three of the four augmented images groups were used for training, while one of the augmented image groups was used for validation. Before feeding the images to the model, all of them were resized to dimensions of 224 $\times$ 224. Given that, 2772 (77 volumes $\times$ 9 slices $\times$ 3 augmentations + 77 volumes $\times$ 9 original slices) images were used for training and 693 images (77 volumes $\times$ 9 slices $\times$ 1 augmentation) were used for validation purposes. In order to test the model, the data augmentation routine was applied again, this time to the validation set, and predictions were made in this new augmented set of 693 images.

Figure 1 depicts the data selection and augmentation procedure for each DBT volume.

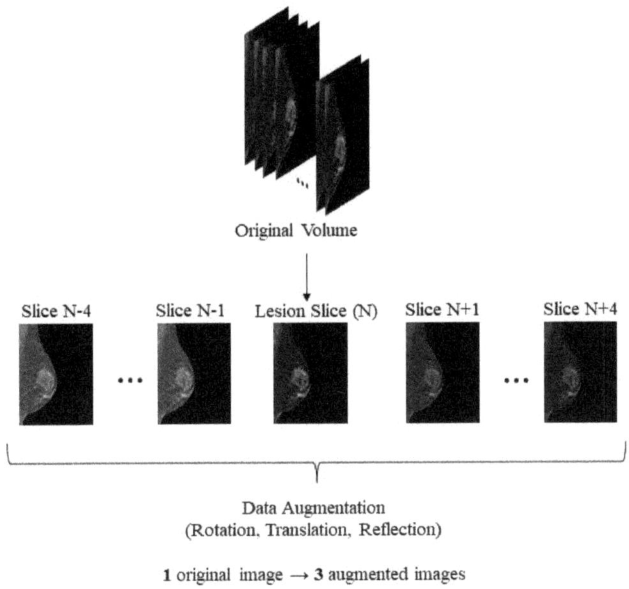

Figure 1. Data selection and augmentation—image contrast modified just for illustration purposes. From each original volume, the most representative slice and the four slices above and below were extracted. The rotation was randomly preformed in a range from −50° to 50°, the translation could be horizontal or vertical, ranging from 1 pixel to 5% of the overall image size, and the reflection could either be vertical or horizontal.

2.3. The Algorithm

As it was discussed during the introduction section, CNNs are widely used for the classification of medical imaging and, in particular, of DBT images. For that reason, a CNN model, inspired by the work of Muduli et al. [18], was developed from scratch. The created model was composed of five different layers: a convolution layer, an activation layer, a pooling layer, a fully connected layer, and a batch normalization layer. The convolutional layer is the base foundation of CNNs. These layers can be considered feature extractors that combine both linear operations (convolution) and nonlinear calculations (activation function). The convolution operation [24] is performed between a random array of numbers—kernel—and the input. Thus, given that the aim is to extract characteristics from the input, it can be said that the use of different kernels results in different feature extractors [25]. The result of this operation is, therefore, a feature map. When applying a convolution layer, one needs to predefine three convolution parameters: the kernel size, the number of filters, and the stride. The first one defines how much of the input the kernel sees at each convolution step; the second one has a self-explanatory name and defines the feature map depth; the stride, on the other hand, defines the space between two successive kernel positions during convolution. Moreover, it is known that the usual convolution operation results in an output feature map with a reduced size in comparison to the input. This is a problem if one wants to stack several sequential convolution layers—the size could become too small. Padding a neighbourhood of the image with pixels of intensity zero allows one to perform convolution operations while obtaining an output with the same size as the input.

In order to increase the relevance of the learned features, a nonlinear operation is applied to the feature map—an activation function. There are several activation functions but one of the most used is the rectified linear unit (ReLU) which takes in the input and outputs the maximum between the input and zero.

The pooling layer aims to reduce the dimensionality of the learnable parameters by summarizing the features present in a specific region of the feature map. This strategy besides introducing the capability of learning invariant features, is also important for reducing the computational cost associated with a high number of parameters. For that reason, training time can also be reduced using pooling strategies [25].

A batch normalization layer, as the name implies, normalizes its input while aiming to maintain the activation values as stable as possible, so that it not only speeds up training but also provides good results. In addition, this layer also helps to avoid problems with overfitting [18].

Finally, fully connected layers usually receive as input the feature map derived from the last layer and aim to map this input to the last output of the network. It is possible to use several fully connected stacked layers and since in these layers, every input is connected to every output, the number of learning parameters can be highly increased. The output of the last fully connected layer is given as the input to the last activation function layer where the chosen activation function needs to match the task that it is being aimed at; a softmax function is usually used.

Our model was inspired by the one developed by Muduli et al., so it was composed of four convolution–batch normalization–ReLU–max pooling blocks. The max pooling parameters were always the same with a (2,2) pooling size and a stride of 2, and a padding routine that maintained the dimensions of the input in the output. On the other hand, the convolution layers increased the number of filters as the network deepened: 16, 32, 64, and 128 filters, respectively. The kernel size, however, decreased along the network: $9 \times 9, 7 \times 7, 5 \times 5$, and 3×3, respectively. After these blocks, two fully connected layers finalized the network, one with 128 units, and another one with 2—the number of classes of our dataset. The outputs of this last layer were mapped to a softmax layer. For learning, the Adam Optimizer [26] was used, with an initial learning rate of 0.001, the loss metric used was the sparse categorical cross-entropy, and the evaluation metric during training was the accuracy. As said, at each epoch, the model was evaluated using the previously created validation set; the training procedure continued until a maximum of 500 epochs unless the performance of the model achieved a baseline accuracy value (0.905) that did not further increase. In that case, the model stopped training before getting to the 500 epochs. In order to avoid overfitting, L2 (factor = 0.001) and dropout (factor = 0.2) regularizations were used Between the fully connected layers the dropout factor was 0.5. Figure 2 generally depicts our model, while Table 1 describes each layer in the model.

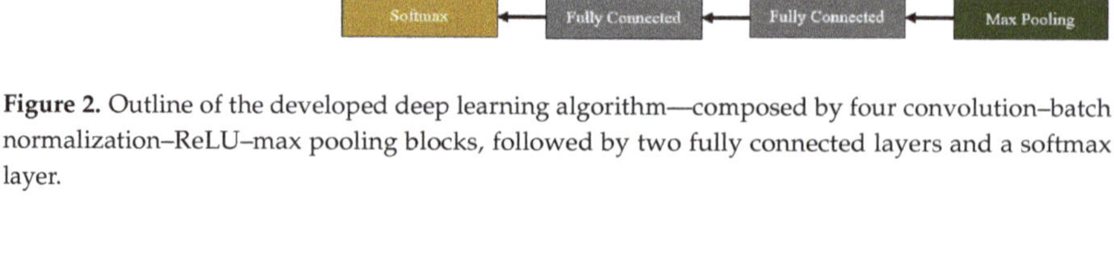

Figure 2. Outline of the developed deep learning algorithm—composed by four convolution–batch normalization–ReLU–max pooling blocks, followed by two fully connected layers and a softmax layer.

Table 1. Layer-by-layer description for the developed model. As the model deepened, the number of filter increased (16-32-64-128) and the kernel size diminished (9-7-5-3). Max pooling had a stride and kernel size of 2 across the network. The numbers of units in the fully connected layers were 128 and 2.

Layer	#Filters (Conv) /Units (FC)	Kernel Size	Stride	Padding
Convolution—1	16	9	1	None
Batch Normalization	-	-	-	-
ReLU	-	-	-	-
Max Pooling	-	2	2	Same
Convolution—2	32	7	1	None
Batch Normalization	-	-	-	-
ReLU	-	-	-	-
Max Pooling	-	2	2	Same
Convolution—3	64	5	1	None
Batch Normalization	-	-	-	-
ReLU	-	-	-	-
Max Pooling	-	2	2	Same
Convolution—4	128	3	-	None
Batch Normalization	-	-	-	-
ReLU	-	-	-	-
Max Pooling	-	2	2	Same
Fully Connected	128	-	-	-
Fully Connected	2	-	-	-
Softmax	-	-	-	-

3. Results

Our methodology started with the preparation of the images before feeding them to the model. First, in Figure 3, the overall results of the image adjustment in the entire image are shown, so that the effect of this methodology is fully noticed. As it can be perceived, the lesion becomes more visible, which is of extreme importance in the task of differentiating benign and malignant lesions.

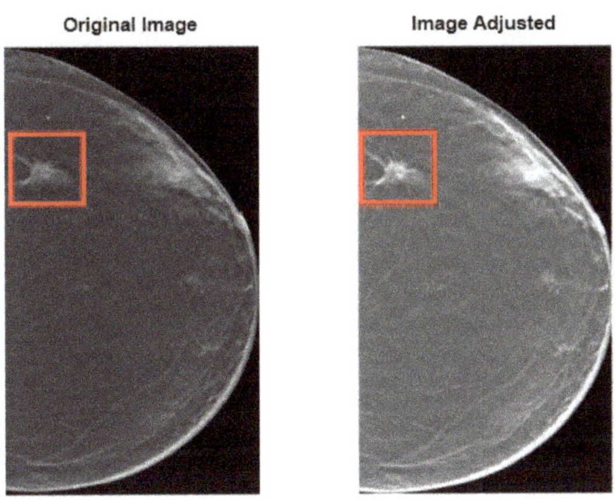

Figure 3. Results of the image adjustment procedure. On the left, the original image is shown with a bounding box encapsulating the lesion. On the right, the same can be seen but for the image after adjusting. It can be perceived how the lesion becomes more visible.

However, as said in the methodology section, rather than using the entire image of the breast, which could be a problem if the image had more than one lesion, ROIs were defined in a way that encompassed the lesion within their borders. Figure 4 shows two examples of defined ROIs, one for each of the considered classes.

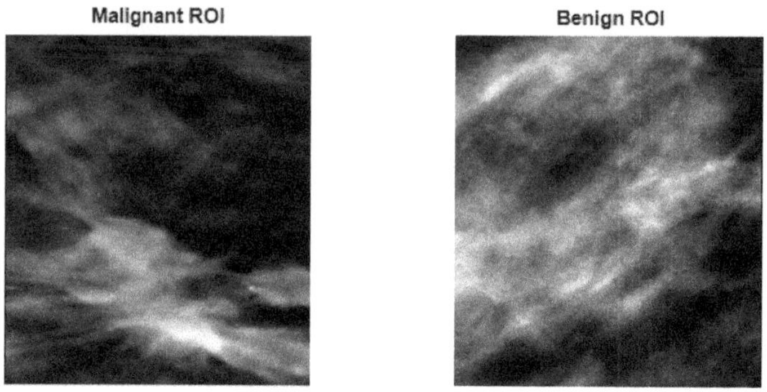

Figure 4. ROI definition to encapsulate malignant (**on the left**) and benign (**on the right**) lesions.

After the image preparation, and given the sparse number of available images to train the model, data augmentation techniques that consisted of random rotation, translations, and mirroring were performed. Figure 5 compares the two previously shown original ROIs with a version of the said images after a random transformation.

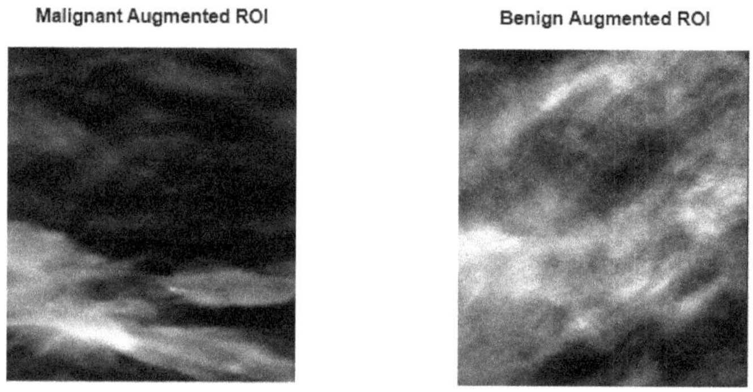

Figure 5. Data augmentation procedure applied to the previously seen ROIs. Random transformations that include rotation and/or translation and/or mirroring were applied to each ROI. As it can be perceived, the images are substantially different from what can be observed in Figure 4, while still maintaining the lesion within the image limits.

As it can be perceived, the images represent a clear distortion of the original images, while maintaining the lesions within the limits of the ROI. This fact means that the data augmentation technique used allowed us not only to increase the number and the variability of images available for training, but also to keep the lesions inside the defined ROI, hence not compromising the ground-truth labels given to the images. Table 2 specifies the number of images used for training, validation, and testing, and their division into the respective classes.

Table 2. Data division. Number of planar images used for training, validation, and testing.

	Training	Validation	Testing
Benign	1404	351	351
Malignant	1368	342	342

The model was evaluated through classical performance metrics: accuracy, sensitivity, specificity, precision, and F1 score. Equations (2)–(6) show how to compute each one of these metrics. In order to compute them, the confusion matrix of both the validation set and the testing set is presented (Figures 6 and 7). For the validation set, used to tune the model, the confusion matrix is seen in Figure 6. With those values, it was possible to compute the accuracy (90.7%) achieved after 494 epochs. Sensitivity, specificity, precision, and F1-score values were 92%, 89%, 89%, and 0.89 Regarding the testing set, the accuracy value was 93.2%. In terms of the previously mentioned metrics, their values in the testing set were 92%, 94%, 94%, and 0.94, respectively.

$$Accuracy = \frac{Correct\ Cases}{All\ Cases} \tag{2}$$

$$Sensitivity = \frac{True\ Positive}{True\ Positive + False\ Negative} \tag{3}$$

$$Specificity = \frac{True\ Negative}{True\ Negative + False\ Positive} \tag{4}$$

$$Precision = \frac{True\ Positive}{True\ Positive + False\ Positive} \tag{5}$$

$$F1\ Score = \frac{2 * Precision * Sensitivity}{Precision + Sensitivity} \tag{6}$$

Figure 6. Confusion matrix obtained with the validation set. From the total 693 instances, the model was capable of correctly classify 629 of them. It can also be noted that approximately 92% of the benign lesions were correctly classified, while for the malignant lesions, the same happened for approximately 89% of the instances.

		Real Labels	
		0	1
Predicted Labels	0	331	20
	1	27	315

Figure 7. Confusion matrix obtained with the test set. From the total 693 instances, the model was capable of correctly classify 646 of them. It can also be noted that approximately 92% of the benign lesions were correctly classified, while for the malignant lesions, the same happened for approximately 94% of them.

4. Discussion

The goal of our study was to develop a CNN model that was able to differentiate malignant from benign lesions. In order to do that, ROIs that encapsulated lesions from the two classes were defined. These ROIs were then passed through an image enhancement algorithm that improved the contrast of the images with the goal of giving more visibility to the lesions, which was important in the context of differentiating the nature of the said lesions. As it was seen in the results section, the goal was achieved, as when comparing the original image of the breast with the image after the adjustment, the improvement of the image contrast was patent, specifically concerning the highlight of the lesions and their borders. After the ROI definition and image adjustment, data augmentation methodologies were employed with the aim of improving both the number of instances for model training and their variability. To do that, to each image in the dataset was applied a random transformation (or transformations) that consisted of rotations and/or translations and/or mirroring. Given the wide range of possible transformations and looking at the results shown in Figure 5, it is safe to say that the goal was accomplished. While the number of images was highly increased, having three variations for each original image on the dataset being used for training, the variability was also increased in a way that allowed the lesions to remain within the margins of the defined ROI. Having achieved the proposed goals in terms of image preparation, it was possible to look correctly at the results obtained in terms of training, validation, and testing of the proposed model. This model was trained from scratch, so, opposite to several of the papers reviewed in the introduction section, no transfer learning methodology was followed. The main metric used to assess model performance during training/validation was accuracy because it was the metric used by Muduli et al., a work close to what was aimed at here, with a very similar network. That being said, the results of the several metrics calculated both for the validation and the testing set were very positive and gave confidence on the robustness of the developed model. From the ten computed metrics, eight of them were above 90% and the remaining ones were not lower than 89%, which was extremely good. However, a good accuracy value by itself does not guarantee a good model. A model that arbitrarily defines all instances as zero can achieve a good accuracy in a scenario where only negative data are given to it. It was not the case in this study as the class balance was taken into account during dataset construction. The remaining metrics gave a higher sense of the robustness of the model; however, it would be interesting to have a metric that gives a clear indication of how much the predictions of the model agreed with the ground-truth labels. With that in mind, Cohen's Kappa coefficient of agreement, that translates into a number of how much the agreement between two raters or readers is, was calculated. In the case of this paper, the "raters" were the ground-truth labels and the predictions made by the model. The calculation of this coefficient was done using Equation (7), where p_0 is the observational probability of agreement, and where p_e is the expected probability of agreement [27]. Table 3 shows how to interpret this factor [27].

$$kappa = \frac{p_o - p_e}{1 - p_e} \tag{7}$$

Table 3. Interpretation of kappa coefficient [27].

Coefficient Value	Strength of Agreement
<0.00	Poor
0.00–0.20	Slight
0.21–0.40	Fair
0.41–0.60	Moderate
0.61–0.80	Substantial
0.81–1.00	Almost perfect

The results for the validation and testing sets were of approximately 0.82 and 0.86, respectively, which according to Table 3 indicated an almost perfect agreement. All these results combined gave us confidence in the model that was able to correctly differentiate benign and malignant images from single slices of DBT images.

Most of the works reviewed in the introduction section used AUC as an evaluation metric, so a comparison between these studies and the work developed here is not fair. However, when looking at the obtained results, it is possible to say that our model achieved a good discriminatory capacity, which was proven by the F1 score and the kappa coefficient value. Thus, in that sense, it compared to most of the state-of-the-art papers reviewed in the introduction. For example, in the work presented in [12], it was found that the AlexNet models with transfer learning were the ones that presented better performance either for mammography (AUC = 0.7274) or DBT (AUC = 0.6632). As it can be perceived, a better performance was obtained with mammography than with DBT, which contradicted what was expected in theory; the authors pointed out that this may have been due to the fact the DBT volume was not entirely used and some of the discarded slices might have had relevant information. Our work, besides achieving a very good performance, tackled some of the problems that that research faced. On one hand, it used information present in the DBT volume from the most relevant slices and on the other hand, we used a 2D image that avoided the tissue overlapping present in mammography.

The authors of [16] compared the performance of their model with a classic AlexNet model but also compared, within their model, the use of a single imaging modality with the use of both modalities (FFDM + DBT). Their model, despite the use of a single or both modalities, outperformed 2D and 3D AlexNet. On the other hand, the use of an ensemble of imaging modalities outperformed the use of just DBT (AUCs of 0.97 and 0.89, respectively). A fair comparison between the work by these authors and the work done here is with their model using only DBT: for the four different backbone architectures, the accuracy values were of 81% (AlexNet), 79% (ResNet), 85% (DenseNet), and 79% (SqueezeNet). Our accuracy value on the testing set was 93.2%, which outperformed any of the models (the same happened for their models trained only on mammography). As for the F1-scores obtained by the authors, they ranged from 78 to 85%, which was lower than the value obtained in this work, 94%.

Zhang et al. [17] found that any of the variations of their proposed methodology (either late fusion of features, or a dynamic image) outperformed the classical architectures (3D convolution), which achieved 0.66 as the maximum AUC value. With their approach, they found out that the extraction of features from each slice and further pooling them into a final feature map produced better results than using the dynamic image, with maximum AUC values for each approach of 0.854 and 0.792, respectively, both using AlexNet architectures. Given the discrepancy in the metrics used, a fair comparison could not be made between their work and ours. However, their conclusions were interesting in that it showed the extraction of features from single slices not only outperformed the use of dynamic image but also classical 3D convolution approaches for DBT volumes.

On the other hand, a fairer comparison could be made between the developed work and the one presented by Muduli et al. These authors developed a model that could differentiate malignant and benign lesions on mammograms and ultrasound images. The architecture of the model was very similar to the one used here, differing in the number of fully connected layers and in the parameters (and type) of regularization. The comparison between the work done here and the one developed by Muduli's team is important because it allows us to understand how this novel way of learning breast characteristics through a single slice of DBT can compare to the use of mammograms.

The three datasets of mammograms used by the authors achieved a performance, in terms of accuracy, of 96.55%, 90.68%, and 91.28%, on each of the three mammography datasets used. Considering our testing set, the model developed in this work reached an accuracy of 93%, which outperformed the model of Muduli in two of the three datasets used. In terms of sensitivity, our model was outperformed in two of the datasets used and was comparable to the other (92% vs. 97.28%, 92.72%, 99.43%). Finally, in terms of specificity, it is our model that outperformed Muduli's work two out of three times (94% vs. 95.92%, 88.21%, 83.13%). As a result, overall, the model developed here was comparable and outperformed the use of mammograms, which indicates the potential that the use of single-slice DBT has in the field of AI applied to breast imaging.

Getting back to the works presented in the introduction section, the work proposed by Sun et al. [19] which consisted of an hybrid method with both 2D and 3D convolutions achieved an accuracy of nearly 80% while their F1 score was 83.54%. Both the results of this methodology, which used 2D convolutions for hierarchical feature representation and 3D convolutions for structural lesion information extraction, were outperformed by our methodology. A similar comparison could be made with the work of Xiao et al. [20], which had the same goal as the work of Sun et al. There, the best accuracy result achieved was 82%, while the F1 score obtained was 85.71%. As it happened with other research works, it was not fair to make a comparison between our work and the one done by [21] since different metrics were used; however, this study showed that the use of single DBT slices could yield very promising results.

However, it is important to analyse some flaws of the used methodology in order to improve future work or even to make it suitable for a real-life application. Originally, the dataset had 78 different 3D images before any strategy to augment the data. On the other hand, the datasets used by Muduli's team had 326, 1500, and 410 different images. This discrepancy in original data, if overcome by the single-slice DBT annotated images, might result in outperforming the classical research based on the use of mammograms.

On the other hand, the use of this classic data augmentation should also be looked upon. While random rotations, translations, noise adding, or contrast variation help to increase variability and the number of instances used for training, it is important to understand, or to think about, how helpful they might be from the clinical point of view. Two transformed images from the same women are in fact different and surely contribute to the increase of variability in the dataset. However, these images are much closer to one another in terms of breast patterns than what happens in clinical practice for two images from different women. While the use of these augmentation techniques is important for the development of novel and better diagnostic models, different methods for increasing the quantity and the variability of data that are closer to real-world scenarios should be followed, such as the use of GANs [28].

Finally, there are several studies [12] that show how much transfer learning can help to improve the performance of the developed models while reducing the computational burden associated with training models. Here, we aimed to train the model completely from scratch, which not only resulted in an increased training time but might also negatively affect the performance of the model.

5. Conclusions

The work done throughout this study aimed to develop a CNN model that could classify breast medical images into two categories: malignant and benign lesions. While the used algorithm was inspired by another research work [18], we introduced the novel use of single-slice DBT for the classification. After extracting ROIs from the images on lesion locations and applying preprocessing methodologies to improve image quality, data augmentation techniques were performed to increase the size of the dataset while also increasing its variability. The results obtained in the several computed metrics showed that the performance of the developed model compared with state-of-the-art research and even overcame it in some cases. This fact suggests that the use of single slice DBT—which overcomes the problem with overlapping tissue present in mammograms—could play a major role once some of the identified flaws of this work are tackled. The information present in single slices is also present in the 3D volume that constitutes DBT. The rationale behind the use of a single-slice approach was based on a literary review where most of the architectures that used 2D convolutions outperformed the models that used a 3D approach. Therefore, it was with that in mind and considering the computational burden of developing from scratch a model using 3D convolution, that the single-slice approach was chosen in this work. While the reviewed works were either based on dynamic images or latent representations, we aimed to look for slices where the lesion was visible because, in one way, it had the planar characteristics of mammograms (allowing the use of 2D convolutions) and, on the other hand, the slice was so thin that it did not present the problem of overlapping tissue present in mammograms. This fact allowed us to diminish the possibilities of either lesion masking or the appearance of false positives, allowing the model to have a better grip on the proposed task. With that in mind, the following steps should be taken in future work: developing more real-life data augmentation techniques, inspired by what is seen in clinical practice in terms of breast configuration variability, and classifying whole mammograms/DBT instead of just one region of interest since it may help to find lesions that were overlooked. Furthermore, there is a great potential in transfer learning approaches, not only because it diminishes the computational burden of training a model (compared to a "from scratch approach"), but because it may also allow one to increase the performance of the developed models. The work developed in this paper served the purpose of showing how promising the use of single slices is for the differentiation between benign and malignant lesions. With that being established, further work should be focused on improving the performance of the models that are based on this approach. Transfer learning can play a major role in this task.

Author Contributions: Conceptualization, J.M., N.M., and N.G.; methodology, J.M., N.M., and N.G.; writing, J.M., N.M., and N.G. All authors have read and agreed to the published version of the manuscript.

Funding: This research was funded by Fundação para a Ciência e Tecnologia–Portugal (FCT-IBEB Strategic Project UIDB/00645/2020), (EXPL/CCI-COM/0656/2021), (UIDB/00408/2020) and (UIDP/00408/2020), and Bolsa de Investigação para Doutoramento (2022.12271.BD).

Institutional Review Board Statement: Not applicable.

Informed Consent Statement: Not applicable.

Data Availability Statement: The dataset analyzed in this study can be found at https://wiki.cancerimagingarchive.net/pages/viewpage.action?pageId=64685580.

Conflicts of Interest: The authors declare no conflict of interest.

References

1. Deo, S.; Sharma, J.; Kumar, S. GLOBOCAN 2020 report on global cancer burden: Challenges and opportunities for surgica oncologists. *Ann. Surg. Oncol.* **2022**, *29*, 6497–6500. [CrossRef] [PubMed]
2. Arnold, M.; Morgan, E.; Rumgay, H.; Mafra, A.; Singh, D.; Laversanne, M.; Vignat, J.; Gralow, J.R.; Cardoso, F.; Siesling, S.; et al Current and future burden of breast cancer: Global statistics for 2020 and 2040. *The Breast* **2022**, *66*, 15–23. [CrossRef] [PubMed]
3. Li, H.; Giger, M.L.; Lan, L.; Bancroft Brown, J.; MacMahon, A.; Mussman, M.; Olopade, O.I.; Sennett, C. Computerized analysis o mammographic parenchymal patterns on a large clinical dataset of full-field digital mammograms: Robustness study with two high-risk datasets. *J. Digit. Imaging* **2012**, *25*, 591–598. [CrossRef] [PubMed]
4. Barba, D.; León-Sosa, A.; Lugo, P.; Suquillo, D.; Torres, F.; Surre, F.; Trojman, L.; Caicedo, A. Breast cancer, screening and diagnostic tools: All you need to know. *Crit. Rev. Oncol.* **2021**, *157*, 103174. [CrossRef]
5. Boyd, N.F.; Martin, L.J.; Yaffe, M.J.; Minkin, S. Mammographic density and breast cancer risk: Current understanding and future prospects. *Breast Cancer Res.* **2011**, *13*, 223. [CrossRef]
6. Løberg, M.; Lousdal, M.L.; Bretthauer, M.; Kalager, M. Benefits and harms of mammography screening. *Breast Cancer Res.* **2015** *17*, 63. [CrossRef]
7. Giess, C.S.; Frost, E.P.; Birdwell, R.L. Interpreting one-view mammographic findings: Minimizing callbacks while maximizing cancer detection. *RadioGraphics* **2014**, *34*, 928–940. [CrossRef]
8. Baker, J.A.; Lo, J.Y. Breast tomosynthesis: State-of-the-art and review of the literature. *Acad. Radiol.* **2011**, *18*, 1298–1310. [CrossRef]
9. Conant, E.F.; Beaber, E.F.; Sprague, B.L.; Herschorn, S.D.; Weaver, D.L.; Onega, T.; Tosteson, A.N.; McCarthy, A.M.; Poplack, S.P. Haas, J.S.; et al. Breast cancer screening using tomosynthesis in combination with digital mammography compared to digital mammography alone: A cohort study within the PROSPR consortium. *Breast Cancer Res. Treat.* **2016**, *156*, 109–116. [CrossRef]
10. Chong, A.; Weinstein, S.P.; McDonald, E.S.; Conant, E.F. Digital breast tomosynthesis: Concepts and clinical practice. *Radiology* **2019**, *292*, 1. [CrossRef]
11. Mendes, J.; Domingues, J.; Aidos, H.; Garcia, N.; Matela, N. AI in Breast Cancer Imaging: A Survey of Different Applications. *J Imaging* **2022**, *8*, 228. [CrossRef] [PubMed]
12. Zhang, X.; Zhang, Y.; Han, E.Y.; Jacobs, N.; Han, Q.; Wang, X.; Liu, J. Classification of whole mammogram and tomosynthesis images using deep convolutional neural networks. *IEEE Trans. Nanobioscience* **2018**, *17*, 237–242. [CrossRef] [PubMed]
13. Krizhevsky, A.; Sutskever, I.; Hinton, G.E. Imagenet classification with deep convolutional neural networks. *Commun. ACM* **2017** *60*, 84–90. [CrossRef]
14. He, K.; Zhang, X.; Ren, S.; Sun, J. Deep residual learning for image recognition. In Proceedings of the IEEE Conference on Computer Vision and Pattern Recognition, Las Vegas, NV, USA, 27–30 June 2016; pp. 770–778.
15. Weiss, K.; Khoshgoftaar, T.M.; Wang, D. A survey of transfer learning. *J. Big Data* **2016**, *3*, 1–40. [CrossRef]
16. Liang, G.; Wang, X.; Zhang, Y.; Xing, X.; Blanton, H.; Salem, T.; Jacobs, N. Joint 2d-3d breast cancer classification. In Proceedings of the 2019 IEEE International Conference on Bioinformatics and Biomedicine (BIBM), San Diego, CA, USA, 18–21 November 2019; pp. 692–696.
17. Zhang, Y.; Wang, X.; Blanton, H.; Liang, G.; Xing, X.; Jacobs, N. 2d convolutional neural networks for 3d digital breast tomosynthesis classification. In Proceedings of the 2019 IEEE International Conference on Bioinformatics and Biomedicine (BIBM) San Diego, CA, USA, 18–21 November 2019; pp. 1013–1017.
18. Muduli, D.; Dash, R.; Majhi, B. Automated diagnosis of breast cancer using multi-modal datasets: A deep convolution neural network based approach. *Biomed. Signal Process. Control.* **2022**, *71*, 102825. [CrossRef]
19. Sun, H.; Wu, S.; Chen, X.; Li, M.; Kong, L.; Yang, X.; Meng, Y.; Chen, S.; Zheng, J. SAH-NET: Structure-Aware Hierarchical Network for Clustered Microcalcification Classification in Digital Breast Tomosynthesis. *IEEE TRansactions Cybern.* **2022**, 1–13 [CrossRef]
20. Xiao, B.; Sun, H.; Meng, Y.; Peng, Y.; Yang, X.; Chen, S.; Yan, Z.; Zheng, J. Classification of microcalcification clusters in digital breast tomosynthesis using ensemble convolutional neural network. *BioMed. Eng. OnLine* **2021**, *20*, 1–20. [CrossRef]
21. Mota, A.M.; Clarkson, M.J.; Almeida, P.; Matela, N. Automatic Classification of Simulated Breast Tomosynthesis Whole Images for the Presence of Microcalcification Clusters Using Deep CNNs. *J. Imaging* **2022**, *8*, 231. [CrossRef]
22. Bilello, E. Breast Cancer Screening—Digital Breast Tomosynthesis (BCS-DBT). Available online: https://wiki.cancerimagingarchive net/pages/viewpage.action?pageId=64685580 (accessed on 20 December 2022).
23. Maciej A. Mazurowski, Ph.D. Available online: https://sites.duke.edu/mazurowski/resources/digital-breast-tomosynthesis-database/ (accessed on 20 December 2022).
24. Yao, X.; Wang, X.; Wang, S.H.; Zhang, Y.D. A comprehensive survey on convolutional neural network in medical image analysis In *Multimedia Tools and Applications*; Springer: Berlin/Heidelberg, Germany, 2020; pp. 1–45.
25. Yamashita, R.; Nishio, M.; Do, R.K.G.; Togashi, K. Convolutional neural networks: An overview and application in radiology *Insights Into Imaging* **2018**, *9*, 611–629. [CrossRef]
26. Kingma, D.P.; Ba, J. Adam: A method for stochastic optimization. *arXiv* **2014**, arXiv:1412.6980.

27. Landis, J.R.; Koch, G.G. The measurement of observer agreement for categorical data. *Biometrics* **1977**, *33*, 159–174. [CrossRef] [PubMed]
28. Dhivya, S.; Mohanavalli, S.; Karthika, S.; Shivani, S.; Mageswari, R. GAN based data augmentation for enhanced tumor classification. In Proceedings of the 2020 4th International Conference on Computer, Communication and Signal Processing (ICCCSP), Chennai, India, 22–23 April 2020; pp. 1–5.

cancers

MDPI

Article

A Comparative Study of Automated Deep Learning Segmentation Models for Prostate MRI

Nuno M. Rodrigues [1,2,*], Sara Silva [1], Leonardo Vanneschi [3] and Nickolas Papanikolaou [2]

1 LASIGE, Faculty of Sciences, University of Lisbon, 1749-016 Lisbon, Portugal
2 Champalimaud Foundation, Centre for the Unknown, 1400-038 Lisbon, Portugal
3 NOVA Information Management School (NOVA IMS), Campus de Campolide, Universidade Nova de Lisboa, 1070-312 Lisboa, Portugal
* Correspondence: nmrodrigues@fc.ul.pt

Simple Summary: Prostate cancer represents a highly prevalent form of cancer worldwide, with timely detection and treatment being crucial for achieving a high survival rate. Manual segmentation, which is the process of manually identifying different anatomical structures or tissues within an image, is the most prevalent detection method. However, it is a time-consuming and subjective task constrained by the radiologists' expertise, which underpins the demand for automated segmentation methods. In this study, we conduct a comprehensive and rigorous comparison of multiple prevalent deep learning-based automatic segmentation models for the prostate gland and both peripheral and transition zones, using multi-parametric MRI data.

Abstract: Prostate cancer is one of the most common forms of cancer globally, affecting roughly one in every eight men according to the American Cancer Society. Although the survival rate for prostate cancer is significantly high given the very high incidence rate, there is an urgent need to improve and develop new clinical aid systems to help detect and treat prostate cancer in a timely manner. In this retrospective study, our contributions are twofold: First, we perform a comparative unified study of different commonly used segmentation models for prostate gland and zone (peripheral and transition) segmentation. Second, we present and evaluate an additional research question regarding the effectiveness of using an object detector as a pre-processing step to aid in the segmentation process. We perform a thorough evaluation of the deep learning models on two public datasets, where one is used for cross-validation and the other as an external test set. Overall, the results reveal that the choice of model is relatively inconsequential, as the majority produce non-significantly different scores, apart from nnU-Net which consistently outperforms others, and that the models trained on data cropped by the object detector often generalize better, despite performing worse during cross-validation.

Keywords: prostate cancer; prostate segmentation; prostate detection; deep learning

Citation: Rodrigues, N.M.; Silva, S.; Vanneschi, L.; Papanikolaou, N. A Comparative Study of Automated Deep Learning Segmentation Models for Prostate MRI. *Cancers* **2023**, *15*, 1467. https://doi.org/10.3390/cancers15051467

Academic Editors: Michał Strzelecki, Adam Piorkowski and Rafał Obuchowicz

Received: 3 January 2023
Revised: 17 February 2023
Accepted: 20 February 2023
Published: 25 February 2023

1. Introduction

Prostate cancer is one of the most common cancer types in the world, affecting roughly one in every eight men according to the American Cancer Society. Despite its high survival rate (5-year relative rates of ≈100% for localized and regional, 30% for distant), it was the third most prominent cancer in 2020 [1]. Therefore, there is an urgent need to develop methods that may aid in early detection and better characterization of disease aggressiveness [2,3]. The latter will make possible the avoidance of over-treatment in patients with non-aggressive disease and the amplification of treatment in patients with aggressive disease. Manual segmentation is still the most common practice, a time-consuming task also limited by the subjectiveness of the radiologists' expertise, resulting in high interobserver variability.

Most automatic segmentation processes developed for clinical applications are based on convolutional neural networks (CNNs), many being variations of the classic Unet architecture [4]. For prostate segmentation, the proposed models continue to improve as shown by the mean dice score used for the following results.Guo et al. [5] presented a two-step pipeline where a stacked sparse auto-encoder was used to learn deep features, and then a sparse patch matching method used those features to infer prostate likelihood, obtaining a Dice score of 87.8 $\pm$ 4.0. Milletari et al. [6] presented Vnet, a volumetric adaptation of unet, achieving a Dice score of 86.9 $\pm$ 3.3. Zhu et al. [7] used deep supervised layers on Unet with additional 1 $\times$ 1 convolutions, obtaining 88.5 Dice. Dai et al. [8] used the object detection and segmentation model Mask-RCNN, achieving 88 $\pm$ 4 and 64 $\pm$ 11 Dice in the segmentation of prostatic gland and intraprostatic lesions, respectively. Zavala-Romero et al. [9] presented a multistream 3D model that employed the three standard planes of an MRI volume as inputs and was trained with data from different scanners, achieving Dice scores of 90.5 $\pm$ 2.7 for the full gland and 79.9 $\pm$ 9.4 for the peripheral zone when using data from one scanner, and 89.2 $\pm$ 3.6 and 81.1 $\pm$ 7.9 when combining data from both scanners. Aldoj et al. [10] presented a Unet variation with two stacked dense blocks in each level of the downsampling and upsampling paths, obtaining Dice scores of 91.2 $\pm$ 0.8, 76.4 $\pm$ 2 and 89.2 $\pm$ 0.8 for full gland, peripheral and central zones, respectively. Duran et al. [11] presented an attention Unet [12] variation, where an additional decoder was added to perform both gland segmentation as well as multi-class lesion segmentation, achieving a Dice score of 87.5 $\pm$ 1.3. Recently, Hung et al. [13] proposed a new take on skip connections, using a transformer to capture the cross-slice information at multiple levels, with the main advantage being that this can be incorporated into most architectures, such as the nnU-Net model.

As of late, more complex and novel architectures and paradigms have been adopted for biomedical imaging segmentation. When dealing with large amounts of data, vision transformer architectures have shown great promise for biomedical segmentation [14–18]. These visions transformer models have the advantage of using self-attention mechanisms, allowing them to learn complex patterns, and to also capture the global spatial dependencies of the images, which is highly advantageous for segmentations. As previously mentioned, since manual segmentation suffers from several limitations, most data end up being unlabeled. Self-Supervised learning strategies have been used to leverage these unlabeled data to improve the performance of segmentation models by first pre-training them on pretext tasks [19–21]. Lastly, another paradigm that has been recently used for biomedical image segmentation is knowledge distillation, where models are first trained on a larger task (e.g., multi-centre data) and are then used to help train other models for more domain-specific tasks [22] (e.g., single-centre data). This technique has already been adopted by some researchers for prostate segmentation [23–25] and is shown to improve overall model performance.

One major problem that presents with most previously addressed Unet-based automatic prostate segmentation works are the different evaluation conditions presented, where the same model produces different results depending on the new architecture being proposed and what it needs to outperform the previous models, regardless of the correctness of the experimental settings. What we propose is not a new method but a comparative study of several of the most common, as well as some new some variations, of Unet-based segmentation models for prostate gland and zone segmentation. Additionally, we present and evaluate another research question, regarding the impact of using an object detector model as a pre-processing step in order to crop the MRI volumes around the prostate gland, reducing computational strain and improving segmentation quality by reducing the redundant area of the volumes without simply resizing the data. We examined this question by comparing the results of the different models in both full and cropped T2W MRI volumes, during cross-validation and in an additional external dataset.

2. Materials and Methods

2.1. Data

We used two publicly available datasets containing retrospective data, one to train both the object detection and the segmentation models (ProstateX), the other to serve as an external test set to assess the quality of the segmentation models (Medical Decathlon prostate dataset). The ProstateX dataset (https://wiki.cancerimagingarchive.net/pages/viewpage.action?pageId=23691656SPIE-AAPM-NCIPROSTATExchallenge, accessed on 2 January 2023) is a collection of prostate MRI volumes that include T2W, DWI and ADC modalities. These volumes were obtained by the Prostate MR Reference Center — Radboud University Medical Centre (Radboudumc) in the Netherlands, using two Siemens 3T MR scanners (MAGNETOM Trio and Skyra). Regarding the acquisition of the images, the following description was provided by the challenge's organizers: "T2-weighted images were acquired using a turbo spin echo sequence and had a resolution of around 0.5 mm in plane and a slice thickness of 3.6 mm. The DWI series were acquired with a single shot echo planar imaging sequence with a resolution of 2-mm in-plane and 3.6-mm slice thickness and with diffusion-encoding gradients in three directions. A total of three b values were acquired (50, 400, and 800), and subsequently, the apparent diffusion coefficient (ADC) map was calculated by the scanner software. All images were acquired without an endorectal coil".

Regarding the segmentations, the prostate gland segmentations were performed by a senior radiologist from the Champalimaud Foundation, while the transition and peripheral zone segmentations were obtained from the public dataset repository. We applied bias field correction, using N4ITK ([26]), to the T2W volumes, and all the masks were resampled to be on the same space and have the same orientation and spacing as the volumes. A total of 153 volumes were used for the gland segmentation, while for the peripheral and transition zone segmentation, a total of 139 volumes was used.

As an external test dataset, we used the publicly available prostate segmentation dataset from the Medical Segmentation Decathlon ([27]), which was acquired at the Radboud University Medical Centre (Radboudumc) in the Netherlands. This dataset consists of 32 MRI volumes of coregistered T2W and ADC modalities, along with segmentation masks with distinct classes for the transition and peripheral zones. We extracted the T2W volumes and, similarly to the ProstateX, performed bias field correction and resampled the masks and made new masks by joining both classes to obtain a prostate mask. The model of the scanner used to take these MRIs is not disclosed.

2.2. Prostate Detection

To prepare the data for object detection, the volumes were converted into 2D 16-bit PNG images (we chose to use 16-bit images so we keep the notion of depth while working in 2D), where the empty slices were discarded. The edge coordinates of the ground truth bounding boxes were obtained from the prostate gland masks. We used these images to train a variation of the YoLo-v4 Open source code available here: https://github.com/ultralytics/yolov5, accessed on 2 January 2023, [28] object detection architecture to locate the prostate gland and accurately draw a bounding box around it (Figure 3). After predicting the bounding boxes, we extracted the coordinates of the edges of the largest bounding box in the volumes, added padding of 40 pixels in each direction, and cropped the entire volume with that box. Using the largest bounding box plus an additional padding, we ensure that all slices contain the entire prostate gland and some additional area. Figure 1 provides a comparison between a standard and cropped slice of a T2W volume.

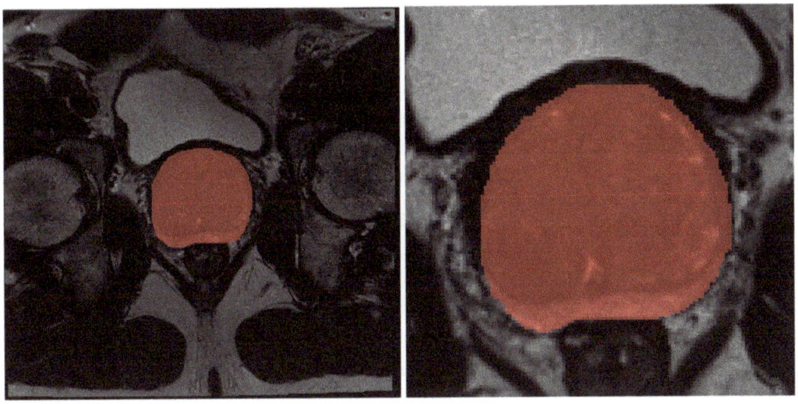

Figure 1. Comparison between original and cropped volumes. Both images correspond to the middle slice of the volumes. The image on the left represents the original full size slice, while the image on the right represents the cropped version.

2.3. T2W Pre-Processing and Augmentation

Before feeding the T2W volumes to the segmentation models, the following pre-processing techniques were applied: Orientation transformation to ensure the models were in RAS+ orientation; Intensity rescaling transformation to ensure the voxel intensity of the volumes was in $[-1, 1]$; Z normalization transformation; Cropping the images into a smaller $160 \times 160 \times 32$ size. This last transformation was done to reduce the total unused area, making it less computationally expensive, and to ensure that the volumes had an appropriate number of slices for the segmentation models. It was only applied to the volumes that had not been previously cropped by the object detector. In addition, the following augmentation transformations were applied: Random affine transformations, including rescaling, translation and rotation; Random changes to the contrast of the volumes by raising their values to the power of γ, where $\gamma = 0.5$; Application of random MRI motion artifacts, and random bias field artifacts. Examples of volumes after pre-processing and augmentation are shown in Figure 2.

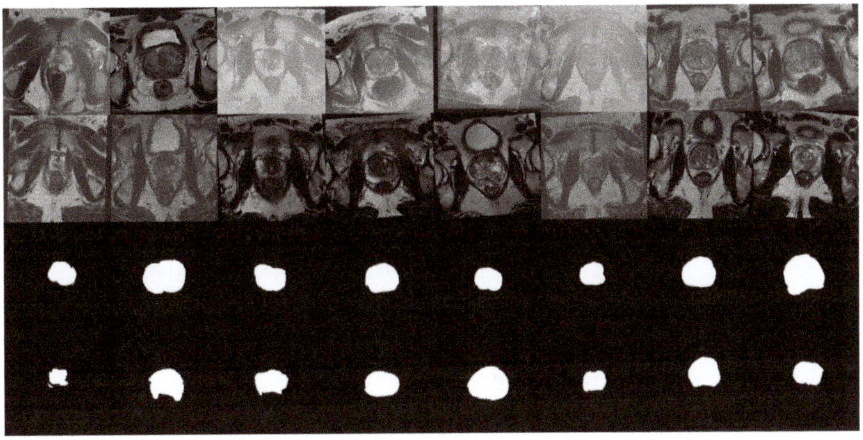

Figure 2. Example of a batch of 16 images, and their respective segmentation masks, from the ProstateX dataset after being pre-processed and augmented. Each image corresponds to the tenth slice of the volume. The images show the effect of the affine transformations as well as the random bias field augmentations.

2.4. Segmentation Models

In this study, we conducted an extensive analysis of various popular Unet-based models from the literature that either had a publicly available implementation or provided enough detail to be reproducible. These models utilized mechanisms falling into three distinct categories: Dense blocks, Recurrent connections, and Attention mechanisms. The inclusion of a diverse range of mechanisms facilitated a thorough investigation of the utility of each mechanism, given that most Unet variations exhibit only minor differences between one another. Additionally, we introduced new networks that build upon the previously published models by incorporating additional combinations of mechanisms. This enabled us to evaluate the performance of these models against established ones, leading to more reliable and informative results.

In total, 13 segmentation models were compared, including: Unet, Unet++ [29], Residual Unet (ResNet) , Attention Unet (aunet) [12], Dense Attention Unet (daunet), Dense-2 Unet (d2unet) [10], Dense-2 Attention Unet (d2aunet), Recurrent Residual Unet (r2unet) [30], Recurrent Residual Attention Unet (r2aunet), nnU-Net [31], Vnet [6], highResNet [32], SegResNet [33].

For the Unets, the standard convolution blocks in the downwards path are composed of two convolution operators with a kernel size 3 and stride of 1, followed by a batch normalization, a ReLU activation function and lastly a maxpool operation for dimensionality reduction. The convolution blocks on the upwards path are composed of an upsample operation with scale 2 to double the size of the previous input, and a convolution of kernel size 2 and stride of 1, followed by a batch normalization and ReLU activation.

For the dense unets, the dense blocks are composed of four blocks, with each one having two convolution operations with kernel size 1 and stride of 1, where after each convolution there is a batch normalization and a ReLU activation function. For the transition blocks also used in the dense unets, we perform an upsample operation with scale 2 and a convolution with kernel size 3 and stride of 1, followed by batch normalization and ReLU activation. Regarding the Dense-2 Unet, our implementation differs from the one presented in the original article. As this article does not include enough information to fully replicate the model, we chose the parameters for the dense blocks and convolution blocks to be similar to the ones of the remaining networks.

The recurrent residual blocks are equal to the ones described in Zahangir et al. [30], and are composed of two residual operation and two convolution blocks, each one having one convolution operation with kernel size 3 and stride of 1, followed by a batch normalization and ReLU activation.

The attention mechanisms are equal to those described in Oktay et al. [12], where we calculate W_g, W_x and ψ, each composed of a convolution of kernel size 1 and stride of 1, and then compute $\sigma(\psi(ReLU(W_g + W_x)))$.

At the start of each Unet, a single convolution block, composed of a convolution of kernel size 3 and stride of 1, followed by a batch normalization and ReLU activation, is applied to double the channels from 32 to 64, so we end up with 1024 channels at the bottleneck area of the unets.

Regarding the Vnet, SegResNet and highResNet, we used the models made available by the MONAI package [34]. We used the 3D full resolution version of the nnunet, which is equal to the one described in [31], and publicly available at https://github.com/MIC-DKFZ/nnUNet, accessed on 2 January 2023. All other models were implemented in Pytorch [35].

2.5. Training and Evaluation

2.5.1. Detection

The detection model was trained with the ProstateX volumes, which differed in size (320 × 320, 384 × 384, 640 × 640) with a varying number of slices (minimum 19, maximum 27).

All volumes were normalized to zero mean and one unit of standard deviation, resized to 256 × 256, and separated in 2D slices. The slices with no prostate were removed, ending with a total of 2801 images. The remaining 2801 images were split into 70% training and 30% validation. To avoid data leakage, we ensured all slices belonging to a volume were only present in one of the sets.

To set the initial parameters for the object detector model, we used the genetic algorithm based hyperparameter evolution included in the package, that ran for 100 generations, with 90% mutation and 10% crossover probabilities. The fitness function used to evaluate this evolution is the weighted average between the mean average precision with a threshold of 0.5 (mAP@0.5), which contributed with 10%, and the mean average precision with different thresholds, from 0.5 to 0.95 in steps of 0.05 (mAP@0.5:0.95), which contributed with the remaining 90%. Then, after having set the initial values for the hyperparameters, the model was trained for 350 epochs.

2.5.2. Segmentation

The datasets were split with 5-fold cross-validation. All models except the nnU-Net were trained for a maximum number of 1500 epochs, using early stopping with patience 30. The optimizer was Weighted Adam (AdamW) with a starting learning rate of 1×10^{-4}, Cosine Annealing learning rate decay, and weight decay of 4×10^{-5}. For nnU-Net, we used the default parameters, training for 1000 epochs with the Ranger optimizer [36]. All models were trained using Pytorch-Lightning.

To assess the quality of the models, we use Mean Dice score (MDS) with 95% confidence interval (CI), Mean Hausdorf distance (MHD) and Mean surface distance (MSD).

The Dice score is a widely used metric for segmentation tasks, and it measures the overlap between the ground truth and predicted masks. The higher the score, the better the segmentation.

$$DS(X,Y) = \frac{2|X \cap Y|}{|X| + |Y|}$$

The Hausdorff distance (HD) measures how far two points are in two images. In this case, how far a point from the predicted mask if from its nearest point in the ground truth mask, essentially indicating the largest segmentation error present in the predicted mask. Low values mean small errors.

$$HD(X,Y) = \max_{x \in X} \min_{y \in Y} ||x - y||$$

The Surface distance (SD)measures the difference between the surface of the predicted mask and the ground truth mask. Low values mean small differences.

$$SD(X,Y) = \min_{x \in X, y \in Y} ||y - x||$$

2.6. Loss Function

Initially, our loss function was the standard averaged sum of Dice (Equation (2)) and Crossentropy (Equation (1)), called Dice CE loss (Equation (3)). It produced good results for the mid regions of the prostate, but struggled with the small and irregular shapes present in both apex and base. Therefore, we included both Focal (Equation (4)) ([37]) and Tversky (Equation (5)) losses, as they mitigate this issue by focusing on the hard negative examples, reducing both false negatives and false positives. The final loss function was named Focal Tversky Dice Crossentropy loss (FTDCEL) (Equation (6)), an averaged sum of all previously mentioned loss functions, all having the same weight (FTDCEL) (Equation (6)). All of these loss functions were implemented using the MONAI package [34].

$$CE = \begin{cases} -log(p) & \text{if } y = 1 \\ -log(1-p) & \text{otherwise} \end{cases} \quad (1)$$

$$DL(p,\hat{p}) = \frac{2p\hat{p}+1}{p+\hat{p}+1} \quad (2)$$

$$DCE(p,\hat{p}) = \frac{DL(p,\hat{p}) + CE(p,\hat{p})}{2} \quad (3)$$

$$\text{rewriting CE as} \quad p_t = \begin{cases} p & \text{if } y = 1 \\ 1-p & \text{otherwise} \end{cases} \quad (4)$$
$$FL(p_t) = -\alpha_t(1-p_t)^{\gamma}log(p_t)$$

$$TL(p,\hat{p}) = 1 - \frac{1+p\hat{p}}{1+p\hat{p}+\beta(1-p)\hat{p}+(1-\beta)p(1-\hat{p})} \quad (5)$$

$$FTDCEL(p,\hat{p}) = \frac{DCE(p,\hat{p}) + FL(p,\hat{p}) + TL(p,\hat{p})}{3} \quad (6)$$

3. Results

3.1. Object Detection

To evaluate the object detection model we followed the procedure described in Section 2.5.1 for four different variations of the yolo model, a small, medium, large and extra-large. Figure A1 shows the results obtained on the medium model which is the one that produced the best results.

As shown in Figure A1, the results obtained were very accurate, with high confidence values on all prostate sections, including the apex, achieving a Precision, Recall, mAP@0.5 and mAP@0.5:0.95 of 0.9709, 0.9534, 0.9653 and 0.6965, respectively. As for the loss values which represent the error meaning lower values are better, this model obtained a bounding box loss (Box loss), which is the measure of how well the predicted bounding box covers the target, of 0.0227 and an object confidence loss (Obj loss), which measures the probability that the target exists inside the predicted bounding box, of 0.0034, on the validation data Figure 3 (bottom row) illustrates a batch of the obtained results. As shown, both the apex, base and middle area of the prostate is properly detected with a high degree of confidence

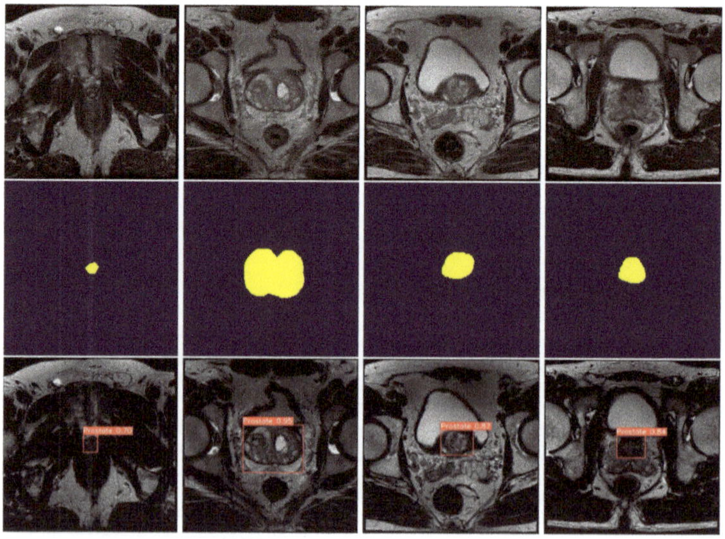

Figure 3. Comparison between original prostate images, masks and predicted volumes. The top row consists of random slices from different volumes, the second row consists of the respective prostate masks, and the third row consists of the respective predicted bounding boxes, with confidence value of the prostate on each of the images.

3.2. Gland Segmentation

Regarding the segmentation of the prostate gland (Table 1A and Figure 4), when working with the full volumes we can see that the nnunet is significantly better than all other models, by ≈2%, achieving a mean Dice score (MDS) of 0.9289 ± 0.0046. This is further corroborated when looking at the boxplots as we can see that most models have a high degree of dispersion, as opposed to the nnunet. A mean Hausdorff distance (MHD) of 5.7155 is considerably better than the ones obtained by the remaining models. It can also be observed that despite small differences, there is no statistical difference between the Dice score results obtained by the other models, although being discernible that some, such as daunet, d2uNet, Vnet and especially the highresnet, have substantially greater MHD values. From Figure 1 right we can see that nnunet is the model that better generalizes, achieving a MDS of 0.8678 on the external test set, a result far greater than most other models, along with a MHD of 10.0231, which is the smallest by far between all models.

When segmenting the prostate gland using the volumes cropped by the object detection model (Figure 1 left and Figure 4), the results are quite different. There is no single significantly best model, there are however four models that under-perform, and nine which have similar scores. The best, although not significantly better, Dice score is obtained by the d2auNet, with a value of 0.9158 ± 0.0068. Looking at the boxplots we can see that the majority of the models, apart from the r2Net and highResNet, present a low degree of dispersion. Looking at the MHD and average surface distance (ASD), we can further see that the choice of the model is mostly inconsequential, with the exception of highresnet. These results still hold for the external test set (Figure 1 right), where we can see that all models, again with the exception of highresnet, have a very similar generalization capability with a MDS of ±0.85. Comparing the performance of full and cropped volume models, looking at Table A1 it is possible see that for 11 of the 13 models there is no significative difference, with the only two exceptions being the Vnet, where it does in fact improve the Dice scores, and the nnU-Net which has the opposite effect, worsening model performance. Looking at the boxplots in Figure 4, we can see that on all metrics the results show less dispersion, which is a good indication. However, when observing the difference between the results on the external test set, we can see that the vast majority of models do improve their generalization capability using cropped data, achieving higher MDS scores and lower MHD scores.

Figure 5 shows a comparison between full volume and cropped volume segmentations of both nnU-Net and d2aunet, as they were the models with the highest Dice score in each task. In this case, despite having a lower MDS, the segmentations provided by the cropped models are arguably better than those provided by the full volumes, with the difference being explained by the difference in size when calculating this metric. In smaller volumes, a wrongly calculated pixel will have more impact in the calculation of the MDS than on larger volumes, which may explain why a model with a slightly lower MDS (0.014 difference) can provide more accurate segmentations.

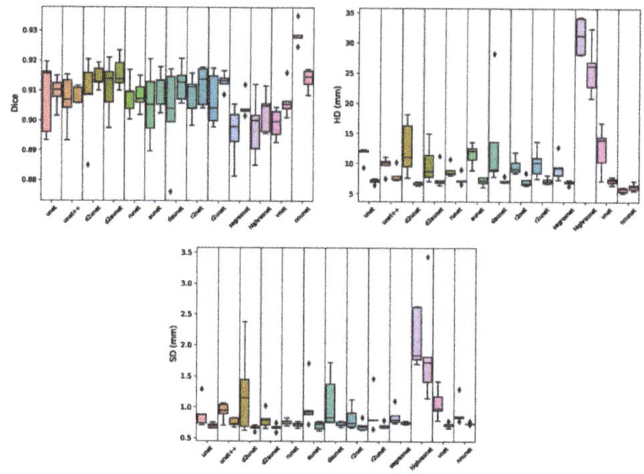

Figure 4. Boxplots showing the distribution of Dice, Hausdorff distance (HD) and Surface distance scores per model during cross-validation for the gland segmentation task. Each segment contains a pair of boxplots, where the left one corresponds to the results of the model on the full data, and the right one on the cropped data.

Table 1. (**A**) Results obtained by the segmentation models on prostate gland segmentation task. The results presented are the mean scores from the validation data over the five-fold models.The best results (based on Table A1) for both the full and cropped volumes are presented in green. (**B**) Results obtained on the external test set by the segmentation models trained on the prostate gland segmentation task. The results presented are the mean scores of the 5-fold models over all the samples on the external test set. The best results for both the full and cropped volumes are presented in red.

(**A**)

	Gland				Cropped Gland			
	MDS	CI	MHD (mm)	ASD (mm)	MDS	CI	MHD (mm)	ASD (mm)
unet	0.9082	0.0154	11.5690	0.8732	0.9094	0.0064	7.0381	0.6990
unet++	0.9066	0.0108	9.7262	0.9289	0.9081	0.0037	7.9955	0.7533
runet	0.9081	0.0078	8.8370	0.7610	0.9083	0.0062	7.3480	0.7186
aunet	0.9052	0.0151	11.5549	1.0364	0.9105	0.0079	7.0760	0.7047
daunet	0.9023	0.0203	13.5009	1.0892	0.9123	0.0076	7.1538	0.7265
d2unet	0.9078	0.0170	12.5156	1.2563	0.9145	0.0047	6.6846	0.6664
d2aunet	0.9110	0.0115	9.9763	0.7997	0.9158	0.0068	7.6787	0.6715
r2unet	0.9084	0.0087	9.6704	0.8259	0.9117	0.0081	7.0649	0.6994
r2aunet	0.9069	0.0110	10.1465	0.8990	0.9130	0.0036	7.1827	0.7096
vnet	0.8991	0.0062	12.5448	1.0621	0.9067	0.0070	7.2024	0.7220
segresnet	0.8960	0.0117	9.3760	0.8540	0.9048	0.0051	6.9018	0.7490
highresnet	0.8980	0.0130	31.1897	2.1118	0.9029	0.0084	25.8385	0.7533
nnunet	0.9289	0.0046	5.7155	0.9218	0.9139	0.0044	6.2565	0.7406

(**B**)

	Gland			Cropped Gland		
	MDS	MHD (mm)	ASD (mm)	MDS	MHD (mm)	ASD (mm)
unet	0.8474	16.9045	2.3177	0.8595	9.9819	1.2044
unet++	0.8544	25.7024	2.0092	0.8472	9.0465	1.3394
runet	0.8501	19.8740	1.9686	0.8547	9.9693	1.2591
aunet	0.8291	25.3664	2.5814	0.8521	9.9022	1.3618
daunet	0.8277	21.4430	2.1456	0.8579	10.6797	1.2818
d2unet	0.8528	15.2098	1.7646	0.8567	9.1642	1.3421
d2aunet	0.8464	13.6260	1.9367	0.8562	8.2389	1.3334
r2unet	0.8517	14.9619	1.7727	0.8546	8.7872	1.3819
r2aunet	0.8514	16.2630	1.9737	0.8573	10.3648	1.7792
vnet	0.8368	21.3305	2.7043	0.8548	9.6195	1.3882
segresnet	0.8479	17.4913	1.7528	0.8468	9.7496	1.3192
highresnet	0.7448	69.6783	11.5506	0.8092	35.0037	3.6740
nnunet	0.8678	10.0231	3.3704	0.8558	9.2565	1.6221

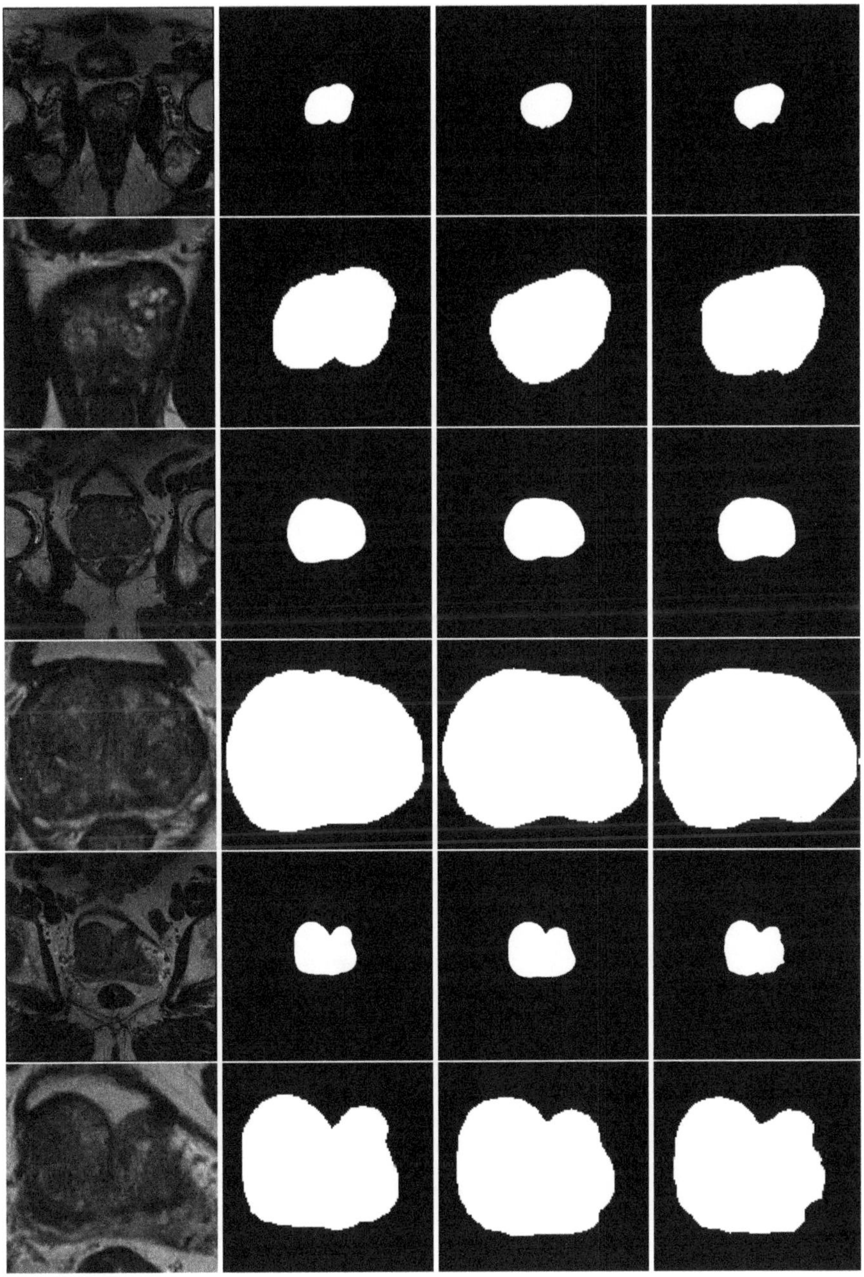

Figure 5. Segmentations of the prostate gland. Column (**A**) contains the volumes, column (**B**) contains the ground truth, column (**C**) contains the nnU-Net segmentations and column (**D**) contains the d2aunet segmentations. Rows are interleaved, showing a full volume and a model cropped by the object detection model, respectively.

3.3. Zone Segmentation

Starting with the transition zone (TZ) segmentation (Table 2A and Figure 6), we can see that the nnU-Net outperformed all the other models, both on the full volume and cropped volume modalities. On the full volume data, it achieved an MDS of 0.8760 $\pm$ 0.0099, which is $\approx$3% better than all other models, while on the cropped data it further increased the distance to the remaining models, by achieving a mean Dice score of 0.8561 $\pm$ 0.0133, $\approx$7% better than the others. Regarding the maximum error, for full and cropped we obtained an MHD distance of 8.7049 and 10.2390, respectively, which are notably smaller than the errors obtained by the remaining models, $\pm$6 and $\pm$4 mm smaller for the full and cropped volumes respectively. However, when the models are evaluated on an external test set (Figure 2) we can see that these results do not hold. For the full volume data the nnU-Net falls short dropping $\approx$15%, while most other variations of the Unet dropped far less, with the regular Unet being the model with the best MDS of 0.77, despite having a larger maximum error. On the cropped volumes, the nnU-Net remained the best performing model, with an MDS of 0.7540.

When comparing the performance of full and cropped volume models, looking at Tables A2 and 2, it is clear that the performance of all models drops significantly when using the cropped data. One interesting particularity to note is that while in the majority of the models, the dispersion of the results on the cropped data is similar to the dispersion on the full data, when observing d2aunet we can see that dispersion of values for both the Hausdorff and surface distances is greatly reduced (Figure 6). Figure 7 shows a comparison between full volume and cropped volume segmentations of both nnU-Net and aunet, as they were the models with the highest Dice scores during cross-validation.

Table 2. (**A**) Results obtained on the external test set by the segmentation models trained on the transitional zone segmentation task. The results presented are the mean scores of the 5-fold models over all the samples on the external test set. The best results (based on Table A2) for both the full and cropped volumes are presented in green. (**B**) Results obtained by the segmentation models on transitional zone segmentation task. The results presented are the mean scores from the validation data over the five folds. The best results for both the full and cropped volumes are presented in red.

(A)

	Tz				Cropped Tz			
	MDS	CI	MHD (mm)	ASD (mm)	MDS	CI	MHD (mm)	ASD (mm)
unet	0.8423	0.0243	14.0153	1.0092	0.7790	0.0250	14.0153	1.3668
unet++	0.8457	0.0254	14.0649	1.0042	0.7715	0.0278	14.0649	1.3840
runet	0.8469	0.0186	12.9054	0.9450	0.7895	0.0432	12.9054	1.3410
Aunet	0.8509	0.0164	14.8667	0.9109	0.7543	0.1045	14.8667	1.4419
daunet	0.8481	0.0155	14.6603	0.9105	0.7703	0.0311	14.6603	1.3604
d2unet	0.8447	0.0144	13.9448	0.9676	0.7790	0.0483	13.9448	1.4247
d2Aunet	0.8328	0.0456	14.5523	1.9611	0.7724	0.0351	14.5523	1.3411
r2unet	0.8390	0.0366	14.4749	1.0748	0.7788	0.0579	14.1360	1.3977
r2Aunet	0.8456	0.0152	13.1392	1.0938	0.7799	0.0365	13.1392	1.3406
vnet	0.8389	0.0172	12.4145	1.0663	0.7998	0.0282	12.4145	1.2589
segresnet	0.8234	0.0199	13.7308	1.2383	0.7729	0.0347	13.7308	1.3995
highresnet	0.8457	0.0186	15.0525	1.5373	0.8083	0.0212	15.0525	1.3386
nnunet	0.8760	0.0099	8.7049	1.2928	0.8561	0.0133	10.2390	1.1226

Table 2. *Cont.*

(B)						
	Tz			Cropped Tz		
	MDS	MHD (mm)	ASD (mm)	MDS	MHD (mm)	ASD (mm)
unet	0.7700	22.5822	2.4379	0.6424	16.8615	2.6838
unet++	0.7681	19.8427	2.3727	0.6536	16.9354	2.4165
runet	0.7671	19.9993	2.0027	0.6477	17.0520	2.5169
Aunet	0.7680	21.7833	2.4354	0.6577	16.0525	2.5444
daunet	0.7688	19.6406	2.4507	0.6471	17.4325	2.5915
d2unet	0.7627	21.2574	2.4447	0.6581	17.3590	2.7397
d2Aunet	0.7638	22.9369	2.3869	0.6581	16.7426	2.5895
r2unet	0.7631	19.8119	2.5560	0.6523	17.3374	2.8883
r2Aunet	0.7672	21.2282	2.1350	0.6553	14.3813	2.6186
vnet	0.7643	20.3670	2.8237	0.6603	18.4141	2.8833
segresnet	0.7381	19.6969	2.6089	0.6482	18.0439	2.7591
highresnet	0.7209	46.6032	6.4525	0.6381	22.3236	3.6603
nnunet	0.7300	16.9107	7.4144	0.7540	14.4592	2.1459

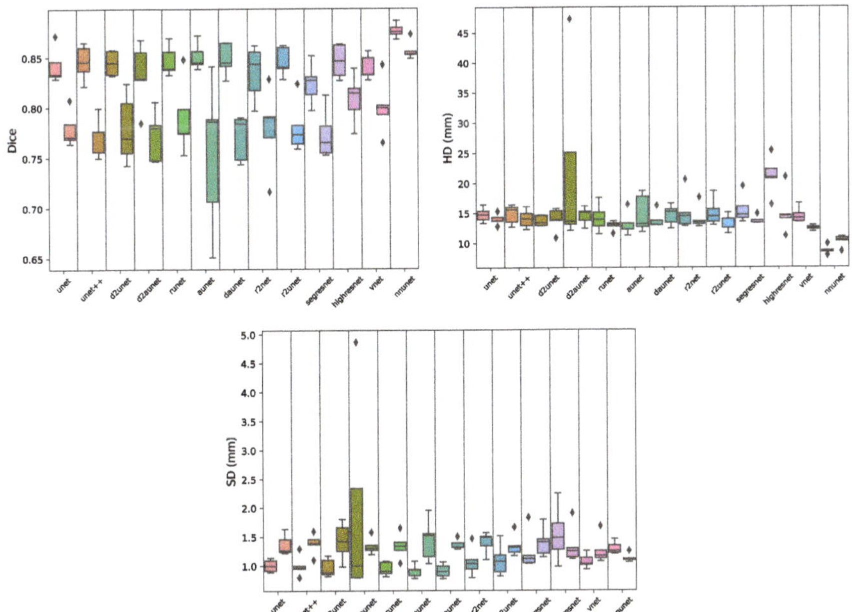

Figure 6. Boxplots showing the distribution of Dice, Hausdorff distance (HD) and Surface distance scores per model during cross-validation for the transition segmentation task. Each segment contains a pair of boxplots, where the left one corresponds to the results of the model on the full data, and the right one on the cropped data.

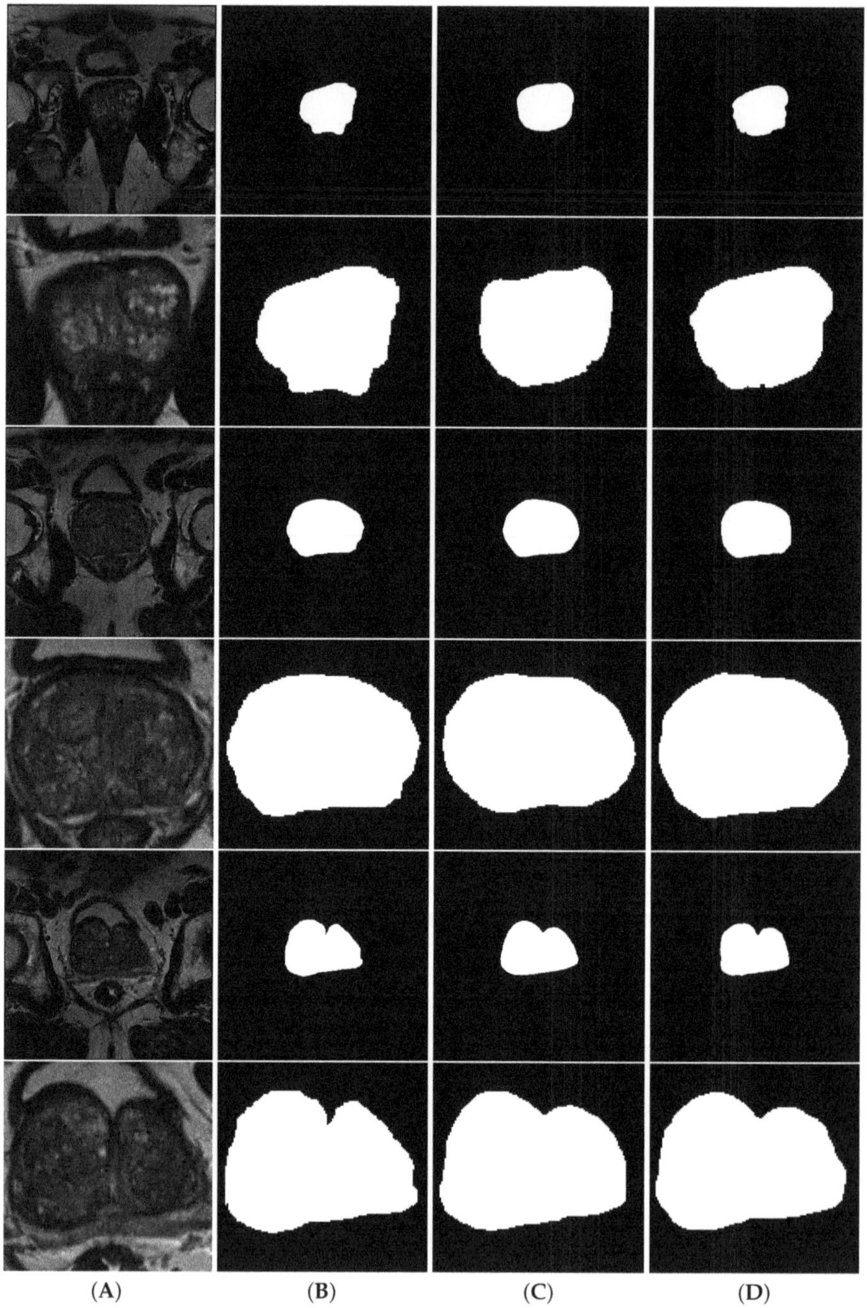

Figure 7. Segmentations of the transition zone. Column (**A**) contains the volumes, column (**B**) contains the ground truth, column (**C**) contains the nnunet segmentations and column (**D**) contains the aunet segmentations. Rows are interleaved, showing a full volume and a model cropped by the object detection model, respectively.

Regarding the peripheral zone (PZ) segmentation (Figures 3 and 8), when working with the full volumes we can see that the nnU-Net outperformed all other models, achieving an MDS of 0.8029 $\pm$ 0.0063, $\approx$5% better than the remaining models, but also the smallest error, with an MHD of 9.8693, less than half of the average of most other models. When working with the cropped volumes, we can see that there is no statistically significant difference between any of the models. They all produce very similar values for all evaluated metrics (MDS, MHD, ASD). Looking at Figure 8, we can also see that the distribution and dispersion are very similar, with only a few outliers such as the aunet when measuring the HD and the d2unet when measuring the SD. When testing the models on the external test set (Figure 3), we can see that while using the full volumes the results hold, with nnU-Net still being the top performing model with an MDS of 0.6835 and having the lowest MHD of only 13.4527. For the cropped volumes, the d2unet was the best-performing model, with an MDS of 0.6387, and the lowest MHD of 16.3358.

When comparing the performance of full and cropped volume models, looking at Tables A3 and 3, we can see that there is no statistical difference between using full or cropped data when segmenting this zone, apart from the nnU-Net that similarly to what was shown in the two previous zones performs significantly worse. However, similarly to what was observable on the gland analysis, we can see that many of the models show small improvements on the external test set, most noticeably when examining the MHD values, showing again that cropping may improve the generalization capabilities. Figure 9 shows a comparison between full volume and cropped volume segmentations of both nnU-Net and highResNet, as they were the models with the highest Dice score during cross-validation. When looking at the full volume segmentations, we can observe a large difference between the nnU-Net and the highResNet. While the first produces smooth and well-defined segmentations, the latter shows clear signs of poorly defined edges and obvious noise in some cases.

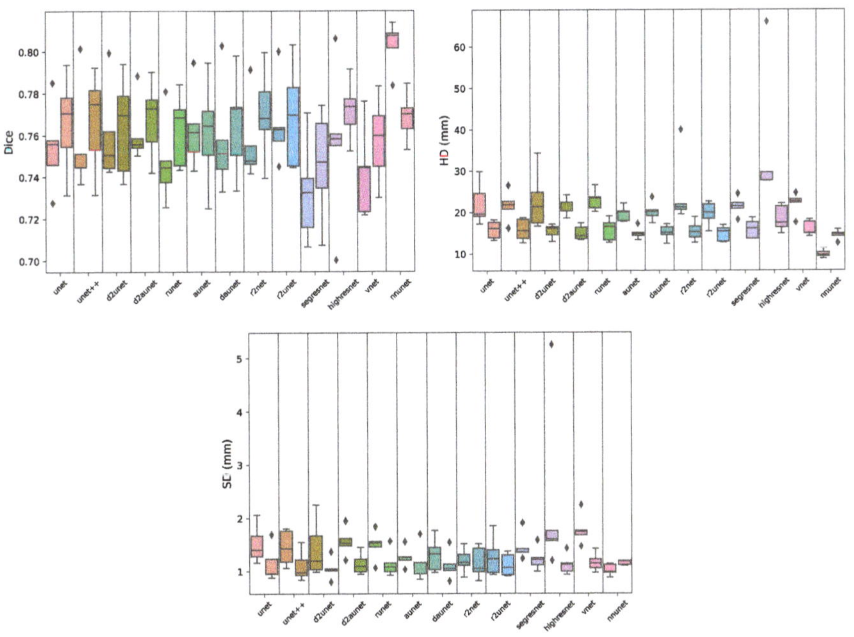

Figure 8. Boxplots showing the distribution of Dice, Hausdorff distance (HD) and Surface distance scores per model during cross-validation for the peripheral segmentation task. Each segment contains a pair of boxplots, where the left one corresponds to the results of the model on the full data, and the right one on the cropped data.

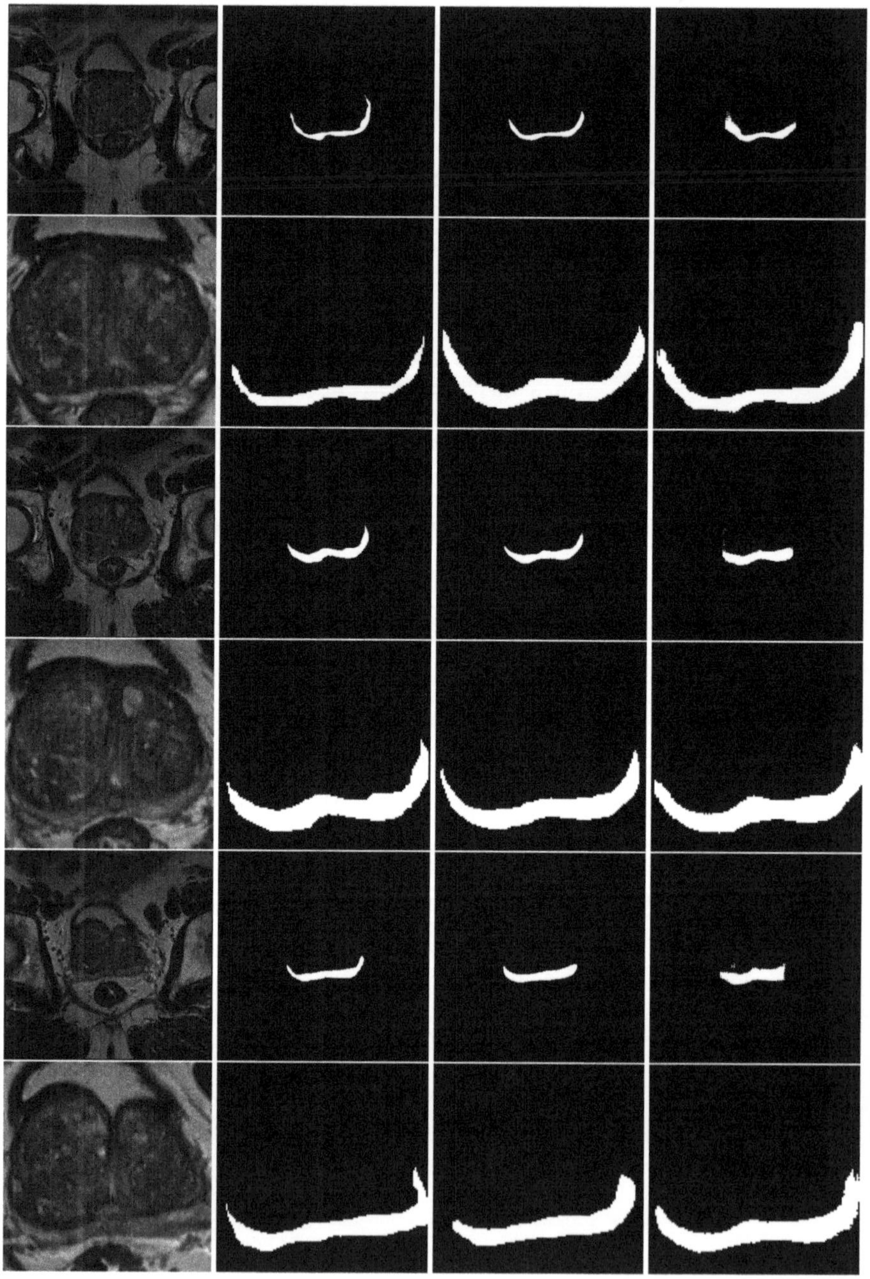

(A) (B) (C) (D)

Figure 9. Segmentations of the peripheral zone. Column (**A**) contains the volumes, column (**B**) contains the ground truth, column (**C**) contains the nnU-Net segmentations and column (**D**) contains the highResNet segmentations. Rows are interleaved, showing a full volume and a model cropped by the object detection model, respectively.

Table 3. (**A**) Results obtained on the external test set by the segmentation models trained on the peripheral zone segmentation task. The results presented are the mean scores of the 5-fold models over all the samples on the external test set. The best results (based on Table A3) for both the full and cropped volumes are presented in green. (**B**) Results obtained by the segmentation models on the peripheral zone segmentation task. The results presented are the mean scores from the validation data over the five folds. The best results for both the full and cropped volumes are presented in red.

(A)	Pz				Cropped Pz			
	MDS	CI	MHD (mm)	ASD (mm)	MDS	CI	MHD (mm)	ASD (mm)
unet	0.7545	0.0300	22.1193	1.5215	0.7656	0.0326	15.8711	1.1443
unet++	0.7557	0.0363	21.6545	1.4484	0.7666	0.0328	15.8122	1.1031
runet	0.7472	0.0296	23.0029	1.4938	0.7627	0.0243	15.8278	1.1482
aunet	0.7632	0.0280	19.6521	1.2725	0.7611	0.0360	15.0016	1.1684
daunet	0.7576	0.0384	20.1854	1.3109	0.7655	0.0349	15.1258	1.1131
d2unet	0.7597	0.0336	22.9687	1.4304	0.7644	0.0327	15.5292	1.0531
d2aunet	0.7614	0.0219	21.2204	1.5646	0.7677	0.0258	15.0999	1.1449
r2unet	0.7563	0.0280	24.6377	1.2007	0.7700	0.0307	15.4505	1.1609
r2aunet	0.7656	0.0294	19.6629	1.2778	0.7689	0.0343	14.8106	1.1195
vnet	0.7420	0.0315	22.0089	1.7818	0.7573	0.0297	15.9604	1.1659
segresnet	0.7329	0.0285	21.4516	1.4579	0.7456	0.0380	15.9265	1.2347
highresnet	0.7560	0.0301	35.7066	2.2781	0.7719	0.0203	18.4239	1.1279
nnunet	0.8029	0.0063	9.8693	1.0210	0.7686	0.0110	14.4054	1.1633

(B)	Pz			Cropped Pz		
	MDS	MHD (mm)	ASD (mm)	MDS	MHD (mm)	ASD (mm)
unet	0.6160	26.9297	2.7719	0.6217	21.3145	2.0339
unet++	0.6129	37.0787	3.5482	0.6064	19.8930	2.3919
runet	0.5828	19.5994	2.8543	0.6104	22.1387	2.4339
aunet	0.6224	20.8184	2.5188	0.6141	20.3642	2.5048
daunet	0.6282	23.0450	3.0015	0.6252	20.2531	2.0680
d2unet	0.6222	18.9975	2.5521	0.6387	16.3358	1.9674
d2aunet	0.6218	26.7974	2.6237	0.6372	17.8484	2.1356
r2unet	0.6273	27.9654	2.5934	0.6308	21.7236	2.2581
r2aunet	0.6228	21.5421	2.7400	0.6300	17.1794	2.0054
vnet	0.5848	40.8916	4.7720	0.5853	23.1373	2.5561
segresnet	0.6106	21.5214	2.5409	0.5339	41.5031	4.4359
highresnet	0.5669	60.2008	5.0504	0.6075	19.7673	2.8573
nnunet	0.6835	13.4527	1.8224	0.6038	17.6085	2.5589

4. Discussion

We conducted an extensive analysis of several commonly used segmentation models for prostate gland and zone segmentation on a unified pipeline with the same settings for all. In addition, we answered the research question regarding the effectiveness of using an object detector to crop the MRI volumes around the prostate gland to aid during the segmentation process.

First, we trained an object detector model to perform bounding box detection on the prostate gland, in order to later crop the images. We employed a variation of the Yolo-v4 model to detect the prostate on 2D slices, which provided very good results, with box and confidence validation losses of 0.0227 and 0.0034, respectively. These results show that it is fairly easy to train a robust prostate detection model, raising the hypothesis for a feature work of leveraging the learned local representations of such a model and using them to aid in a segmentation task.

Using the results obtained from the object detection, we produced a new version of the ProstateX dataset where the volumes were cropped around the gland, reducing the overall size of the image and computational power required. We then trained 13 different commonly used segmentation models, with some additional new variations, on both the full and cropped volumes. The models were subsequently validated on the prostate dataset from the medical imaging decathlon, both on the original and on a cropped version.

When comparing the performance of the models on the full volume segmentation tasks, two main conclusions can be drawn. The first being that the nnU-Net is the overall best model, outperforming all other during cross-validation on the three different segmentation tasks, while still being the model that better generalized on the external test set, on two out of three tasks. Interestingly, despite not being the worst performing model on the external transitional data, it was the one with the highest ASD value, meaning that while it did not make the largest mistakes, it did make the most mistakes out of all models. The second conclusion is that, excluding the nnU-Net, when choosing between the other models, the decision is almost inconsequential. While SegResNet and Vnet produced results significantly worse than some other models, the remaining show no significant differences between each other for the three segmentation tasks.

When comparing the performance of the models on the cropped volume segmentation tasks, the results are not as clear as when using the full volumes. One common aspect among all three segmentation tasks is that the nnU-Net is either the best-performing model, or is at least one of the best. While for the transitional segmentation task, the nnU-Net is the clear winner, for both the gland and peripheral segmentation tasks, the choice of the model is almost inconsequential, with the exception of only four models on the gland task, which perform significantly worse. Regarding the performance on the external test set, there is no clear indication of an overall better model. While for the transition segmentation task, the nnU-Net is clearly better than the other models, and for the remaining two tasks, the results are very similar.

Lastly, comparing the performance of the models when using the two types of data, it is observable that overall the model performance during cross-validation is either the same, or statistically significantly worse. The clearest case being the nnU-Net, where it consistently hinders performance. This is most likely due to the way nnU-Net processes the data, since it is based on a set of heuristics that are applied based on the characteristics of the data, characteristics that are changed when the volumes are cropped. However, in both the gland and peripheral zone segmentation task, it is also observable that the cropped models generalize better, despite their poorer performance during cross-validation, hinting that it is worthwhile to further explore this approach, ideally on a larger dataset and with data to rule out any hypothesis regarding data quantity.

5. Conclusions

In this paper, we perform a comparative study using several of the latest, commonest, and certain variations of Unet-based segmentation models for prostate gland and zone segmentation. We answer the research question regarding the impact of using an object detector model as a pre-processing step in order to crop the MRI volumes around the prostate gland. Regarding the comparison of the different architectures, it is clear that overall there are statistically significant differences between the vast majority of models, regardless of the mechanisms they employ, apart from the nnU-Net model [31], which consistently outperforms all other models. Concerning the object detector, it is shown that despite being straightforward to train and obtain good results for prostate detection, overall the segmentation results are the same or statistically significantly worse.

Author Contributions: Conceptualization, N.M.R. and N.P.; methodology, N.M.R., S.S. and N.P.; software, N.M.R.; validation, N.M.R.; formal analysis, N.M.R. and S.S.; investigation, N.M.R.; resources, S.S. and N.P.; data curation, N.M.R.; writing—original draft preparation, N.M.R.; writing—review and editing, N.M.R., L.V., S.S. and N.P.; visualization, N.M.R.; supervision, L.V., S.S. and N.P.; project administration, N.M.R. and N.P.; funding acquisition, N.M.R., L.V., S.S. and N.P. All authors have read and agreed to the published version of the manuscript.

Funding: The research leading to these results has received funding from the European Union's Horizon 2020 research and innovation programme under grant agreement No 952159 (ProCAncer-I). This work was partially supported by FCT, Portugal, through funding of the LASIGE Research Unit (UIDB/00408/2020 and UIDP/00408/2020), and under the project UIDB/04152/2020—Centro de

Investigação em Gestão de Informação (MagIC)/NOVA IMS. Nuno Rodrigues was supported by PhD Grant 2021/05322/BD.

Institutional Review Board Statement: Ethical review and approval were waived for this study, due to the public nature of the data.

Informed Consent Statement: Patient consent was waived due to the public nature of the data.

Data Availability Statement: Data are available upon reasonable request.

Conflicts of Interest: The authors declare no conflict of interest. The funders had no role in the design of the study; in the collection, analyses, or interpretation of data; in the writing of the manuscript; or in the decision to publish the results.

Appendix A

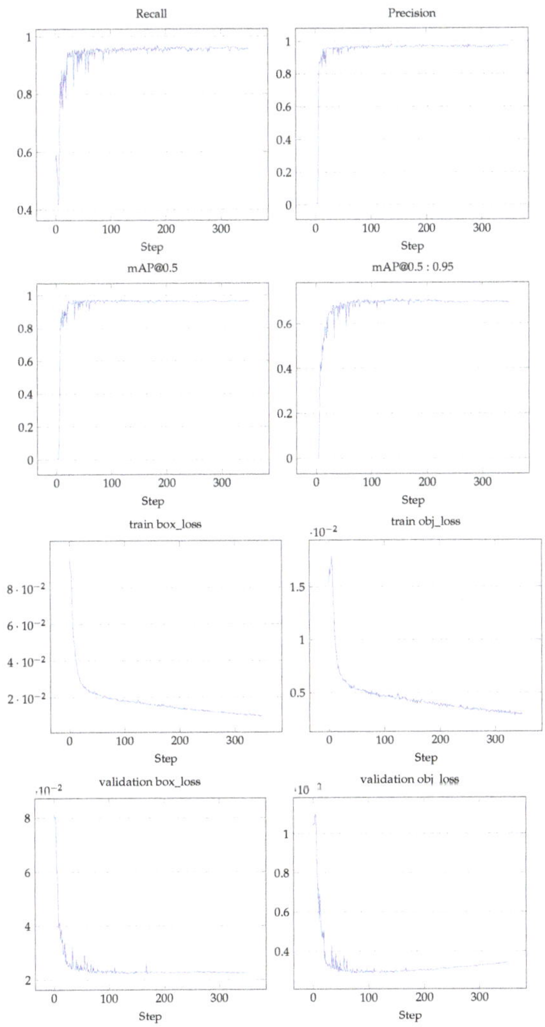

Figure A1. Evolution of the different metrics during the training of the prostate detection model. The figure contains the Recall, Precision, mAP@0., mAP@0.5 : 0.95, and both the object and box loss for the training and validation sets.

Table A1. *p*-values using Kruskal–Wallis for the dice scores obtained in the cross-validation of the gland segmentation task. On the top area we compare the performance of the different architectures on the full data, and on the bottom on the cropped data. On the diagonal we present the comparison of the same architecture on both data types. Significant values ($p < 0.05$) are highlighted in green.

Gland	unet	unet++	d2unet	d2aunet	runet	aunet	daunet	r2unet	r2aunet	segresnet	highresnet	vnet	nnunet	
unet	0.6015	0.3472	0.9168	0.7540	0.9168	0.7540	0.6015	0.7540	0.9168	0.1745	0.1745	0.2506	0.009	
unet++	0.4647	0.9168	0.4647	0.3472	0.9168	0.7540	0.7540	0.9168	0.9168	0.0758	0.1172	0.1172	0.009	
d2unet	0.1172	0.0283	0.3472	0.7540	0.9168	0.7540	0.4647	0.9168	0.6015	0.0758	0.2506	0.1172	0.009	
d2aunet	0.1172	0.0283	0.9168	0.6015	0.6015	0.2506	0.4647	0.4647	0.6015	0.0472	0.0758	0.0472	0.009	
runet	0.7540	0.6015	0.0758	0.0758	0.9168	0.7540	0.7540	0.7540	0.7540	0.0472	0.1172	0.0472	0.009	
aunet	0.6015	0.6015	0.6015	0.2506	0.6015	0.3472	0.9168	0.7540	0.7540	0.3472	0.3472	0.3472	0.009	
daunet	0.6015	0.2506	0.6015	0.4647	0.4647	0.6015	0.2506	0.7540	0.6015	0.3472	0.4647	0.4647	0.009	Full
r2unet	0.4647	0.6015	0.7540	0.3472	0.4647	0.7540	0.7540	0.2506	0.9168	0.0472	0.1745	0.0758	0.009	
r2aunet	0.2506	0.0283	0.6015	0.4647	0.1172	0.6015	0.7540	0.9168	0.4647	0.1745	0.2506	0.1745	0.009	
segresnet	0.2506	0.1172	0.0163	0.0163	0.2506	0.1172	0.0283	0.0283	0.0163	0.1172	0.9168	0.7540	0.009	
highresnet	0.1172	0.0758	0.0163	0.0163	0.1745	0.1172	0.0283	0.1172	0.0163	0.9168	0.4647	0.7540	0.009	
vnet	0.3472	0.3472	0.0472	0.0472	0.4647	0.4647	0.1172	0.2506	0.0758	0.3472	0.4647	0.0472	0.009	
nnunet	0.1745	0.0283	0.6015	0.6015	0.0758	0.4647	0.6015	0.9168	0.7540	0.0163	0.0163	0.0472	0.009	
							Cropped							full vs. cropped

Table A2. *p*-values using Kruskal–Wallis for the dice scores obtained in the cross-validation of the transition segmentation task. On the top area we compare the performance of the different architectures on the full data, and on the bottom on the cropped data. On the diagonal we present the comparison of the same architecture on both data types. Significant values ($p < 0.05$) are highlighted in green.

TZ	unet	unet++	d2unet	d2aunet	runet	aunet	daunet	r2unet	r2aunet	segresnet	highresnet	vnet	nnunet	
unet	0.009	0.7540	0.7540	0.4647	0.4647	0.2506	0.7540	0.7540	0.7540	0.1172	0.9168	0.7540	0.0163	
unet++	0.7540	0.009	0.6015	0.6015	0.9168	0.7540	0.9168	0.4647	0.9168	0.1172	0.9168	0.4647	0.009	
d2unet	0.7540	0.6015	0.009	0.3472	0.9168	0.6015	0.7540	0.6015	0.7540	0.0472	0.9168	0.3472	0.009	
d2aunet	0.4647	0.6015	0.3472	0.0163	0.2506	0.2506	0.6015	0.9168	0.6015	0.4647	0.6015	0.7540	0.0163	
runet	0.4647	0.9168	0.9168	0.2506	0.0472	0.3472	0.7540	0.6015	0.7540	0.0472	0.7540	0.1745	0.0163	
aunet	0.2506	0.7540	0.6015	0.2506	0.3472	0.0163	0.7540	0.3472	0.6015	0.0472	0.7540	0.1745	0.0163	
daunet	0.7540	0.9168	0.7540	0.6015	0.7540	0.7540	0.009	0.3472	0.4647	0.1172	0.7540	0.4647	0.009	Full
r2unet	0.7540	0.4647	0.6015	0.9168	0.6015	0.3472	0.3472	0.0283	0.6015	0.4647	0.3472	0.9168	0.009	
r2aunet	0.7540	0.9168	0.7540	0.6015	0.7540	0.6015	0.4647	0.6015	0.009	0.0758	0.7540	0.3472	0.009	
segresnet	0.1172	0.1172	0.0472	0.4647	0.0472	0.0472	0.1172	0.4647	0.0758	0.0163	0.1172	0.1745	0.009	
highresnet	0.9168	0.9168	0.9168	0.6015	0.7540	0.7540	0.7540	0.3472	0.7540	0.1172	0.0283	0.7540	0.009	
vnet	0.7540	0.4647	0.3472	0.7540	0.1745	0.1745	0.4647	0.9168	0.3472	0.1745	0.7540	0.0472	0.009	
nnunet	0.0163	0.009	0.009	0.0163	0.0163	0.0163	0.009	0.009	0.009	0.009	0.009	0.009	0.0283	
							Cropped							full vs. cropped

Table A3. *p*-values using Kruskal–Wallis for the dice scores obtained in the cross-validation of the peripheral segmentation task. On the top area we compare the performance of the different architectures on the full data, and on the bottom on the cropped data. On the diagonal we present the comparison of the same architecture on both data types. Significant values ($p < 0.05$) are highlighted in green.

PZ	unet	unet++	d2unet	d2aunet	runet	aunet	daunet	r2unet	r2aunet	segresnet	highresnet	vnet	nnunet	
unet	0.4647	0.6015	0.9168	0.6015	0.3472	0.4647	0.9168	0.9168	0.3472	0.1745	0.6015	0.1745	0.0163	
unet++	0.9168	0.4647	0.9168	0.1745	0.4647	0.3472	0.7540	0.6015	0.1745	0.1172	0.3472	0.2506	0.0283	
d2unet	0.9168	0.9168	0.9168	0.6015	0.3472	0.6015	0.9168	0.7540	0.2506	0.0758	0.7540	0.2506	0.0163	
d2aunet	0.9168	0.9168	0.9168	0.4647	0.0758	0.7540	0.4647	0.2506	0.4647	0.0758	0.7540	0.0758	0.0163	
runet	0.7540	0.6015	0.9168	0.6015	0.2506	0.1745	0.4647	0.3472	0.1172	0.2506	0.3472	0.4647	0.009	
aunet	0.7540	0.6015	0.9168	0.6015	0.9168	0.9168	0.4647	0.3472	0.7540	0.0758	0.7540	0.1745	0.0163	
daunet	0.9168	0.7540	0.9168	0.7540	0.6015	0.6015	0.6015	0.9168	0.3472	0.1172	0.4647	0.3472	0.0283	Full
r2unet	0.7540	0.9168	0.7540	0.9168	0.9168	0.6015	0.7540	0.3472	0.2506	0.0758	0.3472	0.1745	0.0163	
r2aunet	0.9168	0.9168	0.6015	0.9168	0.7540	0.7540	0.9168	0.7540	0.7540	0.0758	0.6015	0.0758	0.0163	
segresnet	0.2506	0.1745	0.2506	0.1745	0.3472	0.4647	0.3472	0.1745	0.2506	0.3472	0.3472	0.3472	0.009	
highresnet	0.9168	0.9168	0.7540	0.6015	0.3472	0.3472	0.4647	0.9168	0.7540	0.1172	0.3472	0.3472	0.0283	
vnet	0.4647	0.4647	0.6015	0.4647	0.6015	0.7540	0.4647	0.4647	0.4647	0.6015	0.2506	0.1745	0.009	
nnunet	0.9168	0.7540	0.7540	0.7540	0.4647	0.6015	0.9168	0.9168	0.7540	0.1745	0.6015	0.2506	0.0163	
							Cropped							full vs. cropped

References

1. Sung, H.; Ferlay, J.; Siegel, R.; Laversanne, M.; Soerjomataram, I.; Jemal, A.; Bray, F. Global Cancer Statistics 2020: GLOBOCAN Estimates of Incidence and Mortality Worldwide for 36 Cancers in 185 Countries. *CA Cancer J. Clin.* **2021**, *71*, 209–249. [CrossRef] [PubMed]
2. Ramamurthy, K.; Varikuti, A.R.; Gupta, B.; Aswani, N. A deep learning network for Gleason grading of prostate biopsies using EfficientNet. *Biomed. Tech.* **2022** . [CrossRef] [PubMed]
3. Rodrigues, A.; Santinha, J.; Galvão, B.; Matos, C.; Couto, F.M.; Papanikolaou, N. Prediction of Prostate Cancer Disease Aggressiveness Using Bi-Parametric Mri Radiomics. *Cancers* **2021**, *13*, 6065. [CrossRef] [PubMed]
4. Ronneberger, O.; Fischer, P.; Brox, T. U-Net: Convolutional Networks for Biomedical Image Segmentation. *arXiv* **2015**, arXiv:1505.04597.
5. Guo, Y.; Gao, Y.; Shen, D. Deformable MR Prostate Segmentation via Deep Feature Learning and Sparse Patch Matching. *IEEE Trans. Med. Imaging* **2016**, *35*, 1077–1089. [CrossRef]
6. Milletari, F.; Navab, N.; Ahmadi, S.A. V-Net: Fully Convolutional Neural Networks for Volumetric Medical Image Segmentation. In Proceedings of the 2016 Fourth International Conference on 3D Vision (3DV), Stanford, CA, USA, 25–28 October 2016; pp. 565–571.
7. Zhu, Q.; Du, B.; Turkbey, B.; Choyke, P.; Yan, P. Deeply-supervised CNN for prostate segmentation. In Proceedings of the 2017 International Joint Conference on Neural Networks (IJCNN), Anchorage, Alaska, USA, 14–19 May 2017; pp. 178–184.
8. Dai, Z.; Carver, E.; Liu, C.; Lee, J.; Feldman, A.; Zong, W.; Pantelic, M.; Elshaikh, M.; Wen, N. Segmentation of the Prostatic Gland and the Intraprostatic Lesions on Multiparametic Magnetic Resonance Imaging Using Mask Region-Based Convolutional Neural Networks. *Adv. Radiat. Oncol.* **2020**, *5*, 473–481. [CrossRef]
9. Zavala-Romero, O.; Breto, A.L.; Xu, I.R.; Chang, Y.C.C.; Gautney, N.; Dal Pra, A.; Abramowitz, M.C.; Pollack, A.; Stoyanova, R. Segmentation of prostate and prostate zones using deep learning: A multi-MRI vendor analysis. *Strahlenther. Onkol.* **2020**, *196*, 932–942. [CrossRef]
10. Aldoj, N.; Biavati, F.; Michallek, F.; Stober, S.; Dewey, M. Automatic prostate and prostate zones segmentation of magnetic resonance images using DenseNet-like U-net. *Sci. Rep.* **2020**, *10*, 14315. [CrossRef]
11. Duran, A.; Dussert, G.; Rouvière, O.; Jaouen, T.; Jodoin, P.M.; Lartizien, C. ProstAttention-Net: A deep attention model for prostate cancer segmentation by aggressiveness in MRI scans. *Med. Image Anal.* **2022**, *77*, 102347. [CrossRef]
12. Oktay, O.; Schlemper, J.; Folgoc, L.L.; Lee, M.J.; Heinrich, M.P.; Misawa, K.; Mori, K.; McDonagh, S.G.; Hammerla, N.Y.; Kainz, B.; et al. Attention U-Net: Learning Where to Look for the Pancreas. *arXiv* **2018**, arXiv:1804.03999.
13. Hung, A.L.Y.; Zheng, H.; Miao, Q.; Raman, S.S.; Terzopoulos, D.; Sung, K. CAT-Net: A Cross-Slice Attention Transformer Model for Prostate Zonal Segmentation in MRI. *arXiv* **2022**, arXiv:2203.15163.
14. Petit, O.; Thome, N.; Rambour, C.; Soler, L. U-Net Transformer: Self and Cross Attention for Medical Image Segmentation. *arXiv* **2021**, arXiv:2103.06104.
15. Gao, Y.; Zhou, M.; Liu, D.; Yan, Z.; Zhang, S.; Metaxas, D.N. A Data-scalable Transformer for Medical Image Segmentation: Architecture, Model Efficiency, and Benchmark. *arXiv* **2022**, arXiv:2203.00131.
16. Lin, A.; Chen, B.; Xu, J.; Zhang, Z.; Lu, G. DS-TransUNet:Dual Swin Transformer U-Net for Medical Image Segmentation. *arXiv* **2021**, arXiv:2106.06716.
17. Hatamizadeh, A.; Nath, V.; Tang, Y.; Yang, D.; Roth, H.R.; Xu, D. Swin UNETR: Swin Transformers for Semantic Segmentation of Brain Tumors in MRI Images. In Proceedings of the BrainLes@MICCAI, Virtual Event, 27 September 2021.
18. Shamshad, F.; Khan, S.; Zamir, S.W.; Khan, M.H.; Hayat, M.; Khan, F.S.; Fu, H. Transformers in Medical Imaging: A Survey. *arXiv* **2022**, arXiv:2201.09873.
19. Kalapos, A.; Gyires-T'oth, B. Self-Supervised Pretraining for 2D Medical Image Segmentation. *arXiv* **2022**, arXiv:2209.00314.
20. Azizi, S.; Mustafa, B.; Ryan, F.; Beaver, Z.; von Freyberg, J.; Deaton, J.; Loh, A.; Karthikesalingam, A.; Kornblith, S.; Chen, T.; et al. Big Self-Supervised Models Advance Medical Image Classification. In Proceedings of the 2021 IEEE/CVF International Conference on Computer Vision (ICCV), Montreal, QC, Canada, 11–17 October 2021; pp. 3458–3468.
21. Albelwi, S. Survey on Self-Supervised Learning: Auxiliary Pretext Tasks and Contrastive Learning Methods in Imaging. *Entropy* **2022**, *24*, 551. [CrossRef]
22. Meng, H.; Lin, Z.; Yang, F.; Xu, Y.; Cui, L. Knowledge Distillation In Medical Data Mining: A Survey. In Proceedings of the 5th International Conference on Crowd Science and Engineering, Jinan, China, 16–18 October 2021 ; ICCSE '21; Association for Computing Machinery: New York, NY, USA, 2022; pp. 175–182. [CrossRef]
23. Hassan, T.; Shafay, M.; Hassan, B.; Akram, M.U.; ElBaz, A.; Werghi, N. Knowledge distillation driven instance segmentation for grading prostate cancer. *Comput. Biol. Med.* **2022**, *150*, 106124. [CrossRef]
24. Vesal, S.; Gayo, I.; Bhattacharya, I.; Natarajan, S.; Marks, L.S.; Barratt, D.C.; Fan, R.E.; Hu, Y.; Sonn, G.A.; Rusu, M. Domain Generalization for Prostate Segmentation in Transrectal Ultrasound Images: A Multi-center Study. *arXiv* **2022**, arXiv:2209.02126.
25. Liu, X.; Hu, B.; Huang, W.; Zhang, Y.; Xiong, Z. Efficient Biomedical Instance Segmentation via Knowledge Distillation. In Proceedings of the International Conference on Medical Image Computing and Computer-Assisted Intervention, Singapore, 18–22 September 2022.

26. Tustison, N.; Avants, B.; Cook, P.; Zheng, Y.; Egan, A.; Yushkevich, P.; Gee, J. N4ITK: Improved N3 Bias Correction. *IEEE Trans. Med. Imaging* **2010**, *29*, 1310–1320. [CrossRef]
27. Simpson, A.; Antonelli, M.; Bakas, S.; Bilello, M.; Farahani, K.; Ginneken, B.; Kopp-Schneider, A.; Landman, B.; Litjens, G.; Menze, B.; et al. A large annotated medical image dataset for the development and evaluation of segmentation algorithms. *arXiv* **2019**, arXiv:1902.09063.
28. Bochkovskiy, A.; Wang, C.Y.; Liao, H. YOLOv4: Optimal Speed and Accuracy of Object Detection. *arXiv* **2020**, arXiv:2004.10934.
29. Zhou, Z.; Siddiquee, M.M.R.; Tajbakhsh, N.; Liang, J. UNet++: A Nested U-Net Architecture for Medical Image Segmentation. *arXiv* **2018**, arXiv:1807.10165.
30. Alom, M.Z.; Hasan, M.; Yakopcic, C.; Taha, T.M.; Asari, V.K. Recurrent Residual Convolutional Neural Network based on U-Net (R2U-Net) for Medical Image Segmentation. *arXiv* **2018**, arXiv:1802.06955.
31. Isensee, F.; Jaeger, P.F.; Kohl, S.A.A.; Petersen, J.; Maier-Hein, K.H. nnU-Net: A self-configuring method for deep learning-based biomedical image segmentation. *Nat. Methods* **2021**, *18*, 203–211. [CrossRef]
32. Li, W.; Wang, G.; Fidon, L.; Ourselin, S.; Cardoso, M.J.; Vercauteren, T.K.M. On the Compactness, Efficiency, and Representation of 3D Convolutional Networks: Brain Parcellation as a Pretext Task. *arXiv* **2017**, arXiv:1707.01992.
33. Myronenko, A. 3D MRI brain tumor segmentation using autoencoder regularization. In Proceedings of the BrainLes@MICCAI, Granada, Spain, 16 September 2018.
34. Consortium, T.M. Project MONAI 2020. Available online: https://zenodo.org/record/4323059#.Y_msOR9BxPY (accessed on 2 January 2023).
35. Pytorch Lightning. GitHub. 2019, 3. Available online: https://github.com/PyTorchLightning/pytorch-lightning (accessed on 2 January 2023).
36. Wright, L. Ranger—A Synergistic Optimizer. 2019. Available online: https://github.com/lessw2020/Ranger-Deep-Learning-Optimizer (accessed on 2 January 2023).
37. Lin, T.Y.; Goyal, P.; Girshick, R.B.; He, K.; Dollár, P. Focal Loss for Dense Object Detection. In Proceedings of the 2017 IEEE International Conference on Computer Vision (ICCV), Venice, Italy, 22–29 October 2017; pp. 2999–3007.

diagnostics

Article

DBE-Net: Dual Boundary-Guided Attention Exploration Network for Polyp Segmentation

Haichao Ma [1,2], Chao Xu [1,2,*], Chao Nie [1,2], Jubao Han [1,2], Yingjie Li [1,2] and Chuanxu Liu [1,2]

1 School of Integrated Circuits, Anhui University, Hefei 230601, China
2 Anhui Engineering Laboratory of Agro-Ecological Big Data, Hefei 230601, China
* Correspondence: xchao@ahu.edu.cn; Tel.: +86-133-3919-9368

Abstract: Automatic segmentation of polyps during colonoscopy can help doctors accurately find the polyp area and remove abnormal tissues in time to reduce the possibility of polyps transforming into cancer. However, the current polyp segmentation research still has the following problems: blurry polyp boundaries, multi-scale adaptability of polyps, and close resemblances between polyps and nearby normal tissues. To tackle these issues, this paper proposes a dual boundary-guided attention exploration network (DBE-Net) for polyp segmentation. Firstly, we propose a dual boundary-guided attention exploration module to solve the boundary-blurring problem. This module uses a coarse-to-fine strategy to progressively approximate the real polyp boundary. Secondly, a multi-scale context aggregation enhancement module is introduced to accommodate the multi-scale variation of polyps. Finally, we propose a low-level detail enhancement module, which can extract more low-level details and promote the performance of the overall network. Extensive experiments on five polyp segmentation benchmark datasets show that our method achieves superior performance and stronger generalization ability than state-of-the-art methods. Especially for CVC-ColonDB and ETIS, two challenging datasets among the five datasets, our method achieves excellent results of 82.4% and 80.6% in terms of mDice (mean dice similarity coefficient) and improves by 5.1% and 5.9% compared to the state-of-the-art methods.

Keywords: polyp segmentation; colorectal cancer; boundary exploration; medical image analysis; deep learning; colonoscopy

Citation: Ma, H.; Xu, C.; Nie, C.; Han, J.; Li, Y.; Liu, C. DBE-Net: Dual Boundary-Guided Attention Exploration Network for Polyp Segmentation. *Diagnostics* **2023**, *13*, 896. https://doi.org/10.3390/diagnostics13050896

Academic Editors: Michał Strzelecki, Adam Piorkowski and Rafał Obuchowicz

Received: 1 February 2023
Revised: 22 February 2023
Accepted: 23 February 2023
Published: 27 February 2023

1. Introduction

Colorectal cancer (CRC) is a common digestive tract malignant tumor, which seriously endangers human health, and its incidence ranks third among all cancers [1]. As one of the most important precursors of colorectal cancer, polyps can easily transform into malignant tumors if not treated in time. A colonoscopy is an effective method to detect colon lesions, which can provide doctors with precise positioning information so that doctors can remove them in time before they become cancerous. However, colonoscopy occasionally overlooks cancer-causing polyps due to the inexperience of physicians. Therefore, automatic and accurate segmentation of polyps in colonoscopy images is of great significance for the clinical prevention of colorectal cancer.

At present, the existing methods of polyp segmentation can be divided into two categories, one is traditional methods, and the other is based on deep learning. Traditional polyp segmentation methods usually rely on handcrafted feature-based [2–7] methods to identify polyps, such as texture analysis, color distribution, geometric features, and intensity distribution. Though traditional methods have made considerable progress, these handcrafted feature-based methods still have low polyp segmentation accuracy and poor generalization ability, which cannot meet the requirements of clinical practice.

In recent years, with the continuous development of deep learning, many deep learning-based polyp segmentation methods [8–19] have been proven to outperform traditional handcrafted feature-based methods. With the proposal of the fully convolutional network FCN, the problem of image segmentation is introduced to the pixel level. Brandao et al. [20] and Vazquez et al. [21] used FCN to segment polyps from colonoscopy images, which achieved better segmentation accuracy than traditional methods. Subsequently, Ronneberger et al. [22] proposed a U-shaped network for biomedical image segmentation, namely U-Net, which is the most widely used network in medical image segmentation. Currently, U-Net++ [23], ResUnet++ [24], and other variant models based on the U-Net network architecture are applied to polyp segmentation, and this encoding–decoding network has become the mainstream network architecture in the medical field.

Recently, attention mechanisms have been increasingly applied to medical segmentation, especially polyp segmentation, which is often used to enhance the blurred boundaries of polyps [8–10] or to extract global and local features [25–28]. PraNet [8] uses an inverse attention mechanism to mine boundary information with a global feature map, which is generated by a parallel decoder that aggregates high-level feature information together. ACSNet [16] designed a spatial attention scheme to pay more attention to uncertain regions, but it did not explicitly model the relationship between ambiguous regions and boundary regions. Guo et al. [29] designed a ternary guidance module. This module suppresses background regions through a ternary guidance mask, thereby mining more polyp regions in uncertain regions without interference from tissues in the background. However, due to its limited ability to explore a wide range of uncertain regions, it does not fully utilize boundary information, and it is difficult to detect fine boundaries. Lai et al. [30] proposed a boundary-guided attention module that can guide the network to adaptively learn multi-scale boundary features for accurate polyp segmentation. However, it cannot explore more detailed information in uncertain regions, resulting in polyp segment segmentation and false positive segmentation. Inspired by the above, we propose a dual bounding-guided attention exploration module DBE, which can not only explore more uncertain regions in a wider range but also explore more detailed information with boundary masks.

For the multi-scale feature adaptation of polyps, there are some related studies dealing with the variation of polyp size and shape. MCNet [31] proposes a multi-scale context-guided deep network for automatic lesion segmentation with endoscopy images of the gastrointestinal tract, where both global and local contexts are captured as guidance for model training. MSNet [15] proposed a multi-scale subtraction network, which uses a subtraction unit (SU) to generate differential features between adjacent layers of the network and equips perceptual units at different levels with different perceptual fields in a feature pyramid manner, thus obtaining richer multi-scale difference information. Sun et al. [32] applied a dilated convolution at the end of the encoder to expand the receptive field to improve the performance of polyps, but it is difficult to capture more contextual information with only a dilated convolution. In view of the above, in our network, we propose a multi-scale context aggregation enhancement module CAM, which first aggregates the context information of different scales, and then adopts the channel feature pyramid to obtain the most extensive features from different receptive regions through multiple dilated convolutions with different dilation rates.

The transformer was originally proposed for the field of natural language processing (NLP). Vaswani et al. [33] first proposed a transformer for machine translation tasks. With the great success of transformers in the NLP field, Dosovitskiy et al. [34] proposed a vision transformer (ViT), which extended the transformer model architecture to the field of computer vision and was the first pure transformer for image classification. ViT can well replace the convolution operation and can still achieve good results in image classification tasks without relying on convolution. Chen et al. proposed a segmentation network TransUnet, that combines CNN and Transformer. TransUnet not only encodes image features as sequences through the transformer to enhance global context but also makes good use of low-level CNN features through a U-shaped hybrid structure design and sets a new record in CT multi-organ segmentation task. Zhang et al. [25] adopted a two-branch parallel architecture TransFuse, which runs a CNN-based encoder and a

Transformer-based segmentation network in a parallel manner. Then, the multi-level features of the two branches are efficiently fused by using the proposed fusion module. Wang et al. [26] proposed a simple CNN-free backbone pyramid vision transformer (PVT) for dense prediction tasks. Compared with the design of ViT for image classification, PVT introduces the pyramid structure into the transformer so that it can be applied to various downstream dense prediction tasks. Compared with CNN, PVT performs better in detection and segmentation tasks. This paper adopts PVT as our encoder and designs a new polyp segmentation framework based on it.

In summary, the current polyp segmentation methods based on deep learning have greatly improved the segmentation accuracy and generalization ability compared with traditional methods, but these methods still have shortcomings when facing the challenge of polyp segmentation. In Figure 1, we provide several representative contrast images as an illustration. The first is the problem of blurred polyp boundaries, as shown in Figure 1a. When the colonoscope moves in the intestine, it will cause problems such as motion blur and reflections, which cause blurred boundaries of polyp images and increase the difficulty of polyp segmentation. The second is the multi-scale adaptability of polyps, as shown in Figure 1a–c. Due to the variable size and shape of polyp tissue, the current polyp segmentation methods still have certain limitations in the ability to extract multi-scale features. The third is the close resemblances between polyps and nearby normal tissues, as shown in Figure 1b,c. Polyps have low contrast to the background and high similarity in texture and color to surrounding tissue, which makes accurate identification of polyp areas difficult.

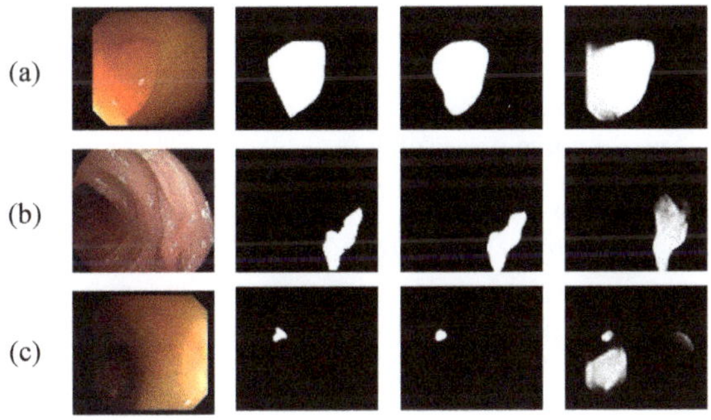

Figure 1. The segmentation examples of our model and CaraNet [18] with different challenge cases: (**a**) blurred boundaries of polyps, (**a**–**c**) multi-scale adaptability of polyps, (**b**,**c**) close resemblances between polyps and nearby normal tissues. These images are from unseen datasets CVC-ColonDB [35] and ETIS [21], which show that our model has better generalization ability.

To solve the above problems, this paper proposes a novel and efficient dual boundary-guided attention exploration network (DBE-Net) for accurate polyp segmentation. This network uses the cooperative exploration of uncertainty regions and border regions as the primary driving force for polyp segmentation. First, to deal with the problem of blurred polyp boundaries, we design a dual-boundary guided attention exploration module (DBE), which gradually approximates real polyp boundaries through a coarse-to-fine strategy. DBE is composed of a ternary mask submodule (TM) and a boundary mask submodule (BM). First, we use TM to suppress the background area, balance the foreground area and enhance the uncertainty area, which can explore more polyp areas in a wider uncertainty area and reduce the interference of normal tissues in the background. Second, we utilize BM to guide the network to adaptively learn the perceptual features of the boundary region to achieve re-enhancement of the polyp boundary. Meanwhile, to address the multi-scale feature adaptation problem, we propose a multi-scale context aggregation augmentation module (CAM). This module first aggregates the features of two adjacent layers to obtain the feature information of different levels. Then, the feature maps are convolved with dilated convolutions with different dilation rates through the channel feature pyramid to accommodate the variation in polyp size and shape. In addition, low-level features usually contain rich, detailed information such as texture, color, and edge information. To address the close resemblances between polyps and nearby normal tissues, we propose a low-level detail enhancement module (LDE). This module can extract more low-level details and provide them with more high-level features for fusion after up-sampling operation to promote the overall performance of the network. Meanwhile, this module uses the combination of channel attention and spatial attention to capture polyp details in different dimensions.

Our main contributions can be highlighted as follows:

- We propose a new dual boundary-guided attention exploration network termed DBE-Net. Different from existing CNN-based methods, this network adapts the pyramid vision transformer as the encoder to obtain more robust features and is combined with the partial decoder PD as the base framework.
- We propose a dual boundary-guided attention exploration module, termed DBE. This module adopts a coarse-to-fine strategy to gradually approximate the real polyp boundary from coarse uncertain regions to fine boundary regions through ternary masks and boundary masks according to the prediction results of previous scales to obtain a clearer polyp segmentation boundary.
- We propose a multi-scale context aggregation augmentation module termed CAM. To address the multi-scale feature adaptation problem, this module adopts multiple dilated convolutions with different dilation rates and obtains the most extensive features from different receptive regions to adapt to the multi-scale changes of polyps.
- To solve the problem of high similarity between polyps and surrounding tissues, we propose a low-level detail enhancement module, termed LDE. This module can extract more low-level details to improve the overall network performance and achieve more accurate polyp segmentation.

2. Materials and Methods

2.1. Proposed Network Structure

The purpose of polyp segmentation is to identify polyp regions from normal tissues in colonoscopy images and to segment polyps robustly. However, the polyp boundaries are irregular, even somewhat blurred, and disturbed by normal tissues in the background, which brings great challenges to improving polyp segmentation performance. Therefore, we propose a dual boundary-guided attention exploration network DBE-Net, which will be used for accurate polyp segmentation.

As shown in Figure 2, the proposed network mainly consists of four parts, including a pyramid vision transformer (PVT) encoder, dual boundary-guided attention exploration module (DBE), multi-scale context aggregation enhancement module (MCA), and low-level detail enhancement module (LDE). In the encoder stage, to obtain more powerful backbone features and provide more foreground information for subsequent decoding stages, PVTv2 is adopted as our encoder. Compared with traditional CNN methods, which pay more attention to local information, PVTv2 exhibits stronger global information extraction ability and better robustness to input disturbances. In the decoder stage, considering that the transformer is good at providing global information but lacks detailed local information, it makes it difficult to solve the multi-scale feature adaptation problem, including size and shape changes. To cope with this problem, MAC can obtain a wider range of features from different receptive regions by aggregating and enhancing the features of each stage to adapt to the multi-scale changes of polyps, thereby obtaining more abundant local and global features. Then, a partial decoder (PD) obtains preliminary rough segmentation results by aggregating features from the outputs of the MAC modules at different stages. However, the preliminary results obtained by the PD module can only obtain significant foreground regions, and it is difficult to judge accurate polyp boundaries for fuzzy boundaries. Therefore, the DBE adopts a coarse-to-fine strategy to segment polyps. According to the preliminary prediction results obtained by PD, the real polyp boundaries are gradually approximated from rough uncertain regions to fine boundary regions by the ternary mask sub-module and the boundary mask sub-module. Low-level features usually contain rich detail information, while the detail fusion sub-module (DF) in LDE can extract more low-level detail information and provide it to high-level features through up-sampling. At the same time, the detail extraction sub-module (DE) can capture the detailed appearance information to promote the overall performance of the network.

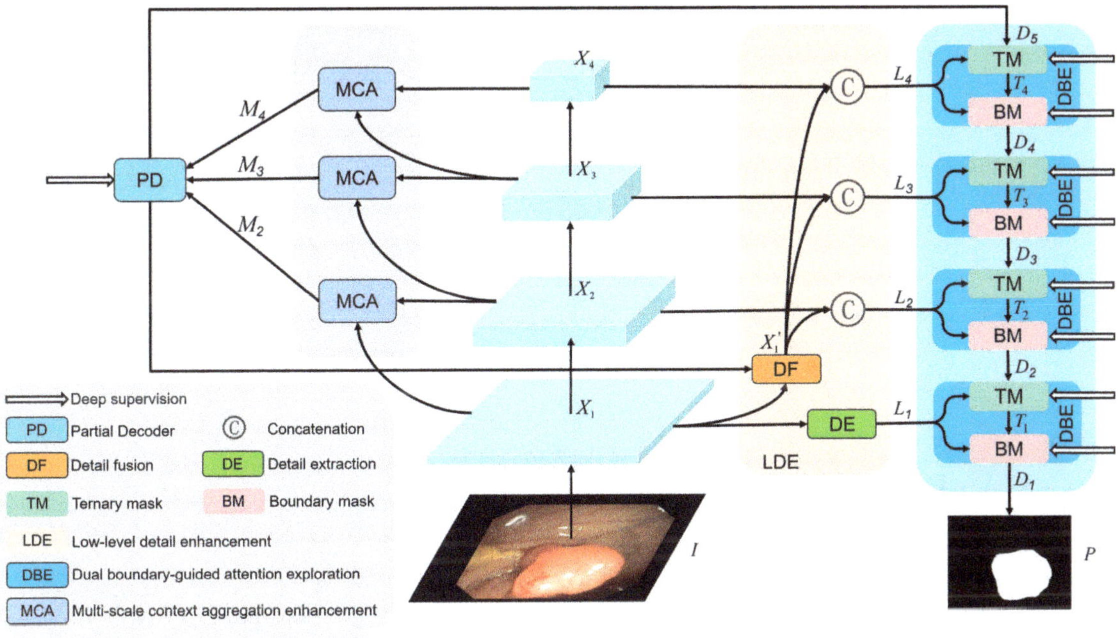

Figure 2. Overall architecture of the proposed DBE-Net, which consists of a pyramid vision transformer (PVT) encoder, dual boundary-guided attention exploration module (DBE), multi-scale context aggregation enhancement module (MCA), and low-level detail enhancement module (LDE).

In our network model, we use PVTv2 as the backbone network to extract four multi-scale pyramid features $\boldsymbol{X}_i$ from the input image $\boldsymbol{I}$, where $i \in \{1, 2, 3, 4\}$. Then, every two adjacent features of these four features are sent to MCA for multi-scale context aggregation enhancement, leading to three output features $\boldsymbol{M}_2$, $\boldsymbol{M}_3$, and $\boldsymbol{M}_4$. Next, we feed the output features $\boldsymbol{M}_2$, $\boldsymbol{M}_3$, and $\boldsymbol{M}_4$ to PD to fuse, resulting in preliminary rough segmentation results $\boldsymbol{D}_5$. At the same time, the coarse segmentation result $\boldsymbol{D}_5$ from PD and low-level features $\boldsymbol{X}_1$ are fed to LDE for detail enhancement to obtain $\boldsymbol{L}_i$, where $I \in \{1, 2, 3, 4\}$. Then the PD features $\boldsymbol{D}_5$ are fed into the DBE, which is combined with the detail enhancement features $\boldsymbol{L}_4$ from the LDE output to obtain a clearer boundary segmentation through two stages of boundary exploration from coarse to fine. Meanwhile, the output features $\boldsymbol{D}_4$ and $\boldsymbol{T}_4$ of the two stages serve as side supervision. Then, the enhanced features $\boldsymbol{D}_4$ are up-sampled and fed to the next DBE. There they are again combined with the detail enhancement features $\boldsymbol{L}_3$ to further obtain a clearer boundary segmentation and output two-stage side supervision $\boldsymbol{T}_3$ and $\boldsymbol{D}_3$. This process is repeated in detail enhancement features $\boldsymbol{L}_2$ and $\boldsymbol{L}_1$ while obtaining four side supervisions, $\boldsymbol{T}_2$, $\boldsymbol{D}_2$, $\boldsymbol{T}_1$, and $\boldsymbol{D}_1$. Finally, through four stepwise boundary explorations, the final predicted output $\boldsymbol{D}_1$ is obtained.

Our overall process is as follows:

$$X_i = T(I)(i = 1, 2, 3, 4) \tag{1}$$

$$M_i = M(X_i, X_{i-1})(i = 2,3,4) \quad (2)$$

$$D_5 = P(M_2, M_3, M_4) \quad (3)$$

$$L_i = L(D_5, X_1, X_2, X_3, X_4)(i = 1,2,3,4) \quad (4)$$

$$D_1 = D(L_1, D(L_2, D(L_3, D(L_4, D_5)))) \quad (5)$$

Here: (1) $\boldsymbol{I}$ represents the input image; (2) $\boldsymbol{T}, \boldsymbol{M}, \boldsymbol{P}, \boldsymbol{L}$, and $\boldsymbol{D}$ represent the functions of PVTv2, MAC, PD, LDE, and DBE, respectively.

2.2. Dual Boundary-Guided Attention Exploration Module

When the colonoscope moves in the intestine, it may produce problems such as motion blur and reflection, which causes blurred boundaries in the polyp image. To pay more attention to the pixels of the blurred boundary and achieve a clearer polyp segmentation boundary, we design a dual boundary-guided attention exploration module DBE, which adopts a coarse-to-fine gradual approximation strategy. First, according to the prediction results of the previous scale and detail enhancement features, more ground-truth polyp regions are obtained from rough uncertain regions through a ternary mask sub-module. Then, the boundary mask sub-module explores the finer polyp boundaries from the fine boundary regions. Finally, multiple DBE operations are used to gradually approach the true polyp boundary. Figure 3 shows the detailed structure of the DBE. This module consists of two sub-modules, namely the ternary mask sub-module TM and the boundary mask sub-module BM.

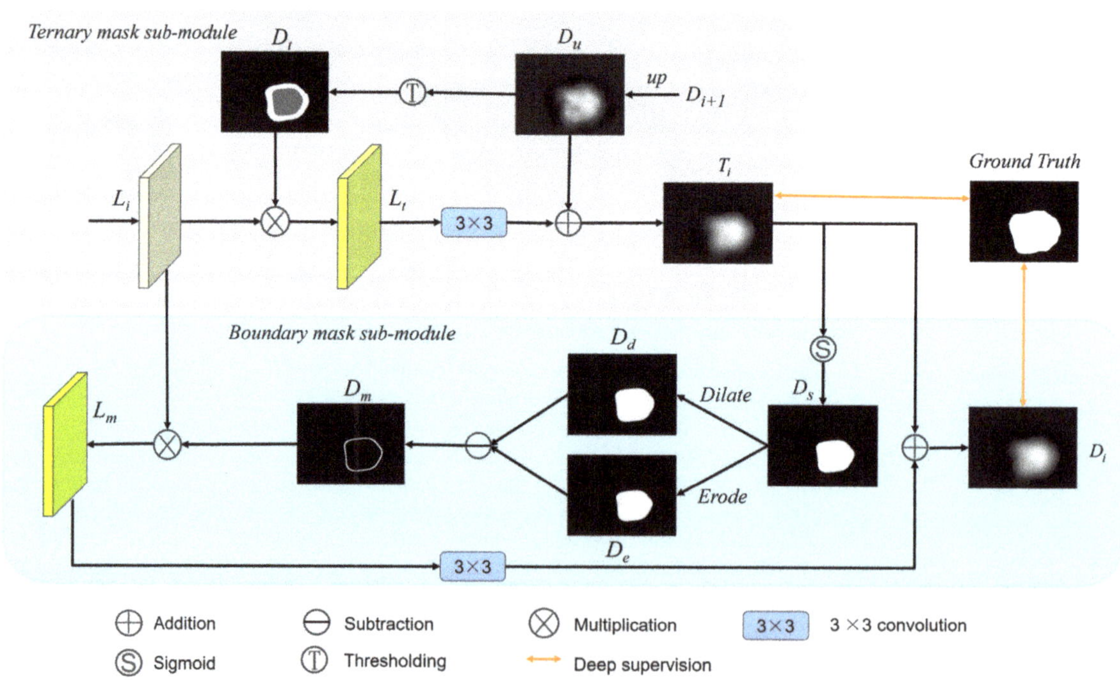

Figure 3. The structure of DBE, which consists of two sub-modules, including the ternary mask sub-module and the boundary mask sub-module. It adopts a coarse-to-fine strategy to gradually approximate the real polyp boundary.

For the TM sub-module, it divides the feature map into three regions, namely foreground, background, and uncertainty regions. The foreground represents polyp regions with high response, the background represents non-polyp regions with low response, and uncertainty regions refer to fuzzy regions with intermediate responses. The uncertain area includes foreground and background objects and is a mixture of the two. The purpose of the TM sub-module is to mine as many foreground objects as possible from the uncertain region and exclude background objects. Specifically, the ternary mask assigns different weight values to different regions. To emphasize the uncertain area, the pixels in the uncertain area are set to the highest weight of 1. Pixels in the foreground area are set to 0 to balance the highly responsive areas. To suppress the interference of the background area, the pixels in the background area are set to the lowest weight of −1. The specific process is as follows:

As shown in the gray rounded rectangle box in Figure 3, the low-resolution output feature map $\boldsymbol{D}_{i+1}$ by the previous stage is up-sampled to obtain $\boldsymbol{D}_u$.

$$D_u = U(D_{i+1}) \tag{6}$$

where $\boldsymbol{U}$ represents the up-sampling operation.

Then, a ternary mask $\boldsymbol{D}_t$ (Equation (7)) is generated based on two thresholds, $\boldsymbol{\alpha}_l$ and $\boldsymbol{\alpha}_h$. In Equation (7), the value of i-th pixel of $\boldsymbol{D}_u$ is greater than $\boldsymbol{\alpha}_h$, which means it belongs to the foreground region and sets the weight to 0. If it is less than $\boldsymbol{\alpha}_l$, it belongs to the background area and sets the weight to −1. When it is between $\boldsymbol{\alpha}_l$ and $\boldsymbol{\alpha}_h$, it belongs to the uncertainty region and sets the weight to 1.

$$D_t^i = \begin{cases} -1, & if\ D_u^i \leq \alpha_l \\ 1, & if\ \alpha_l \leq D_u^i \leq \alpha_h \\ 0, & if\ D_u^i \geq \alpha_h \end{cases} \tag{7}$$

Next, the generated ternary mask $\boldsymbol{D}_t$ is element-wise multiplied by the features $\boldsymbol{L}_i$ from the current stage to obtain weighted feature maps $\boldsymbol{L}_t$. Then, $\boldsymbol{L}_t$ is 3 × 3 convoluted and combined with $\boldsymbol{D}_u$ through element-wise addition to generating the output feature maps $\boldsymbol{T}_i$ of the TM sub-module. In particular, the feature maps $\boldsymbol{T}_i$ is used both as side supervision and as the input of the BM sub-module.

$$T_i = f_i(L_i \otimes D_t) + D_u \tag{8}$$

where ⊗ represents element-wise multiplication. $\boldsymbol{f}_i$ denotes 3 × 3 convolution.

For the BM sub-module, it can extract richer and more detailed boundary field information, which can help to explore more accurate polyp edges. To achieve the goal of enhancing boundary features, we extract boundary information from the output feature maps $\boldsymbol{T}_i$ of the previous DM sub-module through morphological operations. As shown in the cyan rounded rectangular box in Figure 3, the output feature maps $\boldsymbol{T}_i$ of the previous sub-module DM is converted into a binary mask $\boldsymbol{D}_s$ with a threshold of 0.5. If the probability of a pixel belonging to a polyp is greater than 0.5, the pixel is classified as a foreground polyp pixel; otherwise, it is classified as a background pixel. Then, we use morphological operations on the binary mask $\boldsymbol{D}_s$:

$$D_m = Dilate(D_s, E) - Erode(D_s, E) \tag{9}$$

where ***Dilate(D_s, E)*** denotes the morphological dilation operation on the binary mask $\boldsymbol{D}_s$ using multi-structure elements $\boldsymbol{E}$ to generate a dilated mask $\boldsymbol{D}_d$. ***Erode(D_s, E)*** denotes the morphological erosion operation on the binary mask $\boldsymbol{D}_s$ using multi-structure elements $\boldsymbol{E}$ to generate an erosion mask $\boldsymbol{D}_e$. The morphological dilation operation dilates the polyp border outward, thereby expanding the foreground area. The morphological erosion operation shrinks the polyp boundary, thereby reducing the foreground area.

Similar to [36], we also consider multi-structure elements. Let (m, n) be the center of the binary mask $\boldsymbol{D}_s$, and a structural element in the $(2N + 1) \times (2N + 1)$ square window can be represented as:

$$E_i = \{M(m + m_0, n + n_0), \theta_i = i \times \alpha | -N \leq m_0, n_0 \leq N\} \tag{10}$$

where $i = 0, 1, \ldots, 4N - 1$, $\alpha = 180°/4N$, and θ_i is the direction angle of the structure element.

In this paper, for the two DBE modules at higher levels, we choose $N = 2$, and then in a 5×5 square window, the direction angles of all structure elements are 0°, 22.5°, 45°, 67.5°, 90°, 112.5°, 135°, and 157.5°. For the two DBE modules at the lower level, we choose $N = 3$, and then in the 7×7 square window, the direction angles of all structure elements are 0°, 15°, 30°, 45°, 60°, 75°, 90°, 105°, 120°, 135°, 150°, and 165°. They contain most of the directions in which the lines extend in the image. In morphological dilation, each structural element $\boldsymbol{E}_i$ in different directions was used to dilate the image, and the averaged result is taken as the dilated mask. Morphological erosion operation is similar. Then, we perform an element-wise subtraction operation on $\boldsymbol{D}_d$ and $\boldsymbol{D}_e$ to generate the boundary mask $\boldsymbol{D}_m$.

Next, the generated ternary mask $\boldsymbol{D}_m$ is element-wise multiplied by the input features $\boldsymbol{L}_i$ from the previous TM sub-module to obtain boundary-enhanced feature maps $\boldsymbol{L}_m$. Then, $\boldsymbol{L}_m$ is 3×3 convoluted and combined with $\boldsymbol{T}_i$ through element-wise addition to generating the output feature maps $\boldsymbol{D}_i$ of the DM sub-module. Meanwhile, $\boldsymbol{D}_i$ also serves as the final output of the DBE module. In particular, the feature maps $\boldsymbol{D}_i$ is used both as side supervision and as the input of the next DBE module.

$$D_i = f_i(L_i \otimes D_m) + T_i \tag{11}$$

2.3. *Multi-Scale Context Aggregation Enhancement Module*

Due to the variable size and shape of polyp tissue, current polyp segmentation methods still have certain limitations in their ability to extract multi-scale features. To address the multi-scale feature adaptation problem of polyps, we introduce a multi-scale context aggregation augmentation module MCA, which can obtain more detailed local and global feature information, as shown in Figure 4. This module complements the location and spatial information through the interaction of adjacent high- and low-layer feature information. At the same time, this module uses the channel feature pyramid (CFP) [37] module to obtain the most extensive multi-scale feature information from different receptive regions to adapt to the multi-scale variation of polyps.

Specifically, we use 1×1 convolution on two feature maps from different layers to reduce feature channels to 32 to reduce computational resources. The features from lower layers are down-sampled and connected with the features from higher layers as the input of CFP. The CFP module has four branches, and the dilate rates of different branches are set to (1, 2, 4, 8) to obtain more contextual information. Each branch contains three groups of convolution blocks. Each group of convolution blocks decomposes the standard convolution into 3×1 convolution and then performs 1×3 convolution. Then, with this factorized convolution [37], 33% of the parameters can be saved with the same number of filters. Finally, the features of the four branches are concatenated together as the output of the MCA module.

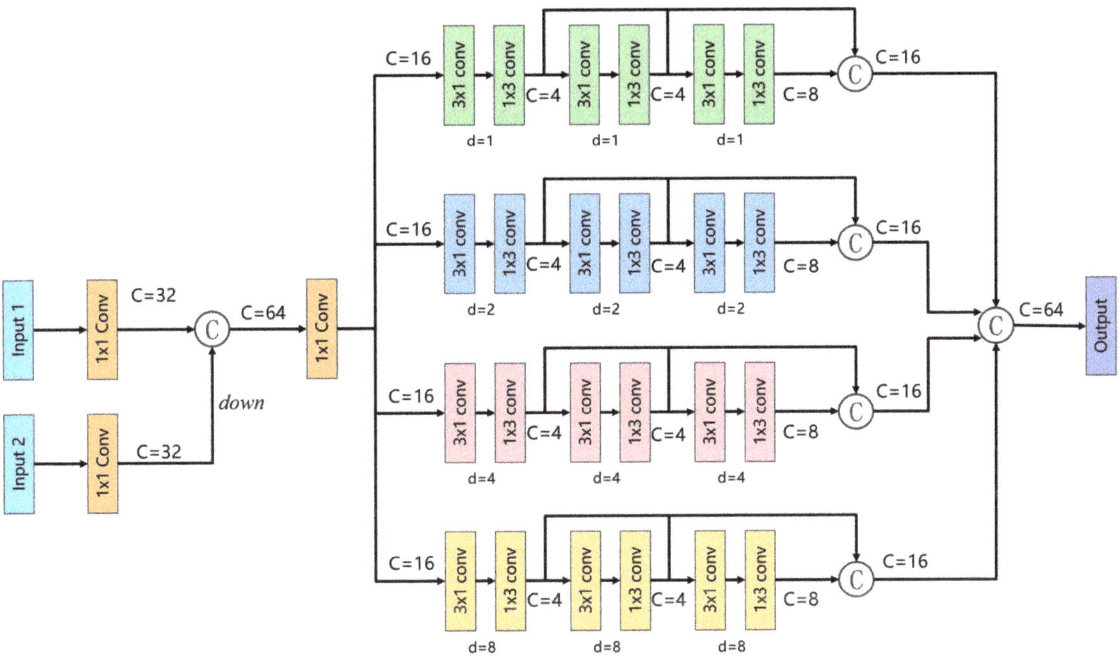

Figure 4. The structure of MCA. It integrates the feature information of adjacent high and low layers and adopts four parallel dilated convolution branches to extract multi-scale features.

2.4. Low-Level Detail Enhancement Module

The appearance of polyps is usually very similar to the surrounding normal tissue; however, low-level feature maps usually contain more detailed information, such as texture, borders, and color. Therefore, we design a low-level detail enhancement module LDE to extract richer detailed information. As shown in Figure 5, the LDE module consists of two parts, namely the detail fusion sub-module DF and the detail extraction sub-module DE.

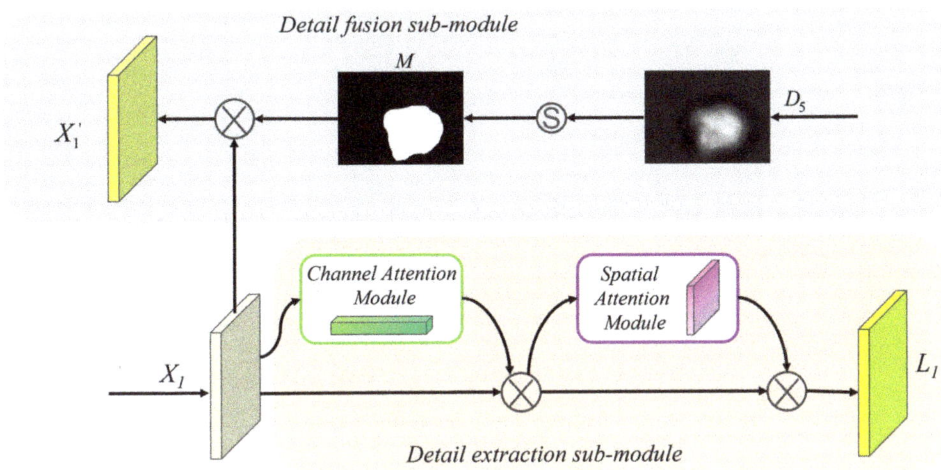

Figure 5. The structure of LDE, which consists of two sub-modules, including the detail fusion sub-module and the detail extraction sub-module. It is used to extract richer low-level details.

First, the DF sub-module reduces the interference of background information by the preliminary segmentation result $\boldsymbol{D}_5$ from the PD module. Then, more detailed information is provided on high-level features for fusion through down-sampling to improve the performance of the overall network and improve the accuracy of polyp segmentation. As shown in the grey rounded rectangle box in Figure 5, the features $\boldsymbol{D}_5$ from the PD module are transformed into a binary mask $\boldsymbol{M}$ with a threshold α. If the probability of belonging to a polyp pixel is less than α, it is classified as a background pixel with a weight of 0. Where α is set to 0.2.

$$M^i = \begin{cases} 0, & if\ X_1^i \leq \alpha \\ 1, & if\ X_1^i \geq \alpha \end{cases} \tag{12}$$

Next, the generated mask $\boldsymbol{M}$ is element-wise multiplied by the low-level features $\boldsymbol{X}_1$ to obtain feature maps $\boldsymbol{X}_1{}'$ with reduced background noise. Then, $\boldsymbol{X}_1{}'$ is up-sampled and concatenated with $\boldsymbol{X}_2$, $\boldsymbol{X}_3$, and $\boldsymbol{X}_4$, respectively, to generate detail enhancement feature maps $\boldsymbol{L}_2$, $\boldsymbol{L}_3$, and $\boldsymbol{L}_4$.

$$X_1{}' = X_1 \otimes M \tag{13}$$

For the DE sub-module, we mainly adopt the tandem use of channel attention operation $\boldsymbol{A}_c(\cdot)$ and spatial attention operation $\boldsymbol{A}_s(\cdot)$ to capture detailed information on polyps from different dimensions.

$$L_1 = A_s(A_c(X_1)) \tag{14}$$

The channel attention module formula is as follows:

$$A_c(F) = \sigma(MLP(P_{avg}(F)) + MLP(P_{max}(F))) \otimes F \tag{15}$$

where $\boldsymbol{P}_{avg}$ and $\boldsymbol{P}_{max}$ represent the max pooling and average pooling functions, respectively. For the input feature map F, the $\boldsymbol{P}_{avg}$ and $\boldsymbol{P}_{max}$ functions are first applied at each spatial location, resulting in two C × 1 × 1 feature maps. Then each is fed into a shared *MLP* containing two fully connected layers. Finally, the channel attention map is obtained by adding pixels and passing through an activation function.

The spatial attention module formula is as follows:

$$A_s(F) = \sigma(f^{7\times7}(Concat(P_{avg}(F), P_{max}(F)))) \otimes F \tag{16}$$

where the feature maps output by the channel attention operation is taken as the input of the spatial attention operation. First, the $\boldsymbol{P}_{avg}$ and $\boldsymbol{P}_{max}$ functions are applied at each channel, resulting in two C × 1 × 1 feature maps. Then, they are connected by channel and fed into a 7 × 7 convolutional layer. Finally, the spatial attention map is obtained through the activation function.

2.5. Loss Function

The DBE-Net is trained via a deep supervision manner. There are a total of 9 aspects of supervision across all stages of the network (Figure 1), including $\boldsymbol{D}_i$ (i = 1, 2, 3, 4, 5) and $\boldsymbol{T}_i$ (i = 1, 2, 3, 4, 5). In the experiments, we use the weighted binary cross entropy ($\boldsymbol{L}_{wbce}$) loss and the weighted intersection over union ($\boldsymbol{L}_{wiou}$) loss to supervise each $\boldsymbol{D}_i$ and $\boldsymbol{T}_i$. These two losses restrict the prediction map from the global object perspective and local pixel perspective, respectively. The total loss can be expressed as:

$$L = L_{main} + L_{aux} \tag{17}$$

where $\boldsymbol{L}_{main}$ and $\boldsymbol{L}_{aux}$ represent the main loss and auxiliary loss, respectively. The main loss $\boldsymbol{L}_{main}$ is to calculate the loss between the feature map $\boldsymbol{D}_i$ and the ground truth $\boldsymbol{G}$, which can be expressed as follows:

$$L_{main} = \sum_{i-1}^{5} (L_{wbce}(D_i, G) + L_{wiou}(D_i, G)) \tag{18}$$

The auxiliary loss L_{aux} is to calculate the loss between the feature map T_i and the ground truth G, which can be expressed as follows:

$$L_{aux} = \sum_{i=1}^{4} (L_{wbce}(T_i, G) + L_{wiou}(T_i, G)) \tag{19}$$

2.6. Datasets

To evaluate the segmentation performance of our proposed method, we conduct experiments on five polyp segmentation data, including Kvasir-SEG [38], CVC-ClinicDB [39], CVC-ColonDB [35], ETIS [21] and CVC-T [5]. The description of these datasets is as follows.

(1) **Kvasir-SEG**. The dataset consists of 1000 polyp images and their corresponding ground truth polyp masks annotated by endoscopy experts. The resolution of the images contained in Kvasir-SEG varies from 332 × 487 to 1920 × 1072 pixels.
(2) **CVC-ClinicDB**. The dataset, also known as *CVC-612*, contains 612 open-access images from 25 colonoscopy videos with a resolution of 384 × 288.
(3) **CVC-ColonDB**. The dataset consists of 380 images from 15 colonoscopy videos with an image resolution of 574 × 500.
(4) **ETIS**. The dataset contains 196 images captured from 34 colonoscopy videos with an image resolution of 1225 × 996.
(5) **CVC-T**. This dataset is a subset of EndoScene and contains 60 images from 44 colonoscopy sequences of 36 patients with an image resolution of 574 × 500.

For a fair comparison, we use the same training and test dataset splits as [8,17,18], as shown in Table 1. A total of 1450 training images are used in the training set, of which 900 images are from Kvasir-SEG, and 550 images are from CVC-ClinicDB. A total of 798 test images are used in the test set, of which 100 images are from Kvasir-SEG, and 62 images are from CVC-ClinicDB. Meanwhile, the other three datasets are used for testing, including 380, 196, and 62 images in CVC-ColonBD, ETIS, and CVC-T, respectively.

Table 1. The colorectal polyp datasets used in our experiments.

Dataset	Image Size	Image Number	Number of Train Samples	Number of Test Samples
Kvasir-SEG	Variable	1000	900	100
CVC-ClinicDB	384 × 288	612	550	62
CVC-ColonDB	574 × 500	380	0	380
ETIS	1225 × 966	196	0	196
CVC-T	574 × 500	60	0	60

3. Results

In this section, we employ three kinds of experiments to validate the performance of DBENet, including qualitative and quantitative experiments and an ablation study. In the experiments, we compare our method with six different methods, including state-of-the-art methods, including UNet [22], UNet++ [23], SFA [40], PraNet [8], SANet [17] and CaraNet [18]. Meanwhile, we analyze and compare three evaluation metrics on five different polyp segmentation datasets. For a fair comparison, the segmentation maps for these methods are published by the authors or directly generated from the original code provided by the authors.

3.1. Training Settings and Evaluation Metrics

Our proposed model is implemented in Pytorch with NVIDIA RTX3090 for GPU acceleration. Before feeding into the network, we resize the input image to 352 × 352. Instead of data augmentation, we use a multi-scale training strategy {0.75, 1.0, 1.25}. We start the training with a batch size of 16 and adopt the Adam optimization method to optimize the model parameters. The initial learning rate is set to 0.0001, and the maximum epoch is set to 100. Thresholds α_l and α_h are set to 0.3 and 0.7, respectively.

In this paper, the mean dice similarity coefficient (*mDice*), the mean intersection over union (*mIoU*), and the mean absolute error (*MAE*) were used to quantitatively evaluate the segmentation performance of polyps. For these evaluation metrics, higher values of mDice and *mIoU* indicate better segmentation results, while lower values of *MAE* indicate better segmentation results.

The Dice coefficient is an ensemble similarity measure and a standard measure for comparing pixel-level classification results between predicted masks and ground truth values. The *mDice* represents the average of the dice similarity coefficients between all predicted images and the ground truth, and the formula is as follows.

$$mDice = \frac{1}{n+1}\sum_{i=1}^{n}\frac{2TP}{FP+2TP+FN} \tag{20}$$

where *TP*, *FP*, *FN*, and n denote true positive, false positive, and false negative predictions and the number of test images, respectively.

The IoU calculates the overlap ratio between the predicted mask and its corresponding ground truth value. The *mIoU* represents the average of the IoU values between all predicted images and the ground truth, and the formula is as follows.

$$mIoU = \frac{1}{n+1}\sum_{i=1}^{n}\frac{TP}{FP+TP+FN} \tag{21}$$

MAE is used to evaluate pixel-level accuracy. We evaluate pixel-level accuracy by computing the mean absolute error between the model's predicted and true values, and the formula is as follows.

$$MAE = \frac{1}{n}\sum_{i=1}^{n}|\hat{p}_i - p_i| \tag{22}$$

where $\hat{p}_i$ and p_i denote the prediction and the corresponding ground truth for the i-th pixel out of the total n pixels.

3.2. Qualitative Analysis

Figure 6 shows a visual comparison between our model DBE-Net and different competing methods. Apparently, our method can segment polyp regions more accurately and outperform several challenging aspects compared to other competing methods. First, DBE-Net can obtain clearer polyp boundaries and reduce the interference of complex backgrounds such as intestinal folds and dim areas. For example, intestinal folds were mistakenly missed by other methods in Figure 6e, and polyps in dim regions were missed or misjudged by other methods in Figure 6a. In both cases, DBE-Net can accurately detect polyps, which may be attributed to the role of the DBE module. This module gradually enhances the fuzzy edge information through a coarse-to-fine strategy to achieve the purpose of approximating the real polyp boundary. Second, DBE-Net can effectively deal with the multi-scale changes of polyps and identify large polyps and small polyps of different shapes, especially to avoid misjudgment of small polyps. As shown in Figure 6a–c, only our method can accurately detect small polyps and reduce false positives. Third, DBE-Net can also better solve the problem of high similarity between polyps and surrounding tissues. As shown in Figure 6d,e, only our method can segment the polyp region more completely.

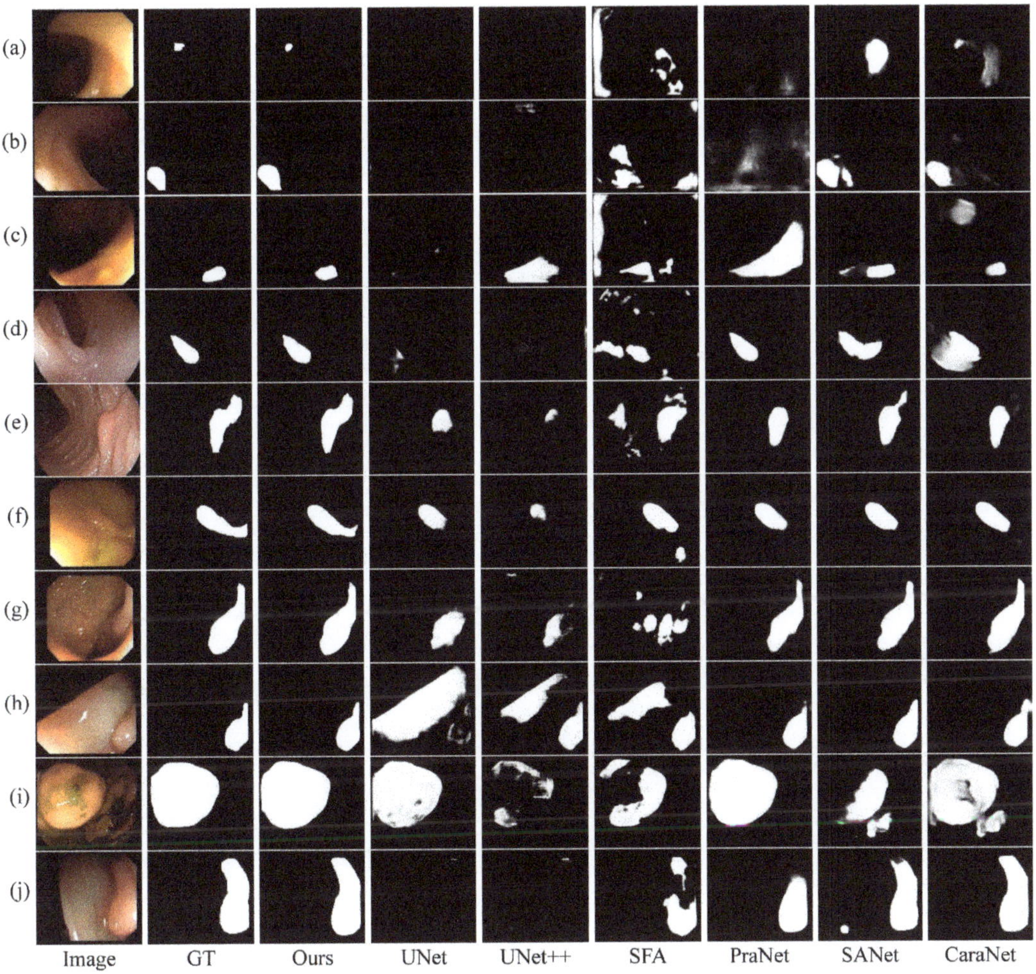

Figure 6. Visualization comparison between the proposed method and six state-of-the-art ones.

3.3. Quantitative Experiments

Tables 2 and 3 present the quantitative results of the statistical comparison of DBE-Net with six different competing methods on five different datasets. For ease of viewing, the best results for each evaluation metric are highlighted in bold. Considering that the training set is selected from CVC-ClinicDB and Kvasir-SEG, the learning ability of the proposed method is first analyzed quantitatively. As shown in Table 2, our proposed method is optimal on various metrics on both datasets. The mDice of our method achieves excellent results of 93.9% and 92.2% in CVC-ClinicDB and Kvasir-SEG.

To verify the generalization abilities of the model, we test it on three unseen datasets (CVC-ColonDB, ETIS, and CVC-T). As shown in Table 3, the results of the proposed method are still optimal and achieve good generalization performance compared to the six competing methods. Especially, as shown in Figure 7, the generalization ability of our method achieves a significant improvement on the challenging datasets CVC-ColonDB and ETIS. On the CVC-ColonDB dataset, the mDice of our method outperforms CaraNet and SANet by 5.1% and 7.1%, respectively. In terms of the ETIS dataset, the mDice of our method outperforms CaraNet and SANet by 5.9% and 5.6%, respectively. On the CVC-T dataset, the mDice of our method surpasses SANet by 1.5%.

Table 2. Quantitative comparison of the learning abilities on CVC-ClinicDB and Kvasir-SEG. The best results are shown in bold.

Method	CVC-ClinicDB			Kvasir-SEG		
	mDice	mIou	MAE	mDice	mIou	MAE
U-Net	0.823	0.755	0.019	0.818	0.746	0.055
UNet++	0.794	0.729	0.022	0.821	0.743	0.048
SFA	0.700	0.607	0.042	0.723	0.611	0.075
PraNet	0.899	0.849	0.009	0.898	0.840	0.030
SANet	0.916	0.859	0.012	0.904	0.847	0.028
CaraNet	0.936	0.887	0.007	0.918	0.865	0.023
Ours	**0.939**	**0.895**	**0.006**	**0.922**	**0.871**	**0.022**

Table 3. Quantitative comparison of the generalization abilities on CVC-ColonDB, ETIS, and CVC-T. The best results are shown in bold.

Method	CVC-ColonDB			ETIS			CVC-T		
	mDice	mIou	MAE	mDice	mIou	MAE	mDice	mIou	MAE
U-Net	0.512	0.444	0.061	0.398	0.335	0.036	0.710	0.627	0.022
UNet++	0.483	0.410	0.064	0.401	0.344	0.035	0.707	0.624	0.018
SFA	0.469	0.347	0.094	0.297	0.217	0.109	0.467	0.329	0.065
PraNet	0.712	0.640	0.045	0.628	0.567	0.031	0.871	0.797	0.010
SANet	0.753	0.670	0.043	0.750	0.654	0.015	0.888	0.815	0.008
CaraNet	0.773	0.689	0.042	0.747	0.672	0.017	**0.903**	**0.838**	0.007
Ours	**0.824**	**0.743**	**0.027**	**0.806**	**0.725**	**0.015**	**0.903**	0.837	**0.006**

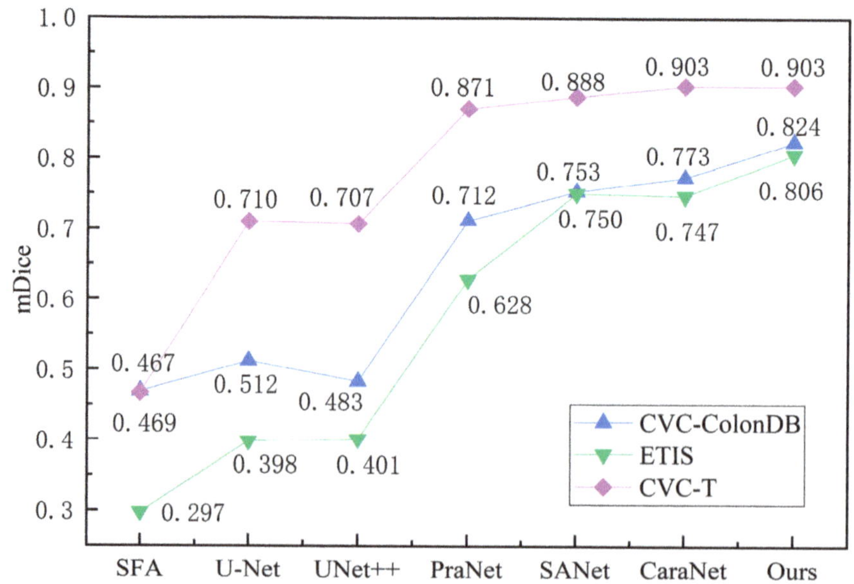

Figure 7. Evaluation of model generalization ability. We provide the mDice results on CVC-ColonDB, ETIS, and CVC-T.

3.4. Ablation Studies

To verify the effectiveness of each component in the model, we conduct ablation experiments on the dual boundary-guided attention exploration module (DBE), multi-scale context aggregation enhancement module (MCA), and low-level detail enhancement module (LDE). The baseline of our backbone network is composed of PVTv2 and PD, and the standard model is composed of "Baseline + MCA + LDE + DBE". We evaluate the effectiveness of different modules by removing or changing them from the standard model. "w/o MCA", "w/o LDE," and "w/o DBE" represent the removal of MCA, LDE, or DBE, respectively, from the standard model.

Ablation studies of DBE: To investigate the effectiveness of the DBE module, we first visualize the output characteristics of the DBE module at different stages. As shown in Figure 8, the redder the area, the more attention the network needs to give. We can observe that it gradually refines the polyp boundary and explores more polyp information as the stage progresses. This shows that the coarse-to-fine strategy of the DBE module can effectively enhance the fuzzy edge information. To quantitatively analyze the effectiveness of the DBE module, we train a version "w/o DBE" that lacks the DBE module. We removed the DBE module altogether and replaced it with an element-wise addition operation. The experimental results are shown in Table 4. Compared to the standard model, the performance of the model without DBE dropped sharply on the five datasets. As shown in the dumbbell diagram in Figure 9, on CVC-ClinicDB, Kvasir-SEG, CVC-ColonDB, ETIS, and CVC-T, compared to the performance of the standard model DBENet, the absence of DBE reduces mDice by 0.8%, 0.6%, 1.7%, 1.6%, and 2.1%, respectively.

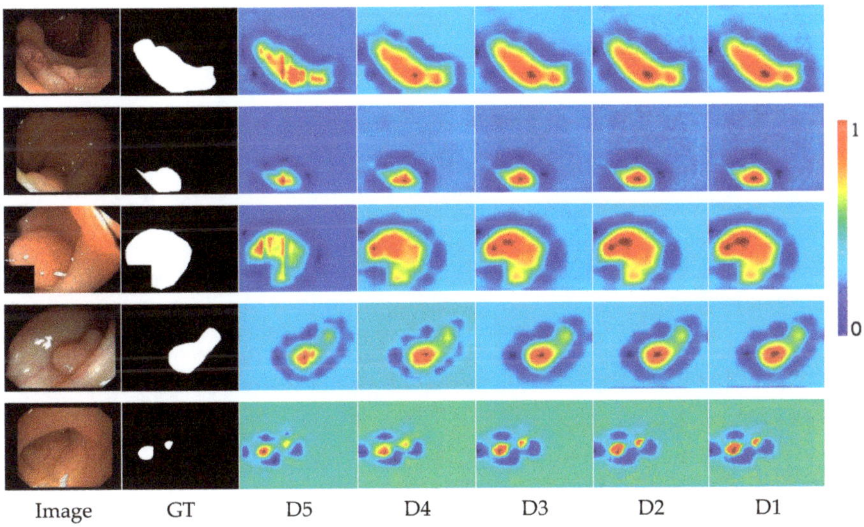

Figure 8. Visualization of output features of the DBE module at different stages. The redder the region, the more attention the network needs to pay.

Table 4. Quantitative results of ablation studies of our method under different configurations.

Dataset	CVC-ClinicDB		Kvasir-SEG		CVC-ColonDB		ETIS		CVC-T	
Metric	**mDice**	**mIou**	**mDice**	**mIou**	**mDice**	**mIou**	**mDice**	**mIou**	**mDice**	**mIou**
Baseline	0.929	0.879	0.913	0.858	0.798	0.716	0.792	0.709	0.891	0.820
w/o MCA	**0.939**	0.892	**0.932**	**0.885**	0.814	0.732	0.801	0.723	0.903	0.836
w/o LDE	0.934	0.885	0.918	0.867	0.818	0.735	0.805	0.723	0.893	0.825
w/o DBE	0.931	0.883	0.916	0.865	0.807	0.726	0.790	0.704	0.882	0.812
Baseline + MCA + LDE + DBE	**0.939**	**0.895**	0.922	0.871	**0.824**	**0.743**	**0.806**	**0.725**	**0.903**	**0.837**

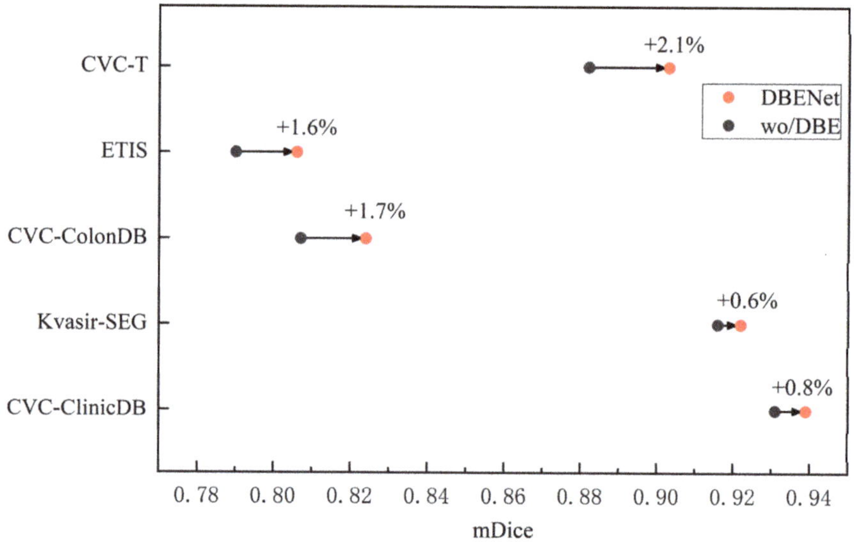

Figure 9. Comparison of mDice results between standard model DBE-Net and "wo/DBE" lacking the DBE module.

Ablation studies of MCA: Similarly, to quantitatively analyze the effectiveness of the MCA module, we train a version "w/o MCA" that lacks the MCA module. We remove all MCA modules and directly fuse the context information of the three high layers (i.e., X_2, X_3, and X_4). The results in Table 4 show that compared with the standard model, the model lacking MCA has a 1.0% and 0.5% drop in mDice on the CVC-ColonDB and ETIS datasets, respectively. In particular, the CVC-ColonDB and ETIS datasets are challenging unseen datasets, which suggests that MCA is helpful for multi-scale generalization.

Ablation studies of LDE: To demonstrate the power of LDE, we train a version "w/o LDE" that lacks the LDE module. As shown in Table 4, removing the LDE results in a slight drop in performance on all five datasets compared to the standard model. On the five datasets, CVC-ClinicDB, Kvasir-SEG, CVC-ColonDB, ETIS, and CVC-T, removing LDE can reduce mDice by 0.5%, 0.4%, 0.6%, 0.1%, and 1.0%, respectively. Therefore, the LDE module plays a role in improving the overall segmentation ability of the model.

4. Discussion

Colonoscopy is an effective way to detect colon lesions. Automatic segmentation of polyp regions can help doctors improve the accuracy of diagnosis. The current polyp segmentation research based on deep learning still has the following problems: blurry polyp boundaries, multi-scale adaptability of polyps, and close resemblances between polyps and nearby normal tissues. For the fuzzy polyp boundary problem, we use a coarse-to-fine strategy to gradually enhance the true polyp boundary. As can be seen from Figure 8, as the stage goes on, the polyp boundary becomes clearer, and the background interference is more suppressed, which shows that our strategy can effectively enhance the blurred edge information. For the multi-scale adaptability of polyps, the proposed MCA module first supplements position and spatial information by fusing adjacent high- and low-layer feature information and then uses dilated convolutions with different dilation rates to obtain the most extensive multi-scale feature information from different receptive areas to accommodate multi-scale changes in polyps. As shown in Figure 6, our model can effectively identify polyps of different shapes and sizes, especially for the misjudgment of small polyps. For the similarity between polyps and adjacent normal tissues, the proposed LDE module makes full use of the low-level features that usually contain more detailed information, such as texture, borders, and color. We reduce the interference of background information in the low-level features through the initial segmentation result D_5 and then capture the detailed information of polyps from different dimensions through the tandem use of channel attention operation and spatial attention operation. From the results of the ablation experiments in Table 4, it can be concluded that the LDE module promotes the overall performance of the model.

The proposed DBE-Net model achieves superior results compared to state-of-the-art methods on all five public polyp segmentation benchmark datasets. From the qualitative results, as shown in Figure 6, it is obvious that the proposed model can obtain sharper segmentation masks while reducing the complex background interference. For polyps of different shapes and sizes, our model can still identify them effectively. Compared with CaraNet, the proposed model improved the mDice by 0.3% and 0.4% on CVC-ClinicDB and Kvasir-SEG datasets, respectively. Test results on three unseen datasets show that our model exhibits a stronger generalization ability. Especially for CVC-ColonDB and ETIS, two challenging datasets among the five datasets, our method achieves excellent results of 82.4% and 80.6% on mDice and improves by 5.1% and 5.9% compared with CaraNet, respectively. This result is significantly ahead of the current state-of-the-art methods.

In this paper, the presented study still has some limitations. We speculate that more data enhancement processing of the original dataset could further improve the efficiency of the model, including increasing the size of larger datasets, applying more image enhancement techniques, and image post-processing steps. The DBE-Net model allows us to design a very deep and clever neural network architecture to achieve more accurate polyp segmentation. However, DBE-Net should not only be limited to the polyp segmentation domain but also can be extended to other medical image segmentation, which will help us to further promote the generalization ability of the proposed model. Furthermore, the proposed model can also be extended to natural image segmentation and other pixel classification tasks. We used all our experience and knowledge to optimize the model, but there may be further optimizations, which may affect the performance of the method.

5. Conclusions

In this paper, we propose a new polyp segmentation framework, DBE-Net, to help address three major challenges in the field of polyp segmentation: blurry polyp boundaries, multi-scale adaptability of polyps, and close resemblances between polyps and nearby normal tissues. To address the blurred boundary problem, we introduce the dual boundary-guided attention exploration module DBE. This module mainly adopts a coarse-to-fine strategy to gradually approach the real polyp boundary. Meanwhile, the proposed network introduces a multi-scale context aggregation enhancement module MCA and a low-level detail enhancement module LDE to help address the problem of polyp multi-scale adaptability and the similarity of polyps to surrounding tissues. Extensive qualitative and quantitative experiments demonstrate that DBE-Net outperforms current state-of-the-art models. For performance analysis on unseen datasets, it is found that the proposed network has a higher generalization ability than other state-of-the-art methods. Especially for the unseen CVC-ColonDB dataset and ETIS dataset, the mDice of the proposed model reaches 82.4% and 80.6%. This result is significantly ahead of current state-of-the-art models. We can conclude that the proposed DBE-Net model can accurately segment polyps and can be considered a powerful polyp segmentation method. Further, we can develop a stable and clinically applicable system for polyp segmentation from colonoscopy video frames.

Author Contributions: Conceptualization, H.M. and C.X.; methodology, H.M.; software, H.M. and C.N.; validation, J.H., Y.L. and H.M.; data curation, C.L. and J.H.; writing—original draft preparation, H.M.; writing—review and editing, C.X.; visualization, C.N.; supervision, C.X.; project administration, C.X.; funding acquisition, C.X. All authors have read and agreed to the published version of the manuscript.

Funding: This research was funded by the National Key Research and Development Program of China (No. 2019YFC0117800).

Institutional Review Board Statement: Not applicable.

Informed Consent Statement: Not applicable.

Data Availability Statement: The datasets used in this research are all benchmark datasets and are publicly available for researchers.

Conflicts of Interest: The authors declare no conflict of interest.

References

1. Siegel, R.L.; Miller, K.D.; Goding Sauer, A.; Fedewa, S.A.; Butterly, L.F.; Anderson, J.C.; Cercek, A.; Smith, R.A.; Jemal, A. Colorectal cancer statistics, 2020. *CA Cancer J. Clin.* **2020**, *70*, 145–164. [CrossRef]
2. Alexandre, L.A.; Nobre, N.; Casteleiro, J. Color and position versus texture features for endoscopic polyp detection. In Proceedings of the 2008 International Conference on BioMedical Engineering and Informatics, Sanya, China, 27–30 May 2008; pp. 38–42.
3. Iwahori, Y.; Shinohara, T.; Hattori, A.; Woodham, R.J.; Fukui, S.; Bhuyan, M.K.; Kasugai, K. Automatic Polyp Detection in Endoscope Images Using a Hessian Filter. In *Proceedings of the MVA*; IEEE: New York, NY, USA, 2013; pp. 21–24.
4. Mamonov, A.V.; Figueiredo, I.N.; Figueiredo, P.N.; Tsai, Y.-H.R. Automated polyp detection in colon capsule endoscopy. *IEEE Trans. Med. Imaging* **2014**, *33*, 1488–1502. [CrossRef]
5. Silva, J.; Histace, A.; Romain, O.; Dray, X.; Granado, B. Toward embedded detection of polyps in wce images for early diagnosis of colorectal cancer. *Int. J. Comput. Assist. Radiol. Surg.* **2014**, *9*, 283–293. [CrossRef]
6. Fiori, M.; Musé, P.; Sapiro, G. A complete system for candidate polyps detection in virtual colonoscopy. *Int. J. Pattern. Recognit Artif. Intell.* **2014**, *28*, 1460014. [CrossRef]
7. Nawarathna, R.; Oh, J.; Muthukudage, J.; Tavanapong, W.; Wong, J.; De Groen, P.C.; Tang, S.J. Abnormal image detection in endoscopy videos using a filter bank and local binary patterns. *Neurocomputing* **2014**, *144*, 70–91. [CrossRef]
8. Fan, D.-P.; Ji, G.-P.; Zhou, T.; Chen, G.; Fu, H.; Shen, J.; Shao, L. Pranet: Parallel reverse attention network for polyp segmentation. In Proceedings of the International Conference on Medical Image Computing and Computer-Assisted Intervention, Lima, Peru, 4–8 October 2020; Springer International Publishing: Berlin/Heidelberg, Germany, 2020; pp. 263–273.
9. Kim, T.; Lee, H.; Kim, D. UACANet: Uncertainty Augmented Context Attention for Polyp Segmentation. In Proceedings of the 29th ACM International Conference on Multimedia, Chendu, China, 20–24 October 2021; pp. 2167–2175.
10. Guo, X.; Yang, C.; Liu, Y.; Yuan, Y. Learn to Threshold: ThresholdNet With Confidence-Guided Manifold Mixup for Polyp Segmentation. *IEEE Trans. Med. Imaging* **2021**, *40*, 1134–1146. [CrossRef]
11. Wu, H.; Zhao, Z.; Zhong, J.; Wang, W.; Wen, Z.; Qin, J. PolypSeg+: A Lightweight Context-Aware Network for Real-Time Polyp Segmentation. *IEEE Trans. Cybern* **2022**, 1–12. [CrossRef]
12. Ta, N.; Chen, H.; Lyu, Y.; Wu, T. BLE-Net: Boundary learning and enhancement network for polyp segmentation. *Multimed. Syst.* **2022**. [CrossRef]
13. Alam, S.; Tomar, N.K.; Thakur, A.; Jha, D.; Rauniyar, A. Automatic polyp segmentation using u-net-resnet50. *arXiv* **2020**. [CrossRef]
14. Nguyen, N.-Q.; Lee, S.-W. Robust Boundary Segmentation in Medical Images Using a Consecutive Deep Encoder-Decoder Network. *IEEE Access* **2019**, *7*, 33795–33808. [CrossRef]
15. Zhao, X.; Zhang, L.; Lu, H. Automatic polyp segmentation via multi-scale subtraction network. In Proceedings of the International Conference on Medical Image Computing and Computer-Assisted Intervention, Strasbourg, France, 27 September–1 October 2021; Springer International Publishing: Berlin/Heidelberg, Germany, 2021; pp. 120–130.
16. Zhang, R.; Li, G.; Li, Z.; Cui, S.; Qian, D.; Yu, Y. Adaptive Context Selection for Polyp Segmentation. In *Medical Image Computing and Computer Assisted Intervention—MICCAI 2020*; Lecture Notes in Computer Science; Springer International Publishing: Berlin/Heidelberg, Germany, 2020; pp. 253–262.
17. Wei, J.; Hu, Y.; Zhang, R.; Li, Z.; Zhou, S.K.; Cui, S. Shallow attention network for polyp segmentation. In Proceedings of the International Conference on Medical Image Computing and Computer-Assisted Intervention, Strasbourg, France, 27 September–1 October 2021; Springer International Publishing: Berlin/Heidelberg, Germany, 2021; pp. 699–708.
18. Lou, A.; Guan, S.; Loew, M. Caranet: Context axial reverse attention network for segmentation of small medical objects. *arXiv* **2021**. [CrossRef]
19. Guo, X.; Chen, Z.; Liu, J.; Yuan, Y. Non-equivalent images and pixels: Confidence-aware resampling with meta-learning mixup for polyp segmentation. *Med. Image Anal.* **2022**, *78*, 102394. [CrossRef] [PubMed]
20. Armato, S.G.; Petrick, N.A.; Brandao, P.; Mazomenos, E.; Ciuti, G.; Caliò, R.; Bianchi, F.; Menciassi, A.; Dario, P.; Koulaouzidis, A.; et al. Fully convolutional neural networks for polyp segmentation in colonoscopy. In *Proceedings of the Medical Imaging 2017: Computer-Aided Diagnosis*; SPIE: Bellingham, WA, USA, 2017.
21. Vazquez, D.; Bernal, J.; Sanchez, F.J.; Fernandez-Esparrach, G.; Lopez, A.M.; Romero, A.; Drozdzal, M.; Courville, A. A Benchmark for Endoluminal Scene Segmentation of Colonoscopy Images. *J. Health Eng.* **2017**, *2017*, 4037190. [CrossRef] [PubMed]

22. Ronneberger, O.; Fischer, P.; Brox, T. U-Net: Convolutional Networks for Biomedical Image Segmentation. In *Medical Image Computing and Computer-Assisted Intervention—MICCAI 2015*; Lecture Notes in Computer Science; Springer International Publishing: Berlin/Heidelberg, Germany, 2015; pp. 234–241.
23. Zhou, Z.; Siddiquee, M.M.R.; Tajbakhsh, N.; Liang, J. UNet++: A Nested U-Net Architecture for Medical Image Segmentation. In *Deep Learning in Medical Image Analysis and Multimodal Learning for Clinical Decision Support. DLMIA ML-CDS 2018 2018. Lecture Notes in Computer Science*; Springer: Cham, Switzerland, 2018; Volume 11045, pp. 3–11. [CrossRef]
24. Jha, D.; Smedsrud, P.H.; Riegler, M.A.; Johansen, D.; De Lange, T.; Halvorsen, P.; Johansen, H.D. Resunet++: An advanced architecture for medical image segmentation. In Proceedings of the 2019 IEEE International Symposium on Multimedia (ISM), San Diego, CA, USA, 9–11 December 2019; pp. 225–2255.
25. Zhang, Y.; Liu, H.; Hu, Q. Transfuse: Fusing transformers and cnns for medical image segmentation. In Proceedings of the International Conference on Medical Image Computing and Computer-Assisted Intervention, Strasbourg, France, 27 September–1 October 2021; Springer International Publishing: Berlin/Heidelberg, Germany, 2021; pp. 14–24.
26. Dong, B.; Wang, W.; Fan, D.-P.; Li, J.; Fu, H.; Shao, L. Polyp-pvt: Polyp segmentation with pyramid vision transformers. *arXiv* **2021**. [CrossRef]
27. Li, S.; Sui, X.; Luo, X.; Xu, X.; Liu, Y.; Goh, R. Medical image segmentation using squeeze-and-expansion transformers. *arXiv* **2021**. [CrossRef]
28. Wang, J.; Huang, Q.; Tang, F.; Meng, J.; Su, J.; Song, S. Stepwise Feature Fusion: Local Guides Global. *arXiv* **2022**. [CrossRef]
29. Guo, Q.; Fang, X.; Wang, L.; Zhang, E. Polyp Segmentation of Colonoscopy Images by Exploring the Uncertain Areas. *IEEE Access* **2022**, *10*, 52971–52981. [CrossRef]
30. Lai, H.; Luo, Y.; Zhang, G.; Shen, X.; Li, B.; Lu, J. Toward accurate polyp segmentation with cascade boundary-guided attention. *Vis. Comput.* **2022**. [CrossRef]
31. Wang, S.; Cong, Y.; Zhu, H.; Chen, X.; Qu, L.; Fan, H.; Zhang, Q.; Liu, M. Multi-Scale Context-Guided Deep Network for Automated Lesion Segmentation With Endoscopy Images of Gastrointestinal Tract. *IEEE J. Biomed. Health Inf.* **2021**, *25*, 514–525. [CrossRef]
32. Sun, X.; Zhang, P.; Wang, D.; Cao, Y.; Liu, B. Colorectal polyp segmentation by u-net with dilation convolution. In Proceedings of the 2019 18th IEEE International Conference On Machine Learning And Applications (ICMLA), Boca Raton, FL, USA, 16–19 December 2019; pp. 851–858.
33. Vaswani, A.; Shazeer, N.; Parmar, N.; Uszkoreit, J.; Jones, L.; Gomez, A.N.; Kaiser, Ł.; Polosukhin, I. Attention is all you need. In Proceedings of the 31st Conference on Neural Information Processing Systems (NIPS 2017), Long Beach, CA, USA, 4–9 December 2017.
34. Dosovitskiy, A.; Beyer, L.; Kolesnikov, A.; Weissenborn, D.; Zhai, X.; Unterthiner, T.; Dehghani, M.; Minderer, M.; Heigold, G.; Gelly, S. An image is worth 16x16 words: Transformers for image recognition at scale. *arXiv* **2020**. [CrossRef]
35. Tajbakhsh, N.; Gurudu, S.R.; Liang, J. Automated polyp detection in colonoscopy videos using shape and context information. *IEEE Trans. Med. Imaging* **2015**, *35*, 630–644. [CrossRef] [PubMed]
36. Lin, H.; Lu, H.; Zhang, L.J.U. 6th world congress on intelligent control and automation. 2006, 10, 9988–9991. *IEEE* **2006**, *10*, 9988–9991.
37. Lou, A.; Loew, M. Cfpnet: Channel-wise feature pyramid for real-time semantic segmentation. In Proceedings of the 2021 IEEE International Conference on Image Processing (ICIP), Anchorage, AK, USA, 19–22 September 2021; pp. 1894–1898.
38. Jha, D.; Smedsrud, P.H.; Riegler, M.A.; Halvorsen, P.; Lange, T.d.; Johansen, D.; Johansen, H.D. Kvasir-seg: A segmented polyp dataset. In Proceedings of the International Conference on Multimedia Modeling, Daejeon, Republic of Korea, 5–8 January 2020; Springer International Publishing: Berlin/Heidelberg, Germany, 2020; pp. 451–462.
39. Bernal, J.; Sánchez, F.J.; Fernández-Esparrach, G.; Gil, D.; Rodríguez, C.; Vilariño, F.; graphics. WM-DOVA maps for accurate polyp highlighting in colonoscopy: Validation vs. saliency maps from physicians. *Comput. Med. Imaging Graph.* **2015**, *43*, 99–111. [CrossRef] [PubMed]
40. Fang, Y.; Chen, C.; Yuan, Y.; Tong, K.-Y. Selective feature aggregation network with area-boundary constraints for polyp segmentation. In Proceedings of the International Conference on Medical Image Computing and Computer-Assisted Intervention, Shenzhen, China, 13–17 October 2019; Springer International Publishing: Berlin/Heidelberg, Germany, 2019; pp. 302–310.

diagnostics

MDPI

Article

A Deep Learning Radiomics Nomogram to Predict Response to Neoadjuvant Chemotherapy for Locally Advanced Cervical Cancer: A Two-Center Study

Yajiao Zhang [1], Chao Wu [2], Zhibo Xiao [2], Furong Lv [2] and Yanbing Liu [1,*]

1 College of Medical Informatics, Chongqing Medical University, Chongqing 400016, China; 18502310801@163.com
2 Department of Radiology, The First Affiliated Hospital of Chongqing Medical University, Chongqing 400016, China
* Correspondence: x1220295030@163.com

Abstract: Purpose: This study aimed to establish a deep learning radiomics nomogram (DLRN) based on multiparametric MR images for predicting the response to neoadjuvant chemotherapy (NACT) in patients with locally advanced cervical cancer (LACC). **Methods:** Patients with LACC (FIGO stage IB-IIIB) who underwent preoperative NACT were enrolled from center 1 (220 cases) and center 2 (independent external validation dataset, 65 cases). Handcrafted and deep learning-based radiomics features were extracted from T2WI, DWI and contrast-enhanced (CE)-T1WI, and radiomics signatures were built based on the optimal features. Two types of radiomics signatures and clinical features were integrated into the DLRN for prediction. The AUC, calibration curve and decision curve analysis (DCA) were employed to illustrate the performance of these models and their clinical utility. In addition, disease-free survival (DFS) was assessed by Kaplan–Meier survival curves based on the DLRN. **Results:** The DLRN showed favorable predictive values in differentiating responders from nonresponders to NACT with AUCs of 0.963, 0.940 and 0.910 in the three datasets, with good calibration (all $p > 0.05$). Furthermore, the DLRN performed better than the clinical model and handcrafted radiomics signature in all datasets (all $p < 0.05$) and slightly higher than the DL-based radiomics signature in the internal validation dataset ($p = 0.251$). DCA indicated that the DLRN has potential in clinical applications. Furthermore, the DLRN was strongly correlated with the DFS of LACC patients (HR = 0.223; $p = 0.004$). **Conclusion:** The DLRN performed well in preoperatively predicting the therapeutic response in LACC and could provide valuable information for individualized treatment.

Keywords: deep learning; radiomics nomogram; locally advanced cervical cancer; neoadjuvant chemotherapy

Citation: Zhang, Y.; Wu, C.; Xiao, Z.; Lv, F.; Liu, Y. A Deep Learning Radiomics Nomogram to Predict Response to Neoadjuvant Chemotherapy for Locally Advanced Cervical Cancer: A Two-Center Study. *Diagnostics* **2023**, *13*, 1073. https://doi.org/10.3390/diagnostics13061073

Academic Editors: Michał Strzelecki, Adam Piorkowski and Rafał Obuchowicz

Received: 15 February 2023
Revised: 9 March 2023
Accepted: 9 March 2023
Published: 11 March 2023

1. Introduction

Cervical cancer is the most frequently diagnosed cancer among women, with over 500,000 new diagnoses and 300,000 deaths per year [1,2]. In the past, for the treatment of locally advanced cervical cancer (LACC), radical hysterectomy (RH) and chemotherapy were suggested; however, the likelihood of treatment failure is high [3,4]. Since the 1990s, neoadjuvant chemotherapy (NACT), a short-course chemotherapy administered before surgery, has been used as a substitute treatment option for cervical cancer [5]. Although the five-year overall and disease-specific survival rates are similar between these treatment modalities, they demonstrate disparate toxicity profiles [6]. An increasing number of findings have suggested that NACT followed by radical hysterectomy or radiotherapy can improve survival. A meta-analysis from 21 randomized studies found that NACT substantially minimized the risk of death in LACC participants [7], while another analysis of 6 randomized controlled clinical trials demonstrated that NACT significantly improved both overall survival and progression-free survival [8]. The potential advantages of NACT have been reinforced by its ability to reduce tumor volume, eradicate subclinical distant metastasis, allow better delivery of radiation and improve prognosis [9,10]. Current research should focus on how to best identify patients who will benefit the most from this method. Ideally, radiotherapy should be avoided to the greatest extent in young women to preserve ovarian and sexual function. Only individuals with excellent reaction and negative lymph nodes should be subjected to radical surgery. This will decrease the requirement for postoperative chemoradiation and the increased morbidity associated with this approach when combined with radical surgery.

MRI has wide utility in pelvic evaluation due to its unique benefits in soft tissue imaging; it is effective in detecting primary tumor size, local diffusion and lymph node metastasis and is considered the gold standard for monitoring the response of cervical tumors to chemotherapy and determining eligibility for eventual excision [11]. In recent years, MRI-based radiomics has received increasing attention due to the advantages of extracting a large number of quantitative features from imaging phenotypes and offering meaningful biologic information [12,13]. Recent efforts have highlighted that MRI-based radiomics features may hold potential in the prediction of the response to NACT in LACC preoperatively [14,15], confirming that radiomics can distinguish patients who are suitable for NACT from those who are not. While the performance of these radiomics models for the prediction of NACT response is within acceptable ranges, there are still some limitations. Handcrafted radiomics features are restricted to the current knowledge of medical imaging [16] and operators' prior knowledge and hence may be insufficiently representative. Because generic features might lead to poor classification model performance, most tasks typically require distinguishing features depending on individual context to produce better outcomes. However, such distinguishing characteristics may not be applicable to other activities [17].

Deep learning (DL) performs well in medical image classification and prediction, extracting and learning deep features directly in a data-driven manner [18–20], and has shown excellent performance in precision cancer diagnosis, therapeutic decision making and survival analysis [21–23]. Since DL provides more abstract and difficult-to-mine radiomics features, several researchers have combined handcrafted and DL-based radiomics features to develop prediction models in cancers [24–26]. Although radiomics is capable of quantifying interpretable tumor image features, DL-based features primarily focus on effective semantic information, making them suitable for large-scale image analysis [27,28]. Moreover, preoperative clinical parameters offer additional useful data for predicting the NACT response [29]. Combining clinical characteristics with the two sets of radiomics features can predict NACT effectiveness from multiple perspectives, providing a more comprehensive assessment.

Nevertheless, investigations on the development of a DL radiomic model for predicting the therapeutic response to NACT in patients with LACC are still lacking. Notably, the pathological response was an independent prognostic variable for disease-free survival (DFS) and overall survival in patients with stage Ib2-IIb disease who underwent radical hysterectomy combined with NACT. Patients who are nonresponsive to NACT have a higher risk of recurrence than those who do respond [30]. However, the application of deep learning radiomics to predict the pathological response to NACT has been infrequently described in LACC. In addition, it is essential to determine radiomics and clinical predictors of the NACT response to stratify cervical cancer patients based on relevant therapy.

In this study, we first built and validated a new preoperative deep learning radiomics nomogram (DLRN) that combines handcrafted and DL-based radiomics features and clinical predictors for the prediction of the clinical response to NACT in cervical cancers.

2. Materials and Methods

2.1. Patients

A retrospective study was performed after receiving approval from the Ethics Committee of our institution, who waived the need for informed consent from the participants.

The study was conducted in accordance with the Declaration of Helsinki. All procedures performed during this study complied with the 1964 Helsinki declaration. The data of patients with cervical cancer with a confirmed histopathology from two centers between January 2015 and June 2021 were enrolled (Figure 1 and Supplementary S1). The inclusion criteria were as follows: (1) conventional pelvic MR scanning examinations performed before NACT, including T2-weighted imaging (T2WI), diffusion-weighted imaging (DWI) and contrast-enhanced (CE)-T1WI, in patients with pathologically confirmed cervical cancer; (2) FIGO stage (2018) from IB to IIIB; (3) use of the same NACT therapy; and (4) the presence of clinicopathological data. The exclusion criteria were as follows: (1) radiation therapy before NACT; (2) incomplete MR sequence or clinicopathological information; and (3) complications with other malignant tumors. Ultimately, a total of 220 patients from center 1 and 65 patients from center 2 were enrolled according to the selection criteria. The overall radiomics pipeline of this study is shown in Figure 2.

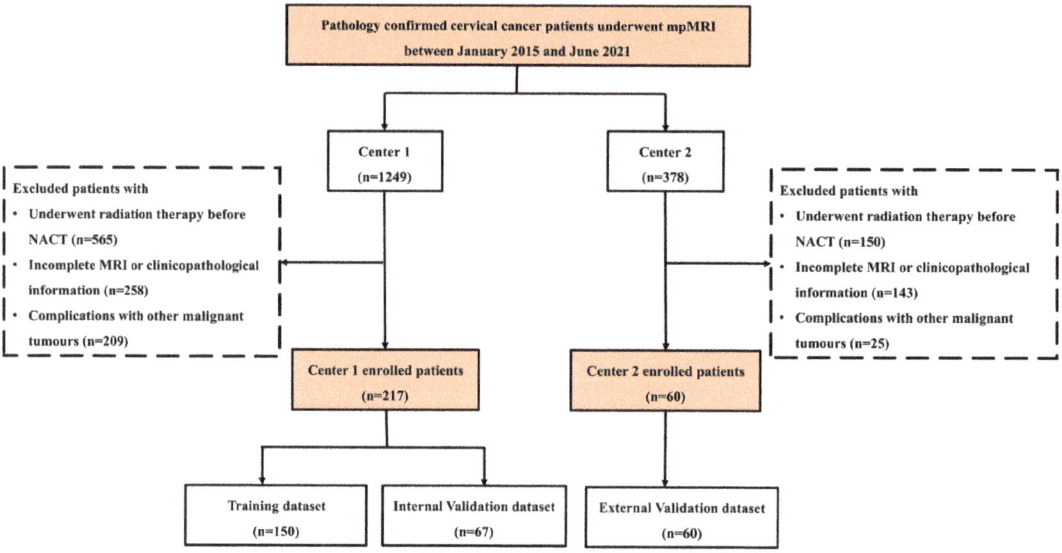

Figure 1. Flowchart of locally advanced cervical cancer (LACC) patients for inclusion in the developmental dataset and three external test datasets.

Regular follow-up was performed every 3–6 months for the first 2 years following treatment, every 6 months for the next 3–5 years and then once a year after that. The period from the date of surgery to the date of the first locoregional recurrence, distant metastasis, death, or the last visit in follow-up, which was defined as DFS, was the endpoint of our study. Gynecological examination and imaging modalities such as CT, MRI and positron emission tomography-computed tomography were applied to screen for locoregional recurrence and distant metastases.

2.2. Assessment of the NACT Response

All participants conducted one or two cycles of NACT and hysterectomies within 2 weeks. The short-term response was measured by the change in tumor size, which was calculated using the Response Evaluation Criteria In Solid Tumors (RECIST v. 1.1) [31]. Before surgery, the clinical response was assessed and categorized as follows: patients who had complete response (CR) or partial response (PR) were assigned to the pathological good responder (pGR) group, while those who had stable disease (SD) or progressive disease (PD) were assigned to the non-pGR group.

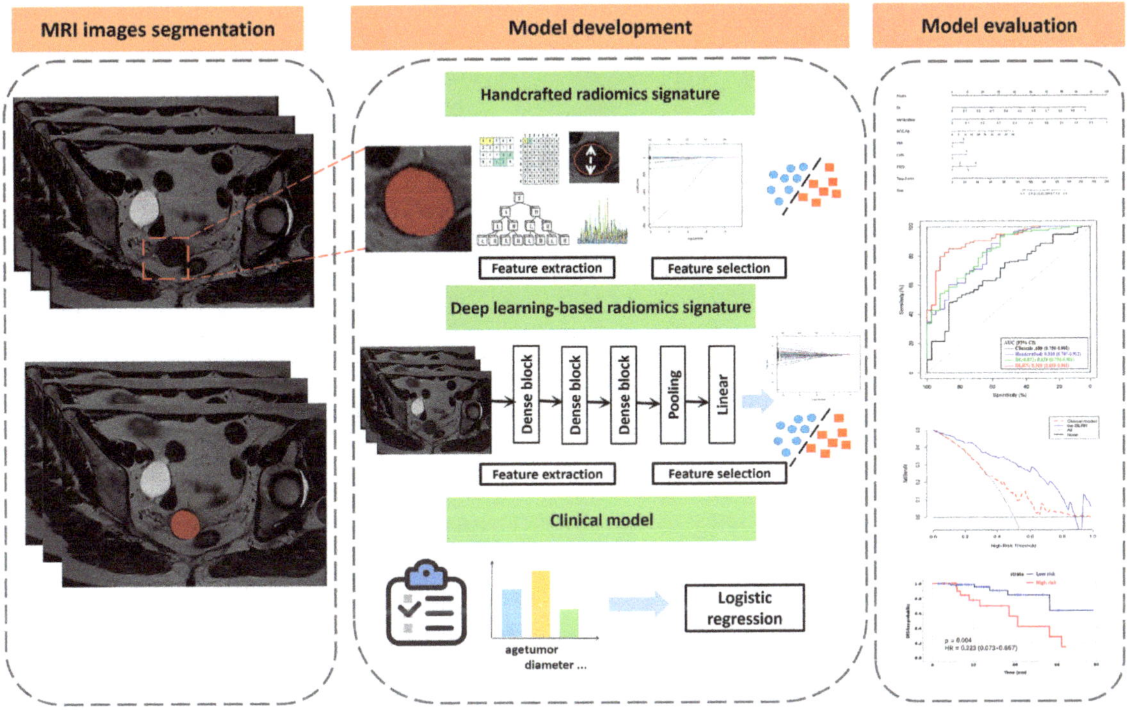

Figure 2. Workflow of this study, including step 1 MR segmentation; step 2 model development: (a) handcrafted radiomics signature, (b) deep learning-based radiomics signature, (c) clinical model; step 3 model evaluation. Receiver operating characteristic (ROC), area under the curve (AUC), calibration, and decision curve analysis (DCA) was used to confirm the performance and clinical utility of model.

2.3. MRI Acquisition

All patients received a systematic pelvic MR examination with a 3.0 T MR scanner at two distinct time periods. The pre- and post-treatment scanning periods were set to be within one week of the initiation of NACT and one week after the completion of NACT, respectively. The MRI parameters of the axial T2WI, DWI and CE-T1WI sequences are detailed in Table S1.

2.4. Image Segmentation and Processing

A radiologist with 9 years of experience in abdominal MRI diagnosis delineated regions of interest (ROIs) for use in radiomics analysis slice by slice along the tumor boundary manually by ITK-SNAP software (version 3.8.0, open-source software available for download at http://www.itksnap.org, accessed on 1 March 2022). All segmentations underwent quality control for proper annotation by an expert pelvic radiologist (with 15 years of experience in abdominal MRI diagnosis). After three months, 40 patients in the training set were selected randomly and resegmented by the radiologists to confirm the stability of radiomics characteristics.

For DL feature extraction, MR images were cropped by defining a rectangular ROI that encompassed the labeled tumor. Then, to satisfy the input size requirement of the pretrained convolutional neural network (CNN) model, the tumor patch was adjusted to a 224 × 224 square, and the intensity of the tumor patch images was standardized to the color range of 0–255 for training. To improve feature discrimination, many preprocessing procedures were applied before extracting quantitative features. Nonlinear intensity normalization and gray-level quantization were used to transform MR images to standardized intensity ranges and map the whole intensity range of the tumor location to distinct gray levels [32,33]. Finally, images were resampled to an isotropic pixel size using bilinear interpolation. Supplementary S2 contains detailed information about image processing.

2.5. Feature Extraction

Two types of features, handcrafted and DL-based radiomics features, were extracted. The open-source Python (v.3.10; https://www.python.org/, accessed on 1 May 2022) toolbox PyRadiomics was used for all handcrafted radiomics features. In total, 1223 handcrafted radiomics features were extracted from each MR sequence, including shape, first-order statistical, texture and transformation features. The details of the feature extraction are shown in Supplementary S3. In this study, DL-based features were extracted using DenseNet-121, which was pretrained on the natural picture dataset ImageNet. All convolutional layers were produced with a constant kernel size of 3×3 with stride and padding set to 1. The network consists of 5 max-pooling layers with a window size of 2×2 and a nonlinear activation function of rectified linear units, with the final fully connected layer having 1001 features with SoftMax enabled.

2.6. Feature Selection and Radiomics Signature Construction

Handcrafted radiomics and DL-based features were normalized individually using Z scores to achieve a standard normal distribution of feature intensity. The following steps were conducted in the training dataset for feature selection and signature establishment. First, the intraclass correlation coefficient (ICC) was determined to confirm the stability of radiomics characteristics, and features with an ICC < 0.85 were removed. Second, univariate analysis was applied to select handcrafted and DL-based radiomics features that were substantially different between the pGR and non-pGR groups using the Mann–Whitney U test. Third, considering that unbalanced classification (pGR group to non-pGR group: 113:35 = 3.23) may lead to inaccurate results, the synthetic minority oversampling technique (SMOTE) was used to balance the minority class in the training dataset so that the ratio of the two groups of patients was set to 1.10:1 (118 patients in the pGR group and 107 in the non-pGR group) in the SMOTE training dataset [34]. Further details of the SMOTE process are shown in Supplementary S4. The SMOTE algorithm was applied to create synthetic samples according to k-nearest neighbour of each minority class sample to balance the number of two-class sample [35,36]. Fourth, to perform feature selection, least absolute shrinkage and selection operator (LASSO) regression with 10-fold cross-validation was utilized. The handcrafted and DL-based radiomics features derived by feature selection were passed through a support vector machine (SVM) classifier to develop the handcrafted radiomics signature and the DL-based radiomics signature, respectively, for predicting the response to NACT.

2.7. Clinical Model and DLRN Development and Validation

To investigate the relationship between clinical variables and NACT response, differences in clinical characteristics were evaluated using the chi-square test or Fisher's exact test for categorical variables and the t test or Mann–Whitney U test for continuous data as applicable. The significant clinical characteristics were then subjected to stepwise backward multivariate logistic regression to determine the independent predictors of the NACT response. The criterion for variable inclusion was $p < 0.1$ for main effects and $p < 0.05$ for interactions. The clinical model comprising only the significant clinical information was constructed.

In this study, as a hybrid model, the DLRN was developed to investigate the potential combined utility of the clinical model, handcrafted radiomics signature, and DL-based radiomics signature by multivariate logistic regression. Given the considerable interpretability of logistic regression, the effect of different models on final prediction outcomes based on model weights may be explained. It is also appropriate for binary classification applications and effective training. The DLRN not only predicts whether the patient would respond to NACT but also gives an approximate probability prediction.

2.8. Statistical Analysis

The receiver operating characteristic (ROC) curve was drawn to assess the performances of the handcrafted and DL-based radiomics signatures, clinical model and the DLRN in training and validation cohorts. DeLong's test was used to compare the AUCs estimation between these prediction models. Performance measurements included sensitivity, specificity and accuracy. Calibration curves and the Hosmer–Lemeshow (H-L) test, which investigate how similar the models' predictions are to the observed outcome, were implemented to verify calibration performance. To assess the clinical value of models, a decision curve analysis (DCA) was performed by quantifying the net benefits to a range of threshold probabilities in both internal and external validation datasets. To investigate the association between the DLRN score and DFS, the Kaplan–Meier curve was employed, and patients were classified into high-risk or low-risk groups. The DFS prediction model was designed using both univariate and multivariate analyses, as well as Cox proportional hazards regression. The scaled Schoenfeld residual test was used to validate the models' proportional hazards assumption. Statistical analyses were conducted using R studio software (R Studio Inc., version 4.0.0, R Foundation for Statistical Computing, Vienna, Austria) and SPSS 24.0 software (SPSS Inc., Chicago, IL, USA). The R software packages are provided in the Supplementary S5. A two-sided $p < 0.05$ was considered statistically significant.

3. Results

3.1. Patient Characteristics

From January 2015 to June 2021, 220 patients from center 1 were included and split into training ($n = 148$; mean age, 52.3 ± 9.54) and validation datasets ($n = 72$; mean age, 51.2 ± 10.52). In addition, 65 patients (mean age, 53.3 ± 10.72) were enrolled from another center as an external validation dataset. The NACT response rate was consistent throughout the three groups of patients, with 76.35% for the training and 63.89% and 72.31% for the internal and external validation datasets, respectively. The baseline clinical characteristics are summarized in Table 1. No significant difference was observed in menopausal status, maximum tumor diameter, pathological type, neutrophil-to-lymphocyte ratio (NLR) or platelet-to-lymphocyte ratio (PLR) between the pGR and non-pGR groups in the datasets ($p > 0.05$). Moreover, age, FIGO stage, serum SCC-Ag (squamous cell carcinoma antigen) level, lymphovascular space invasion (LVSI) status, lymph node metastasis and parametrial invasion (PMI) showed significant differences between the two groups in the training dataset ($p < 0.05$).

3.2. Feature Selection and Radiomics Signature Development

In this study, two types of radiomics features were extracted, including handcrafted radiomics and DL-based radiomics features. For each patient, 3669 handcrafted radiomics features were derived. By univariate analysis, 989 handcrafted radiomics features were found to be significantly different between patients in the pGR and non-pGR groups in the training dataset. As shown in Figure 3a–c, the LASSO classifier selected 10 features (presented in Table 2), including four T2WI features, two DWI features and four CE-T1WI features.

A total of 3000 DL-based features were extracted using DenseNet-121 (Figure 4 and Supplementary Table S2), and 7 optimal DL-based features were acquired through feature selection, including three T2WI features, two DWI features, and two CE-T1WI features (Figure 3d,e and Table 2). Then, we developed a DL-based radiomics signature to differentiate responders from nonresponders in the training data.

Table 1. Clinicopathological characteristics of patients in the training and validation datasets.

Characteristics	Training Dataset (148)		p	Internal Validation Dataset (72)		p	External Validation Dataset (65)		p
	pGR	Non-pGR		pGR	Non-pGR		pGR	Non-pGR	
Number	113	35		46	26		47	18	
Age, mean ± SD, years	53.7 ± 10.64	51.9 ± 11.85	0.003	51.6 ± 10.21	50.7 ± 9.65	0.009	53.9 ± 11.26	51.3 ± 10.93	0.002
Menopausal status									
No	73 (64.60%)	19 (54.29%)		34 (73.91%)	18 (69.23%)		36 (76.60%)	10 (55.56%)	
Yes	40 (35.40%)	16 (45.71%)	0.272	12 (26.09%)	8 (30.77%)	0.670	11 (23.40%)	8 (44.44%)	0.095
FIGO stage									
IB2	23 (20.35%)	8 (22.86%)		8 (17.39%)	5 (19.23%)		12 (25.53%)	6 (33.33%)	
IIA-IIA2	61 (53.98%)	11 (31.43%)		21 (45.65%)	6 (23.08%)		21 (44.68%)	5 (27.78%)	
IIB-IIIB	29 (25.67%)	16 (45.71%)	0.041	17 (36.96%)	15 (57.69%)	0.143	14 (29.79%)	7 (38.89%)	0.461
Maximum tumor diameter	4.46 ± 1.52	5.02 ± 1.24	0.413	4.92 ± 1.39	4.84 ± 1.28	0.934	4.37 ± 1.72	5.29 ± 1.91	0.845
Pathologic type									
Squamous cell carcinoma	93 (82.30%)	24 (68.57%)		36 (78.26%)	23 (88.46%)		34 (72.34%)	15 (83.33%)	
Adenocarcinoma	20 (17.70%)	11 (31.43%)	0.081	9 (19.57%)	3 (11.54%)	0.557	13 (27.66%)	3 (16.67%)	0.549
SCC-Ag	5.20 (1.90, 9.90)	3.40 (0.80, 10.50)	<0.001	4.45 (1.25, 13.50)	3.65 (0.95, 6.65)	<0.001	5.00 (0.85, 13.65)	3.65 (1.15, 11.55)	<0.001
NLR	4.57 ± 2.94	3.23 ± 2.01	<0.001	4.92 ± 1.96	3.01 ± 2.08	<0.001	3.98 ± 2.29	2.94 ± 2.66	<0.001
PLR	214.83 ± 61.93	201.32 ± 72.93	0.119	193.83 ± 73.84	210.83 ± 73.93	0.081	209.14 ± 64.39	219.67 ± 62.92	0.063
LVSI									
Positive	31 (27.43%)	24 (68.57%)		16 (34.78%)	19 (73.08%)		13 (27.66%)	10 (55.56%)	
Negative	82 (72.57%)	11 (31.43%)	<0.001	30 (65.22%)	7 (26.92%)	0.002	34 (72.34%)	8 (44.44%)	0.035
LNM									
Positive	15 (13.27%)	11 (31.43%)		8 (17.39%)	6 (23.08%)		9 (19.15%)	5 (27.78%)	
Negative	98 (86.73%)	24 (68.57%)	0.014	38 (82.61%)	20 (76.92%)	0.558	38 (80.85%)	13 (72.22%)	0.449

Table 1. *Cont.*

Characteristics	Training Dataset (148)		*p*	Internal Validation Dataset (72)		*p*	External Validation Dataset (65)		*p*
	pGR	Non-pGR		pGR	Non-pGR		pGR	Non-pGR	
PMI									
Positive	29 (25.66%)	19 (54.29%)		14 (30.43%)	15 (57.69%)		19 (40.43%)	10 (55.56%)	
Negative	84 (74.34%)	16 (45.71%)	0.002	32 (69.57%)	11 (42.31%)	0.024	28 (59.57%)	8 (44.44%)	0.272

pGR = pathological good responder; FIGO = International Federation of Gynecology and Obstetrics; SCC-Ag = squamous cell carcinoma antigen; NLR = neutrophil-to-lymphocyte ratio; PLR = platelet-to-lymphocyte ratio; LVSI = lymphovascular space invasion; LNM = lymph node metastasis; PMI = parametrial invasion.

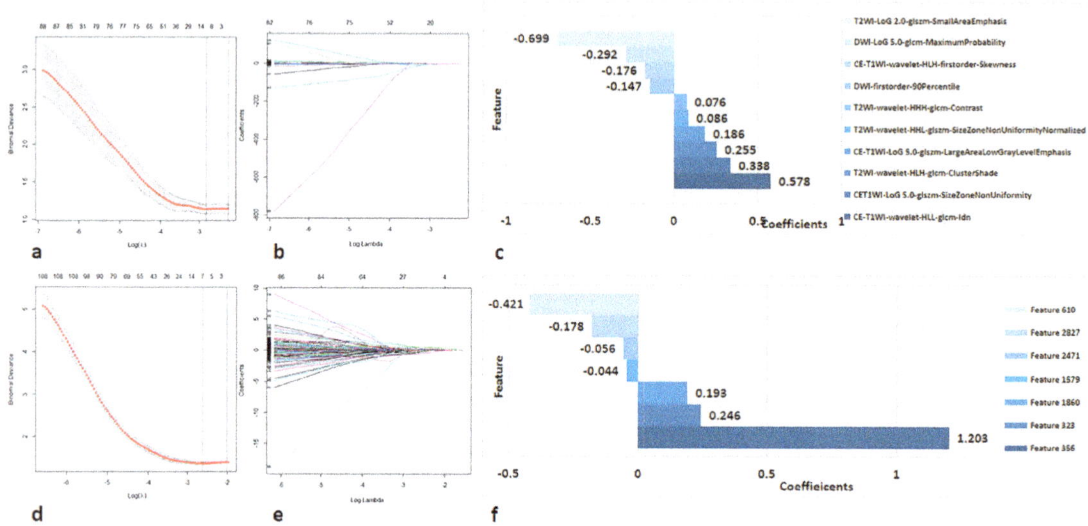

Figure 3. The lasso plots for radiomics feature selection: (**a**,**b**) for handcrafted radiomics features; (**c**) 10 features and corresponding coefficients; (**d**,**e**) for DL-based radiomics features; (**f**) 7 features and corresponding coefficients.

Table 2. The selected handcrafted and DL-based radiomics features and coefficients for radiomics signatures.

	Sequence	Selected Features	Coefficients
Handcrafted radiomics features	T2WI	LoG 2.0_glszm_SmallAreaEmphasis	0.578
	T2WI	Wavelet-HHL_glszm_SizeZoneNonUniformityNormalized	0.076
	T2WI	Wavelet-HHH_glcm_Contrast	0.086
	T2WI	Wavelet-HLH_glcm_ClusterShade	−0.176
	DWI	Firstorder_90Percentile	0.186
	DWI	LoG 5.0_glcm_MaximumProbability	0.338
	CE-T1WI	LoG 5.0_glszm_LargeAreaLowGrayLevelEmphasis	−0.147
	CE-T1WI	Wavelet-HLH_firstorder_Skewness	0.255
	CET1WI	LoG 5.0_glszm_SizeZoneNonUniformity	−0.292
	CE-T1WI	Wavelet-HLL_glcm_Idn	−0.699
DL-based radiomics features	T2WI	Feature 323	−0.178
	T2WI	Feature 356	−0.421
	T2WI	Feature 610	1.203
	DWI	Feature 1579	−0.044
	DWI	Feature 1860	−0.056
	CE-T1WI	Feature 2471	0.193
	CE-T1WI	Feature 2827	0.246

3.3. *Performance of Handcrafted Radiomics Signatures*

The handcrafted radiomics signature achieved an AUC of 0.884 (95% CI, 0.827–0.942) in the training set and 0.858 (95% CI, 0.763–0.953) and 0.810 (95% CI, 0.707–0.912) in the internal and external validation datasets, respectively. The resulting radiomics signature performed stably in all three datasets, and the inaccuracy was within acceptable limits. The internal validation dataset had an accuracy, sensitivity and specificity of 0.781, 0.846 and 0.619, respectively, whereas the corresponding values in the external validation dataset were 0.764, 0.824 and 0.619, respectively.

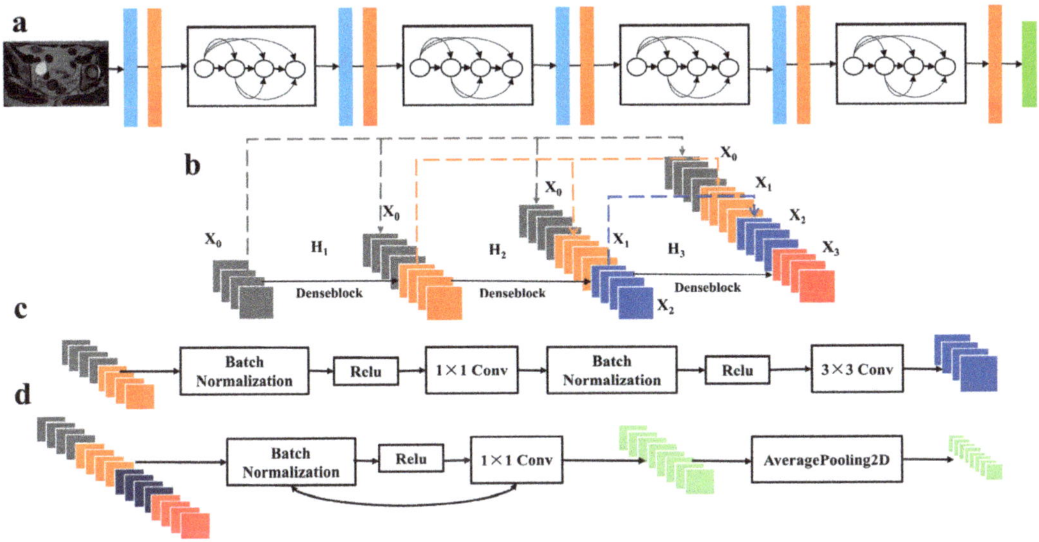

Figure 4. The framework of Densenet-121 model for deep learning-based radiomics feature extraction (**a**). The structures of Dense block (**b**), convolutional block (**c**), and transition layer (**d**) are shown separately.

3.4. Performance of the DL-Based Radiomics Signatures

The DL-based radiomics signature yielded AUCs of 0.871 (95% CI 0.756–0.901) in the training dataset, and 0.893 (95% CI 0.814–0.972) and 0.829 (95% CI 0.680–0.936) in these two validation datasets, respectively. The prediction performance of the DL-based signature was slightly higher than that of the handcrafted radiomics signature in the validation dataset.

3.5. Performance of the Clinical Model

FIGO stage, serum SCC-Ag level, LVSI status and PMI were identified as independent predictors based on univariate and multivariate logistic regression analyses ($p < 0.05$) in the training dataset and were used to build the clinical model (Table 3). The prediction efficiency was determined for the clinical model in the training dataset (AUC = 0.711; 95% CI: 0.620–0.801), internal validation dataset (AUC = 0.665; 95% CI: 0.538–0.792) and external validation dataset (AUC = 0.689; 95% CI 0.750–0.901).

3.6. Performance of the DLRN and Model Comparison

The handcrafted and DL-based radiomics signatures and the abovementioned independent predictors were incorporated to build the DLRN (Figure 5a and Table 4), which showed good performance for NACT response prediction, achieving AUCs of 0.963 (95% CI 0.932–0.995), 0.940 (95% CI 0.877–1.000) and 0.910 (95% CI: 0.859–0.961) in the training, internal validation and external validation datasets, respectively (Table 5 and Figure 6). The calibration curve showed good agreement between the DLRN predicted response and the actual response to NACT (Figure 5b,c), which was also demonstrated by the Hosmer–Lemeshow test. Compared with the clinical model and the handcrafted and DL-based radiomics signatures, the DLRN provided the best performance in predicting the response to NACT ($p < 0.05$) and was externally validated. The patients who had CR and SD to NACT are shown in Figure 7. Compared with the clinical model, the handcrafted radiomics signature and the DL-based signature, the prediction efficiency of DLRN is more accurate, by considering the three models above. This suggests that the DLRN built in this study was more generalizable and allows for more personalized clinical decision making.

Table 3. Results of the univariate and multivariate logistic regression analyses based on the training cohort.

Variables	Univariate Logistic Regression Analysis		Multivariate Logistic Regression Analysis	
	OR (95% CI)	*p*	OR (95% CI)	*p*
Age	1.051 (1.009, 1.086)	<0.05 *	-	-
Menopausal status	-	-	-	-
FIGO stage	1.139 (0.796, 1.650)	<0.05 *	0.263 (0.113, 0.611)	<0.05 *
Maximum tumor diameter	-	-	-	-
Pathologic type	-	-	-	-
CA-125	-	-	-	-
Serum SCC-Ag level	1.021 (0.985, 1.058)	<0.05 *	1.018 (0.978, 1.060)	<0.05 *
NLR	-	-	-	-
PLR	-	-	-	-
LVSI	0.211 (0.103, 0.432)	<0.05 *	0.263 (0.113, 0.611)	<0.05 *
LNM	0.973 (0.387, 2.443)	<0.05 *	-	-
PMI	0.357 (0.176, 0.728)	<0.05 *	0.617 (0.257, 1.485)	<0.05 *

NLR = neutrophil-to-lymphocyte ratio; PLR = platelet-to-lymphocyte ratio. * Indicates the $p < 0.05$.

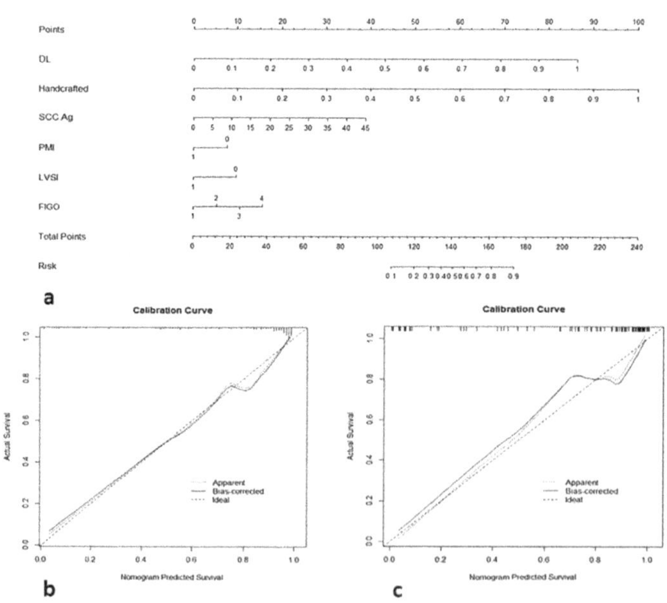

Figure 5. Deep learning radiomics nomogram (DLRN) developed for prediction of response to NACT. (**a**) DLRN combined with the handcrafted and DL-based radiomics signatures and clinical model. Calibration curve of the nomogram in the internal validation (**b**) dataset and external validation (**c**) dataset.

Table 4. Multivariate logistic regression of the DLRN.

Index	β	Odds Ratio (95% CI)	Multivariate *p* Value
Clinical model probability	1.297	3.660 (1.109–12.082)	0.033 *
Handcrafted radiomics signature probability	1.087	2.965 (1.796–4.893)	<0.001 *
DL-based radiomics signature probability	1.157	3.181 (1.845–5.486)	<0.001 *
Intercept	−0.547	-	0.078

Combined model probability = sigmoid (1.297 × Clinical model probability + 1.087 × Handcrafted radiomics signature probability + 1.157 × DL-based radiomics signature probability − 0.547). * Indicates the β is the regression coefficient; the $p < 0.05$.

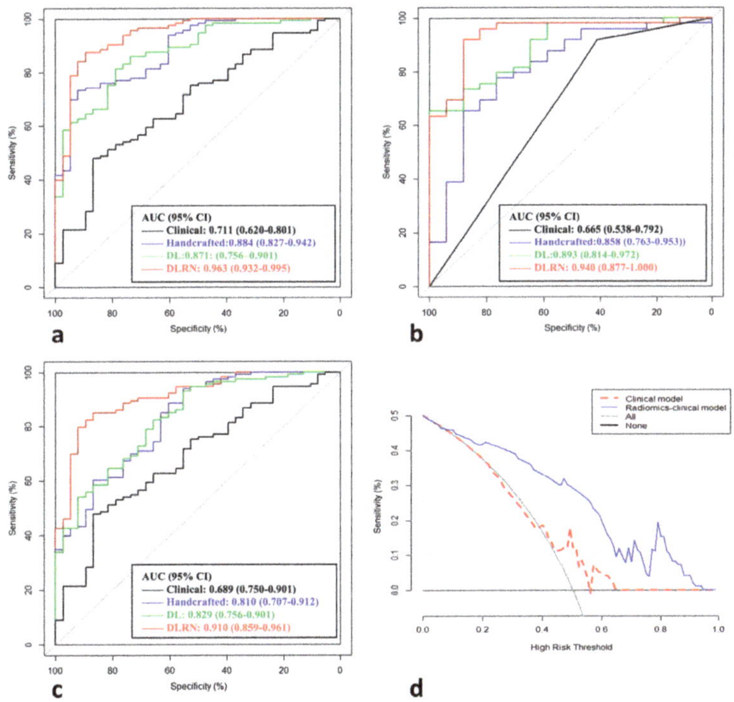

Figure 6. Performance of handcrafted and DL-based radiomics signatures, clinical model and the DLRN. Receiver operating characteristic (ROC) curves of the four models in the training (**a**) dataset, internal validation (**b**) dataset and external validation (**c**) dataset and (**d**) decision curve analysis.

Furthermore, we looked into the prognosis of DLRN in 101 patients with LACC. The median duration of the follow-up was 29 (range, 16–40) months. As shown in Figure 8, higher DLRN scores were demonstrated to be substantially related to improved DFS (hazard ratio (HR), 0.223; 95% CI, 0.073–0.667, log-rank test, $p = 0.004$).

Table 5. Predictive performances of clinical model, radiomics model based on mpMRI and radiomics clinical nomogram.

	Training Dataset				Internal Validation Dataset				External Validation Dataset			
	AUC	Acc [a]	Sen [a]	Spe [a]	AUC	Acc [a]	Sen [a]	Spec [a]	AUC	Acc [a]	Sen [a]	Spe [a]
Handcrafted signature	0.884 (0.827–0.942)	0.795	0.858	0.605	0.858 (0.763–0.953)	0.781	0.846	0.619	0.810 (0.707–0.91)	0.764	0.824	0.619
DL-based signature	0.871 (0.756–0.901)	0.821	0.859	0.711	0.893 (0.814–0.972)	0.833	0.898	0.710	0.829 (0.756–0.90)	0.788	0.867	0.625
Clinical model	0.711 (0.620–0.801)	0.730	0.877	0.389	0.665 (0.538–0.792)	0.788	0.918	0.412	0.689 (0.750–0.90)	0.728	0.885	0.363
DLRN	0.963 (0.932–0.995)	0.927	0.946	0.868	0.940 (0.877–1.000)	0.909	0.918	0.882	0.910 (0.859–0.96)	0.854	0.903	0.830
Handcrafted vs. DL signature	0.376				0.238				0.303			
Handcrafted vs. Clinical model	<0.001				0.014				0.022			
DLRN vs. Handcrafted	<0.001				0.042				0.047			
DLRN vs. Clinical model	<0.001				<0.001				<0.001			
DLRN vs. DL signature	0.003				0.251				0.016			

The DLRN indicates deep learning radiomics nomogram, which integrated handcrafted signature, DL-based signature and clinical model; [a] Sen, Spe, Acc and AUC indicate the sensitivity, specificity, accuracy and area under the curve of the receiver operating characteristic curve, respectively.

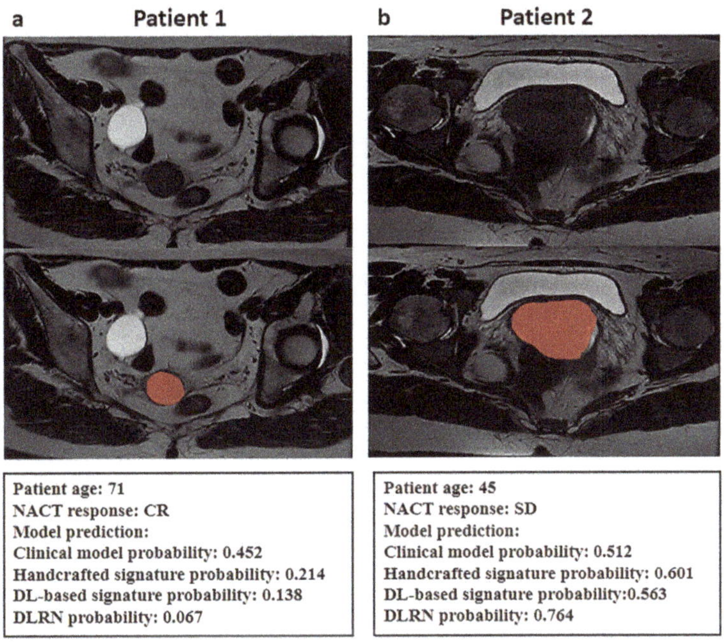

Figure 7. Two typical cases with MR images, clinical characteristics, and predicted results of response to NACT by different models. The red area represented the tumor issue. A 71-year-old woman (**a**) with complete response (CR). The probability of chemoresistance predicted by clinical model, handcrafted radiomics signature and DL-based signature was 0.452, 0.214 and 0.138, and that of chemoresistance predicted by the DLRN was 0.067. A 45-year-old woman (**b**) with stable disease (SD). The probability of chemoresistance predicted by clinical model, handcrafted radiomics signature and DL-based signature was 0.512, 0.601 and 0.5133, and that of chemoresistance predicted by the DLRN was 0.764. T2WI = T2-weighted imaging; DWI = diffusion-weighted imaging; CE-T1WI = contrast-enhanced-T1 weighted imaging.

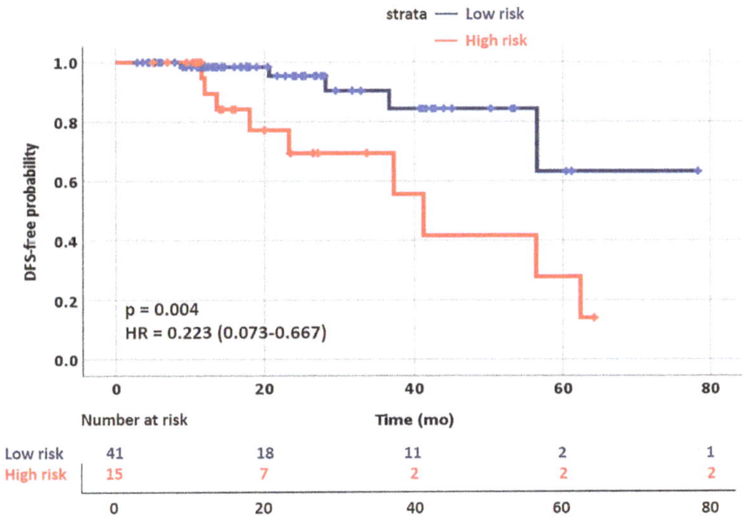

Figure 8. Kaplan–Meier curves of disease-free survival (DFS) on the follow-up LACC dataset.

4. Discussion

In this study, two sets of noninvasive radiomics signatures were developed by handcrafted and DL-based radiomics features based on multiparametric MRI to preoperatively predict the clinical response to NACT in patients with LACC. Furthermore, we incorporated these radiomics signatures with clinical independent predictors, namely, serum SCC-Ag level, PMI, LVSI status and FIGO stage, to build a DL radiomics model, which was presented in the form of a nomogram (DLRN). To the best of our knowledge, this is the first study to explore the predictive value of three ensemble models: clinical model, handcrafted radiomics signature and DL-based radiomics signature. The nomogram might be utilized for multidisciplinary consultation between radiologists and oncologists to predict the NACT response preoperatively and propose appropriate tailored therapy alternatives to these patients, rather than depending exclusively on clinical assessment. Furthermore, the DLRN score was substantially correlated with DFS in patients with LACC, yielding essential supplemental information for prognosis.

Predicting the response to NACT before treatment would be extremely beneficial in terms of preventing unnecessary chemotherapy before surgery, as well as for providing treatment options, individualization and effective patient stratification. An increasing number of methods have been shown to predict the response to NACT in patients with LACC. Tian et al. [15] found that CT-based machine learning analysis could be employed to identify whether NACT is effective for patients with LACC, and their hybrid model, which included radiomics features and clinical features, achieved discriminatory performance with AUCs of 0.803 and 0.821 in the training and validation datasets, respectively. However, when compared to CT, MR is a more effective technique for pelvic examinations in terms of tracking disease progression and the effectiveness of treatment [37,38]. Sun et al. [14] showed that a model based on MRI data could yield excellent accuracy in determining the response to NACT by radiomics analysis. Their model achieved a high AUC (0.999 [95% CI 0.997–1]) and specificity (100% [95% CI 99–100%]) in the validation cohort. The results were higher than the result of the handcrafted radiomics signature in our study. The multicenter statistics in their analysis improved the model's stability, which is worth exploring in future studies.

Previous studies on quantitative image-based prediction have shown that DL-based features, featuring in-depth information in the hidden layers of neural networks without the need for specified features, are complementary to radiomics features [39,40]. Our study demonstrates that this remains true when using DL-based features extracted by Resnet-121, with AUCs that were slightly higher than those of the handcrafted radiomics signature in terms of discrimination for all three datasets. These features have the potential to improve diagnostic value beyond the basic quantification of information currently available in MR images, since they are not being limited to previously established image properties [41,42]. DL-based features provide abundant information on tumor geographic heterogeneity and the tumor microenvironment, both of which are linked to tumor chemical sensitivity, and have utility for cervical cancer screening, detection and prognosis [43,44].

In our study, we extracted radiomics features from MRI, identified the optimal radiomics features, and created a radiomics signature. Likewise, we extracted deep semantic features that can well characterize the tumor tissue to build a DL signature. Finally, the DLRN was developed by incorporating both radiomics signatures with clinical predictors, and its effectiveness was considerably greater than that of the clinical model in all datasets. Earlier studies have found association between FIGO stage and serum SCC-Ag level and NACT effectiveness [29,45]. Our findings are in line with those of Ou et al. [29], who showed that PMI may have an effect on NACT efficacy in LACC. The number of invasive tumor cells is higher in cervical cancer with PMI, which is frequently exposed to strong genetic stress under hypoxic conditions, leading to genetic instability and drug-resistance induction via various mechanisms. Additionally, since cancer stem cells coexist with invasive tumor cells, invasion into surrounding tissue is a significant contribution to therapy response [46,47]. Similarly, the DLRN outperformed the handcrafted and DL-based radiomics signatures in prediction ability in most datasets. Solely DL classification algorithms are prone to overfitting, which may lead to instability; therefore, well-defined handcrafted features were introduced into our processing. The combination of handcrafted and DL-based radiomics signature has the potential to provide higher investment possibilities.

Intriguingly, the score of the DLRN created in this study was strongly related to the DFS of LACC patients. Several studies shown that the response to NACT is a predictor of prognosis and that patients who achieve an NACT response have a better prognosis. Individuals at low risk of an NACT response had a substantially lower chance of favorable long-term results than those at high risk, even after NACT followed by surgery. Patients with higher DLRN scores had a considerably improved DFS. Patients with low DLRN scores tended to have unsatisfactory outcomes; therefore, other treatment options should be provided in a timely manner to minimize excessive toxicity and improve patients' survival prospects. The high-precision online DLRN will not only aid in the evaluation of treatment options and administration of individualized treatment but also enable clinicians to examine patients remotely.

There are some limitations in our study. First, the number of patients enrolled in this research was still limited, and the patients were all Chinese. Therefore, while data used for modeling and validation were obtained from two centers, additional large-scale worldwide validation is required before the model may be used in patients of other ethnicities. Second, only LACC patients with NACT followed by radical hysterectomy were included; those with NACT followed by radiotherapy were excluded, and all those patients will be enrolled for further analysis for predicting the NACT response. Third, image segmentation was performed manually by radiologists in this study, which was time consuming and laborious. An automatic segmentation model of cervical cancer based on multiple MR sequences should be established. Finally, information included in pathological images and genomes may be valuable for prognostic prediction tasks, implying that larger well-designed prospective studies with numerous data points should be conducted in the future.

5. Conclusions

In summary, we developed and validated an MRI-based radiomics nomogram that integrated handcrafted and DL-based radiomics signatures and a clinical model for predicting the response to NACT for patients with LACC preoperatively, which has the potential to become an automatic tool for enabling individualized treatment. Additional validation is needed to generate stronger evidence. The proposed DLRN, incorporating MR-based radiomics signatures and clinical predictors, exhibited promising performance for predicting response and clinical outcomes, and provided valuable information for individualized treatment in LACC. Nevertheless, future prospective investigations with multicenter and large sample sizes are needed to prove the clinical applicability of this DLRN, and radiomics feature explanation at a biological level should be performed in future radiogenomics research.

Supplementary Materials: The following supporting information can be downloaded at: https://www.mdpi.com/article/10.3390/diagnostics13061073/s1.

Author Contributions: Conceptualization: Y.Z. and Y.L.; methodology: Y.Z.; software: C.W.; validation: Z.X., F.L. and Y.L.; formal analysis: F.L.; investigation: Y.Z. and C.W.; data curation: Y.Z. and C.W.; writing—original draft preparation: Y.Z.; writing—review and editing: C.W., Z.X., F.L. and Y.L.; supervision: Y.L.; project administration: Y.Z. and Y.L.; funding acquisition: Y.Z. All authors have read and agreed to the published version of the manuscript.

Funding: This work was funded by the project supported by the Cultivation plan for the 'top-notch' postgraduate of Chongqing Medical University in 2021(BJRC202111), and the program of Intelligence Medicine Special Research and Development of Chongqing Medical University in 2021(YJSZHYX202101).

Institutional Review Board Statement: The study was conducted in accordance with the Declaration of Helsinki, and approved by the Ethics Committee of The First Affiliated Hospital of Chongqing Medical Universtiy (protocol code: 2021-394 and date of approval: 31 Auguest 2021).

Informed Consent Statement: Patient consent was waived due to the retrospective nature of the study.

Data Availability Statement: The data presented in this study are available on request from the corresponding author. The data are not publicly available due to privacy restrictions.

Acknowledgments: The authors would like to thank all staff from the College of Informatics and the Department of Radiology and Gynecology for their hard work and invaluable support for this study. The authors also express gratitude to all the patients for their contribution.

Conflicts of Interest: The authors declare no conflict of interest.

References

1. Cohen, P.A.; Jhingran, A.; Oaknin, A.; Denny, L. Cervical cancer. *Lancet* **2019**, *393*, 169–182. [CrossRef]
2. Sung, H.; Ferlay, J.; Siegel, R.L.; Laversanne, M.; Soerjomataram, I.; Jemal, A.; Bray, F. Global Cancer Statistics 2020: GLOBOCAN Estimates of Incidence and Mortality Worldwide for 36 Cancers in 185 Countries. *CA Cancer J. Clin.* **2021**, *71*, 209–249. [CrossRef]
3. Bhatla, N.; Aoki, D.; Sharma, D.N.; Sankaranarayanan, R. Cancer of the cervix uteri. *Int. J. Gynaecol. Obstet.* **2018**, *143*, 22–36. [CrossRef] [PubMed]
4. Elit, L.; Fyles, A.W.; Devries, M.C.; Oliver, T.K.; Fung-Kee-Fung, M. Follow-up for women after treatment for cervical cancer: A systematic review. *Gynecol. Oncol.* **2009**, *114*, 528–535. [CrossRef] [PubMed]
5. Ye, Q.; Yuan, H.-X.; Chen, H.-L. Responsiveness of neoadjuvant chemotherapy before surgery predicts favorable prognosis for cervical cancer patients: A meta-analysis. *J. Cancer Res. Clin. Oncol.* **2013**, *139*, 1887–1898. [CrossRef] [PubMed]
6. Kim, H.S.; Sardi, J.E.; Katsumata, N.; Ryu, H.S.; Nam, J.H.; Chung, H.H.; Park, N.H.; Song, Y.S.; Behtash, N.; Kamura, T.; et al. Efficacy of neoadjuvant chemotherapy in patients with FIGO stage IB1 to IIA cervical cancer: An international collaborative meta-analysis. *Eur. J. Surg. Oncol. (EJSO)* **2013**, *39*, 115–124. [CrossRef]
7. Jayne, T. Neoadjuvant chemotherapy for locally advanced cervical cancer: A systematic review and meta-analysis of individual patient data from 21 randomised trials. *Eur. J. Cancer* **2003**, *39*, 2470–2486.
8. Rydzewska, L.; Tierney, J.; Vale, C.L.; Symonds, P.R. Neoadjuvant chemotherapy plus surgery versus surgery for cervical cancer. *Cochrane Database Syst. Rev.* **2012**, *12*, CD007406. [CrossRef] [PubMed]
9. Sardi, J.E.; Giaroli, A.; Sananes, C.; Ferreira, M.; Soderini, A.; Bermudez, A.; Snaidas, L.; Vighi, S.; Gomez Rueda, N.; Di Paola, G. Long-Term Follow-up of the First Randomized Trial Using Neoadjuvant Chemotherapy in Stage Ib Squamous Carcinoma of the Cervix: The Final Results. *Gynecol. Oncol.* **1997**, *67*, 61–69. [CrossRef]
10. Benedetti-Panici, P.; Greggi, S.; Scambia, G.; Amoroso, M.; Salerno, M.; Maneschi, F.; Cutillo, G.; Paratore, M.; Scorpiglione, N.; Mancuso, S. Long-term survival following neoadjuvant chemotherapy and radical surgery in locally advanced cervical cancer. *Eur. J. Cancer* **1998**, *34*, 341–346. [CrossRef]
11. Tanderup, K.; Fokdal, L.U.; Sturdza, A.; Haie-Meder, C.; Mazeron, R.; van Limbergen, E.; Jürgenliemk-Schulz, I.; Petric, P.; Hoskin, P.; Dörr, W.; et al. Effect of tumor dose, volume and overall treatment time on local control after radiochemotherapy including MRI guided brachytherapy of locally advanced cervical cancer. *Radiother. Oncol.* **2016**, *120*, 441–446. [CrossRef] [PubMed]
12. Lambin, P.; Leijenaar, R.T.H.; Deist, T.M.; Peerlings, J.; de Jong, E.E.C.; van Timmeren, J.; Sanduleanu, S.; Larue, R.T.H.M.; Even, A.J.G.; Jochems, A.; et al. Radiomics: The bridge between medical imaging and personalized medicine. *Nat. Rev. Clin. Oncol.* **2017**, *14*, 749–762. [CrossRef] [PubMed]
13. Gillies, R.J.; Kinahan, P.E.; Hricak, H. Radiomics: Images Are More than Pictures, They Are Data. *Radiology* **2016**, *278*, 563–577. [CrossRef]
14. Sun, C.; Tian, X.; Liu, Z.; Li, W.; Li, P.; Chen, J.; Zhang, W.; Fang, Z.; Du, P.; Duan, H.; et al. Radiomic analysis for pretreatment prediction of response to neoadjuvant chemotherapy in locally advanced cervical cancer: A multicentre study. *Ebiomedicine* **2019**, *46*, 160–169. [CrossRef] [PubMed]
15. Tian, X.; Sun, C.; Liu, Z.; Li, W.; Duan, H.; Wang, L.; Fan, H.; Li, M.; Li, P.; Wang, L.; et al. Prediction of Response to Preoperative Neoadjuvant Chemotherapy in Locally Advanced Cervical Cancer Using Multicenter CT-Based Radiomic Analysis. *Front. Oncol.* **2020**, *10*, 77. [CrossRef]
16. Lao, J.; Chen, Y.; Li, Z.-C.; Li, Q.; Zhang, J.; Liu, J.; Zhai, G. A Deep Learning-Based Radiomics Model for Prediction of Survival in Glioblastoma Multiforme. *Sci. Rep.* **2017**, *7*, 1–8. [CrossRef] [PubMed]
17. Wu, G.; Chen, Y.; Wang, Y.; Yu, J.; Lv, X.; Ju, X.; Shi, Z.; Chen, L.; Chen, Z. Sparse representation-based radiomics for the diagnosis of brain tumors. *IEEE.Trans. Med. Imaging* **2018**, *37*, 893–905. [CrossRef]
18. Simonyan, K.; Zisserman, A. Very deep convolutional networks for large-scale image recognition. Computer vision and pattern recognition (cs.CV). *arXiv* **2014**, arXiv:1409.1556v6.
19. Huang, G.; Liu, Z.; Van Der Maaten, L.; Weinberger, K.Q. Densely Connected Convolutional Networks. In Proceedings of the 2017 IEEE Conference on Computer Vision and Pattern Recognition (CVPR), Honolulu, HI, USA, 21–26 July 2017.

20. Tan, M.; Q, V.-L. EfficientNet: Rethinking model scaling for convolutional neural networks. International Conference on Machine Learning. *arXiv* **2019**, arXiv:1905.11946v5.
21. Dong, D.; Fang, M.-J.; Tang, L.; Shan, X.-H.; Gao, J.-B.; Giganti, F.; Wang, R.-P.; Chen, X.; Wang, X.-X.; Palumbo, D.; et al. Deep learning radiomic nomogram can predict the number of lymph node metastasis in locally advanced gastric cancer: An international multicenter study. *Ann. Oncol.* **2020**, *31*, 912–920. [CrossRef]
22. Zhang, L.; Dong, D.; Zhang, W.; Hao, X.; Fang, M.; Wang, S.; Li, W.; Liu, Z.; Wang, R.; Zhou, J.; et al. A deep learning risk prediction model for overall survival in patients with gastric cancer: A multicenter study. *Radiother. Oncol.* **2020**, *150*, 73–80. [CrossRef]
23. Coppola, F.; Faggioni, L.; Gabelloni, M.; De Vietro, F.; Mendola, V.; Cattabriga, A.; Cocozza, M.A.; Vara, G.; Piccinino, A.; Monaco, S.L.; et al. Human, all too human? an all-around appraisal of the "artificial intelligence revolution" in medical imaging. *Front. Psychol.* **2021**, *28*, 710982. [CrossRef] [PubMed]
24. Shi, J.; Cui, L.; Wang, H.; Dong, Y.; Yu, T.; Yang, H.; Wang, X.; Liu, G.; Jiang, W.; Luo, Y.; et al. MRI-based intratumoral and peritumoral radiomics on prediction of lymph-vascular space invasion in cervical cancer: A multi-center study. *Biomed. Signal Process. Control.* **2021**, *72*, 103373. [CrossRef]
25. Xuan, R.; Li, T.; Wang, Y.; Xu, J.; Jin, W. Prenatal prediction and typing of placental invasion using MRI deep and radiomic features. *Biomed. Eng. Online* **2021**, *20*, 1–18. [CrossRef] [PubMed]
26. Cui, Y.; Zhang, J.; Li, Z.; Wei, K.; Lei, Y.; Ren, J.; Wu, L.; Shi, Z.; Meng, X.; Yang, X.; et al. A CT-based deep learning radiomics nomogram for predicting the response to neoadjuvant chemotherapy in patients with locally advanced gastric cancer: A multicenter cohort study. *Eclinicalmedicine* **2022**, *46*. [CrossRef]
27. Hua, W.; Xiao, T.; Jiang, X.; Liu, Z.; Wang, M.; Zheng, H.; Wang, S. Lymph-vascular space invasion prediction in cervical cancer: Exploring radiomics and deep learning multilevel features of tumor and peritumor tissue on multiparametric MRI. *Biomed. Signal Process. Control.* **2020**, *58*, 101869. [CrossRef]
28. Jiang, X.; Li, J.; Kan, Y.; Yu, T.; Chang, S.; Sha, X.; Zheng, H.; Luo, Y.; Wang, S. MRI Based Radiomics Approach With Deep Learning for Prediction of Vessel Invasion in Early-Stage Cervical Cancer. *IEEE/ACM Trans. Comput. Biol. Bioinform.* **2020**, *18*, 995–1002. [CrossRef]
29. Ou, Z.; Zhao, D.; Li, B.; Wang, Y.; Liu, S.; Zhang, Y. A Preoperative Nomogram for Predicting Chemoresistance to Neoadjuvant Chemotherapy in Patients with Locally Advanced Cervical Squamous Carcinoma Treated with Radical Hysterectomy. *Cancer Res. Treat.* **2021**, *53*, 233–242. [CrossRef]
30. Gadducci, A.; Sartori, E.; Maggino, T.; Zola, P.; Cosio, S.; Zizioli, V.; Lapresa, M.; Piovano, E.; Landoni, F. Pathological response on surgical samples is an independent prognostic variable for patients with Stage Ib2-IIb cervical cancer treated with neoadjuvant chemotherapy and radical hysterectomy: An Italian multicenterretrospective study (CTFStudy). *Gynecol. Oncol.* **2013**, *131*, 640–644. [CrossRef]
31. Eisenhauer, E.A.; Therasse, P.; Bogaerts, J.; Schwartz, L.H.; Sargent, D.; Ford, R.; Dancey, J.; Arbuck, S.; Gwyther, S.; Mooney, M.; et al. New response evaluation criteria in solid tumours: Revised RECIST guideline (version 1.1). *Eur. J. Cancer.* **2009**, *45*, 228–247. [CrossRef]
32. Collewet, G.; Strzelecki, M.; Mariette, F. Influence of MRI acquisition protocols and image intensity normalization methods on texture classification. *Magn. Reson. Imaging.* **2004**, *22*, 81–91. [CrossRef] [PubMed]
33. Gibbs, P.; Turnbull, L.-W. Textural analysis of contrast-enhanced MR images of the breast. *Magn. Reson. Med.* **2003**, *50*, 92–98. [CrossRef]
34. Chawla, N.V.; Bowyer, K.W.; Hall, L.O.; Kegelmeyer, W.P. SMOTE: Synthetic Minority Over-sampling Technique. *J. Artif. Intell. Res.* **2002**, *16*, 321–357. [CrossRef]
35. Agazzi, G.M.; Ravanelli, M.; Roca, E.; Medicina, D.; Balzarini, P.; Pessina, C.; Vermi, W.; Berruti, A.; Maroldi, R.; Farina, D. CT texture analysis for prediction of EGFR mutational status and ALK rearrangement in patients with non-small cell lung cancer. *Radiol. Med.* **2021**, *126*, 786–794. [CrossRef]
36. Wu, H.; Wu, C.; Zheng, H.; Wang, L.; Guan, W.; Duan, S.; Wang, D. Radiogenomics of neuroblastoma in pediatric patients: CT-based radiomics signature in predicting MYCN amplification. *Eur. Radiol.* **2021**, *31*, 3080–3089. [CrossRef]
37. Balcacer, P.; Shergill, A.; Litkouhi, B. MRI of cervical cancer with a surgical perspective: Staging, prognostic implications and pitfalls. *Abdom. Imaging* **2019**, *44*, 2557–2571. [CrossRef]
38. Liu, L.; Wang, S.; Yu, T.; Bai, H.; Liu, J.; Wang, D.; Luo, Y. Value of diffusion-weighted imaging in preoperative evaluation and prediction of postoperative supplementary therapy for patients with cervical cancer. *Ann. Transl. Med.* **2022**, *10*, 120. [CrossRef] [PubMed]
39. Paul, R.; Hawkins, S.H.; Hall, L.O.; Goldgof, D.B.; Gillies, R.J. Combining deep neural network and traditional image features to improve survival prediction accuracy for lung cancer patients from diagnostic CT. In Proceedings of the IEEE International Conference on Systems, Man, and Cybernetics, Budapest, Hungary, 9–12 October 2016. [CrossRef]
40. Ahmed, K.B.; Hall, L.O.; Goldgof, D.B.; Liu, R.; Gatenby, R.A. Fine-tuning convolutional deep features for MRI based brain tumor classification. Proceedings of Medical Imaging 2017: Computer-Aided Diagnosis, Orlando, FL, USA, 11–16 February 2017. [CrossRef]
41. Han, W.; Qin, L.; Bay, C.; Chen, X.; Yu, K.-H.; Miskin, N.; Li, A.; Xu, X.; Young, G. Deep Transfer Learning and Radiomics Feature Prediction of Survival of Patients with High-Grade Gliomas. *Am. J. Neuroradiol.* **2020**, *41*, 40–48. [CrossRef] [PubMed]

42. Gadducci, A.; Cosio, S. Neoadjuvant chemotherapy in locally advanced cervical cancer: Review of the literature and perspectives of clinical research. *Anticancer Res.* **2020**, *40*, 4819–4828. [CrossRef] [PubMed]
43. Kanavati, F.; Hirose, N.; Ishii, T.; Fukuda, A.; Ichihara, S.; Tsuneki, M. A Deep Learning Model for Cervical Cancer Screening on Liquid-Based Cytology Specimens in Whole Slide Images. *Cancers* **2022**, *14*, 1159. [CrossRef]
44. Wu, Q.; Wang, S.; Zhang, S.; Wang, M.; Ding, Y.; Fang, J.; Wu, Q.; Qian, W.; Liu, Z.; Sun, S.; et al. Development of a Deep Learning Model to Identify Lymph Node Metastasis on Magnetic Resonance Imaging in Patients With Cervical Cancer. *JAMA Netw. Open* **2020**, *3*, e2011625. [CrossRef] [PubMed]
45. Scambia, G.; Panici, P.B.; Foti, E.; Amoroso, M.; Salerno, G.; Ferrandina, G.; Battaglia, F.; Greggi, S.; De Gaetano, A.; Puglia, G.; et al. Squamous cell carcinoma antigen: Prognostic significance and role in the monitoring of neoadjuvant chemotherapy response in cervical cancer. *J. Clin. Oncol.* **1994**, *12*, 2309–2316. [CrossRef] [PubMed]
46. Rein, D.T.; Kurbacher, C.M. The role of chemotherapy in invasive cancer of the cervix uteri: Current standards and future prospects. *Anti-Cancer Drugs* **2001**, *12*, 787–795. [CrossRef] [PubMed]
47. Jiapaer, S.; Furuta, T.; Tanaka, S.; Kitabayashi, T.; Nakada, M. Potential Strategies Overcoming the Temozolomide Resistance for Glioblastoma. *Neurol. Medico-Chirurgica* **2018**, *58*, 405–421. [CrossRef] [PubMed]

Journal of
Clinical Medicine

MDPI

Article

Convolutional Neural Networks to Classify Alzheimer's Disease Severity Based on SPECT Images: A Comparative Study

Wei-Chih Lien [1,2,*], Chung-Hsing Yeh [3], Chun-Yang Chang [4], Chien-Hsiang Chang [4], Wei-Ming Wang [5], Chien-Hsu Chen [4] and Yang-Cheng Lin [4,*]

1 Department of Physical Medicine and Rehabilitation, National Cheng Kung University Hospital, College of Medicine, National Cheng Kung University, Tainan 704, Taiwan
2 Department of Physical Medicine and Rehabilitation, College of Medicine, National Cheng Kung University, Tainan 701, Taiwan
3 Faculty of Information Technology, Monash University, Victoria 3800, Australia
4 Department of Industrial Design, National Cheng Kung University, Tainan 701, Taiwan
5 Department of Statistics, College of Management, National Cheng Kung University, Tainan 701, Taiwan
* Correspondence: lwclwhab@ms8.hinet.net (W.-C.L.); lyc0914@mail.ncku.edu.tw (Y.-C.L.)

Abstract: Image recognition and neuroimaging are increasingly being used to understand the progression of Alzheimer's disease (AD). However, image data from single-photon emission computed tomography (SPECT) are limited. Medical image analysis requires large, labeled training datasets. Therefore, studies have focused on overcoming this problem. In this study, the detection performance of five convolutional neural network (CNN) models (MobileNet V2 and NASNetMobile (lightweight models); VGG16, Inception V3, and ResNet (heavier weight models)) on medical images was compared to establish a classification model for epidemiological research. Brain scan image data were collected from 99 subjects, and 4711 images were used. Demographic data were compared using the chi-squared test and one-way analysis of variance with Bonferroni's post hoc test. Accuracy and loss functions were used to evaluate the performance of CNN models. The cognitive abilities screening instrument and mini mental state exam scores of subjects with a clinical dementia rating (CDR) of 2 were considerably lower than those of subjects with a CDR of 1 or 0.5. This study analyzed the classification performance of various CNN models for medical images and proved the effectiveness of transfer learning in identifying the mild cognitive impairment, mild AD, and moderate AD scoring based on SPECT images.

Keywords: convolutional neural network (CNN); Alzheimer's disease (AD); single-photon emission computed tomography (SPECT); transfer learning; image recognition

Citation: Lien, W.-C.; Yeh, C.-H.; Chang, C.-Y.; Chang, C.-H.; Wang, W.-M.; Chen, C.-H.; Lin, Y.-C. Convolutional Neural Networks to Classify Alzheimer's Disease Severity Based on SPECT Images: A Comparative Study. *J. Clin. Med.* **2023**, *12*, 2218. https://doi.org/10.3390/jcm12062218

Academic Editors: Rafał Obuchowicz, Michał Strzelecki and Adam Piorkowski

Received: 26 January 2023
Revised: 10 March 2023
Accepted: 10 March 2023
Published: 13 March 2023

1. Introduction

With a rapidly aging society, the number of people with Alzheimer's disease (AD) is increasing globally. Approximately 9.9 million new cases of dementia are reported annually, which implies that a new patient is diagnosed with the disease every 3.2 s. The number of people with AD is expected to exceed 100 million by 2050. However, an effective medical treatment for the disease is yet to be devised. Neuroimaging has been used for understanding the progression of AD, and considerable progress has been achieved in the use of deep learning for medical imaging in research and clinical medicine [1]. The macroscopic findings of AD have revealed diffuse brain atrophy [2]. The classification accuracy of AD versus healthy control using deep learning of magnetic resonance imaging (MRI) is 91.4%, and that of mild cognitive impairment (MCI) versus AD is 70.1% [3].

The economic evaluation of dynamic susceptibility MRI compared with nonenhanced computer-assisted tomography (CT) is USD 479,500 per quality-adjusted life year, whereas the corresponding comparison for single-photon emission computed tomography (SPECT) with CT is better (higher effectiveness and lower cost) [4,5]. Regional cerebral blood flow (rCBF) is related to brain metabolism; therefore, changes in CBF reflect variations in neuronal metabolism. From the perspective of neuropathology, subjects with very mild AD typically exhibit abnormal metabolic and rCBF patterns, even at the preclinical stage [6]. A decreased rCBF already occurs in individuals with MCI before they transition to AD [7]. A disrupted cerebral perfusion may cause impaired vascular clearance ability, which promotes the deposition of beta-amyloid and neurofibrillary tangles. Clinical studies have revealed that rCBF alterations are involved in AD pathogenesis. Even before the accumulation of beta-amyloid, subjects with high risk exhibit changes in cerebral blood flow [8]. In addition to clinical manifestations, rCBF SPECT, such as voxel-as-features, has frequently been used by physicians as a diagnostic tool [9,10].

Unlike general natural image recognition tasks, large, labeled training datasets are yet to be devised for medical image analysis [11]. These inadequate labeled data for supervised machine learning using electronic health records are the primary bottleneck in the model development [12]. Therefore, many models based on transfer learning-based methods have been proposed to address this concern [13]. In transfer learning, a previously trained model is applied to another field to improve the learning method for a few labels or small number of datasets in the target data. Pretrained convolutional neural networks (CNNs) (OverFeat) have been used in transfer learning to identify and detect vertical pathologies using X-ray and CT modalities [14]. In previous studies, the accuracy of pretrained CNNs using MRI to detect AD was 0.40 [15]. A fully convolutional network has been used with transfer learning for identifying malignant breast lesions [16] and retinal blood vessel segmentation [17]. Overfitting can easily occur when a small dataset is directly used to train deep learning networks. Transfer learning can improve the initial ability to extract features to alleviate this risk [13]. In this case, transfer learning between task domains is desirable. Furthermore, a fine-tuned CNN after transfer learning should always be the preferred option, regardless of the size of the available training sets. Additionally, the fine-tuned CNN model after transfer learning can quickly attain the maximum performance [18]. By contrast, CNNs trained from scratch require extensive training to achieve the highest performance. Therefore, transfer learning was used in this study to conduct experiments. Such a method can be executed even with limited training data.

In previous studies, rCBF SPECT for the diagnosis of AD revealed a sensitivity of 86%, specificity of 73%, and accuracy of 82% [19]. Brain perfusion SPECT has been proven to be a sensitive tool for assessing functional deficits in the early stages of AD [20]. MCI, referred to with a clinical dementia rating (CDR) of 0.5, with a pooled sensitivity and specificity of 93% and 97%, respectively [21], is an emerging tool for early detection and intervention. A study [9] used voxel-as-features with k-nearest neighbor classification to develop a set of diagnostic models for SPECT imaging. To highlight the benefits of the proposed approach in the early diagnosis of AD, 180 SPECT images with Tc-99m ethyl cysteinate dimer (ECD) as the tracer were used, including 43 normal participants and 30 participants with possible AD. The accuracy of classification of patients with possible AD and normal controls was 71.67%, which indicated acceptable accuracy of the conventional machine learning method. To the best of our knowledge, rCBF SPECT has not been used to classify MCI and mild and moderate AD. Specifically, few studies have comprehensively analyzed the various types of CNN models and discussed their applications in medical imaging [13,22,23].

This study compared various methods based on CNN models for the detection and medical diagnosis of MCI and AD. Specifically, the detection performances of the lightweight CNN models MobileNet V2 [24] and NASNetMobile [25] and the heavier weight CNN models VGG16 [26], Inception V3 [27], and ResNet [28] were compared. These five CNN models are the most widely used in transfer learning for disease diagnosis using medical imaging [29]. The findings of this study can assist physicians in the early stages of clinical diagnosis and reduce the occurrence of misdiagnoses.

2. Materials and Methods

This study was approved by the National Cheng Kung University Human Research Ethics Committee (NCKU HREC-E-108-282-2) and was conducted in accordance with the principles of the Declaration of Helsinki. All participants provided written informed consent. The experimental steps for training the CNN model were divided into two stages. This study compared the accuracies of various architectures of CNN models in predicting SPECT images. In terms of the analysis strategy, adjustments were made to three influencing factors, namely optimizer, fully connected layers, and model parameters, to optimize the CNN model performance. The two-stage experimental process is illustrated in Figure S1.

2.1. Participants

Inclusion criteria: (1) Patients who were over 60 years and sought evaluation for cognitive function decline in the Department of Neurology, National Cheng Kung University Hospital from January 2017 to December 2019. (2) Patients who completed comprehensive cognitive function tests, with results including mini mental state exam (MMSE) score, cognitive abilities screening instrument (CASI) score, CDR score, and sum of box (SOB) and brain SPECT imaging data with Tc-99m ECD as a tracer. Clinical history, physical and neurological examinations, laboratory, and instrumental investigations (including routine laboratory tests, thyroid hormone status, vitamin B12, folate levels, EEG, and CT or MRI brain scans) were performed to exclude secondary dementia. Experienced neurologists evaluated these image data for the possibility of AD and the detailed diagnosis, including the education level, the MMSE score, CASI score, CDR score, and SOB. Brain SPECT image data were collected from 99 subjects, 36 men and 63 women. Detailed demographic and diagnostic data are presented in Table 1.

Table 1. Subjects' demographic and clinical information.

	CDR (0.5)	CDR (1)	CDR (2)	*p* Value
N	52	39	8	
Age (years)	72.08 ± 7.96	76.15 ± 8.91	80.11 ± 3.22	0.01 * (CDR (0.5) < CDR (2))
Gender (female/male)	27/25	31/8	5/3	0.03 *
Education (years)	8.43 ± 4.40	7.03 ± 4.13	7.00 ± 6.08	n.s.
SOB	1.30 ± 0.72	4.73 ± 1.31	10.39 ± 0.93	<0.01 ** (CDR (0.5) < CDR (1); CDR (1) < CDR (2))
CASI	76.96 ± 11.80	63.26 ± 13.68	48.89 ± 14.91	<0.01 ** (CDR (0.5) > CDR (1); CDR (1) > CDR (2))
MMSE	22.92 ± 4.04	19.18 ± 3.91	14.78 ± 3.96	<0.01 ** (CDR (0.5) > CDR (1); CDR (1) > CDR (2))

Note: Values are numbers or mean ± standard deviation (range). Abbreviations: CDR = Clinical Dementia Rating Scale; MMSE = Mini Mental State Exam; N = Number; CASI = Cognitive Abilities Screening Instrument; SOB = Sum of Box. Data were compared using the chi-squared test and one-way analysis of variance with Bonferroni's post hoc test. * $p < 0.05$, ** $p < 0.01$, $p > 0.05$; no significant difference (n.s.).

2.2. SPECT Image Dataset

Generally, the collection of image data suitable for developing a brain imaging recognition system is one of the hardest steps in the procedure. In this study, SPECT imaging data with Tc-99m ECD as a tracer were used to test the effectiveness of the proposed method. To avoid deviations between different hospital sources, the images used were obtained from a single hospital; specifically, the archives of the Department of Neurology, National Cheng Kung University Hospital (Figures 1 and 2).

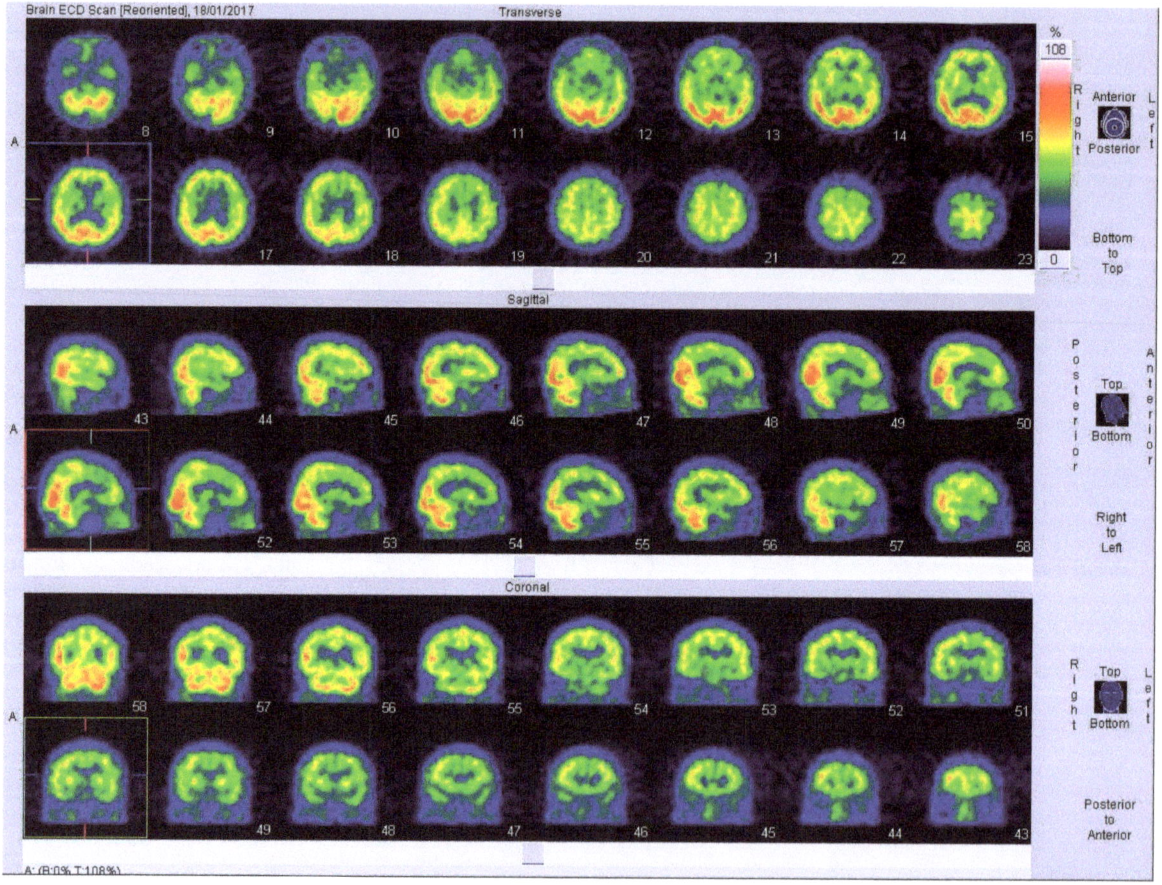

Figure 1. Single-photon emission computed tomography (SPECT) images of one participant. A total of 16 SPECT images were obtained in one section of one patient. Next, 48 SPECT images were obtained in three sections of this participant. A total of 4752 (48 × 99) SPECT images of all 99 participants were used. Of the 4752 images, 41 images beyond the area of brain were deleted. Therefore, 4711 images were used for further analysis. The colors in the figure represent the ranges of blood flow with red being the most blood flow and violet the least.

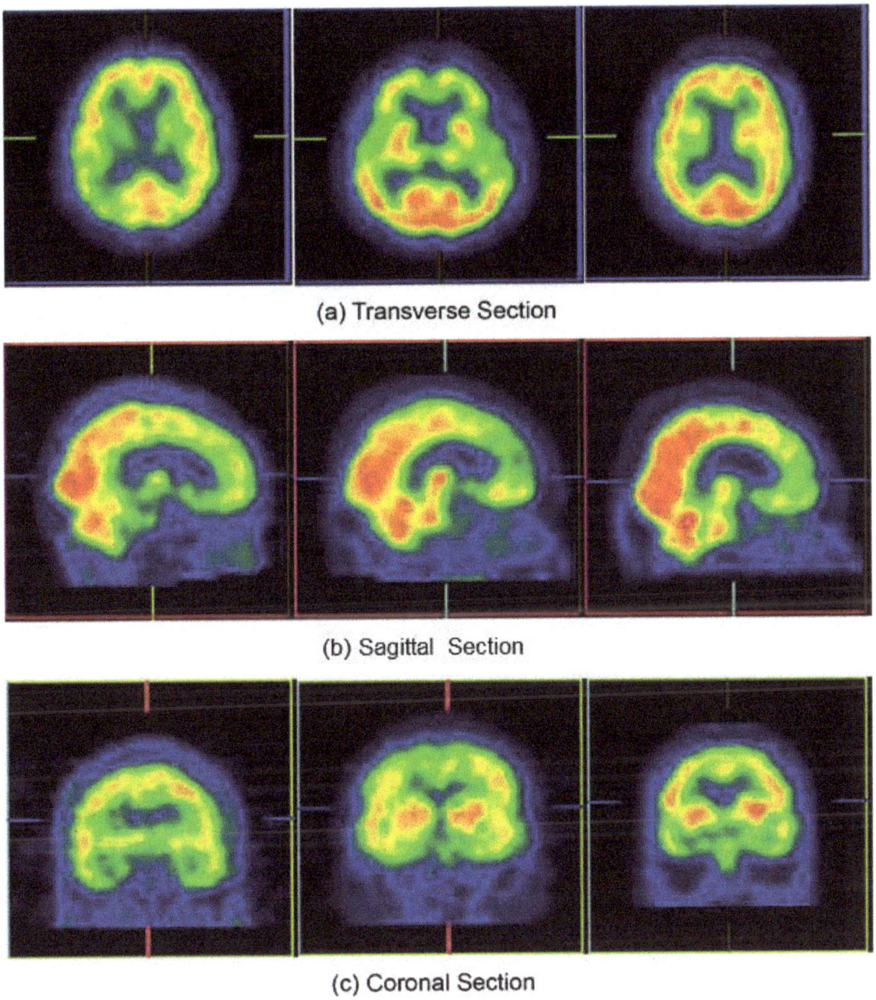

Figure 2. Three sections of SPECT images. (**a**) Transverse section; (**b**) Sagittal Section; (**c**) Coronal section. The colors in the figure represent the ranges of blood flow with red being the most blood flow and violet the least.

To reduce unnecessary information in the images and reduce model training errors, images of all 99 participants were cropped to remove excess background information [30]. After the preprocessing operation, 4711 images of individual slices (as displayed in Figure 3) were obtained. Each cropped SPECT image was a JPG file with a size of 124 × 120 pixels. Of the 99 patients, 52 patients were categorized as having questionable dementia (CDR = 0.5), 39 had MCI (CDR = 1), and 8 had moderate cognitive impairment (CDR = 2). The 4711 images were classified into three categories based on the CDR: 4461 were used as training data (80%) and verification data (20%) to perform all the experiments, and the remainder were manually used as test data (250 images) as suggested in previous studies [31].

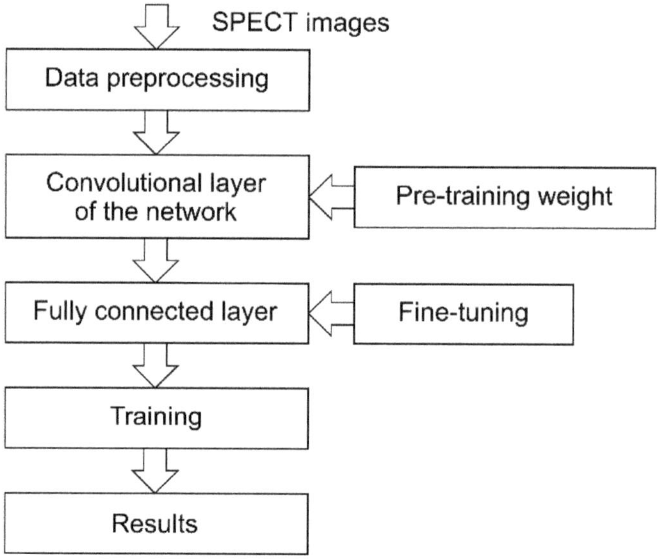

Figure 3. Process and architecture of CNN models.

2.3. Architecture of Five CNN Models after Transfer Learning

After obtaining the SPECT images (Figure 1), the voxel intensities were directly used as features. In the feature set, graphics were output in the transverse/sagittal/coronal sections of three intelligent identifications. In machine learning classification, a neural network was used to classify the three feature sets of CDR 0.5, 1, and 2. The neural network-like architecture is displayed in Figure 3, in the order of the graphics input layer (the size is 124 × 120 × 3), convolution layer, pooling layer, and fully connected layer. The fully connected layer groups and classifies the features extracted by the previous convolutional layer. Furthermore, to verify the accuracy of the prediction, each image is analyzed through an activation function and an output probability value is obtained (Figure 3 and Table 2).

Table 2. Layer settings after transfer learning and preprocessing for data enhancement.

(a) MobileNet V2	Layer	Type
Layer Setting	global_average_pooling2d_1 batch_normalization_1 dense_1 batch_normalization_2 dropout_1	GlobalAveragePooling2D BatchNormalization Dense BatchNormalization Dropout
Data Enhancement	height_shift_range = 0.2 width_shift_range = 0.2 shear_range = 0.2 zoom_range = 0.2 horizontal_flip = True fill_mode = nearest	
(b) NASNetMobile	Layer	Type
Layer Setting	global_average_pooling2d_1 dense_1 dropout_1	GlobalAveragePooling2D Dense Dropout
Data Enhancement	height_shift_range = 0.2 shear_range = 0.2 horizontal_flip = True fill_mode = nearest	

Table 2. *Cont.*

(c) VGG16	Layer	Type
Layer Setting	flatten dense_1 dropout_1 dense_2	Flatten Dense Dropout Dense
Data Enhancement	rotation_range = 40 height_shift_range = 0.2 shear_range = 0.2 zoom_range = 0.2 horizontal_flip = True fill_mode = nearest	
(d) Inception V3	Layer	Type
Layer Setting	global_average_pooling2d_1 dense_1	GlobalAveragePooling2D Dense
Data Enhancement	height_shift_range = 0.2 horizontal_flip = True fill_mode = nearest	
(e) ResNet	Layer	Type
Layer Setting	flatten batch_normalization_1 dense_1 dropout_1 dense_2	Flatten BatchNormalization Dense Dropout Dense
Data Enhancement	height_shift_range = 0.2 width_shift_range = 0.2 horizontal_flip = True fill_mode = nearest	

This study used transfer learning to load the weights of the pretrained model on the new network structure and then trained the network to recognize SPECT images. Five CNNs, namely MobileNet V2, NASNetMobile, VGG16, Inception V3, and ResNet, were used. The structure of the pretrained model was modified through fine-tuning and was then used as the initial model for the SPECT image recognition task. First, we preprocessed the SPECT image data and then used the original convolutional layer in the network structure to extract bottleneck features. Second, we connected the fine-tuned fully connected layer to form a new network structure. The experimental process is illustrated in Figure 3. The fine-tuning process of the fully connected layers in various CNN models is presented in the following subsections. The details of the two lightweight CNN models, namely, MobileNet V2 and NASNetMobile, are presented in the Supplementary Materials.

2.4. VGG16

2.4.1. Fine-Tuning

First, a flattened layer was added to dimensionalize the output of the previous convolutional layer into a two-dimensional matrix. This method reduced the image size without affecting important image features. The output dimension of DenseNet was set to 1024. Finally, the dropout layer was added, and the ratio was set to 0.5 to improve model generalization and avoid excessive reliance on certain regional features (see Table 2).

2.4.2. Experimental Setup

The dataset was normalized to improve data integrity and ensure the similarity of appearance and reading methods of all image data in the records. SoftMax was used as the resulting classifier. For VGG16, the experimental settings (ADAM optimization method, learning rate, exponential decay rate, and attenuation value) were the same as those of MobileNet V2. ReLU was used as the activation function. To increase the amount of training data, we performed rotation, shearing strength, horizontal flip, random scaling, filling, and other processes on the original image (see Table 2).

2.5. Inception V3

2.5.1. Fine-Tuning

To prevent excessive model parameters from causing overfitting problems, Global Average Pooling 2D was added at the end of the model to replace the fully connected layer, and the average value of each feature map was selected as the output. Subsequently, a dense layer was added, and the output dimension was set to 2048 (see Table 2 for details).

2.5.2. Experimental Setup

The input image size of Inception V3 is 299 × 299 pixels. Therefore, rescale = 1/255 was added in the preprocessing stage of the image data, and each pixel value of the SPECT image with an original size of 124 × 120 pixels was multiplied by the scaling factor to facilitate model convergence. To present the classification results as percentages, SoftMax was added as the resulting classifier, and the classification category was set to 3. Similar to the NASNetMobile model, SGD was selected as the optimizer of the model to increase the training speed. The learning rate was set to 10^{-5}, momentum was 0.9, loss function was categorical cross-entropy, batch size was 64, and epoch number was 50. Table 2 details the horizontal flip, random zoom, and original image for data enhancement.

2.6. ResNet

2.6.1. Fine-Tuning

The flattened layer was added to dimensionalize the output of the previous convolutional layer into a two-dimensional matrix, and batch normalization was added. A dense layer was included, and the output dimension was set to 256. Finally, the dropout layer was set to 0.5 to avoid too much reliance on certain regional features. Table 2 details layers added at the end of the ResNet after fine-tuning. Figure 4 illustrates the architecture of the ResNet model.

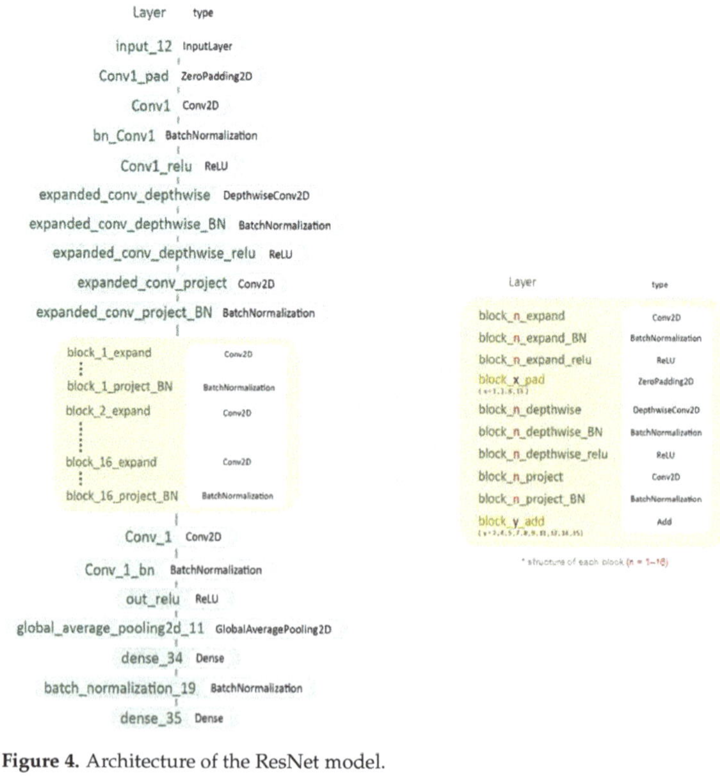

Figure 4. Architecture of the ResNet model.

2.6.2. Experimental Setup

As described for NASNetMobile, rescale = 1/255 was added in the preprocessing stage of the image data to facilitate model convergence. To present the classification results as percentages, SoftMax was selected as the result classifier and SGD was the optimization method. The learning rate was set to 10^{-4}, momentum was 0.9, loss function was categorical cross-entropy, batch size was set to 32, epoch number was set to 50, and activation function was ReLU. Before data analysis, the data enhancement method was used to perform position shift, horizontal flip, and fill processing on the original image (Table 2).

2.7. Statistical Analysis

All 4711 image data were randomly categorized into three separate data frames, namely training, validation, and test sets. First, 250 image data were manually extracted as test data. Subsequently, the random train–validation split was used to divide all remaining data into training (80%) and validation (20%) datasets. The training dataset was used to fit the model, and the validation dataset was used to validate the generalization ability of the model during the training process. The validation and training datasets remained unchanged to avoid the training–serving skew. The demographic data were expressed as the mean ± standard deviation and compared using the chi-squared test and one-way analysis of variance with Bonferroni's post hoc test. The performance of CNN models was evaluated using the learning curves of model accuracy and loss [32]. Learning curves are widely used in machine learning for models that optimize their internal parameters incrementally over training cycles (epochs). The metric used to evaluate learning could be maximizing, which revealed that better scores of classification accuracy indicate more learning. Using a score that is minimizing, such as loss, is preferable, where better scores (smaller loss) indicate more learning, and a value of 0 indicates that the training dataset was learned perfectly, and no errors were made. The confusion matrix was also calculated. A p value of <0.05 was considered statistically significant.

3. Results

Brain scan image data were collected from 36 men and 63 women. The demographic characteristics of the subjects are presented in Table 1.

Subjects with CDRs of 2 were significantly older than those with CDRs of 0.5. The SOB scores of subjects with CDR of 2 were significantly higher than those of subjects with CDRs of 1 and 0.5. The CASI and MMSE scores of subjects with CDR of 2 were significantly lower than those of subjects with CDRs of 1 and 0.5 (Table 1). The scores of the SOB, CASI, and MMSE validated the severity of cognitive impairments classified using the CDR score.

Following the evaluation of the severity of cognitive impairments, SPECT images were analyzed using two lightweight and three heavier weight CNN models to distinguish the severity of AD in patients based on the CDR scores. We used two evaluation indicators, namely accuracy and loss, to evaluate the detection performance of the model.

We first examined the effect of the various sections of brain images (i.e., the transverse, sagittal, and coronal sections) on the model to improve its performance in detecting the severity of AD by detecting different sections of SPECT images. A total of 1602 transverse images, 1584 sagittal images, and 1525 coronal images were obtained. Subsequently, the SPECT image data of the three cross-sections were mixed for model identification experiments.

Table 3 lists the performance indicators of the five CNN models that identify different CDR scores from a single brain section and a mixed section in the validation and testing data. Based on the classification results for a single brain section presented in Table 3, the ResNet model was the best performer among the five CNN models. The validation accuracy rates in the transverse, sagittal, and coronal section image data were 67.23%, 65.37%, and 68.51%, respectively. Inception V3 was the worst-performing CNN model with validation accuracy rates of 56.77%, 53.09%, and 52.78%. The classification validation accuracy of MobileNet V2, NASNetMobile, and VGG16 was approximately 60–69%.

Table 3. Detection performance of the two lightweight and three heavier weight CNN models.

Model	Types of Data	Validation Accuracy (%)	Test Accuracy (%)
MobileNet V2	Transverse section	60.57	60.01
	Sagittal section	58.69	50.45
	Coronal section	60.89	57.77
	Mixed (three kinds of section)	61.87	58.43
NASNetMobile	Transverse section	63.87	61.25
	Sagittal section	57.12	57.03
	Coronal section	63.03	55.49
	Mixed (three kinds of section)	59.89	58.77
VGG16	Transverse section	66.03	64.58
	Sagittal section	64.20	61.89
	Coronal section	58.66	53.01
	Mixed (three kinds of section)	69.45	67.53
Inception V3	Transverse section	56.77	54.22
	Sagittal section	53.09	50.98
	Coronal section	52.78	47.55
	Mixed (three kinds of section)	54.03	52.13
ResNet	Transverse section	67.23	61.20
	Sagittal section	65.37	63.37
	Coronal section	68.51	65.28
	Mixed (three kinds of section)	72.39	68.80

For the classification results of the mixed section, Table 3 reveals that the best model was ResNet, with a validation accuracy rate of 72.39% and a test accuracy rate of 68.8% (confusion matrix in Figure S2; the precision, recall, and F1 score for each class of CDR scores in Table S1), which is sufficient for determining the performance of various CNN models for medical images. The second highest accuracy rate was 69.45% (VGG16). Inception V3 was the worst-performing model, with a validation accuracy rate of only 56.77%. Based on the experimental results, the lightweight CNN models (MobileNet V2 and NASNetMobile) were unsatisfactory. For the image data obtained by mixing the three sections, the best validation accuracies of MobileNet V2 and NASNetMobile were only 61.87% and 59.89%, respectively, indicating that the two lightweight CNN models can be improved to match SPECT medical images. The epoch of each CNN model after transfer learning was set to 50, and the learning curves of model accuracy and loss of mixed data using each CNN model are displayed in Figures S3–S6. The loss of the ResNet model after transfer learning was the smallest, which indicated that it learned the most (Figure S7). The values of validation accuracy were higher than those of training accuracy in the NASNetMobile, VGG16, Inception V3, and ResNet models after approximately 40 epochs. This phenomenon indicated that validation improved the classification accuracy in these four models after a longer training duration (Figures S4–S6).

4. Discussion

Neuroimaging has become a useful tool for understanding AD pathogenesis, and the use of deep learning techniques in medical imaging has achieved considerable progress in research and clinical care.

In this study, five CNN models were trained in the transfer learning process, and performance evaluations and comparisons were performed. The primary objective of this study was to compare the detection performance of various CNN structures for medical images, which was confirmed based on the results. Moreover, compared with using a single cross-sectional image as the input, the use of mixed data from the three cross-sections as the model input produced excellent results.

The validation and test accuracy records in Table 3 indicate that the performance of ResNet was superior to those of MobileNet V2, NASNetMobile, VGG16, and Inception V3. In this type of medical imaging dataset, the heavier weight networks performed better than the lightweight networks. Furthermore, ResNet can effectively avoid gradient explosion, disappearance, and network degradation. Therefore, ResNet can be used to develop an AI expert system that can analyze AD severity using SPECT images, as displayed in Figure 5. The AI expert system does not replace physicians but helps them achieve clearer decisions on disease classification and more confident diagnosis based on systematic objective information to reduce the uncertainty in disease classification diagnosis.

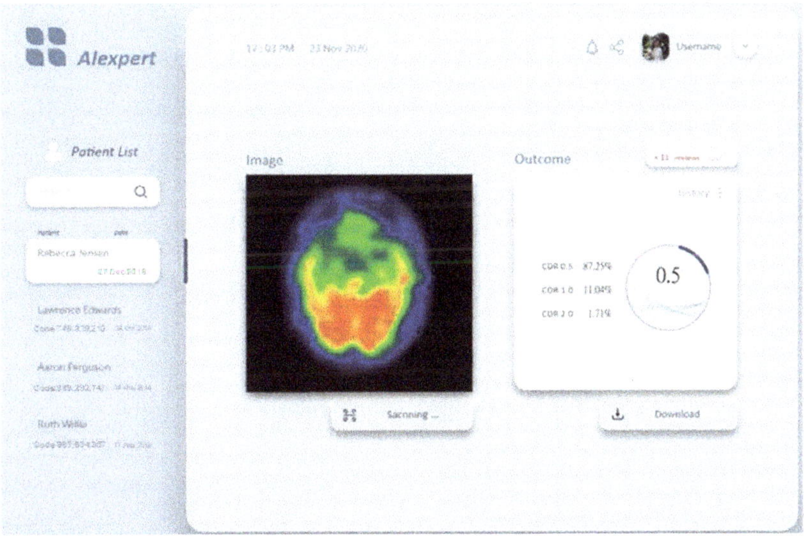

Figure 5. AI expert system based on the ResNet model. Physicians can import the SPECT image (left part), and the AI expert system will show the outcome (here, of 0.5) (right part) to help them make decisions on disease classification.

The performance of CNN models, such as ResNet, varies depending on the type of study, field of study, data used, and imbalance in the sample [33]. In neuroimaging research, image processing and feature recognition have been applied to AD classification. Among the various deep learning methods, ResNet has been widely used for the classification and diagnosis of AD. Amin-Naji et al. [34] used a residual structure in each branch of a CNN. The OASIS dataset was used to evaluate the effectiveness of the model. Finally, an accuracy of 98.72% was obtained in the classification of old patients with AD and normal controls using MRI, which is the best result compared with those in other previous studies on the same database.

Abrol et al. [35] evaluated the ability of a residual structure to learn from structural MRI data using neuroimaging and revealed that it is binary classification or heap classification. Their results showed that the recognition performance of the ResNet architecture is better than that of the SVM and SAE methods. Karasawa and Ohwada [36] proposed a ResNet-based structure for classifying MRI data from the Alzheimer's Disease Neuroimaging Initiative database. The experimental results indicated that the proposed 39-layer residual structure had the highest accuracy compared with those of VGGNet and ResNet-50. Therefore, related research has revealed that the residual structure exhibited a certain effect on the applicability of image recognition in neuroimaging. This finding revealed that the ResNet structure used in this study is effective for AD classification. In previous studies, the performance of the total and subscale scores of the Mattis Dementia Rating Scale for discriminating MCI from NC, MCI from mild AD, and mild AD from moderate AD revealed an accuracy from 61% to 85% [37], which was comparable with the performance of our rCBF SPECT CNN model.

5. Conclusions

Although this study revealed that deep learning technology can achieve excellent performance in SPECT image classification, performance can be improved. The proposed method provides important implications for image recognition and deep learning in the development of mobile applications in AI and medical treatment and exhibits considerable potential in other biomedical fields. The proposed method may open novel avenues for medical image analysis and provide a potentially accurate CNN architecture for researchers and physicians to predict new data. In the future, we hope to collect more clinical data from various hospitals to increase the depth of the training dataset. Moreover, neuroimaging information can be combined with cognitive scale functional information to obtain a superior machine learning model for the improved classification of AD severity.

Supplementary Materials: The following supporting information can be downloaded at: https://www.mdpi.com/article/10.3390/jcm12062218/s1, Table S1: Precision, recall, and F1 score of the ResNet model using the mixed test data of the SPECT brain images for each class of CDR scores; Figure S1: Two-stage experimental process utilized in this study; Figure S2: Confusion matrix of the ResNet model using the mixed test data of the SPECT brain images. 0.5: Clinical Dementia Rating Scale (CDR) = 0.5; 1: CDR = 1; 2: CDR = 2.; Figure S3: Learning curves of model accuracy and loss of mixed data using the MobileNetV2 model. The epoch of the MobileNetV2 model after transfer learning was set to 50, and the SPECT brain images were classified; Figure S4: Learning curves of model accuracy and loss of mixed data using the NasNetMobile model. The epoch of the NasNetMobile model after transfer learning was set to 50 and SPECT brain images were classified; Figure S5: Learning curves of model accuracy and loss of mixed data using the VGG 16 model. The epoch of the VGG 16 model after transfer learning was set to 50 and the SPECT brain images were classified; Figure S6: Learning curves of the model accuracy and loss of mixed data using the Inception V3 model. The epoch of the Inception V3 model after transfer learning was set to 50 and the SPECT brain images were classified; Figure S7: Learning curves of the model accuracy and loss of mixed data using the ResNet model. The epoch of the ResNet model after transfer learning was set to 50 and the SPECT brain images were classified.

Author Contributions: Conceptualization, W.-C.L., C.-H.C. (Chien-Hsu Chen), and Y.-C.L.; Writing—Original Draft, W.-C.L., C.-Y.C., and C.-H.C. (Chien-Hsiang Chang); Methodology, W.-C.L., C.-H.Y., C.-Y.C., C.-H.C. (Chien-Hsiang Chang), C.-H.C. (Chien-Hsu Chen), and Y.-C.L.; Formal Analysis, W.-C.L., C.-Y.C., C.-H.C. (Chien-Hsiang Chang), and W.-M.W.; Validation, W.-C.L., C.-Y.C., and Y.-C.L.; Investigation, W.-C.L., C.-H.Y., C.-Y.C., C.-H.C. (Chien-Hsiang Chang), and W.-M.W.; Funding Acquisition, W.-C.L. and Y.-C.L.; Writing—Review and Editing, C.-H.Y., W.-M.W., C.-H.C. (Chien-Hsu Chen), and Y.-C.L.; Resources, C.-H.Y. and Y.-C.L.; Data Curation, C.-Y.C.; Software, C.-H.C. (Chien-Hsu Chen) and Y.-C.L.; Project Administration, Y.-C.L. All authors have read and agreed to the published version of the manuscript.

Funding: This study was supported by the Ministry of Science and Technology, Taiwan (Grant MOST108-2628-E-006-MY3 for the 3-year project, MOST111-2923-E-006-002-MY3 for the 3-year project, MOST110-2314-B-006-096 for the 1-year project, and MOST111-2314-B-006-064-MY3 for the 3-year project). The funders played no role in the study design, data collection and analysis, interpretation of data, or writing of the manuscript.

Institutional Review Board Statement: This study was approved by the National Cheng Kung University Human Research Ethics Committee (NCKU HREC-E-108-282-2) and conformed to the principles of the Declaration of Helsinki. All participants provided written informed consent prior to participating in the study.

Informed Consent Statement: Informed consent was obtained from all subjects involved in the study.

Data Availability Statement: The data presented in this study are available on request from the corresponding author. The data are not publicly available due to privacy concerns of research participants.

Acknowledgments: The authors are grateful to Yu-Yi Pan, Department of Statistics, College of Management, National Cheng Kung University, for providing statistical consulting services.

Conflicts of Interest: The authors declare that the research was conducted in the absence of any commercial or financial relationships that could be construed as a potential conflict of interest.

References

1. Chetelat, G.A.; Baron, J.-C. Early diagnosis of Alzheimer's disease: Contribution of structural neuroimaging. *Neuroimage* **2003**, *18*, 525–541. [CrossRef]
2. Hashizume, Y. Macroscopic findings of brain with dementia. *Neuropathology* **2022**, *42*, 353–366. [CrossRef] [PubMed]
3. Li, F.; Tran, L.; Thung, K.H.; Ji, S.; Shen, D.; Li, J. A robust deep model for improved classification of AD/MCI patients. *IEEE J. Biomed. Health Inform.* **2015**, *19*, 1610–1616. [CrossRef]
4. Leung, G.M.; Yeung, R.Y.; Chi, I.; Chu, L.W. The economics of Alzheimer disease. *Dement. Geriatr. Cogn. Disord.* **2003**, *15*, 34–43. [CrossRef]
5. Lien, W.-C.; Wang, W.-M.; Wang, F.; Wang, J.-D. Savings of loss-of-life expectancy and lifetime medical costs from prevention of spinal cord injuries: Analysis of nationwide data followed for 17 years. *Inj. Prev.* **2021**, *27*, 567–573. [CrossRef] [PubMed]
6. Binnewijzend, M.A.A.; Benedictus, M.R.; Kuijer, J.P.A.; van der Flier, W.M.; Teunissen, C.E.; Prins, N.D.; Wattjes, M.P.; van Berckel, B.N.M.; Scheltens, P.; Barkhof, F. Cerebral perfusion in the predementia stages of Alzheimer's disease. *Eur. Radiol.* **2016**, *26*, 506–514. [CrossRef] [PubMed]
7. Binnewijzend, M.A.; Kuijer, J.P.A.; Benedictus, M.R.; van der Flier, W.M.; Wink, A.M.; Wattjes, M.P.; van Berckel, B.N.M.; Scheltens, P.; Barkhof, F. Cerebral blood flow measured with 3D pseudocontinuous arterial spin-labeling MR imaging in Alzheimer disease and mild cognitive impairment: A marker for disease severity. *Radiology* **2013**, *267*, 221–230. [CrossRef] [PubMed]
8. Chao, L.L.; Buckley, S.T.; Kornak, J.; Schuff, N.; Madison, C.; Yaffe, K.; Miller, B.L.; Kramer, J.H.; Weiner, M.W. ASL perfusion MRI predicts cognitive decline and conversion from MCI to dementia. *Alzheimer Dis. Assoc. Disord.* **2010**, *24*, 19–27. [CrossRef]
9. Górriz, J.M.; Segovia, F.; Ramírez, J.; Lassl, A.; Salas-Gonzalez, D. GMM based SPECT image classification for the diagnosis of Alzheimer's disease. *Appl. Soft Comput.* **2011**, *11*, 2313–2325. [CrossRef]
10. Lien, W.-C.; Wang, W.-M.; Wang, H.-M.D.; Lin, F.-H.; Yao, F.-Z. Environmental barriers and functional outcomes in patients with schizophrenia in Taiwan: The capacity-performance discrepancy. *Int. J. Environ. Res. Public Health.* **2021**, *19*, 315. [CrossRef]
11. Litjens, G.; Kooi, T.; Bejnordi, B.E.; Setio, A.A.A.; Ciompi, F.; Ghafoorian, M.; van der Laak, J.A.W.M.; van Ginneken, B.; Sánchez, C.I. A survey on deep learning in medical image analysis. *Med. Image Anal.* **2017**, *42*, 60–88. [CrossRef] [PubMed]
12. Lien, W.-C.; Ching, C.T.-S.; Lai, Z. W.; Wang, H.-M.D.; Lin, J.-S.; Huang, Y.-C.; Lin, F.-H.; Wang, W.-F. Intelligent fall-risk assessment based on gait stability and symmetry among older adults using tri-axial accelerometry. *Front. Bioeng. Biotechnol.* **2022**, *10*, 887269. [CrossRef]
13. Raghu, M.; Zhang, C.; Kleinberg, J.; Bengio, S. Transfusion: Understanding transfer learning for medical imaging. *Adv. Neural Inf. Process. Syst.* **2019**, *32*, 3347–3357.
14. van Ginneken, B.; Setio, A.A.; Jacobs, C.; Ciompi, F. Off-the-shelf convolutional neural network features for pulmonary nodule detection in computed tomography scans. In Proceedings of the 2015 IEEE 12th International Symposium on Biomedical Imaging ISBI, Brooklyn, NY, USA, 16–19 April 2015; pp. 286–289.
15. Simon, B.C.; Baskar, D.; Jayanthi, V.S. Alzheimer's disease classification using deep convolutional neural network. In Proceedings of the 2019 9th International Conference on Advances in Computing and Communication (ICACC), Kochi, India, 6–8 November 2019.
16. Sun, Y.; Ma, S.; Sun, S.; Liu, P.; Zhang, L.; Ouyang, J.; Ni, X. Partial discharge pattern recognition of transformers based on MobileNets convolutional neural network. *Appl. Sci.* **2021**, *11*, 6984. [CrossRef]

17. Jiang, Z.; Zhang, H.; Wang, Y.; Ko, S.B. Retinal blood vessel segmentation using fully convolutional network with transfer learning. *Comput. Med. Imaging. Graph.* **2018**, *68*, 1–15. [CrossRef] [PubMed]
18. Shin, H.-C.; Roth, H.R.; Gao, M.; Lu, L.; Xu, Z.; Nogues, I.; Yao, J.; Mollura, D.; Summers, R.M. Deep convolutional neural networks for computer-aided detection: CNN architectures, dataset characteristics and transfer learning. *IEEE Trans. Med. Imaging* **2016**, *35*, 1285–1298. [CrossRef] [PubMed]
19. Bonte, F.J.; Weiner, M.F.; Bigio, E.H.; White, C.L. 3rd. Brain blood flow in the dementias: SPECT with histopathologic correlation in 54 patients. *Radiology* **1997**, *202*, 793–797. [CrossRef]
20. Arnáiz, E.; Almkvist, O. Neuropsychological features of mild cognitive impairment and preclinical Alzheimer's disease. *Acta Neurol. Scand.* **2003**, *107*, 34–41. [CrossRef]
21. Huang, H.-C.; Tseng, Y.-M.; Chen, Y.-C.; Chen, P.-Y.; Chiu, H.-Y. Diagnostic accuracy of the Clinical Dementia Rating scale for detecting mild cognitive impairment and dementia: A bivariate meta-analysis. *Int. J. Geriatr. Psychiatry* **2021**, *36*, 239–251. [CrossRef]
22. Gajbhiye, G.O.; Nandedkar, A.V.; Faye, I. Translating medical image to radiological report: Adaptive multilevel multi-attention approach. *Comput. Methods Programs Biomed.* **2022**, *221*, 106853. [CrossRef]
23. Huynh, B.Q.; Li, H.; Giger, M.L. Digital mammographic tumor classification using transfer learning from deep convolutional neural networks. *J. Med. Imaging* **2016**, *3*, 034501. [CrossRef] [PubMed]
24. Sandler, M.; Howard, A.; Zhu, M.; Zhmoginov, A.; Chen, L.C. Mobilenetv2: Inverted residuals and linear bottlenecks. In Proceedings of the IEEE Conference on Computer Vision and Pattern Recognition, Salt Lake City, UT, USA, 18–23 June 2018; pp. 4510–4520.
25. Zoph, B.; Vasudevan, V.; Shlens, J.; Le, Q.V. Learning transferable architectures for scalable image recognition. In Proceedings of the IEEE Conference on Computer Vision and Pattern Recognition, Salt Lake City, UT, USA, 18–23 June 2018; pp. 8697–8710.
26. Simonyan, K.; Zisserman, A. Very Very deep convolutional networks for large-scale image recognition. *arXiv* **2014**, arXiv:1409.1556.
27. Szegedy, C.; Vanhoucke, V.; Ioffe, S.; Shlens, J.; Wojna, Z. Rethinking the inception architecture for computer vision. In Proceedings of the IEEE Conference on Computer Vision and Pattern Recognition, Las Vegas, NV, USA, 27–30 June 2016; pp. 2818–2826.
28. Targ, S.; Almeida, D.; Lyman, K. ResNet in ResNet: Generalizing residual architectures. *arXiv* **2016**, arXiv:1603.08029.
29. Yu, X.; Wang, J.; Hong, Q.-Q.; Teku, R.; Wang, S.-H.; Zhang, Y.-D. Transfer learning for medical images analyses: A survey. *Neurocomputing* **2022**, *489*, 230–254. [CrossRef]
30. Lien, W.-C.; Zhou, X.-R.; Liang, Y.-J.; Ching, C.T.-S.; Wang, C.-Y.; Lu, F.-I.; Chang, H.-C.; Lin, F.-H.; Wang, H.-M.D. Therapeutic potential of nanoceria pretreatment in preventing the development of urological chronic pelvic pain syndrome: Immunomodulation via reactive oxygen species scavenging and SerpinB2 downregulation. *Bioeng. Transl. Med.* **2023**, *8*, e10346. [CrossRef] [PubMed]
31. Chen, H.; Zhang, K.; Lyu, P.; Li, H.; Zhang, L.; Wu, J.; Lee, C.-H. A deep learning approach to automatic teeth detection and numbering based on object detection in dental periapical films. *Sci. Rep.* **2019**, *9*, 3840. [CrossRef] [PubMed]
32. de Bruijne, M. Machine learning approaches in medical image analysis: From detection to diagnosis. *Med. Image Anal.* **2016**, *33*, 94–97. [CrossRef] [PubMed]
33. Johnson, J.M.; Khoshgoftaar, T.M. Survey on deep learning with class imbalance. *J. Big Data* **2019**, *6*, 27. [CrossRef]
34. Amin-Naji, M.; Mahdavinataj, H.; Aghagolzadeh, A. Alzheimer's disease diagnosis from structural MRI using Siamese convolutional neural network. In Proceedings of the 2019 4th International Conference on Pattern Recognition and Image Analysis IPRIA, Tehran, Iran, 6–7 March 2019; pp. 75–79.
35. Abrol, A.; Bhattarai, M.; Fedorov, A.; Du, Y.; Plis, S.; Calhoun, V.; for the Alzheimer's Disease Neuroimaging Initiative. Deep residual learning for neuroimaging: An application to predict progression to Alzheimer's disease. *J. Neurosci. Methods* **2020**, *339*, 108701. [CrossRef] [PubMed]
36. Karasawa, H.; Liu, C.-L.; Ohwada, H. Deep 3D convolutional neural network architectures for Alzheimer's disease diagnosis. In *Intelligent Information and Database Systems*; Springer: Cham, Switzerland, 2018; pp. 287–296.
37. Qian, S.; Chen, K.; Guan, Q.; Guo, Q. Performance of Mattis dementia rating scale-Chinese version in patients with mild cognitive impairment and Alzheimer's disease. *BMC Neurol.* **2021**, *21*, 172. [CrossRef] [PubMed]

tomography

Article

Using Deep-Learning-Based Artificial Intelligence Technique to Automatically Evaluate the Collateral Status of Multiphase CTA in Acute Ischemic Stroke

Chun-Chao Huang [1,2], Hsin-Fan Chiang [1,2,3], Cheng-Chih Hsieh [1,2,3], Chao-Liang Chou [2,4], Zong-Yi Jhou [5], Ting-Yi Hou [5] and Jin-Siang Shaw [5,*]

1 Department of Radiology, MacKay Memorial Hospital, Taipei 104217, Taiwan; hcc.5306@mmh.org.tw (C.-C.H.)
2 Department of Medicine, MacKay Medical College, New Taipei City 252005, Taiwan
3 Mackay Junior College of Medicine, Nursing, and Management, Taipei 112021, Taiwan
4 Department of Neurology, MacKay Memorial Hospital, Taipei 104217, Taiwan
5 Institute of Mechatronic Engineering, National Taipei University of Technology, Taipei 106344, Taiwan
* Correspondence: jshaw@ntut.edu.tw; Tel.: +886-2-27712171 (ext. 2046); Fax: +886-27317191

Abstract: Background: Collateral status is an important predictor for the outcome of acute ischemic stroke with large vessel occlusion. Multiphase computed-tomography angiography (mCTA) is useful to evaluate the collateral status, but visual evaluation of this examination is time-consuming. This study aims to use an artificial intelligence (AI) technique to develop an automatic AI prediction model for the collateral status of mCTA. Methods: This retrospective study enrolled subjects with acute ischemic stroke receiving endovascular thrombectomy between January 2015 and June 2020 in a tertiary referral hospital. The demographic data and images of mCTA were collected. The collateral status of all mCTA was visually evaluated. Images at the basal ganglion and supraganglion levels of mCTA were selected to produce AI models using the convolutional neural network (CNN) technique to automatically predict the collateral status of mCTA. Results: A total of 82 subjects were enrolled. There were 57 cases randomly selected for the training group and 25 cases for the validation group. In the training group, there were 40 cases with a positive collateral result (good or intermediate) and 17 cases with a negative collateral result (poor). In the validation group, there were 21 cases with a positive collateral result and 4 cases with a negative collateral result. During training for the CNN prediction model, the accuracy of the training group could reach 0.999 ± 0.015, whereas the prediction model had a performance of 0.746 ± 0.008 accuracy on the validation group. The area under the ROC curve was 0.7. Conclusions: This study suggests that the application of the AI model derived from mCTA images to automatically evaluate the collateral status is feasible.

Keywords: multiphase CTA; collateral status; artificial intelligence; convolutional neural network; acute ischemic stroke

Citation: Huang, C.-C.; Chiang, H.-F.; Hsieh, C.-C.; Chou, C.-L.; Jhou, Z.-Y.; Hou, T.-Y.; Shaw, J.-S. Using Deep-Learning-Based Artificial Intelligence Technique to Automatically Evaluate the Collateral Status of Multiphase CTA in Acute Ischemic Stroke. *Tomography* **2023**, *9*, 647–656. https://doi.org/10.3390/tomography9020052

Academic Editors: Rafał Obuchowicz, Michał Strzelecki and Adam Piorkowski

Received: 15 January 2023
Revised: 11 March 2023
Accepted: 13 March 2023
Published: 16 March 2023

1. Introduction

Stroke is a complex disease and is one of the major causes of long-term disability and death. Because of aggressive treatment of underlying risk factors, such as hypertension and dyslipidemia, the mortality due to stroke declines [1]. In acute ischemic stroke with large vessel occlusion, endovascular thrombectomy (EVT) has been proved to be beneficial and is able to improve functional outcome. Roughly 46.0% of patients receiving EVT can achieve a modified Rankin scale (mRS) score of 0 to 2 within 3 months, as compared with only 26.5% of patients not receiving this procedure [2]. The two basic criteria to select candidates for EVT are the time window and image evidence of large vessel occlusion. Collateral status is an additional important imaging result to identify good candidates for EVT, especially at the borderline or in an extended time window. Therefore, pre procedural evaluation, including imaging studies, should be completed as soon as possible.

Multiphase computed-tomography angiography (mCTA) is a useful imaging tool to identify good candidates for EVT. mCTA includes an arterial phase and 2 subsequent venous phases, also known as 2 delayed arterial phases, and there will be three post-contrast images in mCTA. As compared with traditional single-phase CTA, mCTA is advantageous in the detection of arterial occlusion, with higher interrater reliability and in the evaluation of collateral status [3]. Another useful tool is computed-tomography perfusion (CTP), which can calculate the volumes of infarction core and ischemic penumbra, and these measurements can be automatically performed by some automated software [4]. These two techniques are both widely used in acute ischemic stroke. The strengths of CTP include higher correlation with clinical outcome and availability of automated postprocessing software. In contrast, mCTA does not require postprocessing software and is recommended by guidelines. However, the weak point of mCTA is that the evaluation relies on visual assessment, which might still compromise interrater and intrarater agreement and require additional time to evaluate the imaging results [5]. The management of acute ischemic stroke is as soon as possible to reverse blood flow for the ischemic brain in order to rescue these salvageable brain regions. If the results of mCTA could be evaluated automatically as CTP, the pre-EVT evaluation could be more efficient, which would be significantly beneficial for patients.

Artificial intelligence (AI) dramatically improves the performance of clinical radiology. The basic clinical application of AI is computer-aided detection, such as lesion detection in chest plain films or mammography [6,7]. The advanced clinical application includes disease classification and grading and prediction of prognosis [8–10]. Automatic calculation of the results of CTP is also a typical demonstration of the AI technique. If the AI technique can be used to automatically measure the collateral status of mCTA, the weak points of using mCTA in acute ischemic stroke, including time-consuming visual evaluation and discrepancy of interrater or intrarater agreement, might be solved.

This study aims to use the AI technique to develop an AI model, which can automatically evaluate the collateral status of mCTA in acute ischemic stroke with large vessel occlusion.

2. Methods

2.1. Subjects

This was a retrospective study, which had been approved by the Institutional Review Board in Mackay Memorial Hospital (20MMHIS361e). Subjects who had acute ischemic stroke with large vessel occlusion and received EVT between January 2015 and June 2020 were enrolled in this study. Other inclusion criteria included age at least 20 years old, a National Institute of Health Stroke Scale (NIHSS) at onset time of at least 8 scores, having mCTA images, having mRS at 3 months after the procedure of EVT, and having an Alberta Stroke Program Early CT Score (ASPECTS) of at least 5 points on the initial nonenhanced brain CT images. Exclusion criteria were loss of required clinical data, marked imaging motion artifact, severe underlying diseases such as malignancy, severe trauma, and aortic dissection, and acute ischemic infarction due to posterior circulation, including occlusion of vertebral artery or basilar artery.

2.2. Imaging Studies

In the enrolled subjects, all the mCTA of these subjects were reviewed and graded by an experienced neuroradiologist. The collateral status of mCTA was graded based on the original concept of the development of mCTA [11]. The pial arterial filling score within the symptomatic ischemic territory using mCTA ranged from 0 to 5. A score of 5 meant good collateral status, 4 meant intermediate collateral status, and 0 to 3 meant poor collateral status. In a complete mCTA scan, there were a nonenhanced CT scan from head to neck, an arterial phase CT scan from head to neck, an 8 s delay arterial phase CT scan for the whole head, and a 16 s delay arterial phase CT scan for the whole head. The raw images were 1.5 mm on a Siemens SOMATOM Definition AS CT Scanner (Siemens Healthcare

GmbH, Erlangen, Germany) and 1 mm on a Canon Aquilion PRIME CT Scanner (Toshiba Medical Systems, Nasu, Japan). Reconstruction with the maximum intensity projection (MIP) technique was applied on the first arterial phase CT images to produce axial, coronal and sagittal MIP images, as well as on the second and third arterial phase CT images, to produce axial MIP images. The settings of MIP were a 20 mm slice thickness with a 10 mm increment for axial reconstructed MIP images and a 10 mm slice thickness with a 4 mm increment for coronal and sagittal reconstructed MIP images. Two levels of the axial images were identified as the basal ganglion level and the supraganglion level. The basal ganglion level was the image when the size of the basal ganglion was visually measured at the maximum size. The supraganglion level meant the first image level when both lateral ventricles just disappeared on the axial images examined in a caudal to cranial sequence. There were 12 images selected from the mCTA study of each subject for further analysis using the AI technique. The 12 images included the raw image and the MIP image at the basal ganglion level and the supraganglion level from all three arterial phase images (Figure 1).

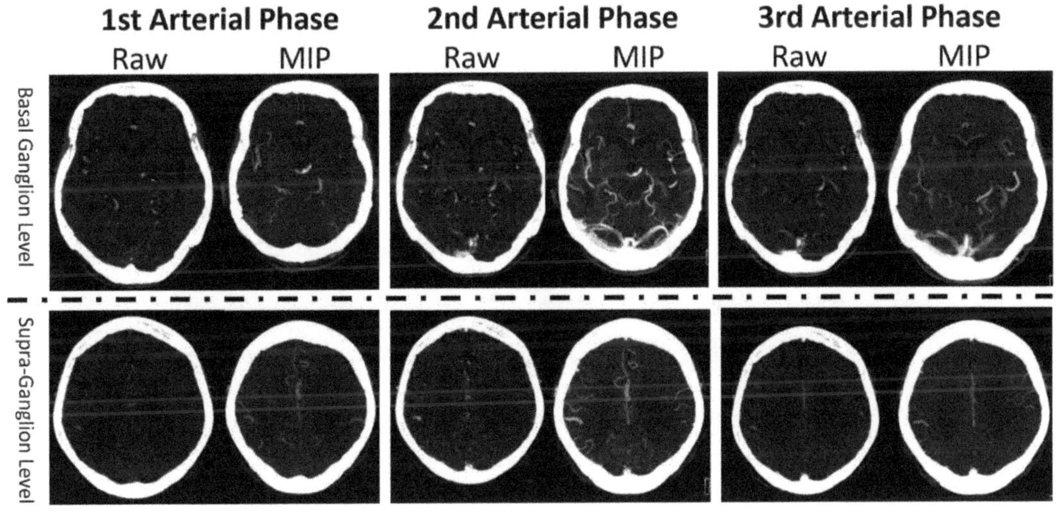

Figure 1. Demonstration of 12 images selected from each multiphase CTA study.

2.3. Data Preprocessing and Normalization

In this study, we mainly used Keras and Scikit-learn libraries in Python to perform data preprocessing, normalization and model establishment. Before establishing deep learning models, data preprocessing and normalization were conducted. The data in this study were images, so the data preprocessing and normalization included the denoising and adjusting contrast of images and resizing images to be identical in size.

2.4. Convolutional Neural Network

The collateral status graded as good or intermediate collateral status was defined as the positive collateral result, while poor collateral status as the negative collateral result (Figure 2). From all the enrolled subjects, 70% subjects were randomly selected for the training group, and the remaining 30% of the subjects were placed in the validation group.

The Convolutional Neural Network (CNN) technique was a deep learning technique and was first described in 1980 [12]. This technique was further consolidated by Yann Lecun [13]. CNN could be used to analyze visual imagery with little preprocessing to perform image classification. In this study, the CNN technique was used to develop a prediction model based on the 12 selected mCTA images to predict collateral status as a positive or negative collateral result. Specifically, the structure of the CNN prediction model is shown in Figure 3, where only 5 convolution layers were employed. It was experimentally evaluated to have a comparable performance as that of the renowned ResNet50 [14] trained using transfer learning for this binary classification problem for collateral status, but with a much-improved convergence rate.

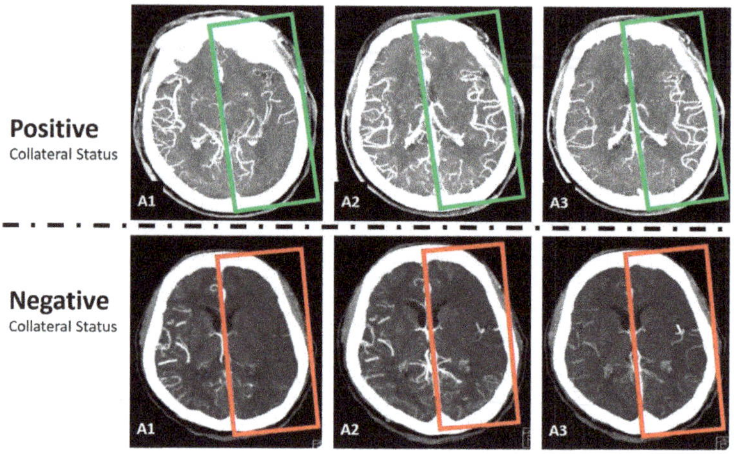

Figure 2. Demonstration of positive and negative collateral status on multiphase CTA. The A1, A2, and A3 indicate the three arterial phases. In the positive collateral status (upper row), the left side brain (green frame) is the ischemic side. When the enhancement of the arterial branches in the ischemic side can achieve similar appearance to the A1 contralateral normal side in A1 or A2 (this case), the collateral status is good or intermediate, respectively, and the result is defined as the positive collateral status. On the contrary, in the negative collateral status (lower row), the arterial branches in the ischemic side (red frame) cannot be enhanced like the A1 contralateral normal side in A1 or A2, suggesting that the collateral status is poor and the result is defined as the negative collateral status.

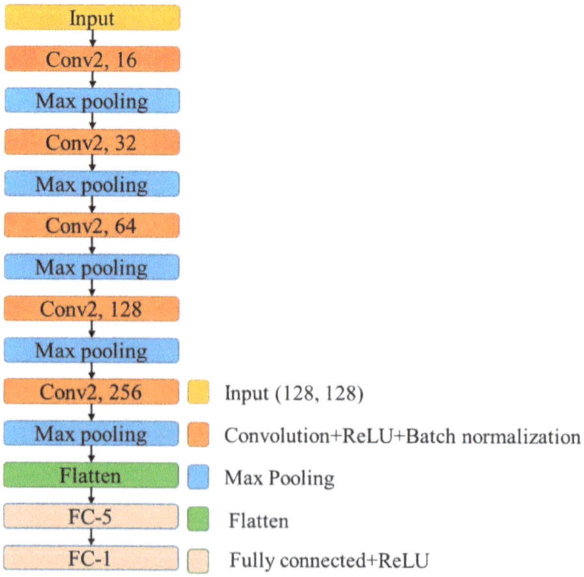

Figure 3. Structure of the CNN model.

3. Results

There were 82 enrolled subjects in this study. The sex ratio was 1:1. The mean age was 70.67 ± 14.09 years old, ranging from 35 to 98 years old. The median of initial NIHSS was 18, ranging from 8 to 31, and the Glasgow Coma Scale was 13.5, ranging from 5 to 15. The median ASPECTS score on the initial nonenhanced brain CT was 8, ranging from 5 to 10. The location of acute arterial occlusion was located at the internal carotid artery in 28 (34.1%) subjects, the first segment of the middle cerebral artery in 41 (50.0%) subjects, and the second segment of the middle cerebral artery in 13 (15.9%) subjects. The procedure results of EVT were 9 (11.0%) with modified treatment in cerebral infarction (mTICI) 0, 4 (4.9%) with mTICI 1, 11 (13.4%) with mTICI 2a, 38 (46.3%) with mTICI 2b, and 20 (24.4%) with mTICI 3. The mean time from onset to reperfusion or end of the procedure was 316.66 ± 86.11 min. Symptomatic intracranial hemorrhage occurred in 14 (17.1%) subjects during the hospitalization course after EVT. The functional outcomes in 3 months were 3 (3.7%) cases with mRS 0, 9 (11.0%) cases with mRS 1, 9 (11.0%) cases with mRS 2, 17 (20.7%) cases with mRS 3, 22 (26.8%) cases with mRS 4, 4 (4.9%) cases with mRS 5, and 18 (22.0%) cases with mRS 6. The demographic data of these subjects with a positive or negative collateral status are separately detailed in Table 1.

Table 1. Demographic data of the enrolled subjects with a positive or negative collateral status.

Collateral Status	Positive (n = 61)	Negative (n = 21)
	Mean ± SD (Range) n (%) Median (Range)	
Age (years)	72.5 ± 13.5 (43–98)	65.4 ± 14.8 (35–96)
Sex (male)	28 (45.9)	13 (61.9)
National Institute of Health Stroke Scale (NIHSS)	17 (8–31)	19 (13–23)
Glasgow Coma Scale	14 (5–15)	11 (7–15)
Alberta Stroke Program Early CT Score (ASPECTS)	8 (5–10)	8 (6–10)
Occlusion site	ICA: 20 (32.8) M1: 31 (50.8) M2: 10 (16.4)	ICA: 8 (38.1) M1: 10 (47.6) M2: 3 (14.3)
Modified treatment in cerebral infarction (mTICI)	0–2a: 16 (26.2) 2b–3: 45 (73.8)	0–2a: 8 (38.1) 2b–3: 13 (61.9)
Onset to Reperfusion time (minutes)	316.3 ± 83.5 (152–528)	317.6 ± 95.4 (189–519)
Symptomatic intracranial hemorrhage	9 (14.8)	5 (23.8)
Modified Rankin Scale	0–2: 18 (29.5) 3–6: 43 (70.5)	0–2: 3 (14.3) 3–6: 18 (85.7)

Abbreviations: ICA: internal carotid artery; M1: the first segment of the middle cerebral artery; M2: the second segment of the middle cerebral artery; SD: standard deviation.

There were 57 cases (69.5%) randomly selected for the training group and 25 cases (30.5%) for the validation group. In the training group, there were 40 cases (70.2%) with a positive collateral result (good or intermediate collateral) and 17 cases (29.8%) with a negative collateral result (poor collateral). In the validation group, there were 21 cases (84.0%) with a positive collateral result and 4 cases (16.0%) with a negative collateral result. The confusion matrix of the validation group is displayed in Figure 4. During training for the CNN prediction model, the accuracy of the training group could reach 0.999 ± 0.015, whereas the prediction model had a performance of 0.746 ± 0.008 accuracy on the validation group (Figure 5). The ROC curve of this model demonstrated an area under the curve (AUC) of 0.7 (Figure 6).

Accuracy = 0.7533 ((222+4)/300)		Ground truth	
		Positive	Negative
Prediction	Positive	222	44
	Negative	30	4

Figure 4. The confusion matrix of the validation group.

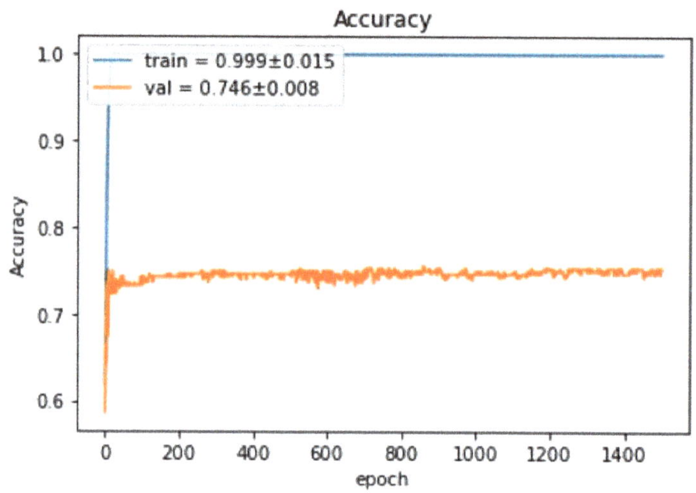

Figure 5. Performance of the prediction model from the selected 12 images of multiphase CTA on the prediction of collateral status.

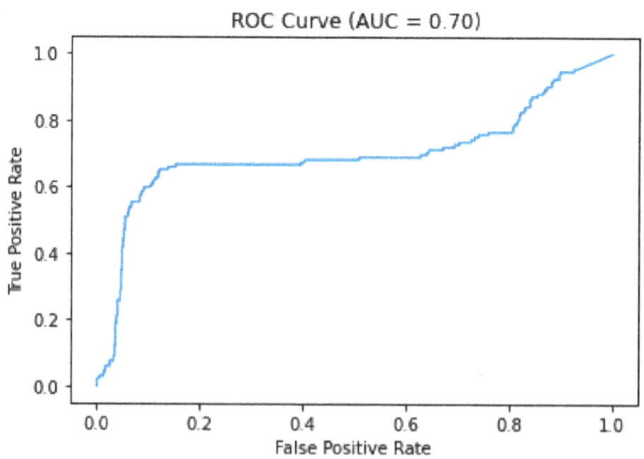

Figure 6. ROC curve of the prediction model with an area under the curve (AUC) of 0.7.

4. Discussion

By using the CNN technique to produce a prediction model from the selected 12 images of mCTA, the model demonstrated a performance of 0.746 ± 0.008 accuracy and a 0.7 AUC on the prediction of the collateral status of mCTA.

The collateral status has been found to be an independent predictor of final infarct volume in patients receiving EVT for acute anterior circulation ischemic stroke due to large vessel occlusion. A good collateral status helps to prevent the middle cerebral arterial territory from infarction [15]. On the contrary, poor collateral circulation is a risk factor for symptomatic intracranial hemorrhage after EVT, and therefore results in a poor functional outcome [16,17]. Furthermore, collateral status sometimes might be helpful to determine eligibility for EVT in some cases [18]. There are many proposed methods to evaluate the collateral status by using CT images, including single-phase CTA, mCTA and CTP. The single-phase CTA tends to underestimate the collateral status due to limited information only from the spatial extent of collateral enhancement in one series of images [19]. CTP-based parameters, such as time-to-maximum, cerebral blood flow, hypoperfusion intensity ratio and cerebral blood volume index, have been used to evaluate the collateral status [20–22]. By using some automated software, these CTP-based parameters could be calculated automatically and quickly without interobserver bias [19]. However, the availability of an automated software might require an additional expense and several factors including bolus shape, scanner protocol, different software, and head motion might affect CTP analysis [19,23]. In addition, a limited brain coverage of the CT scanner will cause inadequate coverage of the ischemic lesion; therefore, the requirement of a wide-coverage CT scanner is also important to perform a better CTP [24]. On the contrary, collateral evaluation based on mCTA might solve some weak points of CTP-based techniques, but mCTA also has its disadvantages, including reduced temporal resolution and time-consuming visual assessment [19]. Though CTP-based collateral evaluation might be more attractive for clinical practitioners because of faster and fully automatic calculation and more reliable results, mCTA sometimes is the only choice due to the limitation of inaccessible postprocessing software or wide-coverage CT scanners. Recently, the rapid development of AI has significantly contributed to automatic imaging analysis in lesion detection and classification. Thus, the development of automatic analysis on mCTA to predict the collateral status might be possible, and the prediction model would be very useful for hospitals where the mCTA technique is the only choice to evaluate patients with acute ischemic stroke.

A previous study produced an AI model to differentiate good and poor collateral status in mCTA and achieved an accuracy of 0.852 ± 0.045. However, the mCTA was extracted from whole-brain 4D CTA/CTP images rather than the traditional mCTA [25]. Another study used single-phase CTA to produce an AI model and achieved accuracy of 0.85 ± 0.01 to differentiate whether there was more than 50% filling [26]. In an article reviewing recent automatic collateral scoring using CTA, the 4D CTA is the most popular one to be used in the research. Other methods are based on single-phase CTA or even nonenhanced CT [27]. The overall accuracy or area under the curve of these methods could be above 70% or even up to 90%. However, an AI model based on the traditional mCTA to evaluate collateral status is less frequently described in the literature, but our preliminary result suggests that the performance of the AI model based on the traditional mCTA could also achieve more than 70% accuracy with an AUC of 0.7. Based on previous studies, CTP-based ischemic core volume shows a better prediction on the 3-month functional outcome for patients with acute ischemic stroke undergoing EVT than the mCTA collateral score [5]. Fully automated commercial software for CTP analysis are also available. However, CTP requires a higher coverage CT scanner and an additional expense for the commercial software, which sometimes makes CTP an impossible choice even with a CT scanner. The concept of 4D CTA is close to that of the traditional mCTA, though the images of different phases in 4D CTA are extracted from a pool of several volume images, while the images of different phases in the traditional mCTA are directly scanned. Again, 4D CTA also requires a wide-coverage CT scanner. As compared with single-phase CTA and nonenhanced CT, mCTA includes much more imaging information and data. Furthermore, mCTA is found to be highly correlated with the CTP result, and the physiological role might be more important than single-phase CTA [28]. For the abovementioned points, mCTA might be a more reliable resource to produce an AI prediction model for the collateral status because of the similarity to 4D CTA and being more correlated with CTP. Although previous studies mainly adopted 4D CTA, single-phase CTA or nonenhanced CT to produce AI prediction models for the collateral status, our study points out the possibility of automatic evaluation of the collateral status from the traditional mCTA, which might sometimes be the best CT collateral evaluation for acute ischemic stroke in hospitals with limited resources, such as a lack of postprocessing software or wide-coverage CT scanners.

There are several limitations in this study. First, the retrospective design of this study has some diverse settings of scan protocol and many uncontrolled parameters. Second, there are two different CT scanners used in this study, and the heterogeneous quality of images might affect AI analysis. The data only enrolled 82 cases and only included 12 images from each mCTA study. In the future, more enrolled cases and included images might be able to improve the performance of the AI model, but the imaging postprocessing time might increase markedly if more images of mCTA studies are included to evaluate the collateral status. In the confusion matrix of this study, there was a very high false-positive rate (91.7%) and a very low true-negative predictive value (8.3%). Possible reasons for this poor result include limited enrolled subjects and images, relatively more enrolled cases with a positive collateral status, and the complexity of collateral evaluation on mCTA. Because poor collateral status is considered as a negative predictor of functional outcome, cases with poor collateral status are not highly suggested to be aggressively treated with EVT, and, thereby, the cases with poor collateral status are fewer than good or intermediate collateral status in this study. To enroll many more cases with a poor collateral status might be helpful to overcome the problem of a high false-positive rate in the future. There are some challenges we encountered during the AI model training, including the limited number of images, overfitting of models, and poor learning efficiency. There was an important problem about image augmentation we did not expect before this study. Image augmentation is a common technique to create more images from existing ones. Initially, we planned to use image augmentation to overcome the problem of a limited number of images in this study. However, the images of mCTA from a single subject had some internal relationship, because they were from the same subject or from the same scan sequence. When we used image augmentation by flipping or rotating images, the result of model training became worse, probably because of breaking the internal relationship of those images from the same subject. Thus, we were focused on using different AI models or changing the structures of AI models to improve the prediction accuracy.

5. Conclusions

This study suggests that even in a hospital with limited resources where only very basic traditional mCTA can be performed, the AI model proposed by this study can still automatically evaluate the collateral status in acute ischemic stroke with large-vessel occlusion, which is promising to facilitate the patient selection for EVT and to decrease inter-observer bias. Since the data required for this AI model only include very basic traditional mCTA images, this AI model can be applied in many more hospitals than other techniques, such as CTP or 4D CTA.

Author Contributions: Conceptualization, C.-C.H. (Chun-Chao Huang), H.-F.C. and C.-C.H. (Cheng-Chih Hsieh); methodology, Z.-Y.J. and T.-Y.H.; software, T.-Y.H.; validation, C.-C.H. (Chun-Chao Huang), H.-F.C. and C.-C.H. (Cheng-Chih Hsieh); formal analysis, C.-C.H. (Chun-Chao Huang) and T.-Y.H.; investigation, C.-L.C.; resources, C.-L.C.; data curation, C.-L.C.; writing—original draft preparation, C.-C.H. (Chun-Chao Huang); writing—review and editing, J.-S.S.; visualization, C.-C.H. (Chun-Chao Huang) and Z.-Y.J.; supervision, J.-S.S.; project administration, C.-C.H. (Chun-Chao Huang); funding acquisition, C.-C.H. (Chun-Chao Huang) and J.-S.S. All authors have read and agreed to the published version of the manuscript.

Funding: This study was funded by the Mackay Memorial Hospital and National Taipei University of Technology (MMH-TT-11003, NTUT-MMH-110-03).

Institutional Review Board Statement: The study was approved by the Institutional Review Board of MacKay Memorial Hospital (ref. 20MMHIS361e).

Informed Consent Statement: Patient consent was waived due to the retrospective nature of the work.

Data Availability Statement: The data presented in this study are available on reasonable request from the corresponding author.

Conflicts of Interest: All authors declare no competing financial interests.

References

1. Cipolla, M.J.; Liebeskind, D.S.; Chan, S.L. The importance of comorbidities in ischemic stroke: Impact of hypertension on the cerebral circulation. *J. Cereb. Blood Flow Metab.* **2018**, *38*, 2129–2149. [CrossRef] [PubMed]
2. Huang, C.; Chou, C.; Huang, W.; Jhou, Z.; Hwang, Y.; Lin, H.; Tsai, Y.; Lin, C. Imaging findings of multiphase ct angiography of acute internal carotid artery occlusion within 6-hour time-window after thrombectomy and its clinical implication. *Iran. J. Radiol.* **2020**, *17*, e102948. [CrossRef]
3. Dundamadappa, S.; Iyer, K.; Agrawal, A.; Choi, D.J. Multiphase ct angiography: A useful technique in acute stroke imaging-collaterals and beyond. *AJNR Am. J. Neuroradiol.* **2021**, *42*, 221–227. [CrossRef] [PubMed]
4. Nogueira, R.G.; Jadhav, A.P.; Haussen, D.C.; Bonafe, A.; Budzik, R.F.; Bhuva, P.; Yavagal, D.R.; Ribo, M.; Cognard, C.; Hanel, R.A.; et al. Thrombectomy 6 to 24 hours after stroke with a mismatch between deficit and infarct. *N. Engl. J. Med.* **2018**, *378*, 11–21. [CrossRef]
5. Tan, Z.; Parsons, M.; Bivard, A.; Sharma, G.; Mitchell, P.; Dowling, R.; Bush, S.; Churilov, L.; Xu, A.; Yan, B. Comparison of computed tomography perfusion and multiphase computed tomography angiogram in predicting clinical outcomes in endovascular thrombectomy. *Stroke* **2022**, *53*, 2926–2934. [CrossRef] [PubMed]
6. Rajpurkar, P.; Irvin, J.; Ball, R.L.; Zhu, K.; Yang, B.; Mehta, H.; Duan, T.; Ding, D.; Bagul, A.; Langlotz, C.P.; et al. Deep learning for chest radiograph diagnosis: A retrospective comparison of the chexnext algorithm to practicing radiologists. *PLoS Med.* **2018**, *15*, e1002686. [CrossRef] [PubMed]
7. Ayer, T.; Chen, Q.; Burnside, E.S. Artificial neural networks in mammography interpretation and diagnostic decision making. *Comput. Math. Methods Med.* **2013**, *2013*, 832509. [CrossRef]
8. Smyser, C.D.; Dosenbach, N.U.; Smyser, T.A.; Snyder, A.Z.; Rogers, C.E.; Inder, T.E.; Schlaggar, B.L.; Neil, J.J. Prediction of brain maturity in infants using machine-learning algorithms. *Neuroimage* **2016**, *136*, 1–9. [CrossRef]
9. Fartaria, M.J.; Bonnier, G.; Roche, A.; Kober, T.; Meuli, R.; Rotzinger, D.; Frackowiak, R.; Schluep, M.; Du Pasquier, R.; Thiran, J.P.; et al. Automated detection of white matter and cortical lesions in early stages of multiple sclerosis. *J. Magn. Reson. Imaging* **2016**, *43*, 1445–1454. [CrossRef]
10. Zhang, X.; Yan, L.F.; Hu, Y.C.; Li, G.; Yang, Y.; Han, Y.; Sun, Y.Z.; Liu, Z.C.; Tian, Q.; Han, Z.Y.; et al. Optimizing a machine learning based glioma grading system using multi-parametric mri histogram and texture features. *Oncotarget* **2017**, *8*, 47816–47830. [CrossRef]
11. Menon, B.K.; d'Esterre, C.D.; Qazi, E.M.; Almekhlafi, M.; Hahn, L.; Demchuk, A.M.; Goyal, M. Multiphase ct angiography: A new tool for the imaging triage of patients with acute ischemic stroke. *Radiology* **2015**, *275*, 510–520. [CrossRef] [PubMed]
12. Fukushima, K. Neocognitron: A self organizing neural network model for a mechanism of pattern recognition unaffected by shift in position. *Biol. Cybern.* **1980**, *36*, 193–202. [CrossRef]
13. LeCun, Y.; Bottou, L.; Bengio, Y.; Haffner, P. Gradient-based learning applied to document recognition. *Proc. IEEE* **1998**, *86*, 2278–2324. [CrossRef]
14. He, K.; Zhang, X.; Ren, S.; Sun, J. Deep residual learning for image recognition. In Proceedings of the 2016 IEEE Conference on Computer Vision and Pattern Recognition (CVPR), Las Vegas, NV, USA, 27–30 June 2016; pp. 770–778.
15. Al-Dasuqi, K.; Payabvash, S.; Torres-Flores, G.A.; Strander, S.M.; Nguyen, C.K.; Peshwe, K.U.; Kodali, S.; Silverman, A.; Malhotra, A.; Johnson, M.H.; et al. Effects of collateral status on infarct distribution following endovascular therapy in large vessel occlusion stroke. *Stroke* **2020**, *51*, e193–e202. [CrossRef] [PubMed]
16. Bang, O.Y.; Saver, J.L.; Kim, S.J.; Kim, G.M.; Chung, C.S.; Ovbiagele, B.; Lee, K.H.; Liebeskind, D.S.; Collaborators, U.C.-S.S. Collateral flow averts hemorrhagic transformation after endovascular therapy for acute ischemic stroke. *Stroke* **2011**, *42*, 2235–2239. [CrossRef]
17. Leng, X.; Fang, H.; Leung, T.W.; Mao, C.; Miao, Z.; Liu, L.; Wong, K.S.; Liebeskind, D.S. Impact of collaterals on the efficacy and safety of endovascular treatment in acute ischaemic stroke: A systematic review and meta-analysis. *J. Neurol. Neurosurg. Psychiatry* **2016**, *87*, 537–544. [CrossRef]
18. Powers, W.J.; Rabinstein, A.A.; Ackerson, T.; Adeoye, O.M.; Bambakidis, N.C.; Becker, K.; Biller, J.; Brown, M.; Demaerschalk, B.M.; Hoh, B.; et al. Guidelines for the early management of patients with acute ischemic stroke: 2019 update to the 2018 guidelines for the early management of acute ischemic stroke: A guideline for healthcare professionals from the american heart association/american stroke association. *Stroke* **2019**, *50*, e344–e418. [PubMed]

19. Potreck, A.; Scheidecker, E.; Weyland, C.S.; Neuberger, U.; Herweh, C.; Mohlenbruch, M.A.; Chen, M.; Nagel, S.; Bendszus, M. Seker, F. Rapid ct perfusion-based relative cbf identifies good collateral status better than hypoperfusion intensity ratio, cbv-index and time-to-maximum in anterior circulation stroke. *AJNR Am. J. Neuroradiol.* **2022**, *43*, 960–965. [CrossRef] [PubMed]
20. Seker, F.; Pfaff, J.; Potreck, A.; Mundiyanapurath, S.; Ringleb, P.A.; Bendszus, M.; Mohlenbruch, M.A. Correlation of t(max) volumes with clinical outcome in anterior circulation stroke. *Brain Behav.* **2017**, *7*, e00772. [CrossRef]
21. Guenego, A.; Fahed, R.; Albers, G.W.; Kuraitis, G.; Sussman, E.S.; Martin, B.W.; Marcellus, D.G.; Olivot, J.M.; Marks, M.P. Lansberg, M.G.; et al. Hypoperfusion intensity ratio correlates with angiographic collaterals in acute ischaemic stroke with m1 occlusion. *Eur. J. Neurol.* **2020**, *27*, 864–870. [CrossRef]
22. Arenillas, J.F.; Cortijo, E.; Garcia-Bermejo, P.; Levy, E.I.; Jahan, R.; Liebeskind, D.; Goyal, M.; Saver, J.L.; Albers, G.W. Relative cerebral blood volume is associated with collateral status and infarct growth in stroke patients in swift prime. *J. Cereb. Blood Flow Metab.* **2018**, *38*, 1839–1847. [CrossRef] [PubMed]
23. Potreck, A.; Seker, F.; Mutke, M.A.; Weyland, C.S.; Herweh, C.; Heiland, S.; Bendszus, M.; Mohlenbruch, M. What is the impact of head movement on automated ct perfusion mismatch evaluation in acute ischemic stroke? *J. Neurointerv. Surg.* **2022**, *14*, 628–633 [CrossRef] [PubMed]
24. Shankar, J.J.; Lum, C. Whole brain ct perfusion on a 320-slice ct scanner. *Indian J. Radiol. Imaging* **2011**, *21*, 209–214. [CrossRef] [PubMed]
25. Wang, J.; Tan, D.; Liu, J.; Wu, J.; Huang, F.; Xiong, H.; Luo, T.; Chen, S.; Li, Y. Merging multiphase cta images and training them simultaneously with a deep learning algorithm could improve the efficacy of ai models for lateral circulation assessment in ischemic stroke. *Diagnostics* **2022**, *12*, 1562. [CrossRef]
26. Rava, R.A.; Seymour, S.E.; Snyder, K.V.; Waqas, M.; Davies, J.M.; Levy, E.I.; Siddiqui, A.H.; Ionita, C.N. Automated collateral flow assessment in patients with acute ischemic stroke using computed tomography with artificial intelligence algorithms. *World Neurosurg.* **2021**, *155*, e748–e760. [CrossRef]
27. Aktar, M.; Tampieri, D.; Rivaz, H.; Kersten-Oertel, M.; Xiao, Y. Automatic collateral circulation scoring in ischemic stroke using 4d ct angiography with low-rank and sparse matrix decomposition. *Int. J. Comput. Assist. Radiol. Surg.* **2020**, *15*, 1501–1511. [CrossRef]
28. Gupta, A.; Garg, P.K.; Khera, P.S.; Panda, S.; Bohra, G.K.; Yadav, T.; Garg, M.K.; Tiwari, S. Multiphase computed tomography angiography (mcta) derived source images in acute ischemic stroke: Beyond collaterals. Can it obviate the need for computed tomography perfusion (ctp)? *Clin. Neurol. Neurosurg.* **2022**, *222*, 107421. [CrossRef]

cancers

MDPI

Article

Radiologic versus Segmentation Measurements to Quantify Wilms Tumor Volume on MRI in Pediatric Patients

Myrthe A. D. Buser [1], Alida F. W. van der Steeg [1,*], Marc H. W. A. Wijnen [1], Matthijs Fitski [1], Harm van Tinteren [1], Marry M. van den Heuvel-Eibrink [1,2], Annemieke S. Littooij [1,2] and Bas H. M. van der Velden [3]

1 Princess Máxima Center for Pediatric Oncology, 3584 CS Utrecht, The Netherlands
2 Wilhelmina Children's Hospital, University Medical Center Utrecht, 3584 EA Utrecht, The Netherlands
3 Image Sciences Institute, University Medical Center Utrecht, Utrecht University, 3584 CX Utrecht, The Netherlands
* Correspondence: a.f.w.vandersteeg@prinsesmaximacentrum.nl; Tel.: +31-88-972-7272

Simple Summary: Volume measurements are important in tumor evaluations of children with a Wilms tumor. Current volume measurements might not be accurate. Our study had two aims. Our first aim was to assess whether manual segmentation of MRI can accurately quantify the volume of Wilms tumors. Our second aim was to show if manual segmentation can be automated using deep learning. We compared radiological-based and manual segmentation-based measurements. Next, we developed an automated segmentation method. Radiological measurements underestimate the actual tumor volume by about 10% irrespective of tumor size. Deep learning can potentially be used to replace manual segmentations in volume measurements.

Abstract: Wilms tumor is a common pediatric solid tumor. To evaluate tumor response to chemotherapy and decide whether nephron-sparing surgery is possible, tumor volume measurements based on magnetic resonance imaging (MRI) are important. Currently, radiological volume measurements are based on measuring tumor dimensions in three directions. Manual segmentation-based volume measurements might be more accurate, but this process is time-consuming and user-dependent. The aim of this study was to investigate whether manual segmentation-based volume measurements are more accurate and to explore whether these segmentations can be automated using deep learning. We included the MRI images of 45 Wilms tumor patients (age 0–18 years). First, we compared radiological tumor volumes with manual segmentation-based tumor volume measurements. Next, we created an automated segmentation method by training a nnU-Net in a five-fold cross-validation. Segmentation quality was validated by comparing the automated segmentation with the manually created ground truth segmentations, using Dice scores and the 95th percentile of the Hausdorff distances (HD95). On average, manual tumor segmentations result in larger tumor volumes. For automated segmentation, the median dice was 0.90. The median HD95 was 7.2 mm. We showed that radiological volume measurements underestimated tumor volume by about 10% when compared to manual segmentation-based volume measurements. Deep learning can potentially be used to replace manual segmentation to benefit from accurate volume measurements without time and observer constraints.

Keywords: Wilms tumor; pediatric oncology; volume measurements; MRI; deep learning

Citation: Buser, M.A.D.; van der Steeg, A.F.W.; Wijnen, M.H.W.A.; Fitski, M.; van Tinteren, H.; van den Heuvel-Eibrink, M.M.; Littooij, A.S.; van der Velden, B.H.M. Radiologic versus Segmentation Measurements to Quantify Wilms Tumor Volume on MRI in Pediatric Patients. *Cancers* **2023**, *15*, 2115. https://doi.org/10.3390/cancers15072115

Academic Editors: Rafał Obuchowicz, Dominik T. Schneider, Michał Strzelecki and Adam Piorkowski

Received: 22 February 2023
Revised: 28 March 2023
Accepted: 30 March 2023
Published: 1 April 2023

1. Introduction

Wilms tumor is one of the most prevalent malignant solid tumors in pediatric patients. With a combination of neoadjuvant chemotherapy, surgery, and post-operative chemotherapy survival up to 90% can be achieved [1,2]. Throughout the treatment process, quantifying tumor volume on magnetic resonance imaging (MRI) or computed tomography (CT) is important for several reasons. In the International Society of Paediatric Oncology–Renal

Study Tumor Group (SIOP-RTSG) UMBRELLA guidelines MRI is preferred because of the absence of a radiation dose and the excellent soft tissue contrast [3]. First, tumor volume can be used as an indication for therapy response and risk stratification, as a tumor volume of >500 mL after neoadjuvant chemotherapy for specific Wilms tumor types places these patients in a higher risk group (see also Supplementary Table S1) [3]. Next, the UMBRELLA guidelines indicate nephron-sparing surgery (NSS) only for tumors <300 mL at diagnosis in patients with a unilateral tumor without a predisposing (tumor) syndrome [4,5]. Last, tumor volume has been suggested to be predictive of patient outcomes [5–7]. Hence, tumor volume measurements are important for Wilms tumor management.

In current good clinical practice, review radiologists assess tumor volume on MRI or CT by measuring the largest diameter of the tumor in three directions. This method is based on the assumption that the tumor is an ellipsoid shape, which does not always reflect the true tumor shape [8–10]. Volume measurement by manual delineation of the tumor on MRI (i.e., segmentation) does not require this assumption. The extent of error in volume measurements in Wilms tumor introduced by the assumption of ellipsoid tumor shape measurements remains unknown.

Although manual segmentation might be more accurate for volume measurements, the process is time-consuming and intra- and interobserver-dependent [11]. Automated segmentation, e.g., using deep learning, has the potential of utilizing more precise manual segmentation-based volume measurements while limiting the time and observer constraints [12]. The aim of this study was twofold. Our first aim was to assess whether manual segmentation of MRI can accurately quantify the volume of Wilms tumors. Our second aim was to prove if manual segmentation can be automated using deep learning.

2. Materials and Methods

2.1. Patients

A total of 45 consecutive Wilms tumor patients, 0–18 years of age and treated in the Princess Máxima Center for pediatric oncology during the period of July 2018 to November 2020 were retrospectively included. All patients signed an informed consent form. Patients were excluded when imaging was incomplete ($n = 4$) or when no informed consent for the SIOP-UMBRELLA study was present ($n = 1$).

All but five patients were treated with pre-operative chemotherapy according to the SIOP-RTSG 2001 or SIOP-RTSG 2016-UMBRELLA protocol [3]. Tumor type and stage were assessed on the resection specimen using hematoxylin and eosin (H&E) staining, and additional immunohistochemistry (IHC), according to international SIOP-RTSG standards [13]. All cases underwent rapid pathology review.

2.2. Magnetic Resonance Imaging

Standard clinical MRI scans were performed and centrally reviewed at diagnosis and before surgery in accordance with the SIOP-RTSG 2016-UMBRELLA protocol [5]. The imaging was performed on a 1.5T scanner (Ingenia; Philips Medical Systems, Best, The Netherlands), and included a 3D T_2-weighted scan and a fat-suppressed T_1-weighted scan with and without intravenous contrast (Gadovist, Bayer Pharma, Berlin, Germany, 0.1 mmol/kg body weight) (Table 1).

Table 1. MRI imaging parameters.

Parameters	T_1-Weighted with Fat Suppression	3D T_2-Weighted
Sequence type	Gradient Echo	Turbo Spin Echo
Repetition time (ms)	5.5	459
Echo time (ms)	2.7	90
Flip angle	10°	90°
Slice thickness (mm)	3.0	1.15
Voxel spacing (mm)	0.74×0.74 mm^2	0.83×0.83 mm^2

2.3. *Tumor Volume Measurement*

We used manual segmentation as reference volume measurements as it is the closest approximation for true tumor volume on MRI. Subsequently, we compared the radiological tumor volume measurement with the manual segmentation-based tumor volume measurement. Next, we compared the deep learning-based segmentations with the manually created ground truth segmentation.

2.3.1. Radiological Tumor Volume Measurements

The largest tumor diameters were measured in three directions by the pediatric radiologist, from which the volume of an ellipsoid was calculated (i.e., the radiological tumor volume).

2.3.2. Manual Segmentation-Based Tumor Volume Measurements

Tumors were manually delineated (i.e., segmented) on the images by three independent observers under supervision of one pediatric radiologist (AL, 13 years of experience with pediatric MRI). Patients were randomly assigned to each observer. A segmentation protocol was developed together with the reviewing radiologist. The manual delineation was performed in 3DSlicer (version 4.11) using a preset protocol [14]:

1. Manual indication of tumor and background on each fourth slice of the postcontrast T_1-weighted scan, due to the best tumor and kidney contrast in this sequence.
2. Initial tumor segmentation using the 3DSlicer algorithm "grow from seeds", which is a 3D volume growing algorithm. After this step, each pixel was assigned either the label tumor or background.
3. Because of the difference in in-slice resolution and slice thickness, the segmentation was reformatted from the T_1-weighted image to the T_2-weighted image using 3DSlicer's inbuild function. These labels were extensively checked and manually corrected if needed.

Tumor volume was calculated by multiplying the number of voxels in the tumor segmentation with the voxel size in mL.

2.3.3. Deep Learning-Based Tumor Volume Measurements

Tumor segmentation was automated using the deep learning application nnU-Net, which is a self-adapting framework for U-net-based segmentation [8,15]. Input to nnU-Net were full 3D T_2-weighted images, output was binary segmentation maps. Patients were randomly split into 5 groups for 5-fold cross-validation [16]. This means that every patient is used as an independent test once and as part of the training data five times. For each fold, a separate nnU-Net training was performed, after which the results of each test group were combined for the final analysis. Differences in baseline characteristics between folds were tested using the chi-squared test to confirm an unbiased split.

A connected component analysis was performed on the output of the nnU-Net segmentation. In unilateral cases, only the biggest connected component was used as tumor segmentation in the next analysis steps. In bilateral cases, the components were visually assessed and manually coupled with the correct tumor label. Figure 1 shows an overview of the deep learning workflow.

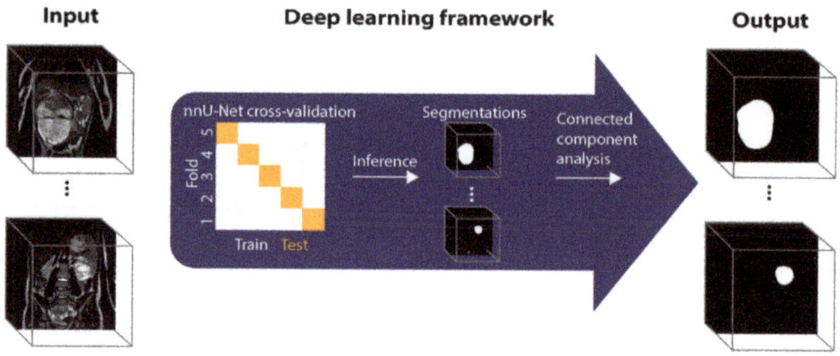

Figure 1. Schematic overview of the deep learning image analysis workflow. All analyses are performed in 3D. Images in the figure are 2D schematic representations of 3D volumes.

2.4. Statistical Analysis

To compare the radiological tumor volume measurement with the manual segmentation based tumor volume measurement, each tumor was analyzed separately. Agreement was assessed using Bland–Altman analyses [17,18]. Tumors were divided into three groups to compare the absolute difference between the groups and the percentage-wise differences.

To assess the performance of the automated deep learning-based tumor volume measurement, automated segmentations were compared to the manual reference segmentations using the Dice similarity coefficient (Dice) and the 95th percentile of the Hausdorff distance (HD95) [19]. The Dice score corresponds with overlay of two segmentation methods. A Dice score of 1.0 is a perfect overlay. The HD95 represents the 95th percentile of the maximum distances between points in border of the two segmentations of the same tumor. A higher HD95 represents a worse comparison. The performance was assessed for all the five independent test sets of the 5-fold cross-validation combined. Differences in segmentation outcomes between different subgroups were tested for statistical significance using Kruskal–Wallis and Mann–Whitney tests. p-values < 0.05 were considered statistically significant. Statistical analyses were performed using Python 3.8 and SciPy 1.7.1.

3. Results

3.1. Patients

Of the 45 patients, 22 were male (49%), and 23 were female (51%). The median age at diagnosis was 39 months (range: 7–109 months) (Table 2). In six patients (13%), bilateral disease was present. One patient (2.2%) had two tumors in one kidney. In total, 52 tumors were separately analyzed (Table 3). The most common histological subtypes consisted of mixed type (n = 13) or regressive (n = 14) type Wilms tumor. Five tumors, all in bilateral patients, were not surgically resected. Four of them were not resected because the lesion was a radiological suspect for a nephrogenic rest. The fifth lesion was not resected because the patient died before surgery of the second kidney was pursued. For these tumors, histological tumor type could not be determined.

Table 2. Baseline characteristics (n = 45). Numbers are count (percentage) unless otherwise specified.

Characteristic		Value
Median age at diagnosis in months (min–max)		39 (7–109)
Gender	Male	22 (49%)
	Female	23 (51%)
Tumor localization	Bilateral	6 (13%)
	Left	16 (36%)
	Right	23 (51%)

Table 3. Overview of analyzed tumors (*n* = 52). Numbers are count (percentage) unless otherwise specified. For 5 tumors, histological type was unknown because these tumors were not resected.

Characteristics		Value
Histological tumor type	Regressive	14 (27%)
	Non-regressive	
	- Mixed	13 (25%)
	- Stromal	9 (17%)
	- Epithelial	2 (4%)
	- Blastemal	1 (2%)
	Diffuse anaplasia	5 (10%)
	Completely necrotic	2 (4%)
	Nephrogenic rest	1 (2%)
	Unknown	5 (10%)
Median radiological volume [mL] (range)		215 (0.68–1774)

3.1.1. Tumor Volume Measurements

The median tumor volume for manual segmentation-based measurements was 200 mL (range: 0.78–1757 mL (Figure 2). The median tumor volume in patients with bilateral disease was 8 mL (range: 0.78–1105 mL), in contrast with a median tumor volume of 307 mL (range: 3–1757 mL) for unilateral tumors.

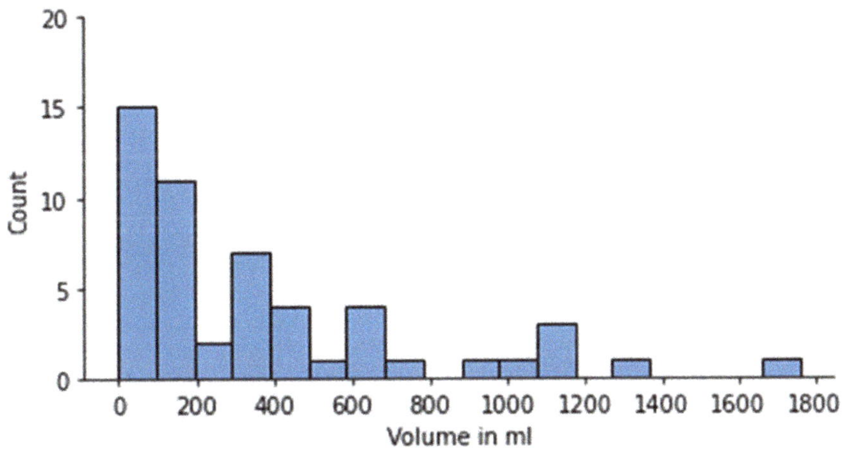

Figure 2. Volumes as assessed by review radiologist.

3.1.2. Radiological versus Manual Segmentation-Based Tumor Volume Measurements

The median difference between radiological tumor volume measurements and segmentation-based tumor volume measurements was 14 mL (range: −81 mL–+175 mL) (Figure 3). This corresponds to a median underestimation of 10% (range: −28.6%–+52.6%). In 38 of 52 tumors (73%), the clinically used radiological tumor volume measurements had a smaller volume compared to the manual segmentation-based volume measurements. For two tumors (4%), the volume measurements yielded, the same result when rounded to the nearest mL. In nineteen tumors (23%), radiological tumor volume measurements had a larger tumor volume.

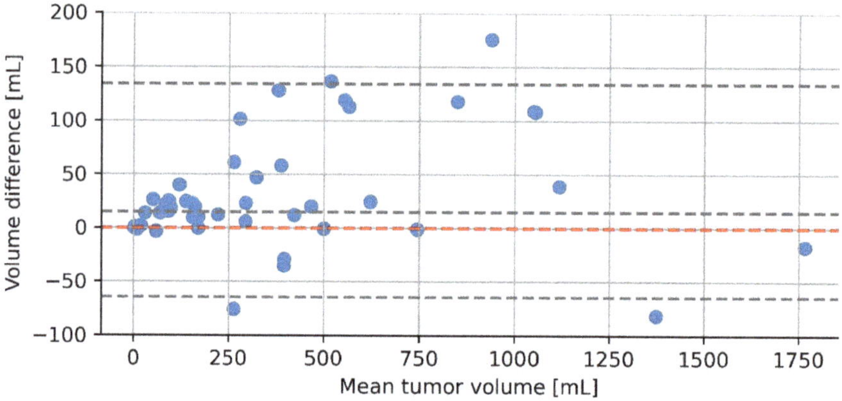

Figure 3. Radiological vs. manual-based tumor volumes. The volume difference plotted on the y-axis was calculated by extracting the radiological tumor volumes from the manually measured reference volume. The red line corresponds to no difference with the manually determined reference volume. Because radiological tumor volume was subtracted from the reference volume, a point above the red line presents an underestimation of tumor volume by the radiologist while points below the line represent an overestimation of tumor volume. In grey, the median and 2.5–97.5 percentile are plotted.

When the tumors were divided into three groups based on their manual volume (0–300, 300–500 mL, and >500 mL) there was a significant increase in the absolute difference between the radiological and manual segmentation volume. However, the difference in the percentage of the volume was the same for the three groups (please refer to Table 4).

Table 4. Differences between radiological and manual segmentation volume in absolute numbers and percentages.

Volume Tumor	Absolute Difference (Mean)	*p*-Value	Percentage Difference (Mean)	*p*-Value
0–300 mL	5.6	0.01	11.9	0.95
300–500 mL	21.5		9.1	
>500 mL	70.2		9.2	

3.1.3. Deep Learning-Based Segmentation

When comparing the manual and automated segmentation for each segmented tumor ($n = 52$), the median non-zero Dice was 0.90 (range: 0.03–0.97) (0.89 for fold 1, 0.9 for fold 2, 0.90 for fold 3, 0.95 for fold 4, 0.89 for fold 5, folds did not differ significantly). The median Hausdorff distance was 7.2 mm (5.5 for fold 1, 7.2 for fold 2, 11.3 for fold 3, 7.0 for fold 4, and 8.0 for fold 5. There was no significant difference between the different folds). Segmentations for the patient with both the highest and lowest Dice are depicted in Figure 4. The median HD95 was 7.2 mm (range: 1.7–27 mm).

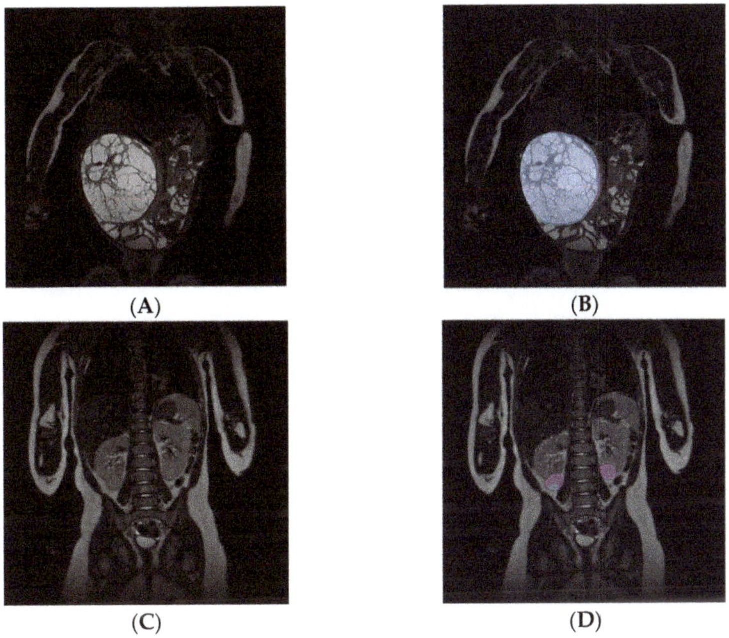

Figure 4. Examples of automated Wilms tumor segmentation. Patients with the highest Dice score (top row (A): MRI, (B): MRI with measurements; Dice score = 0.97) and the lowest overall Dice score (bottom row (C): MRI, (D) MRI with measurements; Dice score = 0.26 for the right kidney tumor, Dice score = 0 for the left kidney tumor as it was missed). The reference manual segmentation is shown in purple, and the deep learning segmentation is projected on the manual segmentation in blue. Note that in (B) these regions have an almost complete overlap.

Deep learning-based segmentation failed to segment three tumors (6%). Each of the missed tumors belonged to the subset of patients with bilateral disease. The missed tumors were relatively small: the manual segmentation-based volumes of the missed tumors were 1.2, 1.9, and 5.7 mL.

Dice scores were significantly lower in patients with bilateral disease. In these patients, the median Dice was 0.79, whereas this was 0.91 for patients with unilateral disease. The median HD95 was 9.3 mm compared with 7.0 mm, which was not significant ($p = 0.43$).

Tumors smaller than 300 mL had a significantly lower median Dice of 0.75 compared with tumors between 300 and 500 mL with a median Dice of 0.90 and above 500 mL with a median Dice of 0.92. The Dice did not significantly differ between tumors of 300 to 500 mL and those above 500 mL ($p = 0.19$). HD95 did not depend on tumor volume ($p = 0.99$).

No significant difference was observed among the various histological subtypes of Wilms tumors in Dice scores ($p = 0.11$) and HD95 ($p = 0.85$). In addition, no difference was found in the Dice score ($p = 0.49$) and HD95 ($p = 0.55$) between patients who received neoadjuvant chemotherapy and those who did not, nor between patients with an age higher or lower than the median age (Dice $p = 0.057$, HD95 $p = 0.27$).

4. Discussion

Our results indicate that conventional radiological tumor volume assessment underestimates tumor volume in the majority of cases when compared to manual segmentation. We further showed that deep learning can be used to replace time-consuming manual segmentation.

Radiological tumor volume measurements assume that the shape of the tumor is an ellipsoid, whereas manual segmentation-based volume measurements are based on the number of tumor voxels. This is a direct method of measuring without assumptions concerning the shape of the tumor. We indeed found an average underestimation of 10% for radiological tumor volume measurements when compared to manual segmentation-based volume measurements.

The literature in adult oncology supports our findings that geometry-based measurements are less accurate. For example, Colvin et al. showed that ellipsoid volume measurements overestimate total prostate volume [10]. Furthermore, Tiruman et al. showed that ellipsoid-based volume measurements were incorrect when the shape of a tumor was irregular [8]. Müller et al. found an average underestimation of 22% in Wilms tumor volumes when measured as an ellipsoid compared to human expert annotations (n = 17 patients) [11]. Our paper, which is the largest to date regarding the number of Wilms tumor patients, also shows an underestimation and provides a deep learning framework to accurately quantify Wilms tumor volume. Together, this inaccuracy of geometry-based tumor volume measurements is consistent with our findings.

Clinically, two volume thresholds are of importance. First, the Society of Paediatric Oncology–Renal Study Tumor Group (SIOP-RTSG) UMBRELLA guidelines consider localized intermediate risk (non-stromal and non-epithelial) stage II and III tumor types high risk when the measured pre-operative volume after chemotherapy is >500 mL [5]. In our cohort, two patients (4.5%) had radiological volume measurements of <500 mL, but >500 mL when using manual segmentation-based volume measurements, which would place these patients in a higher risk group. Our research suggests that if segmentation-based volume measurements become clinically implemented, children can be treated more accurately. Obviously, the prognostic value of volume-based assessment by manual or automated assessment needs to be confirmed based on post hoc analyses before implementation in clinical practice.

The other category of interest is the children eligible for nephron-sparing surgery (NSS), which is currently only considered to be safe for patients with a unilateral tumor with a volume of <300 mL on the diagnostic MRI scan. Two patients (4.5%) were eligible for NSS when measured by the radiologist but crossed the threshold of 300 mL tumor volume when measured manually segmentation-based. Although these results were based on post-chemo MRI scans, we expect that the same applies to diagnostic MRI scans. Together, these results show that the method of tumor volume measurement has potential clinical consequences in clinical decision-making.

In addition to volume measurements, other factors are of importance in staging and risk stratification of Wilms tumor. Ma et al. showed that they could accurately predict WT stage 1 on CT research on the automatic staging of WT on MRI is needed [20].

Although manual segmentation-based volume measurements seem to be more accurate, this process is time-consuming, ranging from one to multiple hours, and is observer dependent [11,21,22]. Automating segmentation using deep learning would limit these constraints. Deep learning-based segmentation was accurate in most of the cases, with a median Dice of 0.90. The highest Dice was 0.97, an almost perfect overlap between the manual and automated segmentation. In contrast, the lowest overall Dice score was 0.26 for one side and 0 for the other side, which is insufficient for future clinical applications. Figure 4 shows both of these patients together. The first thing that stands out is the big difference in tumor volume between these patients. Our analysis indeed showed a significantly lower Dice score for tumors below <300 mL. This can partially be explained by the Dice being a volume-dependent measurement [23], but the lower Dice scores were not merely an artifact of the evaluation metric. Several explanations may exist: first, bilateral tumors were typically smaller (median 8 mL) than unilateral tumors (median 307 mL) and the number of bilateral tumors was smaller in our training set (n = 6). Therefore, our segmentation method might be biased toward large unilateral tumors. We will investigate this in future studies in a larger dataset with an increased number of small unilateral and bilateral tumors separately. Second, the appearance of the small tumors can be visually different, which is exemplified in Figure 4. Both patients in Figure 4 had a tumor of mixed type, but visually these tumors are not the same. The unilateral tumor with the high Dice score is big, hyperintense, and clearly separable from the surrounding kidney tissue. In contrast, the two tumors in the bilateral case are small and of roughly the same intensity as the surrounding kidney tissue. For all other clinical characteristics, we did not find a significant difference in the Dice score or the 95th percentile of the Hausdorff distances.

The differences between manual segmentation and radiological volume measurements may apply to many different types of tumors. We decided to conduct this research on WT patients first because of the importance of volumetric assessment during treatment. Furthermore, in our hospital the solid tumor diagnosed the most is WT. So, this new method of volume measurements will have clinical implications rather quickly. Although the automated segmentation method was developed specifically for Wilms tumor patients, the general methodology can easily be extrapolated to other tumor types. We hypothesize that our conclusion about the radiological volume measurements holds across tumor types. The developed automated segmentation method can easily be retrained for other tumor types, both in the pediatric and adult populations. In future research, we will extend our framework to other pediatric tumor types such as neuroblastoma. This study had some limitations. First, we had a limited sample size. This is, however, the largest unselected national MRI cohort to date including all pediatric Wilms tumor patients in the Netherlands in the period of 2018–2020 [2,24]. Second, although we divided the tumors over three observers for segmentation, the population of tumors and the techniques are derived from a single-institution study. We plan to validate our findings in a large prospective multicenter cohort study.

We showed that clinically used volume measurements in Wilms tumor are not accurate. Changing the method of volume measurements might impact clinical decision-making. Further research has to be conducted to investigate the clinical impact of our findings.

5. Conclusions

Our results show that currently used geometry-based radiological tumor volume measurements by review radiologists underestimate tumor volume in Wilms tumors compared to manual segmentation. This underestimation was about 10%, independent of tumor size. A change in volume measurement technique may have consequences for the stratification of treatment in individual children with a Wilms tumor and thus have clinical implications.

Deep learning shows the potential to replace manual segmentation in this process to utilize the benefits of segmentation-based volume measurements while limiting time and observer constraints. This method of volume measurements using manual or semi-automated segmentation may be extrapolated to other (pediatric) tumor types.

Supplementary Materials: The following supporting information can be downloaded at: https://www.mdpi.com/article/10.3390/cancers15072115/s1, Table S1: Review of postoperative treatment for localized Wilms tumor in UMBRELLA SIOP-RTSG 2016.

Author Contributions: Conceptualization, M.A.D.B., A.F.W.v.d.S., M.H.W.A.W. and B.H.M.v.d.V.; methodology, M.A.D.B., A.F.W.v.d.S. and B.H.M.v.d.V.; software, M.A.D.B. and M.F.; validation, M.A.D.B., A.S.L. and B.H.M.v.d.V.; formal analysis, M.A.D.B. and H.v.T.; investigation, M.A.D.B. and M.F.; resources, M.H.W.A.W., M.M.v.d.H.-E. and A.S.L.; data curation, M.A.D.B. and M.F.; writing—original draft preparation, M.A.D.B.; writing—review and editing, A.F.W.v.d.S., M.H.W.A.W., M.F., H.v.T., M.M.v.d.H.-E., A.S.L. and B.H.M.v.d.V.; visualization, A.F.W.v.d.S., M.H.W.A.W. and B.H.M.v.d.V.; supervision, A.F.W.v.d.S., M.H.W.A.W. and B.H.M.v.d.V. All authors have read and agreed to the published version of the manuscript.

Funding: This research received no external funding.

Institutional Review Board Statement: The study was conducted in accordance with the Declaration of Helsinki. Ethical review and approval were waived for this study due to no direct involvement of patients in this research.

Informed Consent Statement: Informed consent was obtained from all subjects involved in the study.

Data Availability Statement: The datasets analysed during the current study are not publicly available due to patient privacy. Upon reasonable request to the corresponding author the raw data can be made available, however the imaging itself will not be shared.

Acknowledgments: We would like to thank Heleen van den Nieuwenhof for her help with the manual segmentation.

Conflicts of Interest: The authors declare no conflict of interest.

References

1. Spreafico, F.; Bellani, F.F. Wilms' tumor: Past, present and (possibly) future. *Expert Rev. Anticancer Ther.* **2006**, *6*, 249–258. [CrossRef] [PubMed]
2. Roy, P.; van Peer, S.E.; de Witte, M.M.; Tytgat, G.A.M.; Karim-Kos, H.E.; van Grotel, M.; van de Ven, C.P.; Mavinkurve-Groothuis, A.M.C.; Merks, J.H.M.; Kuiper, R.P.; et al. Characteristics and outcome of children with renal tumors in the Netherlands: The first five-year's experience of national centralization. *PLoS ONE* **2022**, *17*, e0261729. [CrossRef] [PubMed]
3. SIOP Renal Tumour Study Group. Paediatric renal tumours: Perspectives from the SIOP-RTSG. *Nat. Rev. Urol.* **2017**, *14*, 3–4. [CrossRef] [PubMed]
4. Vujanić, G.M.; Gessler, M.; Ooms, A.H.A.G.; Collini, P.; Coulomb-l'Hermine, A.; D'Hooghe, E.; de Krijger, R.R.; Perotti, D.; Pritchard-Jones, K.; Vokuhl, C.; et al. The UMBRELLA SIOP–RTSG 2016 Wilms tumour pathology and molecular biology protocol. *Nat. Rev. Urol.* **2018**, *15*, 693–701. [CrossRef]
5. Van den Heuvel-Eibrink, M.M.; Hol, J.A.; Pritchard-Jones, K.; van Tinteren, H.; Furtwängler, R.; Verschuur, A.C.; Vujanic, G.M.; Leuschner, I.; Brok, J.; Rübe, C.; et al. Rationale for the treatment of Wilms tumour in the UMBRELLA SIOP–RTSG 2016 protocol. *Nat. Rev. Urol.* **2017**, *14*, 743–752. [CrossRef]
6. Provenzi, V.O.; Rosa, R.F.M.; Rosa, R.C.M.; Roehe, A.V.; dos Santos, P.P.A.; Faulhaber, F.R.S.; de Oliveira, C.A.V.; Zen, P.R.G. Tumor size and prognosis in patients with Wilms tumor. *Rev. Paul. Pediatr.* **2015**, *33*, 82–87. [CrossRef]
7. Joseph, L.L.; Boddu, D.; Srinivasan, H.N.; Regi, S.S.; Antonisamy, B.; Matthew, L.G.; Totadri, S. Postchemotherapy tumor volume as a prognostic indicator in Wilms tumor: A single-center experience from South India. *Pediatr. Blood Cancer* **2022**, *69*, e29454. [CrossRef]
8. Tirumani, S.H.; Shinagare, A.B.; O'Neill, A.C.; Nishino, M.; Rosenthal, M.H.; Ramaiya, N.H. Accuracy and Feasibility of Estimated Tumour Volumetry in Primary Gastric Gastrointestinal Stromal Tumours: Validation Using Semi-automated Technique in 127 Patients. *Eur. Radiol.* **2016**, *26*, 286–295. [CrossRef]
9. Kim, H.J.; Kim, W. Method of tumor volume evaluation using magnetic resonance imaging for outcome prediction in cervical cancer treated with concurrent chemotherapy and radiotherapy. *Radiat. Oncol. J.* **2012**, *30*, 70–77. [CrossRef]
10. Colvin, R.; Walker, D.; Hafron, J.; Seifman, B.; Nandalur, S.; Gangwish, D.; Nandalur, K.R. Which measurement method should be used for prostate volume for PI-RADS? A comparison of ellipsoid and segmentation methods. *Clin. Imaging* **2021**, *80*, 454–458. [CrossRef]
11. Müller, S.; Farag, I.; Weickert, J.; Braun, Y.; Lollert, A.; Dobberstein, J.; Hötker, A.; Graf, N. Benchmarking Wilms' tumor in multisequence MRI data: Why does current clinical practice fail? Which popular segmentation algorithms perform well? *J. Med. Imaging* **2019**, *6*, 034001. [CrossRef] [PubMed]
12. Litjens, G.; Kooi, T.; Bejnordi, B.E.; Setio, A.A.A.; Ciompi, F.; Ghafoorian, M.; van der Laak, J.A.W.M.; van Ginneken, B.; Sánchez, C.I. A survey on deep learning in medical image analysis. *Med. Image Anal.* **2017**, *42*, 60–88. [CrossRef] [PubMed]
13. Vujanić, G.M.; Sandstedt, B.; Kelsey, A.; Sebire, N.J. Central pathology review in multicenter trials and studies. *Cancer* **2009**, *115*, 1977–1983. [CrossRef] [PubMed]
14. Fedorov, A.; Beichel, R.; Kalpathy-Cramer, J.; Finet, J.; Fillion-Robin, J.-C.; Pujol, S.; Bauer, C.; Jennings, D.; Fennessy, F.; Sonka, M.; et al. 3D Slicer as an image computing platform for the Quantitative Imaging Network. *Magn. Reson. Imaging* **2012**, *30*, 1323–1341. [CrossRef]
15. Ronneberger, O.; Fischer, P.; Brox, T. U-Net: Convolutional Networks for Biomedical Image Segmentation. In *Medical Image Computing and Computer-Assisted Intervention–MICCAI 2015: 18th International Conference, Munich, Germany, 5–9 October 2015, Proceedings, Part III 18*; Springer: Berlin/Heidelberg, Germany, 2015; Volume 9351, pp. 234–241.
16. Berrar, D. Cross-Validation. *Ref. Modul. Life Sci.* **2019**, *1*, 542–545. [CrossRef]
17. Altman, D.G.; Bland, J.M. Measurement in Medicine: The Analysis of Method Comparison Studies. *J. R. Stat. Soc. Ser. Stat.* **1983**, *32*, 307–317. [CrossRef]
18. Dewitte, K.; Fierens, C.; Stöckl, D.; Thienpont, L.M. Application of the Bland–Altman Plot for Interpretation of Method-Comparison Studies: A Critical Investigation of Its Practice. *Clin. Chem.* **2002**, *48*, 799–801. [CrossRef]
19. Taha, A.A.; Hanbury, A. Metrics for evaluating 3D medical image segmentation: Analysis, selection, and tool. *BMC Med. Imaging* **2015**, *15*, 29. [CrossRef]
20. Ma, X.H.; Shu, L.; Jia, X.; Zhou, H.C.; Liu, T.T.; Liang, J.W.; Ding, Y.S.; He, M.; Shu, Q. Machine Learning-Based CT Radiomics Method for Identifying the Stage of Wilms Tumor in Children. *Front Pediatr.* **2022**, *10*, 873035. [CrossRef]

21. Fitski, M.; Meulstee, J.W.; Littooij, A.S.; van der Ven, C.P.; van der Steeg, A.F.W.; Wijnen, M.H.W.A. MRI-Based 3-Dimensional Visualization Workflow for the Preoperative Planning of Nephron-Sparing Surgery in Wilms' Tumor Surgery: A Pilot Study. *J. Healthc. Eng.* **2020**, *2020*, 8899049. [CrossRef]
22. Chaussy, Y.; Vieille, L.; Lacroix, E.; Lenoir, M.; Marie, F.; Corbat, L.; Henriet, J.; Auber, F. 3D reconstruction of Wilms' tumor and kidneys in children: Variability, usefulness and constraints. *J. Pediatr. Urol.* **2020**, *16*, 830.e1–830.e8. [CrossRef] [PubMed]
23. Reinke, A.; Tizabi, M.D.; Sudre, C.; Eisenmann, M.; Rädsch, T.; Baumgartner, M.; Acion, L.; Antonelli, M.; Arbel, T.; Bakas, S.; et al. Common Limitations of Image Processing Metrics: A Picture Story. *arXiv* **2021**, arXiv:2104.05642.
24. Van Peer, S.E.; Hol, J.A.; van der Steeg, A.F.W.; van Grotel, M.; Tytgat, G.A.M.; Mavinkurve-Groothuis, A.M.C.; Janssens, G.O.R.; Littooij, A.S.; de Krijger, R.R.; Jongemans, M.C.J.; et al. Bilateral Renal Tumors in Children: The First 5 Years' Experience of National Centralization in The Netherlands and a Narrative Review of the Literature. *J. Clin. Med.* **2021**, *10*, 5558. [CrossRef] [PubMed]

Journal of
Clinical Medicine

MDPI

Article

Textural Features of MR Images Correlate with an Increased Risk of Clinically Significant Cancer in Patients with High PSA Levels

Sebastian Gibala [1], Rafal Obuchowicz [2,*], Julia Lasek [3], Zofia Schneider [3], Adam Piorkowski [4], Elżbieta Pociask [4] and Karolina Nurzynska [5]

1 Urology Department, Ultragen Medical Center, 31-572 Krakow, Poland
2 Department of Diagnostic Imaging, Jagiellonian University Medical College, 31-501 Krakow, Poland
3 Faculty of Geology, Geophysics and Environmental Protection, AGH University of Science and Technology, 30-059 Krakow, Poland
4 Department of Biocybernetics and Biomedical Engineering, AGH University of Science and Technology, 30-059 Krakow, Poland
5 Department of Algorithmics and Software, Silesian University of Technology, 44-100 Gliwice, Poland
* Correspondence: rafalobuchowicz@su.krakow.pl

Abstract: Background: Prostate cancer, which is associated with gland biology and also with environmental risks, is a serious clinical problem in the male population worldwide. Important progress has been made in the diagnostic and clinical setups designed for the detection of prostate cancer, with a multiparametric magnetic resonance diagnostic process based on the PIRADS protocol playing a key role. This method relies on image evaluation by an imaging specialist. The medical community has expressed its desire for image analysis techniques that can detect important image features that may indicate cancer risk. Methods: Anonymized scans of 41 patients with laboratory diagnosed PSA levels who were routinely scanned for prostate cancer were used. The peripheral and central zones of the prostate were depicted manually with demarcation of suspected tumor foci under medical supervision. More than 7000 textural features in the marked regions were calculated using MaZda software. Then, these 7000 features were used to perform region parameterization. Statistical analyses were performed to find correlations with PSA-level-based diagnosis that might be used to distinguish suspected (different) lesions. Further multiparametrical analysis using MIL-SVM machine learning was used to obtain greater accuracy. Results: Multiparametric classification using MIL-SVM allowed us to reach 92% accuracy. Conclusions: There is an important correlation between the textural parameters of MRI prostate images made using the PIRADS MR protocol with PSA levels > 4 mg/mL. The correlations found express dependence between image features with high cancer markers and hence the cancer risk.

Keywords: prostate cancer; MRI; PSA; textural analysis; multiple-instance learning; support vector machine

Citation: Gibala, S.; Obuchowicz, R.; Lasek, J.; Schneider, Z.; Piorkowski, A.; Pociask, E.; Nurzynska, K. Textural Features of MR Images Correlate with an Increased Risk of Clinically Significant Cancer in Patients with High PSA Levels. *J. Clin. Med.* **2023**, *12*, 2836. https://doi.org/10.3390/jcm12082836

Academic Editor: Enrico G. Caiani

Received: 17 March 2023
Revised: 6 April 2023
Accepted: 11 April 2023
Published: 12 April 2023

1. Introduction

Prostate cancer is the second most frequently diagnosed cancer in men, with approximately 1.4 million diagnoses worldwide in 2020 [1,2]. The basic methods used in the early diagnosis of prostate cancer are laboratory tests, and, in particular, the determination of the PSA tumor marker. Since its discovery in 1979 and through its clinical use in the 1980s and 1990s, the prostate-specific antigen (PSA) has evolved into an invaluable tool for the discovery, assessment, and monitoring of prostate cancer in men. Increased serum PSA levels are likely a product of disturbed cellular architecture in the prostate gland [3]. PSA is believed to correlate highly with prostate disease; it is produced by many glands but is also associated with non-malignant diseases [4]. Moreover, the influence of hormonal therapy influences the detection of prostate cancer [5]. After the introduction of MR scanners to medical practice, the first attempt to diagnose prostate glands was made by Damadian in the early 1970s; ten years later, it was recognized as a valuable tool for the process of diagnosing prostate diseases [6,7]. However, the cost of these examinations has raised some doubts [8] as to overdiagnosis [9]; the role of MR is well established in the process of diagnosing prostate diseases [10], especially after the introduction of diffusion-weighted techniques and contrast enhancement to the protocol [11,12]. The experience gained over the years of using of MRI techniques in prostate imaging, especially where correlated with biopsies [13], resulted in the formation of imaging standards, PIRADS 1 and PIRADS 2, where the PIRADS 4 and PIRADS 5 stages are linked to a high probability of the presence of cancer and associated with high PSA levels [14]. MRI has taken on an increasingly significant role in the diagnostic process with the development of computed systems for image segmentation and analysis, and algorithms for prostate tumor detection have been proposed [15].

However, manual analysis is laborious and dependent on well-trained specialists in the field of diagnostic imaging. Therefore, efforts have been made to develop a computer-aided diagnosis (CAD) tool. It has been proven that automatic approaches improve the specificity, as compared to moderately experienced radiologists, for more difficult tumors [16] and allow for global standardization across radiological centers [17]. The described CAD systems use various modalities, from T2-weighted MRI (T2W-MRI) [18] to prostate biomarkers, and diffusion-weighted MRI (dw-MRI) supported with level-sets methods for prostate delineation [19].

Grayscale MR images are the sum of pixels of different brightness and grayscale levels. Patterns formed by sub-sets of similar pixels can be differentiated with the use of texture maps where pixel distribution changes can be characterized with the application of mathematical formulas [20–22]. A plethora of textural features that cannot be recognized by the human eye can be characterized on the basis of mathematical analyses. This encoded information can be revealed by texture feature maps [22].

Magnetic resonance imaging (MRI) is a non-invasive method of inspecting internal bodily structures. The high quality of the gathered data allows specialists to determine whether the observed tissue corresponds to clinically significant or non-significant prostate cancer [23,24].

Over time, the use of artificial intelligence (AI) methods to enable the creation of a CAD system has been widely investigated. In their review, Booven et al. [25] focus on artificial neural networks (ANNs) as a method of analyzing a variety of data connected with prostate cancer detection.

Starting with prostate-specific antigen (PSA) value analysis, rather than other medical markers, and developing an understanding of MRI images, biomarker diagnosis, histopathological data utilization, etc., the authors conclude that deriving textural features is the most promising approach for prostate cancer description.

Several other reviews support this claim. For instance, Harmon et al. [26] show that AI enables a correlation between pathological and radiological information about the patient when multiparameter MRI (mp-MRI) is considered. Describing the mp-MRI with radiomics (quantitative imaging of features that are poorly recognized by the human eye, but are understandable for computers) supported with AI algorithms allows for the prediction of cancer aggressiveness [27] or general prostate cancer analysis [28]. The influence of deep learning models, which have recently gained popularity, on healthcare professionals working with MRI has been investigated by Liu et al. [29]; meanwhile, Wildeboer et al. [30] concentrated on building a CAD system with this technology.

A significant amount of research addresses the problem of determining the Gleason score using the textural features calculated for various modalities of MRI. Wibmer et al. [31] build a binary classifier using Haralick textural features [32] for the determination of cancerous and non-cancerous prostate images. A similar approach shows that textural features derived from T2W-MRI, dw-MRI, and dynamic contrast-enhanced MRI (DCE) effectively distinguish the Gleason scores [33–35]. More recent research that is still based on first- and second-order histogram features distinguishes up to five grades in the Gleason Grade Group [36,37].

Radiomics features are used to determine the aggressiveness of prostate tumors [38,39]. When calculated for types of mp-MRI including T2W-MRI, dw-MRI, apparent diffusion coefficient imaging (ADC), and DCE, its performance is comparable with the Prostate-Imaging Reporting and Data System (PIRADS) [40–42]. On the other hand, when radiomics features are supported by additional medical risk factors, a model based on multivariate logistic regression allows for the classification of clinically significant and non-significant prostate cancers [43,44]. Using mp-MRI data for feature calculation is justified in previous research where support vector machine (SVM) models based on mp-MRI outperformed those constructed using T2W-MRI or ADC data only.

Recently, deep learning approaches have become a popular method of designing models for building a CAD system [45], detecting and grading prostate cancer [46], and differentiating clinically significant and non-significant prostate cancers [47,48]. In the last case, the performance was comparable to that of the PIRADS system.

There are also methods for automated cancer detection based on T2W-MRI data only [49,50], which perform the automatic grading of prostate biopsy specimens in correlation with Gleason grades based on SVM models [51]. There are also solutions based on textural features derived from mp-MRI for detecting transition zones in prostate tumors [52], and models that use bi-parametric MRI texture analysis for detecting and evaluating high-grade prostate cancer [53].

2. Materials and Methods

2.1. Method Overview

The study protocol was developed according to the Declaration of Helsinki and the Declaration of Good Clinical Practice [54]. All images were anonymized prior to processing to ensure the security of personal data. In addition, written consent from the Local Ethics Committee was obtained to conduct this study, 1072.6120.21.22 (23 February 2022). The study used data from 125 patients aged 27 to 87 years. MRI images were acquired (1.5 T Siemens Avanto, Enlargen, Germany) during normal diagnostic procedures, which met the standard of the PIRADS protocol with the presence of T2-weighted axial sequences with 2 mm slices and a distance factor of 0 and diffusion-weighted sequences with the use of a single shot echo planar sequence (EPI) made with b value equal to 0–800–1500 and a distance factor of 0. Finally, T1-weighted sequences were designed for post-contrast evaluation. The parameters of the T2-weighted sequence utilized in the present study are summarized in Table 1. Images were converted from 12 bits to 8 bits by linearly scaling from min to a range of 0–255. Then, images that met the stable (repeatable) conditions were selected for the analysis of the texture features. After analyzing clinical data, information on PSA value was obtained for 41 patients (values 0.39–19 ng/mL) (see Appendix A).

Table 1. Applied MR protocol in line with PIRADS guidance.

	TE	TR	FA	MX	VOX	FOV	CON	AV	PAT	DF	T
T2 (axial)	105	3320	160	256 × 320	0.6 × 0.6 × 2	200	2	4	2	0	300

TE—time echo, TR—relaxation time, FA—flip angle, MX—imaging matrix, VOX—voxel, FOV—field of view, CON—concentrations, AV—averages, PAT—Parallel acquisition, DF—distance factor, T—overall sequence time.

In the present work, we aimed to verify whether it is possible to distinguish with a high probability between patients with prostate cancer and healthy patients. The content of T2W-MRI scans was analyzed, while the PSA was the reference. For the MRI scans, the textural features were calculated using the MaZda software. For the data processing, manually annotated masks were used. As a result, the texture features were calculated for three different regions: the inner prostate only, the outer prostate only, and the inner and outer prostate regions treated as a whole. In order to create a binary classification model, multiple-instance learning (MIL) with a support vector machine (SVM) was adopted. This approach corresponds well with the characteristics of the collected dataset, where one

PSA value is known for a patient described with several MRI scans. At the same time, the cancerous tissue may be visible only in one part. It was assumed that PSA = 1, 2, 3 correspond to a healthy person (label = 0) and PSA $\geq$ 4 is for an ill one (label 1). Finally, the leave-one-out methodology was applied.

2.2. Prostate MRI Dataset

From the initial dataset, 326 MRI images of prostates collected from 41 patients were selected, in which the images were free of imaging artefacts and segmentation and visualization of the lesions were possible despite the rigorous PIRADS protocol. The number of projections per patient, which include the prostate, varies between 4 and 11 scans. Figure 1 shows the histogram of the number of MRI scans per patient. Each scan is supported with a manual annotation marking the inner (blue) and outer (orange) regions of the prostate, as depicted in Figure 2. For one patient, the outer prostate region was not annotated because this anatomical structure was absent. The mean age was 64.6 $\pm$ 9.8 with median age 65. The mean level of PSA in the cohort was 6.1 $\pm$ 3.9 with the median at 5.14. More detailed information on the cohort is given in Appendix A.

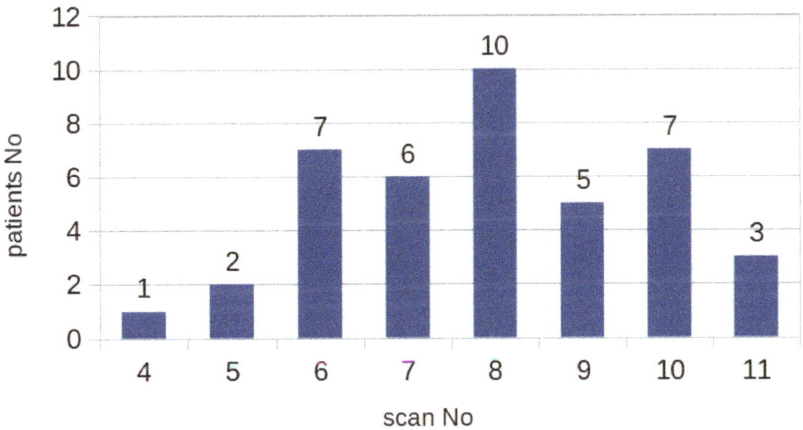

Figure 1. Histogram of the number of patients with similar numbers of MRI scans. The number of analyzed images—slices from different patients varied due to gland size and zone margins. Figure presents groups of patients with different number of slices considered in the study.

The image data with the corresponding segmentations are all publicly available on the website Zenodo [55].

2.3. Textural Features

The MaZda software was used to calculate the textural features for the annotated regions in the MRI scans. This approach was chosen due to the ease of setting the number of significant bits per image as a method parameter. This approach is not straightforward, necessitating the implementation of the *radiomics* Python library, which is also often used for textural feature determination. As we had access to several methods for texture description that may be preprocessed and parameterized in several ways, around 7000 features were calculated to describe a region. A brief overview of the applied methods is presented in this section. Since those methods are well known in the domain of image analysis and are also frequently used as MRI scan descriptors, we refer the reader to the MaZda manual for more details and formulas.

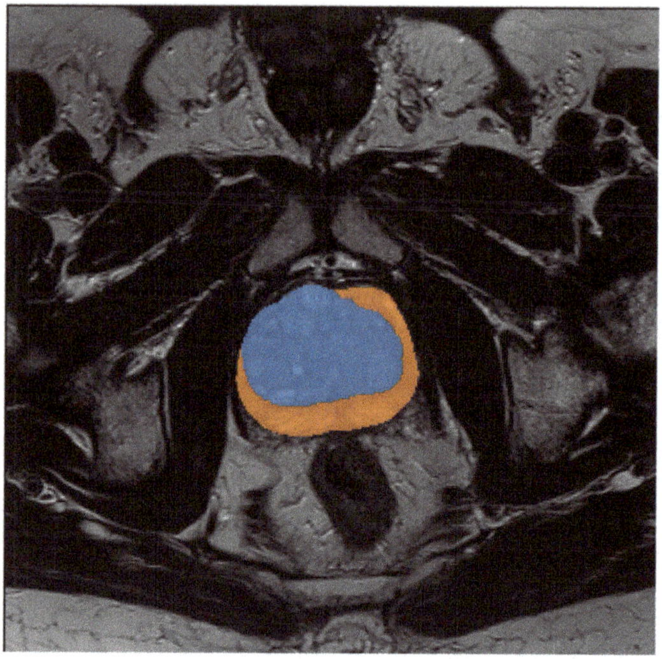

Figure 2. Visualization of the manual annotations prepared for each prostate MRI scan. The inner regions are depicted in blue, while orange shows the outer prostate region.

Before the textural features were calculated, the image color space was converted to YUV. Next, data normalization and quantization were undertaken if necessary. This step diminishes the brightness differences between the analyzed images, which could arise due to differences in the acquisition hardware. The system supports several normalization techniques, the details of which are given in Table 1, and also enables gray-level coding to a chosen number of bits. The preprocessed images were normalized in the region of interest (ROI) according to the methods shown in Table 2; the textural features were calculated using the following methods. The descriptors are derived only for the selected region: the inner prostate, outer prostate, or inner and outer prostate.

Table 2. Normalization methods accessible in MaZda software.

Normalization Method	Description
D	No normalization applied.
S	Gray levels normalized in the range $<\mu \pm 3\sigma>$, where μ—mean gray level value, σ—standard deviation.
M	Linear rescaling of the range to the minimal and maximal values.
N	Linear rescaling of the range defined by the 1st and 99th percentiles of the gray-level histogram.

An image brightness histogram (Hist) is the simplest statistical description of the image content. It allows for the calculation of the following textural features: area, mean, variance, skewness, kurtosis, and percentiles. The gradient map features (Grad) describe the rapid illumination changes in small neighborhoods. This feature map is then transformed into a histogram, allowing us to obtain the following textural features: mean, variance, skewness, kurtosis, and non-zero elements. An autoregressive model (Arm) investigates the influence of neighboring pixels on each other. The method searches for the optimal solution by returning information about directional influence described by four theta parameters and an additional sigma parameter that conveys information about the error standard deviation.

The gray-level co-occurrence matrix method [32] is a more complex approach that concentrates on local neighborhoods. This matrix holds information about the counts of co-occurrence of pixel gray levels next to each other. Then, from this information, several textural features are computed: the angular second moment, contrast, correlation, sum of squares, invariance deformation moment, sum of averages, sum variance, entropy, difference of variance, and difference of entropy. The method acronym is GLCM followed by one letter indicating the direction of runs supplied by the value indicating the distance between pixels considered as neighbors: H—horizontal, V—vertical, Z—diagonal, as shown by the middle part of the letter, N—the other diagonal.

Unlike previously used approaches, the gray-level run-length matrix [56] describes the content by considering a larger region. Here, the image analysis approach reflects the way humans perceive images. The high-quality images have rapid color changes (short runs), while keeping the same colors next to each other (long runs) is characteristic of images of low quality. In this work, the GRLM acronym followed by one letter indicating the direction of runs (H, V, Z, or N) is given. The information about the run length for pixel gray values creates a base matrix from which several textural features are derived: area, short-run emphasis, long-run emphasis, gray-level non-uniformity, mean gray-level non-uniformity, run-length non-uniformity, mean run-length non-uniformity, and fraction.

In local binary patterns (LBPs) [57], the neighborhood of each pixel also influences its understanding. The pixel information is coded considering a circular neighborhood of 4, 8, or 12 pixels. A histogram is prepared from the generated codes, and each of its bins becomes a textural feature described by the LBP or circularity analysis method: over-complete, transition, center-symmetric. This is followed by the number of neighbors.

Another popular method for image description uses histograms of oriented gradients (HOGs) [58]. This technique divides the image into small blocks, for which the gradients are calculated. Their orientations are binned into histograms. For cells (regions composed of several blocks), the histograms are concatenated and normalized to become a feature vector. Using the MaZda software, the HOG method is calculated for the following numbers of bins: 4, 8, 16, and 32.

The frequency components in a local neighborhood are also analyzed using a Gabor transform (Gab). This method is parameterized with the Gaussian envelope, orientation, the period of the sinewave, and the magnitude. On the other hand, the discrete wavelet transform application is also used to derive image information. In this case, the Haar wavelet is used and its energies in the sub-bands become features.

Considering both the large number of methods that can be used to describe the content of MRI prostate scans and the ease of its parameterization, a set of around 7000 features was calculated for each image. Please see the feature naming convention given in Appendix B.

2.4. Multiple-Instance Learning with a Support Vector Machine

In the present problem, several scans are described by one label. Therefore, multiple-instance learning seems to be the best approach to classification. This is a type of supervised learning; however, instead of having a separate label for each sample, all samples with one label are analyzed together and referred to as a bag. Then, the classifier decides that a bag of samples is negative when all samples are negative, and that it is positive when at least one sample within a bag is positive. When applying this method to our problem, the MIL approach returns a "healthy" label if all the MRI scans in the bag are negative, while it returns an "ill" label when at least one MRI scan depicts cancerous tissue. This approach reflects the fact that cancer might be noticed only in one scan, while the other parts are not infected. This approach works with the support vector machine as a binary classifier. For our purposes, the "mil" Python library was used. Figure 3 shows an overview of the presented approach.

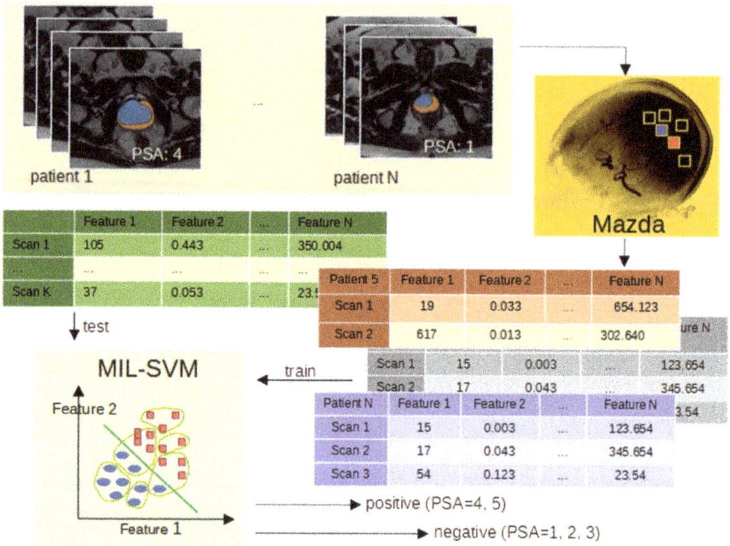

Figure 3. The overview of the presented approach.

3. Results

In this study, three sets of experiments were performed. Each of them was concerned with textural features calculated for different prostate regions annotated on the MRI scans, namely, the inner, outer, and inner and outer regions. Some of the textural features could not be calculated for all scans and have therefore been disregarded. For the rest of the scans, the procedure was as follows. First, the Pearson correlation between pairs of textural features was evaluated. Since the number of features is large (6678 features for the prostate inner region, 4898 for the prostate outer region, and 6718 for the total prostate region) in comparison to the number of samples (41, 40, and 40, respectively), it was decided to prepare the MIL-SVM models using only a pair of non-correlated features. Using a smaller number of representative features is also beneficial because it improves the SVM convergence. We assumed that pairs of features with a correlation higher than 50% should be disregarded (around 10% of all possible combinations). Then, the leave-one-out procedure was adopted to verify the model's generality for each pair of textural features (giving around 60,000/54,000/64,000 possible pairs). This methodology assumes that the model is trained with n-1 samples, using one sample for testing, and that the experiment is repeated n times. The performance of the model was evaluated according to its accuracy, specificity, recall, and F1 score metrics. Both the leave-one-out procedure and SVM had a weight-balanced flag set, since there were 28 positive and 13 negative samples. For SVM, the linear kernel was applied with the regularization factor c = 10.

Tables 3–5 gather the 20 best results recorded for each type of experiment; the outcomes for the remaining pairs of features are attached as Supplementary Materials (Table S1). The highest scores are reached when the inner prostate region is considered (Table 3), with slightly worse outcomes for the whole prostate (Table 2); meanwhile, the outer region of the prostate seems to be slightly less significant (Table 4). However, these differences are very small. It is interesting to note that, depending on the region, different textural features play significant roles. This effect might be induced by the shape of the region of interest, as it is circular in the case of the inner prostate and whole prostate and elongated when only the outer part is analyzed. The elongation with a slight rotation of objects on MRI scans must have influenced the calculation of descriptors, as the computation of many of them depends on the chosen angle. Despite these differences and obstacles, very high F1 scores were recorded: 92.86%, 91.23%, and 89.80% for the inner, whole, and outer prostate regions, respectively.

Table 3. The best 20 pairs of features when the whole prostate region is used to calculate features.

Accuracy	Precision	Recall	F1 Score	Feature 1	Feature 2
0.878049	0.928571	0.896552	0.912281	YN6Gab16H8Mag	YM8HistPerc01
0.878049	0.928571	0.896552	0.912281	YN5Gab16H8Mag	YM8HistPerc01
0.878049	0.928571	0.896552	0.912281	YN8Gab16H8Mag	YM8HistPerc01
0.878049	0.928571	0.896552	0.912281	YN7Gab16H8Mag	YM8HistPerc01
0.878049	0.928571	0.896552	0.912281	YN4Gab16H8Mag	YM8HistPerc01
0.878049	0.892857	0.925926	0.909091	YN6Gab8N4Mag	YS8Gab24N12Mag
0.878049	0.892857	0.925926	0.909091	YS4ArmTeta2	YD8Gab24N12Mag
0.853659	0.857143	0.923077	0.888889	YLbpCs8n15	YM8HistPerc01
0.853659	0.892857	0.892857	0.892857	YM8HistPerc01	YS8Gab12Z6Mag
0.853659	0.892857	0.892857	0.892857	YN5ArmTeta4	YD8GradSkewness
0.853659	0.892857	0.892857	0.892857	YN7ArmTeta4	YD8GradSkewness
0.853659	0.892857	0.892857	0.892857	YN8ArmTeta4	YD8GradSkewness
0.853659	0.928571	0.866667	0.896552	YM4DwtHaarS2HL	YS8HistKurtosis
0.853659	0.857143	0.923077	0.888889	YN5Gab8N4Mag	YS8Gab24N12Mag
0.853659	0.892857	0.892857	0.892857	YS8GlcmZ4SumEntrp	YD8GradSkewness
0.853659	0.892857	0.892857	0.892857	YN8Gab8N4Mag	YS8Gab24N12Mag
0.853659	0.892857	0.892857	0.892857	YN7Gab8N4Mag	YS8Gab24N12Mag
0.853659	0.821429	0.958333	0.884615	YS8ArmTeta2	YD8Gab24N12Mag
0.853659	0.857143	0.923077	0.888889	YS6ArmTeta2	YD8Gab24N12Mag
0.853659	0.857143	0.923077	0.888889	YS5ArmTeta2	YD8Gab24N12Mag

Table 4. The best 20 pairs of features when the inner prostate region is used to calculate features.

Accuracy	Precision	Recall	F1 Score	Feature1	Feature2
0.902439	0.928571	0.928571	0.928571	YM6GlcmH5SumVarnc	YD8Gab24H12Mag
0.878049	0.892857	0.925926	0.909091	YN7Gab12N6Mag	YD8Gab24H12Mag
0.878049	0.892857	0.925926	0.909091	YM5GlcmZ4SumVarnc	YD8Gab24H12Mag
0.878049	0.892857	0.925926	0.909091	YM4GlcmZ4SumVarnc	YD8Gab24H12Mag
0.878049	0.892857	0.925926	0.909091	YM7GlcmZ4SumVarnc	YD8Gab24H12Mag
0.878049	0.892857	0.925926	0.909091	YM6GlcmZ4SumVarnc	YD8Gab24H12Mag
0.878049	0.892857	0.925926	0.909091	YM8GlcmZ4SumVarnc	YD8Gab24H12Mag
0.878049	0.892857	0.925926	0.909091	YM4GlcmH5SumVarnc	YD8Gab24H12Mag
0.878049	0.892857	0.925926	0.909091	YM5GlcmH5SumVarnc	YD8Gab24H12Mag
0.878049	0.892857	0.925926	0.909091	YM7GlcmH5SumVarnc	YD8Gab24H12Mag
0.853659	0.785714	1.000000	0.880000	YM8Gab12H6Mag	YS8HistDomn01
0.853659	0.857143	0.923077	0.888889	YN8Gab12N6Mag	YD8Gab24H12Mag
0.853659	0.857143	0.923077	0.888889	YM8GlcmZ5InvDfMom	YD8Gab24Z12Mag
0.853659	0.821429	0.958333	0.884615	YM4GlcmZ5SumVarnc	YD8Gab24H12Mag
0.853659	0.892857	0.892857	0.892857	YM4GlcmN2SumVarnc	YD8Gab24H12Mag
0.853659	0.857143	0.923077	0.888889	YM4GlcmH4SumVarnc	YD8Gab24H12Mag
0.853659	0.857143	0.923077	0.888889	YM5GlcmH4SumVarnc	YD8Gab24H12Mag
0.853659	0.857143	0.923077	0.888889	YM7GlcmH4SumVarnc	YD8Gab24H12Mag
0.853659	0.857143	0.923077	0.888889	YM6GlcmH4SumVarnc	YD8Gab24H12Mag
0.853659	0.857143	0.923077	0.888889	YM8GlcmH4SumVarnc	YD8Gab24H12Mag

As we can see from Table 3, the best differentiation between clinically significant and non-significant prostate cancers is achieved when the gray-level co-occurrence matrix summed variance feature is used along with Gabor wavelet magnitude. For the first textural feature, slightly better outcomes (92.86%) are attained when the image is preprocessed with min–max normalization (M) using only six significant bits to calculate the matrix in the horizontal direction (H) with stride equal to five. Changing the direction to diagonal (Z) with a slightly varying stride (4, 5), as well as using a different number of bits (4–8), diminishes the performance to 90.91%, while recall is maintained at the same level (92%) and the precision deteriorates from 92% to 89%. In all cases, the second textural feature is calculated for the original image using the Gabor wavelet with constant Gaussian envelope (24) in the horizontal direction and a sinewave period equal to 12. Since the dataset is highly imbalanced, the AUC was not calculated, as it works well only with balanced data and is too optimistic in other cases. Applying two textural features and concentrating on Gabor frequencies (YN7Gab12N6Mag and YD8Gab24H12Mag or YN8Gab12N6Mag and YD8Gab24H12Mag) also give very high F1 scores: 91% and 89%, respectively. The combination of the Gabor magnitude feature (YM8Gab12H6Mag) with histogram parameters (YS8HistDomn01) and other metrics calculated from the gray-level co-occurrence matrix (YM8GlcmZ5InvDfMom) also performs very well, with an F1 score around 88%.

Table 5. The best 20 pairs of features when the outer prostate region is used to calculate features.

Accuracy	Precision	Recall	F1 Score	Feature1	Feature2
0.875000	0.814815	1.000000	0.897959	YM7GlcmN2Correlat	YS8Gab24Z12Mag
0.875000	0.814815	1.000000	0.897959	YM8GlcmN2Correlat	YS8Gab24Z12Mag
0.875000	0.814815	1.000000	0.897959	YM6GlcmN2Correlat	YS8Gab24Z12Mag
0.875000	0.814815	1.000000	0.897959	YM5GlcmN2Correlat	YS8Gab24Z12Mag
0.850000	0.888889	0.888889	0.888889	YS7HogO8b2	YD8GradNonZeros
0.850000	0.888889	0.888889	0.888889	YN6Gab24N12Mag	YD8GradSkewness
0.850000	0.888889	0.888889	0.888889	YN7Gab24N12Mag	YD8GradSkewness
0.850000	0.888889	0.888889	0.888889	YN8Gab24N12Mag	YD8GradSkewness
0.850000	0.888889	0.888889	0.888889	YS4Gab24Z12Mag	YD8Gab16Z8Mag
0.850000	0.888889	0.888889	0.888889	YS6Gab24Z12Mag	YD8Gab16Z8Mag
0.850000	0.888889	0.888889	0.888889	YS7Gab24Z12Mag	YD8Gab16Z8Mag
0.850000	0.888889	0.888889	0.888889	YS8Gab24Z12Mag	YD8Gab16Z8Mag
0.850000	0.888889	0.888889	0.888889	YN7HistMaxm01	YS8Gab12Z6Mag
0.850000	0.777778	1.000000	0.875000	YM4GlcmN2Correlat	YS8Gab24Z12Mag
0.850000	0.814815	0.956522	0.880000	YM6GlcmH4InvDfMom	YD8Gab24N12Mag
0.850000	0.777778	1.000000	0.875000	YM7GlcmH4InvDfMom	YD8Gab24N12Mag
0.850000	0.888889	0.888889	0.888889	YN6GradNonZeros	YS8Gab16H8Mag
0.850000	0.814815	0.956522	0.880000	YN5GlcmV2Entropy	YD8HistKurtosis
0.825000	0.888889	0.857143	0.872727	YS8HogO8b2	YD8GradNonZeros
0.825000	0.814815	0.916667	0.862745	YN4Gab24V12Mag	YS8Gab12Z6Mag

When analyzing the whole prostate region, the best performance was recorded when combining Gabor frequency features with histogram percentiles, reaching an F1 score of 91.12% with recall of 89.66% and precision of 92.86%. Since there are similar outcomes, we can assume that using quantization (diminishing the number of input bits) in the case of Gabor wavelet calculation does not influence the performance, as similar results are achieved for data using from 4 up to 8 bits. Here, the horizontal orientation of a sinewave with period equal to 8 and Gaussian envelope equal to 16 was applied. The data were normalized using min–max transformation to calculate the histogram percentile using 8-bit data. The following results are derived mostly for pairs of features that focus on the image frequency description and not on some textural features, in an effort to express the human perception of the visual information.

The best outcomes for the outer prostate region are recorded when the correlation feature from the gray-level co-occurrence matrix is supported with the Gabor wavelet feature. Next, the combination of two features is used to describe the following texture frequencies. In the best case, the GLCM method is applied to min–max-normalized images; again, changing the number of significant bits (from 5 to 8) does not influence the F1 score—89.80%—with recall of 100% and precision of 81.48%. The second textural feature is derived from a normalized 8-bit image transformed with a Gabor filter with a Gaussian envelope of 24 radios, calculated in a diagonal orientation (Z) with a sinewave amplitude equal to 12.

4. Discussion

Also known as human kallikrein peptidase 3 (hK3), PSA is a member of the kallikrein gene family [59]. Ectopic PSA expression has been found at lower concentrations in malignant breast tissue, normal breast tissue, breast milk, and adrenal and kidney cancer. PSA is highly organ specific, as it is mainly produced by prostate epithelial cells. As evidenced by its imperfect performance as a diagnostic biomarker, PSA is not cancer specific and its values can be elevated in men with benign and malignant prostate diseases [60]. An increase in PSA is secondary to the loss of the barrier provided by the basal layer and the basal membranes of the prostate, which is a mandatory condition for the hormone to enter the circulation. However, loss of the barrier may occur in the case of different prostate diseases (BPH, prostatitis, prostate cancer) and in connection with prostate manipulation of the prostate (prostate massage, prostate biopsy) [61].

Screening for prostate cancer with the use of multiparametric MRI (mp-MRI) is still one of the most controversial topics in the urological literature [62]. Screening results reveal a significantly increased diagnosis of prostate cancer with the use of mp-MRI with detection of less advanced prostate cancer, however, with no overall survival benefit was observed [63]. National USA recommendations against PSA-based screening resulted in a reduction in the use of PSA for early detection and were associated with higher rates of advanced disease [64]. The inclusion of mp-MRI can improve a screening protocol, as it reduces the number of men who undergo biopsies while detecting more high- and intermediate-grade prostate cancer [65,66]. The IP1-PROSTAGRAM study (PSA > 3 ng/mL; MRI Prostate Imaging Reporting and Data System (PIRADS) > 2) showed the highest detection of prostate cancer by MRI compared to transrectal ultrasound-guided prostate biopsy (TRUS) in a population screening setting [65]. Moreover, standard TRUS is not reliable in detecting prostate cancer and the diagnostic yield of additional biopsies performed on hypoechoic lesions is negligible [67]. New sonographic modalities such as micro-Doppler, sonoelastography, or contrast-enhanced US provided promising preliminary findings, alone or combined into the so-called "multiparametric US" [68,69]. In the multiparametric US vs. multiparametric MRI to diagnose prostate cancer (CADMUS) trial, multiparametric US detected 4.3% fewer prostate cancer cases while submitting 11.1% more patients to biopsy than MRI [70]. The 68 m labeled prostate-specific antigen (PSA) positron emission tomography (PET PSMA) method is gaining increasing attention in the medical community as a method with increasing clinical potential [71,72]. It was proven that the diagnostic performance of PET PSMA is better than multiparametric magnetic resonance (mp-MRI) presented in clinical studies of intermediate-risk cancer, with diagnostic confidence reaching 80% in comparison to 60% for mp-MRI [73]. For advanced cancer, reported confidence was up to 99% in comparison to 87–97% reported for mp-MRI [74]. As PET PSMA gains increasing attention and influences treatment change in patients previously diagnosed with MR [75,76], there is a need to improve diagnostic techniques based on mp-MRI, especially in regard to the performance of 1.5 T scanners, as obtained results are less significant compared to those presented by 3T scanners [77,78].

Therefore, in asymptomatic patients with PSA values of 2–10 ng/mL, additional methods should be used to decide whether to refer the patient for a prostate biopsy, the result of which determines the treatment when cancer is diagnosed. These methods include cancer risk calculators as well as imaging-based diagnostics [79]. If PSA level is sufficiently high in correlation with the clinical picture (especially palpation methods), the margin for misleading diagnoses is very narrow [80]. However, in the many cases with intermediate levels of PSA, other diagnostic techniques, including diagnostic imaging, are needed to provide further support.

Magnetic resonance imaging (MRI) has been used for the non-invasive evaluation of the prostate and surrounding tissues since the 1980s. Advances in technology (both in software and hardware) have led to the development of multiparameter MRI (mp-MRI), which combines T2W anatomical imaging with functional and physiological assessments, including diffusion-based imaging (DWI) and its derivatives diffusion coefficient map (ADC) and dynamic MRI with contrast enhancement (DCE), and sometimes other techniques, such as in vivo MR proton spectroscopy [81].

MRI is currently a highly important step in the process of diagnosing suspected prostate cancer [23]. On the basis of over 30 years of experience gained in MR prostate studies, classification systems were created and are constantly updated [82]. However, it is widely acknowledged that a great deal of training is required to determine the image patterns behind prostate cancer, and there are still unclear images where diagnosis is not certain [83]. Therefore, clinical correlation is highly important, especially in relation to levels of PSA antigens [84].

Applying a pair of textural features to create an artificial intelligence model for the automatic detection of patients with high ($\geq$4) scores of PSA as a cutoff value for determination of possible malignancy of the lesion is a promising approach, with an F1 score equal to 92%. In our experiments, we verified whether increasing the number of features could improve the outcome. In order to decrease the number of search possibilities, we concentrated on features that participated in generating models with F1 scores higher than 80%. For this approach, all those features were used as input, but principal component analysis was applied to select the most discriminative data. Nonetheless, these results did not outperform the presented ones, so are not discussed in detail.

5. Conclusions

The results presented are very promising, as the analysis of the MR images has shown a correlation with the biochemical levels of the cancer marker. This is an important step towards making the diagnostic imaging process more objective. The results obtained will be used in the future to develop the analysis in a larger group of patients. This could serve as a solid basis for the development of protocols for automated analysis of prostate magnetic resonance imaging to improve the assessment of prostate cancer in daily clinical practice and to reduce possible diagnostic errors.

5.1. Study Strengths

- Finding the correlation between MR image textural features (in accordance with PIRADS guidance) with proven levels of PSA and hence the risk of prostate cancer.
- Using a novel and unique approach in comparison to formerly applied methods of image analysis, with possible usefulness for automatic, computer-based systems of lesion detection.
- Evaluating magnetic resonance imaging of the prostate with the use of textural analysis; this constitutes another step towards the creation of fully automated protocols for a fast and effective diagnostic path for prostate cancer.

5.2. Study Weakness

- Some limitations of the present work concern the relatively small group of patients analyzed and incomplete data; resolving these issues would allow us to draw even more conclusions.

There is a correlation between the presence of increased PSA level and probability of cancer being identified with the aid of the results of diagnostic imaging analyses. Low PSA in patients with changes in the prostate image was found and the recognized findings in the MR images with no reflection in the increased PSA level might be misleading in practical settings.

Supplementary Materials: The following supporting information can be downloaded at: https://www.mdpi.com/article/10.3390/jcm12082836/s1, Table S1. Prostaty results.

Author Contributions: Conceptualization, S.G., R.O., A.P. and K.N.; methodology, S.G., R.O., J.L., A.P. and K.N.; software, S.G., R.O., J.L., Z.S., A.P. and K.N.; validation, S.G., R.O., J.L., E.P. and K.N.; formal analysis, S.G., R.O., J.L. and K.N.; investigation, S.G., R.O., J.L. and K.N.; resources, S.G., R.O., J.L. and K.N.; data curation, S.G., R.O., J.L., Z.S., A.P., E.P. and K.N.; writing—original draft preparation, S.G., R.O., A.P. and K.N.; writing—review and editing, S.G., R.O., A.P. and K.N.; visualization, S.G., R.O., J.L., Z.S. and K.N.; supervision, S.G., R.O., A.P. and K.N.; project administration, S.G., R.O., J.L., A.P. and K.N.; funding acquisition, S.G., R.O. and K.N. All authors have read and agreed to the published version of the manuscript.

Funding: This research received no external funding.

Institutional Review Board Statement: The study was conducted in accordance with the Declaration of Helsinki and approved by the Institutional Review Board of Jagiellonian University (no 1072.6120.21.22, 23 February 2022).

Informed Consent Statement: This study is retrospective in nature; anonymized images acquired during routine examinations were used with the permission of the Bioethics Committee.

Data Availability Statement: One hundred and fourteen T2-weighted MRI 3D images of the prostate with corresponding segmentations of transition and peripheral zones are available on the Zenodo website, https://doi.org/10.5281/zenodo.7676958.

Conflicts of Interest: The authors declare no conflict of interest.

Abbreviations

ADC	Apparent diffusion coefficient imaging
AI	Artificial intelligence
ANN	Artificial neural network
Arm	Autoregressive model
AUC	Area under curve
AV	Averages
BW	Bandwidth
CAD	Computer-aided diagnosis
CON	Concentrations
DCE	Dynamic contrast-enhanced MRI
DWI	Diffusion-based imaging
dw-MRI	Diffusion-weighted MRI
FA	Flip angle
FOV	Field of view
TF	Turbo factor
PAT	Parallel acquisition
DF	Distance factor
Gab	Gabor transform
GLCM	Grey-level co-occurrence matrix
Grad	Gradient map features
Hist	Brightness histogram
HOG	Histogram of oriented gradients
LBP	Local binary pattern
MIL	Multiple-instance learning
mp-MRI	Multiparameter MRI
MR	Magnetic resonance
MRI	Magnetic resonance imaging
MX	Imaging matrix
PIRADS	Prostate Imaging Reporting and Data System
PSA	Prostate-specific antigen
ROI	Region of interest
SVM	Support vector machine
T	Time
TE	Time echo
TF	Turbo factor
TR	Relaxation time
T2W-MRI	T2-weighted MRI
VOX	Voxel
YUV	Color model, Y—luminance, U—blue projection, V—red projection
TRUS	Transrectal ultrasound
US	Ultrasonography
CADMUS	Cancer Detection by Multiparametric Ultrasound of the Prostate
PET PSMA	PSA labeled positron emission tomography

Appendix A. Dataset Detailed Description

Table A1 presents the basic information about the statistics of the patients' age, PSA level, etc. in the dataset, while Table A1 gives detailed information for each patient.

Table A1. Basic dataset characteristics.

Patient count	41
Age (in years)	
Average	64.6 ± 9.8
Median	65
Range	39–85
PSA level	
Average	6.1 ± 3.9
Median	5.14
Range	0.39–19
Side	
Left	9 patients (22%)
Right	13 patients (32%)
Both	19 patients (46%)
Prostatitis	8 patients (20%)
BPH	23 patients (56%)

Table A2. Detailed description of the medical parameters of the examined cohort.

Patient No	Age	PSA Level	Side	Prostatitis	BPH
1	61	3.50	Left	Yes	Yes
2	62	1.75	Both	Yes	Yes
3	85	5.20	Right	No	Yes
4	63	7.50	Right	No	Yes
5	63	5.60	Right	No	Yes
6	69	6.70	Right	No	Yes
7	67	5.14	Right	No	No
8	78	19.00	Both	No	No
9	68	2.40	Left	Yes	Yes
10	44	3.15	Both	Yes	No
11	62	8.00	Both	No	No
12	62	8.10	Both	No	Yes
13	57	8.60	Both	No	Yes
14	61	8.80	Right	No	No
15	67	4.30	Both	No	No
16	51	9.90	Both	No	No
17	66	17.00	Left	No	No
18	70	5.90	Both	No	Yes
19	65	4.8	Right	No	No
20	70	4.9	Right	No	Yes
21	48	3.3	Both	No	No
22	64	3.4	Left	No	No
23	58	4.5	Both	Yes	No
24	66	2.7	Left	No	No
25	81	4.45	Left	No	Yes
26	62	17.00	Right	Yes	Yes
27	74	5.30	Both	No	Yes
28	58	0.39	Both	Yes	Yes
29	70	6.7	Right	No	Yes
30	77	8.1	Both	No	Yes
31	77	7	Both	No	Yes
32	64	5.2	Left	No	Yes
33	65	6.95	Both	No	Yes
34	55	3.9	Right	Yes	No
35	69	9.00	Both	No	Yes
36	62	3.60	Right	No	Yes
37	39	3.16	Both	No	No
38	50	3.85	Right	No	No
39	73	3.3	Left	No	Yes
40	81	4.5	Left	No	No
41	66	4	Both	No	No

Appendix B. Feature-Naming Conventions

The textural feature names follow a convention whereby the description contains all of the information about the applied method and its parameters and preprocessing steps; the convention also ensures that the names are unique. The unique textural feature identities comprise the following:

- color component—Y always stands for the brightness component from the YUV color space;
- information on normalization, see Table 1;
- the number of bits used for quantization purposes;
- sliding window shape and size (if applied);
- the applied texture feature method;
- an acronym of the parameter;
- and the feature name abbreviation.

For instance, the feature name YM6GlcmH5SumVariance refers to an image converted to the YUV color space where the brightness component is analyzed. It is normalized using the minimum and maximum range of gray-scale pixel values (M) and coded on 6 bits. The gray-level co-occurrence matrix (GLCM) was applied to calculate the matrix in a horizontal (H) fashion, with the distance between adjoining pixels set to 5; finally, the sum of variance is the final feature in the acronym.

References

1. Culp, M.B.; Soerjomataram, I.; Efstathiou, J.A.; Bray, F.; Jemal, A. Recent Global Patterns in Prostate Cancer Incidence and Mortality Rates. *Eur. Urol.* **2020**, *77*, 38–52. [CrossRef]
2. Wang, L.; Lu, B.; He, M.; Wang, Y.; Wang, Z.; Du, L. Prostate Cancer Incidence and Mortality: Global Status and Temporal Trends in 89 Countries From 2000 to 2019. *Front. Public Health* **2022**, *10*, 811044. [CrossRef]
3. Mejak, S.L.; Bayliss, J.; Hanks, S.D. Long Distance Bicycle Riding Causes Prostate-Specific Antigen to Increase in Men Aged 50 Years and Over. *PLoS ONE* **2013**, *8*, e56030. [CrossRef]
4. Ankerst, D.P.; Thompson, I.M. Sensitivity and Specificity of Prostate-Specific Antigen for Prostate Cancer Detection with High Rates of Biopsy Verification. *Arch. Ital. Urol. Androl. Organo Uff.* **2006**, *78*, 125–129.
5. Thompson, I.M.; Chi, C.; Ankerst, D.P.; Goodman, P.J.; Tangen, C.M.; Lippman, S.M.; Lucia, M.S.; Parnes, H.L.; Coltman, C.A. Effect of Finasteride on the Sensitivity of PSA for Detecting Prostate Cancer. *J. Natl. Cancer Inst.* **2006**, *98*, 1128–1133. [CrossRef] [PubMed]
6. Damadian, R. Tumor Detection by Nuclear Magnetic Resonance. *Science* **1971**, *171*, 1151–1153. [CrossRef] [PubMed]
7. Steyn, J.H.; Smith, F.W. Nuclear Magnetic Resonance Imaging of the Prostate. *Br. J. Urol.* **1982**, *54*, 726–728. [CrossRef] [PubMed]
8. Langlotz, C.P. Benefits and Costs of MR Imaging of Prostate Cancer. *Magn. Reson. Imaging Clin. N. Am.* **1996**, *4*, 533–544. [CrossRef]
9. Potosky, A.L. The Role of Increasing Detection in the Rising Incidence of Prostate Cancer. *JAMA* **1995**, *273*, 548. [CrossRef]
10. Murphy, G.; Haider, M.; Ghai, S.; Sreeharsha, B. The Expanding Role of MRI in Prostate Cancer. *Am. J. Roentgenol.* **2013**, *201*, 1229–1238. [CrossRef]
11. Haider, M.A.; van der Kwast, T.H.; Tanguay, J.; Evans, A.J.; Hashmi, A.-T.; Lockwood, G.; Trachtenberg, J. Combined T2-Weighted and Diffusion-Weighted MRI for Localization of Prostate Cancer. *Am. J. Roentgenol.* **2007**, *189*, 323–328. [CrossRef]
12. Verma, S.; Turkbey, B.; Muradyan, N.; Rajesh, A.; Cornud, F.; Haider, M.A.; Choyke, P.L.; Harisinghani, M. Overview of Dynamic Contrast-Enhanced MRI in Prostate Cancer Diagnosis and Management. *Am. J. Roentgenol.* **2012**, *198*, 1277–1288. [CrossRef] [PubMed]
13. Cauni, V.M.; Stanescu, D.; Tanase, F.; Mihai, B.; Persu, C. Magnetic Resonance/Ultrasound Fusion Targeted Biopsy of the Prostate Can Be Improved by Adding Systematic Biopsy. *Med. Ultrason.* **2021**, *23*, 277–282. [CrossRef]
14. Frisbie, J.W.; Van Besien, A.J.; Lee, A.; Xu, L.; Wang, S.; Choksi, A.; Afzal, M.A.; Naslund, M.J.; Lane, B.; Wong, J.; et al. PSA Density Is Complementary to Prostate MP-MRI PI-RADS Scoring System for Risk Stratification of Clinically Significant Prostate Cancer. *Prostate Cancer Prostatic Dis.* **2022**. [CrossRef]
15. Wong, T.; Schieda, N.; Sathiadoss, P.; Haroon, M.; Abreu-Gomez, J.; Ukwatta, E. Fully Automated Detection of Prostate Transition Zone Tumors on T2-Weighted and Apparent Diffusion Coefficient (ADC) Map MR Images Using U-Net Ensemble. *Med. Phys.* **2021**, *48*, 6889–6900. [CrossRef]
16. Gaur, S.; Lay, N.; Harmon, S.A.; Doddakashi, S.; Mehralivand, S.; Argun, B.; Barrett, T.; Bednarova, S.; Girometti, R.; Karaarslan, E.; et al. Can Computer-Aided Diagnosis Assist in the Identification of Prostate Cancer on Prostate MRI? A Multi-Center, Multi-Reader Investigation. *Oncotarget* **2018**, *9*, 33804–33817. [CrossRef]
17. Ishioka, J.; Matsuoka, Y.; Uehara, S.; Yasuda, Y.; Kijima, T.; Yoshida, S.; Yokoyama, M.; Saito, K.; Kihara, K.; Numao, N.; et al. Computer-Aided Diagnosis of Prostate Cancer on Magnetic Resonance Imaging Using a Convolutional Neural Network Algorithm. *BJU Int.* **2018**, *122*, 411–417. [CrossRef]
18. Hambrock, T.; Vos, P.C.; Hulsbergen-van de Kaa, C.A.; Barentsz, J.O.; Huisman, H.J. Prostate Cancer: Computer-Aided Di-agnosis with Multiparametric 3-T MR Imaging—Effect on Observer Performance. *Radiology* **2013**, *266*, 521–530. [CrossRef] [PubMed]
19. Reda, I.; Khalil, A.; Elmogy, M.; Abou El-Fetouh, A.; Shalaby, A.; Abou El-Ghar, M.; Elmaghraby, A.; Ghazal, M.; El-Baz, A. Deep Learning Role in Early Diagnosis of Prostate Cancer. Technol. *Cancer Res. Treat.* **2018**, *17*, 1–11. [CrossRef]
20. Haralick, R.M. Statistical and Structural Approaches to Texture. *Proc. IEEE* **1979**, *67*, 786–804. [CrossRef]

21. Kociołek, M.; Strzelecki, M.; Klepaczko, A. Functional Kidney Analysis Based on Textured DCE-MRI Images. In *Information Technology in Biomedicine*; Pietka, E., Badura, P., Kawa, J., Wieclawek, W., Eds.; Advances in Intelligent Systems and Computing; Springer International Publishing: Cham, Switzerland, 2019; Volume 1011, pp. 38–49. ISBN 978-3-030-23761-5.
22. Szczypiński, P.M.; Strzelecki, M.; Materka, A.; Klepaczko, A. MaZda—The Software Package for Textural Analysis of Bio-medical Images. In *Computers in Medical Activity*; Kącki, E., Rudnicki, M., Stempczyńska, J., Eds.; Advances in Soft Computing; Springer: Berlin/Heidelberg, Germany, 2009; Volume 65, pp. 73–84. ISBN 978-3-642-04461-8.
23. de Rooij, M.; Israël, B.; Tummers, M.; Ahmed, H.U.; Barrett, T.; Giganti, F.; Hamm, B.; Løgager, V.; Padhani, A.; Panebianco, V.; et al. SUR/ESUI consensus statements on multi-parametric MRI for the detection of clinically significant prostate cancer: Quality requirements for image acquisition, interpretation and radiologists' training. *Eur. Radiol.* **2020**, *30*, 5404–5416. [CrossRef]
24. Boesen, L. Multiparametric MRI in Detection and Staging of Prostate Cancer. *Dan. Med. Bull.* **2017**, *64–72*, B5327.
25. Van Booven, D.J.; Kuchakulla, M.; Pai, R.; Frech, F.S.; Ramasahayam, R.; Reddy, P.; Parmar, M.; Ramasamy, R.; Arora, H. A Systematic Review of Artificial Intelligence in Prostate Cancer. *Res. Rep. Urol.* **2021**, *13*, 31–39. [CrossRef] [PubMed]
26. Harmon, S.A.; Tuncer, S.; Sanford, T.; Choyke, P.L.; Türkbey, B. Artificial Intelligence at the Intersection of Pathology and Radiology in Prostate Cancer. *Diagn. Interv. Radiol. Ank. Turk.* **2019**, *25*, 183–188. [CrossRef] [PubMed]
27. Telecan, T.; Andras, I.; Crisan, N.; Giurgiu, L.; Căta, E.D.; Caraiani, C.; Lebovici, A.; Boca, B.; Balint, Z.; Diosan, L.; et al. More than Meets the Eye: Using Textural Analysis and Artificial Intelligence as Decision Support Tools in Prostate Cancer Diag-nosis—A Systematic Review. *J. Pers. Med.* **2022**, *12*, 983. [CrossRef] [PubMed]
28. Patel, N.; Henry, A.; Scarsbrook, A. The Value of MR Textural Analysis in Prostate Cancer. *Clin. Radiol.* **2019**, *74*, 876–885. [CrossRef]
29. Liu, X.; Faes, L.; Kale, A.U.; Wagner, S.K.; Fu, D.J.; Bruynseels, A.; Mahendiran, T.; Moraes, G.; Shamdas, M.; Kern, C.; et al. A Comparison of Deep Learning Performance against Health-Care Professionals in Detecting Diseases from Medical Imaging: A Systematic Review and Meta-Analysis. *Lancet Digit. Health* **2019**, *1*, e271–e297. [CrossRef]
30. Wildeboer, R.R.; van Sloun, R.J.G.; Wijkstra, H.; Mischi, M. Artificial Intelligence in Multiparametric Prostate Cancer Imaging with Focus on Deep-Learning Methods. *Comput. Methods Programs Biomed.* **2020**, *189*, 105316. [CrossRef]
31. Wibmer, A.; Hricak, H.; Gondo, T.; Matsumoto, K.; Veeraraghavan, H.; Fehr, D.; Zheng, J.; Goldman, D.; Moskowitz, C.; Fine, S.W.; et al. Haralick Texture Analysis of Prostate MRI: Utility for Differentiating Non-Cancerous Prostate from Prostate Cancer and Differentiating Prostate Cancers with Different Gleason Scores. *Eur. Radiol.* **2015**, *25*, 2840–2850. [CrossRef]
32. Haralick, R.M.; Shanmugam, K.; Dinstein, I. Textural Features for Image Classification. *IEEE Trans. Syst. Man Cybern.* **1973**, *3*, 610–621. [CrossRef]
33. Nketiah, G.; Elschot, M.; Kim, E.; Teruel, J.R.; Scheenen, T.W.; Bathen, T.F.; Selnæs, K.M. T2-Weighted MRI-Derived Textural Features Reflect Prostate Cancer Aggressiveness: Preliminary Results. *Eur. Radiol.* **2017**, *27*, 3050–3059. [CrossRef]
34. Gnep, K.; Fargeas, A.; Gutiérrez-Carvajal, R.E.; Commandeur, F.; Mathieu, R.; Ospina, J.D.; Rolland, Y.; Rohou, T.; Vincendeau, S.; Hatt, M.; et al. Haralick Textural Features on T2 -Weighted MRI Are Associated with Biochemical Recurrence Following Radiotherapy for Peripheral Zone Prostate Cancer. *J. Magn. Reson. Imaging* **2017**, *45*, 103–117. [CrossRef] [PubMed]
35. Baek, T.W.; Kim, S.H.; Park, S.J.; Park, E.J. Texture Analysis on Bi-Parametric MRI for Evaluation of Aggressiveness in Patients with Prostate Cancer. *Abdom. Radiol. N. Y.* **2020**, *45*, 4214–4222. [CrossRef]
36. Xiong, H.; He, X.; Guo, D. Value of MRI Texture Analysis for Predicting High-Grade Prostate Cancer. *Clin. Imaging* **2021**, *72*, 168–174. [CrossRef]
37. He, X.; Xiong, H.; Zhang, H.; Liu, X.; Zhou, J.; Guo, D. Value of MRI Texture Analysis for Predicting New Gleason Grade Group. *Br. J. Radiol.* **2021**, *94*, 20210005. [CrossRef]
38. Damascelli, A.; Gallivanone, F.; Cristel, G.; Cava, C.; Interlenghi, M.; Esposito, A.; Brembilla, G.; Briganti, A.; Montorsi, F.; Castiglioni, I.; et al. Advanced Imaging Analysis in Prostate MRI: Building a Radiomic Signature to Predict Tumor Aggres-siveness. *Diagnostics* **2021**, *11*, 594. [CrossRef] [PubMed]
39. Giannini, V.; Mazzetti, S.; Defeudis, A.; Stranieri, G.; Calandri, M.; Bollito, E.; Bosco, M.; Porpiglia, F.; Manfredi, M.; De Pascale, A.; et al. A Fully Automatic Artificial Intelligence System Able to Detect and Characterize Prostate Cancer Using Mul-tiparametric MRI: Multicenter and Multi-Scanner Validation. *Front. Oncol.* **2021**, *11*, 718155. [CrossRef]
40. Li, M.; Yang, L.; Yue, Y.; Xu, J.; Huang, C.; Song, B. Use of Radiomics to Improve Diagnostic Performance of PI-RADS v2.1 in Prostate Cancer. *Front. Oncol.* **2020**, *10*, 631831. [CrossRef] [PubMed]
41. Bonekamp, D.; Kohl, S.; Wiesenfarth, M.; Schelb, P.; Radtke, J.P.; Götz, M.; Kickingereder, P.; Yaqubi, K.; Hitthaler, B.; Gählert, N.; et al. Radiomic Machine Learning for Characterization of Prostate Lesions with MRI: Comparison to ADC Values. *Radiology* **2018**, *289*, 128–137. [CrossRef]
42. Wang, J.; Wu, C.-J.; Bao, M.-L.; Zhang, J.; Wang, X.-N.; Zhang, Y.-D. Machine Learning-Based Analysis of MR Radiomics Can Help to Improve the Diagnostic Performance of PI-RADS v2 in Clinically Relevant Prostate Cancer. *Eur. Radiol.* **2017**, *27*, 4082–4090. [CrossRef]
43. Min, X.; Li, M.; Dong, D.; Feng, Z.; Zhang, P.; Ke, Z.; You, H.; Han, F.; Ma, H.; Tian, J.; et al. Multi-Parametric MRI-Based Radiomics Signature for Discriminating between Clinically Significant and Insignificant Prostate Cancer: Cross-Validation of a Machine Learning Method. *Eur. J. Radiol.* **2019**, *115*, 16–21. [CrossRef] [PubMed]

44. Zhang, Y.; Chen, W.; Yue, X.; Shen, J.; Gao, C.; Pang, P.; Cui, F.; Xu, M. Development of a Novel, Multi-Parametric, MRI-Based Radiomic Nomogram for Differentiating between Clinically Significant and Insignificant Prostate Cancer. *Front. Oncol.* **2020**, *10*, 888. [CrossRef]
45. Song, Y.; Zhang, Y.-D.; Yan, X.; Liu, H.; Zhou, M.; Hu, B.; Yang, G. Computer-Aided Diagnosis of Prostate Cancer Using a Deep Convolutional Neural Network from Multiparametric MRI. *J. Magn. Reson. Imaging* **2018**, *48*, 1570–1577. [CrossRef]
46. de Vente, C.; Vos, P.; Hosseinzadeh, M.; Pluim, J.; Veta, M. Deep Learning Regression for Prostate Cancer Detection and Grading in Bi-Parametric MRI. *IEEE Trans. Biomed. Eng.* **2021**, *68*, 374–383. [CrossRef] [PubMed]
47. Liu, Y.; Zheng, H.; Liang, Z.; Miao, Q.; Brisbane, W.G.; Marks, L.S.; Raman, S.S.; Reiter, R.E.; Yang, G.; Sung, K. Textured-Based Deep Learning in Prostate Cancer Classification with 3T Multiparametric MRI: Comparison with PI-RADS-Based Classifica-tion. *Diagnostics* **2021**, *11*, 1785. [CrossRef]
48. Schelb, P.; Kohl, S.; Radtke, J.P.; Wiesenfarth, M.; Kickingereder, P.; Bickelhaupt, S.; Kuder, T.A.; Stenzinger, A.; Hohenfellner, M.; Schlemmer, H.-P.; et al. Classification of Cancer at Prostate MRI: Deep Learning versus Clinical PI-RADS Assessment. *Radiology* **2019**, *293*, 607–617. [CrossRef] [PubMed]
49. Kwak, J.T.; Xu, S.; Wood, B.J.; Turkbey, B.; Choyke, P.L.; Pinto, P.A.; Wang, S.; Summers, R.M. Automated Prostate Cancer Detection Using T2-Weighted and High-b-Value Diffusion-Weighted Magnetic Resonance Imaging. *Med. Phys.* **2015**, *42*, 2368–2378. [CrossRef] [PubMed]
50. Zhao, K.; Wang, C.; Hu, J.; Yang, X.; Wang, H.; Li, F.; Zhang, X.; Zhang, J.; Wang, X. Prostate Cancer Identification: Quantitative Analysis of T2-Weighted MR Images Based on a Back Propagation Artificial Neural Network Model. *Sci. China Life Sci.* **2015**, *58*, 666–673. [CrossRef] [PubMed]
51. Doyle, S.; Hwang, M.; Shah, K.; Madabhushi, A.; Feldman, M.; Tomaszeweski, J. Automated Grading of Prostate Cancer Using Architectural and Textural Image Features. In Proceedings of the 2007 4th IEEE International Symposium on Biomedical Imaging: From Nano to Macro, Arlington, VA, USA, 12–15 April 2007; pp. 1284–1287.
52. Sidhu, H.S.; Benigno, S.; Ganeshan, B.; Dikaios, N.; Johnston, E.W.; Allen, C.; Kirkham, A.; Groves, A.M.; Ahmed, H.U.; Em-berton, M.; et al. Textural Analysis of Multiparametric MRI Detects Transition Zone Prostate Cancer. *Eur. Radiol.* **2017**, *27*, 2348–2358. [CrossRef] [PubMed]
53. Niu, X.-K.; Chen, Z.-F.; Chen, L.; Li, J.; Peng, T.; Li, X. Clinical Application of Biparametric MRI Texture Analysis for Detection and Evaluation of High-Grade Prostate Cancer in Zone-Specific Regions. *Am. J. Roentgenol.* **2018**, *210*, 549–556. [CrossRef] [PubMed]
54. World Medical Association. Issue Information-Declaration of Helsinki. *J. Bone Miner. Res.* **2019**, *34–51*, jbmr.3492. [CrossRef]
55. Gibala, S.; Obuchowicz, R.; Lasek, J.; Schneider, Z.; Piorkowski, A.; Pociask, E.; Nurzynska, K. Prostate MRI T2-Weighted Images with Peripherial and Trasition Zone Segmentations Including Corresponding PIRADS and PSA Values. 2023. Available online: https://zenodo.org/record/7676958#.ZDesiPZByUl (accessed on 2 March 2023).
56. Galloway, M.M. Texture Analysis Using Gray Level Run Lengths. *Comput. Graph. Image Process.* **1975**, *4*, 172–179. [CrossRef]
57. Ojala, T.; Pietikainen, M.; Maenpaa, T. Multiresolution Gray-Scale and Rotation Invariant Texture Classification with Local Binary Patterns. IEEE Trans. *Pattern Anal. Mach. Intell.* **2002**, *24*, 971–987. [CrossRef]
58. Dalal, N.; Triggs, B. Histograms of Oriented Gradients for Human Detection. In Proceedings of the 2005 IEEE Computer Society Conference on Computer Vision and Pattern Recognition (CVPR'05), San Diego, CA, USA, 20–25 June 2005; Volume 1, pp. 886–893.
59. McCormack, R.T.; Wang, T.J.; Rittenhouse, H.G.; Wolfert, R.L.; Finlay, J.A.; Lilja, H.; Okoloff, R.L.; Oesterling, J.E. Molecular Forms of Prostate-Specific Antigen and the Human Kallikrein Gene Family: A New Era. *Urology* **1995**, *45*, 729–744. [CrossRef] [PubMed]
60. Partin, A.W.; Carter, H.B.; Chan, D.W.; Epstein, J.I.; Oesterling, J.E.; Rock, R.C.; Weber, J.P.; Walsh, P.C. Prostate Specific Antigen in the Staging of Localized Prostate Cancer: Influence of Tumor Differentiation, Tumor Volume and Benign Hy-perplasia. *J. Urol.* **1990**, *143*, 747–752. [CrossRef]
61. Morote Robles, J.; Ruibal Morell, A.; Palou Redorta, J.; de Torres Mateos, J.A.; Soler Roselló, A. Clinical Behavior of Prostatic Specific Antigen and Prostatic Acid Phosphatase: A Comparative Study. *Eur. Urol.* **1988**, *14*, 360–366. [CrossRef] [PubMed]
62. Etzioni, R.; Gulati, R.; Cooperberg, M.R.; Penson, D.M.; Weiss, N.S.; Thompson, I.M. Limitations of basing screening policies on screening trials: The US Preventive Services Task Force and Prostate Cancer Screening. *Med. Care* **2013**, *51*, 295. [CrossRef] [PubMed]
63. Ilic, D.; Djulbegovic, M.; Jung, J.H.; Hwang, E.C.; Zhou, Q.; Cleves, A.; Agoritsas, T.; Dahm, P. Prostate cancer screening with prostate-specific antigen (PSA) test: A systematic review and meta-analysis. *BMJ* **2018**, *362*, k3519. [CrossRef] [PubMed]
64. Eklund, M.; Jäderling, F.; Discacciati, A.; Bergman, M.; Annerstedt, M.; Aly, M.; Glaessgen, A.; Carlsson, S.; Grönberg, H.; Nordström, T. MRI-Targeted or Standard Biopsy in Prostate Cancer Screening. *N. Engl. J. Med.* **2021**, *385*, 908. [CrossRef]
65. Eldred-Evans, D.; Ahmed, H.U. Population-Based Prostate Cancer Screening with Magnetic Resonance Imaging or Ultrasonography: The IP1-PROSTAGRAM Study. *JAMA Oncol.* **2021**, *7*, 395. [CrossRef]
66. Van Poppel, H.; Hogenhout, R.; Albers, P.; Bergh, R.C.V.D.; Barentsz, J.O.; Roobol, M.J. Early Detection of Prostate Cancer in 2020 and Beyond: Facts and Recommendations for the European Union and the European Commission. *Eur. Urol.* **2021**, *79*, 327. [CrossRef]
67. Rouviere, O.; Puech, P.; Renard-Penna, R.; Claudon, M.; Roy, C.; Mège-Lechevallier, F.; Decaussin-Petrucci, M.; Dubreuil-Chambardel, M.; Magaud, L.; Remontet, L.; et al. Use of prostate systematic and targeted biopsy on the basis of multiparametric

MRI in biopsy-naive patients (MRI-FIRST): A prospective, multicentre, paired diagnostic study. *Lancet Oncol.* **2019**, *20*, 100. [CrossRef] [PubMed]
68. Correas, J.M.; Halpern, E.J.; Barr, R.G.; Ghai, S.; Walz, J.; Bodard, S.; Dariane, C.; de la Rosette, J. Advanced ultrasound in the diagnosis of prostate cancer. *World J. Urol.* **2021**, *39*, 661. [CrossRef]
69. Mannaerts, C.K.; Engelbrecht, M.R.; Postema, A.W.; van Kollenburg, R.A.; Hoeks, C.M.; Savci-Heijink, C.D.; Van Sloun, R.J.G.; Wildeboer, R.R.; De Reijke, T.M.; Mischi, M.; et al. Detection of clinically significant prostate cancer in biopsy-naïve men: Direct comparison of systematic biopsy, multiparametric MRI- and contrast-ultrasound-dispersion imaging-targeted biopsy. *BJU Int.* **2020**, *126*, 481. [CrossRef]
70. Grey, A.D.R.; Scott, R.; Shah, B.; Acher, P.; Liyanage, S.; Pavlou, M.; Omar, R.; Chinegwundoh, F.; Patki, P.; Shah, T.T.; et al. Multiparametric ultrasound versus multiparametric MRI to diagnose prostate cancer (CADMUS): A prospective, multicentre, paired-cohort, confirmatory study. *Lancet Oncol.* **2022**, *23*, 428. [CrossRef] [PubMed]
71. Zschaeck, S.; Andela, S.B.; Amthauer, H.; Furth, C.; Rogasch, J.M.; Beck, M.; Hofheinz, F.; Huang, K. Correlation Between Quantitative PSMA PET Parameters and Clinical Risk Factors in Non-Metastatic Primary Prostate Cancer Patients. *Front. Oncol.* **2022**, *12*, 879089. [CrossRef]
72. Lisney, A.R.; Leitsmann, C.; Strauß, A.; Meller, B.; Bucerius, J.A.; Sahlmann, C.O. The Role of PSMA PET/CT in the Primary Diagnosis and Follow-Up of Prostate Cancer-A Practical Clinical Review. *Cancers* **2022**, *14*, 3638. [CrossRef] [PubMed]
73. Zhou, C.; Tang, Y.; Deng, Z.; Yang, J.; Zhou, M.; Wang, L.; Hu, S. Comparison of 68Ga-PSMA PET/CT and multiparametric MRI for the detection of low- and intermediate-risk prostate cancer. *EJNMMI Res.* **2022**, *12*, 10. [CrossRef]
74. Regmi, S.K.; Sathianathen, N.; Stout, T.E.; Konety, B.R. MRI/PET Imaging in elevated PSA and localized prostate cancer: A narrative review. *Transl. Androl. Urol.* **2021**, *10*, 3117–3129. [CrossRef]
75. Perera, M.; Papa, N.; Christidis, D.; Wetherell, W.; Hofman, M.S.; Murphy, D.G.; Bolton, D.; Lawrentschuk, N. Sensitivity, Specificity, and Predictors of Positive 68Ga-Prostate-specific Membrane Antigen Positron Emission Tomography in Advanced Prostate Cancer: A Systematic Review and Meta-analysis. *Eur. Urol.* **2016**, *70*, 926–937. [CrossRef]
76. van Kalmthout, L.W.M.; van Melick, H.H.E.; Lavalaye, J.; Meijer, R.P.; Kooistra, A.; de Klerk, J.M.H.; Braat, A.J.A.T.; Kaldeway, H.P.; de Bruin, P.C.; de Keizer, B.; et al. Prospective Validation of Gallium-68 Prostate Specific Membrane Antigen-Positron Emission Tomography/Computerized Tomography for Primary Staging of Prostate Cancer. *J. Urol.* **2020**, *203*, 537–545. [CrossRef]
77. Beyersdorff, D.; Taymoorian, K.; Knösel, T.; Schnorr, D.; Felix, R.; Hamm, B.; Bruhn, H. MRI of prostate cancer at 1.5 and 3.0 T: Comparison of image quality in tumor detection and staging. *Am. J. Roentgenol.* **2005**, *185*, 1214–1220. [CrossRef]
78. Virarkar, M.; Szklaruk, J.; Diab, R.; Bassett, R.; Bhosale, P. Diagnostic value of 3.0 T versus 1.5 T MRI in staging prostate cancer: Systematic review and meta-analysis. *Pol. J. Radiol.* **2022**, *87*, e421–e429. [CrossRef] [PubMed]
79. EAU Pocket Guidelines. In *Proceedings of the EAU Annual Congress, Milan, Italy, 10–13 March 2023*; EAU Guidelines Office: Arnhem, The Netherlands, 2022; ISBN 978-94-92671-16-5.
80. Omri, N.; Kamil, M.; Alexander, K.; Alexander, K.; Edmond, S.; Ariel, Z.; David, K.; Gilad, A.E.; Azik, H. Association between PSA Density and Pathologically Significant Prostate Cancer: The Impact of Prostate Volume. *Prostate* **2020**, *80*, 1444–1449. [CrossRef]
81. American College of Radiology. ACR Appropriateness Criteria®. Available online: https://acsearch.acr.org/list (accessed on 2 March 2023).
82. Beyer, T.; Schlemmer, H.-P.; Weber, M.-A.; Thierfelder, K.M. PI-RADS 2.1—Image Interpretation: The Most Important Updates and Their Clinical Implications. ROFO. *Fortschr. Geb. Rontgenstr. Nuklearmed.* **2021**, *193*, 787–796. [CrossRef] [PubMed]
83. Turkbey, B.; Oto, A. Factors Impacting Performance and Reproducibility of PI-RADS. *Can. Assoc. Radiol. J.* **2021**, *72*, 337–338. [CrossRef]
84. Gündoğdu, E.; Emekli, E.; Kebapçı, M. Evaluation of Relationships between the Final Gleason Score, PI-RADS v2 Score, ADC Value, PSA Level, and Tumor Diameter in Patients That Underwent Radical Prostatectomy Due to Prostate Cancer. *Radiol. Med.* **2020**, *125*, 827–837. [CrossRef] [PubMed]

diagnostics

MDPI

Article

Application of Deep Learning Methods in a Moroccan Ophthalmic Center: Analysis and Discussion

Zineb Farahat [1,2,*], Nabila Zrira [1], Nissrine Souissi [1], Safia Benamar [3,4], Mohammed Belmekki [3,4], Mohamed Nabil Ngote [1,4] and Kawtar Megdiche [2]

1 LISTD Laboratory, Ecole Nationale Supérieure des Mines de Rabat, Rabat 10000, Morocco
2 Cheikh Zaïd Foundation Medical Simulation Center, Rabat 10000, Morocco
3 Cheikh Zaïd Ophthalmic Center, Cheikh Zaïd International University Hospital, Rabat 10000, Morocco
4 Institut Supérieur d'Ingénierie et Technologies de Santé/Faculté de Médecine Abulcasis, Université Internationale Abulcasis des Sciences de la Santé, Rabat 10000, Morocco
* Correspondence: zineb.farahat@enim.ac.ma; Tel.: +212-672-387-783

Abstract: Diabetic retinopathy (DR) remains one of the world's frequent eye illnesses, leading to vision loss among working-aged individuals. Hemorrhages and exudates are examples of signs of DR. However, artificial intelligence (AI), particularly deep learning (DL), is poised to impact nearly every aspect of human life and gradually transform medical practice. Insight into the condition of the retina is becoming more accessible thanks to major advancements in diagnostic technology. AI approaches can be used to assess lots of morphological datasets derived from digital images in a rapid and noninvasive manner. Computer-aided diagnosis tools for automatic detection of DR early-stage signs will ease the pressure on clinicians. In this work, we apply two methods to the color fundus images taken on-site at the Cheikh Zaïd Foundation's Ophthalmic Center in Rabat to detect both exudates and hemorrhages. First, we apply the U-Net method to segment exudates and hemorrhages into red and green colors, respectively. Second, the You Look Only Once Version 5 (YOLOv5) method identifies the presence of hemorrhages and exudates in an image and predicts a probability for each bounding box. The segmentation proposed method obtained a specificity of 85%, a sensitivity of 85%, and a Dice score of 85%. The detection software successfully detected 100% of diabetic retinopathy signs, the expert doctor detected 99% of DR signs, and the resident doctor detected 84%.

Keywords: artificial intelligence; deep learning; diabetic retinopathy; hemorrhages; exudates; automatic screening; method; U-Net; YOLOv5

Citation: Farahat, Z.; Zrira, N.; Souissi, N.; Benamar, S.; Belmekki, M.; Ngote, M.N.; Megdiche, K. Application of Deep Learning Methods in a Moroccan Ophthalmic Center: Analysis and Discussion. *Diagnostics* **2023**, *13*, 1694. https://doi.org/10.3390/diagnostics13101694

Academic Editors: Michał Strzelecki, Adam Piorkowski and Rafał Obuchowicz

Received: 21 February 2023
Revised: 17 March 2023
Accepted: 20 March 2023
Published: 10 May 2023

1. Introduction

Diabetic retinopathy is the primary cause of vision loss in working aged individuals around the world. Its silent progression makes it a sight threatening condition. By 2030, the number of patients with vision-threatening diabetic retinopathy (VTDR) is predicted to jump from 37.3 million to 56.3 million [1]. On the micro scale, according to the Moroccan Society of Ophthalmology [2], the prevalence of DR in Morocco is 35%, approaching 10% of legal blindness.

Diabetic retinopathy is a small-vessel consequence of diabetes and continues to be the leading cause of preventable ocular morbidity amongst working aged individuals. Diabetes currently affects 422 million people globally, with 600 million people expected to be impacted by 2040, primarily in low- and middle-income nations [3]. It is found in one-third of diabetics and it is related to a higher risk of life-threatening systemic vascular problems such as small vessels heart disease, cardiac failure, and stroke [4]. Prevention is the key factor to reduce the risk of diabetic retinopathy evolution by keeping blood pressure, blood glucose, and blood lipids under control. We distinguish many signs of DR, such as retinal hemorrhages, cottony nodules, and exudates. Diabetic retinopathy is a type of retinal microangiopathy. It entails changes in the vascular wall in addition to modifications in the rheological blood's characteristics. Although timely laser therapy can help with both macular oedema and proliferative retinopathy, its potential to reverse vision loss is limited. Endo-ocular surgery may be required in rare cases of advanced retinopathy. Common signs and symptoms of Diabetic Retinopathy are blurred vision, sudden onset of double vision, dryness of the eyes, difficulty perceiving colors, floating bodies, and difficulty seeing in the dark.

Using different methods of measurement, several studies have found that the progression of diabetes to diabetic retinopathy (DR) is associated with changes in hemodynamics or measurable vascular geometry. Possible geometric alterations in the retina might indicate the existence of a systemic disease [5]. However, many parameters, mainly venous, showed a significant change in the development of DR, including an early change two years before the start of DR [6].

Furthermore, many studies demonstrated that the genetic polymorphism of histone ethyltransferases, which are responsible for elevated expression of key proinflammatory factors implicated in vascular injury, can be considered as predictors of the risk for micro- and macrovascular diabetic complications [7].

The Diabetes Control and Complications Trial (DCCT) specified that intensive glycaemic control in type 1 diabetes minimized the risk of development of diabetic retinopathy (primary prevention) and slowed its progression in a group with mild retinopathy at baseline (secondary prevention) [8].

Early detection of DR has demonstrated a significant decrease in the risk of vision impairment. Screening programs are conducted within the framework of a healthcare policy for blindness prevention [9]. In Morocco, very few caravans are organized to alleviate the burden on the health system. Nevertheless, these efforts are diluted by the ongoing need for diagnosis, treatment, and further monitoring.

Other human and geographic challenges are encountered linked to the shortage of trained ophthalmologists and retinal specialists (a ratio of ophthalmologists per capita is 1/68,000), as well as the presence of secluded regions with poor access to medical facilities, as well as uneven scattering between large cities and the countryside [10].

In this context, telemedicine and artificial intelligence are on the rise to respond to the pressing demand in healthcare, increasing productivity and efficiency in care delivery. These methods have the ability to be used in every aspect of medical practice, thanks to recent breakthroughs in digital data acquisition. They are moving into fields that were previously regarded as being entirely the domain of humans, and ophthalmology is no exception [3]. Every year, medical caravans visit many remote areas of Morocco to treat patients. Due to the huge inequity between the number of ophthalmologists and the growing number of diabetic populations, we seek to emphasize a feasible solution in this paper, which is relevant to circumstances where computers can be of a great help to health practitioners. We assume that an intelligent system capable of automatically detecting fundus irregularities and DR signs would be beneficial. It could meet this need and pave the way for a true national DR screening program in the future. In summary, our contributions are five-fold:

1. We collected 1000 retinography images from the Cheikh Zaïd Foundation's Eye Care Center in Rabat and annotated them;
2. We carried out diabetic retinopathy segmentation using U-Net;
3. We carried out diabetic retinopathy detection using Yolov5;
4. We compared the advantages and weaknesses of segmentation and detection methods.
5. We created a software that automatically detects diabetic retinopathy signs.

The remainder of the paper is laid out as follows. Section 2 presents an overview of the existing ophthalmic artificial intelligence solutions. Section 3 is devoted to methods and the dataset. Section 4 deals with experiments and results. Discussion is reported in Section 5. Finally, Section 6 depicts conclusions.

2. Related Work

Using the IDRiD dataset [11], Xu et al. [12] worked on a segmentation model called FFU-Net (Feature Fusion U-Net), which improves U-Net architecture. To begin, the network's pooling layer is replaced with a convolutional layer to minimize the spatial loss of the retinal image. Then, they integrated multiscale feature fusion (MSFF) blocks into the encoders, one that helps the network learn multiscale features and enriches the information provided with skip connections and lower resolution decoders by fusing contextual channel attention (CCA) models. At last, the authors proposed a balanced focal loss function to address misclassification and data imbalance issues.

Kou et al. [13] proposed an enhanced residual U-Net (ERU-Net) for segmenting microaneurysms (MAs) and exudates (EXs). They evaluated ERU-NET's performance for MAs and EXs segmentation on three public datasets: IDRiD, DDR, and E-Ophtha. On these three datasets, the used architecture achieves AUC values of 0.9956, 0.9962, 0.9801, 0.9866, 0.9679, and 0.9609 for microaneurysm and exudates segmentation, which are higher than the original U-Net values.

Li et al. [14] presented MAU-Net, which is a retinal image segmentation method based on the U-Net structure, to segment retinal blood vessels. The authors used DRIVE, STARE, and CHASEDB1 to validate their method.

Zhang et al. [15] proposed a CNN architecture that incorporated the Inception-Res module, as well as densely connected convolutional modules into the U-Net model. The author tested their model on vessel segmentation from retinal images, MRI brain neoplasm segmentation datasets from MICCAI BraTS 2017, and lung CT scan segmentation data from Kaggle datasets. The results of the lung segmentation achieved an average Dice score of 0.9857. The results for brain tumor segmentation achieved a Dice score of 0.9867. The results for vessel segmentation achieved an average Dice score of 0.9582.

Dai et al. [16] developed a DL solution called DeepDR, which allows users to detect different stages of DR. An amount of 466,247 color fundus images were used for training. The detection results of different DR signs, such as microaneurysms, cotton spots, hard exudates, and hemorrhages were 0.901, 0.941, 0.954, and 0.967, respectively. DR classification into mild, moderate, severe, and proliferative achieved an area under the curves of 0.943, 0.955, 0.960, and 0.972, respectively.

Sambya et al. [17] used a U-NET model based on a residual network with sub-pixel convolution initialized to the nearest convolution. The suggested architecture was trained and validated on two publicly accessible datasets, IDRiD and e-ophtha, for microaneurysm and hard exudate segmentation. On the IDRiD dataset, the network obtains a Dice score of 0.9998, as well as 99.88% accuracy, 99.85% sensitivity, 99.95% specificity, for microaneurysms and exudates.

Yaday et al. [18] proposed a U-Net based approach for retinal vessel segmentation. Before starting the segmentation procedure, some preprocessing approaches are used to improve the image's impacted region. Then, a discrete double-tree Ridgelet transform (DT-DRT) is applied to the dataset to extract the features of the region of interest. The proposed segmentation achieved an accuracy of 96.01% in CHASE DB1, 97.65% in DRIVE, and 98.61% in STARE.

Toufique A. Soomro et al. [19] first used preprocessing steps to make the training process more efficient. They implemented the CNN model based on a variational autoencoder (VAE), which is a modified version of U-Net. Their main contribution to the CNN implementation is to replace all pooling layers with progressive convolution and deeper layers. The proposed model generates segmented vessels image. The authors used both DRIVE and STARE datasets to train and test their model. It gave a sensitivity of 0.739 and an accuracy of 0.948 on the DRIVE database. Additionally, a sensitivity of 0.748 and an accuracy of 0.947 are observed for the STARE database.

Swarup et al. [20] compared different architectures (UNET, TLCNN, PCNSVM, and rSVMbCNN). Then, they presented the selected retinal image segmentation method based on a Ranking Support Vector Machine (rSVM) with a convolutional neural network in deep learning for the detection of diabetic retinopathy. They started by computing the pixel-by-pixel score with rSVM, and they then designed a deep convolutional neural network for retinal image segmentation followed by automatic anomaly detection using morphological operations. As a result, they achieved a segmentation accuracy of 96.4%, 97%, and 98.2% for three different databases—STARE, DIARETDB0, and DIARETDB1.

Many AI devices have been developed to revolutionize DR screening [21]. Pal et al. [22] applied the You Only Look Once version 3 (YOLOv3) algorithm to automatically detect hemorrhages in fundus images. The YOLOv3 algorithm recognized all red spots and surrounded them with multiple boxes. It identifies bounding boxes using a CNN-based model named Darknet53 and a squared error loss function, as well as logistic regression to determine an object's confidence score. Finally, non-max suppression was used to delete anything other than the best-fit bounding boxes. In order to train their model, the authors used the MESSIDOR dataset. It is made up of 1200 RGB Fundus images. Only 742 people were chosen out of a total of 1200, with 572 going through training and 170 going through validation. The average precision value of test data was 83.3%.

The results of a YOLO model were published by Rohan Akut [23]. It entails detecting microaneurysms and identifying their location on retinal images. He developed this algorithm based on 85,000 fundus images used for training. The dataset was divided into a ratio of 90–10%, with 90% going to training and 10% going to testing. An amount of 10% of the training dataset was used for validation. The model used enables the creation of a green bounding box around each microaneurysm.

Ming et al. [24] evaluated EyeWisdom in real world, which is an AI solution based on the YOLO detection system. It was created using 25,297 retinography (3785 from Peking Union Medical College Hospital and Henan Eye Hospital and 21,512 from the Kaggle dataset) [25]. The sensitivity was 90.4%, and the specificity was 95.2%.

Yang et al. [26] presented a collaborative learning framework for robust DR grading that integrates patch level lesion and image level grading features (CLPI). They used the IDRiD dataset as a lesion dataset, which contains 81 color fundus images (54 for training and 27 for testing) with pixel level annotations of lesions such as exudates, MAs, and hemorrhages. The authors also used image level datasets such as Messidor-1 [27], Messidor-2, LIQ-EyePACs [28], and other private datasets. They demonstrated that the proposed CLPI outperforms senior ophthalmologists, as well as SOTA algorithms. The authors demonstrated the reliability of CLPI by evaluating the DR grading methods in real world scenarios. The findings demonstrated the effectiveness of the lesion attention scheme, as well as the benefits of CLPI's end-to-end collaborative learning.

Table 1 below summarizes the results of the literature review.

Table 1. Summary of literature review.

Authors	Database	Lesion	Method	Metrics
Xu et al. [8]	Idrid	Microaneurysms (MAs) and exudates (EXs)	FFU-Net	SEN = 11.97%, IOU = 10.68%, and DICE = 5.79%
Kou et al. [9]	E-Ophtha, idrid and DDR		ERU-Net	Idrid: AUC = 0.9956 DDR, and E-Ophtha
Li et al. [10]	DRIVE, STARE, and CHASEDB1	Retinal blood vessels	MAU-Net	DRIVE: ACC = 0.9557 STARE: ACC = 0.9581 CHASEDB1: ACC = 0.9620
Zhang et al. [11]	Kaggle	Retinal blood vessels	U-Net model	Dice score = 0.9582

Table 1. *Cont.*

Authors	Database	Lesion	Method	Metrics
Dai et al. [12]	Local fundus images	Microaneurysms, cotton spots, hard exudates, and hemorrhages	Deepdr developed based on resnet41 and Mask-RCNN57	0.901, 0.941, 0.954, and 0.967
Sambya et al. [13]	Idrid and e-ophtha	Microaneurysms and hard exudates	U-NET	Dice score = 0.9998, accuracy = 99.88%, sensitivity = 99.85%, specificity = 99.95%
Yaday et al. [14]	CHASE DB1, DRIVE and STARE	Retinal blood vessels	U-NET	-CHASE DB1: Accuracy = 96.01%-DRIVE: Accuracy = 97.65% STARE: Accuracy= 98.61%.
Toufique A. Soomro et al. [15]	DRIVE and STARE	Retinal blood vessels	U-Net	-DRIVE: Accuracy = 0.948 Sensitivity = 0.739 STARE: Accuracy = 0.947 Sensitivity = 0.748
Swarup et al. [16]	STARE, DIARETDB0 and DIARETDB1	Exudates	Ranking Support Vector Machine (rsvm) with a convolutional neural network	Accuracy of 96.4%, 97% and 98.2%
Pal et al. [18]	MESSIDOR	Hemorrhages	YOLO	Precision = 83.3%
Rohan Akut [19]	Eyepacs	Microaneurysms	YOLO	Precision = 86.7%
Yang et al. [22]	Idrid, Messidor-1, Messidor-2, LIQ-eyepacs, and a private datasets	Exudates, MAs, and hemorrhages.	CLPI	MESSIDOR: AUC = 0.946 LIQ-EYEPACS: AUC = 0.916 Private Database: AUC = 0.983

The novelty of this paper lies in creating new software that can automatically detect all types of hemorrhages and exudates. Furthermore, the developed software can recognize laser marks, which are the results of a specific retina therapy and are not pathological signs and differentiate them from both hemorrhages and exudates even if they look similar.

3. Methods and Dataset

3.1. Methods

3.1.1. Diabetic Retinopathy Segmentation

In computer vision and deep learning, image segmentation is the process of partitioning an image into segments or highlighted groups of pixels that are considered as meaningful entities [29]. To segment our color fundus images, we trained and tested a U-Net multi-class segmentation model on 200 labeled retinal images (Figure 1). U-Net [30] has become a widely known medical image segmentation technique and has demonstrated excellent performance as a fully convolutional neural network [3,30]. The U-Net architecture contains two "paths." First, the contraction path, known as the encoder, is used to capture an image's context. In fact, it is a combination of convolution and "max pooling" that not only reduces the image size, but it also generates a feature map, thereby decreasing the number of network parameters. The symmetric expansion path, also known as the decoder, is the second path. Due to the transposed convolution, it also provides precise localization. What characterizes the U-Net architecture is the shortened connections between the layers of equal resolution from the analysis path to the expansion path. These connections provide important high-resolution features for the deconvolution layers.

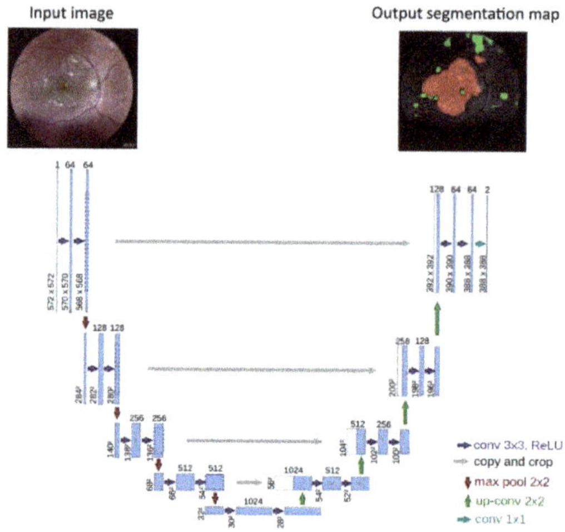

Figure 1. Retinal fundus images with DR signs.

3.1.2. Diabetic Retinopathy Detection

Object detection is an important branch of both image processing and computer vision. It is the process of detecting occurrences of a particular type of object in videos and images [31]. Deep learning has given a lot of attention to object detection algorithms [32]. The recent YOLO series of algorithms, known for their high speed and precision, have been used in different detection tasks. "YOLO" is an abbreviation for "You Only Look Once". In YOLO, a CNN architecture is used only once on the entire image to find the objects. The YOLO system calculates a large number of image features and detects all of the objects. This improves the model's computational efficiency and makes it more suitable for real-time applications. In our paper, we trained a fifth generation of YOLO model, also known as YOLOv5 architecture, to detect both hemorrhages and exudates (Figure 2). The used model processes the entire image with a single neural network, then divides it into parts and predicts probabilities and bounding boxes for each object. The expected probability is used to weight these bounding boxes. An appropriate selection of the activation function is mandatory for DL networks. Leaky ReLU is used in the middle/hidden layer of YOLOv5, and the sigmoid function is used in the last detection layer [33].

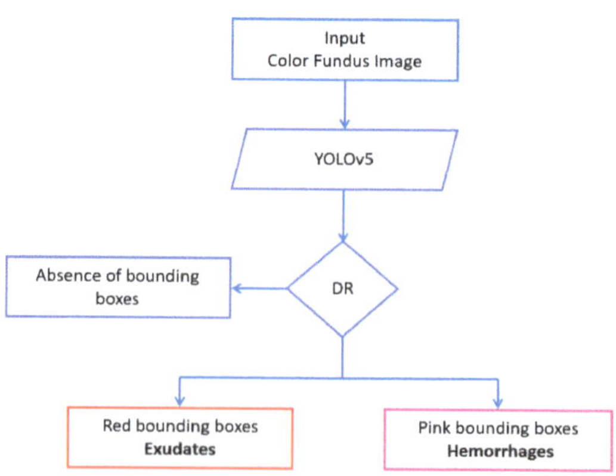

Figure 2. YOLOv5 DR detection.

3.2. Material

Care EIDON is a non-mydriatic retinal camera (Figure 3). It is one of the first TrueColor confocal systems to set new quality standards in retinal imaging by combining the best features of scanning laser ophthalmoscopy (SLO) systems with those of basic fundus imaging. iCare EIDON is a retinal imaging system that provides high image quality, as well as a confocal view in a non-dilating procedure, as well as a wide field and ultra-high resolution imaging. Furthermore, it distinguishes itself by offering users a variety of imaging modalities, including blue, TrueColor, and red-free and red confocal images, as well as infrared. In addition to this, it enables users to work in both fully manual and fully automated modes and to image through cataract and media opacities.

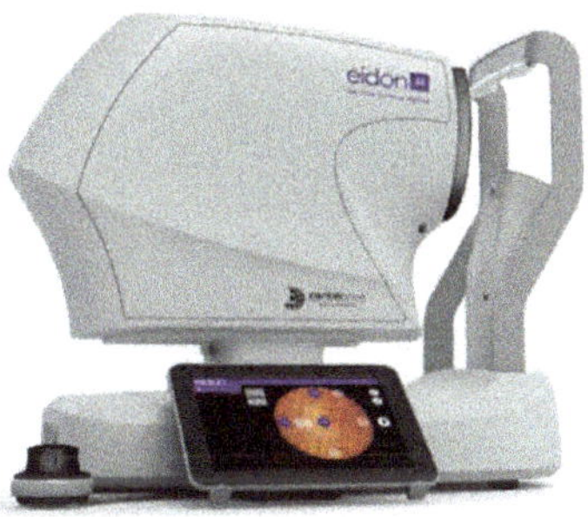

Figure 3. iCare EIDON widefield TrueColor confocal fundus imaging system.

Besides, a personal computer is connected to the iCare Eidon in order to facilitate the display, as well as the management and printing of the color fundus images. This computer will be used for the deployment of the proposed software that analyzes images.

3.3. Dataset

Our study is approved by the ethics committee of the Cheikh Zaid International University Hospital, and patient consent was obtained.

In order to test both the segmentation and detection methods, 1000 color fundus images were collected from the Foundation Ophthalmic Center of Rabat. These images were taken by the EIDON retinograph, which can automatically produce composite images that allow an overview of the retina of the patient.

3.3.1. Mask Generation for Segmentation

An amount of 200 color images were taken from the Cheikh Zaïd International University in Rabat and then used for training and testing the U-Net algorithm. The image implementation procedure is still in progress. More masks are being manually created in order to improve the model dataset. Gimp software has been used to create masks. After being validated by expert ophthalmologists, all of them were saved as JPEG folders and divided into four folders (hard exudates, soft exudates, hemorrhages, and red small dots) (Figure 4). These four folders were then merged into two folders to keep only two classes, which are hemorrhages (small red dots and hemorrhages) and exudates (hard and soft exudates).

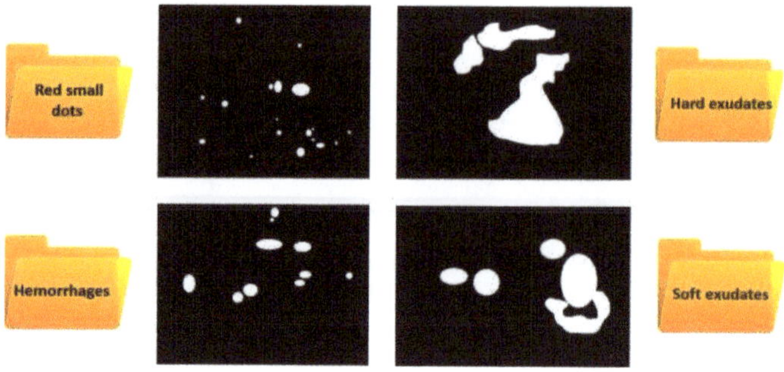

Figure 4. Masks folders of DR signs.

3.3.2. Annotation Generation for Detection

For image detection, a dataset of 1000 local color fundus images taken from the Cheikh Zaïd International University Ophthalmic Center has been used following the approval of the Cheikh Zaïd International University hospital ethics committee. A population with an age range of 39 to 75, a rate of severe pre-proliferative DR of 12%, a rate of complicated DR of 1%, and a rate of legal blindness of 3% was selected. An amount of 70% of our images went for training, 20% went for testing, and 10% went for validation.

Annotations were created manually using LabelImg (Figure 5). Bounding boxes were drawn manually around both hemorrhages and exudates, and then they were saved as ".txt" files.

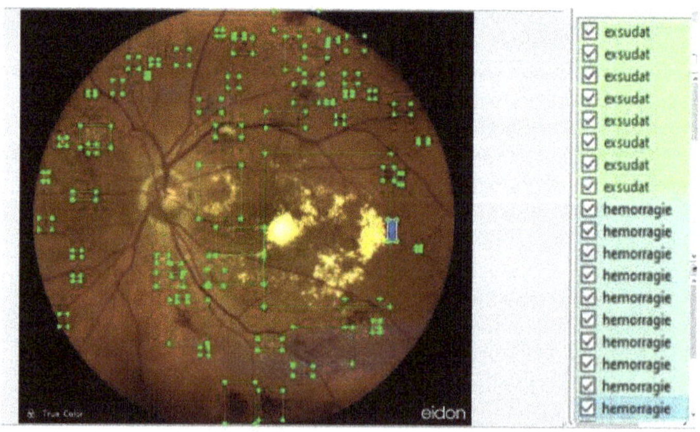

Figure 5. Annotation generation using LabelImg.

The CNN model starts by resizing the input images to 448 × 448. A convolution is then performed through the horizontal and vertical application of several filters, with the aim of extracting the image features. Finally, to improve the accuracy of the bounding boxes of the detected signs, a removal of the non-max is performed (Figure 6).

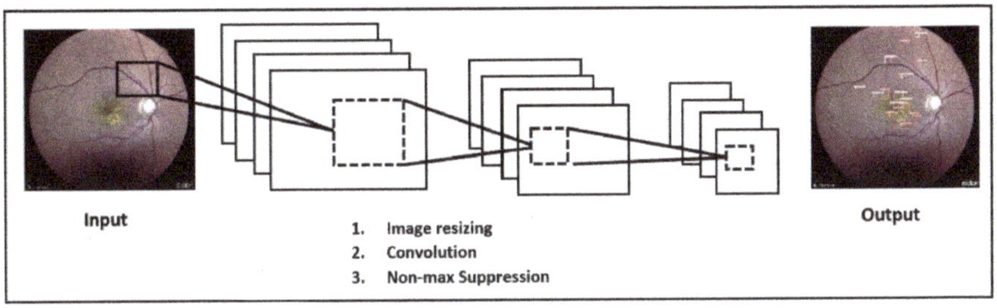

Figure 6. CNN architecture used for detection.

4. Experiments and Results

We run our notebook on a Colab Pro GPU Tesla P100-PCIE-16GB with high RAM. We also used Pytorch for YOLOv5 and Keras with Tensorflow as the backend for U-Net.

4.1. Diabetic Retinopathy Segmentation

As preliminary results of the solution under development by our research team, we obtained an output of either the absence or the presence of DR. As shown in Figure 7, the segmentation enables separating DR signs from the retinal image background to make them recognizable by highlighting exudates in red and hemorrhages in green. The use of this technique, however, necessitates the manual creation of training and validation labels.

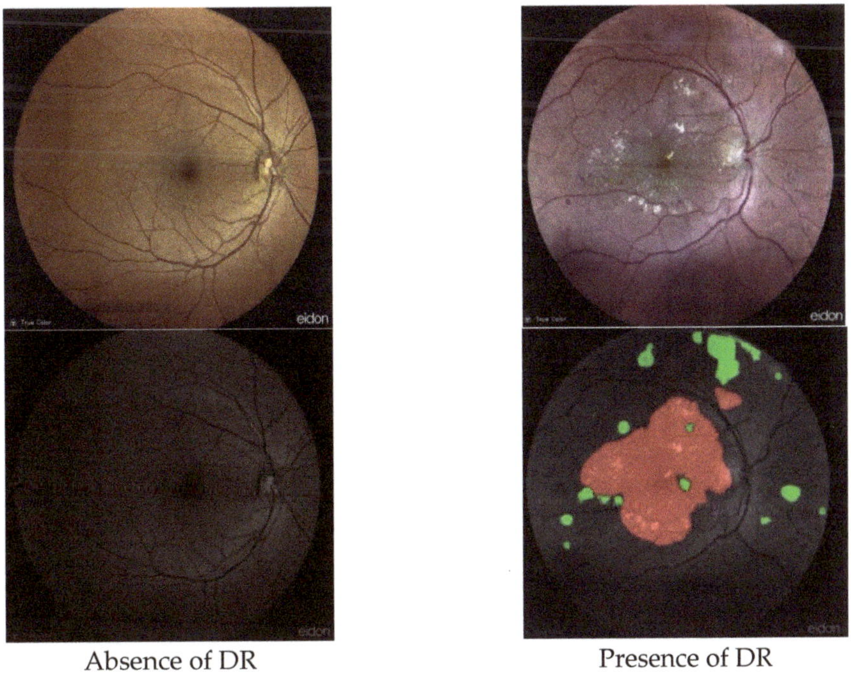

Figure 7. Segmentation results using U-Net.

To assess the adequacy of the proposed solution, evaluation metrics (Table 2) were calculated as follows:

$$J(A,B) = \frac{|A \cup B - |A \cap B|}{|A \cup B|}$$

$$Sensitivity = (True\ Positive)/(True\ Positive + False\ Negative)$$

$$Specificity = (True\ Negative)/(True\ Negative + False\ Positive)$$

$$Dice = \frac{2 \times True\ Positive}{2 \times True\ Positive + False\ Negative + False\ Positive}$$

$$IOU = \frac{A \cap B}{A \cup B}$$

Table 2. Segmentation metrics.

Metric	Value
Jaccard	0.6652
Sensitivity	0.8509
Specificity	0.8506
Dice	0.8506
IOU	0.7503

The proposed segmentation method obtained specificity of 85%, a sensitivity of 85%, a Dice score of 85%, and a Jaccard score of 66% for hemorrhages and exudates segmentation.

4.2. Diabetic Retinopathy Detection

The fifth version of YOLO was used to detect hemorrhages and exudates in color fundus images. Ultralytics introduced it in June 2020, and it is now known as an advanced object detection algorithm. YOLOv5 provides four versions of the object detection network: YOLOv5s, YOLOv5m, YOLOv5l, and YOLOv5x. In this work, the YOLOv5s model was used. In the YOLOv5 series, YOLOv5s is the network with the smallest depth and width. The following are the specifics of training the YOLOv5 model.

Image size: 640;

Batch size: 16;

Data description: coco128.yaml;

Yolo model: YOLOv5s.yaml;

Configuration (cfg): models are described in YAML model configuration files in "model" directory. There are four versions of models of different sizes. "YOLOv5s.yaml" has been used to train.

There are four versions of the model with different sizes. "YOLOv5s.yaml" has been used to train. YOLOv5 was provided with two optimizers: Adam and Stochastic Gradient Descent (SGD). The default one is SGD. For this paper, we used the Adam optimizer, which is a better choice for smaller datasets. After being trained, our detection model allowed us to show diabetic retinopathy signs by drawing red bounding boxes around exudates and pink bounding boxes around hemorrhages. However, in order to assess the detection model's performance in real-life situations, we tested it on recent color fundus images in order to extend the model in real-world conditions. As depicted in Figure 8, our model was able to differentiate between laser impacts and exudates that look alike.

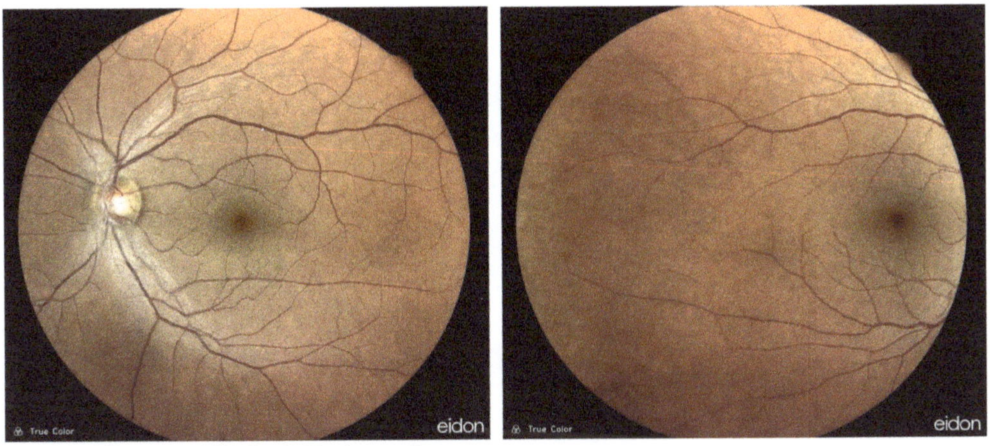

Case 1: Absence of Diabetic Retinopathy

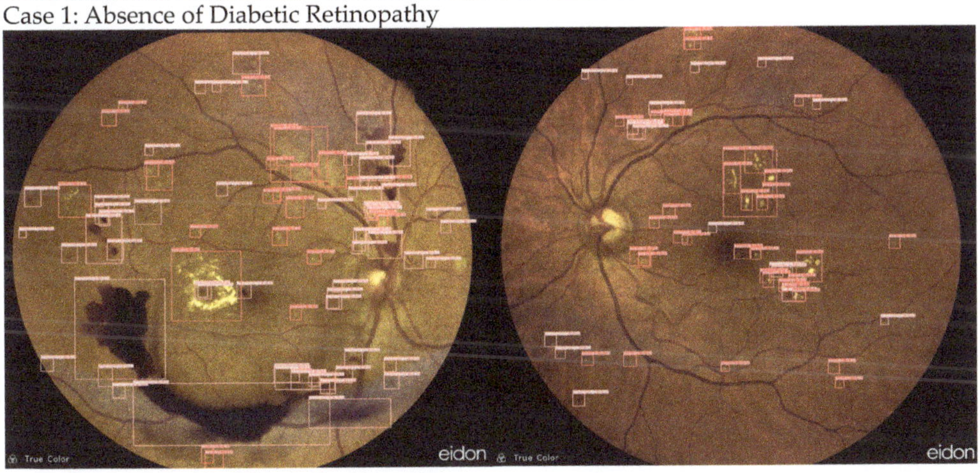

Case 2: Presence of Diabetic Retinopathy

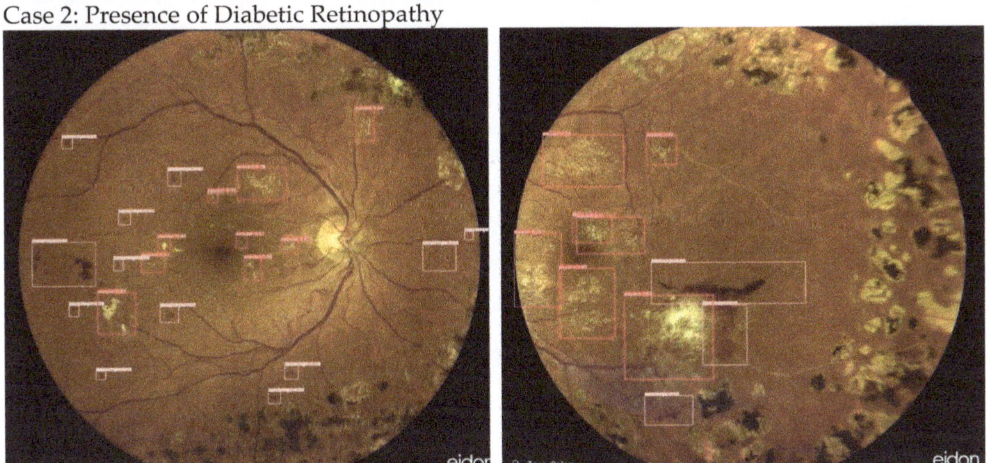

Case 3: Presence of Diabetic Retinopathy and laser impacts where exudates are framed in red and hemorrhages are framed in pink

Figure 8. Detection results using YOLOv5.

4.3. *Abulcasis DR-AI Detection Software*

In order to facilitate the use of the detection code, a graphical interface was created using QT Python and FPDF Python.

The first interface allows users to fill in fields with different information, such as patient's name and date, and they can also add comments, as shown in Figure 9.

Figure 9. Login interface.

A "Validate" button allows the user to validate the filled fields and then to choose the color fundus image path, as well as to switch to the second interface.

The second interface permits us to select a specific image to analyze (Figure 10). Once the "Detection" button is pressed, the third interface is displayed. A "Back" button allows one to go back to the first interface.

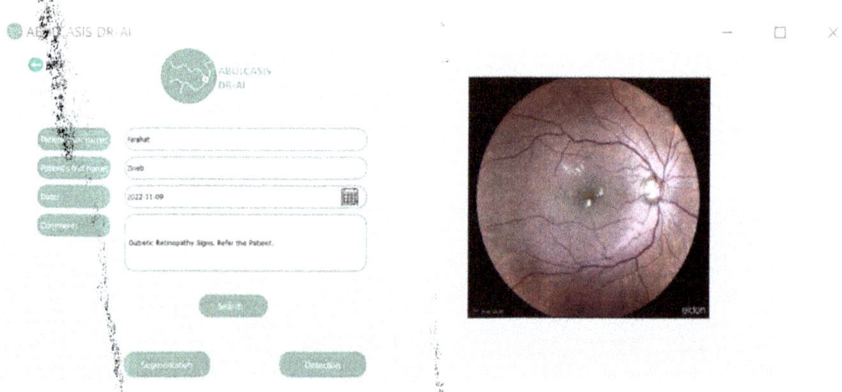

Figure 10. Image choice interface.

The detection result is shown after that, and the user can also enlarge the resulting image in order to see different details of the detected signs of DR.

Furthermore, it is possible to add comments as shown on Figure 11, so as to describe the patient's case.

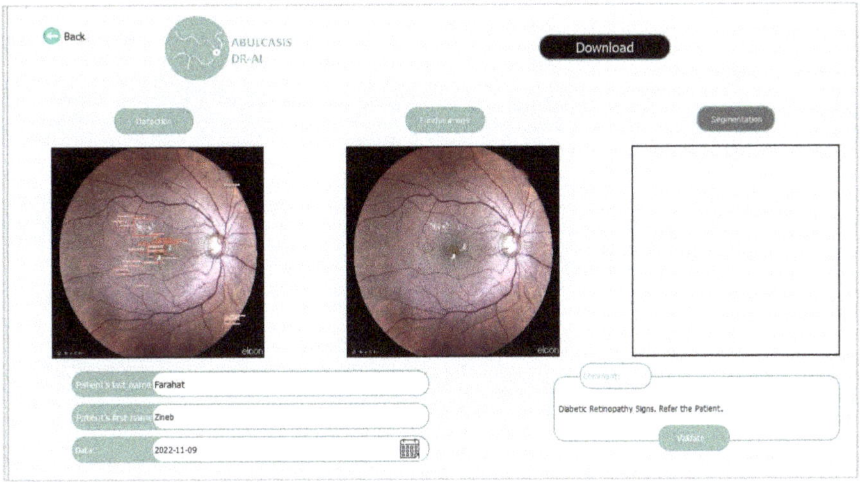

Figure 11. Detection result interface.

By clicking on validate a full report is shown containing the implemented information, as well as the resulting color fundus image (Figure 12).

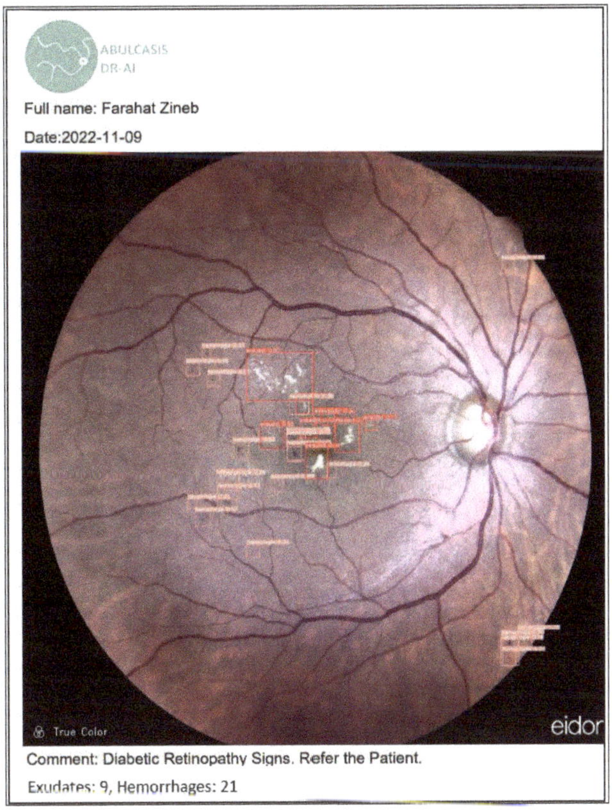

Figure 12. Detection full report.

A blind test was performed using different color fundus real images taken directly from the Icare Eidon modality. The Table 3 below shows clinical characteristics of 20 patients.

Table 3. Clinical characteristics of the patients.

Age	Percentage
20–40	10%
40–60	65%
60–80	25%
Sex	
Male	30%
Female	70%
Diabetic Retinopathy	
Presence	55%
Absence	45%

Comparison results of detection software were summarized and presented in Table 4 below.

Table 4. Detection comparison results.

	Expert Doctor	Resident Doctor	Abulcasis DR-AI	Correct Answer
Total Exudates	98.2%	84.4%	100%	100%
Total Hemorrhages	100%	83.6%	100%	100%
Mean Detection	99%	84%	100%	100%

Figure 13 below represents precision and recall curves obtained using YOLOV5.

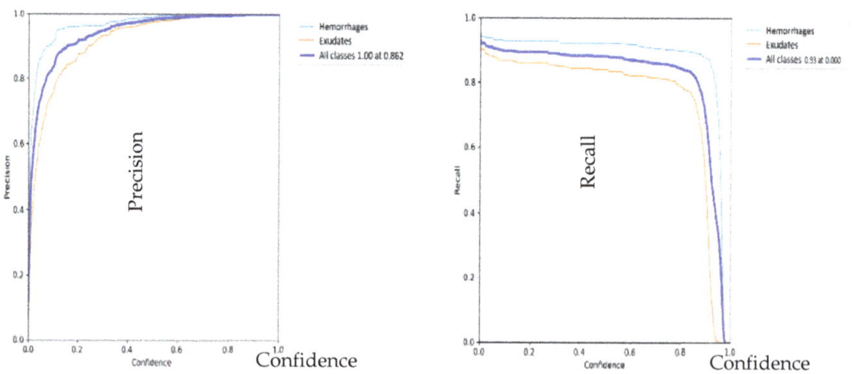

Figure 13. Precision and recall curves.

Furthermore, a public dataset named DeepDRID was used in order to test Abulcasis DR-AI software.

The proposed software detected all hemorrhages and exudates, as presented in Figure 14. It also recognized laser impacts on the example (d) and did not consider them as pathological signs.

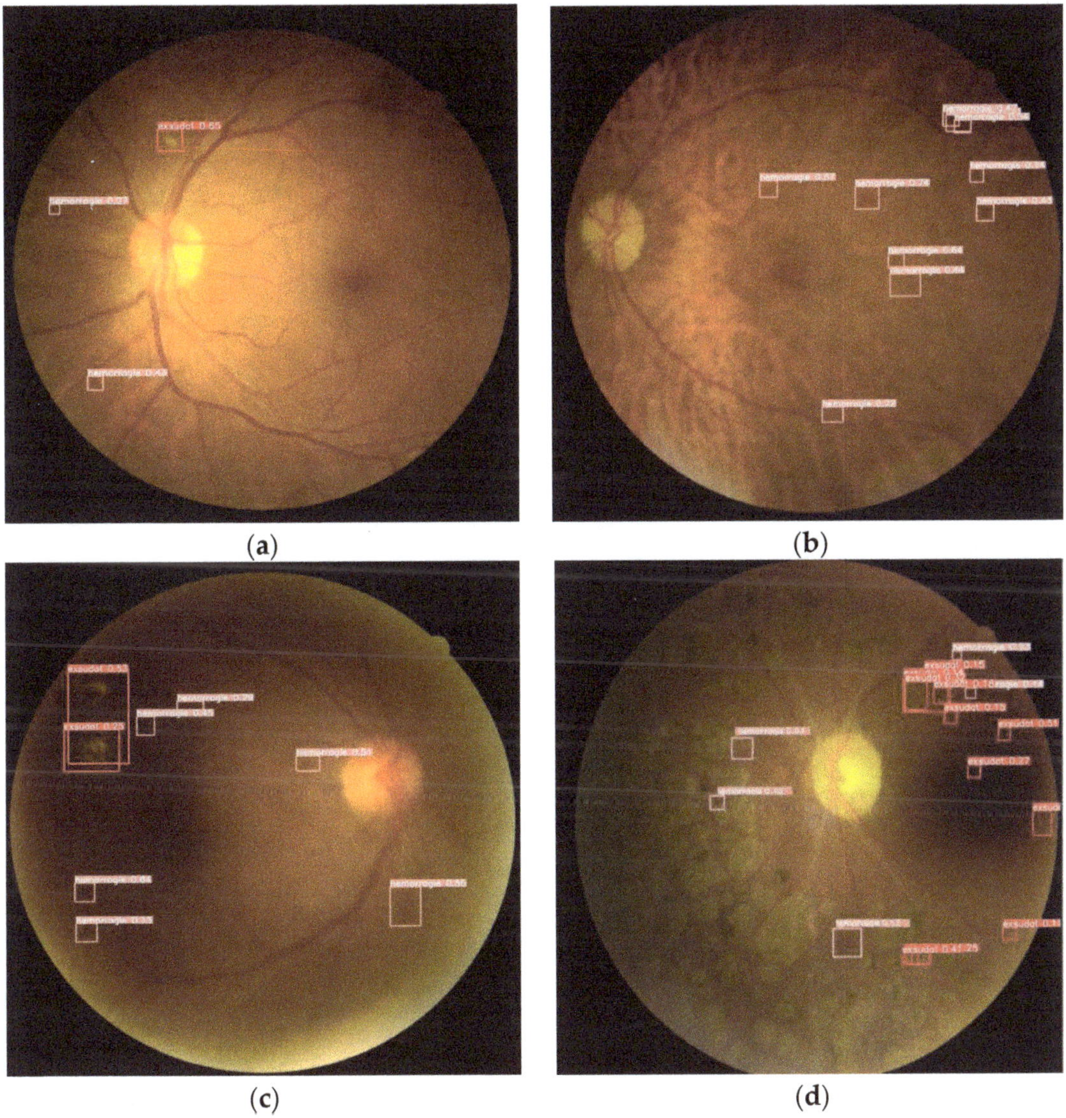

Figure 14. Examples of DR signs detection on DeepDRID public dataset. (**a**) Color fundus image with small hemorrhages and exudates; (**b**) Color fundus image with hemorrhages; (**c**) Color fundus image with hemorrhages and exudates; (**d**) Color fundus image with hemorrhages, exudates and laser impacts.

5. Discussion

As shown in Table 5, each of the two studied solutions has advantages and limitations. What characterized segmentation is the fact that it allowed us to assign a label to every pixel in our color fundus images. Detection enables users to insert colored bounding boxes around the objects in order to make them recognizable.

Table 5. Advantages and limitations of both segmentation and detection approaches.

Approach	Segmentation	Detection
Advantages	-Highlights the segmented parts of the image with different colors, which makes them easily recognizable. -Highlights the segmented parts of the image with different colors, which makes them easily recognizable. -Detects, classifies, and segments every object by assigning a label to each pixel of the image.	-Inserts a bounding colored box around each object.
Limitations	-Requires the creation of learning masks. -The grey color of the output image may hide unsegmented signs.	-Requires the creation of learning masks.

However, since we aim to use this solution for early detection of DR signs, even the smallest exudates and hemorrhages must be taken into consideration and treated to prevent later complications. However, the grey color of the output image may hide unsegmented signs and details. Moreover, the detection method showed better results during the performed blind test.

The proposed software, named Abulcasis DR-AI, allows users to automatically detect DR signs (Figure 15). Differently from the papers presented before, it can detect both types of exudates (hard and soft exudates) and hemorrhages. Abulcasis DR-AI has the capacity to recognize laser impact and does not consider them as DR signs. It has been tested on fundus images, and it has successfully detected 100% of diabetic retinopathy signs, including those not easily detectable with the naked eye. The expert doctor detected 99% of DR signs, and the resident doctor detected 84%. In performing the tests, no false positives were detected.

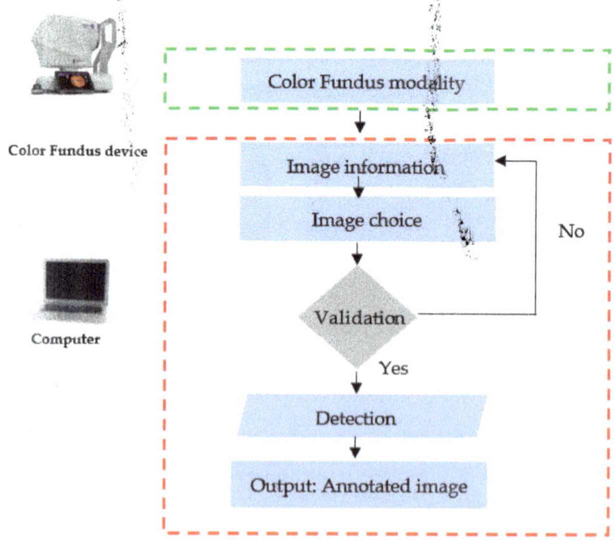

Figure 15. Image detection flowchart.

Artificial intelligence is the future of early screening of diabetic retinopathy and blindness reduction. The solution we have developed could be used in medical caravans organized in regions with low density of hospital structures, with the aim of optimizing the flow of patients within the hospital structures, and thus it will allow people who do not have access to health care to benefit from medical assistance and to save travel costs for those who do not need an examination by a specialist.

6. Conclusions

With a growing diabetic population and a widening gap between demand and the number of trained resources, we believe that early screening and management of DR must be taken into account. Automation of DR screening would be beneficial in Morocco, where there are few doctors for a fast expanding patient population. Artificial intelligence is the future of early screening to address this health issue and help reduce blindness. The U-Net and YOLOv5 models were used and compared to detect hemorrhages and exudates in retinography images. Therefore, we aim to utilize the proposed method in medical caravans arranged in low-density areas with the goal of improving tele-screening in ophthalmology, optimizing patient flow within healthcare buildings, and allowing people who do not require a specialist checkup to save travel expenses. In upcoming projects, the creation of more labeled retinal images is mandatory for further training of our software in order to improve the metrics and make it able to recognize rare cases of DR. Nevertheless, Abulcasis DR-AI software still has limitations that should be improved, and more investigations should be done in order to create a grading tool of DR severity.

Author Contributions: Methodology, N.Z., N.S., M.N.N. and K.M.; Validation, S.B., M.B. and K.M.; Investigation, Z.F.; Resources, M.B.; Data curation, Z.F., S.B. and M.B.; Writing—original draft, Z.F.; Writing—review & editing, N.Z., N.S. and M.N.N.; Supervision, N.S., M.B. and M.N.N.; Project administration, N.S. and M.N.N. All authors have read and agreed to the published version of the manuscript.

Funding: This research received no external funding.

Institutional Review Board Statement: The study was conducted in accordance with the Declaration of Helsinki, and approved by the Ethics Committe of ethics committee of the Cheikh Zaid International University Hospital (protocol code CEFCZ/PR/09/01/23).

Informed Consent Statement: Informed consent was obtained from all subjects involved in the study.

Data Availability Statement: A local dataset was created for training and testing the software, nevertheless it is unavailable due to privacy and ethical restrictions. The software was also tested using a public dataset called DeepDRID: github.com/mutual-ai/Deep-Diabetic-Retinopathy-Image-Dataset-DeepDRiD-, accessed on 20 February 2023.

Acknowledgments: The software presented in this paper is the result of a collaboration between the LISTD Laboratory of the ENSMR and the ophthalmological center of the Cheikh Zaid Foundation. We express our gratitude to President of Rabat's Abulcasis International University of Health Sciences and also the director of Cheikh Zaïd International University Hospital. We thank the Cheikh Zaïd Foundation Ophthalmic Center team for their assistance in conducting this investigation.

Conflicts of Interest: The authors declare no conflict of interest.

References

1. Zheng, Y.; He, M.; Congdon, N. The worldwide epidemic of diabetic retinopathy. *Indian J. Ophthalmol.* **2012**, *60*, 428. [PubMed]
2. Razzak, A.; Kassimi, A.; Mchachi, A.; Benhmidoun, L.; Rachid, R.; Elbelhadji, M. Aspects epidemiologiques de la rétinopathie diabétique à casablanca. *J. De La Société Maroc. D'ophtalmologie* **2019**, *28*, 58–65.
3. Farahat, Z.; Souissi, N.; Belmekki, M.; Megdiche, K.; Benamar, S.; Bennani, Y.; Bencherif, S.; Ngote, N. Diabetic retinopathy: New perspectives with artificial intelligence. In Proceedings of the 2021 Fifth International Conference On Intelligent Computing in Data Sciences (ICDS), Fez, Morocco, 20–22 October 2021; pp. 1–7.
4. Benamar, S.; Bennani, Y.; Bencherif, S.; Farahat, Z.; Souissi, N.; Ngote, N.; Megdiche, K.; Belmekki, M. Diabetic retinopathy screening and management in morocco: Challenges and possible solutions. In Proceedings of the 2021 Fifth International Conference On Intelligent Computing in Data Sciences (ICDS), Fez, Morocco, 20–22 October 2021; pp. 1–8.
5. Leontidis, G.; Al-Diri, B.; Hunter, A. Diabetic retinopathy: Current and future methods for early screening from a retinal hemodynamic and geometric approach. *Expert Rev. Ophthalmol.* **2014**, *9*, 431–442. [CrossRef]
6. Calivá, F.; Leontidis, G.; Chudzik, P.; Hunter, A.; Antiga, L.; Al-Diri, B. Hemodynamics in the retinal vasculature during the progression of diabetic retinopathy. *J. Model. Ophthalmol.* **2017**, *1*, 6–15. [CrossRef]
7. Kang, Q.; Yang, C. Oxidative stress and diabetic retinopathy: Molecular mechanisms, pathogenetic role and therapeutic implications. *Redox Biol.* **2020**, *37*, 101799. [CrossRef]
8. Heng, L.Z.; Comyn, O.; Peto, T.; Tadros, C.; Ng, E.; Sivaprasad, S.; Hykin, P.G. Diabetic retinopathy: Pathogenesis, clinical grading, management and future developments. *Diabet. Med.* **2013**, *30*, 640–650. [CrossRef]

9. Grauslund, J. Diabetic retinopathy screening in the emerging era of artificial intelligence. *Diabetologia* **2022**, *65*, 1415–1423. [CrossRef]
10. Boudry, C.; Al Hajj, H.; Arnould, L.; Mouriaux, F. Analysis of international publication trends in artificial intelligence in ophthalmology. *Graefe's Arch. Clin. Exp. Ophthalmol.* **2022**, *260*, 1779–1788. [CrossRef]
11. Porwal, P.; Pachade, S.; Kamble, R.; Kokare, M.; Deshmukh, G.; Sahasrabuddhe, V.; Meriaudeau, F. Indian diabetic retinopathy image dataset (idrid): A database for diabetic retinopathy screening research. *Data* **2018**, *3*, 25. [CrossRef]
12. Xu, Y.; Zhou, Z.; Li, X.; Zhang, N.; Zhang, M.; Wei, P. Ffu-net: Feature fusion u-net for lesion segmentation of diabetic retinopathy. *BioMed Res. Int.* **2021**, *2021*, 6644071. [CrossRef]
13. Kou, C.; Li, W.; Yu, Z.; Yuan, L. An enhanced residual u-net for microa- neurysms and exudates segmentation in fundus images. *IEEE Access* **2020**, *8*, 185514–185525. [CrossRef]
14. Li, H.; Wang, Y.; Wan, C.; Shen, J.; Chen, Z.; Ye, H.; Yu, Q. Mau-net: A retinal vessels segmentation method. In Proceedings of the 2020 42nd Annual International Conference of the IEEE Engineering in Medicine & Biology Society (EMBC), Montreal, QC, Canada, 20–24 July 2020; pp. 1958–1961.
15. Zhang, Z.; Wu, C.; Coleman, S.; Kerr, D. Dense-inception u-net for medical image segmentation. *Comput. Methods Programs Biomed.* **2020**, *192*, 105395. [CrossRef] [PubMed]
16. Dai, L.; Wu, L.; Li, H.; Cai, C.; Wu, Q.; Kong, H.; Jia, W. A deep learning system for detecting diabetic retinopathy across the disease spectrum. *Nat. Commun.* **2021**, *12*, 3242. [CrossRef] [PubMed]
17. Sambyal, N.; Saini, P.; Syal, R.; Gupta, V. Modified U-Net architecture for semantic segmentation of diabetic retinopathy images. *Biocybern. Biomed. Eng.* **2020**, *40*, 1094–1109. [CrossRef]
18. Yadav, N. A deep data-driven approach for enhanced segmentation of blood vessel for diabetic retinopathy. *Int. J. Imaging Syst. Technol.* **2020**, *32*, 1696–1708. [CrossRef]
19. Soomro, T.A.; Hellwich, O.; Afifi, A.J.; Paul, M.; Gao, J.; Zheng, L. Strided U-Net model: Retinal vessels segmentation using dice loss. In Proceedings of the Digital Image Computing: Techniques and Applications, Canberra, ACT, Australia, 10–13 December 2018; pp. 1–8.
20. Ghosh, S.K.; Ghosh, A. A novel retinal image segmentation using rSVM boosted convolutional neural network for exudates detection. *Biomed. Signal Process. Control* **2021**, *68*, 102785. [CrossRef]
21. Lee, A.Y.; Yanagihara, R.T.; Lee, C.S.; Blazes, M.; Jung, H.C.; Chee, Y.E.; Gencarella, M.D.; Gee, H.; Maa, A.Y.; Cockerham, G.C.; et al. Multicenter, head-to-head, real-world validation study of seven automated artificial intelligence diabetic retinopathy screening systems. *Diabetes Care* **2021**, *44*, 1168–1175. [CrossRef]
22. Pal, P.; Kundu, S.; Dhara, A.K. Detection of red lesions in retinal fundus images using yolo v3. *Curr. Indian Eye Res. J. Ophthalmic Res. Group* **2020**, *7*, 49.
23. Akut, R.R. Film: Finding the location of microaneurysms on the retina. *Biomed. Eng. Lett.* **2019**, *9*, 497–506. [CrossRef]
24. Ming, S.; Xie, K.; Lei, X.; Yang, Y.; Zhao, Z.; Li, S.; Jin, X.; Lei, B. Evaluation of a novel artificial intelligence-based screening system for dia- betic retinopathy in community of china: A real-world study. *Int. Ophthalmol.* **2021**, *41*, 1291–1299. [CrossRef]
25. Gao, S.; Jin, X.; Zhao, Z.; Yu, W.; Chen, Y.; Sun, Y.; Ding, D. Validation and application of an artificial intelligence robot assisted diag- nosis system for diabetic retinopathy. *Chin. J. Exp. Ophthalmol.* **2019**, *12*, 669–673.
26. Yang, Y.; Shang, F.; Wu, B.; Yang, D.; Wang, L.; Xu, Y.; Zhang, W.; Zhang, T. Robust collaborative learning of patch-level and image-level annotations for diabetic retinopathy grading from fundus image. *IEEE Trans. Cybern.* **2021**, *52*, 11407–11417. [CrossRef] [PubMed]
27. Decencière, E.; Zhang, X.; Cazuguel, G.; Lay, B.; Cochener, B.; Trone, C.; Gain, P.; Ordonez, R.; Massin, P.; Erginay, A.; et al. Feedback on a pub-licly distributed image database: The messidor database. *Image Anal. Stereol.* **2014**, *33*, 231–234. [CrossRef]
28. Cuadros, J.; Bresnick, G. Eyepacs: An adaptable telemedicine system for diabetic retinopathy screening. *J. Diabetes Sci. Technol.* **2009**, *3*, 509–516. [CrossRef]
29. Schmidt-Erfurth, U.; Sadeghipour, A.; Gerendas, B.S.; Waldstein, S.M.; Bogunović, H. Artificial intelligence in retina. *Prog. Retin. Eye Res.* **2018**, *67*, 1–29. [CrossRef]
30. Ronneberger, O.; Fischer, P.; Brox, T. U-net: Convolutional networks for biomedical image segmentation. In Proceedings of the International Conference on Medical Image Computing and Computer-Assisted Intervention, Munich, Germany, 5–9 October 2015; pp. 234–241.
31. Pan, C.; Li, X.; Yan, W.Q. A learning-based positive feedback approach in salient object detection. In Proceedings of the 2018 International Conference on Image and Vision Computing New Zealand (IVCNZ), Auckland, New Zealand, 19–21 November 2018; pp. 1–6.
32. Jubayer, F.; Soeb, J.A.; Mojumder, A.N.; Paul, M.K.; Barua, P.; Kayshar, S.; Akter, S.S.; Rahman, M.; Islam, A. Detection of mold on the food surface using yolov5. *Curr. Res. Food Sci.* **2021**, *4*, 724–728. [CrossRef]
33. Wang, L.; Yan, W.Q. Tree leaves detection based on deep learning. In Proceedings of the Geometry and Vision: First International Symposium, ISGV 2021, Auckland, New Zealand, 28–29 January 2021; Revised Selected Papers 1; pp. 26–38.

tomography MDPI

Review

Extended Reality in Diagnostic Imaging—A Literature Review

Paulina Kukla [1], Karolina Maciejewska [1], Iga Strojna [1], Małgorzata Zapał [1,2], Grzegorz Zwierzchowski [1,3] and Bartosz Bąk [1,4,*]

1 Department of Electroradiology, Poznan University of Medical Sciences, 61-866 Poznan, Poland
2 Department of Adult Neurology, Medical University of Gdansk, 80-210 Gdansk, Poland
3 Department of Medical Physics, Greater Poland Cancer Centre, 61-866 Poznan, Poland
4 Department of Radiotherapy II, Greater Poland Cancer Centre, 61-866 Poznan, Poland
* Correspondence: bartosz.bak@wco.pl

Abstract: The utilization of extended reality (ER) has been increasingly explored in the medical field over the past ten years. A comprehensive analysis of scientific publications was conducted to assess the applications of ER in the field of diagnostic imaging, including ultrasound, interventional radiology, and computed tomography. The study also evaluated the use of ER in patient positioning and medical education. Additionally, we explored the potential of ER as a replacement for anesthesia and sedation during examinations. The use of ER technologies in medical education has received increased attention in recent years. This technology allows for a more interactive and engaging educational experience, particularly in anatomy and patient positioning, although the question may be asked: is the technology and maintenance cost worth the investment? The results of the analyzed studies suggest that implementing augmented reality in clinical practice is a positive phenomenon that expands the diagnostic capabilities of imaging studies, education, and positioning. The results suggest that ER has significant potential to improve diagnostic imaging procedures' accuracy and efficiency and enhance the patient experience through increased visualization and understanding of medical conditions. Despite these promising advancements, further research is needed to fully realize the potential of ER in the medical field and to address the challenges and limitations associated with its integration into clinical practice.

Keywords: extended reality; diagnostic imaging; virtual reality; radiology

Citation: Kukla, P.; Maciejewska, K.; Strojna, I.; Zapał, M.; Zwierzchowski, G.; Bąk, B. Extended Reality in Diagnostic Imaging—A Literature Review. *Tomography* **2023**, *9*, 1071–1082. https://doi.org/10.3390/tomography9030088

Academic Editors: Rafał Obuchowicz, Michał Strzelecki and Adam Piorkowski

Received: 7 May 2023
Revised: 21 May 2023
Accepted: 22 May 2023
Published: 24 May 2023

1. Introduction

Extended reality (ER) is a rapidly developing technology with the potential to revolutionize medicine. It allows for virtual and real-world environments to integrate, creating a new level of interaction (immersion) and user engagement [1]. In the field of diagnostic imaging (DI), it has the potential to streamline the diagnostic process, improve patient outcomes, and reduce healthcare costs [2]. ER is an umbrella term that encompasses various simulated reality technologies, including virtual reality (VR), augmented reality (AR), and mixed reality (MR). These technologies provide stereoscopic and three-dimensional (3D) immersion within an environment, as in VR, or are overlaid onto a real-world background, as in AR or MR [2,3].

Extended reality (ER) is an example of rapidly developing technology with the potential to revolutionize various industries, including medicine. It is an umbrella term for simulating objects beyond the real world. It encompasses a range of technologies, including: (1) augmented reality (AR), (2) mixed reality (MR), and (3) virtual reality (VR). These technologies differ in their degree of immersion and interaction with the simulated environment. A graphic of the mix of these technologies is shown in Figure 1.

According to the analysis of the existing literature, there is a need for more standardization in the extended reality terminology, likely due to its ongoing and rapid advancements and potential uses. ER can be divided into three types: augmented reality (AR), mixed reality (MR), and virtual reality (VR). These technologies differ regarding the user's ability to interact with the simulated environment and the degree of reality enhancement. The first type, augmented reality (1), is described as adding digital information to the real world, which the user can still see and interact with. Smartphone games, for example, commonly use this technology, such as in "Pokemon Go" (Niantic, Inc., San Francisco, CA, USA) or "The Witcher: Monster Slayer" (Spokko Inc., Warsaw, PL, USA). In medicine, AR can be used for guidance during procedures by superimposing both virtual and real images into the environment in real time [4,5]. The second type, mixed reality (2), combines digital and natural elements to create a new environment with which the user can interact in real time [6]. Virtual reality (3), the most well-known form of ER, creates an entirely immersive digital environment that replaces the real world [7]. It can be also used to provide a platform for remote training activities and scientific gatherings.

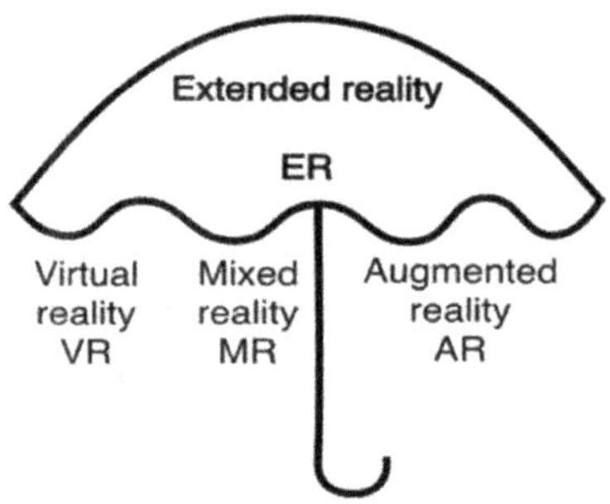

Figure 1. A graphical illustration of the classification of extended reality (ER) into subtypes of technologies such as: augmented reality (AR), mixed reality (MR), and virtual reality (VR). Self-developed illustration.

To create an extended reality environment, two main components are necessary. The first is a device that allows tracking of the user's head and eye position, which provides information about the location of structures in the virtual environment. The second is the visual display of virtual elements from the user's perspective. A smartphone, tablet, or and specialized glasses can still be effective for using AR technology, as they can provide a visual display of virtual elements from the user's perspective. To experience VR and MR, a head-mounted device (HMD) or cave automatic virtual environment (CAVE) is required, allowing for complete immersion in the virtual environment by blocking out the real world and displaying visualizations [8,9]. This division can commonly be found in ER studies published after 2020 [10,11].

However, the implementation of ER technology in medicine is still in its early stages, and many challenges need to be addressed before it can be fully integrated with clinical practice. Diagnostic imaging is the cornerstone of diagnostics in modern medicine; using various techniques capable of "reading" the human body, helping in emergencies, and diagnosing diseases, its continuous development is important to improve the efficiency of diagnosing patients with cancer or cardiovascular diseases [11–14]. Extended reality is one of the fastest-growing technologies used in DI. In recent years, there has been a growing interest in these methods and an increasing number of attempts to implement ER elements in daily life and in other fields, such as medicine and education. The components of extended reality are virtual reality, augmented reality, and mixed reality, which create their own adequate environments [4–8]. This study aims to analyze scientific publications from the last ten years documenting the issues and applications of extended reality, including virtual reality (VR) and augmented reality (AR), in diagnosing and educating patients or medical students. The main objective was to present the possibilities of implementing ER in diagnostic imaging, including ultrasound [15–22], interventional radiology [23–27], computed tomography [28–30], positioning [31–33], and education [34–41], and also to present the role of the patient as a user during the examination or as a replacement in anesthesia and sedation procedures [42–44]. In summary, extended reality is a rapidly developing technology with a vast potential for use in various fields, including medicine, education, and entertainment. Further research is needed to fully explore and understand this technology's potential and to create new and innovative ways of using it.

2. Materials and Methods

This review is based on the available literature from the PubMed and Scopus databases from the last 10 years (2013–2023). The searches included the following keywords and terms applied to titles and abstracts of full-length papers: [Extended reality] AND (virtual reality), [Journal] AND (mixed reality), [Journal] AND (augmented reality), and [Journal] OR "diagnostic imaging" [AllFields] OR "ultrasound" [All Fields]. We included articles and discussions that analyzed the role of augmented reality in diagnostic imaging. The search was limited to full-text articles, including electronic publications before printing. All the work collected was limited to human studies. The last PubMed search update for all sites was conducted on 24 January 2023. Papers in languages other than English and studies only available in abstract form were excluded from this analysis. Studies had to fulfill the following eligibility criteria to be selected for this review:

- Human research and English language only;
- Full articles and reviews;
- Using ER in diagnostic imaging.

According to a PRISMA [45–47] analysis, a total of 495 papers with eligible records were found in PubMed and Scopus, and 10 additional papers were extracted from their reference lists. After removing duplicates, 12,483 papers were screened for further analysis. We excluded 422 studies, as there were: 10 conference papers; 59 papers focused on surgery, radiotherapy, or neurology; 21 papers focused on virtual reconstruction; 5 papers focused on ER in other areas than diagnostic imaging, ultrasound, or education; and 5 papers about deep or machine learning. A detailed analysis is presented in Table 1. Finally, 57 studies could be included in the analysis for this review. All details are listed in the PRISMA workflow (Figure 2).

Table 1. Number of papers excluded from analysis.

Types of Research Excluded	N
Surgery	31
Virtual recon	21
Radiotherapy	12
Neurology	11
Conference paper	10
Forensics	7
Radiation oncology	5
Diagnostic without ER	5
Deep learning	4
Debate, abstract	3
Non-human	3

2.1. *Extended Reality (ER) as a Replacement for Anesthesia and Sedation*

Virtual reality (VR) has been proposed as a potential alternative to anesthesia and sedation in medical procedures such as burn treatment [42], interventional radiology [23], interventional oncology [43], or gynecology [44]. This approach, known as digital sedation, aims to use non-pharmacological and non-invasive VR tools to distract patients and reduce their perception of pain during medical procedures. Some studies have shown that patients perceive VR based distraction methods as more engaging and immersive than other forms of distraction and report less pain during procedures. In their study, Yi-Ling Wang et al. [44] suggest that virtual sedation may also be applied in hysterosalpingography, a solution proposed as a protocol developed by the author. However, in some studies, no significant differences in pain intensity were found between patients using VR distraction and those who did not. Furthermore, more research is needed to fully understand the potential of VR as a replacement for anesthesia and sedation. Additionally, VR distraction may also be a solution to some of the current problems with anesthesia, such as high costs, emissions of nitrous oxide and other gases that can contribute to global warming, or contamination of operating rooms with these compounds [48].

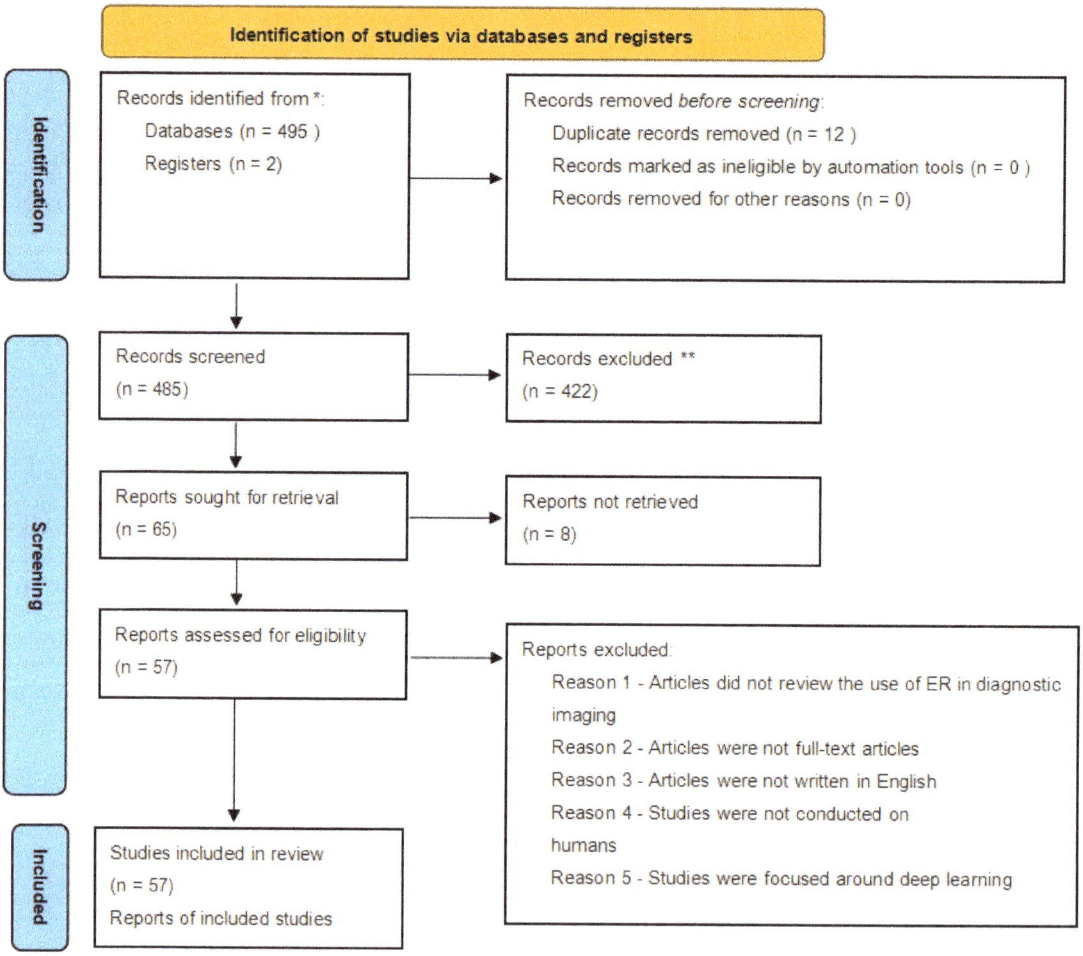

Figure 2. PRISMA analysis. * Records identified through PubMed and Scopus, ** Records excluded for not directly relevant.

2.2. Patient as a User during the Procedure

Recent studies have explored the use of virtual reality (VR) technology as an alternative to improve the patient experience during medical procedures. Nakarada-Kordic et al. [49] studied the differences between using a virtual simulation and a mock magnetic resonance imaging (MRI) simulation to reduce anxiety during the MRI procedure, including the level of comfort, the immersion of the patient in the simulation, and the patient's mood during the simulation. VR in clinical practice can help reduce anxiety symptoms in patients, especially in imaging diagnostics. However, the study did not find statistically significant differences, but participants still found the VR simulation was very helpful in preparing for the examination. Another study by Vu et al. [50] suggests that computer-generated VR environments can safely and effectively provide simple, quick, quantitative assessments of movement dysfunction in patients being evaluated for Parkinson's disease and can serve as an adjunct to brain imaging studies, such as [123I]FP-CIT SPECT/CT. Data on the patient's movement limitations were collected and analyzed by simulating daily activities. VR devices such as controllers and head-mounted displays (HMDs) can provide accurate evaluations of parameters related to head and limb movements. Continuous data analysis using controllers can determine the position of the patient's limbs and may be the basis for a new quantitative scale for evaluating patients with movement disorders.

In summary, recent research has shown that the use of virtual reality technology has the potential to improve the patient experience during medical procedures, especially in reducing anxiety during DI procedures and providing a quantitative assessment of movement dysfunction in patients with Parkinson's disease. Studies have demonstrated that VR simulations can be helpful in preparing for examinations, and devices such as controllers and HMDs can provide accurate evaluations of patients' movements. However, it should be noted that some studies did not find statistically significant differences between the use of VR and non-VR methods.

2.3. Positioning of the Patient in Medicine

Patient positioning for diagnostic examination is one of the essential and very crucial processes for obtaining a suitable image for further evaluation. VR technology using extended reality can be a valuable tool for proper patient positioning and preparation for diagnostics procedures [11,33–36,51].

The Oculus Rift system (Menlo Park, CA, USA) is a device for teaching positioning. In one of their studies, Sapkaroski et al. [48] showed that students who used VR in the process of learning positioning could better position the patient's hand for radiographic examination compared to those who practiced using traditional methods. The advantage of learning positioning using VR is the ability to control bone alignment through the "layer tearing" option, which allows for examination of the positioned part of the body and alignment correction. Additionally, VR provided students with the ability to evaluate each step in the positioning process, which is considered a more effective and valuable form of education. Students' survey responses were analyzed in another study comparing hand positioning learners using VR technology with the learning based on conventional patient positioning methods. One of the groups studied improved their skills using VR technology, while the other practiced traditional forms of learning. Students who did not use VR methods in learning required more time to improve their practical skills [12]. Many non-commercial and freely available virtual libraries, such as gVirtualXray (gVXR), (Bangor, UK) [52], are also used to develop positioning skills. Using these tools allows the user to see both correct and incorrect positioning, which minimizes the risk of exposure [33]. Overall, using VR technology in medical education can be a valuable tool for teaching positioning techniques and providing students with a more hands-on, interactive learning experience. It is worth noting that the use of VR technology in medical education is still in the early stages. Future research will be needed to confirm the benefits of this technology in the long term.

2.4. Education

For medical students and personnel, augmented reality methods such as VR and AR can provide immersive and interactive learning experiences to learn anatomy and patient positioning. With this approach, skills and knowledge in these areas can be effectively and efficiently improved without the need for actual patient interactions—that can ease the anxiety of a learner during the early stages of clinical education [38,40,41].

The DIVA system (Pasteur Institute and Institute Curie, Paris, FR), developed by the Mohamed El Beheiry uses VR and AR to visualize CT scans for medical education and training in diagnosing craniofacial trauma. The software allows the visualization of the scans in 3D and VR environments. According to a study by Bouaoud et al. [34], 92% of surveyed students were satisfied with using DIVA, and 83% noted its ease of use. The DIVA software also allows for detecting structures that would not be visible on a standard 2D image. Another system is the Magic Mirror (Guildford, UK), which is based on AR technology. It enables the learning of anatomy by superimposing 3D images of anatomical structures on the user's body, creating the illusion of being able to see inside the body. A team developed the system at the University of Southern California. It consists of a head-mounted display worn by the user and a motion tracker to track the user's movements. However, one of the main drawbacks of this system is its cost and availability in a commercial setting [35]. In a study undertaken by Weeks et al. [36], augmented reality technology was found to be particularly useful in learning anatomy, particularly the anatomy of the smaller body parts that can be difficult to learn due to their small size. Another example of an educational tool that uses AR technology is Second Life (SL) (San Francisco, CA, USA). SL is a virtual world that allows for e-learning and participation in meetings with users from different locations. Its unique features have made it an attractive tool for education and entertainment, while also raising concerns about addiction and questionable behavior. The platform is designed to simulate a fully realized virtual society, with users being able to interact with each other and participate in various events and activities. Virtual worlds can provide valuable and accessible educational tools, enhancing the learning experience in fields such as radiology [36,51]. Virtual reality simulators have been utilized for skill acquisition in robotic surgery. There have been attempts to enhance surgical robotic skills, and the preliminary outcomes are promising. Nevertheless, only one study involving actual patients has been conducted. Further investigation is necessary [39].

2.5. The Use of ER in Diagnostic Imaging

2.5.1. Ultrasound (US)

The analyzed articles (n = 8) discussed using VR in gynecological, thoracic, and lung ultrasound applications. In ultrasound simulators in clinical settings, two different image generation methods are used: interpolation, in which 2D ultrasound images are generated based on patient data, and generative, in which computer models are generated manually [15]. Reijnders et al. [16] focused on developing a method for assessing the volume of uterine–placental vessels before conception and in the first trimester of pregnancy. The study involved 35 women. They used 3D Power Doppler ultrasounds in two VR systems—Barco I-Space and VR desktop. Petersma C. S. et al. [17] used 3D ultrasound datasets to detect fetal abnormalities in the first trimester of pregnancy. They identified women in the first trimester with a high risk for the fetus. In the study group, a traditional 2D ultrasound was performed in addition to a 3D VR ultrasound. The control group consisted of women who underwent a conventional examination.

M. Bazelmans et al. [18] highlighted the usefulness of the Barco I-space system in detecting additional renal arteries in the fetus. The system allows for a detailed evaluation of the course of additional arteries, their length and width, the number of branches, and their relationship with surrounding structures, which is limited with conventional ultrasound examinations. In another study, Pietersen et al. [19] presented the results of using a simulator for lung ultrasounds. Such studies have also been conducted to evaluate ultrasound simulators for transvaginal ultrasounds, including using the ScanTrainer (Intelligent Ultrasound, Cardiff, UK) as a training tool [20]. Jensen et al. [21] evaluated the FAST ultrasound simulator. The researchers pointed out the possibility of analyzing FAST results in simulated conditions while maintaining appropriate imaging standards.

There is growing interest in using virtual reality technology in diagnostic and ultrasound processes. As educational tools, VR methods can improve personnel confidence in performing the examinations. Additionally, ultrasound simulators can be valuable educational tools for students, residents, and sonographers [22,53,54].

2.5.2. Computed Tomography

Using VR with diagnostic imaging techniques, such as computed tomography (CT), is becoming one of the main areas of interest nowadays. Mirhosseini et al. [28] proposed using VR as a tool for virtual colonoscopies (VCs) as a non-invasive and cost-effective screening procedure for colon cancer in which a fully integrated system with VR glasses allows for the capture of the pathology on a 3D image of the colon. The authors presented an immersive analytics system for VCs which enhances and improves the traditional desktop VC through the use of VR technology. Additionally, in another study, Kang et al. [11] used VR to analyze and evaluate heart abnormalities, specifically in the case of different types of defects; a double-outlet right ventricle converted the CT datasets of 12 heart specimens to stereoscopic images. The images were viewed and evaluated using MR goggles containing two lenses connected to a True 3D (Echopixel; Hewlett-Packard, Mountain View, CA, USA) using the mixed-reality system. The morphological features identified on the stereoscopic models were compared with findings at the macroscopic examination of the actual heart specimens. These studies' results showed a high accuracy level; the average agreement coefficient between the pathological and imaging studies was 97%, as was the agreement between virtual and actual pathological samples. Furthermore, Sun et al. [30] demonstrated the usefulness of VR in minimizing errors in local lung segmentation. Studies conducted on data from eighteen sets of CT scans, in which errors occurred during the automatic segmentation, showed lower error rates than automatically generated segmentations (2.54 ± 0.75 mm before refinement vs. 1.11 ± 0.43 mm post-refinement, $p < 0.001$). The average user interaction time with the system was about 2 min.

2.5.3. Interventional Radiology

In interventional radiology, diagnostic procedures (angiography, cholangiography, and phlebography) and therapeutic procedures (balloon angioplasty, thrombectomy, embolization, etc.) are performed. One advantage of interventional radiology is the ability to perform most procedures under local instead of general anesthesia, which shortens hospitalization times and reduces postoperative complications. However, a limitation of utilizing interventional radiology in medical procedures is the overexposure of the healthcare personnel to ionizing radiation, which is considerably higher than in other radiology departments. That is one reason why radiation protection specialists are increasingly seeking ways to reduce personnel exposure to ionizing radiation. Some authors believe virtual reality has the potential to visualize ionizing radiation doses in daily clinical practice through mixed reality and Monte Carlo simulations [51] or through virtual reality [27]. For example, a study by Yi Guo, Li Mao, et al. [55] estimated the dose delivered to a patient by designing a holographic visualization device which also displayed dose distributions in a room on HoloLens glasses or a monitor. A similar issue was investigated by Takeshi Takata, Susumu Nakabayashi, et al. [6].

Interventional radiology treatment rooms could be a particular application for VR, especially regarding virtual sedation, as discussed in a study by Cornelis, et al. [23]. The authors emphasized VR as an effective tool for reducing acute pain, anxiety, and discomfort. However, more research is needed to fully understand its neuropsychological basis and validate its effectiveness. Virtual sedation uses VR immersion to distract the patient from reality, creating a state of relaxation through enhanced sensory experiences.

Another practical application of VR was presented in a study by Matsuzaki S. et al. [24], who presented an example of virtual reality in interventional radiology to train medical staff in correctly using radiation shielding. The authors presented an application that allows the visualization of scattered radiation in an extended reality, which is difficult to visualize using traditional methods. A study of 33 healthcare workers showed that the application reduced the head exposure of the leading operator by 22–73%. A survey measuring satisfaction, confidence, attention, and accuracy showed high reliability of positive feedback. A study by Popovic B. et al. [25] focused on simulation training for coronary angiography and was conducted with 20 doctors. The results showed that the training participants had better results in radiation protection (collimation and reduced distance between the device and patient), shorter procedure times, and reduced contrast administration. There were no increased post-operative complications. These results suggest that using VR in simulation training is an effective way to improve skills. One of the key advantages of VR training is the ability to simulate real-life scenarios in a controlled environment. In a study by Jensen et al. [26], the researchers allowed trainees to practice the procedure in a safe and risk-free setting before performing it on real patients. The transfer effect of VR training to real-life catheterization labs has also been studied. Several studies have shown that trainees who underwent VR training performed better in real-life procedures compared to those who received traditional training methods. In addition, VR training has been shown to be particularly effective in improving trainees' confidence levels and reducing their anxiety when performing procedures in the catheterization lab. Overall, VR training in coronary angiography has been shown to be an effective and valuable tool for medical professionals. It has the potential to improve patient outcomes, reduce medical errors, and increase the confidence and skills of trainees.

3. Discussion

In recent years, the potential of virtual and augmented reality technologies in medical education has received much attention [56]. They offer more interactive and engaging learning experiences, particularly in anatomy and patient positioning. The DIVA system is a successful tool for facial trauma diagnosis that uses VR to visualize data [34]. The Magic Mirror system uses AR to project 3D anatomy images onto the body for a more realistic understanding [35,57]. However, improvements in cost and accessibility are still needed [35]. AR technology has also proven helpful in teaching the complex anatomy of the head and neck [36]. In addition, as an e-learning tool, SL is used for virtual meetings and distance learning to enhance student engagement, in addition to VR and AR.

In procedures using ionizing radiation, one of the most valuable assets that ER provides is minimizing the doses used during procedures [24,25,55–60], with benefits to the patients and the staff. On the other hand, evaluating the cost-effectiveness of VR and AR in medical education is a crucial question when analyzing these technologies. While VR and AR provide a more interactive and engaging learning experience, the cost of the technology and maintenance must be considered. Based on the available literature, the effectiveness of using AR in ultrasound training, particularly in gynecology or breast applications, is a potential topic of discussion. Reijnders and others [16] discovered that the use of VR technology, specifically the Barco I-Space system, was effective in measuring the volume of maternal and fetal blood vessels before and during the first trimester of pregnancy. Similarly, incorporating 3D VR ultrasound data into traditional 2D ultrasound scans improved the ability to detect fetal abnormalities according to Petersma C. S et al. [17]. Although these VR and AR technology applications are focused on ultrasound, the potential exists to apply them to other medical fields as well. However, more research is needed to determine these technologies' cost-effectiveness and overall effectiveness in the education and training of medical professionals. Another aspect worthy of discussion is the limitations and challenges in the implementation of VR and AR technology in ultrasound training. In addition, there is a call for the standardization of the VR and AR systems used for this purpose to facilitate comparison and to ensure the uniformity of the training. On the other hand, the potential of VR and AR to improve patient care through ultrasound imaging is also remarkable. For example, the DIVA system enables the visualization of CT scans in a 3D virtual environment, potentially helping healthcare providers better identify and diagnose injuries [34].

Augmented reality technologies such as AR, VR, and MR have gained attention in diagnostic imaging over the past years. Based on this review of 12 studies, researchers have shown interest in developing these technologies for DI purposes. However, there is a need for the standardization of concepts and knowledge about the benefits of AR, VR, and MR. Currently, there are varying definitions and understandings of what constitutes augmented reality.

In addition, when using VR technology in the medical field, the frequent occurrence of "cybersickness" is a significant concern. Symptoms of cybersickness, such as eye fatigue, nausea, and vomiting, can reduce the operator's ability to concentrate and feel comfortable during procedures. Further research is needed to address these challenges [6,28]. Based on clinical studies, combining VR/MR technology with CT scanning also increases diagnostic potential. The risk of error and patient discomfort performing screening exams such as virtual colonoscopies could be significantly reduced by implementing these technologies in daily clinical practice. Virtual colonoscopy has many advantages over conventional colonoscopy; these include low cost, improved safety, shorter examination time, and appropriate measurement tools. Furthermore, the use of simulators of diagnostic devices, especially for MRI, in VR technology, may be beneficial for healthcare facilities that cannot install mockups of these devices to prepare patients for exams without exposing them to environmental risks. However, it should be noted that in all the aspects discussed in this review, the use of ER still needs data on the effectiveness of implementing virtual reality as a therapeutic or clinical modality, and there is an insufficient number of clinical studies using augmented reality.

The results of this analysis and their conclusions highlight the importance of integrating advanced medicine and technology. The use of virtual and augmented reality in diagnostic imaging can improve the accuracy and efficiency of examinations while simultaneously increasing patients' comfort during these procedures. Further studies in this area will pave the way for creating innovative techniques and tools that can be applied in clinical practice, improving diagnostic processes and quality of care. This could increase productivity while reducing costs, which could be attractive to healthcare facilities.

4. Conclusions

The results of the modalities discussed indicate that integrating extended reality modalities such as AR, MR, and VR into clinical practice is beneficial. They expand the diagnostic capabilities of imaging examinations, education, and positioning. Currently, this research area has yet to be fully explored. Further research in this area will modernize diagnostic procedures and improve healthcare by creating new techniques and tools that can be used in clinical practice [58–61]. In conclusion, virtual and augmented reality technologies have the potential to transform medical education and patient care in diagnostic imaging.

Author Contributions: Conceptualization, P.K. and K.M.; methodology, P.K and K.M.; data curation, P.K.; writing—original draft preparation, P.K., K.M., I.S. and M.Z.; writing—review and editing, B.B. and G.Z.; supervision, B.B.; project administration, P.K. and B.B. All authors have read and agreed to the published version of the manuscript.

Funding: This research received no external funding.

Institutional Review Board Statement: The study did not require ethical approval.

Informed Consent Statement: Patient consent was waived.

Data Availability Statement: The datasets analyzed in the current study are available from the corresponding author upon request.

Acknowledgments: Not applicable.

Conflicts of Interest: The authors declare no conflict of interest.

References

1. Uppot, R.N.; Laguna, B.; McCarthy, C.J.; De Novi, G.; Phelps, A.; Siegel, E.; Courtier, J. Implementing Virtual and Augmented Reality Tools for Radiology Education and Training, Communication, and Clinical Care. *Radiology* **2019**, *291*, 570–580. [CrossRef]
2. Jung, C.; Wolff, G.; Wernly, B.; Bruno, R.R.; Franz, M.; Schulze, P.C.; Silva, J.N.A.; Silva, J.R.; Bhatt, D.L.; Kelm, M. Virtual and Augmented Reality in Cardiovascular Care: State-of-the-Art and Future Perspectives. *JACC Cardiovasc. Imaging* **2022**, *15*, 519–532. [CrossRef]

3. Park, B.J.; Hunt, S.J.; Martin, C., 3rd; Nadolski, G.J.; Wood, B.J.; Gade, T.P. Augmented and Mixed Reality: Technologies for Enhancing the Future of IR. *J. Vasc. Interv. Radiol.* **2020**, *31*, 1074–1082. [CrossRef] [PubMed]
4. Solbiati, L.A. Augmented Reality: Thrilling Future for Interventional Oncology? *Cardiovasc. Interv. Radiol.* **2021**, *44*, 782–783. [CrossRef] [PubMed]
5. Goo, H.W.; Park, S.J.; Yoo, S.J. Advanced Medical Use of Three-Dimensional Imaging in Congenital Heart Disease: Augmented Reality, Mixed Reality, Virtual Reality, and Three-Dimensional Printing. *Korean J. Radiol.* **2020**, *21*, 133–145. [CrossRef] [PubMed]
6. Takata, T.; Nakabayashi, S.; Kondo, H.; Yamamoto, M.; Furui, S.; Shiraishi, K.; Kobayashi, T.; Oba, H.; Okamoto, T.; Kotoku, J. Mixed Reality Visualization of Radiation Dose for Health Professionals and Patients in Interventional Radiology. *J. Med. Syst.* **2021**, *45*, 38. [CrossRef]
7. Werner, H.; Lopes Dos Santos, J.R.; Ribeiro, G.; Araujo Júnior, E. Prenatal Phenotype of Down Syndrome Using 3-D Virtual Reality. *J. Obstet. Gynaecol. Can.* **2017**, *39*, 886–889. [CrossRef]
8. Sutherland, J.; Belec, J.; Sheikh, A.; Chepelev, L.; Althobaity, W.; Chow, B.J.W.; Mitsouras, D.; Christensen, A.; La Russa, D.J. Applying Modern Virtual and Augmented Reality Technologies to Medical Images and Models. *J. Digit. Imaging* **2018**, *32*, 38–53. [CrossRef]
9. Elsayed, M.; Kadom, N.; Ghobadi, C.; Strauss, B.; Al Dandan, O.; Aggarwal, A.; Anzai, Y.; Griffith, B.; Lazarow, F.; Straus, C.M.; et al. Virtual and augmented reality: Potential applications in radiology. *Acta Radiol.* **2020**, *61*, 1258–1265. [CrossRef]
10. López-Ojeda, W.; Hurley, R.A. Extended-Reality Technologies: An Overview of Emerging Applications in Medical Education and Clinical Care. *J. Neuropsychiatry Clin. Neurosci.* **2021**, *33*, A4-177. [CrossRef]
11. Kang, S.L.; Shkumat, N.; Dragulescu, A.; Guerra, V.; Padfield, N.; Krutikov, K.; Chiasson, D.A.; Chaturvedi, R.R.; Yoo, S.J.; Benson, L.N. Mixed-reality view of cardiac specimens: A new approach to understanding complex intracardiac congenital lesions. *Pediatr. Radiol.* **2020**, *50*, 1610–1616. [CrossRef]
12. Scatliff, J.H.; Morris, P.J. From Roentgen to magnetic resonance imaging: The history of medical imaging. *N. C. Med. J.* **2014**, *75*, 111–113. [CrossRef]
13. Nielsen, M.B.; Søgaard, S.B.; Bech Andersen, S.; Skjoldbye, B.; Hansen, K.L.; Rafaelsen, S.; Nørgaard, N.; Carlsen, J.F. Highlights of the development in ultrasound during the last 70 years: A historical review. *Acta Radiol.* **2021**, *62*, 1499–1514. [CrossRef]
14. Seibert, J.A. One hundred years of medical diagnostic imaging technology. *Health Phys.* **1995**, *69*, 695–720. [CrossRef] [PubMed]
15. Konge, L.; Albrecht-Beste, E.; Nielsen, M.B. Virtual-reality simulation-based training in ultrasound. *Ultraschall Med.* **2014**, *35*, 95–97. [CrossRef] [PubMed]
16. Reijnders, I.F.; Mulders, A.G.M.G.J.; Koster, M.P.H.; Koning, A.H.J.; Frudiger, A.; Willemsen, S.P.; Jauniaux, E.; Burton, G.J.; Steegers-Theunissen, R.P.M.; Steegers, E.A.P. New imaging markers for preconceptional and first-trimester utero-placental vascularization. *Placenta* **2018**, *61*, 96–102. [CrossRef]
17. Pietersma, C.S.; Mulders, A.G.M.G.J.; Moolenaar, L.M.; Hunink, M.G.M.; Koning, A.H.J.; Willemsen, S.P.; Go, A.T.J.I.; Steegers, E.A.P.; Rousian, M. First trimester anomaly scan using virtual reality (VR FETUS study): Study protocol for a randomized clinical trial. *BMC Pregnancy Childbirth* **2020**, *20*, 515. [CrossRef] [PubMed]
18. Bazelmans, M.; Peters, N.C.J.; Koning, A.H.; Eggink., A.J.; Cohen-Overbeek, T.E. Power Doppler rendering of fetal bilateral accessory renal arteries in virtual reality. *Ultrasound Obstet. Gynecol.* **2014**, *44*, 375–376. [CrossRef]
19. Pietersen, P.I.; Konge, L.; Graumann, O.; Nielsen, B.U.; Laursen, C.B. Developing and Gathering Validity Evidence for a Simulation-Based Test of Competencies in Lung Ultrasound. *Respiration* **2019**, *97*, 329–336. [CrossRef]
20. Almestehi, M.; Alomaim, W.; Rainford, L.; Stokes, D.; Stanton, M.; Moran, M. Role of the virtual reality simulator (ScanTrainer) as a multidisciplinary training tool in transvaginal ultrasound: A systematic review and narrative synthesis. *Radiography* **2019**, *25*, 260–268. [CrossRef] [PubMed]
21. Jensen, J.K.; Dyre, L.; Jørgensen, M.E.; Andreasen, L.A.; Tolsgaard, M.G. Collecting Validity Evidence for Simulation-Based Assessment of Point-of-Care Ultrasound Skills. *J. Ultrasound Med.* **2017**, *36*, 2475–2483. [CrossRef] [PubMed]
22. Rosenfeldt Nielsen, M.; Kristensen, E.Q.; Jensen, R.O.; Mollerup, A.M.; Pfeiffer, T.; Graumann, O. Clinical Ultrasound Education for Medical Students: Virtual Reality Versus e-Learning, a Randomized Controlled Pilot Trial. *Ultrasound, Q.* **2021**, *37*, 292–296. [CrossRef]
23. Cornelis, F.H.; Monard, E.; Moulin, M.A.; Vignaud, E.; Laveissiere, F.; Ben Ammar, M.; Nouri-Neuville, M.; Barral, M.; Lombart, B. Sedation and analgesia in interventional radiology: Where do we stand, where are we heading and why does it matter? *Diagn. Interv. Imaging* **2019**, *100*, 753–762. [CrossRef] [PubMed]
24. Matsuzaki, S.; Moritake, T.; Morota, K.; Nagamoto, K.; Nakagami, K.; Kuriyama, T.; Kunugita, N. Development and assessment of an educational application for the proper use of ceiling-suspended radiation shielding screens in angiography rooms using augmented reality technology. *Eur. J. Radiol.* **2021**, *143*, 109925. [CrossRef]
25. Popovic, B.; Pinelli, S.; Albuisson, E.; Metzdorf, P.A.; Mourer, B.; Tran, N.; Camenzind, E. The Simulation Training in Coronary Angiography and Its Impact on Real Life Conduct in the Catheterization Laboratory. *Am. J. Cardiol.* **2019**, *123*, 1208–1213. [CrossRef]
26. Jensen, U.J.; Jensen, J.; Ahlberg, G.; Tornvall, P. Virtual reality training in coronary angiography and its transfer effect to real-life catheterisation lab. *EuroIntervention* **2016**, *11*, 1503–1510. [CrossRef] [PubMed]
27. Fujibuchi, T. Radiation protection education using virtual reality for the visualisation of scattered distributions during radiological examinations. *J. Radiol. Prot.* **2021**, *41*, S317. [CrossRef]
28. Mirhosseini, S.; Gutenko, I.; Ojal, S.; Marino, J.; Kaufman, A. Immersive Virtual Colonoscopy. *IEEE Trans. Vis. Comput. Graph.* **2019**, *25*, 2011–2021. [CrossRef] [PubMed]

29. MacDougall, R.D.; Scherrer, B.; Don, S. Development of a tool to aid the radiologic technologist using augmented reality and computer vision. *Pediatr. Radiol.* **2018**, *48*, 141–145. [CrossRef]
30. Sun, S.; Sonka, M.; Beichel, R.R. Lung segmentation refinement based on optimal surface finding utilizing a hybrid desktop/virtual reality user interface. *Comput. Med. Imaging Graph.* **2013**, *37*, 15–27. [CrossRef]
31. Sapkaroski, D.; Baird, M.; Mundy, M.; Dimmock, M.R. Quantification of Student Radiographic Patient Positioning Using an Immersive Virtual Reality Simulation. *Simul. Healthc.* **2019**, *14*, 258–263. [CrossRef]
32. Sapkaroski, D.; Mundy, M.; Dimmock, M.R. Virtual reality versus conventional clinical role-play for radiographic positioning training: A students' perception study. *Radiography* **2020**, *26*, 57–62. [CrossRef]
33. Sujar, A.; Kelly, G.; García, M.; Vidal, F.P. Interactive teaching environment for diagnostic radiography with real-time X-ray simulation and patient positioning. *Int. J. Comput. Assist. Radiol. Surg.* **2022**, *17*, 85–95. [CrossRef]
34. Bouaoud, J.; El Beheiry, M.; Jablon, E.; Schouman, T.; Bertolus, C.; Picard, A.; Masson, J.B.; Khonsari, R.H. DIVA, a 3D virtual reality platform, improves undergraduate craniofacial trauma education. *J. Stomatol. Oral. Maxillofac. Surg.* **2021**, *122*, 367–371. [CrossRef] [PubMed]
35. Bork, F.; Stratmann, L.; Enssle, S.; Eck, U.; Navab, N.; Waschke, J.; Kugelmann, D. The Benefits of an Augmented Reality Magic Mirror System for Integrated Radiology Teaching in Gross Anatomy. *Anat. Sci. Educ.* **2019**, *12*, 585–598. [CrossRef]
36. Weeks, J.K.; Pakpoor, J.; Park, B.J.; Robinson, N.J.; Rubinstein, N.A.; Prouty, S.M.; Nachiappan, A.C. Harnessing Augmented Reality and CT to Teach First-Year Medical Students Head and Neck Anatomy. *Acad. Radiol.* **2021**, *28*, 871–876. [CrossRef]
37. Lorenzo-Alvarez, R.; Pavia-Molina, J.; Sendra-Portero, F. Exploring the Potential of Undergraduate Radiology Education in the Virtual World Second Life with First-cycle and Second-cycle Medical Students. *Acad. Radiol.* **2018**, *25*, 1087–1096. [CrossRef] [PubMed]
38. Ryan, E.; Poole, C. Impact of Virtual Learning Environment on Students' Satisfaction, Engagement, Recall, and Retention. *J. Med. Imaging Radiat. Sci.* **2019**, *50*, 408–415. [CrossRef]
39. Moglia, A.; Ferrari, V.; Morelli, L.; Ferrari, M.; Mosca, F.; Cuschieri, A. A Systematic Review of Virtual Reality Simulators for Robot-assisted Surgery. *Eur. Urol.* **2016**, *69*, 1065–1080. [CrossRef] [PubMed]
40. O'Connor, M.; Stowe, J.; Potocnik, J.; Giannotti, N.; Murphy, S.; Rainford, L. 3D virtual reality simulation in radiography education: The students' experience. *Radiography* **2021**, *27*, 208–214. [CrossRef]
41. Lorenzo-Alvarez, R.; Rudolphi-Solero, T.; Ruiz-Gomez, M.J.; Sendra-Portero, F. Medical Student Education for Abdominal Radiographs in a 3D Virtual Classroom Versus Traditional Classroom: A Randomized Controlled Trial. *Am. J. Roentgenol.* **2019**, *213*, 644–650. [CrossRef]
42. Bermo, M.S.; Zacharias, C.; Patterson, D.; Wilson, A.; Sharar, S.; Minoshima, S.; Hoffman, H.; Lewis, D.H. The Role of the Cerebellum in Pain Perception: A Brain SPECT Perfusion Study in Patients with Burn Injuries. *J. Neuroimaging.* **2020**, *30*, 815–821. [CrossRef] [PubMed]
43. Cornelis, F.H., Najdawi, M.; Ammar, M.B.; Nouri-Neuville, M.; Lombart, B.; Lotz, J.P.; Cadranel, J.; Barral, M. Integrative Medicine in Interventional Oncology: A Virtuous Alliance. *Medicina* **2020**, *56*, 35. [CrossRef]
44. Wang, Y.L.; Gao, H.X.; Wang, J.S.; Wang, J.H.; Bo, L.; Zhang, T.T.; Dai, Y.L.; Gao, L.L.; Liu, Q.; Zhang, J.J.; et al. Immersive virtual reality as analgesia for women during hysterosalpingography: Study protocol for a randomized controlled trial. *Trials* **2020**, *21*, 102. [CrossRef]
45. Page, M.J.; McKenzie, J.E.; Bossuyt, P.M.; Boutron, I.; Hoffmann, T.C.; Mulrow, C.D.; Shamseer, L.; Tetzlaff, J.M.; Akl, E.A.; Brennan, S.E. The PRISMA 2020 statement: An updated guideline for reporting systematic reviews. *PLoS Med.* **2021**, *18*, e1003583. [CrossRef]
46. Page, M.J.; Moher, D.; Bossuyt, P.M.; Boutron, I.; Hoffmann, T.C.; Mulrow, C.D.; Shamseer, L.; Tetzlaff, J.M.; Akl, E.A.; Brennan, S.E.; et al. PRISMA 2020 explanation and elaboration: Updated guidance and exemplars for reporting systematic reviews. *BMJ* **2021**, *372*, n160. [CrossRef]
47. Wang, X.; Chen, Y.; Liu, Y.; Yao, L.; Estill, J.; Bian, Z.; Wu, T.; Shang, H.; Lee, M.S.; Wei, D.; et al. Reporting items for systematic reviews and meta-analyses of acupuncture: The PRISMA for acupuncture checklist. *BMC Complement. Altern. Med.* **2019**, *19*, 208. [CrossRef] [PubMed]
48. Gordon, D. Sustainability in the Operating Room: Reducing Our Impact on the Planet. *Anesth. Clin.* **2020**, *38*, 679–692. [CrossRef]
49. Nakarada-Kordic, I.; Reay, S.; Bennett, G.; Kruse, J.; Lydon, A.M.; Sim, J. Can virtual reality simulation prepare patients for an MRI experience? *Radiography* **2020**, *26*, 205–213. [CrossRef]
50. Vu, J.P.; Yamin, G.; Reyes, Z.; Shin, A.; Young, A.; Litvan, I.; Xie, P.; Obrzut, S. Assessment of Motor Dysfunction with Virtual Reality in Patients Undergoing [123I]FP-CIT SPECT/CT Brain Imaging. *Tomography* **2021**, *26*, 95–106. [CrossRef]
51. Schroeder, A.N.; Kruse, R.C. The Future of Virtual Sports Ultrasound Education and Collaboration. *Curr. Sport. Med. Rep.* **2021**, *20*, 57–61. [CrossRef] [PubMed]
52. Vidal, F.P.; Villard, P.-F. Development and validation of real-time simulation of X-ray imaging with respiratory motion. *Comput. Med. Imaging Graph.* **2016**, *49*, 1–15. [CrossRef]
53. Chao, C.; Chalouhi, G.E.; Bouhanna, P.; Ville, Y.; Dommergues, M. Randomized Clinical Trial of Virtual Reality Simulation Training for Transvaginal Gynecologic Ultrasound Skills. *J. Ultrasound Med.* **2015**, *34*, 1663–1667. [CrossRef] [PubMed]
54. Jensen, J.K.; Dyre, L.; Jørgensen, M.E.; Andreasen, L.A.; Tolsgaard, M.G. Simulation-based point-of-care ultrasound training: A matter of competency rather than volume. *Acta Anaesthesiol. Scand.* **2018**, *62*, 811–819. [CrossRef]

55. Guo, Y.; Mao, L.; Zhang, G.; Chen, Z.; Pei, X.; Xu, X.G. Conceptual design and preliminary results of a vr-based radiation safety training system for interventional radiologists. *Radiat. Prot. Dosim.* **2020**, *190*, 58–65. [CrossRef]
56. Hayre, C.M.; Kilgour, A. Diagnostic radiography education amidst the COVID-19 pandemic: Current and future use of virtual reality (VR). *J. Med. Imaging Radiat. Sci.* **2021**, *52*, S20–S23. [CrossRef]
57. Ma, M.; Fallavollita, P.; Seelbach, I.; Von Der Heide, A.M.; Euler, E.; Waschke, J.; Navab, N. Personalized augmented reality for anatomy education. *Clin. Anat.* **2016**, *29*, 446–453. [CrossRef] [PubMed]
58. Nishi, K.; Fujibuchi, T.; Yoshinaga, T. Development and evaluation of the effectiveness of educational material for radiological protection that uses augmented reality and virtual reality to visualise the behaviour of scattered radiation. *J. Radiol. Prot.* **2022**, *42*, 011506. [CrossRef]
59. Mohammed, M.A.A.; Khalaf, M.H.; Kesselman, A.; Wang, D.S.; Kothary, N. A Role for Virtual Reality in Planning Endovascular Procedures. *J. Vasc. Interv. Radiol.* **2018**, *29*, 971–974. [CrossRef]
60. Troville, J.; Rudin, S.; Bednarek, D.R. A Prototype Software System for Intra-procedural Staff Dose Monitoring and Virtual Reality Training for Fluoroscopically Guided Interventional Procedures. *J. Digit. Imaging.* **2023**. [CrossRef]
61. Yeung, A.W.K.; Tosevska, A.; Klager, E.; Eibensteiner, F.; Laxar, D.; Stoyanov, J.; Glisic, M.; Zeiner, S.; Kulnik, S.T.; Crutzen, R.; et al. Virtual and Augmented Reality Applications in Medicine: Analysis of the Scientific Literature. *J. Med. Internet Res.* **2021**, *23*, e25499. [CrossRef]

cancers

Article

Deep Learning- and Expert Knowledge-Based Feature Extraction and Performance Evaluation in Breast Histopathology Images

Hepseeba Kode [1] and Buket D. Barkana [2,*]

1 Computer Science and Engineering Department, University of Bridgeport, Bridgeport, CT 06604, USA; hkode@my.bridgeport.edu
2 Electrical Engineering Department, University of Bridgeport, Bridgeport, CT 06604, USA
* Correspondence: bbarkana@bridgeport.edu

Simple Summary: Breast cancer is one of the leading causes of cancer death among women. Developing machine learning-based diagnosis models receives great attention from researchers and scientists using histopathology images. Deep learning (DL) algorithms automatically extract features from raw data through convolutional operations. The generalization of the DL models' results relies on large datasets, although they eliminate the expert knowledge in the feature extraction stage. This work aimed to compare the performance of the features extracted via deep learning and a knowledge-based approach in breast cancer detection from histopathology images.

Abstract: Cancer develops when a single or a group of cells grows and spreads uncontrollably. Histopathology images are used in cancer diagnosis since they show tissue and cell structures under a microscope. Knowledge-based and deep learning-based computer-aided detection is an ongoing research field in cancer diagnosis using histopathology images. Feature extraction is vital in both approaches since the feature set is fed to a classifier and determines the performance. This paper evaluates three feature extraction methods and their performance in breast cancer diagnosis. Features are extracted by (1) a Convolutional Neural Network, (2) a transfer learning architecture VGG16, and (3) a knowledge-based system. The feature sets are tested by seven classifiers, including Neural Network (64 units), Random Forest, Multilayer Perceptron, Decision Tree, Support Vector Machines, K-Nearest Neighbors, and Narrow Neural Network (10 units) on the BreakHis 400× image dataset. The CNN achieved up to 85% for the Neural Network and Random Forest, the VGG16 method achieved up to 86% for the Neural Network, and the knowledge-based features achieved up to 98% for Neural Network, Random Forest, Multilayer Perceptron classifiers.

Keywords: CNNs; VGG16; breast cancer; histopathology; knowledge-based; feature extraction

Citation: Kode, H.; Barkana, B.D. Deep Learning- and Expert Knowledge-Based Feature Extraction and Performance Evaluation in Breast Histopathology Images. *Cancers* **2023**, *15*, 3075. https://doi.org/10.3390/cancers15123075

Academic Editors: Rafał Obuchowicz, Michał Strzelecki and Adam Piorkowski

Received: 24 April 2023
Revised: 31 May 2023
Accepted: 3 June 2023
Published: 6 June 2023

1. Introduction

Cancer is one of the most prevalent diseases worldwide. Many reports have shown that breast cancer is the most common cancer among women in the US and worldwide excluding nonmelanoma skin cancer and the second leading cause of cancer-related mortalities after lung cancer. Breast cancer statistics in 2022 reported that the breast cancer rate increased by 0.5% annually but the death rate declined for women except for American Indians and Alaska Natives [1]. The diagnosis can be made by regular mammography or ultrasound checkups. If the imaging indicates the possibility of malignant tissues, a breast tissue biopsy is performed, allowing pathologists to assess the microscopic structure and elements of the tissue histologically. Pathologists perform analysis, which is a challenging task, on nuclei location, mitosis, and glands in classifying benign and malignant tissues [2]. Figure 1 shows normal and ductal carcinoma images from the Human Protein Atlas [3].

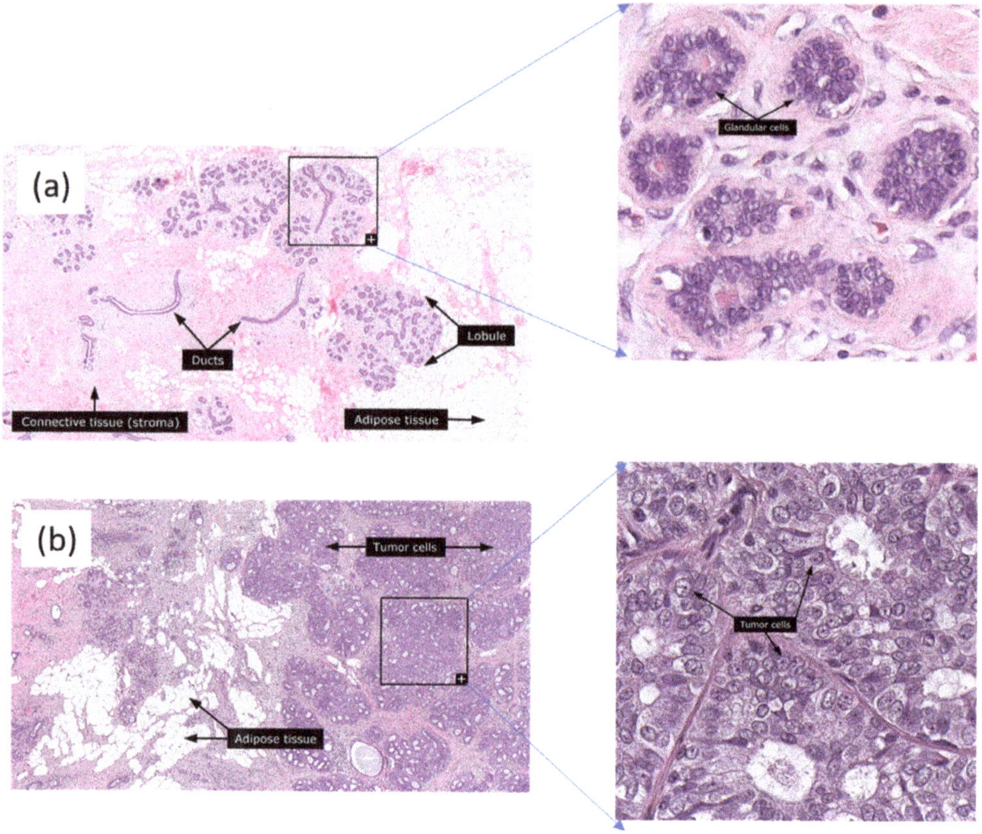

Figure 1. Normal and ductal carcinoma breast histopathology images from the Huma Protein Atlas Tumor cells, adipose tissue, and glandular cells are marked. (**a**) Normal, (**b**) Ductal Carcinoma, 68 years old female, Elston-Ellis score is 4. The images were used with the permission of The Human Protein Atlas; this work met their conditions for the use of the above images.

At the same time, it is time-consuming, expensive, and dependent on the ability and experience of the pathologists. The appearance of cells in a histopathology image can provide important information for disease diagnosis and prognosis. The characteristics of malignant tissues include inconsistent tissue and cell morphology, uneven color distribution, cell overlapping, and variability in the appearance of stained histological sections, creating difficulties in classification [4]. The characteristics of histopathology images play an essential role in diagnosing cancer. Cell morphology in histopathology images refers to individual cells' size, shape, and structure of a tissue sample [5]. For example, the size and shape of cancer cells can be different from normal cells. The analysis of cell morphology in histopathology images can also provide information about the activity of specific cellular pathways and processes, such as cell division and apoptosis (programmed cell death). The study of cell morphology in histopathology images is an integral part of precision medicine, where treatment is tailored to the specific characteristics of a disease.

In contrast, uneven color distribution can occur when there are variations in the staining intensity of different regions in the tissue sample. Figure 2 shows various color distributions. For example, in some cancer tissues, certain areas may appear more intensely stained than others, indicating the presence of abnormal cells or tissues. Similarly, cell overlapping in histopathology images refers to the phenomenon where cells in the tissue sample appear on top of one another [6], making it difficult to distinguish individual cells.

It can occur when cells have undergone abnormal changes, such as increased cell growth or decreased cell death, leading to tissue overcrowding. Cell overlapping can also be a sign of tissue invasion by cancer cells. All these characteristics of histopathology images can provide valuable information for diagnosis and prognosis.

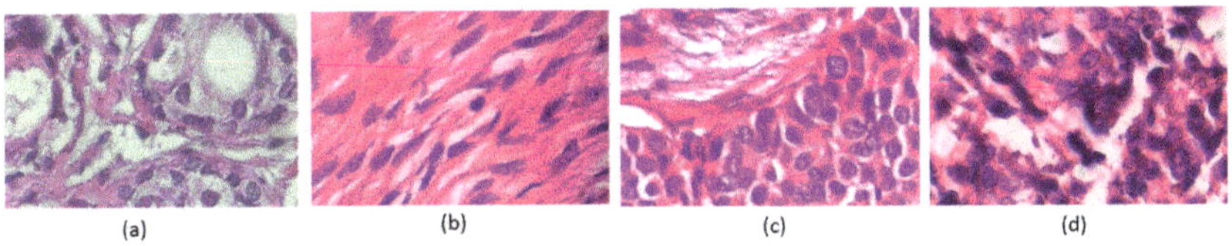

Figure 2. Breast cancer histopathology images from the BreakHis 400× image dataset. The images in (**a**,**b**) are classified as benign, and the images shown in (**c**,**d**) are classified as malignant by pathologists.

Histopathology image biomarkers are features extracted from images of tissue samples to diagnose diseases. These features can include morphological changes in cells and tissues and changes in the expression of a specific protein. Preprocessing histopathology images involves a series of steps to improve the quality of raw images to make them suitable for further analysis and interpretation. Typical preprocessing steps are artifact removal, normalization, segmentation, and registration. Normalization ensures that the image data have the same contrast, brightness, and color balance; segmentation separates the ROI from the background, which can be done using techniques such as thresholding, region-based segmentation, or edge detection [5]. Registration aligns the images to a common coordinate system to make them comparable. Image enhancement improves the visual quality of the images using techniques such as contrast stretching, histogram equalization, or denoising. After the preprocessing stage, the next step is extracting the relevant information from the preprocessed images, called features. In machine learning, features can be extracted manually or automatically. Manually extracted features are obtained from Scale-Invariant Feature Transform (SIFT), Speeded Up Robust Features (SURF), Oriented FAST and rotated BRIEF (ORB), image segmentation, and thresholding techniques. Automatic feature extraction can be done by Deep Convolutional Neural Networks (CNN) and pre-trained networks such as ResNet50, VGG16, AlexNet, etc.

2. Related Works

Nowadays, artificial intelligence (AI) plays an important role in image classification and designing Computer-Aided Diagnosis (CAD) systems to assist medical professionals in increasing and improving the accuracy of diagnosis and reducing the cost [4,7,8]. CAD systems use machine learning (ML) algorithms to classify benign or malignant images. Image classification is fundamental in computer vision, with numerous medical, biology, and robotics applications. Traditionally, image classification relies on handcrafted features derived from the raw pixel values of the images using domain-specific knowledge and expertise [9,10]. These features capture distinctive characteristics of the region of interest (ROI), such as shape, texture, or color, and are used as input to classifiers such as K-Nearest Neighbors, Support Vector Machines, Decision Tree, etc.

In the last decade, however, deep learning (DL) has emerged as a powerful approach to image classification, providing an alternative to handcrafted features. DL consists of neural networks with multiple layers, which can extract features from the raw pixel values of the images automatically without the need for manual design. Convolutional Neural Networks (CNNs) have been particularly successful and have achieved state-of-the-art performance on many image classification benchmarks [11]. Despite the impressive performance of DL on image classification tasks, it remains debatable which approach is best for a given application, especially in medical fields. On the one hand, handcrafted features can be designed specifically for a given domain and may capture essential characteristics of the ROIs that CNNs do not. On the other hand, CNNs can be more robust and generalizable, as they do not rely on domain-specific knowledge and can learn from large amounts of data.

A recent study defined three sets of features to represent the characteristics of the nuclei cell in breast cancer histopathology images to diagnose malignant cases [9]. The work extracted 33 features based on geometric, intensity, and direction information of nuclei and tested them using the BreakHis histopathology image dataset. Four ML algorithms were designed for diagnosis, Decision Tree, a Narrow Neural Network, a K-Nearest Neighbor, and Support Vector Machines. They acquired the best accuracy for a Narrow Neural Network (NNN) of 96.9%. Another research paper [12] has an improved Autoencoder (AE) network using a Siamese framework designed to learn the compelling features of histopathology images. First, they processed the image features using a Gaussian Pyramid to get the multi-scale, and at the feature extraction stage, a Siamese network was used to constrain the pre-trained AE. The experimental results for their paper showed an accuracy of 97.8% on the BreakHis dataset. Further, researchers have used pre-trained networks such as VGG16, VGG19, AlexNet, and Google Net to classify the histopathology image data. One of the studies proposed a framework called the AlexNet-BC model [13], a pre-trained model using the ImageNet dataset and fine-tuned using an augmented dataset. Further, they used an improved cross-entropy function to make the predictions suitable for uniform distributions, and their approach reached an accuracy of 86.31% on the BreakHis dataset.

The work in [14] demonstrated transfer learning ability for the fully connected network on the histopathology images based on three pre-trained networks of VGG16, VGG19, and ResNet50. Among these, their experimental results gave the highest accuracy of 92.60% for the pre-trained VGG16 network with logistic regression classifier. In contrast, AlexNet and VGG16 models were used in another study [15] that used the BreakHis dataset and fine-tuned the parameters with AlexNet. Then, the obtained features were fed to the SVM classifier. Their evaluation showed that the transfer learning produced better results than the deep feature extraction and SVM classifier. A similar study [16] proposed two different architectures to classify breast cancer from histopathology images. One is single-task CNN architecture to predict malignancy which acquired an average accuracy of 83.25%, and another is a multi-task CNN to predict malignancy and image magnification level, which received an accuracy of 82.13%.

Albashish et al. [17] used VGG16 architecture to extract the features from the breast cancer histopathology images. They used different classifiers for classification purposes and acquired the best accuracy of 96% for the RBF-SVM classifier. Another research paper [18] proposed a dual-stream high-order network named DsHoNewith with six convolutional layers. The first stream uses batch normalization (BN) to keep the original data, and the ghost module is used in another stream to extract high-quality features using linear transformations. Li et al. [19] proposed an embedded fusion mutual learning (EFML) method with two parallel heterogeneous networks for feature extraction and two classifiers, adaptive feature fusion and ensemble classifier, embedded into EFML for classification purposes. Atban et al. [20] used the Resnet18 model for deep feature extraction then they applied Practice Swarm Optimization (PSO), Atom Search Optimization (ASO), and Equilibrium Optimizer (EO) algorithms to histopathology images to obtain more valuable features. For this method, they acquired an accuracy of 97.75% for ResNet18-EO for the SVM classifier with Gaussian and radial-based functions (RBF).

Image classification is a widely studied problem in computer vision, with numerous applications in object recognition, scene understanding, and medical image analysis [21]. One of the key challenges in image classification is how to represent the region of interest in the input images distinctively. It is traditionally done through a process called feature extraction, which involves identifying and extracting relevant features from the images. There are two main approaches to extracting features. (1) a Convolutional Neural Network (CNN) and (2) knowledge-based systems based on domain expertise [9,22].

The CNN-based approach has gained popularity recently due to the success of deep learning models in various tasks [13,23,24]. A CNN is an artificial neural network that processes data with grid-like topologies, such as images [25]. It consists of multiple layers of convolutional and pooling operations, which are used to extract features from the images hierarchically [25]. Later, the extracted features are fed to the classifier for the final prediction. In contrast, knowledge-based feature extraction involves expert knowledge in identifying and extracting distinctive features from the images based on prior knowledge or domain expertise [22]. This process can be time-intensive; sometimes, finding the expertise to design the feature extraction model can be difficult.

One of the main areas of machine learning is called deep learning. The system preprocesses the data to extract features using convolutional layers. Later the network feeds the obtained features to the classifier that predicts whether the given image belongs to the same class. Different techniques are involved in preprocessing the data in each process step. Overfitting data, outliers, dropout, and updating weights and biases are some tasks performed to reduce the loss. Currently, deep learning methods achieved remarkable performance in image processing and computer vision applications. These architectures have automatic feature extraction and representation ability. It has been widely adopted for histopathology image classification tasks [26]. The training process in CNN requires constant parameter adjustments to ensure equivalent learning of all the layers due to overfitting and convergence issues [14]. Some researchers say that transfer learning may be a good substitute for training to overcome the above problems [14]. In transfer learning, the network is trained on different datasets, and its parameters are fine-tuned on the desired dataset. Transfer learning can be used as a baseline or feature generator [27].

Our paper compared the performance of a CNN, transfer learning (VGG16), and knowledge-based feature sets [9] using the BreakHis 400× image dataset. The BreakHis dataset consists of histopathology images of breast tissue samples annotated with 'benign' and 'malignant' labels. The obtained features were fed to three classifiers: a neural network, Multilayer Perceptron (MLP), and Random Forest (RF). Their performances were compared by using various evaluation metrics.

3. Dataset

The BreakHis dataset is a publicly available dataset of histopathological images of breast tissues available at https://www.kaggle.com/datasets/ambarish/breakhis, accessed on 1 September 2022. It has been built in collaboration with the P&D Laboratory-Pathological Anatomy and Cytopathology, Parana, Brazil. It consists of 9109 microscopic images of breast biopsy samples divided into benign and malignant. Each image is labeled with information about tissue type, whether benign or malignant, and the diagnosis given by pathologists. The images are further divided into four magnification levels (40×, 100×, 200×, and 400×) to allow for analysis at different levels of detail. The BreakHis dataset has been widely used for developing and evaluating computer-aided diagnosis (CAD) systems for breast cancer detection. It is considered a benchmark dataset for this task, and several research studies have reported high accuracy rates in detecting breast cancer using the BreakHis dataset. In our research, we are using BreakHis 400× magnification level dataset containing two classes with 547 benign and 1146 malignant images labeled accordingly. Each image is a size of 700 × 460 pixels with 3-channel (RGB) having a depth of 8-bit in each channel and PNG format. BreakHis400× is available at https://www.kaggle.com/datasets/scipygaurav/breakhis-400x-breast-cancer-dataset, accessed on 1 September 2022. This work uses 400× magnification since the knowledge-based method mentioned here [9] extracts features from nuclei such as mitotic count, shape, and texture. Magnifications of 40×, 100×, and 200× do not show these characteristics clearly.

4. Methodology

In this paper, we used three approaches to extract the features from the images: (1) a basic CNN architecture, (2) transfer learning, VGG16, and (3) a knowledge-based approach. CNN and transfer learning approaches extract the feature automatically. The origin of the features is not known. Whereas the knowledge-based approach has mathematically defined features, which were driven by expert input. Seven supervised classifiers—Neural Network, Multilayer Perceptron, Random Forest, Decision Tree, SVM, KNN, and NNN-were trained using the feature sets, and their performances were evaluated and compared. Knowledge-based features were obtained from [9], and the details were presented in Section 4.3.

4.1. Basic Convolutional Neural Network (CNN) Based Feature Extraction

A Convolutional Neural Network is used for image recognition and preprocessing tasks. Figure 3a,b shows its architecture. It has multiple layers, including convolutional and pooling layers. The idea behind the convolutional layer is to apply a set of filters known as kernels to the input image, resulting in a set of output feature maps. Each filter is applied to a small section of the input image, and the resulting values are summed up to produce a single output value in the feature map. The convolutional layer can detect various features in the input image, such as edges, corners, and blobs using different filters. The max pooling layer is used to reduce the dimensionality of the output feature maps. It works by dividing the feature map into small, non-overlapping regions and taking the maximum value within each region. It helps reduce the feature maps' spatial size while retaining the most important information.

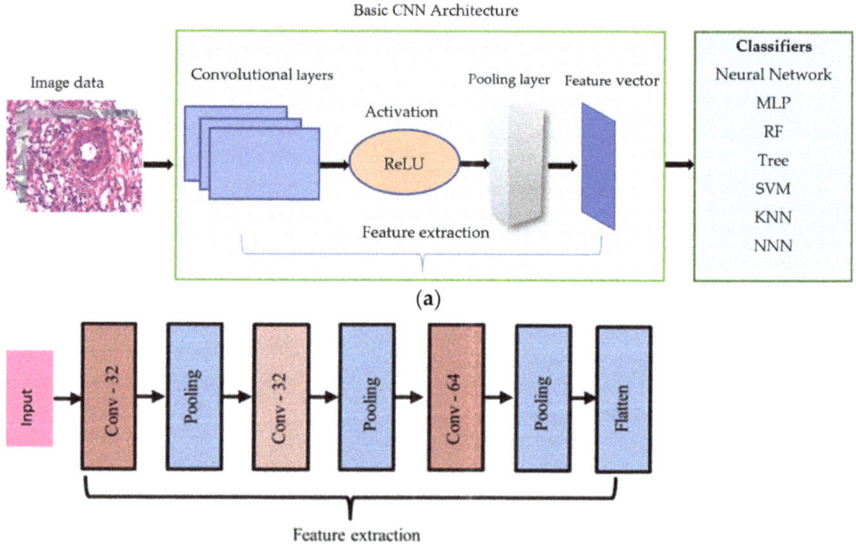

Figure 3. (**a**) The architecture of extracting the features by basic CNN model and feeding them to the three classifiers. (**b**) The details of the architecture.

In our model, we used three convolutional layers along with max pooling layers. The first layer has a set of 32 filters applied to the input image with a kernel size of 3×3, activation function ReLU with an input size of 512×512, including 3-channel RGB. Then, the max pooling layer is applied with a pooling size of 2×2, reducing the spatial size of the output feature maps by taking the maximum value within small, non-overlapping regions. The second layer has the same parameters that the first layers have. The final layer has 64 filters with kernel size 3×3, activation function ReLU applied to the output feature maps from the previous layers and produces a final set of output feature maps. A max pooling layer with size 2×2 is applied further to reduce the spatial size of the output feature maps.

Furthermore, the dropout rate of 0.25 is applied before flattening to effectively drop out entire channels of feature maps rather than individual neurons. It helps to preserve the spatial information in the feature maps and encourages the network to learn more diverse and robust features. After flattening, the feature vector will be fed to classifiers, as shown in Figure 3a.

4.2. Transfer Learning (VGG16) Based Feature Extraction

The VGG16 [25] model consists of 16 layers of convolutional and pooling operations and has been widely used for image classification tasks. It is characterized by using small 3 × 3 filters throughout the network, which leads to a deeper network with smaller receptive fields. We extracted features from the images using the VGG16 CNN model, which was trained on the ImageNet dataset [28]. First, the pre-trained VGG16 model was loaded and did not include the top layers, as these were the layers responsible for the final classification. Then, a new model was created, which took the output of the last layers of VGG16 as input. Next, the preprocessed BreakHis image dataset was fed to the model to extract the features from the images. These extracted features were then fed to the classifiers. Figure 4 illustrates the transfer learning-based feature extraction steps.

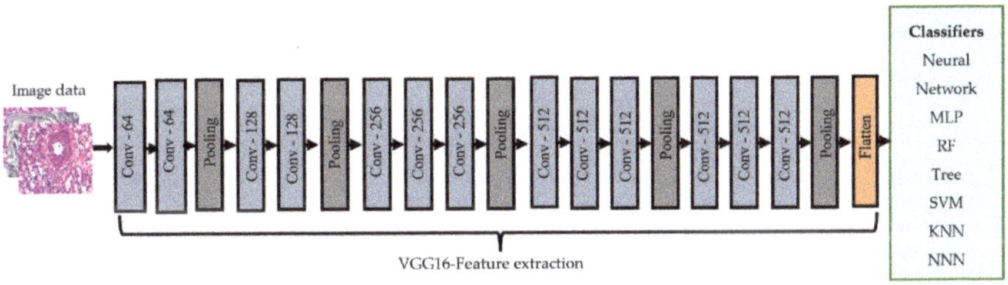

Figure 4. The VGG-16 model was used in our approach.

The below shows the implementation steps of the VGG16 model used in this study.

Steps: Feature Extraction from Input Images Using the VGG16 Model.	
Input:	BreakHis 400× training image data
Preprocessing:	Resizing the images to 512 × 512 Normalizing pixel values to between 0 and 1
Output:	Initialize the VGG16 model. Import VGG16 model Remove the fully connected layers. Load the VGG16 model for feature extraction by removing the FC layers as they were designed for ImageNet classification tasks. VGG_model = VGG16 (weights = 'Imagenet', include_top = False, input_shape = (512, 512, 3) 3-is the RGB channel since we are using color images Pass the input images through the model: Forward pass the preprocessed images through the VGG16 model to obtain the output feature maps. Retrieve the output of the last convolutional layer as they capture the high-level abstract features. Flatten the features. Flatten(img): Convert the 3-D extracted features to the 1-D feature vector Store the extracted features with corresponding image labels for classification tasks Features(img) = Flatten(img)

4.3. Knowledge-Based Feature Extraction

For this approach, the features were calculated based on domain expertise. It involved analyzing the images and identifying specific characteristics or patterns indicative of the class label. In this case, the features include the unique characteristics of cell nuclei representing malignant cases. We adopted this knowledge-based feature set from [9]. Figure 5 illustrates the overall structure of the knowledge-based system.

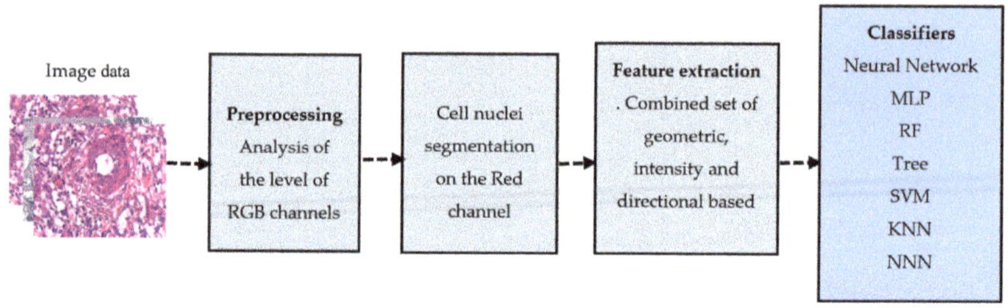

Figure 5. Knowledge-based computer-aided model.

Three sets of features were calculated: Geometric, Directional, and Intensity which include 33 features together to define the characteristics of the cell nuclei for each image. Here, we summarize these feature sets.

4.3.1. Geometrical Features

In [9], the geometric features have been calculated by the parameters such as area (A), perimeter (P), roundness (R), and the number of segmented cells (L). A is the area of the nucleus, where x_k is the number of segmented pixels of each nucleus, P is the perimeter of the nucleus where P_k is the number of pixels at the cell border of each nucleus, $\frac{A}{P}$ is the ratio of the area to the perimeter, and R is the roundness of the ROI calculated from the area and perimeter.

$$f_1 : L \quad (1)$$

$$f_2 : A = \frac{1}{L}\sum_{k=1}^{L} x_k \text{ for } k = 1, 2, \ldots, L \quad (2)$$

$$f_3 : P = \frac{1}{L}\sum_{k=1}^{L} P_k \quad (3)$$

$$f_4 : \frac{A}{P} \quad (4)$$

$$f_5 : R = \frac{4A\pi}{(P)^2} \quad (5)$$

4.3.2. Directional Features

It is known that malignant cells have irregular shapes, whereas benign cells have more definite and round forms. Spatial distance in every eight directions, East (E), West (W), North (N), South (S), Northwest (NW), Northeast (NE), Southwest (SW), and Southeast (SE), are calculated to determine the shape and state of the segmented nuclei.

$$d_N = \sqrt{(m - m')^2 + (n - n')^2} \quad (6)$$

$$f_6 \to f_{13} : \bar{m}_D = \frac{1}{L}\sum_{k=1}^{L} d_{D_k} \quad (7)$$

$$D = N, S, E, W, NW, NE, SW, SE$$

$\bar{m}_D$ is mean for a particular direction of the segmented region. The standard deviation represents the variation in the eight directions (D). σ_D is the standard deviation for each cardinal direction.

$f_{14} \to f_{21}$:

$$\sigma_D = \sqrt{\frac{\sum_{k=1}^{L}\left(d_{D_k} - \bar{m}_D\right)^2}{L}} \text{ for } D = N, S, E, W, NW, NE, SW, SE \tag{8}$$

The range is calculated as the difference between the eight directions' maximum and minimum spatial distance values. This feature set will make up of features $f_{22} \rightarrow f_{29}$.
$f_{22} \rightarrow f_{29}$:

$$range(D_k) = \max\left(d_{D_k}\right) - \min\left(d_{D_k}\right)$$

$$\text{for } k = 1, 2, \ldots, L \text{ and } D = N, S, E, W, NW, NE, SW, SE \tag{9}$$

The ratio of the segmented region to the area of the fitting rectangle (A_R) is defined to represent the directional feature set.

$$f_{30} : AR_{ratio} = \frac{1}{L}\sum_{k=1}^{L}\frac{A_k}{A_{R_k}} \tag{10}$$

4.3.3. Intensity-Based Features

Intensity-based features capture mitotic cell-related details. The intensity-based features include mean, standard deviation, and range measurements.

$$f_{31} : \bar{I} = \frac{1}{L}\sum_{k=1}^{L} I_k \tag{11}$$

Here, I is the average intensity of a nucleus.

$$f_{32} : \sigma_I = \frac{1}{L}\sum_{k=1}^{L} \sigma_{Ik} \tag{12}$$

$$\sigma_{I_k} = \sqrt{\frac{\sum_{i=1}^{x_k}(J_k(i) - I_k)^2}{x_k}} \tag{13}$$

R_I is the intensity range that is calculated by the average value of the difference between the maximum and minimum pixel intensity values of nuclei.

$$f_{33} : R_I = \frac{1}{L}\sum_{k=1}^{L} R_{Ik} \tag{14}$$

$$R_{I_k} = \max(J_k) - \min(J_k) \text{ for } k = 1, 2, \ldots L \tag{15}$$

The calculated features were analyzed in terms of their efficiency in separating benign and malignant cases. We ranked the 33 features using the minimum redundancy maximum relevance (MRMR) algorithm. Figure 6 illustrates the feature ranking. The drop in the importance score represents the confidence in the selection algorithm. There is a large drop between the first two and the remaining features. The mean intensity value of the intensity-based set was found to be the second most significant feature. The ranking order was as follows: Feature number: 27, 31, 18, 24, 30, 29, 8, 21, 23, 15, 26, 25, 28, 12, 20, 3, 5, 6, 22, 11, 9, 17, 16, 14, 4, 19, 10, 13, 7, 32, 1, 2, 33. However, overall, directional features were found to be more significant than the geometrical and intensity-based features as 14 out of the 15 features in the ranking belonged to the directional features that were developed to represent the irregular border shape of the segmented cells in an image. Details of the features can be found in [9]. Figures 7 and 8 plot some of the features versus each other to show the benign and malignant cases. As seen in the figures, the represented directional and intensity features distinguish two classes with some overlapping. All features were fed to the classier in this work.

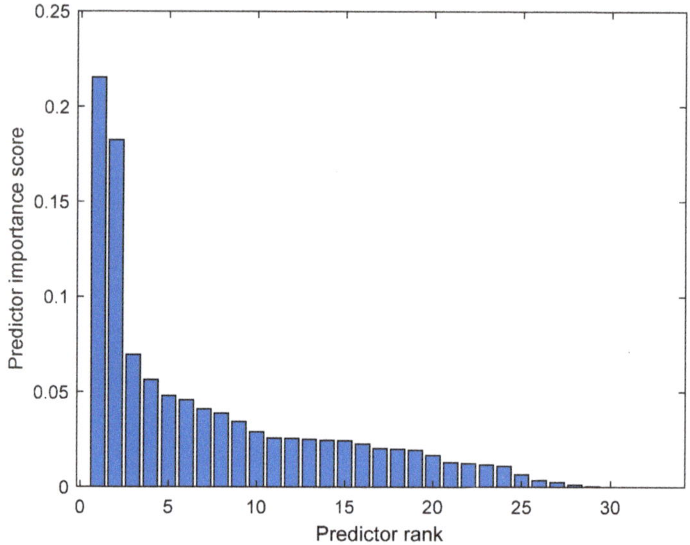

Figure 6. Feature ranking using minimum redundancy maximum relevance (MRMR) algorithm.

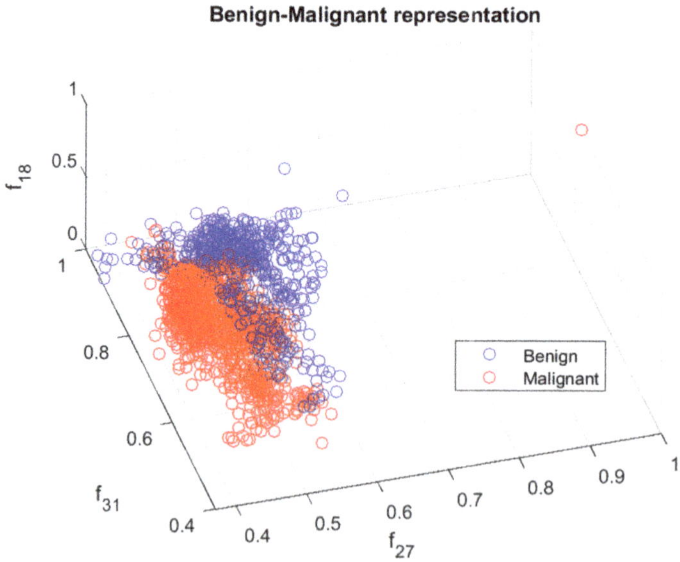

Figure 7. Benign and malignant characterization by directional features f_{18}, f_{27} and intensity-based feature f_{31}. Benign and malignant cases are represented in blue and red, respectively.

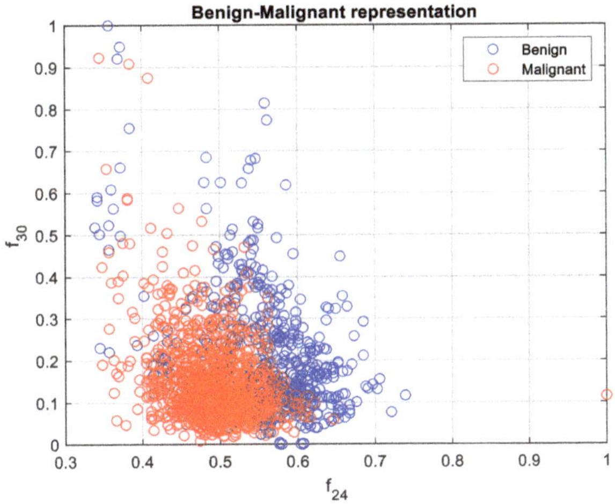

Figure 8. Benign and malignant characterization by two directional features f_{24} and f_{30}. Benign and malignant cases are represented in blue and red, respectively.

5. Classification and Performance Evaluation

We evaluated the performance of the feature sets using seven different classifiers: a Neural Network (NN) (64 units), a Random Forest (RF), a Multilayer Perceptron (MLP), Decision Tree, Support Vector Machine (SVM), KNN, Narrow NN (10 units).

5.1. Neural Network (NN)

The Neural Network (NN) classifier is widely used for image classification tasks. It consists of multiple layers of interconnected "neurons" that are used to process and analyze the input data. The NN is trained using a large dataset of labeled images and adjusts the weights and biases of the neurons to optimize the prediction performance. In our experiment, we designed an NN where a dense layer with 64 units and a 'ReLu' activation function is added, followed by a final dense layer with one unit and a 'Sigmoid' activation function. The model was compiled with the 'Adam' optimizer and binary cross entropy loss function and trained with extracted features and corresponding labels.

5.2. Multilayer Perceptron (MLP)

This type of neural network classifier consists of multiple layers of neurons with a feedforward architecture. It is used for image classification tasks due to its ability to learn complex patterns in datasets. The MLP classifier is trained by adjusting the weights and biases of the neurons through an optimization process, such as backpropagation. In our study, we designed MLP classifier with 'hidden_layer_sizes = 100', 'max_iter = 500', and a 'ReLu' activation function.

5.3. Random Forest (RF)

Random Forest classifiers use an ensemble learning method that combines multiple Decision Trees and use the average prediction of all the trees to make a final prediction. It is widely used for image classification tasks due to its ability to handle high-dimensional data. A bootstrap sample is taken from the original data set with a replacement for each tree in the forest. It means that some data points may be included multiple times in a given bootstrap sample. From this sample, a Decision Tree is constructed using a random subset of features at each node. It helps to reduce the correlation between trees and improves the overall performance of the forest. Once all trees have been constructed, the Random Forest predicts a given input by averaging the predictions of all the trees. In our work, we imported a Random Forest classifier from the scikit-learn library and used the parameter 'n_estimators = 100' for all three feature extraction approaches.

5.4. Decision Tree

The Decision Tree is one of the popular supervised classifiers. It can be defined by decision nodes and leaves, where the leaves are the decisions. In this work, the fine tree classifier was set up with the same parameters used in [9]. The maximum number of splits: 100. Split criterion: Gini's diversity index and Surrogate decision splits: off.

5.5. Support Vector Machines (SVM)

We set the SVM parameters as follows. The kernel function was set as linear, and the kernel scale was set to automatic. The box constraint level was set to 1, and the multiclass method was set to One-v-One. The value for the standardized data was set to true.

5.6. KNN

The Fine KNN classifier was used in our work as the number of neighbors was set to 1, and the distance metric was set to Euclidean. The distance weight and the standardized data values were set to Equal and true.

5.7. Narrow Neural Networks (NNN)

The Narrow Neural Network had one fully connected layer with a size of 10. The activation used was ReLU with an iteration limit of 1000. The regularization strength (Lambda) was set to 0.

5.8. BreakHist Dataset and Evaluation Metrics

We used the BreakHis 400× Histopathology image dataset, which consists of 547 benign and 1146 malignant images Spanhol et al. 2016 [29]. For this dataset, the images were resized to (512 × 512) and split the data into a training set and testing set of 80% and 20%, respectively. Performance evaluations were done by accuracy, precision, recall, F1 score, and ROC curves. All these metrics are based on confusion matrices. The metric parameters are computed using true negative (TN), true positive (TP), false positive (FP), and false negative (FN). TP refers to the cases where the model predicted the positive class and the actual class is positive; similarly, TN refers to the cases where the model predicted the negative class and the actual class is also negative. Whereas FP refers to the cases where the model predicted the positive class, the actual class is negative. It is known as the Type 1 error. Similarly, FN refers to the cases where the model predicted the negative class, but the actual class is positive. It is known as the Type 2 error. A confusion matrix evaluates the performance of a classification model by providing a detailed picture of how well the model is performing. In medical diagnosis, minimizing FN is often more important. The accuracy is calculated as

$$\text{Accuracy} = \frac{TN + TP}{TN + TP + FN + FP} \tag{16}$$

Precision and Recall are other widely used metrics to evaluate the performance of the testing data. Precision and Recall can be calculated as follows:

$$\text{Precision} = \frac{TP}{TP + FP} \tag{17}$$

$$\text{Recall} = \frac{TP}{TP + FN} \tag{18}$$

The other evaluation metric is F1-score, which is commonly used for statistical measures. The values of this F1-score range from 0 to 1. It can be calculated as follows:

$$\text{F1-score} = 2 * \frac{\text{Precision} * \text{Recall}}{\text{Precision} + \text{Recall}} \tag{19}$$

6. Experimental Results

Table 1 presents the performance of the proposed methodology. The presented values represent the macro average with ±0.02 tolerance, which supervised classifiers' performances slightly change depending on the training data. As can be seen, the performance of the knowledge-based feature extractor consistently outperformed the CNN and VGG16-based feature extractors using three classifiers on the BreaKHis 400× dataset.

Table 1. Macro average performances of the proposed approaches (±0.02)—VGG16, CNN, and knowledge-based—for binary classification using Neural Networks (64 units), Multilayer Perceptron, Random Forest, Decision Tree, Support Vector Machines, K-Nearest Neighbor, and Narrow Neural Network (10 units) classifiers on the BreakHis 400× dataset.

Feature Extractor	Classifier	Accuracy	Precision	Recall	F1 Score
Basic CNN	Neural Network (64 units)	0.85	0.84	0.82	0.83
	MLP	0.84	0.84	0.79	0.81
	RF	0.85	0.85	0.79	0.81
	Tree	0.75	0.72	0.73	0.73
	SVM	0.79	0.76	0.75	0.76
	KNN	0.79	0.77	0.73	0.75
	NNN (10 units)	0.84	0.83	0.79	0.8
VGG16	Neural Network (64 units)	0.86	0.87	0.81	0.83
	MLP	0.82	0.83	0.76	0.88
	RF	0.74	0.84	0.62	0.61
	Tree	0.73	0.7	0.7	0.7
	SVM	0.87	0.86	0.83	0.84
	KNN	0.73	0.71	0.63	0.64
	NNN (10 units)	0.8	0.86	0.67	0.69
Knowledge-based extraction	Neural Network (64 units)	0.98	0.98	0.96	0.97
	MLP	0.98	0.98	0.97	0.98
	RF	0.98	0.98	0.96	0.98
	Tree [9]	0.85	0.9	0.88	0.89
	SVM [9]	0.96	0.9	0.88	0.89
	KNN [9]	0.94	0.98	0.94	0.96
	NNN [9]	0.97	0.98	0.98	0.98

We employed the BreakHis 400× dataset for a couple of reasons. (1) Our work compares the knowledge-based features with CNN and VGG16-based features. Knowledge-based features are defined for histopathology images with 400× magnification since it shows the nuclei characteristics and segmentation of nuclei, which is still challenging due to overlapping and clustered nuclei. (2) The dataset provides binary classification, and our work proposed three models for binary classification.

The ROC (Receiving Operating Characteristic) curve is a graphical representation of the performance of a binary classifier at a different classification threshold. It is created by plotting the *TP* rate against the *FP* rate for various threshold values. The TPR is the proportion of actual positives correctly identified as such, while the FPR is the proportion of actual negatives incorrectly identified as positives. The ROC curve is a useful tool for evaluating and comparing the performance of binary classifiers. The closer the ROC curve is to the top left corner, the better the performance of the classifier. AUC (Area Under the Curve) is a metric that measures the overall performance of a binary classifier.

It represents the area under the ROC curve and is interpreted as the probability that a randomly chosen positive example will be ranked higher than a randomly chosen negative example. The AUC value ranges from 0 to 1, with a value of 0.5 indicating a random classifier and a value of 1 indicating a perfect classifier. The higher AUC value indicates a better overall performance of the classifier. The ROC and AUC curves provide a graphical and numerical measure of the classifier's performance, making them helpful in evaluating and comparing different classification models. Figures 9–11 show the ROC and AUC of the CNN, VGG16, and knowledge-based feature extraction approaches. We present the convergence performance of the employed models in Figure 12.

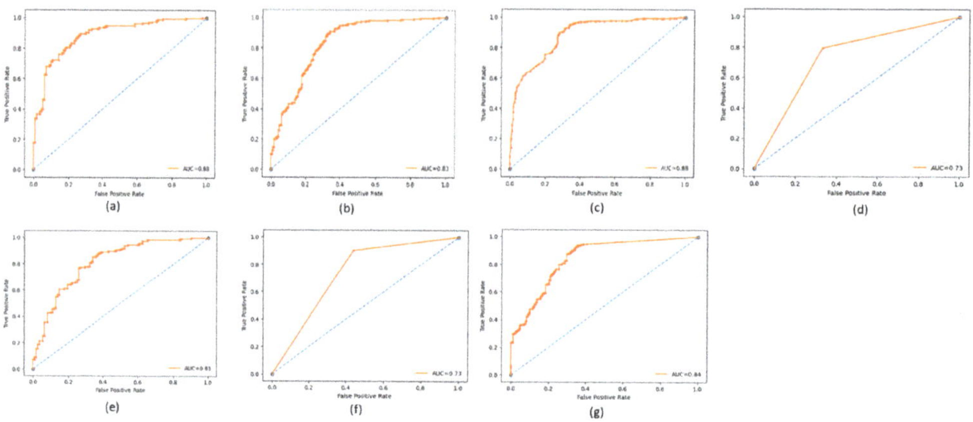

Figure 9. ROC and AUC graphical representations of the CNN feature extraction approach using the classifiers. (**a**) NN (64 units), (**b**) MLP, (**c**) RF, (**d**) Decision tree, (**e**) SVM, (**f**) KNN, and (**g**) NNN (10 units).

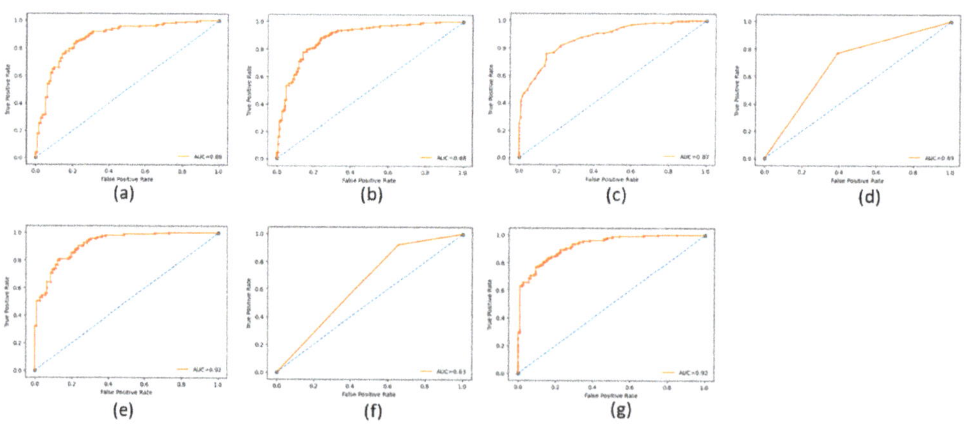

Figure 10. ROC and AUC graphical representation of VGG16 feature extraction approach using the classifiers. (**a**) NN (64 units), (**b**) MLP, (**c**) RF, (**d**) Decision tree, (**e**) SVM, (**f**) KNN, and (**g**) NNN (10 units).

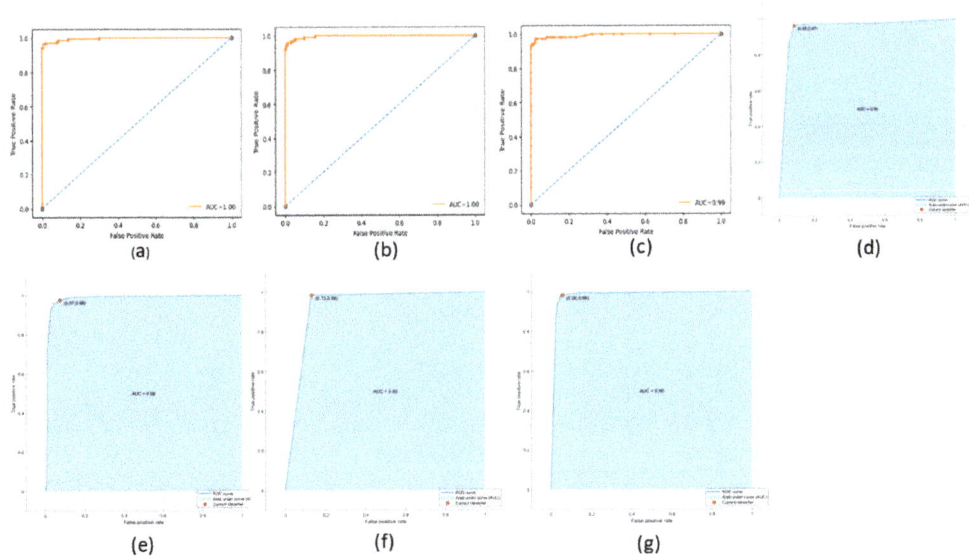

Figure 11. ROC and AUC graphical representation of knowledge-based feature extraction approach using the classifiers. (**a**) NN (64 units), (**b**) MLP, (**c**) RF, (**d**) Decision tree, (**e**) SVM, (**f**) KNN, and (**g**) NNN (10 units).

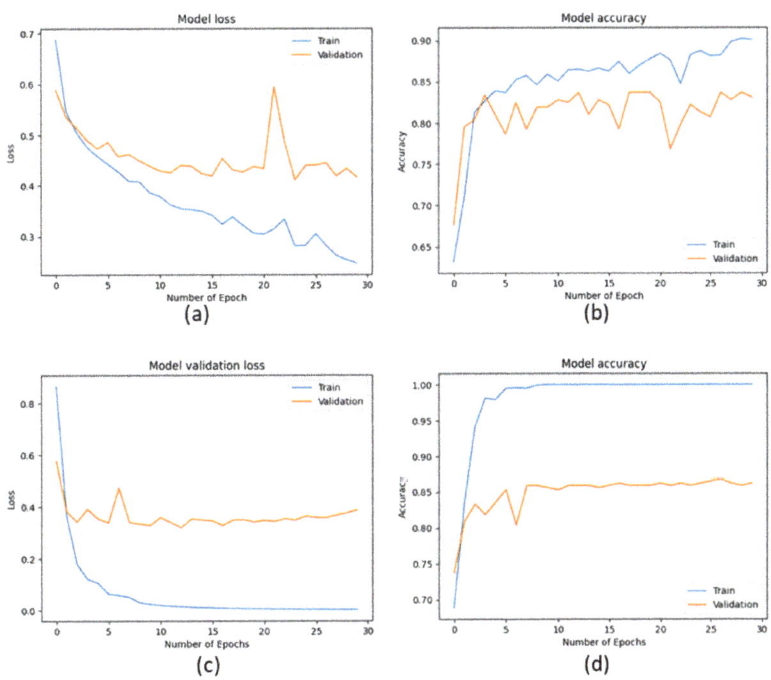

Figure 12. Model loss and accuracy performance for training and validation data. (**a**,**b**) CNN (**c**,**d**) VGG16.

Table 2 compares the performance of the proposed method with the previously proposed methods in literature on the BreaKHis 400× dataset. The results show that the evaluated knowledge base feature set outperformed the feature sets extracted by the ML-based features.

Table 2. Comparison of the state-of-art works on 400× BreakHis dataset with binary class classification.

Method	Feature Extraction	Classifier	Accuracy	Precision	Recall	F1 Score
Labrada et al., 2022 [9]	Knowledge-based [9]	Decision Tree SVM KNN NNN	0.85 0.96 0.94 0.97	0.9 0.9 0.98 0.98	0.88 0.88 0.94 0.97	0.89 0.89 0.96 0.98
Sharma and Mehra 2020 [10]	VGG16	RF SVM LR	0.69 0.92 0.85	0.7 0.92 0.86	0.69 0.91 0.86	0.69 0.91 0.86
Sharma and Mehra, 2020 [10]	Hu moment, Colored histogram, and Haralick texture	RF SVM	0.86 0.83	-	-	-
Gupta et al., 2020 [11]	ResNet50	Linear Regression SVM	0.93 0.93	-	-	-
Deniz et al., 2018 [15]	VGG16+Alexnet	Fully connected layers (FC6)	0.87	-	-	-
Albashish et al., 2021 [17]	VGG16	RBF-SVM NN	0.96 0.9	-	-	-
Zou et al., 2022 [18]	DsHoNet	-	0.98	0.98	0.98	0.98
Li et al., 2023 [19]	EFML	-	0.99	0.99	0.99	0.99
Atban et al., 2023 [20]	ResNet18	KNN SVM Decision Tree	0.9 0.95 0.75	0.9 0.95 0.75	0.9 0.95 0.77	0.9 0.95 0.76
Spanhol et al., 2016 [29]	PFTAS	SVM RF	0.82 0.81	-	-	-
Zhang et al., 2018 [30]	CNN	SVM RF KNN	0.78 0.75 0.75	-	-	-
Spanhol et al., 2017 [31]	CaffeNet	Neural network (Fully connected layers (fc7))	0.82	-	-	0.87
Fabio Alexandre Spanhol et al., 2016 [32]	CNN	Neural network (Fully connected layers)	0.8	-	-	-
Adeshina et al. nd. [33]	Deep CNN	Neural network (Fully connected layers	0.91	0.63	0.77	-
Gour et al., 2020 [34]	ResHist features	RF SVM	0.86 0.86	-	-	-
Gour et al., 2020 [34]	VGG16	Simple Neural network (with fully connected layers)	0.81	-	-	-
Togacar et al., 2020 [35]	BreastNet	ML classifier	0.96	0.95	0.95	0.95
Zhu et al., 2019 [36]	Hybrid CNN model	-	0.81	0.82	0.93	0.87
Zou et al., 2021 [37]	AHoNet	-	0.96	0.95	0.96	0.95
This work	**CNN** **VGG16**	**Neural network with FC layers**	**0.85** **0.86**	**0.84** **0.87**	**0.82** **0.81**	**0.83** **0.83**
This work	**Knowledge-based**	**Neural Network** **MLP** **Random forest**	**0.98** **0.98** **0.98**	**0.98** **0.98** **0.98**	**0.96** **0.97** **0.96**	**0.97** **0.98** **0.98**

The results of our evaluation demonstrate the high capability of the knowledge-based approach to feature extraction for image classification tasks. It is time intensive, relies on expert knowledge for feature definition and selection design, and extracts high-quality features. In contrast, automatically extracted features by CNN and VGG16 were less effective than the knowledge-based features.

One possible reason for the superior performance of the knowledge-based approach is identifying each characteristic of the cell nuclei based on expert opinion and designing the model to extract that information from the cell nuclei, which determines whether the cell is malignant. Another reason is robustness to noise and variability. Histopathology images can be subject to noise and variability, such as differences in staining techniques or tissue preparation. Knowledge-based approaches can help identify features robust to these variations, leading to more accurate and consistent results. These features are often based on biological structures and processes, which can provide greater interpretability and insight into the underlying biology of the tissue. It is particularly important since understanding the biological basis of the disease is crucial for developing effective treatments.

The size of the dataset does not affect the knowledge-based feature calculations. The features' mathematical model was developed based on content knowledge. The classifier's performance may change because of the type of classifier design and specifications, such as Support Vector Machines, KNN, ANN, CNN, etc. However, we can employ the knowledge-based features to design a rule-based classifier such as a Fuzzy classifier. In contrast, dataset size affects the features calculated via CNN and VGG16. It is the major disadvantage of these two models. To highlight the limitation of such models, our study compares the capabilities of the three feature sets coming from traditional CNN, transfer learning, and knowledge-based.

The CNN models performed well on large datasets of images and learned features well-suited for image classification tasks, as shown in previous studies [25]. Some research reported that CNN-based and VGG16-based feature extraction methods were more effective than knowledge-based feature extraction. It could be because their handcrafted features were not as unique and informative as those learned by CNN and VGG16 models [38]. Here, we used three knowledge-based feature sets that were initially developed by Labrada in 2022. The 33 features represent the cell nuclei's morphological, directional, and orientation [9]. The VGG16 feature extraction method also performed better when used with a simple Neural Network than MLP and RF.

In contrast, it performed poorly compared with knowledge-based feature extraction. It could be because the VGG16 model was not explicitly developed for histopathology image classification tasks. Our results are consistent with previous studies showing that fine-tuning a pre-trained CNN model can improve performance on image classification tasks [39].

The works in [18,19] employed DsHoNet and EFML methods, which achieved high classification performance. The research approach of using a dual-stream high-order network (DsHoNet) for breast cancer classification can be considered efficient in terms of its potential to improve accuracy and performance. By employing a dual-stream architecture, the model can leverage multiple sources of information or features extracted from histopathology images, enhancing the classification process. Additionally, incorporating high-order network components allows for capturing complex relationships and patterns within the data, potentially leading to improved classification results. However, the actual efficiency of this research would depend on factors such as the size and diversity of the dataset, the training methodology, and the evaluation metrics used to assess the performance of the classification model.

However, DsHoNet can generally be considered a relatively complex model compared to more straightforward approaches. The use of a dual-stream architecture implies the presence of two separate pathways or streams within the network, each handling different types of input or extracting different types of features. It introduces additional complexity compared to single-stream models.

Furthermore, incorporating high-order network components suggests the model captures higher-order interactions or dependencies among features. It typically involves modeling complex relationships and may require more computational resources and advanced techniques.

Overall, DsHoNet is likely to have a higher complexity level than traditional models, as it aims to leverage multiple streams of information and capture more intricate relationships within the data.

EFML involves designing a network architecture that integrates and learns from multiple modalities or feature sources. It typically requires the development of specialized network modules or fusion strategies to combine information from different sources. The complexity of the architecture depends on factors such as the number and type of modalities, the fusion approach used (early fusion, late fusion, or intermediate fusion), and the interactions between modalities within the network.

EFML involves training a model with multiple modalities or feature sources, which increases the computational requirements compared to single-modal approaches. The training process may include backpropagation and optimization across multiple branches or streams, potentially leading to increased training time and memory usage. Additionally, the computational complexity can increase if the modalities have different resolutions or require separate preprocessing steps. Our work's models have less complexity than DsHoNet and EFML.

Work in [11] used ResNet50, achieved an accuracy of 93%, and did not report other performance metrics. [20] used ResNet18 and achieved accuracies of 90%, 95%, and 75 using KNN, SVM, and Decision Tree classifiers, respectively. ResNet50 is undoubtedly an advanced model with remarkable results in many applications. Although many studies proved that the Resnet50 model performs better than VGG16, detecting errors becomes difficult due to its deeper network. Furthermore, if the network is shallow, the learning might be inefficient. Medical image datasets need to build efficient learning models to predict the classes. In our research, we wanted to use VGG16 due to its very small receptive fields instead of massive fields, such as Resnet50, for the BreakHis dataset. VGG16 has its own merits, making it a suitable choice for our research. VGG16 is a deep Convolutional Neural Network, and its simplicity and straightforward design make it easier to understand and implement, which was beneficial for our research objectives. Additionally, VGG16 has demonstrated strong performance across various image classification benchmarks, and its pre-trained weights are readily available, allowing for efficient transfer learning.

Several directions for future work would build on the results of this work. One possible way is to explore using different transfer learning and CNN models for feature extraction. Many CNN models are available, each with unique and various characteristics and capabilities. It would be an exciting topic to explore and compare the performance of these models on different image classification methods to determine which is most effective. Another work can examine the impact of other preprocessing techniques on the performance of the CNN- based approach. The preprocessing stage impacts the extracted features' quality and the classifier's overall performance. Therefore, it would be helpful to study the effect of different preprocessing techniques on the performance of the CNN-based approach. Finally, it would be interesting to study the effectiveness of CNN-based and knowledge-based feature extraction and their variations in the training data. It would be helpful to evaluate the performance of the approaches when the data are limited or biased in some way. It could provide insights into the generalization capabilities of the CNN- and knowledge-based systems.

We are implementing a Vision transformer (VIT) for breast cancer diagnosis from the histopathology images as an ongoing work.

Among different image-based datasets, histopathology images are one where deep neural network techniques perform well in diagnosing benign and malignant tissues. However, the present mainstream techniques face challenges. One of them is that the high-resolution characteristics of histopathological images have not been fully utilized to improve classification accuracy. One reason is that the current patch-based method does not adequately integrate these patches to make the classification result from a complete histopathology image [4].

7. Conclusions

In this study, we designed CNNs, VGG16, and knowledge-based models for feature extraction. We evaluated their performance on the BreakHis 400× dataset using Neural Networks (64 units), Multilayer Perceptron, Random Forest, Decision Tree, Support Vector Machines, K-Nearest Neighbor, and Narrow Neural Network (10 units) classifiers. The basic CNN-based features achieved an accuracy of up to 85%. The VGG16-based features attained an accuracy of up to 86%. Knowledge-based features outperformed the CNN and VGG16 features significantly. It reached an accuracy of 98% for all classifiers mentioned above. Without depending on the size of the dataset and training, the features were calculated by mathematical equations that were defined to capture expert knowledge. Overall, our work reports that the knowledge-based CAD system produced a robust high performance in benign/malignant classification using histopathology images.

Author Contributions: Conceptualization B.D.B. and H.K.; methodology H.K. and B.D.B.; software H.K.; validation H.K. and B.D.B.; formal analysis B.D.B. and H.K.; investigation H.K. and B.D.B.; resources H.K. and B.D.B.; data curation H.K. and B.D.B.; writing—original draft preparation H.K. and B.D.B.; writing—review and editing, H.K. and B.D.B.; visualization H.K. and B.D.B.; supervision B.D.B.; project administration B.D.B.; funding acquisition None. All authors have read and agreed to the published version of the manuscript.

Funding: This research received no external funding.

Institutional Review Board Statement: Not applicable.

Informed Consent Statement: Not applicable.

Data Availability Statement: Our work does not have a publicly archived dataset.

Conflicts of Interest: The authors declare no conflict of interest.

References

1. Giaquinto, A.N.; Sung, H.; Miller, K.D.; Kramer, J.L.; Newman, L.A.; Minihan, A.; Jemal, A.; Siegel, R.L. Breast Cancer Statistics, 2022. *CA Cancer J. Clin.* **2022**, *72*, 524–541. [CrossRef] [PubMed]
2. Uhlén, M.; Björling, E.; Agaton, C.; Szigyarto, C.A.-K.; Amini, B.; Andersen, E.; Andersson, A.-C.; Angelidou, P.; Asplund, A.; Asplund, C.; et al. A Human Protein Atlas for Normal and Cancer Tissues Based on Antibody Proteomics. *Mol. Cell. Proteom. MCP* **2005**, *4*, 1920–1932. [CrossRef] [PubMed]
3. Available online: https://v15.proteinatlas.org/learn/dictionary/cancer/breast+cancer+1 (accessed on 17 May 2023).
4. Yan, R.; Ren, F.; Wang, Z.; Wang, L.; Zhang, T.; Liu, Y.; Rao, X.; Zheng, C.; Zhang, F. Breast cancer histopathological image classification using a hybrid deep neural network. *Methods* **2020**, *173*, 52–60. [CrossRef]
5. Zeng, Z.; Xie, W.; Zhang, Y.; Lu, Y. RIC-Unet: An Improved Neural Network Based on Unet for Nuclei Segmentation in Histology Images. *IEEE Access* **2019**, *7*, 21420–21428. [CrossRef]
6. Zhang, X.; Zhu, X.; Tang, K.; Zhao, Y.; Lu, Z.; Feng, Q. DDTNet: A dense dual-task network for tumor-infiltrating lymphocyte detection and segmentation in histopathological images of breast cancer. *Med. Image Anal.* **2022**, *78*, 102415. [CrossRef]
7. Barkana, B.D.; El-Sayed, A.; Khaled, R.H.; Helal, M.; Khaled, H.; Deeb, R.; Pitcher, M.; Pfeiffer, R.; Roubidoux, M.; Schairer, C.; et al. Imaging Modalities in Inflammatory Breast Cancer (IBC) Diagnosis: A Computer-Aided Diagnosis System Using Bilateral Mammography Images. *Sensors* **2023**, *23*, 64. [CrossRef]
8. Barkana, B.D.; Sariçiçek, I. Classification of Breast Masses in Mammograms Using 2D Homomorphic Transform Features and Supervised Classifiers. *J. Med. Imaging Health Inform.* **2017**, *7*, 1566–1571. [CrossRef]
9. Labrada, A.; Barkana, B.D. Breast cancer diagnosis from histopathology images using supervised algorithms. In Proceedings of the IEEE Symposium on Computer-Based Medical Systems, Shenzen, China, 21–23 July 2022; pp. 102–107. [CrossRef]

10. Sharma, S.; Mehra, R. Conventional Machine Learning and Deep Learning Approach for Multi-Classification of Breast Cancer Histopathology Images—A Comparative Insight. *J. Digit. Imaging* **2020**, *33*, 632–654. [CrossRef]
11. Gupta, K.; Chawla, N. Analysis of Histopathological Images for Prediction of Breast Cancer Using Traditional Classifiers with Pre-Trained CNN. In *Procedia Computer Science*; Elsevier: Amsterdam, The Netherlands, 2020; Volume 167, pp. 878–889. [CrossRef]
12. Liu, M.; He, Y.; Wu, M.; Zeng, C. Breast Histopathological Image Classification Method Based on Autoencoder and Siamese Framework. *Information* **2022**, *13*, 107. [CrossRef]
13. Liu, M.; Hu, L.; Tang, Y.; Wang, C.; He, Y.; Zeng, C.; Lin, K.; He, Z.; Huo, W. A Deep Learning Method for Breast Cancer Classification in the Pathology Images. *IEEE J. Biomed. Health Inform.* **2022**, *26*, 5025–5032. [CrossRef]
14. Shallu; Mehra, R. Breast cancer histology images classification: Training from scratch or transfer learning? *ICT Express* **2018**, *4*, 247–254. [CrossRef]
15. Deniz, E.; Şengür, A.; Kadiroğlu, Z.; Guo, Y.; Bajaj, V.; Budak, Ü. Transfer learning based histopathologic image classification for breast cancer detection. *Health Inf. Sci. Syst.* **2018**, *6*, 18. [CrossRef] [PubMed]
16. Bayramoglu, N.; Kannala, J.; Heikkila, J. Deep learning for magnification independent breast cancer histopathology image classification. In Proceedings of the 23rd International Conference on Pattern Recognition (ICPR), Cancun, Mexico, 4–8 December 2016; pp. 2440–2445. [CrossRef]
17. Albashish, D.; Al-Sayyed, R.; Abdullah, A.; Ryalat, M.H.; Almansour, N.A. Deep CNN model based on VGG16 for breast cancer classification. In Proceedings of the International Conference on Information Technology, ICIT 2021, Amman, Jordan, 14–15 July 2021; pp. 805–810. [CrossRef]
18. Zou, Y.; Chen, S.; Che, C.; Zhang, J.; Zhang, Q. Breast cancer histopathology image classification based on dual-stream high-order network. *Biomed. Signal Process. Control.* **2022**, *78*, 104007. [CrossRef]
19. Li, G.; Wu, G.; Xu, G.; Li, C.; Zhu, Z.; Ye, Y.; Zhang, H. Pathological image classification via embedded fusion mutual learning. *Biomed. Signal Process. Control.* **2023**, *79*, 104181. [CrossRef]
20. Atban, F.; Ekinci, E.; Garip, Z. Traditional machine learning algorithms for breast cancer image classification with optimized deep features. *Biomed. Signal Process. Control.* **2023**, *81*, 104534. [CrossRef]
21. LeCun, Y.; Bengio, Y.; Hinton, G. Deep learning. *Nature* **2015**, *521*, 436–444. [CrossRef]
22. Joseph, A.A.; Abdullahi, M.; Junaidu, S.B.; Ibrahim, H.; Chiroma, H. Improved multi-classification of breast cancer histopathological images using handcrafted features and deep neural network (dense layer). *Intell. Syst. Appl.* **2022**, *14*, 66. [CrossRef]
23. Ahmad, N.; Asghar, S.; Gillani, S.A. Transfer learning-assisted multi-resolution breast cancer histopathological images classification. *Vis. Comput.* **2022**, *38*, 2751–2770. [CrossRef]
24. Egriboz, E.; Gokcen, B.; Bilgin, G. Classification of Breast Cancer Histopathology Images by Cell-Centered Deep Learning Approach. In Proceedings of the 28th Signal Processing and Communications Applications Conference, Gaziantep, Turkey, 5–7 October 2020; pp. 1–4. [CrossRef]
25. Simonyan, K.; Zisserman, A. Very Deep Convolutional Networks for Large-Scale Image Recognition. *arXiv* **2014**, arXiv:1409.1556.
26. Shen, D.; Wu, G.; Suk, H.-I. Deep Learning in Medical Image Analysis. *Annu. Rev. Biomed. Eng.* **2017**, *19*, 221–248. [CrossRef]
27. Gopalakrishnan, K.; Khaitan, S.K.; Choudhary, A.; Agrawal, A. Deep Convolutional Neural Networks with transfer learning for computer vision-based data-driven pavement distress detection. *Constr. Build. Mater.* **2017**, *157*, 322–330. [CrossRef]
28. Deng, J.; Dong, W.; Socher, R.; Li, L.-J.; Li, K.; Fei-Fei, L. ImageNet: A large-scale hierarchical image database. In Proceedings of the 2009 IEEE Conference on Computer Vision and Pattern Recognition, Miami, FL, USA, 20–25 June 2009; pp. 248–255. [CrossRef]
29. Spanhol, F.A.; Oliveira, L.S.; Petitjean, C.; Heutte, L. A Dataset for Breast Cancer Histopathological Image Classification. *IEEE Trans. Biomed. Eng.* **2016**, *63*, 1455–1462. [CrossRef] [PubMed]
30. Bardou, D.; Zhang, K.; Ahmad, S.M. Classification of Breast Cancer Based on Histology Images Using Convolutional Neural Networks. *IEEE Access* **2018**, *6*, 24680–24693. [CrossRef]
31. Spanhol, F.A.; Oliveira, L.S.; Cavalin, P.R.; Petitjean, C.; Heutte, L. Deep features for breast cancer histopathological image classification. In Proceedings of the IEEE International Conference on Systems, Man, and Cybernetics (SMC), Banff, AB, Canada, 5–8 October 2017; pp. 1868–1873. [CrossRef]
32. Spanhol, F.A.; Oliveira, L.S.; Petitjean, C.; Heutte, L. Breast cancer histopathological image classification using Convolutional Neural Networks. In Proceedings of the 2016 International Joint Conference on Neural Networks (IJCNN), Vancouver, BC, Canada, 24–29 July 2016; pp. 2560–2567.
33. Adeshina, S.A.; Adedigba, A.P.; Adeniyi, A.A.; Aibinu, A.M. Breast Cancer Histopathology Image Classification with Deep Convolutional Neural Networks. In Proceedings of the 14th International Conference on Electronics Computer and Computation (ICECCO), Kaskelen, Kazakhstan, 29 November–1 December 2018; pp. 206–212. [CrossRef]
34. Gour, M.; Jain, S.; Kumar, T.S. Residual learning based CNN for breast cancer histopathological image classification. *Int. J. Imaging Syst. Technol.* **2020**, *30*, 621–635. [CrossRef]
35. Toğaçar, M.; Özkurt, K.B.; Ergen, B.; Cömert, Z. BreastNet: A novel convolutional neural network model through histopathological images for the diagnosis of breast cancer. *Phys. A: Stat. Mech. its Appl.* **2020**, *545*, 123592. [CrossRef]
36. Zhu, C.; Song, F.; Wang, Y.; Dong, H.; Guo, Y.; Liu, J. Breast cancer histopathology image classification through assembling multiple compact CNNs. *BMC Med. Inform. Decis. Mak.* **2019**, *19*, 198. [CrossRef]

37. Zou, Y.; Zhang, J.; Huang, S.; Liu, B. Breast cancer histopathological image classification using attention high-order deep network. *Int. J. Imaging Syst. Technol.* **2022**, *32*, 266–279. [CrossRef]
38. Krizhevsky, A.; Sutskever, I.; Hinton, G.E. Imagenet classification with deep convolutional neural networks. *Commun. ACM* **2017**, *60*, 84–90. [CrossRef]
39. Alzubaidi, L.; Santamaría, J.; Manoufali, M.; Mohammed, B.; Fadhel, M.A.; Zhang, J.; Al-Timemy, A.H.; Al-Shamma, O.; Duan, Y. MedNet: Pre-trained Convolutional Neural Network Model for the Medical Imaging Tasks. *arXiv* **2021**, arXiv:2110.06512.

cancers

MDPI

Article

'Earlier than Early' Detection of Breast Cancer in Israeli *BRCA* Mutation Carriers Applying AI-Based Analysis to Consecutive MRI Scans

Debbie Anaby [1,2,*], David Shavin [1], Gali Zimmerman-Moreno [1], Noam Nissan [1,2], Eitan Friedman [2,3] and Miri Sklair-Levy [1,2,3]

1 Department of Diagnostic Imaging, Sheba Medical Center, Ramat Gan 52621, Israel
2 Sackler Faculty of Medicine, Tel-Aviv University, Tel-Aviv 6910201, Israel
3 Meirav High Risk Center, Sheba Medical Center, Ramat Gan 52621, Israel
* Correspondence: debbie.anaby@sheba.health.gov.il

Simple Summary: Women who have a genetic mutation of BRCA1 or BRCA2 are at a significantly higher risk for developing breast cancer. Early detection is crucial for an improved prognosis, therefore they are offered an intensive follow-up program, including a yearly MRI scan. Although MRI is the most sensitive imaging modality for breast cancer detection, it was found that a significant number of tumors are overlooked or misinterpreted, leading to a delayed diagnosis. Aiming to improve breast cancer diagnosis at early stages, we developed an artificial-intelligence based tool that is shown to classify correctly ~65% of the tumors at an early time point. These tumors were not suspected/diagnosed by the radiologists at that time point, but only at the next MRI scan. We believe that such an AI-system could serve as an aid to radiologists, improve their decision-making and achieve an 'earlier than early' diagnosis of breast cancer in BRCA carriers.

Abstract: Female *BRCA1/BRCA2* (=*BRCA*) pathogenic variants (PVs) carriers are at a substantially higher risk for developing breast cancer (BC) compared with the average risk population. Detection of BC at an early stage significantly improves prognosis. To facilitate early BC detection, a surveillance scheme is offered to *BRCA* PV carriers from age 25–30 years that includes annual MRI based breast imaging. Indeed, adherence to the recommended scheme has been shown to be associated with earlier disease stages at BC diagnosis, more in-situ pathology, smaller tumors, and less axillary involvement. While MRI is the most sensitive modality for BC detection in *BRCA* PV carriers, there are a significant number of overlooked or misinterpreted radiological lesions (mostly enhancing foci), leading to a delayed BC diagnosis at a more advanced stage. In this study we developed an artificial intelligence (AI)-network, aimed at a more accurate classification of enhancing foci, in MRIs of *BRCA* PV carriers, thus reducing false-negative interpretations. Retrospectively identified foci in prior MRIs that were either diagnosed as BC or benign/normal in a subsequent MRI were manually segmented and served as input for a convolutional network architecture. The model was successful in classification of 65% of the cancerous foci, most of them triple-negative BC. If validated, applying this scheme routinely may facilitate 'earlier than early' BC diagnosis in *BRCA* PV carriers.

Keywords: BRCA; early diagnosis; artificial intelligence; breast cancer; MRI

Citation: Anaby, D.; Shavin, D.; Zimmerman-Moreno, G.; Nissan, N.; Friedman, E.; Sklair-Levy, M. 'Earlier than Early' Detection of Breast Cancer in Israeli *BRCA* Mutation Carriers Applying AI-Based Analysis to Consecutive MRI Scans. *Cancers* **2023**, *15*, 3120. https://doi.org/10.3390/cancers15123120

Academic Editors: Rafał Obuchowicz, Michał Strzelecki and Adam Piorkowski

Received: 24 April 2023
Revised: 30 May 2023
Accepted: 2 June 2023
Published: 8 June 2023

1. Introduction

BRCA1 and *BRCA2* (heretofore *BRCA*) germline pathogenic variant (PVs) carriers are at a significantly higher risk for breast cancer (BC), with estimated cumulative lifetime risk of 55% to 65% and 45% to 47%, respectively [1,2]. *BRCA1* carriers commonly develop aggressive cancers (triple-negative with high nuclear grade), with peak incidence in the 41–50 age-group, while *BRCA2* carriers commonly develop cancers that are less aggressive (estrogen receptor (ER)-positive) with peak incidence in the 51–60 age-group [3,4]. Tumor volume doubling time in invasive BCs is reportedly twice as high in *BRCA* carriers (46 and 52 days in *BRCA1* and *BRCA2*, respectively) compared with age adjusted *BRCA* wild type high-risk women [5]. Thus, within 1 year, a tumor may double its volume by 4–5 times, implying a doubling of tumor diameter, a fact that may adversely affect tumor stage at diagnosis and consequent prognosis. Therefore, early detection of BC is crucial for diagnosis of curable non-metastatic disease, at a stage in which the cancer cells have not yet gained the ability to metastasize and become resistant to targeted therapies [6].

The current screening guidelines for female *BRCA* PVs carriers, including bi-annual breast imaging [7], have been shown to increase the rate of early diagnosis with more in-situ pathology, smaller tumors, and less axillary involvement compared with carriers who do not adhere to the suggested surveillance scheme [7–10]. MRI is considered the most sensitive modality for BC detection in *BRCA* PV carriers, with sensitivity rates of up to 96% [10–13]. Yet several studies have shown that, in retrospective analyses, in a significant number of MRI detected BCs (50–74%), localized radiological abnormalities were present in the prior MRI, which was performed approximately one year before actual BC diagnosis [12,14–17]. Pages et al. [18] and Seo et al. [19] reported that in ~50% of the reviewed cases (from a total of 56 and 72 consecutive pairs of MR imaging studies, respectively) a suspicious radiological finding of ~1 cm in size could be retrospectively visualized in a previously performed MRI. Notably, Gubern-Merida et al. [20] reported rates of 60% (24/40 pairs of MRIs) displaying misclassified pre BC diagnosis MRI abnormal findings, and Bilocq-Lacoste et al. [17] reported even higher rates of 74% (57/77 pairs of MRIs). The false negative pre-BC diagnosed MRIs in these studies were attributed to several factors: resemblance of the cancerous lesion to physiologic enhancement, small lesion size, stability in size and location of a non-mass in a postsurgical area [18,19]. Even when small lesions (usually 5 mm in size) are detected by MRI, their classification may be challenging because their morphologic and kinetic characteristics are commonly benign [21].

Recently, artificial intelligence (AI) algorithms are being assessed as additional tools for BC screening and diagnosis, mainly based on mammography captured data analyses [22–29]. MRI-based AI algorithms are also being evaluated for BC diagnosis, improved specificity, treatment response assessment, and anatomic segmentations of fibro-glandular tissue (FGT), allowing the quantification of background parenchymal enhancement (BPE) [30–34]. One of the major challenges of MRI breast imaging analyses is that of characterizing sub-centimeter lesions, which often display benign appearing features, and differentiating them from abundant types of enhancing foci, for either benign etiologies or focal background parenchymal enhancement (BPE). With that in mind, we developed and clinically-tested an AI network, trained to correctly classify enhancing foci detected on consecutive breast MRI studies of *BRCA* PV carriers, in order to facilitate 'earlier than early' BC diagnosis.

2. Materials and Methods

2.1. Study Population

This study was approved by the institutional review board and the need to obtain a written informed consent was waived, given the study outline and retrospective nature.

A retrospective search was performed in the dataset of the Meirav high-risk Clinic repository at Sheba Medical Center for all *BRCA* PV carriers with MRI detected BC and a prior MRI examination in the preceding 18 months, between 2012–2021. Diagnosis was based on biopsy obtained pathological report and the location of the tumor was determined by MRI. Cancer-free *BRCA* PV carriers seen at the same clinic during the study period with consecutive MRIs and with at least one year clinical and radiological follow-up were retrieved and served as controls.

Breast MRI was performed at 1.5 Tesla (Signa Excite HDX, GE Healthcare, Chicago, IL, USA) using a dedicated double breast coil equipped with eight channels. Dynamic contrast enhanced (DCE)-MRI protocol was obtained via axial vibrant multiphase 3D DCE T1 weighted sequence, prior and four times after an automated injection of contrast agent bolus (0.1 mL/kg at 2 mL/s Dotarem, (gadoterate meglumine, Guebet)), followed by a 20 mL saline flush. The first post-contrast images were centered at 1:25 min after injection and the rest were centered every 2 min following the first images, such that the delayed images were centered at 7:35 min after injection. The parameters used for the DCE-MRI were as follows: echo time (TE) = 2.6 ms, repetition time (TR) = 5.4 ms, flip angle = 15°, bandwidth = 83.3 kHz, matrix = 512 × 364, field of view (FOV) = 340 mm and slice thickness = 2 mm.

The pre-contrast images were subtracted from the 2nd time-point post-contrast images to create sub-DCE images. These were collected from the prior MRIs of all patients.

2.2. MR Images, Clinical, Radiological and Pathological Data Analysis

Patients' consecutive MRIs; 'prior' (~1 year before diagnosis) and 'diagnosis' were collected along with their clinical data, pathology of the diagnosed tumors and image radiology features, including age, time between consecutive scans, exact *BRCA* PV, breast imaging reporting and data system (BI-RADS) score, and BPE grade in prior scans, tumor size at diagnosis, tumor histopathological type, histological grade and immunohistochemistry results for hormonal receptor expression. Lesions at diagnosis and abnormalities in the prior scans (if existed) were retrospectively morphologically characterized (focus, mass, non-mass) according to the BI-RADS lexicon (acr.org/birads) by an experienced breast radiologist (MSL).

2.3. Lesion Segmentation and Morphological/Kinetic Assessment

The prior MRIs were used for analysis and the diagnosis MRIs and biopsies served as ground-truth for cancer/cancer-free labeling and for identification of the tumor area and its delineation.

Manual segmentation was performed on picture archiving and communication system (PACS). In the cancer patients, the central slice of the tumor was manually marked on the sub-DCE image of the diagnosis MRI by DA in agreement with radiologist MSL. Manual co-registration between the two consecutive scans in each patient was performed in PACS and then the corresponding region of the known cancerous tumor was identified in the prior scan based on anatomical landmarks. In cases in which the region comprised of an enhancing abnormality, its central slice was manually marked. In the cancer-free women, if the prior MRIs had a reported lesion/abnormality, it was marked for control. Otherwise, a prominent enhancing focus (regarded as BPE) was chosen and marked. The corresponding regions in the following MRI scans were also carefully identified and marked, if existed. In all cases, the segmentation was copied from the subtraction image to the raw data including the pre-contrast, the first and the fourth post-contrast time point images. Then, kinetic features were calculated according to:

$$Initial\ enhancement = \frac{(SE\ in\ 1st\ time\ point) - (SE\ in\ pre\ contrast)}{SE\ in\ pre\ contrast} * 100 \quad (1)$$

$$Delayed\ phase = \frac{(SE\ in\ 4th\ time\ point) - (SE\ in\ 1st\ time\ point)}{SE\ in\ 1st\ time\ point} * 100 \quad (2)$$

where SE stands for signal enhancement. Initial enhancement was categorized as 'slow' when smaller than 50%, 'medium' when between 50–100% and 'fast' when larger than 100%. Delayed phase was defined as 'persistent' for a larger increase than 10%, 'plateau' when between a decrease of 10% and an increase of 10%, and 'washout' for a decrease of more than 10% [35].

Segmented lesions and abnormalities in both prior and diagnosis MRIs were morphologically characterized into 3 categories: focus, mass and non-mass, according to BI-RADS lexicon (acr.org/birads). Lesion size was manually measured in PACS based on the largest diameter.

2.4. Convolutional Neural Network (CNN) Architecture

The model was categorized between tumor and non-tumor ROIs. Two patches per abnormality or BPE regions served as input: (1) the regions immediately around the abnormality or BPE (resized to 24 × 24) and (2) a larger field of view (FOV) region including the abnormality's or BPE's (128 × 128) surroundings. This allowed capture of both global spatial information and smaller local details [36]. The proposed network architecture, illustrated in Figure 1, included a repeated application of a 3 × 3 convolution, followed by a rectified linear activation function (ReLU) and a 2 × 2 max pooling operation with stride 2 for down-sampling for feature extraction [37]. This process was repeated separately for both inputs described above. Then, the feature vectors were flattened and concatenated into a single feature vector. This then served as input for a multi-layer-perceptron classifier and finally pushed the output through a soft-max activation to provide a valid distribution. The training was performed using an Adam optimizer with a learning rate of 1×10^{-4} that minimizes a binary cross entropy loss function. The feature extractors were pre-trained on a Cifar-19 dataset for an improved initialization of the model's weights. Cross-validation was performed with a leave-one-out procedure.

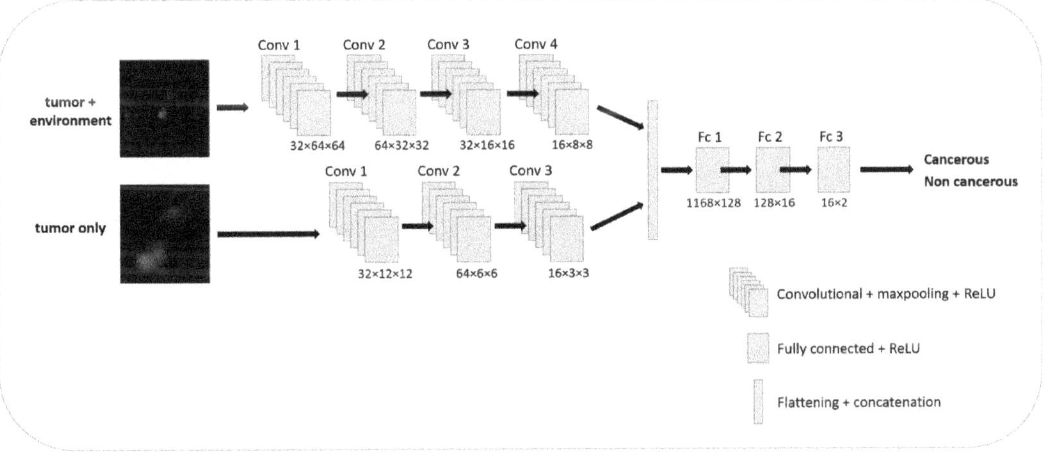

Figure 1. The proposed network architecture with input of 2 ROIs per lesion from a sub-DCE 2D image.

2.5. Statistics

The data were statistically analyzed using SPSS 29.0 (Chicago, IL, USA). *T*-tests or Chi-square tests were used to assess differences between independent groups. Repeated measures ANOVA with a Greenhouse-Geisser correction was used to test the change in lesion characteristics between prior and diagnosis scans in the cancerous and non-cancerous groups. Time of scan was used as the within-subject factor and group as the between-subjects factor. Paired t-tests were applied as post-hoc analysis in the cases where a significant interaction between time and group was found. Significance was set at $p < 0.05$.

3. Results

Overall, the study group encompassed 53 biopsy-proven BCs in *BRCA* PV carriers with a pre diagnosis MRI available for analysis (367.6 days between scans on average). As controls, 53 cancer-free *BRCA* PV carriers with two consecutive MRIs performed during the study period were analyzed.

In retrospective visualization, a radiological abnormality could be detected in previous MRIs of 32/53 BC patients (60.4%) in the same anatomical region where a tumor was subsequently located. Figure 2 presents consecutive MRIs of three representative BC cases

(Figure 2A–C) in whom an abnormality could be visualized in their prior MRI (Figure 2D–F). The time difference between the two scans was 357, 418, and 387 days, respectively.

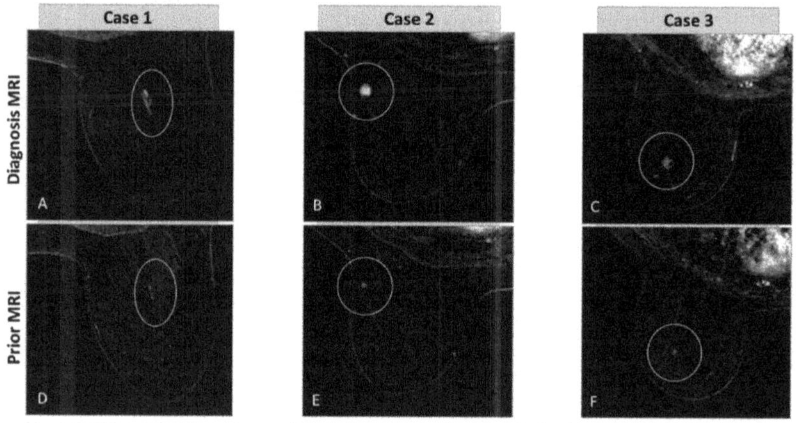

Features	Diagnosis MRI	Prior MRI	Diagnosis MRI	Prior MRI	Diagnosis MRI	Prior MRI
Morphology	Linear non-mass	Linear non-mass	round mass	focus	round mass	focus
Initial enhancement	fast	fast	fast	fast	fast	fast
Delayed enhancement	slow	plateau	plateau	plateau	plateau	plateau
Tumor size [mm]	10	6	6	3	6	3

Figure 2. Diagnosed BCs in three patients (top row) and their corresponding radiological abnormalities as appearing in the previous MRIs (bottom row). Tumors and abnormalities are marked by a white circle. The morphological and kinetic features are shown below the images. Case 1–74 years old *BRCA1* PV carrier, diagnosed with grade 3 triple negative ductal carcinoma in situ (DCIS) (**A**). Linear non mass enhancement at the prior MRI (**D**) was unchanged throughout several previous MRIs (not presented), and therefore reported as BI-RADS 2. Case 2–64 years old *BRCA1* PV carrier, diagnosed with grade 3 triple negative invasive ductal carcinoma (IDC) (**B**). Enhancing focus at the prior MRI was misinterpreted as intramammary lymph node and reported as BI-RADS 2 (**E**). Case 3–67 years old *BRCA1* PV carrier, diagnosed with grade 3 triple negative IDC (**C**). Enhancing focus at the prior MRI was not detected by the radiologist and reported as BI-RADS 2 (**F**).

Radiological features and tumor pathological characteristics of all 53 BC cases are shown in Table 1. Of 32 patients in whom a radiological abnormality was present in previous MRI imaging, five (15.6%) cases were reported as BI-RADS 0 due to that abnormality or due to dense breasts, indicating that more imaging information is required (mammography and ultrasound) in order to make an informed medical decision. Of the five patients, three followed the recommendations, and the findings were either determined as post-surgical changes or were not detected, thus were not diagnosed at that time point. 8/32 (25%) patients were reported as BI-RADS 3 due to unchanged findings, post-surgical changes, dense breasts or benign-looking lesions. The rest (59.4%) were reported as radiologically negative (BI-RADS 1 or 2 on prior scan). Age at diagnosis, distribution of *BRCA* PV, days between the two scans, BI-RADS and BPE in the prior scan were similar between the case group (those with BC) and the cancer-free group.

Table 1. Radiological features and pathological characteristics of study participants.

Characteristics	*All Cancer Patients*	*Cancer Patients with Abnormality*	*Cancer Patients no Abnormality*	*p Value Cancer with/without Abnormality*	*All Cancer-Free Patients*	*p Value Cancer/Cancer-Free*
Number of patients	53	32 (60.4)	21 (39.6)		53	
Age at diagnosis (range-years)	52 ± 14 (32–80)	51.4 ± 13.6 (33–78)	51.9 ± 14.4 (32–80)	0.96	50 ± 15.9 (23–78)	0.71
Mutated gene				0.70		**0.015**
BRCA1	39 (73.6)	25 (78.1)	14 (66.7)		25 (47.2)	
BRCA2	14 (26.4)	7 (21.9)	7 (33.3)		26 (49.1)	
unknown					2 (3.8)	
Days between scans (range-days)	367.6 ± 130.2 (177–938)	361 ± 118.6 (177–779)	380 ± 148.2 (182–938)	0.68	373 ± 78.5 (177–604)	0.78
BIRADS on prior scan				0.44		0.28
0	7 (13.2)	5 (15.6)	2 (9.5)		4 (7.6)	
1	2 (3.8)	2 (6.3)	–		5 (9.4)	
2	33 (62.3)	17 (53.1)	16 (76.2)		33 (62.3)	
3	11 (2.8)	8 (25)	3 (14.3)		8 (15.1)	
4	–	–	–		3 (5.7)	
BPE on prior scan				0.08		0.09
minimal-mild	31 (58.5)	16 (50)	15 (71.4)		40 (75.5)	
moderate-marked	22 (41.5)	16 (50)	6 (28.6)		13 (24.5)	
Tumor size at diagnosis [mm]	10.8 ± 7.3 (2–35)	12.8 ± 8.4 (2–35)	7.8 ± 4 (3–16)	**0.01**		
Tumor type				0.77		
IDC	33 (62.3)	18 (56.3)	15 (71.4)			
DCIS	17 (32.1)	12 (37.5)	5 (23.8)			
IDC+DCIS	2 (3.8)	1 (3.1)	1 (4.8)			
unknown	1 (1.9)	1 (3.1)	–			
Histological grade				0.36		
Low	2 (3.8)	1 (3.1)	2 (9.5)			
Intermediate	14 (26.4)	8 (25)	6 (28.6)			
High	33 (62.3)	19 (59.4)	14 (66.7)			
unknown	4 (7.6)	4 (12.5)	–			
Luminal type				0.35		
HR+/HER2−	18 (33.9)	10 (31.3)	8 (38.1)			
HR+/HER2+	3 (5.7)	2 (6.3)	1 (4.8)			
HR−/HER2+	4 (7.6)	1 (3.1)	1 (4.8)			
Triple negative	30 (56.6)	19 (59.4)	11 (52.4)			

Continuous data: mean ± SD, with ranges in parenthesis. Categorical data: number of patients with percentages in parenthesis. IDC = invasive ductal carcinoma, DCIS = ductal carcinoma in situ, HER2 = human epidermal growth factor receptor, HR = hormone receptor.

Invasive ductal carcinoma (IDC) tumors were more common in both groups compared with ductal carcinoma in-situ (DCIS), while a large number were high grade triple negative (44% and 60% of patients with abnormality and without abnormality in the prior MRI, respectively). Tumor size at diagnosis was significantly different between the groups (student's *t*-test, $p = 0.01$), while patients who had an abnormality in the early scan presented with larger tumor sizes than those who did not at the diagnosis MRI (12.8 mm vs. 7.8 mm on average, respectively). The distribution of BPE in the prior scans was different between the groups: in patients who had an abnormality on the prior scan, BPE was distributed evenly between minimal-mild and moderate-marked whereas, in those who did not show an abnormality more MRIs showed a minimal-mild BPE score than moderate-marked (71.4% vs. 28.6%, respectively). Yet this difference was not statistically significant ($p = 0.08$).

Age of the individuals diagnosed with BC and age at analysis for the cancer-free group, days between the consecutive scans, BIRADS and BPE distributions were similar between these two groups. Mutated *BRCA* gene distribution was statistically significantly different ($p = 0.015$): most cancer patients were *BRCA1* PV carriers (73.6%) whereas in the cancer-free cohort an almost even representation was seen for both genes.

Lesion characteristics were compared between the two consecutive scans in the cancer and cancer-free patients (Table 2). The overall mean difference of the lesion size over time was statistically significant ($F(1,59) = 23.742$, $p < 0.001$), as was the interaction time*group ($F(1,59) = 25.990$, $p < 0.001$). Additionally, the overall mean difference of the morphology characteristics over time (focus, mass, non-mass) was statistically significant ($F(1,63) = 5.968$, $p = 0.017$) as was the interaction ($F(1,63) = 13.871$, $p < 0.001$). A paired *t*-test revealed significance between these features in the cancer group, only showing smaller lesion size in the prior scans compared with the diagnosis scans (6.1 mm vs. 10.8 mm on average, $p < 0.0001$), most of which were determined to be 'focus' (59.3%). The morphology type distribution was significantly different between the prior and diagnosis scans in the cancer group ($p = 0.0001$), showing mostly 'mass' lesions at diagnosis (56.3%).

In the cancer-free cohort, almost 40% of the enhancements that were detected in the prior scans did not appear in the follow-up MRI. In cases where lesions were visualized in both consecutive MRIs, lesion size and morphology were similar.

Kinetic characteristics were similar between the consecutive scans in both the cancer and cancer-free groups.

Analysis based on an AI network architecture was used to achieve early detection of the abnormalities on prior MRIs. The network was based on a total of 85 MRI scans: prior MRIs of the 32 cancer patients who had an abnormality and prior scans of the 53 cancer-free patients. The network successfully classified 21/32 cancer cases (65.6%) as cancerous and 47/53 cancer-free cases (88.7%) as non-cancerous, as shown in Figure 3.

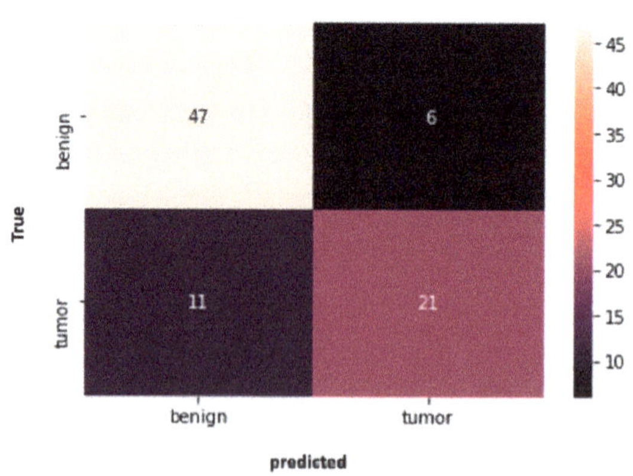

Figure 3. Confusion matrix of the performance of the network for classification of very early lesions/abnormalities in prior MRI scans to diagnosis scans of breast cancer.

Table 2. Lesion characteristics at diagnosis and in the prior scans.

		Cancerous Lesions		*Non-Cancerous Lesions/BPE*		*Time*		*Interaction Time × Group*		*Paired t-Test*	
		Prior Scan (N = 32)	***At Diagnosis (N = 32)***	***Prior Scan (N = 53)***	***Follow-Up Scan (N = 33)***	***F***	***p***	***F***	***p***	***Cancerous Lesions p***	***Non-Cancerous Lesions p***
Lesion size [mm]		6.1 ± 4.2 (1.5–17)	10.8 ± 7.3 (2–35)	7.4 ± 4.2 (1.7–27.3)	7.1 ± 5.1 (2.3–28.9)	F(1,59) = 23.74	**<0.0001**	F(1,59) = 25.99	**<0.001**	**<0.0001**	0.69
Morphology						F(1,63) = 5.96	**0.02**	F(1,63) = 13.87	**<0.001**	**0.0001**	0.37
	Focus	19 (59.3)	6 (18.8)	12 (22.6)	12 (36.4)						
	Mass	6 (18.8)	18 (56.3)	17 (32.1)	10 (30.3)						
	Non-mass	7 (21.8)	8 (25)	25 (47.2)	11 (33.3)						
Kinetics											
	CAD	9 (28.1)	14 (43.8)	4 (7.6)	1 (3)	F(1,61) = 0.93	0.34	F(1,61) = 4.78	**0.03**	0.09	0.16
Initial phase						F(1,61) = 0.41	0.53	F(1,61) = 1.55	0.22		
	Slow	8 (25)	5 (15.6)	24 (45.3)	16 (48.5)						
	Medium	13 (40.6)	13 (40.6)	15 (28.3)	8 (24.2)						
	Fast	11 (34.4)	14 (43.8)	14 (26.4)	9 (27.3)						
Delayed phase						F(1,62) = 2.11	0.15	F(1,62) = 0.24	0.75		
	Persistent	23 (71.9)	21 (65.6)	45 (84.9)	24 (72.7)						
	Plateau	8 (25)	10 (31.3)	7 (13.2)	7 (21.2)						
	Washout	1 (3.1)	1 (3.1)	1 (1.9)	2 (6.1)						

Continuous data: means, with ranges in parenthesis. Categorical data: numbers of patients with percentages in parenthesis.

MR scan characteristics and tumor pathology and morphology were observed for the 21 successfully classified cancer cases, compared with the 11 unsuccessful classified cases, by the network (Tables 3 and 4).

Table 3. Radiological and pathological characteristics of successfully and unsuccessfully classified abnormalities in prior scans of cancer patients.

Characteristics		AI Success (21)	AI Failure (11)	*p*-Value
Age at diagnosis (range-years)		52.5 ± 13.4 (34–77)	49.3 ± 14.3 (33–78)	0.53
Mutated gene				0.39
	BRCA1	18 (85.7)	7 (63.6)	
	BRCA2	3 (14.3)	4 (36.4)	
Days between scans (range-days)		362.7 ± 89.3 (196–511)	357.7 ± 166.1 (177–779)	0.85
BIRADS on prior scan				1
	0	4 (19.1)	1 (10)	
	1	1 (4.8)	1 (10)	
	2	11 (52.4)	6 (60)	
	3	5 (23.8)	2 (20)	
BPE on previous scan				0.57
	Minimal to mild	12 (57.1)	5 (45.5)	
	Moderate to marked	9 (42.9)	6 (54.5)	
Tumor size [mm]		10.6 ± 8 (2–28)	16.7 ± 7.8 (7–35)	**0.05**
Tumor type				0.42
	IDC	13 (61.9)	5 (45.5)	
	DCIS	6 (28.6)	6 (54.5)	
	IDC+DCIS	1 (4.8)	--	
	unknown	1 (4.8)	--	
Histological grade				0.68
	Low	0	--	
	Intermediate	4 (19.1)	4(40)	
	High	14 (66.7)	5(50)	
	unknown	3 (14.3)	1(10)	
Molecular subtype				**0.016**
	HR+/HER2-	5 (23.8)	5 (50)	
	HR+/HER2+	1 (4.8)	--	
	HR-/HER2+	2 (9.5)	2 (20)	
	Triple negative	13 (61.9)	3 (30)	

Table 4. Lesion characteristics of successfully and unsuccessfully classified abnormalities in prior scans of cancer patients.

Characteristics of Early Scan		AI Success (21)	AI Failure (11)	*p*-Value
Morphology				1
	focus	12 (57.1)	7 (63.6)	
	mass	4 (19.1)	2 (18.2)	
	non-mass	5 (23.8)	2 (18.2)	
Initial enhancement				0.71
	Slow	5 (23.8)	2 (18.2)	
	Medium	8 (38.1)	6 (54.5)	
	fast	8 (38.1)	3 (27.3)	
Delayed phase				0.47
	Persistent	15 (71.4)	8 (72.7)	
	Plateau	6 (28.6)	2 (18.2)	
	washout	--	1 (9)	
CAD				1
	Positive	6 (28.6)	3 (27.3)	
	negative	13 (61.9)	8 (72.7)	
	unknown	1 (4.8)		

Abbreviation: CAD, computer aided detection.

Interestingly, the molecular subtype distribution was significantly different between the groups ($p = 0.016$). In the successfully classified group of patients, most tumors were triple negative (61.9%), while in the unsuccessfully classified patients, most tumors were HR+/HER2− (50%). The tumor size showed a trend p-value of 0.05, while the correctly classified abnormalities in the prior scans had a smaller diameter compared with those which were not classified correctly (10.6 mm vs. 16.7 mm, respectively).

Morphology and kinetic characteristics were similar in both the successfully and unsuccessfully classified groups of patients. In both groups, most abnormalities were 'focus', with a medium/fast initial enhancement and persistent delayed phase. Computer aided detection (CAD) was negative in most cases.

Representative cases of successful and unsuccessful classification of the AI algorithm in cancerous and benign lesions are shown in Figures 4–6 (distinct from the cases presented in Figure 1). In Figure 4, four successful classifications of cancer tumors are seen, showing the abnormalities that appeared in the prior MRIs (top row) and the detected tumors in the diagnostic MRIs (bottom row). In Figure 4A,B, the abnormalities were detected at the time of the prior MRIs by the radiologist, but were said to be post-surgical changes and given BI-RADS 3. Both were diagnosed ~1 year later by the 'diagnostic' MRI as IDCs. In Figure 4C,D, there was no suspicious radiological finding at the time of the prior MRIs. The abnormalities were noted only retrospectively. Both were diagnosed by the 'diagnostic' MRIs as in-situ carcinomas. Figure 5 shows three examples of unsuccessful classifications of cancerous tumors. All were given BI-RADS 2 in the prior MRIs (Figure 5A–D) and the radiological abnormalities were only retrospectively detected. Note that, in Figure 5A,C, the abnormalities are mildly conspicuous and particularly difficult to differentiate from the enhancing background.

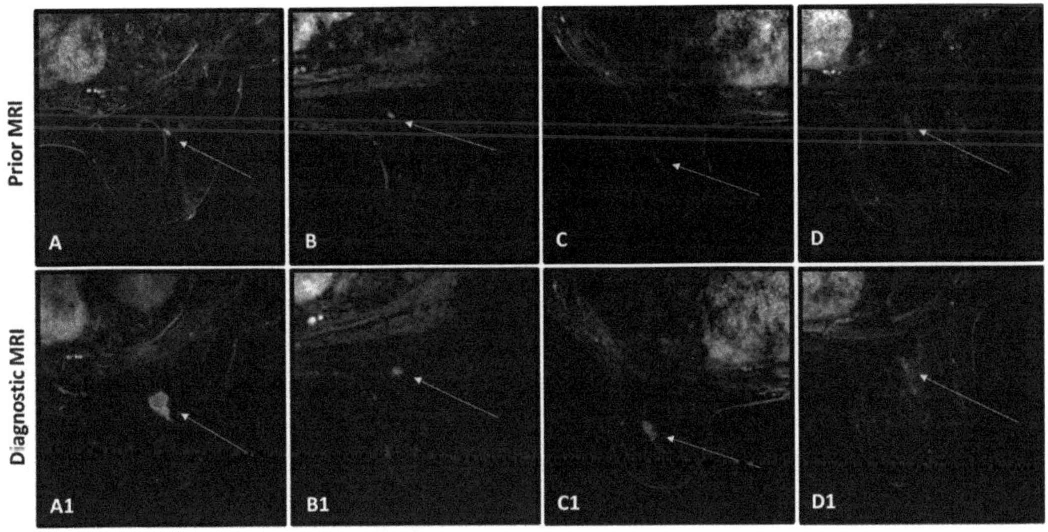

Figure 4. Four cancer cases of a successful classification of the AI network based on the prior MRIs. In all cases, an abnormality was retrospectively found in the prior scans but was not suspected at the time. Top row shows the prior MRIs of the patients (**A–D**), arrow pointing to an enhancing abnormality. Bottom row shows the 'diagnostic' MRIs of the patients (**A1–D1**), arrows pointing to the diagnosed cancerous tumors.

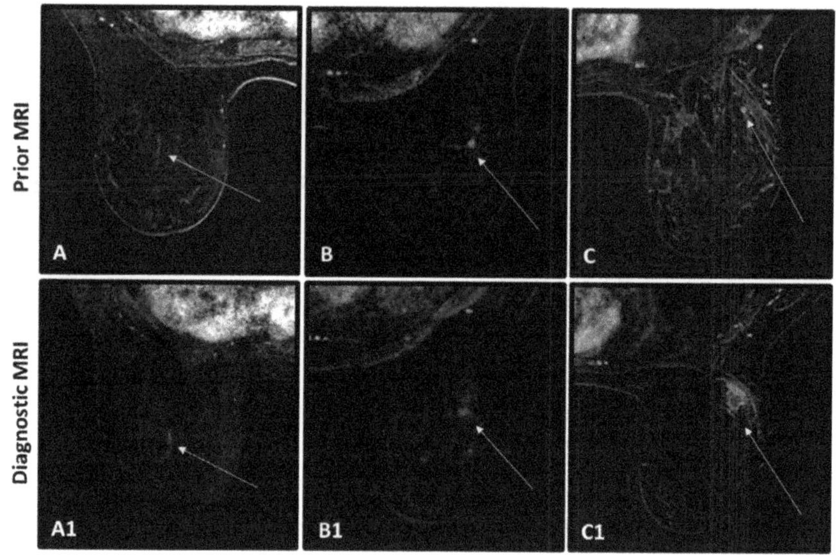

Figure 5. Three cancer cases of an unsuccessful classification of the AI network. In all cases, an abnormality was retrospectively found in the prior scans but was not suspected at the time by the radiologist. Top row shows the prior MRIs of the patients (**A–C**), arrow pointing to an enhancing abnormality. Bottom row shows the 'diagnostic' MRIs of the patients (**A1–C1**), arrows pointing to the diagnosed cancerous tumors.

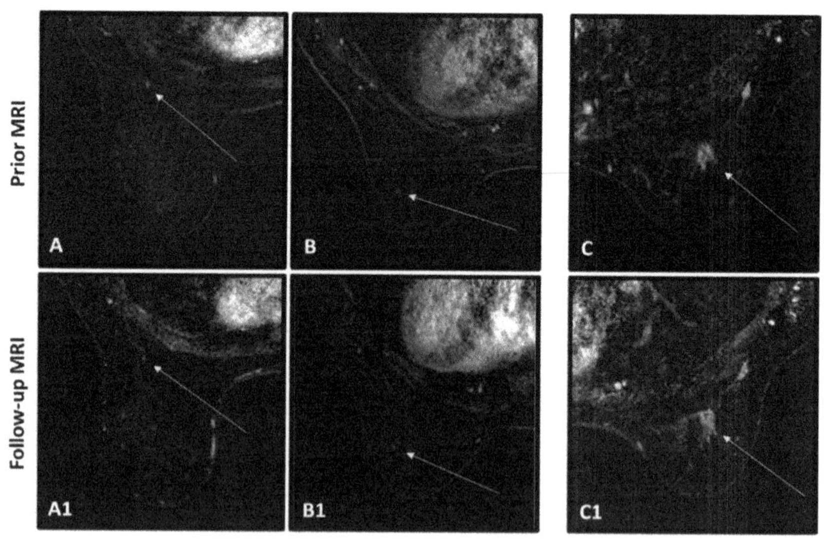

Figure 6. Two benign cases of a successful classification of the AI network (**A**,**A1**,**B**,**B1**) and one benign case of an unsuccessful classification of the AI network (**C**,**C1**). In all cases, an abnormality was retrospectively found in the prior scans. Top row shows the prior MRIs of the patients, arrow pointing to an enhancing abnormality. Bottom row shows the follow-up MRIs of the patients, arrows pointing to the benign tumors.

Figure 6 shows three benign cases. Two of them were correctly classified by the AI algorithm and one was not. The correctly classified lesions were radiologically detected in the prior MRIs, one was given BI-RADS 0 and the patient's previous MRI was requested to observe whether the lesion had already been present (Figure 6A), and the other was given BI-RADS 2 because the lesion was known and seemed unchanged (Figure 6B). The following MRIs of both cases were given BI-RADS 2. The incorrectly classified case (Figure 6C) was given BI-RADS 4 due to non-mass enhancement that appeared in the prior scan. The patient was recommended to complete more imaging tests and a biopsy, but the lesion was not detected under targeted ultrasound (US) nor a following MRI, therefore was finally determined as parenchymal enhancement. A follow-up MRI, almost 1 year later, showed sporadic enhancement areas which were defined as BPE, and the MRI was given BI-RADS 2.

4. Discussion

In this study, we challenged an AI-based network by investigating its ability to classify sub-centimeter breast abnormalities that were not originally suspected by the radiologist. As shown herein, breast abnormalities that are localized to the same anatomical area where the tumor was diagnosed on subsequent MRI can be visualized on initial MRIin ~60% of the cases. These results are in line with previously published studies [17–20]. Korhonen et al. [15] summarized the main reasons for false-negative errors: (1) technical—patient motion, artifacts etc. (2) perceptual—poor lesion conspicuity, subtle appearance of lesion etc. (3) cognitive—wrong interpretation. Misinterpretation of lesions was also found to increase the false-negative errors, mainly due to the presence of multiple breast lesions, prior biopsy or surgery and stability in size [17,18]. Bilocq-Lacoste et al. [17] reported that cancers that were undiagnosed had no specific MRI characteristics, receptor status, or risk factors, such as gene mutation, chest radiation, family history and site of previous biopsy. Most of the undiagnosed cancers (51 from 77, 66%) were overlooked due to their small size and high BPE. In agreement with Bilocq-Lacoste et al. [17], in the current study, the abnormalities in the prior scans were significantly smaller than the diagnosed tumors.

Clauser et al. [38] reported that the rate of small foci (<0.5 cm lesions) identified by MRI in a high-risk population was 31.3% (from a total of 166 patients). An automated approach based on CAD was reported in 2016 [20], where 71% of prior visible lesions and 31% of prior minimally visible lesions were detected in a group of 40 high risk cases. Several studies evaluated additional tools for an improved characterization of sub-centimeter breast lesions detected by breast imaging. Gibbs et al. [39] showed that radiomic analysis of small breast lesions is feasible; they showed significant differences between benign and malignant lesions for 53/133 calculated features, with high negative (>89%) and positive (>83%) predictive values. Lo Gullo et al. [40] showed that the combination of radiomics and machine learning improves the differentiation between benign and malignant small breast lesions in *BRCA* PV carriers compared with the BI-RADS classification by the radiologists. The integration of AI algorithms with the Internet-of-Things (IoT) has the potential to drive substantial advancement in the diagnosis of early stage breast cancer. In recent years, IoT has emerged as a revolutionary technology in healthcare. For example, it has been reported as a breakthrough in the surgical field [41]. Its ability to securely transmit, store and analyze data in real-time may assist radiologists in detection and characterization of breast lesions with increased efficiency and accuracy. Peta et al. [42] proposed an IoT-based model using data from bio-thermal sensors for the classification of breast cancer. The model achieved an accuracy of above 90%. Majji et al. [43] used mammograms and IoT for classification of breast lesions. Here too, the model achieved a high accuracy of above 90%. In this study, an AI-based network could accurately re-classify more than half of cancerous breast MRI abnormalities in *BRCA* PV carriers with a low rate of false-positives. Notably, triple-negative tumors were more frequently successfully classified as cancerous compared with other histological tumor types. This may be related, in all likelihood, to the fact that *BRCA1* mutated cases represented 85.7% of the successfully classified group and to the well-established association of triple-negative breast tumors with specific MRI features [44–46]. In a study by Motta et al. [45], the majority of the triple-negative tumors appeared as regular shaped mass enhancements (round or oval) with circumscribed margins by MRI. A large portion of triple-negative breast cancers also presented with a rim enhancement which was shown to be a positive predictor of this subtype. Importantly, triple-negative is associated with an increased angiogenesis and with a higher recurrence rate [45]. Therefore, the ability to detect triple-negative breast cancer at such an early stage, while abnormalities are mostly foci with no rim-enhancement, may reduce the clinical burden of the disease.

There were several limitations in this study that need to be acknowledged: a small cohort size, the use of one 2D central tumor slice instead of the whole 3D volume and no age-matched cancer-free control patients. Yet the limited spectrum of germline *BRCA* PV, the minimal inter-observer variability (due to the fact that all were carried out in a single medical center using the same staff and equipment), and the availability of longitudinal data on study participants are notable advantages.

5. Conclusions

In this study, we presented the ability to accurately classify sub-centimeter MRI-detected breast abnormalities that were not suspected by the radiologist and were subsequently diagnosed as cancer at an average of 1 year later. Future work includes applying such a classification on larger groups of patients in a prospective manner and broadening the AI-network to an automatic detection of suspicious abnormalities. If successful in these additional studies, such an approach can potentially facilitate earlier BC detection in high-risk women.

Author Contributions: Conceptualization, D.A., N.N. and M.S.-L.; methodology, D.A., D.S. and G.Z.-M.; investigation, D.A. and M.S.-L.; data curation. D.A.; visualization, D.A.; software, D.S. and G.Z.-M.; writing—original draft preparation, D.A.; writing—review and editing, N.N., E.F., M.S.-L., G.Z.-M. and D.A. Funding acquisition, D.A. All authors have read and agreed to the published version of the manuscript.

Funding: The authors are grateful for the support of this research by the Earlier.org—Friends For an Earlier Breast Cancer Test foundation and the Dahlia Greidinger Anti-Cancer Fund.

Institutional Review Board Statement: The study was conducted in accordance with the Declaration of Helsinki, and approved by the Institutional Review Board of The Sheba Medical Center (protocol code SMC-9219-22, approved at 8/11/2022).

Informed Consent Statement: Patient consent was waived due to the retrospective nature of the study.

Data Availability Statement: The data that support the findings of this study are available from the corresponding author, D.A., upon reasonable request.

Conflicts of Interest: The authors have no declared conflict of interest.

References

1. Antoniou, A.; Pharoah, P.D.P.; Narod, S.; Risch, H.A.; Eyfjord, J.E.; Hopper, J.L.; Loman, N.; Olsson, H.; Johannsson, O.; Borg, A.; et al. Average Risks of Breast and Ovarian Cancer Associated with *BRCA1* or *BRCA2* Mutations Detected in Case Series Unselected for Family History: A Combined Analysis of 22 Studies. *Am. J. Hum. Genet.* **2003**, *72*, 1117–1130. [CrossRef]
2. Chen, S.; Parmigiani, G. Meta-Analysis of BRCA1 and BRCA2 Penetrance. *J. Clin. Oncol.* **2007**, *25*, 1329–1333. [CrossRef]
3. Kuchenbaecker, K.B.; Hopper, J.L.; Barnes, D.R.; Phillips, K.A.; Mooij, T.M.; Roos-Blom, M.J.; Jervis, S.; Van Leeuwen, F.E.; Milne, R.L.; Andrieu, N.; et al. Risks of Breast, Ovarian, and Contralateral Breast Cancer for BRCA1 and BRCA2 Mutation Carriers. *JAMA* **2017**, *317*, 2402–2416. [CrossRef]
4. You, C.; Xiao, Q.; Zhu, X.; Sun, Y.; Di, G.; Liu, G.; Hou, Y.; Chen, C.; Wu, J.; Shao, Z.; et al. The Clinicopathological and MRI Features of Patients with BRCA1/2 Mutations in Familial Breast Cancer. *Gland. Surg.* **2021**, *10*, 262–272. [CrossRef]
5. Tilanus-linthorst, M.M.A.; Obdeijn, I.; Hop, W.C.J.; Causer, P.A.; Leach, M.O.; Pointon, L.; Hill, K.; Klijn, J.G.M.; Warren, R.M.L.; Gilbert, F.J. BRCA1Mutation and Young Age Predict Fast Breast Cancer Growth in the Dutch, United Kingdom, and Canadian Magnetic Resonance Imaging Screening TRials. *Clin. Cancer Res.* **2007**, *13*, 7357–7362. [CrossRef]
6. Greaves, M.; Maley, C.C. Clonal Evolution in Cancer. *Nature* **2012**, *481*, 306–313. [CrossRef]
7. Elezaby, M.; Lees, B.; Maturen, K.E.; Barroilhet, L.; Wisinski, K.B.; Schrager, S.; Wilke, L.G.; Sadowski, E. BRCA Mutation Carriers: Breast and Ovarian Cancer Screening Guidelines and Imaging Considerations. *Radiology* **2019**, *291*, 554–569. [CrossRef]

8. Bernstein-molho, R.; Kaufman, B.; Ben, M.A.; Sklair-levy, M.; Madoursky, D.; Zippel, D.; Laitman, Y.; Friedman, E. Breast Cancer Surveillance for BRCA1/2 Mutation Carriers—Is "Early Detection" Early Enough? *Breast* **2020**, *49*, 81–86. [CrossRef]
9. Guindalini, R.S.; Zheng, Y.; Abe, H.; Whitaker, K.; Toshio, F.; Walsh, T.; Schacht, D.; Kulkarni, K.; Sheth, D.; Verp, M.S.; et al. Intensive Surveillance with Biannual Dynamic Contrast-Enhanced Magnetic Resonance Imaging Downstages Breast Cancer in BRCA1 Mutation Carriers. *Clin. Cancer Res.* **2020**, *25*, 1786–1794. [CrossRef]
10. Shraga, S.; Grinshpun, A.; Zick, A.; Kadouri, L.; Cohen, Y.; Maimon, O.; Adler-Levy, Y.; Zeltzer, G.; Granit, A.; Maly, B.; et al. High-Risk Breast Cancer Screening in BRCA1/2 Carriers Leads to Early Detection and Improved Survival After a Breast Cancer Diagnosis. *Front. Oncol.* **2021**, *11*, 683656. [CrossRef]
11. Kriege, M.; Brekelmans, C.T.M.; Boetes, C.; Besnard, P.; Zonderland, H.M.; Obdeijn, I.; Manoliu, R.; Kok, T.; Peterse, H.L.; Tilanus-linthorst, M.M.; et al. Efficacy of MRI and mammography for breast-cancer screening in women with a familial or genetic predisposition. *N. Engl. J. Med.* **2004**, *351*, 427–437. [CrossRef]
12. Lo, G.; Scaranelo, A.M.; Aboras, H.; Ghai, S.; Kulkarni, S.; Fleming, R.; Bukhanov, K.; Crystal, P. Evaluation of the Utility of Screening Mammography for High-Risk Women Undergoing Screening Breast MR Imaging. *Radiology* **2017**, *285*, 36–43. [CrossRef]
13. Warner, E.; Hill, K.; Causer, P.; Plewes, D.; Jong, R.; Yaffe, M.; Foulkes, W.D.; Ghadirian, P.; Lynch, H.; Couch, F.; et al. Prospective Study of Breast Cancer Incidence in Women With a BRCA1 or BRCA2 Mutation Under Surveillance With and Without Magnetic Resonance Imaging. *J. Clin. Oncol.* **2011**, *29*, 1664. [CrossRef]
14. Maxwell, A.J.; Lim, Y.Y.; Hurley, E.; Evans, D.G.; Howell, A.; Gadde, S. False-Negative MRI Breast Screening in High-Risk Women. *Clin. Radiol.* **2017**, *72*, 207–216. [CrossRef]
15. Korhonen, K.E.; Samantha, P.; Susan, P.; Tobey, J.; Birnbaum, J.A.; Mcdonald, E.S. Breast MRI: False-Negative Results and Missed Opportunities. *RadioGraphics* **2021**, *41*, 10–15. [CrossRef]
16. Gao, Y.; Reig, B.; Heacock, L.; Bennett, D.L.; Heller, S.L.; Moy, L.; Louis, S. Magnetic Resonance Imaging in Screening of Breast Cancer. *Radiol. Clin. N. Am.* **2021**, *59*, 85–98. [CrossRef]
17. Bilocq-Lacoste, J.; Ferre, R.; Kuling, G.; Martel, A.L.; Tyrrell, P.N.; Li, S.; Wang, G.; Curpen, B. Missed Breast Cancers on MRI in High-Risk Patients: A Retrospective Case–Control Study. *Tomography* **2022**, *8*, 27. [CrossRef]
18. Pages, E.B.; Millet, I.; Hoa, D.; Doyon, F.C. Undiagnosed Breast Cancer at MR imaging: Analysis of causes. *Radiology* **2012**, *264*, 40–50. [CrossRef]
19. Seo, M.; Cho, N.; Bae, M.S.; Koo, H.R.; Kim, W.H.; Lee, S.H.; Chu, A. Features of Undiagnosed Breast Cancers at Screening Breast MR Imaging and Potential Utility of Computer-Aided Evaluation. *Korean J. Radiol.* **2016**, *17*, 59–68. [CrossRef]
20. Gubern-mérida, A.; Vreemann, S.; Martí, R.; Melendez, J.; Lardenoije, S.; Mann, R.M.; Karssemeijer, N.; Platel, B. Automated Detection of Breast Cancer in False-Negative Screening MRI Studies from Women at Increased Risk. *Eur. J. Radiol.* **2016**, *85*, 472–479. [CrossRef]
21. Meissnitzer, M.; Dershaw, D.D.; Feigin, K.; Bernard-davila, B.; Barra, F.; Morris, E.A. MRI Appearance of Invasive Subcentimetre Breast Carcinoma: Benign Characteristics Are Common. *Br. J. Radiol.* **2017**, *90*, 20170102. [CrossRef]
22. Lång, K.; Dustler, M.; Dahlblom, V.; Åkesson, A.; Andersson, I.; Zackrisson, S. Identifying Normal Mammograms in a Large Screening Population Using Artificial Intelligence. *Eur. Radiol.* **2021**, *31*, 1687–1692. [CrossRef]
23. Rodriguez-Ruiz, A.; Krupinski, E.; Mordang, J.; Schilling, K.; Heywand-Kobrunner, S.H.; Sechopoulos, I.; Mann, R.M. Detection of Breast Cancer with Mammography: Effect of an Artificial Intelligence Support System. *Radiology* **2019**, *290*, 305–314. [CrossRef]
24. Rodriguez-Ruiz, A.; Lång, K.; Gubern-Merida, A.; Broeders, M.; Gennaro, G.; Clauser, P.; Helbich, T.H.; Chevalier, M.; Tan, T.; Mertelmeier, T.; et al. Stand-Alone Artificial Intelligence for Breast Cancer Detection in Mammography: Comparison With 101 Radiologists. *J. Natl. Cancer Inst.* **2019**, *111*, 916–922. [CrossRef]
25. Dembrower, K.; Wåhlin, E.; Liu, Y.; Salim, M.; Smith, K.; Lindholm, P.; Eklund, M.; Strand, F. Effect of Artificial Intelligence-Based Triaging of Breast Cancer Screening Mammograms on Cancer Detection and Radiologist Workload: A Retrospective Simulation Study. *Lancet Digit. Health* **2020**, *2*, e468–e474. [CrossRef]
26. McKinney, S.M.; Sieniek, M.; Godbole, V.; Godwin, J.; Antropova, N.; Ashrafian, H.; Back, T.; Chesus, M.; Corrado, G.C.; Darzi, A.; et al. International Evaluation of an AI System for Breast Cancer Screening. *Nature* **2020**, *577*, 89–94. [CrossRef]
27. Lotter, W.; Diab, A.R.; Haslam, B.; Kim, J.G.; Grisot, G.; Wu, E.; Wu, K.; Onieva, J.O.; Boyer, Y.; Boxerman, J.L.; et al. Robust Breast Cancer Detection in Mammography and Digital Breast Tomosynthesis Using an Annotation-Efficient Deep Learning Approach. *Nat. Med.* **2021**, *27*, 244–249. [CrossRef]
28. Raya-Povedano, J.L.; Romero-Martín, S.; Elías-Cabot, E.; Gubern-Mérida, A.; Rodríguez-Ruiz, A.; Álvarez-Benito, M. AI-Based Strategies to Reduce Workload in Breast Cancer Screening with Mammography and Tomosynthesis: A Retrospective Evaluation. *Radiology* **2021**, *300*, 57–65. [CrossRef]
29. Larsen, M.; Aglen, C.F.; Lee, C.I.; Hoff, S.R.; Lund-Hanssen, H.; Lång, K.; Nygård, J.F.; Ursin, G.; Hofvind, S. Artificial Intelligence Evaluation of 122969 Mammography Examinations from a Population-Based Screening Program. *Radiology* **2022**, *303*, 502–511. [CrossRef]
30. Xu, X.; Fu, L.; Chen, Y.; Larsson, R.; Zhang, D.; Suo, S.; Hua, J.; Member, Z. Breast Region Segmentation Using Convolutional Neural Network in Dynamic Contrast Enhanced MRI. In Proceedings of the 40th Annual International Conference of the EMBC, Honolulu, HI, USA, 18–21 July 2018; pp. 750–753.

31. Ha, R.; Chang, P.; Mema, E.; Mutasa, S.; Karcich, J.; Wynn, R.T.; Liu, M.Z. Fully Automated Convolutional Neural Network Method for Quantification of Breast MRI Fibroglandular Tissue and Background Parenchymal Enhancement. *J. Digit. Imaging* **2019**, *32*, 141–147. [CrossRef]
32. Comes, M.C.; Fanizzi, A.; Bove, S.; Didonna, V.; Diotaiuti, S.; La Forgia, D.; Latorre, A.; Martinelli, E.; Mencattini, A.; Nardone, A.; et al. Early Prediction of Neoadjuvant Chemotherapy Response by Exploiting a Transfer Learning Approach on Breast DCE-MRIs. *Sci. Rep.* **2021**, *11*, 14123. [CrossRef] [PubMed]
33. Verburg, E.; Van Gils, C.H.; Van Der Velden, B.H.M.; Bakker, M.F. Deep Learning for Automated Triaging of 4581 Breast MRI Examinations from the DENSE Trial. *Radiology* **2022**, *302*, 29–36. [CrossRef] [PubMed]
34. Abe, H.; Giger, M.L. Use of Clinical MRI Maximum Intensity Projections for Improved Breast Lesion Classification with Deep Convolutional Neural Networks. *J. Med. Imaging* **2023**, *5*, 014503. [CrossRef]
35. Kim, H.; Ko, E.Y.; Kim, K.E.; Kim, M.K.; Choi, J.S.; Ko, E.S.; Han, B.K. Assessment of Enhancement Kinetics Improves the Specificity of Abbreviated Breast MRI: Performance in an Enriched Cohort. *Diagnostics* **2023**, *13*, 136. [CrossRef]
36. Frid-Adar, M.; Diamant, I.; Klang, E.; Amitai, M.; Goldberger, J.; Greenspan, H. Modeling the Intra-Class Variability for Liver Lesion Detection Using a Multi-Class Patch-Based CNN. *arXiv* **2017**, arXiv:1707.06053v2.
37. Alahmer, H.; Ahmed, A. Hierarchical Classification of Liver Tumor from CT Images Based on Difference-of-Features (DOF). In Proceedings of the International Conference of Signal and Engineering, London, UK, 29 June–1 July 2016.
38. Clauser, P.; Cassano, E.; De Nicolò, A.; Rotili, A.; Bonanni, B.; Bazzocchi, M.; Zuiani, C. Foci on Breast Magnetic Resonance Imaging in High-Risk Women: Cancer or Not ? *Radiol. Med.* **2016**, *121*, 611–617. [CrossRef]
39. Gibbs, P.; Onishi, N.; Sadinski, M.; Gallagher, K.M.; Hughes, M.; Martinez, D.F.; Morris, E.A.; Sutton, E.J. Characterization of Sub-1 Cm Breast Lesions Using Radiomics Analysis. *J. Magn. Reson. Imaging* **2019**, *50*, 1468–1477. [CrossRef]
40. Lo Gullo, R.; Daimiel, I.; Saccarelli, C.R.; Bitencourt, A.; Gibbs, P.; Fox, M.J.; Thakur, S.B.; Martinez, D.F.; Jochelson, M.S.; Morris, E.A.; et al. Improved Characterization of Sub-Centimeter Enhancing Breast Masses on MRI with Radiomics and Machine Learning in BRCA Mutation Carriers. *Eur. Radiol.* **2020**, *30*, 6721–6731. [CrossRef]
41. Mulita, F.; Verras, G.; Anagnostopoulos, C.; Kotis, K. A Smarter Health through the Internet of Surgical Things. *Sensors* **2022**, *22*, 4577. [CrossRef]
42. Peta, J.; Koppu, S. An IoT-Based Framework and Ensemble Optimized Deep Maxout Network Model for Breast Cancer Classification. *Electronics* **2022**, *11*, 4137. [CrossRef]
43. Majji, R.P.G.; Prakash, O.; Rajeswari, R.; Cristin, R. Smart IoT in Breast Cancer Detection Using Optimal Deep Learning. *J. Digit. Imaging* **2023**. [CrossRef] [PubMed]
44. Li, J.; Han, X. Research and Progress in Magnetic Resonance Imaging of Triple-Negative Breast Cancer. *Magn. Reson. Imaging* **2014**, *32*, 392–396. [CrossRef] [PubMed]
45. Moffa, G.; Galati, F.; Collalunga, E.; Rizzo, V.; Amati, G.D.; Pediconi, F.; Kripa, E. Can MRI Biomarkers Predict Triple-Negative Breast Cancer? *Diagnostics* **2020**, *15*, 1090. [CrossRef] [PubMed]
46. Chen, I.E.; Lee-Felker, S. Triple-Negative Breast Cancer: Multimodality Appearance. *Curr. Radiol. Rep.* **2022**, *11*, 53–59. [CrossRef]

cancers MDPI

Article

Deep Learning Algorithm for Differentiating Patients with a Healthy Liver from Patients with Liver Lesions Based on MR Images

Maciej Skwirczyński [1], Zbisław Tabor [2], Julia Lasek [3], Zofia Schneider [3], Sebastian Gibała [4], Iwona Kucybała [5], Andrzej Urbanik [5] and Rafał Obuchowicz [5,*]

1 Faculty of Mathematics and Computer Science, Jagiellonian University, 30-348 Krakow, Poland
2 Faculty of Electrical Engineering, Automatics, Computer Science, and Biomedical Engineering, AGH University of Science and Technology, 30-059 Krakow, Poland
3 Faculty of Geology, Geophysics, and Environmental Protection, AGH University of Science and Technology, 30-059 Krakow, Poland
4 Ultragen Medical Center, 31-572 Krakow, Poland
5 Department of Diagnostic Imaging, Jagiellonian University Medical College, 31-501 Krakow, Poland
* Correspondence: r.obuchowicz@gmail.com

Simple Summary: We propose an automated system that handles numerous problems encountered when diagnosing the presence of numerous types of lesions in the liver based on multiparametric magnetic resonance (MR) images. Using a properly built and processed dataset and deep-learning segmentation algorithms, we devised a method for screening MR images for the presence of focal lesions in the liver.

Abstract: The problems in diagnosing the state of a vital organ such as the liver are complex and remain unresolved. These problems are underscored by frequently published studies on this issue. At the same time, demand for imaging diagnostics, preferably using a method that can detect the disease at the earliest possible stage, is constantly increasing. In this paper, we present liver diseases in the context of diagnosis, diagnostic problems, and possible elimination. We discuss the dataset and methods and present the stages of the pipeline we developed, leading to multiclass segmentation of the liver in multiparametric MR image into lesions and normal tissue. Finally, based on the processing results, each case is classified as either a healthy liver or a liver with lesions. For the training set, the AUC ROC is 0.925 (standard error 0.013 and a p-value less than 0.001), and for the test set, the AUC ROC is 0.852 (standard error 0.039 and a p-value less than 0.001). Further refinements to the proposed pipeline are also discussed. The proposed approach could be used in the detection of focal lesions in the liver and the description of liver tumors. Practical application of the developed multi-class segmentation method represents a key step toward standardizing the medical evaluation of focal lesions in the liver.

Keywords: liver tumor segmentation; classification; deep learning; focal lesions; hepatocellular carcinoma; adenoma; focal nodular hyperplasia; magnetic resonance imaging (MRI)

Citation: Skwirczyński, M.; Tabor, Z.; Lasek, J.; Schneider, Z.; Gibała, S.; Kucybała, I.; Urbanik, A.; Obuchowicz, R. Deep Learning Algorithm for Differentiating Patients with a Healthy Liver from Patients with Liver Lesions Based on MR Images. *Cancers* **2023**, *15*, 3142. https://doi.org/10.3390/cancers15123142

Academic Editor: Antonio Grieco

Received: 26 April 2023
Revised: 5 June 2023
Accepted: 9 June 2023
Published: 11 June 2023

1. Introduction

The liver is a vital organ unique for metabolism, in which numerous biochemical pathways are present [1]. Due to the liver's complexity and unique functional role, liver health and preservation of the liver parenchyma represent major causes of concern. The liver, as the only place where certain processes occur (such as detoxification), is vulnerable to damage caused by various toxins, with alcohol being the most prevalent among the population [2,3]. Prolonged exposure to toxic substances leads to cirrhosis with a high incidence of hepatocellular carcinoma [4]. Hepatocellular carcinoma is the most common primary liver cancer [5]. The probability of developing HCC is more than 20% in the cirrhotic liver, and outcome of the tumor is fatal [6]. Due to its involvement in the hormonal pathways, the liver is also subject to the influence of the both endogenous and exogenous hormones, which are related to the formation of lesions such as FNH [7,8]. Hormonal simulation can also induce the dimensional growth of hemangiomas [9]. Finally, the liver's anatomy itself, due to its many tubular structures filled with fluids, promotes the formation of cysts. Consequently, benign cysts are the most frequent lesions [10]. The importance of the liver itself and probability of different lesions being formed makes the liver a point of interest in evaluations using diagnostic imaging procedures. Magnetic resonance is the most important modality in the diagnosis of the most common primary form of liver cancer: hepatocellular carcinoma. The diagnostic accuracy of MRI in the detection of liver tumors is high compared to that of tumors in other organs [11–13]. According to international guidelines, MRI is the first step of the diagnostic process, and histopathological analysis is required for liver nodules with an atypical imaging presentation [14,15]. However, a liver biopsy is not always feasible, especially in the case of very small liver nodules; furthermore, this procedure is not free from risks and is limited by an elevated rate of false negatives (about 30%) [16]. Therefore, the implementation of radiomics and machine learning software in clinical practice appears to be a promising and interesting new field of investigation.

Sensitivity and specificity for MRI-supported diagnosis are over 90% in hemangioma, cyst, and FNH detection [17–20]. With this high detection sensitivity, MRI outperforms CT in the detection of lesions with diameters less than 2 cm [21]. Notably, MRI is also advantageous in the detection and characterization of intracellular compounds, which may indicate the state of disease or even level of advancement of morbidity. The use of hepatobiliary contrast agents and the introduction of diffusion-weighted imaging sequences in the protocol supported the recognition of premalignant lesions helps to assess risk in atypical or indeterminate nodules [22,23]. Importantly, MR requires no radiation exposure, which is especially important in abdominal CT, for which the dose is relatively large, ranging from 15 to 24 mSv and corresponding to a dose equivalent to 220 to 350 single chest X-rays. This exposure may influence lifetime attributable cancer risk [24]. Although MRI efficiency is reported as remarkably high, it must be noted that those numbers represent the best human experience in the detection of lesions reported in diagnostic processes. Human perception of the patterns of textures present in the analyzed image plays a key role in the diagnostic process [25]. Detection relies on differences in the lesion textures compared to the background. Recognition is a process based on a comparison of the perceived textures against the "gold standard" lesion attributes as specified in, e.g., atlas recommendations. This process is subjective and prone to errors and intra-reader variability, which makes computer-based supporting systems highly anticipated by the medical community.

In the past years, an enormous increase in the number of medical imaging examinations performed has been observed [24]. This trend is connected to an increase in demand for radiological opinions in clinical decision-making processes [26]. Increasing the number of imaging exams introduces large amounts of data, which cannot be analyzed in the time expected by patients. Therefore, the concept of automated image analysis is growing in popularity as a promising approach for obtaining results faster and more effectively in order to meet clinical needs [27]. Among the problems related to the automated analysis of medical images, the task of screening for the presence of cancer mass is emerging and continues to gain popularity, especially in the fields of lung cancer and brain lesions but also in liver cancer detection [28–30]. The use of automated analysis methods such as radiomics and machine learning improves cancer detection in early stages, which is connected to an increased survival rate [31]. This increase is often connected to the detection of incipient cancer, which in the worst case scenario, could potentially be overlooked by a human [32].

Recently, studies have presented many automatic and semi-automatic methods based on classical methods of image analysis and deep learning (DL) techniques, aimed at creating systems for the reliable diagnosis and detection of liver tumors. The proposed approaches are mostly applied to CT images of livers and based on available public liver CT databases. Given that the liver can undergo various pathologies such as hepatic fat, iron deposits, fibrosis, and tumors, which can modify the liver's density and signal intensity or simply distort its shape, existing methods suffer from inaccuracies. Another important aspect is the location of the liver, whose immediate surroundings, due to having a similar intensity, add to the difficulty of corresponding problems. Liver segmentation is one of the key steps in various clinical applications such as tumor detection and is often preceded by the preprocessing required to achieve high accuracy in later stages. Preprocessing can include applying windowing to the raw image [33]; histogram equalization, the most widely used being contrast limited adaptive histogram equalization [34]; data normalization [35]; image denoising, e.g., median filter [36]; MR bias field correction, e.g., N4 bias correction [37]; or the joint removal of MRI bias [38].

MR images provide much better differentiation between healthy tissues and lesions. Nevertheless, automated methods for the analysis of these images are not common, especially when it comes to the analysis of multiparametric MR images of the liver. In light of these facts, in our study, we analyze T1 (pre- and post-contrast), T2 TSE, and DWI. We propose a DL-based processing pipeline aimed at screening MR liver examinations for the presence of lesions. Because multiparametric MR liver databases are not available in the public domain, as a part of this study, an imaging database was collected, which includes both MR abdominal examinations and reference manual segmentation of the liver and lesions.

2. Materials and Methods

2.1. Materials

The study protocol was developed based on the guidelines of the Helsinki Declaration and the Declaration of Good Clinical Practice. All images were anonymized prior to processing to ensure the security of personal data. In addition, written consent from the Local Ethics Committee was obtained to conduct this study 1072.6120.22.22 (23 February 2022).

Images were selected in the Hospital Data Reporting System—Hospital Information System (HIS). To select the group of interest, 6500 records were reviewed by data scientists and the radiology team. From this dataset, 1520 records were selected for further evaluation and examined in the hospital PACS for the presence of lesions (hepatocellular carcinoma, hemangioma, liver cysts, and focal nodular hyperplasia). The control group, with no lesions confirmed, was also selected from the database. In total, 500 records that met the quality criteria (free of artefacts and completeness of the desired sequences) were exported for further analyses. Meticulous scanning and verification of the lesions revealed 145 cystic lesions (C), 126 hemangiomas (HM), 28 hepatocellular carcinomas (HCC), and 78 focal nodular hyperplasia (FNH) cases. For some patients, more than one type of lesion was present in the liver. The distribution of particular lesions is given in Figure 1.

In this study, we used data from the selected 500 patients scanned in a 1.5 T MR system. MRI images were acquired using the standard diagnostic protocol for patients referred for an abdominal evaluation in MRI. DWI, single-shot echo-planar imaging, EPI, T2 turbo-spin echo, T1 turbo-spin echo, post-contrast early (T1 + C), and T1 turbo-spin echo post-contrast (non-hepatospecific) delayed (T1 Delayed) were used in in this study. Diffusion weighted sequences (EPI) were produced with 50-400-800 B using 2, 4, and 8 averages, with three diffusion weightings. In all applied sequences, the signal/noise ratios were equal to 1. The protocol outline of sequences used for the evaluation is presented in Table 1.

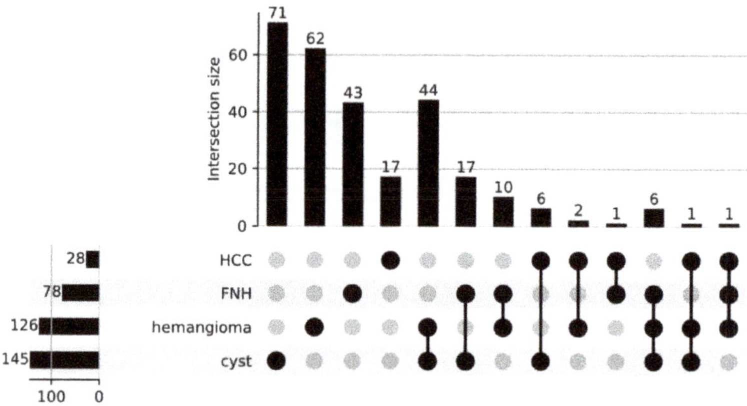

Figure 1. Distribution of the focal lesions in the examined set of approved images.

Table 1. MR protocol applied in standard abdominal imaging.

	TE	TR	FA	MX	VOX	FOV	CON	AV	PAT	DF	T
T2 (AX)	100	4000	180	182 × 320	1.3 × 1.3 × 4	200	4	1	2	30	1.38
T1 (AX)	2.58	215	70	154 × 256	1.6 × 1.6 × 4	400	4	1	2	30	0.44
DWI (AX)	71	7800	0	118 × 192	2.1 × 2.1 × 4	400	4	2,4,8	2	30	5.53

TE—time echo, TR—relaxation time, FA—flip angle, MX—imaging matrix, VOX—voxel size in mm^3, FOV—field of view (mm), BW—bandwidth, AV—averages, CON—concentrations, TF–turbo factor, T—overall sequence time (minutes).

The reference manual segmentation of the liver and lesions was prepared with the 3D Slicer Imaging Computing Platform (https://www.slicer.org/, accessed on 1 February 2023). For each patient, the liver was segmented in each MR sequence. For each lesion, the lesion's presence was confirmed, and its outlines were drawn for each slice in which the lesion was visible. Each lesion was outlined for a single MR sequence where the lesion was most prominent, as determined using a visual assessment. The liver and lesion outlines were prepared by two members of the research team. The initial outlines were then verified by a group of three radiologists (with 40, 15, and 6 years of experience) who applied an intrareader multistep consensus procedure to arrive at the final outlines. Agreements over 92% were accepted for further image analysis. Lower agreement resulted in corrections of the initial outlines. The workflow is presented in Figure 2.

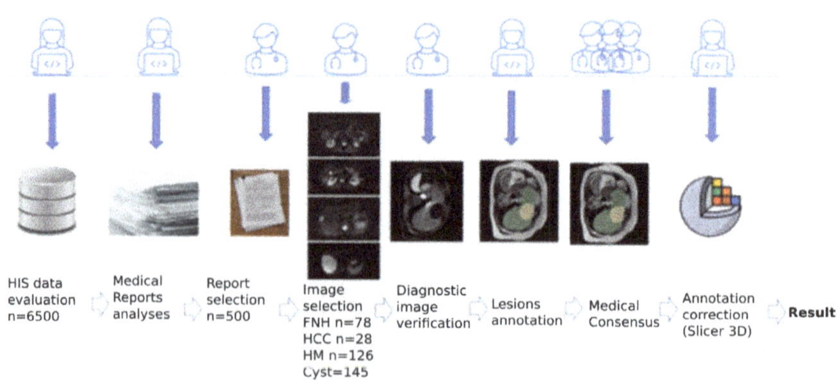

Figure 2. Workflow for collecting from the image database (designed as a "Result" in the figure) used to develop deep-learning algorithms.

2.2. Methods

The dataset of 500 MR examinations was randomly split into training and testing portions. Stratified sampling was applied to the training and test sets when selecting examinations: approximately 20% of examinations were selected for the test set from each group of examinations (healthy, with HM, with HCC, with cyst, with cyst and FNH, etc.). The training set consisted of 405 MR examinations, including 178 examinations in which no lesions in the liver were visible and 227 examinations in which lesions were present. The test set consisted of 95 MR examinations, including 43 examinations in which lesions in the liver were not visible and 52 examinations in which changes were present.

To complete the task of patient classification, we devised a pipeline consisting of the steps described below. Firstly, the four diagnostic images, denoted T2, T1 + C, T1 Delayed, and DWI in the following, were loaded from the corresponding DICOM sequences and entered into a spatial mesh with unified spatial resolution based on linear interpolation and information stored in appropriate DICOM tags (pixel size and slice thickness equal to inter-slice distance).

In the next step, the liver in each of the four images was segmented. For this process, we used U-Net, as implemented in nnU-Net framework [39]. The nnU-Net implementation of U-Net differs slightly from that of the vanilla U-Net [40]. Most notably, in nnU-Net, stride convolutions were used instead of max-pooling, and Leaky ReLU was used instead of plain ReLU. There was also a difference in data preparation. The authors of nnU-Net implemented a universal framework, which means that this framework can be used for very different datasets, often with better results than neural networks solutions aimed at operating on only one concrete dataset. Network topology details are also established based on datasets. This process allows for the appropriate aggregation of spatial information and is also influenced by hardware parameters, such as GPU memory. To account for these factors, we started with the initial patch size and verified that the size would fit within the GPU memory. If so, the batch size was increased; if not, the patch size was decreased. In addition to input patch size and number, pooling operations per axis were set, which also implicitly set the number of convolutional layers [41]. For the above-mentioned neural network, we chose training in a fivefold cross-validation scheme, where the training set was randomly divided into five parts. Four of these parts were used as the actual training set and one as the validation set. The validation set was selected in five possible ways by running five training sessions and obtaining five models of liver segmentation. These five models were used in the prediction stage by averaging their responses.

Data augmentation was applied during the dataset preparation stage and was implemented using the Batchgenerators framework [42] with the following augmentations [39] (the value of the probability of applying each augmentation is given in parenthesis): Spatial: rotation, scaling (0.4)—in the case of isotropic patches, the angles of rotation around each axis, given in degrees, were sampled from U (−30, 30), and for anisotropic patches, from U (−180, 180); the scaling factor was sampled from U (0.7, 1.4). Gaussian noise (0.15), if applied, was added to each voxel independently with the variance of the noise drawn from U (0, 0.1). For Gaussian blur (0.2), the width of the Gaussian kernel, expressed in voxels, was taken from U (0.5, 1.5), and the blur (0.5) was applied to each modality independently. For brightness (0.15), the intensity of each voxel was multiplied by the value drawn from U (0.7, 1.3). For contrast (0.15), the intensity of each voxel was multiplied by the value drawn from U (0.7, 1.3); the latter was subsequently trimmed to the original range. A simulation of low resolution (0.25) was achieved using nearest neighbor interpolation to downsample the modalities by a U (1, 2) factor and then resample the modalities back to their original size using cubic interpolation. The gamma intensities (0.15, and the voxel intensities were inverted beforehand) were scaled to a factor [0, 1] of their respective range of values, a nonlinear transformation of the intensities per voxel was applied, and then the intensities were scaled back to their original range of values. Mirroring (0.5) was performed with a given probability along all axes. After liver segmentation, we removed all but the largest connected components. This process was found to be beneficial for our task.

Then, we performed registration between T2 and the remaining diagnostic images. The registration step was necessary because, typically, the effects of a patient's movements between acquisition of the four sequences were clearly visible. We first found three transformations between the liver segmentations in T2 images, which we set as fixed images. These transformations were, respectively, T1 + C, T1 Delayed, and DWI liver segmentations and were set as moving images. We used a method that allows for two-stage registration, where each stage is characterized by different transforms. At the beginning of the first stage, we set a similarity 3D transform with rotation, uniform scaling around a fixed center, and translation. To ascertain the agreement between the two registered images, we used negative normalized cross correlation with the sampling strategy set to random (sample image voxels with replacements using a uniform distribution) and the sampling percentage set to 1%. At this stage, we used a linear interpolator and regular step gradient descent optimizer. The second stage was implemented in an analogous way. We changed the transform to B-spline and the optimizer to Limited memory Broyden, Fletcher, Goldfarb, Shanno, and Bound Constrained (LBFGSB) [43,44]. Subsequently, we applied the found transformations to the MR images obtained by combining all sequences under one spatial mesh. After the registration step, we cropped the images to the region of liver segmentation.

We then fed the data into nnU-Net, which was trained in the same way as before, only this time for simultaneous liver and liver lesion segmentations. The output returned multiclass segmentation of the liver and lesions.

The ultimate screening task was solved based on the results of the multiclass segmentation. Whenever the segmentation indicated that the number of voxels labeled as being contained within a lesion was larger than a certain decision threshold, the MR examination was labeled as containing a lesion. ROC curves were then obtained by varying the decision threshold in a range from 0 to 100,000 voxels. Additionally, the liver and lesion segmentation quality were measured by means of two standard metrics: the Dice coefficient and robust Hausdorff distance (with tolerance set to 0.99). Computation of the segmentation quality metrics was accomplished using the surface–distance library (https://github.com/deepmind/surface-distance, accessed on 1 February 2023). The processing workflow is shown in Figure 3.

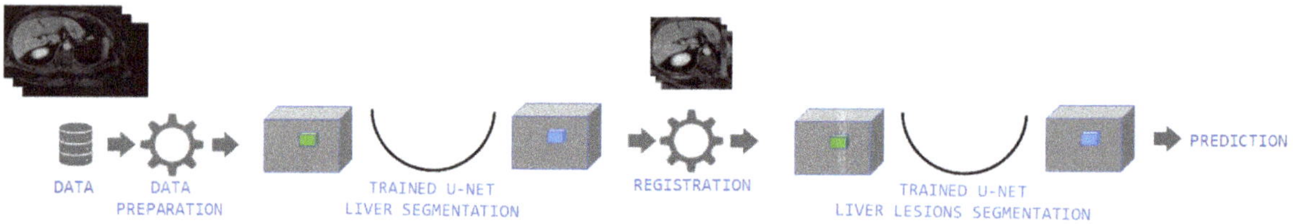

Figure 3. MR image processing workflow.

3. Results

In Figure 4, we show plots of the loss function for the task of liver segmentation in T2 images. The U-Net models were trained until a plateau of the loss function was reached for a validation set. During training, the best models for the validation set were saved and subsequently used for predictions. The training and validation curves for other segmentation tasks looked similar to the examples shown in Figure 4. In Figure 5, we show the results of registration between T2 (fixed) images and the remaining diagnostic images. Clearly, the effect of patient movement was very prominent, as demonstrated by relatively low values of the Dice coefficient before registration. Registration compensated for patient movement to some extent, as the Dice coefficient after registration was around 96%. Figure 6 shows the ROC curves for the training and test sets in the screening task. The ROC AUC value was found to be substantially higher for the training set than for the test set. Given the number of true positive and true negative cases in the training and test sets, the ROC AUC values equal to 0.925 (standard error 0.013) and 0.852 (standard error 0.039) for the training and test sets, respectively, were statistically significant at a *p*-value less than 0.001. After fixing specificity (SPE) at 0.79 for the test set, the sensitivity (SEN) of detection for patients with liver lesions was equal to 0.79. For the test set and decision threshold equal to 100 voxels, the SEN, SPE, and accuracy (ACC) of classification were equal to 0.79, and the positive predictive value (PPV) was equal to 0.82. The corresponding values for the training set were 0.85 and 0.88. For the test set and the decision threshold equal to 300 voxels, the ACC was equal to 0.83, SEN was equal to 0.73, SPE was equal to 0.95, and PPV was equal to 0.95. The respective values for the training set were 0.82, 0.75, 0.91, and 0.91. Figure 7 shows box plots for the liver segmentation task. The values of the Dice coefficient were found to be quite high at about 0.98. In contrast, the task of lesion segmentation is not solved equally well, as demonstrated by the box plots for the liver lesion segmentation task (Figure 8). Figure 9 demonstrates that the accuracy of hepatic lesion segmentation, measured by the Dice coefficient, was strongly dependent on the lesion volume measured by the number of voxels, with the segmentation of larger lesions being more accurate. These conclusions regarding segmentation quality were further supported by the values of robust Hausdorff Distance (rHD). For the task of liver segmentation, the median rHD in the training set was equal to 1.0 voxels, standard deviation was equal to 2.5 voxels, and the 95% quantile was equal to 4.0 voxels. The respective values for the test set were 1.4, 5.5, and 15.2 voxels. For the task of lesion segmentation, the median rHD in the training set was equal to 23.2 voxels, the standard deviation was equal to 42.9 voxels, and the 95% quantile was equal to 193.5 voxels. The corresponding values for the test set were 25.9, 53.8, and 168.7 voxels.

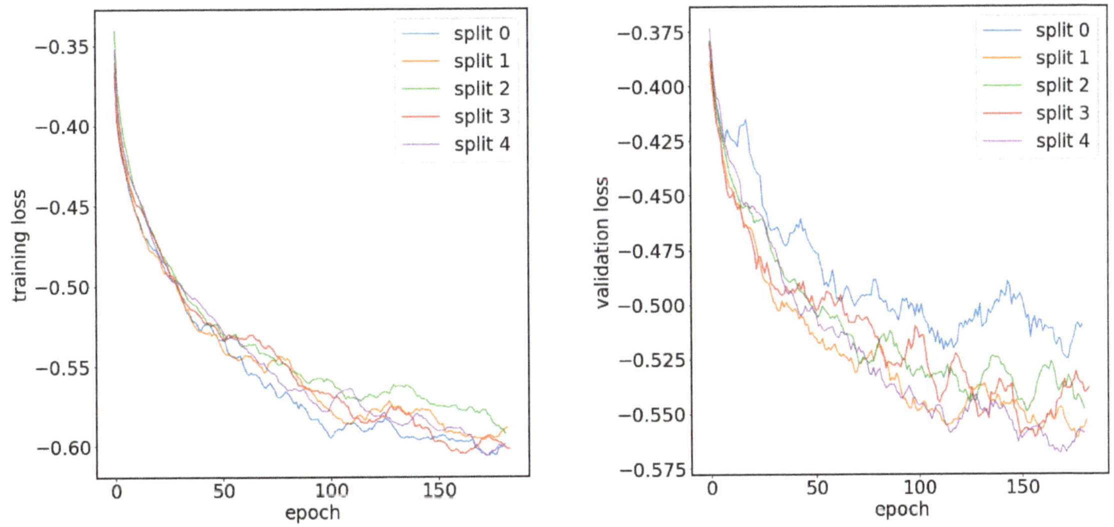

Figure 4. Plots of the loss function for the task of liver segmentation in T2 images for the training sets (left) and validation sets (right).

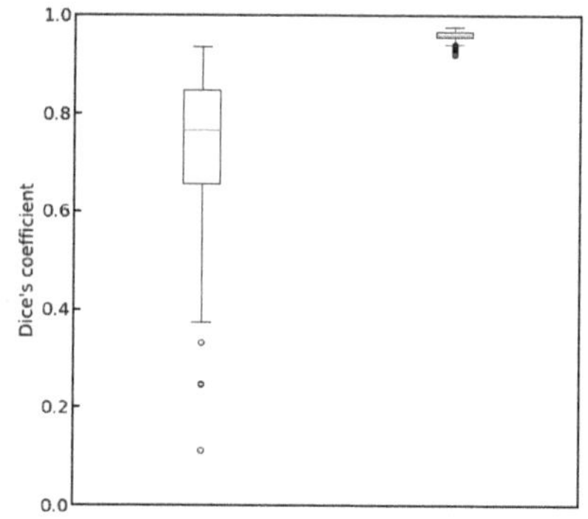

(a)

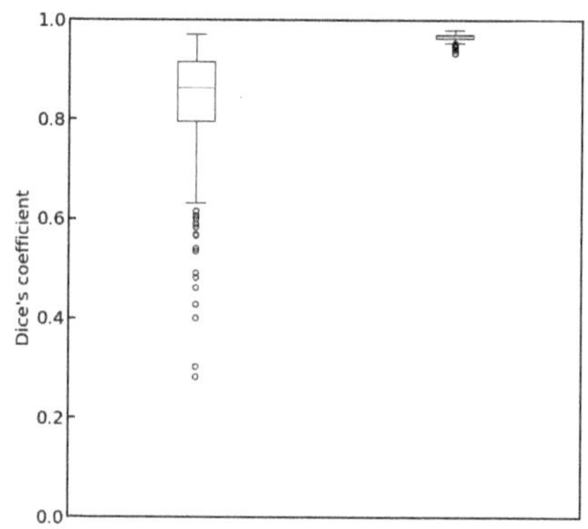

(b)

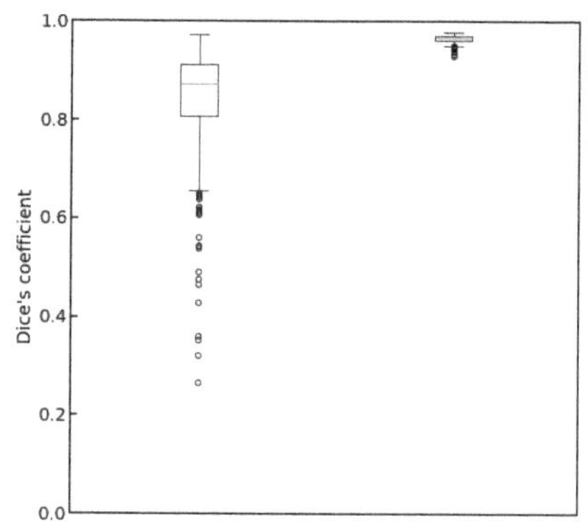

(c)

Figure 5. The results of registering moving images: (**a**) DWI, (**b**) T1 + C, and (**c**) T1 delayed against the fixed image T2. In each plot, the Dice coefficients between liver segmentation in moving and fixed images are shown before registration (left box plots) and after registration (right box plots).

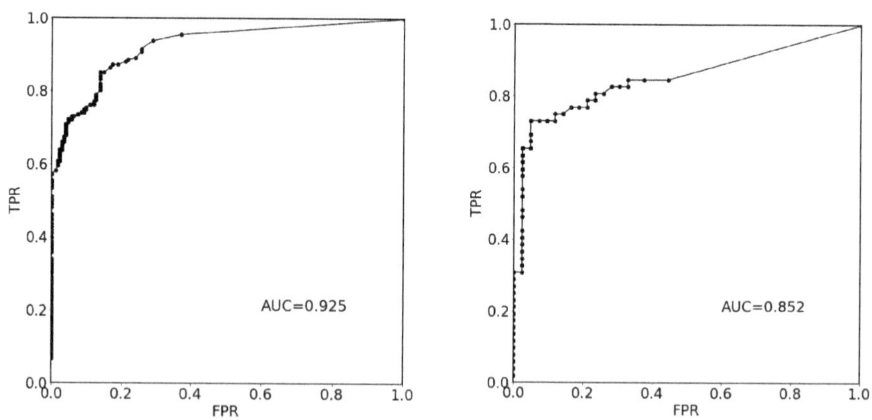

Figure 6. ROC curve for the "healthy liver/liver with a lesion present" classification task. On the left, the figure presents results for the training set. Here, the area under the ROC (AUC ROC) is equal to 0.925 (standard error 0.013, *p*-value less than 0.001). On the right, the results for the test set are presented. Here, the AUC ROC is equal to 0.852 (standard error 0.039, *p*-value less than 0.001). TPR: true positive rate; FPR: false positive rate.

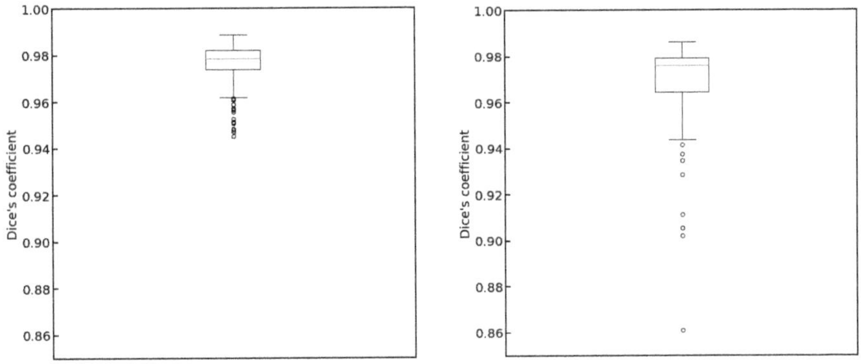

Figure 7. Box plots for the liver segmentation task. On the left, the results for the training set are presented. On the right, the results for the test set are presented.

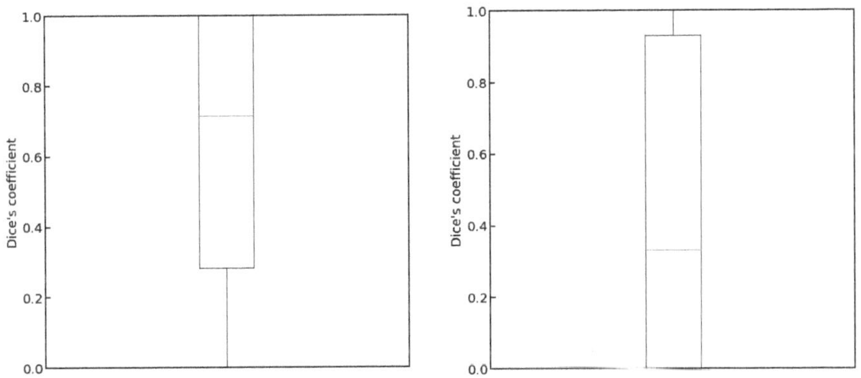

Figure 8. Box plots for the liver lesion segmentation task. On the left, the results for the training set are presented. On the right, the results for the test set are presented.

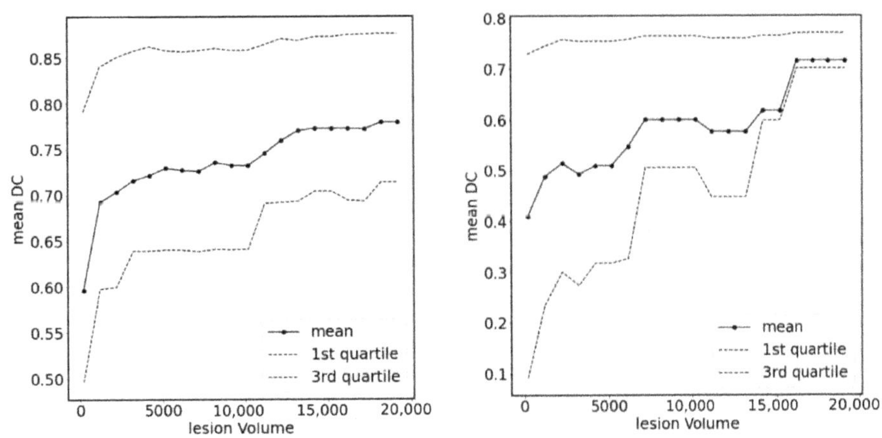

Figure 9. Accuracy of hepatic lesion segmentation measured by the Dice coefficient, presented according to lesion volume (measured by number of voxels). On the left, the results for the training set are presented. On the right, the results for the test set are presented.

4. Discussion

There are numerous possible lesions in the liver. These lesions can be aggressive in terms of growth and also represent a danger in replacing healthy parenchyma [45,46]. The impact of liver dysfunction on the health status of an individual is enormous, so there is a strong need for accurate diagnoses. Contemporary radiology with MR is the most effective modality used for the detection of soft tissue lesions. One of the most important limitations of this process is the ability of the human eye to find differences between lesions and the liver parenchyma in the most effective and repeatable way. The field training of imaging specialists, psychosomatic conditions, and overall light condition in the examination room are all recognized factors that can influence lesion detection [47,48]. The application of different sequences where various constituents of the lesions can be visualized aids in recognition of such lesions. However, this effect might be alleviated with lesion polymorphism [49]. Therefore, the first step of lesion recognition—a border setting between normal parenchyma and the lesion—is a segmentation task run based on training with multi-step verified man-made descriptions. The second task—recognition of the patterns related to lesions—relies on associated processes and requires significance experience from the reader, including the ability to engage in comparative interpretation [50]. Both steps are part of daily practice in radiology, with special importance in oncology, where lesion dimensions and associated change are of particular importance in treatment assessment and further planning [51].

Deep learning methods are promising in the field of lesion detection and could help optimize the detection of liver lesions [52]. Some studies describing DL methodologies explored combined liver segmentation and liver tumors, but many focused only on liver segmentation. The 3D Deep Supervised Network described by Dou et al. [53] is based on a fully convolutional network architecture. The presented framework was evaluated on the MICCAI SLiver07 CT data ASD set. For this architecture, the total inference time for a single subject was 5 s for the 3D Deep Supervised Network and 87 s for conditional random fields, which corresponds to approximately one and a half minutes in total. Another algorithm using deep supervision was presented by Lv et al. [54]. The authors replaced the encoder's standard convolution with a residual block and provided an atrous inception module to connect the encoder and decoder, allowing the model to obtain multi-scale features; the authors also incorporated a deep supervision mechanism into the decoder. In addition, the authors employed the Tversky loss function to balance segmented and nonsegmented regions and performed further refinements with a dense conditional random field. This method was validated on three public CT databases: LiTS17, 3DIRCADb, and SLiver07. As a drawback, the authors noted that the large amount of 3D network parameters hindered and slowed the training process. The authors also noted the susceptibility of the method to errors when considering livers with borderline tumors. Additionally, since this work was aimed at segmenting the liver, it may not be suitable for segmenting liver tumors. A two-step method was presented by Lu et al. [55]. In the first step, this method uses a 3D convolutional neural network to segment the liver. In the second step, segmentation accuracy is improved using cutting charts and maps. The model was tested on 40 CT volumes derived from the publicly available datasets Sliver07 and 3DIRCADb. The authors concluded that this method performs reasonably well compared to the manual method due to the latter's long processing time and erroneous results. An evaluation of three different convolutional neural network (CNN) architectures was performed by Hoang et al. [56]. These architectures were a Fully Convolutional Network with Conditional Random Fields, Deep retinal image understanding, and the V-net model. The authors used contrast-enhanced images from the different datasets of LiTS, mayo, EMC_LD, EMC_NC_LD, which were collected from different medical centers. The results showed that all three CNNs achieved a mean Dice score of over 90% in liver segmentation with typical contrast-enhanced CT images of the liver, but these three architectures did not perform well for non-contrast-enhanced CT and low dose images. An automatic methods based on CNNs using 30 enhanced CT volumes were compared by Li et al. [57]. A Dice score of 80%

was achieved in this study, demonstrating that the CNN methods performed better than the classical machine learning methods AdaBoost, Random Forests, and support vector machine. Li et al. [58] proposed a deep attention neural network including a high-resolution branch that can maintain input image resolution and thus preserve spatial details, as well as multiscale feature aggregation for cascading liver and tumor segmentation from CT images. This model achieved a Dice score of 76.3% for lesions and 96.0% for the liver when LiTS datasets were used for the evaluation.

For liver tumor segmentation in CT, the task is more difficult than liver segmentation. This process requires the patient to be injected with a contrast agent that enhances the contrast between the healthy liver and cancerous tissues of the liver. Even then, the variety of tumors makes this task difficult. However, the main reason segmentation of liver tumors is not frequently studied compared to whole liver segmentation is caused by the unavailability of public datasets with ground truth masks for liver tumors, which is due to the fact that manually annotating liver and liver tumor data is time consuming and prone to errors.

Magnetic resonance imaging has become the golden standard radiological modality for liver mass evaluation [59] due to the absence of ionizing radiation and good differentiation of morphological features of different tumor classes under this method [60]. With increasing clinical demand for MR-based diagnosis, there is a need to develop fast and comprehensive clinical protocols for liver mass detection and differentiation [61]. Proposed protocols consist of the sequences important for the detection of malignancies with less of a possible time expense [62]. Studies presenting different approaches showed the importance of T1-weighted pre- and post-contrast images but also T2-weighted images and DWI sequences, as summarized in the work of Brunsing [63]. These protocols called AMRI (Abbreviated MRI) protocols, enable the rejection of time consuming and costly abdominal procedures. The sensitivity and specificity of AMRI protocols were found to exceed 84% and 92%, respectively [64].

While the results presented in this study are promising, further research in this area is warranted to improve the quality of both screening and lesion segmentation. The differences between the results obtained for the training and test sets show that the trained models do not generalize very well. Since methods preventing overfitting (data augmentation) were intensively used during training, it can be concluded that in order to further improve the quality of the models, it is necessary to increase the training set. The relatively weak performance of the models in the task of lesion segmentation could also be a consequence of the heterogeneity of the dataset, which contains both very small and very large (e.g., 1/3 of the whole liver) lesions.

Further improvements are possible in the designed pipeline. The first step of liver segmentation in the diagnostic sequences was quite accurate (with a Dice coefficient above 0.98). While marginal improvements to liver segmentation are possible, it is unlikely that such improvements would improve the quality of the end-to-end process. The registration was also quite accurate (with a Dice coefficient around 0.96). However, because all four diagnostic sequences were used to segment the lesions, some small inaccuracies in the registration may lead to large inaccuracies in the segmentation of small lesions. For this reason, further research on high-quality inter-modality registration is required to improve end-to-end processing. Besides classic elastic registration, other methods were also tested (including ones based on deep learning, particularly the DeepReg framework; https://github.com/DeepRegNet/DeepReg, accessed on 1 February 2023), but no better results were obtained. Finally, after improving the inter-modality registration step and increasing the image database, keeping in mind the heterogeneity of the data, multi-class segmentation, which works well for large lesions, could be supplemented by object detector tools to detect small lesions. Deep-learning-based object detectors (MONAI library: https://monai.io/, accessed on 1 February 2023, nnDetection framework: https://github.com/MIC-DKFZ/nnDetection, accessed on 1 February 2023) were already applied to the current dataset, given the current registration algorithm, but the results were not sufficient to improve the end-to-end quality of the screening task. Today, there are tremendous efforts being undertaken to develop solutions based on radiomics analysis aimed at increasing the diagnostic utility of MR in the task of diagnosing liver lesions. Different approaches have been used to detect and classify liver lesions with promising results [65]. In the near future, it is expected that the expansion of computer-assisted classification systems developed in the field of liver malignancies will improve diagnostic accuracy to the level where making accurate diagnoses will be possible without expanding the process to invasive procedures. This measure will markedly increase patient comfort and safety.

The present study also has limitations, first of which is lack of adenomas and metastases in our database. Hemangiomas were rare in our material (over 30 cases) and were difficult to differentiate in the stage of manual segmentation preparation. Therefore, these hemangiomas were not ultimately used in the present study. The limited incidence of Hemangiomas is a consequence of the fact that these lesions are rare. Some studies reported a frequency of up to 20% in the general population [66], but in other studies, a frequency as low as 0.4% was found [67]. Metastases were not present in our material due to profile of the medical unit where the data were collected. Another limitation is that the analyses were based on four sequences only. The inclusion of more sequences may have improved the performance of the detection model. On the other hand, the inclusion of more sequences should be preceded by improving the inter-sequence registration algorithm. Otherwise, the benefits of using more data may be hindered by inaccuracies in the registration. The final limitation of our study is the size of the dataset, which, despite including 500 cases, was found to be too small based on a comparison of the results found for the training and test sets.

5. Conclusions

In this study, we designed a pipeline for liver screening. This method involves classifying examined patients into one of two groups: patients with healthy livers and patients with liver lesions. The quality of this method was found to be promising, with the AUC ROC for the testing set equal to 0.85 and sensitivity and specificity equal to 0.79. Further improvements to this pipeline and increasing the volume of the training set would likely further improve the quality of the results. The proposed system could aid practical radiological routines in the detection and differentiation of lesions of different types. We believe that further implementation of the proposed approach will improve the diagnostic process, for which the accurate recognition of lesions is mandatory but remains challenging.

Author Contributions: Conceptualization, M.S., Z.T., R.O., I.K. and A.U.; methodology M.S., Z.T., J.L., Z.S., I.K. and R.O.; software, M.S., Z.T., J.L. and Z.S.; validation, M.S., Z.T., J.L., Z.S., I.K., A.U. and R.O.; formal analysis, M.S., Z.T., J.L., Z.S., I.K. and R.O.; investigation, M.S., Z.T., J.L., Z.S., I.K. and R.O.; resources, M.S., Z.T., J.L., Z.S., I.K., A.U. and R.O.; data curation, M.S., Z.T., J.L., Z.S., I.K., A.U. and R.O.; writing—original draft preparation, M.S., Z.T., S.G. and R.O. writing—review and editing, M.S., Z.T., S.G. and R.O.; visualization, M.S., Z.T., J.L., Z.S., I.K., A.U. and R.O.; supervision, Z.T. and R.O. All authors have read and agreed to the published version of the manuscript.

Funding: This research was funded by the National Center for Research and Development (NCBR), Poland, grant number POIR.01.01.01-00-1666/20.

Institutional Review Board Statement: The study protocol was developed according to the Declaration of Helsinki and the Declaration of Good Clinical Practice. All images were anonymized prior to processing to ensure the security of personal data. In addition, written consent from the Jagiellonian University Committee was obtained to conduct this study, 1072.6120.22.22 (23 February 2022).

Informed Consent Statement: Patient consent was waived due to retrospective nature of this study. All images used were obtained during standard patient examinations, and strict anonymization procedures were implemented in this study.

Data Availability Statement: Data can be made available on request by contacting the corresponding author.

Acknowledgments: This work was supported by the PLGrid Infrastructure (Academic Computer Centre Cyfronet, AGH University of Science and Technology.

Conflicts of Interest: The authors declare no conflict of interest.

References

1. Giannitrapani, L.; Soresi, M.; La Spada, E.; Cervello, M.; D'Alessandro, N.; Montalto, G. Sex Hormones and Risk of Liver Tumor. *Ann. N. Y. Acad. Sci.* **2006**, *1089*, 228–236. [CrossRef] [PubMed]
2. Ye, F.; Zhai, M.; Long, J.; Gong, Y.; Ren, C.; Zhang, D.; Lin, X.; Liu, S. The Burden of Liver Cirrhosis in Mortality: Results from the Global Burden of Disease Study. *Front. Public Health* **2022**, *10*, 909455. [CrossRef]
3. Roerecke, M.; Vafaei, A.; Hasan, O.S.M.; Chrystoja, B.R.; Cruz, M.; Lee, R.; Neuman, M.G.; Rehm, J. Alcohol Consumption and Risk of Liver Cirrhosis: A Systematic Review and Meta-Analysis. *Am. J. Gastroenterol.* **2019**, *114*, 1574–1586. [CrossRef] [PubMed]
4. Jiang, H.-Y.; Chen, J.; Xia, C.-C.; Cao, L.-K.; Duan, T.; Song, B. Noninvasive Imaging of Hepatocellular Carcinoma: From Diagnosis to Prognosis. *World J. Gastroenterol.* **2018**, *24*, 2348–2362. [CrossRef] [PubMed]
5. Souto, E.; Gores, G.J. When Should a Liver Mass Suspected of Being a Hepatocellular Carcinoma Be Biopsied? *Liver Transpl.* **2000**, *6*, 73–75. [CrossRef] [PubMed]
6. Luna, A.; Pahwa, S.; Bonini, C.; Alcalá-Mata, L.; Wright, K.L.; Gulani, V. Multiparametric MR Imaging in Abdominal Malignancies. *Magn. Reson. Imaging Clin.* **2016**, *24*, 157–186. [CrossRef]
7. Taseva, A.; Tasev, V.; Bulanov, D.; Dimitrov, K.; Popov, V.; Zivkov, E.; Dimitrova, V. Diagnosis of liver hemangioma. *Khirurgiia* **2013**, *3*, 8–13.
8. Caturelli, E.; Pompili, M.; Bartolucci, F.; Siena, D.A.; Sperandeo, M.; Andriulli, A.; Bisceglia, M. Hemangioma-like Lesions in Chronic Liver Disease: Diagnostic Evaluation in Patients. *Radiology* **2001**, *220*, 337–342. [CrossRef]
9. Glinkova, V.; Shevah, O.; Boaz, M.; Levine, A.; Shirin, H. Hepatic Haemangiomas: Possible Association with Female Sex Hormones. *Gut* **2004**, *53*, 1352–1355. [CrossRef]
10. Lantinga, M.A.; Gevers, T.J.; Drenth, J.P. Evaluation of Hepatic Cystic Lesions. *World J. Gastroenterol.* **2013**, *19*, 3543–3554. [CrossRef]
11. Albiin, N. MRI of Focal Liver Lesions. *Curr. Med. Imaging Rev.* **2012**, *8*, 107–116. [CrossRef]
12. Renzulli, M.; Brandi, N.; Pecorelli, A.; Pastore, L.V.; Granito, A.; Martinese, G.; Tovoli, F.; Simonetti, M.; Dajti, E.; Colecchia, A.; et al. Segmental Distribution of Hepatocellular Carcinoma in Cirrhotic Livers. *Diagnostics* **2022**, *12*, 834. [CrossRef]
13. Pecorelli, A.; Franceschi, P.; Braccischi, L.; Izzo, F.; Renzulli, M.; Golfieri, R. MRI Appearance of Focal Lesions in Liver Iron Overload. *Diagnostics* **2022**, *12*, 891. [CrossRef] [PubMed]
14. Bowlus, C.L.; Arrivé, L.; Bergquist, A.; Deneau, M.; Forman, L.; Ilyas, S.I.; Lunsford, K.E.; Martinez, M.; Sapisochin, G.; Shroff, R.; et al. AASLD Practice Guidance on Primary Sclerosing Cholangitis and Cholangiocarcinoma. *Hepatology* **2023**, *77*, 659–702. [CrossRef]
15. European Association for the Study of the Liver. Electronic address: Easloffice@easloffice.eu; European Association for the Study of the Liver EASL-ILCA Clinical Practice Guidelines on Intrahepatic Cholangiocarcinoma. *J. Hepatol.* **2023**, *69*, 182–236. [CrossRef]
16. Renzulli, M.; Pecorelli, A.; Brandi, N.; Brocchi, S.; Tovoli, F.; Granito, A.; Carrafiello, G.; Ierardi, A.M.; Golfieri, R. The Feasibility of Liver Biopsy for Undefined Nodules in Patients under Surveillance for Hepatocellular Carcinoma: Is Biopsy Really a Useful Tool? *J. Clin. Med.* **2022**, *11*, 4399. [CrossRef] [PubMed]
17. Blonski, W.; Reddy, K.R. Evaluation of Nonmalignant Liver Masses. *Curr. Gastroenterol. Rep.* **2006**, *8*, 38–45. [CrossRef]
18. Zviniene, K.; Zaboriene, I.; Basevicius, A.; Jurkiene, N.; Barauskas, G.; Pundzius, J. Comparative Diagnostic Value of Contrast-Enhanced Ultrasonography, Computed Tomography, and Magnetic Resonance Imaging in Diagnosis of Hepatic Hemangiomas. *Medicina* **2010**, *46*, 329–335. [CrossRef] [PubMed]
19. Vilgrain, V. Focal Nodular Hyperplasia. *Eur. J. Radiol.* **2006**, *58*, 236–245. [CrossRef]
20. Mortelé, K.J.; Praet, M.; Van Vlierberghe, H.; de Hemptinne, B.; Zou, K.; Ros, P.R. Focal Nodular Hyperplasia of the Liver: Detection and Characterization with Plain and Dynamic-Enhanced MRI. *Abdom. Imaging* **2002**, *27*, 700–707. [CrossRef]
21. Lee, Y.J.; Lee, J.M.; Lee, J.S.; Lee, H.Y.; Park, B.H.; Kim, Y.H.; Han, J.K.; Choi, B.I. Hepatocellular Carcinoma: Diagnostic Performance of Multidetector CT and MR Imaging-a Systematic Review and Meta-Analysis. *Radiology* **2015**, *275*, 97–109. [CrossRef] [PubMed]
22. Renzulli, M.; Braccischi, L.; D'Errico, A.; Pecorelli, A.; Brandi, N.; Golfieri, R.; Albertini, E.; Vasuri, F. State-of-the-Art Review on the Correlations between Pathological and Magnetic Resonance Features of Cirrhotic Nodules. *Histol. Histopathol.* **2022**, *37*, 1151–1165. [CrossRef] [PubMed]
23. Orlacchio, A.; Chegai, F.; Fabiano, S.; Merolla, S.; Funel, V.; Di Giuliano, F.; Manuelli, M.; Tisone, G.; Francioso, S.; Angelico, M.; et al. Role of MRI with Hepatospecific Contrast Agent in the Identification and Characterization of Focal Liver Lesions: Pathological Correlation in Explanted Livers. *Radiol. Med.* **2016**, *121*, 588–596. [CrossRef] [PubMed]
24. Smith-Bindman, R.; Miglioretti, D.L.; Larson, E.B. Rising Use of Diagnostic Medical Imaging in a Large Integrated Health System. *Health Aff.* **2008**, *27*, 1491–1502. [CrossRef]

25. Abbey, C.K.; Barrett, H.H. Human- and Model-Observer Performance in Ramp-Spectrum Noise: Effects of Regularization and Object Variability. *J. Opt. Soc. Am. A Opt. Image Sci. Vis.* **2001**, *18*, 473–488. [CrossRef]
26. Ciarrapico, A.M.; Ugenti, R.; Di Minco, L.; Santori, E.; Altobelli, S.; Coco, I.; D'Onofrio, S.; Simonetti, G. Diagnostic Imaging and Spending Review: Extreme Problems Call for Extreme Measures. *Radiol. Med.* **2017**, *122*, 288–293. [CrossRef]
27. Van Timmeren, J.E.; Cester, D.; Tanadini-Lang, S.; Alkadhi, H.; Baessler, B. Radiomics in Medical Imaging—"How-to" Guide and Critical Reflection. *Insights Imaging* **2020**, *11*, 91. [CrossRef]
28. Binczyk, F.; Prazuch, W.; Bozek, P.; Polanska, J. Radiomics and Artificial Intelligence in Lung Cancer Screening. *Transl. Lung Cancer Res.* **2021**, *10*, 1186. [CrossRef]
29. Ozkara, B.B.; Chen, M.M.; Federau, C.; Karabacak, M.; Briere, T.M.; Li, J.; Wintermark, M. Deep Learning for Detecting Brain Metastases on MRI: A Systematic Review and Meta-Analysis. *Cancers* **2023**, *15*, 334. [CrossRef]
30. Dreher, C.; Linde, P.; Boda-Heggemann, J.; Baessler, B. Radiomics for Liver Tumours. *Strahlenther. Onkol.* **2020**, *196*, 888–899. [CrossRef]
31. Gillies, R.J.; Schabath, M.B. Radiomics Improves Cancer Screening and Early Detection. *Cancer Epidemiol. Biomark. Prev.* **2020**, *29*, 2556–2567. [CrossRef] [PubMed]
32. Freitas, P.S.; Janicas, C.; Veiga, J.; Matos, A.P.; Herédia, V.; Ramalho, M. Imaging Evaluation of the Liver in Oncology Patients: A Comparison of Techniques. *World J. Hepatol.* **2021**, *13*, 1936–1955. [CrossRef]
33. Kim, K.; Chun, J. A New Hyper Parameter of Hounsfield Unit Range in Liver Segmentation. *J. Internet Comput. Serv.* **2020**, *21*, 103–111.
34. Stimper, V.; Bauer, S.; Ernstorfer, R.; Scholkopf, B.; Xian, R.P. Multidimensional Contrast Limited Adaptive Histogram Equalization. *IEEE Access* **2019**, *7*, 165437–165447. [CrossRef]
35. García, S.; Luengo, J.; Herrera, F. *Data Preprocessing in Data Mining*; Intelligent Systems Reference Library; Springer International Publishing: Cham, Switzerland, 2015; Volume 72, ISBN 978-3-319-10246-7.
36. Dinc, I.; Dinc, S.; Sigdel, M.; Sigdel, M.; Aygun, R.; Pusey, M. Chapter 12—DT-Binarize: A Decision Tree Based Binarization for Protein Crystal Images. In *Emerging Trends in Image Processing, Computer Vision and Pattern Recognition*; Morgan Kaufmann: Burlington, MA, USA, 2015; pp. 183–199.
37. Tustison, N.J.; Avants, B.B.; Cook, P.A.; Zheng, Y.; Egan, A.; Yushkevich, P.A.; Gee, J.C. N4ITK: Improved N3 Bias Correction. *IEEE Trans. Med. Imaging* **2010**, *29*, 1310–1320. [CrossRef]
38. Learned-Miller, E.G.; Jain, V. Many Heads Are Better Than One: Jointly Removing Bias from Multiple MRIs Using Nonparametric Maximum Likelihood. In *Proceedings of the Information Processing in Medical Imaging*; Christensen, G.E., Sonka, M., Eds.; Springer: Berlin/Heidelberg, Germany, 2005; pp. 615–626.
39. Isensee, F.; Petersen, J.; Klein, A.; Zimmerer, D.; Jaeger, P.F.; Kohl, S.; Wasserthal, J.; Koehler, G.; Norajitra, T.; Wirkert, S.; et al. NnU-Net: Self-Adapting Framework for U-Net-Based Medical Image Segmentation. *arXiv* **2018**, arXiv:1809.10486.
40. Ronneberger, O.; Fischer, P.; Brox, T. U-Net: Convolutional Networks for Biomedical Image Segmentation. In Proceedings of the Medical Image Computing and Computer-Assisted Intervention—MICCAI 2015, Munich, Germany, 5–9 October 2015; Navab, N., Hornegger, J., Wells, W.M., Frangi, A.F., Eds.; Springer International Publishing: Cham, Switzerland, 2015; pp. 234–241.
41. Isensee, F.; Jaeger, P.; Kohl, S.; Petersen, J.; Maier-Hein, K. NnU-Net: A Self-Configuring Method for Deep Learning-Based Biomedical Image Segmentation. *Nat. Methods* **2021**, *18*, 1–9. [CrossRef] [PubMed]
42. Isensee, F.; Jäger, P.; Wasserthal, J.; Zimmerer, D.; Petersen, J.; Kohl, S.; Schock, J.; Klein, A.; Roß, T.; Wirkert, S.; et al. *Batchgenerators—a Python Framework for Data Augmentation*, Version 0.19.6; Zenodo: Geneva, Switzerland, 2020. [CrossRef]
43. Byrd, R.H.; Lu, P.; Nocedal, J.; Zhu, C. A Limited Memory Algorithm for Bound Constrained Optimization. *SIAM J. Sci. Comput.* **1995**, *16*, 1190–1208. [CrossRef]
44. Zhu, C.; Byrd, R.H.; Lu, P.; Nocedal, J. Algorithm 778: L-BFGS-B: Fortran Subroutines for Large-Scale Bound-Constrained Optimization. *ACM Trans. Math. Softw.* **1997**, *23*, 550–560. [CrossRef]
45. Assy, N.; Nasser, G.; Djibre, A.; Beniashvili, Z.; Elias, S.; Zidan, J. Characteristics of Common Solid Liver Lesions and Recommendations for Diagnostic Workup. *World J. Gastroenterol* **2009**, *15*, 3217–3227. [CrossRef] [PubMed]
46. Grazioli, L.; Ambrosini, R.; Frittoli, B.; Grazioli, M.; Morone, M. Primary Benign Liver Lesions. *Eur. J. Radiol.* **2017**, *95*, 378–398. [CrossRef]
47. Krupa, K.; Bekiesińska-Figatowska, M. Artifacts in Magnetic Resonance Imaging. *Pol. J. Radiol.* **2015**, *80*, 93–106. [CrossRef]
48. Kundel, H.L. Reader Error, Object Recognition, and Visual Search. In Proceedings of the Medical Imaging 2004: Image Perception, Observer Performance, and Technology Assessment. SPIE, San Diego, CA, USA, 17–19 February 2004; pp. 1–11. [CrossRef]
49. Manning, D.J.; Gale, A.; Krupinski, E.A. Perception Research in Medical Imaging. *Br. J. Radiol.* **2005**, *78*, 683–685. [CrossRef]
50. Krupinski, E.A. Current Perspectives in Medical Image Perception. *Atten. Percept. Psychophys.* **2010**, *72*, 1205–1217. [CrossRef]
51. Sung, W. Quantitative Imaging in Oncology. *Tomography* **2022**, *8*, 1676–1677. [CrossRef]
52. Park, H.J.; Park, B.; Lee, S.S. Radiomics and Deep Learning: Hepatic Applications. *Korean J. Radiol.* **2020**, *21*, 387–401. [CrossRef]
53. Dou, Q.; Chen, H.; Jin, Y.; Yu, L.; Qin, J.; Heng, P.-A. 3D Deeply Supervised Network for Automatic Liver Segmentation from CT Volumes. In Proceedings of the Medical Image Computing and Computer-Assisted Intervention–MICCAI 2016: 19th International Conference, Athens, Greece, 17–21 October 2016; Springer International Publishing: Cham, Switzerland, 2016; pp. 149–157.
54. Lv, P.; Wang, J.; Zhang, X.; Shi, C. Deep Supervision and Atrous Inception-Based U-Net Combining CRF for Automatic Liver Segmentation from CT. *Sci. Rep.* **2022**, *12*, 16995. [CrossRef]

55. Lu, F.; Wu, F.; Hu, P.; Peng, Z.; Kong, D. Automatic 3D Liver Location and Segmentation via Convolutional Neural Network and Graph Cut. *Int. J. Comput. Assist. Radiol. Surg.* **2017**, *12*, 171–182. [CrossRef] [PubMed]
56. Hoang, H.; Pham, C.; Franklin, D.; Walsum, T.; Luu, M. An Evaluation of CNN-Based Liver Segmentation Methods Using Multi-Types of CT Abdominal Images from Multiple Medical Centers. In Proceedings of the 19th International Symposium on Communications and Information Technologies (ISCIT), Ho Chi Minh City, Vietnam, 25–27 September 2019; pp. 20–25.
57. Li, W.; Jia, F.; Hu, Q. Automatic Segmentation of Liver Tumor in CT Images with Deep Convolutional Neural Networks. *J. Comput. Commun.* **2015**, *3*, 146–151. [CrossRef]
58. Li, Y.; Zou, B.; Liu, Q. A Deep Attention Network via High-Resolution Representation for Liver and Liver Tumor Segmentation. *J. Appl. Biomed.* **2021**, *41*, 1518–1532. [CrossRef]
59. Taouli, B.; Ehman, R.L.; Reeder, S.B. Advanced MRI Methods for Assessment of Chronic Liver Disease. *AJR Am. J. Roentgenol.* **2009**, *193*, 14–27. [CrossRef] [PubMed]
60. Kitao, A.; Matsui, O.; Yoneda, N.; Kita, R.; Kozaka, K.; Kobayashi, S.; Gabata, T. Differentiation Between Hepatocellular Carcinoma Showing Hyperintensity on the Hepatobiliary Phase of Gadoxetic Acid-Enhanced MRI and Focal Nodular Hyperplasia by CT and MRI. *AJR Am. J. Roentgenol.* **2018**, *211*, 347–357. [CrossRef] [PubMed]
61. Carbonell, G.; Taouli, B. Abbreviated MR Protocols for Chronic Liver Disease and Liver Cancer. *Magn. Reson. Imaging Clin.* **2021**, *29*, 321–327. [CrossRef] [PubMed]
62. Tillman, B.G.; Gorman, J.D.; Hru, J.M.; Lee, M.H.; King, M.C.; Sirlin, C.B.; Marks, R.M. Diagnostic Per-Lesion Performance of a Simulated Gadoxetate Disodium-Enhanced Abbreviated MRI Protocol for Hepatocellular Carcinoma Screening. *Clin. Radiol.* **2018**, *73*, 485–493. [CrossRef]
63. Brunsing, R.L.; Fowler, K.J.; Yokoo, T.; Cunha, G.M.; Sirlin, C.B.; Marks, R.M. Alternative Approach of Hepatocellular Carcinoma Surveillance: Abbreviated MRI. *Hepatoma Res.* **2020**, *6*, 59. [CrossRef]
64. Chan, M.V.; McDonald, S.J.; Ong, Y.-Y.; Mastrocostas, K.; Ho, E.; Huo, Y.R.; Santhakumar, C.; Lee, A.U.; Yang, J. HCC Screening: Assessment of an Abbreviated Non-Contrast MRI Protocol. *Eur. Radiol. Exp.* **2019**, *3*, 49. [CrossRef]
65. Renzulli, M.; Mottola, M.; Coppola, F.; Cocozza, M.A.; Malavasi, S.; Cattabriga, A.; Vara, G.; Ravaioli, M.; Cescon, M.; Vasuri, F.; et al. Automatically Extracted Machine Learning Features from Preoperative CT to Early Predict Microvascular Invasion in HCC: The Role of the Zone of Transition (ZOT). *Cancers* **2022**, *14*, 1816. [CrossRef]
66. Mogahed, M.M.; Zytoon, A.A.; Essa, B.; Abdellatif, W.; Ghanem, N.; ElWakeel, B. Natural History of Hepatic Hemangiomas as a Guide for Surgical Indication. *Egypt. Liver J.* **2020**, *10*, 56. [CrossRef]
67. Bajenaru, N.; Balaban, V.; Săvulescu, F.; Campeanu, I.; Patrascu, T. Hepatic Hemangioma-Review. *J. Med. Life* **2015**, *8*, 4–11.

tomography

MDPI

Article

Sinogram Inpainting with Generative Adversarial Networks and Shape Priors

Emilien Valat [1,*], Katayoun Farrahi [2] and Thomas Blumensath [3]

[1] Cambridge Image Analysis Group, Department of Applied Mathematics and Theoretical Physics, Centre for Mathematical Sciences, University of Cambridge, Wilberforce Rd., Cambridge CB3 0WA, UK
[2] Vision, Learning and Control Group, Department of Electronics and Computer Science, University of Southampton, University Rd., Southampton SO17 1BJ, UK
[3] Institute of Sound and Vibration Research, Department of Engineering and the Environment, University of Southampton, University Rd., Southampton SO17 1BJ, UK
* Correspondence: ev373@cam.ac.uk

Abstract: X-ray computed tomography is a widely used, non-destructive imaging technique that computes cross-sectional images of an object from a set of X-ray absorption profiles (the so-called sinogram). The computation of the image from the sinogram is an ill-posed inverse problem, which becomes underdetermined when we are only able to collect insufficiently many X-ray measurements. We are here interested in solving X-ray tomography image reconstruction problems where we are unable to scan the object from all directions, but where we have prior information about the object's shape. We thus propose a method that reduces image artefacts due to limited tomographic measurements by inferring missing measurements using shape priors. Our method uses a Generative Adversarial Network that combines limited acquisition data and shape information. While most existing methods focus on evenly spaced missing scanning angles, we propose an approach that infers a substantial number of consecutive missing acquisitions. We show that our method consistently improves image quality compared to images reconstructed using the previous state-of-the-art sinogram-inpainting techniques. In particular, we demonstrate a 7 dB Peak Signal-to-Noise Ratio improvement compared to other methods.

Keywords: X-ray computed tomography; machine-learning; computer assisted design data; Generative Adversarial Network

Citation: Valat, E.; Farrahi, K.; Blumensath, T. Sinogram Inpainting with Generative Adversarial Networks and Shape Priors. *Tomography* **2023**, *9*, 1137–1152. https://doi.org/10.3390/tomography9030094

Academic Editors: Rafał Obuchowicz, Michał Strzelecki and Adam Piorkowski

Received: 6 May 2023
Revised: 31 May 2023
Accepted: 1 June 2023
Published: 13 June 2023

1. Introduction

X-ray Computed Tomography (XCT) is a versatile 3D imaging technique that allows the estimation of volumetric X-ray attenuation profiles. The technique is used routinely in medical diagnosis but is also increasingly popular in scientific investigations and non-destructive testing [1,2]. However, in many applications, we can only collect limited X-ray measurements. For example, in industrial applications, object geometry might prevent full object rotation and thus prevent the use of standard XCT scanning strategies. The method might also be applicable to high-resolution medical scanning, where only a low-resolution full scan is possible to provide overall shape information, whilst a high-resolution scan can only be performed over a limited angular range.

Standard XCT reconstruction methods rely on measuring X-ray absorption along sufficiently many paths through an object. For example, for parallel beam tomography, measurements in a plane orthogonal to the rotation axis can be described as

$$R(\theta,r) = \int_x \int_y f(x,y)\delta(x\cos(\theta) + y\sin(\theta) - r)\,dx\,dy, \quad (1)$$

where r is the distance of the measurement along the detector, θ the angle at which the measurement is performed, (x,y) the coordinates of the object and f the X-ray attenuation of the object. $R(\theta, r)$ is known as the sinogram of the image $f(x, y)$ and the tomographic reconstruction problem can be posed as estimating $f(x, y)$ from a set of measurements $R(\theta, r)$. This 2D scanning setting can also be extended to 3D scanning. For example, in industrial settings, XCT systems often use a 3D cone beam scan trajectory; here, working directly on the full 3D scan data is crucial. Alternatively, where X-ray scattering is a problem, collimated fan-beam systems are also often used. Whilst these require longer scan times, they lead to decoupled inverse problems where each 2D image can be recovered independently.

Unfortunately, the tomographic inverse problem is severely ill-posed, even if enough angular measurements are available. Regularising the inverse problem is thus crucial, especially working in settings where only limited projections can be acquired. Standard image reconstruction in XCT relies on analytical methods based on the implicitly regularised Filtered Back-Projection (FBP) framework, initially developed for parallel beam tomography, but easily extendable to 2D fan-beam scan geometries by rearranging and interpolating the measurements. For 3D cone beam scans, the so-called FDK [3] algorithm provides a fast approximate analytical solution.

An alternative is to use iterative reconstruction methods. Whilst these offer more flexibility in terms of the constraints and regularisation approaches that can be used, given the size of the inverse problem, these methods remain computationally too costly for many practical applications. For example, many XCT systems routinely collect billions of measurements to reconstruct a 3D volumetric image with billions of unknown parameters. Fast and memory-efficient analytic approaches thus remain popular in practice.

When working with missing data, iterative regularised frameworks are tempting, as they often produce better images and allow the easy use of additional prior information. However, these are mainly applied to parallel or fan-beam scans, though advanced GPU accelerated methods are slowly allowing increased use of these approaches for 3D cone-beam data [4]. These methods solve the ill-posed or underdetermined linear inverse problem $Ax = y$ using additional regularisation terms such as the Total Variation constraint [5] or more XCT-specific regularisers [6–8]. Unfortunately, for real 3D cone beam CT applications, these methods remain too slow to be of practical significance. Thus, there remains significant interest in developing efficient approaches that can estimate the missing information required by analytical reconstruction methods [9,10].

In order to develop efficient methods for XCT image reconstruction, we here develop an approach that tries to estimate missing measurements. We infer the missing information in the sinogram and perform the inference using additional shape information. Then, standard analytical methods can be used to estimate the image. Whilst it would also be possible to incorporate shape priors into an iterative image recovery framework, for example, using masks that set attenuation values outside of objects to zero, for many realistic applications, this approach is still too slow for many problems of interest.

1.1. Related Work

To the best of our knowledge, there are currently no known published methods that use shape information to infer missing data for XCT reconstruction. There are, however, several techniques that are related to our sinogram estimation approach but that do not utilise shape information.

Earliest attempts on 2D sinogram data encode the sinogram as an image and missing measurements as zero-columns and apply various models, such as diffusion [9] or grid-interpolation [11], to fill out the missing values in the image. That is, given a sinogram displayed as an image, interpolating the missing pixel values in the image using a diffusion model or a grid-interpolation model. Tackling the problem of inferring measurements from that perspective is similar to the one of inpainting an image, that is, inferring arbitrarily large regions in images based on image semantics.

From a similar perspective, recent approaches [12,13] use data-driven techniques to improve sinograms. Their approach relies on initial deterministic interpolation to estimate the missing measurements. Then, they patch the sinogram into fixed blocks and train a Convolutional Neural Network (CNN) to change all the true and interpolated values. Refs. [14,15] use a Unet [16] instead of a simple CNN. One must stress that these approaches aim at inferring measurements in scenarios where missing measurements are evenly spaced, which is different from the problem we chose to tackle. Other methods using Generative Adversarial Networks (GANs) [17] try to encode a latent representation of the few-view sinogram [18,19]. This method does not scale to reasonable image sizes (no more than 128 $\times$ 128 pixels). A GAN architecture has also previously been used for sinogram synthesis [20], with an optimisation procedure to perform the inpainting as in [21], though again without the use of shape priors.

Our proposed approach is based on semantic image inpainting, which is a constrained image generation problem [21]. Missing parts of an image are inpainted using a generative network and solving an optimisation problem. GAN and their fully-convolutional version Deep-Convolutional GAN (DCGAN) [22] are adapted to this task: they were used for image [23] and sinogram [20] inpainting. The optimisation relies on finding the "closest" encoding in the latent space of the GAN distribution by minimising a penalty function that encompasses contextual and conditional information. This method has two limitations. First, walking the latent space of the distribution can only yield specific improvements in the image generated by the GAN: ref. [24] shows that images can only be transformed to some degree (brightness, zoom, rotation). In this context, walking refers to applying small perturbations to the data's encoding in the GAN's latent space. Not only is the transformation corresponding to inpainting not defined, but it has no certainty to be achievable by applying some transformation to the latent representation, mainly when guided by a generic loss function rather than by a supervised walk. Second, the optimisation function of the inpainting procedure adds computational time and hyperparameters. In addition to training the generative model, the optimisation process is time-consuming and requires fine-tuning the learning rate and the number of iterations.

We propose an approach that uses shape information to side-step the complicated optimisation process of GAN-based inpainting methods.

1.2. Utilising Object Shape Information

Shape information is often available in medical and industrial imaging but rarely used in XCT reconstruction. For instance, in medical imaging, refs. [25,26] demonstrate the potential use of shape models of a generic human body to minimise X-ray exposure. Moreover, recent advances in anatomic modelling such as [27] demonstrate the growing availability of computational, yet accurate, priors. We want to stress that a generic shape prior would likely have limited benefits, so a patient-specific prior might be required. Such information could, for example, be obtained from faster, lower-resolution CT scans, such as in [28,29], or from different modalities, such as MRI.

In the field of non-destructive testing for high value manufacturing, Computer Assisted Design (CAD) drawings are often available, providing strong constraints on object shape. Frequently, when using CT scans to ensure the highest quality of manufacturing, the commercial entity that commissions the scan also possesses the CAD data related to the object that is scanned.

2. Proposed Approach

In an XCT system, measurements are taken at different rotation angles. We here address the problem of limited angle tomography, where X-ray measurements are not available from a (possibly extensive) range of consecutive angular directions(note that the limited angle setting here is different from a limited measurement setting where reduced numbers of measurements are collected at evenly or randomly distributed angles). The problem of inferring these missing measurements can thus be cast as an image inpainting problem [21]. If additional knowledge about the object's shape is available, the problem becomes an image inpainting with edge information task, which has already been studied in [30]. We thus here use the pix2pix Generative Adversarial Network (GAN) [31] architecture proposed in [30]. Whilst this generic inpainting approach has been proposed and demonstrated for image inpainting with additional shape information, it has never been used for XCT sinogram estimation. The generic idea is shown in Figure 1.

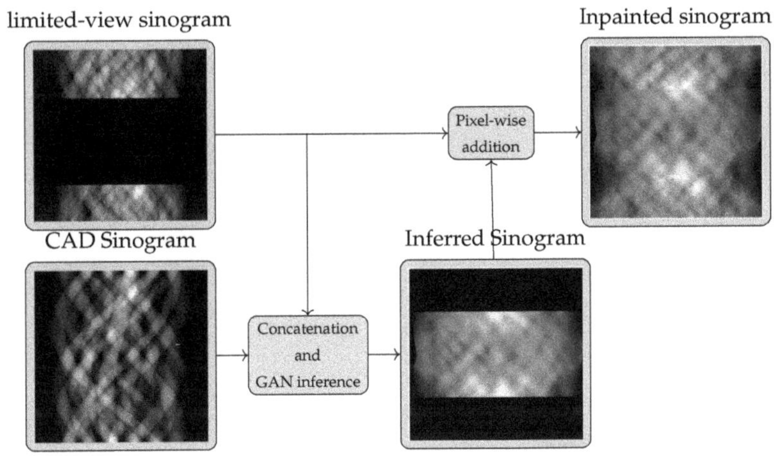

Figure 1. The pix2pix architecture using a generator and a discriminator. The generator is trained to generate missing sinogram information from a multi-channel input. One channel is the sinogram with the missing information, one is a sinogram encoding the object shape and one is a mask indicating where the CAD expects missing information to be located. The discriminator (or detector) is a classification network that tries to detect whether the multi-channel input is an actual full sinogram or is a sinogram where the generator estimates missing information. GAN training is done iteratively, with the generator trying to fool the discriminator better and the discriminator trying to improve the detection of estimated data.

We propose using the pix2pix network architecture to incorporate CAD information into the XCT inpainting processes. We develop the method using a 2D/3D parallel beam setting, which can easily be extended to 2D fan-beam scans. For 3D cone-beam data, the same approach can be used in theory on the entire set of projections, though here a block-based approach might be required to reduce computational requirements. We show how measurement information from limited angles can improve the initial shape prior.

2.1. Generative Adversarial Networks for Sinogram Inpainting with Edge Information

The pix2pix architecture uses a GAN approach. GANs are generative models that try to learn complex probability distributions by providing an explicit method to generate samples from that distribution. GANs comprise two networks that compete against each other. A generator G generates new data that follow the same distribution p_{data} as the training samples by transforming a noise vector drawn from a distribution p_z. The discriminator D tries to discriminate between real training samples and those generated by G. They are a powerful alternative to other deep generative models and density estimators.

GAN optimisation is done by alternatively solving the minimax problem:

$$\min_G \max_D \mathcal{L}(G,D) = \mathbb{E}_{x\sim p_{data}} \log(D(x)) + \tag{2}$$

$$\mathbb{E}_{z\sim p_z} \log(1 - D(G(z))), \tag{3}$$

where $\mathbb{E}$ is the expectation over the training data.

2.1.1. Deep Convolutional GAN and U-Net—The pix2pix Architecture

A problem commonly found in GANs is instability during training. To address this problem and use convolutions to scale to larger data sizes, the Deep Convolutional GAN

(DCGAN) architecture [22] was proposed. Other changes proposed include the replacement of the maxout [32] activation by ReLu and Tanh in the generator and LeakyReLu in the discriminator, the inclusion of batch-normalisation [33] and the replacement of pooling units with learnt sampling units. The pix2pix architecture uses a U-net with several input channels as the generator to include prior information. In our setting, one channel is the measured sinogram, where missing information is set to zero, a second channel encodes the shape information and a third channel provides a mask to tell the network which information is to be estimated and which information is given. An example of a target sinogram and the three different input sinograms to our generator model are shown in Figure 2 for one of the datasets introduced below. U-nets were initially designed for image segmentation and are fully convolutional auto-encoders which use the feature channels inferred from a down-sampling step to a latent representation to up-sample this latent representation back to the original space. They are convenient architectures designed to map a multi-channel input image to a one-channel output image or vice versa. U-nets have been shown repeatedly to outperform other architecture on many tasks and have thus developed as a de facto standard in many image processing tasks. The training procedure is detailed in Figure 3.

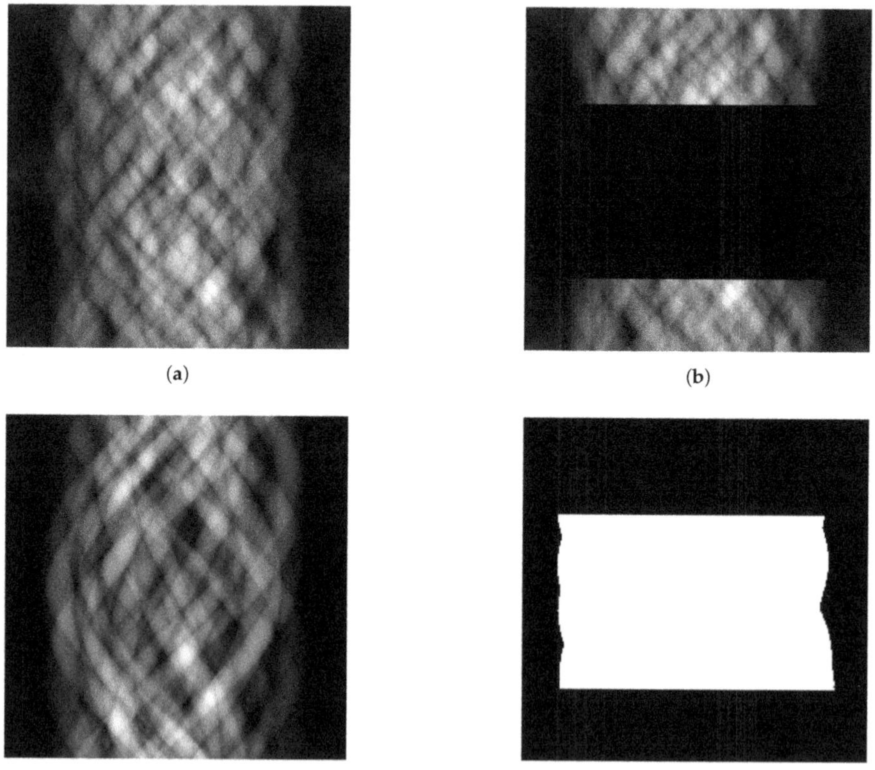

(**a**) (**b**) (**c**) (**d**)

Figure 2. Sample of the input data given to the generator. We estimate the full sinogram (**a**) from the sinogram with missing data (**b**), the sinogram of the shape prior (**c**) and the sinogram's mask indicating the location of missing projections (**d**).

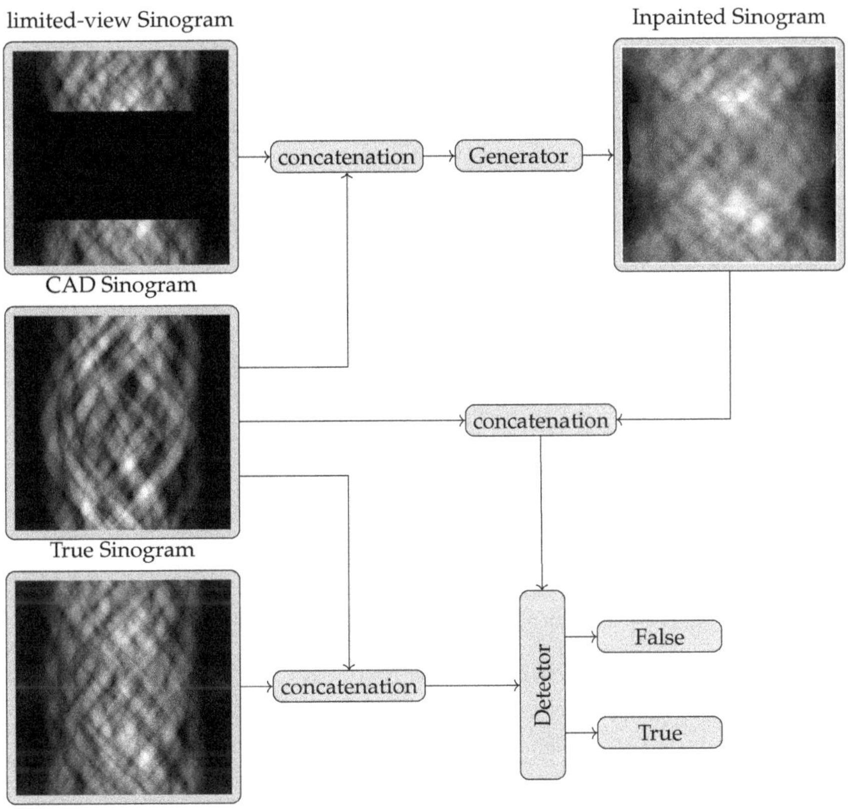

Figure 3. Explanation of the training procedure. The CAD sinogram is added to both target and limited-view sinograms.

2.1.2. Adapted Loss Function

The standard GAN loss function in Equation (2) does not account for conditional modalities. Furthermore, in order to encourage the generator to produce images close to their target value, an L1-loss is often used [21] and is also a good choice for image quality when Poisson noise is present, as is the case in XCT images [34]. The pix2pix architecture is thus trained using the loss function:

$$\min_G \max_D \mathcal{L}(G,D) = \mathbb{E}_{x\sim p_{true}} \log(D(x)) + \quad (4)$$

$$\mathbb{E}_{\tilde{x}\sim p_{impaired}} \log(1 - D(G(\tilde{x},y))) + \quad (5)$$

$$\lambda L_1(G(\tilde{x},y),x), \quad (6)$$

In our case, x and $\tilde{x}$ are the target sinograms with all data and the input sinogram without the missing data, respectively. y is an image encoding the prior shape and λ is a weighting parameter.

The problem that the discriminator has to solve is as follows: given a full sinogram and its associated shape prior, classify the sinogram as either generated by the GAN or drawn from the authentic database. Given a sinogram with missing data, the generator has to generate missing acquisitions that are close to the ground-truth acquisitions in terms of L1-loss [35].

2.2. Encoding the Shape Prior

There are several ways to encode the object's shape information for use in the GAN. We work in the sinogram domain, encoding shape priors in the same domain. This will be done by projecting images containing shape information into the sinogram domain. One way to generate the shape image would be to generate images where the boundary pixels of an object are set to a non-zero value. Alternatively, we can set the object boundary and interior pixels to a single value before projection, using the same value for all objects or choosing a random value for each object. We here chose the latter approach as it allows a greater range of different densities to be encoded and was found to work better in initial experiments. An example shape prior encoding is shown in Figure 4, where we show the original image and the shape prior.

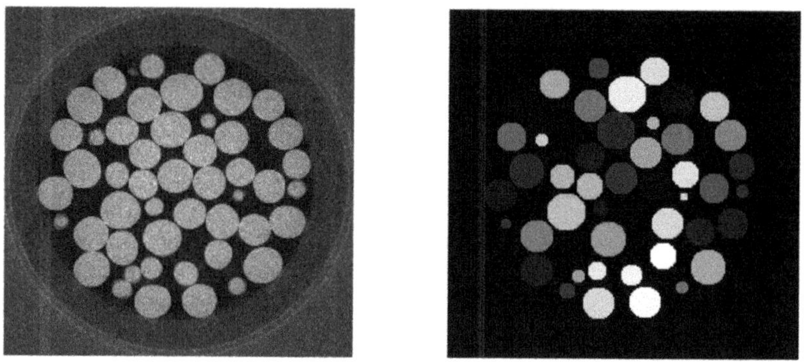

(**a**) (**b**)

Figure 4. Real training samples. A 2D slice from the SophiaBeads dataset, showing the full slice we want to estimate (**a**) and the image that encodes the shape prior (**b**). Here the shape prior indicates each of the locations for each sphere in the image. However, as the attenuation value of each sphere is unknown, random attenuation values are assigned.

To further enhance the encoding of the prior, we add a further pre-processing step that uses a property of sinograms. For XCT scans for which the entire object remains in the field of view, at any acquisition angle, when summing all attenuation values, the total attenuation should be constant, independent of projection angle:

$$\int_r R(\theta, r) = \int_y \int_x f(x, y)\, dx\, dy = \text{const.} \tag{7}$$

We use this property to scale the shape encoding values for each sinogram angle. For region-of-interest scans where objects can move outside the field of view, interpolation could be used to estimate this value for each missing angle, though we do not explore this here.

Architectural Details

The architecture of our generator follows the guidelines of the pix2pix architecture, and our description follows the same notation. Let C_k be a Convolution-BatchNorm-ReLU layer with k output channels. Let CD_k be a Convolution-BatchNorm-Dropout-ReLU layer with a dropout rate of 50% [36]. All convolutions are four-by-four spatial filters applied with a stride of 2. Convolutions in the encoder down-sample by a factor of 2, whereas in the decoder, they up-sample by a factor of 2. After the last layer in the decoder, a convolution is applied to map to the number of output channels (which in turn depends on how we chose to encode data), followed by a Tanh function and an element-wise multiplication with a mask, as explained in Section 2.2. As an exception to the above notation, BatchNorm is not applied to the first C64 layer in the encoder. All ReLUs [37] in the encoder are leaky [38], with slope 0.2, while ReLUs in the decoder are not leaky. This results in the following generator:

Encoder:

C_{64}-C_{128}-C_{256}-C_{512}-C_{512}-C_{512}-C_{512}-C_{512}.

Decoder:

CD_{512}-CD_{1024}-CD_{1024}-C_{1024}-C_{1024}-C_{512}-C_{256}-C_{128}.

The discriminator works on image patches of size 16 × 16. For the discriminator, after its last layer and after extracting the patch, a convolution is applied to map to a one-dimensional output, followed by a Sigmoid function. This results in the following discriminator:

Discriminator:

C_{64}-C_{128}-C_{256}-C_{512}.

3. Datasets

Deep-Learning is a data-driven paradigm that requires large datasets for training. However, databases of XCT acquisitions with associated shape priors do not exist. We, therefore, adapted the SophiaBeads [39] dataset to report the performance of our process on actual data.

3.1. Preparing the SophiaBeads Training Data

Initially designed to provide data for researchers into XCT reconstruction from limited measurements, this dataset is a collection of six cone beam XCT acquisitions of a plastic tube filled with soda-lime glass beads. As the beads are approximately cylindrical, each slice through the volume contains only circular objects. The difference between each dataset is the number of projections used per volume.

To generate our training data with both complete and missing projections as well as shape information, we reconstructed the two datasets that had 512 and 1024 projections using the approach in [4,40], generating 256 × 256 × 200 volumes, using the FDK algorithm for cone-beam tomography. The other four datasets needed more projections to provide suitable ground-truth training data and were thus discarded. We used circle-finding methods to identify object boundaries in the full reconstructions, which were used as shape priors. Training data were then generated from individual 2D slices using a parallel beam geometry for experimental convenience and speed.

Thus, our datasets consisted of 400 sinograms (200 generated from each of the two volumes) and the associated CAD data. Each volume is split into 180 training slices and 20 test slices. A slice of the volume reconstructed with 512 projections and the identified beads, from which the object boundaries can be inferred, is shown in Figure 4a and Figure 4b, respectively, whilst the associated sinograms were those previously shown in Figure 2. To test our method's robustness, we chose not to capture the plastic container's outline in the CAD data, and several small circles were also not included. Doing so ensured the images had features not represented in the CAD data.

We generate full sinograms from each slice, simulating a detector with 256 pixels and measuring from 256 equally spaced angles in a 180-degree arc. We use the implementation of the radon transform proposed in [4]. The sinograms corresponding to Figure 4a and Figure 4b are shown in Figure 2a and Figure 2c, respectively. Sinograms with missing projections were generated by setting between 5 and 95% of consecutive angular measurements to 0. Figure 2b shows an example of a sinogram with 50 per cent missing data.

3.2. Synthetic Training Data

Having a synthetic dataset allows experimentation with the data model. In real applications, the shape prior will only sometimes precisely match the object shape. There could be boundary deviations, or objects could have unknown inclusions and internal defects. Therefore, we also generate a separate dataset with additional internal defects in the objects not captured by the shape prior.

In order to simulate such defects, we create a synthetic dataset to mirror the SophiaBeads data, as shown in Figure 5a,b. Each slice is a 256 by 256 image of several circles generated with the circle function from the scikit-image.draw Python module with different radii and positions and uniform densities with added Gaussian noise. Object attenuation values are set to 25, while the image background is assumed to have an X-ray attenuation much lower than the object and is set to 1. We generate the shape prior associated with a slice using the same function with the same radii and position for each circle but with different densities and no Gaussian noise. Whilst we chose this model to replicate the SophiaBeads dataset, similar data have previously been used in [41–43]. Figure 6 displays an example image slice.

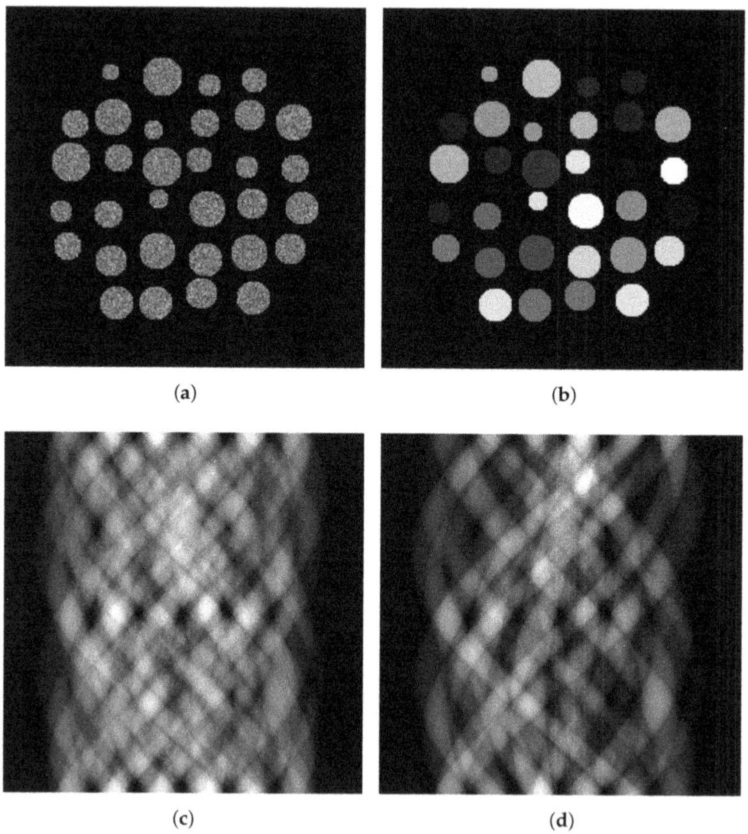

Figure 5. Slice samples from the synthetic dataset. The ground-truth reconstruction (**a**), and the shape prior (**b**) showing each object with a different randomly assigned grey value, in which no plastic container is visible. In an attempt to reproduce noise in the image, Gaussian noise is added to the material of uniform densities. (**b**) Sinogram samples from the synthetic dataset, from the ground-truth reconstruction (**c**) and from the shape prior (**d**). The observed sinograms are obtained using the forward radon transform with a parallel beam geometry.

3.3. GAN Training

We train the pix2pix networks on the SophiaBeads training dataset reconstructed with 1024 projections using a hundred epochs. We then evaluate them on the dataset reconstructed from 512 projections. The three-channel input data are re-scaled to have values between 0 and 1. We used a batch size of 8 and the Adam optimiser [44]. We set the learning rate at 0.0002 and λ to 100. We used label-smoothing and noise addition in the discriminator, as suggested in [31]. We implemented the top-k [45] training procedure which discards 6/8 of the generated samples when updating the G loss.

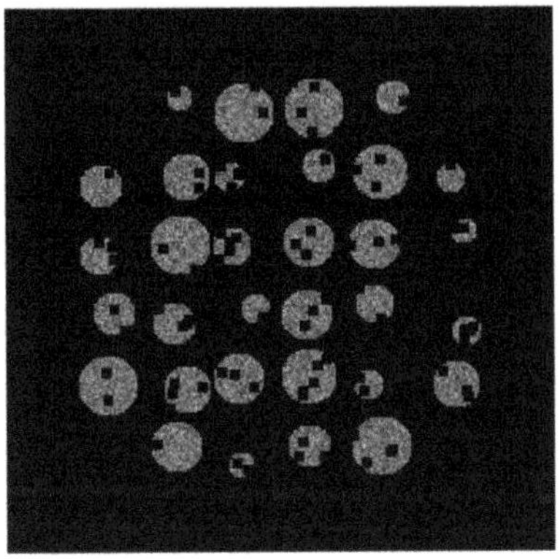

Figure 6. Simulated XCT images with objects with internal defects that are not encoded in shape prior.

We perform training on the SophiaBeads dataset with the priors and on a synthetic dataset that models internal defects separately. This means two copies of the same architecture were trained separately on each dataset but with the same hyperparameters.

4. Results

Whilst no known methods incorporate shape information into XCT reconstruction, we propose to compare our approach to a set of techniques that address the problem of inferring missing measurements in a sinogram.

First, we propose one straightforward method given CAD data: replacing the missing projections with the ones expected by the CAD. We also implement scaling the CAD data values developed in Section 2.2. These two CAD-only approaches are reported in rows one and two of Table 1. Then, we also implement a linear interpolation method. This straightforward technique is used as a pre-processing step in many interpolation procedures, such as in [13]. In row three of Table 1, we refer to this method as "Linear Interpolation".

Table 1. Performance comparison between the different methods regarding the difference between the estimated and true sinogram.

Missing Angles	**30°**	**60°**	**90°**
(1) Replacing missing projections with CAD data	22.5	19.2	17.1
(2) Replacing missing projections with CAD data (and with attenuation values scaling)	23.1	21.3	19.7
(3) Linear Interpolation	23.6	22.6	20.0
(4) Interpolation using a Unet	9.9	8.4	7.7
(5) Interpolation using a Unet and CAD data	16.8	12.9	11.0
(6) Interpolation using a GAN	7.0	3.82	2.0
(7) Interpolation using pix2pix and CAD data	**29.1**	**27.8**	**27.0**

In order to compare the difference between generative models and their non-generative counterparts, we implement two methods based on the Unet architecture. To begin with, we train a Unet to infer missing projections from the impaired sinogram only. Then we train a Unet with the same architecture to tackle the same goal, but we provide it with shape information. This comparison will also hint at the importance of shape priors for tackling the problem of inferring missing projections in a sinogram. These two methods are referred to as "Interpolation using a Unet" and "Interpolation using a Unet and CAD data" in Table 1. They are given in rows four and five, respectively.

We also compare our method to the GAN-only approach developed in [20]. We refer to this method as "Interpolation using a GAN" in Table 1, in row six. Unfortunately, we did not manage to replicate their results on our dataset. We suspect this might be due to an increased sinogram size. Finally, we report the results of our method based on the pix2pix network and refer to this method as "Interpolation using pix2pix and CAD data" in row seven of Table 1.

We report the results using the Peak Signal-to-Noise Ratio (PSNR).

4.1. *Estimation Performance on the SophiaBeads Dataset*

We have two volumes of 200 slices each. We split each volume into 90 per cent for training and 10 per cent for testing. We train using only the data from one volume, with convergence monitored on the same volume's test data. We then compute the performance results presented below by using the test split from the other volume.

We start by comparing the approaches over a range of misusing angles, from 30 degrees (17% missing data) to 90 degrees (50% missing data). In particular, we compare the method that estimates missing information from the CAD data with and without scaling, simple linear interpolation of the sinogram, interpolation followed by a U-net, interpolation followed by a U-net that also sees the CAD data, GAN-based inpainting without CAD data and finally the pix2pix U-net-based GAN that also uses CAD data. We evaluate the performance by comparing the estimated sinogram to the target sinograms.

Table 1 shows the results, where we can see that the pix2pix architecture combined with our CAD prior outperforms all other approaches in terms of PSNR. It is interesting to note that using the CAD prior significantly enhances the performance of the U-net. It is also worth mentioning that the scaling operation slightly improves the quality of the CAD replacement. The GAN method was hard to train. Hence, the inpainting process has limitations that we will detail in Section 5.

Sinograms contain more low-frequency information than images, so the error measure in the sinogram domain differs from the error in the image domain. To evaluate the performance of the different methods once we have recovered the image, we compute a reconstruction using the SIRT algorithm as implemented in the TIGRE Python toolbox [4].

Figure 7 shows the reconstructions associated with each method when 50% of the data is missing. As can be seen in Table 2, not only does the pix2pix architecture combined with the CAD prior outperform all other approaches in the image space, but it is the only one that yields an improvement compared to the reconstruction made from the original sinogram as shown here in the first row of the table. It is also important to underline that methods performing well in the sinogram space do not necessarily improve the image space.

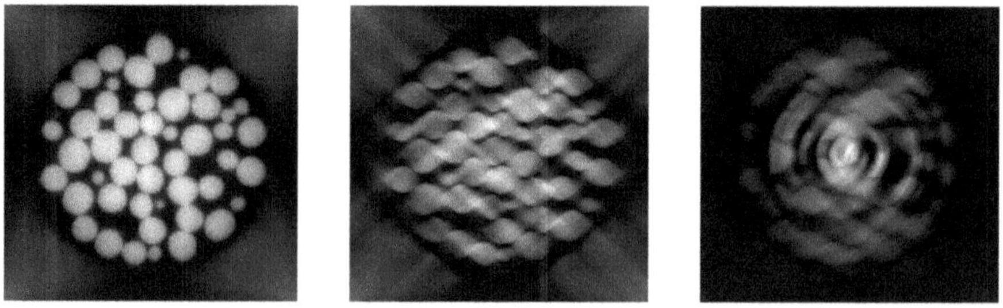

(a) (b) (c)

Figure 7. *Cont.*

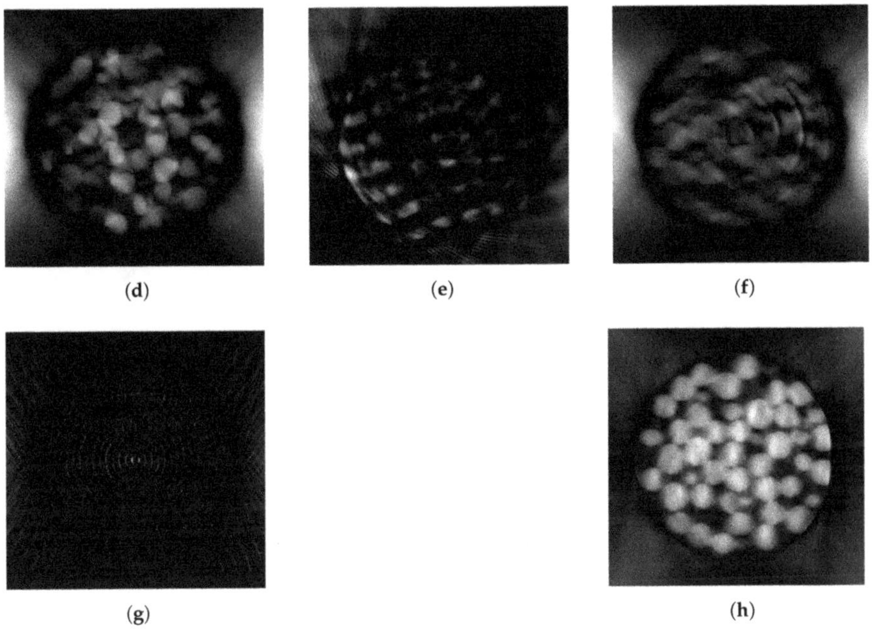

Figure 7. Reconstructions of a slice from one of the SophiaBeads test datasets. First, the target image (**a**), reconstructed from 256 measurements. Then, the image reconstructed from half of the measurements (**b**), the linear interpolation (**c**), the replacement of the missing measurements by the ones expected by the CAD (**d**), the measurements inferred by the Unet method without CAD data (**e**), the measurements inferred by the Unet method with CAD data (**f**), the measurements inferred by the GAN without CAD data (**g**) and the measurements inferred by our method (**h**).

Table 2. Performance of the different methods in terms of the quality of the reconstructed image.

Missing Angles	**30 °**	**60°**	**90°**
Reconstruction from sinograms without treatment	69.1	66.4	63.8
Reconstruction corresponding to Table 1 (2)	61.3	58.2	56.3
Reconstruction corresponding to Table 1 (3)	61.2	61.8	59.4
Reconstruction corresponding to Table 1 (4)	52.7	51.4	51.4
Reconstruction corresponding to Table 1 (5)	60.4	56.5	55.1
Reconstruction corresponding to Table 1 (6)	53.4	51.4	50.9
Reconstruction corresponding to Table 1 (7)	**69.6**	**67.9**	**66.5**

4.2. Effect of Inconsistencies between the Two Modalities on the Synthetic Dataset

We finally also use the data with simulated defects. The advantage of our method here is less pronounced, and a qualitative inspection shows that the network fails at inferring the hole pattern, thus showing a limited ability to recover defects. This can be seen in Figure 8, where we show four examples of the difference between the image reconstructed with our method and the target image for data with 33% per cent missing projections. This suggests that the GAN might fail to learn the fine structure of the object attenuation.

One reason for this might be that we are estimating a considerable number of missing projections, which is highly challenging. Moreover, the nature of holes makes the choice of loss function delicate, as the defects are high-frequency image components whilst low-frequency information dominates. In a future study, we plan to modify the loss function to provide more appropriate weighting to the cost function. The image reconstruction process treats measured and inferred data similarly, which might not be optimal if one or the other is known to have a more significant error.

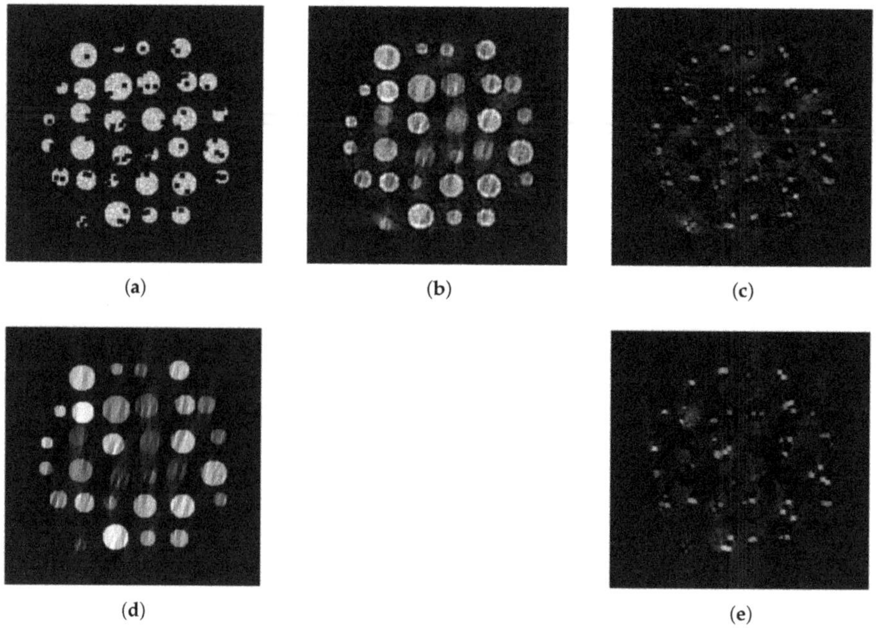

Figure 8. Reconstructions of the synthetic samples using various methods. First, the one from the sinogram 256 measurements (**a**), then from the sinogram interpolated with our method (**b**), the difference between the latter and the target (**c**), the one interpolated with the shape prior (**d**) and the difference between the latter and the target (**e**). When reconstructing data with defects using shape priors, the method can identify the objects' location, but the defect's fine details are more difficult to identify.

5. Discussion and Conclusions

In this paper, we have shown how to use a GAN to exploit shape prior information to address the image inpainting with edge information problem in XCT. Our experiments show the significant advantage that our method offers over state-of-the-art methods that do not include shape prior information for inferring a large number of consecutive missing acquisitions. We also demonstrated that the CAD prior facilitates GAN training.

The main limitation of our approach is the failure to extrapolate faults that are not encoded in the shape priors. One reason might be that our cost function does not sufficiently account for high-frequency image information. Moreover, whilst the image reconstruction process is not studied here, weighting the ground truth and the generated acquisitions could improve the final image quality.

Comparing our method to the GAN-only approach of [20], it is worth stressing that the sinograms that are measured in industrial applications are much larger than those used in their report and so scaling of the approach has to be considered. The method in [20] was directly inspired by [23], where results are achieved on 64 $\times$ 64 images, and learning is based on contextual as well as discriminator losses. This approach failed because it did not directly scale to the increased data sizes used here.

In a future study, we will investigate the mitigation of the impact of the discrepancies between the CAD and the actual scan. To do so, we plan to add random image perturbations to the CAD before sampling. We hope that this approach will help the model learn more robust image features.

Author Contributions: Conceptualization, E.V. and T.B.; methodology, E.V. and T.B.; software, E.V. and T.B.; validation, E.V.; formal analysis, E.V. and T.B.; investigation, E.V. and T.B.; resources, T.B.; data curation, E.V. and T.B.; writing—original draft preparation, E.V.; writing—review and editing, E.V., K.F. and T.B.; visualization, E.V.; supervision, K.F. and T.B.; project administration, T.B.; funding acquisition, T.B. All authors have read and agreed to the published version of the manuscript.

Funding: This research was funded by Anglo-French DSTL-AID Joint-PhD program.

Institutional Review Board Statement: Not applicable.

Informed Consent Statement: Not applicable.

Data Availability Statement: The data used in this study is the SophiaBeads dataset, that can be downloaded from Zenodo. We refer the reader to [39] for the dataset and to [40] for the code to reconstruct the tomographic images.

Conflicts of Interest: The authors declare no conflict of interest.

Abbreviations

The following abbreviations are used in this manuscript:

XCT	X-ray Computed Tomography
CAD	Computer Assisted Design
(DC)GAN	(Deep-Convolutionnal) Generative Adversarial Network
CNN	Convolutionnal Neural Network

References

1. Hall, E.J.; Brenner, D.J. Cancer risks from diagnostic radiology. *Br. J. Radiol.* **2008**, *81*, 362–378. . [CrossRef] [PubMed]
2. du Plessis, A.; le Roux, S.G.; Guelpa, A. Comparison of medical and industrial X-ray computed tomography for non-destructive testing. *Case Stud. Nondestruct. Test. Eval.* **2016**, *6*, 17–25. [CrossRef]
3. Feldkamp, L.A.; Davis, L.C.; Kress, J.W. *Practical Cone-Beam Algorithm*; Technical Report 6; Ford Motor Company: Dearborn, MI, USA, 1984.
4. Biguri, A.; Dosanjh, M.; Hancock, S.; Soleimani, M. TIGRE: A MATLAB-GPU toolbox for CBCT image reconstruction. *Biomed. Phys. Eng. Express* **2016**, *2*, 055010. [CrossRef]
5. Candès, E.J.; Romberg, J.; Tao, T. Robust uncertainty principles: Exact signal reconstruction from highly incomplete frequency information. *IEEE Trans. Inf. Theory* **2006**, *52*, 489–509. [CrossRef]
6. Xu, Q.; Yu, H.Y.; Mou, X.Q.; Zhang, L.; Hsieh, J.; Wang, G. Low-dose X-ray CT reconstruction via dictionary learning. *IEEE Trans. Med. Imaging* **2012**, *31*, 1682–1697. [CrossRef] [PubMed]
7. Wu, D.; Kim, K.; El Fakhri, G.; Li, Q. Iterative low-dose CT reconstruction with priors trained by artificial neural network. *IEEE Trans. Med. Imaging* **2017**, *36*, 2479–2486. [CrossRef] [PubMed]
8. Karimi, D.; Ward, R. Reducing streak artifacts in computed tomography via sparse representation in coupled dictionaries. *Med. Phys.* **2016**, *43*, 1473–1486. [CrossRef] [PubMed]
9. Köstler, H.; Prümmer, M.; Rüde, U.; Hornegger, J. Adaptive variational sinogram interpolation of sparsely sampled CT data. In Proceedings of the International Conference on Pattern Recognition, Hong Kong, China, 20–24 August 2006; Volume 3, pp. 778–781. [CrossRef]
10. Pohlmann, M.; Berger, M.; Maier, A.; Hornegger, J.; Fahrig, R. *Estimation of Missing Fan-Beam Projections Using Frequency Consistency Conditions*; Technical Report; Friedrich Alexander Universitat Erlangen-Nurnberg: Nurnberg, Germany, 2014.
11. Benammar, A. Sinogram interpolation method for limited angle tomography. In Proceedings of the 2019 International Conference on Advanced Electrical Engineering (ICAEE), Algiers, Algeria, 19–21 November 2019; pp. 1–5.
12. Lee, H.; Lee, J.; Cho, S. View-interpolation of sparsely sampled sinogram using convolutional neural network. In Proceedings of the Medical Imaging 2017, Orlando, FL, USA, 13–16 February 2017; Volume 10133, p. 1013328. . [CrossRef]
13. Lee, H.; Cho, S.; Lee, J. Sinogram synthesis using convolutional-neural-network for sparsely view-sampled CT. In Proceedings of the Medical Imaging 2018: Image Processing, Houston, TX, USA, 11–13 February 2018; pp. 592–601. [CrossRef]
14. Lee, H.; Lee, J.; Kim, H.; Cho, B.; Cho, S. Deep-neural-network-based sinogram synthesis for sparse-view CT image reconstruction. *IEEE Trans. Radiat. Plasma Med. Sci.* **2019**, *3*, 109–119. [CrossRef]
15. Yuan, H.; Jia, J.; Zhu, Z. SIPID: A deep learning framework for sinogram interpolation and image denoising in low-dose CT reconstruction. In Proceedings of the International Symposium on Biomedical Imaging, Washington, DC, USA, 4–7 April 2018; pp. 1521–1524. [CrossRef]
16. Ronneberger, O.; Fischer, P.; Brox, T. U-Net: Convolutional Networks for Biomedical Image Segmentation. In Proceedings of the International Conference on Medical Image Computing and Computer-Assisted Intervention, Munich, Germany, 5–9 October 2015; pp. 234–241.

17. Goodfellow, I.J.; Pouget-Abadie, J.; Mirza, M.; Xu, B.; Warde-Farley, D.; Ozair, S.; Courville, A.; Bengio, Y. Generative Adversarial Networks. *Commun. ACM* **2020**, *63*, 139–144. [CrossRef]
18. Anirudh, R.; Kim, H.; Thiagarajan, J.J.; Mohan, K.A.; Champley, K.; Bremer, T. Lose the Views: Limited Angle CT Reconstruction via Implicit Sinogram Completion. In Proceedings of the IEEE Computer Society Conference on Computer Vision and Pattern Recognition, Salt Lake City, UT, USA, 18–23 June 2018; pp. 6343–6352. [CrossRef]
19. Kim, H.; Anirudh, R.; Mohan, K.A.; Champley, K. *Extreme Few-View CT Reconstruction Using Deep Inference*; Technical Report; Lawrence Livermore National Laboratory: Livermore, CA, USA, 2019.
20. Yoo, S.; Yang, X.; Wolfman, M.; Gursoy, D.; Katsaggelos, A.K. Sinogram Image Completion for Limited Angle Tomography with Generative Adversarial Networks. In Proceedings of the International Conference on Image Processing, ICIP, Taipei, Taiwan, 22–25 September 2019; pp. 1252–1256. [CrossRef]
21. Pathak, D.; Krahenbuhl, P.; Donahue, J.; Darrell, T.; Efros, A.A. Context Encoders: Feature Learning by Inpainting. In Proceedings of the IEEE Conference on Computer Vision and Pattern Recognition, Vegas, NV, USA, 27–30 June 2016; pp. 2536–2544.
22. Radford, A.; Metz, L.; Chintala, S. *Unsupervised Representation Learning with Deep Convolutional Generative Adversarial Networks*; Technical Report: Indico Research: Boston, MA, USA, 2015.
23. Yeh, R.A.; Chen, C.; Lim, T.Y.; Schwing, A.G.; Hasegawa-Johnson, M.; Do, M.N. Semantic Image Inpainting with Deep Generative Models. In Proceedings of the IEEE Conference on Computer Vision and Pattern Recognition, Vegas, NV, USA, 27–30 June 2016; pp. 5485–5493.
24. Jahanian, A.; Chai, L.; Isola, P. *On the "Steerability" of Generative Adversarial Networks*; Technical Report; Massachussetts Institute of Technology: Cambridge, MA, USA, 2019.
25. Ackerman, M.J. The visible human project. *Proc. IEEE* **1998**, *86*, 504–511. [CrossRef]
26. Xu, X.G.; Taranenko, V.; Zhang, J.; Shi, C. A boundary-representation method for designing whole-body radiation dosimetry models: Pregnant females at the ends of three gestational periods—RPI-P3, -P6 and -P9. *Phys. Med. Biol.* **2007**, *52*, 7023–7044. [CrossRef]
27. Kervyn, G. *The First Open Source 3D Atlas of Human Anatomy*; Technical Report; Felix Meritis: Amsterdam, The Netherlands, 2022.
28. Huang, J.; Zhang, Y.; Ma, J.; Zeng, D.; Bian, Z.; Niu, S.; Feng, Q.; Liang, Z.; Chen, W. Iterative image reconstruction for sparse-view CT using normal-dose image induced total variation prior. *PLoS ONE* **2013**, *8*, e0079709. [CrossRef] [PubMed]
29. Abbas, S.; Min, J.; Cho, S. Super-sparsely view-sampled cone-beam CT by incorporating prior data. *J. X-ray Sci. Technol.* **2013**, *21*, 71–83. [CrossRef] [PubMed]
30. Isola, P.; Zhu, J.Y.; Zhou, T.; Efros, A.A. Image-to-Image Translation with Conditional Adversarial Networks. In Proceedings of the IEEE Conference on Computer Vision and Pattern Recognition, Honolulu, HI, USA, 21–26 July 2017; pp. 1125–1134.
31. Goodfellow, I. *NIPS 2016 Tutorial: Generative Adversarial Networks*; Technical Report; OpenAI: San Francisco, CA, USA, 2016.
32. Goodfellow, I.J.; Warde-Farley, D.; Mirza, M.; Courville, A.; Bengio, Y. Maxout Networks. In Proceedings of the International Conference on Machine Learning, Atlanta, GA, USA, 16–21 June 2013; pp. 1319–1327.
33. Ioffe, S.; Szegedy, C. Batch Normalization: Accelerating Deep Network Training by Reducing Internal Covariate Shift. In Proceedings of the International Conference on Machine Learning, Lille, France, 6–11 July 2015; pp. 448–456.
34. Dupé, F.X.; Fadili, J.; Starck, J.L. Inverse Problems with Poisson noise: Primal and Primal-Dual Splitting. In Proceedings of the 2011 18th IEEE International Conference on Image Processing, Brussels, Belgium, 11–14 September 2011; pp. 1901–1904.
35. Larsen, A.B.L.; Sønderby, S.K.; Larochelle, H.; Winther, O. Autoencoding beyond pixels using a learned similarity metric. In Proceedings of the International Conference on Machine Learning, Lille, France, 6–11 July 2015; pp. 1558–1566.
36. Srivastava, N.; Hinton, G.; Krizhevsky, A.; Salakhutdinov, R. Dropout: A Simple Way to Prevent Neural Networks from Overfitting. *J. Mach. Learn. Res.* **2014**, *15*, 1929–1958.
37. Nair, V.; Hinton, G.E. Rectified Linear Units Improve Restricted Boltzmann Machines. In Proceedings of the 27th International Conference on Machine Learning, Haifa, Israel , 21–24 June 2010; pp. 807–814.
38. Maas, A.L.; Hannun, A.Y.; Ng, A.Y. Rectifier Nonlinearities Improve Neural Network Acoustic Models. In Proceedings of the International Conference on Machine Learning, Atlanta, GA, USA, 16–21 June 2013; p. 3.
39. Coban, S. *SophiaBeads Dataset Project [Data Set]* ; Technical Report; University of Manchester: Manchester, UK, 2015.
40. Coban, S. *SophiaBeads Dataset Project Codes (v1.0)*; Technical Report; University of Manchester: Manchester, UK, 2015.
41. Jin, K.H.; McCann, M.T.; Froustey, E.; Unser, M. Deep Convolutional Neural Network for Inverse Problems in Imaging. *IEEE Trans. Image Process.* **2017**, *26*, 4509–4522. [CrossRef] [PubMed]
42. Adler, J.; Ringh, A.; Öktem, O.; Karlsson, J. *Learning to Solve Inverse Problems Using Wasserstein Loss*; Technical Report; KTH Royal Institute of Technology, Deptartment of Mathematics: Stockholm, Sweden, 2017.
43. Kelly, B.; Matthews, T.P.; Anastasio, M.A. *Deep Learning-Guided Image Reconstruction from Incomplete Data*; Technical Report; Washington University, Department of Computer Science: Saint Louis, MO, USA, 2017.

44. Kingma, D.P.; Ba, J. *Adam: A Method for Stochastic Optimization*; Technical Report; Google Brain: Mountain View, CA, USA, 2014.
45. Sinha, S.; Zhao, Z.; Goyal, A.; Raffel, C.; Odena, A. Top-k Training of GANs: Improving GAN Performance by Throwing Away Bad Samples. In Proceedings of the Advances in Neural Information Processing Systems, Online, 6–12 December 2020; pp. 14638–14649.

Journal of
Clinical Medicine

MDPI

Article

Use of Automated Machine Learning for Classifying Hemoperitoneum on Ultrasonographic Images of Morrison's Pouch: A Multicenter Retrospective Study

Dongkil Jeong [1], Wonjoon Jeong [2], Ji Han Lee [3] and Sin-Youl Park [4,*] on behalf of the Society of Emergency & Critical Care Imaging (SECCI)

1 Department of Emergency Medicine, College of Medicine, Soonchunhyang University, Cheonan 31151, Republic of Korea; 397012@naver.com
2 Department of Emergency Medicine, School of Medicine, Chungnam National University, Daejeon 35015, Republic of Korea; gardenjun@hanmail.net
3 Division of Emergency Medicine, Department of Medicine, The Catholic University of Korea, Seoul 11765, Republic of Korea; capcloud80@hanmail.net
4 Department of Emergency Medicine, College of Medicine, Yeungnam University, Daegu 42415, Republic of Korea
* Correspondence: dryuri@naver.com; Tel.: +82-053-620-4322

Abstract: This study evaluated automated machine learning (AutoML) in classifying the presence or absence of hemoperitoneum in ultrasonography (USG) images of Morrison's pouch. In this multicenter, retrospective study, 864 trauma patients from trauma and emergency medical centers in South Korea were included. In all, 2200 USG images (1100 hemoperitoneum and 1100 normal) were collected. Of these, 1800 images were used for training and 200 were used for the internal validation of AutoML. External validation was performed using 100 hemoperitoneum images and 100 normal images collected separately from a trauma center that were not included in the training and internal validation sets. Google's open-source AutoML was used to train the algorithm in classifying hemoperitoneum in USG images, followed by internal and external validation. In the internal validation, the sensitivity, specificity, and area under the receiver operating characteristic (AUROC) curve were 95%, 99%, and 0.97, respectively. In the external validation, the sensitivity, specificity, and AUROC were 94%, 99%, and 0.97, respectively. The performances of AutoML in the internal and external validation were not statistically different ($p = 0.78$). A publicly available, general-purpose AutoML can accurately classify the presence or absence of hemoperitoneum in USG images of the Morrison's pouch of real-world trauma patients.

Keywords: ultrasonography; automated machine learning; emergency medicine; trauma; hemoperitoneum

Citation: Jeong, D.; Jeong, W.; Lee, J.H.; Park, S.-Y., on behalf of the Society of Emergency & Critical Care Imaging (SECCI). Use of Automated Machine Learning for Classifying Hemoperitoneum on Ultrasonographic Images of Morrison's Pouch: A Multicenter Retrospective Study. *J. Clin. Med.* **2023**, *12*, 4043. https://doi.org/10.3390/jcm12124043

Academic Editors: Michał Strzelecki, Adam Piorkowski and Rafał Obuchowicz

Received: 8 March 2023
Revised: 9 June 2023
Accepted: 11 June 2023
Published: 14 June 2023

1. Introduction

Uncontrolled massive bleeding is a leading cause of potentially preventable death in major trauma [1]. Massive bleeding from an organ and vascular injury in the chest or abdomen are associated with approximately 20–40% of trauma deaths that occur after hospital arrival, and the early recognition and control of the bleeding can reduce these mortality rates [2,3]. Therefore, the early recognition of massive bleeding in major trauma patients and the rapid control of bleeding while initiating resuscitation are critical steps in the emergency department [4].

Computed tomography (CT) is an imaging modality that is considered the gold standard for assessing hemothorax and hemoperitoneum and the presence of internal organ and vascular injury; however, CT can be difficult to obtain in hemodynamically unstable patients because it requires transport of the patient from the emergency department [5].

Focused assessment with sonography for trauma (FAST) is currently the initial assessment tool of choice for hemodynamically unstable trauma patients and has the advantage of being a simple, non-invasive method with reasonable sensitivity, specificity, and accuracy that can be performed safely, quickly, and repeatedly at the bedside [6,7].

However, the disadvantages of ultrasonography (USG), such as the low clarity of USG images, interobserver variability of the findings, and low quality of acquired images depending on the scanning skill of the operator, can lead to the poor interpretation accuracy of the USG images if the operator is not sufficiently trained in USG or has limited experience in performing it, and poor sensitivity in detecting hemoperitoneum on USG images, especially when the amount of hemoperitoneum is insignificant [8–10]. The accuracy of FAST hemoperitoneum diagnosis varies widely, from 61.3% to 100%, depending on the operator [11–13].

In the field of medical imaging, deep learning (DL) algorithms can not only reduce the workload of pre-screening but also mark areas of suspected lesions, thereby increasing the reliability of clinicians' diagnosis from medical images and reducing the likelihood of diagnostic errors [14,15]. In the field of emergency medicine, where clinical decisions must be made within minutes of arrival at the emergency department for resuscitation and stabilization, the help of DL algorithms in classifying these medical images is highly anticipated [16,17].

Recent studies of DL algorithms in radiography and CT have shown that preliminary classification by DL algorithms can help clinicians improve the accuracy of imaging diagnoses [18–20]. Based on this observation, we hypothesized that if a DL algorithm can provide accurate medical assistance in classifying hemoperitoneum on trauma USG, it can improve the reliability of hemoperitoneum diagnosis for clinicians who are less experienced in interpreting trauma findings on USG and reduce errors to improve clinical decision making.

Building DL algorithms in USG imaging requires high-performance computing equipment and collaboration with DL coding experts to develop non-sparse DL algorithms, which is a major barrier to DL research by clinicians without the support of equipment and technology [21,22]. Recently, global information technology companies have offered cloud-computing-based automated machine learning (AutoML) platforms that can perform DL without the need for high-performance computing equipment and specialized programming skills. If these AutoML platforms can be applied to medical imaging, the equipment and technical barriers that clinicians face in applying DL algorithms to clinical research can be reduced. As a preliminary investigation into the application of AutoML in classifying hemoperitoneum in trauma USG, this study aimed to evaluate the performance of AutoML in classifying the presence or absence of hemoperitoneum in Morrison's pouch USG images of trauma patients.

2. Materials and Methods

2.1. Study Design and Setting

We retrospectively collected USG images of adult trauma patients from multiple medical centers in the Republic of Korea to build and evaluate an AutoML system capable of classifying the presence or absence of hemoperitoneum on USG images of Morrison's pouch.

2.2. Study Population

For the development cohort, adult trauma patients who presented to one level I trauma center, two level I emergency medical centers, and one level II emergency medical center between 1 June 2019 and 31 July 2021, and who initially underwent trauma USG were included in the analysis. Patients who did not undergo USG imaging, whose USG images of Morrison's pouch were not available, and those who were younger than 18 years were excluded.

2.3. Data Collection

An emergency medicine (EM) specialist with at least 10 years of experience performing trauma USG and at least 5 years of experience teaching trauma USG at USG-related academic organizations collected USG images at his emergency medical center using a photo archiving and communication system (PACS). Among the various trauma USG images, only Morrison's pouch USG images were included in this study. Morrison's pouch refers to the hepatorenal recess, which is the space between the liver and the right kidney; only USG images in which this area was clearly visible were selected. If hypoechoic blood was observed in Morrison's pouch, it was classified as hemoperitoneum, and if only hyperechoic lines were observed, it was classified as normal (Figure 1).

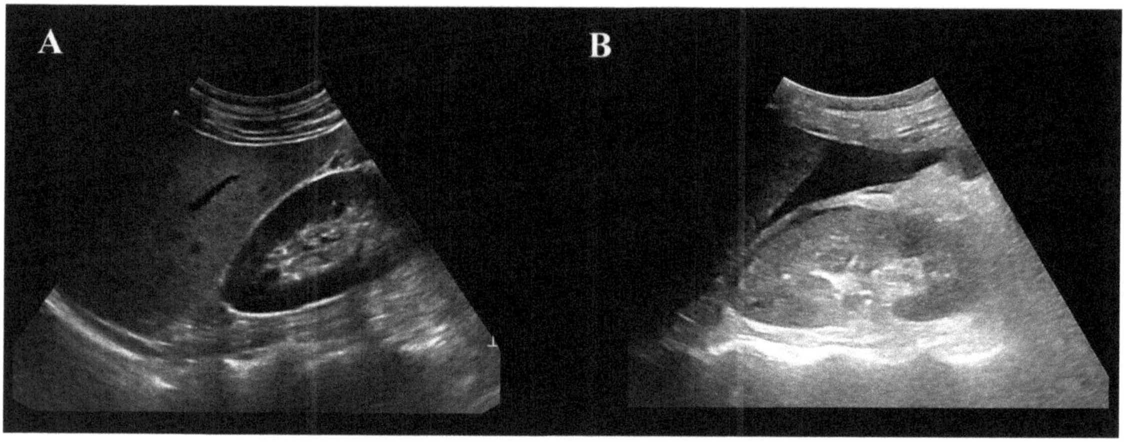

Figure 1. Ultrasonographic image of Morrison's pouch. (**A**) Normal ultrasonographic image of Morrison's pouch. Morrison's pouch is observed as hyperechoic lines. (**B**) Ultrasonographic image with hemoperitoneum in Morrison's pouch. Hypoechoic fluid collection in Morrison's pouch can be observed.

The selected USG images were collected as JPG files at a size of 1024 × 768 pixels. All USG images were anonymized and sent to the lead researcher, who first checked the quality and anonymity of the USG images.

The lead researcher checked the anonymity, clarity, and size of the USG images collected from each hospital. Then, after randomly shuffling the collected USG images so that the classification of normal and hemoperitoneum could not be confirmed, three separate emergency medicine experts with more than 10 years of FAST clinical experience were recruited and asked to reclassify the presence or absence of hemoperitoneum in the Morrison's pouch USG images. If there was disagreement on the classification of the presence or absence of hematoperitoneum, the classification was determined by consensus among the three experts.

2.4. USG Image Distribution

A total of 2200 USG images from 864 patients were collected from one level 1 trauma center, two level 1 urgent care centers, and one level 2 urgent care center.

Training and internal validation of the AutoML were performed with a total of 2000 USG images (1000 hemoperitoneum USG images and 1000 normal Morrison's pouch USG images) collected from 782 patients at the three emergency medical centers, of which 1800 randomly selected images (900 hemoperitoneum USG images and 900 normal USG images) were used for training the AutoML. The remaining 200 images (100 hemoperitoneum USG images and 100 normal USG images) were used for internal validation.

External validation of AutoML was performed using 200 USG images (100 hemoperitoneum images and 100 normal images) collected separately from a Level 1 trauma center that was not included in the training and internal validation. The training, internal validation, and external validation of AutoML and the distribution of USG images are shown in Figure 2.

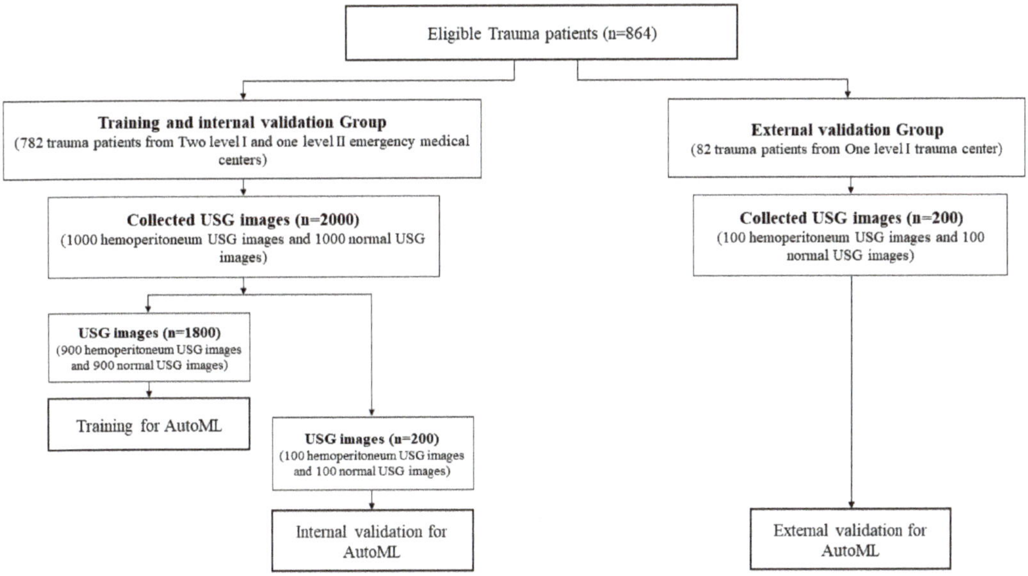

Figure 2. Distribution of ultrasonography images from eligible trauma patients for training, internal validation, and external validation of automated machine learning.

2.5. USG Image Analysis with AutoML

To train the AutoML for classifying the presence or absence of hemoperitoneum in trauma USG images, we used the open-source DL provided by Google (https://teachablemachine.withgoogle.co, accessed on 1 March 2023) (Figure 3).

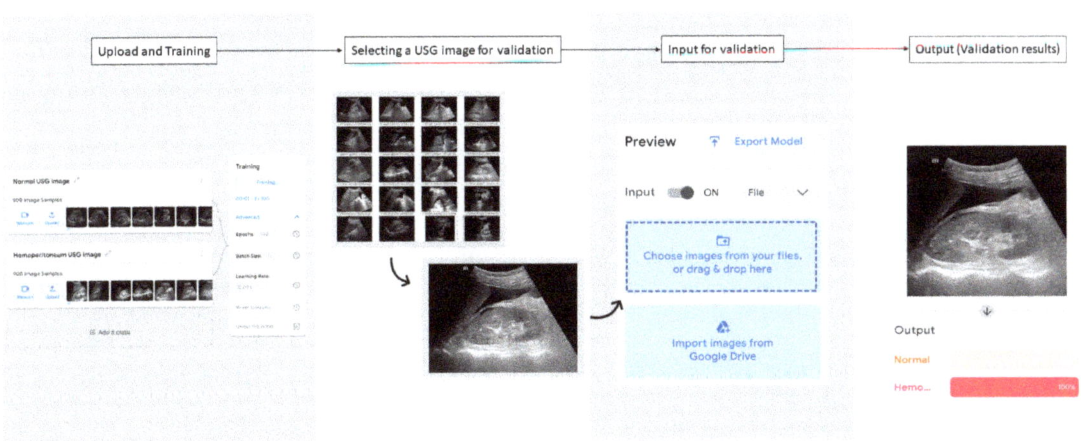

Figure 3. Training and validation of automated machine learning.

AutoML, popularized this time, is an online platform that can classify images, actions, sounds, and more. The professional work involved in DL coding is already built in, so users can train AutoML by simply uploading images or videos by class after deciding which class to classify. Detailed training settings, such as epoch, batch size, and training speed, can be selected by the user. By clicking the "Training" button, the custom DL is trained; training takes a few minutes depending on the training settings and the number of images. Once the AutoML is trained, users can validate it by uploading individual images or videos to the "Input" section and see the results in real time in the "Output" section of the AutoML. The results in the "Output" section are displayed as the size of the color bar graph and the percentage of the class to which the video and image you uploaded for validation belong.

The training and validation process of AutoML for classifying the presence or absence of hemoperitoneum in trauma USG images in this study is shown in Figure 3. Two classes, "normal" and "hemoperitoneum", were created for training AutoML in this study. For training, 900 normal USG images were dragged and dropped into the "normal" class, and similarly, 900 hemoperitoneum USG images were dragged and dropped into the "hemoperitoneum" class. The training settings used 100 epochs, batch size of 16, and learning rate of 0.001. Training took approximately 1 min.

Then, the trained AutoML was validated using 100 normal USG images and 100 hemoperitoneum USG images. After randomly mixing the 200 USG images prepared for internal validation, the USG images were dragged and dropped one at a time into the validation input field, and the results were checked and recorded. In the "Output" section, the groups represented by the size and percentage of the color bar graph were defined as "AutoML's Classification". If both groups were represented in the color bar graph, the group with the higher percentage was defined as "AutoML's Classification". The results of the 200 internal validations were compared to the classifications of an emergency physician with 10 years of experience, categorized as negative and positive, and a confusion matrix based on the results was created.

External validation of AutoML was performed using 200 separately collected USG images that were not used in the training and internal validation. After randomly mixing 100 normal USG images and 100 hemoperitoneum images, we uploaded the USG images one at a time by dragging and dropping them into the validation "Input" section and recorded the results. We defined the AutoML classification results in the same way as the internal validation, and the results were compared with the standard reference results to create a confusion matrix.

2.6. Statistical Analysis

The Shapiro–Wilk test was used to test for normality of the continuous variables. When the data did not follow normal distribution, medians (95% confidence interval [CI]) were used, and the Mann–Whitney U test was performed to test for significance. Categorical variables were expressed as frequency (percentage), and statistical significance was assessed using either the chi-squared or Fisher's exact test. The discriminatory performance of AutoML in recognizing hemoperitoneum in Morrison's pouch during internal and external validation was evaluated in terms of sensitivity, specificity, positive predictive values (PPV), and negative predictive values (NPV). The performance of AutoML in both validation groups was compared with that of three EM experts using the McNemar test for paired proportions. The statistical differences between the internal and external validation groups in terms of sensitivity, specificity, PPV, and NPV were evaluated using the two-proportions test. The performance of the AutoML was evaluated by comparing the area under the receiver operating characteristic (AUROC) curves of the internal and external validation groups. Statistical tests were conducted using MedCalc (version 19.4.1, MedCalc Software, Ostend, Belgium), and all tests were two-sided, with p-values considered significant at 0.05.

3. Results

3.1. Clinical Characteristics of Enrolled Trauma Patients

A total of 2200 USG images were collected from USG still frames and video clips of 864 patients. Of these, 2000 USG images were collected from 782 patients at three hospitals and were used for training (1800 images) and internal validation (200 images). The remaining 200 USG images were separately collected from 82 patients at one hospital and were used for external validation. The sample population included 596 men (68.98%) and 268 women (31.01%), with a median age of 58 years (95% CI, 56.0–60.0). Of the enrolled patients, 429 (49.65%) had hemoperitoneum. Demographic data of the patients in the training and internal and external validation groups are presented in Table 1; no statistically significant differences were found between the two groups.

Table 1. Clinical characteristics of enrolled trauma patients.

	Entire Sample (n = 864)	Training and Internal Validation Group (n = 782)	External Validation Group (n = 82)	p
Sex (n, %)				0.17
Male	596 (68.98)	534 (68.27)	62 (75.61)	
Female	268 (31.02)	248 (31.71)	20 (24.39)	
Median age (yr) (95% CI *)	58.0 (56.0–60.0)	58 (56.0–60.0)	57 (51.7–62.3)	0.81
Hemoperitoneum (n, %)	429 (49.65)	388 (49.62)	41 (50.00)	0.95

* CI: confidence interval.

3.2. Performance of AutoML in Classifying Hemoperitoneum in Morrison's Pouch USG Images: The Internal Validation Group

Table 2 shows the internal validation results of AutoML classifying the presence or absence of hemoperitoneum in USG images of Morrison's pouch. The performance of AutoML in the internal validation showed a sensitivity of 95% (95% CI, 88.72–98.36%), a specificity of 99% (95% CI, 94.55–99.97%), a PPV of 98.96% (95% CI, 93.11–99.85%), and an NPV of 95.19% (95% CI, 89.39–97.90%). The accuracy was 97.00% (93.58–98.89%).

Table 2. Classification performance of AutoML in classifying hemoperitoneum in Morrison's pouch USG images in the internal validation group.

	Internal Validation Group	Difference (95% CI) with Standard Reference *	p
Sensitivity	95.00% (88.72–98.36%)	2.00% (−0.38–4.38%)	0.22
Specificity	99.00% (94.55–99.97%)		
PPV	98.96% (93.11–99.85%)		
NPV	95.19% (89.39–97.90%)		
Accuracy	97.00% (93.58–98.89%)		

AutoML: Automated machine learning; CI: confidence interval; PPV: positive predictive value; and NPV: negative predictive value. * Standard reference: decision of three emergency medicine experts.

When comparing the internal validation results of AutoML with the classification results of the three EM experts, the statistical difference was found to be 2.00% (95% CI, −0.38–4.38%). There was no statistically significant difference between the performance of the EM experts and the AutoML in classifying the presence or absence of hemoperitoneum in the USG images of Morrison's pouch (p = 0.22).

3.3. Performance of AutoML in Classifying Hemoperitoneum in Morrison's Pouch USG Images: The External Validation Group

In the external validation, the performance of AutoML showed a sensitivity of 94% (87.40–97.77%), a specificity of 99% (95% CI, 94.55–99.97%), a PPV of 98.95% (95% CI, 93.04–99.85%), an NPV of 94.29% (95% CI, 88.36–97.29%), and an accuracy of 96.50% (92.92–98.58%) (Table 3). The statistical difference between the performance of AutoML in the external validation and the classification by EM experts was 2.50% (95% CI, −0.07–5.07%). There was no statistically significant difference between the performance of the EM experts and the AutoML in classifying the presence or absence of hemoperitoneum in USG images (p = 0.13).

Table 3. Classification performance of AutoML in classifying hemoperitoneum in Morrison's pouch ultrasonography images in the external validation group.

	External Validation Group	Difference (95% CI) with Standard Reference *	p
Sensitivity	94.00% (87.40–97.77%)	2.50% (−0.07–5.07%)	0.13
Specificity	99.00% (94.55–99.97%)		
PPV	98.95% (93.04–99.85%)		
NPV	94.29% (88.36–97.29%)		
Accuracy	96.50% (92.92–98.58%)		

AutoML: Automated machine learning; CI: confidence interval; PPV: positive predictive value; and NPV: negative predictive value. * Standard reference: decision of three emergency medicine experts.

3.4. Comparison of Internal and External Validation of the Performance of AutoML in Classifying the Presence or Absence of Hemoperitoneum on USG Images of Morrison's Pouch

Table 4 shows the results of a statistical comparison of the internal and external validation of the performance of AutoML in classifying the presence or absence of hemoperitoneum on USG images of Morrison's pouch. The difference in the performances between internal and external validation indicated a sensitivity of 1% (95% CI, −3.70–5.76%), a specificity of 0% (95% CI, −2.67–2.67%), a PPV of 0.01% (95% CI, −2.69–2.71%), and an NPV of 0.90% (95% CI, −3.72–5.58%). There were no statistically significant differences in the performance of the two groups in terms of sensitivity, specificity, PPV, and NPV.

Table 4. Comparison of the performance of AutoML in internal and external validation in classifying hemoperitoneum on ultrasonographic images of Morrison's pouch.

	Difference (95% CI) between Internal and External Validation	p
Sensitivity	1% (−3.70–5.76%)	0.66
Specificity	0% (−2.67–2.67%)	1.0
PPV	0.01% (−2.69–2.71%)	0.99
NPV	0.90% (−3.72–5.58%)	0.69

AutoML: Automated machine learning; CI: confidence interval. PPV: positive predictive value; and NPV: negative predictive value.

3.5. ROC Curve of AutoML in Classifying the Presence or Absence of Hemoperitoneum in Morrison's Pouch USG Images

Figure 4 shows the confusion matrix and ROC curve for the performance of AutoML in classifying hemoperitoneum in USG images in the internal and external validation groups, respectively. The AUROC values for the internal and external validation groups were 0.97 (95% CI, 0.94–0.99) and 0.97 (95% CI, 0.93–0.99), respectively. The statistical difference in AUROC values between the internal and external validation groups was 0.01 (standard error, 0.02), and there was no statistically significant difference between the two groups (p = 0.78).

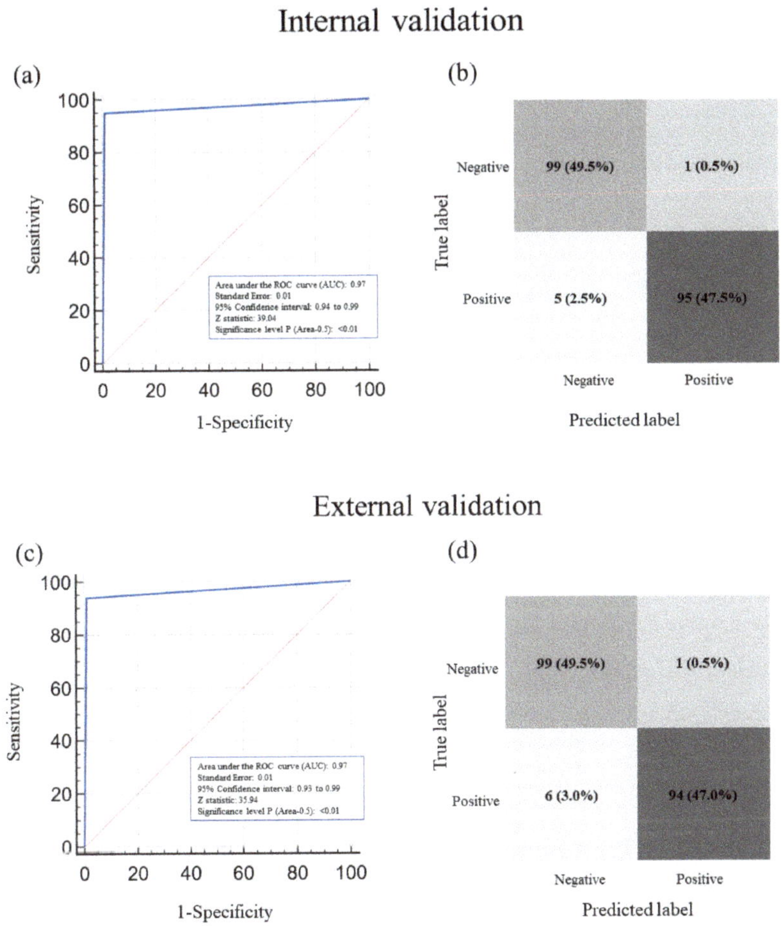

Figure 4. Receiver operating characteristic (ROC) curves and confusion matrix for the performance of AutoML in classifying the presence or absence of hemoperitoneum in Morrison's pouch USG image. (**a**) ROC curve of automated ML classification of hemoperitoneum in Morrison's pouch USG image in internal validation. (**b**) Confusion matrix of AutoML classification of hemoperitoneum in Morrison's pouch USG image in internal validation. (**c**) ROC curve of AutoML classification of hemoperitoneum in Morrison's pouch USG image in external validation. (**d**) Confusion matrix of AutoML classification of hemoperitoneum in Morrison's pouch USG image in external validation. The statistical difference in the AUROC between the internal and external validation groups was 0.01 (standard error, 0.02), and the difference in the AUROC between the two groups was not statistically significant (p = 0.78). USG, ultrasonography; and AutoML, automated machine learning.

4. Discussion

The purpose of this study was to evaluate the performance of AutoML in classifying the presence or absence of hemoperitoneum in USG images of Morrison's pouch using the AutoML platform. To our knowledge, this is the first study to evaluate the performance of a publicly available, general-purpose AutoML platform that requires no technical support or equipment to classify USG images of real-world trauma patients.

Independent external validation with separate USG images showed that a publicly available, general-purpose AutoML can classify the presence or absence of hemoperitoneum in Morrison's pouch USG images of real trauma patients with a high accuracy and performance comparable to that of USG-trained emergency physicians.

The utility of USG in the primary and secondary examination in trauma patients, allowing early bedside interpretation of results in real time while making informed medical decisions about treatment and care, is well-recognized. However, for untrained or inexperienced novice USG users, obtaining good-quality USG images can be challenging, and the accuracy or speed of determining the presence or absence of hemoperitoneum from acquired images can be compromised. This can negatively impact medical decision making for trauma patients.

Recently, DL, one of the most widely used artificial intelligence technologies in medical imaging, has been shown to be useful in various medical fields. In particular, with regard to radiography and computed tomography, DL algorithms have shown a high accuracy in classifying the presence or absence of lesions, comparable to that of related specialists, and DL algorithms can contribute to improving the diagnostic performance of physicians, especially novice physicians, by providing accurate medical assistance [23–27].

In a study by Jin et al., an artificial intelligence solution using chest radiographs showed acceptable diagnostic accuracy in a respiratory outpatient clinic, and the use of the solution improved the diagnostic performance of physicians in judging chest radiographs [28]. Although research on DL with USG images has been less active than that with radiographs and CT and MR images, recent improvements in USG technology and image quality have led to attempts to classify lesions using DL algorithms with a variety of USG images.

Raimondo et al. found that in the diagnosis of adenomyosis using USG images, DL models had a low accuracy but high specificity of diagnosis compared to that of residents with intermediate skills (fourth-year residents), which demonstrated that DL algorithms can be useful in assisting novice physicians with clinical decision making [29]. In addition, a DL algorithm built to support lesion detection in breast USG images was able to discriminate between benign and malignant breast nodules with a diagnostic accuracy equivalent to that of experts, demonstrating that DL algorithms can classify the presence or absence of lesions in USG images [30].

Furthermore, Sjogren et al. examined the use of DL algorithms in the field of trauma USG; they evaluated the performance of ML in determining the presence or absence of hemoperitoneum in 20 USG images in the perihepatic view of trauma patients [31]. They found that ML could distinguish between the presence and absence of hemoperitoneum with a sensitivity of 100% and specificity of 90% and concluded that such DL algorithms can help clinicians interpret trauma USG results. A study applying a customized DL algorithm to USG still images and video clips of pediatric trauma patients showed that, compared to experts, the DL algorithm was able to detect hemoperitoneum in Morrison's pouch USG images with a sensitivity of 91.9% (95% CI, 91.7–92.1%) and a specificity of 97.9% (95% CI, 97.8–98.0%) [32]. The authors showed that the accurate classification of USG images via DL algorithms is critical for ensuring the quality and feasibility of multi-level DL FAST models and improving the assessment of injured children.

Recently, in a study performed to detect hemoperitoneum in Morrison's pouch on USG images using a bespoke DL model, the model was able to detect the presence or absence of hemoperitoneum in Morrison's pouch with a sensitivity and specificity of 96.7% and 98.5%, respectively; hence, a DL algorithm with an automated feedback and training system could help improve the reading and image acquisition skills of inexperienced sonographers [33]. Unlike previous studies that used custom-built ML or DL algorithms for trauma USG images, this study evaluated the performance of the algorithm in classifying USG images of real trauma patients using the public AutoML platform, which does not require professional coding or large-scale equipment. We found that AutoML could classify the presence or absence of hemoperitoneum with a high accuracy; its results were comparable to the classification made by relevant experts and consistent with the results of previous studies using bespoke ML and DL algorithms.

The use of AutoML in classifying the presence of hemoperitoneum in USG images of Morrison's pouch is expected to help improve the accuracy of the assessment by healthcare providers who are unfamiliar or inexperienced in examining or reading trauma USG images.

However, to apply AutoML to real-world clinical patients, it must be convenient in addition to being accurate. This is especially true for USG, which lacks the clarity of other medical imaging modalities, and USG images tend to vary in quality, depending on the skills of the operator. Therefore, for the practical clinical application of AutoML, it is important that AutoML demonstrates high accuracy in classifying not only selected USG images with high clarity in a defined area but also trauma USG images with varying clarity levels in different areas. The clinical value of AutoML as a potential decision support tool needs to be evaluated in a large, multi-center, prospective study of trauma USG images of varying clarity levels at different sites and in different practice models. In addition, to receive real-time medical advice using AutoML-based point of care technologies, technological advances that allow AutoML to be embedded in the USG machine or integrated with a PACS or electronic medical record system are required. However, the advice of AutoML is not perfect; in a real-world clinical setting, clinicians should not rely solely on the advice of AutoML but should consider it in conjunction with other tests and results to make a final medical decision [34]. Nevertheless, medical advice of AutoML is expected to help clinicians improve their clinical decisions when identifying hemoperitoneum on trauma USG images.

Limitations

Given the retrospective nature of this study, the possibility of bias cannot be excluded. This study also had the following limitations:

First, per the FAST protocol, the left and right upper and lower abdomen are scanned consecutively, and then the images are evaluated collectively to assess the presence or absence of hemoperitoneum. While Morrison's pouch is a highly specific area for recognizing peritoneum, USG images of this organ alone are not sufficient to assess the presence or absence of hemoperitoneum [35,36]. Because this study was conducted with images of one of three sites within the peritoneal cavity imaged under FAST, the results of this study cannot be directly compared to the performance of FAST by healthcare providers. The results of this study are also limited by the fact that AutoML could only distinguish between the presence and absence of hemoperitoneum on USG images of Morrison's pouch. Further studies are needed to evaluate the classification performance of AutoML using USG images of the lower and left upper abdomen.

Second, although this study was conducted with USG images of real trauma patients, only high-quality images of Morrison's pouch with high clarity were selected. It is necessary to evaluate the presence or absence of hemoperitoneum on USG images with varying levels of clarity in clinical practice. Therefore, further AutoML studies involving USG images with varying levels of clarity are necessary.

Third, in this study, we used still-frame images. As the concept of "point-of-care USG" is gaining popularity, free fluid in clinical situations is being identified via a comprehensive evaluation of continuous dynamic images rather than by the evaluation of still images. Additionally, the performance of customized DL algorithms in classifying USG images may be better with video clips than with still frames [32]. Therefore, further research on the performance of AutoML in analyzing USG images from video clips is needed.

5. Conclusions

A publicly available, general-purpose AutoML was able to accurately classify the presence or absence of hemoperitoneum in Morrison's pouch USG images of real-world trauma patients. The assistance provided by AutoML in classifying hemoperitoneum in Morrison's pouch USG images is expected to improve the accuracy and speed of hemoperitoneum diagnosis using trauma USG images by clinicians unfamiliar or inexperienced in examining and reading trauma USG images.

Author Contributions: All authors were involved in drafting the article or revising it critically for important intellectual content. Study conceptualization and design, S.-Y.P., W.J., J.H.L. and D.J.; acquisition and analysis of data, S.-Y.P., W.J., J.H.L. and D.J.; discussions and data interpretation, S.-Y.P. and W.J.; and drafting of the article, S.-Y.P. and D.J. All authors have read and agreed to the published version of the manuscript.

Funding: This work was supported by the Yeungnam University Research Grant.

Institutional Review Board Statement: The institutional review board (IRB) of the principal hospital responsible for the study reviewed and approved the study (IRB registration number: 2021-12-016).

Informed Consent Statement: The requirement for informed consent was waived owing to the retrospective nature of this study.

Data Availability Statement: The data used in this study are not publicly accessible due to image management policy issues at each hospital but are available from the corresponding author upon reasonable request.

Acknowledgments: SECCI and its members have been very supportive and instrumental in the conduct of this study.

Conflicts of Interest: The authors declare no conflict of interest.

References

1. Harvin, J.A.; Maxim, T.; Inaba, K.; Martinez-Aguilar, M.A.; King, D.R.; Choudhry, A.J.; Zielinski, M.D.; Akinyeye, S.; Todd, S.R.; Griffin, R.L. Mortality following emergent trauma laparotomy: A multicenter, retrospective study: Mortality after emergent trauma laparotomy. *J. Trauma Acute Care Surg.* **2017**, *83*, 464. [CrossRef]
2. Eastridge, B.J.; Holcomb, J.B.; Shackelford, S. Outcomes of traumatic hemorrhagic shock and the epidemiology of preventable death from injury. *Transfusion* **2019**, *59*, 1423–1428. [CrossRef]
3. Holcomb, J.B.; Pati, S. Optimal trauma resuscitation with plasma as the primary resuscitative fluid: The surgeon's perspective. *Hematol. Am. Soc. Hematol. Educ. Program* **2013**, *2013*, 656–659. [CrossRef]
4. Murao, S.; Yamakawa, K.; Kabata, D.; Kinoshita, T.; Umemura, Y.; Shintani, A.; Fujimi, S. Effect of earlier door-to-CT and door-to-bleeding control in severe blunt trauma: A retrospective cohort study. *J. Clin. Med.* **2021**, *10*, 1522. [CrossRef]
5. Lateef, A.U.; Khan, A.A.; Rana, M.M. Comparison of efficacy of FAST and CT scan in patients with blunt abdominal trauma. *Ann. Punjab Med. Coll. (APMC)* **2019**, *13*, 10–13.
6. Kumar, S.; Bansal, V.K.; Muduly, D.K.; Sharma, P.; Misra, M.C.; Chumber, S.; Singh, S.; Bhardwaj, D. Accuracy of focused assessment with sonography for trauma (fast) in blunt trauma abdomen—A prospective study. *Indian J. Surg.* **2015**, *77*, 393–397. [CrossRef] [PubMed]
7. Ingeman, J.E.; Plewa, M.C.; Okasinski, R.E.; King, R.W.; Knotts, F.B. Emergency physician use of ultrasonography in blunt abdominal trauma. *Acad. Emerg. Med.* **1996**, *3*, 931–937. [CrossRef] [PubMed]
8. Basnet, S.; Shrestha, S.K.; Pradhan, A.; Shrestha, R.; Shrestha, A.P.; Sharma, G.; Bade, S.; Giri, L. Diagnostic performance of the extended focused assessment with sonography for trauma (EFAST) patients in a tertiary care hospital of Nepal. *Trauma Surg. Acute Care Open* **2020**, *5*, e000438. [CrossRef] [PubMed]
9. Shanmuganathan, K.; Mirvis, S.E.; Sherbourne, C.D.; Chiu, W.C.; Rodriguez, A. Hemoperitoneum as the sole indicator of abdominal visceral injuries: A potential limitation of screening abdominal US for trauma. *Radiology* **1999**, *212*, 423–430. [CrossRef]
10. Emery, K.H.; McAneney, C.M.; Racadio, J.M.; Johnson, N.D.; Evora, D.K.; Garcia, V.F. Absent peritoneal fluid on screening trauma ultrasonography in children: A prospective comparison with computed tomography. *J. Pediatr. Surg.* **2001**, *36*, 565–569. [CrossRef]
11. Körner, M.; Krötz, M.M.; Degenhart, C.; Pfeifer, K.J.; Reiser, M.F.; Linsenmaier, U. Current role of emergency US in patients with major trauma. *RadioGraphics* **2008**, *28*, 225–242. [CrossRef]
12. Boulanger, B.R.; McLellan, B.A.; Brenneman, F.D.; Wherrett, L.; Rizoli, S.B.; Culhane, J.; Hamilton, P. Emergent abdominal sonography as a screening test in a new diagnostic algorithm for blunt trauma. *J. Trauma Acute Care Surg.* **1996**, *40*, 867–874. [CrossRef]
13. Netherton, S.; Milenkovic, V.; Taylor, M.; Davis, P.J. Diagnostic accuracy of eFAST in the trauma patient: A systematic review and meta-analysis. *Can. J. Emerg. Med.* **2019**, *21*, 727–738. [CrossRef]
14. Currie, G.; Hawk, K.E.; Rohren, E.; Vial, A.; Klein, R. Machine learning and deep learning in medical imaging: Intelligent imaging. *J. Med. Imaging Radiat. Sci.* **2019**, *50*, 477–487. [CrossRef]
15. Zhou, S.K.; Greenspan, H.; Davatzikos, C.; Duncan, J.S.; Van Ginneken, B.; Madabhushi, A.; Prince, J.L.; Rueckert, D.; Summers, R.M. A review of deep learning in medical imaging: Imaging traits, technology trends, case studies with progress highlights, and future promises. *Proc. IEEE* **2021**, *109*, 820–838. [CrossRef]
16. Stewart, J.; Sprivulis, P.; Dwivedi, G. Artificial intelligence and machine learning in emergency medicine. *Emerg. Med. Australas.* **2018**, *30*, 870–874. [CrossRef]

17. Tang, K.J.W.; Ang, C.K.E.; Constantinides, T.; Rajinikanth, V.; Acharya, U.R.; Cheong, K.H. Artificial intelligence and machine learning in emergency medicine. *Biocybern. Biomed. Eng.* **2021**, *41*, 156–172. [CrossRef]
18. McBee, M.P.; Awan, O.A.; Colucci, A.T.; Ghobadi, C.W.; Kadom, N.; Kansagra, A.P.; Tridandapani, S.; Auffermann, W.F. Deep learning in radiology. *Acad. Radiol.* **2018**, *25*, 1472–1480. [CrossRef]
19. Jang, S.B.; Lee, S.H.; Lee, D.E.; Park, S.-Y.; Kim, J.K.; Cho, J.W.; Cho, J.; Kim, K.B.; Park, B.; Park, J. Deep-learning algorithms for the interpretation of chest radiographs to aid in the triage of COVID-19 patients: A multicenter retrospective study. *PLoS ONE* **2020**, *15*, e0242759. [CrossRef]
20. Meng, X.H.; Wu, D.J.; Wang, Z.; Ma, X.L.; Dong, X.M.; Liu, A.E.; Chen, L. A fully automated rib fracture detection system on chest CT images and its impact on radiologist performance. *Skelet. Radiol.* **2021**, *50*, 1821–1828. [CrossRef]
21. Alom, M.Z.; Taha, T.M.; Yakopcic, C.; Westberg, S.; Sidike, P.; Nasrin, M.S.; Hasan, M.; Van Essen, B.C.; Awwal, A.A.S.; Asari, V.K. A state-of-the-art survey on deep learning theory and architectures. *Electronics* **2019**, *8*, 292. [CrossRef]
22. Misra, R.; O'Byrne, C.; Keane, P. Code-Free Deep Learning: A Step into the Future of Ophthalmology. 2022. Available online: https://www.eyenews.uk.com/features/ophthalmology/post/code-free-deep-learning-a-step-into-the-future-of-ophthalmology (accessed on 1 April 2023).
23. Soares, T.R.; de Oliveira, R.D.; Liu, Y.E.; da Silva Santos, A.; dos Santos, P.C.P.; Monte, L.R.S.; de Oliveira, L.M.; Park, C.M.; Hwang, E.J.; Andrews, J.R.; et al. Evaluation of chest X-ray with automated interpretation algorithms for mass tuberculosis screening in prisons: A cross-sectional study. *Lancet Reg. Health Am.* **2023**, *17*, 100388. [CrossRef] [PubMed]
24. Yoo, H.; Lee, S.H.; Arru, C.D.; Doda Khera, R.; Singh, R.; Siebert, S.; Kim, D.; Lee, Y.; Park, J.H.; Eom, H.J.; et al. AI-based improvement in lung cancer detection on chest radiographs: Results of a multi-reader study in NLST dataset. *Eur. Radiol.* **2021**, *31*, 9664–9674. [CrossRef] [PubMed]
25. van Beek, E.; Ahn, J.; Kim, M.; Murchison, J. Validation study of machine-learning chest radiograph software in primary and emergency medicine. *Clin. Radiol.* **2023**, *78*, 1–7. [CrossRef] [PubMed]
26. Singh, R.; Kalra, M.K.; Nitiwarangkul, C.; Patti, J.A.; Homayounieh, F.; Padole, A.; Rao, P.; Putha, P.; Muse, V.V.; Sharma, A.; et al. Deep learning in chest radiography: Detection of findings and presence of change. *PLoS ONE* **2018**, *13*, e0204155. [CrossRef]
27. Wong, D.J.; Gandomkar, Z.; Wu, W.J.; Zhang, G.; Gao, W.; He, X.; Wang, Y.; Reed, W. Artificial intelligence and convolution neural networks assessing mammographic images: A narrative literature review. *J. Med. Radiat. Sci.* **2020**, *67*, 134–142. [CrossRef]
28. Jin, K.N.; Kim, E.Y.; Kim, Y.J.; Lee, G.P.; Kim, H.; Oh, S.; Kim, Y.S.; Han, J.H.; Cho, Y.J. Diagnostic effect of artificial intelligence solution for referable thoracic abnormalities on chest radiography: A multicenter respiratory outpatient diagnostic cohort study. *Eur. Radiol.* **2022**, *32*, 3469–3479. [CrossRef]
29. Raimondo, D.; Raffone, A.; Aru, A.C.; Giorgi, M.; Giaquinto, I.; Spagnolo, E.; Travaglino, A.; Galatolo, F.A.; Cimino, M.G.C.A.; Lenzi, J. Application of deep learning model in the sonographic diagnosis of uterine adenomyosis. *Int. J. Environ. Health Res.* **2023**, *20*, 1724. [CrossRef]
30. Li, J.; Bu, Y.; Lu, S.; Pang, H.; Luo, C.; Liu, Y.; Qian, L. Development of a deep learning–based model for diagnosing breast nodules with ultrasound. *J. Ultrasound Med.* **2021**, *40*, 513–520. [CrossRef]
31. Sjogren, A.R.; Leo, M.M.; Feldman, J.; Gwin, J.T. Image segmentation and machine learning for detection of abdominal free fluid in focused assessment with sonography for trauma examinations: A pilot study. *J. Ultrasound Med.* **2016**, *35*, 2501–2509. [CrossRef]
32. Kornblith, A.E.; Addo, N.; Dong, R.; Rogers, R.; Grupp-Phelan, J.; Butte, A.; Gupta, P.; Callcut, R.A.; Arnaout, R. Development and validation of a deep learning model for automated view classification of pediatric focused assessment with sonography for trauma (FAST). *MedRxiv* **2020**. [CrossRef]
33. Cheng, C.Y.; Chiu, I.M.; Hsu, M.Y.; Pan, H.Y.; Tsai, C.M.; Lin, C.-H.R. Deep learning assisted detection of abdominal free fluid in Morison's pouch during focused assessment with sonography in trauma. *Front. Med.* **2021**, *8*, 707437. [CrossRef]
34. Shokoohi, H.; LeSaux, M.A.; Roohani, Y.H.; Liteplo, A.; Huang, C.; Blaivas, M. Enhanced point-of-care ultrasound applications by integrating automated feature-learning systems using deep learning. *J. Ultrasound Med.* **2019**, *38*, 1887–1897. [CrossRef]
35. Rozycki, G.S.; Ochsner, M.G.; Feliciano, D.V.; Thomas, B.; Boulanger, B.R.; Davis, F.E.; Falcone, R.E.; Schmidt, J.A. Early detection of hemoperitoneum by USG examination of the right upper quadrant. A multicenter study. *J. Trauma.* **1998**, *45*, 878–883. [CrossRef]
36. Ma, O.J.; Kefer, M.P.; Mateer, J.R.; Thoma, B. Evaluation of hemoperitoneum using a single-vs multiple-view ultrasonographic examination. *Acad. Emerg. Med.* **1995**, *2*, 581–586. [CrossRef]

tomography

MDPI

Article

A Deep Learning Approach for Rapid and Generalizable Denoising of Photon-Counting Micro-CT Images

Rohan Nadkarni, Darin P. Clark, Alex J. Allphin and Cristian T. Badea *

Quantitative Imaging and Analysis Lab, Department of Radiology, Duke University Medical Center, Durham, NC 27710, USA; rohan.nadkarni@duke.edu (R.N.); darin.clark@duke.edu (D.P.C.); alex.allphin@duke.edu (A.J.A.)

* Correspondence: cristian.badea@duke.edu

Abstract: Photon-counting CT (PCCT) is powerful for spectral imaging and material decomposition but produces noisy weighted filtered backprojection (wFBP) reconstructions. Although iterative reconstruction effectively denoises these images, it requires extensive computation time. To overcome this limitation, we propose a deep learning (DL) model, UnetU, which quickly estimates iterative reconstruction from wFBP. Utilizing a 2D U-net convolutional neural network (CNN) with a custom loss function and transformation of wFBP, UnetU promotes accurate material decomposition across various photon-counting detector (PCD) energy threshold settings. UnetU outperformed multi-energy non-local means (ME NLM) and a conventional denoising CNN called UnetwFBP in terms of root mean square error (RMSE) in test set reconstructions and their respective matrix inversion material decompositions. Qualitative results in reconstruction and material decomposition domains revealed that UnetU is the best approximation of iterative reconstruction. In reconstructions with varying undersampling factors from a high dose ex vivo scan, UnetU consistently gave higher structural similarity (SSIM) and peak signal-to-noise ratio (PSNR) to the fully sampled iterative reconstruction than ME NLM and UnetwFBP. This research demonstrates UnetU's potential as a fast (i.e., 15 times faster than iterative reconstruction) and generalizable approach for PCCT denoising, holding promise for advancing preclinical PCCT research.

Keywords: denoising; deep learning; preclinical; micro-CT; photon-counting CT; contrast agents

Citation: Nadkarni, R.; Clark, D.P.; Allphin, A.J.; Badea, C.T. A Deep Learning Approach for Rapid and Generalizable Denoising of Photon-Counting Micro-CT Images. *Tomography* **2023**, *9*, 1286–1302. https://doi.org/10.3390/tomography9040102

Academic Editors: Rafał Obuchowicz, Michał Strzelecki and Adam Piorkowski

Received: 9 June 2023
Revised: 27 June 2023
Accepted: 30 June 2023
Published: 2 July 2023

1. Introduction

Photon-counting detectors (PCDs) offer the potential to improve performance of CT imaging at both clinical and preclinical levels and to extend the use of CT to novel applications [1]. PCDs acquire images at two or more energies simultaneously by detecting individual x-ray photons and sorting them into bins based on energy thresholds, making photon-counting CT (PCCT) a promising tool for spectral imaging tasks that require material decomposition. Moreover, PCD energy thresholds can be adjusted near the K-edges of various contrast agents, enabling their spectral separation in novel imaging procedures [2,3].

However, PCCT dose constraints and photon binning result in high noise levels, particularly in high-energy bins, reducing image quality and the accuracy of material decomposition. Multi-energy CT can be denoised using various post-reconstruction algorithms such as non-local means (NLM) filtering [4,5]. Alternatively, lower noise can be achieved via iterative reconstruction [6–8] or deep learning (DL)-based methods [9]. In our own work, iterative reconstruction algorithms such as those using rank-sparse kernel regression (RSKR) [10] are effective at producing low-bias denoised PCCT reconstructions that result in highly accurate material decomposition. However, such iterative approaches are computationally expensive.

To address this limitation, DL models can be used to approximate iterative reconstruction, providing robust denoising with reduced computation time. Commonly, DL methods use convolutional neural networks (CNNs) to learn the mapping between noisy and clean images directly from data, eliminating the need for explicit knowledge of the underlying statistical model. Previous studies have demonstrated that DL-based denoising can achieve comparable performance to iterative reconstruction, with significantly reduced computational time [11]. Various DL architectures have been used for denoising CT images, including U-nets [12], residual encoder–decoder (RED) CNN [13], generative adversarial networks (GANs) [14], and vision transformers [15]. U-nets are a popular choice for CT image denoising [16–20] because they offer several advantages. These include fast computation at test time due to a relatively small number of parameters and simple architecture, an encoder–decoder structure that allows for efficient feature extraction and information propagation, and skip connections between encoder and decoder that allow the network to retain and use both high- and low-level features during the denoising process [21]. Nevertheless, challenges remain in DL-based denoising, such as the risk of overfitting to training data, leading to limited generalization performance. Generalizability is crucial when denoising PCCT images, given that PCD energy thresholds change based on the contrast agents used, affecting the noise level in each energy channel. Furthermore, there is a need for robust evaluations that can accurately measure the performance of DL-based denoising algorithms.

In this study, we introduce a supervised learning approach to develop a denoising CNN that adapts to various PCD threshold settings. Our goal is to perform fast and reliable denoising of weighted filtered backprojection (wFBP) reconstructions at the level of iterative reconstruction in scans acquired on our in-house developed preclinical PCCT imaging system [22] with four energy thresholds. We trained UnetU, which is a U-net CNN with a unique preprocessing method and loss function intended to promote generalizability and accurate material decomposition. Then, we compared UnetU to a traditional U-net denoising approach called UnetwFBP as well as a non-data-driven filtering method called multi-energy non-local means (ME NLM). Instead of proposing a novel CNN architecture, our study contributes to a growing body of DL methods for denoising micro-CT data by introducing a unique input preprocessing approach and loss function that enable a well-known CNN such as the U-net to produce accurate reconstructions and material separation at various PCD energy threshold settings.

2. Materials and Methods

2.1. Image Acquisition

For this study, we acquired all PCD image data using a dual-source micro-CT imaging system consisting of an energy-integrating detector (EID) and a PCD (Dectris Santis 1604, 150 μm pixel size, 4 energy thresholds) [22]. We used a G-297 x-ray tube (Varian Medical Systems, Palo Alto, CA, USA) with a 0.3 mm focal spot size and a 0.1 mm Cu filter and utilized an Epsilon high-frequency x-ray generator (EMD Technologies, Saint-Eustache, QC, Canada). We used only PCD-based acquisitions for the scans in this study.

The PCCT scans used in this work are summarized in Table 1. Our CNN training and validation sets consisted of data from 10 different helical scans of mice (9 ex vivo, 1 in vivo) that provided 4 distinct PCD threshold settings and 5 different contrast agents: Iodine (I), Gadolinium (Gd), Tantalum (Ta), Bismuth (Bi), and Gold (Au). The testing set included 5 different scans: 2 scans of phantoms containing vials of known material concentrations, 1 ex vivo mouse, and 1 in vivo mouse to provide sets at each of the threshold settings included in training, and an additional in vivo scan of a mouse with thresholds 28, 34, 39, 45 keV, and contrast agents I and Barium (Ba) to determine how our CNNs generalized to threshold settings not included in training. PCD energy thresholds were chosen to be slightly above the K-edges of contrast agents that were used as basis materials in the decomposition of the PCCT reconstruction. Each scan was acquired using a helical trajectory with 1070 degrees rotation. The number of projection angles for the helical scans varied between 900 and 1500, and vertical translation varied between 1.25 and 2.5 cm.

Table 1. Summary of scan protocols for training and testing sets. Samples scanned for this study included in vivo mice, ex vivo mice, and phantoms containing vials of known material concentration.

PCD Thresholds (keV)	Training Sets	Validation Sets	Test Sets	Contrast Materials	kVp	mA	Dose (mGy)	Types of Samples Scanned
25, 34, 50, 60	3	0	1	I, Gd	80	4	36	in vivo, ex vivo
25, 34, 50, 80	2	1	1	I, Au	125	2.5	38	ex vivo
25, 28, 34, 40	2	0	1	I	80	4	36	ex vivo, phantom
50, 65, 80, 90	2	0	1	Gd, Ta, Bi	125	2.5	38	ex vivo, phantom
28, 34, 39, 45	0	0	1	I, Ba	80	4	36	in vivo

2.2. Image Reconstruction

All PCD micro-CT scan data underwent reconstruction using the weighted filtered backprojection (wFBP) algorithm [23] with an isotropic voxel size of 125 microns. Additionally, the PCCT data were jointly reconstructed using a multi-channel iterative algorithm to create training labels and for comparison with ME NLM, UnetwFBP, and UnetU denoising in the test set. Specifically, we applied the split Bregman method with the add-residual-back strategy [24] and rank-sparse kernel regression (RSKR) regularization [10]. Reconstruction iteratively solved the following equation:

$$X = \underset{X}{arg\ min}\frac{1}{2}\sum_{e}||\ RX(e) - Y(e)\ ||_2^2 + \lambda||\ X\ ||_{BTV}. \tag{1}$$

The reconstructed data (X) at each energy (e) minimizes the reprojection error (R representing the system projection matrix) relative to log-transformed projection data (Y). To reduce noise in the reconstructed results, this data fidelity term is minimized subject to the bilateral total variation (BTV). BTV measured within and between energies is reduced via RSKR.

2.3. Material Decomposition

The method of material decomposition was applied to the wFBP and iterative RSKR reconstructions, as well as ME NLM, UnetwFBP, and UnetU denoised reconstructions. We performed image-based material decomposition using the method of Alvarez and Macovski [25] extended to include K-edge contrast materials. Thus, we performed a post-reconstruction spectral decomposition with H_2O, Calcium (Ca), and N K-edge materials as basis functions, M_i:

$$X(e) = C_{H2O}M_{H2O}(e) + C_{Ca}M_{Ca}(e) + \sum_{i=1}^{N} C_iM_i(e). \tag{2}$$

In this formulation, M is a matrix of material sensitivities at each energy. We computed these values by fitting the slope of attenuation measurements taken in vials of known material concentrations. C represents density in g/mL for water and concentration in mg/mL for all other materials. Finally, X is the attenuation coefficient of the voxel under consideration at energy e. Material decomposition was performed by matrix inversion, solving the following linear system at each voxel:

$$C = XM^{-1} \tag{3}$$

Orthogonal subspace projection was used to prevent negative concentrations [10]. Post decomposition, the material maps were assigned to colors and combined and visualized in ImageJ.

2.4. Network Training and Testing

In Section 2.1, we allocated ten sets of reconstructions for training and validation, and reserved five sets for testing. We implemented two 2D U-net [26] CNNs (with four levels and two convolutions per level) using the PyTorch library [27] and compared their performance. Figure 1 illustrates the implementation of these U-nets.

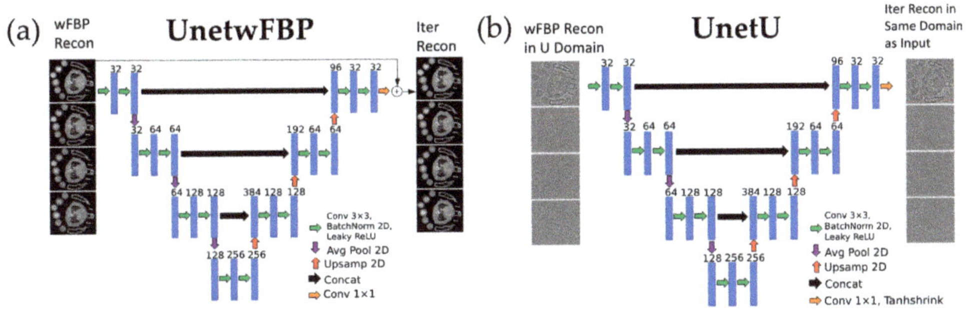

Figure 1. Diagram of CNN architecture and training procedure for each denoising U-net. (**a**) UnetwFBP, a denoising U-net approach with conventional choice of network input and labels. (**b**) UnetU, a denoising U-net approach augmented by our novel input preprocessing method.

Our first U-net, named UnetwFBP, represents a conventional U-net denoising approach that serves as a baseline for comparison. UnetwFBP uses wFBP reconstructions as input and iterative RSKR reconstructions as its training labels. Given the non-zero-centered nature of the input data, we included a residual connection between the input and output in UnetwFBP to address intensity bias. We trained UnetwFBP using a mean squared error (MSE) loss function.

The second U-net, named UnetU, is a novel approach that improves the ability of a standard U-net CNN to perform generalizable PCCT denoising through our unique preprocessing of the wFBP input. First, we divide each energy channel of the wFBP image by its noise variance measured as the Median Absolute Deviation (MAD) of image intensity gradients [28] in the central axial slice. Then, we compute the singular value decomposition (SVD) along the energy dimension. The noise variance weighted SVD helps to compensate for the noise level of each channel to promote generalizability, while creating a representation with less redundancy between channels than the wFBP reconstruction. Next, each left singular vector image from the SVD is high-pass-filtered by subtracting a copy of itself that has been blurred with a 3D Gaussian filter of kernel size $37 \times 37 \times 37$ and full width at half maximum (FWHM) of 10 voxels. This filtering helps accelerate training convergence by making the network input zero-centered to match the non-linear activation functions and initialization of batch normalization in the U-net. Furthermore, it promotes generalization because high-frequency information in PCCT images is more consistent across different energy threshold settings than low-frequency information. Finally, each singular vector is divided by its absolute 99.95 percentile intensity value to create the images (U) that are used as the input to UnetU. This division brings the extremely small variance of each high-pass-filtered singular vector much closer to the variance of 1 that is assumed during initialization of U-net parameters. We used the following output activation function for UnetU:

$$Tanhshrink(x) = x - \tanh(x) \tag{4}$$

to help reduce the bias of network predictions for large negative and positive values in the U images. Although our transformation is applied globally for each 4D image volume, our loss function exploited the fact that it can be inverted for individual axial slices at all 4 energies. We trained UnetU using the following four-term loss function, with labels derived from iterative RSKR reconstruction:

$$Loss = \frac{1}{N} \| X - X_{label} \|_F^2 + \lambda_1 \frac{1}{N} \| U - U_{label} \|_F^2 + \lambda_2 \frac{1}{N} \| XM^{-1} - X_{label} M^{-1} \|_F^2 + \frac{1}{N} \| X_{blur} - X_{blur,\,label} \|_F^2 \tag{5}$$

where N is the number of voxels, X and X_{label} are the prediction and label in the reconstruction domain, U and U_{label} are the prediction and label in the normalized, high-pass-filtered left singular vector domain of the wFBP, M is the sensitivity matrix used for material decomposition, and X_{blur} and $X_{blur,label}$ are the prediction and label in the reconstruction domain after blurring axial slices with a 2D Gaussian filter of kernel size 19 × 19 and FWHM 5. Although the first term in the loss function already penalizes error in the reconstruction domain, the fourth term further increases penalization of bias in attenuation values. To ensure that the material decomposition loss term penalizes each material map at a similar intensity scale, the sensitivity matrices (M) used in the loss function were rescaled via multiplication of each row by the corresponding calibration vial concentration. To avoid problems such as excessively long training time per epoch, discrepancies between the magnitudes of cost function terms, and vanishing gradients associated with thresholding, the decompositions in the loss function did not incorporate a non-negativity constraint. We used $\lambda_1 = 15$ and $\lambda_2 = 0.2$ to give the first 3 loss terms similar magnitude. In addition to training UnetU with the custom loss function in Equation (5), we also trained UnetU with only the first term of Equation (5) (MSE loss) to allow us to assess how the custom loss function affects test set performance.

The UnetwFBP and UnetU models both take 2D axial image slices of size 400 × 400 voxels with 4 channels as the input. Both networks have 4 input channels and 4 output channels to match the number of PCD energy thresholds. We held out one set of matching wFBP and iterative reconstruction volumes as a validation set, with the remaining nine sets used for training. Then, we randomly shuffled the order of the slices in both the training and validation sets. Data augmentation for both UnetwFBP and UnetU consisted of randomly flipping input image slices horizontally and vertically with 50% probability for each flip. For UnetU, we randomly multiplied both the left and right singular vectors by –1 with 50% probability as an additional form of data augmentation. All U-nets were trained to run for a maximum of 2000 epochs with a batch size of 8, Adam optimizer [29], an initial learning rate of 2×10^{-4}, a learning rate scheduler, and early stopping. The learning rate scheduler cuts the learning rate in half every time the validation loss fails to decrease for 10 consecutive epochs, and early stopping terminates training when the validation loss fails to decrease for 20 consecutive epochs. UnetwFBP, UnetU with MSE loss, and UnetU with custom loss were trained on a computer running Linux with four NVIDIA RTX 6000 Ada GPUs. After training was completed, all three models were used to predict estimates of iterative RSKR reconstruction for the test sets described in Section 2.1.

2.5. Performance Evaluation

We conducted a thorough evaluation of the performance of the UnetwFBP and UnetU approaches, considering computation time, quantitative metrics, qualitative results, and performance at varying radiation doses. For all analyses, UnetwFBP and UnetU were compared both to each other and to the ME NLM algorithm. As suggested in the original paper [5], our ME NLM implementation uses a search window $W_\Omega = 11^3$ voxels, patch size $W_p = 3^3$ voxels, filtering strength h = 1.2, and noise standard deviation values at each energy derived from homogeneous region of interest (ROI) measurements (from a water vial). Each time we call the ME NLM function, we pass in 31 adjacent axial slices of the wFBP image and take the filtered version of the 16th axial slice as the output.

To assess computational burden, we measured the time required for iterative RSKR reconstruction, ME NLM denoising, UnetwFBP prediction, and UnetU prediction on one of the test sets.

Our quantitative metrics included average bias in Hounsfield Units (HU) relative to wFBP for each denoising method (iterative RSKR, ME NLM, UnetwFBP, UnetU with MSE loss, UnetU with custom loss) and average noise level in HU for wFBP and each denoising method. For each of the 5 test sets, we chose an axial slice with a water vial, defined an ROI within the vial, and measured the mean and standard deviation in this ROI for wFBP and each denoising method at all four energy thresholds. To determine average bias for each denoising method at each energy threshold, we computed the absolute difference between mean value in the water vial ROI from denoised reconstruction and from wFBP in each test set and took the mean of the 5 absolute difference values. The average noise level for wFBP and for each denoising method at each energy threshold was calculated by taking the mean of the 5 standard deviation values from water vial ROIs in the test sets. In addition to computing average bias and noise metrics in the reconstructions, we compared the average root mean square error (RMSE) in HU relative to iterative RSKR reconstruction across the 5 test sets for wFBP reconstruction, ME NLM, UnetwFBP, UnetU with MSE loss and UnetU with custom loss.

For each test set, we used the sensitivity matrix derived from calibration vial measurements in the iterative RSKR reconstruction to convert reconstructions from wFBP and each denoising method into material decompositions via sensitivity matrix inversion with the approach described in Section 2.3. Then, we computed the average test set RMSE of wFBP, ME NLM, UnetwFBP, UnetU with MSE loss, and UnetU with custom loss in the material maps relative to the decompositions from iterative reconstruction. Since quantitative comparisons of bias, noise level, and RMSE in reconstructions and material maps are sufficient to demonstrate the effect of loss function selection on UnetU, the remaining analyses only include UnetU with the custom loss function.

For qualitative comparison, we show reconstructions and corresponding matrix inversion material decompositions from iterative RSKR, wFBP, ME NLM, UnetwFBP, and UnetU on a test set scan of a phantom containing vials of water and known concentrations of the K-edge contrast materials Gd, Ta, and Bi. This set was chosen because of the very low photon counts and resulting high noise level associated with the high energy thresholds corresponding to the K-edges of its contrast agents (e.g., K-edge at ~90 keV for Bi). In addition, we show reconstructions and the corresponding matrix inversion decompositions on a test set in vivo scan of a mouse with tumors that was injected both with I and Ba-based contrast agents to demonstrate how our U-net denoising generalizes to threshold settings not included in training. Like our RMSE analysis, these qualitative comparisons treat iterative RSKR reconstruction and its matrix inversion decomposition as the ground truth. Therefore, we also display absolute subtraction maps from the iterative reconstruction for ME NLM, UnetwFBP, and UnetU in order to make their artifacts more visible. The conclusions we draw from these qualitative analyses are based on the assessment of small animal imagers on our team (CTB, DPC) who have more than a decade of experience.

To evaluate each denoising method at different dose levels, we acquired a helical scan with I and Gd K-edge threshold settings of an ex vivo mouse with 2700 projections, which is about 3 times the dose level of a typical preclinical in vivo helical scan (using 900 projections with dose 36 mGy). We performed wFBP reconstruction of the full set of 2700 projections and projection data generated by undersampling factors of 2, 3, 4, 5, and 6 to simulate a wide range of radiation dose levels. For each of the 6 wFBP reconstructions, we used iterative RSKR reconstruction, ME NLM, UnetwFBP, and UnetU to denoise the images. We show wFBP and denoised images from the highest energy threshold (60 keV) for the fully sampled, 3x undersampled, and 6x undersampled cases in an ROI from a coronal slice. We also plot structural similarity (SSIM) and peak signal-to-noise ratio (PSNR) values in this slice for each denoising method at each undersampling factor relative to the iterative RSKR reconstruction of the complete set of 2700 projections.

3. Results

3.1. Loss Curves

Figure 2 shows the training and validation loss curves for UnetwFBP, UnetU with MSE loss, and UnetU with custom loss. All three networks completed training within 300 epochs due to the early stopping implementation. Training took 8 h and 42 min for UnetwFBP, 11 h and 9 min for UnetU with MSE loss, and 16 h and 56 min for UnetU with custom loss.

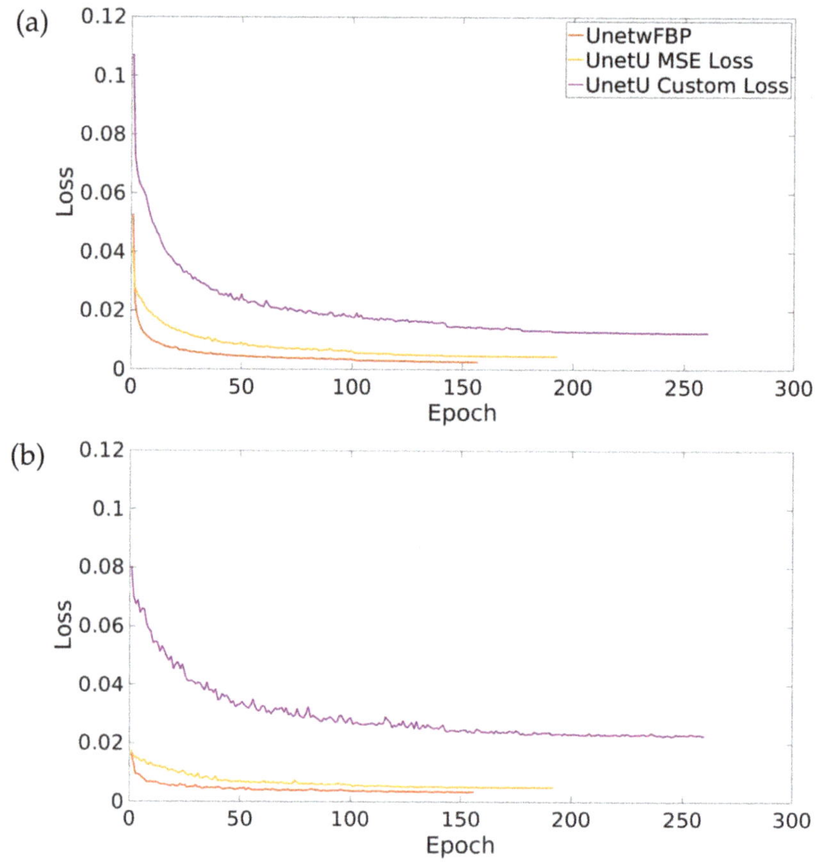

Figure 2. Training (**a**) and validation (**b**) loss curves for all three denoising U-nets.

3.2. Computation Time

On the same computer used to train the U-nets, iterative RSKR reconstruction took 915 s on a test set with dimensions of 400 × 400 × 400 voxels and 4 energy thresholds, while ME NLM took 854 s, UnetwFBP took 36 s, and UnetU took 61 s to denoise the same test set. Please note that our ME NLM implementation is not optimized for speed, and it is likely that greater reduction in computation time can be achieved with ME NLM.

3.3. Quantitative Analyses

Figure 3 shows the average bias relative to wFBP as well as the average noise level of wFBP and the denoising methods at each of the four PCD thresholds in water vial measurements. All denoising approaches except UnetwFBP have an average bias of less than 11 HU at all four energy thresholds. Iterative RSKR reconstruction, ME NLM, and all three U-nets give an average noise level of less than 35 HU at all four thresholds, while the average noise level of wFBP reconstruction is above 70 HU at all four thresholds. ME NLM has the lowest bias as well as the lowest noise level. Compared to UnetU with MSE loss, UnetU with custom loss shows lower bias for three out of four PCD thresholds and lower noise level for all four thresholds.

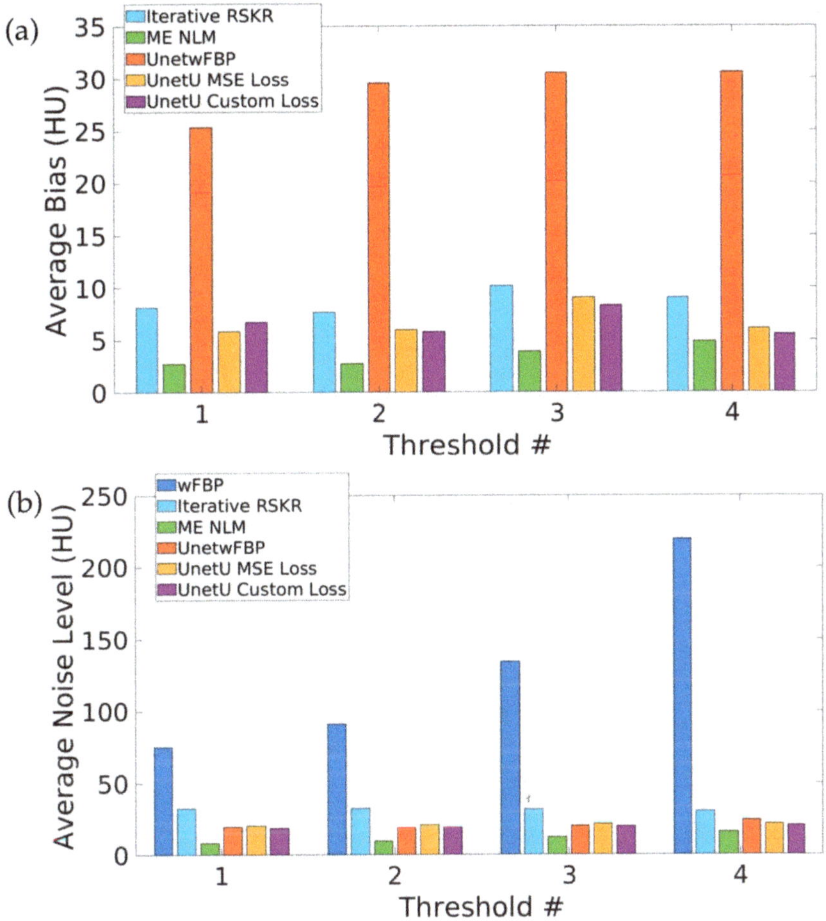

Figure 3. Average bias (**a**) and noise level (**b**) in water vial measurements from test set images.

Figure 4 shows the average test set RMSE of wFBP, ME NLM, and the denoising U-nets relative to iterative RSKR reconstruction and its matrix inversion decompositions. ME NLM and all three U-nets provide dramatically lower error than the wFBP reconstruction and its decomposition. UnetU with custom loss is most accurate on average in the reconstruction (38.631 $\pm$ 9.270 HU) and consistently provides the lowest RMSE in the material maps (0.200 $\pm$ 0.026 mg/mL for Iodine, 1.445 $\pm$ 0.436 mg/mL for Calcium, 0.042 $\pm$ 0.018 g/mL for Water, 0.399 $\pm$ 0.010 mg/mL for Gadolinium, 0.537 mg/mL for Tantalum, 0.491 mg/mL for Bismuth, 0.959 mg/mL for Gold, 0.297 mg/mL for Barium).

3.4. Vials Phantom

Figure 5 compares ME NLM, UnetwFBP, and UnetU with custom loss to iterative RSKR reconstruction in a test set scan of a vials phantom with threshold settings for the K-edges of Gd, Ta, and Bi. ME NLM (RMSE = 80.271 HU) denoises with low bias but causes blurring at edges, as shown in Figure 5d. Both U-net predictions show less edge blurring. However, UnetU with custom loss (RMSE = 50.605 HU) introduces much less bias in the vials than UnetwFBP (RMSE = 91.178 HU), as shown in Figure 5f,h.

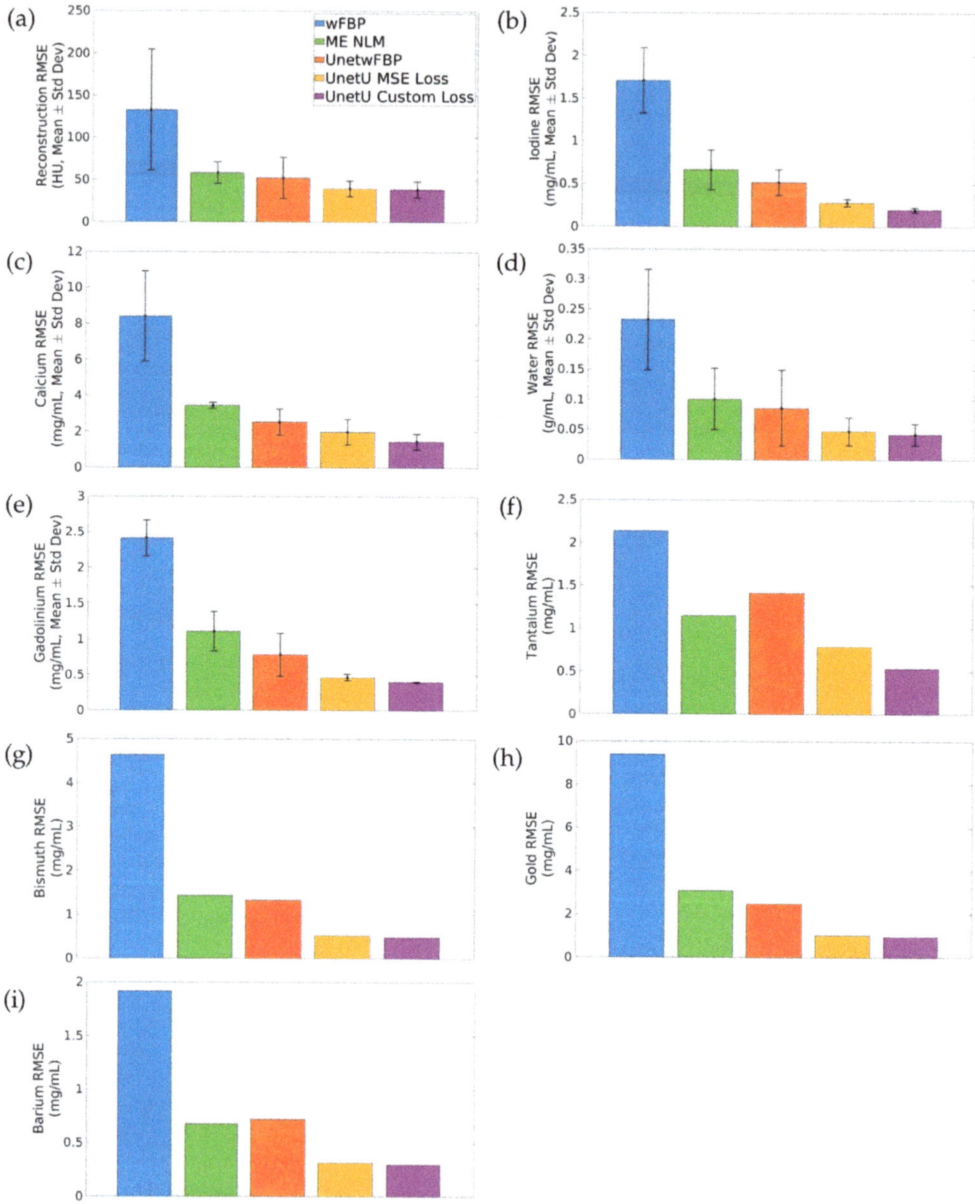

Figure 4. Average root mean square error (RMSE) of wFBP and each denoising method relative to iterative RSKR reconstruction and its matrix inversion decompositions. (**a**) Reconstruction; (**b**) Iodine; (**c**) Calcium; (**d**) Water; (**e**) Gadolinium; (**f**) Tantalum; (**g**) Bismuth; (**h**) Gold; (**i**) Barium. Error bars indicate standard deviation. Error bars are not included in the plots for Tantalum, Bismuth, Gold, or Barium because these materials are only present in the decomposition of one test set.

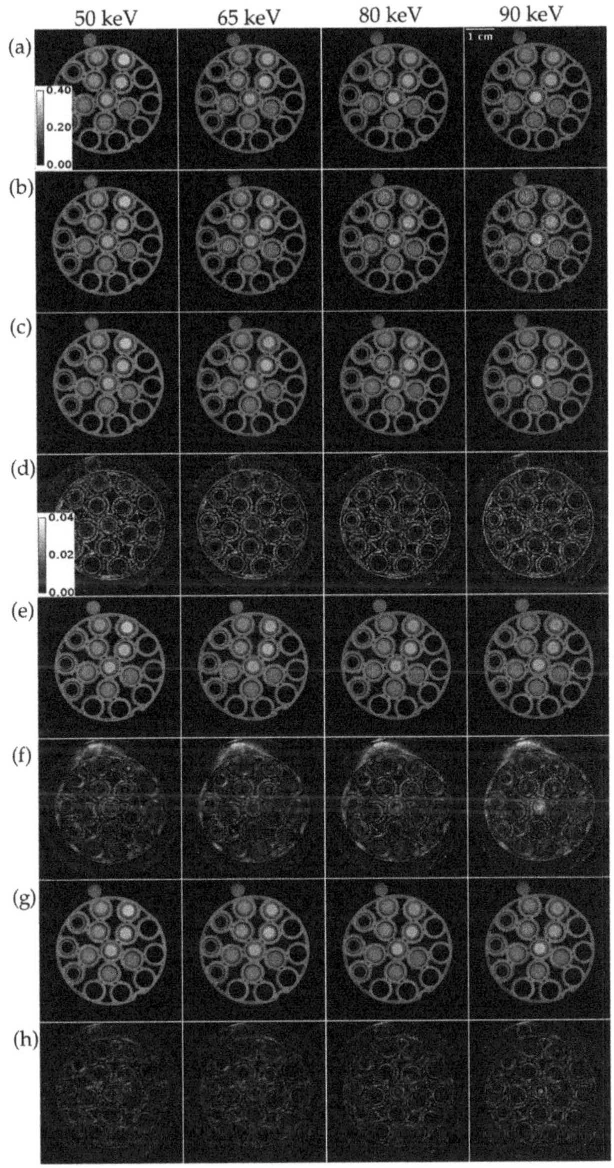

Figure 5. Reconstructions of a vials phantom scan with threshold settings at the Gd, Ta, and Bi K-edges. (**a**) Iterative RSKR reconstruction; (**b**) wFBP reconstruction; (**c**) ME NLM denoised reconstruction; (**d**) Absolute difference between (**c**,**a**); (**e**) UnetwFBP prediction; (**f**) Absolute difference between (**e**,**a**); (**g**) UnetU with custom loss prediction; (**h**) Absolute difference between (**g**,**a**). Calibration bars show the display window range of reconstructions and absolute difference maps in units of cm^{-1}.

Figure 6 shows material decompositions of the reconstructions from Figure 5. The basis materials are the three contrast agents in the phantom (Gd, Ta, Bi) and water. We show both individual material maps in different colors and their composite. The decomposition from UnetwFBP prediction still has a lot of background noise, while both ME NLM and UnetU images result in low noise decompositions. However, the decomposition from ME

NLM shows more artifacts and erroneous enhancement of the background due to excessive blurring of the reconstruction.

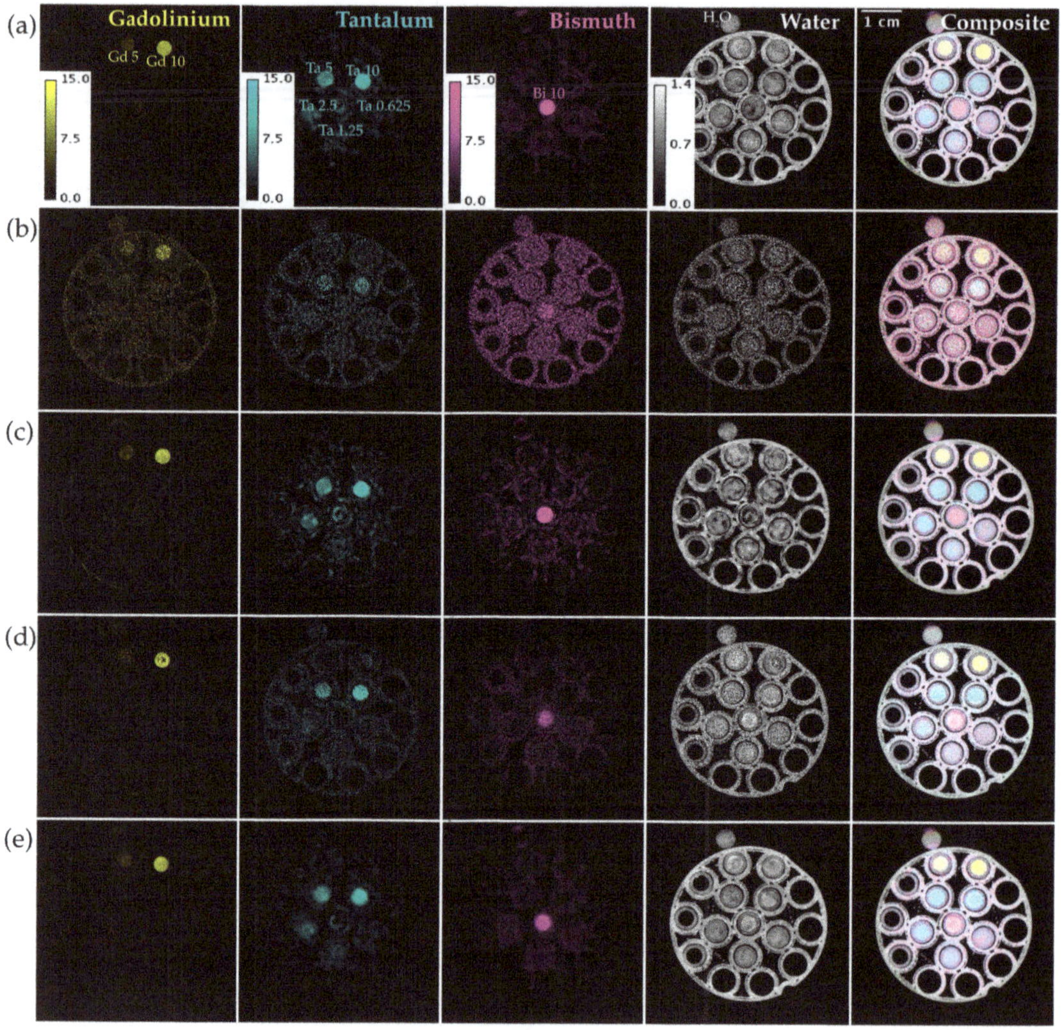

Figure 6. Material decomposition of a vials phantom scan with threshold settings at the Gd, Ta, and Bi K-edges. (**a**) From iterative RSKR reconstruction; (**b**) From wFBP reconstruction; (**c**) From ME NLM denoised reconstruction; (**d**) From UnetwFBP prediction; (**e**) From UnetU with custom loss prediction. Labels on the material maps in (**a**) indicate the locations of the water vial and vials with known concentrations (in mg/mL) of Gd, Ta, and Bi. Calibration bars show the display window range in units of g/mL for water and mg/mL for all other materials.

3.5. In Vivo Mouse Scan

Figure 7 compares ME NLM, UnetwFBP, and UnetU with custom loss to iterative RSKR reconstruction in a test set scan of an in vivo mouse with a tumor injected with I and Ba that was acquired with threshold settings not included in the training set. While ME NLM and both U-nets show some bias in high intensity regions such as the bone, UnetU with custom loss (RMSE = 41.524 HU) has lower error than both ME NLM (RMSE = 47.948 HU) and UnetwFBP (RMSE = 46.199 HU), due to superior noise reduction and preservation of edges.

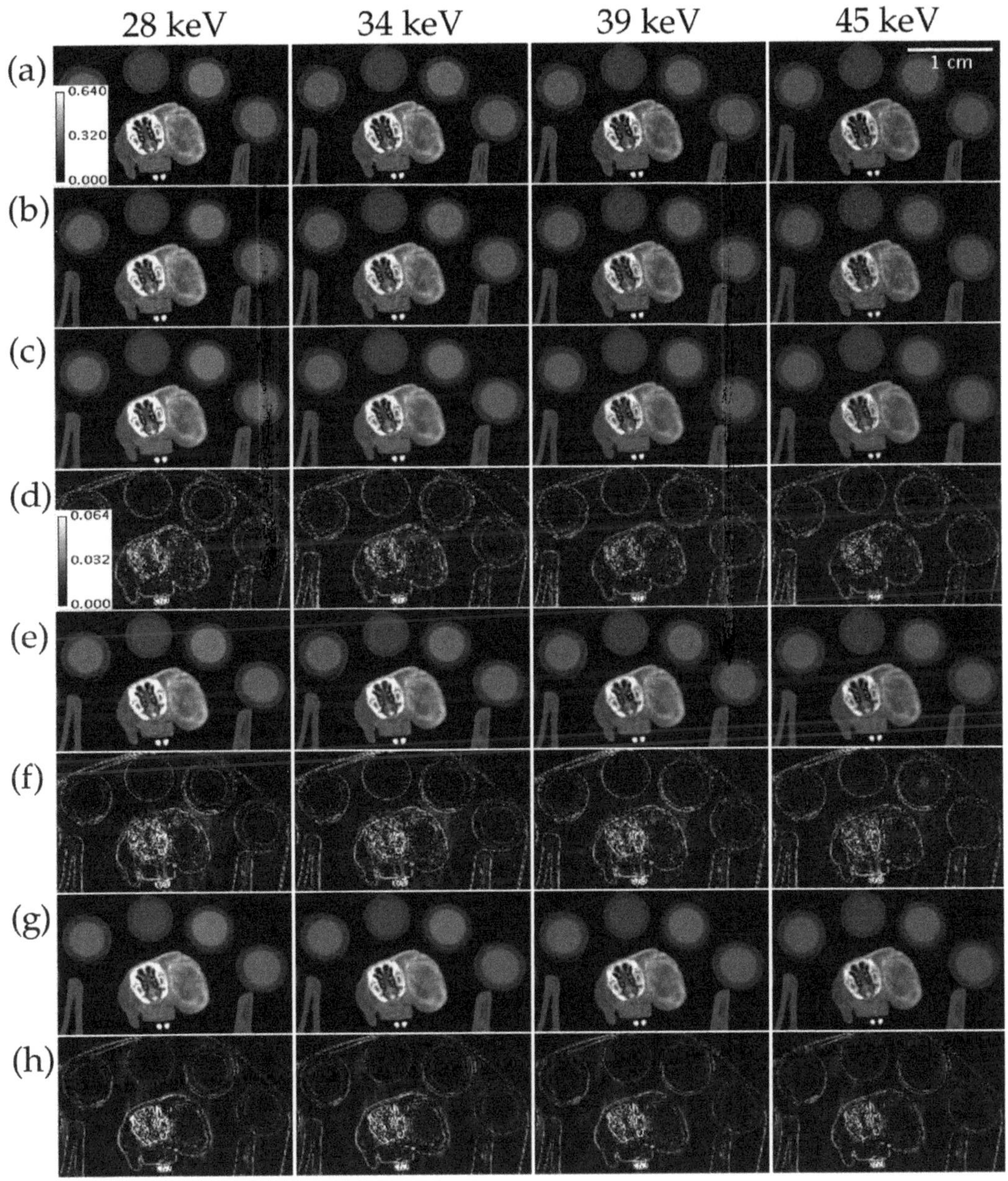

Figure 7. Reconstructions of an in vivo mouse scan with threshold settings at the I and Ba K-edges. (**a**) Iterative RSKR reconstruction; (**b**) wFBP reconstruction; (**c**) ME NLM denoised reconstruction; (**d**) Absolute difference between (**c**,**a**); (**e**) UnetwFBP prediction; (**f**) Absolute difference between (**e**,**a**); (**g**) UnetU with custom loss prediction; (**h**) Absolute difference between (**g**,**a**). Calibration bars show the display window range of reconstructions and absolute difference maps in units of cm^{-1}.

Figure 8 shows material decompositions of the reconstructions from Figure 7. The decomposition from ME NLM has some noisy regions in the body of the mouse. The decomposition from UnetwFBP shows noticeable bias in the bone, tumor, and Calcium vial in all the material maps. The decomposition from UnetU appears to show some blurring, but comes closest to reproducing the decomposition from iterative RSKR reconstruction.

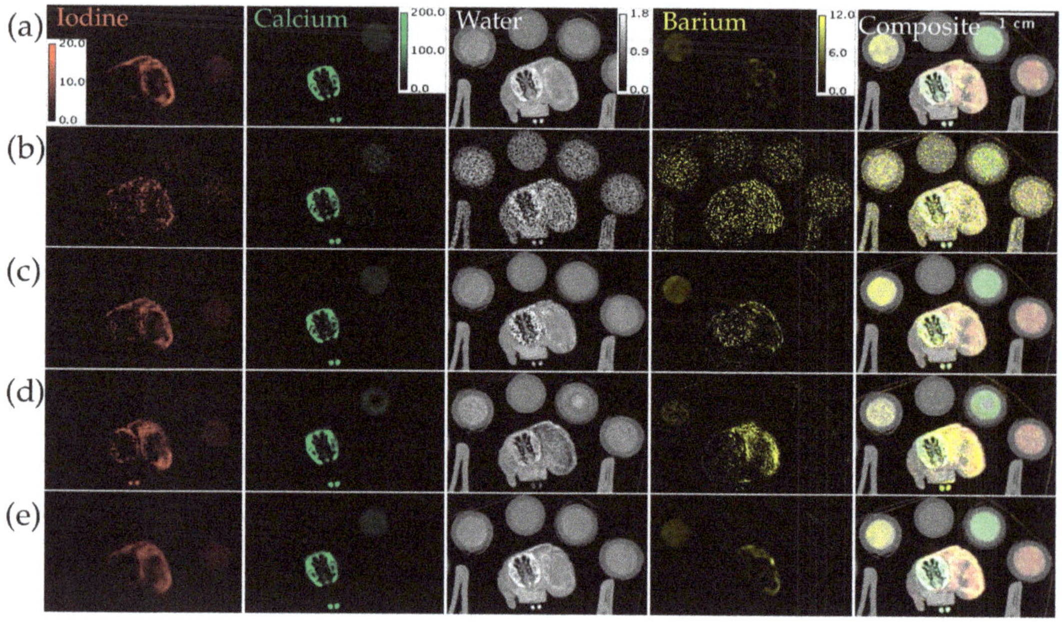

Figure 8. Material decomposition of an in vivo mouse scan with threshold settings at the I and Ba K-edges. (**a**) From iterative RSKR reconstruction; (**b**) From wFBP reconstruction; (**c**) From ME NLM denoised reconstruction; (**d**) From UnetwFBP prediction; (**e**) From UnetU with custom loss prediction. Calibration bars show the display window range in units of g/mL for water and mg/mL for all other materials.

3.6. Performance at Different Dose Levels

Figure 9 displays coronal slices from the chest of an ex vivo mouse scan, reconstructed using varying numbers of projections (specifically, 2700, 900, and 450), thereby altering radiation dose. These images serve to exemplify the performance of wFBP, iterative RSKR reconstruction, ME NLM, UnetwFBP, and UnetU with custom loss at the highest energy threshold of 60 keV. ME NLM denoising significantly blurs the image, particularly at lower dose levels. UnetwFBP does not blur as much as ME NLM, but it is unable to provide sufficient noise reduction at lower dose. UnetU with custom loss and iterative RSKR reconstruction provide a much more optimal balance between noise reduction and spatial resolution. This qualitative observation is supported by quantitative results in part (f) of Figure 9, which shows that iterative reconstruction gives the highest SSIM and PSNR at all dose levels, while UnetU with custom loss gives the 2nd highest values for all dose levels except the full set of 2700 projections. It is worth noting that UnetU predictions have high SSIM and PSNR values in a coronal slice even though the network denoises axial slices.

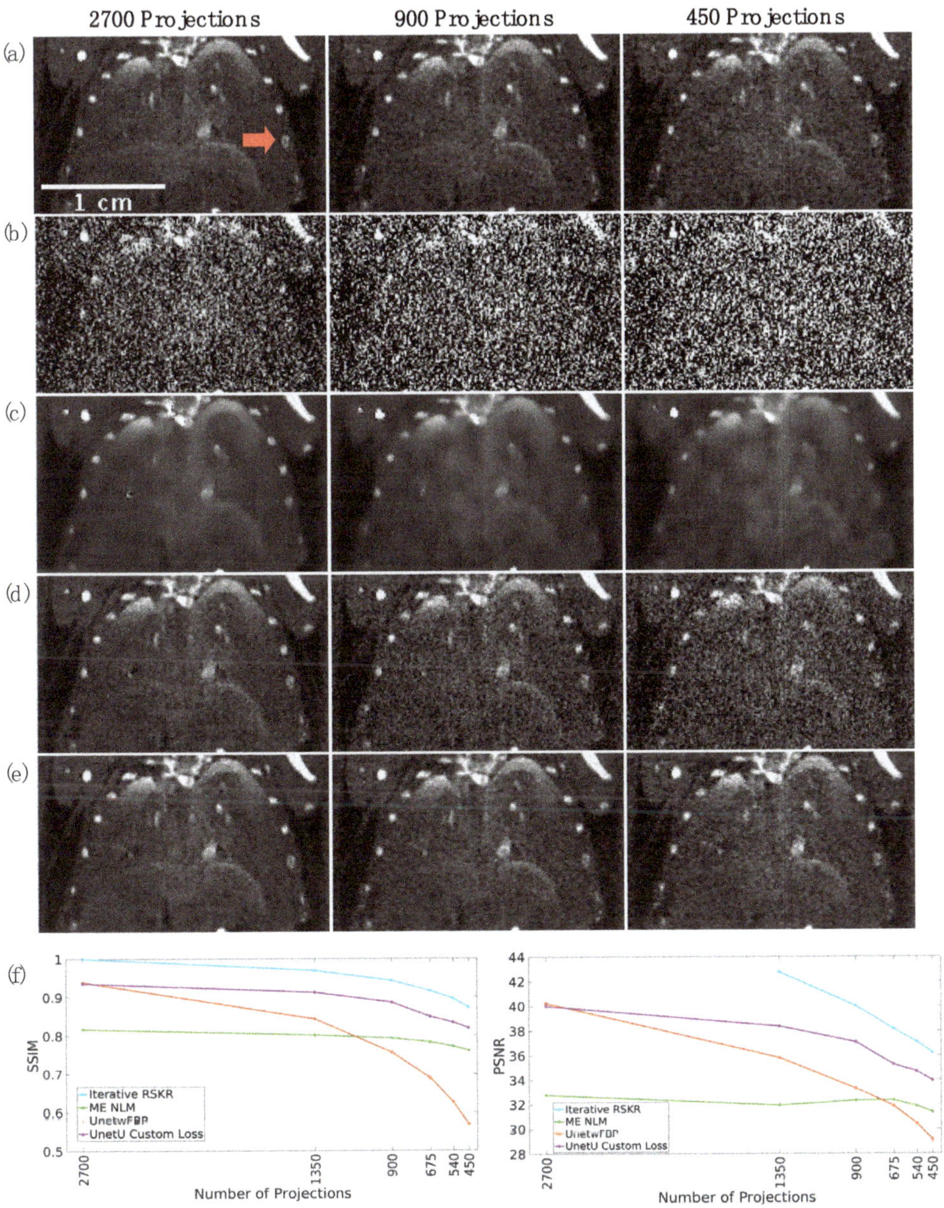

Figure 9. Performance of each denoising method at different dose levels in the highest energy threshold (60 keV) of an ex vivo mouse scan. (**a**) Iterative RSKR reconstructions; (**b**) wFBP reconstructions; (**c**) ME NLM denoised reconstructions; (**d**) UnetwFBP predictions; (**e**) UnetU with custom loss predictions; (**f**) SSIM (left) and PSNR (right) values for each denoising method at each dose level relative to the iterative reconstruction with 2700 projections. Since SSIM and PSNR values for iterative RSKR reconstruction at 2700 projections compare the reference image to itself, they have values of 1 and $+\infty$, respectively. The calibration bar shows the display window range for reconstructions in units of cm^{-1}. The red arrow in (**a**) indicates a rib bone that is blurred in ME NLM images and corrupted by noise in UnetwFBP predictions but preserved in iterative RSKR reconstructions and UnetU predictions.

4. Discussion

This work presents UnetU as a fast and adaptable denoising method for PCCT. Remarkably, UnetU can denoise a standard four-energy PCCT reconstruction 15 times faster than iterative reconstruction with RSKR. Results from the test set showed that UnetU with custom loss function outputs denoised images with higher spatial resolution than ME NLM and lower bias than UnetwFBP. In addition, UnetU with custom loss consistently has greater accuracy in the material decomposition than both ME NLM and UnetwFBP. Assessment of denoising performance at different dose levels (see Figure 9) showed that UnetwFBP has inadequate noise reduction and ME NLM severely degrades spatial resolution at low dose, while UnetU with custom loss can provide noise reduction at a wide range of dose levels with much less blurring than ME NLM. These results demonstrate that UnetU is a generalizable DL method that can quickly predict a good approximation of iterative RSKR reconstruction for a variety of acquisition settings with the same PCD.

There are a few limitations that merit further discussion. Like all PCDs, the detector used to acquire our training and testing sets is affected by spectral distortions such as charge sharing and pulse pileup [2]. The iterative RSKR reconstruction algorithm used to generate our training labels does not correct for these effects. Future work could further improve the accuracy of our denoising approach by using PCCT simulations, a model-based correction, or EID scans to make iterative RSKR reconstructions without spectral distortions as training labels. In recent work, we have improved PCCT material decomposition accuracy by training a U-net with material maps derived from iterative reconstruction of high-dose, multi-energy scans with an EID [30]. In a similar fashion, we can create training labels for PCCT denoising by converting multi-energy EID material decompositions into PCD energy threshold images by using material basis functions. Although our UnetU method results in greater quantitative accuracy than ME NLM and UnetwFBP in both the reconstruction and material decomposition domains, we acknowledge that UnetU predictions still have some bias in high-intensity regions of the reconstructions. Additional refinements to the input preprocessing approach, loss function, or CNN architecture could help address this issue.

In addition to these specific limitations, there are more general challenges associated with a DL approach that should be considered. The availability of training data poses a significant challenge, as DL-based denoising methods typically require high-quality data, which in CT often necessitates higher-dose acquisitions. However, our approach leverages iterative RSKR reconstructions as training labels, eliminating the need for additional high-dose images and simplifying the data acquisition process. DL models also demand substantial computational resources for training and inference, which can preclude their use in resource-limited settings. In our study, we were fortunate to have access to high-performance, multi-GPU computers to facilitate our DL training and evaluation. DL models' inherent complexity can impede interpretability, making it challenging to understand the decision-making process. However, the use of a U-net CNN architecture in our approach mitigates this challenge to some extent, since it has far fewer weights than other more complicated networks. This allows insights into the denoising process and contributes to transparency. Finally, the robustness and adaptability of DL models to different data acquisition protocols needs careful investigation. By training and testing our U-net models on diverse datasets encompassing several PCCT energy threshold settings, we aimed to assess their ability to handle variations encountered in practical scenarios. However, a large test set with systematic variation in multiple acquisition parameters would allow a more complete understanding of the extent to which DL approaches such as UnetU are generalizable.

Nevertheless, our work establishes an effective supervised DL approach that enables fast approximation of iterative RSKR reconstruction and accurate material decomposition for a wide variety of PCD scan protocols. A generalizable denoising approach such as UnetU assists in imaging studies involving novel contrast agents, opening new avenues for research in spectral CT.

5. Conclusions

In this work, we presented UnetU, a deep learning approach for fast PCCT denoising that combines a standard U-net CNN architecture with novel input preprocessing and a complex loss function to provide accurate denoising and material decomposition at a variety of PCD energy threshold settings. We determined that UnetU denoises a typical wFBP reconstruction 15 times faster than iterative RSKR reconstruction. UnetU gave lower average test set RMSE (relative to iterative RSKR reconstruction) than UnetwFBP and ME NLM in reconstructions and material maps. This quantitative result was supported by qualitative results, indicating that UnetU provides the best approximation of iterative RSKR reconstruction in both the reconstruction and material decomposition domains.

Author Contributions: Conceptualization, R.N. and C.T.B.; methodology, R.N., D.P.C. and C.T.B.; software, R.N. and D.P.C.; validation, R.N.; formal analysis, R.N.; investigation, R.N. and A.J.A.; resources, C.T.B.; data curation, R.N. and A.J.A.; writing—original draft preparation, R.N. and C.T.B.; writing—review and editing, R.N., D.P.C., A.J.A. and C.T.B.; visualization, R.N.; supervision, C.T.B.; project administration, C.T.B.; funding acquisition, C.T.B. All authors have read and agreed to the published version of the manuscript.

Funding: This research was performed at the Duke Quantitative Imaging and Analysis Lab funded by NIH grants number U24 CA220245 and RF1AG070149-01.

Institutional Review Board Statement: The in vivo animal work was conducted under a protocol approved by the Duke University Institutional Animal Care and Use Committee (IACUC).

Informed Consent Statement: Not applicable.

Data Availability Statement: The data and code will be shared on https://gitlab.oit.duke.edu/rohan.nadkarni/unetu_pcctdenoising/ (accessed on 1 July 2023).

Acknowledgments: We acknowledge the help of Yi Qi with the animal experiments.

Conflicts of Interest: The authors declare no conflict of interest.

References

1. Taguchi, K.; Iwanczyk, J.S. Vision 20/20: Single photon counting X-ray detectors in medical imaging. *Med. Phys.* **2013**, *40*, 100901. [CrossRef]
2. Leng, S.; Bruesewitz, M.; Tao, S.; Rajendran, K.; Halaweish, A.F.; Campeau, N.; Fletcher, J.G.; McCollough, C.H. Photon-counting Detector CT: System Design and Clinical Applications of an Emerging Technology. *RadioGraphics* **2019**, *39*, 609–912. [CrossRef] [PubMed]
3. Allphin, A.J.; Clark, D.P.; Thuering, T.; Bhandari, P.; Ghaghada, K.B.; Badea, C.T. Micro-CT imaging of multiple K-edge elements using GaAs and CdTe photon counting detectors. *Phys. Med. Biol.* **2023**, *68*, 085023. [CrossRef] [PubMed]
4. Li, Z.; Yu, L.; Trzasko, J.D.; Lake, D.S.; Blezek, D.J.; Fletcher, J.G.; McCollough, C.H.; Manduca, A. Adaptive nonlocal means filtering based on local noise level for CT denoising. *Med. Phys.* **2014**, *41*, 011908. [CrossRef]
5. Li, Z.; Leng, S.; Yu, L.; Manduca, A.; McCollough, C.H. An effective noise reduction method for multi-energy CT images that exploit spatio-spectral features. *Med. Phys.* **2017**, *44*, 1610–1623. [CrossRef] [PubMed]
6. Mechlem, K.; Ehn, S.; Sellerer, T.; Braig, E.; Munzel, D.; Pfeiffer, F.; Noel, P.B. Joint Statistical Iterative Material Image Reconstruction for Spectral Computed Tomography Using a Semi-Empirical Forward Model. *IEEE Trans. Med. Imaging* **2018**, *37*, 68–80. [CrossRef]
7. Niu, S.; Zhang, Y.; Zhong, Y.; Liu, G.; Lu, S.; Zhang, X.; Hu, S.; Wang, T.; Yu, G.; Wang, J. Iterative reconstruction for photon-counting CT using prior image constrained total generalized variation. *Comput. Biol. Med.* **2018**, *103*, 167–182. [CrossRef] [PubMed]
8. Yu, Z.; Leng, S.; Li, Z.; McCollough, C.H. Spectral prior image constrained compressed sensing (spectral PICCS) for photon-counting computed tomography. *Phys. Med. Biol.* **2016**, *61*, 6707–6732. [CrossRef]
9. Baffour, F.I.; Huber, N.R.; Ferrero, A.; Rajendran, K.; Glazebrook, K.N.; Larson, N.B.; Kumar, S.; Cook, J.M.; Leng, S.; Shanblatt, E.R.; et al. Photon-counting Detector CT with Deep Learning Noise Reduction to Detect Multiple Myeloma. *Radiology* **2023**, *306*, 229–236. [CrossRef]
10. Clark, D.P.; Badea, C.T. Hybrid spectral CT reconstruction. *PLoS ONE* **2017**, *12*, e0180324. [CrossRef]
11. Jin, K.H.; McCann, M.T.; Froustey, E.; Unser, M. Deep Convolutional Neural Network for Inverse Problems in Imaging. *IEEE Trans. Image Process.* **2017**, *26*, 4509–4522. [CrossRef] [PubMed]
12. Wang, S.; Yang, Y.; Yin, Z.; Wang, A. Noise2Noise for denoising photon counting CT images: Generating training data from existing scans. Proceedings of SPIE Medical Imaging 2023: Physics of Medical Imaging, San Diego, CA, USA.

13. Chen, H.; Zhang, Y.; Kalra, M.K.; Lin, F.; Chen, Y.; Liao, P.; Zhou, J.; Wang, G. Low-Dose CT With a Residual Encoder-Decoder Convolutional Neural Network. *IEEE Trans. Med. Imaging* **2017**, *36*, 2524–2535. [CrossRef]
14. Yang, Q.; Yan, P.; Zhang, Y.; Yu, H.; Shi, Y.; Mou, X.; Kalra, M.K.; Zhang, Y.; Sun, L.; Wang, G. Low-Dose CT Image Denoising Using a Generative Adversarial Network With Wasserstein Distance and Perceptual Loss. *IEEE Trans. Med. Imaging* **2018**, *37*, 1348–1357. [CrossRef]
15. Wang, D.; Fan, F.; Wu, Z.; Liu, R.; Wang, F.; Yu, H. CTformer: Convolution-free Token2Token dilated vision transformer for low-dose CT denoising. *Phys. Med. Biol.* **2023**, *68*, 065012. [CrossRef] [PubMed]
16. Huber, N.R.; Ferrero, A.; Rajendran, K.; Baffour, F.; Glazebrook, K.N.; Diehn, F.E.; Inoue, A.; Fletcher, J.G.; Yu, L.; Leng, S.; et al. Dedicated convolutional neural network for noise reduction in ultra-high-resolution photon-counting detector computed tomography. *Phys. Med. Biol.* **2022**, *67*, 175014. [CrossRef] [PubMed]
17. Heinrich, M.P.; Stille, M.; Buzug, T.M. Residual U-Net Convolutional Neural Network Architecture for Low-Dose CT Denoising. *Curr. Dir. Biomed. Eng.* **2018**, *4*, 297–300. [CrossRef]
18. Abascal, J.F.P.J.; Bussod, S.; Ducros, N.; Si-Mohamed, S.; Douek, P.; Chappard, C.; Peyrin, F. A residual U-Net network with image prior for 3D image denoising. In Proceedings of the 2020 28th European Signal Processing Conference (EUSIPCO), Amsterdam, The Netherlands, 18–21 January 2021; pp. 1264–1268.
19. Yuan, N.; Zhou, J.; Qi, J. Half2Half: Deep neural network based CT image denoising without independent reference data. *Phys. Med. Biol.* **2020**, *65*, 215020. [CrossRef]
20. Wu, W.; Hu, D.; Niu, C.; Vanden Broeke, L.; Butler, A.P.H.; Cao, P.; Atlas, J.; Chernoglazov, A.; Vardhanabhuti, V.; Wang, G. Deep learning based spectral CT imaging. *Neural Netw.* **2021**, *144*, 342–358. [CrossRef]
21. Komatsu, R.; Gonsalves, T. Comparing U-Net Based Models for Denoising Color Images. *AI* **2020**, *1*, 465–486. [CrossRef]
22. Holbrook, M.D.; Clark, D.P.; Badea, C.T. Dual source hybrid spectral micro-CT using an energy-integrating and a photon-counting detector. *Phys. Med. Biol.* **2020**, *65*, 205012. [CrossRef]
23. Stierstorfer, K.; Rauscher, A.; Boese, J.; Bruder, H.; Schaller, S.; Flohr, T. Weighted FBP—A simple approximate 3D FBP algorithm for multislice spiral CT with good dose usage for arbitrary pitch. *Phys. Med. Biol.* **2004**, *49*, 2209–2218. [CrossRef] [PubMed]
24. Gao, H.; Yu, H.; Osher, S.; Wang, G. Multi-energy CT based on a prior rank, intensity and sparsity model (PRISM). *Inverse Probl.* **2011**, *27*, 115012. [CrossRef] [PubMed]
25. Alvarez, R.E.; Macovski, A. Energy-selective reconstructions in X-ray computerised tomography. *Phys. Med. Biol.* **1976**, *21*, 733–744. [CrossRef]
26. Ronneberger, O.; Fischer, P.; Brox, T. U-Net: Convolutional Networks for Biomedical Image Segmentation. In Proceedings of the Medical Image Computing and Computer-Assisted Intervention–MICCAI 2015, 18th International Conference, Munich, Germany, 5–9 October 2015; Springer: Cham, Switzerland, 2015; pp. 234–241.
27. Paszke, A.; Gross, S.; Massa, F.; Lerer, A.; Bradbury, J.; Chanan, G.; Killeen, T.; Lin, Z.M.; Gimelshein, N.; Antiga, L.; et al. PyTorch: An Imperative Style, High-Performance Deep Learning Library. *Adv. Neural Inf. Process. Syst.* **2019**, *32*, 721.
28. Donoho, D.L.; Johnstone, I.M. Adapting to Unknown Smoothness via Wavelet Shrinkage. *J. Am. Stat. Assoc.* **1995**, *90*, 1200–1224. [CrossRef]
29. Kingma, D.P.; Ba, J.L. Adam: A Method for Stochastic Optimization. In Proceedings of the 3rd International Conference on Learning Representations, ICLR 2015, San Diego, CA, USA, 7–9 May 2015.
30. Nadkarni, R.; Allphin, A.; Clark, D.P.; Badea, C.T. Material decomposition from photon-counting CT using a convolutional neural network and energy-integrating CT training labels. *Phys. Med. Biol.* **2022**, *67*, 155003. [CrossRef] [PubMed]

diagnostics

MDPI

Article

Segmentation of Portal Vein in Multiphase CTA Image Based on Unsupervised Domain Transfer and Pseudo Label

Genshen Song [1,2,†], Ziyue Xie [1,2,†], Haoran Wang [1,2], Shiman Li [1,2], Demin Yao [1,2], Shiyao Chen [3,*] and Yonghong Shi [1,2,4,*]

1 Digital Medical Research Center, School of Basic Medical Sciences, Fudan University, Shanghai 200032, China; gssong20@fudan.edu.cn (G.S.); zyxie22@m.fudan.edu.cn (Z.X.); hrwang20@fudan.edu.cn (H.W.); smli21@m.fudan.edu.cn (S.L.); rh386@sina.com (D.Y.)
2 Shanghai Key Laboratory of Medical Imaging Computing and Computer Assisted Intervention, Shanghai 200032, China
3 Department of Gastroenterology and Hepatology, Zhongshan Hospital, Fudan University, Shanghai 200032, China
4 Academy for Engineering & Technology, Fudan University, Shanghai 200433, China
* Correspondence: chen.shiyao@zs-hospital.sh.cn (S.C.); yonghong.shi@fudan.edu.cn (Y.S.)
† These authors contributed equally to this work.

Abstract: Background: Clinically, physicians diagnose portal vein diseases on abdominal CT angiography (CTA) images scanned in the hepatic arterial phase (H-phase), portal vein phase (P-phase) and equilibrium phase (E-phase) simultaneously. However, existing studies typically segment the portal vein on P-phase images without considering other phase images. Method: We propose a method for segmenting portal veins on multiphase images based on unsupervised domain transfer and pseudo labels by using annotated P-phase images. Firstly, unsupervised domain transfer is performed to make the H-phase and E-phase images of the same patient approach the P-phase image in style, reducing the image differences caused by contrast media. Secondly, the H-phase (or E-phase) image and its style transferred image are input into the segmentation module together with the P-phase image. Under the constraints of pseudo labels, accurate prediction results are obtained. Results: This method was evaluated on the multiphase CTA images of 169 patients. The portal vein segmented from the H-phase and E-phase images achieved DSC values of 0.76 and 0.86 and Jaccard values of 0.61 and 0.76, respectively. Conclusion: The method can automatically segment the portal vein on H-phase and E-phase images when only the portal vein on the P-phase CTA image is annotated, which greatly assists in clinical diagnosis.

Keywords: portal vein segmentation; multiphase CTA image; pseudo label; unsupervised domain transfer

Citation: Song, G.; Xie, Z.; Wang, H.; Li, S.; Yao, D.; Chen, S.; Shi, Y. Segmentation of Portal Vein in Multiphase CTA Image Based on Unsupervised Domain Transfer and Pseudo Label. *Diagnostics* **2023**, *13*, 2250. https://doi.org/10.3390/diagnostics13132250

Academic Editors: Michał Strzelecki, Adam Piorkowski and Rafał Obuchowicz

Received: 5 May 2023
Revised: 27 June 2023
Accepted: 29 June 2023
Published: 3 July 2023

1. Introduction

The portal vein enters the liver and supplies hepatic blood, accounting for approximately 75% of the total blood supply to the liver [1]. Portal vein thrombosis or cavernous transformation of the portal vein causes obstruction of portal blood flow and reduces hepatic blood supply, which aggravates liver injuries, leads to increased portal vein pressure, and further increases the risk of complications associated with portal hypertension such as ascites, esophageal and gastric varices bleeding, thus affecting the prognosis and quality of life of patients [2,3]. Many studies aim to use machine learning and deep learning methods to segment the liver and related vessels in different modal medical images to promote disease diagnosis, prognosis and surgical planning [4–6].

Clinically, multiphase computer tomography angiography (CTA) images are obtained by scanning at different times of peak venous flow. Due to the complementary vessel information on these images presenting a comprehensive status of the collateral circulation, they are routinely used for disease diagnosis and patient prognosis prediction [7]. For example, the diagnosis of portal vein disease often depends on observing the portal vein on the abdominal CTA images of the hepatic artery phase (H-phase), the portal vein phase (P-phase) and the equilibrium phase (E-phase) simultaneously, and intuitively fusing CTA images of different phases, so as to complement vessel information and implement effective treatment [8]. Therefore, it is crucial to accurately segment portal veins on multiphase CTA images.

Due to the different distribution of contrast media between portal veins and surrounding tissues in different phase CTA images, the contrast between the portal vein and the surrounding tissues in the P-phase is the most obvious, which is beneficial for doctors to manually label and construct labeled datasets. Therefore, the existing deep learning methods almost only label the portal veins in P-phase images. Segmenting portal veins on P-phase CTA images is a supervised problem. Supervised methods based on deep learning can generate a best-fit internal representation of the vessel on the image. For example, deep learning has been combined with anatomy for portal vein segmentation in 2D images for liver SBRT planning [9]. As 3D medical images can retain more contextual information, most vessel segmentation algorithms are implemented and improved on the basis of classic 3D segmentation models such as nnUNet and 3D U-Net [10]. For example, Kitrungrotsakul et al. [11] proposed a deep convolutional neural network with multiple pathways called VesselNet for robust segmentation of liver vessels. Yu et al. [12] constructed a novel 3D residual U-Net framework for portal vein segmentation from abdominal CT images. Xu et al. [13] proposed a mean-teacher-assisted confidence learning method for vessel segmentation.

In order to segment the portal veins from unlabeled H-phase and E-phase images by using the label of P-phase, the existing methods overlap the ground truth of the portal veins in the P-phase images to the corresponding images by simple image alignment for supervised learning. For example, liver segmentation based on EM clustering and GVF level set was performed based on the registration of the multiphase CT dataset [14], and automatic landmark detection and TPS deformation were used for non-rigid registration of multiphase liver CT data [15]. However, these methods cannot accurately reflect the status of the portal vein in different phase images. Specifically, Figure 1 shows the slices of H-phase, P-phase and E-phase CTA images of a subject scanned on the same device, as well as the corresponding portal vein annotation. The orange boxes highlight the portal vein in different phases, while the red curves show the main portal vein. It can be seen that the portal vein is a short and thick main vein [16], which is a smaller target compared to larger organs such as the liver and spleen, making it relatively difficult for physicians to locate the portal vein during observation, especially in H-phase and E-phase images. In addition, the concentration of contrast media in the portal vein varies in the different phases: the contrast between the portal vein and the adjacent tissues is the lowest in H-phase images, the highest in P-phase images and decreases again in E-phase images. To make matters worse, vessels and tissues undergo a certain degree of displacement and distortion between phases. As can be seen, there is an inter-phase shift in the annotation results of the vessels, and the complete portal vein trunk cannot be obtained directly on each phase image by simple alignment.

The above research shows that the existing supervised vascular segmentation deep learning methods mainly focus on changing the network structure, segmenting single-phase images and not exploring cross-phase information. The existing exploration of interphase information utilization, that is, the simple alignment method, ignores the contrast and morphological changes of intertemporal blood vessels. Therefore, we believe that the key to portal vein segmentation in H-phase and E-phase CTA images is to solve the unsupervised domain transfer problem, which is limited by vessel labeling.

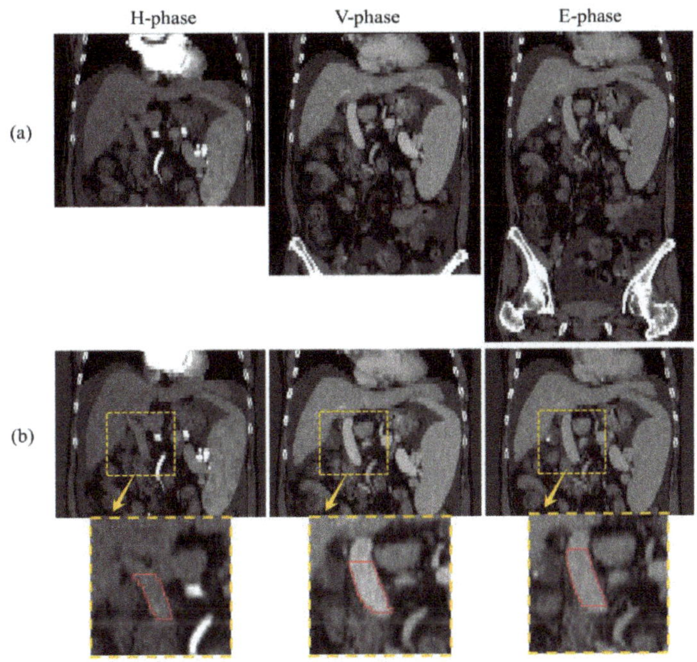

Figure 1. Illustration of the original and aligned CTA images of H-phase, P-phase and E-phase, as well as the magnified portal vein region (orange box) and main portal vein (red curve) in the corresponding images. (**a**) Original image. (**b**) Aligned image.

It is known that Generative Adversarial Network (GAN) [17] provides a way to learn deep representations without a large amount of annotated training data. And it can generalize the models trained on the source domain (annotated training datasets) to the target domain (test datasets) through transfer learning [18] and visual domain transfer techniques [19]. In particular, the technological breakthrough around GAN, such as CycleGAN [20], can generate high-quality images under unsupervised conditions and is often applied to multimodality image annotation problems. Recently, Xu et al. [13] used noise-labeled data for the challenging task of liver vessel segmentation. Jiang et al. [21] used unpaired CT and MRI images for domain adaptive transformations and guided the student CT networks with the help of an informative teacher MRI network to extract features indicating foreground and background differences. Other studies have explored domain transfer segmentation on CT and MRI images of the heart [22,23]. In general, obtaining the similarity between different images by domain transfer is a feasible annotation method.

In addition, the lack of sufficient public datasets has led to an increasing interest in the study of unsupervised and semi-supervised deep learning [24,25]. Medical images are usually presented in multiple modalities, such as MRI and CT, so unsupervised domain adaptive learning using annotations of one modality to obtain annotations of another modality is a frequently explored topic. For example, Raju et al. [26] proposed the co-heterogeneous and adaptive segmentation (CHASe) method based on liver-enhanced CT images, which requires only a small amount of annotated monophasic data to adapt to any unannotated heterogeneous multiphase data and is highly labor-saving. Qu et al. [27] conducted a similar study on pancreatic-enhanced CT images using two-phase scanning, and their proposed framework can integrate multi-scale and multi-view information for multi-phase pancreas segmentation. This study has a strong reference value for the segmentation of portal veins on the multiphase images. Currently, most studies on unsupervised domain adaptation for image segmentation focus on simple tasks such as organs and tumors. Few unsupervised cross-modality studies have been applied to vessel segmentation due to the inconsistent size and a number of vessels observed on CTA images of different phases.

Therefore, for the multiphase CTA dataset of patients with portal-related diseases, based on the annotated P-phase images, this paper proposes a method to segment portal veins on the H-phase and E-phase images based on unsupervised domain transfer and pseudo labels. Firstly, a style transfer network is used to implement modal transformation from H-phase (or E-phase) images to P-phase images, narrowing the style differences between different phases and achieving higher apparent similarity. At the same time, the limitation of insufficient contrast of the portal vein in the original images of the different phases is suppressed on the corresponding ones after style transfer, which is conducive to constructing the pseudo labels. Secondly, the H-phase (or E-phase) images and the style transferred ones that will be input into the 3D U-Net network along with the P-phase images. When training the segmentation model, under the constraint of the ground truths of the portal vein of the P-phase images, the predicted results of the H-phase (or E-phase) images are used to jointly construct pseudo labels with confidence for the corresponding phase. This not only reduces the impact of vessel shift between different phases but also achieves accurate segmentation of portal veins in corresponding phases. Finally, in the case of only one phase image annotation, this method can obtain segmented portal veins on the other two phase images, respectively.

2. Materials and Methods

2.1. Data Description

The dataset included 169 patients with liver cirrhosis and portal vein disease who visited Zhongshan Hospital, affiliated with Fudan University in Shanghai, China, between January 2016 and May 2020. These patients were scanned for abdominal CTA images in the H-phase, the P-phase and the E-phase. The CTA images were acquired with a delayed scan time of 25 to 30 s for the H-phase, 60 to 70 s for the P-phase and 85 to 90 s for E-phase after contrast media injection. The imaging parameters were as follows: in-plane resolution is between 0.610 mm $\times$ 0.610 mm and 0.881 mm $\times$ 0.881 mm; slice thickness is 5.0 mm; and the size of the image acquisition matrix is 512 $\times$ 512.

Finally, an experienced physician annotated the portal vein trunk on all images.

2.2. Image Preprocessing

The preprocessing of CTA images includes removing the CT beds, aligning images of different phases of the same patient, and extracting and normalizing the portal vein blocks.

Removing the CT beds. Selecting a threshold of −300 Hounsfield units to binarize the image can remove the CT bed from all images completely and preserve the largest body contour in the images.

Aligning images of different phases. Figure 1a shows coronal views of H-phase, P-phase and E-phase CTA images of the same patient. The images of different phases contain different numbers of axial slices, indicating that even for the same patient, it is not possible to determine the location of portal veins in other phases solely based on portal vein annotation in P-phase. According to the body contour extracted by binarization, the slice locations with the largest intersection of the binary images of the three phases were taken so that the images of different phases of the same patient were aligned axially. Figure 1b shows the results of the H-phase, P-phase and E-phase CTA images after the removal of the CT bed and alignment of the images of the three phases.

Extracting and normalizing the portal vein blocks. The portal vein occupies a relatively small volume in the whole CTA image and has an irregular shape, especially the weak contrast between the portal vein and the surrounding tissues in H-phase and E-phase images, making it difficult to segment directly in the original image. For this purpose, this study utilizes portal vein annotation in the P-phase image to automatically crop image blocks containing portal veins from the aligned H-phase and E-phase images, with a size of 32 $\times$ 128 $\times$ 128.

2.3. Multiphase Segmentation Network

Figure 2 depicts the portal vein segmentation network (called as PVSegNet) on the multiphase images, which consists of two parts: a style transfer module and a segmentation module. The inputs to the network are the H-phase (or E-phase) image, I^A, and the P-phase image, I^V. The ground truth of the portal vein in the P-phase image is denoted as $True^V$. The output of the network is the prediction results of the portal vein on the H-phase (or E-phase) image, P^A. In the style transfer module, the H-phase (or E-phase) image, I^A, is transformed to construct the image, $I^{A\to V}$, with a similar style to the P-phase image, I^V. Similarly, the P-phase image, I^V, itself can also be transferred to the image, $I^{V\to V}$, with a consistent style. Then, in the segmentation module, these style-transferred images, $I^{A\to V}$ and $I^{V\to V}$, and their original images, I^A and I^V are trained only under the constraint of the ground truth of the portal vein in the P-phase image, $True^V$, and the corresponding prediction results are generated to construct pseudo labels for further optimizing the segmentation. Finally, the portal vein on the H-phase (or E-phase) images is segmented automatically and accurately.

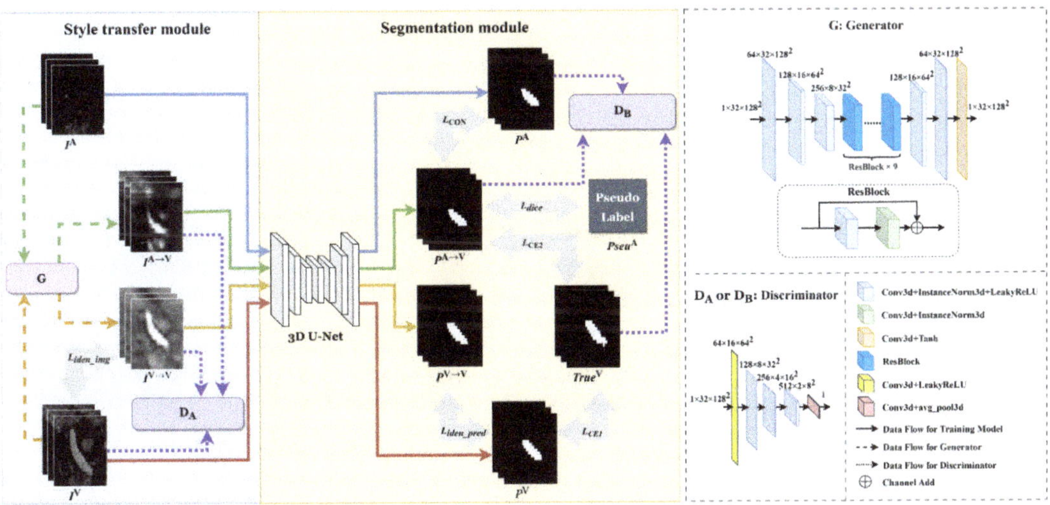

Figure 2. Portal vein segmentation network on multiphase images.

2.3.1. Style Transfer Module

Traditional style transfer neural networks are capable of converting information from one representation to another. In order to enable the H-phase (or E-phase) image to have similar characteristics to the P-phase image and segment the portal vein on the corresponding images, inspired by the idea of the domain transfer of the CycleGan network [20], this paper builds the style transfer module composed of a generator and a discriminator, as shown in Figure 2.

First, I^A is input into the generator G to generate an image $I^{A\to V}$ with a similar style to I^V. Since there is a high similarity of organs and tissues between the two-phase images, in order to ensure that the semantic characteristics similar to I^V in I^A are not changed by the generator and can be preserved, I^V also goes through the generator to get $I^{V\to V}$. The image block input to the generator is represented by a tensor with size $1 \times 32 \times 128 \times 128$ (representing channel $\times$ depth $\times$ height $\times$ width). These images are sequentially processed through three encoding convolution layers, nine residual structures (ResBlock) and two decoding convolution layers, and the output is the result of style transfer transformation with the same size as the input image. Here, the convolutional kernels of the encoding and decoding layers are connected with InstanceNorm3d and LeakyReLU to form the basic structural module. The size of the convolution kernel is $3 \times 3 \times 3$, and the step sizes of the convolution kernels are all 2.

Secondly, whether the style transfer result $I^{A \to V}$ is similar to I^{V} in terms of style characteristics will be determined by the discriminator D_A shown in Figure 2. At the same time, the discriminator also evaluates the transferred image, $I^{V \to V}$, to enhance the ability of the generator to preserve the transformation P-phase image style characteristics. The image block input to the discriminator is represented by a tensor with a size of $1 \times 32 \times 128 \times 128$. And these images are sequentially processed through five layers of convolution, and finally, the value 1 or 0 is output by the average pooling (avg_pool3d), which indicates whether the style transfer result is similar to I^{V}. The step size of the convolutional kernel in the discriminator network is 2.

It should be noted that in Figure 2, I^{V} is an annotated P-phase image, and I^{A} may be an H-phase or E-phase image to be segmented. This study separately trains the style transfer module on the H-phase or E-phase images and its subsequent segmentation module on the corresponding phase under the guidance of the P-phase images.

2.3.2. Segmentation Module

When the H-phase (or E-phase) images undergo the style transfer transformation and have a similar style as the P-phase image, the segmentation module can segment both the P-phase image I^{V} and its style-transferred image $I^{V \to V}$, as well as the H-phase (or E-phase) image I^{A} and its corresponding style-transferred image $I^{A \to V}$, and output the corresponding prediction results, respectively. The segmentation module consists of a 3D U-Net [10] segmentation network and a discriminator, D_B, as shown in Figure 2.

Figure 3 shows the 3D U-Net network. The image represented by a tensor with size $1 \times 32 \times 128 \times 128$ sequentially undergoes four downsamplings (max-pooling) operations with a step size of 2, and four upsampling (ConvTranspose) operations with a step size of 2. And each level's upsampling layer is directly concatenated with the same level's downsampling layer through skip connections. The final output is the binary segmentation result of the portal vein on the image. Before each downsampling and upsampling operation, the input image or intermediate calculation results are sequentially extracted at pixel level by two basic structures formed by connecting the convolutional layers, InstanceNorm3d and LeakyReLU.

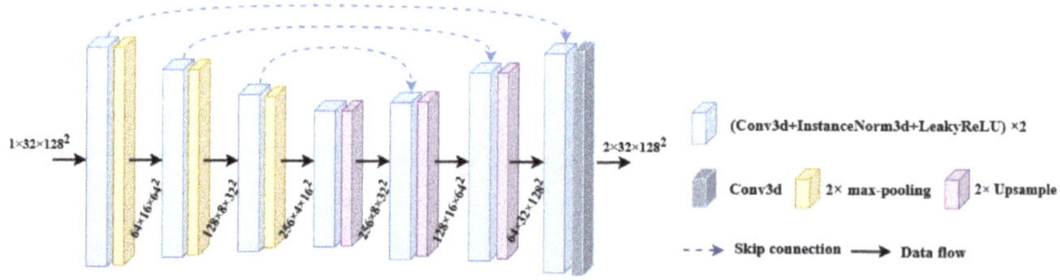

Figure 3. 3D U-Net segmentation module.

The discriminator D_B of the segmentation module is used to judge whether the output of the 3D U-net network is similar to the ground truth $True^{V}$ on the P-phase image, which is an image-level judgment and more suitable as a soft constraint of the segmentation module. The discriminator D_B judges the authenticity of the prediction results, $P^{A \to V}$ and P^{A}, enhancing the constraint of the network on the final prediction result P^{A}. Here, the structure of the discriminator D_B is the same as that of D_A, as shown in Figure 2.

2.3.3. Construction of Pseudo Label

During the training process, when using the ground truth $True^{V}$ of the P-phase image to guide the segmentation network in training the H-phase (or E-phase) images, it is common to learn incorrect information due to the vessel shift on different phase images. Therefore, this study proposes to use the ground truth of the portal vein in the P-phase image and the prediction results of the segmentation network to jointly construct pseudo labels and guide network training through pseudo labels. Figure 4 shows the process of constructing pseudo labels.

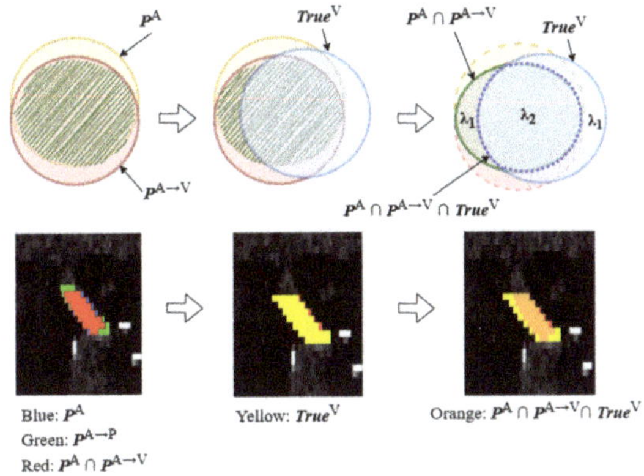

Figure 4. Construction of pseudo label.

The original H-phase (or E-phase) image I^A and its style transferred image $I^{A \to V}$ are input into the 3D U-Net network, and the predicted results P^A and $P^{A \to V}$ are obtained. Although they are similar, there are differences between the two. In particular, the intersection area, $P^A \cap P^{A \to V}$, may belong to the possible label area. This intersection area of $P^A \cap P^{A \to V}$ further intersects with the ground truth $True^V$ of the P-phase image I^V, and the resulting intersection $P^A \cap P^{A \to V} \cap True^V$ can be used as the correct label area on image I^A. The remaining part of $True^V$ is also considered as a possible label area for image $I^{A \to V}$. Then, set weights $\lambda_2 - \lambda_1$ and λ_1 for the correct and possible label regions separately, and the pseudo labels are calculated as Equation (1):

$$Pseu^A = \lambda_1 P^A \cap P^{A \to V} + \lambda_1 True^V + (\lambda_2 - 2\lambda_1)\left(P^A \cap P^{A \to V} \cap True^V\right) \tag{1}$$

Here, referring to the overlapping area between the predicted results and the ground truths in Figure 4, λ_1 and λ_2 are set as 0.1 and 1 for the H-phase image, 0.7 and 1 for the E-phase image and 0 for the remaining areas. In addition, Dice loss is set between the pseudo label $Pseu^A$ and the network prediction result $P^{A \to V}$ for guiding the PVSegNet network to better train and segment $I^{A \to V}$.

2.3.4. Loss Function

The cost function optimized by this method, Loss, is calculated in Equation (2) and consists of five types of the loss function, including Identity loss, L_{iden_img} and L_{iden_pred}, Consistency loss, L_{CON}, Cross entropy loss, L_{CE1} and L_{CE2}, Dice Loss, L_{dice}, as well as the Discrimination loss, L_{G_img} and L_{G_pred}.

$$Loss = \omega_1 L_{iden\_img} + \omega_2 L_{iden\_pred} + \omega_3 L_{G\_img} + \omega_4 L_{G\_pred} + \omega_5 L_{CON} + \omega_6 L_{CE1} + \omega_7 L_{dice} + \omega_8 L_{CE2} \tag{2}$$

where $\omega_1, \omega_2, \ldots, \omega_8$ represent the weights of the loss functions, respectively, and in this study, $\omega_1 = \omega_2 = \omega_7 = \omega_8 = 5$, $\omega_3 = \omega_4 = 0.01$, $\omega_5 = 1$, $\omega_6 = 10$.

Identity loss. It is generally believed that images should retain their own features after style transfer [20]. This study uses Identity loss to constrain the image I^V in order to preserve its own features after transformation by the generator, as shown in Equation (3). L_{iden_img} measures the difference between the style-transferred image $I^{V \to V}$ and the original image I^V. In addition, using the same segmentation network to predict the same images should result in the same results. That is to say, the predicted results of images I^V and

$I^{V\to V}$, P^{V} and $P^{V\to V}$, should also be subject to the constraint of Identity loss, as shown in Equation (4).

$$L_{iden\_img} = \begin{cases} \sum_{i=1}^{H\times W\times D}(I_i^V - I_i^{V\to V})^2 & if\ |I_i^V - I_i^{V\to V}| < 1 \\ \sum_{i=1}^{H\times W\times D}\left|I_i^V - I_i^{V\to V}\right| - 0.5 & otherwise \end{cases} \tag{3}$$

$$L_{iden\_pred} = \sum_{c=1}^{C} \begin{cases} \sum_{i=1}^{H\times W\times D}\left(P^{V_i^c} - P^{V\to V_i^c}\right)^2 & if\ \left|P^{V_i^c} - P^{V\to V_i^c}\right| < 1 \\ \sum_{i=1}^{H\times W\times D}\left|P^{V_i^c} - P^{V\to V_i^c}\right| - 0.5 & otherwise \end{cases} \tag{4}$$

where $C = 2$ denotes the portal vein and the background. H, W and D represent the height, width and depth of the image, respectively.

Discrimination loss. The discriminator network D_A in the style transfer module used the binary cross entropy loss function to separate the original P-phase image from the generated P-phase image. When training discriminator D_A, set the discriminator D_A to judge $I^{A\to V}$ as 0, and judge I^V and $I^{V\to V}$ as 1.

The discriminator network D_B in the segmentation module also optimized the binary cross entropy loss function to discriminate the ground truth $True^V$ from the predicted results $P^{A\to V}$ or P^A. When training the discriminator, set the discriminator D_B to judge the predicted results $P^{A\to V}$ and P^A as 0 and determine that the annotation $True^V$ of the P-phase image is 1.

In order to ensure that the image $I^{A\to V}$ generated by the generator predicts sufficiently realistic results through the segmentation network so as to deceive the discriminator D_A which judges that the generated image $I^{A\to V}$ is 1, and at the same time, makes the discriminator D_B determines that the predicted image $P^{A\to V}$ is 1. Therefore, the Discrimination loss functions, i.e., L_{G_img} and L_{G_pred}, are set respectively in the training process to enhance the effectiveness of the generator and segmentation network, as shown in Equations (5) and (6).

$$L_{G\_img} = -\ log\left(D\left(I^{A\to V}\right)\right) \tag{5}$$

$$L_{G\_pred} = -log\left(D\left(P^{A\to V}\right)\right) \tag{6}$$

where $D(I^{A\to V})$ and $D(P^{A\to V})$ represent the binary cross entropy loss results of $I^{A\to V}$ and $P^{A\to V}$ input to the corresponding discriminator, respectively.

Consistency loss. In order to form constraints between the original H-phase (or E-phase) image and its style-transferred image and make the segmentation network learn useful knowledge lost due to style transfer, Consistency loss L_{CON} is set between the corresponding prediction results P^A and $P^{A\to V}$, as shown in Equation (7).

$$L_{CON} = \sum_{c=1}^{C} \frac{1}{H\times W\times D} \sum_{i=1}^{H\times W\times D} \left(P^{A_i^c} - P^{A\to V_i^c}\right)^2 \tag{7}$$

Cross entropy loss. As shown in Equation (8), Cross entropy loss L_{CE1} is set between the segmentation result P^V on the P-phase image and the ground truth $True^V$ to guide the correct convergence of the segmentation network. Multiphase images have similarities, and to some extent, there is information that can be gleaned from one another. Therefore, the Cross entropy loss L_{CE2} is also set to calculate the similarity between the prediction result $P^{A\to V}$ and the ground truth $True^V$, as shown in Equation (9).

$$L_{CE1} = -\sum_{c=1}^{C} \sum_{i=1}^{H\times W\times D} P^{V_i^c} log\left(\frac{e^{Ture^{V_i^c}}}{\sum_{j=1}^{C} e^j}\right) \tag{8}$$

$$L_{CE2} = -\sum_{c=1}^{C} \sum_{i=1}^{H\times W\times D} P^{A\to V_i^c} log\left(\frac{e^{Ture^{V_i^c}}}{\sum_{j=1}^{C} e^j}\right) \tag{9}$$

Dice loss. In addition to the Cross entropy loss, the Dice loss is also set in the segmentation network to reduce the impact caused by the vessel shift between different phases, as shown in Equation (10), which is used to calculate the loss between the prediction result $P^{A \to V}$ and the pseudo label $Pseu^{A}$.

$$L_{dice} = \sum_{c=1}^{C} \left(1 - \frac{2\sum_{i=1}^{H\times W\times D} P^{A\to V}{}_i^c Pseu^{A}{}_i^c}{\sum_{i=1}^{H\times W\times D} \left(P^{A\to V}{}_i^c\right)^2 + \sum_{i=1}^{H\times W\times D} \left(Pseu^{A}{}_i^c\right)^2} \right) \tag{10}$$

3. Experiments

3.1. Experimental Setup and Evaluation Index

The experiments were conducted on Ubuntu 18.04 operating system and PyTorch framework, configured with Intel® Core™ i5-9600K (3.70 GHz × 6 CPUs), 64 GB RAM and one RTX 3090 GPU.

All the experiments are performed using three-fold cross-validation. During the training stage, the H-phase and E-phase training datasets were trained for 50 and 40 epochs to obtain the model with the best performance for testing on the respective test dataset, respectively. The training was performed using the Adam optimizer with an initial learning rate of 0.001.

The evaluation metrics of the segmentation results are standard metrics based on the pixel-level confusion matrix calculation, including the dice similarity coefficient (DSC) [28] and the Jaccard coefficient (Jaccard), as shown in Equations (11) and (12), respectively.

$$\text{DSC} = \frac{2TP_c}{2TP_c + FP_c + FN_c} \tag{11}$$

$$\text{Jaccard} = \frac{TN_c + TP_c}{TN_c + TP_c + FP_c + FN_c} \tag{12}$$

where c denotes a category region. TP_c and TN_c denote the numbers of the true positive and the true negative pixels in the cth region, while FP_c and FN_c are the numbers of the false positive and the false negative pixels in that category, respectively. Two-tailed student t-test was used to test the p-value between the prediction results of PVSegNet and other methods. And a p-value less than 0.05 indicates a significant difference between the predicted results of these two methods. The specific p-value results are shown in Tables A1–A3.

3.2. Comparison of Experimental Results

To evaluate the effectiveness of the proposed network architecture, nnUNet, 3D U-Net and CycleGan are used for comparison experiments. The supervised experiments use the annotations of the corresponding portal vein in the H-phase or E-phase images to train the 3D U-Net network. In the other comparison experiments, the portal vein annotation of the P-phase images will be used directly as pseudo labels for training the network on the H-phase or E-phase images. And the same portal vein training dataset was used for all methods. Table 1 shows the performance of different methods in segmenting the portal vein on the H-phase and E-phase images in the test dataset. For fair comparisons, no data enhancement was performed for all experiments in this paper.

In the segmentation results of the portal vein on the H-phase images, the supervised segmentation results of nnUNet can reach 0.832 for DSC and 0.724 for Jaccard, respectively. The supervised segmentation results of 3D U-Net can reach 0.724 for DSC and 0.581 for Jaccard, respectively. Using images I^{A} or I^{V} and annotations $True^{V}$ in the P-phase, 3D U-Net can reach 0.590 and 0.117 for DSC and 0.581 and 0.079 for Jaccard, respectively. The results show that the model trained with only the P-phase image I^{V} and $True^{V}$ has poor generalization results in the H-phase and can barely recognize the features of the portal vein. The results of the model trained with image I^{A} as well as $True^{V}$ show that useful information can be learned from H-phase images and ground truths, but the lack of contrast has an impact on H-phase image segmentation. According to the segmentation results of CycleGan, the effect of using only style transfer can reach 0.633 for DSC and 0.482 for Jaccard, respectively. Finally, the DSC value is 0.689 and Jaccard is 0.546 for the PVSegNet method proposed in this study, which is better than the segmentation results in the comparative experiments that also used portal vein annotations $True^V$.

Table 1. Performance comparison of different methods in segmenting the portal vein on H-phase and E-phase images in the test set. Inputs with "✔" are used to train the models.

Methods	I^A	I^V	$True^A$	$True^V$	H-Phase		E-Phase	
					DSC	Jaccard	DSC	Jaccard
nnUNet (Supervised)	✔		✔		0.832 ± 0.001	0.724 ± 0.001	0.894 ± 0.000	0.816 ± 0.001
3D U-Net (Supervised)	✔		✔		0.724 ± 0.000	0.581 ± 0.000	0.832 ± 0.000	0.723 ± 0.001
3D U-Net	✔			✔	0.590 ± 0.001	0.437 ± 0.001	0.650 ± 0.001	0.500 ± 0.001
3D U-Net		✔		✔	0.117 ± 0.000	0.079 ± 0.000	0.801 ± 0.001	0.680 ± 0.001
CycleGan	✔	✔		✔	0.633 ± 0.001	0.482 ± 0.001	0.643 ± 0.001	0.493 ± 0.001
PVSegNet	✔	✔		✔	**0.689 ± 0.001**	**0.546 ± 0.001**	**0.826 ± 0.001**	**0.712 ± 0.001**

In the results of portal vein segmentation on the E-phase images, the supervised segmentation results of nnUNet can reach 0.894 for DSC and 0.816 for Jaccard and the supervised segmentation results of 3D U-Net can reach 0.832 for DSC and 0.723 for Jaccard, respectively. Three-dimensional U-Net can reach 0.650 and 0.801 for DSC and 0.500 and 0.680 for Jaccard using images I^A or I^V and $True^V$ in the P-phase, respectively. The results show that the E-phase image is more similar to the P-phase image than the H-phase image, and the model trained on the P-phase images has higher generalization in the E-phase. The results trained on image I^A as well as $True^V$ show that different phase vessel shifts still exist when using P-phase ground truth $True^V$ as pseudo labels. In addition, the segmentation results using only style transfer of CycleGan can reach 0.643 for DSC and 0.493 for Jaccard, respectively. And DSC value is 0.828 and Jaccard is 0.712 for the PVSegNet method proposed in this study, which is superior to the results in other comparative experiments trained with the ground truths of P-phase images.

3.3. *Ablation Experimental Results*

In this section, ablation experiments were conducted on two parts of the proposed PVSegNet network, analyzing the effect of each module of our method and the pseudo label weights on the experimental results.

(1) Effect of each module on experimental results

Table 2 shows the impact of each module of our method on the experimental results when using the ground truths of the P-phase images to guide the portal vein segmentation on the H-phase or E-phase images. The first row shows the results of removing the style transfer module, discriminator D_B and pseudo labels; the second row shows the results of removing the pseudo labels; the third row shows the results of removing the style transfer module, i.e., removing the generator G and discriminator D_A; and the fourth row shows the results of removing the discriminator D_B. The experimental results in Table 2 show that the results all decrease after removing these structures. And our proposed PVSegNet network contains each structure and achieves the best segmentation performance.

(2) Effect of pseudo label weights on experimental results

Table 2. Effect of each module of PVSegNet on the experimental results. Modules with "✔" are included.

	#	D_B	G&D_A	PL	H-Phase		E-Phase	
					DSC	Jaccard	DSC	Jaccard
PVSegNet based	1				0.643 ± 0.000	0.486 ± 0.000	0.758 ± 0.003	0.672 ± 0.007
	2	✔	✔		0.666 ± 0.000	0.485 ± 0.002	0.819 ± 0.000	0.702 ± 0.001
	3	✔		✔	0.648 ± 0.000	0.493 ± 0.000	0.759 ± 0.003	0.624 ± 0.004
	4		✔	✔	0.676 ± 0.002	0.534 ± 0.002	0.823 ± 0.001	0.709 ± 0.002
PVSegNet	5	✔	✔	✔	**0.689 ± 0.001**	**0.546 ± 0.001**	**0.826 ± 0.001**	**0.712 ± 0.001**

Notes: #: Method number; D_B: Discriminator D_B; G&D_A: Generator G and Discriminator D_A; PL: Pseudo labels.

Table 3 shows that when different weights are set to construct pseudo labels, the network can obtain different portal vein segmentation results for H-phase and E-phase images. According to Table 3, when λ_1 is set as the weights of 0.1 or 0.7, the constructed pseudo labels can achieve the best results when segmenting the portal vein on H-phase or E-phase images, respectively.

Table 3. Effect of setting different weights for the pseudo labels on the experimental results.

λ_1	H-Phase		E-Phase	
	DSC	Jaccard	DSC	Jaccard
0.1	**0.689 ± 0.001**	**0.546 ± 0.001**	0.818 ± 0.000	0.708 ± 0.000
0.3	0.665 ± 0.001	0.529 ± 0.001	0.815 ± 0.001	0.698 ± 0.002
0.5	0.681 ± 0.002	0.520 ± 0.002	0.819 ± 0.001	0.703 ± 0.001
0.7	0.621 ± 0.002	0.474 ± 0.002	**0.826 ± 0.001**	**0.712 ± 0.001**
0.9	0.656 ± 0.000	0.498 ± 0.000	0.824 ± 0.001	0.708 ± 0.002

3.4. Visualization Results

Figure 5 shows the images of two cases from the sagittal plane. Each case was visualized with its original three-phase images and style transferred images generated by the generator G. The transferred P-phase image retains the features of the original portal vein after going through the generator G. There are contrast media on both the E-phase and P-phase images and the portal vein is clear. And after these images are transferred by the generator G, the transferred images also retain most of the information used to segment the portal vein even. The portal vein is not clear in the H-phase image, but after being transferred by generator G, the portal vein is clearer compared to the original image, and the global position information of the portal vein can be extracted despite the loss of some edge information.

To observe whether the proposed PVSegNet method achieves an effective segmentation of the portal vein, Figure 6 shows the segmentation results on the H-phase and E-phase images, respectively, with the green line indicating the contours of the ground truth and the red line indicating the contours of the segmentation results. From Figure 6, it can be seen that the networks trained directly with the annotations of portal vein images as pseudo-labels, i.e., 3D U-Net and CycleGan, have obviously missed the segmentation phenomenon and cannot segment the complete portal vein, and the segmentation results are slightly worse than those of PVSegNet. In the visualization results, the segmentation results in the E-phase are better relative to those in the H-phase because the vessels in the E-phase are more clearly defined.

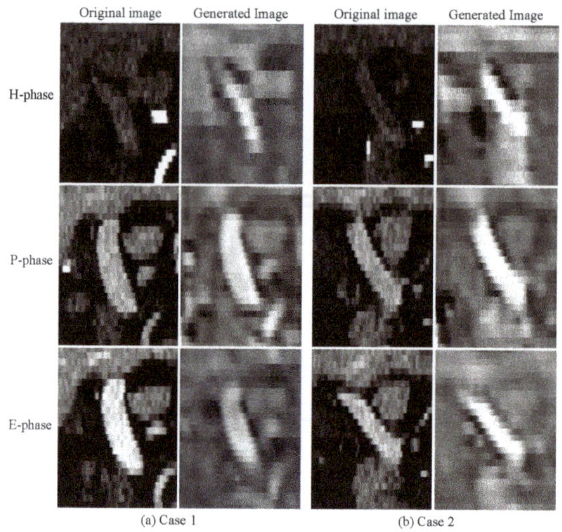

Figure 5. Images generated by the generator G in the style transfer module with similar styles to their respective P-phase image of case 1 (**a**) and case 2 (**b**).

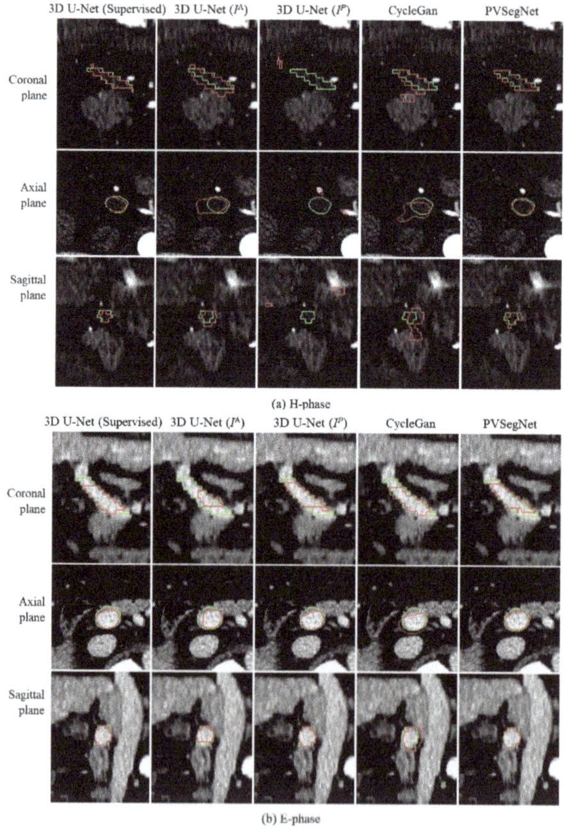

Figure 6. Visualization results of the comparison experiment in the coronal, axial and sagittal planes for the H-phase (**a**) and E-phase (**b**) images, respectively. The green line is the outline of the ground truth and the red line is the outline of segmentation results.

4. Discussion

Due to the labor-intensive and time-consuming manual annotation of the portal vein on multiphase abdominal CTA images, as well as the challenge of manually annotating the portal vein on H-phase images, we propose a multiphase portal vein segmentation method based on style transfer and pseudo labels. This method can automatically segment the portal vein on H-phase or E-phase images when only P-phase images have portal vein annotations.

The portal vein segmentation network, PVSegNet, mainly consists of a style transfer module and a segmentation module. The style transfer module can convert the H-phase and E-phase images of the same patient to the P-phase images, which makes the segmentation network more easily adaptable to both H-phase and E-phase images. The segmentation module consists of a 3D U-Net network and a discriminator. The discriminator is used to determine whether the segmentation results are similar to the P-phase ground truth, forming a soft constraint on the segmentation network. When using only the annotation of the P-phase images as the pseudo labels to guide the training of the network on the H-phase, as well as the E-phase images, the wrong information is often learned due to the vessel shift. In this paper, we use the annotation of P-phase images and the prediction results obtained from the segmentation network to jointly construct the pseudo labels of the corresponding phase during the training process so as to reduce the effect of vessel shift and obtain more accurate segmentation results. From Table 1, it can be concluded that the proposed method outperforms the model trained directly with the P-phase annotations and H-phase images in the H-phase test dataset. And the segmentation effect in the E-phase images is even better.

However, there is still room for improvement in the study. Firstly, the portal vein segmentation performance in the H-phase images needs further enhancement for the task. The next work will make full use of the annotations of P-phase images to improve the construction of pseudo labels and set strong constraints in the style transfer to achieve the improvement of the algorithm. Secondly, the deep learning segmentation method proposed in this study has been evaluated only on a CT dataset from one hospital, and the future plan is to apply the results of this study to multi-center data for validating the reliability of the algorithm through more clinical applications. Finally, there is a lack of diagnostic studies. Although the portal vein has been segmented, this paper did not use the portal vein to predict portal pressure, and the clinical study will be carried out after collecting more data from relevant patients.

5. Conclusions

In this paper, we propose a multiphase segmentation algorithm for portal veins based on style transfer and pseudo labels. Under the condition that only the P-phase portal vein is annotated, this method achieves automatic and effective segmentation of the portal vein in H-phase and E-phase CTA images. The portal vein segmentation network PVSegNet mainly consists of a style transfer module and a segmentation module. The style transfer module converts the E-phase and H-phase images of the same patient to the P-phase images with the same style. The segmentation module consists of pseudo labels guided 3D U-Net network and a discriminator with a soft constraint function. Since the pseudo labels are constructed based on three mutually constrained components, which are annotations on the P-phase image, prediction results obtained from the segmentation network, and soft constraints of the discriminator on the segmentation network, the method can reduce the effect of learned misinformation from vessel shift and distortion. And the method obtains more accurate portal vein segmentation results on both H-phase and E-phase images. This greatly aids clinical practice.

Author Contributions: Conceptualization, G.S. and Z.X.; methodology, G.S.; software, D.Y.; validation, G.S., Z.X. and H.W.; formal analysis, G.S.; investigation, G.S. and Y.S.; resources, S.C.; data curation, Z.X. and S.C.; writing—original draft preparation, G.S., Z.X. and S.L.; writing—review and editing, Y.S.; visualization, G.S.; supervision, Y.S.; project administration, Y.S.; funding acquisition, Y.S. All authors have read and agreed to the published version of the manuscript.

Funding: This work was supported by the medical-industrial integration project of Fudan University under grant XM03211181. This work was also supported by the National Natural Science Foundation of China under grant 82072021.

Institutional Review Board Statement: Not applicable.

Informed Consent Statement: Written informed consent has been obtained from the patient(s) to publish this paper.

Data Availability Statement: Not applicable.

Conflicts of Interest: The authors declare no conflict of interest.

Appendix A

Table A1. The *p*-value obtained by comparing the DSC and Jaccard of PVSegNet with other methods using a two-tailed student *t*-test. Inputs with "✔" are used to train the models.

Methods (vs. PVSegNet)	I^A	I^V	$True^A$	$True^V$	H-Phase		E-Phase	
					DSC	Jaccard	DSC	Jaccard
nnUNet (Supervised)	✔		✔		** 4.91×10^{-22}	** 9.09×10^{-46}	** 1.15×10^{-26}	2.73×10^{-1}
3D U-Net (Supervised)	✔		✔		** 4.02×10^{-3}	** 4.98×10^{-3}	2.40×10^{-1}	1.50×10^{-1}
3D U-Net	✔			✔	** 1.24×10^{-9}	** 1.38×10^{-11}	** 1.52×10^{-33}	** 5.89×10^{-40}
3D U-Net		✔		✔	** 1.75×10^{-75}	** 1.36×10^{-74}	** 1.06×10^{-5}	** 1.37×10^{-5}
CycleGan	✔	✔		✔	** 4.15×10^{-4}	** 3.66×10^{-5}	** 6.48×10^{-34}	** 5.15×10^{-41}

Notes: **: *p*-value < 0.01, significant statistical difference.

Table A2. The *p*-value obtained by comparing the DSC and Jaccard of PVSegNet with PVSegNet-based methods using a two-tailed student *t*-test. Modules with "✔" are included.

	#	D_B	$G\&D_A$	PL	H-Phase		E-Phase	
					DSC	Jaccard	DSC	Jaccard
PVSegNet Based vs. PVSegNet	1				** 1.59×10^{-4}	** 4.15×10^{-6}	** 2.98×10^{-20}	* 9.82×10^{-22}
	2	✔	✔		* 1.72×10^{-2}	** 1.05×10^{-3}	9.76×10^{-2}	5.92×10^{-2}
	3	✔		✔	** 3.47×10^{-4}	** 2.58×10^{-6}	** 3.93×10^{-18}	* 1.61×10^{-20}
	4		✔	✔	1.06×10^{-1}	1.10×10^{-1}	4.31×10^{-1}	6.63×10^{-1}

Notes: #: Method number; D_B: Discriminator D_B; $G\&D_A$: Generator G and Discriminator D_A; PL: Pseudo labels; *: *p*-value < 0.05, statistical difference; **: *p*-value < 0.01, significant statistical difference.

Table A3. The *p*-value obtained by comparing the DSC and Jaccard of different λ_1 using a two-tailed student *t*-test. For the H-phase, compare the values of $\lambda_1 = 0.1$ with other λ_1 values, and for the E-phase, compare the values of $\lambda_1 = 0.7$ with other λ_1 values.

λ_1	H-Phase		E-Phase	
	DSC	Jaccard	DSC	Jaccard
0.1			* 4.65×10^{-2}	* 2.77×10^{-2}
0.3	** 8.89×10^{-3}	** 5.24×10^{-3}	** 6.81×10^{-3}	** 8.92×10^{-3}
0.5	** 8.79×10^{-3}	** 1.58×10^{-3}	8.66×10^{-2}	9.18×10^{-2}
0.7	** 4.60×10^{-10}	** 3.09×10^{-11}		
0.9	** 8.12×10^{-4}	** 1.99×10^{-4}	5.49×10^{-1}	4.70×10^{-1}

Notes: *: *p*-value < 0.05, statistical difference; **: *p*-value < 0.01, significant statistical difference.

References

1. Carneiro, C.; Brito, J.; Bilreiro, C.; Barros, M.; Bahia, C.; Santiago, I.; Caseiro-Alves, F. All about portal vein: A pictorial display to anatomy, variants and physiopathology. *Insights Into Imaging* **2019**, *10*, 38. [CrossRef]
2. Gallego, C.; Velasco, M.; Marcuello, P.; Tejedor, D.; De Campo, L.; Friera, A. Congenital and acquired anomalies of the portal venous system. *Radiographics* **2002**, *22*, 141–159. [CrossRef]
3. Neil Granger, D.; Holm, L.; Kvietys, P. The gastrointestinal circulation: Physiology and pathophysiology. *Compr. Physiol.* **2011**, *5*, 1541–1583.
4. Lebre, M.A.; Vacavant, A.; Grand-Brochier, M.; Rositi, H.; Abergel, A.; Chabrot, P.; Magnin, B. Automatic segmentation methods for liver and hepatic vessels from CT and MRI volumes, applied to the Couinaud scheme. *Comput. Biol. Med.* **2019**, *110*, 42–51. [CrossRef] [PubMed]
5. Han, X.; Wu, X.; Wang, S.; Xu, L.; Xu, H.; Zheng, D.; Yu, N.; Hong, Y.; Yu, Z.; Yang, D.; et al. Automated segmentation of liver segment on portal venous phase MR images using a 3D convolutional neural network. *Insights Imaging* **2022**, *13*, 26. [CrossRef]
6. Guo, Q.; Song, H.; Fan, J.; Ai, D.; Gao, Y.; Yu, X.; Yang, J. Portal Vein and Hepatic Vein Segmentation in Multi-Phase MR Images Using Flow-Guided Change Detection. *IEEE Trans. Image Process.* **2022**, *31*, 2503–2517. [CrossRef] [PubMed]
7. Huang, A.; Cheng, W.H.; Lee, C.W.; Yang, C.Y.; Liu, H.M. Multiphase computed tomographic angiography with bone subtraction using 3D multichannel convolution neural networks. In Proceedings of the 2020 42nd Annual International Conference of the IEEE Engineering in Medicine & Biology Society (EMBC), Montreal, QC, Canada, 20–24 July 2020; pp. 1274–1277.
8. Bharati, S.; Mondal, M.; Podder, P.; Prasath, V. Deep learning for medical image registration: A comprehensive review. *arXiv* **2022**, arXiv:2204.11341.
9. Ibragimov, B.; Toesca, D.; Chang, D.; Koong, A.; Xing, L. Combining deep learning with anatomical analysis for segmentation of the portal vein for liver SBRT planning. *Phys. Med. Biol.* **2017**, *62*, 8943. [CrossRef]
10. Çiçek, Ö.; Abdulkadir, A.; Lienkamp, S.S.; Brox, T.; Ronneberger, O. 3D U-Net: Learning dense volumetric segmentation from sparse annotation. In Proceedings of the Medical Image Computing and Computer-Assisted Intervention–MICCAI 2016: 19th International Conference, Athens, Greece, 17–21 October 2016; Part II 19. Springer: Cham, Switzerland, 2016; pp. 424–432.
11. Kitrungrotsakul, T.; Han, X.H.; Iwamoto, Y.; Lin, L.; Foruzan, A.H.; Xiong, W.; Chen, Y.W. VesselNet: A deep convolutional neural network with multi pathways for robust hepatic vessel segmentation. *Comput. Med. Imaging Graph.* **2019**, *75*, 74–83. [CrossRef]
12. Yu, W.; Fang, B.; Liu, Y.; Gao, M.; Zheng, S.; Wang, Y. Liver vessels segmentation based on 3d residual U-NET. In Proceedings of the 2019 IEEE International Conference on Image Processing (ICIP), Taipei, Taiwan, 22–25 September 2019; pp. 250–254.
13. Xu, Z.; Lu, D.; Wang, Y.; Luo, J.; Jayender, J.; Ma, K.; Zheng, Y.; Li, X. Noisy labels are treasure: Mean-teacher-assisted confident learning for hepatic vessel segmentation. In Proceedings of the Medical Image Computing and Computer Assisted Intervention–MICCAI 2021: 24th International Conference, Strasbourg, France, 27 September–1 October 2021; Proceedings, Part I 24. Springer: Cham, Switzerland, 2021; pp. 3–13.
14. Tao, G.; Singh, A.; Bidaut, L. Liver segmentation from registered multiphase CT data sets with EM clustering and GVF level set. In Proceedings of the Medical Imaging 2010: Image Processing, San Diego, CA, USA, 14–16 February 2010; Volume 7623, pp. 300–308.
15. Zhang, X.; Tan, X.; Gao, X.; Wu, D.; Zhou, X.; Fujita, H. Non-rigid registration of multi-phase liver CT data using fully automated landmark detection and TPS deformation. *Clust. Comput.* **2019**, *22*, 15305–15319. [CrossRef]
16. Rameshbabu, C.; Wani, Z.A.; Rai, P.; Abdulqader, A.; Garg, S.; Sharma, M. Standard imaging techniques for assessment of portal venous system and its tributaries by linear endoscopic ultrasound: A pictorial essay. *Endosc. Ultrasound* **2013**, *2*, 16. [PubMed]
17. Goodfellow, I.; Pouget-Abadie, J.; Mirza, M.; Xu, B.; Warde-Farley, D.; Ozair, S.; Courville, A.; Bengio, Y. Generative adversarial networks. *Commun. ACM* **2020**, *63*, 139–144. [CrossRef]
18. Weiss, K.; Khoshgoftaar, T.M.; Wang, D. A survey of transfer learning. *J. Big Data* **2016**, *3*, 9. [CrossRef]
19. Patel, V.M.; Gopalan, R.; Li, R.; Chellappa, R. Visual domain adaptation: A survey of recent advances. *IEEE Signal Process. Mag.* **2015**, *32*, 53–69. [CrossRef]
20. Zhu, J.Y.; Park, T.; Isola, P.; Efros, A.A. Unpaired image-to-image translation using cycle-consistent adversarial networks. In Proceedings of the IEEE International Conference on Computer Vision, Venice, Italy, 22–29 October 2017; pp. 2223–2232.
21. Jiang, J.; Rimner, A.; Deasy, J.O.; Veeraraghavan, H. Unpaired cross-modality educed distillation (CMEDL) for medical image segmentation. *IEEE Trans. Med. Imaging* **2021**, *41*, 1057–1068. [CrossRef] [PubMed]
22. Zhao, Z.; Xu, K.; Li, S.; Zeng, Z.; Guan, C. Mt-uda: Towards unsupervised cross-modality medical image segmentation with limited source labels. In Proceedings of the Medical Image Computing and Computer Assisted Intervention–MICCAI 2021: 24th International Conference, Strasbourg, France, 27 September–1 October 2021; Proceedings, Part I 24. Springer: Cham, Switzerland, 2021; pp. 293–303.
23. Chen, C.; Dou, Q.; Chen, H.; Qin, J.; Heng, P.A. Unsupervised bidirectional cross-modality adaptation via deeply synergistic image and feature alignment for medical image segmentation. *IEEE Trans. Med. Imaging* **2020**, *39*, 2494–2505. [CrossRef] [PubMed]
24. Weston, J.; Ratle, F.; Collobert, R. Deep learning via semi-supervised embedding. In Proceedings of the 25th International Conference on Machine Learning, Helsinki, Finland, 5–9 July 2008; pp. 1168–1175.
25. Qu, L.; Liu, S.; Liu, X.; Wang, M.; Song, Z. Towards label-efficient automatic diagnosis and analysis: A comprehensive survey of advanced deep learning-based weakly-supervised, semi-supervised and self-supervised techniques in histopathological image analysis. *Phys. Med. Biol.* **2022**, *67*, 20. [CrossRef] [PubMed]

26. Raju, A.; Cheng, C.T.; Huo, Y.; Cai, J.; Huang, J.; Xiao, J.; Lu, L.; Liao, C.; Harrison, A.P. Co-heterogeneous and adaptive segmentation from multi-source and multi-phase CT imaging data: A study on pathological liver and lesion segmentation. In Proceedings of the Computer Vision–ECCV 2020: 16th European Conference, Glasgow, UK, 23–28 August 2020; Proceedings, Part XXIII. Springer: Cham, Switzerland, 2020; pp. 448–465.
27. Qu, T.; Wang, X.; Fang, C.; Mao, L.; Li, J.; Li, P.; Qu, J.; Li, X.; Xue, H.; Yu, Y.; et al. M3Net: A multi-scale multi-view framework for multi-phase pancreas segmentation based on cross-phase non-local attention. *Med. Image Anal.* **2022**, *75*, 102232. [CrossRef] [PubMed]
28. Zou, K.H.; Warfield, S.K.; Bharatha, A.; Tempany, C.M.; Kaus, M.R.; Haker, S.J.; Wells, W.M., III; Jolesz, F.A.; Kikinis, R. Statistical validation of image segmentation quality based on a spatial overlap index1: Scientific reports. *Acad. Radiol.* **2004**, *11*, 178–189. [CrossRef] [PubMed]

Journal of
Clinical Medicine

MDPI

Article

Performance of Fully Automated Algorithm Detecting Bone Marrow Edema in Sacroiliac Joints

Joanna Ożga [1,*], Michał Wyka [1], Agata Raczko [1], Zbisław Tabor [2], Zuzanna Oleniacz [1], Michał Korman [1] and Wadim Wojciechowski [1]

1 Department of Radiology, Jagiellonian University Medical College, ul. Botaniczna 3, 31-503 Krakow, Poland; michal.wyka@student.uj.edu.pl (M.W.); agata.raczko@student.uj.edu.pl (A.R.); zuzanna.oleniacz@student.uj.edu.pl (Z.O.); michal.korman@student.uj.edu.pl (M.K.); wadim.wojciechowski@uj.edu.pl (W.W.)

2 Department of Biocybernetics and Biomedical Engineering, AGH University of Science and Technology, al. Adama Mickiewicza 30, 30-059 Krakow, Poland; ztabor@agh.edu.pl

* Correspondence: joannakarolina.ozga@student.uj.edu.pl; Tel.: +48-600-298-383

Abstract: This study evaluates the performance of a fully automated algorithm to detect active inflammation in the form of bone marrow edema (BME) in iliac and sacral bones, depending on the quality of the coronal oblique plane in patients with axial spondyloarthritis (axSpA). The results were assessed based on the technical correctness of MRI examination of the sacroiliac joints (SIJs). A total of 173 patients with suspected axSpA were included in the study. In order to verify the correctness of the MRI, a deviation angle was measured on the slice acquired in the sagittal plane in the T2-weighted sequence. This angle was located between the line drawn between the posterior edges of S1 and S2 vertebrae and the line that marks the actual plane in which the slices were acquired in T1 and STIR sequences. All examinations were divided into quartiles according to the deviation angle measured in degrees as follows: 1st group [0; 2.2], 2nd group (2.2; 5.7], 3rd group (5.7; 10] and 4th group (10; 29.2]. Segmentations of the sacral and iliac bones were acquired manually and automatically using the fully automated algorithm on the T1 sequence. The Dice coefficient for automated bone segmentations with respect to reference manual segmentations was 0.9820 (95% CI [0.9804, 0.9835]). Examinations of BME lesions were assessed using the SPARCC scale (in 68 cases SPARCC > 0). Manual and automatic segmentations of the lesions were performed on STIR sequences and compared. The sensitivity of detection of BME ranged from 0.58 (group 1) to 0.83 (group 2) versus 0.76 (total), while the specificity was equal to 0.97 in each group. The study indicates that the performance of the algorithm is satisfactory regardless of the deviation angle.

Keywords: deep learning; axial spondyloarthritis; sacroiliac joint; fully automated algorithm; coronal oblique plane

Citation: Ożga, J.; Wyka, M.; Raczko, A.; Tabor, Z.; Oleniacz, Z.; Korman, M.; Wojciechowski, W. Performance of Fully Automated Algorithm Detecting Bone Marrow Edema in Sacroiliac Joints. *J. Clin. Med.* **2023**, *12*, 4852. https://doi.org/10.3390/jcm12144852

Academic Editors: Rafał Obuchowicz, Michał Strzelecki, Adam Piorkowski and Giuseppe Ingravallo

Received: 9 May 2023
Revised: 18 June 2023
Accepted: 21 July 2023
Published: 24 July 2023

1. Introduction

Axial spondyloarthritis (axSpA) is a group of rheumatic inflammatory conditions that affect the sacroiliac, intervertebral and facet joints [1]. The prevalence of axSpA ranges from nine to 30 per 10,000 people [2]. It is approximately two- to three-fold higher in men than in women [3]. In general, axSpA initially presents during the third decade of life [4]. The classification criteria for axial spondyloarthritis, developed by the Assessment in SpondyloArthritis international Society (ASAS), include the presence of sacroiliitis detected by radiography or by magnetic resonance imaging (MRI) with the addition of at least one SpA feature ("imaging arm"), or the presence of HLA-B27 with the addition of at least two SpA features ("clinical arm") [5]. AxSpA results in persistent inflammation of the sacroiliac joints (SIJs), causing chronic back pain, stiffness, skeletal and postural changes. This condition limits the ability to perform daily living activities, which contributes to a negative effect that axSpA has on HRQoL (health-related quality of life) [5]. However, appropriate treatment, physical therapy, and psychosocial interventions can improve HRQoL [6].

Currently, magnetic resonance imaging (MRI) of sacroiliac joints is considered a first-line method of choice for early axSpA diagnostics [7], although other imaging modalities can also be used to diagnose abnormal conditions in the joints [8] This examination is performed only for specific indications, including bone, joint or soft tissue infection, inflammation of SIJ, SIJ dysfunction, ankylosing spondylitis, degenerative arthritis, SIJ pain or stiffness, avascular necrosis, rheumatoid arthritis, osteoarthritis, sacroiliitis, psoriasis and gout [9]. MRI is, therefore, a diagnostic procedure that enables the assessment of active inflammation within SIJs (sacroiliitis) [10,11].

This type of examination allows the evaluation of bone marrow edema (BME), erosions, fat lesions, sclerosis or ankylosis (bone formation) [12]. According to the ASAS classification criteria, active sacroiliitis in MRI is defined by the following:

1. Bone marrow edema (collection of fluid between bone marrow cells due to inflammation) visible on "fluid-sensitive" STIR sequence (Short Tau Inversion Recovery) (also known as T2-weighted sensitive to water sequence) as areas hyperintense to the sacral interforaminal bone marrow (brighter areas).
2. Bone marrow enhancement (osteitis), which is observed on T1-weighted sequence after contrast media administration [9,13,14].

To determine the presence of sacroiliitis in MRI, such changes must be clearly present, located in a typical anatomical area (subchondral bone), and apparent on at least two consecutive slices of examination, or more than one lesion must occur on the same slice [9,10,15]. The SPARCC (The Spondyloarthritis Research Consortium of Canada) scale is used for evaluating inflammatory lesions in the SIJs, measuring the presence, depth, and intensity of bone marrow inflammation [16].

Early diagnosis of axSpA is crucial to treatment results and the future quality of life of affected patients [17]. However, there is often perennial delay between the onset of symptoms and diagnosis [18,19].

Using the criteria of the European Spondyloarthropathy Study Group (ESSG) or Amor [20], patients who do not fulfill the criteria for AS but present with clinical features of SpA can be classified as undifferentiated SpA. Nevertheless, long-term studies suggest that more than 40% of patients in this group may not develop AS even after 10 years of follow-up. This clearly demonstrates the importance of early diagnosis, especially in the preradiographic phase when sacroiliac arthritis is not yet visible on radiographs [21], necessitating the use of other imaging techniques. Hence, the crucial role of using MRI in the early diagnosis of axSpA [22,23]. The main benefits of accelerating the diagnostic process are improved results of biological treatment [24] and prevention or delay of the onset progression in the form of structural damage [25].

Artificial intelligence (AI) can be defined as the ability of a computer system to perform tasks that would typically or traditionally require human intelligence [26]. Machine learning is a branch of artificial intelligence that allows the extraction of meaningful patterns from practical examples [27,28]. MRI analysis of patients with axSpA is a time-consuming process, requiring radiological specialized knowledge and evaluation experience. Streamlining this process can be possible through the use of artificial intelligence [29,30]. Up to now, only two algorithms for semi-automated detection and quantification of axSpA-related sacroiliitis have been developed—the first by Zarco et al. [31] and the other one by Kucybała et al. [12]. In the study of Rzecki et al. [30], a fully automated algorithm for BME evaluation was described. This fully automated algorithm was compared with other approaches in [32], further demonstrating its potential for clinical use. The following year, another algorithm for analyzing MR images of the SIJs in axial spondyloarthritis was developed by Bressem et al. [33]. In contrast to the algorithm described in [30], the algorithm of Bressem et al. [33] focuses only on classifying whole MR images as either normal or abnormal; that is, the location and the volume of the lesion are not determined using the algorithm [33].

The algorithm described in [30] was, however, developed and tested using a carefully selected dataset. During the selection, attention was paid to the high technical quality of the MR examination, which do not necessarily always meet everyday clinical conditions. Before the algorithm of [30] can be implemented in clinical settings, it must be demonstrated that it is not sensitive to typical variations in the MR acquisition protocol. For this reason, the aim of this study is to validate the performance of the fully automated algorithm [30] designed for BME evaluation depending on the technical correctness of the MRI examinations of the SIJs.

2. Materials and Methods

2.1. Material

The study obtained approval from the Institutional Bioethics Committee (No. Of approval: 1072.6120.332. 2022, date of approval: 18 January 2023). This is a retrospective study of 173 patients who underwent MRI examinations of the SIJs. In this group, 47% (n = 82) of patients were male, while 53% (n = 91) of patients were female. The median age of the patients was 31, with an IQR of 23 years (range 18–86 years). All of them underwent sacroiliac joint MRI due to suspected axSpA. Therefore, the patients population is very diverse and among them there are people without confirmed axSpA, people in the early stages of the axSpA, and those who were found to have irreversible changes after examination. Random selection of patients was crucial in order to evaluate the performance of the algorithm, as it best reflects the population that the algorithm will deal with if implemented as a tool used in daily practice by radiologists or other specialists. The same applies to the verification of the algorithm with regard to the technical correctness of the MRI examinations of the SIJs.

All analyses were based on processing T1-weighted, T2-weighted, and STIR sequence images. T2-weighted sequences were acquired in the sagittal plane, while T1-weighted and STIR sequences were in the coronal oblique plane, and the position of the patient remained unchanged during their acquisition time. The examinations involved in the present research were performed using a 3.0 Tesla MRI scanner (Achieva, Philips Healthcare, Amsterdam, The Netherlands) and an 8-channel phased-array XL-torso body matrix coil.

Detailed scan parameters were as follows:

- For T1-weighted turbo spin echo (TSE) sequence—TR 500 ms, TE 14 ms, flip angle 90, NEX 1, slice thickness 3 mm, matrix 560 × 560, FOV 240 × 240 × 71,
- For STIR TSE sequence—TR 5239 ms, TE 30 ms, inversion time 190 ms, flip angle 90, NEX 2, slice thickness 3 mm, matrix 400 × 400, FOV 240 × 240 × 71.

In each MRI examination, the number of slices in each series varies between 12 and 24.

2.2. Methods

In order to properly evaluate the performance of the algorithm depending on the technical correctness of the MRI examination of the sacroiliac joints, this study consisted of the following steps:

1. Assessment of the correctness of the alignment of the MRI sections of the SIJs.
2. Enhancement of the pre-existing algorithm in the form of post-processing adjustments.
3. Manual and automatic segmentation of the sacrum and iliac bones.
4. Evaluation of BME using the SPARCC scale.
5. Manual and automatic segmentation of BME.
6. Statistical analysis of the results.

2.2.1. Assessment of the Correctness of the Alignment of the MRI Section of the SIJs

In order to standardize the assessment, the researchers decided to create an angle, which will determine the technical correctness of the entire examination. This angle is called the deviation angle and is located between the fully correct line drawn between the posterior edges of the S1 and S2 vertebrae and the line that marks the actual plane in which the slices were acquired in the T1 and STIR sequences. Owing to this method, the researchers could verify how the fully automated algorithm performs in real-life conditions. The greater the deviation angle measured by the investigators, the lower the quality of the slices in T1-weighted and STIR sequences. Slices obtained in an inaccurate plane do not fully visualize the sacroiliac joints, which impairs their assessment.

The deviation angles were measured on T2-weighted sequence slices obtained in the sagittal plane, as shown in Figure 1. Figure 2, presented below, shows the interrelationship of the three sequences of the same examination (as shown in Figure 1) that were used in this study. The RadiAnt DICOM Viewer was used to take measurements. All examinations were analyzed by two independent researchers in a blind study. The final results of the measurements were averaged to obtain a single deviation angle corresponding to each examination.

Furthermore, all examinations have been divided into four groups based on the quartile values of the angles measured in degrees, ensuring that the groups have similar sizes. The criteria for dividing the examinations into groups are listed below:

1. Angle dimensions [0; 2.2].
2. Angle dimensions (2.2; 5.7].
3. Angle dimensions (5.7; 10].
4. Angle dimensions (10; 29.2].

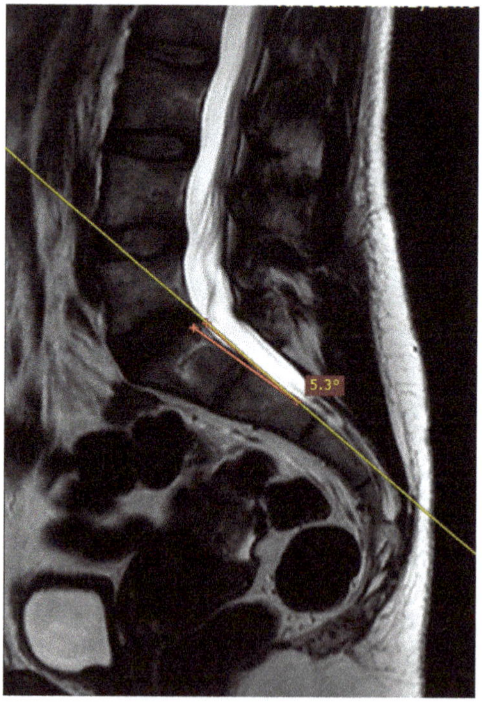

Figure 1. The measured deviation angle equals 5.3 degrees (group 2). The red line is the fully correct line drawn between the posterior edges of the S1 and S2 vertebrae. The yellow line marks the actual plane in which the slices were acquired.

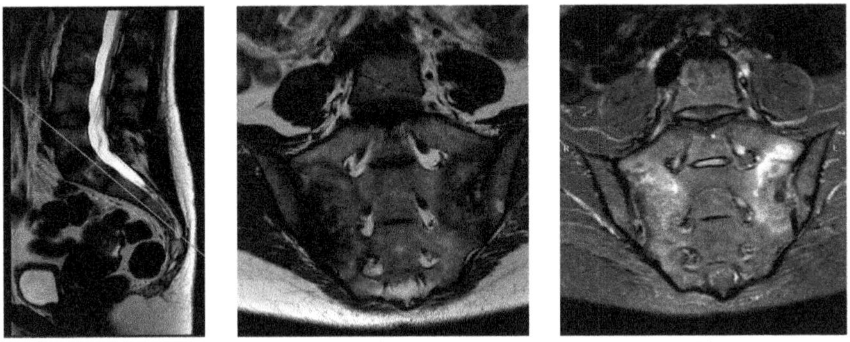

Figure 2. (**a**) Example of sagittal T2 series. Yellow line indicates the position of coronal projections (**b**,**c**). (**b**) Example of coronal T2 series. (**c**) Example of coronal STIR series.

2.2.2. Enhancement of the Pre-Existing Algorithm in the Form of Post-Processing Adjustments

To perform the detection of the first sacrum and iliac bones, and then the BME, the fully automated algorithm described previously was used [12,30]. The algorithm, described in detail in the study of Kucybała et al. [12] and Rzecki et al. [30], creates segmentations of three regions corresponding to both the iliac bones and sacrum in T1-weighted sequences, and then these segmentations are superimposed on STIR sequence so that the algorithm can perform segmentations of the BME. The transfer of bone segmentations from T1-weighted to STIR-weighted sequences is feasible due to the assumption that the patient's position was unchanged during the acquisition of both sequences.

Post-processing adjustments were implemented to adapt the performance of the algorithm [30] to the guidelines [9,10,15], which state that an inflammatory lesion can be identified if it is present on two consecutive MRI slices or if there is more than one lesion on a single slice. In addition, the inflammatory lesion must be located in typical anatomical areas, such as subchondral or periarticular bone marrow [9].

2.2.3. Manual and Automatic Segmentation of the Sacrum and Iliac Bones

In order to compare the performance of the improved algorithm depending on the technical correctness of the MRI of the sacroiliac joints, it is necessary to perform both manual and automatic bone segmentations. Therefore, it was decided to manually segment the iliac and sacral bones and then evaluate their agreement with the segmentations created by the algorithm. The area of the right and left iliac bones, as well as the sacrum, were manually labeled on consecutive slices of the T1-weighted sequence by one researcher. The segmentations were prepared manually with the 3D Slicer (version 4.11.20210226) program and then verified by two other members of the research team. After the verification, it was stated that the created segmentations cannot be more accurate and can serve as a reference for segmentations performed by the algorithm. The establishment of reference segmentation considerably simplifies the process of comparing algorithm performance, as it provides more tools that can assess the compatibility of human and artificial intelligence. The compatibility of both segmentations was evaluated by several means. Initially, to have all bone segmentations manually reviewed one by one by the researchers, a proprietary scale was introduced to assess compliance. The scale is shown in Figure 3 and was based on the SPARCC scale. In this visual scale, each examination can receive a maximum score of 48 points. Only six slices that represent the largest surface of the SIJs are selected and afterwards evaluated. Each SIJ is divided into quadrants. In each quadrant, the correspondence of manual and fully automated segmentation is evaluated; if both segmentations correspond, one point is granted. Secondly, the Dice coefficient (DC) is calculated as follows:

$$DC = \frac{2|M \cap A|}{|M| + |A|}$$

where M and A stand for manually and automatically segmented bone regions, respectively, while $|X|$ for some region X denotes the volume (number of voxels) of X.

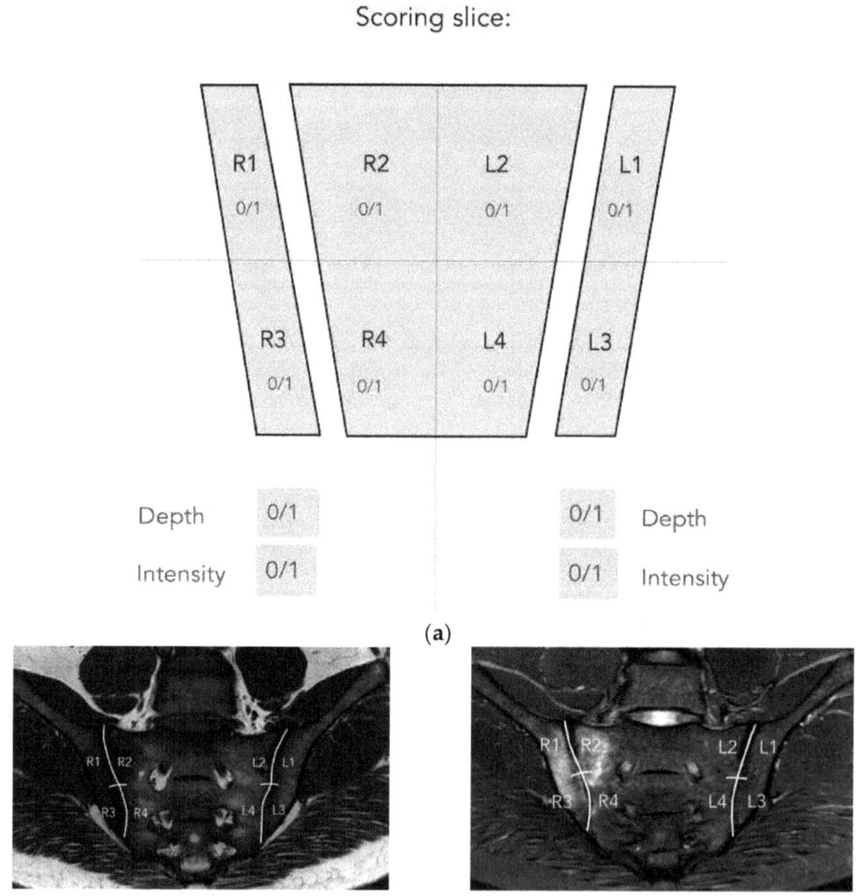

Figure 3. Scoring scale used in the study. The evaluation involves six slices of each examination. (**a**) Graphic illustrating the division of the SIJ into quadrants: R1—right iliac (upper), R2—right sacral (upper), R3—right iliac (lower), R4—right sacral (lower), L1—left iliac (upper), L2—left sacral (upper), L3—left iliac (lower), L4—left sacral (lower). Depth and intensity rating details (SPARCC scale) are provided below the graphic. (**b**) Quadrant division illustrated on the T1 sequence. (**c**) Corresponding quadrant division illustrated on the STIR sequence.

2.2.4. Evaluation of BME Using the SPARCC Scale

To assess the presence, depth, and intensity of BME, the SPARCC scale was used. Only six selected slices representing the largest proportion of the SIJ are evaluated. Points are assigned dichotomously: present—1, absent—0. Each SIJ is divided into four quadrants: 1 upper iliac, 2 lower iliac, 3 upper sacral, 4 lower sacral (Figure 3). The presence of increased signal in each quadrant is recorded. The maximum score for two joints in each slice is eight. The maximum score for the whole examination is 48. All lesions presented in the iliac bone are evaluated, whereas in the sacrum, lesions medially extend as far as the lateral border of the sacral foramina. A score for "intensity" can be granted separately for each of the SIJs on each slice. The reference for awarding a point for intensity is high signal from slow-flowing venous blood within presacral veins. One point is awarded when the "intense" signal is present in at least one quadrant of the joint on a single slice. The maximum score for one slice is two, therefore for six slices it is 12. A score for "depth" can be granted separately for each of the SIJs on each slice. A lesion is classified as "deep" if there is a homogeneous and unambiguous increase in signal extending over a depth of at least 1 cm from the articular surface. One point is awarded when the "deep" signal is present in at least one quadrant of the joint on a single slice. The maximum score for one slice is two, therefore for six slices it is 12 [14].

The total maximum score is 72:

1. Presence of bone marrow edema = 48.
2. Presence of intense edema = 12.
3. Presence of deep edema = 12.

All examinations were analyzed by two researchers who did not contact each other during the SPARCC scale completion. After obtaining the scores, the results of the two researchers were compared, and examinations that were rated unequally were submitted to a consortium composed of all of members of the research team, who collectively decided on the ultimate scores.

2.2.5. Manual and Automatic Segmentation of BME

The manual segmentations of BME based on SPARCC scale scores were completed for each examination using the 3D Slicer (version 4.11.20210226) program. Segmentations were obtained on all slices of the STIR sequence of each examination. Once all the manual segmentations were produced, each segmentation was further verified by three researchers for complete accuracy and consistency. This method of creating reference segmentations was adopted because the evaluation of lesions is subjective (meaning that the ground truth is burdened with uncertainty, making the problem difficult). Therefore, it was decided that the review performed by a total of three researchers can indeed be considered reliable. The resulting segmentations were consensually recognized as reference segmentations to be subsequently used for verification of automatic segmentations. The evaluation of segmentations performed manually and those performed by the automated algorithm has also been verified in several ways. The first step was a semi-quantitative assessment based on a scale created by the researchers (Figure 3), similar to the scale used previously to evaluate the correspondence of the manual and automated segmentation of sacral and iliac bones. The maximum score was 48 points, with one point granted for each quadrant of both joints if the manual reference and automated segmentations equally detected or excluded inflammatory lesion. Due to this scale, all segmentations of inflammatory lesions were required to be viewed by the researchers.

In addition to this visual scale, both versions of segmentations were further compared in a quantitative manner. The automated segmentation of lesions assigned a real number to every voxel within the volume of interest, which can be interpreted as a probability that the voxel is within a lesion region. Then, if the probability is larger than some user-selected decision threshold, the voxel receives a lesion labeled. Otherwise, it receives a label of healthy tissue. Based on this labeling, the evaluation is further based on counting the voxels of true positive (TP), true negative (TN), false positive (FP) and false negative (FN). Voxels corresponding to areas of inflammatory lesions marked by both manual and automatic segmentations were considered true positive. Truly negative areas were those where the inflammatory lesions were not covered by any segmentation. On the basis of the above data, specificity, accuracy and sensitivity were calculated.

The selection of the decision threshold mentioned above is usually based on the analysis of the ROC (Receiver Operating Characteristic). To generate such curves, the decision threshold is varied from some low to some high value. Then, TP, TN, FP and FN are counted, and from these numbers, the false positive rate (FPR) and true positive rate (TPR) are calculated. The ROC curve is the plot of TPR vs. FPR. The area under the ROC curve (AUC ROC) is a measure of automated classification accuracy, with AUC ROC equal to one for a hypothetical ideal classification algorithm. The optimal decision threshold can be derived from the ROC curve, for example, as the one that maximizes the sum of specificity and sensitivity [12].

2.2.6. Statistical Analysis of the Results

Since the studies were divided into groups on the basis of the mean measured deviation angle of the two raters, the intraclass correlation coefficient (ICC) was calculated to test the absolute agreement of those raters. The ICC (2, 2) estimate and its 95% confidence interval were calculated based on the mean rating (k = 2), absolute agreement, and the two-way random effects model.

Statistical analysis was then applied to the bone and BME segmentations, visual scale results, and SPARCC scores. In particular, they were checked for differences in groups based on the deviation angle value using the Kruskal-Wallis H test due to the rank character of those scales. Additionally, the Dice coefficient (DC) was calculated on the basis of comparison between algorithm prediction and the manually created mask (ground truth) and analyzed in terms of group differences also using the Kruskal-Wallis H test. However, this coefficient was calculated only for bone segmentations due to its limitations when describing small structures [34]. Algorithm predictions for BME were evaluated with the use of ROC. ROC curves were calculated for all the samples and separately for the samples assigned to the four groups differing by the technical correctness of the MRI examination. Additionally, specificity, sensitivity and accuracy based on total number of voxels in algorithm predictions indicated as TP, TN, FP or FN when comparing with manual segmentations of BME were calculated, assuming the decision threshold equal to 0.5. Sensitivity based on the count of single lesions detected by the algorithm was also calculated.

Statistical analysis was performed using the Stata/SE 17.0 package with the assumption of a significance level at 0.05.

3. Results

3.1. Assessment of the Correctness of the Alignment of the MRI Section of the SIJs

Overall, 20% (n = 35) of the examinations included in the study were performed perfectly correctly considering the technical aspect of the MRI procedure. The examination was defined as fully correct if the deviation angle measured by all of the researchers equaled 0 degrees (Figure 4). The median deviation angle was equal to 5.7 degrees. The maximal deviation angle measured was 29.2 degrees (Figure 5). The ICC (2, 2) was computed as 0.9960 (95% CI [0.9945, 0.9970]), which indicates excellent reliability between the researchers when measuring the deviation angle. Additionally, Figure 6 represents a Bland-Altman diagram with the limits of 95% agreement stated. The results of the division into quartiles due to the deviation angle measurements of all the examinations are shown in Table 1. The obtained division was considered to be valid for the whole study, and on its basis the analysis of the subsequent parts of the study was performed.

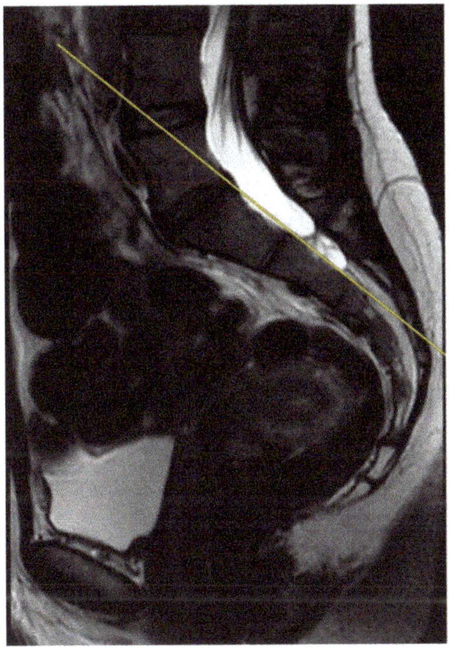

Figure 4. The measured deviation angle equals 0 degrees, properly performed MRI examination of the sacroiliac joints, T2-weighted sequence. The yellow line marks the actual plane in which the slices were acquired, while it is also the fully correct line drawn between the posterior edges of the S1 and S2 vertebrae.

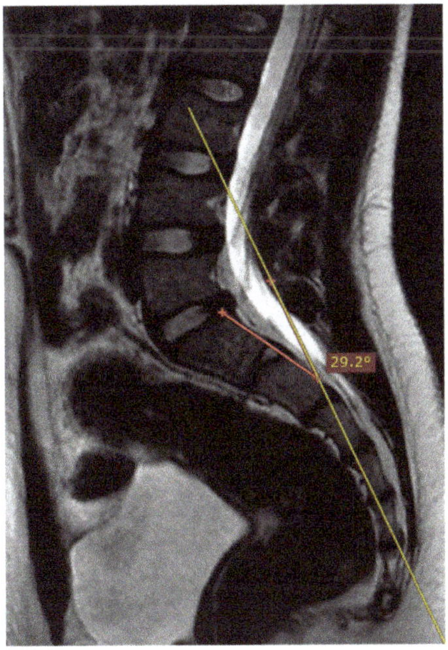

Figure 5. The measured deviation angle equals 29.2 degrees (group 4), T2-weighted sequence. The red line is the fully correct line drawn between the posterior edges of the S1 and S2 vertebrae. The yellow line marks the actual plane in which the slices were acquired.

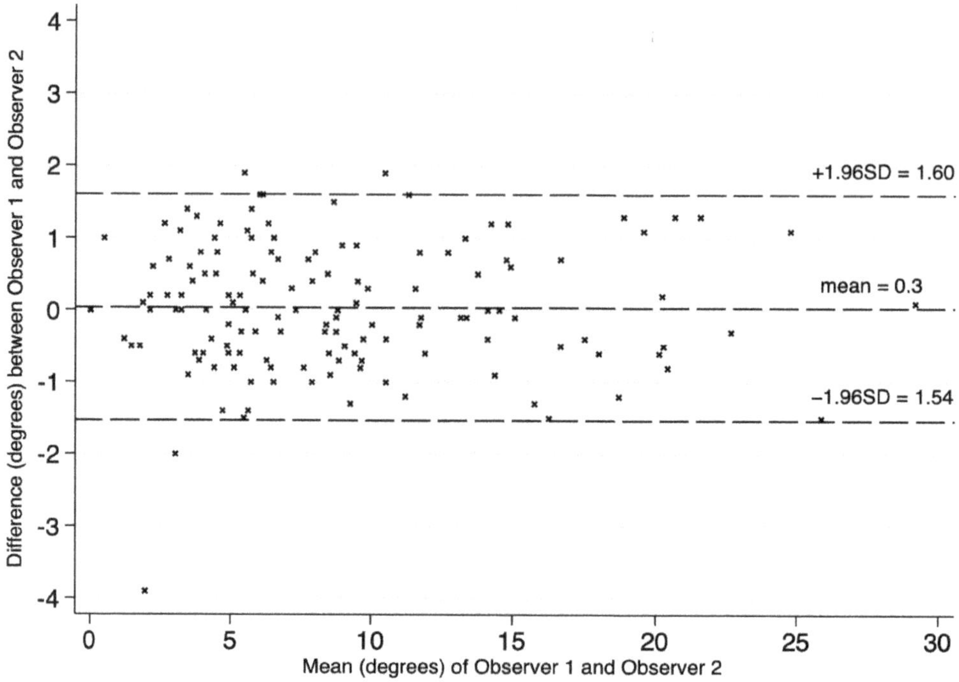

Figure 6. Difference between measurements of deviation angle (degrees) in 173 studies taken by Observer 1 and Observer 2 plotted against their mean.

Table 1. The results of the division of all examinations into quartiles.

		Age			Deviation Angle (Degrees)		
		Median (IQR)	**Min**	**Max**	**Median (IQR)**	**Min**	**Max**
Total	Females (n = 91)	33 (18)	10	86	5.9 (9.8)	0	25.9
	Males (n = 82)	29 (22)	8	64	5.65 (6.6)	0	29.2
	n = 173	31 (23)	8	86	5.7 (7.8)	0	29.2
Group 1	Females (n = 25)	36 (18)	10	86	0 (0)	0	2.1
	Males (n = 19)	36 (28)	8	63	0 (0)	0	2.2
	n = 44	36 (20)	8	86	0 (0)	0	2.2
Group 2	Females (n = 19)	41 (22)	15	67	4.1 (1)	2.7	5.7
	Males (n = 24)	28 (16.5)	10	57	4.9 (1.8)	2.6	5.7
	n = 43	30 (25)	10	67	4.4 (1.8)	2.6	5.7

Table 1. *Cont.*

		Age			Deviation Angle (Degrees)		
		Median (IQR)	**Min**	**Max**	**Median (IQR)**	**Min**	**Max**
Group 3	Females ($n = 20$)	29.5 (19.5)	12	55	7.85 (2.7)	5.8	9.9
	Males ($n = 23$)	29 (22)	11	64	8.4 (2.4)	6.0	10.0
	$n = 43$	29 (21)	11	64	8.4 (2.6)	5.8	10.0
Group 4	Females ($n = 27$)	30 (21)	13	68	14.9 (5.5)	10.5	25.9
	Males ($n = 16$)	27 (26)	11	52	14.45 (8.35)	11.2	29.2
	$n = 43$	28 (22)	11	68	14.8 (6.9)	10.5	29.2

3.2. Manual and Automatic Segmentation of the Sacrum and Iliac Bones

The results of the segmentation of the sacrum and iliac bones are shown in the Table 2. The score awarded by the researcher was calculated for each examination and ranged from 0 to 48 points, with 48 points were awarded in the absence of uncertainty about the complete validity of the automatic segmentation. Fourteen percent ($n = 25$) of examinations were given 48 points, while the minimum score amounted to 26 points. The median of scores was equal to 47 (IQR, 5). Samples of the manual and automatic segmentations of the sacrum and iliac bones are presented in Figure 7. Kruskal–Wallis H test (H(3) = 0.512, $p = 0.9162$) was not statistically significant when comparing the results of the visual scale between groups, which suggests no differences in algorithm performance depending on deviation angle value.

The Dice similarity coefficient for comparison between manual and automatic segmentations, averaged over 173 examinations, ranged from 0.9277 to the maximum of 0.9968 (Table 2). The mean value of DC was 0.9820 (95% CI [0.9804, 0.9835]). As shown in Table 2, the mean value of Dice similarity coefficients varied between the groups from 0.9795 (95% CI [0.9758, 0.9832]) in the 3rd group to 0.9839 (95% CI [0.9807, 0.9871]) in the 1st group. Analysis in terms of group differences of DC with Kruskal–Wallis H test yielded a statistically significant difference between groups, H(3) = 8.362, $p = 0.0391$. Post-hoc comparison with the use of Dunn's test with Bonferroni adjustment revealed a difference between group one and group three, which was considered statistically significant ($p = 0.0166$).

Table 2. The results of the comparison between manual and automatic segmentations of the sacrum and iliac bones with the use of the visual scale (score 0–48) and Dice similarity coefficient.

Groups Based on Deviation Angle	Visual Scale			Dice Similarity Coefficient			
	Median (IQR)	**Min**	**Max**	**Mean (SEM)**	**95% CI**	**Min**	**Max**
Total	47 (5)	26	48	0.9820 (0.0008)	[0.9804, 0.9835]	0.9277	0.9968
Group 1 [0; 2.2]	47 (4)	34	48	0.9839 (0.0016)	[0.9807, 0.9871]	0.9340	0.9968
Group 2 (2.2; 5.7]	47 (4)	26	48	0.9825 (0.0012)	[0.9801, 0.9848)	0.9564	0.9962
Group 3 (5.7; 10]	46 (5)	32	48	0.9795 (0.0019)	[0.9758, 0.9832]	0.9277	0.9958
Group 4 (10; 29.2]	47 (5)	32	48	0.9819 (0.0014)	[0.9793, 0.9846]	0.9584	0.9960

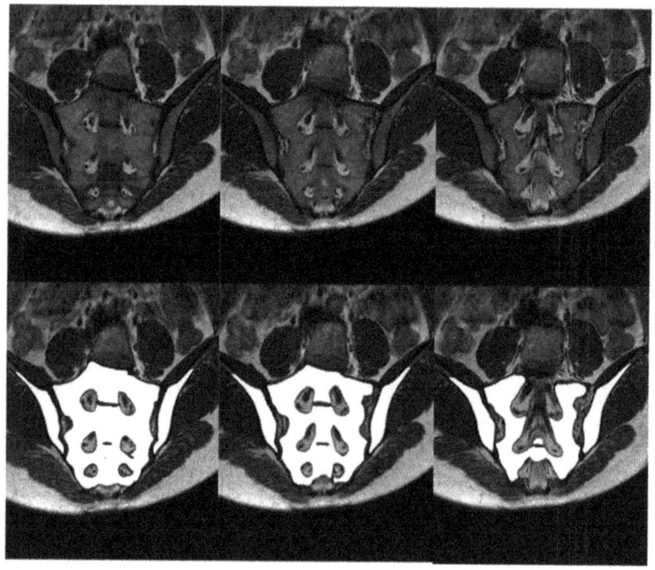

Figure 7. Sample slices of an average case (visual scale equal to 47, DC = 0.98): T1 sequence in the upper row, manual segmentation in the middle row and automated segmentation in the lower row.

3.3. Evaluation of BME Using the SPARCC Scale

The final results of the SPARCC scale assessment are shown in Table 3. It is worth noting that statistical analysis showed no differences between groups in SPARCC scale points, which indicates homogeneity of groups in terms of the advancement of BME. The difference of SPARCC points was also checked using the Kruskal–Wallis H test, and the result was not statistically significant, $H(3) = 0.754$, $p = 0.8604$.

Table 3. The results of SPARCC scale assessment.

Groups Based on Deviation Angle		SPARCC Scale									
		Median (IQR)	Minimal Value	Maximal Value	SPARCC = 0	SPARCC > 0	SPARCC DEPTH = 0	SPARCC DEPTH > 0	SPARCC INTENSITIVITY = 0	SPARCC INTENSITIVITY > 0
Total	Females (n = 91)	0 (5)	0	49	54	37	80	11	61	30
	Males (n = 82)	0 (6)	0	56	51	31	70	12	52	30
	n = 173	0 (5)	0	56	105	68	150	23	113	60
Group 1 [0; 2.2]	Females (n = 25)	0 (3)	0	48	14	11	21	4	15	10
	Males (n = 19)	0 (4)	0	20	10	9	15	4	11	8
	n = 44	0 (6.5)	0	48	24	20	36	8	26	18
Group 2 (2.2; 5.7]	Females (n = 19)	1 (6)	0	45	9	10	17	2	12	7
	Males (n = 24)	0 (6.5)	0	51	17	7	19	5	17	7
	n = 43	0 (6)	0	51	26	17	36	7	29	14
Group 3 (5.7; 10]	Females (n = 20)	0 (4)	0	49	13	7	16	4	14	6
	Males (n = 23)	0 (10)	0	56	15	8	20	3	15	8
	n = 43	0 (9)	0	56	28	15	36	7	29	14
Group 4 (10; 29.2]	Females (n = 27)	0 (5)	0	25	18	9	26	1	20	7
	Males (n = 16)	0 (4)	0	31	9	7	16	0	9	7
	n = 43	0 (4)	0	31	27	16	42	1	29	14

3.4. Manual and Automatic Segmentation of BME

The sample comparison of the performance of manual and automatic segmentations of BME is presented in Figure 8. The results of assessing the compatibility of manual and automatic BME segmentations with the visual scale described previously are provided in Table 4. Differences between the groups were checked using Kruskal–Wallis H test, which resulted in not being statistically significant, $H(3) = 1.125$, $p = 0.7710$. This outcome implies no difference in the algorithm's performance in BME segmentation in relationship to the deviation angle. Due to the limitations of the SPARCC scale, the number of examinations which points were awarded (SPARCC > 0, $n = 68$) differs from the number of examinations with BME recognized when preparing manual segmentations ($n = 83$).

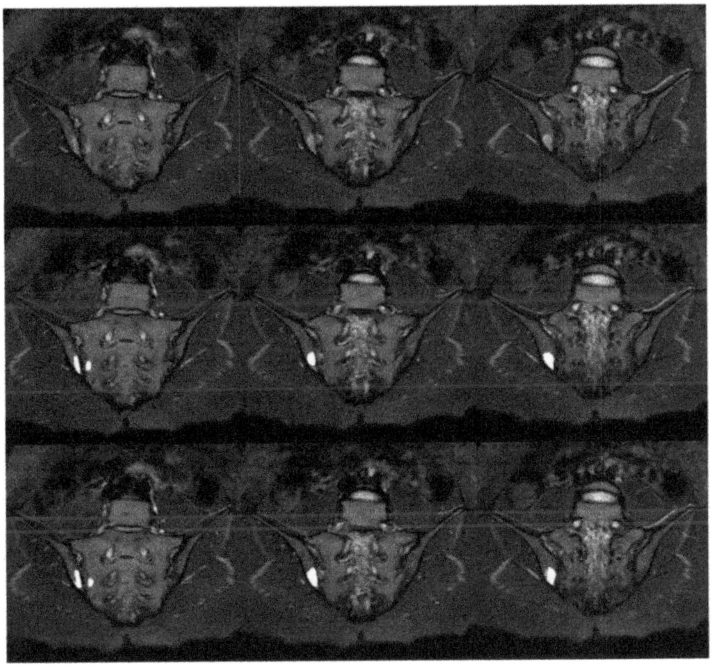

Figure 8. Sample slices with inflammatory lesions (SPARCC = 17): STIR sequence in the upper row, manual segmentation in the middle row and automated segmentation in the lower row.

Table 4. The results of the assessment of the compatibility between manual and automatic BME segmentations.

	Visual Scale		
Groups Based on Deviation Angle	**Median (IQR)**	**Min**	**Max**
Total	43 (7)	22	48
Group 1 [0; 2.2]	42.5 (9)	22	48
Group 2 (2.2; 5.7]	43 (8)	23	48
Group 3 (5.7; 10]	43 (8)	25	48
Group 4 (10; 29.2]	45 (7)	18	48

Table 5 summarizes the results of the visual scoring scale with the division of each joint into quadrants. These results suggest that the algorithm underperforms when searching for lesions in quadrants lying on the sacral bone, due to higher values of interquartile range. Computed Receiver Operating Characteristics (ROC) curves are presented in Figure 9. The area under the curve (AUC) for all examinations is equal to 0.895. In the selected groups, the highest values were in groups 2 (AUC = 0.917) and 3 (AUC = 0.923), while the smallest in group 1 (AUC = 0.823) and group 4 (AUC = 0.887).

Table 5. The summary of the results of visual scale assessment with the division into quadrants of each SIJ.

Groups Based on Deviation Angle	**Median for the Sum of Points Scored on Six Slices in Each Quadrant (IQR—Interquartile Range)**			
Total	Right Joint		Left Joint	
	Upper iliac quadrant	Upper sacrum quadrant	Upper sacrum quadrant	Upper iliac quadrant
	6 (0)	6 (1)	5 (2)	6 (0)
	6 (0)	6 (1)	5 (3)	6 (1)
	Lower iliac quadrant	Lower sacrum quadrant	Lower sacrum quadrant	Lower iliac quadrant
Group 1 [0; 2.2]	Right joint		Left joint	
	Upper iliac quadrant	Upper sacrum quadrant	Upper sacrum quadrant	Upper iliac quadrant
	6 (0)	6 (1)	4.5 (2)	6 (0)
	6 (0)	6 (1.5)	5 (3)	6 (1)
	Lower iliac quadrant	Lower sacrum quadrant	Lower sacrum quadrant	Lower iliac quadrant
Group 2 (2.2; 5.7]	Right joint		Left joint	
	Upper iliac quadrant	Upper sacrum quadrant	Upper sacrum quadrant	Upper iliac quadrant
	6 (0)	6 (2)	5 (2)	6 (0)
	6 (1)	6 (2)	5 (3)	6 (2)
	Lower iliac quadrant	Lower sacrum quadrant	Lower sacrum quadrant	Lower iliac quadrant
Group 3 (5.7; 10]	Right joint		Left joint	
	Upper iliac quadrant	Upper sacrum quadrant	Upper sacrum quadrant	Upper iliac quadrant
	6 (0)	6 (1)	5 (2)	6 (1)
	6 (1)	6 (1)	6 (3)	6 (1)
	Lower iliac quadrant	Lower sacrum quadrant	Lower sacrum quadrant	Lower iliac quadrant
Group 4 (10; 29.2]	Right joint		Left joint	
	Upper iliac quadrant	Upper sacrum quadrant	Upper sacrum quadrant	Upper iliac quadrant
	6 (0)	6 (1)	6 (2)	6 (0)
	6 (0)	6 (2)	5 (3)	6 (1)
	Lower iliac quadrant	Lower sacrum quadrant	Lower sacrum quadrant	Lower iliac quadrant

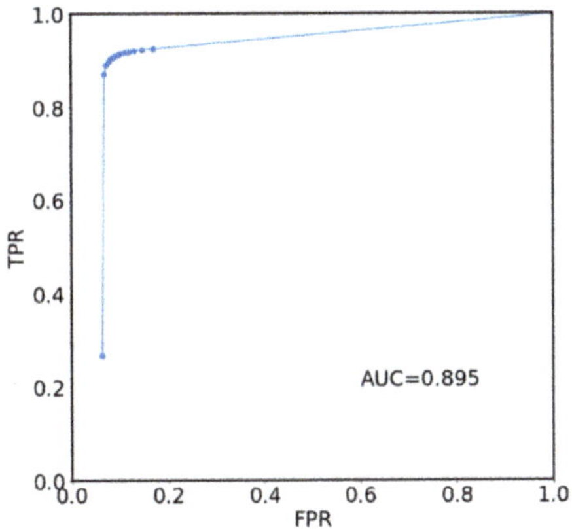

(a)

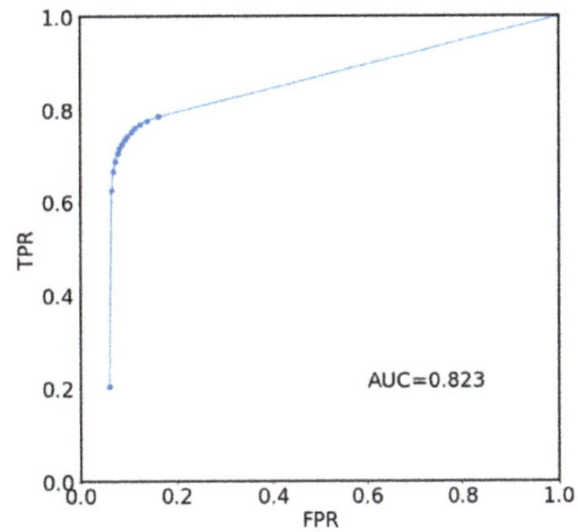

(b)

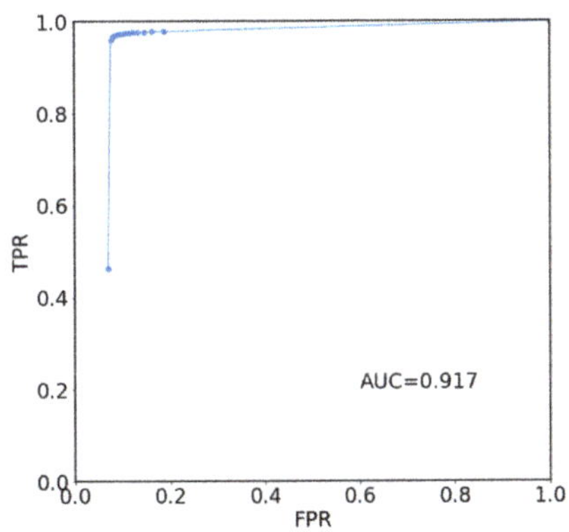

(c)

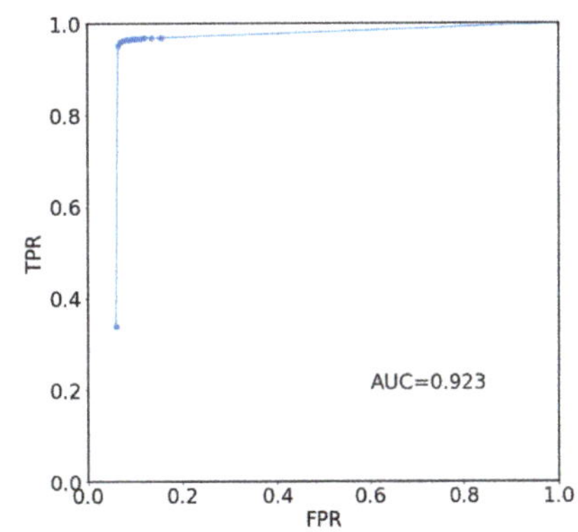

(d)

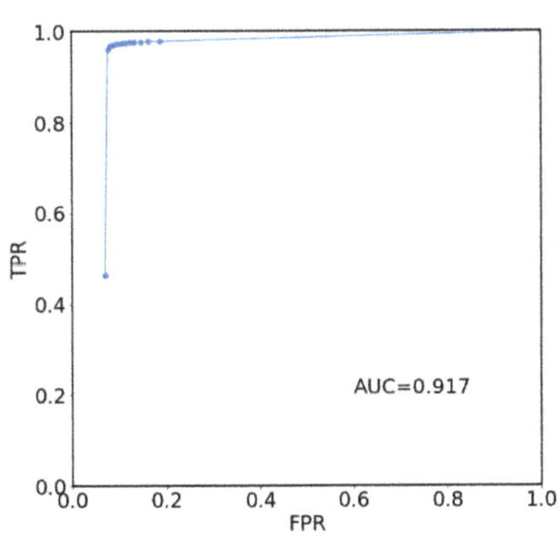

(e)

Figure 9. ROC curves calculated for all the samples (**a**) and separately for the samples assigned to the four groups: (**b**) Group 1; (**c**) Group 2; (**d**) Group 3; (**e**) Group 4.

In order to further evaluate the performance of the algorithm, sensitivity, specificity and accuracy were also calculated basing on the total sum of voxels recognized as TP, TN, FP and FN. Moreover, the sensitivity for recognizing singular lesions was calculated. The results of these calculations are included in Table 6.

Table 6. The results of sensitivity, specificity and accuracy calculated based on the total sum of voxels recognized as TP, TN, FP and FN together with the results of sensitivity calculated from the total amount of lesions recognized as TP and FN by the algorithm.

Groups Based on Deviation Angle		Total Number of Pixels	Total Number of Lesions
Total	Sensitivity	0.76	0.75
	Specificity	0.97	-
	Accuracy	0.97	-
Group 1 [0; 2.2]	Sensitivity	0.58	0.80
	Specificity	0.97	-
	Accuracy	0.96	-
Group 2 (2.2; 5.7]	Sensitivity	0.83	0.78
	Specificity	0.97	-
	Accuracy	0.96	-
Group 3 (5.7; 10]	Sensitivity	0.83	0.77
	Specificity	0.97	-
	Accuracy	0.97	-
Group 4 (10; 29.2]	Sensitivity	0.70	0.66
	Specificity	0.97	-
	Accuracy	0.97	-

3.5. Summary of the Results

The most important data that summarize the performance of the algorithm is presented in Table 7.

Table 7. Summary of the most important results [1].

		Group 1	Group 2	Group 3	Group 4	Differences between Groups
Bone segmentation	Visual scale [Median (IQR)]	47 (4)	47 (4)	46 (5)	47 (5)	No
	Dice similarity coefficient	0.9839 (0.0016)	0.9825 (0.0012)	0.9795 (0.0019)	0.9819 (0.0014)	Yes [2]
BME segmentation	Visual scale [Median (IQR)]	42.5 (9)	43 (8)	43 (8)	45 (7)	No
	SPARCC [Median (IQR)]	0 (6.5)	0 (6)	0 (9)	0 (4)	No
	AUC [Value]	0.823	0.917	0.923	0.917	-

[1] Detailed description is provided in text. [2] Post-hoc analysis showed difference between group 1 and group 3.

4. Discussion

The technical aspect of performing sacroiliac joint MRI is often omitted in research studies. For radiologists to correctly evaluate the examination, it is crucial that the obtained images adequately represent the joints. Due to the anatomy of the pelvic and iliac bones, improper adjustment of the scanning plane can result in obtaining a completely different image than expected. For this reason, when developing algorithms aimed at assisting clinicians in the evaluation of SIJs, it is crucial to also retest them on examinations that were not performed completely appropriately in the technical aspect. The angle of deviation was selected as a method of assessing the technical correctness of the examination, since it determines the visual appearance of the obtained slices in the STIR sequence, which is used to assess BME.

Dividing all examinations into equal groups on the basis of the deviation angle enabled verification of the fully automated algorithm's performance under various parameters. In addition, each group was similar in terms of the number of studies in which SPARCC was 0 or was >0 (Table 3). The results of the work demonstrated that despite the possibility that the algorithm could perform less properly with examinations that are not fully correct, it performs equally satisfactorily regardless of the deviation angle.

None of the previously developed semi-automatic or fully automatic algorithms [12,30,31,33] created for BME detection in the SIJs have not been tested on such a large number of examinations and against a technical correctness of the examination. Precisely for this reason, it makes the improved algorithm presented in this paper superior to previous ones. It is worth noting that machine learning algorithms' performance critically depends on the size of the sample used during training, on the basis of which artificial intelligence "learns" to recognize a particular element. Therefore, it can create a sort of pattern that may eventually serve to distinguish pathology from normality [35]. Hence, the more extensive collection of examinations is assembled, the better the results achieved by the algorithm.

In order to segment BME in general, it is obligatory to correctly perform the bone segmentation. Therefore, it is essential for the fully automated algorithm to first achieve quality bone segmentation results, such as those reached in this study, as the average DC score is 0.9820 (95% CI [0.9804, 0.9835]), and even the minimum DC = 0.9277, which is also satisfactory. Statistical analysis does not disclose significant differences in DC values between groups one and three; however, obtained values of DC in all groups are high enough to assume that the algorithm detects bones accurately irrespective of the deviation angle.

Segmentation of BME is performed by the algorithm only on the region of bones marked in the previous segmentation. For this reason, if bone segmentation is performed correctly regardless the technical correctness of the examination, segmentation of BME as well will be performed within the relevant regions.

In order to compare manual and automatic segmentation of inflammatory lesions, not only a visual scale based on the SPARCC scale was used but also a mathematical evaluation of voxels, which eliminated the limitation of the visual scale resulting from the selection of only six slices in each examination. However, it is important to remember that the visual evaluation of the segmentations is essential since as of now, the radiologists' expertise is considered to be the ground truth.

Compared to the algorithm created by Bressem [33], the algorithm developed by our research team allows not only to globally determine whether a lesion is present in a given study but also to locate it in a specific quadrant of SIJs. In addition, what distinguishes our algorithm from the other one [33] is the ability to measure the size of the BME. This may prove useful in further studies on the correlation of inflammatory lesions with symptom severity in patients with axial spondyloarthritis.

With this study, the fully automated algorithm for detecting BME was successfully validated on a number of examinations at least five times greater than the number of examinations on which it was initially created. It is worth mentioning that the age of the patients included in the study ranged from 18 to 86 years, which allows to conclude that the algorithm may have clinical application in the detection of BME in adults.

In conclusion, the fully automated algorithm remains competent to satisfactorily evaluate inflammatory changes in MRI examinations of the sacroiliac joints regardless of the deviation angle differences described. In addition, the algorithm was demonstrated to be useful for the exclusion of BME.

Author Contributions: Conceptualization, W.W.; methodology, J.O., M.W., A.R., Z.T., M.K. and W.W.; software, Z.T.; validation, J.O., M.W., A.R., M.K. and W.W.; formal analysis, M.W.; investigation, J.O., Z.O. and M.K.; resources, W.W.; data curation, J.O., M.W. and W.W.; writing—original draft preparation, J.O., M.W., A.R., Z.T. and Z.O.; writing—review and editing, J.O., M.W., A.R., Z.T., Z.O., M.K. and W.W.; project administration, W.W. All authors have read and agreed to the published version of the manuscript.

Funding: This research received no external funding.

Institutional Review Board Statement: The study was conducted in accordance with the Declaration of Helsinki and approved by the Institutional Review Board (or Ethics Committee) of Jagiellonian University Bioethics Committee (1072.6120.332.2022, 18 January 2023).

Informed Consent Statement: Patient consent was waived due to the fact that the study was a retrospective, fully anonymized study and was based on an analysis of MRI examinations already performed on patients. In addition, consent was obtained from the institutions from where the data used in the study were obtained. It must be indicated that all examinations used in the study were anonymized by institutions providing them before the researchers received them.

Data Availability Statement: Not applicable.

Conflicts of Interest: The authors declare no conflict of interest.

References

1. Sieper, J.; Poddubnyy, D. Axial Spondyloarthritis. *Lancet* **2017**, *390*, 73–84. [CrossRef]
2. Wang, R.; Ward, M.M. Epidemiology of Axial Spondyloarthritis: An Update. *Curr. Opin. Rheumatol.* **2018**, *30*, 137–143. [CrossRef]
3. Wright, G.C.; Kaine, J.; Deodhar, A. Understanding Differences between Men and Women with Axial Spondyloarthritis. *Semin. Arthritis Rheum.* **2020**, *50*, 687–694. [CrossRef]
4. Dean, L.E.; Jones, G.T.; MacDonald, A.G.; Downham, C.; Sturrock, R.D.; Macfarlane, G.J. Global Prevalence of Ankylosing Spondylitis. *Rheumatology* **2014**, *53*, 650–657. [CrossRef]
5. Rudwaleit, M.; van der Heijde, D.; Landewé, R.; Listing, J.; Akkoc, N.; Brandt, J.; Braun, J.; Chou, C.T.; Collantes-Estevez, E.; Dougados, M.; et al. The Development of Assessment of SpondyloArthritis International Society Classification Criteria for Axial Spondyloarthritis (Part II): Validation and Final Selection. *Ann. Rheum. Dis.* **2009**, *68*, 777–783. [CrossRef]
6. Strand, V.; Singh, J.A. Patient Burden of Axial Spondyloarthritis. *J. Clin. Rheumatol.* **2017**, *23*, 383–391. [CrossRef]
7. Reijnierse, M. Axial Skeleton Bone Marrow Changes in Inflammatory Rheumatologic Disorders. *Semin. Musculoskelet. Radiol.* **2023**, *27*, 091–102. [CrossRef]
8. Obuchowicz, R.; Bonczar, M. Ultrasonographic Differentiation of Lateral Elbow Pain. *Ultrasound Int. Open* **2016**, *2*, E38–E46. [CrossRef]
9. Sieper, J.; Rudwaleit, M.; Baraliakos, X.; Brandt, J.; Braun, J.; Burgos-Vargas, R.; Dougados, M.; Hermann, K.-G.; Landewé, R.; Maksymowych, W.; et al. The Assessment of SpondyloArthritis International Society (ASAS) Handbook: A Guide to Assess Spondyloarthritis. *Ann. Rheum. Dis.* **2009**, *68* (Suppl. 2), ii1–ii44. [CrossRef]
10. Rudwaleit, M.; Jurik, A.G.; Hermann, K.-G.A.; Landewe, R.; van der Heijde, D.; Baraliakos, X.; Marzo-Ortega, H.; Ostergaard, M.; Braun, J.; Sieper, J. Defining Active Sacroiliitis on Magnetic Resonance Imaging (MRI) for Classification of Axial Spondyloarthritis: A Consensual Approach by the ASAS/OMERACT MRI Group. *Ann. Rheum. Dis.* **2009**, *68*, 1520–1527. [CrossRef]
11. Jurik, A.G. Diagnostics of Sacroiliac Joint Differentials to Axial Spondyloarthritis Changes by Magnetic Resonance Imaging. *J. Clin. Med.* **2023**, *12*, 1039. [CrossRef]
12. Kucybała, I.; Tabor, Z.; Polak, J.; Urbanik, A.; Wojciechowski, W. The Semi-Automated Algorithm for the Detection of Bone Marrow Oedema Lesions in Patients with Axial Spondyloarthritis. *Rheumatol. Int.* **2020**, *40*, 625–633. [CrossRef]
13. Tsoi, C.; Griffith, J.F.; Lee, R.K.L.; Wong, P.C.H.; Tam, L.S. Imaging of Sacroiliitis: Current Status, Limitations and Pitfalls. *Quant. Imaging Med. Surg.* **2019**, *9*, 318–335. [CrossRef]
14. Maksymowych, W.P.; Lambert, R.G.; Baraliakos, X.; Weber, U.; Machado, P.M.; Pedersen, S.J.; de Hooge, M.; Sieper, J.; Wichuk, S.; Poddubnyy, D.; et al. Data-Driven Definitions for Active and Structural MRI Lesions in the Sacroiliac Joint in Spondyloarthritis and Their Predictive Utility. *Rheumatology* **2021**, *60*, 4778–4789. [CrossRef]
15. Maksymowych, W.P.; Inman, R.D.; Salonen, D.; Dhillon, S.S.; Williams, M.; Stone, M.; Conner-Spady, B.; Palsat, J.; Lambert, R.G.W. Spondyloarthritis Research Consortium of Canada Magnetic Resonance Imaging Index for Assessment of Sacroiliac Joint Inflammation in Ankylosing Spondylitis. *Arthritis Rheum.* **2005**, *53*, 703–709. [CrossRef]
16. Wendling, D.; Claudepierre, P.; Prati, C. Early Diagnosis and Management Are Crucial in Spondyloarthritis. *Jt. Bone Spine* **2013**, *80*, 582–585. [CrossRef]
17. Lambert, R.G.W.; Bakker, P.A.C.; van der Heijde, D.; Weber, U.; Rudwaleit, M.; Hermann, K.-G.A.; Sieper, J.; Baraliakos, X.; Bennett, A.; Braun, J.; et al. Defining Active Sacroiliitis on MRI for Classification of Axial Spondyloarthritis: Update by the ASAS MRI Working Group. *Ann. Rheum. Dis.* **2016**, *75*, 1958–1963. [CrossRef]

18. Lapane, K.L.; Dubé, C.; Ferrucci, K.; Khan, S.; Kuhn, K.A.; Yi, E.; Kay, J.; Liu, S.-H. Patient Perspectives on Health Care Provider Practices Leading to an Axial Spondyloarthritis Diagnosis: An Exploratory Qualitative Research Study. *BMC Fam. Pract.* **2021**, *22*, 251. [CrossRef]
19. Hay, C.A.; Packham, J.; Ryan, S.; Mallen, C.D.; Chatzixenitidis, A.; Prior, J.A. Diagnostic Delay in Axial Spondyloarthritis: A Systematic Review. *Clin. Rheumatol.* **2022**, *41*, 1939–1950. [CrossRef]
20. Akgul, O. Classification Criteria for Spondyloarthropathies. *World J. Orthop.* **2011**, *2*, 107. [CrossRef]
21. Colbert, R.A. Early Axial Spondyloarthritis. *Curr. Opin. Rheumatol.* **2010**, *22*, 603–607. [CrossRef]
22. Maksymowych, W.P. The Role of Imaging in the Diagnosis and Management of Axial Spondyloarthritis. *Nat. Rev. Rheumatol.* **2019**, *15*, 657–672. [CrossRef]
23. Ramiro, S.; Nikiphorou, E.; Sepriano, A.; Ortolan, A.; Webers, C.; Baraliakos, X.; Landewé, R.B.M.; Van den Bosch, F.E.; Boteva, B.; Bremander, A.; et al. ASAS-EULAR Recommendations for the Management of Axial Spondyloarthritis: 2022 Update. *Ann. Rheum. Dis.* **2023**, *82*, 19–34. [CrossRef]
24. Tam, L.S.; Wei, J.C.-C.; Aggarwal, A.; Baek, H.J.; Cheung, P.P.; Chiowchanwisawakit, P.; Dans, L.; Gu, J.; Hagino, N.; Kishimoto, M.; et al. 2018 APLAR Axial Spondyloarthritis Treatment Recommendations. *Int. J. Rheum. Dis.* **2019**, *22*, 340–356. [CrossRef]
25. Aouad, K.; Ziade, N.; Baraliakos, X. Structural Progression in Axial Spondyloarthritis. *Jt. Bone Spine* **2020**, *87*, 131–136. [CrossRef]
26. Stoel, B. Use of Artificial Intelligence in Imaging in Rheumatology—Current Status and Future Perspectives. *RMD Open* **2020**, *6*, e001063. [CrossRef]
27. Erickson, B.J.; Korfiatis, P.; Akkus, Z.; Kline, T.L. Machine Learning for Medical Imaging. *Radiographics* **2017**, *37*, 505–515. [CrossRef]
28. Chartrand, G.; Cheng, P.M.; Vorontsov, E.; Drozdzal, M.; Turcotte, S.; Pal, C.J.; Kadoury, S.; Tang, A. Deep Learning: A Primer for Radiologists. *Radiographics* **2017**, *37*, 2113–2131. [CrossRef]
29. Faleiros, M.C.; Nogueira-Barbosa, M.H.; Dalto, V.F.; Júnior, J.R.F.; Tenório, A.P.M.; Luppino-Assad, R.; Louzada-Junior, P.; Rangayyan, R.M.; de Azevedo-Marques, P.M. Machine Learning Techniques for Computer-Aided Classification of Active Inflammatory Sacroiliitis in Magnetic Resonance Imaging. *Adv. Rheumatol.* **2020**, *60*, 25. [CrossRef]
30. Rzecki, K.; Kucybała, I.; Gut, D.; Jarosz, A.; Nabagło, T.; Tabor, Z.; Wojciechowski, W. Fully Automated Algorithm for the Detection of Bone Marrow Oedema Lesions in Patients with Axial Spondyloarthritis—Feasibility Study. *Biocybern. Biomed. Eng.* **2021**, *41*, 833–853. [CrossRef]
31. Zarco, P.; Almodóvar, R.; Bueno, Á.; Molinero, L.M.; SCAISS Study Group. Development and Validation of SCAISS, a Tool for Semi-Automated Quantification of Sacroilitis by Magnetic Resonance in Spondyloarthritis. *Rheumatol. Int.* **2018**, *38*, 1919–1926. [CrossRef]
32. Garrido-González, C.; Pineda, M.L.; Garrido-Castro, J.L.; Tabor, Z.; Kucybała, I.; Wojciechowski, W.; Zarco-Montejo, P.; Almodovar, R. Collantes Estevez POS0958 Responsiveness of conventional, semi-automatic and full-automatic methods to quantify marrow bone edema lesions in MRI of axial spondyloarthritis patients: A pilot study. *Ann. Rheum. Dis.* **2021**, *80*, 743–744. [CrossRef]
33. Bressem, K.K.; Adams, L.C.; Proft, F.; Hermann, K.G.A.; Diekhoff, T.; Spiller, L.; Niehues, S.M.; Makowski, M.R.; Hamm, B.; Protopopov, M.; et al. Deep Learning Detects Changes Indicative of Axial Spondyloarthritis at MRI of Sacroiliac Joints. *Radiology* **2022**, *305*, 655–665. [CrossRef]
34. Reinke, A.; Tizabi, M.D.; Sudre, C.H.; Eisenmann, M.; Rädsch, T.; Baumgartner, M.; Acion, L.; Antonelli, M.; Arbel, T.; Bakas, S.; et al. Common Limitations of Image Processing Metrics: A Picture Story. *arXiv* **2021**, arXiv:2104.05642.
35. Szegedy, C.; Wei, L.; Yangqing, J.; Sermanet, P.; Reed, S.; Anguelov, D. Going Deeper with Convolutions. In Proceedings of the IEEE Conference on Computer Vision and Pattern Recognition (CVPR), Boston, MA, USA, 7–12 June 2015; pp. 1–9.

tomography

MDPI

Article

Effects of Path-Finding Algorithms on the Labeling of the Centerlines of Circle of Willis Arteries

Se-On Kim and Yoon-Chul Kim *

Division of Digital Healthcare, College of Software and Digital Healthcare Convergence, Yonsei University, Wonju 26493, Republic of Korea
* Correspondence: yoonckim@yonsei.ac.kr; Tel.: +82-33-760-2318

Abstract: Quantitative analysis of intracranial vessel segments typically requires the identification of the vessels' centerlines, and a path-finding algorithm can be used to automatically detect vessel segments' centerlines. This study compared the performance of path-finding algorithms for vessel labeling. Three-dimensional (3D) time-of-flight magnetic resonance angiography (MRA) images from the publicly available dataset were considered for this study. After manual annotations of the endpoints of each vessel segment, three path-finding methods were compared: (Method 1) depth-first search algorithm, (Method 2) Dijkstra's algorithm, and (Method 3) A* algorithm. The rate of correctly found paths was quantified and compared among the three methods in each segment of the circle of Willis arteries. In the analysis of 840 vessel segments, Method 2 showed the highest accuracy (97.1%) of correctly found paths, while Method 1 and 3 showed an accuracy of 83.5% and 96.1%, respectively. The AComm artery was highly inaccurately identified in Method 1, with an accuracy of 43.2%. Incorrect paths by Method 2 were noted in the R-ICA, L-ICA, and R-PCA-P1 segments. The Dijkstra and A* algorithms showed similar accuracy in path-finding, and they were comparable in the speed of path-finding in the circle of Willis arterial segments.

Keywords: magnetic resonance angiography; cerebral arteries; vessel segmentation; graph structure; Dijkstra algorithm; A* algorithm; depth first search

Citation: Kim, S.-O.; Kim, Y.-C. Effects of Path-Finding Algorithms on the Labeling of the Centerlines of Circle of Willis Arteries. *Tomography* **2023**, *9*, 1423–1433. https://doi.org/10.3390/tomography9040113

Academic Editors: Michał Strzelecki, Adam Piorkowski and Rafał Obuchowicz

Received: 19 May 2023
Revised: 21 July 2023
Accepted: 21 July 2023
Published: 24 July 2023

1. Introduction

Abnormal morphological characteristics of the cerebral blood vessels are associated with vascular diseases. For example, stenosis of any artery due to intracranial atherosclerosis can cause ischemia in the brain tissue, resulting in stroke and other cognitive brain disorders [1,2]. Development of intracranial aneurysms [3,4] may be associated with high blood pressure on the weakened vessel wall and potentially could result in ruptures and cerebral hemorrhages if left untreated [5]. Three-dimensional (3D) computed tomography angiography (CTA) and 3D time-of-flight MR angiography (TOF-MRA) are commonly used to noninvasively obtain data on the cerebral vessels [6,7]. The geometric information such as vessel tortuosity [8] and distributions of the diameters of the vessels [9] is extracted and evaluated to study correlations between vessel abnormalities and diseases. Quantitative evaluation based on the geometry of the cerebral arteries has been reported in the literature [9–11]. For example, a cumulative distribution function of the cross-sectional diameters in a vessel segment was evaluated [9]. Tortuosity descriptors such as the sum of angle metrics, inflection count metric, triangular index, relative length, and product of angle distance were used to find the relationship between diseases and morphological features [12,13]. These quantities are typically calculated on a vessel segment after identifying the vessel's centerline, and a path-finding algorithm can automate the identification of the centerline.

Labeling of the arteries in the circle of Willis (CoW) has been of interest to researchers since the CoW is essential in maintaining the circulation of blood flow even in the occlusion of an artery and has important bifurcations that anatomically separate the intracranial arteries of interest [14–16]. For example, the internal carotid artery (ICA) branches into the anterior cerebral artery (ACA) and middle cerebral artery (MCA). In the posterior circulation, the basilar artery (BA) is connected to the left and right posterior cerebral arteries (PCA). Tracking of the vessel's centerline can be performed via a path-finding algorithm after manual annotations of two endpoints of a vessel segment. This procedure is relatively simple and straightforward when compared with deep learning-based segmentation of the intracranial vessel segments [17]. The development of deep learning segmentation models requires manual segmentations of individual arterial segments for the generation of training data. The manual segmentation process involves slice-wise manual tracing in a 3D volume consisting of hundreds of slices, and it is thus time-consuming and laborious [18]. Hence, the centerline tracking approaches can improve the efficiency of vessel labeling and its subsequent quantification without the need for deep learning model training data generation. Previous studies have demonstrated methods adopting path-finding algorithms such as Dijkstra's algorithm in analyzing 3D angiography data [19–21]. However, they did not demonstrate detailed comparisons of accuracy among available path-finding algorithms in vessel image analysis.

In this study, we evaluated the performance of three path-finding algorithms in robustly identifying the cerebral arterial segments in the CoW. In terms of path-finding accuracy and computational time, we compared a depth-first search (DFS) based algorithm that does not involve a graph structure to a Dijkstra algorithm and an A* (A star) algorithm, both of which require a graph representation.

2. Materials and Methods

Figure 1 illustrates the flowchart of the presented vessel labeling methods. The preprocessing steps involving Otsu thresholding and 3D seeded region growing have been implemented in Matlab version 9.13 (The Mathworks Inc., Natick, MA, USA). Skeletonization, graph structure generation, path-finding methods, and visualization of the centerlines in 3D were implemented in Python version 3.10 (Python Software Foundation).

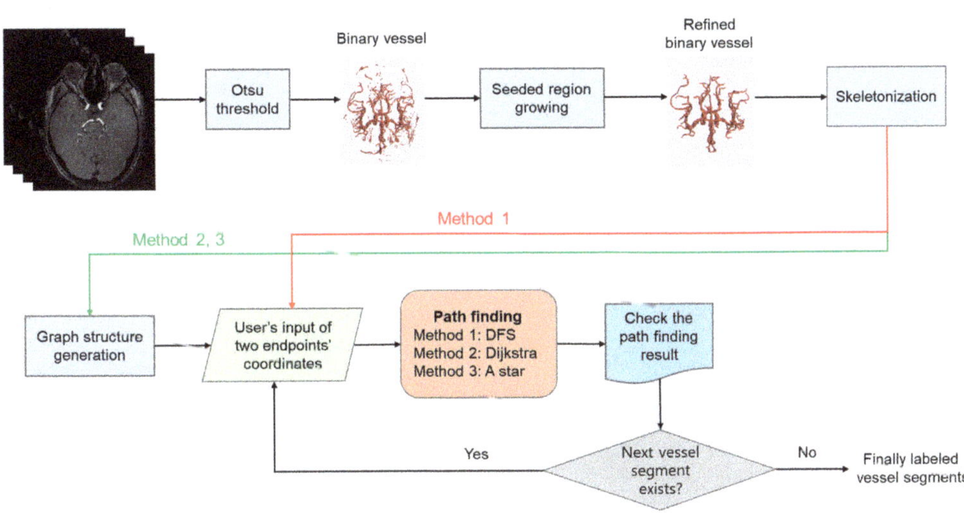

Figure 1. A flowchart of the segmental vessel-labeling algorithm. Three path-finding methods were used to label the segments in the circle of Willis. Notably, Method 1 does not require a graph structure generation process, while Methods 2 and 3 require the graph structure generation process.

2.1. Data

We used publicly available magnetic resonance angiography (MRA) data from the IXI Dataset (https://brain-development.org/ixi-dataset) (accessed on 21 July 2023). Sixty Neuroimaging Informatics Technology Initiative (NIfTI) files were considered to evaluate the performance of path-finding algorithms. The imaging parameters are as follows: Philips Medical Systems Intera 3T, repetition time = 16.7 ms, echo time = 5.8 ms, number of phase-encoding steps = 286, reconstruction diameter = 240 × 240 mm^2, acquisition matrix = 288 × 286, flip angle = 16-deg, in-plane pixel spacing = 0.4–0.6 mm, spacing between slices = 0.8 mm. We applied the bi-cubic interpolation along the slice dimension to generate iso-resolution image data with the same pixel spacing in all three directions. After the interpolation, the final voxel size was isotropic with the 0.6 mm voxel spacing.

2.2. Vessel Segmentation and Skeletonization

The user selected an axial slice that shows three cross-sections of the vessels, which are the right internal carotid artery (ICA), left ICA, and basilar artery (BA). Supplementary Figure S1 shows an example axial slice showing three vessels' cross-sections. A seed point was manually annotated by the mouse click within each of the three cross-sections (Supplementary Figure S1). With each seed point, 3D seeded region growing was performed using the region-growing function [22] available in Matlab. The segmented arteries after the region growing included the region of the CoW. The three segmentation results were obtained in binary masks, and the union set operation was performed to combine the three segmented masks into one final binary mask of the vessels. Skeletonization was performed to find the centerlines of the vessels using the scikit-image Python library [23].

2.3. Annotation of Two Endpoints

Following the anatomical terminology provided by Dumais et al. [17], we identified 14 vessel segments by manually annotating two endpoints in each vessel segment. The vessel segments of interest were (1) AComm (anterior communicating artery), (2) R-A1 (right anterior cerebral artery A1), (3) L-A1 (left anterior cerebral artery A1), (4) R-M1 (right middle cerebral artery M1), (5) L-M1 (left middle cerebral artery M1), (6) R-ICA (right internal carotid artery), (7) L-ICA (left internal carotid artery), (8) R-PComm (right posterior communicating artery), (9) L-PComm (left posterior communicating artery), (10) R-P1 (right posterior cerebral artery P1), (11) L-P1 (left posterior cerebral artery P1), (12) R-P2 (right posterior cerebral artery P2), (13) L-P2 (left posterior cerebral artery P2), and (14) BA (basilar artery). Some endpoints were shared among the vessel segments. For example, the R-A1 and R-M1 segments share a common branch point (Figure 2). The same is true for the BA, R-P1, and L-P1 segments.

The Plotly (Plotly Technologies Inc., Montreal, QC, Canada) Python library (https://plotly.com/python) (accessed on 21 July 2023) was used to visually identify the vessel segments in the 3D space. The mouse hovering on the target endpoint shows its position information in 3D coordinates. For each vessel segment, the position information of the two endpoints was the input to our path-finding algorithms. During manual annotation, we recorded the position information of the two endpoints in vessel segments. If no centerline exists in a certain vessel segment, we recorded '0 0 0' for the vessel segment. The position information was saved in a text file with .txt extension for each subject. The visualization result of the colored vessel segments was saved in a .html file, which was reserved for further investigation to check the accuracy of the vessel annotations.

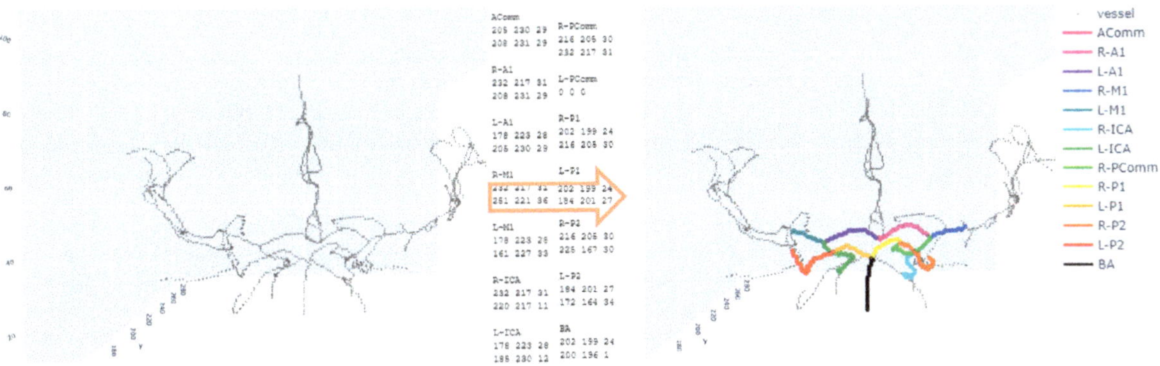

Figure 2. An example of colored labeling of the arteries in the circle of Willis. The **left** figure is the centerlines after skeletonization, and the **right** figure shows the labeled segments in colors after applying a path-finding algorithm. The numbers in the middle text show the 3D coordinates of the two endpoints in each vessel segment along with the vessel segment's name.

2.4. Path-Finding Algorithms

In this study, we implemented three path-finding algorithms for comparisons: (Method 1) DFS-based path-finding algorithm, (Method 2) Dijkstra's algorithm, and (Method 3) A* algorithm.

For Method 1, we implemented a maze-solving algorithm based on DFS (Supplementary Figure S2). Given two endpoints, from the start point, Method 1 attempted to find the next neighbor pixel and record the visited pixel locations using the push operation in a stack. When there is no neighboring pixel anymore, the visited locations were taken out of the stack until the algorithm found the neighboring pixel which was not visited. The algorithm was terminated when the endpoint was reached. The remaining data containing pixel locations in the stack indicate the coordinates along the path that connects the two endpoints. We implemented the method in C++ on the Microsoft Visual Studio environment. PyBind11 (https://github.com/pybind/pybind11) (accessed on 21 July 2023) was used to create Python bindings to the C++-compiled code [24]. Hence, the 3D coordinates along the path were able to be obtained and saved in a variable in Python.

Methods 2 and 3 require the generation of a graph structure from the skeleton vessel image as shown in Figure 1. To generate a graph from the skeleton image, we used the Skan version 0.10.0 Python library (https://skeleton-analysis.org/stable) (accessed on 21 July 2023) [25]. Dijkstra's algorithm is well known as the shortest path-finding algorithm in a graph structure [26]. In our implementation, the pixels along the centerlines represented vertices (or nodes), and the connections with their neighboring pixels represented edges. The weights between the two connected nodes were calculated as the Euclidean distance of the two points. We used the Dijkstra() function provided by the SciPy Python library [27]. The Dijkstra() function is based on the Fibonacci heap implementation, which has the time complexity of $O(E + V \log V)$ (here, E is the number of edges, and V is the number of vertices) and is more time efficient than the list implementation whose time complexity is $O(V^2)$ [28]. A* algorithm is another method for the shortest path search in a graph structure. In contrast to Dijkstra's algorithm, which finds the shortest path from a starting point to all goal points, the A* algorithm only finds the shortest path from a starting point to a destination point, and it introduces heuristic cost values as well as graph weights to find the next node [29]. In our study, we used and modified the code available from the python-astar open-source library (https://github.com/jrialland/python-astar) (accessed on 21 July 2023). The heuristic cost value was calculated by computing the Euclidean distance between the nodes of interest and the destination node.

2.5. Evaluation

We compared the three methods in terms of path-finding accuracy and computational time in Python. For each vessel segment, we provided the two endpoints and let the methods automatically find a path. We counted the number of incorrect paths and the number of correct paths for each vessel segment after building consensus on the path-finding correctness. An incorrect path was defined as a path that connects the two endpoints but is not the path we expected. To measure computational time, we used the time.perf_counter() function. Since we used the already annotated endpoints, during the time measurements there were no manual annotation procedures for recording two endpoints in all the vessel segments. For Method 1, we measured the time interval of the DFS path-finding algorithm taken to find paths in all segments in each subject. For Methods 2 and 3, we first measured the time interval of the graph structure generation from the skeleton image and then measured each time interval of the path-finding algorithm (i.e., SciPy's Dijkstra() function for Method 2 and python-astar's find_path() function for Method 3). For each method of Methods 2 and 3, we summed the time intervals including the graph structure generation and the path finding. These time intervals were compared to the time measured from Method 1 for each subject's data. The evaluation was performed on a Windows PC (13th Generation Intel® Core™ i7-13700K 16-Core Processor Central Processing Unit).

A Fisher's exact test was used to analyze any differences between the path-finding methods in detecting the correct centerlines of the vessel segments. A two-sample unpaired Student's *t*-test was performed to determine if the computational time between the two methods was significantly different. A *p*-value of <0.05 was considered statistically significant.

3. Results

Table 1 summarizes the accuracy of the three algorithms in finding the paths between two endpoints in each vessel segment. Method 1 produced 121 incorrect paths out of 735 paths, and Method 2 produced only 21 incorrect paths out of 735 paths. Method 3 produced 29 incorrect paths. The accuracy (97.1%) of Method 2 was significantly higher than that (83.5%) of Method 1. The accuracy of Method 3 was 96.1%. The AComm is highly inaccurate in Method 1 with an accuracy of 43.2%. Method 2 is superior to Method 1 in all vessel segments except for the R-ICA segment. Method 3 is comparable to Method 2 in every vessel segment, except for the AComm segment. Fisher's exact test resulted in statistically significant associations in most vessel segments in Method 1 and Method 2. As shown in Table 1, AComm, ACA A1, PComm, L-PCA P1, L-PCA P2, and BA showed statistically significant associations, while MCA M1, R-PCA P1, R-PCA P2, and ICA were not statistically significant ($p > 0.05$). This implies that the choice of a path-finding method affected the correct detection of paths in AComm, ACA A1, PComm, L-PCA P1, L-PCA P2, and BA. The same was true in the association of Method 1 with Method 3. However, there was no statistically significant association between Method 2 and Method 3, except for AComm.

The seeded region-growing algorithm was advantageous because it did not segment unwanted vessels such as veins or outer intracranial vessels. The inclusion of the veins or outer intracranial vessels can obscure the visual appearances of the CoW arteries (Supplementary Figure S3) and make it challenging to locate the two endpoints when identifying a vessel segment in the CoW. The undetected paths in Table 1 resulted from a non-existent binary vessel mask possibly due to either under-segmentation of the seeded region-growing algorithm or non-enhancement of the vessel itself, and thus the path-finding algorithms were not able to be applied to the vessel segments. AComm had 16 undetected paths (i.e., 26.7% of the total subjects). R-PComm and L-PComm had 46 and 42 undetected paths (i.e., 76.7% and 70.0% of the total subjects), respectively.

Table 1. Evaluation of the path-finding methods.

		Method 1 (DFS Algorithm) No. of Correct Paths	Method 1 (DFS Algorithm) No. of Incorrect Paths	Method 2 (Dijkstra Algorithm) No. of Correct Paths	Method 2 (Dijkstra Algorithm) No. of Incorrect Paths	Method 3 (A* Algorithm) No. of Correct Paths	Method 3 (A* Algorithm) No. of Incorrect Paths	p-Value *	p-Value †	p-Value ‡	No. of Undetected Paths	Total
AComm		19	25	44	0	36	8	<0.001	<0.001	0.006	16	60
ACA A1	R	51	9	60	0	60	0	0.003	0.003	1	0	60
	L	49	10	59	0	59	0	0.001	0.001	1	1	60
MCA M1	R	55	5	60	0	60	0	0.057	0.057	1	0	60
	L	58	2	60	0	59	1	0.496	1	1	0	60
PComm	R	8	6	14	0	14	0	0.016	0.016	1	46	60
	L	10	8	18	0	18	0	0.003	0.003	1	42	60
PCA P1	R	52	8	57	3	58	2	0.204	0.095	1	0	60
	L	51	9	60	0	60	0	0.003	0.003	1	0	60
PCA P2	R	55	5	60	0	60	0	0.057	0.057	1	0	60
	L	51	9	60	0	60	0	0.003	0.003	1	0	60
BA		53	7	60	0	60	0	0.013	0.013	1	0	60
ICA	R	51	9	50	10	50	10	1	1	1	0	60
	L	51	9	52	8	52	8	1	1	1	0	60
Total		614	121	714	21	706	29				105	840

* Comparison between Method 1 (DFS algorithm) and Method 2 (Dijkstra algorithm). † Comparison between Method 1 (DFS algorithm) and Method 3 (A* algorithm). ‡ Comparison between Method 2 (Dijkstra algorithm) and Method 3 (A* algorithm).

Computational time was compared among the three methods in all subjects' data (Supplementary Table S1). To isolate the path-finding algorithms from the manual annotation, we loaded the text file which contained all the manually annotated endpoints of the vessel segments, and we focused on measuring time on the path-finding procedures subject by subject. Method 1 took an average (±standard deviation) of 1489.4 (±191.4) ms. Method 2 took shorter than Method 1 with an average (±standard deviation) of 458.2 (±63.4) ms. Method 3 took an average (±standard deviation) of 458.0 (±63.4) ms, which was comparable to Method 2. The difference between Method 1 and Method 2 (or Method 3) was statistically significant ($p < 0.0001$), while the difference between Method 2 and Method 3 was not statistically significant ($p > 0.4$). In Methods 2 and 3, the conversion of the skeleton to the graph structure was the main bottleneck and took 98.95% and 99.00% of the total computational time, respectively. Moreover, it is important to note that manual annotation of the endpoints required the user to rotate and zoom in and out of the 3D skeletal vessel for correct identification of the vessel segments and took approximately 10 min per subject.

4. Discussion

Path finding is essential for annotating and quantifying vessel segments and has the potential to be useful in planning invasive procedures such as mechanical thrombectomy. This current study compared the performance of three path-finding algorithms in the total 840 arterial segments in the CoWs. Method 1 does not require a conversion of the skeleton's centerlines to the graph representation and is based on the DFS algorithm. Since it does not construct a graph representation, it is deemed as rather simple, when compared to Methods 2 and 3, which require the construction of a graph and utilize the shortest path-finding Dijkstra and A* algorithms, respectively. Our path-finding results indicate that Method 1 is prone to errors when there is a loop in the vessel's centerlines (Figure 3). The shortest path-finding algorithms can avoid such errors in case of the existence of a loop.

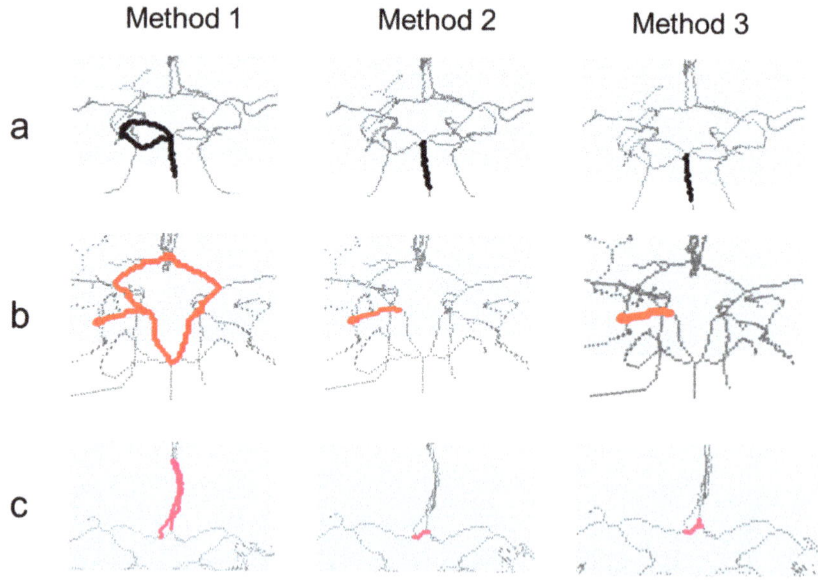

Figure 3. Incorrect path-finding results using Method 1 when compared with correct path-finding results using Methods 2 and 3. (**a**) Basilar artery (BA) segment, (**b**) right posterior cerebral artery P2 (R-P2) segment, and (**c**) anterior communicating artery (AComm) segment. Method 1 finds detoured paths in (**a**–**c**). The colorful lines indicate the specific vessel segments (see Figure 2 for the correspondences between colors and vessel segments).

There are two possible reasons for the incorrect path-finding results. First, when the path forms a loop, Method 1 may not find the shortest path and instead can take a path that leads to the other endpoint regardless of the path's length. The 'loop' path was frequent in the AComm segments, where there are alternative paths that detour via A2 segments. Second, when there are multiple paths between the two endpoints, Method 2 always finds the shortest path, which is sometimes not the correct path (Figure 4). We often detected an incorrect path by comparing it with a symmetrical vessel anatomy located in the other hemisphere of the brain. Additional routes in the main arteries may be attributed to either incorrect segmentation due to noise in the image or errors in the skeletonization process. By improving segmentation performance or denoising the gray-scale TOF-MRA images, the incorrect path-finding problem may be resolved.

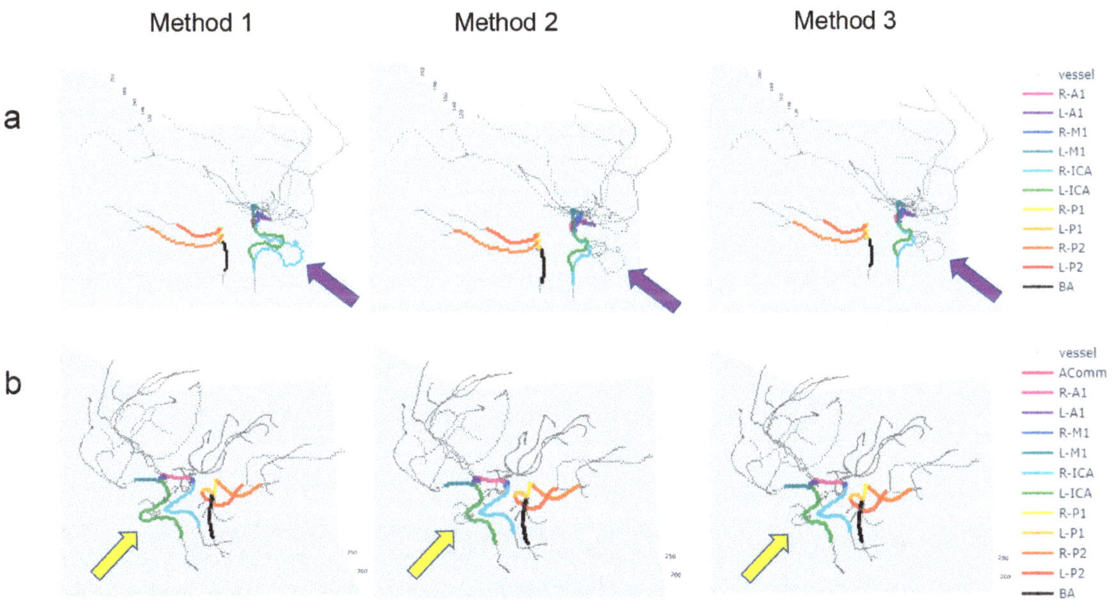

Figure 4. Incorrect path-finding results using Methods 2 and 3 when compared with correct path-finding results using Method 1. (**a**) The R-ICA and L-ICA segments in Method 2 found the shortest paths, but the paths are incorrect, given the nature of tortuous vessels in the ICA in general. Compare the ICA vessels indicated by the purple arrows. (**b**) The L-ICA segment in Methods 2 and 3 found the shortest paths, which is incorrect. Compare the ICA vessels indicated by the yellow arrows.

Compared to the A* algorithm, Dijkstra's algorithm finds the shortest paths for all nodes, although in our case we need only a single shortest path between the initial and destination nodes. Hence, the Dijkstra algorithm may not be the most time-efficient for path finding by nature. Since the A* algorithm only cares about finding the shortest path between the two endpoints, it is theoretically more time-efficient than the Dijkstra algorithm. However, in our evaluation, there were little or no noticeable improvements in the A* algorithm over the Dijkstra algorithm in terms of computational time, which was not expected given the inherent advantage in time efficiency in the A* algorithm.

The seeded region-growing algorithm used in this study provided under-segmentation results. This under-segmentation occurred dominantly in the PComm segments. Since we focused on the evaluation of the path-finding methods, it is a fair comparison, although there was an under-segmentation issue in several vessel segments. However, there is room for improvement in automatic segmentation. One way would be to use advanced segmentation methods based on encoder–decoder deep convolutional neural network architectures [30,31] or a multiscale image analysis approach [32]. Our study was based on MRA images, but it is possible to extract vessel segments from CTA images. Notably, a direct application of our region-growing segmentation to CTA images would be challenging because the region-growing method is based on the similarity of intensity in the blood, and contrast-enhanced arteries may not be distinguished from nearby bone structures in CTA images [33,34]. The errors in vessel segmentation are likely to occur especially in the regions where the arteries and bones are very close (e.g., internal carotid arteries, vertebral arteries, etc.). For CTA images, recent studies proposed automatic deep learning-based vessel segmentation methods to improve the accuracy of vessel segmentation [18,35].

Manual annotation of the endpoints in 3D space is tedious and can be prone to errors if not carefully checked. As such, it would be intriguing to investigate the feasibility of automatic landmark localization methods [36–38] to localize the vessel segments' endpoints which we manually annotated in this study. Landmark localization of the two endpoints may also be helpful in automatically detecting occlusions of the arterial segments in ischemic stroke patients' data.

This study has several limitations. First, the number of subjects used for the analysis is small. However, increasing the number of subjects would not significantly affect the study outcome, as there are already more than 10 segments for each subject's MRA image data. Second, we did not perform intrarater or interrater variability in evaluating the correctness of the paths. Third, the image segmentation quality has imperfections since it showed under-segmentation results, especially in the PComm segments. However, improving the segmentation results may not significantly affect the outcome of the study because all three methods used the same segmented binary vessel masks for the analysis. Fourth, we did not consider disconnected vessel segments, which may result from the occlusions or severe stenoses of arterial segments. The development of a method for identifying disconnected vessel segments would be important for automatically detecting the vessel's occlusion. Last but not least, we only considered arterial segments in the CoW arteries. Extending the analysis to other intracranial vessel segments such as ACA A2 and MCA M2 would be an important venue for further research.

5. Conclusions

Our study indicates that a graph representation of the centerlines of cerebral arteries in the CoW is advantageous in semi-automatically labeling the cerebral arteries when it is used with the shortest path-finding algorithms such as the Dijkstra algorithm and the A* algorithm. Among the three path-finding algorithms, the Dijkstra algorithm resulted in the highest accuracy of 97.1%, and it was slightly higher in accuracy than the A* algorithm. When the DFS approach was used, the incorrect path-finding results mostly occurred for the vessel segments with multiple paths connecting the two endpoints. Path-finding results for the Dijkstra and A* algorithms were incorrect when the paths were visually compared with the paths in the other brain hemisphere. Full automatization of vessel segmentation and landmark localization with deep learning would be desirable in order to increase time efficiency and avoid manual annotation procedures.

Supplementary Materials: The following supporting information can be downloaded at: https://www.mdpi.com/article/10.3390/tomography9040113/s1, Figure S1: A slice used for the selection of three seed points. The cross-sections of the right internal carotid artery (orange), the left internal carotid artery (blue), and the basilar artery (green) are shown. The seed locations for seeded region growing are indicated by the arrows; Figure S2: The pseudo-code of the depth first search (DFS) based maze solving algorithm. This algorithm is referred to as Method 1; Figure S3: Seeded region growing. (a) A vessel segmentation result after the Otsu's thresholding. Disconnected vessels are present, which sometimes obscure the view of the artery of interests. (b) A vessel segmentation result after applying the seeded region growing algorithm to the Otsu-thresholded vessel. Disconnected vessels disappear, and vessel visualization is improved; Table S1: Comparison of computational time measurements.

Author Contributions: Conceptualization, Y.-C.K.; methodology, S.-O.K. and Y.-C.K.; software, S.-O.K. and Y.-C.K.; validation, S.-O.K.; formal analysis, S.-O.K.; investigation, S.-O.K.; resources, S.-O.K.; data curation, S.-O.K. and Y.-C.K.; writing—original draft preparation, Y.-C.K.; writing—review and editing, Y.-C.K.; visualization, S.-O.K. and Y.-C.K.; supervision, Y.-C.K.; project administration, Y.-C.K.; funding acquisition, Y.-C.K. All authors have read and agreed to the published version of the manuscript.

Funding: This study was supported by the "Regional Innovation Strategy (RIS)" through the National Research Foundation of Korea (NRF) funded by the Ministry of Education (MOE) (2022RIS-005).

Institutional Review Board Statement: Ethical review and approval were waived for this study due to the use of publicly available data.

Informed Consent Statement: Patient consent was waived due to the use of publicly available data.

Data Availability Statement: The data presented in this study are available on request from the corresponding author.

Conflicts of Interest: The authors declare no conflict of interest.

References

1. Ritz, K.; Denswil, N.P.; Stam, O.C.; van Lieshout, J.J.; Daemen, M.J. Cause and mechanisms of intracranial atherosclerosis. *Circulation* **2014**, *130*, 1407–1414. [CrossRef]
2. Hu, X.; De Silva, T.M.; Chen, J.; Faraci, F.M. Cerebral Vascular Disease and Neurovascular Injury in Ischemic Stroke. *Circ. Res.* **2017**, *120*, 449–471. [CrossRef] [PubMed]
3. Krzyzewski, R.M.; Klis, K.M.; Kwinta, B.M.; Gackowska, M.; Gasowski, J. Increased tortuosity of ACA might be associated with increased risk of ACoA aneurysm development and less aneurysm dome size: A computer-aided analysis. *Eur. Radiol.* **2019**, *29*, 6309–6318. [CrossRef] [PubMed]
4. Jeong, W.; Rhee, K. Hemodynamics of cerebral aneurysms: Computational analyses of aneurysm progress and treatment. *Comput. Math. Methods Med.* **2012**, *2012*, 782801. [CrossRef]
5. Magid-Bernstein, J.; Girard, R.; Polster, S.; Srinath, A.; Romanos, S.; Awad, I.A.; Sansing, L.H. Cerebral Hemorrhage: Pathophysiology, Treatment, and Future Directions. *Circ. Res.* **2022**, *130*, 1204–1229. [CrossRef]
6. Cirillo, M.; Scomazzoni, F.; Cirillo, L.; Cadioli, M.; Simionato, F.; Iadanza, A.; Kirchin, M.; Righi, C.; Anzalone, N. Comparison of 3D TOF-MRA and 3D CE-MRA at 3T for imaging of intracranial aneurysms. *Eur. J. Radiol.* **2013**, *82*, e853–e859. [CrossRef] [PubMed]
7. Lell, M.M.; Anders, K.; Uder, M.; Klotz, E.; Ditt, H.; Vega-Higuera, F.; Boskamp, T.; Bautz, W.A.; Tomandl, B.F. New techniques in CT angiography. *Radiographics* **2006**, *26* (Suppl. 1), S45–S62. [CrossRef]
8. Han, H.C. Twisted blood vessels: Symptoms, etiology and biomechanical mechanisms. *J. Vasc. Res.* **2012**, *49*, 185–197. [CrossRef] [PubMed]
9. Kandil, H.; Soliman, A.; Ghazal, M.; Mahmoud, A.; Shalaby, A.; Keynton, R.; Elmaghraby, A.; Giridharan, G.; El-Baz, A. A Novel Framework for Early Detection of Hypertension using Magnetic Resonance Angiography. *Sci. Rep.* **2019**, *9*, 11105. [CrossRef]
10. Kim, H.J.; Song, H.N.; Lee, J.E.; Kim, Y.C.; Baek, I.Y.; Kim, Y.S.; Chung, J.W.; Jee, T.K.; Yeon, J.Y.; Bang, O.Y.; et al. How Cerebral Vessel Tortuosity Affects Development and Recurrence of Aneurysm: Outer Curvature versus Bifurcation Type. *J. Stroke* **2021**, *23*, 213–222. [CrossRef]
11. Klis, K.M.; Krzyzewski, R.M.; Kwinta, B.M.; Lasocha, B.; Brzegowy, P.; Stachura, K.; Popiela, T.J.; Borek, R.; Gasowski, J. Increased tortuosity of basilar artery might be associated with higher risk of aneurysm development. *Eur. Radiol.* **2020**, *30*, 5625–5632. [CrossRef] [PubMed]
12. Sodi, A.; Guarducci, M.; Vauthier, L.; Ioannidis, A.S.; Pitz, S.; Abbruzzese, G.; Sofi, F.; Mecocci, A.; Miele, A.; Menchini, U. Computer assisted evaluation of retinal vessels tortuosity in Fabry disease. *Acta Ophthalmol.* **2013**, *91*, e113–e119. [CrossRef] [PubMed]
13. Klis, K.M.; Krzyzewski, R.M.; Kwinta, B.M.; Stachura, K.; Gasowski, J. Tortuosity of the Internal Carotid Artery and Its Clinical Significance in the Development of Aneurysms. *J. Clin. Med.* **2019**, *8*, 237. [CrossRef] [PubMed]
14. Bogunovic, H.; Pozo, J.M.; Cardenes, R.; San Roman, L.; Frangi, A.F. Anatomical labeling of the Circle of Willis using maximum a posteriori probability estimation. *IEEE Trans. Med. Imaging* **2013**, *32*, 1587–1599. [CrossRef] [PubMed]
15. Chen, L.; Mossa-Basha, M.; Balu, N.; Canton, G.; Sun, J.; Pimentel, K.; Hatsukami, T.S.; Hwang, J.N.; Yuan, C. Development of a quantitative intracranial vascular features extraction tool on 3D MRA using semiautomated open-curve active contour vessel tracing. *Magn. Reson. Med.* **2018**, *79*, 3229–3238. [CrossRef]
16. Robben, D.; Turetken, E.; Sunaert, S.; Thijs, V.; Wilms, G.; Fua, P.; Maes, F.; Suetens, P. Simultaneous segmentation and anatomical labeling of the cerebral vasculature. *Med. Image Anal.* **2016**, *32*, 201–215. [CrossRef]
17. Dumais, F.; Caceres, M.P.; Janelle, F.; Seifeldine, K.; Ares-Bruneau, N.; Gutierrez, J.; Bocti, C.; Whittingstall, K. eICAB: A novel deep learning pipeline for Circle of Willis multiclass segmentation and analysis. *Neuroimage* **2022**, *260*, 119425. [CrossRef]

18. Nazir, A.; Cheema, M.N.; Sheng, B.; Li, H.T.; Li, P.; Yang, P.; Jung, Y.Y.; Qin, J.; Kim, J.M.; Feng, D.D. OFF-eNET: An Optimally Fused Fully End-to-End Network for Automatic Dense Volumetric 3D Intracranial Blood Vessels Segmentation. *IEEE Trans. Image Process.* **2020**, *29*, 7192–7202. [CrossRef]
19. Suran, S.; Pattanaik, V.; Malathi, D. Discovering shortest path between points in cerebrovascular system. In Proceedings of the 6th IBM Collaborative Academia Research Exchange Conference (I-CARE) on I-CARE 2014, Bangalore, India, 9–11 October 2014; pp. 1–3.
20. Shen, M.; Wei, J.; Fan, J.; Tan, J.; Wang, Z.; Yang, Z.; Qiao, P.; Liao, F. Automatic cerebral artery system labeling using registration and key points tracking. In Proceedings of the Knowledge Science, Engineering and Management: 13th International Conference, KSEM 2020, Hangzhou, China, 28–30 August 2020; pp. 355–367.
21. Thamm, F.; Jurgens, M.; Taubmann, O.; Thamm, A.; Rist, L.; Ditt, H.; Maier, A. An algorithm for the labeling and interactive visualization of the cerebrovascular system of ischemic strokes. *Biomed. Phys. Eng. Express* **2022**, *8*, 065016. [CrossRef]
22. Kroon, D.-J. *Region Growing, MATLAB Central File Exchange*; The MathWorks, Inc.: Natick, MA, USA, 2021.
23. Van der Walt, S.; Schönberger, J.L.; Nunez-Iglesias, J.; Boulogne, F.; Warner, J.D.; Yager, N.; Gouillart, E.; Yu, T. Scikit-Image: Image processing in Python. *PeerJ* **2014**, *2*, e453. [CrossRef]
24. Kim, Y.C.; Kim, K.R.; Lee, H.; Choe, Y.H. Fast calculation software for modified Look-Locker inversion recovery (MOLLI) T1 mapping. *BMC Med. Imaging* **2021**, *21*, 26. [CrossRef] [PubMed]
25. Nunez-Iglesias, J.; Blanch, A.J.; Looker, O.; Dixon, M.W.; Tilley, L. A new Python library to analyse skeleton images confirms malaria parasite remodelling of the red blood cell membrane skeleton. *PeerJ* **2018**, *6*, e4312. [CrossRef] [PubMed]
26. Lee, K.D.; Lee, K.D.; Steve Hubbard, S.H. *Data Structures and Algorithms with Python*; Springer: Berlin/Heidelberg, Germany, 2015.
27. Virtanen, P.; Gommers, R.; Oliphant, T.E.; Haberland, M.; Reddy, T.; Cournapeau, D.; Burovski, E.; Peterson, P.; Weckesser, W.; Bright, J.; et al. SciPy 1.0: Fundamental algorithms for scientific computing in Python. *Nat. Methods* **2020**, *17*, 261–272. [CrossRef]
28. Schrijver, A. *Combinatorial Optimization: Polyhedra and Efficiency*; Springer: Berlin/Heidelberg, Germany, 2003; Volume 24.
29. Russell, S.J. *Artificial Intelligence a Modern Approach*; Pearson Education, Inc.: London, UK, 2010.
30. Chen, Y.; Jin, D.; Guo, B.; Bai, X. Attention-Assisted Adversarial Model for Cerebrovascular Segmentation in 3D TOF-MRA Volumes. *IEEE Trans. Med. Imaging* **2022**, *41*, 3520–3532. [CrossRef]
31. Livne, M.; Rieger, J.; Aydin, O.U.; Taha, A.A.; Akay, E.M.; Kossen, T.; Sobesky, J.; Kelleher, J.D.; Hildebrand, K.; Frey, D.; et al. A U-Net Deep Learning Framework for High Performance Vessel Segmentation in Patients With Cerebrovascular Disease. *Front. Neurosci.* **2019**, *13*, 97. [CrossRef] [PubMed]
32. Wozniak, T.; Strzelecki, M.; Majos, A.; Stefanczyk, L. 3D vascular tree segmentation using a multiscale vesselness function and a level set approach. *Biocybern. Biomed. Eng.* **2017**, *37*, 66–77. [CrossRef]
33. Lell, M.M.; Ruehm, S.G.; Kramer, M.; Panknin, C.; Habibi, R.; Klotz, E.; Villablanca, P. Cranial computed tomography angiography with automated bone subtraction: A feasibility study. *Investig. Radiol.* **2009**, *44*, 38–43. [CrossRef]
34. van Straten, M.; Venema, H.W.; Streekstra, G.J.; Majoie, C.B.; den Heeten, G.J.; Grimbergen, C.A. Removal of bone in CT angiography of the cervical arteries by piecewise matched mask bone elimination. *Med. Phys.* **2004**, *31*, 2924–2933. [CrossRef]
35. Fu, F.; Wei, J.; Zhang, M.; Yu, F.; Xiao, Y.; Rong, D.; Shan, Y.; Li, Y.; Zhao, C.; Liao, F.; et al. Rapid vessel segmentation and reconstruction of head and neck angiograms using 3D convolutional neural network. *Nat. Commun.* **2020**, *11*, 4829. [CrossRef]
36. Payer, C.; Stern, D.; Bischof, H.; Urschler, M. Integrating spatial configuration into heatmap regression based CNNs for landmark localization. *Med. Image Anal.* **2019**, *54*, 207–219. [CrossRef]
37. Alansary, A.; Oktay, O.; Li, Y.; Folgoc, L.L.; Hou, B.; Vaillant, G.; Kamnitsas, K.; Vlontzos, A.; Glocker, B.; Kainz, B.; et al. Evaluating reinforcement learning agents for anatomical landmark detection. *Med. Image Anal.* **2019**, *53*, 156–164. [CrossRef] [PubMed]
38. Xue, H.; Artico, J.; Fontana, M.; Moon, J.C.; Davies, R.H.; Kellman, P. Landmark Detection in Cardiac MRI by Using a Convolutional Neural Network. *Radiol. Artif. Intell.* **2021**, *3*, e200197. [CrossRef] [PubMed]

tomography

MDPI

Article

Image Quality Improvement in Deep Learning Image Reconstruction of Head Computed Tomography Examination

Michal Pula [1], Emilia Kucharczyk [2], Agata Zdanowicz [3] and Maciej Guzinski [3,*]

1 Lower Silesian Oncology, Pulmonology and Hematology Center, Hirszfelda Square 12, 53-413 Wrocław, Poland; michal.pula97@gmail.com
2 Faculty of Medicine, Wroclaw Medical University, Ludwika Pasteura 1, 50-367 Wrocław, Poland; emilia.kucharczyk@student.umed.wroc.pl
3 Department of General Radiology, Interventional Radiology and Neuroradiology, Wroclaw Medical University, Borowska 213, 50-556 Wrocław, Poland; agatazdanowicz@gmail.com
* Correspondence: maciej.guzinski@umw.edu.pl; Tel.: +48-72-744-26-68

Abstract: In this study, we assess image quality in computed tomography scans reconstructed via DLIR (Deep Learning Image Reconstruction) and compare it with iterative reconstruction ASIR-V (Adaptive Statistical Iterative Reconstruction) in CT (computed tomography) scans of the head. The CT scans of 109 patients were subjected to both objective and subjective evaluation of image quality. The objective evaluation was based on the SNR (signal-to-noise ratio) and CNR (contrast-to-noise ratio) of the brain's gray and white matter. The regions of interest for our study were set in the BGA (basal ganglia area) and PCF (posterior cranial fossa). Simultaneously, a subjective assessment of image quality, based on brain structure visibility, was conducted by experienced radiologists. In the assessed scans, we obtained up to a 54% increase in SNR for gray matter and a 60% increase for white matter using DLIR in comparison to ASIR-V. Moreover, we achieved a CNR increment of 58% in the BGA structures and 50% in the PCF. In the subjective assessment of the obtained images, DLIR had a mean rating score of 2.8, compared to the mean score of 2.6 for ASIR-V images. In conclusion, DLIR shows improved image quality compared to the standard iterative reconstruction of CT images of the head.

Keywords: deep learning reconstruction; deep neural network; head CT; image quality improvement; true fidelity

Citation: Pula, M.; Kucharczyk, E.; Zdanowicz, A.; Guzinski, M. Image Quality Improvement in Deep Learning Image Reconstruction of Head Computed Tomography Examination. *Tomography* **2023**, *9*, 1485–1493. https://doi.org/10.3390/tomography9040118

Academic Editors: Rafał Obuchowicz, Michał Strzelecki and Adam Piorkowski

Received: 10 May 2023
Revised: 14 July 2023
Accepted: 19 July 2023
Published: 9 August 2023

1. Introduction

Computed tomography has traditionally been regarded as a fundamental diagnostic modality, particularly in the evaluation of head structures. As a result of technological advancements in CT (computed tomography) imaging and the expanding range of clinical indications, there has been a significant rise in the annual utilization of CT scans. Hence, there is increasing demand for more precise imaging, which minimizes the need for additional tests. This becomes particularly crucial in situations where an urgent response is necessary for the patient's condition [1,2]. CT scans that are performed without the use of intravenous contrast agents have an inherent limitation in differentiating soft tissues with contrast resolution [3,4]. For instance, in the brain parenchyma, the variation in CT attenuation between gray and white matter is typically minor [5,6]. Identifying specific alterations in attenuation between gray and white matter is necessary for detecting certain intracranial pathologies [4–6]. The aforementioned limitations pose considerable challenges in the interpretation of non-contrast CT scans of the head, particularly in cases of trauma [7].

The ability to visualize anatomic structures and pathologic features in a CT image is influenced by two critical factors: blur and noise. Higher levels of blurring can reduce the visibility of minor objects or image details, while higher levels of visual noise can reduce the visibility of low-contrast objects [8]. Therefore, it is crucial to develop an optimized protocol that considers these physical principles for a specific clinical study and adjusts to achieve a balance between detail, low noise, and patient exposure. As the physical parameters of devices are modified at the expense of others, the role of image reconstruction becomes increasingly significant [9–11].

Recent advancements in imaging technology have spurred the development of novel techniques for processing CT images. Two prominent image reconstruction techniques used in CT are DLIR (Deep Learning Image Reconstruction) and ASiR-V (Adaptive Statistical Iterative Reconstruction). DLIR utilizes advanced machine learning algorithms to reconstruct CT images [6,10,12], whereas ASiR employs statistical models to iteratively refine the image data [7,13,14]. The primary objective of these reconstruction algorithms is to minimize image noise in low-dose CT scans while preserving image quality [6,9,12,15,16]. In clinical practice, the noise reduction capability of Iterative Reconstruction (IR) is often leveraged to reduce radiation dose while maintaining consistent image noise levels [7,11]. However, IR algorithms have limitations, particularly in balancing noise reduction and image texture, as well as overall image quality [5,6,17]. To tackle these challenges, a novel Deep Learning Image Reconstruction (DLIR) algorithm has been introduced. This algorithm utilizes a deep neural network-based reconstruction engine, which aims to overcome the limitations associated with iterative reconstruction algorithms [5–7,17].

The purpose of this study is to compare the quality of images obtained using both of the aforementioned techniques in order to establish a gold standard image reconstruction method for CT of the head. In current research, ASiR remains the benchmark reconstruction model in CT. However, given the constant pursuit of reducing radiation doses while maintaining image quality, and the rising number of parameters subject to evaluation, DLIR appears to be the natural successor to iterative reconstruction (IR).

2. Materials and Methods

A total of 163 consecutive head CT scans were retrospectively evaluated. The scans were initially carried out for clinical purposes, using a non-contrast imaging technique. We included 109 patients in the study, and the selection of a suitable research group was based on the implementation of exclusion criteria. These were: (1) neurosurgical treatment such as clipping, coiling, drainage, etc. (7 patients); (2) misregistration artifacts, which appear as blurring, streaking, or shading—caused by patient movement during a CT scan (28 patients); (3) mass effect induced by hematoma, edema, tumors, etc. (19 patients). The purpose of the exclusion criteria was to prevent the degradation of image quality resulting from the existence of various disruptive factors.

A Revolution Apex, 256-row scanner (GE Healthcare, Chicago, IL, USA) was used for all of the examinations. The assessments were conducted using the axial scanning method. All patients were in a supine position; the patient's head was fixed in the dedicated head holder. The tube voltage was 120 kVp; the mA was set to 320; the beam collimation was 80 mm; the slice thickness was 0.625 mm; the scan range covered the area between the foramen magnum and the top of the head; the scan rotation time was 1 s; and the automatic exposure control (AEC) was activated. DLIR reconstruction was obtained using TrueFidelity, a GE Healthcare deep neural network engine trained on FBP images of high quality, and a DLIR of high strength (DLIR-H) was utilized. The newest generation of iterative reconstruction from GE Healthcare, the Adaptive Statistical Iterative Reconstruction V (ASIR-V), in a combination of 50% ASIR-V and 50% of FBP, was utilized for ASIR-V.

2.1. Subjective Image Quality

Paired images were reviewed independently by a group of 3 radiologists with 3, 8, and 10 years of experience in neuroradiology, respectively. The readers were blinded to the reconstruction method. Images were displayed in a brain window (80 W, 40 L) on a GE workstation AW server 3.2 ext. 4.0. The reviewers were requested to evaluate the sharpness of the image, the differentiation between gray and white matter, as well as the visibility of the basal ganglia, pons, cerebellum, and ventricular structures. A Likert scale-based rating system was implemented, where images of excellent quality received a rating of 3, good-quality images were rated 2, and images of poor quality but that were still usable for evaluation received a rating of 1. The figures were assessed sequentially by the radiologist with the least experience, followed by the more experienced readers, with the most proficient neuroradiologist granting the final rank. A Fleiss' Kappa measurement was utilized to evaluate the agreement between reviewers. The assessment of image quality adhered to the European Guidelines on Quality Criteria for Computed Tomography.

2.2. Objective Image Quality:

The objective assessment of image quality involved analyzing the signal-to-noise (SNR) and contrast-to-noise (CNR) ratios. To maintain consistency, a control variable was introduced in the form of manually set Regions of Interest (ROIs), which are analogous and predefined study areas in each pair of images. The average size of the ROI area was 24 mm^2, with the smallest and the largest areas measuring 20 mm^2 and 29 mm^2, respectively. ROIs were positioned manually on the ASIR-V (Figure 1) reconstruction image, and then, duplicated at identical coordinates on the DLIR (Figure 2) reconstruction, which excluded potential deviations resulting from the lack of homogeneity of the measured tissues. The designation of the slice subject to the assessment was based solely on the anatomical factors of the patient. The first two ROIs were located in the white matter of the posterior limb of the internal capsule and the grey matter of the caudate nucleus, while the other two ROIs were positioned in the white and grey matter of the cerebellum. The fifth ROI was set in the cerebrospinal fluid of the fourth ventricle. To determine the signal level and noise, the mean CT number (mean) within the ROI and the standard deviation (SD) were taken into account.

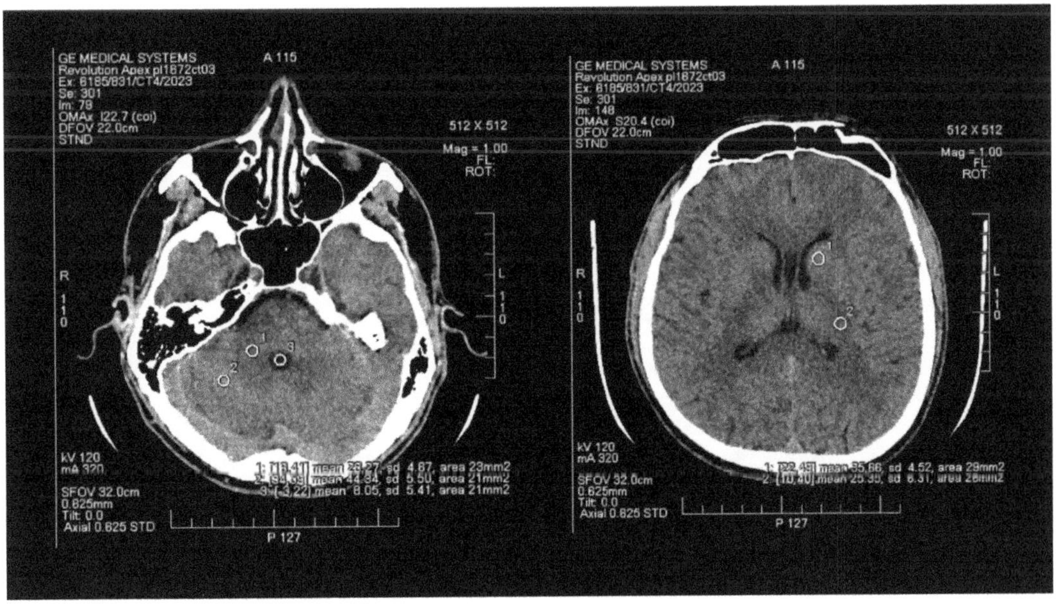

Figure 1. Images of head CT with marked ROIs located in BGA (basal ganglia area) and PCF (posterior cranial fossa) obtained using FBP + ASIR.

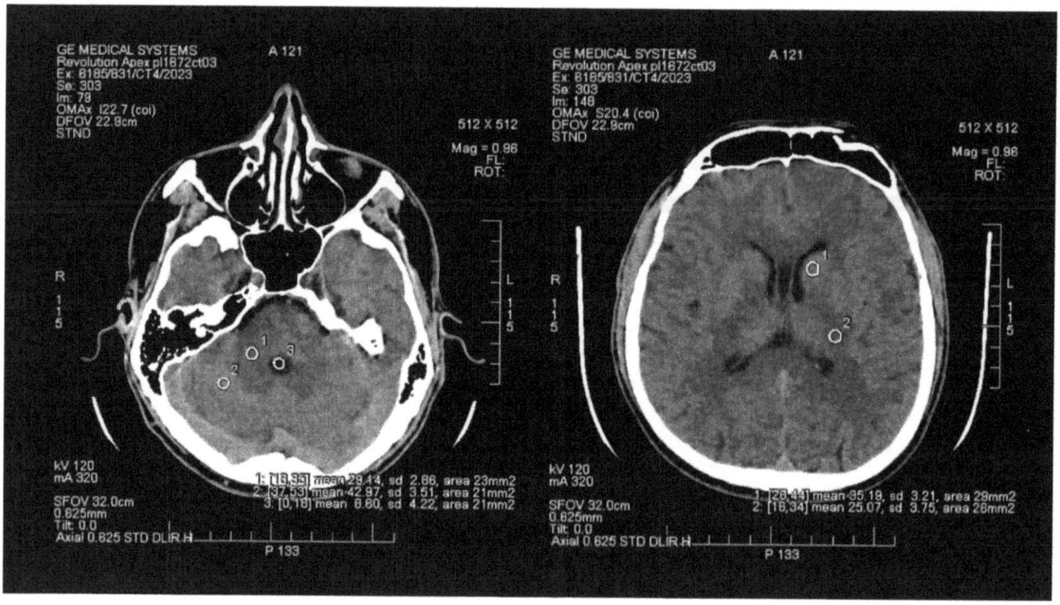

Figure 2. Images of head CT with marked ROIs located in BGA (basal ganglia area) and PCF (posterior cranial fossa) obtained using DLIR.

To facilitate data comparison, the calculation of numerical variables was employed, and specific formulae were developed to determine the contrast-to-noise ratio (CNR) and the signal-to-noise ratio (SNR), incorporating the relevant variables. The SNR was calculated by dividing the mean by the standard deviation (SD). For ROI pairs comprising white and grey matter, the CNR was determined using the following formula: the difference between the means of the two ROIs divided by the square root of the sum of the squared noise in the two ROIs.

$$CNR = \frac{Mean1 - Mean2}{\sqrt{SD1^2 + SD2^2}}$$

2.3. Statistical Analysis

We conducted a statistical analysis using RStudio (version 2022.12.0 + 353) Statistical Software (R Foundation for Statistical Computing, Vienna, Austria) [18]. We performed a paired t-test with Welsh modification for unequal variances to compare the SNR and CNR measurements between image pairs. To reduce the risk of significance due to chance resulting from multiple statistical testing, we applied Bonferroni correction to the *p*-value. A *p*-value of less than 0.01 was considered significant. In order to assess the agreement of the measurements between the standard technique (ASIR) and the novel solution (DLIR), Bland–Altman plots were implemented.

3. Results

The characteristics observed in the Bland–Altman plots for the selected ROIs, namely, uniform dispersion of data points, data points primarily falling within the limits of agreement, no systematic outliners, and a lack of patterns or trends in the data, indicated disparity in the measurements due to the implementation of different techniques (Figure 3, Table 1).

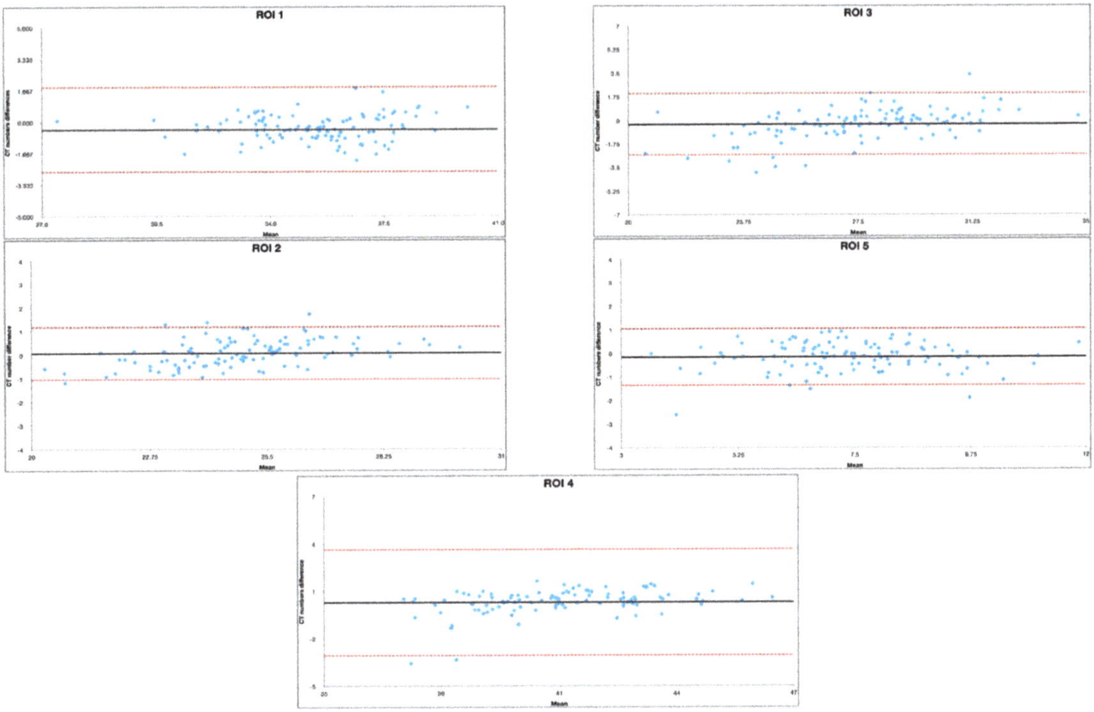

Figure 3. Bland–Altman plot assessment of agreement between DLIR and ASIR in selected ROIs.

Table 1. Limits of agreement and bias for Bland–Altman plot assessment of DLIR and ASIR measurement similarity.

Region	Limits of Agreement		Mean Bias
ROI 1	−2.6483	1.8449	0.4017
ROI 2	−1.0265	1.1848	0.0791
ROI 3	−2.5678	2.0095	−0.2791
ROI 4	−3.0607	3.6421	0.2907
ROI 5	−1.3425	1.0463	−0.1480

Pairwise comparisons of the images acquired using DLIR and ASIR-V in the BGA (basal ganglia area) presented increases in the SNR of 46.7% ($p < 0.01$) and 59.5% ($p < 0.01$) for gray and white matter, respectively. Likewise, in the posterior cranial fossa, DLIR reconstructions accounted for higher SNR values. Specifically, the SNR increased by 54.2% ($p < 0.01$) in the gray matter of the cerebellum and 60.1% ($p < 0.01$) in the white matter. In the assessment of the ventricular system based on the ROI placed in the fourth ventricle of the brain, the SNR values were 51.2% ($p < 0.01$) higher in DLIR reconstruction compared to ASIR-V. The DLIR-acquired images of BGA showed a 57.6% increase in the CNR compared to the ASIR-V scans, while the PCF (posterior cranial fossa) scans achieved a 50.3% increase. The mean SNR and CNR measurements are presented in Table 2. Our analysis revealed a statistically significant ($p < 0.01$) image improvement in both BGA and PCF under DLIR utilization. Amongst the 109 assessed scans, the highest recorded Dose Length Product (DLP) was 1208.77 mGy*cm, while the lowest was 433.27 mGy*cm. The average DLP was 903.97 mGy*cm.

Table 2. Objective image quality evaluation of head CT: DLIR compared to ASIR.

BGA					
	DLIR	**ASIR**	**DLIR vs. ASIR**	**Mean Difference between DLIR and ASIR**	***p*-Value**
GM SNR	9.52 ± 2.12	6.49 ± 1.30	46.68%	3.03 ± 2.32	<0.001
WM SNR	7.48 ± 1.47	4.69 ± 1.03	59.49%	2.78 ± 1.88	<0.001
CNR	2.08 ± 0.58	1.32 ± 0.40	57.58%	0.76 ± 0.62	<0.001
PCF					
	DLIR	**ASIR**	**DLIR vs. ASIR**	**Mean difference between DLIR and ASIR**	***p*-value**
GM SNR	11.67 ± 2.32	7.57 ± 1.45	54.16%	4.10 ± 2.82	<0.001
WM SNR	7.83 ± 1.72	4.89 ± 1.11	60.12%	2.94 ± 2.06	<0.001
CNR	2.57 ± 0.70	1.71 ± 0.50	50.29%	0.86 ± 0.74	<0.001
IV Ventricle					
	DLIR	**ASIR**	**DLIR vs. ASIR**	**Mean difference between DLIR and ASIR**	***p*-value**
SNR	1.89 ± 0.45	1.25 ± 0.32	51.20%	0.63 ± 0.50	<0.001

BGA—basal ganglia area, DLIR—Deep Learning Image Reconstruction, ASIR—Adaptive Statistical Iterative Reconstruction, SNR—signal-to-noise ratio, GM—grey matter, WM—white matter, CNR—contrast-to-noise ratio.

During the subjective assessment of the compared images, the researchers observed that DLIR had a slight advantage over ASIR-V in terms of its ability to visualize brain structures. A total of 89 of the acquired DLIR figures were given the highest possible mark (evaluated as being of excellent quality), compared to only 69 ASIR images. However, both reconstruction techniques faced difficulties in improving images of poor quality, with four of the ASIR, and three of the DLIR datasets rated 1. The average score awarded by the evaluators for the reconstructed images was 2.8 for DLIR and 2.6 for ASIR-V. The reviewers reached near-perfect agreement in the assessment of the images (Fleiss' Kappa 0.88 and 0.91 for DLIR and ASIR-V, respectively). The subjective assessment is presented in Figure 4.

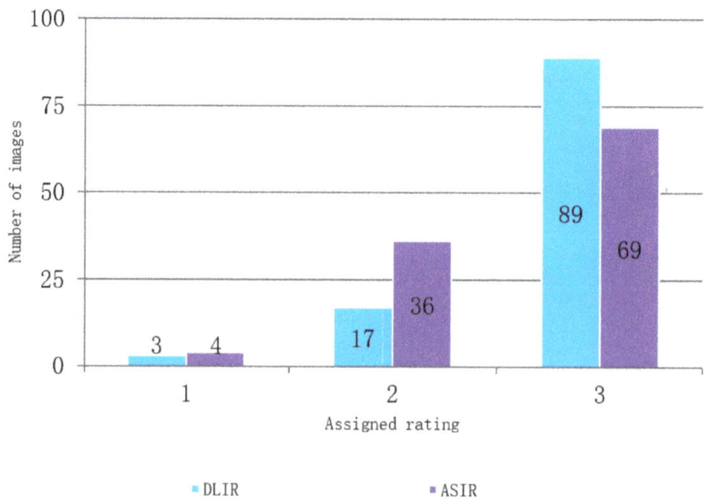

Figure 4. Subjective image quality assessment in DLIR and ASIR.

4. Discussion

The objective of this study was to conduct a comprehensive comparison of clinical brain CT image quality, utilizing both objective and subjective evaluation methods. Two distinct reconstruction algorithms, namely, ASIR-V and DLIR, were employed in this study. The primary advantage of our approach was the collection of average CT values within regions of interest (ROIs) for both protocols during a single scan. This methodology ensured the exclusion of any potential variations arising from changes in brain morphology or the emergence of previously undetected pathological findings. Additionally, obtaining both reconstructions in a single scan enabled the evaluation of image quality enhancement while maintaining a consistent radiation dose.

The results of our study demonstrate that DLIR provided a notable improvement in SNR, with increases of 46.68% up to 60.12% observed in both gray and white matter, as well as CNR increases of 57.58% and 50.29% in the BGA and PCF, respectively. The results of our study are in line with the current literature concerning the selection of patient cohorts, whether the studies implement comparable exclusion criteria, as in Kim et al.'s research [17], or administer no clinically valid exclusion criteria at all, as in the corresponding study of Alagic et al. [7]. However, our study contains a larger patient group. Moreover, in our assessment, a greater focus was placed on establishing PCF image quality improvement, whereas the previous studies aimed mostly to analyze the basal ganglia and cortex regions [7,17].

The utilization of DLIR in non-contrast head CT scans appears to have particular significance as differences in mean CT numbers between the grey and white matter of the brain are very subtle. Moreover, the neuralgic region of the cerebellum, pons, and medulla is rarely free of motion and bone-based artifacts, thus making image improvement and noise canceling in this region of high clinical importance [19,20]. Conventional ASIR-V approaches face challenges due to the increasing number of parameters, which can make convergence difficult. Recent studies have demonstrated that DLIR can handle intricate problems owing to training processes, thus surpassing the limitations of ASIR-V. DLIR was trained with thousands of FBP images that had been taken at both low and high dose levels, obtaining images free of statistical, physical, mechanical, or physiological biases. The high-dose FBP images were set as the target images for the DLIR recon algorithm, while the respective low-dose ones, and more specifically, their sinograms, fed the input layer of the algorithm. The neurons of the DLIR algorithm were finetuned in an iterative process, such that low-dose sinograms managed to achieve high dose level image quality [6].

There are several limitations to our study that should be acknowledged. As a result of the exclusion criteria we stated, this retrospective study had a relatively small sample size. Even though the number of patients in our study was larger compared to recent studies that compared DLIR and ASIR-V [7,17], it is necessary to conduct further research with a larger patient cohort to confirm our findings. Secondly, our study only included patients without clear neuropathological evidence or foreign bodies, and we were not able to assess the improvement in diagnostic accuracy. While our results indicate that DLIR yields superior image quality in individuals with normal-aging brains, additional studies involving patients with different clinical conditions are necessary to assess the impact of DLIR on lesion detection and diagnostic accuracy. Additionally, we only investigated CT images using a standard radiation dose protocol that was adjusted to our institution's clinical requirements. Given that radiation dose reduction can be achieved by adjusting the reconstruction mechanism, it is important to assess the effects of DLIR on low-dose radiation protocols in future studies. Moreover, CNR and SNR are linear metrics, and as such, a more sophisticated image quality assessment in terms of detectability index (d') and MTF (modulation transfer function) or TTF (task-based transfer function)/NPS-based metrics will follow.

In conclusion, our study demonstrates that DLIR surpasses ASIR-V in enhancing image quality and reducing noise in non-contrast head CT examinations. DLIR exhibited superiority in both subjective and objective assessments of image quality improvement.

These findings suggest that DLIR has the potential to become the standard method for reconstructing head CT scans. However, further research is needed to validate these results, consider different patient populations, and explore the potential benefits of DLIR in low-dose radiation protocols using comprehensive image quality metrics.

Author Contributions: Conceptualization, M.P. and M.G.; methodology, E.K. and A.Z.; validation, A.Z. and M.G.; data curation, M.P. and E.K.; writing—original draft preparation, M.P., E.K. and A.Z.; writing—review and editing, M.G.; visualization, M.P. and M.G.; supervision, M.G.; project administration, M.P.; funding acquisition, M.G. All authors have read and agreed to the published version of the manuscript.

Funding: This research was funded by Wroclaw Medical University, Grant SUBZ.C270.23.077.

Institutional Review Board Statement: Institutional Review Board approval was obtained for this research. Approval number: KB-591/2019.

Informed Consent Statement: Patient consent was waived due to retrospective nature of this study and because no sensitive personal information was published.

Data Availability Statement: Restrictions apply to the availability of these data. Data generated or analyzed during this study are available from the corresponding author by request subject to institutional review and a data use agreement.

Conflicts of Interest: The authors declare no conflict of interest.

References

1. Tanaka, S.; Kadoya, N.; Sugai, Y.; Umeda, M.; Ishizawa, M.; Katsuta, Y.; Ito, K.; Takeda, K.; Jingu, K. A deep learning-based radiomics approach to predict head and neck tumor regression for adaptive radiotherapy. *Sci. Rep.* **2022**, *12*, 8899. [CrossRef] [PubMed]
2. Schweitzer, A.D.; Niogi, S.N.; Whitlow, C.J.; Tsiouris, A.J. Traumatic Brain Injury: Imaging Patterns and Complications. *Radiographics* **2019**, *39*, 1571–1595. [CrossRef] [PubMed]
3. Lin, E.; Alessio, A. What are the basic concepts of temporal, contrast, and spatial resolution in cardiac CT? *J. Cardiovasc. Comput. Tomogr.* **2009**, *3*, 403–408. [CrossRef] [PubMed]
4. Bier, G.; Bongers, M.N.; Ditt, H.; Bender, B.; Ernemann, U.; Horger, M. Enhanced gray-white matter differentiation on non-enhanced CT using a frequency selective non-linear blending. *Neuroradiology* **2016**, *58*, 649–655. [CrossRef]
5. Sun, J.; Li, H.; Wang, B.; Li, J.; Li, M.; Zhou, Z.; Peng, Y. Application of a deep learning image reconstruction (DLIR) algorithm in head CT imaging for children to improve image quality and lesion detection. *BMC Med. Imaging* **2021**, *21*, 108. [CrossRef] [PubMed]
6. Hsieh, J.; Liu, E.; Nett, B.; Tang, J.; Thibault, J.-B.; Sahney, S. *New Era of Image Reconstruction: TrueFidelity^{TM} Technical White Paper on Deep Learning Image Reconstruction*; GE Healthcare: Chicago, IL, USA, 2019.
7. Alagic, Z.; Cardenas, J.D.; Halldorsson, K.; Grozman, V.; Wallgren, S.; Suzuki, C.; Helmenkamp, J.; Koskinen, S.K. Deep learning versus iterative image reconstruction algorithm for head CT in trauma. *Emerg. Radiol.* **2022**, *29*, 339–352. [CrossRef] [PubMed]
8. Sprawls, P. AAPM tutorial. CT image detail and noise. *RadioGraphics* **1992**, *12*, 1041–1046. [CrossRef] [PubMed]
9. Nagayama, Y.; Iwashita, K.; Maruyama, N.; Uetani, H.; Goto, M.; Sakabe, D.; Emoto, T.; Nakato, K.; Shigematsu, S.; Kato, Y.; et al. Deep learning-based reconstruction can improve the image quality of low radiation dose head CT. *Eur. Radiol.* **2023**, *33*, 3253–3265. [CrossRef] [PubMed]
10. McLeavy, C.M.; Chunara, M.J.; Gravell, R.J.; Rauf, A.; Cushnie, A.; Staley Talbot, C.; Hawkins, R.M. The future of CT: Deep learning reconstruction. *Clin. Radiol.* **2021**, *76*, 407–415. [CrossRef] [PubMed]
11. Greffier, J.; Hamard, A.; Pereira, F.; Barrau, C.; Pasquier, H.; Beregi, J.P.; Frandon, J. Image quality and dose reduction opportunity of deep learning image reconstruction algorithm for CT: A phantom study. *Eur. Radiol.* **2020**, *30*, 3951–3959. [CrossRef] [PubMed]
12. Lee, N.; Cho, H.-H.; Lee, S.M.; You, S.K. Adaptation of Deep Learning Image Reconstruction for Pediatric Head CT: A Focus on the Image Quality. *J. Korean Soc. Radiol.* **2023**, *84*, 240–252. [CrossRef] [PubMed]
13. Kaul, D.; Kahn, J.; Huizing, L.; Wiener, E.; Grupp, U.; Böning, G.; Ghadjar, P.; Renz, D.M.; Streitparth, F. Reducing Radiation Dose in Adult Head CT using Iterative Reconstruction—A Clinical Study in 177 Patients. *Rofo* **2016**, *188*, 155–162. [CrossRef] [PubMed]
14. Silva, A.C.; Lawder, H.J.; Hara, A.; Kujak, J.; Pavlicek, W. Innovations in CT Dose Reduction Strategy: Application of the Adaptive Statistical Iterative Reconstruction Algorithm. *Am. J. Roentgenol.* **2010**, *194*, 191–199. [CrossRef] [PubMed]
15. Singh, R.; Digumarthy, S.R.; Muse, V.V.; Kambadakone, A.R.; Blake, M.A.; Tabari, A.; Hoi, Y.; Akino, N.; Angel, E.; Madan, R.; et al. Image Quality and Lesion Detection on Deep Learning Reconstruction and Iterative Reconstruction of Submillisievert Chest and Abdominal CT. *Am. J. Roentgenol.* **2020**, *214*, 566–573. [CrossRef] [PubMed]

16. Jensen, C.T.; Liu, X.; Tamm, E.P.; Chandler, A.G.; Sun, J.; Morani, A.C.; Javadi, S.; Wagner-Bartak, N.A. Image Quality Assessment of Abdominal CT by Use of New Deep Learning Image Reconstruction: Initial Experience. *Am. J. Roentgenol.* **2020**, *215*, 50–57. [CrossRef] [PubMed]
17. Kim, I.; Kang, H.; Yoon, H.J.; Chung, B.M.; Shin, N.-Y. Deep learning–based image reconstruction for brain CT: Improved image quality compared with adaptive statistical iterative reconstruction-Veo (ASIR-V). *Neuroradiology* **2021**, *63*, 905–912. [CrossRef] [PubMed]
18. R Core Team. *R: A Language and Environment for Statistical Computing*; R Foundation for Statistical Computing: Vienna, Austria, 2013.
19. Rozeik, C.; Kotterer, O.; Preiss, J.; Schütz, M.; Dingler, W.; Deininger, H.K. Cranial CT Artifacts and Gantry Angulation. *J. Comput. Assist. Tomogr.* **1991**, *15*, 381–386. [CrossRef] [PubMed]
20. Alberico, R.A.; Loud, P.; Pollina, J.; Greco, W.; Patel, M.; Klufas, R. Thick-Section Reformatting of Thinly Collimated Helical CT for Reduction of Skull Base-Related Artifacts. *Am. J. Roentgenol.* **2000**, *175*, 1361–1366. [CrossRef] [PubMed]

cancers MDPI

Article

Deep-Learning-Based Dose Predictor for Glioblastoma–Assessing the Sensitivity and Robustness for Dose Awareness in Contouring

Robert Poel [1,2,*,†], Amith J. Kamath [2,†], Jonas Willmann [3], Nicolaus Andratschke [3], Ekin Ermiş [1], Daniel M. Aebersold [1], Peter Manser [1,4] and Mauricio Reyes [1,2]

1 Department of Radiation Oncology, Inselspital, Bern University Hospital, University of Bern, CH-3010 Bern, Switzerland
2 ARTORG Center for Biomedical Research, University of Bern, CH-3010 Bern, Switzerland
3 Department of Radiation Oncology, University Hospital Zurich, University of Zurich, CH-8091 Zurich, Switzerland
4 Division of Medical Radiation Physics, Inselspital, Bern University Hospital, University of Bern, CH-3010 Bern, Switzerland
* Correspondence: robert.poel@insel.ch
† These authors contributed equally to this work.

Simple Summary: For accurate radiotherapy, a clear definition of the geometric extent of organs and tumor volumes is important. Due to the laborious task of manually drawing contours to define these, automatic segmentation models are becoming increasingly available. These models, however, need to be visually evaluated by radiation oncology experts. This evaluation itself takes up valuable time, therefore making an efficient and clinically relevant validation of auto-segmented results desirable. An accurate 3D dose prediction model can help create dose awareness prior to the actual dose-planning step. It can provide useful information for the quality assurance of the contouring step. In this study, we trained a 3D dose predictor for volumetric modulated arc therapy (VMAT) treatment of glioblastoma patients based on an existing architecture called a cascaded 3D U-Net. We further tested this model's sensitivity and robustness for the purpose of estimating dose changes due to contour variations.

Abstract: External beam radiation therapy requires a sophisticated and laborious planning procedure. To improve the efficiency and quality of this procedure, machine-learning models that predict these dose distributions were introduced. The most recent dose prediction models are based on deep-learning architectures called 3D U-Nets that give good approximations of the dose in 3D almost instantly. Our purpose was to train such a 3D dose prediction model for glioblastoma VMAT treatment and test its robustness and sensitivity for the purpose of quality assurance of automatic contouring. From a cohort of 125 glioblastoma (GBM) patients, VMAT plans were created according to a clinical protocol. The initial model was trained on a cascaded 3D U-Net. A total of 60 cases were used for training, 15 for validation and 20 for testing. The prediction model was tested for sensitivity to dose changes when subject to realistic contour variations. Additionally, the model was tested for robustness by exposing it to a worst-case test set containing out-of-distribution cases. The initially trained prediction model had a dose score of 0.94 Gy and a mean DVH (dose volume histograms) score for all structures of 1.95 Gy. In terms of sensitivity, the model was able to predict the dose changes that occurred due to the contour variations with a mean error of 1.38 Gy. We obtained a 3D VMAT dose prediction model for GBM with limited data, providing good sensitivity to realistic contour variations. We tested and improved the model's robustness by targeted updates to the training set, making it a useful technique for introducing dose awareness in the contouring evaluation and quality assurance process.

Keywords: radiotherapy; dose prediction; deep learning; quality assurance; VMAT; glioblastoma

Citation: Poel, R.; Kamath, A.J.; Willmann, J.; Andratschke, N.; Ermiş, E.; Aebersold, D.M.; Manser, P.; Reyes, M. Deep-Learning-Based Dose Predictor for Glioblastoma–Assessing the Sensitivity and Robustness for Dose Awareness in Contouring. *Cancers* **2023**, *15*, 4226. https://doi.org/10.3390/cancers15174226

Academic Editors: Michał Strzelecki, Adam Piorkowski and Rafał Obuchowicz

Received: 3 August 2023
Revised: 16 August 2023
Accepted: 21 August 2023
Published: 23 August 2023

1. Introduction

Many cancers are currently treated by a combination of local and systemic therapy. Local therapy often consists of a combination of surgery and radiotherapy. The latter requires a sophisticated planning process to guarantee successful treatment. To improve the efficiency and quality of treatment planning in radiotherapy, methods for predicting possible dose distributions or dose volume histogram (DVH) curves have been introduced in previous years [1–7]. Since 2012, knowledge-based methods have been used to predict what is achievable in treatment planning. This was not only used for quality assurance in treatment planning [8–10], but also to speed up the planning process by initializing the treatment plan based on the prediction [11]. By reducing the number of inputs from the user, treatment planning becomes much more consistent, more resource-efficient, and potentially beneficial for treatment quality.

In recent years, three-dimensional dose prediction through neural networks has been shown to be a viable method for this purpose. In 2016, the first study using neural networks for dose predictions was published by Shiraishi et al. [12]. In the following years, approximately 30 more studies that tried to predict the 3D dose distribution with deep learning were published. Most of these were based on treatments with a relatively standard target orientation, with minor anatomical variations from patient to patient, such as prostate and oro/nasopharyngeal cancers [13–16]. However, there are also promising results for models in the brain, breast, and lungs [1,3,17]. In 2020, the Open-Access Knowledge-Based Planning (OpenKBP) Challenge was organized, providing an open-access dataset of head and neck treatment plans to train prediction models and evaluate them on a set of standardized metrics [18]. A total of 195 participants competed in this challenge, where the best-ranked team scored a mean absolute error (MAE) of 2.43 and 1.48 for the dose and DVH scores, respectively (see Section 2.6). Their methodology is publicly available and described as a technical note [19].

Dose prediction models are mainly used for treatment planning. This means that, in practice, in addition to the dose prediction, a second model is required to convert the predicted dose into an actual plan that is executable for the specific treatment technique. In this latter step, a final optimization incorporates individual case properties, physical constraints, and dose delivery hardware [20]. In our case, we want to use the dose prediction model for another purpose, namely contour evaluation.

The contouring of targets and organs at risk (OARs), the step that takes place prior to planning, is also subject to automation to improve efficiency and consistency with respect to the current manual process. With the implementation of artificial intelligence (AI), it is important to have specific tools for quality assurance [21]. To ensure quality, an assessment of the contours is required. Usually, visual inspection is the go-to method; however, this is a time-consuming task. For each target and OAR, every image slice needs to be visually inspected. If necessary, manual adjustments need to be made if a contour is deemed incorrect. Especially for deep-learning-based auto-segmentation models, a lack of robustness could result in unpredictable errors that can happen anywhere within the image volume [22]. There are, thus, good reasons to automate this QA step as well. There have been several attempts to provide an automatic assessment of automatic segmentations. Most of these were based on geometrical prior knowledge [23–25], but they lacked certainty for proper QA [26]. Moreover, more recent deep-learning-based approaches based on uncertainty maps seem to have issues [27]. To complement the current work performed on QA for auto-segmentation, we postulate that it would be beneficial to know the possible clinical impact of a specific segmentation. A deep-learning model that can give an accurate prediction of the dose received by an OAR instantly could provide the required information to assess the clinical impact of contour variations. Such dose awareness, for the evaluation of contours, can improve the efficiency of the process. Moreover, it provides clinically relevant feedback to the evaluator, who can then focus on potential segmentation errors that will have a larger impact on the treatment.

To assess the feasibility of a near-instant dose prediction model in providing dose awareness for the evaluation of auto-segmented contours, we made use of a deep-learning model to predict the dose for glioblastoma cases. Based on the network of Liu et al. [19] that was used in the OpenKBP challenge, a model was trained on a set of curated glioblastoma (GBM) cases. Unlike with current dose prediction algorithms, we wanted to verify the model's performance for contouring quality assessment (QA). This means that, beyond specific accuracy and sensitivity, robust predictions for a broad range of situations are required. To do so, we tested our trained model on specific sets of contour alterations to assess its sensitivity. Furthermore, we stress-tested the model by using a specific worst-case test set, including rare cases where we expected it to fail. This enabled us to determine the robustness of the model and understand where further improvements are required. Subsequently, based on the outcome observed on the worst-case test set, we improved the robustness of the model by augmenting the training set with synthetically generated cases characterizing the observed failure patterns.

2. Materials and Methods

2.1. Data Collection and Preparation

Imaging data from a cohort of 125 GBM patients treated with radiotherapy at the Inselspital University Hospital (Bern, Switzerland) were available. For all patients, the planning target volume (PTV) and the OARs (organs at risk) were curated by a mutual agreement between three radiation oncology professionals. A plan was constructed using a strict dose prescription and standard templates for planning setup and dose optimization initiation for all cases. Of the first 95 cases, 60 were selected for model training, 15 were chosen as validation, and 20 were used as a test set. Of the remaining 30 cases, 10 were used to construct a worst-case test set manually, and the other 20 for improving the training by adding specific out-of-distribution cases. Section 2.5 further details how the worst-case test set and the out-of-distribution cases were designed.

2.2. Dose Planning

All cases were planned according to the clinical dose prescription of 60 Gy in 30 fractions based on the ESTRO-EANO guidelines [28] in the Eclipse treatment planning system (TPS) V15.06.05 (Varian Medical Systems, Palo Alto). All OARs were subject to a dose constraint, which, according to a priority list, could or could not be compromised (Table 1). All plans used a volumetric arc technique (VMAT) with a double full co-planar arc with 6 MV beams containing a flattening filter. The plans were optimized with the photon optimizer, and doses were calculated with the Anisotropic Analytical Algorithm [29]. After dose calculation, the dose was normalized so that 50% of the PTV was covered by 100% of the prescribed dose, according to the institutional clinical guidelines.

Table 1. Clinical dose-planning guidelines for GBM treatment.

OAR	Constraint	Priority
Brain-PTV	• $V_{60\,Gy} \leq 3$ cc	2
Brainstem	• $D_{0.03cc} \leq 60$ Gy (hard constraint) • $D_{0.03cc} < 54$ Gy	1 4
Chiasm	• $D_{0.03cc} \leq 54$ Gy (hard constraint) • $D_{0.03cc} \leq 50$ Gy	1 3
Cochlea (Ipsi-lat)	• $D_{mean} \leq 45$ Gy (<30% hearing loss) • $D_{mean} \leq 32$ Gy (<20% tinnitus)	5 9
Cochlea (Bi-lat)	• $D_{mean} \leq 45$ Gy (<30% hearing loss) • $D_{mean} \leq 32$ Gy (<20% tinnitus)	7 9

Table 1. *Cont.*

OAR	Constraint	Priority
Hippocampus	• $D_{mean} \leq 30$ Gy (<30% IQ loss) • $D_{0.03cc} \leq 30$ Gy • $D_{40\%} \leq 7.3$ Gy (long-term NCF)	8 14 11
Lacrimal Gland	• $D_{mean} \leq 25$ Gy (clinic) (hard constraint)	1
Lens	• $D_{0.03cc} \leq 7$ Gy (<25% cataract)	12
Optic nerves (Ipsi-lat)	• $D_{0.03cc} \leq 54$ Gy (hard constraint) • $D_{0.03cc} \leq 50$ Gy	1 3
Optic nerve (Bi-lat)	• $D_{0.03cc} \leq 54$ Gy (hard constraint) • $D_{0.03cc} \leq 50$ Gy	1 6
Pituitary	• $D_{mean} \leq 45$ Gy (panhypopituitarism) • $D_{mean} \leq 20$ Gy (growth hormone deficiency)	10 13
Retina	• $D_{0.03cc} \leq 45$ Gy (hard constraint)	1
Target	**Objective**	**Priority**
PTV	• $D_{90\%} > 57$ Gy (95%)	1
CTV	• $D_{100\%} > 60$ Gy (100%)	2
PTV	• $D_{0.03cc} < 64$ Gy (107%)	3

2.3. *Training*

The planning CT and the structures were available in DICOM format for each case. All data were converted from DICOM to NIfTI files, using the PyRaDise package [30]. The RTSS files containing the PTV and the OARs were divided into 14 separate 3D binary masks, each containing a single structure. The input files consisted of 16 3D volumes per case: the planning CT, the dose distribution, the PTV binary mask, and 13 OAR binary masks (Figure 1).

We trained a two-level cascaded 3D (C3D) U-Net [31] as the dose prediction network (i.e., the input to the second U-Net is the output of the first, concatenated with the input to the first U-Net). The U-Net is the most commonly used deep-learning neural network for dose prediction. It was first used in this context as a 2D network by Nguyen et al. [32]. Since then, many advances have been made, notably an extension to a 3D U-Net [33]. The C3D model was proposed to incorporate both global and local anatomical features for dose prediction and showed the best results among all the competing model architectures in the OpenKBP challenge [18,31].

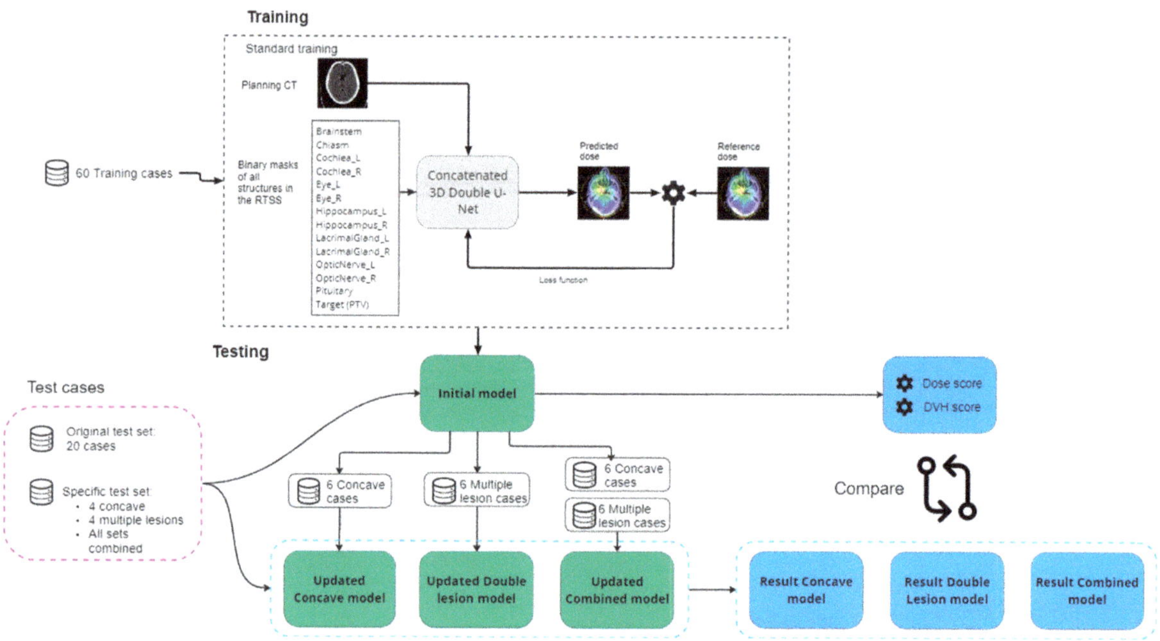

Figure 1. Schematic overview of the training and testing process. The upper block represents the training procedure of the initial model with its inputs and outputs. The initial model is tested on the test cases, resulting in dose and DVH scores for each test set. The initial model (green block) is updated threefold with concave cases, multiple lesion cases, and a combination of the two. The updated models are tested on the same test sets. The results are then compared (blue blocks).

The model input was a normalized CT volume and binary segmentation masks for each of the 13 OARs and target volume. As output, the model predicted a continuous-valued 3D dose (upscaled from [0, 1] to [0, 70 Gy] to normalize to the full range of the dose within the cohort) of the same dimension as the input. The loss was computed as follows:

$$Loss = 0.5 * L1(reference, A) + L1\ (reference, B)$$

where A and B are the outputs of the first and second U-Nets, respectively. In this equation, *reference* indicates the reference dose, and $L1$ refers to the $L1$ loss. All volumes were resampled to 128^3 voxels due to GPU memory constraints. The hyperparameters for training the C3D model were unchanged from the original implementation [9], except that the number of input binary masks was updated to 14 to match the number of structures in our dataset. The model's weights were randomly initialized using the "He" method [34]. The training process ran for 80,000 iterations, and the model with the best validation dose score was saved. All experiments were run with PyTorch1.12 on an NVIDIA RTX A5000 graphics processing unit (GPU). We trained the model five times with the same hyperparameter set but a different random seed initialization to ensure reliable convergence. Each training run took 24 h. A single inference to run the dose prediction model takes about 1–2 min on a standard consumer level PC and about 15 s with a GPU.

2.4. Assessing the Model's Sensitivity

One of the goals of the dose predictor is to provide realistic dosimetry information based on the input contours. It should additionally be able to predict realistic dose changes produced by small and realistic changes to the contours of an organ (i.e., inter-expert variability [35]). To analyze the sensitivity of the dose prediction model to these changes, a specific case was chosen where the GBM target is near the left optic nerve (ONL). In practice, the optic nerves are prone to variability in contours due to the intra- and inter-fractional movement of the eyes which also affect the optic nerves. In this case, small changes in the contour of the ONL would lead to significant dose changes. To simulate this situation, ten alternative contours of the ONL were manually drawn. The dose was re-optimized and recalculated on the TPS for each alternative contour in order to serve as a reference dose. The reference doses were then compared qualitatively and quantitatively to the doses predicted by the model. The dice similarity coefficients (DSCs) for the alternatives were calculated to correlate the dose differences to the geometric discrepancy of the ONL contours.

2.5. Improving the Model–Worst-Case Test Set

To assess the robustness of the model, a worst-case test set was selected. Based on the analysis and evaluation of the dose score results on the standard test set and the statistical analysis of the normal distribution of the training set, a number of test cases were defined where we expected the model to fail or have difficulties. The PTVs of these cases were manually manipulated to simulate rare cases not described by the training dataset (out-of-distribution cases), as well as to present a challenge in terms of the physical limitations of obtaining perfect dose conformity. Among these 10 cases, we included (i) targets of larger and smaller size than those present in the training set; (ii) targets consisting of multiple lesions; (iii) irregular shapes, such as elongated or concave targets; and (iv) targets that present an overlap with the OARs.

According to the worst-case test set results, we gained insight into which situations the model performs poorly in and where it could benefit from additional training. Our observations showed that the prediction model struggled mostly with the physical limitations of conforming the dose according to the targets outlined for specific shapes. Where conformity of the actual dose prediction decreases with concave shapes or multiple targets close to each other, the dose predictor overestimates the dose fall-off in these regions. To increase the overall robustness of the model while also improving the model for these situations, we updated the trained model by including a set of concave-shaped target cases and a set of cases where the target consists of multiple lesions.

The respective new training sets were constructed by manually adjusting the target volume (Figure 2). The ten cases used for both sets are from different patients and were not used in any previous model training. All new cases included dose planning according to the same protocol described above in order to serve as the reference dose.

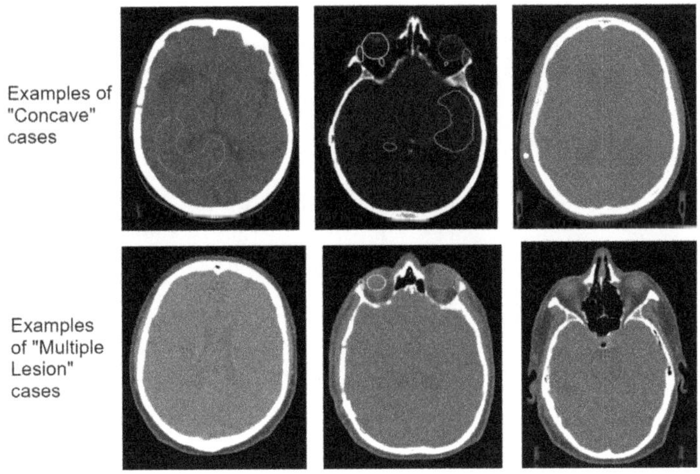

Figure 2. Examples of the additional training cases for the concave targets (**above**) and the multiple targets (**below**). The targets are drawn manually in red and do not represent actual tumor situations. The structures in other colors represent OARs.

First, we trained an updated "Concave Model" with 60 standard cases + 6 concave cases. The remaining 4 concave cases were used as the test set. Second, we trained an updated "Multiple Lesion Model" with the 60 standard cases + 6 multiple lesion cases. Again, 4 cases were used as a test set. Finally, we retrained the initial model with 60 standard cases + 6 concave cases + 6 multiple lesion cases. We tested these models on the standard test set, as well as these specific additional test cases, and compared this with the results of the initial model. An overview of the experimental setup is given in Figure 1.

2.6. Evaluation

The trained models were evaluated on the test set, and the prediction of the dose was compared to the actual planned dose by means of the standardized metrics used by the OpenKBP challenge [18]: the dose score and the DVH score. The dose score measures the mean error over all the voxels between the two 3D volumes. In this case, we used the whole brain to measure the dose score instead of the whole CT volume or whole body volume. Taking a larger volume dilutes the results to a more positive outcome. The DVH score is the mean error over a set of criteria specific to the given volume. For OARs, these criteria are the mean dose (Dmean) and the maximum dose to 0.1cc (D0.1cc). For the target volume, the criteria are the dose received by 1%, 95%, and 99% of the voxels within the volume (D1, D95, and D99). The DVH score is calculated for all OARs used in training (lens and retina are combined within the eye, since overlapping masks were not possible). We report the mean results for the five trained models and use one of them for a subsequent sensitivity analysis.

Additionally, the initial trained model and the updated models were tested on a set of concave target cases, a set of multiple lesion cases, and a combined test set that included both plus the standard test set.

3. Results

Based on the initial training set of 60 cases, the performance of the model was determined based on the standard test set of 20 cases. The mean result over five independently trained models showed a dose score, which was measured over the whole brain volume of 0.94 (standard deviation (SD) = 0.36). The mean DVH score over all OARs and the target was 1.95 (SD = 0.95).

3.1. Results for Sensitivity

An overview of the nine alternative left optic nerve contours is shown in Figure 3. The mean dose to the ONL, based on the treatment planning system, and the mean dose based on the prediction model for the reference and the nine alternative contours are shown in Table 2. There is a reasonable variation in the mean dose among the different alternative contours with respect to the reference contour. In some cases, only minor changes to the mean dose occur, even though the DSC metric shows a significant difference in contour similarity. In other cases, the mean dose change with respect to the reference contour can be as high as 5 to 7 Gy. The difference between the calculated dose and the predicted dose seems to follow a trend and varies under a maximum of 3.50 Gy, with a mean of 1.38 Gy. This shows that the predicted dose is more often overestimated.

The average difference of the calculated mean dose for the alternatives with respect to the calculated reference mean dose was 2.44 Gy (i.e., the difference between the alternatives to the reference dose). For the predicted dose, this difference was 2.32 Gy. This means that the correlation coefficient between reference and predicted dose differences across the contour alternatives was 0.89, while the correlation coefficient with the DSC was −0.42 [36].

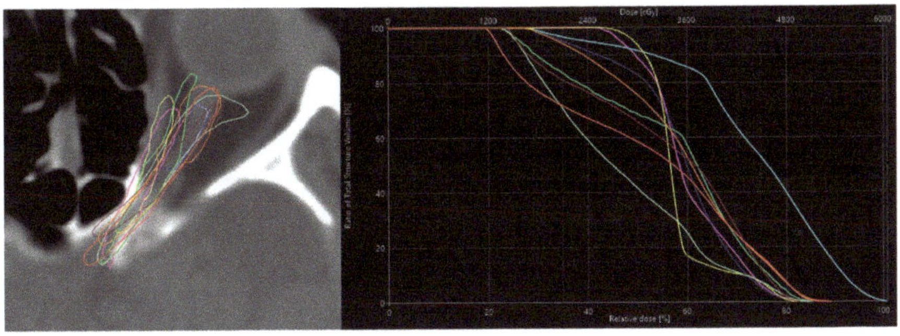

Figure 3. On the left is an overview of all the 9 alternative contours of the ONL. On the right, the dose's respective DVH curves are calculated with the treatment planning system, which shows the variation in the dose these contours have. The colors in the DVH curve correspond to the colors of the contours on the left.

Table 2. Predicted mean doses in Gy for the different optic nerve left contours.

ONL Contour	Calc. Dose	Pred. Dose	Δ Dose Calc-Pred	DSC	Δ to Calc-Ref	Δ to Pred-Ref
Reference	*34.7*	*35.5*	*−0.8*	*n.a.*	*n.a.*	*n.a*
Alternative-1	32.2	35.7	−3.5	0.31	−2.5	0.2
Alternative-2	30.7	32.4	−1.7	0.26	−4	−3.1
Alternative-3	34.2	34.5	−0.3	0.63	−0.5	-1
Alternative-4	31.8	34.1	−2.3	0.59	−2.9	−1.4
Alternative-5	26.9	30.1	−3.2	0.51	−7.8	−5.4
Alternative-6	32.8	36	−3.2	0.20	−1.9	0.5
Alternative-7	41.8	41.2	0.6	0.16	7.1	5.7
Alternative-8	35.3	33.1	2.2	0.58	0.6	−2.4
Alternative-9	34.5	36.1	−1.6	0.05	−0.2	0.6
Mean	33.49	34.87	−1.38	0.37	**Corr. Coeff.:** 0.89	

3.2. *Improving the Model–Worst-Case Test Set*

While analyzing the results of the worst-case test set, we saw flaws, particularly in cases where targets have concave shapes and consist of multiple lesions (see Figure 4). In such cases, the prediction model overestimated the ability to conform the dose to the targets. This is mainly reflected in higher dose scores and less so in the DVH scores of the target since the dose discrepancy occurs just outside of the target structure. We updated our training data with six concave target cases and six multiple lesion target cases and a combination of both. The results for the different test sets are given for the dose score, the DVH score for the OARs, and the DVH score for the targets separately in Table 3.

Based on the standard test set, the updated models scored similarly to the initial model on the dose score and the DVH score for targets. The DVH score for OARs improved for all updated models. The updated model with the concave target cases shows, by far, the best results of all trained models.

Focusing on the concave updated model, we observed improvements in the dose and DVH score on the concave test set, as well as improved results on the multiple test set. This means that, for this this small set of specific cases, the concave updated model scores significantly better than the initial model. For the multiple-lesions updated model, we also observed an improvement, but to a lesser extent. The combined updated model, containing both concave and multiple target lesions, scored the worst of all updated models. For the standard test set, the combined updated model did not show improved scores with respect to the initial model.

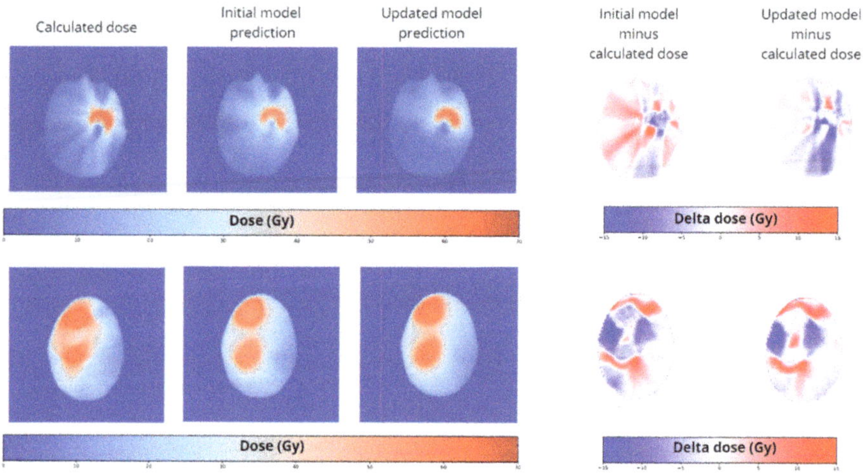

Figure 4. Dosimetric comparison of the calculated dose, the initial prediction model, and the updated model for a concave (**above**) and a multiple lesion case (**below**). The images represent a single axial slice. On the right, the dose difference maps of the corresponding axial slice are given. The difference between the latter shows improvements in the dose prediction. The depicted cases were not used in the training of the initial model or the updated model.

Table 3. Results of the dose score and DVH scores of the initial and the updated dose prediction models. Lower values represent better scores.

Test Set	Initial Model	Concave Updated Model	Multiple Lesion Updated Model	Combined Updated Model
		Dose scores' whole brain volume		
Standard test set	0.94	0.94	0.92	0.98
Concave test set	0.87	0.81	0.81	0.87
Multiple test set	1.30	0.84	1.24	1.02
Combined test set	0.98	0.90	0.95	0.97
		DVH scores' OARs		
Standard test set	2.01	1.73	1.85	1.89
Concave test set	2.11	1.67	1.99	2.08
Multiple test set	3.05	1.86	3.05	2.67
Combined test set	2.18	1.74	2.04	2.03
		DVH scores' targets		
Standard test set	1.19	1.12	1.20	1.26
Concave test set	1.72	1.67	1.51	1.66
Multiple test set	3.62	1.92	3.18	2.91
Combined test set	1.61	1.31	1.53	1.55

For the combined test set, which is a combination of the three previous test sets, the updated models scored consistently better than the initial model.

Qualitatively, we can see an improvement in the spatial distribution of the predicted dose in the updated models at exactly the locations of concern, the concave parts of the target, and the space between multiple lesions, especially in the axial direction.

4. Discussion

By means of an existing dose prediction model that was trained for head and neck cases, we obtained good results for translating the dose prediction model to glioblastoma cases in the brain. For these initial results, only 60 cases were used for training. Compared to the results of the OpenKBP challenge, using their proposed metrics, our initial trained model scored better values than the top ranked participants did. However, we are aware that the anatomies of the evaluated treatment sites are different. It must be noted that different 3D volumes are used, as well as different OAR structures, which will have an influence on the used scores. We do not yet have a benchmark for these metrics in the brain region. On the other hand, whereas head and neck targets are much more similar from case to case, GBM targets vary much more in size, shape, and location. Nonetheless, the model was able to achieve good results on a relatively small training set. The overall dose score was less than 1 Gy. Although this might be relevant in a final dose distribution, for the dose predictor model and the purposes we aim to use it for, we believe this to be of sufficient accuracy. Although the prediction model might have struggled in reproducing the exact streaking patterns of the actual calculated dose, this did not have an impact on the relevant metrics, such as the mean and maximum dose received by the specific OARs. We conclude from this that a cascaded 3D U-Net is capable of predicting the dose when provided with high-quality and well-curated data to train it.

Although some other groups have published on dose prediction in the brain [7,12,17], for both VMAT and conventional intensity modulated radiotherapy (IMRT), they did not specify the nature of the brain tumors. Furthermore, different treatment prescriptions were included in these works, certain tumor locations were excluded, and additional planning parameters were used in the training model. This is the first dose prediction model specified for GBM treatment.

It seems that 3D dose distribution prediction is a great application for deep-learning models. Although the predictions are not perfect, they are useful in the initialization processes of automatic planning. It also provides a near-instant estimation of the dose distribution that outperforms any currently available analytical or mathematical prediction method. Our work shows that a dose prediction model for a specific purpose, i.e., treatment area and modality, is relatively easy to obtain. For our specific purpose, we wanted to predict the dose to OARs. We were able to improve the accuracy of the DVH score for the OARs with some minor additions to the training set.

Considering it is a 3D U-Net, the training times can be relatively long. In our scaled-down version (128 by 128 by 128 voxels), a single training course lasted around 24 h. Adding more training data—for instance, doubling or tripling the amount—would be feasible. In addition, increasing the resolution to 256 voxels would improve dose prediction resolutions at the cost of longer training and inference times. On the other hand, the fast inference time of only seconds on a high-end GPU enables multiple applications where the inference is run in a high-throughput manner. Some examples where a multitude of dose predictions are used are to show the dose effect of atomic surface transformations in OARs in the brain [37] or for a clinically relevant guidance of the loss function in automatic segmentation of targets volumes for GBM [38].

Furthermore, we did not analyze the interpretability of the model. It is, however, feasible to obtain saliency maps for the outcomes of the dose prediction model since it is based on the standard U-Net architecture [39].

To obtain the training data, we used plans that were made according to a strict and standardized protocol. This makes the resulting dose distribution better to predict. It might therefore be one of the reasons for the good results. On the other hand, using such a strict protocol makes the model only valid for treatments following this strict protocol. However, in clinical practice, different approaches are used depending not only on case specifics but also on the individual preference of doctors, planners, and the availability of specific hardware. In this case, we used a VMAT technique. The dose distribution of such plans is more predictable since, in every case, two full 360-degree co-planar arcs are used. In other techniques, such as conventional IMRT, using a set number of beam angles, or in more sophisticated VMAT techniques making use of non-coplanar arcs, the dose might be more difficult to predict. To solve this issue, one needs to make a dose prediction model for every different treatment strategy. Although this seems cumbersome, in clinical reality, this comes down to a few treatment strategies per treatment site for any department. Given our model's results, only a limited number of data are required to obtain a viable model.

The main contribution of this paper was to show the feasibility of training an accurate deep-learning-based dose predictor for GBM treatment. If data are limited for a particular scenario but a demarcated treatment protocol exists, even for non-homogeneous anatomies (i.e., not prostate or head and neck, for which most of the dose prediction methods are proposed), satisfactory results can be obtained compared to currently reported outcomes. Although the main drive behind dose prediction models is the purpose of automation in dose planning, dose prediction models can also be important for many other purposes. Our hypothesis is that they can be useful in the quality management of the radiotherapy steps that take place prior to planning. There are possible positive consequences for the effectiveness of quality assurance, the accuracy of segmentation, and the optimization of treatment planning. By introducing clinically relevant objectives early in the radiotherapy planning process, we can improve automation and enable automated quality assurance. This makes the radiotherapy process faster and also improves the quality.

We tested the obtained dose prediction model for specific criteria important for a dose predictor in quality management, sensitivity, and robustness. We showed that the initially trained model is sensitive enough to detect dose trends on realistic contour variability in a critical organ such as the optic nerve. We also tested the initial model against robustness. Although we found that the model did lack a certain accuracy in specific situations, we showed that with a simple strategy of adding specific cases to the training set, the robustness and the overall accuracy of the model increased. We anticipate that dose prediction models can be more accurate when using larger datasets of carefully curated data. In addition, the models can be tailored to have specific characteristics to fulfil the needs of different tasks in radiotherapy management.

5. Conclusions

This manuscript showed that obtaining a dose prediction model for GBM VMAT treatment that is both robust and sensitive to realistic segmentation variations is possible with less than 100 cases. Although the model is only valid for one specific treatment strategy, the relative ease of implementation and the near-instant inference time make it a useful technique to incorporate dose awareness into the automation processes in radiotherapy prior to planning.

Author Contributions: R.P., conceptualization, methodology, data curation, data analysis, and writing original draft; A.J.K., conceptualization, methodology, training framework, model training, data analysis, and writing—review and editing; J.W., conceptualization and writing—review and editing; N.A., conceptualization and writing—review and editing; E.E., data curation and writing—review and editing; D.M.A., supervision and writing—review and editing; P.M., methodology, supervision, and writing—review and editing; M.R., conceptualization, methodology, supervision, and writing—review and editing. All authors have read and agreed to the published version of the manuscript.

Funding: This work is supported by the Innosuisse Grant 31274.1 IP-LS and by Swiss Cancer Research (KFS-5127-08-2020).

Institutional Review Board Statement: Not applicable.

Informed Consent Statement: Patient consent was waived due to the data used being anonymized imaging data from a previous retrospective study.

Data Availability Statement: The data presented in this study are available upon request from the corresponding author. The data on which the models are trained are not able to be shared due to privacy and ethical considerations. The data of the networks that were trained are available from third party sources, which are referred to in the main text of the manuscript.

Conflicts of Interest: The authors of this manuscript declare no conflict of interest.

References

1. Dahiya, N.; Jhanwar, G.; Yezzi, A.; Zarepisheh, M.; Nadeem, S. Deep Learning 3D Dose Prediction for Conventional Lung IMRT Using Consistent/Unbiased Automated Plans. *arXiv* **2021**, arXiv:2106.03705.
2. Kontaxis, C.; Bol, G.H.; Lagendijk, J.J.W.; Raaymakers, B.W. DeepDose: Towards a fast dose calculation engine for radiation therapy using deep learning. *Phys. Med. Biol.* **2020**, *65*, 075013. [CrossRef]
3. Bakx, N.; Bluemink, H.; Hagelaar, E.; Van Der Sangen, M.; Theuws, J. Development and evaluation of radiotherapy deep learning dose prediction models for breast cancer. *Phys. Imaging Radiat. Oncol.* **2021**, *17*, 65–70. [CrossRef]
4. Ahn, S.H.; Kim, E.; Kim, C.; Cheon, W.; Kim, M.; Lee, S.B.; Lim, Y.K.; Kim, H.; Shin, D.; Kim, D.Y.; et al. Deep learning method for prediction of patient-specific dose distribution in breast cancer. *Radiat. Oncol.* **2021**, *16*, 154. [CrossRef]
5. Liu, Y.; Chen, Z.; Wang, J.; Wang, X.; Qu, B.; Ma, L.; Zhao, W.; Zhang, G.; Xu, S. Dose Prediction Using a Three-Dimensional Convolutional Neural Network for Nasopharyngeal Carcinoma with Tomotherapy. *Front. Oncol.* **2021**, *11*, 752007. [CrossRef]
6. Yue, M.; Xue, X.; Wang, Z.; Lambo, R.L.; Zhao, W.; Xie, Y.; Cai, J.; Qin, W. Dose prediction via distance-guided deep learning: Initial development for nasopharyngeal carcinoma radiotherapy. *Radiother. Oncol.* **2022**, *170*, 198–204. [CrossRef]
7. Yang, J.; Zhao, Y.; Zhang, F.; Liao, M.; Yang, X. Deep learning architecture with transformer and semantic field alignment for voxel-level dose prediction on brain tumors. *Med. Phys.* **2022**, *50*, 1149–1161. [CrossRef]
8. Appenzoller, L.M.; Michalski, J.M.; Thorstad, W.L.; Mutic, S.; Moore, K.L. Predicting dose-volume histograms for organs-at-risk in IMRT planning. *Med. Phys.* **2012**, *39*, 7446–7461. [CrossRef]
9. Tol, J.P.; Dahele, M.; Delaney, A.R.; Slotman, B.J.; Verbakel, W.F.A.R. Can knowledge-based DVH predictions be used for automated, individualized quality assurance of radiotherapy treatment plans? *Radiat. Oncol.* **2015**, *10*, 234. [CrossRef]
10. Gronberg, M.P.; Beadle, B.M.; Garden, A.S.; Skinner, H.; Gay, S.; Netherton, T.; Cao, W.; Cardenas, C.E.; Chung, C.; Fuentes, D.T.; et al. Deep Learning–Based Dose Prediction for Automated, Individualized Quality Assurance of Head and Neck Radiation Therapy Plans. *Pract. Radiat. Oncol.* **2023**, *13*, e282–e291. [CrossRef]
11. McIntosh, C.; Purdie, T.G. Contextual Atlas Regression Forests: Multiple-Atlas-Based Automated Dose Prediction in Radiation Therapy. *IEEE Trans. Med. Imaging* **2016**, *35*, 1000–1012. [CrossRef]
12. Shiraishi, S.; Moore, K.L. Knowledge-based prediction of three-dimensional dose distributions for external beam radiotherapy. *Med. Phys.* **2016**, *43*, 378–387. [CrossRef]
13. Ma, J.; Bai, T.; Nguyen, D.; Folkerts, M.; Jia, X. Individualized 3D Dose Distribution Prediction Using Deep Learning. In *Artificial Intelligence in Radiation Therapy*; Springer Nature: Cham, Switzerland, 2019; Volume 11850, pp. 110–118.
14. Willems, S.; Crijns, W.; Sterpin, E.; Haustermans, K.; Maes, F. Feasibility of CT-Only 3D Dose Prediction for VMAT Prostate Plans Using Deep Learning. In *Artificial Intelligence in Radiation Therapy*; Springer: Cham, Switzerland, 2019; Volume 11850, pp. 10–17.
15. Chen, X.; Men, K.; Li, Y.; Yi, J.; Dai, J. A feasibility study on an automated method to generate patient-specific dose distributions for radiotherapy using deep learning. *Med. Phys.* **2019**, *46*, 56–64. [CrossRef]
16. Chen, X.; Zhu, J.; Yang, B.; Chen, D.; Men, K.; Dai, J. Combining distance and anatomical information for deep-learning based dose distribution predictions for nasopharyngeal cancer radiotherapy planning. *Front. Oncol.* **2023**, *13*, 1041769. [CrossRef]
17. McIntosh, C.; Purdie, T.G. Voxel-based dose prediction with multi-patient atlas selection for automated radiotherapy treatment planning. *Phys. Med. Biol.* **2017**, *62*, 415–431. [CrossRef]
18. Babier, A.; Zhang, B.; Mahmood, R.; Moore, K.L.; Purdie, T.G.; McNiven, A.L.; Chan, T.C.Y. OpenKBP: The open-access knowledge-based planning grand challenge and dataset. *Med. Phys.* **2021**, *48*, 5549–5561. [CrossRef]
19. Wang, J.; Lu, J.; Qin, G.; Shen, L.; Sun, Y.; Ying, H.; Zhang, Z.; Hu, W. Technical Note: A deep learning-based autosegmentation of rectal tumors in MR images. *Med. Phys.* **2018**, *45*, 2560–2564. [CrossRef]
20. Sun, Z.; Xia, X.; Fan, J.; Zhao, J.; Zhang, K.; Wang, J.; Hu, W. A hybrid optimization strategy for deliverable intensity-modulated radiotherapy plan generation using deep learning-based dose prediction. *Med. Phys.* **2022**, *49*, 1344–1356. [CrossRef]
21. Huynh, E.; Hosny, A.; Guthier, C.; Bitterman, D.S.; Petit, S.F.; Haas-Kogan, D.A.; Kann, B.; Aerts, H.J.W.L.; Mak, R.H. Artificial intelligence in radiation oncology. *Nat. Rev. Clin. Oncol.* **2020**, *17*, 771–781. [CrossRef]
22. Poel, R.; Rüfenacht, E.; Ermis, E.; Müller, M.; Fix, M.K.; Aebersold, D.M.; Manser, P.; Reyes, M. Impact of random outliers in auto-segmented targets on radiotherapy treatment plans for glioblastoma. *Radiat. Oncol.* **2022**, *17*, 170. [CrossRef]
23. Altman, M.B.; Wooten, H.O.; Dewees, T.A.; Thorstad, W. A framework for automated contour quality assurance in radiation therapy including adaptive techniques A framework for automated contour quality assurance in radiation therapy including adaptive techniques. *Phys. Med. Biol.* **2015**, *60*, 5199–5209. [CrossRef]
24. Chen, H.-C.; Tan, J.; Dolly, S.; Kavanaugh, J.; Anastasio, M.A.; Low, D.A.; Li, H.H.; Altman, M.; Gay, H.; Thorstad, W.L.; et al. Automated contouring error detection based on supervised geometric attribute distribution models for radiation therapy: A general strategy. *Med. Phys.* **2015**, *42*, 1048–1059. [CrossRef]
25. Zhang, J.; Ates, O.; Li, A. Implementation of a Machine Learning–Based Automatic Contour Quality Assurance Tool for Online Adaptive Radiation Therapy of Prostate Cancer. *Int. J. Radiat. Oncol.* **2016**, *96*, E668. [CrossRef]
26. Isaksson, L.J.; Summers, P.; Bhalerao, A.; Gandini, S.; Raimondi, S.; Pepa, M.; Zaffaroni, M.; Corrao, G.; Mazzola, G.C.; Rotondi, M.; et al. Quality assurance for automatically generated contours with additional deep learning. *Insights Imaging* **2022**, *13*, 137. [CrossRef]
27. Jungo, A.; Meier, R.; Ermis, E.; Herrmann, E.; Reyes, M. Uncertainty-driven Sanity Check: Application to Postoperative Brain Tumor Cavity Segmentation. *arXiv* **2018**, arXiv:1806.03106.

28. Niyazi, M.; Andratschke, N.; Bendszus, M.; Chalmers, A.J.; Erridge, S.C.; Galldiks, N.; Lagerwaard, F.J.; Navarria, P.; Rosenschöld, P.M.A.; Ricardi, U.; et al. ESTRO-EANO guideline on target delineation and radiotherapy details for glioblastoma. *Radiother. Oncol.* **2023**, *184*, 109663. [CrossRef]
29. Van Esch, A.; Tillikainen, L.; Pyykkonen, J.; Tenhunen, M.; Helminen, H.; Siljamäki, S.; Alakuijala, J.; Paiusco, M.; Iori, M.; Huyskens, D.P. Testing of the analytical anisotropic algorithm for photon dose calculation. *Med. Phys.* **2006**, *33*, 4130–4148. [CrossRef]
30. Rüfenacht, E.; Kamath, A.; Suter, Y.; Poel, R.; Ermiş, E.; Scheib, S.; Reyes, M. PyRaDiSe: A Python package for DICOM-RT-based auto-segmentation pipeline construction and DICOM-RT data conversion. *Comput. Methods Programs Biomed.* **2023**, *231*, 107374. [CrossRef]
31. Liu, S.; Zhang, J.; Li, T.; Yan, H.; Liu, J. Technical Note: A cascade 3D U-Net for dose prediction in radiotherapy. *Med. Phys.* **2021**, *48*, 5574–5582. [CrossRef]
32. Nguyen, D.; Long, T.; Jia, X.; Lu, W.; Gu, X.; Iqbal, Z.; Jiang, S. A feasibility study for predicting optimal radiation therapy dose distributions of prostate cancer patients from patient anatomy using deep learning. *Sci. Rep.* **2019**, *9*, 1076. [CrossRef]
33. Kearney, V.; Chan, J.W.; Haaf, S.; Descovich, M.; Solberg, T.D. DoseNet: A volumetric dose prediction algorithm using 3D fully-convolutional neural networks. *Phys. Med. Biol.* **2018**, *63*, 235022. [CrossRef] [PubMed]
34. He, K.; Zhang, X.; Ren, S.; Sun, J. Delving deep into rectifiers: Surpassing human-level performance on imagenet classification. *Proc. IEEE Int. Conf. Comput. Vis.* **2015**, *2015*, 1026–1034.
35. Deeley, M.A.; Chen, A.; Datteri, R.; Noble, J.H.; Cmelak, A.J.; Donnelly, E.F.; Malcolm, A.W.; Moretti, L.; Jaboin, J.; Niermann, K.; et al. Comparison of manual and automatic segmentation methods for brain structures in the presence of space-occupying lesions: A multi-expert study. *Phys. Med. Biol.* **2011**, *56*, 4557–4577. [CrossRef] [PubMed]
36. Poel, R.; Rüfenacht, E.; Hermann, E.; Scheib, S.; Manser, P.; Aebersold, D.M.; Reyes, M. The predictive value of segmentation metrics on dosimetry in organs at risk of the brain. *Med. Image Anal.* **2021**, *73*, 102161. [CrossRef]
37. Kamath, A.; Poel, R.; Willmann, J.; Ermis, E.; Andratschke, N.; Reyes, M. ASTRA: Atomic Surface Transformations for Radiotherapy Quality Assurance. In Proceedings of the International Conference of the IEEE Engineering in Medicine and Biology Society, Sydney, Australia, 24–27 July 2023.
38. Rüfenacht, E.; Kamath, A.; Suter, Y.; Poel, R.; Ermiş, E.; Scheib, S.; Reyes, M. Dose Guidance for Radiotherapy-oriented Deep Learning Segmentation. In Proceedings of the MICCAI Proceedings, Vancouver, BC, Canada, 8–12 October 2023.
39. You, S.; Reyes, M. Influence of contrast and texture based image modifications on the performance and attention shift of U-Net models for brain tissue segmentation. *Front. Neuroimaging* **2022**, *1*, 1012639. [CrossRef]

diagnostics MDPI

Review

Redefining Radiology: A Review of Artificial Intelligence Integration in Medical Imaging

Reabal Najjar

Canberra Health Services, Australian Capital Territory 2605, Australia; reabal.najjar@act.gov.au

Abstract: This comprehensive review unfolds a detailed narrative of Artificial Intelligence (AI) making its foray into radiology, a move that is catalysing transformational shifts in the healthcare landscape. It traces the evolution of radiology, from the initial discovery of X-rays to the application of machine learning and deep learning in modern medical image analysis. The primary focus of this review is to shed light on AI applications in radiology, elucidating their seminal roles in image segmentation, computer-aided diagnosis, predictive analytics, and workflow optimisation. A spotlight is cast on the profound impact of AI on diagnostic processes, personalised medicine, and clinical workflows, with empirical evidence derived from a series of case studies across multiple medical disciplines. However, the integration of AI in radiology is not devoid of challenges. The review ventures into the labyrinth of obstacles that are inherent to AI-driven radiology—data quality, the 'black box' enigma, infrastructural and technical complexities, as well as ethical implications. Peering into the future, the review contends that the road ahead for AI in radiology is paved with promising opportunities. It advocates for continuous research, embracing avant-garde imaging technologies, and fostering robust collaborations between radiologists and AI developers. The conclusion underlines the role of AI as a catalyst for change in radiology, a stance that is firmly rooted in sustained innovation, dynamic partnerships, and a steadfast commitment to ethical responsibility.

Keywords: medical imaging; radiology; artificial intelligence; machine learning; deep learning; convolutional neural networks; computer-aided diagnosis; radiomics

Citation: Najjar, R. Redefining Radiology: A Review of Artificial Intelligence Integration in Medical Imaging. *Diagnostics* **2023**, *13*, 2760. https://doi.org/10.3390/diagnostics13172760

Academic Editors: Michał Strzelecki, Adam Piorkowski and Rafał Obuchowicz

Received: 13 June 2023
Revised: 1 August 2023
Accepted: 10 August 2023
Published: 25 August 2023

1. Introduction

Radiology, since its inception, has experienced a revolutionary journey, punctuating modern medicine with its profound influence. From the discovery of X-rays to the subsequent integration of artificial intelligence (AI) and machine learning (ML), this multifaceted discipline continually evolves, transforming itself and the healthcare ecosystem it underpins.

This comprehensive review scrutinises the interplay of AI and ML in radiology, exploring their foundational principles, historical progression, practical applications, inherent challenges, and ethical dilemmas. By enriching understanding of AI and ML's contributions to radiology, the review aims to foster insightful discussions among clinicians, researchers, and policymakers, ultimately shaping the field's direction and enhancing patient outcomes. The exploration delves into the fundamental constructs of AI and ML, their growing influence in radiology, practical integration strategies, and illustrative case studies across various medical specialties. Further, it addresses challenges such as data quality, ethical concerns, and contemplates potential future directions in AI-driven radiology.

1.1. Radiology in Modern Medicine

Radiology, the medical discipline centred on the utilisation of imaging modalities to diagnose and treat diseases, has emerged as a cornerstone of contemporary medicine, forming an integral part of clinical practice. It extends beyond mere disease detection to encompass treatment guidance and ongoing disease management. Expertise in diagnostic modalities such as computed tomography (CT), magnetic resonance imaging (MRI), positron emission tomography (PET), ultrasound, and X-rays guide immediate clinical interventions, treatment monitoring, and chronicle a visual narrative of a patient's health. The intricate insights into anatomical, physiological, and molecular disease processes provided by medical imaging have a significant impact on patient care, facilitating the tailoring of treatments, thereby improving therapeutic outcomes and minimising adverse effects [1–3].

Radiology serves as a crucial gear in the intricate machinery of interdisciplinary medical teams. Radiologists deliver precise, timely imaging reports, thus enhancing communication among various specialists and shaping crucial decisions, which contributes to a holistic, patient-focused healthcare approach [4]. As valued consultative partners, radiologists offer key insights into the choice and interpretation of suitable imaging studies, playing a significant role in radiation safety and dose management while their expertise elucidates the clinical picture, offering insights that can markedly influence patient management [5,6].

1.2. From Roentgen to Magnetic Fields: A Brief History

The metamorphosis of modern medical imaging technology, from Wilhelm Roentgen's pioneering discovery of X-ray technology in 1895 to contemporary advanced techniques, epitomises the relentless pursuit of scientific advancement and its profound impact on radiology (Figure 1). Roentgen's unprecedented X-ray innovation offered a non-invasive glimpse into the human body, forming the foundation for modern imaging. Despite its initial limitations in 2D representation and soft tissue contrast, this fundamental concept laid the groundwork for more sophisticated, non-invasive imaging modalities [7].

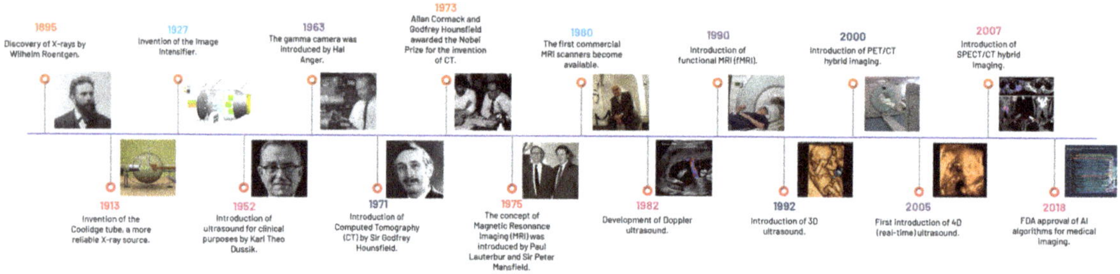

Figure 1. A historical timeline of the key discoveries in medical imaging.

The advent of CT in 1973 by Sir Godfrey Hounsfield and Allan Cormack marked a significant milestone, transcending the limitations of 2D imaging by introducing a three-dimensional (3D) format [8]. CT utilises the basic principle of differential absorption but couples this with the synchronised rotation of X-ray sources and detectors around the patient's body, paired with sophisticated computational algorithms, enabling the reconstruction of 3D volumetric data from the collected 2D images [7].

Introduced in the 20th century, ultrasound imaging marked a departure away from ionising radiation-based technologies by using high-frequency sound waves to create real-time images of internal body structures. Its non-ionising radiation nature, real-time imaging capability, and cost-effectiveness have allowed broad applicability in various clinical fields, such as obstetrics, gynaecology, cardiology, and emergency medicine [9]. As an indispensable tool in emergency and critical care medicine, its transformative role, especially through point-of-care ultrasound (POCUS), has facilitated rapid bedside assessments and hastened clinical decision-making [7].

In the 1970s, Paul Lauterbur and Sir Peter Mansfield led the development of MRI, a technology that employs a robust magnetic field and radio waves to generate exceptionally detailed images of the body, particularly of soft tissue structures [10,11]. The non-ionising nature of MRI, coupled with its unrivalled soft tissue differentiation ability, revolutionised medical imaging. Manipulation of RF pulse sequence timing in MRI further enhanced its diagnostic utility, enabling the acquisition of various image types and discernment of distinct tissues and pathologies [7].

1.3. *From Film to Function: An Evolution in Radiology*

Parallel to the disruptive innovations in imaging modalities, another pivotal shift was occurring in the late 20th century: the evolution from film-based to digital radiography and the introduction of Picture Archiving and Communication Systems (PACS). This transition radically improved the efficiency of image acquisition, storage, and retrieval, while also enabling the seamless sharing and transfer of images within and across healthcare institutions [12].

The transformation in medical imaging technology did not stop with these innovations. Functional imaging techniques such as PET, which is uniquely characterised by its use of radiolabelled biochemical substances, and Single-Photon Emission Computed Tomography (SPECT), employing gamma-emitting radionuclides to trace biological processes, illuminated metabolic and biological processes, opening a window into cellular activity and providing invaluable insights into the functional status of organs [13,14].

3D imaging heralded a significant evolution in medical imaging by providing a more precise understanding of spatial relationships within the body, thereby improving diagnostic accuracy and surgical planning. The subsequent development of four-dimensional (4D) imaging further pushed the boundaries by incorporating the element of time, allowing for real-time monitoring of physiological processes [15].

The confluence of functional and anatomical imaging gave rise to hybrid imaging technologies such as PET/CT and SPECT/CT. These modalities amalgamate the strengths of both techniques, providing comprehensive diagnostic information. For instance, PET/CT combines the metabolic insight of PET with the detailed anatomical context of CT, considerably enhancing the accuracy of lesion localisation and characterisation [16].

Lastly, the emerging field of interventional radiology, which leverages imaging for guidance during minimally invasive procedures, has reshaped the healthcare landscape. By providing real-time visualisation of the target area, these procedures offer enhanced precision, potentially improving patient outcomes and reducing recovery times. For example, image-guided biopsies provide a safer and less invasive alternative to surgical biopsies, leading to fewer complications and shorter hospital stays [17].

1.4. *A Glimpse into the Future: New Frontiers*

Radiology's future promises transformation through the integration of virtual/augmented reality (VR/AR) and AI, heralding a new era of medical imaging. Originating from the gaming and entertainment industries, VR/AR technologies are gradually permeating radiology, providing an immersive environment advantageous for radiology training and clinical practice. In the latter, the technologies can augment imaging data visualisation, thereby enhancing diagnosis and treatment planning [18].

AI, particularly its subset machine learning, is radically improving radiology, strengthening image analysis, and mitigating diagnostic errors. AI algorithms process and interpret data, performing tasks that emulate or even surpass human cognitive capabilities. ML, through exposure to labeled examples, is capable of extracting complex, high-level data, even from unlabelled datasets. By integrating AI into VR/AR technologies, the potential to boost radiological efficiency, improve diagnostic accuracy, and improve treatment planning exponentially exists [19].

Over the last two decades, the radiology community has refined computer-aided diagnosis (CAD) tools based on ML, which are set to bring about an integrated diagnostic service by incorporating radiology, pathology, and genomics data to improve the performance of CAD and enhance the productivity of radiology service by AI-assisted workflow [20].

Notwithstanding, the integration of AI and VR/AR in radiology encounters technical hurdles, notably incorporating AI-derived results into existing workflows. A proposed roadmap, however, advocates for AI-based image analysis algorithm integration, featuring a radiologist-AI feedback loop system for continuous improvement. This is exemplified by a case study demonstrating improved detection of brain metastases with AI application and radiologist feedback [21].

The ethical, legal, and societal implications of these technologies in radiology warrant careful scrutiny. The application of AI in radiology, while promising, can raise ethical issues, particularly related to bias and "black box" issues (i.e., the lack of transparency in AI decision-making processes). Advocacy for ethical AI underscores the need to address possible discriminatory effects and injustices, recommending future radiology-focused AI developments to incorporate social science perspectives [22].

2. The Fundamentals of Artificial Intelligence and Machine Learning

This section offers a journey through complex developments of AI and ML, sketching their historical trajectory and explaining the distinctive yet interconnected terminologies of AI, ML, and Deep Learning (DL). It also sheds light on the key ML algorithms and techniques that have shaped the technological landscape, underscoring the indelible imprint they have left on it.

2.1. *Chronicle of Artificial Intelligence: Milestones and Breakthroughs*

The compelling development of AI, spanning from antiquity—marked by folklore and tales of artificial beings—to our current era of sophisticated AI systems, has roots in the philosophical depictions of human cognition as a mechanistic process (Figure 2). The inception of modern AI concepts is largely accredited to the development of the programmable digital computer in the 1940s, reaching a milestone with the field's official establishment at the 1956 Dartmouth Conference [23].

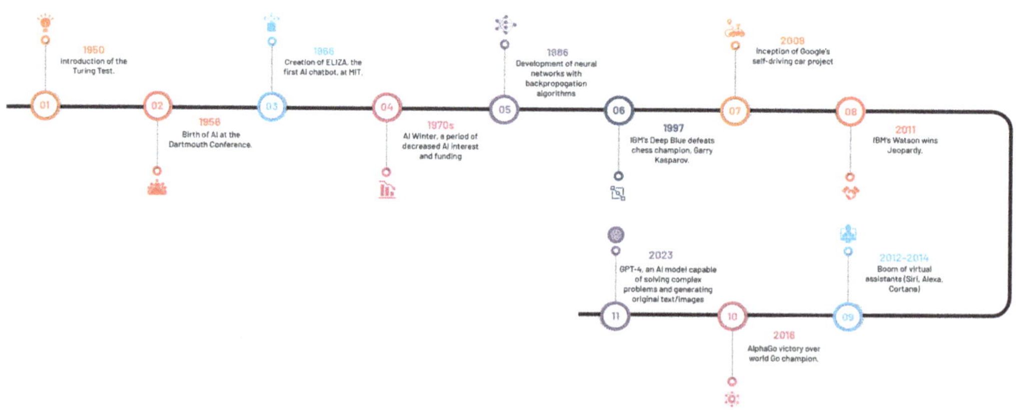

Figure 2. Significant milestones in the evolution of artificial intelligence.

The 1970s saw a significant leap forward with the emergence of rule-based expert systems like MYCIN, a seminal creation by Buchanan and Shortliffe [23,24]. These expert systems, designed to replicate human expertise by leveraging knowledge bases and inference engines, laid the foundation for AI's influential role in medical diagnosis and clinical decision-making.

Machine learning algorithms, which surfaced in subsequent years, spearheaded a novel paradigm for data-driven predictions and classification. Decision trees introduced in 1986, support vector machines in 1995, and neural networks in 1986, collectively broadened the AI horizon in healthcare [25–27]. These algorithms propelled the analysis of extensive datasets, thereby inaugurating a novel epoch of pattern recognition and predictive modelling in healthcare.

The turn of the century witnessed a paradigm shift with the advent of deep learning approaches, particularly convolutional neural networks (CNNs). Surpassing previous methodologies in image recognition tasks, CNNs, architecturally modelled on structure and function of the human brain, precipitated major advancements in medical image classification, segmentation, and detection, due to their capacity to learn hierarchical representations from vast amounts of labelled data [28,29].

The accelerated progression of AI in recent years is a by-product of the synergistic interplay of two factors: the surge in big data and advancements in computational power. The widespread dissemination of electronic health records (EHRs), medical imaging archives, and annotated datasets provide ample training data, while progress in hardware, including graphical processing units (GPUs) and distributed computing, have expedited the deployment of computationally demanding AI algorithms [30].

The rapid advancement of AI language models, such as GPT-4, has led to unique applications and implications across a multitude of discplines, including healthcare and medical imaging [31]. These models, whilst capable of generating human-like text and facilitating communication, have raised salient concerns. The potential for these models to contribute substantially to medical research and patient care is undeniable; however, experts express reservations about their limitations and potential to inadvertently engender inequities or disseminate misinformation, thus emphasising the need for robust strategies to manage these risks responsibly. These strategies include increasing transparency about potential harms, promoting early detection of issues, and implementing regulatory measures and peer reviews [32]. By doing so, the goal is to ensure that AI technologies, including language models, are utilised optimally and ethically, contributing positively to medical imaging and healthcare outcomes.

By recognising these milestones, we can appreciate the progression of AI in healthcare, particularly in medical imaging. The intertwining of rule-based systems, traditional machine learning algorithms, and the transformative influence of deep learning techniques has laid the foundation for the current state-of-the-art AI applications in radiology and other medical specialties.

2.2. *Decoding the Terminology: AI, ML, and DL*

The computational intelligence landscape is an expansive system, constituted of distinct yet interconnected systems, with each sector playing a specific role and interacts uniquely within the data science sphere (Figure 3).

Artificial intelligence is defined as the replication of human intelligence in machines that are programmed to emulate human cognition and actions, encompassing learning, problem-solving, reasoning, and perception. AI can be classified into two major types: narrow AI, which is designed for specific tasks (e.g., facial recognition or voice commands), and general AI, which mimics a broader spectrum of human intellect. AI aims to develop systems with autonomous intelligent functionality, capable of problem-solving, decision-making, and performing tasks typically requiring human intelligence [33].

Machine learning, a subset of AI, is centred around the development of software that learns autonomously from accessed data. The learning process is derived from the analysis of observations or data to identify patterns and make informed future decisions based on these observations. Its fundamental objective is to enable computers to adapt their actions autonomously without human intervention. The primary categories of ML algorithms are supervised learning, which predicts or classifies new data based on examples, and unsupervised learning, which identifies inherent patterns and structures within the data without guidance from pre-established outputs [34].

Deep learning, an ML subset, utilises multilayered artificial neural networks (aptly coined "deep"), enabling DL algorithms to model and understand complex data patterns. These algorithms are especially effective for tasks where manual feature extraction proves challenging, such as image or speech recognition. It is noteworthy that these feature layers are data-learned, not human-engineered, embodying an architecture inspired by the human brain and have proven successful in visual recognition tasks, even outperforming human performance [28].

Recognising the nuanced interrelationships between AI, ML, and DL aids in conceptualising each subfield's contribution and progression within the broader AI narrative. AI lays the foundation for ML and DL. ML amplifies AI's potential by enabling machine data learning, and DL further deepens these capabilities with neural networks that decipher complex data patterns. Each field enriches the wider AI domain, culminating in the modern AI landscape where each layer contributes to the evolution of intelligent systems.

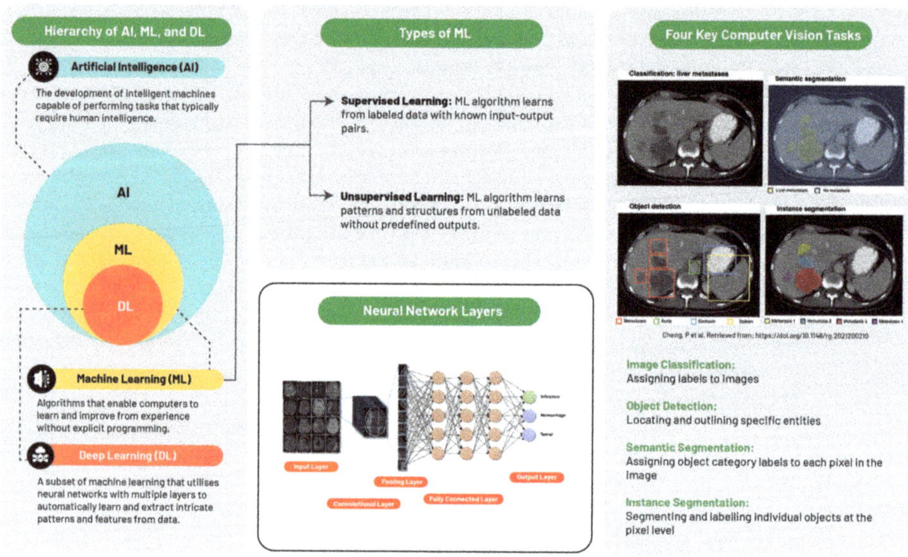

Figure 3. Schematic illustration of artificial intelligence and machine learning frameworks [35].

2.3. *Machine Learning Foundations: Algorithms and Techniques*

Machine learning is anchored by a plethora of algorithms and techniques that enable computers to learn from data. The two primary types of machine learning, supervised and unsupervised learning, lie at the heart of this field.

Supervised learning, a dominant machine learning variant, harnesses pre-established examples or training data, consisting of input-output pairs. The objective is to formulate a function that maps input data to corresponding outputs, enabling accurate predictions or classifications for unfamiliar data. Central to supervised learning are algorithms such as linear regression, logistic regression, and decision trees [36].

In contrast, unsupervised learning navigates the data space autonomously to uncover inherent patterns, structures, or relationships devoid of predefined outputs. Its focus is on discovering intrinsic data structures, thus offering insights that could potentially solve complex problems. Notable unsupervised learning algorithms include clustering techniques like k-means, hierarchical clustering, and dimensionality reduction methods such as principal component analysis (PCA) [37].

Artificial Neural Networks (ANNs) simulate the operational framework of the human brain, executing intricate tasks via a network of interconnected artificial neurons arranged in layers. The backpropagation algorithm, a vital cog in ANN operations, exhibits high fault tolerance, ensuring the system's functionality despite occasional neuronal failures [38]. ANNs facilitate feature extraction and sophisticated pattern recognition, crucial for machine learning, which enhance data representation or class differentiation by aiding raw data pre-processing for feature extraction or selection [39].

Developments in ANNs have given rise to complex structures like deep learning models comprising multilayer neurons. Notably, CNNs employ convolution in lieu of standard matrix multiplication in certain layers. Tailored for processing pixel data, CNNs excel in tasks related to pattern recognition in images, audio, or text, substantially contributing to computer vision and Natural Language Processing (NLP) by simplifying complex patterns into abstract representations through layers of features [40].

3. Integrating AI into Medical Imaging: The Dawn of Radiology 2.0

This section delves into the remarkable role of AI within medical imaging, spotlighting the revolutionary shifts it catalyses in the world of radiology. Its multifaceted potential, from revolutionising image acquisition to reshaping radiological analyses, and from streamlining reporting to crafting personalised medical narratives, positions AI at the epicentre of the ongoing healthcare revolution. Beyond radiology, this transformation extends to other areas of healthcare—such as pathology, cardiology, genomics, drug discovery, and healthcare delivery—where the impactful strides of AI are being increasingly recognised. Concluding this exploration is the emergent paradigm of AI-facilitated personalised medicine, underscoring a more proactive, patient-oriented, and holistic patient care approach.

3.1. A Paradigm Shift in Radiology

AI has instigated a profound metamorphosis in the field of radiology, redefining traditional workflows and elevating the radiologist's role (Figure 4). In the realm of image acquisition, AI augments scanning procedures, optimises image fidelity, and fosters sophisticated image reconstruction across MRI, CT, and PET modalities. Foremost among these advancements, deep learning accelerates MRI scanning, harmonising efficiency and quality, with commensurate progress witnessed in CT and PET image reconstruction [41].

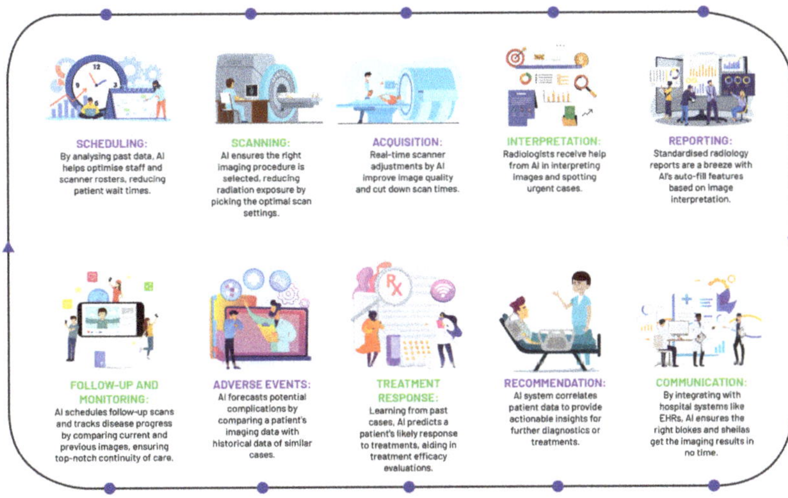

Figure 4. A streamlined workflow diagram illustrating the role of artificial intelligence in radiological practice.

AI significantly streamlines the acquisition of radiologist analyses on chest X-rays, as evidenced by a study wherein an AI system reduced interpretation delivery times from 11.2 days to a mere 2.7 days, reinforcing the potency of automated triaging systems in streamlining healthcare workflows and amplifying patient care standards [42].

As one of the pioneering healthcare specialties to adopt digital technology, radiology has capitalised on machine learning in CAD tools for over two decades, demonstrating robust performance in sensitivity and specificity [43]. Although clinical adoption has been slow due to various challenges, AI is posited as a pivotal tool to surmount these, enhancing CAD performance, streamlining radiology services, and fostering the development of integrated diagnostic services.

The integration of AI into radiology reporting has structured and annotated data to bolster report uniformity and streamlining patient history tracking. These cutting-edge tools generate comprehensive task lists, incorporating pertinent information from the patient's history into EHRs, with the prime objective of enhancing report accessibility and integration into care pathways [44].

Beyond AI's transformative influence on reporting and imaging procedures, these state-of-the-art systems play a pivotal role in maintaining continuity in provider communication and patient care, validating correlations between imaging diagnoses, radiological reports, and treatment plans, alerting providers to any discrepancies. Moreover, AI optimises personnel allocation and scanner usage and reduces radiation exposure, thereby boosting efficiency and quality of care [45]. By virtue of its broad-spectrum capabilities, AI is redefining the landscape of radiology, cementing its indispensable status within the discipline.

3.2. *Beyond Radiology: Broader Applications of AI in Healthcare*

AI has indelibly altered several healthcare areas, demonstrating its capability to enhance clinical practice beyond radiology via improvements in diagnostics, genomics, drug discovery, and healthcare delivery optimisation.

Pathology has witnessed the successful application of AI algorithms in tissue analysis, markedly enhancing diagnostic accuracy and speed. Automated image analysis tools enable pathologists to scrutinise tissues microscopically, identifying subtle histopathological attributes often overlooked by the human eye [46]. AI also accelerates the move towards digital pathology, converting traditional glass slides into digital scans for remote diagnostics and collaborative work—both of which are vital in the digital age of telemedicine [47].

AI shows considerable promise in cardiology, particularly in interpreting electrocardiograms and echocardiograms. Sophisticated ML algorithms detect complex cardiac patterns and abnormalities, accurately predicting conditions such as atrial fibrillation and myocardial infarction [48]. Its exponential growth in echocardiography is evident, with automated algorithms aiding in the interpretation of cardiac structure parameters, mitigating interobserver variability, and enhancing diagnostic precision [49].

The complex nature of genomics renders it an ideal candidate for AI intervention. DL techniques decipher genomic data, aiding in identifying genetic variants linked to disease susceptibility, and opening avenues for tailored treatment strategies based on individual genetic profiles [50].

AI has proven invaluable in the field of drug discovery by expediting the search for potent therapeutic compounds, and subsequently, accelerating the drug development process. For example, AI can predict the pharmacokinetic and pharmacodynamic properties of novel compounds, pinpoint potential drug targets, and simulate clinical trials, substantially reducing both time and costs associated with drug development [51].

Additionally, AI's ability to optimise healthcare delivery is remarkable. AI-driven predictive analytics can enhance hospital workflows, accurately predicting patient admission rates, and optimising resource allocation [52]. AI applications in cost reduction have also emerged, with machine learning algorithms identifying inefficiencies in healthcare systems, thus enabling cost-effective care (SHAH 2021). The overarching aim of these AI applications is to improve patient outcomes by streamlining diagnostic processes and personalising treatment plans.

3.3. A New Era of Personalised Medicine

AI has catalysed a paradigm shift in the arena of personalised medicine. With its unparalleled capacity for processing vast amounts of complex data, AI is broadening the scope of personalised medicine, venturing into previously uncharted territories.

One of the most significant contributions of AI to personalised medicine is its potential to unlock the treasure trove of information embedded in EHRs. By harnessing the power of advanced machine learning algorithms, AI can discern patterns within EHRs, which can provide critical insights into specific disease states or risk factors. This process enables a proactive approach to patient care, allowing healthcare providers to foresee potential health risks and intervene accordingly [4]. In a landmark study by Rajkomar et al. (2018), AI was successfully utilised to predict medical events using data from EHRs, underlining its instrumental role in preventative medicine [53].

By integrating various sources of patient data—ranging from medical imaging and EHRs to genomic data—AI enables a more holistic approach to healthcare. This unification allows for personalised treatment plans, precise disease risk prediction, and improved monitoring of treatment responses. AI systems also enhance clinical decision-making by harmonising patient-specific data with up-to-date scientific discoveries, generating clinically relevant recommendations [54]. However, it is vital to ensure that AI's implementation is patient-centric and ethical considerations such as data privacy, security, and algorithmic bias are rigorously adhered to. This vigilance guarantees the equitable application of AI technologies, respecting the rights and needs of all patients [4].

In conclusion, AI is shaping a new era of truly personalised patient care. However, it is paramount to acknowledge the necessity for continuous research and meticulous validation to ensure the safe and effective deployment of AI in healthcare. Multidisciplinary collaborations, involving specialists from AI, radiology, genomics, and clinical practice, are crucial for fine-tuning AI-driven models and technologies. Such partnerships will help maximise the potential benefits of AI, leading to enhanced patient outcomes and propelling the field of medicine into the future.

4. Practical Applications of AI in Radiology Practice

This section critically examines the practical applications of AI in the field of radiology, elaborating on the novel approaches it brings to imaging techniques, diagnosis, and patient care. By navigating through AI-driven methodologies like DL and CNNs, the section illuminates how AI is redefining the way image segmentation and classification, and diagnostics are conducted. Moreover, it explores the prognostic power of radiomics and predictive potential of AI in optimising workflows. Throughout the discourse, the section also confronts the inherent challenges and bottlenecks in the integration of AI within radiology, underpinning the critical need for interpretability, validation, standardisation, and the need to preserve the human element in healthcare.

4.1. Image Segmentation and Classification

DL has ignited a significant shift within radiology that is particularly noticeable in the domains of image segmentation and classification, where substantial strides have been made. The advancements brought about by these AI-centric methods have amplified the precision and speed of diagnosis, thereby amplifying the competency of radiologists and raising the bar of patient care. Nevertheless, the assimilation of AI brings about numerous challenges that need to be overcome to facilitate its optimal integration and application within radiology.

CNNs, given their inherent ability to learn complex patterns through backpropagation, have emerged as formidable tools for computational visual tasks, including various radiological applications. Their distinct architectural layers, from calculations in convolutional layers to the generation of predictions in fully connected layers, all converge to form a proficient object detection system. These networks have demonstrated remarkable proficiency in object detection tasks, thanks to their integrated capabilities of feature extraction, semantic segmentation, and the handling of multi-scale features. The efficiency and effectiveness of CNNs can be further amplified through the application of transfer learning, which allows for the reuse of pre-existing models. This, in turn, enhances accuracy, facilitates efficient training even with limited datasets, and minimises the need for labour-intensive and error-prone manual segmentation [40,55].

Illustrating CNNs' potential, the segmentation of lung nodules from CT scans using AI has shown superior performance in the early detection and treatment of lung cancer, achieving an area under the receiver operating characteristic curve (AUROC) of 94.4% and outperforming six radiologists in the task [56]. Similarly, CNNs have played a pivotal role in the segmentation of brain tumours from MRI scans and the analysis of retinal images for early symptoms of diabetic retinopathy, further underlining AI's broad applicability and versatility of AI in the sphere of medical imaging [40,55].

Image classification, an equally critical application of AI in radiology, leverages CNNs to differentiate between normal and pathological findings. AI models have been engineered to distinguish between benign and malignant tumours in mammography, achieving performance metrics comparable to human radiologists, thus facilitating early detection of breast cancer [57].

Notwithstanding these significant strides, integrating AI within radiology invites a suite of challenges that need to be addressed. The construction of robust and reliable AI models necessitates access to extensive datasets and thorough validation, both of which can be challenging due to privacy concerns and the heterogeneous nature of medical imaging data. Second, ensuring the interpretability and explicability of the "black-box" dilemma in deep learning models is paramount to garner trust and adoption from both radiologists and patients. Lastly, integrating AI into existing clinical workflows and fostering a harmonious human-AI collaboration are essential to fully realise the potential of AI in radiology and translate these technological advances into tangible enhancements in patient care.

4.2. *Advancing Diagnostics with AI and CAD Systems*

The third wave of AI, notably the integration of deep learning technology into CAD systems, has propelled radiology into a digital era. The emergence of AI-integrated CAD (AI-CAD) systems has revolutionised radiology services, underpinning transformative changes in the landscape of diagnostics. These systems have redefined how radiologists interpret images, increasing diagnostic accuracy, reducing false positives, and significantly improving workflow efficiencies.

AI-CAD systems' merit lies in their substantial reduction of false positives, enhancing dependability in clinical settings. This is corroborated by a study comparing AI-CAD and traditional CAD software, where the AI system outperformed by decreasing the false-positive marks per image (FPPI) by a significant 69%. It specifically excelled in identifying microcalcifications and masses, reducing false positives by 83% and 56% respectively. Such reductions optimise radiologists' efficiency, potentially cutting their case reading time by an estimated 17%, and mitigate socioeconomic issues like patients' unnecessary emotional stress and financial burden [58].

Employing ensemble learning methodologies can hone CAD systems further, as evidenced by a recent study deploying a calibrated ensemble of deep learners for the task of detecting abnormalities in musculoskeletal radiographs. This ensemble model excelled over individual models, even outperforming expert radiologists in three out of the seven upper extremity anatomical regions (with an AUC of 0.93, Accuracy of 0.87, and Precision of 0.93). These results lend compelling support to the utility of the calibrated ensemble approach in identifying abnormalities in musculoskeletal X-rays [59].

AI has also shown remarkable promise in equalling, and in some cases surpassing, the performance of radiologists in breast screening through the deployment of AI algorithms for automated patient triage and predicting treatment outcomes—tasks that extend beyond human capabilities [60]. However, to fully exploit the benefits of AI in routine breast imaging fully, there's a need to ensure sufficient data availability for testing and monitoring

AI algorithms pre and post-integration into healthcare systems [60]. Yet, the dawn of a new AI epoch has presented a promising avenue that is anticipated to spur the next major transformation in radiology: the growth of radiomics—a field seeking to unify data from radiology, pathology, and genomics to offer a comprehensive diagnostic service [20].

4.3. *Prognostics with Radiomics and Predictive Analytics*

Radiomics is an emerging field within medicine, predicated on the extraction of high-dimensional data from radiological images and harbours immense potential for the landscape of medical diagnostics, prognosis, and the evaluation of disease response to treatment. However, radiomics still faces challenges, including the need for standardisation and validation to ensure reliable and reproducible outcomes. The main strength of radiomics lies in its ability to complement traditional clinical practice with precise, quantitative information, thereby revolutionising medical decision-making processes [61].

The rapid exponential growth of medical imaging data has fostered a conducive environment for the application of ML and data-driven science. Radiomics-based decision-support systems for precision diagnosis and treatment are poised to become integral tools in the armamentarium of modern medicine. This, however, is not to say that the journey of radiomics towards full clinical applicability is without challenges. The field currently grapples with a lack of standardisation and validation that are quintessential for ensuring reliable and reproducible outcomes [61,62].

In this milieu, the advent of AI presents promising avenues for overcoming these challenges and unlocking the full potential of radiomics. Through modelling high-dimensional data, AI-driven analytics enable accurate predictions pertaining to disease progression, treatment response, or patient survival. This significant stride offers clinicians an unprecedented wealth of information that vastly transcends the limitations of human perception. Particularly within oncology, radiomics has proven instrumental in identifying molecular phenotypes and lymph node metastases, appraising treatment response, and prognosticating disease survival [61].

It is crucial to acknowledge that the amalgamation of AI into radiomics remains in its infancy. To fully harness its potential in medical imaging, a concerted effort towards research and development is crucial. A key facet of this development will be the facilitation of large-scale data sharing, the establishment of standardised data collection protocols, clear evaluation criteria, and robust reporting guidelines. These elements are fundamental to the maturation and widespread adoption of radiomics as a discipline, opening the door to a new era of precision medicine.

4.4. *Workflow Optimisation Using AI*

AI is gaining traction in radiology, aiming to optimise workflows and enhance non-interpretative tasks' efficacy. When coupled with NLP, AI can automate the triage of imaging studies, prioritising urgent cases based on the retrieval and analysis of key data from patients' EHR. This expedites patient triage, radiology reporting, and managing incidental finding follow-ups [20].

AI significantly enhances the radiology process by automating triage and improving report generation. It swiftly sorts and prioritises radiological studies like CT scans and MRIs based on urgency, highlighting critical cases for immediate review. This automation aids in consistently detecting severe conditions such as stroke, haemorrhage, and malignancy, reducing errors. The use of AI, particularly NLP, in non-interpretative tasks alleviates tedious aspects of the workflow, potentially mitigating radiologist burnout [4].

Furthermore, AI enhances the generation and interpretation of radiology reports. Deep Learning algorithms address the limitations of traditional reporting, including fatigue-induced errors or inconsistencies due to varied expertise levels. They detect and characterise findings to improve consistency, facilitate standardised report creation, and reduce errors. This additional analysis layer streamlines the workflow and augments report clarity, significantly contributing to the quality of radiology services [63].

The incorporation of AI transcends the limits of purely diagnostic capabilities, profoundly amplifying interdisciplinary collaboration and patient-radiologist communication. AI platforms serve as a vital conduit that nurtures a collective understanding of imaging results among disparate healthcare professionals as they possess the potential to demystify the complexity of medical terminologies for patients. This transparency aids in establishing a strong rapport between the patient and the radiologist while also cultivating a higher degree of patient involvement in their personal health [61].

As strides are made towards AI integration in radiology, it's crucial to recognise that as of 2021, only 30% of radiologists reported clinical AI use, with over 70% expressing reluctance to invest in AI. Many perceived AI as offering negligible benefits, hinting at the field of radiology being in the "trough of disillusionment" phase in the AI adoption process [64]. This disillusionment emanates from factors such as scepticism about AI performance and applicability, a perceived lack of necessity, inadequate workflows for efficient AI utilisation, and a dearth of scalable AI-supporting infrastructure.

To transition into the "slope of enlightenment", the field must establish infrastructure that supports optimal AI functionality, involving the redefinition and disruption of existing systems such as image management and PACS for intelligent workflow orchestration [65]. Notwithstanding AI's potential, the importance of preserving the human element in patient care cannot be overstated, underlining that while AI can augment the work of radiologists, it is not a substitute for the nuanced judgement and empathetic communication that are at the heart of patient care.

5. Case Studies: AI across Medical Specialties

5.1. Neuroradiology

The rapidly expanding field of ML, particularly supervised techniques and DL, has proven to be indispensable in managing high-dimensional data within the realm of neuroradiology (Figure 5). This cutting-edge technology has facilitated the early detection of a different stroke subtypes, as evidenced in the study by Yedavalli et al. (2021) [66]. CNNs demonstrate remarkable expertise in various tasks, including detecting infarcts or haemorrhages, segmentation, classification, and identification of large vessel occlusion. The application of CNNs in these areas has significantly influenced the approach to stroke treatment, as expounded by the research carried out by Soun et al. (2021) [67].

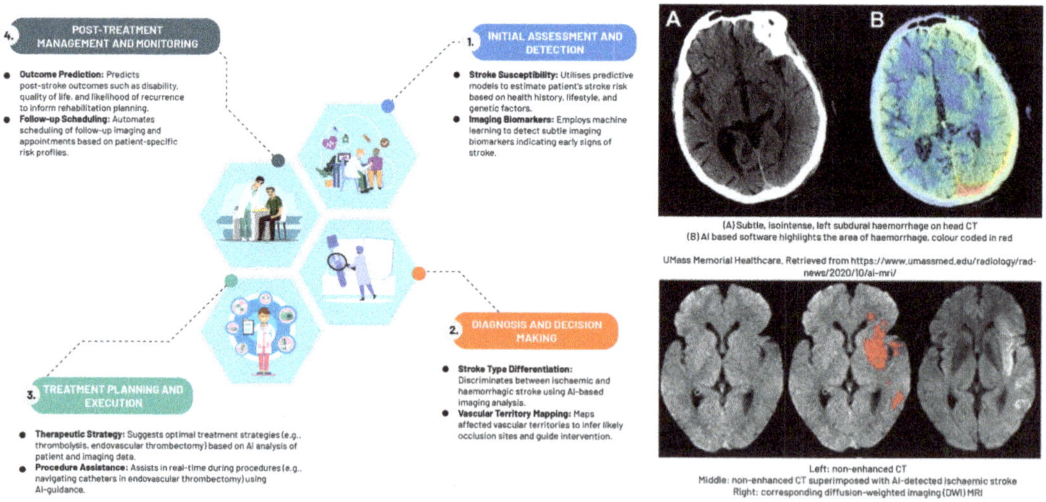

Figure 5. An overview of machine learning driven applications in neuroradiology [68,69].

AI transcends diagnostic boundaries to bolster clinical decision-making, particularly in scenarios marked by substantial inter-rater variability. Its applications span from classifying stroke subtypes and detecting haemorrhages to identifying segmentation and large vessel occlusions. This advancement presents distinct advantages for facilities managing a small number of stroke patients or those functioning as regional hubs [70].

An ever-growing body of research underscores AI's potential in bolstering decisions related to thrombolysis and thrombectomy. Shlobin et al. (2021), for instance, devised an AI model capable of precisely detecting large vessel occlusions using CT imaging, which demonstrates a high level of sensitivity and specificity in identifying patients suitable for timely thrombectomy intervention [71]. Zhu et al. (2022) utilised AI algorithms to predict thrombolysis responses in patients afflicted with acute ischaemic stroke, integrating imaging features with clinical data to support clinicians in formulating the most effective treatment strategies [72].

AI also plays a critical role in the early detection of neurodegenerative disorders, specifically for conditions like Alzheimer's and Parkinson's diseases. Sophisticated AI algorithms have been designed to analyse MR images for detecting specific biomarkers or characteristic patterns associated with these conditions. The task of discerning subtle changes in brain structure or function, which is crucial in diagnosing these diseases, is made more efficient with AI due to its ability to detect refined voxel-level patterns and provide objective, quantitative assessments [73].

Furthermore, AI has demonstrated its potential in predicting postoperative outcomes for brain and spine surgeries. By examining preoperative imaging data, AI models can generate prognostications regarding surgical outcomes, such as the likelihood of complications or the extent of functional improvement, aiding surgeons in treatment planning and managing patient expectations, as documented by Soun et al. (2021) [67].

5.2. Oncological Imaging

AI and ML technologies, fuelled by the surge in high-performance computing, have catalysed significant advancements in oncology, particularly in cancer imaging (Figure 6). In precision oncology, the synergy between AI, superior computing, and deep learning strategies, coupled with the integration of multi-omics data, has streamlined cancer diagnosis, prognosis, and treatment [74,75].

The intrinsically digital nature of oncological imaging lends itself to AI and ML applications. Here, the imaging pipeline from acquisition to interpretation, reporting, and communication thrives within the digital space, enabling efficient data capture for AI and ML analysis. As a result, these technologies are being actively explored and adopted in cancer imaging, which constitutes a significant proportion of the workload in numerous healthcare facilities [74].

In tumour detection and classification, AI is increasingly being used to discern between benign and malignant lesions and various tumour types, particularly in diagnostics for breast, lung, and prostate cancers, with AI-based devices already making their way into clinical practice [75]. Studies reveal that deep learning models and CNNs can classify lung nodules on CT scans and differentiate renal cell carcinoma subtypes on MRI with high accuracy, often rivalling the expertise of experienced radiologists [76,77].

AI algorithms provide an objective, consistent means of assessing changes in tumour size or metabolic activity changes, automating measurements that were traditionally time-consuming and susceptible to inter-observer variability, such as those under the Response Evaluation Criteria In Solid Tumors (RECIST) [78]. AI accomplishes this by harnessing radiomic features—high-dimensional data extracted from radiological images—to build mathematical models adept at detecting subtle changes indicative of treatment response [4]. Furthermore, AI plays a key role in monitoring treatment response by quantifying tumour changes via granular analysis of medical image subunits (pixels/voxels). These smaller elements can be examined by computers to reveal objective mathematical features linked with disease behaviour or outcomes [74]. AI also procures valuable prognostic insights by analysing radiomic signatures, such as texture analysis, to predict survival rates in lung cancer patients from pre-treatment CT images, while radiomic features extracted from MRI scans have shown a correlation with recurrence risk in patients with glioblastoma [79,80]. Consequently, the integration of AI in radiology leads to efficient and precise tracking of tumour progression, significantly enhancing overall treatment assessment and patient care.

Traditional methods of monitoring radiation therapy response, based on manual assessment of changes in tumour size and characteristics, are often subjective and may overlook subtle indicators of treatment efficacy. AI, particularly CNNs, provide an objective means of assessing treatment response, utilising vast datasets of annotated imaging scans to accurately identify and delineate tumours [78]. This automation streamlines the planning process, potentially enhancing treatment outcomes via precise radiation dosing. The effectiveness of this approach is often quantified by the Sørensen–Dice coefficient (DSC), facilitating early and accurate therapy efficacy assessment and timely treatment adjustments, if necessary [81].

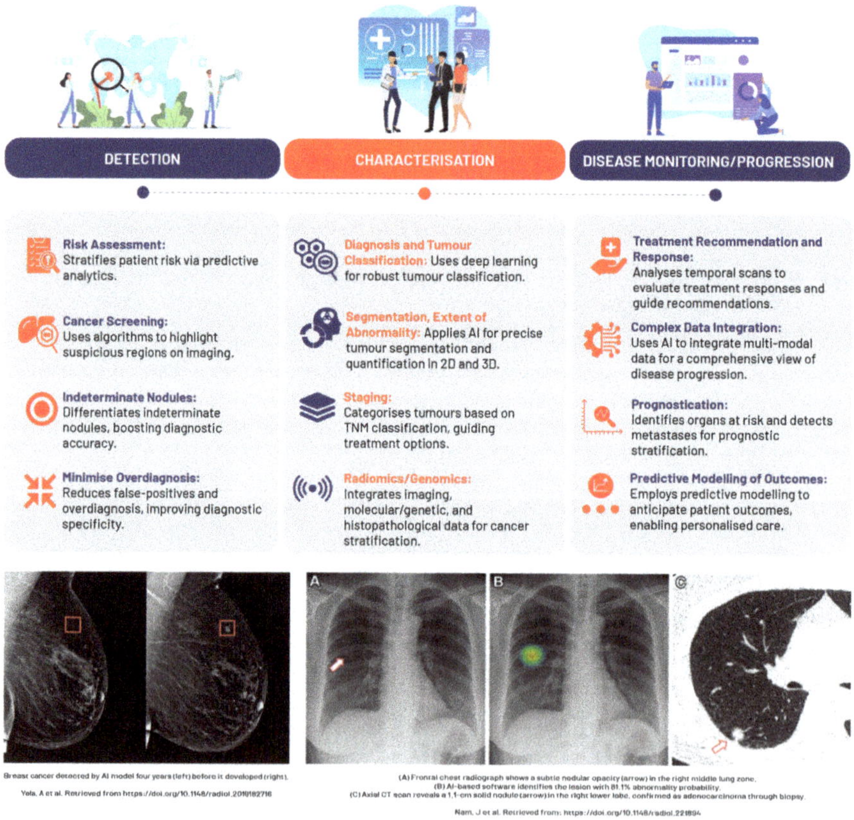

Figure 6. An overview of machine learning applications in oncological imaging [82,83].

5.3. Cardiovascular Imaging

AI has considerably advanced cardiovascular imaging in recent years, enabling enhanced detection and quantification of heart diseases, comprehensive analysis of vascular abnormalities, and the integration of multi-modality imaging data (Figure 7). AI algorithms can efficiently interpret complex imaging data, recognising initial stages of cardiac diseases, such as coronary artery disease and congestive heart failure, through modalities like cardiac CT, MRI, or echocardiography. For example, ML models and CNNs have demonstrated the ability to automatically detect coronary artery calcification and perform automatic segmentation of left ventricular myocardium, respectively, with strong correlation to manual analyses [84,85]. AI technology has also enhanced the functional evaluation of the left ventricle in echocardiographic diagnosis by automating tasks traditionally reliant on visual observation and manual boundary tracing, such as measuring left ventricular ejection fraction using the Simpson method. Such advancements promise to reduce the dependence on physician experience, increasing the repeatability and accuracy of these evaluations [84].

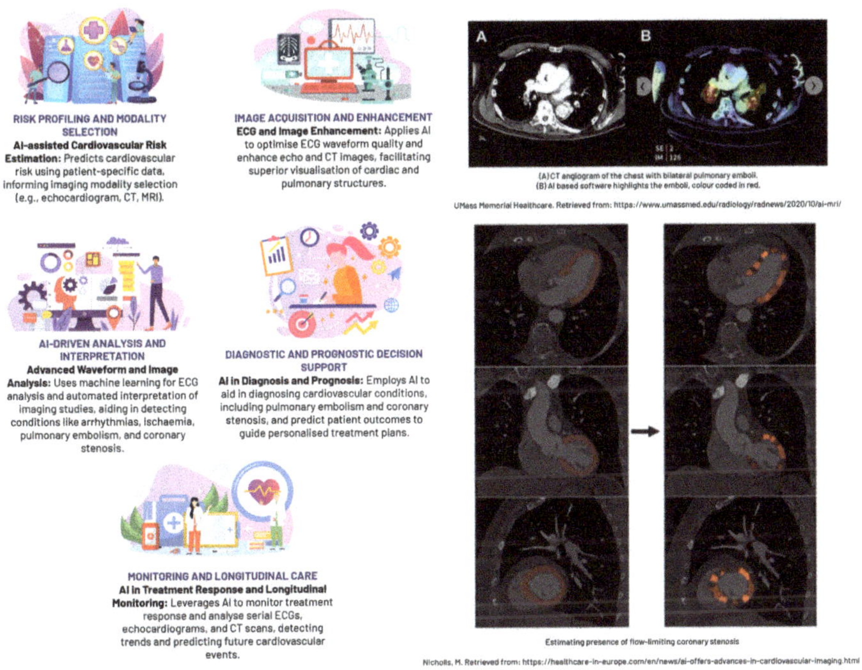

Figure 7. An overview of machine learning applications in chest imaging [68,86].

Beyond cardiac strucuture conditions, AI is also instrumental in the analysis of vascular abnormalities, such as aortic aneurysms or peripheral artery disease, facilitating early intervention and potentially improving patient outcomes. CNNs have proven effective in evaluating abdominal aortic aneurysms from CT images, demonstrating high accuracy in detecting and sizing these potentially life-threatening conditions [87]. Furthermore, AI-assisted standard section recognition has significantly reduced the time required for evaluation, enhanced detection ability, and improved the accuracy of novice practitioners, which has been particularly beneficial in settings with limited resources for training echocardiography physicians [84].

A major breakthrough in AI-based cardiovascular imaging is the integration of data from multi-modality imaging, which combines information from CT, MRI, and echocardiography to provide a holistic representation of cardiac structure and function. This consolidated data is essential for complex assessments, such as detecting ischaemia or planning interventions. For instance, ML algorithms can merge perfusion data from MRI with coronary anatomy from CT, generating sophisticated 3D heart models, thus improving the detection of cardiac ischaemia and facilitate precise procedural planning [88,89].

5.4. Abdominal Imaging

AI has fostered unprecedented advancements in abdominal and pelvic imaging, particularly within the domain of gastrointestinal imaging (Figure 8). AI's contributions have been pivotal in improving detection, diagnosis, and staging of liver and pancreatic diseases. The conceptualisation of diverse AI-based predictive models has broadened the diagnostic spectrum, encompassing gastrointestinal and inflammatory diseases, non-malignant conditions, and the detection of bowel bleeding using cutting-edge technology such as wireless capsule endoscopy [90]. AI has also proven instrumental in detecting hepatic-associated fibrosis by leveraging EHRs to derive meaningful insights into patient health data and medical history. Furthermore, AI's fusion with endoscopic ultrasound technology has substantially improved both the accuracy and speed of diagnosing pancreatic carcinoma, consequently enhancing patient management strategies [90].

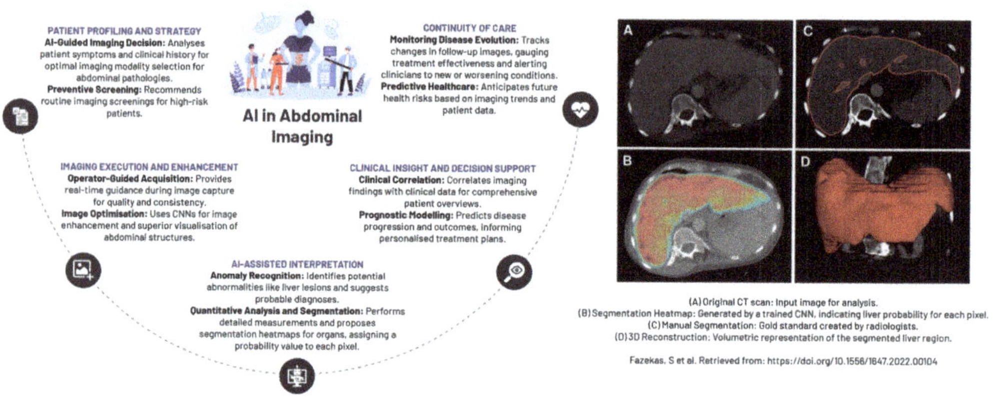

Figure 8. An overview of AI-driven applications in abdominal imaging [91].

In the specialised fields of hepatology and pancreatology, the integration of AI with various imaging techniques has brought about a diagnostic paradigm shift in diagnosing liver and pancreatic diseases. The techniques that have benefited from AI include ultrasound, endoscopic ultrasonography, CT, MR, and PET/CT. AI's influence permeates beyond diagnostics, aiding in the selection of the most appropriate diagnostic test for a patient based on individualised medical profile. Furthermore, AI has been instrumental in optimising image quality, accelerating image acquisition, and predicting patient prognosis as well as their response to treatments [92].

AI has reshaped abdominal and pelvic imaging with its ability to deliver precise and reproducible imaging diagnoses. The technology facilitates automated or semi-automated segmentation and registration of the liver and pancreatic glands, along with their associated lesions, thereby increasing the diagnostic accuracy and treatment efficacy. The integration of radiomics introduces novel quantitative metrics into radiological reports, enriching the detection and characterisation of focal lesions and diffuse diseases of the liver and pancreas, leading to potential enhancements in clinical outcomes [92].

In nephrology, contemporary AI applications display remarkable potential in anticipating the onset of acute kidney injury before significant biochemical changes become evident, potentially allowing for timely interventions to prevent disease progression. Additionally, AI's ability to identify modifiable risk factors for chronic kidney disease progression offers valuable insights for preventative care [93].

Furthermore, in the sphere of renal tumour detection, AI models have demonstrated proficiency matching, or even exceeding human accuracy, in interpreting imaging studies. This impressive feat could augment prognostication and decision-making processes post-renal transplantation. The heightened precision in detecting and diagnosing renal tumours fosters more effective treatment strategies, potentially leading to improved patient outcomes [93].

6. Challenges, Limitations, and Future Directions

The expansion of AI in healthcare, particularly in the field of diagnostic radiology, has presented unprecedented opportunities to enhance the quality and efficiency of patient care. Despite this, the rapid growth comes with a myriad of challenges, including the necessity for adequate data quality and volume, the "black box" dilemma, integration into clinical practice, and ethical considerations. This section dissects these issues and proposes potential resolutions that could facilitate the assimilation and responsible usage of AI in radiology while addressing various technical, infrastructural, regulatory, and human factors (Figure 9).

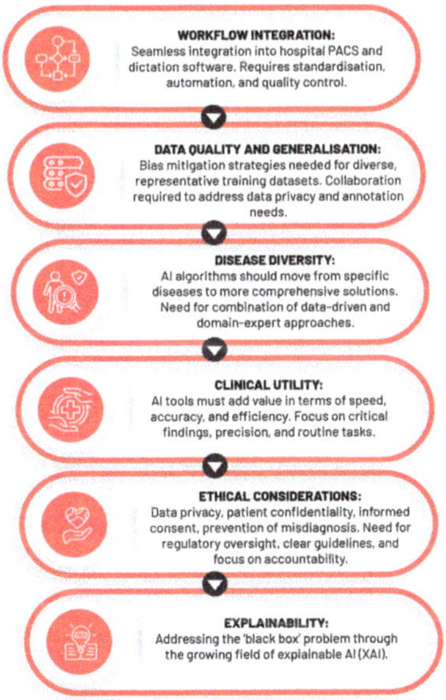

Figure 9. Challenges of AI integration into clinical practice.

6.1. Data Quality, Quantity, and the "Black Box" Problem

The performance of AI algorithms, essentially mathematical mirroring of reality, depends not only on their training datasets and precision but also on calibration in interpreting medical images depends on comprehensive datasets that accurately reflect varied patient demographics, including age, sex, ethnicity, and disease stages [94,95].

The construction of such datasets is frequently obstructed by representation biases due to the utilisation of restricted demographic groups or specific clinical settings [96]. Strategies like data augmentation, oversampling, and undersampling are often employed to remedy data scarcity, ensuring dataset diversity and balanced representation during model training [97].

It is essential to recognise and address potential risks associated with biased or unrepresentative data as mismanagement can inadvertently perpetuate health disparities and yield AI models with subpar performance in certain patient populations. The "black box" problem in AI, denoting the lack of transparency in AI models, complicates error detection and bias identification, thus adversely affecting underrepresented groups and clinical utility [22]. Tackling these challenges requires a concerted effort to diversify data collection, combat bias in AI system design, perform population subgroup-based performance analyses, and use representative population samples for clinical validation [22].

Adding to these strategies, a growing field known as explainable artificial intelligence (XAI) seeks to demystify AI decision-making processes through improving inference reliability and enhancing transparency and interpretability [98]. XAI techniques, such as saliency maps, feature importance, and surrogate models, aid in visualising and explaining the rationale behind AI models' decisions, making them accessible to both experts and lay audiences [99,100].

6.2. Clinical Integration of AI into Radiology Practice

The integration of AI into clinical radiology has been met with both enthusiasm and scepticism, necessitating a clear roadmap to address the multifaceted challenges that arise. Central to these is the prerequisite for robust hardware and reliable software, which are instrumental in managing the vast amounts of data generated by medical imaging systems. The untapped potential in the estimated 97% of unused hospital data can be harnessed by AI-enabled applications, substantially augmenting disease trajectory prediction and treatment regimen modification [101].

The effective implementation of AI demands high-performance hardware capable of executing complex computations in real time, implying the need for significant investment in powerful, reliable hardware infrastructure to circumvent any adverse effects on patient care due to system failures [101]. On the other hand, the software must be robust, intuitive, and designed to integrate within existing radiological systems, necessitating close collaboration between AI developers, radiologists, and other healthcare professionals [102]. AI also holds the promise of alleviating administrative burdens that beleaguer radiologists who, on average, dedicate approximately 16.6% of their working hours (roughly nine hours per week) to administrative tasks [103].

Navigating the hurdles of regulatory approval and ongoing monitoring is equally imperative, as AI tools need to undergo rigorous validation to demonstrate safety and effectiveness before gaining regulatory approval [104]. Once approved, continuous monitoring and evaluation are necessary to track performance and ensure continuous improvement and the delivery of reliable results. Lastly, successful AI integration into radiology hinges on acceptance and adoption by end-users, primarily radiologists [103]. This highlights the importance of providing sufficient training and support to ensure radiologists can effectively use AI tools and comfortably integrate them into daily practice.

6.3. The Ethical Conundrums

The ethical implications of AI integration within radiology unearths several ethical quandaries, namely data privacy and security, patient confidentiality, informed consent, misdiagnosis risks, and the preservation of the human element in patient care.

The intersection of data privacy and security with AI pivots on the governance of patient data, notably when considering public-private partnerships involving large tech corporations. A review of commercial healthcare AI unveiled that a considerable fraction of these technologies are under the control of private entities, raising alarms about potential data misuse. This is epitomised by the DeepMind incident, which involved the transfer of patient data from the United Kingdom to the United States without explicit patient consent [105]. These events call for more stringent regulatory oversight to guarantee that patient data remains within its original jurisdiction and is safeguarded against unauthorised access.

Patient confidentiality and informed consent, inextricably linked to data privacy, require patients to exercise autonomy over their data, comprehending its use, potential risks, and associated benefits. This aspect becomes especially significant within AI-driven healthcare, where the black box nature of learning algorithms can veil the decision-making processes [105]. Consequently, requisite safeguards and transparency procedures need implementation to uphold privacy and secure patient autonomy.

Furthermore, the potential for misdiagnosis, liability, and accountability in AI-assisted radiology raises significant ethical concerns. Although AI offers promising diagnostic capabilities, it is susceptible to errors and biases, which can lead to incorrect diagnoses and patient harm [106]. Addressing this issue necessitates the development and implementation of clear guidelines and policies for AI use in medical settings, with an emphasis on decision-making accountability and delineating the responsibilities of healthcare professionals and AI systems.

With that being said, the risk of over-reliance on AI emphasises the importance of preserving the human element in patient care. While AI can enhance healthcare delivery and complement the skills of healthcare professionals, it should not overshadow the irreplaceable expertise and nuanced decision-making inherent in the practice of medicine, particularly given the complexity and variability of individual cases [22]. As the role of AI in radiology continues to evolve, these ethical considerations must remain at the forefront of discussions, policy-making, and research, ensuring the responsible and equitable application of this transformative technology in healthcare.

6.4. Bridging the Gap: Collaboration between Radiologists and AI Developers

The growing intersection of AI and radiology necessitates a synergistic collaboration between radiologists and AI developers to create advanced tools that drive clinical innovation and improve patient care. This section discusses the fusion of these disparate fields and the importance of interdisciplinary collaboration, its benefits, and strategies to bridge the gap between academia and industry.

Radiologists bring to the table an in-depth understanding of clinical needs, disease processes, and imaging interpretation nuances, while AI developers possess the technical acumen to design, implement, and optimise machine learning algorithms. This diverse expertise is not just complementary but also essential for the development of AI tools that are clinically viable, safe, and effective. Without the input of radiologists, AI tools risk being designed without consideration for real-world clinical workflows, limiting their utility or potentially compromising patient safety. Conversely, without the technical skills of AI developers, radiologists would struggle to leverage the vast potential of AI in imaging analysis [65].

Interdisciplinary research is the key to marrying these distinct skill sets. Collaboration between radiologists and AI developers can result in AI tools more attuned with clinical needs, effectively bridging the gap between theory and practice. Such partnerships also encourage a collective sense of ownership and responsibility, driving the adoption and optimisation of AI tools in clinical settings. A common model for fostering collaboration is the Academia-Industry Collaboration Plan (AICP), which outlines processes and methods for establishing partnerships between academia and industry [107]. This is crucial as universities supply the workforce and innovative ideas, while the industry provides necessary funding for research and innovation.

However, promoting collaboration between academia and industry is not without its challenges. One significant hurdle that can stifle innovation and delay the deployment of AI tools in radiology is the contemporary dichotomy between these two sectors. While academia often focuses on theory and exploration, industry is more concerned with practical applications and market viability [107]. To overcome this, there are several strategies that can be employed. First, joint projects involving both academic and industry partners can facilitate the exchange of ideas and resources, leading to innovative and robust AI solutions. Shared datasets enable a more diverse range of research, enhancing the generalisability of AI tools. Finally, open-source software offers a platform for collaboration, promoting transparency and reproducibility, both of which are fundamental to scientific progress.

6.5. Medical Education and Training in Healthcare

The incorporation of AI into clinical practice necessitates the acquisition of new skills. With AI algorithms capable of processing large volumes of data far beyond human capacities, the emphasis on memorising extensive medical information or developing procedural skills through repetitive practice may diminish. This shift necessitates the need for clinicians to acquire new skills, such as data science, statistics, and AI ethics, which will be crucial for interacting safely and effectively with AI technologies [108]. These skills will empower clinicians to competently input data, interpret algorithmic outputs, and communicate AI-derived treatment plans to patients.

In the face of this shift, the roles and responsibilities of radiologists may extend beyond image interpretation to encompass a broader understanding and application of AI technologies. Radiologists, traditionally concerned primarily with image interpretation, may find themselves more engaged in tasks such as AI model development, validation, or monitoring. For instance, a significant component of AI in radiology involves the labelling of images, a labor-intensive and costly process that demands the expertise of radiologists [20]. As AI technologies advance, radiologists will increasingly interact with decision support systems that can suggest diagnoses, alert test results, and even automate clinical documentation [108].

The professional evolution of radiologists may also coincide with significant scientific and technical advancements in radiomics and pathomics, facilitating the integration of diagnostic services and personalised medicine, especially in resource-poor settings with limited infrastructure where the adoption of AI might be challenging [20].

The Royal Australian and New Zealand College of Radiologists (RANZCR) has adopted a progressive approach to these challenges and opportunities. In their 2023 curriculum, AI-related topics have been incorporated, reflecting the growing recognition of the importance of AI in radiology and the need for radiologists to acquire new competencies. This crucial inclusion of AI ensures that RANZCR remains at the forefront of technology and innovation, solidifying its commitment to equipping radiologists with the most advanced tools and knowledge in the rapidly evolving field of medical imaging [109].

7. Conclusions

This review concludes by summarising the key insights, transformative potential, and future path of the intricate relationship between AI and medical imaging. With AI playing a pivotal role in modern radiology, it provides a plethora of advantages such as improved diagnostic accuracy, workflow efficiency, and personalised patient care. These advancements, including tools for image partitioning, categorisation, computer-aided diagnosis, and innovative diagnostic and prognostic tools driven by radiomics and predictive analytics, herald a promising potential for improving patient outcomes. However, challenges concerning data privacy, security, and the 'black box' nature of AI models remain to be addressed. Despite these hurdles, the future is promising with new algorithms and architectures broadening the scope of medical image analysis. An essential synergy between radiologists and AI developers is needed to foster interdisciplinary research and bridge academia and industry. This need also extends to preparing healthcare professionals for an AI-infused landscape and evolving the role of radiologists in the AI era. Embracing AI's potential in redefining radiology involves not only an unwavering dedication to innovation and development of advanced algorithms, but also nurturing collaborations among radiologists, AI developers, patients, and policy-makers. These joint efforts should aim to meet clinical needs, translate research into practical applications, and ensure ethical AI deployment, always prioritising patient safety, privacy, and dignity. In this light, the forthcoming aeon of AI in radiology, though challenging, unfolds its vast potential in transforming healthcare.

Funding: This research received no external funding.

Institutional Review Board Statement: Not applicable.

Informed Consent Statement: Not applicable.

Data Availability Statement: No new data were created or analysed in this study. Data sharing is not applicable to this article.

Conflicts of Interest: The author declares no conflict of interest.

References

1. Brady, A.P.; Bello, J.A.; Derchi, L.E.; Fuchsjäger, M.; Goergen, S.; Krestin, G.P.; Lee, E.J.Y.; Levin, D.C.; Pressacco, J.; Rao, V.M.; et al. Radiology in the era of value-based healthcare: A multi-society expert statement from the ACR, CAR, ESR, IS3R, RANZCR, and RSNA. *Insights Imaging* **2020**, *11*, 136. [CrossRef]
2. Giardino, A.; Gupta, S.; Olson, E.; Sepulveda, K.; Lenchik, L.; Ivanidze, J.; Rakow-Penner, R.; Patel, M.J.; Subramaniam, R.M.; Ganeshan, D. Role of Imaging in the Era of Precision Medicine. *Acad. Radiol.* **2017**, *24*, 639–649. [CrossRef]
3. Jameson, J.L.; Longo, D.L. Precision medicine–personalized, problematic, and promising. *N. Engl. J. Med.* **2015**, *372*, 2229–2234. [CrossRef]
4. Hosny, A.; Parmar, C.; Quackenbush, J.; Schwartz, L.H.; Aerts, H.J. Artificial intelligence in radiology. *Nat. Rev. Cancer* **2018**, *18*, 500–510. [CrossRef] [PubMed]
5. European Society of Radiology (ESR); European Federation of Radiographer Societies (EFRS). Patient Safety in Medical Imaging: A joint paper of the European Society of Radiology (ESR) and the European Federation of Radiographer Societies (EFRS). *Insights Imaging* **2019**, *10*, 45. [CrossRef] [PubMed]
6. Dreyer, K.J.; Geis, J.R. When Machines Think: Radiology's Next Frontier. *Radiology* **2017**, *285*, 713–718. [CrossRef] [PubMed]
7. Bushberg, J.T.; Seibert, J.A.; Leidholdt, E.M. *The Essential Physics of Medical Imaging*, 4th ed.; Lippincott Williams & Wilkins (LWW): Philadelphia, PA, USA, 2020.
8. Hounsfield, G.N. Computerized transverse axial scanning (tomography): Part 1. Description of system. *Br. J. Radiol.* **1973**, *46*, 1016–1022. [CrossRef]
9. Edler, I.; Hertz, C.H. The use of ultrasonic reflectoscope for the continuous recording of the movements of heart walls. *Kungl Fysiogr Sallsk i Lund Forhandl* **1954**, *24*, 1–19. [CrossRef]
10. Lauterbur, P.C. Image formation by induced local interactions: Examples employing nuclear magnetic resonance. *Nature* **1973**, *242*, 190–191. [CrossRef]
11. Mansfield, P.; Grannell, P.K. NMR 'diffraction' in solids? *J. Phys. C Solid State Phys.* **1977**, *10*, L55–L58. [CrossRef]
12. Huang, H. *PACS and Imaging Informatics: Basic Principles and Applications*; John Wiley & Sons: Hoboken, NJ, USA, 2011.
13. Cherry, S.R.; Jones, T.; Karp, J.S.; Qi, J.; Moses, W.W.; Badawi, R.D. Total-body PET: Maximizing sensitivity to create new opportunities for clinical research and patient care. *J. Nucl. Med.* **2018**, *59*, 3–12. [CrossRef] [PubMed]
14. Hutton, B.; Buvat, I.; Beekman, F. Review and current status of SPECT scatter correction. *Phys. Med. Biol.* **2011**, *56*, R85–R112. [CrossRef] [PubMed]
15. Vannan, M.; Pedrizzetti, G.; Li, P.; Gurudevan, S.; Houle, H.; Main, J.; Jackson, J.; Nanda, N. Effect of cardiac resynchronization therapy on longitudinal and circumferential left ventricular mechanics by velocity vector imaging: Description and initial clinical application of a novel method using high-frame rate B-mode echocardiographic images. *Echocardiography* **2005**, *22*, 826–830. [CrossRef]
16. Delbeke, D.; Coleman, R.; Guiberteau, M.; Brown, M.; Royal, H.; Siegel, B.; Townsend, D.; Berland, L.; Parker, J.; Zubal, G.; et al. Procedure Guideline for SPECT/CT Imaging 1.0. *J. Nucl. Med.* **2006**, *47*, 1227–1234. [PubMed]
17. Lorenz, J. Management of Malignant Biliary Obstruction. *Semin. Interv. Radiol.* **2016**, *33*, 259–267. [CrossRef] [PubMed]
18. Uppot, R.; Laguna, B.; McCarthy, C.; De Novi, G.; Phelps, A.; Siegel, E.; Courtier, J. Implementing Virtual and Augmented Reality Tools for Radiology Education and Training, Communication, and Clinical Care. *Radiology* **2019**, *291*, 570–580. [CrossRef]
19. von Ende, E.; Ryan, S.; Crain, M.; Makary, M. Artificial Intelligence, Augmented Reality, and Virtual Reality Advances and Applications in Interventional Radiology. *Diagnostics* **2023**, *13*, 892. [CrossRef]
20. Mun, S.K.; Wong, K.H.; Lo, S.B.; Li, Y.; Bayarsaikhan, S. Artificial Intelligence for the Future Radiology Diagnostic Service. *Front. Mol. Biosci.* **2021**, *7*, 614258. [CrossRef]
21. Dikici, E.; Bigelow, M.; Prevedello, L.; White, R.; Erdal, B. Integrating AI into radiology workflow: Levels of research, production, and feedback maturity. *J. Med. Imaging* **2020**, *7*, 016502. [CrossRef]
22. Goisauf, M.; Cano Abadía, M. Ethics of AI in Radiology: A Review of Ethical and Societal Implications. *Front. Big Data* **2022**, *5*, 850383. [CrossRef]
23. Buchanan, B.; Shortliffe, E. *Rule Based Expert Systems: The MYCIN Experiments of the Stanford Heuristic Programming Project*; Addison-Wesley Longman Publishing Co., Inc.: Boston, MA, USA, 1984.
24. Shortliffe, E. (Ed.) *Computer-Based Medical Consultations: MYCIN*, 1st ed.; Elsevier: Amsterdam, The Netherlands, 1976.
25. Quinlan, J. Induction of decision trees. *Mach. Learn.* **1986**, *1*, 81–106. [CrossRef]
26. Cortes, C.; Vapnik, V. Support-vector networks. *Mach. Learn.* **1995**, *20*, 273–297. [CrossRef]
27. Rumelhart, D.; Hinton, G.; Williams, R. Learning representations by back-propagating errors. *Nature* **1986**, *323*, 533–536. [CrossRef]
28. LeCun, Y.; Bengio, Y.; Hinton, G. Deep learning. *Nature* **2015**, *521*, 436–444. [CrossRef] [PubMed]
29. Krizhevsky, A.; Sutskever, I.; Hinton, G.E. ImageNet classification with deep convolutional neural networks. *Commun. ACM* **2017**, *60*, 84–90. [CrossRef]
30. Russakovsky, O.; Deng, J.; Su, H.; Krause, J.; Satheesh, S.; Ma, S.; Huang, Z.; Karpathy, A.; Khosla, A.; Bernstein, M.; et al. ImageNet Large Scale Visual Recognition Challenge. *Int. J. Comput. Vis.* **2015**, *115*, 211–252. [CrossRef]

31. OpenAI. ChatGPT: Improving Language Understanding by Generative Pre-Training. 2022. Available online: https://openai.com/research/language-unsupervised (accessed on 19 May 2023).
32. Brown, T.; Mann, B.; Ryder, N.; Subbiah, M.; Kaplan, J.; Dhariwal, P.; Neelakantan, A.; Shyam, P.; Sastry, G.; Askell, A.; et al. Language Models are Few-Shot Learners. *arXiv* **2020**, arXiv:2005.14165.
33. Pennachin, C.; Goertzel, B. (Eds.) Contemporary Approaches to Artificial General Intelligence. In *Artificial General Intelligence*; Cognitive Technologies; Springer: Berlin/Heidelberg, Germany, 2007. [CrossRef]
34. Mitchell, T.M. *Machine Learning*, 1st ed.; McGraw-Hill, Inc.: New York, NY, USA, 1997.
35. Cheng, P.M.; Montagnon, E.; Yamashita, R.; Pan, I.; Cadrin-Chênevert, A.; Perdigón Romero, F.; Chartrand, G.; Kadoury, S.; Tang, A. Deep Learning: An Update for Radiologists. *Radiographics* **2021**, *41*, 1427–1445. [CrossRef]
36. Rashidi, H.H.; Tran, N.K.; Betts, E.V.; Howell, L.P.; Green, R. Artificial Intelligence and Machine Learning in Pathology: The Present Landscape of Supervised Methods. *Acad. Pathol.* **2019**, *6*, 2374289519873088. [CrossRef]
37. Glickenstein, D.; Hamm, K.; Huo, X.; Mei, Y.; Stoll, M. Editorial: Mathematical Fundamentals of Machine Learning. *Front. Appl. Math. Stat.* **2021**, *7*, 674785. [CrossRef]
38. Pires, P.B.; Santos, J.D.; Pereira, I.V. Artificial Neural Networks: History and State of the Art. In *Encyclopedia of Information Science and Technology*, 6th ed.; Khosrow-Pour, M., Ed.; IGI Global: Hershey, PA, USA, 2023. [CrossRef]
39. Ghojogh, B.; Samad, M.N.; Mashhadi, S.A.; Kapoor, T.; Ali, W.; Karray, F.; Crowley, M. Feature Selection and Feature Extraction in Pattern Analysis: A Literature Review. *arXiv* **2019**, arXiv:1905.02845.
40. Salehi, A.W.; Khan, S.; Gupta, G.; Alabduallah, B.I.; Almjally, A.; Alsolai, H.; Siddiqui, T.; Mellit, A. A Study of CNN and Transfer Learning in Medical Imaging: Advantages, Challenges, Future Scope. *Sustainability* **2023**, *15*, 5930. [CrossRef]
41. Litjens, G.; Kooi, T.; Bejnordi, B.E.; Setio, A.A.A.; Ciompi, F.; Ghafoorian, M.; van der Laak, J.A.W.M.; van Ginneken, B.; Sánchez, C.I. A survey on deep learning in medical image analysis. *Med. Image Anal.* **2017**, *42*, 60–88. [CrossRef]
42. Annarumma, M.; Withey, S.J.; Bakewell, R.J.; Pesce, E.; Goh, V.; Montana, G. Automated Triaging of Adult Chest Radiographs with Deep Artificial Neural Networks. *Radiology* **2019**, *291*, 196–202. [CrossRef]
43. Tadavarthi, Y.; Makeeva, V.; Wagstaff, W.; Zhan, H.; Podlasek, A.; Bhatia, N.; Heilbrun, M.; Krupinski, E.; Safdar, N.; Banerjee, I.; et al. Overview of Noninterpretive Artificial Intelligence Models for Safety, Quality, Workflow, and Education Applications in Radiology Practice. *Radiol. Artif. Intell.* **2022**, *4*, e210114. [CrossRef] [PubMed]
44. Bizzo, B.C.; Almeida, R.R.; Alkasab, T.K. Artificial Intelligence Enabling Radiology Reporting. *Radiol. Clin. N. Am.* **2021**, *59*, 1045–1052. [CrossRef] [PubMed]
45. European Society of Radiology (ESR). What the radiologist should know about artificial intelligence—An ESR white paper. *Insights Imaging* **2019**, *10*, 44. [CrossRef] [PubMed]
46. Kumar, N.; Verma, R.; Sharma, S.; Bhargava, S.; Vahadane, A.; Sethi, A. A Dataset and a Technique for Generalized Nuclear Segmentation for Computational Pathology. *IEEE Trans. Med. Imaging* **2017**, *36*, 1550–1560. [CrossRef]
47. Komura, D.; Ishikawa, S. Machine Learning Methods for Histopathological Image Analysis. *Comput. Struct. Biotechnol. J.* **2018**, *16*, 34–42. [CrossRef]
48. Attia, Z.I.; Noseworthy, P.A.; Lopez-Jimenez, F.; Asirvatham, S.J.; Deshmukh, A.J.; Gersh, B.J.; Carter, R.E.; Yao, X.; Rabinstein, A.A.; Erickson, B.J.; et al. An artificial intelligence-enabled ECG algorithm for the identification of patients with atrial fibrillation during sinus rhythm: A retrospective analysis of outcome prediction. *Lancet* **2019**, *394*, 861–867. [CrossRef]
49. Narula, S.; Shameer, K.; Salem Omar, A.M.; Dudley, J.T.; Sengupta, P.P. Machine-Learning Algorithms to Automate Morphological and Functional Assessments in 2D Echocardiography. *J. Am. Coll. Cardiol.* **2016**, *68*, 2287–2295. [CrossRef] [PubMed]
50. Zou, J.; Huss, M.; Abid, A.; Mohammadi, P.; Torkamani, A.; Telenti, A. A primer on deep learning in genomics. *Nat. Genet.* **2019**, *51*, 12–18. [CrossRef] [PubMed]
51. Vamathevan, J.; Clark, D.; Czodrowski, P.; Dunham, I.; Ferran, E.; Lee, G.; Li, B.; Madabhushi, A.; Shah, P.; Spitzer, M.; et al. Applications of machine learning in drug discovery and development. *Nat. Rev. Drug Discov.* **2019**, *18*, 463–477. [CrossRef] [PubMed]
52. Davenport, T.; Kalakota, R. The potential for artificial intelligence in healthcare. *Future Healthc. J.* **2019**, *6*, 94–98. [CrossRef]
53. Rajkomar, A.; Oren, E.; Chen, K.; Dai, A.M.; Hajaj, N.; Hardt, M.; Liu, P.J.; Liu, X.; Marcus, J.; Sun, M.; et al. Scalable and accurate deep learning with electronic health records. *NPJ Digit. Med.* **2018**, *1*, 18. [CrossRef] [PubMed]
54. Esteva, A.; Robicquet, A.; Ramsundar, B.; Kuleshov, V.; De Pristo, M.; Chou, K.; Cui, C.; Corrado, G.; Thrun, S.; Dean, J. A guide to deep learning in healthcare. *Nat. Med.* **2019**, *25*, 24–29. [CrossRef] [PubMed]
55. Qiang, B.; Chen, R.; Zhou, M.; Pang, Y.; Zhai, Y.; Yang, M. Convolutional Neural Networks-Based Object Detection Algorithm by Jointing Semantic Segmentation for Images. *Sensors* **2020**, *20*, 5080. [CrossRef]
56. Ardila, D.; Kiraly, A.P.; Bharadwaj, S.; Choi, B.; Reicher, J.J.; Peng, L.; Tse, D.; Etemadi, M.; Ye, W.; Corrado, G.; et al. End-to-end lung cancer screening with three-dimensional deep learning on low-dose chest computed tomography. *Nat. Med.* **2019**, *25*, 954–961. [CrossRef]
57. Rodriguez-Ruiz, A.; Lång, K.; Gubern-Merida, A.; Broeders, M.; Gennaro, G.; Clauser, P.; Helbich, T.H.; Chevalier, M.; Tan, T.; Mertelmeier, T.; et al. Stand-Alone Artificial Intelligence for Breast Cancer Detection in Mammography: Comparison with 101 Radiologists. *J. Natl. Cancer Inst.* **2019**, *111*, 916–922. [CrossRef]

58. Mayo, R.C.; Kent, D.; Sen, L.C.; Kapoor, M.; Leung, J.W.T.; Watanabe, A.T. Reduction of False-Positive Markings on Mammograms: A Retrospective Comparison Study Using an Artificial Intelligence-Based CAD. *J. Digit. Imaging* **2019**, *32*, 618–624. [CrossRef]
59. He, M.; Wang, X.; Zhao, Y. A calibrated deep learning ensemble for abnormality detection in musculoskeletal radiographs. *Sci. Rep.* **2021**, *11*, 9097. [CrossRef]
60. Hickman, S.E.; Baxter, G.C.; Gilbert, F.J. Adoption of artificial intelligence in breast imaging: Evaluation, ethical constraints and limitations. *Br. J. Cancer* **2021**, *125*, 15–22. [CrossRef]
61. Pesapane, F.; Rotili, A.; Agazzi, G.M.; Botta, F.; Raimondi, S.; Penco, S.; Dominelli, V.; Cremonesi, M.; Jereczek-Fossa, B.A.; Carrafiello, G.; et al. Recent Radiomics Advancements in Breast Cancer: Lessons and Pitfalls for the Next Future. *Curr. Oncol.* **2021**, *28*, 2351–2372. [CrossRef]
62. Lambin, P.; Leijenaar, R.T.H.; Deist, T.M.; Peerlings, J.; de Jong, E.E.C.; van Timmeren, J.; Sanduleanu, S.; Larue, R.T.H.M.; Even, A.J.G.; Jochems, A.; et al. Radiomics: The bridge between medical imaging and personalized medicine. *Nat. Rev. Clin. Oncol.* **2017**, *14*, 749–762. [CrossRef] [PubMed]
63. Lakhani, P.; Sundaram, B. Deep Learning at Chest Radiography: Automated Classification of Pulmonary Tuberculosis by Using Convolutional Neural Networks. *Radiology* **2017**, *284*, 574–582. [CrossRef] [PubMed]
64. Allen, B.; Agarwal, S.; Coombs, L.; Wald, C.; Dreyer, K. 2020 ACR Data Science Institute Artificial Intelligence Survey. *J. Am. Coll. Radiol.* **2021**, *18*, 1153–1159. [CrossRef]
65. Makeeva, V. An Essential Roadmap for AI in Radiology. *Am. Coll. Radiol.* **2022**. Available online: https://www.acr.org/Practice-Management-Quality-Informatics/ACR-Bulletin/Articles/September-2022/An-Essential-Roadmap-for-AI-in-Radiology (accessed on 20 May 2023).
66. Yedavalli, V.S.; Tong, E.; Martin, D.; Yeom, K.W.; Forkert, N.D. Artificial intelligence in stroke imaging: Current and future perspectives. *Clin. Imaging* **2021**, *69*, 246–254. [CrossRef]
67. Soun, J.E.; Chow, D.S.; Nagamine, M.; Takhtawala, R.S.; Filippi, C.G.; Yu, W.; Chang, P.D. Artificial Intelligence and Acute Stroke Imaging. *AJNR Am. J. Neuroradiol.* **2021**, *42*, 2–11. [CrossRef]
68. UMass Chan Medical School Department of Radiology. Artificial Intelligence in MRI. 2020. Available online: https://www.umassmed.edu/radiology/radnews/2020/10/ai-mri/ (accessed on 27 May 2023).
69. Qiu, W.; Kuang, H.; Teleg, E.; Ospel, J.M.; Sohn, S.I.; Almekhlafi, M.; Goyal, M.; Hill, M.D.; Demchuk, A.M.; Menon, B.K. Machine Learning for Detecting Early Infarction in Acute Stroke with Non-Contrast-enhanced CT. *Radiology* **2020**, *294*, 638–644. [CrossRef]
70. Bivard, A.; Churilov, L.; Parsons, M. Artificial intelligence for decision support in acute stroke - current roles and potential. *Nat. Rev. Neurol.* **2020**, *16*, 575–585. [CrossRef] [PubMed]
71. Shlobin, N.; Baig, A.; Waqas, M.; Patel, T.; Dossani, R.; Wilson, M.; Cappuzzo, J.; Siddiqui, A.; Tutino, V.; Levy, E. Artificial Intelligence for Large-Vessel Occlusion Stroke: A Systematic Review. *World Neurosurg.* **2022**, *159*, 207–220.e1. [CrossRef]
72. Zhu, B.; Zhao, J.; Cao, M.; Du, W.; Yang, L.; Su, M.; Tian, Y.; Wu, M.; Wu, T.; Wang, M.; et al. Predicting 1-Hour Thrombolysis Effect of r-tPA in Patients with Acute Ischemic Stroke Using Machine Learning Algorithm. *Front. Pharmacol.* **2022**, *12*, 759782. [CrossRef] [PubMed]
73. Lui, Y.W.; Chang, P.D.; Zaharchuk, G.; Barboriak, D.P.; Flanders, A.E.; Wintermark, M.; Hess, C.P.; Filippi, C.G. Artificial Intelligence in Neuroradiology: Current Status and Future Directions. *AJNR Am. J. Neuroradiol.* **2020**, *41*, E52–E59. [CrossRef] [PubMed]
74. Koh, D.M.; Papanikolaou, N.; Bick, U.; Rowland, I.; Charles, E.K., Jr.; Jayshree, K.-C.; Celso, M.; Luis, M.-B.; Anne, M.; Seong, K.M.; et al. Artificial intelligence and machine learning in cancer imaging. *Commun. Med.* **2022**, *2*, 133. [CrossRef] [PubMed]
75. Luchini, C.; Pea, A.; Scarpa, A. Artificial intelligence in oncology: Current applications and future perspectives. *Br. J. Cancer* **2022**, *126*, 4–9. [CrossRef]
76. Wang, H.; Zhu, H.; Ding, L. Accurate classification of lung nodules on CT images using the TransUnet. *Front. Public Health* **2022**, *10*, 1060798. [CrossRef]
77. Kowalewski, K.F.; Egen, L.; Fischetti, C.E.; Puliatti, S.; Juan, G.R.; Taratkin, M.; Ines, R.B.; Sidoti Abate, M.A., Mühlbauer, J.; Wessels, F.; et al. Artificial intelligence for renal cancer: From imaging to histology and beyond. *Asian J. Urol.* **2022**, *9*, 243–252. [CrossRef]
78. Bhandari, A.; Koppen, J.; Agzarian, M. Convolutional neural networks for brain tumour segmentation. *Insights Imaging* **2020**, *11*, 77. [CrossRef]
79. Starkov, P.; Aguilera, T.A.; Golden, D.I.; Shultz, D.B.; Trakul, N.; Maxim, P.G.; Le, Q.T.; Loo, B.W.; Diehn, M.; Depeursinge, A.; et al. The use of texture-based radiomics CT analysis to predict outcomes in early-stage non-small cell lung cancer treated with stereotactic ablative radiotherapy. *Br. J. Radiol.* **2019**, *92*, 20180228. [CrossRef]
80. Liu, C.; Li, Y.; Xia, X.; Wang, J.; Hu, C. Application of radiomics feature captured from MRI for prediction of recurrence for glioma patients. *J. Cancer* **2022**, *13*, 965–974. [CrossRef]
81. Boldrini, L.; Bibault, J.E.; Masciocchi, C.; Shen, Y.; Bittner, M.I. Deep Learning: A Review for the Radiation Oncologist. *Front. Oncol.* **2019**, *9*, 977. [CrossRef]
82. Yala, A.; Lehman, C.; Schuster, T.; Portnoi, T.; Barzilay, R. A Deep Learning Mammography-based Model for Improved Breast Cancer Risk Prediction. *Radiology* **2019**, *292*, 60–66. [CrossRef]

83. Nam, J.G.; Hwang, E.J.; Kim, J.; Park, N.; Lee, E.H.; Kim, H.J.; Nam, M.; Lee, J.H.; Park, C.M.; Goo, J.M. AI Improves Nodule Detection on Chest Radiographs in a Health Screening Population: A Randomized Controlled Trial. *Radiology* **2023**, *307*, e221894. [CrossRef]
84. Zhou, J.; Du, M.; Chang, S.; Chen, Z. Artificial intelligence in echocardiography: Detection, functional evaluation, and disease diagnosis. *Cardiovasc. Ultrasound* **2021**, *19*, 29. [CrossRef]
85. Guo, F.; Ng, M.; Roifman, I.; Wright, G. Cardiac Magnetic Resonance Left Ventricle Segmentation and Function Evaluation Using a Trained Deep-Learning Model. *Appl. Sci.* **2022**, *12*, 2627. [CrossRef]
86. Nicholls, M. AI Offers Advances in Cardiovascular Imaging. *Healthcare in Europe*. 2023. Available online: https://healthcare-in-europe.com/en/news/ai-offers-advances-in-cardiovascular-imaging.html (accessed on 24 May 2023).
87. Camara, J.R.; Tomihama, R.T.; Pop, A.; Shedd, M.P.; Dobrowski, B.S.; Knox, C.J.; Abou-Zamzam, A.M.J.; Kiang, S.C. Development of a convolutional neural network to detect abdominal aortic aneurysms. *J. Vasc. Surg. Cases Innov. Tech.* **2022**, *8*, 305–311. [CrossRef]
88. Betancur, J.; Otaki, Y.; Motwani, M.; Fish, M.B.; Lemley, M.; Dey, D.; Gransar, H.; Tamarappoo, B.; Germano, G.; Sharir, T.; et al. Prognostic Value of Combined Clinical and Myocardial Perfusion Imaging Data Using Machine Learning. *JACC Cardiovasc. Imaging* **2018**, *11*, 1000–1009. [CrossRef]
89. Al'Aref, S.J.; Anchouche, K.; Singh, G.; Slomka, P.J.; Kolli, K.K.; Kumar, A.; Pandey, M.; Maliakal, G.; van Rosendael, A.R.; Beecy, A.N.; et al. Clinical applications of machine learning in cardiovascular disease and its relevance to cardiac imaging. *Eur. Heart J.* **2019**, *40*, 1975–1986. [CrossRef]
90. Kumar, R.; Khan, F.U.; Sharma, A.; Aziz, I.B.A.; Poddar, N.K. Recent Applications of Artificial Intelligence in the Detection of Gastrointestinal, Hepatic and Pancreatic Diseases. *Curr. Med. Chem.* **2022**, *29*, 66–85. [CrossRef]
91. Fazekas, S.; Budai, B.K.; Stollmayer, R.; Kaposi, P.N.; Bérczi, V. Artificial intelligence and neural networks in radiology – Basics that all radiology residents should know. *Imaging* **2022**, *14*, 73–81. [CrossRef]
92. Berbís, M.A.; Godino, F.P.; Del Val, J.R.; PMata, L.A.; Luna, A. Clinical impact of artificial intelligence-based solutions on imaging of the pancreas and liver. *World J. Gastroenterol.* **2023**, *29*, 1427–1445. [CrossRef]
93. Loftus, T.J.; Shickel, B.; Ozrazgat-Baslanti, T.; Ren, Y.; Glicksberg, B.S.; Cao, J.; Singh, K.; Chan, L.; Nadkarni, G.N.; Bihorac, A. Artificial intelligence-enabled decision support in nephrology. *Nat. Rev. Nephrol.* **2022**, *18*, 452–465. [CrossRef]
94. Jayakumar, S.; Sounderajah, V.; Normahani, P.; Harling, L.; Markar, S.R.; Ashrafian, H.; Darzi, A. Quality assessment standards in artificial intelligence diagnostic accuracy systematic reviews: A meta-research study. *NPJ Digit. Med.* **2022**, *5*, 11. [CrossRef]
95. Roberts, M.; Driggs, D.; Thorpe, M.; Gilbey, J.; Yeung, M.; Ursprung, S.; Aviles-Rivero, A.I.; Etmann, C.; McCague, C.; Beer, L.; et al. Common pitfalls and recommendations for using machine learning to detect and prognosticate for COVID-19 using chest radiographs and CT scans. *Nat. Mach. Intell.* **2021**, *3*, 199–217. [CrossRef]
96. Obermeyer, Z.; Powers, B.; Vogeli, C.; Mullainathan, S. Dissecting racial bias in an algorithm used to manage the health of populations. *Science* **2019**, *366*, 447–453. [CrossRef]
97. Candemir, S.; Nguyen, X.V.; Folio, L.R.; Prevedello, L.M. Training Strategies for Radiology Deep Learning Models in Data-limited Scenarios. *Radiol. Artif. Intell.* **2021**, *3*, 6. [CrossRef]
98. Qian, J.; Li, H.; Wang, J.; He, L. Recent Advances in Explainable Artificial Intelligence for Magnetic Resonance Imaging. *Diagnostics* **2023**, *13*, 1571. [CrossRef]
99. London, A. Artificial Intelligence and Black-Box Medical Decisions: Accuracy versus Explainability. *Hastings Cent. Rep.* **2019**, *49*, 15–21. [CrossRef]
100. Holzinger, A.; Langs, G.; Denk, H.; Zatloukal, K.; Müller, H. Causability and explainability of artificial intelligence in medicine. *Wiley Interdiscip. Rev. Data Min. Knowl. Discov.* **2019**, *9*, 4. [CrossRef]
101. Barragán-Montero, A.; Javaid, U.; Valdés, G.; Nguyen, D.; Desbordes, P.; Macq, B.; Willems, S.; Vandewinckele, L.; Holmström, M.; Löfman, F.; et al. Artificial intelligence and machine learning for medical imaging: A technology review. *Phys. Med.* **2021**, *83*, 242–256. [CrossRef]
102. Wang, G. A Perspective on Deep Imaging. *IEEE Access* **2016**, *4*, 8914–8924. [CrossRef]
103. Khosla, P.; Teterwak, P.; Wang, C.; Sarna, A.; Tian, Y.; Isola, P.; Maschinot, A.; Liu, C.; Krishnan, D. Supervised contrastive learning. In Proceedings of the Advances in Neural Information Processing Systems, Online, 6–12 December 2020; Volume 33, pp. 18661–18673.
104. Topol, E.J. High-performance medicine: The convergence of human and artificial intelligence. *Nat. Med.* **2019**, *25*, 44–56. [CrossRef]
105. Murdoch, B. Privacy and artificial intelligence: Challenges for protecting health information in a new era. *BMC Med. Ethics* **2021**, *22*, 122. [CrossRef]
106. Pianykh, O.S.; Langs, G.; Dewey, M.; Enzmann, D.R.; Herold, C.J.; Schoenberg, S.O.; Brink, J.A. Continuous Learning AI in Radiology: Implementation Principles and Early Applications. *Radiology* **2020**, *297*, 6–14. [CrossRef]
107. Ahmed, F.; Fattani, M.T.; Ali, S.R.; Enam, R.N. Strengthening the Bridge Between Academic and the Industry Through the Academia-Industry Collaboration Plan Design Model. *Front. Psychol.* **2022**, *13*, 875940. [CrossRef]

108. Banerjee, M.; Chiew, D.; Patel, K.T.; Johns, I.; Chappell, D.; Linton, N.; Cole, G.D.; Francis, D.P.; Szram, J.; Ross, J.; et al. The impact of artificial intelligence on clinical education: Perceptions of postgraduate trainee doctors in London (UK) and recommendations for trainers. *BMC Med. Educ.* **2021**, *21*, 429. [CrossRef]
109. Australian, T.R.; New Zealand College of Radiologists. Clinical Radiology Curriculum Learning Outcomes. 2023. Available online: https://www.ranzcr.com/trainees/clinical-radiology-training-program/learning-outcomes-and-handbook (accessed on 30 May 2023).

tomography

MDPI

Article

Deep Learning-Based Versus Iterative Image Reconstruction for Unenhanced Brain CT: A Quantitative Comparison of Image Quality

Andrea Cozzi [1,†], Maurizio Cè [2,†], Giuseppe De Padova [2], Dario Libri [2], Nazarena Caldarelli [2], Fabio Zucconi [3], Giancarlo Oliva [4] and Michaela Cellina [4,*]

[1] Service of Radiology, Imaging Institute of Southern Switzerland (IIMSI), Ente Ospedaliero Cantonale (EOC), Via Tesserete 46, 6900 Lugano, Switzerland; andrea.cozzi@gmail.com
[2] Postgraduation School in Radiodiagnostics, Università degli Studi di Milano, Via Festa del Perdono 7, 20122 Milano, Italy; giuseppe.depadova@unimi.it (G.D.P.); dario.libri@unimi.it (D.L.); nazarena.caldarelli1@unimi.it (N.C.)
[3] Department of Radioprotection, Fatebenefratelli Hospital, ASST Fatebenefratelli Sacco, Milano, Piazza Principessa Clotilde 3, 20121 Milano, Italy; fabio.zucconi@asst-fbf-sacco.it
[4] Radiology Department, Fatebenefratelli Hospital, ASST Fatebenefratelli Sacco, Milano, Piazza Principessa Clotilde 3, 20121 Milano, Italy; linforisonanza@gmail.com
* Correspondence: michaela.cellina@asst-fbf-sacco.it
† These authors contributed equally to this work.

Abstract: This exploratory retrospective study aims to quantitatively compare the image quality of unenhanced brain computed tomography (CT) reconstructed with an iterative (AIDR-3D) and a deep learning-based (AiCE) reconstruction algorithm. After a preliminary phantom study, AIDR-3D and AiCE reconstructions (0.5 mm thickness) of 100 consecutive brain CTs acquired in the emergency setting on the same 320-detector row CT scanner were retrospectively analyzed, calculating image noise reduction attributable to the AiCE algorithm, artifact indexes in the posterior cranial fossa, and contrast-to-noise ratios (CNRs) at the cortical and thalamic levels. In the phantom study, the spatial resolution of the two datasets proved to be comparable; conversely, AIDR-3D reconstructions showed a broader noise pattern. In the human study, median image noise was lower with AiCE compared to AIDR-3D (4.7 vs. 5.3, $p < 0.001$, median 19.6% noise reduction), whereas AIDR-3D yielded a lower artifact index than AiCE (7.5 vs. 8.4, $p < 0.001$). AiCE also showed higher median CNRs at the cortical (2.5 vs. 1.8, $p < 0.001$) and thalamic levels (2.8 vs. 1.7, $p < 0.001$). These results highlight how image quality improvements granted by deep learning-based (AiCE) and iterative (AIDR-3D) image reconstruction algorithms vary according to different brain areas.

Keywords: brain computed tomography; iterative reconstruction algorithms; deep learning-based reconstruction algorithms; image quality; image noise

Citation: Cozzi, A.; Cè, M.; De Padova, G.; Libri, D.; Caldarelli, N.; Zucconi, F.; Oliva, G.; Cellina, M. Deep Learning-Based Versus Iterative Image Reconstruction for Unenhanced Brain CT: A Quantitative Comparison of Image Quality. *Tomography* **2023**, *9*, 1629–1637. https://doi.org/10.3390/tomography9050130

Academic Editor: Benjamin M. Ellingson

Received: 30 June 2023
Revised: 23 August 2023
Accepted: 30 August 2023
Published: 31 August 2023

1. Introduction

Unenhanced brain computed tomography (CT) is a widely used medical imaging technique with a crucial role in the diagnosis, monitoring, and treatment of neurological disorders and injuries [1]. Its ability to provide detailed structural information in a non-invasive and low-cost manner, with quick turnaround times and wide availability, makes it the examination of choice in the emergency setting [2]. However, the acquisition of high-quality CT images is often hindered by various factors, such as beam hardening, streak, and partial volume artifacts observed in the posterior cranial fossa, image noise, and limited variation of Hounsfield units (HU) in subtle pathologic conditions. Moreover, repeated CT examinations are related to cumulative radiation exposure.

Filtered back projection (FBP) is a commonly used reconstruction technique in CT. While FBP has been the go-to method for many years, it does have some limitations that can affect image quality and diagnostic accuracy [3]. For example, FBP does not incorporate advanced noise suppression techniques; as a result, the inherent noise in the raw projection data can be amplified during the reconstruction process, leading to grainy or speckled images. This can obscure subtle anatomical features and reduce overall image quality, particularly when using low-dose CTR protocols. Moreover, FBP lacks sophisticated algorithms for artifact correction; thus, image artifacts can be easily caused by several factors, such as metallic implants, beam hardening, and patient motion.

Iterative reconstruction techniques were introduced to overcome these limitations of FBP. By iteratively refining the reconstructed image using mathematical models and optimization techniques, these algorithms can mitigate the blurring effect caused by the finite size of X-ray beams and improve the visualization of fine anatomical details, achieving higher spatial resolution compared to FBP [4]. Iterative reconstructions incorporate sophisticated noise suppression algorithms, which effectively reduce noise in the reconstructed images. These methods employ statistical models and regularization techniques to differentiate between true anatomical structures and random noise, resulting in cleaner and less grainy images. Improved noise suppression enhances the visibility of subtle structures and increases the signal-to-noise ratio, leading to improved diagnostic confidence, especially with low-dose CT protocols. However, iterative reconstruction techniques still show some limitations, such as longer reconstruction time due to the iterative nature of these algorithms; moreover, when high reconstruction strength levels are combined with low-dose acquisition CT, the resulting images have a waxy and plastic look or a simply unnatural appearance [5].

In recent years, deep learning (DL) has emerged as a transformative technology in medical imaging [6]. DL methods leverage complex neural network architectures to learn intricate patterns and features directly from vast amounts of training data, enabling them to achieve high performances in various imaging-related tasks, one of them indeed being the improvement of image reconstruction [7,8]. Of note, the training process behind DL-based image reconstruction can deal with a fundamental issue of iterative reconstruction, i.e., the loss of algorithm convergence due to the high number of factors that must be balanced during iteration [9].

DL-based reconstructions, with the True Fidelity (GE Healthcare, Chicago, IL, USA) and the Advanced Intelligent Clear-IQ Engine (AiCE, Canon Medical Systems, Otawara, Japan) algorithms, have been extensively investigated in abdominal imaging [8] and have also been the object of pilot studies in unenhanced brain CT for adult and pediatric populations. For example, comparisons of image quality between iterative reconstructions and the True Fidelity algorithm have been reported in adult populations by Kim et al. [10] and Alagic et al. [11] in adult patients, and by Sun et al. [12] and Lee et al. [13] in pediatric patients. In all these studies, DL-based reconstructions significantly reduced image noise and improved image quality, both from a qualitative and quantitative point of view. The AiCE algorithm has been tested in unenhanced brain CT by Oostveen et al. [14]. Their qualitative evaluation found that in four out of five image quality characteristics, DL-based reconstructions outperformed hybrid iterative reconstructions, with significant differences in noise reduction and gray matter–white matter (GM-WM) differentiation. While no statistically significant difference in overall image quality rating was found, DL-based reconstructions outperformed iterative reconstructions in all three analyzed quantitative parameters.

The clinical relevance of image quality improvements granted by DL-based image reconstruction algorithms is particularly high in the emergency setting [15,16]. However, most of the published studies include a heterogeneous case mix; a more homogeneous correspondence with the neuropathological scenario found in the emergency setting has been implemented only in the study by Kim et al. [10], which was, however, solely focused on the True Fidelity algorithm.

Thus, this study aims to explore the application of the AiCE algorithm on a set of consecutive brain CT examinations performed in the emergency setting. In particular, after a preliminary phantom study, the image quality of AiCE reconstructions will be quantitatively compared with that of hybrid iterative reconstructions (AIDR-3D) in terms of image noise values, image noise reduction attributable to the AiCE algorithm, artifact indexes, and contrast-to-noise ratios (CNRs).

2. Materials and Methods

This single-center exploratory retrospective study was conducted at the Radiology Department of Fatebenefratelli Hospital (ASST Fatebenefratelli Sacco, Milano, Italy) after approval from the Internal Review Board (protocol code DLBRI, approved on 30th April 2023). All patients provided their informed consent for CT execution and use of their de-identified images.

A preliminary phantom study was conducted by scanning quality control phantoms to assess the basic image quality performances in a well-known and controlled environment. The phantom used for the determination of the spatial resolution was a Catphan 600 unit, whereas a cylindrical water-filled phantom was used for the analysis and quantification of noise on the scans.

Then, for the in-human study, unenhanced brain CT scans of consecutive patients aged > 18 years requested by the Emergency Department of our institution for suspected stroke or after head trauma and performed between 10 April and 12 April 2023, were retrospectively retrieved from the institutional picture archiving and communication system (PACS).

2.1. *Phantom Study*

2.1.1. Spatial Resolution

The Catphan 600 phantom was scanned on a 320-detector row CT scanner (Aquilion One PRISM edition, Canon Medical Systems) on the CTP 591 section with the following exposure parameters: helical acquisition; tube voltage 120 kV; tube current exposure time product 125 mAs; pitch 0.625; collimation 320 $\times$ 0.5 mm.

The images were reconstructed in two datasets, the first with the iterative AIDR-3D FC26 kernel and the second with the DL-based AiCE Brain LCD kernel. The following reconstruction parameters were kept unmodified in both datasets: slice thickness 0.5 mm; field of view 22 cm.

On each dataset, five images in the center of the CTP 591 section were extracted. The images were analyzed with IQWorks v0.7.2 (IQWorks Project, University of Edinburgh, Edinburgh, UK) and a point spread function analysis was performed for each image centered on the tungsten wire on CTP 591. A task transfer function (TTF) was determined for each image. The five functions were averaged, and the resulting function was taken as the representative TTF of the dataset.

2.1.2. Noise Analysis

The cylindrical water-filled phantom was then scanned with the following exposure parameters: helical acquisition; tube voltage 120 kV; tube current exposure time product 360 mAs. Again, the images were reconstructed in two datasets, the first with the AIDR-3D FC26 kernel and the second with the AiCE Brain LCD kernel, keeping for both datasets the following reconstruction parameters: slice thickness 0.5 mm; field of view 22 cm; 512 $\times$ 512 (rows $\times$ columns) matrix.

On each dataset, twenty images in the middle of the phantom were extracted. A 200 mm^3 ROI was drawn in the center of each image, and the standard deviation of the ROI was registered as a measure of the amount of noise. Each image was then analyzed with the Qadistri plugin of ImageJ (ImageJ, U.S. National Institutes of Health, Bethesda, MD, USA) to determine the noise power spectrum (NPS). The spectra in the same set were averaged to reduce the statistical fluctuations in the data and determine the NPS of the dataset.

The average frequency of the NPS was calculated and registered as a measure of the noise texture appearance.

2.2. In-Human Study

All acquisitions were performed on the same 320-detector row CT scanner (Aquilion One PRISM edition, Canon Medical Systems). All examinations were acquired in helical mode with the following acquisition parameters: tube voltage 120 kVp; tube current 125 mAs; rotation time 1 s; collimation 0.5 × 40 mm; pitch 0.625; field of view 220 mm; 512 × 512 matrix. The effective dose was calculated through the dose length product multiplied by the appropriate conversion coefficient [17].

CT datasets were first reconstructed at 0.5 mm with the iterative reconstruction algorithm (AIDR-3D) and then, always at 0.5 mm, with the DL-based reconstruction algorithm (AiCE). Image datasets were visualized on the institutional PACS (Carestream Vue PACS v12.2, Carestream Health, Rochester, NY, USA). To analyze the GM-WM differentiation, four regions of interest (ROI) of 15.07 mm^2 were delineated: frontal WM and adjacent cortical GM at the level of the centrum semiovale, and thalamic deep GM and WM of the adjacent posterior limb of the internal capsule (Figures 1 and 2). To evaluate artifacts in the posterior cranial fossa, another ROI of 200 mm^2 was drawn in the inter-petrous region of the posterior fossa at the pons level, where the most noticeable artifacts were seen (Figure 3). All ROIs were delineated by two board-certified neuroradiologists and then reviewed by another board-certified neuroradiologist. We defined CT numbers (HU) of the thalamic deep GM and the frontal cortex as CT attenuation values of GM, whereas image noise was defined as the standard deviation (SD) of attenuation values measured in the deep WM at the centrum semiovale level in a ROI of 100 mm^2 (Figure 4).

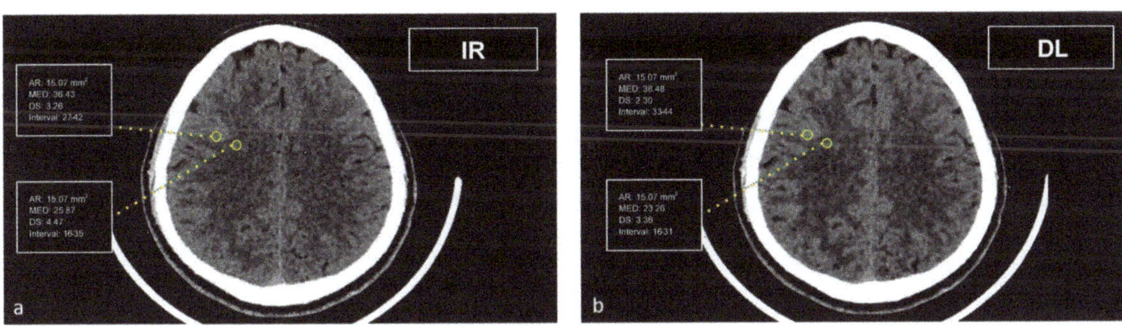

Figure 1. Demonstration of the region of interest placement in the frontal cortex and adjacent WM on iterative reconstructions (**a**) and artificial intelligence-based reconstructions (**b**).

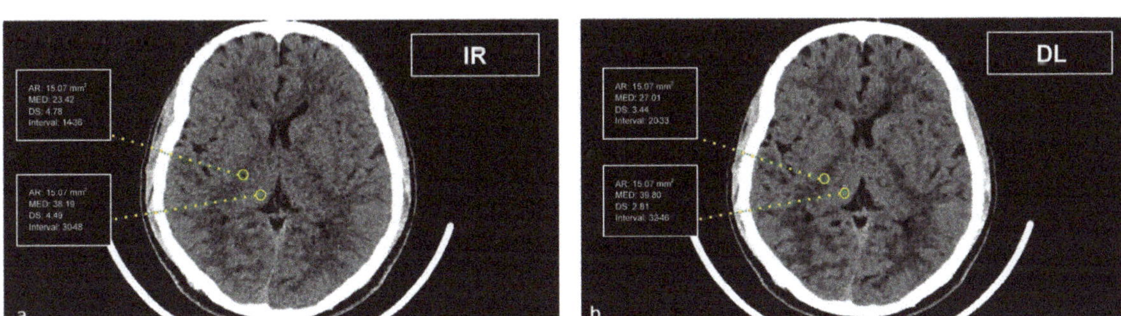

Figure 2. Demonstration of the region of interest placement in the right thalamus and adjacent WM of the internal capsule on iterative reconstructions (**a**) and artificial intelligence-based reconstructions (**b**).

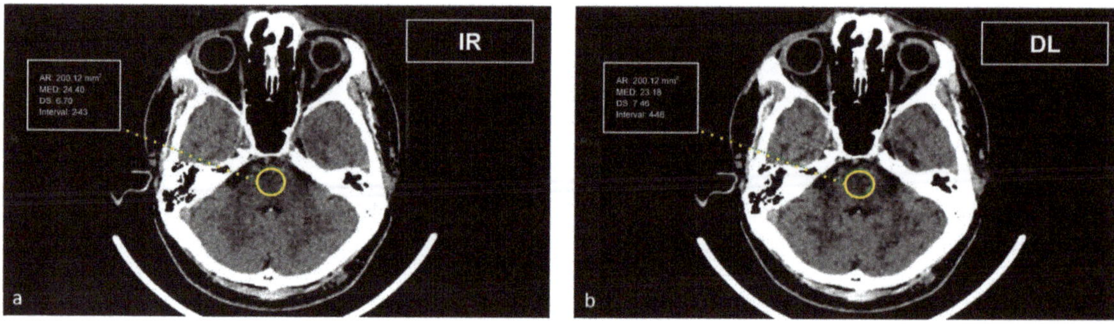

Figure 3. Demonstration of the region of interest placement in the intrapetrous region to assess the image noise in the posterior fossa on iterative reconstructions (**a**) and artificial intelligence-based reconstructions (**b**).

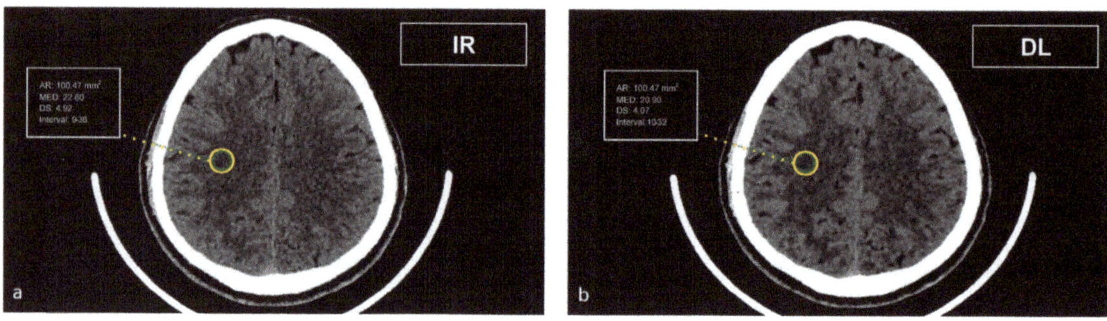

Figure 4. Demonstration of the region of interest placement in the right centrum semiovale on iterative reconstructions (**a**) and artificial intelligence-based reconstructions (**b**) to assess the image noise.

The artifact index was defined as the SD within the ROI of the posterior cranial fossa, which is prone to beam hardening, streak, and/or partial volume artifacts. Therefore, the artifact index may reflect the amount of CT number variations caused by artifacts in addition to the inherent image noise associated with scanner- and patient-related factors.

The CNRs at the centrum semiovale and basal ganglia levels were calculated with the following formula:

$$\frac{Mean\ HU_{GM} - Mean\ HU_{WM}}{\sqrt{Mean\ SD_{HU_{GM}}{}^{2} - Mean\ SD_{HU_{WM}}{}^{2}}}$$

The noise reduction rate of AiCE (%) was calculated as follows:

$$\frac{SD_{AIDR\text{-}3D} - SD_{AiCE}}{SD_{AIDR\text{-}3D}} \times 100$$

2.3. Statistical Analysis

The normality of AIDR-3D and AiCE data distributions obtained from the analysis of human CT images was evaluated with the Shapiro–Wilk test, which confirmed the non-normal distribution of these data. Thus, the non-parametric Wilcoxon test was used for all comparisons between the two sets of measurements, considering their continuous and paired nature. Values of image noise, artifact index, and cortical and thalamic CNRs in the two sets of measurements were expressed as the median and interquartile range (IQR). In order to account for multiple comparisons, the Bonferroni correction was applied considering the four aforementioned tests, with an ensuing $p < 0.013$ threshold for statistical significance. RStudio (with R version 4.2.1, The R Foundation for Statistical Computing, Vienna, Austria) and STATA (version MP 17.1, StataCorp, College Station, TX, USA) were used to perform all analyses.

3. Results

3.1. *Image Analysis in Phantom*

In the phantom study, the spatial resolution of the two datasets resulted in being comparable: the TTF of the AIDR-3D dataset showed a 50% frequency of 4.0/cm, whereas the TTF of the AiCE dataset showed a 50% frequency of 3.9/cm.

Conversely, the AIDR-3D reconstruction algorithm resulted in a broader noise pattern: the analysis of the noise amount resulted in an average noise of 6.0 ± 0.2 for the AIDR-3D dataset, whereas for the AiCE dataset the average noise was 4.1 ± 0.2. The NPS analysis showed that the average noise frequency for the AIDR-3D dataset was 2.3/cm, while for the AiCE dataset the average frequency was 1.8/cm.

3.2. *Image Analysis in Patients*

A total of 100 patients (40 females, overall median age 67 years, IQR 41–80) undergoing unenhanced brain CT for suspected stroke or head trauma were retrieved and included in this study. The mean dose length product was 1028 mGy × cm, whereas the mean effective dose was 2.06 mSv. As shown in Figure 5, in the deep WM of the centrum semiovale the AiCE images had a significantly lower median image noise (4.7, IQR 4.4–5.0) than the AIDR-3D ones (median 5.3 IQR 5.0–5.7, $p < 0.001$), with a median 19.6% noise reduction (IQR 11.7–28.0%, minimum −11.5%, maximum 50.0%). Conversely, AiCE images had a significantly higher median artifact index in the posterior cranial fossa (8.4, IQR 7.3–9.2) compared to AIDR-3D images (median 7.5, IQR 6.9–8.3, $p < 0.001$). However, AiCE images had a significantly higher median CNR both at the cortical level (2.5, IQR 2.0–3.3, $p < 0.001$) and the thalamic level (median 2.8, IQR 2.3–3.2, $p < 0.001$) compared to the median AIDR-3D CNRs of 1.8 (IQR 1.4–2.3, cortical level) and 1.7 (IQR 1.5–1.8, thalamic level).

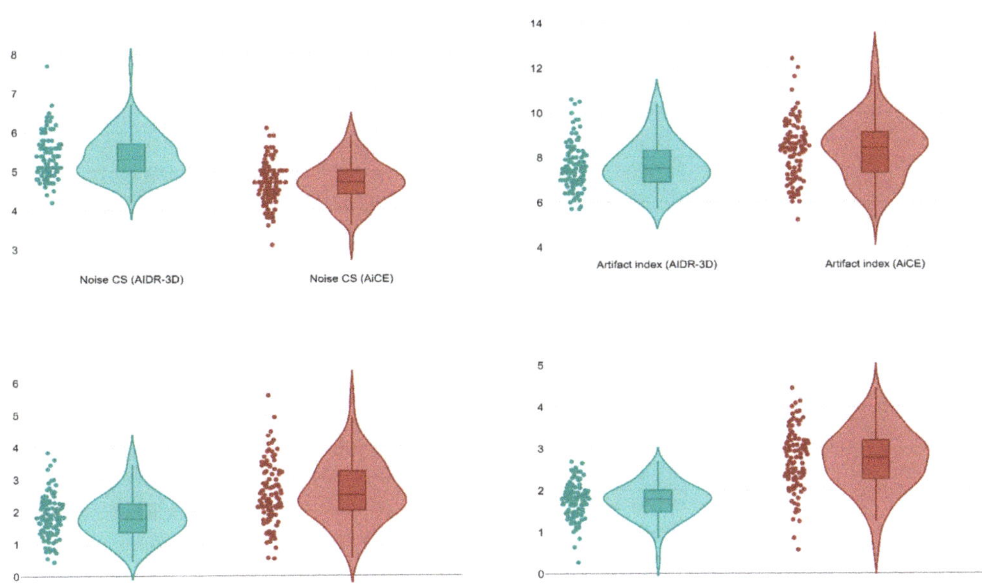

Figure 5. Violin plots comparing image noise in the centrum semiovale, artifact index in the posterior fossa, and CNRs at cortical and thalamic levels between AIDR-3D and AiCE reconstructions.

4. Discussion

This study aimed to quantitatively compare the image quality of unenhanced brain CT images reconstructed with a DL-based (AiCE) and a hybrid iterative reconstruction algorithm (AIDR-3D), two types of reconstruction algorithms that are nowadays widely applied in routine clinical practice.

Our study demonstrated that DL-based reconstructions can significantly improve the CNR both at the frontal lobe and basal ganglia levels and provide a low image noise at the supratentorial level, with a mean noise reduction of 19.6%. These findings are consistent with previous studies evaluating CT image reconstruction using DL-based algorithms in the brain [9–14] and other anatomical regions [7,8,18–21]. These results at the supratentorial level confirm the utility of DL-based reconstructions, especially in the emergency setting, where subtle definition of anatomical structures boosts the diagnostic performance of CT in the detection of ischemic and hemorrhagic alterations.

However, we observed that AiCE reconstructions had a higher median artifact index in the posterior cranial fossa compared to AIDR-3D images. The crucial relevance of the posterior cranial fossa in unenhanced brain CT is based on two different aspects: first, because the presence of several delicate structures (such as the brainstem and the cranial nerves) in a small area implies a considerable diagnostic challenge in the emergency setting for the rule-out of hemorrhagic and ischemic events; second, because the posterior cranial fossa is typically affected by artifacts due to the predominant presence of neighboring bony structures [15,22]. This point deserves careful consideration, as the study by Oostveen et al. [14] using the same DL-based algorithm had found a significant reduction of artifacts in the posterior cranial fossa; however, the ROI positioning chosen by Oostveen et al. [14] (in the fourth ventricle) is much less challenging than the placement in the pons implemented by Kim et al. [10] and also adopted in this study, as this positioning is closer to skull base bones. Notably, the latter option is better suited to mirror clinical necessities, as the emergency setting requires the ability to appropriately evaluate the margins of the posterior cranial fossa, which are a particularly problematic area from a clinical point of view. On the one hand, this different ROI positioning in the posterior cranial fossa might, therefore, explain the higher presence of artifacts observed in our study compared to Oostveen et al. [14]; on the other hand, this indirect observation of a worse performance of the AiCE algorithm compared to other studies with the True Fidelity algorithm and the same ROI positioning in the posterior cranial fossa [10,11]—which had found significant reductions of image artifacts—warrants the need for studies comparing different DL-based reconstruction algorithms.

Overall, the results of this study confirm that DL-based reconstructions are able to quantitatively improve image quality in selected areas of the supratentorial gray and white matter, particularly in regions such as the frontal lobe and basal ganglia, which are common sites of pathological processes encountered in the emergency setting. Conversely, the less encouraging results obtained in equally critical but more challenging areas suggest that iterative reconstructions still have an advantage when we intend to compare some quantitative parameters—such as the artifact index—which are nearer to a qualitative assessment and where the long-standing familiarity of radiologists with images from iterative reconstructions still plays a part. However, the capabilities of DL-based reconstruction algorithms, provided that adequate training is performed, should be well-suited to overcome these limitations. Furthermore, the extreme complexity of the cranial anatomical environment implies different necessities according to different areas: the ductility of DL-based techniques could lead to the development of blended models in which, according to the clinical context, optimal reconstruction performance can be achieved through the selection of reconstruction models according to different anatomical regions.

This study has limitations. First, only thin-slice reconstructions (0.5 mm) were evaluated: the impact of DL-based reconstructions on thicker slices is expected to be similar, albeit with less pronounced noise reduction, since thicker slices inherently have less noise regardless of the reconstruction method. Second, image noise was only measured in terms of standard deviations; this does not account for different noise textures associated with each reconstruction algorithm, which play a role in the perception of "natural" image appearance. Third, our evaluation was limited to adult patients, even though the benefits of DL-based reconstruction algorithms are foreseen to be even higher in pediatric patients, where the application of low radiation dose protocols favors the emergence of imaging artifacts.

In conclusion, data from this study show how the performance of DL-based (AiCE) and iterative (AIDR-3D) image reconstruction algorithms for unenhanced brain CT varies according to different brain areas. Wherever anatomical complexity poses a particular challenge to CT image reconstruction in the emergency setting, the combined opportunistic use of hybrid iterative and DL-based appears as a viable option to achieve optimal image quality until further development of DL-based models leads to high-level image reconstruction in all regions of the brain.

Author Contributions: Conceptualization, M.C. (Michaela Cellina); methodology, A.C., M.C. (Maurizio Cè) and M.C. (Michaela Cellina); validation, M.C. (Michaela Cellina); formal analysis, A.C. and M.C. (Maurizio Cè); investigation, N.C., D.L., G.D.P. and F.Z.; data curation, A.C., M.C. (Maurizio Cè), N.C., D.L. and G.D.P.; writing—original draft preparation, M.C. (Michaela Cellina) and F.Z.; writing—review and editing, A.C. and M.C. (Maurizio Cè); supervision, G.O.; project administration, M.C. (Michaela Cellina). All authors have read and agreed to the published version of the manuscript.

Funding: This research received no external funding.

Institutional Review Board Statement: The study was conducted in accordance with the Declaration of Helsinki and approved by the Institutional Review Board (protocol code DLBRI, approved 30 April 2023).

Informed Consent Statement: Informed consent was obtained from all subjects involved in the study.

Data Availability Statement: All analyzed data are included in this paper.

Conflicts of Interest: The authors declare no conflict of interest.

References

1. Cellina, M.; Panzeri, M.; Floridi, C.; Martinenghi, C.M.A.; Clesceri, G.; Oliva, G. Overuse of computed tomography for minor head injury in young patients: An analysis of promoting factors. *Radiol. Med.* **2018**, *123*, 507–514. [CrossRef] [PubMed]
2. Leone, E.; Ferrari, R.; Trinci, M.; Cingolani, E.; Galluzzo, M. Imaging features of electric scooter trauma: What an emergency radiologist needs to know. *Radiol. Med.* **2022**, *127*, 872–880. [CrossRef] [PubMed]
3. Willemink, M.J.; Noël, P.B. The evolution of image reconstruction for CT—From filtered back projection to artificial intelligence. *Eur. Radiol.* **2019**, *29*, 2185–2195. [CrossRef] [PubMed]
4. Agostini, A.; Borgheresi, A.; Carotti, M.; Ottaviani, L.; Badaloni, M.; Floridi, C.; Giovagnoni, A. Third-generation iterative reconstruction on a dual-source, high-pitch, low-dose chest CT protocol with tin filter for spectral shaping at 100 kV: A study on a small series of COVID-19 patients. *Radiol. Med.* **2021**, *126*, 388–398. [CrossRef] [PubMed]
5. Geyer, L.L.; Schoepf, U.J.; Meinel, F.G.; Nance, J.W.; Bastarrika, G.; Leipsic, J.A.; Paul, N.S.; Rengo, M.; Laghi, A.; De Cecco, C.N. State of the Art: Iterative CT Reconstruction Techniques. *Radiology* **2015**, *276*, 339–357. [CrossRef]
6. Castiglioni, I.; Rundo, L.; Codari, M.; Di Leo, G.; Salvatore, C.; Interlenghi, M.; Gallivanone, F.; Cozzi, A.; D'Amico, N.C.; Sardanelli, F. AI applications to medical images: From machine learning to deep learning. *Phys. Medica* **2021**, *83*, 9–24. [CrossRef]
7. Szczykutowicz, T.P.; Toia, G.V.; Dhanantwari, A.; Nett, B. A Review of Deep Learning CT Reconstruction: Concepts, Limitations, and Promise in Clinical Practice. *Curr. Radiol. Rep.* **2022**, *10*, 101–115. [CrossRef]
8. Shehata, M.A.; Saad, A.M.; Kamel, S.; Stanietzky, N.; Roman-Colon, A.M.; Morani, A.C.; Elsayes, K.M.; Jensen, C.T. Deep-learning CT reconstruction in clinical scans of the abdomen: A systematic review and meta-analysis. *Abdom. Radiol.* **2023**, *48*, 2724–2756. [CrossRef]
9. Altmann, S.; Abello Mercado, M.A.; Ucar, F.A.; Kronfeld, A.; Al-Nawas, B.; Mukhopadhyay, A.; Booz, C.; Brockmann, M.A.; Othman, A.E. Ultra-High-Resolution CT of the Head and Neck with Deep Learning Reconstruction—Assessment of Image Quality and Radiation Exposure and Intraindividual Comparison with Normal-Resolution CT. *Diagnostics* **2023**, *13*, 1534. [CrossRef]
10. Kim, I.; Kang, H.; Yoon, H.J.; Chung, B.M.; Shin, N.-Y. Deep learning–based image reconstruction for brain CT: Improved image quality compared with adaptive statistical iterative reconstruction-Veo (ASIR-V). *Neuroradiology* **2021**, *63*, 905–912. [CrossRef]

11. Alagic, Z.; Diaz Cardenas, J.; Halldorsson, K.; Grozman, V.; Wallgren, S.; Suzuki, C.; Helmenkamp, J.; Koskinen, S.K. Deep learning versus iterative image reconstruction algorithm for head CT in trauma. *Emerg. Radiol.* **2022**, *29*, 339–352. [CrossRef] [PubMed]
12. Sun, J.; Li, H.; Wang, B.; Li, J.; Li, M.; Zhou, Z.; Peng, Y. Application of a deep learning image reconstruction (DLIR) algorithm in head CT imaging for children to improve image quality and lesion detection. *BMC Med. Imaging* **2021**, *21*, 108. [CrossRef]
13. Lee, N.; Cho, H.-H.; Lee, S.M.; You, S.K. Adaptation of Deep Learning Image Reconstruction for Pediatric Head CT: A Focus on the Image Quality. *J. Korean Soc. Radiol.* **2023**, *84*, 240–252. [CrossRef] [PubMed]
14. Oostveen, L.J.; Meijer, F.J.A.; de Lange, F.; Smit, E.J.; Pegge, S.A.; Steens, S.C.A.; van Amerongen, M.J.; Prokop, M.; Sechopoulos, I. Deep learning–based reconstruction may improve non-contrast cerebral CT imaging compared to other current reconstruction algorithms. *Eur. Radiol.* **2021**, *31*, 5498–5506. [CrossRef] [PubMed]
15. Barrett, J.F.; Keat, N. Artifacts in CT: Recognition and Avoidance. *RadioGraphics* **2004**, *24*, 1679–1691. [CrossRef]
16. Cauley, K.A.; Yorks, P.J.; Flora, S.; Fielden, S.W. The effects of the skull on CT imaging of the brain: A skull and brain phantom study. *Br. J. Radiol.* **2021**, *94*, 20200714. [CrossRef]
17. Christner, J.A.; Kofler, J.M.; McCollough, C.H. Estimating Effective Dose for CT Using Dose–Length Product Compared With Using Organ Doses: Consequences of Adopting International Commission on Radiological Protection Publication 103 or Dual-Energy Scanning. *Am. J. Roentgenol.* **2010**, *194*, 881–889. [CrossRef]
18. Cellina, M.; Cè, M.; Rossini, N.; Cacioppa, L.M.; Ascenti, V.; Carrafiello, G.; Floridi, C. Computed Tomography Urography: State of the Art and Beyond. *Tomography* **2023**, *9*, 909–930. [CrossRef]
19. Akagi, M.; Nakamura, Y.; Higaki, T.; Narita, K.; Honda, Y.; Zhou, J.; Yu, Z.; Akino, N.; Awai, K. Deep learning reconstruction improves image quality of abdominal ultra-high-resolution CT. *Eur. Radiol.* **2019**, *29*, 6163–6171. [CrossRef]
20. Tamura, A.; Mukaida, E.; Ota, Y.; Nakamura, I.; Arakita, K.; Yoshioka, K. Deep learning reconstruction allows low-dose imaging while maintaining image quality: Comparison of deep learning reconstruction and hybrid iterative reconstruction in contrast-enhanced abdominal CT. *Quant. Imaging Med. Surg.* **2022**, *12*, 2977–2984. [CrossRef]
21. Tanabe, N.; Sakamoto, R.; Kozawa, S.; Oguma, T.; Shima, H.; Shiraishi, Y.; Koizumi, K.; Sato, S.; Nakamoto, Y.; Hirai, T. Deep learning-based reconstruction of chest ultra-high-resolution computed tomography and quantitative evaluations of smaller airways. *Respir. Investig.* **2022**, *60*, 167–170. [CrossRef]
22. Guziński, M.; Waszczuk, Ł.; Sąsiadek, M.J. Head CT: Image quality improvement of posterior fossa and radiation dose reduction with ASiR—Comparative studies of CT head examinations. *Eur. Radiol.* **2016**, *26*, 3691–3696. [CrossRef]

Journal of
Clinical Medicine

MDPI

Article

Chest X-ray Foreign Objects Detection Using Artificial Intelligence

Jakub Kufel [1,*], Katarzyna Bargieł-Łączek [2,3], Maciej Koźlik [4], Łukasz Czogalik [5], Piotr Dudek [5], Mikołaj Magiera [5], Wiktoria Bartnikowska [2], Anna Lis [6], Iga Paszkiewicz [5], Szymon Kocot [7], Maciej Cebula [8], Katarzyna Gruszczyńska [3] and Zbigniew Nawrat [1,9]

1 Department of Biophysics, Faculty of Medical Sciences in Zabrze, Medical University of Silesia, Jordana 19, 41-808 Zabrze, Poland; znawrat@sum.edu.pl or nawrat@frk.pl
2 Paediatric Radiology Students' Scientific Association at the Division of Diagnostic Imaging, 40-752 Katowice, Poland; katarzyna.bargiel@op.pl (K.B.-Ł.); wiktoriabart96@gmail.com (W.B.)
3 Department of Radiology and Nuclear Medicine, Faculty of Medical Sciences in Katowice, Medical University of Silesia, 40-752 Katowice, Poland; kgruszczynska@sum.edu.pl
4 Division of Cardiology and Structural Heart Disease, Medical University of Silesia, 40-635 Katowice, Poland; kozlik.maciej@gmail.com
5 Professor Zbigniew Religa Student Scientific Association at the Department of Biophysic, Faculty of Medical Sciences in Zabrze, Medical University of Silesia, Jordana 19, 41-808 Zabrze, Poland; lukczog@gmail.com (Ł.C.); piotrekd233@gmail.com (P.D.); mikaczu7422@gmail.com (M.M.); igapaszkiewicz.ip@gmail.com (I.P.)
6 Cardiology Students' Scientific Association at the III Department of Cardiology, Faculty of Medical Sciences in Katowice, Medical University of Silesia, 40-635 Katowice, Poland; lis.anna9898@gmail.com
7 Bright Coders' Factory, Technologiczna 2, 45-839 Opole, Poland; szymko1995@gmail.com
8 Individual Specialist Medical Practice, 40-754 Katowice, Poland; maciejmichalcebula@gmail.com
9 Foundation of Cardiac Surgery Development, 41-800 Zabrze, Poland
* Correspondence: jakubkufel92@gmail.com; Tel.: +48-518530868

Abstract: Diagnostic imaging has become an integral part of the healthcare system. In recent years, scientists around the world have been working on artificial intelligence-based tools that help in achieving better and faster diagnoses. Their accuracy is crucial for successful treatment, especially for imaging diagnostics. This study used a deep convolutional neural network to detect four categories of objects on digital chest X-ray images. The data were obtained from the publicly available National Institutes of Health (NIH) Chest X-ray (CXR) Dataset. In total, 112,120 CXRs from 30,805 patients were manually checked for foreign objects: vascular port, shoulder endoprosthesis, necklace, and implantable cardioverter-defibrillator (ICD). Then, they were annotated with the use of a computer program, and the necessary image preprocessing was performed, such as resizing, normalization, and cropping. The object detection model was trained using the You Only Look Once v8 architecture and the Ultralytics framework. The results showed not only that the obtained average precision of foreign object detection on the CXR was 0.815 but also that the model can be useful in detecting foreign objects on the CXR images. Models of this type may be used as a tool for specialists, in particular, with the growing popularity of radiology comes an increasing workload. We are optimistic that it could accelerate and facilitate the work to provide a faster diagnosis.

Keywords: artificial intelligence; artifacts; chest X-ray; convolutional neural network; foreign body

Citation: Kufel, J.; Bargieł-Łączek, K.; Koźlik, M.; Czogalik, Ł.; Dudek, P.; Magiera, M.; Bartnikowska, W.; Lis, A.; Paszkiewicz, I.; Kocot, S.; et al. Chest X-ray Foreign Objects Detection Using Artificial Intelligence. *J. Clin. Med.* **2023**, *12*, 5841. https://doi.org/10.3390/jcm12185841

Academic Editor: Moritz Wildgruber

Received: 13 August 2023
Revised: 2 September 2023
Accepted: 5 September 2023
Published: 8 September 2023

1. Introduction

Diagnostic imaging has become an essential part of the healthcare system, and its accuracy is crucial for undertaking proper treatment. However, as the number of radiological examinations increases, the workload of specialists increases. Chest X-rays represent 40% of the 3.6 billion imaging procedures performed worldwide every year [1]. In the United States (US), more than 70 million CXR examinations are performed annually [2]. Artificial intelligence (AI) has received considerable attention over the past decade. Deep learning (DL) is a subcategory of AI and is based on deep neural networks (DNNs). The element that differentiates DNNs from artificial neural networks (ANNs) is the use of multiple hidden layers in a network's architecture. This enables the network to understand and mimic more complex behaviors and processes [3]. As an example of the use of DNNs, Ghaderzadeh et al. created a computer-aided detection (CAD) system for COVID-19 on CT scans. The proposed model achieved an accuracy of 99.6% on the test dataset [4]. In other research carried out in another field, Ghaderzadeh et al. developed a CNN-based model able to detect ALL from hematogone cases and assist hematologists and laboratory personnel in diagnosing ALL subtypes. Peripheral blood smear was used as the imaging material in this study. The model based on DenseNet201 accomplished sensitivity, specificity, and accuracy on the level over 99% on the test dataset [5]. Other hematology research conducted by Hosseini et al. established a model and mobile application for screening B-ALL cases from non-B-ALL cases. Accuracy, sensitivity, and specificity of the DNN-based model reached 100% in both test dataset and in laboratory settings [6]. Research by Garavand et al. from the other side of the diagnostic spectrum aimed to review and evaluate machine learning methods in the diagnosis of coronary artery disease. Laboratory and demographic data classification has been used by AI systems such as traditional ML, DL/NN, and ensemble. Of the 54 studies included in the review, an impressive 24 were studies based on the use of DNNs [7].

Motivated by the success of artificial intelligence, especially deep learning in medical image analysis, researchers around the world are working on AI-based tools to help radiologists in the diagnostic process. A big part of this is the error-free nature of artificial intelligence. One of the reasons that make radiological image evaluation difficult is artifacts. These are image fragments that are not anatomical structures caused by hardware defects, including software error, operator error, or foreign bodies (e.g., ICDs [8], ECG leads, jewelry [9], endotracheal tubes, cardiac clips [10], coins [11], or silicone breast implants [12]) inside or outside the patient's body, as well as patient movement during the examination (motion artifacts). A separate radiographic imaging method should be chosen for each foreign body, depending on the material and its depth and location in the body. Metal and glass bodies are easily detectable, unlike wooden ones. In studies using X-rays, ultrasound or CT-scan is recommended if the location is inaccurate or the presence and characteristics of the foreign body are inconclusive [13]. Baram et al. proved that more than 50% of imaging studies with foreign bodies in the trachea and 25% of foreign bodies in the bronchi were not detected on X-rays [14]. Another example is the misinterpretation of ECG electrodes, or a ruptured silicone implant [15], considered to be pulmonary nodules on chest X-rays in AP projection, which was confirmed by a follow-up X-ray in lateral projection [16–18]. Artifacts can also hinder diagnosis by obscuring organs. An example is endotracheal tubes, which, when placed incorrectly, can cause health complications [19], as well as hinder the diagnostic process. This is especially useful in intensive care units where the speed of accurately identifying its placement is often critical for the patient [20]. Neural network-based systems are also now being used to recognize and classify implanted devices such as pacemakers/loop recorders/defibrillators (Implantable cardioverter-defibrillators, ICDs) visible on chest X-rays. This is particularly important in clinical emergencies that require medical personnel to act immediately. The presence of ICD devices on the images can be considered an artifact in the form of an object but can also provide a lot of relevant clinical information [21]. This information can be provided by the patient themself and is also contained in the medical records of the ICD device. If it is not possible to obtain them, specialized personnel can distinguish the ICD model on an X-ray, but the method is not perfect and often poses difficulties [22]. For this reason, artificial intelligence algorithms are eagerly becoming involved in the process, shortening it considerably, with accuracy in identifying the manufacturer, and models reaching 99.6% [8]. Applications for mobile devices are also being developed to further facilitate the process for clinicians [21]. Examples of devices a radiologist may encounter include shoulder joint endoprostheses [23,24], bowel motility markers, gastric banding, abdominal tubes [25], neurostimulator endoprostheses, hemostatic cellulose gauze, vascular bypass [26], intra-aortic balloon pumps [26], ventriculoperitoneal valves [26,27], tracheobronchial stents [26], coronary artery stents [26,27], ventricular assist devices [26–28], endoscopic capsules, endobronchial devices, devices used in prostate brachytherapy, or hernia meshes [25,26]. In the following study, AI was used to determine selected artifacts on X-ray images. The focus was on detection by a neural network of objects such as an ICD, a necklace, a shoulder joint endoprosthesis, and a vascular port. It has been proven that AI using a convolutional neural network (CNN) can recognize pathologies on X-ray images with comparable efficiency to radiologists [29].

2. Materials and Methods

2.1. Radiological Part

For the present study, the NIH Chest X-ray Dataset was used. The dataset contains 112,120 X-ray images of the scans of 30,805 unique individuals [30]. To create these labels, the authors used natural language processing to extract disease classifications via text from relevant radiology reports. The labels are over 90% well-picked and are expected to be suitable for weakly supervised learning. Five researchers using the LabelImg program (Figure 1) marked a total of 9937 foreign objects on chest X-rays, such as vascular port (7648), necklace (1613), ICD (498), and shoulder joint endoprosthesis (178). Markings were made in the form of boxes that covered the area of interest—the foreign objects. The dataset created was validated twice by a specialist radiologist. After the validation process, corrections were also applied twice in order to obtain better quality labels. The corrections were made by people other than those who originally performed the labeling.

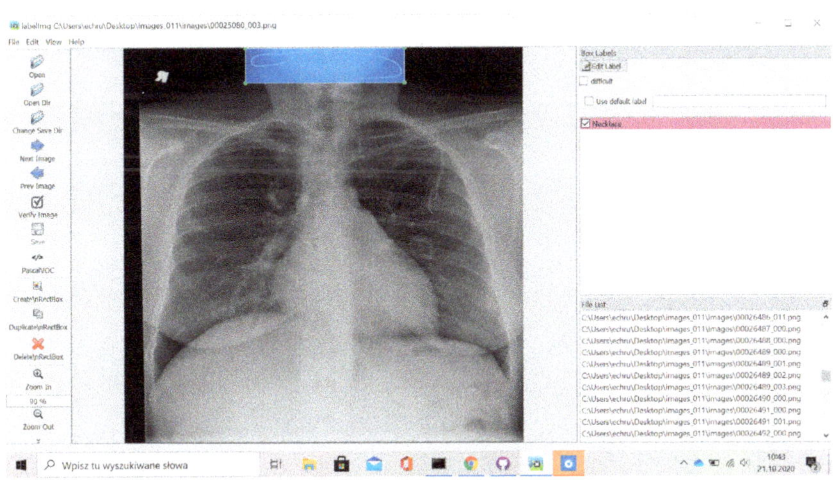

Figure 1. Presentation of the labeling process in the LabelImg program. The image shows a selected necklace (structure in the blue area) with the appropriate color of the label.

2.2. Technical Part

The location of the box on the chest X-ray showing the foreign object was then encoded as coordinates, e.g., [880.0, 435.0, 1004.0, 532.0] (Figure 1). The object detection model was trained using the YOLOv8 architecture and the Ultralytics framework [26]. The dataset used for training and evaluating the models consisted of 8703 images of 1024 $\times$ 1024 pixels with 1 or more foreign objects of the following classes: Endo (shoulder joint endoprosthesis), ICD (implantable cardioverter-defibrillator), Necklace (neck jewelry), and Port (vascular port), which have been divided into training and validation subsets using a 90:10 split. The YOLO v8 architecture is a real-time object detection model. This is the latest version of the popular algorithm for fast detection of objects in images. YOLO v8 was released in 2021 and includes several improvements over previous versions such as better accuracy, faster performance, and increased flexibility. It uses a single convolutional network to predict object boundaries and class probabilities, directly from complete images.

The training was conducted on a single Nvidia A100 GPU using the Google Colab architecture. The models were trained using the Stochastic Gradient Descent (SGD) optimizer with a training rate of 0.01 and a batch size of 64. The training process was stopped after 100 epochs, when the validation loss plateaued. Data augmentation was used to increase the diversity of the training dataset. This included random horizontal flips, random scaling, and random framing. During training, model performance was assessed using the Mean Average Precision (mAP) metric. The final model was selected based on the mAP score on the validation set. The models were implemented in Pytorch and trained using the Ultralytics YOLOv8 library and a nano variant of the model using transfer learning. The model consisted of 225 layers and 3.01 M parameters.

3. Results

For the evaluation of our model, the average precision (AP) for Intersection over Union (IoU) from 0.5 to 0.95 was used. The step size was set as 0.05. The trained object-detection model achieved an impressive AP of 0.815 on the validation set, indicating high object detection accuracy. The model was able to accurately detect and classify a wide range of object classes, including necklaces, vascular ports, ICDs, and shoulder joint replacements. The average AP for the IoU threshold of 50–95% (mAP 50–95) was: Endo—0.859, ICD—0.823, Necklace—0.79, and Port—0.789. The value for all foreign objects was 0.815. The Area under the Precision–Recall curve at 50% IoU threshold: Endo—0.988, ICD—0.995, Necklace—0.971, and Port—0.993. The value for all foreign objects was equal to 0.987 (Figure 2).

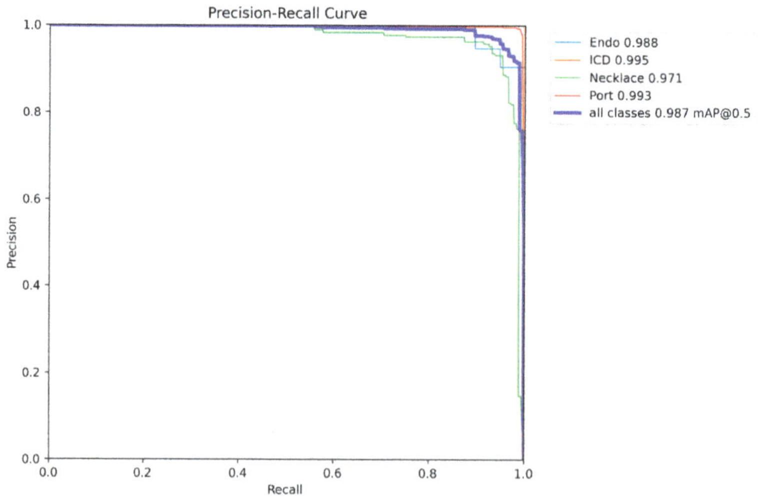

Figure 2. The graph shows the Precision–Recall Curve for each of the foreign objects classes (endo, ICD, necklace, and port) as well as for all the classes. The mean average precision at intersection over union threshold equal to 0.5 was calculated.

A qualitative analysis of the model's predictions showed that the model was able to accurately detect and classify objects even on tests that were particularly difficult to analyze, such as overexposed, shaded, not covering the full examined area or presented incorrectly. Some images have been specially modified so that their quality differs significantly from standard X-rays (Figure 3).

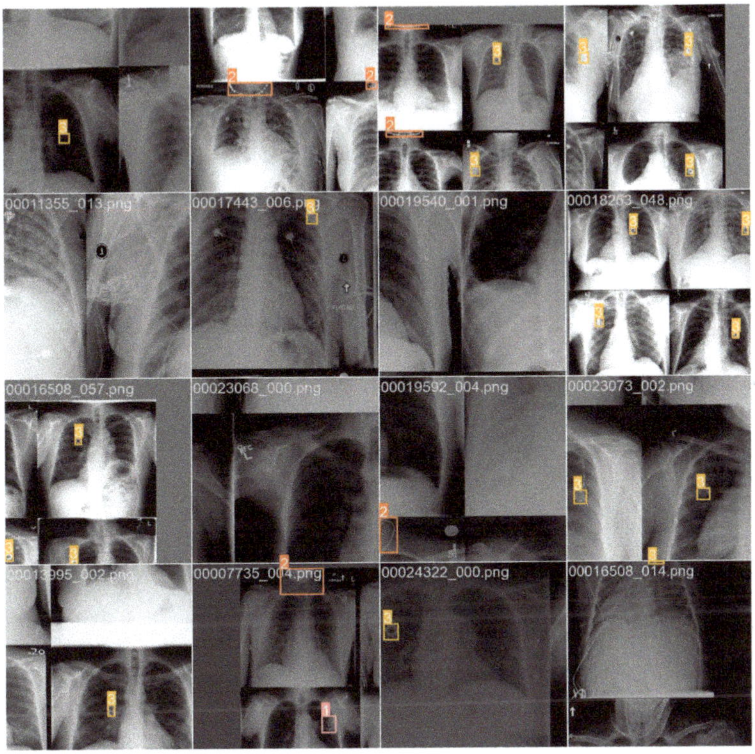

Figure 3. Sample-labeled images used to test the model. Rectangle-shaped and square-shaped areas in yellow, orange, and pink colors are areas with marked foreign objects (bounding boxes).

In addition, the performance of the real-time model was evaluated by measuring the prediction time on the Nvidia A100. The model was able to process 250 images per second. The predictions obtained by the network reflected the actual locations of foreign objects. The comparison of the results of the network and the determinations made by the researchers is shown in (Figure 4).

For verification, images with the neural network's boxes overlaid on chest X-rays were generated to visually assess the correctness of the network (Figure 4). A total of 100 randomly selected chest X-ray images with artifacts were generated and manually checked for the correctness of the type of foreign body and the correctness of the box marking by the AI. All the 100 randomly selected chest X-ray images checked had the type and boxed area correctly applied.

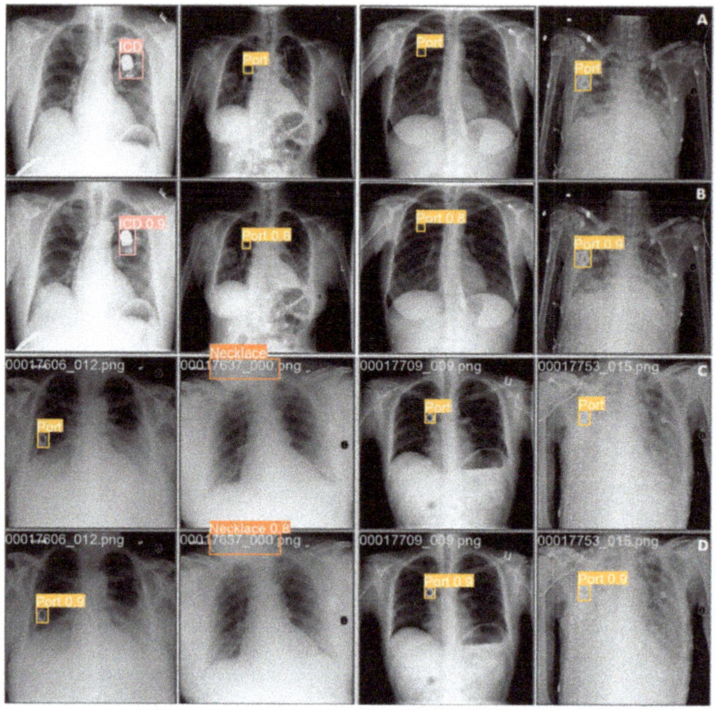

Figure 4. The image shows a comparison of the placement of bounding boxes (rectangular-shaped and square-shaped areas in pink, yellow, and orange colours) labeled by the model and the annotator evaluating the X-ray images. Rows A and C represent images evaluated by the annotator, whereas rows B and D were labeled by the AI.

4. Related Works

The use of a variety of algorithms is gaining increasing application in various fields of medicine, especially radiology. AI-based solutions can be applied, for example, in detecting pathologies or different objects on CXRs. The study from Kara et al. showed high accuracy for both the identification and localization of endotracheal tube tips and carinae on chest X-ray images [20]. Moreover, most common diseases like pneumonia can be detected by an algorithm—Rajpurkar et al. developed an algorithm called CheXNet that can detect pneumonia from chest X-rays at a level exceeding practicing radiologists [29]. There is also a growing interest in creating mobile phone applications that help physicians in everyday work, like in Weinreich et al.'s study on an application for identification of cardiac implanted electrical devices (CIEDs) (e.g., pacemakers and defibrillators) in urgent or emergent settings [21]. Furthermore, as CIEDs become more and more common, it is useful to develop and evaluate a deep learning-based algorithm that performs the detection and characterization of parameters, including MRI safety, of CIEDs on chest X-ray images, like Kim et al. performed [31]. Comparisons of different machine learning algorithms are also available—for example, Chudow et al. compared algorithms for identification of CIEDs using chest radiography: The PacemakerID algorithm, available as a mobile phone application (PIDa) and a web platform (PIDw), and The Pacemaker Identification with Neural Networks (PPMnn), available via web platform [22]. However, AI does not only support diagnostic processes; it can also help when planning and carrying out therapy. Sultan et al. proposed three different deep learning-based frameworks to identify different types of shoulder implants in X-ray scans, mainly an ensemble network called the Inception

Mobile Fully Connected Convolutional Network (IMFC-Net) [23]. Only selected examples are listed here, but they illustrate how broadly AI-based algorithms can be used.

5. Discussion

Over the last decades, many new devices and elements of the instrumentation were developed and have become a permanent part of the diagnostic images described by radiologists. These tools are often implanted in the patient. An example are the ones placed on the patient's body surface in order to perform specific tests—clips, cables, or ECG electrodes [32] or others inside the body, such as endovascular occlusion devices, thrombolysis catheters [27]. The widespread use of X-rays as a diagnostic tool is associated with the equally frequent occurrence of the above-mentioned devices as elements of radiographs.

The model proposed in this study achieved an impressive average precision and high object detection accuracy and was able to classify foreign bodies in four classes. This shows that it can contribute to a decrease in the number of missed foreign bodies on X-ray images. Thus, these detected foreign objects would not be mistaken for pathologies and other artifacts. In the training process, the photos had been specially transformed to imitate images of lower quality and difficult to analyze, such as can be found in everyday clinical practice (for instance bed-side chest X-ray). This clearly proves its usefulness in a clinical environment. Moreover, it was amazingly fast (250 images/s), which would also contribute to clinical practice.

The rapid and efficient detection of foreign bodies is also highly useful for various reasons. The strong magnetic field generated by magnetic resonance imaging (MRI), among other things, poses a danger to the health and life of patients with implanted devices made of metal, such as ICDs and endoprostheses. Performing such an examination on a patient with an implanted metal device can lead to, among other things, abnormal heart rhythms, inhibited pacing, or heating up or even damage to the entire device. Most implanted devices used in cardiology are contraindications for diagnostic tests using magnetic fields. An exception is the modern SureScan devices [33], which are approved for MRI. Hence, the conclusion is that detection of metallic components increases safety when classifying a patient for MRI [34,35] and should be performed with due care by a specialist radiologist before qualifying a patient for MRI. The literature provides studies concerning the use of artificial neural networks to identify different types of foreign bodies on radiographs. Kim et al. proposed a model that detects cardiac implantable electronic devices (CIEDs) and characterizes their parameters. The main aim was to create a tool to help classify patients with such implants for MRI. This is especially important when there are no documentation or problems obtaining device information from the patient, such as when the patient is unconscious [36]. Advances in the refinement and miniaturization of cardiac rhythm management devices have led to the widespread use of safer and less fail-safe implantable leadless electrical devices (LLIEDs) [37]. For most devices of this type, MRI is referred to as "conditional", i.e., the test is safe under certain conditions, as specified by the device manufacturer [38]. However, due to the small size and common factors that interfere with the evaluation of radiographs (such as suboptimal inspection techniques, motion artifacts, or other impermeable objects), the presence of the device can be ignored. The resulting model developed by Kim et al. took less than 7 min to analyze about 8000 images in the training phase, resulting in about 19 CXRs per second [31]. In comparison, the model created for our study analyzes as many as 250 X-ray images per second. A study by Thurston et al. aimed to create a similar tool, which achieved 99.67% accuracy on the test set [39]. Also, models developed by White et al. to support detection and identification of LLIED on CXRs performed prior to scheduled MRI scans achieved accuracies of 94.5% and 95%, respectively. Safety-level categorization was high during both testing (AUC $\geq$ 0.98 and $\geq$0.99, respectively) and trialing (accuracy 98% and 97%, respectively) [36]. Comparatively, our model achieved an accuracy of >99%.

Another type of foreign object for which artificial neural networks have been designed to recognise is shoulder joint endoprostheses. Joint degeneration, recognized by the WHO as a disease of civilization, has contributed to the widespread performance of endoprostheses, which, in Poland, reaches about 90,000 procedures per year, which is associated with the emergence of a large group of patients, with the implant also being an artifact on the X-ray image [40]. Sultan et al. proposed using AI to develop a personalized approach for patients just qualified for shoulder joint prosthesis revision. The standard method usually used in the absence of implant documentation, i.e., the comparison of radiographs of the patient with radiographs of the shoulder prosthesis by a specialist, is time consuming and error prone. The results of the IMFC-Net model they developed were promising—an average accuracy of 89.09%, a precision rate of 89.54%, a recall rate of 86.57%, and an F1 score of 87.94% [23]. However, our model demonstrated the ability to detect shoulder endoprostheses with an accuracy of 98.80%. Another study by Deshpande et al. showed AUC of 0.997 for pacemakers and 0.988 for jewelry (necklaces and earrings). Our results for pacemakers and necklaces were quite similar (0.995 and 0.988, respectively) [41].

The use of foreign body detection models does not end there. It is important to keep in mind the risk of radiologists misinterpreting additional elements present on radiographs and attributing to them a pathological etiology or, on the contrary, blurring the pathological image through the foreign body. In their work, Dick et al. described the case of a patient who had a rupture of one of her breast implants that resulted in migration of silicone, the image of which suggested a nodular lesion in the lung [15]. An entirely different situation has been described by Fitter and Cowie: visible artifacts on radiographs, a consequence of the presence of breast implants, obscuring pathological lesions present in the lung; this is only seen on x-rays after the implant is removed [42].

Research to date has primarily focused on AI's use in identifying objects in X-rays within specific fields. Targeted tools have shown promising results in various sectors. However, the ultimate goal is to develop a comprehensive model capable of detecting and describing multiple foreign elements in a single patient during one examination. Our model takes a significant step towards this by simultaneously detecting four different elements. To improve our neural network's usability, we should use a larger, diverse dataset during training. Including more low-quality radiographs for continuous learning and retraining will enhance the tool's accuracy and reliability. Moreover, adding new foreign objects to the re-trained model opens up the possibility to determine the correctness of the insertion of certain implantable elements like endotracheal tubes or the exact type of endoprosthesis or ICD. Identifying foreign objects on X-rays requires experience, and it is also a time-consuming process. Manual labeling of such a large amount of data can involve the possibility of misidentifications. Most of these are corrected by the verification team. The additional screening can, therefore, reduce the likelihood of worthless data being fed into the network. Another solution is to use an auxiliary network model, verifying the input image. Despite the interest in AI, its direct implication in the medical field is a drawn-out process that is treated with a great deal of distrust. As the technical solutions that are put into clinical use must be of the highest quality, the following tool can serve as an aid to the work of the radiologist, who will still remain responsible for the final diagnosis.

The pursuit of this research article was challenged by several limitations, primarily stemming from the limited availability of CXR featuring foreign bodies. This scarcity precluded the creation of a comprehensive test dataset, although it is noteworthy that this constraint did not adversely impact the ultimate findings of our study. Furthermore, the absence of a sufficient pool of comparable studies in the field posed an additional constraint, restricting our ability to make direct comparisons with other researchers who have explored similar topics.

6. Conclusions

The impressive results obtained by our model suggest its usefulness in the detection of artifacts in X-ray images, both the type of artifact and its location. To further enhance the accuracy and reliability of the discussed tool, it is worth implementing a larger amount of more diverse image data in model training. Nevertheless, the present results suggest the possibility of imminent acceptance of the network in clinical use. Such validation would make it possible to use it as an advisory tool used to signal the occurrence of artifacts without engaging the attention of a medical specialist. The ability to detect objects in low-quality or abnormal images highlights the strength of our network. Automating the artifact detection-process using AI would guarantee the accuracy of object classification, reduce the risk of missing objects and significantly speed up the creation of X-ray descriptions. The time saved on the identification and description of the artifact could be spent on the analysis of other pathologies visible in the image. Further research is necessary to ensure the safe use of the model in clinical practice and its potential to enhance the work of radiology specialists.

Author Contributions: Conceptualization, J.K. and I.P.; Methodology, J.K.; software, S.K.; validation, K.B.-Ł., I.P. and W.B.; formal analysis, K.G.; investigation, Z.N.; resources, M.C.; data curation, Ł.C., P.D. and M.M.; writing—original draft preparation, J.K., I.P., W.B. and A.L.; writing—review and editing, K.B.-Ł., A.L. and M.K.; visualization, Ł.C.; supervision, K.G.; project administration, J.K.; funding acquisition, M.C. and K.G. All authors have read and agreed to the published version of the manuscript.

Funding: We gratefully acknowledge the Polish high-performance computing infrastructure PLGrid (HPC Centers: ACK Cyfronet AGH) for providing computer facilities and support within computational grant no. PLG/2021/015184.

Institutional Review Board Statement: Not applicable.

Informed Consent Statement: Not applicable.

Data Availability Statement: Dataset used: https://www.kaggle.com/datasets/nih-chest-xrays/data, Other data is available from the authors and is made available on request to interested parties.

Conflicts of Interest: The authors declare no conflict of interest.

References

1. Murphy, K. The Global Innovation Index 2019. How Data Will Improve Healthcare Without Adding Staff or Beds. Available online: https://www.wipo.int/edocs/pubdocs/en/wipo_pub_gii_2019-chapter8.pdf (accessed on 12 July 2023).
2. Rogers, M. Routine Admission CXR (RACXR). Core EM. Available online: https://coreem.net/core/routine-admission-cxr-racxr/ (accessed on 13 July 2023).
3. Kufel, J.; Bargieł-Łączek, K.; Kocot, S.; Koźlik, M.; Bartnikowska, W.; Janik, M.; Czogalik, Ł.; Dudek, P.; Magiera, M.; Lis, A.; et al. What Is Machine Learning, Artificial Neural Networks and Deep Learning?—Examples of Practical Applications in Medicine. *Diagnostics* **2023**, *13*, 2582. [CrossRef]
4. Ghaderzadeh, M.; Asadi, F.; Jafari, R.; Bashash, D.; Abolghasemi, H.; Aria, M. Deep Convolutional Neural Network–Based Computer-Aided Detection System for COVID-19 Using Multiple Lung Scans: Design and Implementation Study. *J. Med. Internet Res.* **2021**, *23*, e27468. [CrossRef]
5. Ghaderzadeh, M.; Aria, M.; Hosseini, A.; Asadi, F.; Bashash, D.; Abolghasemi, H. A Fast and Efficient CNN Model for B-ALL Diagnosis and Its Subtypes Classification Using Peripheral Blood Smear Images. *Int. J. Intell. Syst.* **2022**, *37*, 5113–5133. [CrossRef]
6. Hosseini, A.; Eshraghi, M.A.; Taami, T.; Sadeghsalehi, H.; Hoseinzadeh, Z.; Ghaderzadeh, M.; Rafiee, M. A Mobile Application Based on Efficient Lightweight CNN Model for Classification of B-ALL Cancer from Non-Cancerous Cells: A Design and Implementation Study. *Inform. Med. Unlocked* **2023**, *39*, 101244. [CrossRef]
7. Garavand, A.; Behmanesh, A.; Aslani, N.; Sadeghsalehi, H.; Ghaderzadeh, M. Towards Diagnostic Aided Systems in Coronary Artery Disease Detection: A Comprehensive Multiview Survey of the State of the Art. *Int. J. Intell. Syst.* **2023**, *2023*, 6442756. [CrossRef]
8. Howard, J.P.; Fisher, L.; Shun-Shin, M.J.; Keene, D.; Arnold, A.D.; Ahmad, Y.; Cook, C.M.; Moon, J.C.; Manisty, C.H.; Whinnett, Z.I.; et al. Cardiac Rhythm Device Identification Using Neural Networks. *JACC Clin. Electrophysiol.* **2019**, *5*, 576–586. [CrossRef]
9. Jain, R.K.; Vokes, T. Dual-Energy X-ray Absorptiometry. *J. Clin. Densitom.* **2017**, *20*, 291–303. [CrossRef]
10. Morgan, H.; Pettet, G.; Reed, M.; Paul, S.P. Indications for Chest X-rays in Children and How to Obtain and Interpret Them. *Nurs. Child. Young People* **2018**, *30*, 30–37. [CrossRef]
11. Dedhia, K.; Chang, Y.-F.; Leonardis, R.; Chi, D.H. Is There a Need for Repeat Radiologic Examination of Children with Esophageal Coin Foreign Body? *Otolaryngol.–Head Neck Surg.* **2017**, *156*, 173–179. [CrossRef]
12. de Boer, M.; van Middelkoop, M.; Hauptmann, M.; van der Bijl, N.; Bosmans, J.A.W.; Hendriks-Brouwer, N.; Schop, S.J.; de Boer, J.P.; Hijmering, N.J.; Overbeek, L.I.H.; et al. Breast Implant Prevalence in the Dutch Female Population Assessed by Chest Radiographs. *Aesthetic Surg. J.* **2020**, *40*, 156–164. [CrossRef]

13. Javadrashid, R.; Fouladi, D.F.; Golamian, M.; Hajalioghli, P.; Daghighi, M.H.; Shahmorady, Z.; Niknejad, M.T. Visibility of Different Foreign Bodies in the Maxillofacial Region Using Plain Radiography, CT, MRI and Ultrasonography: An In Vitro Study. *Dentomaxillofac. Radiol.* **2015**, *44*, 20140229. [CrossRef] [PubMed]
14. Baram, A.; Sherzad, H.; Saeed, S.; Kakamad, F.H.; Hamawandi, A.M.H. Tracheobronchial Foreign Bodies in Children: The Role of Emergency Rigid Bronchoscopy. *Glob. Pediatr. Health* **2017**, *4*, 2333794X17743663. [CrossRef] [PubMed]
15. Dick, A.C.; Deans, G.T.; Johnston, L.; Spence, R.A. Ruptured Silicone Breast Implant: A Misleading Chest X-ray. *Ulster Med. J.* **1994**, *63*, 238–240. [PubMed]
16. Kim, T.; Messersmith, R.N.; MacMahon, H. Pulmonary Nodule Mimicked by ECG Lead Artifact. *Chest* **1989**, *95*, 237–238. [CrossRef]
17. Gronner, A.T.; Eisenberg, R.L. Pseudonodule of the Lung Caused by Offset DX ECG Electrode. *Am. J. Roentgenol.* **1994**, *162*, 728–729. [CrossRef] [PubMed]
18. Edwards, M.E.; Harris, R.D.; Spiegel, P.K. Offset Electrocardiographic Electrode Simulating a Pulmonary Nodule on Radiographs. *Radiology* **1989**, *172*, 281–282. [CrossRef]
19. Gupta, P.K.; Gupta, K.; Jain, M.; Garg, T. Postprocedural Chest Radiograph: Impact on the Management in Critical Care Unit. *Anesth. Essays Res.* **2014**, *8*, 139. [CrossRef]
20. Kara, S.; Akers, J.Y.; Chang, P.D. Identification and Localization of Endotracheal Tube on Chest Radiographs Using a Cascaded Convolutional Neural Network Approach. *J. Digit. Imaging* **2021**, *34*, 898–904. [CrossRef]
21. Weinreich, M.; Chudow, J.J.; Weinreich, B.; Krumerman, T.; Nag, T.; Rahgozar, K.; Shulman, E.; Fisher, J.; Ferrick, K.J. Development of an Artificially Intelligent Mobile Phone Application to Identify Cardiac Devices on Chest Radiography. *JACC Clin. Electrophysiol.* **2019**, *5*, 1094–1095. [CrossRef]
22. Chudow, J.J.; Jones, D.; Weinreich, M.; Zaremski, L.; Lee, S.; Weinreich, B.; Krumerman, A.; Fisher, J.D.; Ferrick, K.J. A Head-to Head Comparison of Machine Learning Algorithms for Identification of Implanted Cardiac Devices. *Am. J. Cardiol.* **2021**, *144*, 77–82. [CrossRef]
23. Sultan, H.; Owais, M.; Choi, J.; Mahmood, T.; Haider, A.; Ullah, N.; Park, K.R. Artificial Intelligence-Based Solution in Personalized Computer-Aided Arthroscopy of Shoulder Prostheses. *J. Pers. Med.* **2022**, *12*, 109. [CrossRef] [PubMed]
24. Sultan, H.; Owais, M.; Park, C.; Mahmood, T.; Haider, A.; Park, K.R. Artificial Intelligence-Based Recognition of Different Types of Shoulder Implants in X-Ray Scans Based on Dense Residual Ensemble-Network for Personalized Medicine. *J. Pers. Med.* **2021**, *11*, 482. [CrossRef] [PubMed]
25. Jeong, Y.; Lee, J.; Kim, K.A.; Park, C.M. Medical Devices of the Abdomen and Pelvis and Their Complications: A Radiologic Atlas. *Taehan Yongsang Uihakhoe Chi* **2020**, *81*, 863–885. [CrossRef] [PubMed]
26. Mathew, R.P.; Alexander, T.; Patel, V.; Low, G. Chest Radiographs of Cardiac Devices (Part 1): Lines, Tubes, Non-Cardiac Medical Devices and Materials. *S. Afr. J. Radiol.* **2019**, *23*, 1–9. [CrossRef]
27. Sigakis, C.J.G.; Mathai, S.K.; Suby-Long, T.D.; Restauri, N.L.; Ocazionez, D.; Bang, T.J.; Restrepo, C.S.; Sachs, P.B.; Vargas, D. Radiographic Review of Current Therapeutic and Monitoring Devices in the Chest. *RadioGraphics* **2018**, *38*, 1027–1045. [CrossRef]
28. Taljanovic, M.S.; Hunter, T.B.; Freundlich, I.M.; Mar, W.A.; Smyth, S.H.; O'Brien, M.J. Misplaced Devices in the Chest, Abdomen, and Pelvis: Part I. In *Seminars in Ultrasound, CT and MRI*; Saunders: Philadelphia, PA, USA, 2006; Volume 27, pp. 78–97. [CrossRef]
29. Rajpurkar, P.; Irvin, J.; Zhu, K.; Yang, B.; Mehta, H.; Duan, T.; Ding, D.; Bagul, A.; Langlotz, C.; Shpanskaya, K.; et al. CheXNet: Radiologist-Level Pneumonia Detection on Chest X-rays with Deep Learning. *arXiv* **2017**, arXiv:1711.05225. [CrossRef]
30. National Institutes of Health—Clinical Center. CXR8. Available online: https://nihcc.app.box.com/v/ChestXray-NIHCC (accessed on 11 July 2023).
31. Kim, U.-H.; Kim, M.Y.; Park, E.-A.; Lee, W.; Lim, W.-H.; Kim, H.-L.; Oh, S.; Jin, K.N. Deep Learning-Based Algorithm for the Detection and Characterization of MRI Safety of Cardiac Implantable Electronic Devices on Chest Radiographs. *Korean J. Radiol.* **2021**, *22*, 1918–1928. [CrossRef]
32. Graham, L.-J. Chest X-ray—Cardiac Disease—Other Artifacts. Coronary Artery Stent. 12 March 2023. Available online: https://www.radiologymasterclass.co.uk/gallery/chest/cardiac_disease/artifacts (accessed on 22 July 2023).
33. MRI Access Surescan Implantable Cardiac Devices, Medtronic. 2020. Available online: https://www.medtronic.com/us-en/healthcare-professionals/mri-resources/implantable-cardiac-devices.html (accessed on 21 July 2023).
34. Ainslie, M.; Miller, C.; Brown, B.; Schmitt, M. Cardiac MRI of Patients with Implanted Electrical Cardiac Devices. *Heart* **2014**, *100*, 363–369. [CrossRef]
35. Higaki, A.; Kurokawa, T.; Kazatani, T.; Kido, S.; Aono, T.; Matsuda, K.; Tanaka, Y.; Kosaki, T.; Kawamura, G.; Shigematsu, T.; et al. Image Similarity-Based Cardiac Rhythm Device Identification from X-Rays Using Feature Point Matching. *Pacing Clin. Electrophysiol.* **2021**, *44*, 633–640. [CrossRef]
36. White, R.D.; Demirer, M.; Gupta, V.; Sebro, R.A.; Kusumoto, F.M.; Erdal, B.S. Pre-Deployment Assessment of an AI Model to Assist Radiologists in Chest X-Ray Detection and Identification of Lead-Less Implanted Electronic Devices for Pre-MRI Safety Screening: Realized Implementation Needs and Proposed Operational Solutions. *J. Med. Imaging* **2022**, *9*, 054504. [CrossRef]
37. Rav Acha, M.; Soifer, E.; Hasin, T. Cardiac Implantable Electronic Miniaturized and Micro Devices. *Micromachines* **2020**, *11*, 902. [CrossRef] [PubMed]
38. Muthalaly, R.G.; Nerlekar, N.; Ge, Y.; Kwong, R.Y.; Nasis, A. MRI in Patients with Cardiac Implantable Electronic Devices. *Radiology* **2018**, *289*, 281–292. [CrossRef] [PubMed]

39. Thurston, M.D.V.; Kim, D.H.; Wit, H.K. Neural Network Detection of Pacemakers for MRI Safety. *J. Digit. Imaging* **2022**, *35*, 1673–1680. [CrossRef] [PubMed]
40. NIK o Sytuacji Pacjentów Potrzebujących Wszczepienia Stawu Biodrowego Lub Kolanowego—Najwyższa Izba Kontroli. Available online: https://www.nik.gov.pl/aktualnosci/pacjenci-potrzebujacy-wszczepienia-stawu.html (accessed on 29 July 2023).
41. Deshpande, H.; Harder, T.; Saalbach, A.; Sawarkar, A.; Buelow, T. Detection of Foreign Objects In Chest Radiographs Using Deep Learning. In Proceedings of the 2020 IEEE 17th International Symposium on Biomedical Imaging Workshops (ISBI Workshops), Iowa City, IA, USA, 4 April 2020; IEEE: Iowa City, IA, USA, 2020; pp. 1–4.
42. Fitter, A.C.; Cowie, J.C. Minerva. *Br. Med. J.* **1992**, *305*, 842. [CrossRef]

tomography

MDPI

Communication

Generating Synthetic Radiological Images with PySynthMRI: An Open-Source Cross-Platform Tool

Luca Peretti [1,2,3], Graziella Donatelli [2,4], Matteo Cencini [5], Paolo Cecchi [2,6], Guido Buonincontri [1], Mirco Cosottini [6], Michela Tosetti [1,*] and Mauro Costagli [1,7]

1 Laboratory of Medical Physics and Magnetic Resonance, IRCCS Stella Maris, 56128 Pisa, Italy; luca.peretti@fsm.unipi.it (L.P.)
2 Imago 7 Research Foundation, 56128 Pisa, Italy
3 Department of Computer Science, University of Pisa, 56127 Pisa, Italy
4 Department of Diagnostic and Interventional Radiology and Nuclear Medicine, Azienda Ospedaliero-Universitaria Pisana, 56124 Pisa, Italy
5 Italian National Institute of Nuclear Physics (INFN), Section of Pisa, 56127 Pisa, Italy
6 Department of Translational Research and New Technologies in Medicine and Surgery, University of Pisa, 56126 Pisa, Italy
7 Department of Neuroscience, Rehabilitation, Ophthalmology, Genetics, Maternal and Child Sciences (DINOGMI), University of Genoa, 16132 Genoa, Italy
* Correspondence: michela.tosetti@fsm.unipi.it; Tel.: +39-050-886215

Abstract: Synthetic MR Imaging allows for the reconstruction of different image contrasts from a single acquisition, reducing scan times. Commercial products that implement synthetic MRI are used in research. They rely on vendor-specific acquisitions and do not include the possibility of using custom multiparametric imaging techniques. We introduce PySynthMRI, an open-source tool with a user-friendly interface that uses a set of input images to generate synthetic images with diverse radiological contrasts by varying representative parameters of the desired target sequence, including the echo time, repetition time and inversion time(s). PySynthMRI is written in Python 3.6, and it can be executed under Linux, Windows, or MacOS as a python script or an executable. The tool is free and open source and is developed while taking into consideration the possibility of software customization by the end user. PySynthMRI generates synthetic images by calculating the pixelwise signal intensity as a function of a set of input images (e.g., T1 and T2 maps) and simulated scanner parameters chosen by the user via a graphical interface. The distribution provides a set of default synthetic contrasts, including T1w gradient echo, T2w spin echo, FLAIR and Double Inversion Recovery. The synthetic images can be exported in DICOM or NiFTI format. PySynthMRI allows for the fast synthetization of differently weighted MR images based on quantitative maps. Specialists can use the provided signal models to retrospectively generate contrasts and add custom ones. The modular architecture of the tool can be exploited to add new features without impacting the codebase.

Keywords: magnetic resonance imaging; synthetic imaging; software tool

Citation: Peretti, L.; Donatelli, G.; Cencini, M.; Cecchi, P.; Buonincontri, G.; Cosottini, M.; Tosetti, M.; Costagli, M. Generating Synthetic Radiological Images with PySynthMRI: An Open-Source Cross-Platform Tool. *Tomography* **2023**, *9*, 1723–1733. https://doi.org/10.3390/tomography9050137

Academic Editors: Rafał Obuchowicz, Michał Strzelecki and Adam Piorkowski

Received: 17 August 2023
Revised: 7 September 2023
Accepted: 8 September 2023
Published: 11 September 2023

1. Introduction

Magnetic Resonance Imaging (MRI) is a noninvasive modality that allows one to obtain images at a high spatial resolution with a wide range of radiologically meaningful structural soft tissue contrast and functional information. A complete MRI exam typically consists of the acquisition of several sequences, designed to obtain images with different contrasts, which depend on the differences in tissue properties, such as T1 and T2 relaxation times and proton density, and acquisition parameters, such as the times of repetition (TR), echo (TE) and inversion (TI). Each MRI sequence has a typical duration of several minutes and is designed to achieve only one specific tissue contrast. As a consequence, a complete MRI exam involves long acquisition times that can be challenging for children and uncollaborative patients. The recent development of methods for quantitative multiparametric mapping with a single time-efficient acquisition has paved the way for a new imaging approach, known as synthetic MRI, which aims to generate diverse contrast-weighted MR images from a set of tissue parameters [1–4]. This approach has been used and validated for clinical feasibility, and commercial products implementing synthetic MRI, such as SyMRI (SyntheticMR, Linköping, Sweden), are being used in research and clinical exams [5,6]. These products, however, rely on vendor-specific acquisitions and do not include the possibility of using custom implementations of multiparametric imaging techniques, such as MR Fingerprinting [7], MR-STAT [8], QALAS [9] and Quantitative Transient-state Imaging (QTI [10]). To the best of our knowledge, there is currently no open-source software available to perform synthetic MRI.

This article presents PySynthMRI, an open-source tool that uses a set of input images, such as quantitative maps of T1, T2 and proton density (PD), and allows the user to combine them to produce synthetic contrast-weighted images by varying simulated sequence parameters, such as TE, TR and TI(s). Pre-release versions of PySynthMRI have successfully been used to produce synthetic radiological images in the context of clinical research studies, recently presented in abstract form [11,12].

2. Method

The PySynthMRI tool is written in Python 3.6, an open programming language available in all computing platforms. The tool makes use of the open source numpy and open-cv packages, chosen with the aim of maximizing the efficiency in manipulating image matrices and therefore of making the interaction with the graphical interface fluid for the user. The code is organized in a Model-View-Controller (MVC) pattern [13], and its modularity simplifies the tests and makes it possible to add new features without impacting the codebase. PySynthMRI has been developed in continuous collaboration with technicians and clinicians in order to obtain an easily usable interface, shown in Figure 1. PySynthMRI can be executed under Linux, Windows or MacOS as a python script or an executable. The minimum operating system requirements to run the software are Windows 7, Ubuntu 16.04, Red Hat Linux 6.6 or MacOS 10.11, depending on the system used. PySynthMRI can be executed on standard laptops with no special hardware requirements and is guaranteed to run on 2 GHz CPU and 4 GB RAM without a graphic card.

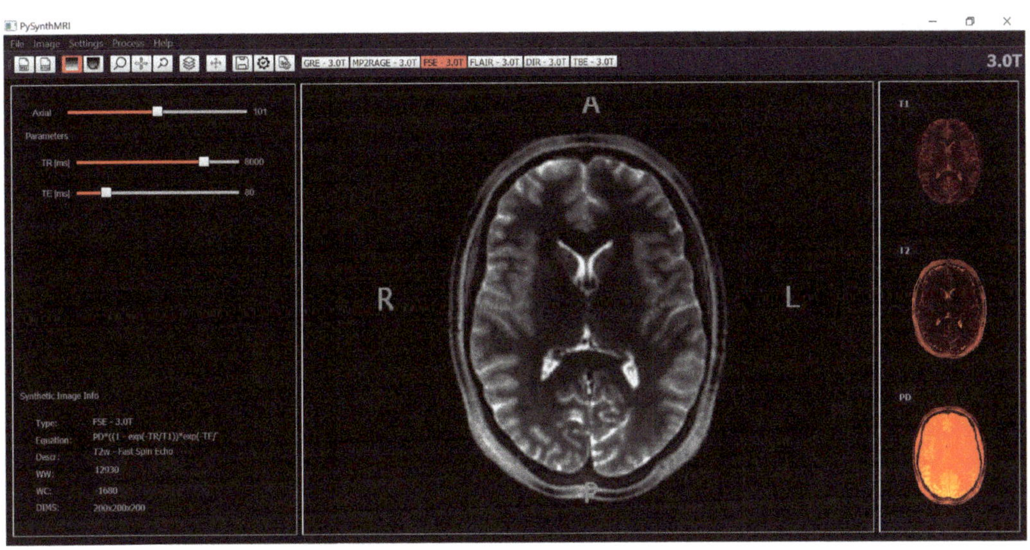

Figure 1. The graphical interface of PySynthMRI. On the left, the provided scanner parameter. In the middle, the synthesized FLAIR image. On the right, the input quantitative maps.

PyInstaller library has been used to bundle the tool and all its dependencies into a single executable package. PySynthMRI software version 1.0.0 with the source code and documentation is released on GitHub under a free GPL software license and can be accessed at: https://github.com/FiRMLAB-Pisa/pySynthMRI.git (accessed on 10 September 2023).

A scheme of its architecture is shown in Figure 2. The specific features of the tool are described below.

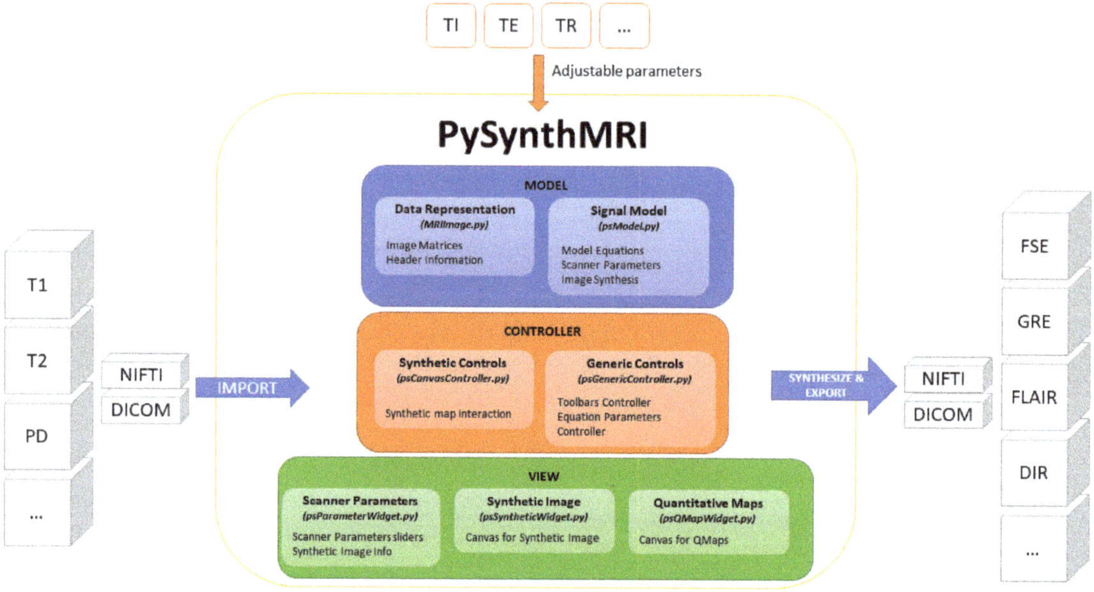

Figure 2. PySynthMRI architecture. The architecture follows the Model View Controller design pattern. PySynthMRI supports both DICOM and NIfTI files.

2.1. Synthetic Image Generation

The core functionality of PySynthMRI implements the generation of contrast images.

The tool interface provides a simple way to modify sequence parameters in order to choose values that generate the requested synthetic image, which is immediately shown to the user using the preferred interpolation for best rendering. The image synthesis is performed by calculating, pixel by pixel, the expected signal intensity as a function of a set of input images (for example, quantitative maps of T1, T2 and proton density) and simulated scanner parameters. The generic signal model that takes into account the effects of inversion recovery, T1 and T2 weighting is:

$$S = abs(PD * (1 - 2 * exp(-TI/T1)) * (1 - exp(-TR/T1)) * exp(-TE/T2)) \quad (1)$$

PySynthMRI is distributed with a pre-defined set of sample presets for the generation of synthetic T1-weighted gradient-recalled echo (GRE), bias-free T1-weighted MPRAGE [14]/MP2RAGE [15], T2-weighted spin echo (SE), FLuid Attenuated Inversion Recovery (FLAIR) [16], Tissue Border Enhancement [17] and Double Inversion Recovery (DIR) [18]. The sample presets, as well as a sample input dataset consisting of quantitative maps of the T1, T2 and proton density of a healthy subject, are meant as a starting point to help the user get acquainted with the software. However, it is worth highlighting that PySynthMRI is designed to enable the full customization of the signal equations, which can be easily expanded to include additional input data, such as quantitative maps of T2* and magnetic susceptibility.

2.2. Data Formats

PySynthMRI provides complete support for the Digital COmmunications in Medicine standard (DICOM) and Neuroimaging Informatics Technology Initiative standard (NIfTI). Both standards can be used for importing the input data and exporting the synthetic images. Although NIfTI images include an extended header to store, amongst others, DICOM tags and attributes, they usually do not contain all tags requested by DICOM viewers [19]. To overcome this issue, PySynthMRI generates DICOM images using a template containing all necessary tags and automatically updates modified parameters, such as the TR, TE, TI, window scale, and series description. The Nibabel python library was used to handle both DICOM and NIfTI formats.

To achieve the proposed data-agnostic model, PySynthMRI maps the file format internally using an ad hoc Python data loader. The design choice allows developers to easily add new data formats if necessary.

2.3. Configuration File

Signal models can be easily defined using a configuration file formatted as a JavaScript Object Notation (JSON) open standard format. The JSON format supports all data types needed by PySynthMRI, and it provides human readability, which is a key feature for easy access. An example of a configuration file is shown in Figure 3. Signal models of synthetic images are grouped in presets in order to facilitate the use of the appropriate set of signal models and parameters depending on the input dataset (for example, to load appropriate simulated scanning parameters for MR scanners operating at different fields). Each signal model is provided as a JSON object consisting of an equation that represents the actual model and the list of scanner parameters, along with their default values. The Configuration File contains the list of quantitative maps that are accepted by PySynthMRI, and it can be adjusted as desired and modified while using the tool: the updates are immediately applied. A validator checks for errors and inconsistencies each time the file is loaded, helping the user with the correct compilation.

2.4. Graphical Interface Interaction

The interface of PySynthMRI is shown in Figure 1. The user can load the desired number of input images, which can be quantitative maps of T1, T2 and PD, as well as other types of maps (including user-defined binary masks). The files can be uploaded so as to provide a path or using the drag-n-drop feature.

The user can immediately evaluate the synthesized images resulting from simulated scanner parameters, which can be modified using sliders and/or mouse/keyboard interactions. The provided image viewer features include zoom, span and window greyscale regulation, as well as slicing and reformatting in the three orthogonal planes. Custom signal models can be added at runtime or using the Configuration File in order to obtain new synthetic contrasts.

2.5. Batch Synthesization

Although PySynthMRI is mainly intended for a direct graphical usage, one secondary feature is the execution of batch synthesization. The batch synthesization consists of automatically generating multiple contrast images from the provided input data of different subjects. The user selects the contrasts to be generated and provides the path containing the input data of different subjects. The tool sequentially synthesizes all the images without the need to export them individually. Batch synthesization drastically reduces the time required to apply synthetic MRI models to multiple acquisitions once the simulated acquisition parameters have been determined.

```
{
    "synthetic_images": {
        "3.0T": {
            "FSE": {···
            },
            "GRE": {···
            },
            "FLAIR": {
                "title": "T2w - Fluid Attenuated Inversion Recovery",
                "equation": "abs(PD*exp(-TSAT/T1)*exp(-TE/T2)*(1-2*exp(-TI/T1)))",
                "series_number": 9104,
                "parameters": {
                    "TE": {
                        "label": "TE [ms]",
                        "value": 80,
                        "min": 1,
                        "max": 100,
                        "step": 1
                    },
                    "TI": {
                        "label": "TI [ms]",
                        "value": 2075,
                        "min": 1000,
                        "max": 3500,
                        "step": 5
                    },
                    "TSAT": {
                        "label": "TSAT [ms]",
                        "value": 1405,
                        "min": 400,
                        "max": 10000,
                        "step": 5
                    }
                },
                "window_center": 1952,
                "window_width": 4876
            },
            "MP2RAGE": {···
            }
        },
        "7.0T": {···
        }
    },

    "quantitative_maps": {···
    },
    "image_interpolation": {···
    }
}
```

Figure 3. Example configuration file of PysynthMRI.

2.6. Testing

PySynthMRI has been tested by using input images representing maps of T1, T2 and PD obtained with different methods [10,20,21] on different scanners operating at 1.5T, 3T and 7T.

For demonstration purposes, a neuroradiologist with little programming skills and nine years of research experience used PySynthMRI to generate six types of synthetic images (T1w GRE and MP2RAGE [15]; T2w FSE and FLAIR [16]; TBE [17]; DIR [18]) from a set of input images consisting of maps of T1, T2 and PD obtained on a 3T scanner (MR750, GE Healthcare, Chicago, IL, USA). The input dataset was obtained using a 3D

QTI sequence [10,20], consisting of an inversion-prepared steady-state free precession acquisition with spiral k-space sampling achieving the simultaneous quantification of PD, T1 and T2 of the whole head at a 1.1 mm isotropic resolution in 7 min. The user empirically edited the sample equations and the simulated scanning parameters to optimize the desired synthetic contrasts.

3. Results and Discussion

Figure 4 displays six types of synthetic images (T1w GRE and MP2RAGE; T2w FSE and FLAIR; DIR; TBE) obtained with PySynthMRI using maps of T1, T2 and PD of one subject (18-year-old male) as input. To obtain these images, the user empirically edited the sample equations and set the simulated scanning parameters.

- Synthetic T1-weighted images simulating GRE (Gradient-recalled Echo) were synthesized with the following equation, incorporating the input maps of PD and T1 and setting the simulated acquisition parameter TR:

$$\begin{aligned} S_{GRE} &= PD * (1 - exp(-TR/T1)) \\ \text{TR} &= 113 \text{ ms.} \end{aligned} \tag{2}$$

- Bias-free T1-weighted MPRAGE was synthesized by taking into consideration that B1 bias in QTI is incorporated in the PD map, which can be excluded from the formula, achieving a similar appearance to MP2RAGE uniform imaging [15]:

$$\begin{aligned} S_{MP2RAGE} &= abs(1 - 2 * exp(-TI/T1)) \\ \text{TI} &= 1816 \text{ ms} \end{aligned} \tag{3}$$

- Synthetic T2-weighted images simulating a SE (Spin Echo) acquisition were obtained as follows:

$$\begin{aligned} S_{SE} &= PD * (1 - exp(-TR/T1)) * exp(-TE/T2) \\ \text{TR} &= 8000 \text{ ms; TE} = 80 \text{ ms} \end{aligned} \tag{4}$$

- Synthetic T2-FLAIR was obtained with a user-modified formula incorporating a coefficient TSAT, which enables the introduction of T1-weighting, which better mimics the one in conventional imaging:

$$\begin{aligned} S_{FLAIR} = abs(PD * exp(-TE/T2) * exp(-TSAT/T1) \\ *(1 - 2 * exp(-TI/T1))) \\ \text{TE} = 80 \text{ ms; TSAT} = 1405 \text{ ms; TI} = 2075 \text{ ms.} \end{aligned} \tag{5}$$

- Synthetic TBE acquisitions were obtained from the generic signal model, with the appropriate parameters TI, TR, TE:

$$\begin{aligned} S_{TBE} = abs(PD * (1 - 2 * exp(-TI/T1)) * (1 - exp(-TR/T1)) * exp(-TE/T2)) \\ \text{TR} = 5020 \text{ ms; TE} = 1 \text{ ms; TI} = 795 \text{ ms} \end{aligned} \tag{6}$$

- Double Inversion Recovery (DIR [18,22]) was obtained by including the two times of the inversion TI1 and TI2 in the signal equation, as follows:

$$\begin{aligned} S_{DIR} = abs(PD * (1 - 2 * exp(-TI2/T1) + 2 \\ *exp(-(TI1 + TI2)/T1) \\ -exp(-TR/T1)) * exp(-TE/T2)) \\ \text{TR} = 6670 \text{ ms; TE} = 80 \text{ ms; TI1} = 2208 \text{ ms; TI2} = 545 \text{ ms} \end{aligned} \tag{7}$$

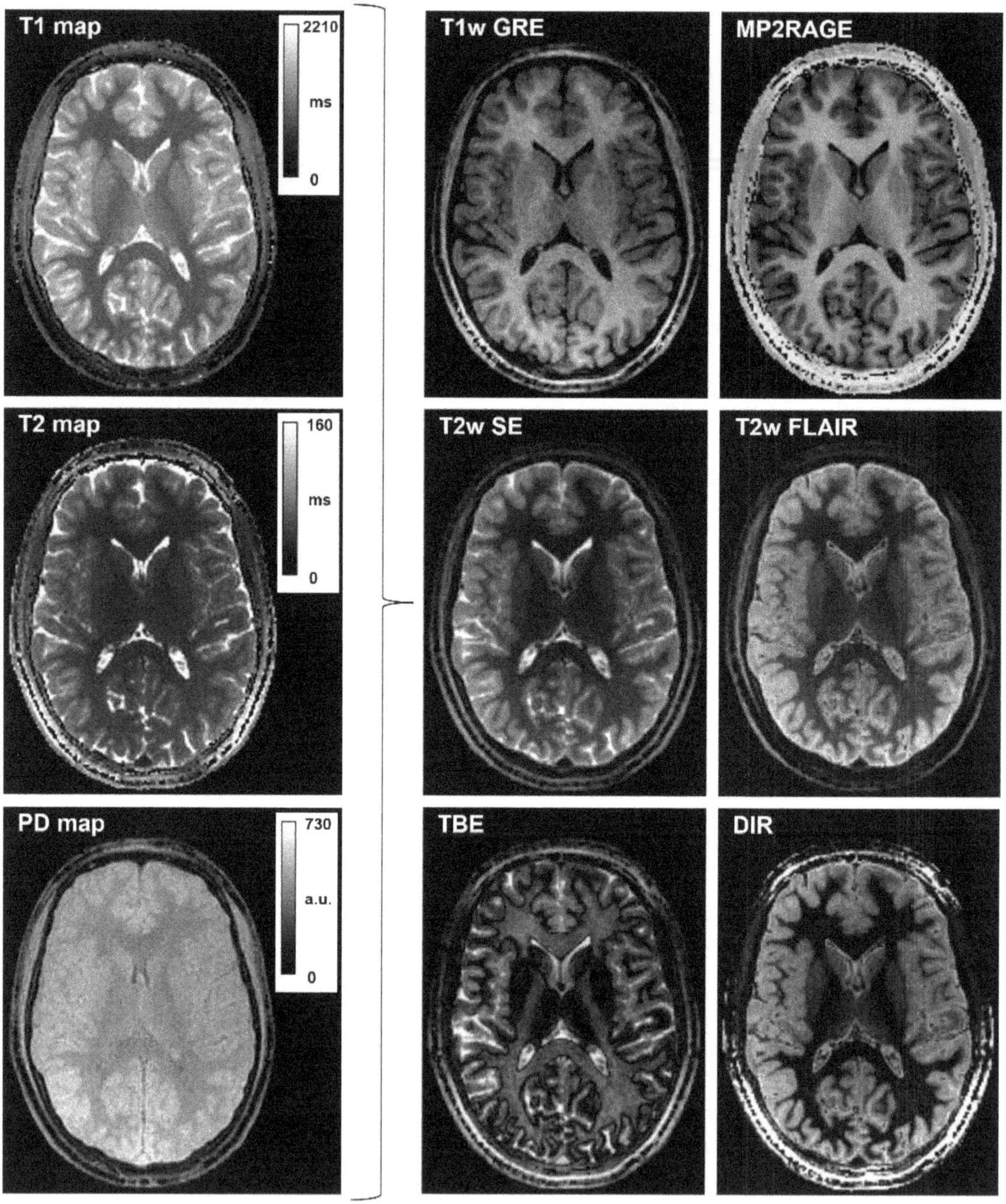

Figure 4. Synthetic T1-weighted and T2-weighted contrasts generated with PySynthMRI. On the left, the input quantitative maps; on the right, the synthesized images.

The user was able to obtain the desired synthetic images without any programming skills.

For a qualitative comparison with images obtained with conventional acquisitions, Figure 5 displays three synthetic image contrasts generated with PySynthMRI (T1-w, T2w and T2-FLAIR) using the quantitative maps obtained with one single QTI acquisition (total acquisition time: 7 min) from a 28-year-old female, compared with the images obtained with conventional 3D FSPGR (TR = 8.1 ms; TE = 3.2 ms; TI = 450 ms; spatial resolution = 1 × 1 × 1 mm^3; scan duration = 4′11″), 2D FSE (TR = 3636 ms; effective TE = 113.3 ms; resolution = 0.57 × 0.57 × 4 mm^3; scan duration = 2′11″) and 3D T2-FLAIR (TR = 7000 ms; effective TE = 113.3 ms; TI = 1945 ms; resolution = 1.16 × 1.16 × 1.2 mm^3; scan duration = 5′25″).

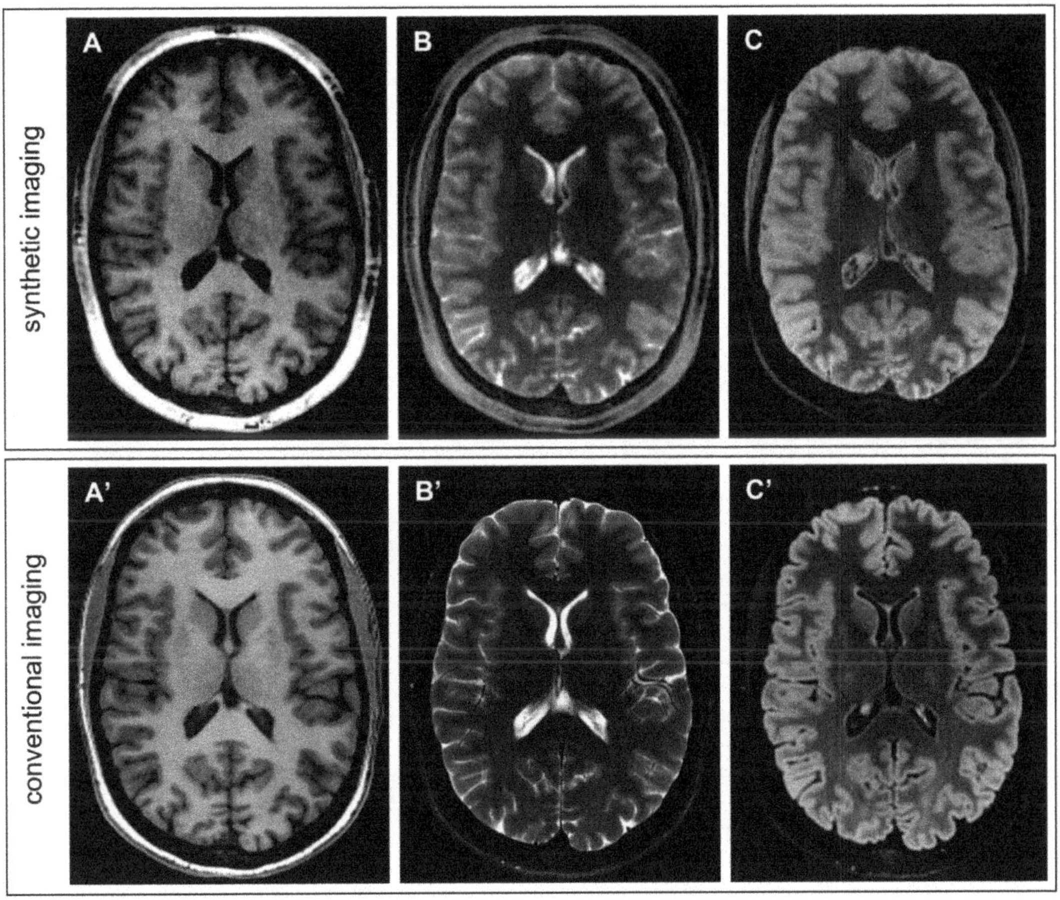

Figure 5. Direct qualitative comparison between synthetic images obtained with PySynthMRI from QTI-derived maps (**top row**) and conventional T1-weighted (**A**,**A′**), T2-weighted (**B**,**B′**) and T2-FLAIR (**C**,**C′**) images of conventional acquisitions (**bottom row**).

The storage size is exactly the same for synthetic and conventional images with the same resolution and coverage.

Synthetic FLAIR image (Figure 5C) fails in fluid suppression and artifacts are visible in the basal ganglia. Those artefacts are not related to PySynthMRI but are instead related to the input maps that were used for testing the software. QTI acquisition is known to provide erroneous relaxation estimates in moving tissue, that is, in particular, in the ventricular cerebrospinal fluid. As previously reported in the literature, the FLAIR artefacts are also caused by partial volume effects at the interface between periventricular tissue and cerebrospinal fluid [23]. One possible way of mitigating the problem of partial voluming is to leverage the software so as to include an additional input map in the signal model (e.g., a water fraction map) that can be produced from the same QTI acquisition used for T1 and T2 mapping. Furthermore, the synthesized FLAIR can be improved using a deep-learning method based on the Generative Adversarial Network [24].

The synthetic images were saved in DICOM and NIfTI formats, suitable for radiological evaluation and storage on conventional PACS as well as for further image processing/analysis using external software. The same presets were applied to a batch of 210 different patients (including patients with neurodegenerative diseases, neurodevelopmental disorders, epilepsy, and brain tumors) in order to provide six radiological contrasts for each patient, resulting in the generation of 210 × 6 = 1260 3D volume images, which were saved in the DICOM format. The processing time was 1.1 s per 3D synthetic image on a Windows machine equipped with an i5-8500 CPU @ 3.0 GHz, 6 cores and 16 GB RAM.

Despite the fact that we tested PySynthMRI on the provided signal models, it is possible to include further input maps, such as B1 field and/or noise models, which may, on the one hand, have limited use in the generation of synthetic images for radiological inspection but may, on the other hand, have a potential use when the software is used for educational purposes.

The development and dissemination of time-efficient quantitative MRI techniques motivate the use of PySynthMRI in clinical research. It allows specialists to easily synthesize, at post-processing time, multiple contrasts for a particular subject. This function can be useful for refining the desired contrasts, as optimal parameters depend on the patient age, region of interest and/or pathology. The synthesization process can be useful in cases where it is desirable to generate contrasts not originally considered at the time of acquisition.

PySynthMRI also shows a high potential in education, as it can be used to show how sequence parameters impact the generation of images with different tissue contrasts.

PySynthMRI complements other types of software whose main purpose is the accurate simulation of the MR signal. In this respect, it is worth referring to simulation tools such as SIMRI [25], JEMRIS [26], MRiLab [27] and others [28–30]. These simulators are primarily intended as tools for MRI physicists to support the development of new sequences and the optimization of acquisition parameters. On the contrary, the primary function of PySynthMRI is that of an interactive viewer tailored for radiologists in a clinical research context, and it is aimed at evaluating and producing synthetic contrasts in real time. Its computation time for one entire 3D volume is in the order of 1 s, while for signal simulators, it is in the order of minutes or hours. The MRI community may take advantage of the future development of tools in which both of these complementary aspects (that is, an accurate MRI signal simulation and the production of synthetic images for radiological purposes in a clinical context) are implemented.

The modular architecture allows developers and contributors to easily extend and customize the tool to incorporate additional parameter maps, such as T2* or MT, or different predefined contrast images. Future releases of PySynthMRI will aim to incorporate further capabilities, including an automatic tissue segmentation software module capable of computing and showing probabilistic maps of white matter (WM), gray matter (GM) and cerebrospinal fluid (CSF) using a pre-built lookup table representing the relationship between partial volume tissue composition and provided quantitative maps [31].

Author Contributions: Conceptualization, L.P. and M.C. (Mauro Costagli); Data curation, L.P., P.C. and M.C. (Mirco Cosottini); Funding acquisition, G.B. and M.C. (Mauro Costagli); Investigation, L.P., G.D., P.C. and M.C. (Mauro Costagli); Methodology, M.C. (Matteo Cencini) and M.C. (Mauro Costagli); Project administration, M.T. and M.C. (Mauro Costagli); Resources, M.C. (Mirco Cosottini) and M.T.; Software, L.P., M.C. (Matteo Cencini) and G.B.; Supervision, M.T. and M.C. (Mauro Costagli); Validation, G.D. and M.C. (Mauro Costagli); Visualization, L.P. and G.D.; Writing—original draft, L.P., M.C. (Matteo Cencini) and M.C. (Mauro Costagli); Writing—review & editing, G.D., M.C. (Matteo Cencini), P.C., G.B., M.C. (Mirco Cosottini), M.T. and M.C. (Mauro Costagli). All authors have read and agreed to the published version of the manuscript.

Funding: This study is part of a research project funded by the Italian Ministry of Health (grant n. GR-2016-02361693). This study was also partially supported by grant RC and the 5 × 1000 voluntary contributions to IRCCS Fondazione Stella Maris, funded by the Italian Ministry of Health.

Institutional Review Board Statement: The study was conducted in accordance with the Declaration of Helsinki, and approved by the Ethics Committee "Comitato Etico Regionale per la Sperimentazione Clinica della Regione Toscana" (protocol code 50/2018, 16 April 2018 and protocol code 9351, 18 February 2019).

Informed Consent Statement: Informed consent was obtained from all subjects involved in the study.

Data Availability Statement: Software and sample data are available at https://github.com/FiRMLAB-Pisa/pySynthMRI.git (access date: 10 September 2023).

Conflicts of Interest: Matteo Cencini and Michela Tosetti received a research grant from GE Healthcare. The other authors (Luca Peretti, Graziella Donatelli, Paolo Cecchi, Guido Buonincontri, Mirco Cosottini, Mauro Costagli) declare no conflict of interest.

References

1. Hagiwara, A.; Hori, M.; Yokoyama, K.; Takemura, M.Y.; Andica, C.; Tabata, T.; Kamagata, K.; Suzuki, M.; Kumamaru, K.K.; Nakazawa, M.; et al. Synthetic MRI in the Detection of Multiple Sclerosis Plaques. *Am. J. Neuroradiol.* **2017**, *38*, 257–263. [CrossRef] [PubMed]
2. Andica, C.; Hagiwara, A.; Hori, M.; Kamagata, K.; Koshino, S.; Maekawa, T.; Suzuki, M.; Fujiwara, H.; Ikeno, M.; Shimizu, T.; et al. Review of synthetic MRI in pediatric brains: Basic principle of MR quantification, its features, clinical applications, and limitations. *J. Neuroradiol.* **2019**, *46*, 268–275. [CrossRef] [PubMed]
3. Ji, S.; Yang, D.; Lee, J.; Choi, S.H.; Kim, H.; Kang, K.M. Synthetic MRI: Technologies and Applications in Neuroradiology. *J. Magn. Reson. Imaging* **2022**, *55*, 1013–1025. [CrossRef] [PubMed]
4. Blystad, I.; Warntjes, J.B.M.; Smedby, O.; Landtblom, A.-M.; Lundberg, P.; Larsson, E.-M. Synthetic Mri of the Brain in a Clinical Setting. *Acta Radiol.* **2012**, *53*, 1158–1163. [CrossRef]
5. Tanenbaum, L.N.; Tsiouris, A.J.; Johnson, A.N.; Naidich, T.P.; DeLano, M.C.; Melhem, E.R.; Quarterman, P.; Parameswaran, S.X.; Shankaranarayanan, A.; Goyen, M.; et al. Synthetic MRI for Clinical Neuroimaging: Results of the Magnetic Resonance Image Compilation (MAGiC) Prospective, Multicenter, Multireader Trial. *Am. J. Neuroradiol.* **2017**, *38*, 1103–1110. [CrossRef]
6. Ryu, K.H.; Baek, H.J.; Moon, J.I.; Choi, B.H.; Park, S.E.; Ha, J.Y.; Jeon, K.N.; Bae, K.; Choi, D.S.; Cho, S.B.; et al. Initial clinical experience of synthetic MRI as a routine neuroimaging protocol in daily practice: A single-center study. *J. Neuroradiol.* **2020**, *47*, 151–160. [CrossRef]
7. Ma, D.; Gulani, V.; Seiberlich, N.; Liu, K.; Sunshine, J.L.; Duerk, J.L.; Griswold, M.A. Magnetic resonance fingerprinting. *Nature* **2013**, *495*, 187–192. [CrossRef]
8. Sbrizzi, A.; van der Heide, O.; Cloos, M.; van der Toorn, A.; Hoogduin, H.; Luijten, P.R.; Berg, C.A.T. van den Fast quantitative MRI as a nonlinear tomography problem. *Magn. Reson. Imaging* **2018**, *46*, 56–63. [CrossRef]
9. Kvernby, S.; Warntjes, M.J.B.; Haraldsson, H.; Carlhäll, C.-J.; Engvall, J.; Ebbers, T. Simultaneous three-dimensional myocardial T1 and T2 mapping in one breath hold with 3D-QALAS. *J. Cardiovasc. Magn. Reson.* **2014**, *16*, 102. [CrossRef]
10. Gómez, P.A.; Cencini, M.; Golbabaee, M.; Schulte, R.F.; Pirkl, C.; Horvath, I.; Fallo, G.; Peretti, L.; Tosetti, M.; Menze, B.H.; et al. Rapid three-dimensional multiparametric MRI with quantitative transient-state imaging. *Sci. Rep.* **2020**, *10*, 13769. [CrossRef]
11. Donatelli, G.; Migaleddu, G.; Cencini, M.; Cecchi, P.; Peretti, L.; D'Amelio, C.; Buonincontri, G.; Tosetti, M.; Cosottini, M.; Costagli, M. Postcontrast 3D MRF-derived synthetic T1-weighted images capture pathological contrast enhancement in brain diseases. In Proceedings of the XIII Annual Meeting of the Italian Association of Magnetic Resonance in Medicine (AIRMM), Pisa, Italy, 23–25 November 2022.
12. Donatelli, G.; Migaleddu, G.; Cencini, M.; Cecchi, P.; Peretti, L.; D'Amelio, C.; Buonincontri, G.; Tosetti, M.; Cosottini, M.; Costagli, M. Pathological contrast enhancement in different brain diseases in synthetic T1-weigthed images derived from 3D quantitative transient-state imaging (QTI). In Proceedings of the Annual Meeting of the International Society of Magnetic Resonance in Medicine (ISMRM), Toronto, ON, Canda, 3–8 June 2023.
13. Krasner, G.E.; Pope, S.T. A description of the model-view-controller user interface paradigm in the smalltalk-80 system. *J. Object Oriented Program.* **1988**, *1*, 26–49.
14. Mugler, J.P.; Brookeman, J.R. Three-dimensional magnetization-prepared rapid gradient-echo imaging (3D MP RAGE). *Magn. Reson. Med.* **1990**, *15*, 152–157. [CrossRef] [PubMed]
15. Marques, J.P.; Kober, T.; Krueger, G.; van der Zwaag, W.; Van de Moortele, P.-F.; Gruetter, R. MP2RAGE, a self bias-field corrected sequence for improved segmentation and T1-mapping at high field. *NeuroImage* **2010**, *49*, 1271–1281. [CrossRef] [PubMed]
16. Hajnal, J.V.; Bryant, D.J.; Kasuboski, L.; Pattany, P.M.; Coene, B.D.; Lewis, P.D.; Pennock, J.M.; Oatridge, A.; Young, I.R.; Bydder, G.M. Use of Fluid Attenuated Inversion Recovery (FLAIR) Pulse Sequences in MRI of the Brain. *J. Comput. Assist. Tomogr.* **1992**, *16*, 841–844. [CrossRef]

17. Costagli, M.; Kelley, D.A.C.; Symms, M.R.; Biagi, L.; Stara, R.; Maggioni, E.; Tiberi, G.; Barba, C.; Guerrini, R.; Cosottini, M.; et al. Tissue Border Enhancement by inversion recovery MRI at 7.0 Tesla. *Neuroradiology* **2014**, *56*, 517–523. [CrossRef]
18. Saranathan, M.; Worters, P.W.; Rettmann, D.W.; Winegar, B.; Becker, J. Physics for clinicians: Fluid-attenuated inversion recovery (FLAIR) and double inversion recovery (DIR) Imaging: FLAIR and DIR Imaging. *J. Magn. Reson. Imaging* **2017**, *46*, 1590–1600. [CrossRef] [PubMed]
19. Mildenberger, P.; Eichelberg, M.; Martin, E. Introduction to the DICOM standard. *Eur. Radiol.* **2002**, *12*, 920–927. [CrossRef] [PubMed]
20. Kurzawski, J.W.; Cencini, M.; Peretti, L.; Gómez, P.A.; Schulte, R.F.; Donatelli, G.; Cosottini, M.; Cecchi, P.; Costagli, M.; Retico, A.; et al. Retrospective rigid motion correction of three-dimensional magnetic resonance fingerprinting of the human brain. *Magn. Reson. Med.* **2020**, *84*, 2606–2615. [CrossRef]
21. Pirkl, C.M.; Cencini, M.; Kurzawski, J.W.; Waldmannstetter, D.; Li, H.; Sekuboyina, A.; Endt, S.; Peretti, L.; Donatelli, G.; Pasquariello, R.; et al. Learning residual motion correction for fast and robust 3D multiparametric MRI. *Med. Image Anal.* **2022**, *77*, 102387. [CrossRef]
22. Redpath, T.W.; Smith, F.W. Use of a double inversion recovery pulse sequence to image selectively grey or white brain matter. *Br. J. Radiol.* **1994**, *67*, 1258–1263. [CrossRef] [PubMed]
23. Hagiwara, A.; Warntjes, M.; Hori, M.; Andica, C.; Nakazawa, M.; Kumamaru, K.K.; Abe, O.; Aoki, S. SyMRI of the Brain: Rapid Quantification of Relaxation Rates and Proton Density, With Synthetic MRI, Automatic Brain Segmentation, and Myelin Measurement. *Invest. Radiol.* **2017**, *52*, 647–657. [CrossRef]
24. Hagiwara, A.; Otsuka, Y.; Hori, M.; Tachibana, Y.; Yokoyama, K.; Fujita, S.; Andica, C.; Kamagata, K.; Irie, R.; Koshino, S.; et al. Improving the Quality of Synthetic FLAIR Images with Deep Learning Using a Conditional Generative Adversarial Network for Pixel-by-Pixel Image Translation. *Am. J. Neuroradiol.* **2019**, *40*, 224–230. [CrossRef]
25. Benoit-Cattin, H.; Collewet, G.; Belaroussi, B.; Saint-Jalmes, H.; Odet, C. The SIMRI project: A versatile and interactive MRI simulator. *J. Magn. Reson.* **2005**, *173*, 97–115. [CrossRef] [PubMed]
26. Stöcker, T.; Vahedipour, K.; Pflugfelder, D.; Shah, N.J. High-performance computing MRI simulations. *Magn. Reson. Med.* **2010**, *64*, 186–193. [CrossRef] [PubMed]
27. Liu, F.; Velikina, J.V.; Block, W.F.; Kijowski, R.; Samsonov, A.A. Fast Realistic MRI Simulations Based on Generalized Multi-Pool Exchange Tissue Model. *IEEE Trans. Med. Imaging* **2017**, *36*, 527–537. [CrossRef] [PubMed]
28. Kwan, R.K.-S.; Evans, A.C.; Pike, G.B. MRI simulation-based evaluation of image-processing and classification methods. *IEEE Trans. Med. Imaging* **1999**, *18*, 1085–1097. [CrossRef]
29. Yoder, D.A.; Zhao, Y.; Paschal, C.B.; Fitzpatrick, J.M. MRI simulator with object-specific field map calculations. *Magn. Reson. Imaging* **2004**, *22*, 315–328. [CrossRef]
30. Klepaczko, A.; Szczypiński, P.; Dwojakowski, G.; Strzelecki, M.; Materka, A. Computer Simulation of Magnetic Resonance Angiography Imaging: Model Description and Validation. *PLoS ONE* **2014**, *9*, e93689. [CrossRef]
31. West, J.; Warntjes, J.B.M.; Lundberg, P. Novel whole brain segmentation and volume estimation using quantitative MRI. *Eur. Radiol.* **2012**, *22*, 998–1007. [CrossRef]

tomography

MDPI

Article

Retrospective Motion Artifact Reduction by Spatial Scaling of Liver Diffusion-Weighted Images

Johannes Raspe [1,2,*], Felix N. Harder [1], Selina Rupp [1], Sean McTavish [1], Johannes M. Peeters [3], Kilian Weiss [4], Marcus R. Makowski [1], Rickmer F. Braren [1], Dimitrios C. Karampinos [1] and Anh T. Van [1]

1 School of Medicine and Health, Technical University of Munich, 81675 Munich, Germany; dimitrios.karampinos@tum.de (D.C.K.); anh.van@tum.de (A.T.V.)
2 School of Natural Sciences, Technical University of Munich, 85748 Garching, Germany
3 Philips Healthcare, 5684 PC Best, The Netherlands
4 Philips GmbH Market DACH, 22335 Hamburg, Germany
* Correspondence: j.raspe@tum.de; Tel.: +49-89-4140-6974

Abstract: Cardiac motion causes unpredictable signal loss in respiratory-triggered diffusion-weighted magnetic resonance imaging (DWI) of the liver, especially inside the left lobe. The left liver lobe may thus be frequently neglected in the clinical evaluation of liver DWI. In this work, a data-driven algorithm that relies on the statistics of the signal in the left liver lobe to mitigate the motion-induced signal loss is presented. The proposed data-driven algorithm utilizes the exclusion of severely corrupted images with subsequent spatially dependent image scaling based on a signal-loss model to correctly combine the multi-average diffusion-weighted images. The signal in the left liver lobe is restored and the liver signal is more homogeneous after applying the proposed algorithm. Furthermore, overestimation of the apparent diffusion coefficient (ADC) in the left liver lobe is reduced. The proposed algorithm can therefore contribute to reduce the motion-induced bias in DWI of the liver and help to increase the diagnostic value of DWI in the left liver lobe.

Keywords: liver; magnetic resonance imaging (MRI); diffusion weighted imaging (DWI); cardiac motion; artifact; intravoxel dephasing induced signal loss; apparent diffusion coefficient (ADC) bias

Citation: Raspe, J.; Harder, F.N.; Rupp, S.; McTavish, S.; Peeters, J.M.; Weiss, K.; Makowski, M.R.; Braren, R.F.; Karampinos, D.C.; Van, A.T. Retrospective Motion Artifact Reduction by Spatial Scaling of Liver Diffusion-Weighted Images. *Tomography* **2023**, *9*, 1839–1856. https://doi.org/10.3390/tomography9050146

Academic Editor: Rafał Obuchowicz, Michał Strzelecki and Adam Piorkowski

Received: 31 August 2023
Revised: 27 September 2023
Accepted: 29 September 2023
Published: 6 October 2023

1. Introduction

Diffusion-weighted (DW) magnetic resonance imaging (MRI) [1,2] depicts tissue microstructure through probing the molecular diffusion process [3]. It has crucial, established applications in brain imaging, both in neuroscience and neurological imaging [4–7]. Translation of diffusion-weighted imaging (DWI) to the body is desirable and has been performed in multiple organs including the liver [8–11]. In the liver, DWI has been shown to be useful for the detection and characterization of focal hepatic lesions [12–14], for the assessment of treatment response [9,15], and for the evaluation of diffuse liver disease [16,17].

DWI is influenced by multiple sources of systematic errors, including eddy current effects [18] and gradient non-linearities [19,20] as well as other higher-order effects, such as the spatial dependence of b-values [21]. In liver applications, DWI is prone to more severe sources of error, including complex motion effects [9,22,23], worse B0 and B1 inhomogeneity [24], difficult fat suppression [25], and shorter tissue T2 relaxation [26,27]. Types of motion that affect the liver include cardiac [9,23,28,29], respiratory [9,28], and gross body motion [9]. In the presence of diffusion encoding gradients, motion leads to spin phase accumulation. At the voxel level, this motion-induced spin phase accumulation results in intra-voxel dephasing and/or a voxel phase. When intra-voxel dephasing occurs due to non-uniform motion at the voxel scale, the affected diffusion-weighted images exhibit non-diffusion-induced signal loss, confounding the estimated diffusion parameters [23,30,31]. When motion leads to an additional voxel phase, the effect is commonly ignored when a single-shot acquisition is employed since only the magnitude images are typically needed.

Many studies have investigated and compared different methods for minimizing the effect of motion, especially respiratory motion, in liver DWI [28,30,32–34]. Breath-holding and respiratory triggering are the most widely used methods in liver DWI for reducing respiratory motion effects [28,33,35]. Breath-holding performance can vary depending on the cooperation and the fitness of the subject and might not be suitable for all patient scans. The performance of respiratory triggering depends on the stability of the breathing pattern and therefore warrants residual motion effects. With respect to cardiac motion, although cardiac gating can mitigate cardiac motion effects, including in the left liver lobe [32,36], the inefficiency of a cardiac-gated scan prevents it from being widely applied [9]. In recent years, motion-compensated diffusion encoding (MCDE) waveforms have emerged as an additional method for minimizing motion in liver DWI [20,37–39]. MCDE can reduce the effect of motion regardless of its sources. However, MCDE waveforms result in a prolonged echo time (TE) that might prevent their application, especially in patients with a short liver T2. The prolonged TE and the accompanied flow-compensation properties of MCDE waveforms also brighten up the vessel signal in liver DWI, confounding estimated apparent diffusion coefficient (ADC) maps and compromising diagnostic value [20,39]. Reduction of the vessel signal requires a reduction in the motion compensation capability [39].

Post-processing techniques have also been proposed to mitigate motion-induced signal loss in liver DWI. Chang et al. [40] implemented an algorithm rejecting outlier data. However, eliminating data might worsen the noise performance of the method if the rejected data are above the noise floor. By employing power-of-four averaging instead of the standard sum of magnitudes (equivalent to power-of-one averaging), Liau et al. [29] showed that the cardiac-motion-induced signal loss and the discrepancy of ADC between the left and right liver lobes are reduced. However, increasing the power of averaging also leads to an increase in noise in the averaged image. Furthermore, the optimum power changes with individual subjects and b-values, and there is no systematic method for choosing the proper power. Weighted averaging was also proposed to reduce signal loss in the left liver lobe in liver DWI [41]. The weights were computed per diffusion encoding direction by dividing the square of individual averages by the sum of squares of all averages. Algorithms similar to all of the aforementioned ones were compared by Führes et al. [42] when using MCDE waveforms. A weighted averaging approach using low-pass-filtered weight maps was shown to increase image quality most in terms of minimizing pulsation artifacts and increasing the lesion contrast-to-noise ratio while maintaining data consistency.

More recently, deep learning approaches have also been introduced to perform weighted averaging of liver DWI. In Gadjimuradov et al. [43], a network was trained to compute the weights for a weighted averaging. The same authors developed another method that uses a sliding window for adaptive weighting based on small patches of multiple repetitions [44]. They also used a trained classifier to predict whether an average contains significant signal dropout to robustly estimate a good reference value to compare to.

The purpose of the present work was to develop a model and algorithm to tackle the problem of motion-induced signal loss in liver DWI. Therefore, the present work proposes a novel algorithm that reduces the motion-induced signal loss in respiratory-triggered liver DWI. It combines the rejection of heavily corrupted data and spatial scaling of the average diffusion-weighted images. It therefore can potentially increase the diagnostic quality of liver diffusion data. Data rejection was performed at the repetition and voxel levels and was incorporated before the averaging. The proposed algorithm was tested on in vivo DWI data from patients receiving an abdominal MRI examination.

2. Materials and Methods

2.1. Implicit Weighted Averaging

To improve the signal-to-noise ratio (SNR) and mitigate motion artifacts, diffusion imaging often acquires multiple repetitions for each b-value and diffusion encoding direction. Let $\{I_{z,i}(x,y)|i = 1, \ldots, N\}$ be the magnitude of the DW image at the in-plane spatial location (x, y) from the i-th repetition of slice z for each b-value. The average DW image is conventionally computed as the sum of the magnitudes of all averages as

$$\mu_{z,\text{standard}}(x, y) = \frac{\sum_{i=1}^{N} I_{z,i}(x, y)}{N}, \quad (1)$$

though other conventions, such as a combination of geometric and arithmetic averages, exist [42].

An example of 18 repetitions of liver DWI at a b-value of $b = 600\ \text{s/mm}^2$ is depicted in Figure 1a, wherein the loss of signal in some repetitions—especially in the left liver lobe—is apparent. The sum of magnitudes of these images as in Equation (1) leads to an average diffusion-weighted image with an inhomogeneous liver, which is depicted in Figure 1b on the left. An ADC map calculated from motion-corrupted images typically manifests overestimation in the left liver, as shown on the right in Figure 1b.

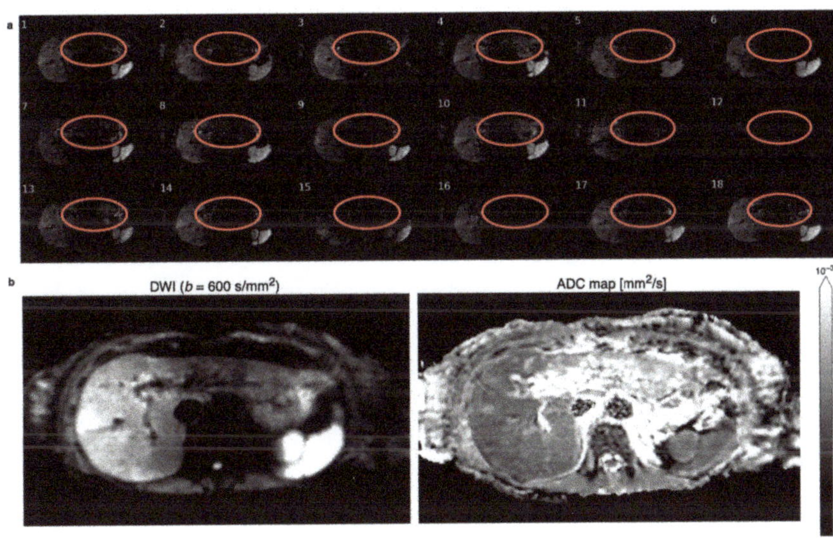

Figure 1. (**a**) Six repetitions of respiratory-triggered diffusion-weighted images (DWIs) with a b-value of $b = 600\ \text{s/mm}^2$ for three different diffusion directions that show different patterns of motion-induced signal-loss artifacts. The location of the left liver lobe is circled in red for each repetition. Some repetitions, like number 12, show global signal loss, while more often, local signal loss in the left liver lobe is observable, e.g., repetition 15. The local signal loss is incoherent and unpredictable between repetitions. (**b**) A sum of magnitudes of the 18 repetitions yields the diffusion-weighted (DW) image on the left that displays an inhomogeneous liver with evidently higher values in the right liver lobe. Apparent diffusion coefficient (ADC) mapping calculated using the motion-affected DWIs results in a inhomogeneous appearance of ADC in the liver.

The motion-induced signal loss as observable in multiple repetitions in Figure 1a can be viewed as repetition- and spatially dependent signal weighting. Thus, the magnitude of the obtained image can be expressed as

$$I_{z,i}(x,y) = w_{z,i}(x,y)\hat{I}_z(x,y) + n_z(x,y), \tag{2}$$

where $w_{z,i}(x,y)$ denotes the weighting factor, $\hat{I}_z(x,y)$ is the image magnitude without motion-induced signal loss that is independent of repetitions, and $n_z(x,y)$ is a noise term. For simplicity, Equation (2) only considers the motion-induced signal fluctuation across repetitions. Other changes across repetitions—induced by misregistration, phase variation, and distortion—are ignored. Substituting Equation (2) into Equation (1) yields

$$\mu_{z,\,\text{standard}}(x,y) = \frac{\sum_{i=1}^{N} w_{z,i}(x,y)\hat{I}_z(x,y) + n_z(x,y)}{N}. \tag{3}$$

Motion-induced signal loss in DWI results in $\sum_{i=1}^{N} w_{z,i}(x,y) < N$ and therefore in incorrectly smaller values of the estimated average DW image. Ignoring noise, if the signal weighting factor $w_{z,i}(x,y)$ is known, the true image magnitude can be estimated as

$$\hat{I}_z(x,y) \approx \frac{\sum_{i=1}^{N} I_{z,i}(x,y)}{W_z(x,y)}, \tag{4}$$

where

$$W_z(x,y) = \sum_{i=1}^{N} w_{z,i}(x,y). \tag{5}$$

2.2. Spatial Scaling for the Reduction of Motion-Induced Signal Loss

The proposed algorithm is an extension of a previously presented conference abstract [45]. It involves two primary processes: first, rejection of repetitions and voxel outliers, and second, spatially dependent magnitude-based scaling of the average DW image. The overall workflow of the algorithm is depicted in Figure 2 and is applied to all repetitions and slices of a single b-value as well as for all b-values of interest. In addition, the isotropic diffusivity assumption of diffusion in the liver parenchyma [46] enables the treatment of diffusion encoding directions as additional repetitions, thereby providing better statistics for the algorithm. Thus, the total number of repetitions N in this work is a product of the number of repetitions per diffusion direction and the number of diffusion encoding directions.

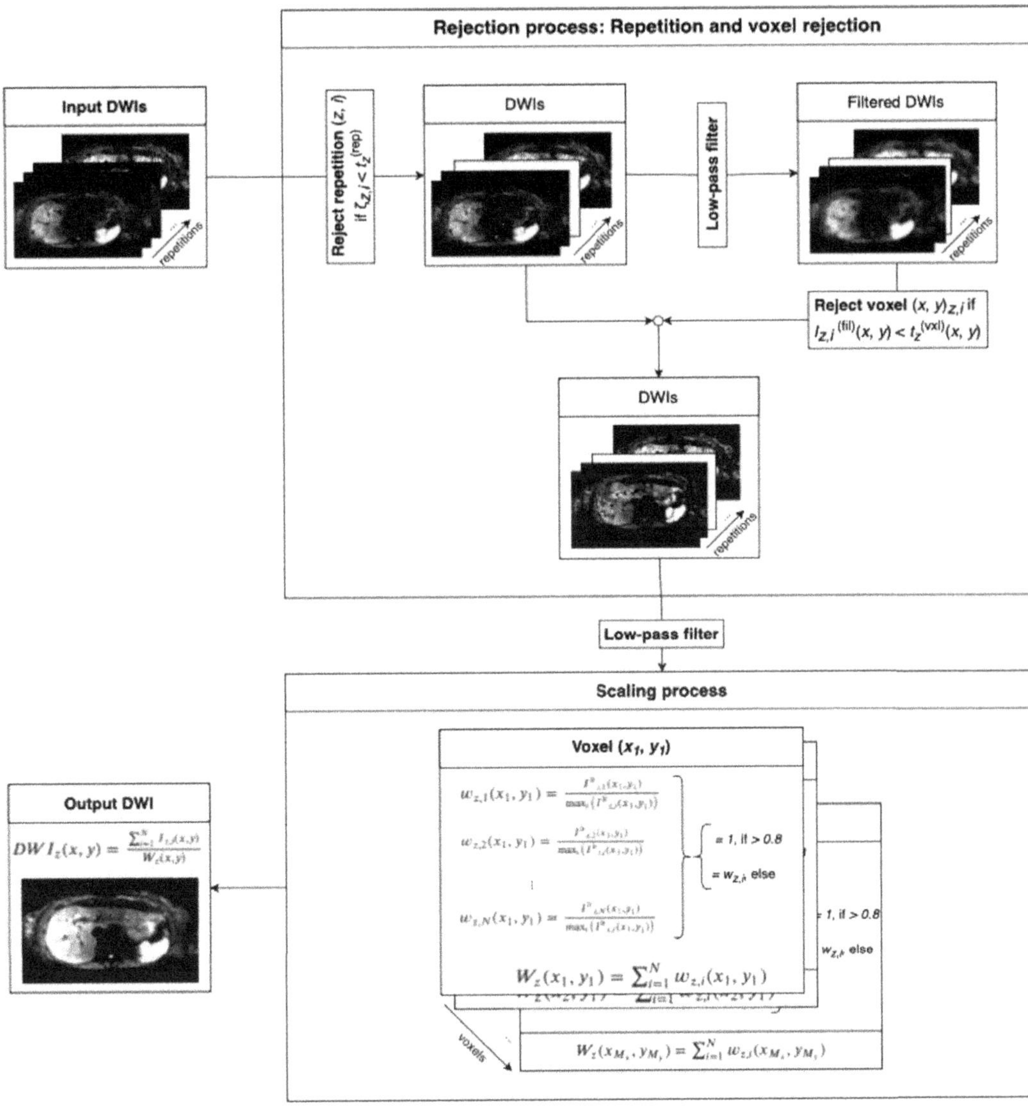

Figure 2. The two-step algorithm takes repetitions of one slice and one b-value as input. DW images in different directions are treated as additional repetitions. The sum of intensities of all voxels in each repetition is compared to a threshold $t_z^{(\text{rep})}$ for repetition rejection. Subsequent low-pass filtering blurs the images to consider that motion artifacts tend to affect patches of the image in the voxel rejection step. Spatially dependent thresholding on the filtered images determines whether a voxel is rejected. The spatially dependent weights $w_{z,i}(x,y)$ are computed on low-pass-filtered data by dividing the intensity of each repetition by the maximum intensity projection across all repetitions. Weights that exceed a value of 0.8 are set to 1. The resulting DW image is found by scaling the sum across repetitions of the magnitude of non-rejected voxels with the spatially dependent scaling factor $W_z(x,y)$ according to Equation (4).

2.2.1. Rejection Process

The rejection process is divided into two steps: repetition rejection and voxel rejection. Complete repetitions are only rejected when they contain little to no valuable information that adds to the resulting DW image. An example of a repetition that should be rejected is repetition number 12 in Figure 1a. Whether the i-th repetition of slice z, $I_{z,i}$, gets rejected is determined by thresholding. $I_{z,i}$ gets rejected if the sum of intensities $\zeta_{z,i} = \sum_{(x,y)} I_{z,i}(x,y)$ of all of its voxels is below the threshold

$$t_z{}^{(\text{rep})} = \underset{i}{\text{med}}(\zeta_z) - \lambda^{(\text{rep})}\underset{i}{\text{mad}}(\zeta_z), \tag{6}$$

where $\zeta_z = \{\zeta_{z,i}\}_i$ is the set of sums of intensities of all repetitions of slice z, $\lambda^{(\text{rep})}$ is a thresholding parameter, med () denotes the median [47], and mad () denotes the median absolute deviation (MAD) [48,49]. The MAD represents a more robust measure of data dispersion as compared to the standard deviation in the presence of outliers [49]. The median absolute deviation is defined by

$$\underset{i}{\text{mad}}(\zeta_z) = \underset{i}{\text{med}}\left(\left\{\left|\zeta_{z,i} - \underset{j}{\text{med}}(\{\zeta_z\})\right|\right\}\right), \tag{7}$$

where i and j are repetition indices and z is the slice index. The variable $\lambda^{(\text{rep})}$ in Equation (6) decides the distance of the threshold from the median value $\text{med}_i\,(\zeta_z)$. Since only repetitions that are totally corrupted and have almost no signal are targeted, $\lambda^{(\text{rep})}$ should be positive: $\lambda^{(\text{rep})} > 0$ implies that at most half of the repetitions from one slice are rejected. In the absence of motion-corrupted repetitions, all repetitions of the slice in consideration are not affected by motion-induced signal loss and their magnitude fluctuation should be caused primarily by noise. By choosing a threshold that is multiple MADs below the median, the possibility of rejecting slices unaffected by signal loss is low. When all repetitions of the slice in consideration are heavily corrupted by motion-induced signal loss, the aforementioned repetition rejection step cannot provide better performance than standard averaging, meaning the slice in consideration is corrupted in the final average DW image. In our implementation, $\lambda^{(\text{rep})}$ is fixed to the empirical value of 5, which means the threshold is five times the MAD below the median. This value marks a conservative compromise between rejection of slices with valuable information and not rejecting slices with total signal loss. "Conservative" in this situation refers to the choice of a threshold parameter rather greater than smaller in order to avoid over-rejection. Rejecting too little data can in part be compensated by the scaling described in Section 2.2.2, while over-rejection will lead to irreversible loss of data.

The second step focuses on rejecting outlier voxels that experience severe signal loss. The voxel rejection step targets localized motion-induced signal loss mainly caused by cardiac motion. Examples of the local signal loss can be seen in Figure 1a, e.g., repetitions number 6 and 15. Despite being localized, the motion-induced signal loss should affect a neighborhood of voxels instead of individual voxels. Thus, a low-pass filter is applied to the image before the voxel rejection step to utilize the fact that motion artifacts tend to affect patches of an image. A voxel $(x,y)_{z,i}$ at slice z of the low-pass-filtered repetition i gets rejected if its magnitude is below the threshold

$$t_z{}^{(\text{vxl})}(x,y) = \underset{i}{\text{med}}(I^{\text{lr}}{}_z(x,y)) - \lambda^{(\text{vxl})}\underset{i}{\text{mad}}(I^{\text{lr}}{}_z(x,y)), \tag{8}$$

where $I^{\text{lr}}{}_z$ is the set of all low-resolution repetitions of slice z and $\lambda^{(\text{vxl})}$ is an adjustable thresholding parameter. It is worth noting that the threshold $t_z{}^{(\text{vxl})}(x,y)$ is spatially dependent. Similar to $\lambda^{(\text{rep})}$, $\lambda^{(\text{vxl})}$ is fixed to an empirical value of 3 in the present algorithm.

2.2.2. Spatial Scaling of Average Diffusion-Weighted Images

The spatial scaling factor $w_{z,i}(x,y)$ in Equation (2) represents the motion-induced signal change map of each repetition as compared to the ground truth image. Since motion during diffusion encoding leads predominantly to signal loss, the maximum intensity projection (MIP) of all repetitions can approximate the ground truth image. In this case, for each slice z, repetition i, and in-plane location (x,y), the weighting factor $w_{z,i}(x,y)$ can be computed as

$$w_{z,i}(x,y) = \frac{I^{\mathrm{lr}}{}_{z,i}(x,y)}{\max\limits_{i}(I^{\mathrm{lr}}{}_{z,i}(x,y))}, \tag{9}$$

where $I^{\mathrm{lr}}{}_{z,i}$ is the i-th repetition of slice z with a low-pass filter applied. This filtering step before the weight computation assures better noise performance in the scaled image. The scaling factors are computed for non-rejected voxels only and are set to 0 for rejected voxels. To reduce the underestimation of the scaling factors and subsequent overestimation of the average image magnitude due to noise, scaling factors that exceed a value of 0.8 are set to 1. The final scaled average DW image of slice z is

$$DWI_z(x,y) = \frac{\sum_{i=1}^{N} I_{z,i}(x,y)}{\sum_{i=1}^{N} w_{z,i}(x,y)} = \frac{\sum_{i=1}^{N} I_{z,i}(x,y)}{W_z(x,y)}. \tag{10}$$

2.3. Data Acquisition

A total of 72 patients receiving an abdominal MRI examination were scanned on the 3 T Ingenia Elition X whole-body scanner by Philips Healthcare (Best, The Netherlands) using a 16-channel torso coil and the build-in-table 12-channel posterior coil. The local ethics committee approved the study. A spin-echo DW sequence with a single-shot echo-planar imaging (EPI) readout was employed with the following acquisition parameters: field of view $FOV = (420 \times 370)\ \mathrm{mm}^2$, voxel size of $(3 \times 3 \times 4)\ \mathrm{mm}^3$; 60 slices with 0.4 mm gap; b-values of $b = (0, 50, 300, 600)\ \mathrm{s/mm}^2$ with 2, 1, 2, and 6 repetitions, respectively; and 3 diffusion encoding directions for $b > 0\ \mathrm{s/mm}^2$. All scans were respiratory triggered with a fixed trigger delay of 200 ms.

The data were analyzed retrospectively for the purpose of the present work. Therefore, this work relies on the analysis of routine clinical DW image data from patients in the institution receiving diagnostic abdominal MRI for a wide range of pathologies.

2.4. Image Reconstruction and Analysis

Reconstructions were performed offline using standard averaging (Equation (1)) and spatially scaled averaging (Equation (10)) with prior outlier rejection. A mono-exponential model was applied on the average DWIs at $b = (50, 300, 600)\ \mathrm{s/mm}^2$ to estimate ADC maps.

The offline reconstruction was performed using MATLAB by MathWorks (Natick, MA, USA) [50] and the ReconFrame framework by GyroTools (Zurich, Switzerland). The proposed algorithm was implemented in Python [51], including the external libraries NumPy [52], SciPy [53], Astropy [54,55], Matplotlib [56], and HDF5 for Python [57].

2.4.1. ADC Comparison

For ADC comparison between standard averaging and the proposed method—composed of outlier rejection and spatially scaled averaging—circular regions of interest (ROIs) were placed in the nine Couinaud liver segments [58]. Care was taken to avoid including pathological regions and big vessels in the ROIs. Segments I, II, III, IVa, and IVb were assigned to the left liver lobe, and the remaining segments V, VI, VII, and VIII were assigned to the right liver lobe [39,58,59].

ADC evaluation was also performed in the paraspinal muscle region—where there is minimal motion-induced signal loss—to evaluate the bias of the proposed method. Since diffusion in the muscle is anisotropic, diffusion encoding directions were not considered as additional averages in the application of the spatially scaled averaging algorithm for the muscle region. Instead, each diffusion encoding direction was treated independently. Two ROIs in the left and right sides of the paraspinal muscle of each patient were drawn.

2.4.2. Clinical Evaluation

A radiologist blindly evaluated and compared average DWIs at a b-value of $b = 600\ \mathrm{s/mm^2}$ from standard averaging and the proposed method. Four non-diagnostic data sets and one that only partly covered the liver were excluded, leaving 67 data sets for evaluation. The comparison criteria were

1. Overall image quality: the gross appearance of the whole image volume;
2. Liver homogeneity: the contrast between the signal of the liver parenchyma of the left lobe versus the right lobe;
3. Perceived signal-to-noise ratio (SNR): the visual perception of the noise performance;
4. Quality of lesion detection: the possibility to distinguish healthy liver parenchyma from lesions.

The scores for each criterion ranged from ranged from 1 to 4, with 4 being the best score. The ADC images were only checked in the case of at least one lesion detected, which occurred in 28 of the 67 patients, and these were not scored.

2.4.3. Statistical Evaluation

Multiple instances of two-sided paired t-tests [60] were performed throughout the analysis. Two data sets are assumed to be significantly different when $p \leq 0.05$ [61]. Linear regression was performed using a least-square fit [62] of a function of the form $y = ax + b$ to the data characterized by the Pearson correlation coefficient r [63,64].

3. Results

3.1. Repetition and Voxel Rejection

Figure 3 shows a representative example of slice rejection. The sums of the intensities of a single slice for 18 repetitions are plotted in Figure 3a with the rejection region indicated by a red background. The slice at repetition 12 is rejected due to its overall low signal intensity, as is also apparent from Figure 3b.

Figure 4a shows the magnitude variation at two voxels across repetitions with their respective rejection thresholds indicated by the red line and background. The arrowheads indicate the corresponding voxel locations in Figure 4b. While both the red and the blue voxels exhibit high magnitude values in repetition number 4, their magnitude values in repetition number 11 (bottom figures) are severely reduced. Therefore, both voxels are rejected (dark red shade) in repetition number 11 and not in repetition number 4. Repetition number 4 does not suffer from signal dropout, and consequently, no voxels inside the liver are rejected. Repetition number 11 suffers from local signal loss effects, and patches of voxels are rejected, as indicated in dark red shade in Figure 4b. Some parts of the left liver lobe were spared from rejection due to localized signal loss in multiple repetitions. Though, the overall weights reflect the signal loss as presented in the next section.

3.2. Spatial Scaling of Average DWI

The weight map $W_z(x, y)$ computed for the same slice as in Figures 3 and 4 is depicted in Figure 5a. The values in the region of the left liver lobe are smaller than those in the right liver, where the scaling factor mostly tends to the maximum value of 18 (the total number of averages). Another example of the weight map shown in Figure 5b also illustrates the concentration of lower weights to the motion-affected left liver.

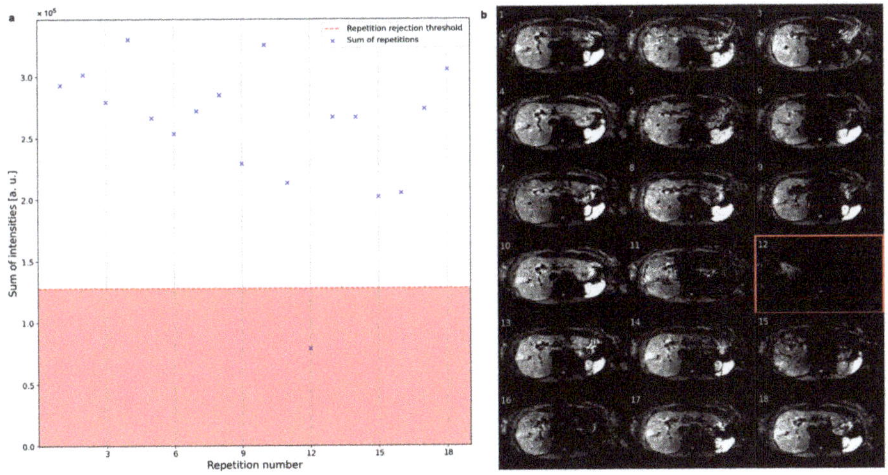

Figure 3. (**a**) The sum of intensities of one slice across all repetitions can vary substantially between repetitions. The red line marks the threshold under which slices get rejected. (**b**) The rejected slice at repetition number 12 suffers from severe global signal loss, while some other repetitions suffer from local signal loss.

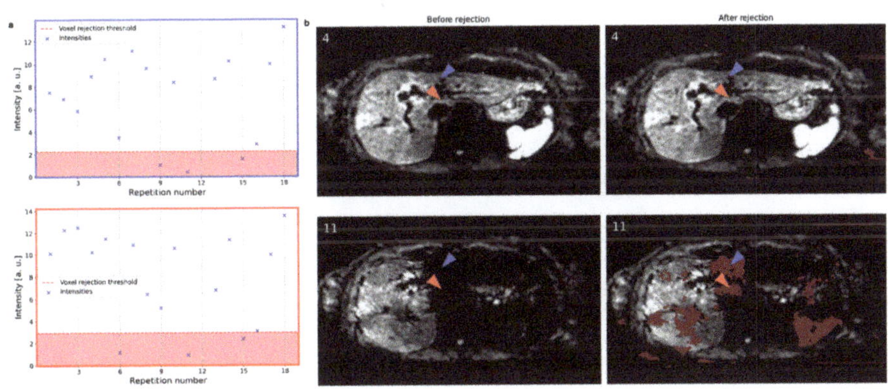

Figure 4. (**a**) The magnitude values of two voxels across all repetitions (excluding the previously rejected slice at repetition number 12) are shown for two exemplary voxels. The rejection threshold indicated by the red line and background depends on the spatial location. (**b**) Slices at repetitions 4 and 11 before and after the voxel rejection are shown with rejected voxels indicated in dark red. The colored arrowheads point to the voxels evaluated in (**a**) with their colors matching the color of the plots in (**a**). Repetition number 4 does not suffer from signal dropout, and consequently, no voxels inside the liver are rejected. Repetition number 11 suffers from signal loss, and patches of voxels are rejected.

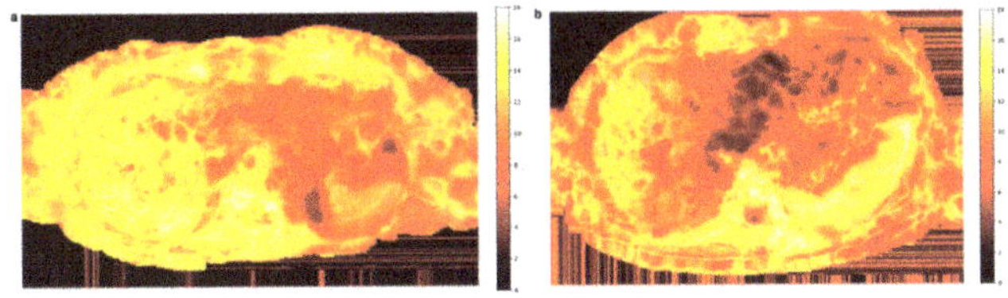

Figure 5. Two representative weight maps calculated with Equation (9) for the b600 diffusion-weighted images. Lower weight values reflect higher signal loss and mainly localize to the left liver lobe.

The diffusion-weighted images at the b-value of $b = 600\ \mathrm{s/mm^2}$ and ADC maps from the standard and proposed methods with the weight map of Figure 5a applied are shown in Figure 6a. The left liver lobe appears brighter, and ADC overestimation is reduced. A second example case in Figure 6b illustrates the algorithm's ability to mitigate the motion-induced signal loss in the left liver. With spatial scaling, the liver is more homogeneous in both the diffusion-weighted image and the ADC map. The reduction in signal loss appearance in the DW image gives better delineation of the lesion inside the left liver lobe.

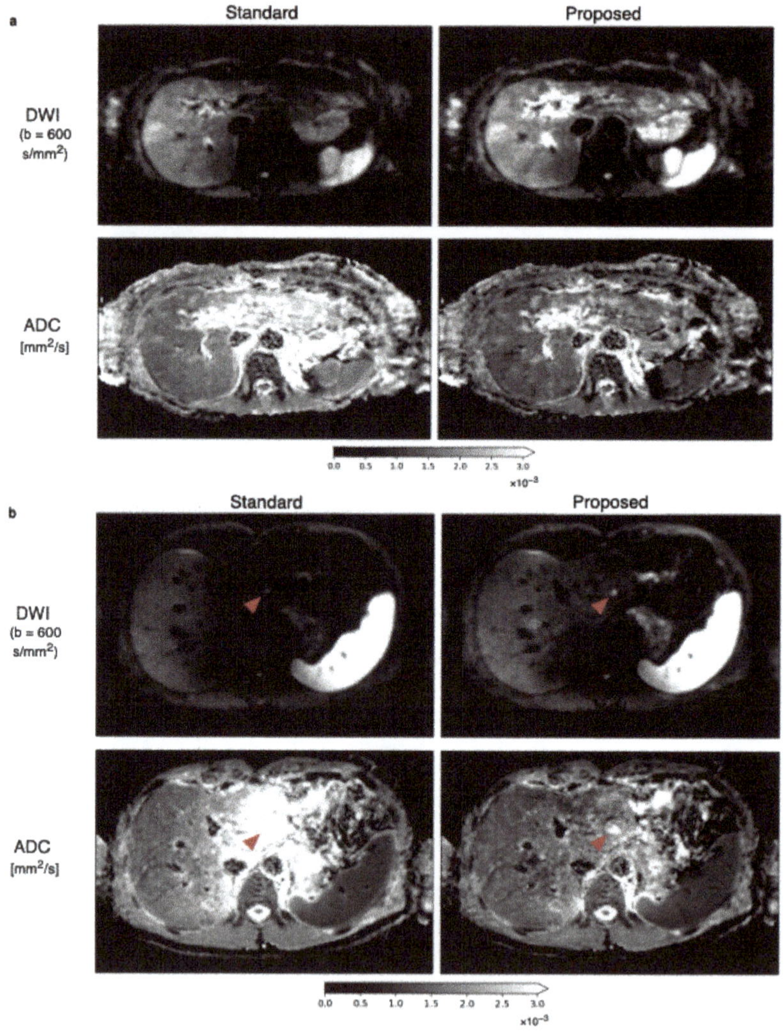

Figure 6. Average DWIs at b-values of $b = 600\ \mathrm{s/mm^2}$ and corresponding ADC maps with the standard averaging and proposed methods. (**a**) An increase of intensity inside the left liver in the DW image in the top row for the proposed method compared to the standard averaging is observable, leading also to a more homogeneous appearance of the liver in the ADC images in the bottom row. (**b**) Another in vivo data set visualizes the ability of the proposed method to reduce signal loss in the left liver lobe in DWIs and to reduce the artificial overestimation of ADC. The lesion on the left side of the liver, which is marked with the red arrow, has improved visibility and is better defined in the DW image and particularly in the ADC image.

3.3. Quantitative ADC Measurement

Means and standard deviations of ADC values across patients are plotted for both averaging methods in Figure 7a. The standard averaging method yields higher ADC values in the left liver lobe than in the rest of the liver. The obtained ADC values within the left liver lobe are more stable across all liver segments when employing the proposed method. Two-sided paired *t*-tests show that the differences between the ADCs obtained from the two averaging methods are significant in all liver segments, with p-values listed in Table 1 next to the mean ADCs of both methods in the Couinaud liver segments.

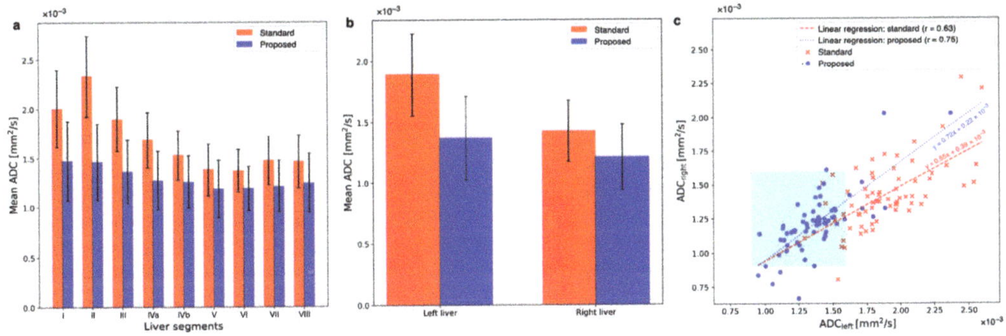

Figure 7. (**a**) A comparison of ADCs in the nine Couinaud liver segments reveals that ADCs in segments I, II, and III are overestimated the most, but they also they show the greatest correction to lower values after using the proposed method. ADCs of the segments of the right liver experience smaller changes toward lower values after using the proposed method, although they are significant. (**b**) The average ADC for all patients in the left liver is reduced by 38.2%, while in the right liver, the reduction is only 17.6%. This results in a more homogeneous liver appearance using the proposed method: with an ADC liver lobe ratio (ALR) of 1.13 ± 0.53 for the proposed method but 1.33 ± 0.47 for the standard method. (**c**) The comparison between left and right liver is plotted for all patients. The cohort of ADCs from the proposed method is notably more dense and shifted towards the lower left corner and inside the cyan square that indicates the range of ADCs of healthy liver parenchyma. Linear regression of both these cohorts yields a steeper slope that is closer to 1 for the ADCs using the proposed method.

Table 1. The mean ADCs over all subjects are reduced in every Couinaud liver segment. The *p*-values of differences in ADCs between standard averaging and the proposed methods in the different liver segments all show statistical significance.

Liver Segment	$ADC_{standard}$ [mm²/s]	$ADC_{proposed}$ [mm²/s]	*p*-Value
I	1.97 ± 0.35	1.71 ± 0.36	9.0×10^{-20}
II	2.25 ± 0.33	1.82 ± 0.33	2.5×10^{-19}
III	1.87 ± 0.27	1.60 ± 0.28	1.3×10^{-12}
IVa	1.70 ± 0.25	1.48 ± 0.25	1.7×10^{-13}
IVb	1.58 ± 0.22	1.43 ± 0.23	1.2×10^{-8}
V	1.43 ± 0.23	1.32 ± 0.24	5.9×10^{-9}
VI	1.42 ± 0.18	1.31 ± 0.18	2.4×10^{-5}
VII	1.54 ± 0.20	1.40 ± 0.21	3.8×10^{-4}
VIII	1.51 ± 0.23	1.38 ± 0.24	8.4×10^{-6}

When averaging the respective segments for the left (segments I–IVb) and the right (segments V–VIII) liver, an overestimation of the ADC in the left liver compared to the right liver can be observed when standard averaging is employed (see Figure 7b). Applying the proposed method reduces the overestimation of the ADC in the left liver. The mean ADC of the left liver is reduced by 38.2% by using the proposed method ($p < 0.01$). With the proposed method, ADC in the right liver is also reduced, but only by 17.6% ($p < 0.001$).

The ratio of ADC of the left liver to ADC of the right liver (ADC liver lobe ratio, ALR) is $ALR_{\text{standard}} = 1.33 \pm 0.47$ for the standard method and $ALR_{\text{proposed}} = 1.13 \pm 0.53$ for the spatially scaled method. The ADCs in the left and the right liver are different with standard averaging ($p = 0.021$) and the proposed method ($p = 0.020$).

Figure 7c compares the ADCs of the left and right liver for all individual patients. Note that the majority of ADC values lie inside the expected range of $(0.9 - 1.6) \times 10^{-3}$ mm^2/s only after applying the proposed method. This range is indicated in the plot by the cyan square. The slope of a linear regression between the left and right liver ADC values is closer to 1 for the proposed method: with 0.72 as compared to 0.55 with the standard method. Also, the Pearson-r correlation coefficient is greater for the ADC maps using the proposed method: with $r = 0.75$ for the proposed and $r = 0.63$ for the standard averaging method. The greater r-value indicates a higher linear correlation between the left and right liver lobes after applying the proposed algorithm. The ALRs of the individual patients with the proposed algorithm applied are significantly different from those of the standard averaged data set ($p < 0.001$).

The bias of the proposed method was tested by comparing ADC maps with the standard and proposed methods in the paraspinal muscle regions where motion is negligible. Figure 8a visualizes the mean ADCs averaged over both ROIs of the standard averaging and proposed methods. The mean ADCs are reduced on average from $1.82 \pm 0.08 \times 10^{-3}$ mm^2/s to $1.80 \pm 0.08 \times 10^{-3}$ mm^2/s. The ADC reduction is not statistically significant ($p = 0.31$). Figure 8b plots the mean ADCs in the paraspinal muscles of all patients with the standard averaging and proposed methods. The linear regression with a slope close to 1 is rotated by a small amount from the $y = x$ line, indicating a slight trend towards smaller ADCs.

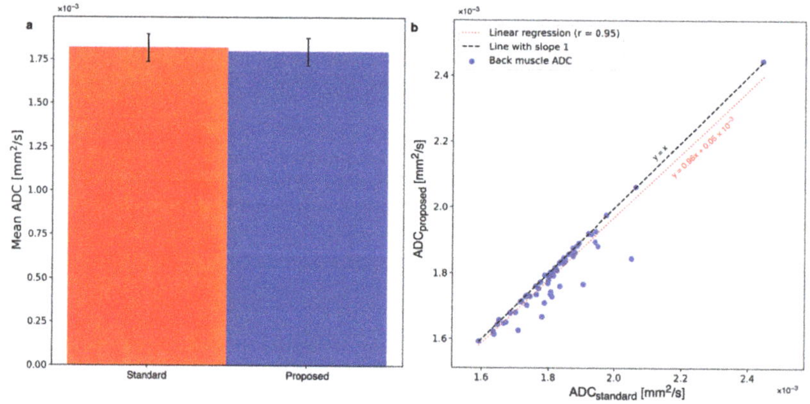

Figure 8. (**a**) The mean ADC in the back muscle changes from $1.82 \pm 0.08 \times 10^{-3}$ mm^2/s to $1.80 \pm 0.08 \times 10^{-3}$ mm^2/s. The change is statistically insignificant. (**b**) The shift towards smaller ADCs can also be observed when comparing the whole cohort of patients with standard averaging and the proposed method. The linear regression of the plot has a slope of 0.96.

3.4. Radiological Reading

The radiologist's scores of the DWIs from standard averaging and the proposed method are compared in Table 2. The rows indicate the number of patient data sets in which the respective category scored higher, the same, or worse for the proposed spatially scaled averaging compared to standard averaging. There is statistically significantly higher liver homogeneity for the proposed method with a mean score of 3.13 ± 0.92 as compared to standard averaging with a mean score of 2.82 ± 0.89 ($p = 0.05$). The improvement in the other categories is found to be not statistically significant for image quality ($p = 0.93$), perceived SNR ($p = 0.51$), and better lesion detection ($p = 0.53$).

Table 2. Radiological reading scores for all patient data sets for the standard and proposed methods in four categories. Liver homogeneity scored significantly better, showing an improvement in 24 out of 67 cases ($p = 0.05$). Improvements in the other three categories were not statistically significant.

	Image Quality	Liver Homogeneity	Perceived SNR	Lesion Detection
Better	1	24	13	3
Same	66	39	48	25
Worse	0	4	6	0

4. Discussion

The present work proposes a fast post-processing method to mitigate motion-induced signal loss in DWI of the liver, especially in the left lobe, relying on the acquisition of typically many repetitions. The two-step method deploys the rejection of slice and voxel outliers based on data-based thresholds before averaging the multi-repetition data deploying a scaled average. The proposed algorithm reduces the motion-induced signal loss in liver DWI and consequently also corrects for ADC overestimation.

Outlier rejection had been proposed earlier [40] using a different method. In the proposed method, the rejection step discards noise-level data with simple thresholding steps on the repetition and voxel levels. Rejected data are excluded from further processing to improve the noise performance of the method. Both the slice and voxel rejection procedures employ tunable parameters controlling the amount of data to be rejected. Optimal rejection parameters can be found by evaluating the trade-off between the reduction in motion-induced signal loss and the SNR of the averaged image. The trade-off stayed stable when the thresholds of the rejections fell below the median of the data. Thus, the rejection parameters in the current study were fixed to empirical values. In another study, a voxel-wise exclusion algorithm was optimized regarding an exclusion threshold [42].

The spatial scaling factor $w_{z,i}(x, y)$ in Equation (9) reflects the motion-induced signal loss at each spatial location in the DWIs. Estimating these spatial scaling factors requires reference DWIs that are minimally affected by motion-induced signal loss. In this work, such reference images were chosen as the MIP across all acquired repetitions. Another choice of reference images such as the sum of squares (or higher power) of all repetitions or the average of manually selected good images can also be used [41,43]. The choice of reference images dictates the trade-off between SNR performance, processing time and manual work, and the residual motion-induced signal loss in the final averaged images. More specifically, the choice of a noisy MIP as reference leads to scaling factors that are affected by noise but reflect the signal loss well. The sum-of-squares images have high SNR but still retain significant signal loss, especially in the case where most of the repetitions are corrupted. While manual selection of good images arguably gives the best possible reference, the selection must be carried out for each and every data set separately and is time-consuming and labor intensive. With the choice of an MIP as the reference, the proposed algorithm dampens the effect of noise by operating on low-pass-filtered images and setting all scaling factors larger than 0.8 to 1.

Current post-processing approaches to mitigate motion-induced signal loss in DWI rely on heuristic methods that penalize the contributions of signals with lower magnitudes to the final combined images. In contrast, the proposed method is based on a mathematical formulation of the effect of motion-induced signal loss on each repetition (Equation (2)), allowing for more flexible ways to combine the repetitions. Future work can find different ways to combine individual repetitions based on the proposed model.

Quantitatively, the proposed algorithm significantly reduced the ADC bias towards elevated values in the left liver lobe, shifting it towards the reported range of healthy liver ADC ($0.9 - 1.6 \times 10^{-3}$ mm^2/s) [34,65,66]. Other studies also showed that reducing the bias in the liver with post-processing methods is possible [29,41,44]. Moreover, for the whole cohort of patient data, a trend towards a more homogeneous liver was observed, as reflected by a slope closer to 1 for the regression comparing left and right liver ADCs. The significant change in ADC in the right liver lobe can be attributed to motion affecting this area, albeit to a lesser extent than the left liver lobe [23]. Thus, the proposed method reduces the signal-loss artifact due to all kinds of motion and not only the cardiac-motion-induced signal loss in the left liver lobe. Additionally, the proposed method never decreases the signal compared to the standard averaging method and is likely to increase it, especially for higher b-values, leading to a corresponding reduction in ADC. However, in the paraspinal muscle, where motion is minimal, the ADC of the proposed method is not significantly different from the ADC of the standard averaging method.

The proposed method was evaluated qualitatively by a radiologist, showing that the proposed method produces more homogeneous liver DWIs compared to the standard averaging method. Improving the homogeneity of liver DWIs is essential for diagnostic accuracy, and this was achieved in 35.8% of all evaluated patient cases using the proposed algorithm. The homogeneous appearance of healthy liver parenchyma is essential for diagnostic accuracy. In cases of motion-induced signal loss, the left liver can be excluded from diagnosis due to the low image quality [67]. The proposed method can effectively mitigate this issue and provide images with improved homogeneity of healthy liver parenchyma. In some patient cases, the proposed method also improved lesion detection, demonstrating its potential value in clinical routine. However, since only 28 out of 67 patients in the present cohort had lesions detected in the liver, and some of the lesions were only present in parts of the liver where the signal loss was less severe, further larger-sized studies with a dedicated cohort of patients with diagnosed hepatic lesions are needed to assess the potential enhancement in lesion detection. While another study was conducted on a similarly sized cohort of patients with hepatic lesions [42], there is no other study to the best of our knowledge that used clinical reading as evaluation of post-processing methods for the reduction in motion-induced signal loss in liver DWI on a similar patient cohort size.

Several additional processing and analysis techniques can further improve the proposed method. In diffusion imaging, motion and eddy currents can cause misregistration across repetitions. Also, there are multiple possible sources for other systematic errors that can be addressed next to the motion-induced signal loss, such as the gradient non-linearity or the spatial distribution of b-values. [18–21,68,69]. Though, the effective ADC bias due to motion observed in this study is substantially greater than the effects of any higher-order sources of bias. Therefore, the proposed method focuses on reducing motion-induced signal loss with the possibility to include higher-order effects in a future study. Including higher-order effects in the correction algorithm can be a direction of future work.

Since motion-induced signal loss is expectedly patch-wise and of low resolution, the rejection of data and the estimation of the scaling factors were performed on low-pass-filtered data. Therefore, the effect of misregistration on the estimated scaling factors is reduced. Nevertheless, registration before further processing might further improve the precision of the scaling factors and reduce blurring in the final averaged image. Furthermore, the proposed method assumes that the only difference across repetitions of each b-value is the motion-induced signal loss. While this assumption is true most of the time, the existence of other types of artifacts that change from repetition to repetition at each b-value would compromise the performance of the method. These artifacts, if any, must be addressed before applying the spatial scaling method. Penultimately, the proposed approach is a data-driven approach and thus does not guarantee the complete removal of motion-induced ADC quantification bias. Finally, the proposed approach would not work in cases where severe motion artifacts appear across all repetitions. In such cases, the use of MCDE waveforms would be preferred.

A complete understanding of the proposed algorithm might benefit from analyzing the SNR performance. While the slice and voxel rejection steps can help with noise reduction, the spatial scaling step might increase noise. The noise performance of the spatial scaling step depends on the choice of reference images. A Monte Carlo simulation can give insights into intermediate and final SNR behavior of the method. The behavior of the rejection steps might change in cases of free-breathing acquisitions because more outliers are expected compared to respiratory-triggered scans. Therefore, the effect of SNR benefit might be greater under free-breathing conditions. However, the robustness has to be evaluated first since too many repetitions containing outliers are likely to interfere with the rejection process.

The study has some limitations that should be noted. First, the significant change in ADC in the right liver lobe after applying the proposed method needs to be fully explained. Pulsation artifacts from vessels could partly explain this observed reduction in ADC. Second, the thresholds for outlier rejection were chosen empirically, which may limit reproducibility and the possibility of applying the method to other datasets, particularly since only data from a single scanner were used. A more thorough investigation of the influence of those parameters and of a more robust way to determine them can be a direction for future work. Adaptive thresholding [40] or optimization of a threshold [42] might be beneficial. Third, ADC values were only compared to the reported range of healthy liver parenchyma ADCs in the literature, and there was no ground truth ADC for comparison to assess the outcome of the proposed method. Last, the clinical reading was performed by only a single radiologist, rendering it impossible to evaluate the quality of scoring or to perform an inter-rater comparison.

5. Conclusions

A spatially scaled averaging approach with prior outlier rejection for multi-repetition liver DWI was developed to mitigate the motion-induced signal loss in repetition-combined DW images. Quantitative and qualitative assessments show the superior performance of the proposed method in achieving more homogeneous liver DWI and reduction of ADC overestimation, especially in the left liver lobe, as compared to the standard averaging method. This can potentially improve clinical diagnostics, as the left liver lobe can be assessed more precisely. The algorithm is easy to implement and has a fast processing time. The proposed model of the effect of motion-induced signal loss on each repetition and on the final combined image can be used as the base for other advanced repetition combination approaches that minimize signal loss.

Author Contributions: Conceptualization, D.C.K. and A.T.V.; data curation, S.M., J.R. and S.R.; formal analysis, F.N.H. and J.R.; funding acquisition, D.C.K.; investigation, J.R. and A.T.V.; methodology, J.R. and A.T.V.; project administration, R.F.B. and D.C.K.; resources, D.C.K. and M.R.M.; software, J.R.; supervision, D.C.K. and A.T.V.; validation, J.M.P., A.T.V. and K.W.; visualization, J.R.; writing—original draft, J.R.; writing—review and editing, D.C.K. and A.T.V. All authors have read and agreed to the published version of the manuscript.

Funding: The present work was supported by Philips Healthcare.

Institutional Review Board Statement: The study was conducted according to the guidelines of the Declaration of Helsinki and was approved by the Institutional Review Board (or Ethics Committee) of Klinikum rechts der Isar, Munich (protocol code 87/18S, approved on 5 March 2018).

Informed Consent Statement: Informed consent was waived by the local ethics committee due to the retrospective nature of the data according to Bavarian state regulations.

Data Availability Statement: Data are available upon request and pending institutional review.

Conflicts of Interest: Johannes M. Peeters and Kilian Weiss are employees of Philips Healthcare. Dimitrios C. Karampinos receives grant support from Philips Healthcare.

Abbreviations

The following abbreviations are used in this manuscript:

ADC	Apparent Diffusion Coefficient
ALR	Apparent diffusion coefficient liver Lobe Ratio
DW	Diffusion-Weighted
DWI	Diffusion-Weighted (magnetic resonance) Imaging
EPI	Echo-Planar Imaging
FOV	Field Of View
MCDE	Motion-Compensated Diffusion Encoding
MRI	Magnetic Resonance Imaging
ROI	Region Of Interest
SNR	Signal-to-Noise Ratio
TE	Echo Time
T2	T2 relaxation time

References

1. Stejskal, E.O.; Tanner, J.E. Spin Diffusion Measurements: Spin Echoes in the Presence of a Time-Dependent Field Gradient. *J. Chem. Phys.* **1965**, *42*, 288–292. [CrossRef]
2. Frahm, J.; Merboldt, K.D.; Hänicke, W.; Haase, A. Stimulated Echo Imaging. *J. Magn. Reson.* **1985**, *64*, 81–93. [CrossRef]
3. Le Bihan, D.; Breton, E.; Lallemand, D.; Aubin, M.L.; Vignaud, J.; Laval-Jeantet, M. Separation of diffusion and perfusion in intravoxel incoherent motion MR imaging. *Radiology* **1988**, *168*, 497–505. [CrossRef]
4. Le Bihan, D. Looking into the functional architecture of the brain with diffusion MRI. *Nat. Rev. Neurosci.* **2003**, *4*, 469–480. [CrossRef] [PubMed]
5. Le Bihan, D. Diffusion, confusion and functional MRI. *NeuroImage* **2012**, *62*, 1131–1136. [CrossRef]
6. Le Bihan, D.; Johansen-Berg, H. Diffusion MRI at 25: Exploring brain tissue structure and function. *Neuroimage* **2012**, *61*, 324–341. [CrossRef]
7. Alexander, D.C.; Dyrby, T.B.; Nilsson, M.; Zhang, H. Imaging brain microstructure with diffusion MRI: Practicality and applications. *NMR Biomed.* **2017**, *32*, e3841. [CrossRef] [PubMed]
8. Koh, D.M.; Collins, D.J. Diffusion-Weighted MRI in the Body: Applications and Challenges in Oncology. *Am. J. Roentgenol.* **2007**, *188*, 1622–1635. [CrossRef]
9. Kele, P.G.; van der Jagt, E.J. Diffusion weighted imaging in the liver. *World J. Gastroentorol.* **2010**, *16*, 1567–1576. [CrossRef]
10. Narquin, S.; Ingrand, P.; Azais, I.; Delwail, V.; Vialle, R.; Boucecbi, S.; Tasu, J.P. Comparison of whole-body diffusion MRI and conventional radiological assessment in the staging of myeloma. *Diagn. Interv. Imaging* **2013**, *94*, 629–636. [CrossRef]
11. Ayuso, C.; Rimola, J.; Vilana, R.; Burrel, M.; Darnell, A.; García-Criado, Á.; Bianchi, L.; Belmonte, E.; Caparroz, C.; Barrufet, M.; et al. Diagnosis and staging of hepatocellular carcinoma (HCC): Current guidelines. *Eur. J. Radiol.* **2018**, *101*, 72–81. [CrossRef] [PubMed]
12. Ichikawa, T.; Haradome, H.; Hachiya, J.; Nitatori, T.; Araki, T. Diffusion-weighted MR imaging with a single-shot echoplanar sequence: Detection and characterization of focal hepatic lesions. *Am. J. Roentgenol.* **1998**, *170*, 397–402. [CrossRef] [PubMed]
13. Parikh, T.; Drew, S.J.; Lee, V.S.; Wong, S.; Hecht, E.M.; Babb, J.S.; Taouli, B. Focal liver lesion detection and characterization with diffusion-weighted MR imaging: Comparison with standard breath-hold T2-weighted imaging. *Radiology* **2008**, *246*, 812–822. [CrossRef]
14. Yoon, J.H.; Lee, J.M.; Yu, M.H.; Kiefer, B.; Han, J.K.; Choi, B.I. Evaluation of hepatic focal lesions using diffusion-weighted MR imaging: Comparison of apparent diffusion coefficient and intravoxel incoherent motion-derived parameters. *J. Magn. Reson. Imaging* **2014**, *39*, 276–285. [CrossRef] [PubMed]
15. Kim, J.H.; Joo, I.; Kim, T.Y.; Han, S.W.; Kim, Y.J.; Lee, J.M.; Han, J.K. Diffusion-Related MRI Parameters for Assessing Early Treatment Response of Liver Metastases to Cytotoxic Therapy in Colorectal Cancer. *Am. J. Roentgenol.* **2016**, *207*, W26–W32. [CrossRef]
16. Koinuma, M.; Ohashi, I.; Hanafusa, K.; Shibuya, H. Apparent diffusion coefficient measurements with diffusion-weighted magnetic resonance imaging for evaluation of hepatic fibrosis. *J. Magn. Reson. Imaging* **2005**, *22*, 80–85. [CrossRef] [PubMed]
17. Amano, Y.; Kumazaki, T.; Ishihara, M. Single-shot diffusion-weighted echo-planar imaging of normal and cirrhotic livers using a phased-array multicoil. *Acta Radiol.* **2010**, *39*, 440–442. [CrossRef]
18. Meier, C.; Zwanger, M.; Feiweier, T.; Porter, D. Concomitant field terms for asymmetric gradient coils: Consequences for diffusion, flow, and echo-planar imaging. *Magn. Reson. Med. Off. J. Int. Soc. Magn. Reson. Med.* **2008**, *60*, 128–134. [CrossRef] [PubMed]
19. Malyarenko, D.I.; Newitt, D.; Wilmes, L.J.; Tudorica, A.; Helmer, K.G.; Arlinghaus, L.R.; Jacobs, M.A.; Jajamovich, G.; Taouli, B.; Yankeelov, T.E.; et al. Demonstration of nonlinearity bias in the measurement of the apparent diffusion coefficient in multicenter trials. *Magn. Reson. Med.* **2016**, *75*, 1312–1323. [CrossRef]

20. McTavish, S.; Van, A.T.; Peeters, J.M.; Weiss, K.; Makowski, M.R.; Braren, R.F.; Karampinos, D.C. Gradient nonlinearity correction in liver DWI using motion-compensated diffusion encoding waveforms. *Magn. Reson. Mater. Phys. Biol. Med.* **2021**, *35*, 827–841. . [CrossRef]
21. Krzyżak, A.T.; Olejniczak, Z. Improving the accuracy of PGSE DTI experiments using the spatial distribution of b matrix. *Magn. Reson. Imaging* **2015**, *33*, 286–295. [CrossRef]
22. Koh, D.M.; Takahara, T.; Imai, Y.; Collins, D.J. Practical Aspects of Assessing Tumors Using Clinical Diffusion-weighted Imaging in the Body. *Magn. Reson. Med. Sci.* **2007**, *6*, 211–224. [CrossRef] [PubMed]
23. Kwee, T.C.; Takahara, T.; Niwa, T.; Ivancevic, M.K.; Herigault, G.; Van Cauteren, M.; Luijten, P.R. Influence of cardiac motion on diffusion-weighted magnetic resonance imaging of the liver. *Magn. Reson. Mater. Phys. Biol. Med.* **2009**, *22*, 319–325. [CrossRef] [PubMed]
24. Roberts, N.T.; Hinshaw, L.A.; Colgan, T.J.; Ii, T.; Hernando, D.; Reeder, S.B. B_0 and B_1 inhomogeneities in the liver at 1.5 T and 3.0 T. *Magn. Reson. Med.* **2021**, *85*, 2212–2220. [CrossRef] [PubMed]
25. Luna, A.; Ribes, R.; Soto, J.A. *Diffusion MRI Outside the Brain: A Cased-Based Review and Clinical Applications*; Springer Science+Business Media: Berlin/Heidelberg, Germany; University of British Columbia: Vancouver, BC, Canada, 2012; Chapter 1. [CrossRef]
26. de Bazelaire, C.M.J.; Duhamel, G.D.; Rofsky, N.M.; Alsop, D.C. MR Imaging Relaxation Times of Abdominal and Pelvic Tissues Measured in Vivo at 3.0 T: Preliminary Results. *Radiology* **2021**, *230*, 652–659. [CrossRef] [PubMed]
27. Kumar, R.; Delshad, S.; Woo, M.A.; Macey, P.M.; Harper, R.M. Age-related regional brain T2-relaxation changes in healthy adults. *J. Magn. Reson. Imaging* **2012**, *35*, 300–308. [CrossRef] [PubMed]
28. Naganawa, S.; Kawai, H.; Fukatsu, H.; Sakurai, Y.; Aoki, I.; Miura, S.; Mimura, T.; Kanazawe, H.; Ishigaki, T. Diffusion-weighted Imaging of the Liver: Technical Challenges and Prospects for the Future. *Magn. Reson. Med. Sci.* **2005**, *4*, 175–186. [CrossRef]
29. Liau, J.; Lee, J.; Schroeder, M.E.; Sirlin, C.B.; Bydder, M. Cardiac motion in diffusion-weighted MRI of the liver: Artifact and a method of correction. *J. Magn. Reson. Imaging* **2011**, *35*, 318–327. [CrossRef]
30. Kwee, T.C.; Takahara, T.; Koh, D.M.; Nievelstein, R.A.J.; Luijten, P.R. Comparison and reproducibility of ADC measurements in breathhold, respiratory triggered, and free-breathing diffusion-weighted MR imaging of the liver. *J. Magn. Reson. Imaging* **2008**, *28*, 1141–1148. [CrossRef]
31. Rauh, S.S.; Riexinger, A.J.; Ohlmeyer, S.; Hammon, M.; Saake, M.; Stemmer, A.; Uder, M.; Hensel, B.; Laun, F.B. A mixed waveform protocol for reduction of the cardiac motion artifact in black-blood diffusion-weighted imaging of the liver. *Magn. Reson. Imaging* **2020**, *67*, 59–68. [CrossRef]
32. Nasu, K.; Kuroki, Y.; Sekiguchi, R.; Kazama, T.; Nakajima, H. Measurement of the apparent diffusion coefficient in the liver: Is it a reliable index for hepatic disease diagnosis? *Radiat. Med.* **2006**, *24*, 438–444. [CrossRef] [PubMed]
33. Kandpal, H.; Sharma, R.; Madhusudhan, K.S.; Kapoor, K.S. Respiratory-Triggered Versus Breath-Hold Diffusion-Weighted MRI of Liver Lesions: Comparison of Image Quality and Apparent Diffusion Coefficient Values. *Am. J. Roentgenol.* **2009**, *192*, 915–922. [CrossRef] [PubMed]
34. Taouli, B.; Sandberg, A.; Stemmer, A.; Parikh, T.; Wong, S.; Xu, J.; Lee, V.S. Diffusion-weighted imaging of the liver: Comparison of navigator triggered and breathhold acquisitions. *J. Magn. Reson. Imaging* **2009**, *30*, 561–568. [CrossRef] [PubMed]
35. Dyvorne, H.A.; Galea, N.; Nevers, T.; Fiel, M.I.; Carpenter, D.; Wong, E.; Orton, M.; de Oliveira, A.; Feiweier, T.; Vachon, M.L.; et al. Diffusion-weighted Imaging of the Liver with Multiple *b* Values: Effect of Diffusion Gradient Polarity and Breathing Acquisition on Image Quality and Intravoxel Incoherent Motion Parameters—A Pilot Study. *Radiology* **2013**, *266*, 920–929. [CrossRef] [PubMed]
36. Mürtz, P.; Flacke, S.; Träber, F.; van den Brink, J.S.; Gieseke, J.; Schild, H.H. Abdomen: Diffusion-weighted MR imaging with pulse-triggered single-shot sequences. *Radiology* **2002**, *224*, 258–264. [CrossRef]
37. Aliotta, E.; Wu, H.H.; Ennis, D.B. Convex optimized diffusion encoding (CODE) gradient waveforms for minimum echo time and bulk motion-compensated diffusion MRI. *Magn. Reson. Med.* **2017**, *77*, 717–729. [CrossRef]
38. Peña-Nogales, Ó.; Zhang, Y.; Wang, X.; de Luis-Garcia, R.; Aja-Fernández, S.; Holmes, J.H.; Hernando, D. Optimized Diffusion-Weighting Gradient Waveform Design (ODGD) formulation for motion compensation and concomitant gradient nulling. *Magn. Reson. Med.* **2019**, *81*, 989–1003. [CrossRef]
39. Zhang, Y.; Peña-Nogales, Ó.; Holmes, J.H.; Hernando, D. Motion-robust and blood-suppressed *M*1-optimized diffusion MR imaging of the liver. *Magn. Reson. Med.* **2019**, *82*, 302–311. [CrossRef]
40. Chang, L.C.; Walker, L.; Pierpaoli, C. Informed RESTORE: A method for robust estimation of diffusion tensor from low redundancy datasets in the presence of physiological noise artifacts. *Magn. Reson. Med.* **2012**, *68*, 1654–1663. [CrossRef]
41. Ichikawa, S.; Motosugi, I.; Tamada, D.; Wakayama, T.; Sato, K.; Funayama, S.; Onishi, H. Improving the Quality of Diffusion-weighted Imaging of the Left Hepatic Lobe Using Weighted Averaging of Signals from Multiple Excitations. *Magn. Reson. Med. Sci.* **2019**, *18*, 225–232. [CrossRef]
42. Führes, T.; Saake, M.; Lorenz, J.; Seuss, H.; Stemmer, A.; Benkert, T.; Uder, M.; Laun, F.B. Reduction of the cardiac pulsation artifact and improvement of lesion conspicuity in flow-compensated diffusion images in the liver—A quantitative evaluation of postprocessing algorithms. *Magn. Reson. Med.* **2023**, *89*, 423–439. [CrossRef] [PubMed]
43. Gadjimuradov, F.; Benkert, T.; Nickel, M.D.; Maier, A. Deep Learning-based Adaptive Image Combination for Signal-Dropout Suppression in Liver DWI. In Proceedings of the 2021 ISMRM & SMRT Annual Meeting and Exhibition, Virtual, 15–20 May 2021.

44. Gadjimuradov, F.; Benkert, T.; Nickel, M.D.; Führes, T.; Saake, M.; Maier, A. Deep learning–guided weighted averaging for signal dropout compensation in DWI of the liver. *Magn. Reson. Med.* **2022**, *88*, 2679–2693. [CrossRef] [PubMed]
45. Raspe, J.; Van, A.T.; Harder, F.; McTavish, S.; Peeters, J.; Weiss, K.; Makowski, M.R.; Braren, R.F.; Karampinos, D.C. Spatial Scaling of Respiratory-Triggered Liver Diffusion Weighted Imaging. In Proceedings of the 2022 ISMRM & SMRT Annual Meeting and Exhibition, London, UK, 7–12 May 2022.
46. Taouli, B.; Vilgrain, V.; Dumont, E.; Daire, J.L.; Fan, B.; Menu, Y. Evaluation of Liver Diffusion Isotropy and Characterization of Focal Hepatic Lesions with Two Single-Shot Echo-Planar MR Imaging Sequences: Prospective Study in 66 Patients. *Radiology* **2003**, *226*, 71–76. [CrossRef] [PubMed]
47. Daintith, J. *Dictionary of Physics*, 6th ed.; Oxford University Press: Oxford, UK, 2009.
48. Gauß, C.F. Bestimmung der Genauigkeit der Beobachtungen. In *Zeitschrift für Astronomie und verwandte Wissenschaften*; J.G. Cotta'sche Buchhandlung: Tübingen, Germany, 1816; Volume 1, pp. 185–197.
49. Leys, C.; Ley, C.; Klein, O.; Bernard, P.; Licata, L. Detecting outliers: Do not use standard deviation around the mean, use absolute deviation around the median. *J. Exp. Soc. Psychol.* **2013**, *49*, 764–766. [CrossRef]
50. *MATLAB, Version 9.10.0.1684407 (R2021a) Update3*; The MathWorks: Natick, MA, USA, 2021.
51. Van Rossum, G.; Drake, F.L. *Python 3 Reference Manual*; CreateSpace: Scotts Valley, CA, USA, 2009.
52. Harris, C.R.; Millman, K.J.; van der Walt, S.J.; Gommers, R.; Virtanen, P.; Cournapeau, D.; Wieser, E.; Taylor, J.; Berg, S.; Smith, N.J.; et al. Array programming with NumPy. *Nature* **2020**, *585*, 357–362. [CrossRef] [PubMed]
53. Virtanen, P.; Gommers, R.; Oliphant, T.E.; Haberland, M.; Reddy, T.; Cournapeau, D.; Burovski, E.; Peterson, P.; Weckesser, W.; Bright, J.; et al. SciPy 1.0: Fundamental Algorithms for Scientific Computing in Python. *Nat. Methods* **2020**, *17*, 261–272. [CrossRef]
54. Astropy Collaboration; Robitaille, T.P.; Tollerud, E.J.; Greenfield, P.; Droettboom, M.; Bray, E.; Aldcroft, T.; Davis, M.; Ginsburg, A.; Price-Whelan, A.M.; et al. Astropy: A community Python package for astronomy. *Astron. Astrophys.* **2013**, *558*, A33. . [CrossRef]
55. Astropy Collaboration; Price-Whelan, A.M.; Sipőcz, B.M.; Günther, H.M.; Lim, P.L.; Crawford, S.M.; Conseil, S.; Shupe, D.L.; Craig, M.W.; Dencheva, N.; et al. The Astropy Project: Building an Open-science Project and Status of the v2.0 Core Package. *Astron. J.* **2018**, *156*, 123. [CrossRef]
56. Hunter, J.D. Matplotlib: A 2D graphics environment. *Comput. Sci. Eng.* **2007**, *9*, 90–95. [CrossRef]
57. Collette, A. *Python and HDF5*; O'Reilly: Sebastopol, CA, USA, 2013.
58. Juza, R.M.; Pauli, E.M. Clinical and surgical anatomy of the liver: A review for clinicians. *Clin. Anat.* **2014**, *27*, 764–769. [CrossRef]
59. Sibulesky, L. Normal liver anatomy. *Clin. Liver Dis.* **2013**, *2*, S1–S3. [CrossRef] [PubMed]
60. Student. The Probable Error of a Mean. *Biometrika* **1908**, *6*, 4–25. [CrossRef]
61. Fisher, R.A. III. The influence of rainfall on the yield of wheat at Rothamsted. *Philos. Trans. R. Soc. Lond. Ser. B Contain. Pap. A Biol. Character* **1925**, *213*, 89–142. [CrossRef]
62. Merriman, M. *A List of Writings Relating to the Method of Least Squares: With Historical and Critical Notes*; Academic Press: Cambridge, MA, USA, 1877; Volume 4.
63. Pearson, K. VII. Note on regression and inheritance in the case of two parents. *Proc. R. Soc. Lond.* **1895**, *58*, 240–242. [CrossRef]
64. Wright, S. Correlation and causation. *J. Agric. Res.* **1921**, *20*, 557–585.
65. Sandrasegaran, K.; Akisik, F.M.; Lin, C.; Tahir, B.; Rajan, J.; Saxena, R.; Aisen, A.M. Value of Diffusion-Weighted MRI for Assessing Liver Fibrosis and Cirrhosis. *Am. J. Roentgenol.* **2009**, *193*, 1556–1560. [CrossRef] [PubMed]
66. Papanikolaou, N.; Gourtsoyianni, S.; Yarmenitis, S.; Maris, T.; Gourtsoyiannis, N. Comparison between two-point and four-point methods for quantification of normal liver parenchyma and focal lesions. Value of normalization with spleen. *Eur. J. Radiol.* **2010**, *73*, 305–309. [CrossRef]
67. Chen, X.; Qin, L.; Pan, D.; Huang, Y.; Yan, L.; Wang, G.; Liu, Y.; Liang, C.; Liu, Z. Liver diffusion-weighted MR imaging: Reproducibility comparison of ADC measurements obtained with multiple breath-hold, free-breathing, respiratory-triggered, and navigator-triggered techniques. *Radiology* **2014**, *271*, 113–125. [CrossRef] [PubMed]
68. Le Bihan, D.; Poupon, C.; Amadon, A.; Lethimonnier, F. Artifacts and Pitfalls in Diffusion MRI. *J. Magn. Reson. Imaging* **2006**, *24*, 478–488. [CrossRef] [PubMed]
69. Borkowski, K.; Krzyżak, A.T. Assessment of the systematic errors caused by diffusion gradient inhomogeneity in DTI-computer simulations. *NMR Biomed.* **2019**, *32*, e4130. [CrossRef]

diagnostics

MDPI

Article

Exploration and Enhancement of Classifiers in the Detection of Lung Cancer from Histopathological Images

Karthikeyan Shanmugam and Harikumar Rajaguru *

Department of Electronics and Communication Engineering, Bannari Amman Institute of Technology, Sathyamangalam 638401, India; ms.karthi.1388@gmail.com
* Correspondence: harikumarrajaguru@gmail.com

Abstract: Lung cancer is a prevalent malignancy that impacts individuals of all genders and is often diagnosed late due to delayed symptoms. To catch it early, researchers are developing algorithms to study lung cancer images. The primary objective of this work is to propose a novel approach for the detection of lung cancer using histopathological images. In this work, the histopathological images underwent preprocessing, followed by segmentation using a modified approach of KFCM-based segmentation and the segmented image intensity values were dimensionally reduced using Particle Swarm Optimization (PSO) and Grey Wolf Optimization (GWO). Algorithms such as KL Divergence and Invasive Weed Optimization (IWO) are used for feature selection. Seven different classifiers such as SVM, KNN, Random Forest, Decision Tree, Softmax Discriminant, Multilayer Perceptron, and BLDC were used to analyze and classify the images as benign or malignant. Results were compared using standard metrics, and kappa analysis assessed classifier agreement. The Decision Tree Classifier with GWO feature extraction achieved good accuracy of 85.01% without feature selection and hyperparameter tuning approaches. Furthermore, we present a methodology to enhance the accuracy of the classifiers by employing hyperparameter tuning algorithms based on Adam and RAdam. By combining features from GWO and IWO, and using the RAdam algorithm, the Decision Tree classifier achieves the commendable accuracy of 91.57%.

Keywords: histopathology; benign; adenocarcinoma; PSO; GWO; KL divergence; IWO; multilayer perceptron; bayesian linear discriminant analysis classifier

Citation: Shanmugam, K.; Rajaguru, H. Exploration and Enhancement of Classifiers in the Detection of Lung Cancer from Histopathological Images. *Diagnostics* **2023**, *13*, 3289. https://doi.org/10.3390/diagnostics13203289

Academic Editors: Rafał Obuchowicz, Michał Strzelecki and Adam Piorkowski

Received: 12 October 2023
Revised: 20 October 2023
Accepted: 21 October 2023
Published: 23 October 2023

1. Introduction

Cancer is increasingly common, and doctors use blood tests, biopsies, and image analysis for its diagnosis. It originates from damaged cells and varies among individuals. Understanding its source helps us comprehend its condition [1]. Lung cancer, often tied to smoking or harmful exposures, is a prevalent cancer type causing rising death tolls globally [2]. It affects both genders and has a low survival rate. Early detection is crucial for better outcomes. The five-year survival rate is approximately 34% for surgically removable early-stage cancer, compared to less than 10% for inoperable cases. Lung cancer treatment depends on histological characteristics, categorized as small cell (SCLC) and non-small cell (NSCLC) types, of which 80% to 85% are NSCLC and the rest are SCLC [3]. NSCLC has subtypes such as benign, adenocarcinoma (ACA), and squamous cell carcinoma (SCC). SCC displays characteristics such as the presence of clusters of polyhedral cells, keratinization, and the formulation of keratin pearls. Once the tissue type is identified, suitable treatments can be selected: either surgery, chemotherapy, radiation, targeted therapy, or immunotherapy.

Early detection and treatment of cancer are vital for better patient outcomes. Traditional diagnostic methods involve clinical assessments, lab tests, imaging, and a procedure called biopsy [4], which is considered the gold standard. During biopsy, tissue samples are taken and examined under a microscope using techniques such as hematoxylin and eosin staining. This histopathological analysis helps identify abnormal tissue growth and cell characteristics. Accurate identification and classification of individual cell nuclei are of utmost importance when evaluating tissue samples for cancer diagnosis. Pathologists inspect these samples at different magnifications, looking for signs of malignancy such as irregular cell shape, dark nuclei, and increased mitotic figures [5], and count to generate reliable results [6]. Manual histopathological examination is a time-consuming process due to the frequent presence of numerous nuclei from diverse categories clustered together in histopathological images, which can result in disagreements among pathologists [7], prompting the development of automated systems. Researchers have utilized image processing, pattern recognition, and machine learning/deep learning techniques to create computer-aided diagnostic (CAD) systems. These systems aim to detect and classify carcinomas quickly and reliably [8]. Machine learning and deep learning algorithms improve CAD performance as they learn from more data. These approaches use either microscopic images or whole slide images (WSIs) and extract features to aid in diagnosis. The challenge is to create a novel, versatile, and fully automated CAD system that can handle both microscopic images and WSIs, regardless of any imaging artifacts. Automated analysis of microscopic images is vital for evaluating digitized specimens, reducing inter-observer variations, and improving objectivity and reproducibility, as emphasized by Foran et al. [9]. This advancement can enable comparative studies of diseases and potentially aid in diagnostic decision-making.

Different imaging techniques, such as ultrasounds, MRIs, CT scans, X-rays, and needle biopsies, are used to diagnose lung cancer. X-ray imaging, considered a fundamental technique for lung examination, possesses restricted resolution and the potential to overlook specific areas of interest [10]. CT scans are commonly used to detect early stages of lung cancer and locate tumors before surgery, but they expose patients to harmful radiation with repeated scans. MRI demonstrates notable sensitivity and specificity, valuable for identifying bone metastases, although it is not advisable for diagnosing lung cancer. Ultrasound, a non-invasive method, proves adept at identifying postoperative lung issues and surpasses X-rays in effectiveness [11]. While image examination aids in diagnosis, staging, treatment evaluation, and prognosis assessment, histopathological examination remains the most reliable method to determine tumor characteristics and clinical stages. Histopathological images offer an intricate view of cellular and tissue-level transformations linked in differentiating between various conditions and cancer types, empowering pathologists to deliver precise and reliable diagnoses. Moreover, they are invaluable for pinpointing distinct biomarkers linked to various cancer types and grades, facilitating tumor classification and subtyping. Histopathological images form a dependable diagnostic framework known for its consistency and reliability in cancer diagnosis [12]. By harnessing extensive datasets of annotated histopathological images, it becomes feasible to create highly dependable algorithms for automated cancer diagnosis. These algorithms effectively streamline the diagnostic process, reducing the necessity for extensive manual examination [13].

The objective of this study is to create a classification framework that can analyze histopathological images data related to lung cancer. The goal is to accurately classify individuals as either having cancer or not, using machine learning techniques and meta-heuristic algorithms for tasks such as feature extraction, feature selection, and classification. The following subsection analyzes various methods for cancer detection and classification using image processing and classification techniques.

Review of Previous Work

In recent times, the research community has shown significant interest in diagnosing Lung Cancer through histopathological images. Numerous methodologies have been explored, utilizing a range of machine learning and deep learning techniques, across diverse datasets to detect instances of lung cancer.

Various strategies have been proposed to identify irregularities in lung-related images, encompassing chest radiographs, CT scans, ultrasound images, histopathological images, and microarray data. Ozekes and Camurcu [14] utilized template matching, while Schilham et al. devised a computer-aided detection (CAD) system that encompasses preprocessing, the identification of candidate nodules, feature extraction, and cancer classification [15]. Wang et al. [16] executed the classification of pathology images concerning lung cancer using a convolutional neural network (CNN) methodology, incorporating cell segmentation. The final layer of the CNN model integrated the Softmax activation function to enhance the classification process. Through the application of the region of interest (ROI) technique as a preliminary step, they focused on cell areas containing relevant tumors. The achieved classification accuracy for the three-class image dataset reached 90.1%. Dehmeshki et al. [17] employed a genetic algorithm based on shapes for template matching, while Suarez-Cuenca et al. used an iris filter for CT image discrimination [18]. Murphy et al. used a K-nearest neighbor (KNN) classifier for nodule detection [19], and Giger et al. used geometric features in their CAD system for CT images.

Wei et al. [20] undertook the categorization of histopathological images depicting six classes of lung cancer utilizing CNNs. They specifically employed ResNet models for their investigation. The ResNet models were integrated with pre-trained approaches from ImageNet and COCO image databases. Prior to the model training phase, the input data underwent preprocessing, which included the application of augmentation techniques. The study's achievement in terms of classification F-score reached a notable 90.4%. Mohammed Al-Jabber et al. [21] employed histopathological images from the LC25000 dataset, employing both ANN and the GoogLeNet and VGG-19 models. This combination yielded an impressive accuracy of 99.64%. Teramoto et al. [22] effectively distinguished histopathological images spanning three types of lung cancer through the application of a deep learning model. They implemented an augmentation approach that involved rotating, flipping, and applying filters to each image. Following this, they employed their developed deep CNN model to carry out the classification process. The outcomes of their classification efforts yielded an accuracy of around 70%. Shapcott et al. [23] conducted their model training by initially subjecting the input data to a preprocessing stage, integrating the augmentation technique. They employed a deep learning methodology for classifying histopathological images related to colon cancer. The dataset encompassed four distinct classes. To facilitate cell identification, a cell patches algorithm was employed on each image. The images were segmented into specific dimensions through segmentation procedures. The classification process was then conducted using the CNN model based on the defined cell patches. The obtained correlation accuracy rates ranged between 90% and 96.9%.

Barker et al. an automated system to classify brain tumors using digital pathology images [24]. Ojansivu et al. explored an automated method for categorizing breast cancer from tissue samples [25]. Ficsor et al. proposed an automated classification method for colon inflammation using digital microscopy images of histological sections [26]. The authors of a study, Mouelhi et al. [27], used various techniques such as Haralick's textures, histogram of oriented gradients (HOG), and color-based statistical moments (CCSM) to extract features from biopsy images and classify cancerous cells. The features included energy, correlation, homogeneity, contrast, GLCM texture features, as well as RGB, gray level, and HSV color components. Huang and Lai [28] focused on histology image analysis, employing texture features and KNN, SVM for image classification and segmentation. Their approach achieved a classification accuracy of 90.07% and 92.8%. Gessert et al. [29] executed the classification procedure employing CNN models based on transfer learning, leveraging microscopic images of colon cancer. Their study employed a dataset that comprised both benign and malignant images. They trained the dataset using various models including Inception, VGG, and DenseNet. Among these, the DenseNet model yielded the most promising classification outcome, achieving a classification accuracy of 91.2%.

Sinha and Ramkrishan [30] studied small biopsy images, analyzing cell characteristics such as shape, size, color, and other properties. Four classification methods were compared: Bayesian, KNN, neural networks, and SVM. The last two methods achieved the highest accuracy rates of 94.1%, while the first two had lower rates of 82.3% and 70.6%.

Kasmin et al. [31] examined microscopic biopsy images, considering characteristics such as cell/nuclei size, cell boundary length, minimum polygon area enclosing a cell, major axis length of an ellipse fitted to a cell, filled cell area, and average cytoplasmic intensity. They used neural networks and achieved classification accuracies of 86% and 92%. Chia-Hung Chen et al. [32] used a convolutional neural network to diagnose endobronchial ultrasound images, achieving an improved accuracy of 85.4% compared to traditional methods. Azka Khoirunnia et al. [33] developed a lung cancer detection system using a combination of CNN and RNN with Microarray data. In their research, CNN achieved 83% accuracy, RNN reached 71%, and the fusion of CNN and RNN (CRNN) attained the highest accuracy at 91%. Shahid Mehmood et al. [34] focused on classifying histopathological images of lung and colon cancers. By using AlexNet along with a technique called Class-Selective Contrast Enhancement, they achieved an impressive accuracy of 98.4%.

This paper is structured as follows: Section 2 focuses on the methodology employed for detecting lung cancer. Section 3 explores the feature extraction techniques including Particle Swarm Optimization and Grey Wolf Optimization whereas Section 4 explores the feature selection techniques, such as KL Divergence, and Invasive Weed Optimization. Section 5 explains the different classifiers used and hyper parameter updating method and its implementation. Section 6 presents the cumulative results, and Section 7 concludes the paper.

The following section deals with the methodology employed for identifying lung cancer through histopathological images.

2. Methodology for Lung Cancer Detection

This study employed lung histopathological images sourced from the LC25000 Dataset, which is available online. Andrew Borkowski and his colleagues from James Hospital Tampa, University of South Florida, and the Moffitt Cancer Center in Florida, USA, worked together to collectively assemble this dataset. The dataset encompasses histopathological images representing lung and colon cancer cases. Excluding colon cancer cases, the collection includes a total of 500 lung tissue images, divided equally between Benign Lung tissue and Lung Adenocarcinomas. These images were originally captured from pathology glass slides and were later resized to square dimensions of 768 × 768 pixels, down from their original size of 1024 × 768 pixels. The dataset underwent augmentation, resulting in an expansion to a comprehensive set of 10,000 lung histopathological color images which are categorized into two classes: Benign (N) and Adenocarcinoma (ACA), each consisting of 5000 images. These images are resized to a standard size of 256 × 256 followed by converting into a grey scale image. Notably, the images portray lung benign tissue characterized by abnormality but not indicative of cancer, while lung adenocarcinoma, the most prevalent form of lung cancer in the United States and notably linked to smoking, forms the second category.

Figure 1 shows the general schematic diagram for identifying and categorizing lung cancer in histopathological images. The input histopathological image will undergo conversion into a linear vector comprising 65,536 elements (due to the image's size of [256 × 256]). The procedure involves image pre-processing and a modified KFCM-based segmentation. During segmentation process approximately as [190 × 190] of the original image (i.e., nearly 36,100 intensity values) are segmented and used for further processing. These values will be directly employed to initialize the positions of birds in the Particle Swarm Optimization (PSO) and grey wolves in the Grey Wolf Optimization (GWO) algorithms. Optimization algorithms such as PSO and GWO are used to obtain a matrix of [512 × 10] dimensionally reduced intensity values from the segmented images. These dimensionally reduced features undergo feature selection techniques such as KL divergence and IWO. The selected features are then inputted into classifiers to evaluate their performance of the classifiers. Furthermore, an enhancement in the accuracy of lung cancer classification across various classifiers including SVM, KNN, Random Forest, Decision Tree, Softmax Discriminant,

Multilayer Perceptron, and BLDC classifiers is achieved through the implementation of a Hyper Parameter Updation algorithm based on the RAdam technique.

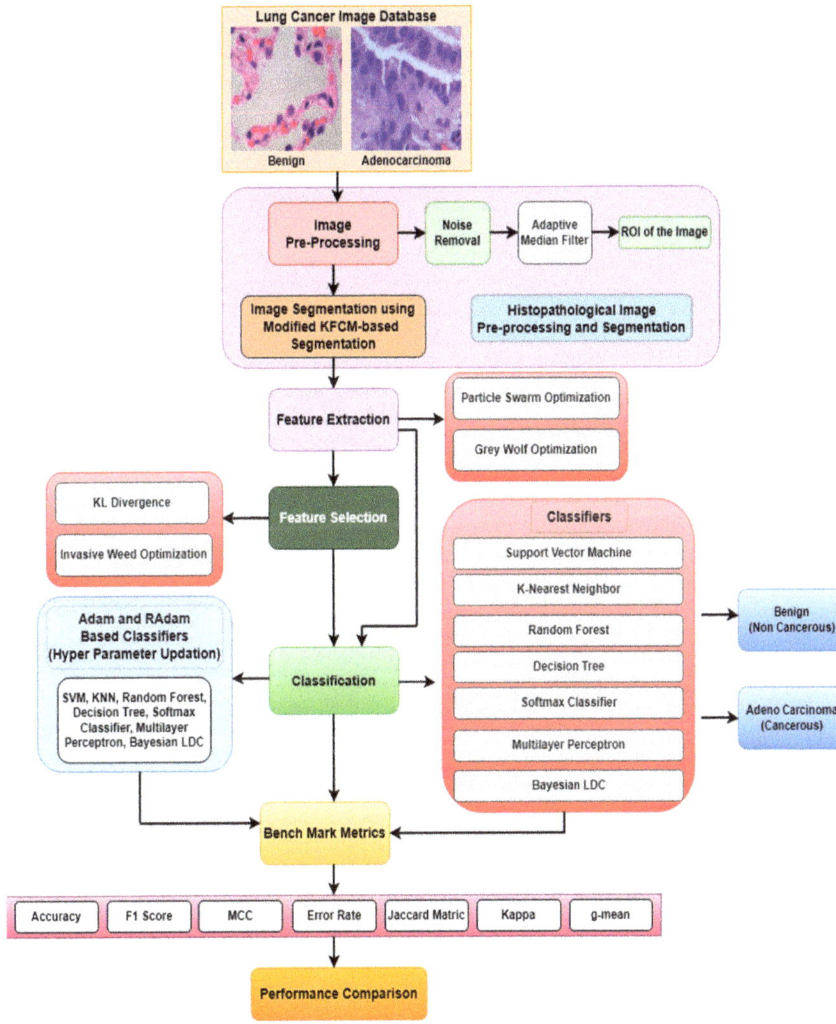

Figure 1. Schematic representation for detecting lung abnormalities from Histopathological Images.

2.1. Histopathological Image Preprocessing

Histopathological analysis serves as the definitive standard for evaluating the quality and clinical staging of tumors [35]. In the realm of diagnosing and treating medical conditions, healthcare professionals heavily rely on histopathological images. These images establish a crucial cornerstone for predicting patient survival rates [36].

As per available reports, histopathological images present several challenges:

- The images exhibit intricate geometric structures and complex textures that arise from the vast diversity in structural morphology [37].
- Notably, histopathological images are susceptible to color inconsistencies and noise due to external factors such as variations in illumination conditions [38].

- Variations in microscope magnification, equipment settings, and other variables contribute to inconsistencies in image sizes and resolutions within histopathological images [39].
- Elements of significance, such as local micro-vessels with distinctive textural characteristics, significantly influence disease diagnosis within histopathological images. Extracting these features is of paramount importance in supporting the classification and diagnosis of lung cancer [40].

Due to these factors, the histopathological images we encounter are frequently not perfect and these images show that image quality is affected by noise during acquisition and artifacts during sample preparation and slide digitization. Preprocessing methods are employed in histopathological images to enhance image quality, rectify anomalies, amplify pertinent characteristics, and establish uniformity, ultimately resulting in heightened precision and dependability of diagnostic outcomes. The study demonstrates that using an efficient adaptive median filter enhances image quality, reduces artifacts, and facilitates accurate diagnosis and analysis. However, when subjected to an adaptive median filter, these images tend to become smoother and exhibit reduced noise, rendering them suitable for our forthcoming investigations. After artifact removal, the filtered histopathological images are used for segmentation. Here, the size of the selected region of interest (ROI) is 256×256 which is the complete original image.

2.2. Histopathological Image Segmentation

A Modified Kernel Fuzzy C-Means methodology is employed to effectively segment normal and abnormal regions in histopathological images even though outliers are encountered. Image segmentation is the process of dividing an image into distinct regions based on certain image characteristics, with the goal of isolating and identifying specific regions within the image [41]. In this scenario, we have an input histopathological image denoted as H, which consists of a set of color images x_i at pixel i $(i = 1, 2, \ldots N)$ and these color images are represented as $X = \{x_1, x_2, \ldots, x_N\} \subset R^k$, in the k-dimensional space. The cluster centers within the histopathological images are represented as $Y = \{y_1, y_2, \ldots, y_c\}$, where c is a positive integer $(2 < c \ll N)$, and m_{ij} represents the membership value for each pixel i in the j-th cluster $(j = 1, 2, \ldots c)$. In the Kernel Fuzzy C-Means algorithm, clusters are formed in the image space by assigning distinct membership values to all pixels. The objective function or general equation for the Kernel Fuzzy C-Means algorithm is expressed as follows in the Equation (1):

$$O_{KFCM} = \sum_{i=1}^{N} \sum_{j=1}^{c} m_{ij}^{n} \|x_i - y_j\|^2,\ 1 \le n \le \infty \tag{1}$$

where n represent an exponent used for regularization, with the condition that, $n > 1$, and $\|x_i - y_j\|^2$ denotes the squared grayscale Euclidean distance between x_i and y_j, which is given in Equation (2):

$$\sum_{j=1}^{c} m_{ij} = 1,\ m_{ij} \in [0,1],\ 0 \le \sum_{i=1}^{N} m_{ij} \le N \tag{2}$$

Using the membership function derived from the alternate optimization approach, the process of iteratively updating the cluster centers is carried out according to the Equations (3) and (4).

$$m_{ij} = \frac{1}{\sum_{k=1}^{c} \left(\|x_i - y_j\|^2 / \|x_i - y_k\|^2 \right)^{1/(n-1)}} \tag{3}$$

$$y_j = \frac{\sum_{i=1}^{N} m_{ij}^n x_i}{\sum_{i=1}^{N} m_{ij}^n} \tag{4}$$

To reduce the impact of noise, the Equation (5) incorporates the spatial information of neighboring pixels,

$$O_{KFCM-S} = \sum_{i=1}^{N}\sum_{j=1}^{c} m_{ij}^n \|x_i - y_j\|^2 + \frac{\alpha}{N_R}\sum_{i=1}^{N}\sum_{j=1}^{c} m_{ij}^n \left(\sum_{r\in N_i} \|x_i - y_j\|^2\right) \tag{5}$$

Here, the spatial information is denoted by α, N_i represents the set of pixels and its cardinality is defined as N_R, the neighborhood function is substituted by $\left\|\acute{x}_i - y_j\right\|^2$, in place of $\frac{1}{N_R}\sum_{r\in N_R} \left\|x_i - y_j\right\|^2$, where, $\acute{x}$ represents a color scale-filtered image, and the Euclidean distance is replaced with the correlation distance measure to avoid the neighborhood function. The updated equation is represented in Equation (6):

$$O_{KFCM-S(1,2)} = \sum_{i=1}^{N}\sum_{j=1}^{c} m_{ij}^n \|x_i - y_j\|^2 + \alpha\sum_{i=1}^{N}\sum_{j=1}^{c} m_{ij}^n \left(\|x_i' - y_j\|^2\right) \tag{6}$$

In this study, a modified version of KFCM computes the parameter η_j for each cluster at every iteration to substitute for α [42]. The calculation of this parameter utilizes the correlation function, as outlined in the Equation (7):

$$\eta_j = \frac{\min_{j'\neq j}\left(1 - C\left(y'_{j'}, y_j\right)\right)}{\max_k\left(1 - C(y_k, x')\right)} \tag{7}$$

Here C represents the correlation function or correlation distance measure. Here, determining the precise characteristics of C typically necessitates a large number of patterns and numerous cluster centers to identify optimal value for η_j. To address this challenge, a solution is devised by integrating spatial context and scale information through the incorporation of fuzzy factor. The objective function of the KFCM, as presented in Equation (8), incorporates the inclusion of the fuzzy factor F_{ij}.

$$O_{C-KFCM} = \sum_{i=1}^{N}\sum_{j=1}^{c}\left[m\,|ij^n|\,\|x_i - y_j\|^2 + F_{ij}\right] \tag{8}$$

Then the altered fuzzy factor F_{ij}' is derived using Equation (9).

$$F_{ij}' = \sum_{C\in N, i\neq k} w_{ik}(1 - m_{ij})^m \tag{9}$$

This adjusted fuzzy factor plays a crucial role in influencing local neighbor relationships and substituting the traditional distance metric with a correlation function. Here, w_{ik} represents the fuzzy factor for cluster i, and $1 - C(x_i - y_j)$ signifies the correlation metric function. Since the histopathological images contain variation in intensities, gradients, and complex backgrounds, it becomes imperative to employ a modified KFCM-based segmentation method to distinguish between the region of interest (ROI) and the background in the image. Figure 2 illustrates the sequence of the original image, the filtered image, the identified ROI within the ACA image, and the segmented image generated using the modified KFCM for the Adenocarcinoma (ACA) class.

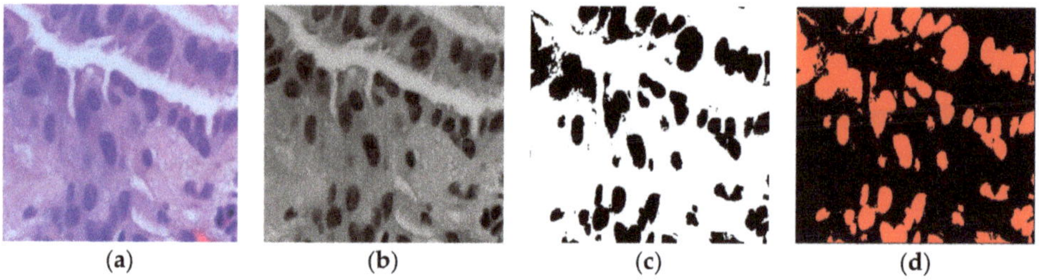

Figure 2. (a) Original ACA image; (b) Filtered ACA image; (c) ROI of the ACA Image; (d) Segmented ACA image.

The following section focuses on the methods utilized for extracting dimensionally reduced image features, aimed at enhancing the classification and detection of lung cancer using histopathological images.

3. Feature Extraction

Feature extraction techniques condense essential information from images into compact feature vectors, enabling the effective classification of complex image datasets using linear algorithms [43]. As the abundant features within histopathological images serve as a fundamental resource for clinicians to conduct diagnoses, the proficient extraction of these image features stands as a pivotal factor in enhancing the precision of computer-aided diagnosis [44]. This study delves into the impact of two distinct feature extraction techniques such as PSO and GWO on the classification of histopathological images related to lung cancer.

3.1. Particle Swarm Optimization (PSO)

Kennedy and Eberhart introduced the PSO algorithm in 1995, which draws inspiration from the hunting behavior of birds. This optimization method relies on a population and leverages the social dynamics of bird flocks. It starts by creating particles and setting key parameters for the optimization process. [45].

Every particle has a unique position that is traced by the following equation:

$$x_i^k = \left(x_{i1}^k, x_{i2}^k, \ldots, x_{iq}^k\right) \quad (10)$$

The velocity is traced by the following equation:

$$y_i^k = \left(y_{i1}^k, y_{i2}^k, \ldots, y_{iq}^k\right) \quad (11)$$

Each particle's velocity is updated as:

$$y_i^{k+1} = w_i y_i^k + c_1 r_1 \left(pbest_i - x_i^k\right) + c_2 r_2 \left(gbest_i - x_i^k\right) \quad (12)$$

Here, r_1 and r_2 represent randomly selected values within the range of 0 to 1. The acceleration coefficients, denoted as c_1 and c_2, play a role in analyzing the motion of particles. The weight function is expressed as:

$$w_i = \frac{w_{max} - w_{min}}{k_{max}} \times k \quad (13)$$

The position of each particle is given by:

$$x_i^{k+1} = x_i^k + y_i^{k+1} \quad (14)$$

The particle that possesses the optimal position progresses to the next level. The best position for an individual particle is represented by the letters "p-best", while the letters "g-best" represent the best position among all particles. The weight parameter "w_i" is chosen between 0.45–0.9, maximum iteration values are 100–1000, both r_1 and r_2 are set to 0.85, cognitive component (c_1) and Social Component (c_2) are chosen between 1.0–2.0. The above values are determined based on the trial-and-error method.

3.2. Grey Wolf Optimization (GWO)

Grey wolves are known for living and hunting in groups called packs [46]. The process of searching and hunting involves plotting to track and approach a target efficiently. This optimization technique, inspired by the search and hunting patterns of gray wolves, employs symbols such as Alpha (α), Beta (β), and Gamma (γ) to represent the best, next best, and third best solutions in mathematical modeling. Lambdas are presumed to be the remaining possible solutions and they guide the alpha, beta, and gamma wolves in searching and surrounding the prey. Three coefficients, A, B, and C are suggested to describe the encircling behavior. The equation of hunting strategy is formulated as follows:

$$D_\alpha = |B_1 \cdot X_\alpha - X(t)| \tag{15}$$

$$D_\beta = |B_2 \cdot X_\beta - X(t)| \tag{16}$$

$$D_\gamma = |B_3 \cdot X_\gamma - X(t)| \tag{17}$$

where D_α, D_β and D_γ denotes the adjusted distance variables from the alpha, beta, and delta positions to the other wolves, B_1, B_2 and B_3 are coefficients that assist in adapting these distance variables, t signifies the ongoing iteration, X indicates the position of the grey wolf and it follows as,

$$X_1 = |X_\alpha - A_1 D_\alpha| \tag{18}$$

$$X_2 = |X_\beta - A_2 D_\beta| \tag{19}$$

$$X_3 = |X_\gamma - A_3 D_\gamma| \tag{20}$$

$$X(t) = \frac{X_1 + X_2 + X_3}{3} \tag{21}$$

The parameters A and B can be mathematically expressed as follows:

$$A = 2i \cdot r_1 - i \tag{22}$$

$$B = 2 \cdot r_2 \tag{23}$$

The control parameter i chases A, which eventually drives the lambda wolves to flee from the dominant wolves such as α, β and γ. When there are multiple dominant wolves ($|A| > 1$), the grey wolves run away from them, allowing lambda wolves to search extensively and explore more during optimization. However, when there are fewer dominant wolves ($|A| < 1$), the grey wolves approach them and follow their guidance in hunting, which is called local search in optimization. During the iterations, the control parameter i is linearly decreased from 2 to 0, and is represented as,

$$i = 2 - (iter) \cdot \frac{2}{max\_iter} \tag{24}$$

where max_iter indicates the maximum iteration, and it is started from the beginning.

In the context of the classification problem, the introduction of randomness through variables r_1 and r_2 leads to heightened fluctuations in the wolves' positions. Consequently, their ability to effectively converge towards the target (prey) becomes hindered. To address this issue, a decision has been made to treat the values of r_1 and r_2 in Equations (9) and (10) as control parameters within a confined range of [0, 1], rather than allowing them to remain purely random. Through empirical experimentation, it has been determined that the optimal performance of the Grey Wolf Optimization (GWO) algorithm is achieved when both r_1 and r_2 are set to 0.8. This adjustment enhances the accuracy of the GWO algorithm in tackling the classification problem.

3.3. Statistical Analysis

To enhance the accuracy of cancer prediction using dimensionally reduced features, it is advisable to calculate statistical parameters from the region of interest. The intensity values, which have been reduced in dimensionality through methods such as PSO (Particle Swarm Optimization) and GWO (Grey Wolf Optimization), are then examined using statistical measures such as Mean, Variance, Skewness, Kurtosis, Pearson Correlation Coefficient (PCC), and CCA (Canonical Correlation Analysis). These statistical parameters help determine whether the outcomes accurately reflect the inherent properties of lung cancer data within the subspace. These attributes were derived for both normal and malignant classes.

The statistical parameters of cancer data, extracted using the PSO and GWO methods, are shown in Table 1. Variance quantifies data spread. Notably, Table 1 reveals lower mean values for normal cases using both PSO and GWO, while higher mean values are evident for malignant cases using both methods. Furthermore, the Malignant group demonstrates greater data spread compared to the Normal group as indicated by Table 1. GWO shows a Pearson correlation coefficient of 1 for both cases, implying strong intra-class correlation. Skewness and kurtosis are highly skewed for both normal and malignant instances. When CCA values exceed 0.5, strong inter-class correlation is present. However, Table 1 indicates that PSO and GWO methods exhibit the lowest inter-class correlation. Consequently, the analysis of these extracted features emphasizes the need for improved classifiers.

Table 1. Statistical Parameters in PSO and GWO for Feature Extraction in Malignant and Normal Data.

Statistical Parameters	PSO		GWO	
	Malignant	Normal	Malignant	Normal
Mean	0.8598080214	0.1109701363	0.01878313748	0.01751341349
Variance	0.05867975074	0.07425036326	0.07492946326	0.07494543857
Skewness	19.87029488	19.83047771	22.52231557	22.56212107
Kurtosis	441.8828416	444.9961882	509.1565306	510.3537192
Pearson CC	0.9019022281	0.9269991469	0.9985202125	0.997858273
CCA	0.12309		0.11291	

In cases where the features exhibit linear separability, a straightforward binary thresholding approach can be employed for the classification of Histopathological Lung images into two distinct classes: N and ACA. The characteristics of malignancy exhibit non-linear and non-Gaussian features that overlap with each other. To analyze these dimensionally reduced values which was obtained from PSO and GWO methods, histogram and scatterplot plots are used as illustrated in Figures 3 and 4.

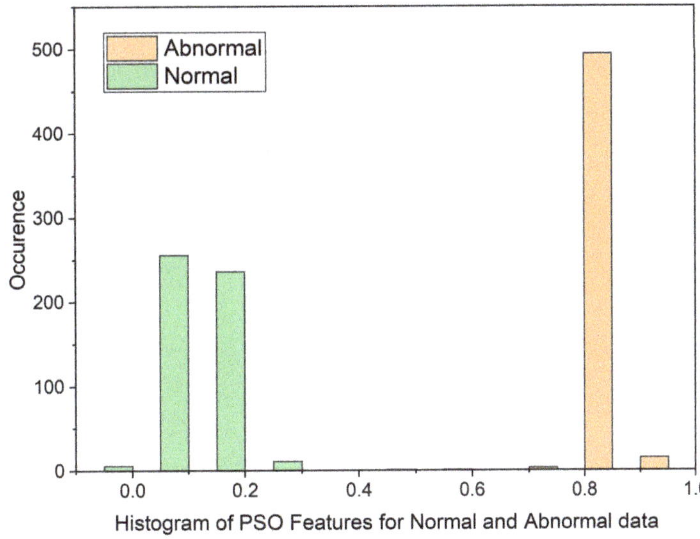

Figure 3. Histogram of PSO Features for Normal and Malignant data.

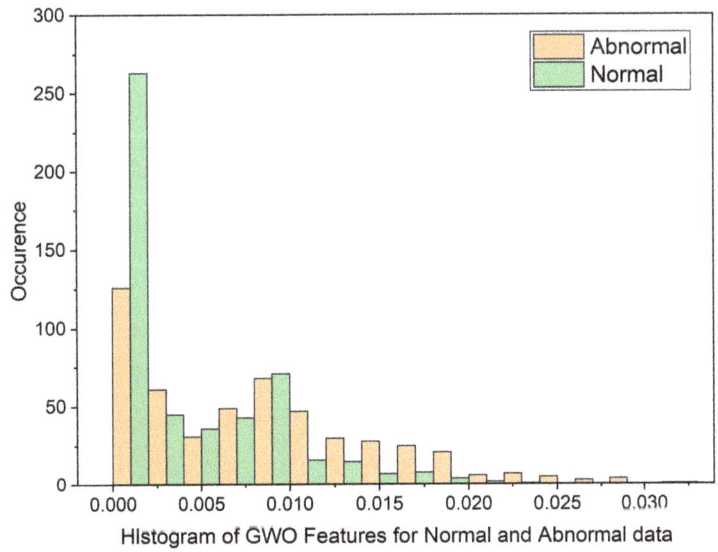

Figure 4. Histogram of GWO Features for Normal and Malignant data.

The histogram plot in Figure 3 illustrates the distribution of PSO feature data for normal and malignant cancer cases. The histogram illustrates PSO features characterized by outliers, substantial gaps, downward trends, and a non-Gaussian distribution. From Table 1, In the PSO-based extraction technique, the Canonical Correlation Coefficient (CCA) value is significantly low at 0.12309, suggesting a non-linear relationship between normal and malignant cases. Figure 4 showcases the histogram plot for GWO feature distribution, indicating skewed Poisson distributed data, and a non-linear nature.

Figures 5 and 6 display scatterplots demonstrating the feature output of normal and malignant cancer data utilizing the PSO and GWO methods. Scatter plots are useful for identifying data clustering, detecting nonlinearity, and overlapping. Both figures indicate the presence of nonlinearity and overlapping in the data. Therefore, from the histogram and scatterplot it is evident to employ accurate classifiers capable of distinguishing between normal and cancer cases in lung data using PSO and GWO features. The next section centers on the techniques applied to choose optimal image features, with the goal of improving the classification and identification of lung cancer in histopathological images.

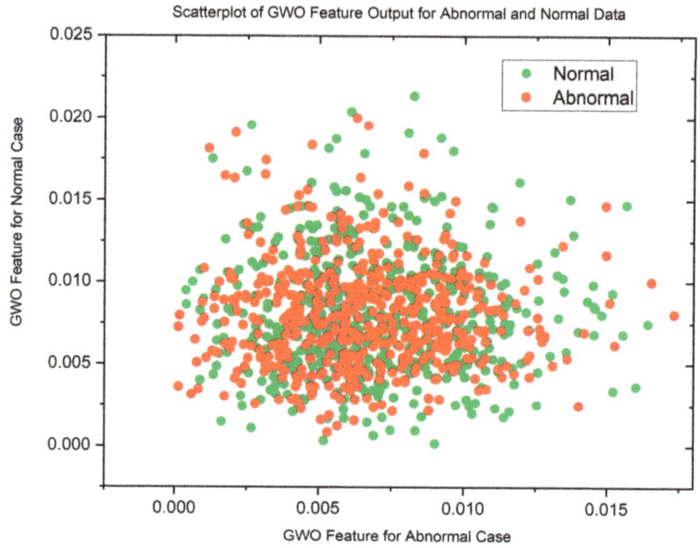

Figure 5. Scatterplot of PSO Features for Normal and Malignant Case.

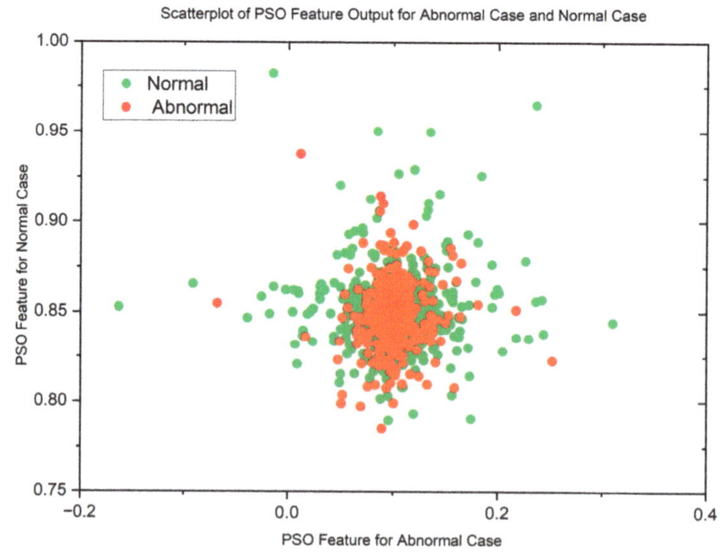

Figure 6. Scatterplot of GWO Features for Normal and Malignant Case.

4. Feature Selection

Feature selection aims to reduce input variables, excluding irrelevant characteristics for a more accurate, less complex, and unbiased model. Optimal feature selection is crucial for creating an effective, accurate machine learning model with high generalization ability [40]. In this paper, Feature Selection is performed using the KL Divergence and Invasive Weed

Optimization (IWO) methods. Following the feature extraction procedures as described in Section 3, which involves Particle Swarm Optimization (PSO) and Grey Wolf Optimization (GWO). After feature extraction method [256 × 256] is dimensionally reduced to [512 × 10] per histopathological image. These [512 × 10] intensity values per image serve as the initial input for the Feature Selection techniques namely KL Divergence and IWO. However, the application of Feature Selection techniques is a further dimensionality reduction method only. After PSO and GWO feature extraction methods through KL Divergence and IWO process [512 × 10] intensity values per image is reduced to [100 × 12] which represents the most relevant intensity values of the images are retained as input given to the classifier for the subsequent classification process. The histopathological images are represented as a relevant intensity value in matrix form as described above.

4.1. KL Divergence

KL Divergence, also known as relative entropy, measures disparities between probability distributions, but in an asymmetric manner. The KL divergence between a probability distribution $q = (q_1, q_2, \ldots, q_n)$ and another distribution $p = (p_1, p_2, \ldots, p_n)$ is defined as,

$$D_{KL}(q \mid\mid p) = \sum_{j=1}^{m} q_j log \frac{q_j}{p_j} \tag{25}$$

The integral form of the *KL* divergence for continuous distributions is expressed as follows:

$$D_{KL}(q \mid\mid p) = \int_{-\infty}^{\infty} q_j log \frac{q_j}{p_j} dx \tag{26}$$

The *KL* divergence exhibits mutual convexity for both discrete and continuous distributions. The following are the properties of the *KL* divergence measure:

$$D_{KL}(q \mid\mid p) = \begin{cases} 0,\ if\ q = p \\ c,\ c > 0,\ if\ partially\ overlapping \\ +\infty,\ if\ non-overlapping \end{cases} \tag{27}$$

From the above equation, it can be observed that when the *KL* divergence is smaller, the two compared distributions are more similar.

4.2. Invasive Weed Optimization

The invasive weed optimization algorithm is a popular population-based metaheuristic approach [47]. The dynamic and versatile characteristics of weed colonies have sparked the creation of an optimization algorithm that imitates their behavior. By leveraging the qualities of weeds, a straightforward and efficient optimization technique can be developed. This method, called the IWO algorithm, incorporates phases such as seeding, growth, and competition. The following are the strategy for simulating weed habitat behavior:

1. Primary Population Initialization: A few seeds are dispersed to start the search.
2. Reproduction process: Seeds have the potential to grow into flowering plants, which then choose and spread the fittest seeds for survival and reproduction. The quantity of grass grain grains decreases in a linear fashion from Y_{max} to Y_{min} as follows:

$$n(weed_j) = \frac{Y_{\max}(\max\_fit - fit(weed_j)) + Y_{\min}(fit(weed_j) - \min\_fit)}{\max\_fit - \min\_fit} \tag{28}$$

3. Spectral Spread Method: The group's seeds are distributed normally with a mean planting position and standard deviation (SD) determined by the equation below.

$$\sigma_t = \left(\frac{N-t}{N}\right)^m \left(\sigma_{intital} - \sigma_{final}\right) + \sigma_{final} \tag{29}$$

4. Competitive Deprivation: If the colony has more grasses than the maximum limit (Smax), the grass with the lowest fitness is eliminated to maintain a consistent number of herbs.
5. The process continues until the maximum iteration is reached, keeping the lowest cost value of the grasses.

The upcoming next section revolves around the utilization of classification methods to categorize lung cancer images within histopathological images.

5. Classifiers for the Detection of Lung Cancer

Classifiers have a crucial role in categorizing data effectively. An optimal classifier is characterized by its ability to achieve high accuracy and low error rates while maintaining manageable computational complexity. Addressing the classification challenge involves constructing a model for the purpose of classifying images and assigning them appropriate class labels. The following sections of this paper delve into the classifiers that were used for this purpose.

5.1. Support Vector Machine

SVM is known for its scalability and classification performance [42]. It aims to create a hyperplane that maximizes class separation by minimizing the cost function. It is given by the following expression:

$$Minimize,\ \frac{1}{2}\ \|w\|^2 + C\sum_{k=1}^{m}\mu_k \tag{30}$$

$$\text{Subject to } \mathrm{z_k}\left(\mathrm{w^T x_k} + \mathrm{f}\right)^3 \geq 1 - \mathrm{\mu_k}, \mathrm{\mu_k} \geq 0$$

where $w^T, x_k \in R^2$ and $f \in R$, $\|w\|^2 = w^T w$.

C represents the trade-off between the margin and the error. The training data's size is represented by $\sum_k$, and the class label for each sample is represented as $\mathrm{z_k}$. SVM is a flexible classifier suitable for linear and nonlinear cases. To handle nonlinear data, we employ Polynomial, RBF, and Sigmoid kernel functions. In this study, we exclusively enhance the classification accuracy by utilizing the SVM-RBF kernel.

5.2. K-Nearest Neighbor

KNN stands as a widely utilized and efficient non-parametric classification technique. In KNN, the symbol 'k' denotes the count of nearest neighbors involved in the voting process. To enhance prediction accuracy, employing an odd value for k is recommended. KNN determines the classification of a test sample by conducting a majority vote among neighboring training samples. Measuring distances between individuals is crucial, and the Euclidean distance is commonly used for this purpose [48]. For example, in the Euclidean space if u and v are the two points and it is assumed that $u = (u_1, u_2, u_3, \ldots, u_n)$ and $v = (v_1, v_2, v_3, \ldots, v_n)$, then the Euclidean distance of line segment can be expressed as follows:

$$Dist(u,y) = \sqrt{(u_1 - v_1)^2 + (u_2 - v_2)^2 + \ldots + (u_n - v_n)^2} = \sqrt{\sum_{i=1}^{n}(u_i - v_i)^2} \tag{31}$$

5.3. Random Forest

This tree-based ensemble learning algorithm is highly accurate and resilient in image classification [49]. It utilizes multiple decision trees that work independently. Two important parameters for the algorithm are the number of decision trees and the number of predictive variables used in each tree's decision-making process. By combining the votes of multiple decision trees, a random forest can accurately predict binary tasks. For a training set × consisting of M samples, each containing N features and a classification label Y. The following steps are involved in the construction of Random Forest.

1. Randomly select M samples from × using the Bootstrap method.
2. Choose n random features (where n < N) to split a decision tree node. Determine the split criterion by selecting the feature with the lowest Gini value. Gini is computed using the formula:
$$Gini = 1 - \sum_{i=1}^{c} (p_i)^2 \quad (32)$$
where p_i represents the relative frequency of dataset features and c represents the number of classes.
3. Generate M decision trees by repeating steps 1 and 2, M times.
4. Create a random forest by combining the decision trees and utilize voting to determine the classification outcome.

5.4. Decision Tree

It is a well-known machine learning algorithm that partitions input data recursively [50]. A decision tree starts with a root node and branches. This work utilizes CART, which splits the data based on its ability to distinguish between groups. The process continues until all data groups have the same label or match the training set. CART uses the Gini impurity measure at each node to determine the best split. The data at node 'd' are divided into two subsets, X-left and X-right, based on the splitting features and a threshold determined by CART and the amount of data X.

At node 'd' the input is computed through impurity measure Gini as $\sum_k p_{dk}(1 - p_{dk})$ with the proportion of class k observation in the node 'd'. Construction time of a decision tree depends on the dataset's size (samples and features). Overfitting can occur if the tree is built using CART and results in few samples per leaf. To prevent overfitting and improve accuracy, a pruning algorithm can be used to simplify the tree, reducing construction time while maintaining performance.

5.5. Softmax Discriminant Classifier

SDC's main objective is to classify a given test specimen [51] by comparing its distance to the training sample within its category. The process entails gauging the distance between training and test samples belonging to the same class to derive the outcome. Supposing, the training set $M = [M1,\ M2, \ldots,\ Mq] \in \mathfrak{R}^{c\times d}$ comes from q distinct classes. $M_q = [M1^q,\ M2^q, \ldots\ Mdq^q] \in \mathfrak{R}^{c\times d_q}$ Indicates dq samples from the qth class where $\sum_{j=1}^{q} d_j = d$. Assuming $w \epsilon \mathfrak{R}^{c\times 1}$ represents the test sample, within the classifier, we employ samples from class q to recognize the test sample, aiming to minimize the reconstruction error. To uphold the principle of SDC, we can enhance the non-linear transformation linking the q class samples and the test sample. Therefore, the SDC can be defined as follows:

$$h(w) = \underset{j}{\text{argmax}} z_w^j \quad (33)$$

$$h(w) = \underset{j}{\text{argmax}} \log\left(\sum_{j=1}^{d_j} \exp\left(-\lambda \left\|w - w_k^j\right\|_2\right)\right) \quad (34)$$

where $h\ (w)$ defines the distance between the jth class and the test sample. The value of λ should be greater than zero, to provide a penalty cost. If w relates to the jth class, then w and w_k^j would have likely same characteristics and so $\left\|w - w_k^j\right\|_2$ is progressing close towards zero and hence maximizing z_w^j can achieve the maximum possible value in an asymptotic manner.

5.6. Multilayer Perceptron

MLP is often used to approximate functions such as regression [52]. It consists of an input layer with n nodes, a hidden layer, and an output layer. The given input and output pairs be denoted as (m_p, n_p), $p = 1, 2, \ldots, m$, where $m_p = (m_{p1}, m_{p2}, \ldots, m_{pn})$ and y_p are the input vector and the corresponding desired output value, respectively. Sigmoid function is commonly used for hidden and output nodes, producing values from 0 to 1.

The kth hidden node in the MLP calculates its output when the input is given. The output value is computed as

$$c_{pk} = f_s\left(\sum_{j=1}^{n} w_{jk}m_{pj} + \theta_k\right) \quad (35)$$

The output value of the output node is determined by the sigmoid function (f_s), along with bias (θ_k), and connection weight (w_{jk}) associated with the corresponding hidden node. Then the final output value is computed as,

$$c_p = f_s\left(\sum_{k=1}^{l} w_k c_{pk} + \theta\right) \quad (36)$$

The number of hidden nodes is denoted by l, the bias to the output node is represented by θ, and w_k signifies the connection weight from the kth hidden node to the output node. This results in a total of $(n+2)l+1$ synaptic connections. To train the Multilayer Perceptron (MLP), the following cost function can be utilized.

$$E = \frac{1}{2}\sum_{j=1}^{t}(n_p - c_p)^2 \quad (37)$$

where t denotes the number of training patterns. In our study, we used a three-layer model, which is known to effectively approximate any continuous function with high accuracy [53].

5.7. Bayesian Linear Discriminant Classifier

The BLDC, or Bayesian Linear Discriminant Classifier, can distinguish between multiple classes. It uses the Fisher linear discriminant and applies the Bayes decision rule to estimate the error probability [54]. Bayesian regression assumes that the target variable y is a linear combination of vector k, and Gaussian noise m. This relationship is expressed as $y = q^T k + m$, where q represents the weight coefficients.

The given expression represents the likelihood function,

$$p\left(\frac{C}{\beta, q}\right) = \left(\frac{\beta}{2\pi}\right)^{m/2} \exp\left(-\frac{\beta}{2}\left\|M^T q - y\right\|^2\right) \quad (38)$$

In the above equation, y is the target values for regression, M is a matrix made by combining the training feature vectors horizontally, and C is the combination of $\{M, y\}$. β represents the noise's inverse variance, and T is the total number of samples in the training set.

5.8. Methods for Updating Hyperparameters in Various Classifiers

The performance of a classifier greatly depends on the values assigned to its hyperparameters [55]. To find the best hyperparameter values, different methods such as Stochastic Gradient Descent (SGD), Grid Search (GS), and Adaptive Moment Estimation Method (ADAM) can be used. This study introduces a new approach called R-Adam, which aims to enhance lung cancer classification accuracy for the Decision Tree classifier and other classifiers. While Adam is a prevalent choice for hyperparameter selection in deep learning networks, this study introduces R-Adam, an adapted version proposed for hyperparameter selection across diverse classifiers. Utilizing controlled randomness, the envisioned R-Adam algorithm aims to discover hyperparameter values in proximity to the optimal values recommended by the Adam method. The investigation assesses the classification performance using both Adam and the newly introduced R-Adam technique.

5.8.1. Adam Approach

The Adam approach involves employing squared gradients and exponential moving averages. The validation of hyperparameters is achieved based on the expressions provided below [56]:

$$x_{t+1} = x_t - \frac{L_r}{\varepsilon + \sqrt{\hat{P}_t}} * \hat{M}_t \quad (39)$$

where x_t represents the previous hyperparameters, x_{t+1} denotes the updated hyperparameters, L_r signifies the learning rate, and ε is a small constant used to avoid division by zero. The constants in the Adam method are Z_1 and Z_2.

$$\hat{M}_t = \frac{m_t}{1 - Z_1^t} \quad (40)$$

$$\hat{P}_t = \frac{p_t}{1 - Z_2^t} \quad (41)$$

$$m_t = Z_1 * m_{t-1} + (1 - Z_1) * \frac{\partial L}{\partial x_t} \quad (42)$$

$$p_t = Z_2 * p_{t-1} + (1 - Z_2) * \left(\frac{\partial L}{\partial x_t}\right)^2 \quad (43)$$

where $\frac{\partial L}{\partial x_t}$ signifies the derivative of the loss function with respect to x. Thus, the mathematical representation of the loss function is as follows:

$$\frac{\partial L}{\partial x_{tr}} = \frac{ER_{tr}}{x_{in}},\ if\ tr = 1 \quad (44)$$

$$\frac{\partial L}{\partial x_{tr}} = \frac{ER_{tr} - ER_{tr-1}}{x_{tr} - x_{tr-1}},\ if\ tr > 1 \quad (45)$$

where ER stands for the error rate, tr indicates the current iteration and $tr - 1$ denotes the previous iteration of in the Adam approach. Algorithm 1 outlines the process of utilizing the Adam optimizer to update hyperparameters in a Decision Tree model, aiming to minimize the error rate, which serves as the loss function. In Decision Tree, the key hyperparameters include maximum depth and criterion. In Decision tree, the hyperparameters are set as maximum depth = 20 and criterion = MSE. The Adam's approach employs specific constants in this work: L_r = 0.001, Z_1 = 0.89, Z_2 = 0.9 and $\varepsilon = 10^{-9}$. Through experimentation, the optimal number of iterations for the Adam approach was determined to be 40. This iterative process aims to uncover the lowest error rate, helping identify the best hyperparameters. Notably, similar approaches involving SVM, KNN, Random Forest, SDC, MLP, and BLDC models could also leverage the Adam optimizer to update hyperparameters in a comparable manner.

Algorithm 1. Adam Approach

1. Initialization: Set initial values for hyperparameters:

Target value, maximum iterations, maximum depth and criterion for Decision Tree, maximum iterations for Adam, L_r, Z_1, Z_2, ε.

2. Hyperparameter Tuning Loop:
   (a) For t_r = 1 to maximum iterations for Adam
      - Calculate maximum depth.
      - Determine the criterion.

   (b) For t = 1 to maximum iterations for Decision Tree
      - Update values for maximum depth
      - Criterion is set to MSE.
      - Determine the optimal values for maximum depth.

   end for.
3. Formulate a confusion matrix and compute the error rate (ER).
4. Compute the loss gradient using Equation (45).
5. Establish new optimal hyperparameter values using Equation (39) through (43).
   end for.

5.8.2. RAdam's Approach

The Randomized Adam (RAdam) technique is tailored to enhance the precision of the Decision Tree classifier. Algorithm 2 presents a methodology for implementing the Decision Tree using the RAdam approach. RAdam amalgamates two core components: the Adam method and controlled randomization. The controlled randomization process is pivotal in elevating classification performance. Within each iteration of the Adam method, hyperparameters are updated. The Adam process, which meticulously refines hyperparameter ranges, is nested within the iterative controlled randomization. This controlled randomization strategy integrates two control parameters—solution considering rate and solution adjusting rate—to fulfill its objective. Constants for R-Adam are defined as follows: bandwidth is set at 0.0098, the maximum number of iterations for randomization is 15, solution considering rate is 0.6, and solution adjusting rate is 0.92. In Algorithm 2, randomization 1, randomization 2, randomization 5, and randomization 6 indicate random values from the range [0, 1], while randomization 3 and randomization 4 correspond to random values within [0, 0.1]. Following this iterative process, the lowest error rate is found, leading to the identification of optimal hyperparameters. Significantly, analogous methodologies that pertain to SVM, KNN, Random Forest, SDC, MLP, and BLDC models could also make use of the Adam optimizer for adjusting hyperparameters in a similar fashion.

The following section pertains to the outcomes derived from employing diverse classification techniques for the categorization of lung cancer images within histopathological images.

Algorithm 2. RAdam's Approach

1. Initialization: Set initial values for hyperparameters:

Target value, maximum iterations, maximum depth and criterion for Decision Tree, maximum iterations for Adam, L_r, Z_1, Z_2, ε, solution considering rate, solution adjusting rate and bandwidth.

2. Hyperparameter Tuning Loop:
   (a) For $t_r = 1$ to maximum iterations for Adam
      - Calculate maximum depth.
      - Determine the criterion.

   (b) For $t = 1$ to maximum iterations for Decision Tree
      - Update values for maximum depth
      - Criterion is set to MSE.
      - Determine the optimal values for maximum depth.

   end for.
3. Formulate a confusion matrix and compute the error rate (ER).
4. Compute the loss gradient using Equation (45).
5. Establish new optimal hyperparameter values using Equation (39) through (43).
   end for.
6. For each iteration: current iterations for randomization = 1 to maximum iterations for randomization
   If randomization 1 < solution considering rate.
   Set r_1 for this iteration as $r_1\prime$
   Set r_2 for this iteration as $r_2\prime$
   If randomization 2 < solution adjusting rate.
   Set r_1 for this iteration as $r_1\prime$+bandwidth$\times$ randomization 3.
   Set r_2 for this iteration as $=r_2\prime$+bandwidth$\times$ randomization 4.
   end if.
   If r_1 for this iteration is less than the lower bound, set it to the lower bound.
   end if.
   If r_2 for this iteration is less than the lower bound, set it to the lower bound.
   end if.
   If r_1 for this iteration is less than the upper bound, set it to the upper bound.
   end if.
   If r_2 for this iteration is less than the upper bound, set it to the upper bound.
   end if.
   Set r_1 for this iteration as lower bound + (bandwidth $\times$ randomization 5).
   Set r_2 for this iteration as lower bound + (bandwidth $\times$ randomization 6).
   end if.
7. Repeat.
   - Calculate maximum depth.
   - Determine the criterion.
8. For each iteration t from 1 to maximum iterations for Decision Tree
   - Update the values of maximum depth.
   - Set the criterion to MSE.
   - Determine the optimal values for maximum depth.

   end for.
9. Formulate a confusion matrix and compute the error rate (ER).
10. Compute the ER using r_1 & r_2 as hyperparameters. end for.

6. Results and Discussion

This section explores the efficacy of different classifiers based on their benchmark parameters. A higher classification accuracy combined with a decreased error rate signifies robust performance of the classifier. As a result, the classifiers underwent training and testing using the extracted and chosen feature values within the Lung Histopathological Image Dataset.

6.1. Training and Testing of the Classifiers

The training and testing of the classifiers constitute crucial phases within classification procedures. Training facilitates the acquisition of patterns linked to the provided dimensionally reduced intensity values of the histopathological images by the classifier. In this study, the entire dataset, comprising histopathological image values related to lung cancer detection and classification, is divided into 10 equal folds. The analysis involves a series of iterations. During each iteration, one-fold is designated as the testing set, while the remaining nine folds are combined and used as the training set. In essence, 10% of the data is reserved for testing in each iteration, and the remaining 90% is utilized for training. Various performance metrics are computed for each iteration. The results obtained from all 10 iterations are collected and aggregated. This aggregation often involves calculating average values for the performance metrics. The conclusion of training and testing for the classifiers was established based on the mean square error (MSE) acting as the termination criterion. The mathematical expression for MSE is given below:

$$MSE = \frac{1}{M}\sum_{i=1}^{M}(O_i - T_k)^2 \tag{46}$$

where O_i signifies the value observed at a definite time; T_k indicates the target value for model k, with "k" ranging from 1 to 15; and the value of M is assumed to be 5000 and indicates the total number of images.

6.2. Selection of the Optimal Parameters for the Classifiers

In this study, seven classifiers were used to categorize images into benign or adenocarcinoma based on the target selection. The target selection for the benign case $\left(T_{benign}\right)$ is represented as follows:

$$\frac{1}{M}\sum_{k=1}^{M}\mu_k \leq T_{benign} \tag{47}$$

The characteristics of the entire set of benign lung data (M) were subjected to normalization, and their average is denoted as μ_k as outlined in Equation (38), applicable for classification purposes.

The average of the normalized features is denoted as μ_k. For benign images, a target value of 0.1 was selected, which falls within the lower end of the 0–1 scale.

The condition for choosing a target in a case of adenocarcinoma (aca) is:

$$\frac{1}{N}\sum_{i=1}^{N}\mu_i \leq T_{aca} \tag{48}$$

The characteristics of the entire set of lung adenocarcinoma data (N) were subjected to normalization, and their average is denoted as μ_i as outlined in Equation (39), applicable for classification purposes.

To enhance adenocarcinoma classification, the target selection should exceed the mean value μ_k, which represents the average of normalized features across N images. Improving classification requires a target value of 0.5 or higher, as specified by the condition:

$$\left(\left\|T_{aca} - T_{benign}\right\| \geq 0.5\right) \tag{49}$$

Depending on the criteria described in Equation (40), the selected targets for this study were set at 0.1 for benign cases and 0.85 for adenocarcinoma cases. The classifiers underwent training using a 10-fold cross validation training and testing approach, with the stopping criterion being an MSE value of $\left(10^{-5}\right)$ or a maximum operation of 1000, whichever was achieved first. The selection of optimal parameters for the classifiers during the training process is outlined in Table 2. In the case of SVM (RBF) classifier the parameters are selected through the trial-and-error method. The classifiers parameters are α, Kernel width parameter (σ), w, and b are selected with the constraint of minimum MSE. In the case of KNN, K value indicates the number of clusters and in this case, it is K = 5 is selected randomly. With Euclidean Distance measure as the cluster coefficient with weight w = 0.5 is selected with the constrain of minimum MSE. In the case of Random Forest, the parameters such as Number of trees, Maximum depth and Bootstrap sample are initialized with random selection. Similarly in the case of Decision Tree, the parameter maximum depth is initialized with random selection. Since it is a binary classification problem, the class weight value for Random Forest is settled at 0.45, whereas for Decision tree, the class weight is settled at 0.4. In the case of Softmax Discriminant Classifier, it is a binary classification problem, so the λ value is settled at 0.5 along with the mean of each class target values as 0.1 and 0.85. In the case of Multilayer Perceptron Classifier, the network is trained using LM (Levenberg-Marquardt) algorithm to minimize the square output error. This error back propagation algorithm is used to calculate the weights updates in each layer of the network. As the number of hidden units gradually increased from its initial value, then there will be a reduction in the minimum Mean Squared Error (MSE) on the testing set. The optimal number of hidden units is the one that results in the lowest MSE. If the number of hidden units is increased beyond this point, the model's performance does not show any further improvement; instead, it often starts to decline. This decline occurs since the neural network becomes unnecessarily complex, exceeding the complexity necessary to solve the problem effectively. The choice of the learning rate as 0.3 is determined based on the distribution of training patterns and their associated MSE. In case of BLDC, the parameters such as prior probability p(x) − 0.5, Class mean μ_x = 0.8 and μ_y = 0.1 are selected with constrain of minimum MSE. The training process demonstrated that the MSE value was attained either as low as 1.0×10^{-10} or after 1000 iterations.

Table 2. Selection of the Optimal Parameters for the Classifiers.

Classifiers	Optimal Parameters of the Classifiers
Support Vector Machine	Kernel—RBF; α—1; Kernel width parameter (σ)—100; w—0.85; b—0.01; Convergence Criterion—MSE.
K-Nearest Neighbor	K—5; Distance Metric—Euclidian; w—0.5; Criterion—MSE.
Random Forest	Number of Trees—200; Maximum Depth—10; Bootstrap Sample—20; Class Weight—0.45.
Decision Tree	Maximum Depth—20; Impurity Criterion—MSE; Class Weight—0.4.
Softmax Discriminant Classifier	λ = 0.5 along with mean of each class target values as 0.1 and 0.85.
Multilayer Perceptron	Learning rate—0.3; Learning Algorithm—LM; Criterion—MSE.
Bayesian Linear Discriminant Classifier	Prior Probability P(x)—0.5; Class mean μ_x = 0.8 and μ_y = 0.1, Criterion = MSE.

6.3. Performance Metrics of the Classifiers

The primary objective of the classifier was to effectively distinguish between cancer cells and normal data samples in the dataset. As this research focuses on binary classification, it is essential to select appropriate performance metrics. In binary classification tasks, one of the key evaluation tools is the confusion matrix. This matrix provides a concise summary of the model's predictions in relation to the actual labels of the dataset. The confusion matrix consists of four elements: True Positive (TP), True Negative (TN), False Positive (FP), and False Negative (FN). TP indicates the presence of lung cancer, while TN indicates its absence, both representing correct classification. FP and FN represent misclassification, where lung cancer is incorrectly predicted as present (FP), or lung cancer is present but wrongly classified as not present (FN).

Table 3 displays TP, TN, FP, FN values, and average MSE for PSO and GWO features along with seven classifiers without employing Feature Selection Methods. Achieving the lowest MSE serves as an indicator for improved classifier performance, while a higher MSE value results in inferior classifier performance, regardless of the employed feature selection methods. PSO features show Decision Tree Classifier with the lowest MSE (3.60×10^{-7}) and Random Forest Classifier with the highest MSE (1.60×10^{-5}). GWO features show Bayesian LDC Classifier with the minimum MSE (2.50×10^{-7}) and KNN Classifier with the maximum MSE (1.44×10^{-5}).

Table 3. Confusion Matrix for Classifiers without Feature Selection.

Feature Extraction	Classifiers	Confusion Matrix				MSE
		TP	TN	FP	FN	
PSO	SVM	3944	4009	991	1056	7.29×10^{-6}
	KNN	4267	3725	1275	733	4.49×10^{-5}
	Random Forest	2692	2933	2067	2308	1.60×10^{-5}
	Decision Tree	3184	3217	1783	1816	3.60×10^{-7}
	Softmax Discriminant	4033	3750	1250	967	4.00×10^{-8}
	Multilayer Perceptron	3425	3675	1325	1575	2.25×10^{-6}
	Bayesian LDC	4367	3975	1025	633	5.63×10^{-5}
GWO	SVM	3617	4175	825	1383	5.76×10^{-6}
	KNN	3500	3725	1275	1500	1.44×10^{-5}
	Random Forest	3967	3817	1183	1033	3.36×10^{-5}
	Decision Tree	4517	3984	1016	483	8.41×10^{-6}
	Softmax Discriminant	4083	4275	725	917	1.96×10^{-4}
	Multilayer Perceptron	4050	4384	616	950	4.84×10^{-4}
	Bayesian LDC	3967	3692	1308	1033	2.50×10^{-7}

The features extracted were given to seven classifiers for performance analysis, following feature selection methods. Table 4 shows the average MSE and confusion matrix for PSO Feature Extraction with KL Divergence and IWO feature selection. The Decision Tree had the lowest MSE (9.00×10^{-6}) using PSO with KL Divergence, while the Bayesian LDC had the highest MSE (1.02×10^{-5}). With PSO and IWO, the Decision Tree had the lowest MSE (7.84×10^{-6}), while the Softmax Discriminant had the highest MSE (1.22×10^{-5}).

Table 5 displays the average MSE and confusion matrix for GWO Feature Extraction with KL Divergence and IWO feature selection methods. The results include SVM, KNN, Random Forest, Decision Tree, Softmax Discriminant, Multilayer Perceptron, and Bayesian LDC classifiers. In the GWO with KL Divergence approach, Bayesian LDC achieves the lowest MSE (1.00×10^{-8}), while the Multilayer Perceptron Classifier has the highest MSE (2.03×10^{-5}). Similarly, in the PSO with IWO approach, SVM achieves the minimum MSE (4.90×10^{-7}), while the Random Forest Classifier has the maximum MSE (1.52×10^{-5}).

Table 6 presents the mean Mean Squared Error (MSE) and confusion matrix outcomes for PSO Feature Extraction using KL Divergence and IWO feature selection techniques in Adam Hyperparameter Tuning. Among these, Bayesian LDC achieved the smallest MSE (8.41×10^{-6}) through PSO with KL Divergence, whereas Random Forest showed the highest MSE (2.72×10^{-4}). When considering PSO and IWO, Random Forest demonstrated the lowest MSE (9.00×10^{-8}), whereas Softmax Discriminant had the highest MSE (4.00×10^{-4}).

Table 4. Confusion Matrix for Classifiers for PSO with KL Divergence and IWO.

Feature Selection	Classifiers	Confusion Matrix				MSE
		TP	TN	FP	FN	
KL Divergence	SVM	3297	2747	2253	1703	3.24×10^{-6}
	KNN	3978	2605	2395	1022	8.41×10^{-6}
	Random Forest	4115	3294	1706	885	2.30×10^{-5}
	Decision Tree	3919	4089	911	1081	9.00×10^{-6}
	Softmax Discriminant	4089	4258	742	911	4.84×10^{-6}
	Multilayer Perceptron	4271	3633	1367	729	2.56×10^{-6}
	Bayesian LDC	3298	3311	1690	1702	1.02×10^{-5}
IWO	SVM	3854	3503	1497	1146	2.21×10^{-5}
	KNN	3490	3985	1016	1510	3.36×10^{-5}
	Random Forest	3574	2757	2243	1426	1.94×10^{-5}
	Decision Tree	2982	2871	2129	2018	7.84×10^{-6}
	Softmax Discriminant	2734	3047	1953	2266	1.22×10^{-5}
	Multilayer Perceptron	3047	2592	2408	1953	1.00×10^{-6}
	Bayesian LDC	2681	2698	2302	2319	1.85×10^{-5}

Table 5. Confusion Matrix for Classifiers for GWO with KL Divergence and IWO.

Feature Selection	Classifiers	Confusion Matrix				MSE
		TP	TN	FP	FN	
KL Divergence	SVM	4029	2742	2258	971	1.00×10^{-6}
	KNN	3789	4147	853	1211	4.90×10^{-5}
	Random Forest	3490	4089	911	1510	6.40×10^{-7}
	Decision Tree	3594	4147	853	1406	2.50×10^{-7}
	Softmax Discriminant	4896	2668	2333	104	1.00×10^{-6}
	Multilayer Perceptron	3737	2982	2018	1263	2.03×10^{-5}
	Bayesian LDC	3460	2767	2233	1540	1.00×10^{-8}
IWO	SVM	4401	3262	1738	599	4.90×10^{-7}
	KNN	3203	3880	1120	1797	1.60×10^{-5}
	Random Forest	4440	2735	2265	560	1.52×10^{-5}
	Decision Tree	4167	2620	2380	833	5.29×10^{-6}
	Softmax Discriminant	4219	2687	2313	781	2.30×10^{-5}
	Multilayer Perceptron	4375	2747	2253	625	9.61×10^{-6}
	Bayesian LDC	3216	2760	2240	1784	6.89×10^{-5}

Table 7 displays the average Mean Squared Error (MSE) and the results of the confusion matrices obtained from GWO Feature Extraction using KL Divergence and IWO feature selection techniques in Adam Hyperparameter Tuning. Among these approaches, Multilayer Perceptron achieved the smallest MSE of 6.40×10^{-7} using GWO with KL Divergence, while SVM exhibited the highest MSE of 1.23×10^{-5}. Considering both GWO and IWO, Decision Tree showcased the lowest MSE of 6.40×10^{-7}, whereas Softmax Discriminant had the highest MSE of 1.04×10^{-4}.

Table 6. Confusion Matrix for Classifiers: PSO with KL Divergence and IWO for Adam Hyperparameter Tuning.

Feature Selection	Classifiers	Confusion Matrix				MSE
		TP	TN	FP	FN	
KL Divergence	SVM	4089	3568	1433	911	6.61×10^{-4}
	KNN	4184	4487	514	817	1.44×10^{-5}
	Random Forest	4555	3520	1480	445	2.72×10^{-4}
	Decision Tree	3815	3809	1191	1185	6.72×10^{-5}
	Softmax Discriminant	4392	3948	1052	608	2.40×10^{-5}
	Multilayer Perceptron	3881	4048	952	1119	1.96×10^{-6}
	Bayesian LDC	4156	3947	1053	844	8.41×10^{-6}
IWO	SVM	3599	4085	915	1401	8.10×10^{-5}
	KNN	4058	4375	625	942	7.23×10^{-5}
	Random Forest	4129	4038	962	871	9.00×10^{-8}
	Decision Tree	3713	4308	692	1288	6.40×10^{-5}
	Softmax Discriminant	4129	4161	839	871	4.00×10^{-4}
	Multilayer Perceptron	4539	4024	976	461	2.50×10^{-5}
	Bayesian LDC	3817	3797	1203	1183	1.44×10^{-5}

Table 7. Confusion Matrix for Classifiers: GWO with KL Divergence and IWO for Adam Hyperparameter Tuning.

Feature Selection	Classifiers	Confusion Matrix				MSE
		TP	TN	FP	FN	
KL Divergence	SVM	3653	4466	534	1347	1.23×10^{-5}
	KNN	4139	4948	52	862	7.23×10^{-5}
	Random Forest	4044	3913	1088	956	1.30×10^{-5}
	Decision Tree	3635	3985	1016	1365	6.89×10^{-5}
	Softmax Discriminant	3565	4297	703	1435	1.37×10^{-5}
	Multilayer Perceptron	3740	4034	966	1260	6.40×10^{-7}
	Bayesian LDC	3775	3987	1013	1225	4.90×10^{-7}
IWO	SVM	4339	4617	383	661	1.94×10^{-5}
	KNN	4129	4321	680	871	5.76×10^{-6}
	Random Forest	4509	4466	534	491	7.57×10^{-5}
	Decision Tree	4617	4390	610	383	6.40×10^{-7}
	Softmax Discriminant	4409	4005	995	592	1.04×10^{-4}
	Multilayer Perceptron	4409	3913	1088	592	4.49×10^{-5}
	Bayesian LDC	4754	3973	1027	246	4.90×10^{-7}

Table 8 presents the average Mean Squared Error (MSE) and the results of confusion matrices obtained by using PSO Feature Extraction with KL Divergence and IWO feature selection techniques during R-Adam Hyperparameter Tuning. Among these methods, SVM achieved the smallest MSE of 6.56×10^{-5} when using GWO with KL Divergence, while Random Forest had the highest MSE of 1.09×10^{-5}. Considering both PSO and IWO,

Random Forest had the lowest MSE of 4.49×10^{-5}, while Softmax Discriminant had the highest MSE of 1.10×10^{-4}.

Table 8. Confusion Matrix for Classifiers: PSO with KL Divergence and IWO for RAdam Hyperparameter Tuning.

Feature Selection	Classifiers	Confusion Matrix				MSE
		TP	TN	FP	FN	
KL Divergence	SVM	4144	3668	1333	856	6.56×10^{-5}
	KNN	4209	4537	464	792	2.92×10^{-5}
	Random Forest	4575	3620	1380	425	1.09×10^{-5}
	Decision Tree	3950	3859	1141	1050	5.93×10^{-5}
	Softmax Discriminant	4417	4098	902	583	1.60×10^{-5}
	Multilayer Perceptron	4011	4198	802	989	3.03×10^{-5}
	Bayesian LDC	4245	4047	953	755	3.97×10^{-5}
IWO	SVM	3710	4235	765	1290	1.37×10^{-5}
	KNN	4208	4375	625	792	4.22×10^{-5}
	Random Forest	4229	4188	812	771	4.49×10^{-5}
	Decision Tree	3813	4408	592	1188	4.36×10^{-5}
	Softmax Discriminant	4229	4211	789	771	1.10×10^{-4}
	Multilayer Perceptron	4558	4074	926	443	2.30×10^{-5}
	Bayesian LDC	3917	3897	1103	1083	3.02×10^{-5}

Table 9 displays the average Mean Squared Error (MSE) and the outcomes of confusion matrices. These were derived using GWO Feature Extraction with KL Divergence and IWO feature selection methods within R-Adam Hyperparameter Tuning. Among the techniques, Bayesian LDC achieved the lowest MSE of 9.61×10^{-6} with GWO and KL Divergence. Conversely, Random Forest had the highest MSE of 1.02×10^{-5}. When considering both GWO and IWO, KNN displayed the smallest MSE of 5.48×10^{-5}, while Random Forest exhibited the highest MSE of 1.90×10^{-4}.

Table 10 presents the metrics used to evaluate the performance of classifiers, including Accuracy, Error Rate, F1 Score, MCC, Jaccard Index, g-Mean, and Kappa. The mathematical expressions for these metrics are also provided.

The lung cancer data are processed using PSO and GWO techniques to extract features from normal and malignant data. These features are then used as inputs for seven classification models. Table 11 shows the performance of the classifiers without Feature Selection. The Decision Tree Classifier stands out with the highest accuracy of 85.01% for GWO features. It also achieves the highest F1 score (85.77%), MCC value (0.70), Jaccard Index (75.08%), g-mean (85.33%), kappa score (0.70), and the lowest error rate (14.99%). In contrast, the Random Forest classifier performs poorly for PSO features, with an accuracy of 56.25%, F1 score of 55.17%, MCC value of 0.13, Jaccard Index of 38.09%, g-mean of 56.26%, kappa value of 0.13, and the highest error rate of 43.75%. Without feature selection, the Decision Tree Classifier with GWO feature extraction method achieves the best accuracy and outperforms other classifiers.

The performance of a random forest model heavily relies on the quality of individual trees and the diversity among them. If a random forest includes subpar or correlated trees, it can result in reduced overall accuracy. Correlation among trees can introduce redundant information, hampering the model's ability to generalize effectively to new data and causing a drop in accuracy.

Table 9. Confusion Matrix for Classifiers: GWO with KL Divergence and IWO for RAdam Hyperparameter Tuning.

Feature Selection	Classifiers	Confusion Matrix				MSE
		TP	TN	FP	FN	
KL Divergence	SVM	3758	4466	534	1242	1.37×10^{-5}
	KNN	4159	4948	52	842	2.40×10^{-5}
	Random Forest	4094	4063	938	906	1.02×10^{-5}
	Decision Tree	3750	3985	1016	1250	1.23×10^{-5}
	Softmax Discriminant	3670	4297	703	1330	4.76×10^{-5}
	Multilayer Perceptron	3860	4084	916	1140	2.12×10^{-5}
	Bayesian LDC	3905	4137	863	1095	9.61×10^{-6}
IWO	SVM	4439	4667	333	561	4.36×10^{-5}
	KNN	4229	4321	680	771	5.48×10^{-5}
	Random Forest	4559	4466	534	441	1.90×10^{-4}
	Decision Tree	4667	4490	510	333	2.40×10^{-5}
	Softmax Discriminant	4459	4055	945	542	5.33×10^{-5}
	Multilayer Perceptron	4459	4063	938	542	5.04×10^{-5}
	Bayesian LDC	4789	4073	927	211	1.09×10^{-5}

Table 10. Standard Benchmark Parameters.

Performance Metrics	Equation	Significance
Accuracy (%)	$Accuracy = \frac{TP+TN}{TP+TN+FP+FN}$	Average positive-to-negative sample ratio.
Error Rate	$Err = \frac{FP+FN}{TP+TN+FP+FN}$	The number of incorrect predictions, based on recorded observations.
F1 Score (%)	$F1 = \frac{2TP}{2TP+FP+FN}$	Average of precision and recall to obtain the classification accuracy of a specific class.
MCC	$MCC = \frac{TN \times TP - FN \times FP}{\sqrt{(TP+FP)(TP+FN)(TN+FP)(TN+FN)}}$	Pearson correlation between the actual output and the achieved output.
Jaccard Index (%)	$Jaccard = \frac{TP}{TP+FP+FN}$	The number of predicted true positives exceeded the number of actual positives, regardless of whether they were real or predicted.
g-mean (%)	$g - mean = \sqrt{\frac{TP}{TP+FN} * \frac{TN}{TN+FP}}$	Combination of sensitivity and specificity into a single value that balances both objectives.
Kappa	$Kappa = \frac{\Pr(a)-\Pr(e)}{1-\Pr(e)}$	Inter-rater agreement measure for assessing agreement between two methods in categorizing cancer cases.

In Table 11, the Random Forest Classifier, when paired with the feature extraction technique of Particle Swarm Optimization (PSO), achieves a lower accuracy of 56.25% compared to the Genetic Wolf Optimization (GWO) approach, which achieves an accuracy of 77.84%. This discrepancy is primarily due to PSO selecting suboptimal intensity values of the segmented image, including less informative or irrelevant intensity values in the random forest model. Effective selection of intensity values is crucial for any classifier to yield better results. If the process of selecting intensity values fails to filter out irrelevant ones, it can negatively impact the performance of the Random Forest model. And also,

PSO selects intensity values that are highly specific to the training dataset, resulting in a model that performs well on the training data but struggles to generalize to new, unseen data. Furthermore, the computational demands of PSO can lead to longer tree construction times, affecting the overall classifier's performance.

Table 11. Performance Analysis of the Classifiers without Feature Selection.

Feature Extraction	Classifiers	Accuracy (%)	Error Rate (%)	F1 Score (%)	MCC	Jaccard Index (%)	g-Mean (%)	Kappa
PSO	SVM	79.53	20.47	79.40	0.59	65.83	79.53	0.59
	KNN	79.92	20.08	80.95	0.60	68.00	80.21	0.60
	Random Forest	56.25	43.75	55.17	0.13	38.09	56.26	0.13
	Decision Tree	64.01	35.99	63.89	0.28	46.94	64.01	0.28
	Softmax Discriminant	77.83	22.17	78.44	0.56	64.53	77.90	0.56
	Multilayer Perceptron	71	29	70.26	0.42	54.15	71.04	0.42
	Bayesian LDC	83.42	16.58	84.05	0.67	72.48	83.59	0.67
GWO	SVM	77.92	22.08	76.62	0.56	62.09	78.21	0.56
	KNN	72.25	27.75	71.61	0.45	55.78	72.29	0.45
	Random Forest	77.84	22.16	78.17	0.56	64.16	77.86	0.56
	Decision Tree	85.01	14.99	85.77	0.70	75.08	85.33	0.70
	Softmax Discriminant	83.58	16.42	83.26	0.67	71.32	83.62	0.67
	Multilayer Perceptron	84.34	15.66	83.80	0.69	72.12	84.46	0.69
	Bayesian LDC	76.59	23.41	77.22	0.53	62.89	76.66	0.53

Table 12 compares the performance of seven classifiers with PSO, KL Divergence, and IWO Feature Selection. The Softmax Discriminant Classifier stands out with superior results for KL Divergence features, achieving an accuracy of 83.47%, the highest F1 score of 83.18%, MCC of 0.67, Jaccard Index of 71.21%, g-mean of 83.50%, Kappa score of 0.67, and the lowest error rate of 16.53%. Conversely, the Bayesian LDC classifier performs poorly with IWO features, obtaining an accuracy of 53.79%, F1 score of 53.71%, MCC of 0.08, Jaccard Index of 36.72%, g-mean of 53.79%, Kappa of 0.08, and the highest error rate of 46.21%. Overall, the Softmax Discriminant Classifier using PSO and KL Divergence Feature selection achieves the highest accuracy and outperforms other classifiers.

Table 13 presents the performance of seven classifiers using GWO features, KL Divergence, and IWO Feature Selection. The KNN Classifier achieves the highest accuracy of 79.36% with KL Divergence features. It also obtains the highest F1 score (78.60%), MCC value (0.59), Jaccard Index (64.74%), g-mean (79.49%), kappa score (0.59), and lowest error rate (20.64%) among all classifiers. However, the Bayesian LDC classifier performs poorly with IWO features, achieving an accuracy of 59.76%, F1 score of 61.52%, MCC value of 0.20, Jaccard Index of 44.42%, g-mean of 59.84%, kappa value of 0.20, and the highest error rate (40.24%). The KNN Classifier with GWO and KL Divergence Feature selection method demonstrates the best accuracy and outperforms other classifiers.

Table 14 presents a comprehensive performance analysis of various classifiers, utilizing PSO with KL Divergence and IWO in combination with Adam Hyperparameter Tuning. The findings highlight that the K-Nearest Neighbors (KNN) Classifier attains the highest accuracy at 86.70% when incorporating KL Divergence features. This classifier also excels in other evaluation metrics, boasting the highest F1 score (86.30%), MCC value (0.74), Jaccard Index (75.96%), geometric mean (g-mean) (86.81%), kappa score (0.73), and displaying the lowest error rate (13.30%) compared to all other classifiers. Conversely, the performance of the Bayesian Linear Discriminant Classifier (LDC) is notably subpar when employing IWO features, achieving an accuracy of 76.14%, an F1 score of 76.19%, an MCC value of 0.52, a Jaccard Index of 61.54%, a g-mean of 76.14%, a kappa value of 0.52, and the highest error rate (23.86%) among the classifiers considered. Overall, the KNN Classifier in conjunction with PSO and the KL Divergence Feature selection method emerges as the standout performer, showcasing superior accuracy and outclassing the other classifiers in the evaluation.

Table 12. Performance Analysis of the Classifiers for PSO with KL Divergence and IWO.

Feature Selection	Classifiers	Accuracy (%)	Error Rate (%)	F1 Score (%)	MCC	Jaccard Index (%)	g-mean (%)	Kappa
KL Divergence	SVM	60.44	39.56	62.51	0.21	45.46	60.56	0.21
	KNN	65.83	34.17	69.96	0.33	53.79	66.96	0.32
	Random Forest	74.09	25.91	76.05	0.49	61.36	74.65	0.48
	Decision Tree	80.08	19.92	79.74	0.60	66.30	80.11	0.60
	Softmax Discriminant	83.47	16.53	83.18	0.67	71.21	83.50	0.67
	Multilayer Perceptron	79.04	20.96	80.30	0.59	67.08	79.43	0.58
	Bayesian LDC	66.08	33.92	66.04	0.32	49.30	66.08	0.32
IWO	SVM	73.57	26.43	74.47	0.47	59.32	73.67	0.47
	KNN	74.74	25.26	73.43	0.50	58.01	74.95	0.49
	Random Forest	63.31	36.69	66.09	0.27	49.35	63.64	0.27
	Decision Tree	58.53	41.47	58.99	0.17	41.83	58.54	0.17
	Softmax Discriminant	57.81	42.19	56.45	0.16	39.32	57.84	0.16
	Multilayer Perceptron	56.39	43.61	58.29	0.13	41.13	56.44	0.13
	Bayesian LDC	53.79	46.21	53.71	0.08	36.72	53.79	0.08

Table 13. Performance Analysis of the Classifiers for GWO with KL Divergence and IWO.

Feature Selection	Classifiers	Accuracy (%)	Error Rate (%)	F1 Score (%)	MCC	Jaccard Index (%)	g-mean (%)	Kappa
KL Divergence	SVM	67.72	32.28	71.40	0.37	55.52	68.80	0.35
	KNN	79.36	20.64	78.60	0.59	64.74	79.49	0.59
	Random Forest	75.78	24.22	74.24	0.52	59.03	76.09	0.52
	Decision Tree	77.41	22.59	76.09	0.55	61.41	77.69	0.55
	Softmax Discriminant	75.64	24.37	80.08	0.57	66.77	80.74	0.51
	Multilayer Perceptron	67.19	32.81	69.49	0.35	53.25	67.54	0.34
	Bayesian LDC	62.27	37.73	64.72	0.25	47.84	62.49	0.25
IWO	SVM	76.63	23.37	79.02	0.55	65.31	77.82	0.53
	KNN	70.83	29.17	68.71	0.42	52.34	71.16	0.42
	Random Forest	71.75	28.25	75.87	0.46	61.12	74.14	0.43
	Decision Tree	67.87	32.13	72.18	0.38	56.47	69.50	0.36
	Softmax Discriminant	69.06	30.94	73.17	0.40	57.69	70.74	0.38
	Multilayer Perceptron	71.22	28.78	75.25	0.45	60.32	73.33	0.42
	Bayesian LDC	59.76	40.24	61.52	0.20	44.42	59.84	0.20

Table 14. Performance Analysis of the Classifiers: PSO with KL Divergence and IWO for Adam Hyperparameter Tuning.

Feature Selection	Classifiers	Accuracy (%)	Error Rate (%)	F1 Score (%)	MCC	Jaccard Index (%)	g-mean (%)	Kappa
KL Divergence	SVM	76.56	23.44	77.72	0.53	63.56	76.80	0.53
	KNN	86.70	13.30	86.30	0.74	75.96	86.81	0.73
	Random Forest	80.75	19.25	82.56	0.63	70.30	81.86	0.62
	Decision Tree	76.24	23.76	76.26	0.52	61.63	76.24	0.53
	Softmax Discriminant	83.40	16.60	84.11	0.67	72.58	83.62	0.67
	Multilayer Perceptron	79.28	20.72	78.93	0.59	65.20	79.31	0.59
	Bayesian LDC	81.03	18.97	81.42	0.62	68.67	81.08	0.62
IWO	SVM	76.84	23.16	75.66	0.54	61.84	77.05	0.54
	KNN	84.33	15.67	83.82	0.69	72.14	84.44	0.69
	Random Forest	81.67	18.33	81.83	0.63	69.25	81.68	0.63
	Decision Tree	80.21	19.79	78.95	0.61	65.23	80.56	0.60
	Softmax Discriminant	82.90	17.10	82.84	0.66	70.71	82.90	0.66
	Multilayer Perceptron	85.64	14.36	86.28	0.72	75.88	85.94	0.71
	Bayesian LDC	76.14	23.86	76.19	0.52	61.54	76.14	0.52

Table 15 provides a comprehensive analysis of classifier performance, utilizing a combination of PSO with KL Divergence and IWO along with R-Adam Hyperparameter Tuning. The results highlight that the K-Nearest Neighbors (KNN) Classifier achieves the highest accuracy of 87.45% when incorporating KL Divergence features. This classifier also excels across various evaluation metrics, including the highest F1 score (87.02%), MCC value (0.75), Jaccard Index (77.03%), geometric mean (g-mean) (87.58%), kappa score (0.75), and the lowest error rate (12.55%) compared to other classifiers. On the other hand, the Decision Tree's performance is notably weaker when using KL Divergence features, with an accuracy of 78.09%, F1 score of 78.19%, MCC value of 0.55, Jaccard Index of 64.19%, g-mean of 78.10%, kappa value of 0.54, and the highest error rate (21.91%) among considered classifiers. In summary, the KNN Classifier, in combination with PSO and the KL Divergence Feature selection method, stands out as the top performer, showcasing exceptional accuracy and surpassing other classifiers in the evaluation.

Table 16 provides a comprehensive analysis of classifier performance, utilizing GWO with KL Divergence and IWO along with Adam Hyperparameter Tuning. The results emphasize that the K-Nearest Neighbors (KNN) Classifier achieves the highest accuracy at 90.87% when incorporating KL Divergence features. This classifier also excels in various evaluation metrics, recording the highest F1 score (90.06%), MCC value (0.83), Jaccard Index (81.92%), geometric mean (g-mean) (91.71%), kappa score (0.82), and demonstrating the lowest error rate (9.14%) compared to alternative classifiers. In contrast, the performance of the Decision Tree is notably below par when utilizing KL Divergence features, attaining an accuracy of 76.20%, an F1 score of 75.33%, an MCC value of 0.53, a Jaccard Index of 60.43%, a g-mean of 76.30%, a kappa score of 0.52, and the highest error rate (23.81%) among the considered classifiers. In summary, the KNN Classifier, combined with GWO and the KL Divergence Feature selection approach, stands out as the top performer, showcasing remarkable accuracy and surpassing the other classifiers in the evaluation.

Table 15. Performance Analysis of the Classifiers: PSO with KL Divergence and IWO for RAdam Hyperparameter Tuning.

Feature Selection	Classifiers	Accuracy (%)	Error Rate (%)	F1 Score (%)	MCC	Jaccard Index (%)	g-mean (%)	Kappa
KL Divergence	SVM	78.11	21.89	79.11	0.56	65.44	78.32	0.56
	KNN	87.45	12.55	87.02	0.75	77.03	87.58	0.75
	Random Forest	81.95	18.05	83.53	0.65	71.71	82.92	0.64
	Decision Tree	78.09	21.91	78.19	0.55	64.19	78.10	0.54
	Softmax Discriminant	85.15	14.85	85.61	0.70	74.84	85.27	0.70
	Multilayer Perceptron	82.09	17.91	81.75	0.64	69.13	82.12	0.64
	Bayesian LDC	82.92	17.08	83.25	0.66	71.31	82.96	0.66
IWO	SVM	79.45	20.55	78.31	0.59	64.35	79.72	0.59
	KNN	85.83	14.17	85.59	0.72	74.81	85.86	0.72
	Random Forest	84.17	15.83	84.23	0.68	72.76	84.17	0.68
	Decision Tree	82.21	17.79	81.08	0.65	68.18	82.58	0.64
	Softmax Discriminant	84.40	15.60	84.43	0.69	73.05	84.40	0.69
	Multilayer Perceptron	86.32	13.68	86.95	0.73	76.91	86.59	0.73
	Bayesian LDC	78.14	21.86	78.29	0.56	64.21	78.14	0.56

Table 16. Performance Analysis of the Classifiers: GWO with KL Divergence and IWO for Adam Hyperparameter Tuning.

Feature Selection	Classifiers	Accuracy (%)	Error Rate (%)	F1 Score (%)	MCC	Jaccard Index (%)	g-mean (%)	Kappa
KL Divergence	SVM	81.19	18.81	79.53	0.63	66.02	81.88	0.62
	KNN	90.87	9.14	90.06	0.83	81.92	91.71	0.82
	Random Forest	79.56	20.44	79.83	0.59	66.43	79.58	0.59
	Decision Tree	76.20	23.81	75.33	0.53	60.43	76.30	0.52
	Softmax Discriminant	78.62	21.38	76.93	0.58	62.51	79.13	0.57
	Multilayer Perceptron	77.74	22.26	77.07	0.56	62.69	77.82	0.55
	Bayesian LDC	77.62	22.38	77.14	0.55	62.78	77.66	0.55
IWO	SVM	89.56	10.44	89.27	0.79	80.61	89.66	0.79
	KNN	84.50	15.51	84.19	0.69	72.70	84.54	0.69
	Random Forest	89.76	10.24	89.80	0.80	81.49	89.76	0.80
	Decision Tree	90.07	9.93	90.29	0.80	81.30	90.13	0.80
	Softmax Discriminant	84.13	15.87	84.75	0.68	73.54	84.31	0.68
	Multilayer Perceptron	83.21	16.79	84.00	0.67	72.42	83.47	0.66
	Bayesian LDC	87.27	12.73	88.19	0.75	78.88	88.00	0.75

Table 17 presents a comprehensive analysis of classifier performance, utilizing GWO with KL Divergence, and IWO alongside R-Adam Hyperparameter Tuning. The outcomes underscore the Decision Tree Classifier's exceptional performance, achieving the highest accuracy at 91.57% when integrating KL Divergence features. This classifier also outperforms others across various evaluation metrics, achieving the highest F1 score (91.71%), MCC value (0.83), Jaccard Index (84.70%), geometric mean (g-mean) (91.87%), kappa score (0.83), and demonstrating the lowest error rate (8.43%) compared to alternative classifiers. In contrast, the performance of the Decision Tree notably drops when utilizing IWO, with an accuracy of 77.35%, an F1 score of 76.80%, an MCC value of 0.55, a Jaccard Index of 62.34%, a g-mean of 77.39%, a kappa score of 0.55, and the highest error rate (22.66%) among the considered classifiers. In recap, the Decision Tree classifier, combined with GWO and the KL Divergence Feature selection approach, emerges as the leading performer, showcasing remarkable accuracy and surpassing the other classifiers in the evaluation.

Table 17. Performance Analysis of the Classifiers: GWO with KL Divergence and IWO for RAdam Hyperparameter Tuning.

Feature Selection	Classifiers	Accuracy (%)	Error Rate (%)	F1 Score (%)	MCC	Jaccard Index (%)	g-mean (%)	Kappa
KL Divergence	SVM	82.24	17.76	80.89	0.65	67.91	82.77	0.64
	KNN	91.07	8.94	90.30	0.82	82.31	91.61	0.82
	Random Forest	81.56	18.44	81.62	0.63	68.95	81.56	0.63
	Decision Tree	77.35	22.66	76.80	0.55	62.34	77.39	0.55
	Softmax Discriminant	79.67	20.33	78.31	0.60	64.35	80.05	0.59
	Multilayer Perceptron	79.44	20.56	78.97	0.59	65.25	79.49	0.59
	Bayesian LDC	80.42	19.58	79.96	0.61	66.61	80.47	0.61
IWO	SVM	91.06	8.94	90.86	0.82	83.24	91.13	0.82
	KNN	85.50	14.51	85.36	0.71	74.46	85.50	0.71
	Random Forest	90.26	9.74	90.35	0.81	82.39	90.27	0.81
	Decision Tree	91.57	8.43	91.71	0.83	84.70	91.87	0.83
	Softmax Discriminant	85.13	14.87	85.71	0.70	75.00	85.32	0.70
	Multilayer Perceptron	85.21	14.79	85.77	0.71	75.09	85.39	0.70
	Bayesian LDC	88.62	11.38	89.38	0.78	80.80	89.25	0.77

Table 18 presents a summary of the performance outcomes for each combination of feature extraction and feature selection using Adam and R-Adam Hyperparameter tuning methods across all seven classifiers. The highest accuracy of 91.57% in the Decision Tree classifier was attained by combining GWO and IWO techniques, utilizing the RAdam Hyperparameter tuning approach.

Figure 7 illustrates the comparative performance of classifiers in relation to Accuracy, both with and without the integration of feature selection. As depicted in the graph, among all of the classifier types, the Decision Tree classifier employing the GWO (Grey Wolf Optimization) feature extraction method outperformed the rest in terms of achieving the highest accuracy. When utilizing the KL Divergence feature selection technique, along with PSO feature extraction technique, the Softmax Discriminant Classifier demonstrated a commendable accuracy of 83.47%. Similarly, when employing the IWO (Invasive Weed Optimization) feature selection technique, along with GWO feature extraction technique, the SVM (Support Vector Machine) classifier exhibited a notable accuracy of 76.63%. In contrast, the Mathematical feature selection approaches yielded comparatively lower accuracy when compared to scenarios where feature selection was not applied.

Figure 8 displays how classifiers perform when hyperparameter tuning methods such as Adam and RAdam are used to enhance accuracy. Even after applying feature selection techniques, there's no significant improvement in classifier accuracy compared to using no feature selection. To address this, hyperparameter update algorithms are introduced. The accuracy achieved through the KL Divergence feature selection method is notably high across all classifiers. However, for the IWO feature selection technique, accuracy seems to be somewhat lower. This prompts the use of hyperparameter update algorithms specifically for the IWO feature selection. As a result of employing these algorithms, there's a substantial accuracy improvement for all classifiers using the IWO feature selection. Notably, the Decision Tree classifier combined with GWO feature extraction and IWO feature selection, along with the RAdam hyperparameter update algorithm, achieves the highest accuracy at 91.57%.

Table 18. Performance Analysis of the classifiers for Maximum Accuracy.

S No	Feature Extraction	Feature Selection	Classifiers	Accuracy (%)
1	PSO	-	Bayesian LDC	83.42%
2	GWO	-	Decision Tree	85.01%
3	PSO	KL Divergence	Softmax Discriminant	83.47%
4	PSO	IWO	KNN	74.74%
5	GWO	KL Divergence	KNN	79.36%
6	GWO	IWO	SVM	76.63%
7	PSO	KL Divergence	KNN with Adam	86.70%
8	PSO	IWO	MLP with Adam	85.64%
9	PSO	KL Divergence	KNN with RAdam	87.45%
10	PSO	IWO	MLP with RAdam	86.32%
11	GWO	KL Divergence	KNN with Adam	90.87%
12	GWO	IWO	Decision Tree with Adam	90.07%
13	GWO	KL Divergence	KNN with RAdam	91.07%
14	GWO	IWO	Decision Tree with RAdam	91.57%

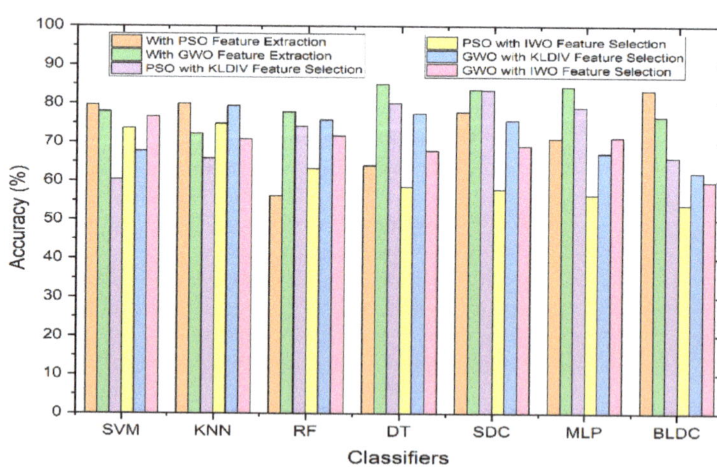

Figure 7. Performance of Classifiers with and without Feature Selection Methods in terms of Accuracy.

Figure 9 illustrates the classifiers' performance by analyzing the Deviation of MCC and Kappa parameters in relation to their mean values. These parameters, MCC and Kappa, serve as benchmarks for evaluating how classifiers respond to different inputs. The study involves two input categories: features extracted using PSO and GWO, followed by feature selection through KL Divergence and IWO. The selected features are then inputted into the classifiers, and their effectiveness is evaluated through the resulting MCC and Kappa values. The average MCC and Kappa values attained from the classifiers are 0.56661 and 0.56256, respectively. A methodology is devised to assess classifier performance by examining the variability of MCC and Kappa values from their respective means. Notably,

Figure 9 depicts a trend where MCC and Kappa values in the graph's third quadrant correspond to non-linear outcomes with lower performance metrics. Conversely, values in the graph's first quadrant indicate improved classifier performance, with MCC and Kappa values surpassing the average. This pattern suggests an enhancement in classifier performance for GWO inputs when coupled with IWO feature selection, particularly within the context of the RAdam hyperparameter tuning approach. Figure 9 is also characterized by a linear curve fitting described by the equation $Y = 1.0017\,X + 4 \times 10^{-06}$, with an R^2 value of 0.998.

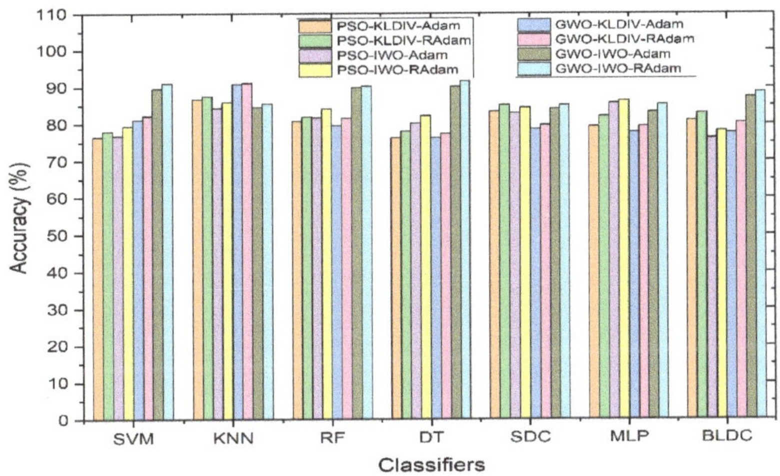

Figure 8. Performance of Classifiers with Hyperparameter Tuning Methods for Adam and RAdam in terms of Accuracy.

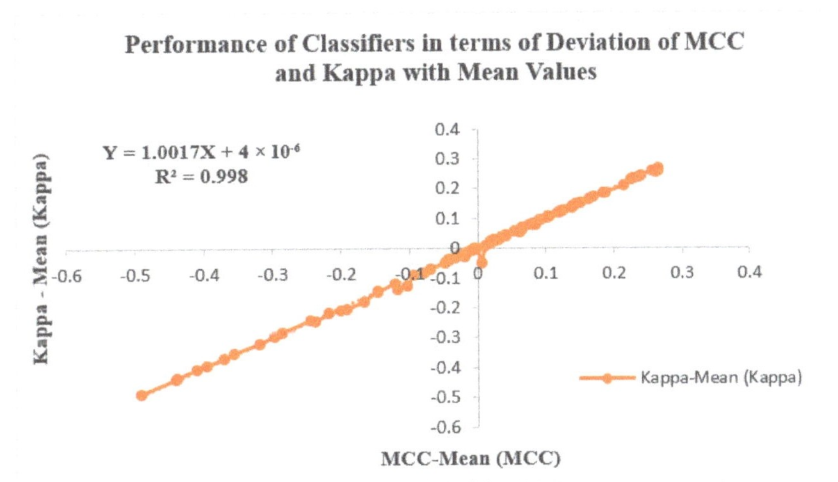

Figure 9. Performance of Classifiers in terms of Deviation of MCC and Kappa Parameters with mean values.

6.4. Computational Complexity Analysis of the Classifiers

Computational complexity also acts as a performance metric for classifiers, encompassing time and space complexities. This study utilizes the Big O notation to characterize the computational complexity of feature extraction, feature selection, and classification methods. The assessment of Computational Complexity involves an input size labelled as 'n'. When the input size is $O(1)$, the computational complexity remains minimal. However, as the input size increases, so does the computational complexity. The relationship between input size and computational complexity is encapsulated by the Big O notation. Specifically, if the complexity grows logarithmically with the increase in 'n', it is represented as $O(\log n)$. The classifiers examined in this research integrate either feature extraction methods, feature selection techniques, or a combination of both. Hence, computational complexity becomes a blend of these hybrid methodologies. Table 19 offers an overview of the computational complexity associated with the classifiers across diverse Feature Extraction and Feature Selection Techniques.

Table 19. Computational Complexity of the classifiers among Feature Extraction, Feature Selection and Hyperparameter Tuning approaches.

S No	Classifiers	Without Feature Extraction	With Feature Extraction		With Feature Selection		With Hyperparameter Tuning of IWO Feature Selection Method	
			PSO	GWO	KL Divergence	IWO	Adam	RAdam
1	SVM	$O\ (2n^2)$	$O\ (2n^5)$	$O\ (2n^5)$	$O\ (2n^6)$	$O\ (2n^6\log n)$	$O\ (2n^2\log n)$	$O\left(4n^7\log 5n\right)$
2	KNN	$O\ (n^2)$	$O\ (n^5)$	$O\ (n^5)$	$O\ (n^6)$	$O\ (n^6\log n)$	$O\ (2n^7\log 2n)$	$O\ (2n^7\log 5n)$
3	RF	$O\ (n\log n)$	$O\ (n^4\log n)$	$O\ (n^4\log n)$	$O\ (n^5\log n)$	$O\ (n^5\log 2n)$	$O\ (2n^6\log 3n)$	$O\ (2n^6\log 6n)$
4	DT	$O\ (\log n)$	$O\ (n^3\log n)$	$O\ (n^3\log n)$	$O\ (n^4\log n)$	$O\ (n^4\log 2n)$	$O\ (2n^5\log 3n)$	$O\ (2n^5\log 6n)$
5	SDC	$O\ (n^2)$	$O\ (n^5)$	$O\ (n^5)$	$O\ (n^6)$	$O\ (n^6\log n)$	$O\ (2n^7\log 2n)$	$O\ (2n^7\log 5n)$
6	MLP	$O\ (n^5)$	$O\ (n^8)$	$O\ (n^8)$	$O\ (n^9)$	$O\ (n^9\log n)$	$O\left(2n^{10}\log 2n\right)$	$O\left(2n^{10}\log 5n\right)$
7	BLDC	$O\ (n^2)$	$O\ (n^5)$	$O\ (n^5)$	$O\ (n^6)$	$O\ (n^6\log n)$	$O\ (2n^7\log 2n)$	$O\ (2n^7\log 5n)$

As evident from Table 19, when feature extraction techniques are not employed, classifiers such as SVM, KNN, Random Forest (RF), Decision Tree (DT), Softmax Discriminant classifier (SDC), Multilayer Perceptron (MLP), and Bayesian LDC (BLDC) exhibit lower levels of computational complexity. When utilizing the GWO feature extraction technique, the Decision Tree classifier stands out with a computational complexity of $O\ (n^3\log n)$ and achieves a high accuracy of 85.01%. However, the BLDC classifier, with a computational complexity of $O\ (n^6\log n)$ for GWO feature extraction, performs poorly when IWO feature selection methods are applied across the classifiers. The observed underperformance is linked to outlier problems present in the GWO features. To improve classifier performance, this study integrates hyperparameter tuning into the IWO feature selection method. Particularly, the Decision Tree classifier demonstrates remarkable performance with accuracies of 90.07% and 91.57% when utilizing GWO feature extraction in conjunction with IWO. These improvements come with a moderate computational complexity represented by $O(2n^5\log 3n)$ for the Adam Hyperparameter tuning approach and $O(2n^5\log 6n)$ for the RAdam Hyperparameter tuning approach.

6.5. Comparison of Previous Works

Comparison charts with different Machine Learning and Deep Learning models along with classifiers are shown in the Table 20 for the different datasets of lung cancer. As noted in the Table 20, that the classifier performance is analyzed for the four different datasets, namely CRAG, LIDC-IDRI, LUNA16 and LC25000. The following classifiers namely Ensemble, ResNet50, KNN, AlexNet and CNN were analyzed. As shown in the Table 20., for the CRAG database, the ResNet50 attains the maximum accuracy of 93.91% whereas for CT image database (LIDC-IDRI, LUNA16), the CNN-ALCDC model attains the maximum accuracy of 97.2%. This is due to the smaller number of CT images as the data set. Similarly, for LC25000 database, for a binary classification problem, 5000 images for each class are taken, which attains a commendable accuracy of 91.57%.

Table 20. Comparison of classifier performance with different datasets.

S No	Authors	Dataset Used	Machine Learning Models/Classifiers	Accuracy (%)
1	Bukhari, S. et al. [57]	CRAG Dataset	ResNet-50	93.91%
2	Jinsa Kuruvilla. et al. [58]	LIDC Dataset (155 Patients—CT images)	Feed Forward Back Propagation Neural Networks	93.3%
3	Dabass, M. et al. [59]	CRAG Dataset	Atrous Convolved Hybrid Seg-Net Architecture	87.63%
4	Supriya, Suresh.et al. [60]	LIDC-IDRI Repository (CT scans)	CNN	93.9%
5	Wadood, Abdul. [61]	LIDC-IDRI Repository (CT scans)	CNN-ALCDC	97.2%
6	Rekka, Mastouri. et al. [62]	LUNA16 Database (3186 CT images)	BCNN [VGG16, VGG19]	91.99%
7	Tasnim, Ahmed. et al. [63]	LUNA16 Database	3D CNN	80%
8	Mesut Toğaçar. et al. [64]	Cancer Imaging Archieve (CT images)	AlexNet and kNN classifier	98.74%
9	Anum, Masood. et al. [65]	Biomedical Datasets–IoT	CNN DFCNet	77.6% 84.58%
10	Wahyudi, Setiawan. et al. [66]	LC25000 Database	CNN	87.16%
11	Manaswini, Pradhan. [67]	LC25000 Database	Without Feature Selection (EGOA)—KNN With Feature Selection (EGOA)–KNN	80.16% 81.59%
12	Phankokkruad, M [68]	LC25000 Database	Ensemble ResNet50V2	91% 90%
13	Karthikeyan Shanmugam, Harikumar Rajaguru	LC25000 Database	Feature Extraction-GWO Feature Selection—IWO Decision tree with RAdam Hyper parameter Updation method	91.57%

7. Conclusions

Early diagnosis of lung cancer enhances patient life expectancy. This paper proposes machine learning techniques to enhance classifier accuracy and enable early identification using histopathological images. The primary aim is to achieve lung cancer classification with high accuracy, while minimizing false positives and false negatives. The study applies adaptive median filtering and a modified KFCM-based segmentation method to obtain the segmented images for better classification results. Feature extraction involves optimization techniques such as PSO and GWO which reduces the dimensionality of the segmented image to [512 $\times$ 10], followed by statistical analysis. Feature selection reduces the number of intensity values to [100 $\times$ 12] for lung cancer classification. Through the utilization of KL Divergence and Invasive Weed Optimization to evaluate the dimensionally reduced features, the datasets undergo classification with various classifiers to achieve better accuracy. The classification process entails seven classifiers, coupled with Hyperparameter selection using Adam and Radam methods, which are compared and analyzed. The Decision Tree Classifier for GWO features without feature selection achieves a better accuracy of 85.01%.

Mathematical feature selection methods had lower accuracy compared to scenarios without feature selection. The results are further enhanced when the Hyperparameter Up-to-date methods are employed which reveal that the combination of GWO-IWO-Decision Tree classifier for RAdam outperforms all other classifiers, achieving an overall accuracy of 91.57% in classifying Benign and Adenocarcinoma classes. Future research directions will explore diverse feature selection techniques, optimization methodologies, and the inclusion of deep learning approaches such as CNN, DNN, and LSTM to further enhance lung cancer classification.

Author Contributions: Conceptualization, K.S.; Methodology, H.R.; Software, K.S.; Validation, H.R.; Formal analysis, K.S.; Investigation, H.R.; Resources, K.S.; Data curation, K.S.; Writing-original draft, K.S.; Writing-review and editing, H.R.; Visualization, H.R.; Supervision, H.R. All authors have read and agreed to the published version of the manuscript.

Funding: This research received no external funding.

Institutional Review Board Statement: Not Applicable.

Informed Consent Statement: Not Applicable.

Data Availability Statement: Not Applicable.

Conflicts of Interest: The authors declare no conflict of interest.

References

1. Prabhakar, S.K.; Lee, S.W. An Integrated Approach for Ovarian Cancer Classification with the Application of Stochastic Optimization. *IEEE Access.* **2020**, *8*, 127866–127882. [CrossRef]
2. Jemal, A.; Ward, E.M.; Johnson, C.J.; Cronin, K.A.; Ma, J.; Ryerson, A.B.; Mariotto, A.; Lake, A.J.; Wilson, R.; Sherman, R.L.; et al. Annual Report to the Nation on the Status of Cancer, 1975–2014, Featuring Survival. *J. Natl. Cancer Inst.* **2017**, *109*, djx030. [CrossRef] [PubMed]
3. Miki, T.; Yano, S.; Hanibuchi, M.; Sone, S. Bone Metastasis Model with Multiorgan Dissemination of Human Small-Cell Lung Cancer (SBC-5) Cells in Natural Killer Cell-Depleted SCID Mice. *Oncol. Res. Featur. Preclin. Clin. Cancer Ther.* **2001**, *12*, 209–217. [CrossRef] [PubMed]
4. Mahmood, H.; Shaban, M.; Indave, B.I.; Santos-Silva, A.R.; Rajpoot, N.; Khurram, S.A. Use of artificial intelligence in diagnosis of head and neck precancerous and cancerous lesions: A systematic review. *Oral Oncol.* **2020**, *110*, 104885. [CrossRef] [PubMed]
5. He, L.; Long, L.R.; Antani, S.; Thoma, G.R. Histology image analysis for carcinoma detection and grading. *Comput. Methods Programs Biomed.* **2012**, *107*, 538–556. [CrossRef] [PubMed]
6. Komura, D.; Ishikawa, S. Machine Learning Methods for Histopathological Image Analysis. *Comput. Struct. Biotechnol. J.* **2018**, *16*, 34–42. [CrossRef]
7. Andreadis, D.A.; Pavlou, A.M.; Panta, P. Biopsy and oral squamous cell carcinoma histopathology. In *Oral Cancer Detection: Novel Strategies and Clinical Impact*; Springer International Publishing: Cham, Switzerland, 2019; pp. 133–151.
8. Gertych, A.; Swiderska-Chadaj, Z.; Ma, Z.; Ing, N.; Markiewicz, T.; Cierniak, S.; Salemi, H.; Guzman, S.; Walts, A.E.; Knudsen, B.S. Convolutional neural networks can accurately distinguish four histologic growth patterns of lung adenocarcinoma in digital slides. *Sci. Rep.* **2019**, *9*, 1483. [CrossRef]
9. Zhang, X.; Xing, F.; Su, H.; Yang, L.; Zhang, S. High-throughput histopathological image analysis via robust cell segmentation and hashing. *Med. Image Anal.* **2015**, *26*, 306–315. [CrossRef]
10. Draelos, R.L.; Dov, D.; Mazurowski, M.A.; Lo, J.Y.; Henao, R.; Rubin, G.D.; Carin, L. Machine-learning-based multiple abnormality prediction with large-scale chest computed tomography volumes. *Med. Image Anal.* **2021**, *67*, 101857. [CrossRef]
11. Szabó, M.; Bozó, A.; Darvas, K.; Soós, S.; Őzse, M.; Iványi, Z.D. The role of ultrasonographic lung aeration score in the prediction of postoperative pulmonary complications: An observational study. *BMC Anesthesiol.* **2021**, *21*, 19. [CrossRef]
12. Toğaçar, M. Disease type detection in lung and colon cancer images using the complement approach of inefficient sets. *Comput. Biol. Med.* **2021**, *137*, 104827. [CrossRef] [PubMed]
13. Talukder, M.A.; Islam, M.M.; Uddin, M.A.; Akhter, A.; Hasan, K.F.; Moni, M.A. Machine Learning-based Lung and Colon Cancer Detection using Deep Feature Extraction and Ensemble Learning. *Expert Syst. Appl.* **2022**, *205*, 117695. [CrossRef]
14. Ozekes, S.; Camurcu, A.Y. Automatic lung nodule detection using template matching. In Proceedings of the 4th International Conference on Advances in Information Systems, ADVIS, Izmir, Turkey, 18–20 October 2006; Volume 4243, pp. 247–253.
15. Schilham, A.M.; van Ginneken, B.; Loog, M. A computer-aided diagnosis system for detection of lung nodules in chest radiographs with an evaluation on a public database. *Med. Image Anal.* **2006**, *10*, 247–258. [CrossRef]
16. Wang, S.; Wang, T.; Yang, L.; Yang, D.M.; Fujimoto, J.; Yi, F.; Luo, X.; Yang, Y.; Yao, B.; Lin, S.; et al. ConvPath: A software tool for lung adenocarcinoma digital pathological image analysis aided by a convolutional neural network. *eBioMedicine* **2019**, *50*, 103–110. [CrossRef]

17. Dehmeshki, J.; Ye, X.; Lin, X.; Valdivieso, M.; Amin, H. Automated detection of lung nodules in CT images using shape-based genetic algorithm. *Comput. Med. Imaging Graph.* **2007**, *31*, 408–417. [CrossRef] [PubMed]
18. Suárez-Cuenca, J.J.; Tahoces, P.G.; Souto, M.; Lado, M.J.; Remy-Jardin, M.; Remy, J.; Vidal, J.J. Application of the iris filter for automatic detection of pulmonary nodules on computed tomography images. *Comput. Biol. Med.* **2009**, *39*, 921–933. [CrossRef] [PubMed]
19. Murphy, K.; van Ginneken, B.; Schilham, A.M.; de Hoop, B.J.; Gietema, H.A.; Prokop, M. A large-scale evaluation of automatic pulmonary nodule detection in chest CT using local image features and k-nearest-neighbour classification. *Med. Image Anal.* **2009**, *13*, 757–770. [CrossRef]
20. Wei, J.W.; Tafe, L.J.; Linnik, Y.A.; Vaickus, L.J.; Tomita, N.; Hassanpour, S. Pathologist-level classification of histologic patterns on resected lung adenocarcinoma slides with deep neural networks. *Sci. Rep.* **2019**, *9*, 3358. [CrossRef]
21. Al-Jabbar, M.; Alshahrani, M.; Senan, E.M.; Ahmed, I.A. Histopathological Analysis for Detecting Lung and Colon Cancer Malignancies Using Hybrid Systems with Fused Features. *Bioengineering* **2023**, *10*, 383. [CrossRef]
22. Teramoto, A.; Tsukamoto, T.; Kiriyama, Y.; Fujita, H. Automated Classification of Lung Cancer Types from Cytological Images Using Deep Convolutional Neural Networks. *Biomed. Res. Int.* **2017**, *2017*, 4067832. [CrossRef]
23. Shapcott, M.; Hewitt, K.J.; Rajpoot, N. Deep Learning with Sampling in Colon Cancer Histology. *Front. Bioeng. Biotechnol.* **2019**, *7*, 52. [CrossRef] [PubMed]
24. Barker, J.; Hoogi, A.; Depeursinge, A.; Rubin, D.L. Automated classification of brain tumor type in whole-slide digital pathology images using local representative tiles. *Med. Image Anal.* **2016**, *30*, 60–71. [CrossRef] [PubMed]
25. Ojansivu, V.; Linder, N.; Rahtu, E.; Pietikäinen, M.; Lundin, M.; Joensuu, H.; Lundin, J. Automated classification of breast cancer morphology in histopathological images. *Diagn. Pathol.* **2013**, *8*, S29. [CrossRef]
26. Ficsor, L.; Varga, V.S.; Tagscherer, A.; Tulassay, Z.; Molnar, B. Automated classification of inflammation in colon histological sections based on digital microscopy and advanced image analysis. *Cytom. Part A* **2008**, *73*, 230–237. [CrossRef]
27. Taheri, M.; Hamer, G.; Son, S.H.; Shin, S.Y. Automated Single and Multi-Breast Tumor Segmentation Using Improved Watershed Technique in 2D MRI Images. In Proceedings of the International Conference on Research in Adaptive and Convergent Systems (RACS '16), Association for Computing Machinery, New York, NY, USA, 11–14 October 2016; pp. 61–66.
28. Huang, P.W.; Lai, Y.H. Effective segmentation and classification for HCC biopsy images. *Pattern Recognit.* **2010**, *43*, 1550–1563. [CrossRef]
29. Gessert, N.; Bengs, M.; Wittig, L.; Drömann, D.; Keck, T.; Schlaefer, A.; Ellebrecht, D.B. Deep transfer learning methods for colon cancer classification in confocal laser microscopy images. *Int. J. Comput. Assist. Radiol. Surg.* **2019**, *14*, 1837–1845. [CrossRef]
30. Sinha, N.; Ramakrishnan, A.G. Automation of differential blood count. In Proceedings of the TENCON 2003, Conference on Convergent Technologies for Asia-Pacific Region, Bangalore, India, 15–17 October 2003; Volume 2, pp. 547–551.
31. Fauziah, K.; Anton, S.P.; Abdullah, A. Detection of leukemia in human blood sample based on microscopic images: A study. *J. Theor. Appl. Inf. Technol.* **2012**, *46*, 579–586.
32. Chen, C.H.; Lee, Y.W.; Huang, Y.S.; Lan, W.R.; Chang, R.F.; Tu, C.Y.; Chen, C.Y.; Liao, W.C. Computer-aided diagnosis of endobronchial ultrasound images using convolutional neural network. *Comput. Methods Programs Biomed.* **2019**, *177*, 175–182. [CrossRef]
33. Khoirunnisa, A.; Adytia, D. Implementation of CRNN Method for Lung Cancer Detection based on Microarray Data. *Int. J. Inform. Vis.* **2023**, *7*, 2. [CrossRef]
34. Mehmood, S.; Ghazal, T.M.; Khan, M.A.; Zubair, M.; Naseem, M.T.; Faiz, T.; Ahmad, M. Malignancy Detection in Lung and Colon Histopathology Images Using Transfer Learning with Class Selective Image Processing. *IEEE Access* **2022**, *10*, 25657–25668. [CrossRef]
35. Boumaraf, S.; Liu, X.; Zheng, Z.; Ma, X.; Ferkous, C. A new transfer learning based approach to magnification dependent and independent classification of breast cancer in histopathological images. *Biomed. Signal Process. Control.* **2021**, *63*, 102192. [CrossRef]
36. Rodríguez Fernández, A.; Gómez Río, M.; Llamas Elvira, J.M.; Sánchez-Palencia Ramos, A.; Bellón Guardia, M.; Ramos Font, C.; Torné Poyatos, P.; Pedraza Muriel, V. Diagnosis efficacy of structural (CT) and functional (FDG-PET) imaging methods in the thoracic and extrathoracic staging of non-small cell lung cancer. *Clin. Transl. Oncol.* **2007**, *9*, 32–39. [CrossRef] [PubMed]
37. Spanhol, F.A.; Oliveira, L.; Cavalin, P.R.; Petitjean, C.; Heutte, L. Deep features for breast cancer histopathological image classification. In Proceedings of the IEEE International Conference on Systems, Man, and Cybernetics (SMC), Banff, AB, Canada, 5–8 October 2017; pp. 1868–1873.
38. Ma, Z.; Tavares, J.M.R.; Jorge, R.N. A review on the current segmentation algorithms for medical images. In Proceedings of the 1st International Conference on Imaging Theory and Applications (IMAGAPP), Lisboa, Portugal, 5–8 February 2009.
39. Ding, Y.; Fu, X. Kernel-based fuzzy c-means clustering algorithm based on genetic algorithm. *Neurocomputing* **2016**, *188*, 233–238. [CrossRef]
40. Ciompi, F.; Geessink, O.G.; Bejnordi, B.E.; Souza, G.S.; Baidoshvili, A.; Litjens, G.J.; Ginneken, B.V.; Nagtegaal, I.D.; Laak, J.V. The importance of stain normalization in colorectal tissue classification with convolutional networks. In Proceedings of the 2017 IEEE 14th International Symposium on Biomedical Imaging (ISBI 2017), Melbourne, Australia, 18–21 April 2017; pp. 160–163.
41. Liu, X.; Wang, C.; Bai, J.; Liao, G. Fine-tuning Pre-trained Convolutional Neural Networks for Gastric Precancerous Disease Classification on Magnification Narrow-band Imaging Images. *Neurocomputing* **2020**, *392*, 253–267. [CrossRef]

42. Sun, C.; Xu, A.; Liu, D.; Xiong, Z.; Zhao, F.; Ding, W. Deep Learning-Based Classification of Liver Cancer Histopathology Images Using Only Global Labels. *IEEE J. Biomed. Health Inform.* **2020**, *24*, 1643–1651. [CrossRef]
43. Alinsaif, S.; Lang, J. Texture features in the Shearlet domain for histopathological image classification. *BMC Med. Inform. Decis. Mak.* **2020**, *20* (Suppl. S14), 312. [CrossRef]
44. Sujin, P.R.; Prakash, T.R.; Linda, M.M. Particle Swarm Optimization Based Reactive Power Optimization. *ArXiv* **2010**, arXiv:1001.3491.
45. Saremi, S.; Mirjalili, S.Z.; Mirjalili, S.M. Evolutionary population dynamics and grey wolf optimizer. *Neural Comput. Appl.* **2015**, *26*, 1257–1263. [CrossRef]
46. Prabhakar, S.K.; Rajaguru, H. Alcoholic EEG signal classification with Correlation Dimension based distance metrics approach and Modified Adaboost classification. *Heliyon* **2020**, *6*, e05689. [CrossRef]
47. Mallahzadeh, A.R.; Oraizi, H.; Davoodi-Rad, Z. Application of The Invasive Weed Optimization Technique for Antenna Configurations. *Prog. Electromagn. Res.* **2008**, *79*, 425–428. [CrossRef]
48. Xia, L.; Huang, Q.; Wu, D. Decision Tree-Based Contextual Location Prediction from Mobile Device Logs. *Mob. Inf. Syst.* **2018**, *2018*, 1852861. [CrossRef]
49. Kumar, M.; Rath, N.K.; Swain, A.; Rath, S.K. Feature Selection and Classification of Microarray Data using MapReduce based ANOVA and K-Nearest Neighbor. *Procedia Comput. Sci.* **2015**, *54*, 301–310. [CrossRef]
50. Gibson, R.; Danaher, T.; Hehir, W.; Collins, L. A remote sensing approach to mapping fire severity in south-eastern Australia using sentinel 2 and random forest. *Remote Sens. Environ.* **2020**, *240*, 111702–111715. [CrossRef]
51. Zang, F.; Zhang, J. Softmax Discriminant Classifier. In Proceedings of the Third International Conference on Multimedia Information Networking and Security, Shanghai, China, 4–6 November 2011; pp. 16–19.
52. Smith, K.A.; Gupta, J.N.D. Neural networks in business: Techniques and applications for the operations researcher. *Comput. Oper. Res.* **2000**, *27*, 1023–1044. [CrossRef]
53. Cybenko, G.V. Approximation by superpositions of a sigmoidal function. *Math. Control. Signals Syst.* **1989**, *2*, 303–314. [CrossRef]
54. Fonseca, P.; Den Teuling, N.; Long, X.; Aarts, R.M. Cardiorespiratory Sleep Stage Detection Using Conditional Random Fields. *IEEE J. Biomed. Health Inform.* **2017**, *21*, 956–966. [CrossRef]
55. Claesen, M.; Moor, B.D. Hyperparameter Search in Machine Learning. *arXiv* **2015**, arXiv:1502.02127.
56. Sen, S.Y.; Ozkurt, N. Convolutional Neural Network Hyperparameter Tuning with Adam Optimizer for ECG Classification. In Proceedings of the 2020 Innovations in Intelligent Systems and Applications Conference, ASYU, Istanbul, Turkey, 15–17 October 2020; Institute of Electrical and Electronics Engineers Inc.: New York, NY, USA, 2020.
57. Bukhari, S.U.K.; Syed, A.; Bokhari, S.K.A.; Hussain, S.S.; Armaghan, S.U.; Shah, S.S.H. The histological diagnosis of colonic adenocarcinoma by applying partial self supervised learning. *medRxiv* **2020**. [CrossRef]
58. Kuruvilla, J.; Gunavathi, K. Lung cancer classification using neural networks for CT images. *Comput. Methods Programs Biomed.* **2014**, *113*, 202–209. [CrossRef]
59. Dabass, M.; Dabass, J. An Atrous Convolved Hybrid Seg-Net Model with residual and attention mechanism for gland detection and segmentation in histopathological images. *Comput. Biol. Med.* **2023**, *155*, 106690. [CrossRef]
60. Suresh, S.; Mohan, S. ROI-based feature learning for efficient true positive prediction using convolutional neural network for lung cancer diagnosis. *Neural Comput. Appl.* **2020**, *32*, 15989–16009. [CrossRef]
61. Abdul, W. An Automatic Lung Cancer Detection and Classification (ALCDC) System Using Convolutional Neural Network. In Proceedings of the 2020 13th International Conference on Developments in eSystems Engineering (DeSE), Liverpool, UK, 14–17 December 2020; pp. 443–446.
62. Mastouri, R.; Khlifa, N.; Neji, H.; Hantous-Zannad, S. A bilinear convolutional neural network for lung nodules classification on CT images. *Int. J. Comput. Assist. Radiol. Surg.* **2021**, *16*, 91–101. [CrossRef] [PubMed]
63. Ahmed, T.; Parvin, M.S.; Haque, M.R.; Uddin, M.S. Lung Cancer Detection Using CT Image Based on 3D Convolutional Neural Network. *J. Comput. Chem.* **2020**, *8*, 35–42. [CrossRef]
64. Toğaçar, M.; Ergen, B.; Cömert, Z. Detection of lung cancer on chest CT images using minimum redundancy maximum relevance feature selection method with convolutional neural networks. *Biocybern. Biomed. Eng.* **2020**, *40*, 23–39. [CrossRef]
65. Masood, A.; Sheng, B.; Li, P.; Hou, X.; Wei, X.; Qin, J.; Feng, D. Computer-Assisted Decision Support System in Pulmonary Cancer detection and stage classification on CT images. *J. Biomed. Inform.* **2018**, *79*, 117–128. [CrossRef]
66. Setiawan, W.; Suhadi, M.M.; Husni; Pramudita, Y.D. Histopathology of Lung Cancer Classification Using Convolutional Neural Network with Gamma Correction. Commun. *Math. Biol. Neurosci.* **2022**, *2022*, 81.
67. Pradhan, M.; Bhuiyan, A.; Mishra, S.; Thieu, T.; Coman, I.L. Histopathological Lung cancer Detection using Enhanced Grasshopper Optimization Algorithm with Random Forest. *Int. J. Intell. Eng. Syst.* **2022**, *15*, 11–20.
68. Phankokkruad, M. Ensemble Transfer Learning for Lung Cancer Detection. In Proceedings of the 4th International Conferenceon Data Science and Information Technology, Shanghai, China, 23–25 July 2021; pp. 438–442.

cancers

MDPI

Article

Towards Realistic 3D Models of Tumor Vascular Networks

Max C. Lindemann [1,†], Lukas Glänzer [1,†], Anjali A. Roeth [2,3], Thomas Schmitz-Rode [1] and Ioana Slabu [1,*]

1 Institute of Applied Medical Engineering, Helmholtz Institute, Medical Faculty, RWTH Aachen University, Pauwelsstraße 20, 52074 Aachen, Germany; glaenzer@ame.rwth-aachen.de (L.G.); smiro@ame.rwth-aachen.de (T.S.-R.)
2 Department of General, Visceral and Transplant Surgery, RWTH Aachen University Hospital, Pauwelsstrasse 30, 52074 Aachen, Germany
3 Department of Surgery, Maastricht University, P. Debyelaan 25, 6229 HX Maastricht, The Netherlands
* Correspondence: slabu@ame.rwth-aachen.de; Tel.: +49-241-80-89102
† These authors contributed equally to this work.

Simple Summary: Three-dimensional models of tumor vascular networks are of significant importance for in vitro and in silico investigations of, for example, the efficiency of anti-cancer drugs in an early stage of clinical transition and can be potentially used for the development of in vitro systems as 3D-printable vascular networks to facilitate personalized medicine and randomized controlled clinical trials. In this work, histologic slices of a human pancreatic tumor are used as examples to establish an algorithm-based method that enables the reconstruction of a 3D vascular network model. The advantages of this method are high resolution and accuracy concerning the characteristics of the vascular network (e.g., density, trajectory of vessels).

Abstract: For reliable in silico or in vitro investigations in, for example, biosensing and drug delivery applications, accurate models of tumor vascular networks down to the capillary size are essential. Compared to images acquired with conventional medical imaging techniques, digitalized histological tumor slices have a higher resolution, enabling the delineation of capillaries. Volume rendering procedures can then be used to generate a 3D model. However, the preparation of such slices leads to misalignments in relative slice orientation between consecutive slices. Thus, image registration algorithms are necessary to re-align the slices. Here, we present an algorithm for the registration and reconstruction of a vascular network from histologic slices applied to 169 tumor slices. The registration includes two steps. First, consecutive images are incrementally pre-aligned using feature- and area-based transformations. Second, using the previous transformations, parallel registration for all images is enabled. Combining intensity- and color-based thresholds along with heuristic analysis, vascular structures are segmented. A 3D interpolation technique is used for volume rendering. This results in a 3D vascular network with approximately 400–450 vessels with diameters down to 25–30 µm. A delineation of vessel structures with close distance was limited in areas of high structural density. Improvement can be achieved by using images with higher resolution and or machine learning techniques.

Keywords: image registration; segmentation; vascular network model; tumor; reconstruction; image processing; histologic images

Citation: Lindemann, M.C.; Glänzer, L.; Roeth, A.A.; Schmitz-Rode, T.; Slabu, I. Towards Realistic 3D Models of Tumor Vascular Networks. *Cancers* **2023**, *15*, 5352. https://doi.org/10.3390/cancers15225352

Academic Editors: Stephen T. C. Wong, Rafał Obuchowicz, Michał Strzelecki and Adam Piorkowski

Received: 28 August 2023
Revised: 3 November 2023
Accepted: 3 November 2023
Published: 9 November 2023

1. Introduction

Three-dimensional models of vascular networks have the potential to improve the accuracy of computational and experimental investigations in many biomedical fields. Some of these are biosensing, organ-on-a-chip (OOAC), tissue engineering and scaffolding, and development of anti-cancer drugs (ACDs) in cancer therapy. In biosensing, vasculature models are necessary for evaluation of the sensors' efficiency both in silico [1] and in vitro [2]. However, accurate vasculature models would increase the reliability of these results. Similarly, for OOACs, microfluidic systems are built to simulate the physiology of human organs [3]. Current OOACs allow for the use of 3D structures [4] for which, in general, vascularization is still needed and part of ongoing research [5]. For the development of ACDs, accurate tumor models are needed to evaluate their effectiveness in early-stage development [6]. The ACDs' development also includes approaches such as magnetic drug targeting to realize more efficient drug delivery [7,8]. This approach can be combined with other tumor therapies to increase their effectiveness [9–13].

Animal models deliver the necessary tissue function but also have limitations (i.e., the comparatively small number of specimens that can be investigated and high variability in the experimental conditions, e.g., due to biodynamic signatures) [14]. In addition, it is recommended to avoid animal experiments whenever possible [15]; however, reliable alternatives exist only for macroscopic structures [16] that do not mimic human tissue function. In tissue engineering, 3D models of vascular networks are already used for the design of scaffolds [17]. The models are usually based on artificially generated data [18] or on medical imaging data (e.g., from micro-computed tomography (μCT)) [19]. The artificially generated models include approaches that allow reverse-engineering; however, they still need rigorous experimental validation. Methods such as CT, μCT, and two-photon microscopy allow imaging of vasculature without destroying the tissue, and spatial encoding is available. As these methods are well established, the model acquisition is relatively fast. Furthermore, these methods allow imaging in vivo. Explantation of the tumor is not necessary. However, the resolution of CT and μCT images is too low (CT: approx. 1 mm, μCT: approx. 5 μm) to sufficiently delineate capillaries [19–21]. For two-photon microscopy, the resolution is around 100 nm [22], but the tissue penetration depth is comparatively low (approx. 1 mm). This method was also used for the investigation of in vivo drug delivery [23].

In clinical practice, reliable treatment planning is of paramount importance. Hence, to further clinical translation of the above-mentioned approaches, accurate in silico investigation based on realistic vascular network models plays a key role. In this work, histologic slices of a human pancreatic tumor are used as examples to establish a new approach of building a virtual 3D vascular network model. The advantages of this method compared to the ones mentioned above are high resolution, accuracy concerning the characteristics of the vascular network (e.g., density, trajectory of vessels), and a potential use for in vitro systems as 3D-printable vascular networks to facilitate personalized medicine and randomized controlled clinical trials. Such 3D networks can enable precise simulations of blood flow dynamics, aiding in the development of targeted treatments and predicting patient-specific responses.

This workflow for the generation of a vascular network model of a tumor using histological images presented in this work consists of several steps: slicing a tumor, fixation of slices on slides, staining the structures of interest, digitization of slides, registration of the digitalized slides (i.e., alignment to the same coordinate system), image segmentation of the structures of interest (i.e., vessels), and reconstruction of the registered and segmented images to a 3D model (cf. Figure 1).

The major challenges for the generation of a 3D model from histologic images compared to state-of-the-art procedures are the registration of high-resolution and highly repetitive features, segmentation robust to artifacts from sample preparation, and implementation that allows for upscaling of the number of images used. To overcome these challenges, this work uses a combination of area- and feature-based registration, and segmentation including intensity- and color-based thresholds and heuristic analysis. Here, we present as an example the registration and reconstruction of a vascular network from histologic slices applied to 169 tumor slices. The algorithms are discussed, and results are compared to state-of-the-art registration (see Table 2), segmentation, and reconstruction approaches in Section 3.

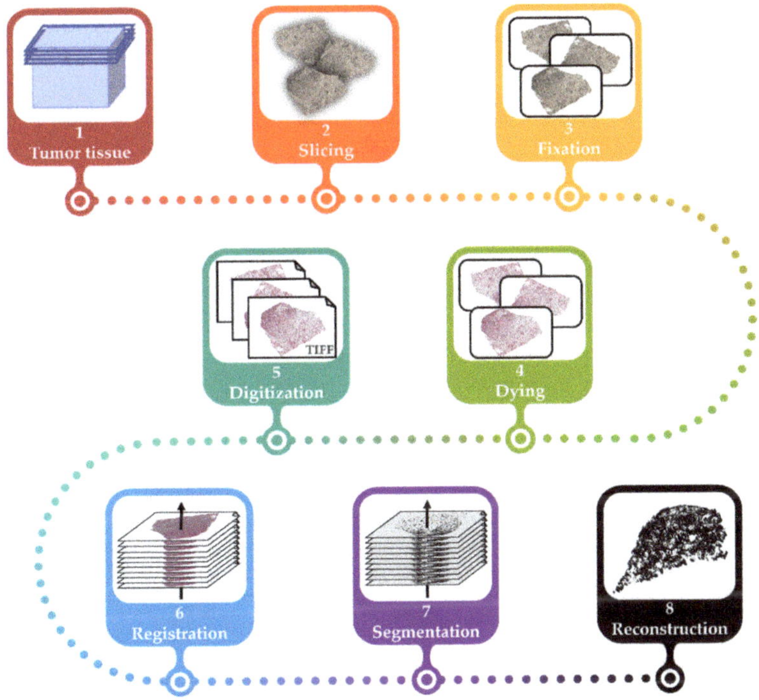

Figure 1. The general workflow for the reconstruction of the vascular network. The tumor tissue is explanted and embedded in paraffin wax (1), sliced into 2.5 µm slices (2), fixated in 3.5% formaldehyde, and placed on slides (3). The tissue on each slide is then immunohistochemically stained (4) and then digitized (5). The resulting images are registered (6). Using segmentation, vascular structures are extracted (7) and reconstructed for the vascular model of the initial tumor (8).

2. Materials and Methods

The general workflow for the generation of the 3D vascular network is shown in Figure 1. Steps (1) to (5) are described in Section 2.1, step (6) is described in Sections 2.2 and 2.3, step (7) in Section 2.4, and step (8) in Section 2.5.

2.1. Sample Preparation

After informed consent of the patient and with permission of the local IRB (EK206/09), a piece of a human pancreatic ductal adenocarcinoma was used after resection. It was fixed in a 3.5% formaldehyde solution (Otto-Fischar GmbH, Saarbrücken, Germany). The tumor was embedded in paraffin wax and cut in 169 slices of 2.5 µm thickness each using a microtome (HM 340 E, Thermo Fisher Scientific Inc., Waltham, MA, USA). Slices were manually placed on slides (Starforst, Engelbrecht Medizin- und Labortechnik, Edermünde, Germany) and dried overnight at 56 °C. The slices were immunohistochemically stained (Figure 2) with the antibody "von Willebrandt-factor" (DakoCytomation, Glostrup, Denmark) using the ZytoChem Plus AP Polymer System (Mouse/Rabbit) (Zytomed Systems, Berlin, Germany) and Hematoxylin (Zytomed Systems, Berlin, Germany). Then, the slides were covered with Vitro-Clud (R. Langenbrinck GmbH, Emmendingen, Germany).

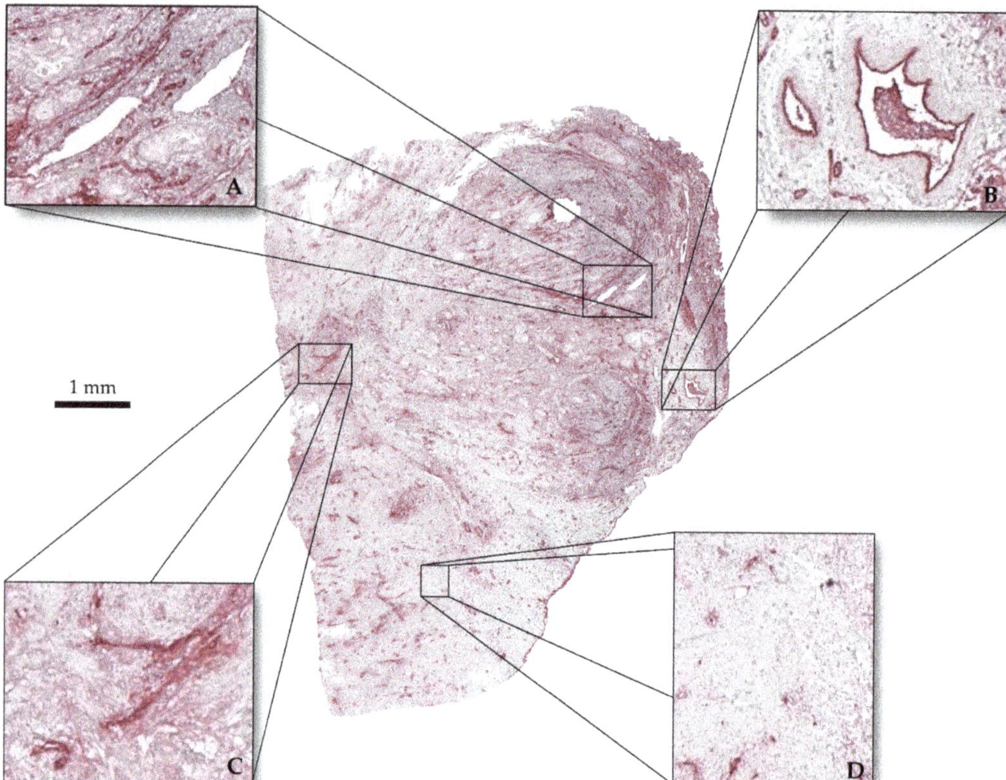

Figure 2. Tumor slice immunohistochemically stained. (**A**) Vessels with clear outlines and homogeneous interiors. (**B**) Vessels with clear outline and perturbed interiors. (**C**) Vessels with perturbed outlines and no interiors. (**D**) Stained cell nuclei.

The slides were digitalized using a microscope (Axio Imager.Z2, Carl Zeiss Microscopy Deutschland GmbH, Jena, Germany). The slide images were converted and cropped to a size of 13020×10422 px^2, resulting in images with a pixel size of 0.5 µm.

2.2. Preprocessing

Prior to the registration, the images were preprocessed to achieve better registration results. First, the background of the images (i.e., the area of the image not showing any tissue) was replaced with white. This is necessary, as the registration algorithm (see Section 2.3) requires a white background. Additionally, a white margin (12.5% of the image size) was added to each image. This made the image registration more robust against losing image information through rotation (cf. Section 2.3 on registration). Second, the imregtform function of Matlab (The MathWorks, Inc., Natick, MA, USA) was used to perform a so-called similarity transformation between consecutive images. This similarity transformation included rotation, translation, reflection, and scaling of the images. However, it did not include elastic transformation, which was performed in a later step (see Section 2.3).

One image preprocessing took only few seconds and could be computed in parallel.

All computations were performed with computing resources granted by RWTH Aachen University. The RWTH Aachen University High-Performance Cluster provides access to computational nodes using 2 Intel Xeon Platinum 8160 Processors "SkyLake" with 2.1 GHz, 24 cores each. Each node hast 48 cores and 192 GB of main memory.

2.3. Registration

Because of the slides' preparation (see chapter 2.1), a misalignment of the images was inevitable. Therefore, an image registration was required, which transfers the images into the same coordinate system. For the registration, the open-source software package Fiji [24] was used.

The registration algorithm presented here is based on the works of Wang et al. [25,26], which was available as a plug-in for Fiji (version 1.45) called ImproveCWR. We updated the bUnwarpJ version 3 [27,28] within this plugin in order to process images with larger file sizes. We also adjusted the ImproveCWR plugin to be able to save the transformation parameters of each registration.

As registering one image after the other would have been very time consuming when working with large numbers of images, we chose a two-step process that parallelized the registration, as described in Figure 3.

Figure 3. The registration proceeds in two steps. In the first step, each image is aligned to the previous one (i.e., 2 to 1, 3 to 2, etc.). Each of these registrations yield a transformation that can be saved to file (A, B, etc.). In the second step, each image is aligned to the first one in the set using concatenations of the transformations saved in the first step. All calculations within a step are performed in parallel.

The adjusted ImproveCWR registration algorithm was validated using the training datasets of the ANHIR challenge [29,30]. This provides 230 histological image pairs with annotated image features (i.e., SIFT features). Using these registered images, the relative target registration error (*rTRE*) for every annotated image feature *l* in every image pair (*i,j*) was calculated as follows:

$$rTRE_l^{ij} = \frac{\left\| \hat{x}_l^j - x_l^i \right\|_2}{d_j} \qquad (1)$$

with feature position x, the transformed feature position $\hat{x}$, the diagonal of the image d_j, and the Euclidian norm $\|\cdot\|_2$. Then, the median value of all features in a pair of images and the average of all medians were calculated.

The robustness R_i is the number of improved features compared to the total number of features in the set L_i in the image i. To calculate R_i, the relative initial registration error (*rIRE*) was determined as follows:

$$rIRE_l^{ij} = \frac{\left\| x_l^j - x_l^i \right\|_2}{d_j}. \tag{2}$$

Using *rTRE* and *rIRE*, R_i can be calculated:

$$R_i = \frac{1}{\|L_i\|} \sum_{l \in L_i} k_l, \quad k_l = \begin{cases} 1, & rTRE_l^{ij} < rIRE_l^{ij} \\ 0, & rTRE_l^{ij} \geq rIRE_l^{ij} \end{cases} \tag{3}$$

The average R_i over all images was calculated.

As the dataset of the ANHIR challenge includes a variety of tissues, we created a subset of images including only images that are similar to the tissue used in this work and performed a separate validation as described above (Equations (1)–(3)). We calculated the average R_i and average median *rTRE* for the subset of 90 image pairs and the complete set of 230 image pairs.

2.4. Segmentation

The images were converted to grayscale as the segmentation algorithm proposed in this section only relies on the pixel intensities. The brightness for each image was homogenized using the contrast-limited adaptive histogram equalization (CLAHE) algorithm [31]. Then, a median filter was used to reduce noise (from the cell nuclei).

Starting from here, the segmentation was performed as described in Figure 4, consisting of five steps: (i) delineation of vessel interiors, (ii) delineation of vessel contours, (iii) removal of unwanted structures, (iv) fusing contour and interior information, and (v) filling of remaining gaps.

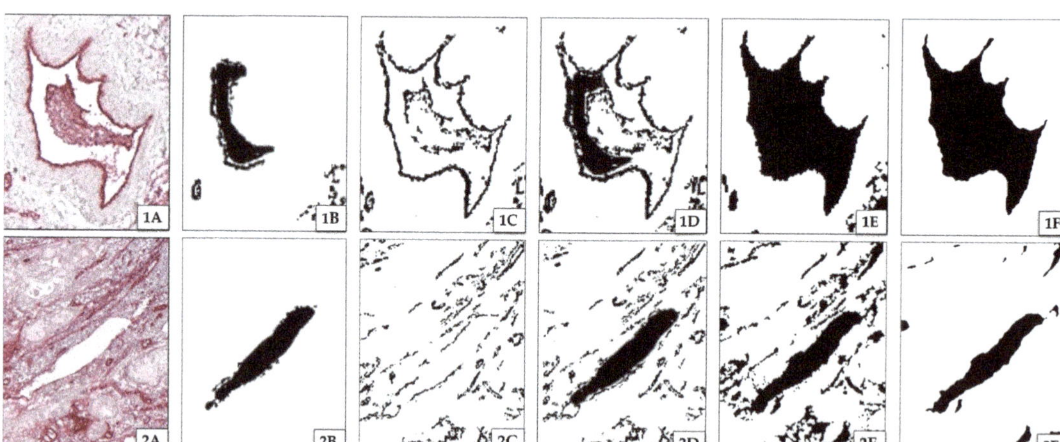

Figure 4. Results of segmentation steps for two image excerpts 1 and 2. (**A**) shows excerpts with one vascular structure each. (**1A**) has a well-defined contour with a perturbed interior, whereas (**2A**) has a clear interior with a perturbed contour. (**B**) shows the results of extracted interiors of vessels, and (**C**) shows the results of their extracted contours. (**D**) is a fusion of (**B**) and (**C**), indicating significant differences in possibly detected interiors and contours (white spaces between interior and contours). (**E**) shows the result of the segmentation after dilating and filling the detected vessels to their contours (see step (v)). Images (**F**) show the final segmentation results with applied smoothing.

(i) The first step of the segmentation process was the delineation of vessel interiors (see Figure 4B, black area). These are prominent due to their bright and plain white appearance. To isolate these areas, a threshold with a cut-off value of 82.35% color intensity is used. This threshold was determined using Otsu's method [32]. Because of the preprocessing, one cut-off value was sufficient for all images. Figure 4(1B) shows that the lumen was only partially delineated due to presence of other structures, such as cells or debris, within the vessel. However, these missed parts of the lumen will be gained by reconstruction (see below step (iii)).

(ii) The second step was the delineation of the contours of the vessels in the preprocessed image (see Figure 4C). This includes the contour of vessel interiors (delineated in step one) and of collapsed vessels that do not have an interior. Because of the staining, the latter are darker compared to the surrounding tissue. Hence, the intensity of the vessel contours varies throughout the image. As the segmentation is based on thresholding, the image was rasterized and a local cut-off value $I_{cut-off}$ was calculated for every rasterized part as follows:

$$I_{cut-off} = \mu + x \cdot \sigma \,, \quad (4)$$

where μ is the mean intensity, σ is the standard deviation, and x is an empirical value. The results of both segmentation steps were binarized. In this way, two types of images were created: one with delineated vessel interiors and the other with delineated vessel walls (contours). In both image types, damage artifacts that occurred during the slicing process and real vessel structures were not differentiated (for example, there was no difference between delineated fissures and vessel interiors).

(iii) In the third step, unwanted structures (such as cell nuclei, megakaryocytes, or damaged tissue) were removed from the images that resulted after processing according the two previous steps. This was performed as follows:

1. Pixels that were not related to vessels (bright or dark spots in the image) were removed by application of a sliding window algorithm. This algorithm removes every structure that completely fits into a region of the image with a size of $n \times n$ pixels, where n is an experimental value. If n is too big, vessel structures are deleted. If n is too small, the unrelated pixels might not be removed effectively. The size of n depends on the $I_{cut-off}$ and the magnification used.
2. Structures in the image of damaged tissue (induced by the manual preparation of the slides) are removed by using the Moore–Neighbor tracing algorithm [33] with Jacob's stopping criterion [34]. Usually, after segmentation, vessel interiors (delineated in step (i), Figure 4B) show continuous contours of intensely stained tissue (delineated in step (ii), Figure 4C). These contours are dilated by five pixels. Only vessel interiors that have at least 45% of a corresponding contour length are kept.
3. Further removal of damaged tissue was performed by comparing vessel interiors of three consecutive images using the Moore–Neighbor tracing algorithm. Repeating structures in the images were kept, the other ones were discarded (Figure 5, red circles). Further, interiors not found in step (i) were added when they appeared in two of three images (Figure 5, green circles). To increase accuracy, these removal and addition procedures can be extended to a higher number of consecutive images that are analyzed. For both removed and added structures, a small offset up to 10 pixels of the position of the structure between two consecutive images is allowed to compensate for errors in the registration.

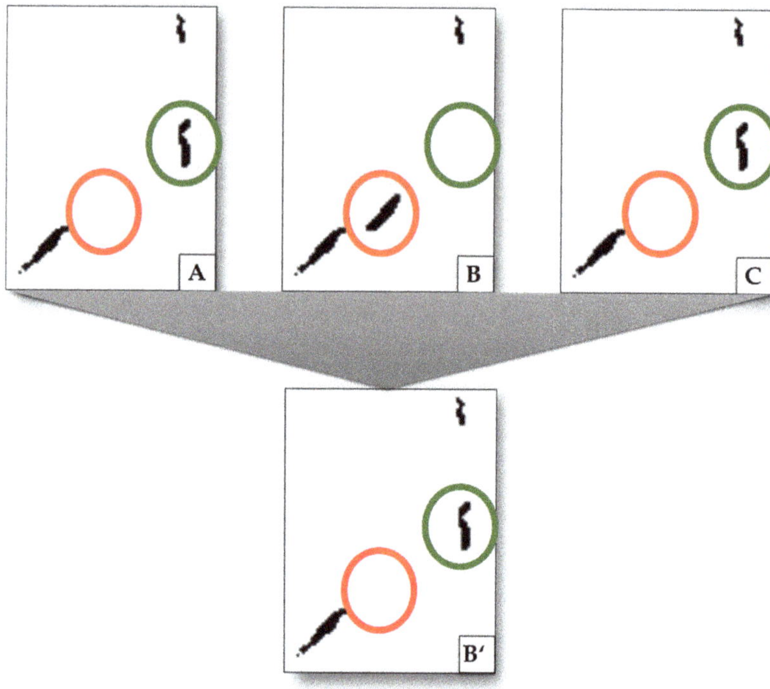

Figure 5. Exemplary procedure to remove structures that are related to damage artifacts and not to vessels as well as to add missing structures by interpolation. Images (**A–C**) show excerpts from consecutive images, (**B′**) is the result after (**B**) is compared to its next neighbors. Highlighted in green is a structure detected in image (**A**) and (**C**), but not in (**B**). By interpolation, this structure is added to image (**B′**). Highlighted in red is a structure that only appears in (**B**) but not in (**A**) or (**C**). This is interpreted as damage artifacts and is removed (does not appear in (**B′**)).

(iv) The fourth step of the segmentation process was the fusion of images with the information on vessel contours with those of the vessel interiors (see Figure 4D). The interiors are a bit smaller than the contours, due to different cut-off values (cf. steps (i) and (ii)) in the respective segmentations.

(v) The fifth step was the closing of open ends of contours to form closed contours and filling of the gaps between these closed contours and vessel interiors (see Figure 4E). The algorithm uses binary dilation [35] with a disk of radius 5 px on every pixel. If any other pixel is within the radius, the gap is closed. Filled structures were discarded if the filled area was larger than the area of the biggest vessel in the respective image. This procedure was repeated, and the maximum gap-size was doubled for each iteration. If a filling was discarded for the first time, the disc size of the last iteration was increased by 5 px for every following iteration when the filling was discarded for the second time.

For further filling of remaining gaps, contours were dilated, then filled and eroded afterwards. To prevent clustering in areas of high vessel density, the image was rasterized, and the dilation was adapted to the current raster's vessel density.

Further refinement was performed in the 3D reconstruction.

2.5. 3D Reconstruction

Three-dimensional reconstruction was performed by interpolation of the 2D segmented images. For interpolation, every unset voxel was examined. If there were 15 voxels set in its 3 × 3 × 10 voxel neighborhood, the examined voxel was also set. Looking 10 voxels in z-direction enables the bridging of larger gaps resulting from missing image information. To close discontinued vessels resulting from inaccuracies in registration or segmentation, for every voxel its 3D neighborhood of voxels was checked again. This neighborhood is defined as a 3 × 3 × 3 voxel neighborhood around the examined voxel in the original stack. If there were less than 10 voxels identified as vessel in the original voxel neighborhood, the voxel was discarded. Both the size of the neighborhood and the value of 10 voxels are empirical values.

To achieve a smoother and in this way more natural look of vessel structures, a Gaussian filter in 3D was applied. This was followed by thresholding to receive binary images again. The smoothing also led to a more robust algorithm as remaining discontinuities in a vessel's trajectory were eliminated.

The reconstructed 3D vessel network was qualitatively evaluated by an experienced radiologist.

3. Results and Discussion

In this work, we developed an algorithm to reconstruct a 3D vascular network from histological slices. The algorithm consists of a two-step registration based on the ImproveCWR algorithm by Wang et al. [25,26] (see Section 2.3), a five-step segmentation (see Section 2.4), and a 3D reconstruction (see Section 2.5).

The computing time for the registration, segmentation, and reconstruction of 169 consecutive slices of an example pancreatic ductal adenocarcinoma is shown in Table 1.

Table 1. Computation time of the different tasks to reconstruct a 3D vascular network from histologic slices.

Task	Time for Completion (in Minutes)
registration (step 1): calculation of transformation of each slice to its unregistered predecessor	11.93
registration (step 2): application of all predeceasing transformations to each slice	58.78
complete registration (step 1 + step 2)	70.71
segmentation	72.28
reconstruction	54.38
total computation time	197.37

During the registration, one computational node was allocated, which then used all 48 cores, and 187 GB of main memory was allocated. For the segmentation, five computational nodes were allocated, using in total 158 cores and 750 GB of main memory. For the reconstruction, one computational node was allocated using one core and 4 GB of main memory. The total computation time of 197 min shows the benefits of the parallelization of the registration steps. If the registration were to be computed linearly on a single CPU, the runtime could be approximated to last over 66 days for the 169 slices. The total of 197 min is therefore within a reasonable time frame in a research context for the development of reliable in vitro models. This would be, however, not appropriate for, as an example, tissue analysis in a clinical setting, in which decisions are time critical.

Figure 6 shows unregistered and registered images, demonstrating the performance of the registration algorithm.

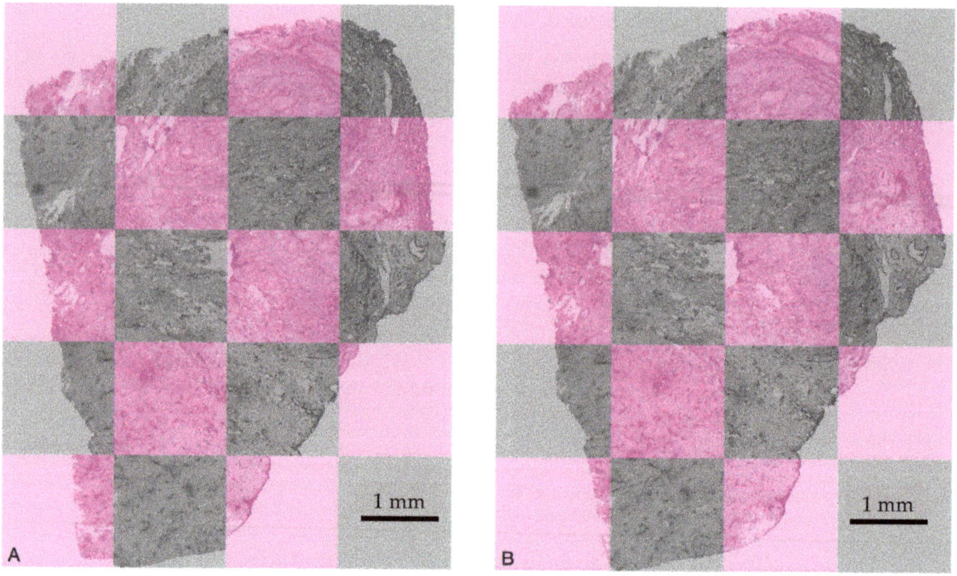

Figure 6. Two superimposed images of slices displayed in grey and in pink, respectively: (**A**) without registration and (**B**) after registration. In (**A**), a clear mismatch between the outlines of the slices displayed in grey and in pink is observed. In (**B**), the outlines of the slices coincide.

Due to manual handling of the slices for sample preparation and digitization, image artifacts were created. Therefore, registration of the digitalized slides was necessary. Table 2 summarizes registration methods reported in the literature and their applicability for high-resolution histologic slices of tumors. For registration, physical markers can be used, which are applied to the tissue before slicing [36]. This technique has the disadvantage of damaging the tissue, which can no longer be evaluated at the position of the markers. To overcome this problem, Schwier et al. [37] introduced a method in which vessels are used as markers to register the slices as they are re-occurring features on every slice. Another approach is the scale-invariant feature transform (SIFT) by Lowe [38], which extracts distinctive areas, called features, from unregistered images that can be matched to one another. This SIFT algorithm and the similar SURF algorithm [39] are used in a couple of other registration approaches [40–47]. However, feature extraction alone in histologic images of a tumor is not sufficient because of the highly repetitive patterns that are not sufficiently distinctive. Kugler et al. [48,49] also applied the extraction of features technique using normalized cross-correlation and smoothening procedures. Rigid image registration algorithms [50,51] are frequently used, enabling the rotation, displacement, and scaling of digitalized slices. However, these algorithms are not sufficient for the registration of typical histological slices, which show further damages gained during preparation (deformations, fissures, etc.). Consequently, most existing registration methods show shortcomings in registering high-resolution images of tumor slices as they cannot handle extensive damage of tissue such as shearing [40,51,52], ruptures [53], or highly repetitive patterns [38,39,41,45,47,54,55]. Although the images of the example tumor investigated in this work consist of highly repetitive patterns, which limits exclusively feature-based approaches such as SIFT and SURF [38–41,44,47,55,56], the algorithm used is suitable to register whole slide images. This is due to the combination of area- and feature-based registration. Furthermore, it does not rely on previously segmented images [54,57].

Table 2. Registration methods reported in the literature, application field, and limitations of the method.

Focused Methods	Reconstruction-Related Application	Limitations
	Fiducial markers for point matching	
Rigid registration [36]	Protocol for inducing fiducial markers	Markers locally destroy tissue and vessels
Affine registration [58]	3D reconstruction specifically for muscle fibers	
	Feature extraction and matching methods	
Finding scale-invariant feature transformation SIFT/SURF: local feature descriptors [38–41,44]	Detection and matching of distinctive features invariant to the transformation	Unsuitable for highly repetitive patterns of histologic samples
Object tracing using SIFT [47]	Registration for images with high deformation, artifacts, and missing tissue, incorporating quality assurance	
Point matching improving RANSAC [56]	Faster and more robust point matching of features	
Non-rigid feature matching [55]	Matching and handling large deformations	
	Rigid registration methods	
Sparse-feature-based registration [42]	Handling registration of objects with different orders of magnitudes in structure size	Rigid registration is not sufficient for samples deformed by shearing
Rigid registration [51]	Co-registering re-stained histological images	
Trajectory tracing [49]	Registration and reconstruction of 2D histological images with different staining	Mainly captures vessels along z-axis
	Non-rigid registration methods	
Rigid and elastic registration [57]	Automatic sectioning, segmentation, registration, 3D reconstruction	Registration of already segmented images
Intensity-based registration, curvature flow [54]	Smoothed 3D reconstruction of distinct object boundaries	
Feature- and area-based registration [25]	Different image types including highly repetitive histological images	Computationally expensive transformations
B-spline [59]	3D reconstruction preserving tissue microstructures	Not capable of handling large distortions
Affine registration, mutual information (MI), matrix exponential neural network (MINE) [53]	Unsupervised registration for mono- and multi-modal images	
Registering multi-resolution scales for ROIs [45]	Registration of histological images dealing with major artifacts	No whole-section registration
Intensity-based stochastic gradient descent method (SGDM), region-based convolutional neural networks [60]	Registration, segmentation, and reconstruction of multi-modal histological images, focused on nerval specimen	Unsuitable for highly repetitive patterns of histologic samples
	Non-automatic methods	
Rigid registration [50]	Manual rigid registration	Unfeasible for big data
Manual segmentation, rigid registration, mesh generation [52]	Direct 3D mesh generation from images	

The results of the validation (see Section 2.3) are shown in Table 3. The quality of the presented method ranks in the middle of the earlier published methods for the complete set of images of the ANHIR challenge and for images similar to those presented in this work at the very front [30].

Table 3. Results of the registration validation. The registration error (*r*TRE) and the average robustness R_i are calculated as described in Section 2.3 and compared to results from literature.

	Complete Set of Images	Subset of Similar Images	Range of Top 10 Algorithms in ANHIR Leaderboard [61]
average median rTRE	0.0044	0.002163	0.00067–0.00250
average robustness R_i	0.85	0.986152	0.99276–0.88992

It can be concluded that the results referring to the investigation of the complete set of images are in the middle of the bulk of those published in the ANHIR challenge [30] or compared to the current leaderboard [61]. However, for the image subset, the results rank within the top ranks published. Parallelization of different steps in the image registration (s. Figure 3) resulted in time-efficient computation (cf. Table 1), making the algorithm suitable for the investigation of a large number of images.

Regarding the visual quality of the segmentation, the results are presented as an overlay of the original slice and the segmented areas in Figure 7. It can be observed that clustered vessels occur in areas with a high structure density. The delineation of individual structures with a very small space of a few pixels between them gets lost in the dilation step. This is a limitation of the algorithm. A delineation of single vessels within these clusters could be achieved for images with a higher resolution. However, this leads to an increase in computation time.

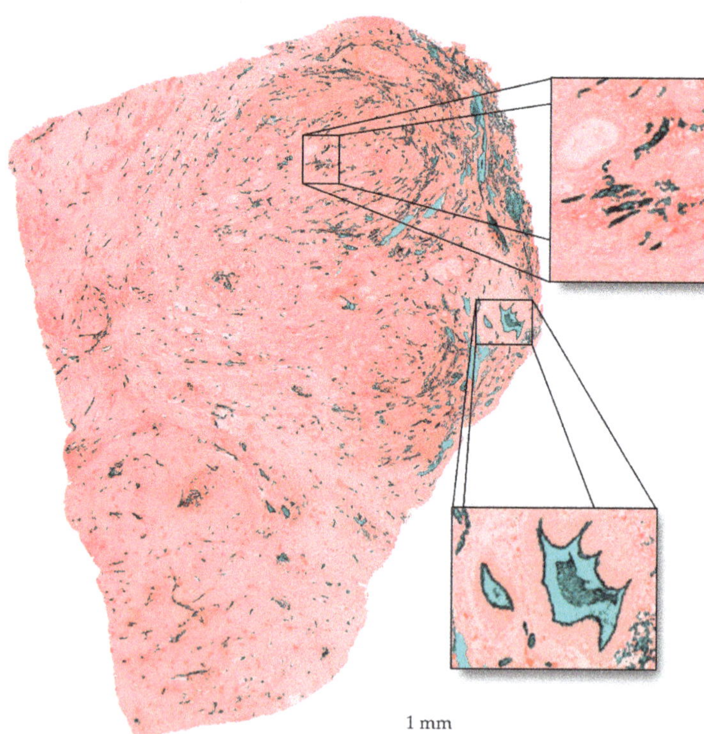

Figure 7. Overlay of original image and segmentation. Vessel interiors are marked in blue, vessel contours in black, discarded tissue in red.

To develop a 3D vascular network from registered images of slices, vessels need to be segmented from the surrounding tissue. With regard to segmentation, for some of the published works, commercial software was used that is not available anymore [50,62] Other approaches were specifically developed for certain tissue types, such as fascicles [60], and are not applicable to histologic images of tumor tissue. Most segmentation algorithms are designed to delineate cancerous tissue regions from non-cancerous surroundings [63–66]. These algorithms identify cancerous regions by analysis of misshaped structures, color variations in staining, or different textures within the image. Recent approaches to segment histologic images use standard computer vision techniques such as color-based segmentation with thresholding [67,68] or Markov random fields [69], pattern recognition [65], or clustering such as k-means or mean-shift [70–72]. Histological images usually show very high resolutions at large image sizes due to their high level of detail, which makes the above-mentioned approaches very expensive in terms of computing time [70]. With the developments in computational power, the utilization of machine learning in medical image analysis has experienced a huge increase. Many approaches to identify cancerous regions have been developed and have greatly improved the field of medical image analysis [73–79]. However, identifying cancerous regions is a completely different task compared to delineating small structures such as capillaries in a tumor. The main differences refer to the level of detail of highly repetitive small features that need to be delineated.

Reyes-Aldasoro et al. [62] developed an algorithm for automated segmentation of vascular structures from histologic images. A region-growing algorithm detects vascular structures that are then—using characteristics of continuous vessels—joined, separated, or closed to align with a certain definition of a realistic pathway for vessels. Schwier et al. [80] used the software MeVisLab [81] to reconstruct a whole vessel tree from murine liver sections. The reported procedure combines color thresholds and object-based analysis to include spatial information and assumed characteristics of vessel trajectories such as continuity. A broad segmentation approach is Trainable Weka Segmentation [82]. It uses machine learning to distinguish between different areas and must manually be trained by a user. As the method presented by Schwier et al. [80] shows promising results for the segmentation of consecutive tissue slices, we tested this method with our images and compared it to the method described above. The results are described in the Supplementary Materials S1, Table S1.

Another approach for segmenting vascular structures is vesselness functions [83–85]. These functions enhance the appearance of blood vessels in medical images [86,87] and use a vesselness measure to obtain probability masks that state for each pixel if it is likely to belong to a vessel or not. One of the primary challenges here would be the heterogeneity in staining intensities and colors within the histological dataset. Vesselness functions are typically designed to detect structures with distinct intensity or contrast characteristics [83,88], which are likely to be obscured by the irregular and heterogeneous staining. As a result, these functions would exhibit inconsistent and unreliable responses, making the accurate identification of vessels difficult. Additionally, the arbitrary shapes of vessels within the tumor slices are challenging. Vesselness functions rely on detecting tubular or vessel-like shapes [87,88]. In the present data, vessels often exhibited irregular geometries, which do not conform to typical vessel shapes. Consequently, the vesselness functions are likely to fail to effectively capture these diverse structures.

The comparison of the resulting segmentation showed that for the images used in this work, our segmentation is more robust to distortions such as ruptures and tissue located in vessels (see Supplementary Materials S1, Table S1).

The complete reconstruction of the vascular network from 169 histologic slices is shown in Figure 8. Single vessels and bifurcations are clearly visible.

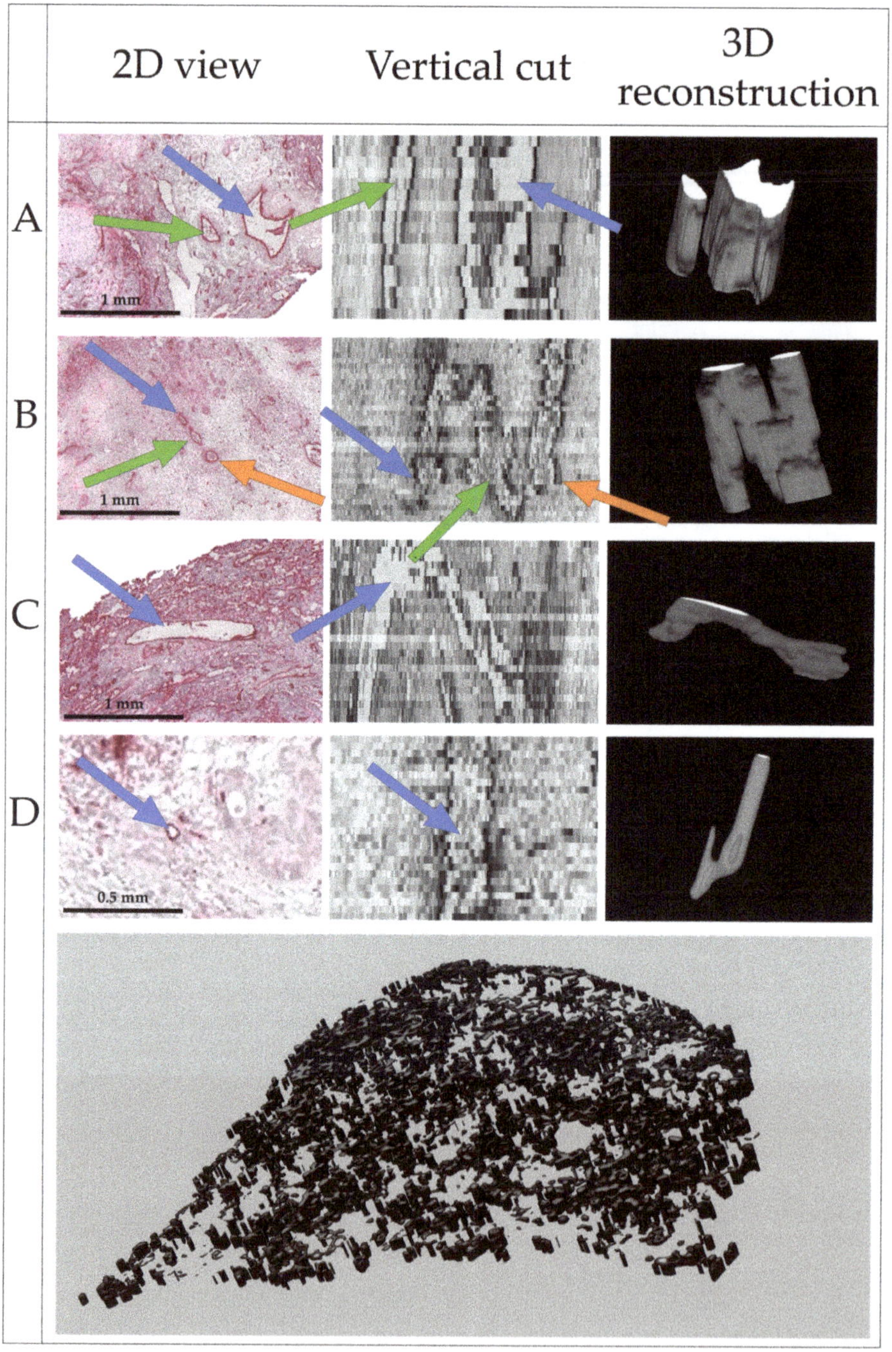

Figure 8. Final 3D reconstruction of the 169 slices with close-up views on four distinct vessels, where (**A**) shows two parallel vessels; (**B**) shows two bifurcations, one join and one split of two vessels each; (**C**) shows a vessel perpendicular to the image plane; and (**D**) shows a very small vessel. Within one row, arrows of the same color point to the same structures.

As no ground truth is available, the quality of the segmentation and 3D reconstruction was evaluated visually by an experienced radiologist (see Acknowledgements). Based on the results shown in Figure 8, the radiologist estimated a realistic delineation of vessels parallel and perpendicular to the image plane of the histologic slices. Some artifacts were present, but the general structure of the vessels was well represented if the vessels were visible in the histopathological source images.

A validation of the continuity of the vessels was not possible with the available dataset. As the histologic slices have a thickness of 2.5 µm, the reconstruction of the 169 slices represents a vessel network with a thickness of 422.5 µm. For an assessment of vessel continuity, a larger dataset is necessary. Therefore, the evaluation of vessel continuity is subject to future research.

Reconstruction of 3D structures based on histological images has been performed before for different medical applications: Feuerstein et al. [89] used Markov random fields with deformable stack registration. Fonyad et al. [90] reconstructed a coronary model of mice using the Voloom software (former µCore by microDimensions GmbH, Munich, Germany). Onozato et al. [91] reconstructed adenocarcinomas of the lung using a former version of the CaseViewer software (3DHistech Ltd., Budapest, Hungary). Another predecessor of CaseViewer was used by Wu et al. [92] to reconstruct colorectal biopsies. Tovbis et al. [60] additionally segmented the fascicles in the reconstructed tissue. The approach presented here focuses on the reconstruction of capillaries on a much smaller scale than previous works.

4. Conclusions

With this work, we developed an algorithm that enables the reconstruction of a 3D vascular network of a human tumor based on histologic images. The algorithm consists of a two-step registration approach, feature- and area-based registration, and is suitable for the investigation of a large number of high-resolution images. It was validated with the ANHIR dataset and competes with the top ranks for similar tissue types as used in this work. It further uses a five-step segmentation, which separately delineates vessel interiors and contours, and fuses this information with that from consecutive images, to extract the vessels from the surrounding tissue. The reconstruction algorithm refines the registration and segmentation results to achieve a natural-looking model of a vascular network. The resulting 3D model was qualitatively evaluated by an experienced radiologist and considered as promising.

Supplementary Materials: The following are available online at https://www.mdpi.com/article/10.3390/cancers15225352/s1, Chapter S1: Comparison to alternative segmentation algorithm, Table S1: Comparison of the segmentation algorithms. Reference [80] is cited in the Supplementary Materials.

Author Contributions: Conceptualization, M.C.L. and I.S.; methodology, M.C.L.; software, L.G.; validation, M.C.L. and L.G.; formal analysis, M.C.L., L.G. and I.S.; resources, A.A.R., I.S. and T.S.-R.; data curation, M.C.L. and I.S.; writing—original draft preparation, M.C.L.; writing—review and editing, M.C.L., L.G. and I.S.; visualization, M.C.L. and L.G.; supervision, I.S.; project administration, I.S.; funding acquisition, I.S. All authors have read and agreed to the published version of the manuscript.

Funding: Funded by the Federal Ministry of Education and Research (BMBF) and the Ministry of Culture and Science of the German State of North Rhine-Westphalia (MKW) under the Excellence Strategy of the Federal Government and the Länder.

Institutional Review Board Statement: The study was conducted according to the guidelines of the Declaration of Helsinki and approved by the Institutional Review Board of the RWTH Aachen University Hospital (EK206/09, 05.01.2010).

Informed Consent Statement: Not applicable.

Data Availability Statement: Data are available in a pseudonymized manner upon request due to ethical restrictions (patient confidentially). Therefore, the data are not available publicly.

Acknowledgments: We thank Daniel Truhn from the Department of Radiology of the University Hospital RWTH Aachen for his evaluation in assessing the segmentation and registration results. Simulations were performed with computing resources granted by RWTH Aachen University under project rwth0481.

Conflicts of Interest: The authors declare no conflict of interest. The funders had no role in the design of the study; in the collection, analyses, or interpretation of data; in the writing of the manuscript, or in the decision to publish the results.

References

1. Chen, Y.; Ali, M.; Shi, S.; Cheang, U.K. Biosensing-by-Learning Direct Targeting Strategy for Enhanced Tumor Sensitization. *IEEE Trans. Nanobioscience* **2019**, *18*, 498–509. [CrossRef] [PubMed]
2. Haun, J.B.; Yoon, T.-J.; Lee, H.; Weissleder, R. Magnetic nanoparticle biosensors. *Wiley Interdiscip. Rev. Nanomed. Nanobiotechnol.* **2010**, *2*, 291–304. [CrossRef] [PubMed]
3. Bhatia, S.N.; Ingber, D.E. Microfluidic organs-on-chips. *Nat. Biotechnol.* **2014**, *32*, 760–772. [CrossRef] [PubMed]
4. Koyilot, M.C.; Natarajan, P.; Hunt, C.R.; Sivarajkumar, S.; Roy, R.; Joglekar, S.; Pandita, S.; Tong, C.W.; Marakkar, S.; Subramanian, L.; et al. Breakthroughs and Applications of Organ-on-a-Chip Technology. *Cells* **2022**, *11*, 1828. [CrossRef] [PubMed]
5. Bongio, M.; Lopa, S.; Gilardi, M.; Bersini, S.; Moretti, M. A 3D vascularized bone remodeling model combining osteoblasts and osteoclasts in a CaP nanoparticle-enriched matrix. *Nanomedicine* **2016**, *11*, 1073–1091. [CrossRef]
6. Unger, C.; Kramer, N.; Walzl, A.; Scherzer, M.; Hengstschläger, M.; Dolznig, H. Modeling human carcinomas: Physiologically relevant 3D models to improve anti-cancer drug development. *Adv. Drug Deliv. Rev.* **2014**, *79–80*, 50–67. [CrossRef]
7. Abd Elrahman, A.A.; Mansour, F.R. Targeted magnetic iron oxide nanoparticles: Preparation, functionalization and biomedical application: Preparation, functionalization and biomedical application. *J. Drug Deliv. Sci. Technol.* **2019**, *52*, 702–712. [CrossRef]
8. Dadfar, S.M.; Roemhild, K.; Drude, N.I.; von Stillfried, S.; Knüchel, R.; Kiessling, F.; Lammers, T. Iron oxide nanoparticles: Diagnostic, therapeutic and theranostic applications. *Adv. Drug Deliv. Rev.* **2019**, *138*, 302–325. [CrossRef] [PubMed]
9. Chawla, A.; Ferrone, C.R. Neoadjuvant Therapy for Resectable Pancreatic Cancer: An Evolving Paradigm Shift. *Front. Oncol.* **2019**, *9*, 1085. [CrossRef]
10. Roeth, A.A.; Slabu, I.; Baumann, M.; Alizai, P.H.; Schmeding, M.; Guentherodt, G.; Schmitz-Rode, T.; Neumann, U.P. Establishment of a biophysical model to optimize endoscopic targeting of magnetic nanoparticles for cancer treatment. *Int. J. Nanomed.* **2017**, *12*, 5933–5940. [CrossRef]
11. Lindemann, M.C.; Luttke, T.; Nottrodt, N.; Schmitz-Rode, T.; Slabu, I. FEM based simulation of magnetic drug targeting in a multibranched vessel model. *Comput. Methods Programs Biomed.* **2021**, *210*, 106354. [CrossRef]
12. Singh, M.; Ma, R.; Zhu, L. Quantitative evaluation of effects of coupled temperature elevation, thermal damage, and enlarged porosity on nanoparticle migration in tumors during magnetic nanoparticle hyperthermia. *Int. Commun. Heat Mass Transf.* **2021**, *126*, 105393. [CrossRef]
13. Singh, M.; Singh, T.; Soni, S. Pre-operative Assessment of Ablation Margins for Variable Blood Perfusion Metrics in a Magnetic Resonance Imaging Based Complex Breast Tumour Anatomy: Simulation Paradigms in Thermal Therapies. *Comput. Methods Programs Biomed.* **2021**, *198*, 105781. [CrossRef] [PubMed]
14. Van der Meer, A.D.; van den Berg, A. Organs-on-chips: Breaking the in vitro impasse. *Integr. Biol.* **2012**, *4*, 461–470. [CrossRef] [PubMed]
15. Russell, W.; Burch, R.L. *The Principles of Humane Experimental Technique;* Methuen Publishing: London, UK, 1959.
16. Roeth, A.A.; Garretson, I.; Beltz, M.; Herbold, T.; Schulze-Hagen, M.; Quaisser, S.; Georgens, A.; Reith, D.; Slabu, I.; Klink, C.D.; et al. 3D-Printed Replica and Porcine Explants for Pre-Clinical Optimization of Endoscopic Tumor Treatment by Magnetic Targeting. *Cancers* **2021**, *13*, 5496. [CrossRef] [PubMed]
17. Wang, P.; Sun, Y.; Shi, X.; Shen, H.; Ning, H.; Liu, H. 3D printing of tissue engineering scaffolds: A focus on vascular regeneration. *Bio-Des. Manuf.* **2021**, *4*, 344–378. [CrossRef] [PubMed]
18. Szafron, J.M.; Ramachandra, A.B.; Breuer, C.K.; Marsden, A.L.; Humphrey, J.D. Optimization of Tissue-Engineered Vascular Graft Design Using Computational Modeling. *Tissue Eng. Part C Methods* **2019**, *25*, 561–570. [CrossRef] [PubMed]
19. Tu, S.; Hu, F.; Cai, W.; Xiao, L.; Zhang, L.; Zheng, H.; Jiang, Q.; Chen, L. Visualizing polymeric bioresorbable scaffolds with three-dimensional image reconstruction using contrast-enhanced micro-computed tomography. *Int. J. Cardiovasc. Imaging* **2017**, *33*, 731–737. [CrossRef] [PubMed]
20. Ghaghada, K.B.; Sato, A.F.; Starosolski, Z.A.; Berg, J.; Vail, D.M. Computed Tomography Imaging of Solid Tumors Using a Liposomal-Iodine Contrast Agent in Companion Dogs with Naturally Occurring Cancer. *PLoS ONE* **2016**, *11*, e0152718. [CrossRef] [PubMed]
21. Ghani, M.U.; Zhou, Z.; Ren, L.; Li, Y.; Zheng, B.; Yang, K.; Liu, H. Investigation of spatial resolution characteristics of an in vivo micro computed tomography system. *Nucl. Instrum. Methods Phys. Res. A* **2016**, *807*, 129–136. [CrossRef]
22. McDonald, D.M.; Choyke, P.L. Imaging of angiogenesis: From microscope to clinic. *Nat. Med.* **2003**, *9*, 713–725. [CrossRef]

23. Pena, A.-M.; Chen, X.; Pence, I.J.; Bornschlögl, T.; Jeong, S.; Grégoire, S.; Luengo, G.S.; Hallegot, P.; Obeidy, P.; Feizpour, A.; et al. Imaging and quantifying drug delivery in skin—Part 2: Fluorescence andvibrational spectroscopic imaging methods. *Adv. Drug Deliv. Rev.* **2020**, *153*, 147–168. [CrossRef] [PubMed]
24. Schindelin, J.; Arganda-Carreras, I.; Frise, E.; Kaynig, V.; Longair, M.; Pietzsch, T.; Preibisch, S.; Rueden, C.; Saalfeld, S.; Schmid, B.; et al. Fiji: An open-source platform for biological-image analysis. *Nat. Methods* **2012**, *9*, 676–682. [CrossRef]
25. Wang, C.-W.; Chen, H.-C. Improved image alignment method in application to X-ray images and biological images. *Bioinformatics* **2013**, *29*, 1879–1887. [CrossRef] [PubMed]
26. Wang, C.-W.; Ka, S.-M.; Chen, A. Robust image registration of biological microscopic images. *Sci. Rep.* **2014**, *4*, 6050. [CrossRef] [PubMed]
27. Arganda-Carreras, I.; Sorzano, C.O.S.; Marabini, R.; Carazo, J.M.; Ortiz-De-Solorzano, C.; Kybic, J. Consistent and Elastic Registration of Histological Sections Using Vector-Spline Regularization. In *Computer Vision Approaches to Medical Image Analysis*; Hutchison, D., Kanade, T., Kittler, J., Kleinberg, J.M., Mattern, F., Mitchell, J.C., Naor, M., Nierstrasz, O., Pandu Rangan, C., et al., Eds.; Springer: Berlin/Heidelberg, Germany, 2006; pp. 85–95. ISBN 978-3-540-46257-6.
28. Sorzano, C.O.S.; Thévenaz, P.; Unser, M. Elastic registration of biological images using vector-spline regularization. *IEEE Trans. Biomed. Eng.* **2005**, *52*, 652–663. [CrossRef] [PubMed]
29. Borovec, J.; Munoz-Barrutia, A.; Kybic, J. Benchmarking of Image Registration Methods for Differently Stained Histological Slides. In Proceedings of the 2018 25th IEEE International Conference on Image Processing (ICIP), Athens, Greece, 7–10 October 2018; pp. 3368–3372, ISBN 978-1-4799-7061-2.
30. Borovec, J.; Kybic, J.; Arganda-Carreras, I.; Sorokin, D.V.; Bueno, G.; Khvostikov, A.V.; Bakas, S.; Chang, E.I.-C.; Heldmann, S.; Kartasalo, K.; et al. ANHIR: Automatic Non-Rigid Histological Image Registration Challenge. *IEEE Trans. Med. Imaging* **2020**, *39*, 3042–3052. [CrossRef]
31. Zuiderveld, K. Contrast limited adaptive histogram equilization. In *Graphics Gems IV*; Heckbert, P.S., Ed.; Academic Press: San Diego, CA, USA, 1994; pp. 474–485. ISBN 0123361559.
32. Otsu, N. A Threshold Selection Method from Gray-Level Histograms. *IEEE Trans. Syst., Man Cybern.* **1979**, *9*, 62–66. [CrossRef]
33. Gonzalez, R.C.; Woods, R.E.; Eddins, S.L. *Digital Image Processing Using MATLAB*; Pearson Prentice Hall: Upper Saddle River, NJ, USA, 2004; ISBN 0-13-008519-7.
34. Reddy, P.R.; Amarnadh, V.; Bhaskar, M. Evaluation of Stopping Criterion in Contour Tracing Algorithms. *Int. J. Comput. Sci. Inf. Technol.* **2012**, *3*, 3888–3894.
35. Haralick, R.M.; Shapiro, L.G. *Computer and Robot Vision, 2*; Addison-Wesley: Reading, MA, USA, 1993; ISBN 0201569434.
36. Hughes, C.; Rouvière, O.; Mege-Lechevallier, F.; Souchon, R.; Prost, R. Robust alignment of prostate histology slices with quantified accuracy. *IEEE Trans. Biomed. Eng.* **2013**, *60*, 281–291. [CrossRef]
37. Schwier, M.; Böhler, T.; Hahn, H.K.; Dahmen, U.; Dirsch, O. Registration of histological whole slide images guided by vessel structures. *J. Pathol. Inform.* **2013**, *4*, S10. [CrossRef] [PubMed]
38. Lowe, D.G. Distinctive Image Features from Scale-Invariant Keypoints. *Int. J. Comput. Vis.* **2004**, *60*, 91–110. [CrossRef]
39. Bay, H.; Ess, A.; Tuytelaars, T.; van Gool, L. Speeded-Up Robust Features (SURF). *Comput. Vis. Image Underst.* **2008**, *110*, 346–359. [CrossRef]
40. Kupfer, B.; Netanyahu, N.S.; Shimshoni, I. An Efficient SIFT-Based Mode-Seeking Algorithm for Sub-Pixel Registration of Remotely Sensed Images. *IEEE Geosci. Remote Sens. Lett.* **2015**, *12*, 379–383. [CrossRef]
41. Liu, Q.; Zhao, G.; Deng, J.; Xue, Q.; Hou, W.; He, Y. Image Registration Algorithm for Sequence Pathology Slices of Pulmonary Nodule. In Proceedings of the 2019 8th International Symposium on Next Generation Electronics (ISNE), Zhengzhou, China, 9–10 October 2019; pp. 1–3, ISBN 978-1-7281-2062-1.
42. Lobachev, O.; Ulrich, C.; Steiniger, B.S.; Wilhelmi, V.; Stachniss, V.; Guthe, M. Feature-based multi-resolution registration of immunostained serial sections. *Med. Image Anal.* **2017**, *35*, 288–302. [CrossRef]
43. Saalfeld, S.; Fetter, R.; Cardona, A.; Tomancak, P. Elastic volume reconstruction from series of ultra-thin microscopy sections. *Nat. Methods* **2012**, *9*, 717–720. [CrossRef] [PubMed]
44. Hermann, J.; Brehmer, K.; Jankowski, V.; Lellig, M.; Hohl, M.; Mahfoud, F.; Speer, T.; Schunk, S.J., Tschernig, T.; Thiele, H.; et al. Registration of Image Modalities for Analyses of Tissue Samples Using 3D Image Modelling. *Proteom. Clin. Appl.* **2020**, *15*, e1900143. [CrossRef] [PubMed]
45. Paknezhad, M.; Loh, S.Y.M.; Choudhury, Y.; Koh, V.K.C.; Yong, T.T.K.; Tan, H.S.; Kanesvaran, R.; Tan, P.H.; Peng, J.Y.S.; Yu, W.; et al. Regional registration of whole slide image stacks containing major histological artifacts. *BMC Bioinform.* **2020**, *21*, 558. [CrossRef]
46. Zhang, J.; Li, Z.; Yu, Q. Point-Based Registration for Multi-stained Histology Images. In Proceedings of the 2020 IEEE 5th International Conference on Image, Vision and Computing (ICIVC), Beijing, China, 10–12 July 2020; pp. 92–96, ISBN 978-1-7281-6661-2.
47. Deng, R.; Yang, H.; Jha, A.; Lu, Y.; Chu, P.; Fogo, A.B.; Huo, Y. Map3D: Registration-Based Multi-Object Tracking on 3D Serial Whole Slide Images. *IEEE Trans. Med. Imaging* **2021**, *40*, 1924–1933. [CrossRef] [PubMed]

48. Kugler, M.; Goto, Y.; Kawamura, N.; Kobayashi, H.; Yokota, T.; Iwamoto, C.; Ohuchida, K.; Hashizume, M.; Hontani, H. Accurate 3D Reconstruction of a Whole Pancreatic Cancer Tumor from Pathology Images with Different Stains. In *Computational Pathology and Ophthalmic Medical Image Analysis: First International Workshop, COMPAY 2018, and 5th International Workshop, OMIA 2018, held in Conjunction with MICCAI 2018, Granada, Spain, 16–20 September 2018: Proceedings*; Stoyanov, D., Taylor, Z., Ciompi, F., Xu, Y., Martel, A., Maier-Hein, L., Rajpoot, N., van der Laak, J., Veta, M., McKenna, S., et al., Eds.; Springer: Cham, Switzerland, 2018; pp. 35–43. ISBN 9783030009496.
49. Kugler, M.; Goto, Y.; Tamura, Y.; Kawamura, N.; Kobayashi, H.; Yokota, T.; Iwamoto, C.; Ohuchida, K.; Hashizume, M.; Shimizu, A.; et al. Robust 3D image reconstruction of pancreatic cancer tumors from histopathological images with different stains and its quantitative performance evaluation. *Int. J. Comput. Assist. Radiol. Surg.* **2019**, *14*, 2047–2055. [CrossRef]
50. Paish, E.C.; Green, A.R.; Rakha, E.A.; Macmillan, R.D.; Maddison, J.R.; Ellis, I.O. Three-dimensional reconstruction of sentinel lymph nodes with metastatic breast cancer indicates three distinct patterns of tumour growth. *J. Clin. Pathol.* **2009**, *62*, 617–623. [CrossRef] [PubMed]
51. Jiang, J.; Larson, N.B.; Prodduturi, N.; Flotte, T.J.; Hart, S.N. Robust hierarchical density estimation and regression for re-stained histological whole slide image co-registration. *PLoS ONE* **2019**, *14*, e0220074. [CrossRef] [PubMed]
52. Niemann, A.; Weigand, S.; Hoffmann, T.; Skalej, M.; Tulamo, R.; Preim, B.; Saalfeld, S. Interactive exploration of a 3D intracranial aneurysm wall model extracted from histologic slices. *Int. J. Comput. Assist. Radiol. Surg.* **2020**, *15*, 99–107. [CrossRef] [PubMed]
53. Nan, A.; Tennant, M.; Rubin, U.; Ray, N. DRMIME: Differentiable Mutual Information and Matrix Exponential for Multi-Resolution Image Registration. In Proceedings of the Third Conference on Medical Imaging with Deep Learning, Montreal, QC, Canada, 6–8 July 2020.
54. Cifor, A.; Bai, L.; Pitiot, A. Smoothness-guided 3-D reconstruction of 2-D histological images. *Neuroimage* **2011**, *56*, 197–211. [CrossRef] [PubMed]
55. Cai, N.; Chen, H.; Li, Y.; Peng, Y.; Li, J.; Li, X. Reducing non-realistic deformations in registration using precise and reliable landmark correspondences. *Comput. Biol. Med.* **2019**, *115*, 103515. [CrossRef] [PubMed]
56. Wu, Y.; Ma, W.; Gong, M.; Su, L.; Jiao, L. A Novel Point-Matching Algorithm Based on Fast Sample Consensus for Image Registration. *IEEE Geosci. Remote Sens. Lett.* **2015**, *12*, 43–47. [CrossRef]
57. Arganda-Carreras, I.; Fernández-González, R.; Muñoz-Barrutia, A.; Ortiz-De-Solorzano, C. 3D reconstruction of histological sections: Application to mammary gland tissue. *Microsc. Res. Tech.* **2010**, *73*, 1019–1029. [CrossRef] [PubMed]
58. Kuravi, R.; Leichsenring, K.; Böl, M.; Ehret, A.E. 3D finite element models from serial section histology of skeletal muscle tissue —The role of micro-architecture on mechanical behaviour. *J. Mech. Behav. Biomed. Mater.* **2020**, *113*, 104109. [CrossRef]
59. Liu, J.; Wu, X.; Xu, C.; Ma, M.; Zhao, J.; Li, M.; Yu, Q.; Hao, X.; Wang, G.; Wei, B.; et al. A Novel Method for Observing Tumor Margin in Hepatoblastoma Based on Microstructure 3D Reconstruction. *Fetal Pediatr. Pathol.* **2020**, *41*, 371–380. [CrossRef] [PubMed]
60. Tovbis, D.; Agur, A.; Mogk, J.P.M.; Zariffa, J. Automatic three-dimensional reconstruction of fascicles in peripheral nerves from histological images. *PLoS ONE* **2020**, *15*, e0233028. [CrossRef]
61. Borovec, J.; Kybic, J.; Barrutia, A.M. Leaderboard—Grand Challenge. Available online: https://anhir.grand-challenge.org/evaluation/challenge/leaderboard/ (accessed on 16 May 2022).
62. Reyes-Aldasoro, C.C.; Williams, L.J.; Akerman, S.; Kanthou, C.; Tozer, G.M. An automatic algorithm for the segmentation and morphological analysis of microvessels in immunostained histological tumour sections. *J. Microsc.* **2011**, *242*, 262–278. [CrossRef]
63. Li, C.; Chen, H.; Li, X.; Xu, N.; Hu, Z.; Xue, D.; Qi, S.; Ma, H.; Zhang, L.; Sun, H. A review for cervical histopathology image analysis using machine vision approaches. *Artif. Intell. Rev.* **2020**, *53*, 4821–4862. [CrossRef]
64. Azevedo Tosta, T.A.; Neves, L.A.; do Nascimento, M.Z. Segmentation methods of H&E-stained histological images of lymphoma: A review. *Inform. Med. Unlocked* **2017**, *9*, 35–43. [CrossRef]
65. Tabesh, A.; Teverovskiy, M.; Pang, H.-Y.; Kumar, V.P.; Verbel, D.; Kotsianti, A.; Saidi, O. Multifeature prostate cancer diagnosis and Gleason grading of histological images. *IEEE Trans. Med. Imaging* **2007**, *26*, 1366–1378. [CrossRef] [PubMed]
66. Gurcan, M.N.; Boucheron, L.E.; Can, A.; Madabhushi, A.; Rajpoot, N.M.; Yener, B. Histopathological image analysis: A review. *IEEE Rev. Biomed. Eng.* **2009**, *2*, 147–171. [CrossRef] [PubMed]
67. Goddard, J.C.; Sutton, C.D.; Furness, P.N.; Kockelbergh, R.C.; O'Byrne, K.J. A computer image analysis system for microvessel density measurement in solid tumours. *Angiogenesis* **2002**, *5*, 15–20. [CrossRef] [PubMed]
68. Steiniger, B.; Bette, M.; Schwarzbach, H. The open microcirculation in human spleens: A three-dimensional approach. *J. Histochem. Cytochem.* **2011**, *59*, 639–648. [CrossRef] [PubMed]
69. Meas-Yedid, V.; Tilie, S.; Olivo-Marin, J.-C. Color image segmentation based on Markov random field clustering for histological image analysis. In Proceedings of the 16th International Conference on Pattern Recognition, Quebec City, QC, Canada, 11–15 August 2002; pp. 796–799.
70. Wu, G.; Zhao, X.; Luo, S.; Shi, H. Histological image segmentation using fast mean shift clustering method. *Biomed. Eng. Online* **2015**, *14*, 24. [CrossRef]
71. Yang, L.; Meer, P.; Foran, D.J. Unsupervised segmentation based on robust estimation and color active contour models. *IEEE Trans. Inf. Technol. Biomed.* **2005**, *9*, 475–486. [CrossRef] [PubMed]
72. He, L.; Long, L.R.; Antani, S.; Thoma, G. Distribution fitting-based pixel labeling for histology image segmentation. *Med. Imaging 2011 Comput.-Aided Diagn.* **2011**, *7963*, 79633D. [CrossRef]

73. Xu, Y.; Zhu, J.-Y.; Chang, E.I.-C.; Lai, M.; Tu, Z. Weakly supervised histopathology cancer image segmentation and classification. *Med. Image Anal.* **2014**, *18*, 591–604. [CrossRef]
74. Rodríguez, R. Blood Vessel Segmentation via Neural Network in Histological Images. *J. Intell. Robot. Syst.* **2003**, *36*, 451–465. [CrossRef]
75. Kothari, S.; Phan, J.H.; Moffitt, R.A.; Stokes, T.H.; Hassberger, S.E.; Chaudry, Q.; Young, A.N.; Wang, M.D. Automatic batch-invariant color segmentation of histological cancer images. *Proc. IEEE Int. Symp. Biomed. Imaging* **2011**, *2011*, 657–660. [CrossRef] [PubMed]
76. Jia, Z.; Huang, X.; Chang, E.I.-C.; Xu, Y. Constrained Deep Weak Supervision for Histopathology Image Segmentation. *IEEE Trans. Med. Imaging* **2017**, *36*, 2376–2388. [CrossRef] [PubMed]
77. Xu, G.; Song, Z.; Sun, Z.; Ku, C.; Yang, Z.; Liu, C.; Wang, S.; Ma, J.; Xu, W. CAMEL: A Weakly Supervised Learning Framework for Histopathology Image Segmentation. In Proceedings of the 2019 IEEE/CVF International Conference on Computer Vision (ICCV), Seoul, Republic of Korea, 27 October–2 November 2019. [CrossRef]
78. Sun, C.; Li, C.; Zhang, J.; Rahaman, M.M.; Ai, S.; Chen, H.; Kulwa, F.; Li, Y.; Li, X.; Jiang, T. Gastric histopathology image segmentation using a hierarchical conditional random field. *Biocybern. Biomed. Eng.* **2020**, *40*, 1535–1555. [CrossRef]
79. Janssens, T.; Antanas, L.; Derde, S.; Vanhorebeek, I.; van den Berghe, G.; Güiza Grandas, F. Charisma: An integrated approach to automatic H&E-stained skeletal muscle cell segmentation using supervised learning and novel robust clump splitting. *Med. Image Anal.* **2013**, *17*, 1206–1219. [CrossRef] [PubMed]
80. Schwier, M.; Hahn, H.K.; Dahmen, U.; Dirsch, O. Reconstruction of vessel structures from serial whole slide sections of murine liver samples. In Proceedings of the SPIE Medical Imaging, Lake Buena Vista, FL, USA, 9–14 February 2013; Gurcan, M.N., Madabhushi, A., Eds.; SPIE: Washington, DC, USA, 2013; p. 86760D. [CrossRef]
81. Ritter, F.; Boskamp, T.; Homeyer, A.; Laue, H.; Schwier, M.; Link, F.; Peitgen, H.-O. Medical image analysis. *IEEE Pulse* **2011**, *2*, 60–70. [CrossRef]
82. Arganda-Carreras, I.; Kaynig, V.; Rueden, C.; Eliceiri, K.W.; Schindelin, J.; Cardona, A.; Sebastian Seung, H. Trainable Weka Segmentation: A machine learning tool for microscopy pixel classification. *Bioinformatics* **2017**, *33*, 2424–2426. [CrossRef]
83. Drechsler, K.; Oyarzun Laura, C. Comparison of vesselness functions for multiscale analysis of the liver vasculature. In Proceedings of the 2010 10th IEEE International Conference on Information Technology and Applications in Biomedicine (ITAB 2010), Corfu, Greece, 3–5 November 2010; pp. 1–5, ISBN 978-1-4244-6559-0.
84. Tetteh, G.; Efremov, V.; Forkert, N.D.; Schneider, M.; Kirschke, J.; Weber, B.; Zimmer, C.; Piraud, M.; Menze, B.H. DeepVesselNet: Vessel Segmentation, Centerline Prediction, and Bifurcation Detection in 3-D Angiographic Volumes. *Front. Neurosci.* **2020**, *14*, 592352. [CrossRef] [PubMed]
85. Woźniak, T.; Strzelecki, M.; Majos, A.; Stefańczyk, L. 3D vascular tree segmentation using a multiscale vesselness function and a level set approach. *Biocybern. Biomed. Eng.* **2017**, *37*, 66–77. [CrossRef]
86. Survarachakan, S.; Pelanis, E.; Khan, Z.A.; Kumar, R.P.; Edwin, B.; Lindseth, F. Effects of Enhancement on Deep Learning Based Hepatic Vessel Segmentation. *Electronics* **2021**, *10*, 1165. [CrossRef]
87. Frangi, A.F.; Niessen, W.J.; Vincken, K.L.; Viergever, M.A. Multiscale vessel enhancement filtering. In *Medical Image Computing and Computer-Assisted Interventation—MICCAI'98*; Wells, W.M., Colchester, A., Delp, S., Eds.; Springer Nature: Berlin/Heidelberg, Germany, 1998; pp. 130–137. ISBN 978-3-540-65136-9.
88. Lamy, J.; Merveille, O.; Kerautret, B.; Passat, N.; Vacavant, A. Vesselness Filters: A Survey with Benchmarks Applied to Liver Imaging. In Proceedings of the 2020 25th International Conference on Pattern Recognition (ICPR), Milan, Italy, 10–15 January 2021; pp. 3528–3535, ISBN 978-1-7281-8808-9.
89. Feuerstein, M.; Heibel, H.; Gardiazabal, J.; Navab, N.; Groher, M. Reconstruction of 3-D histology images by simultaneous deformable registration. *Med. Image Comput. Comput. Assist. Interv.* **2011**, *14*, 582–589. [CrossRef] [PubMed]
90. Fónyad, L.; Shinoda, K.; Farkash, E.A.; Groher, M.; Sebastian, D.P.; Szász, A.M.; Colvin, R.B.; Yagi, Y. 3-dimensional digital reconstruction of the murine coronary system for the evaluation of chronic allograft vasculopathy. *Diagn. Pathol.* **2015**, *10*, 16. [CrossRef] [PubMed]
91. Onozato, M.L.; Klepeis, V.E.; Yagi, Y.; Mino-Kenudson, M. A role of three-dimensional (3D) reconstruction in the classification of lung adenocarcinoma. *Stud. Health Technol. Inform.* **2012**, *179*, 250–256. [CrossRef] [PubMed]
92. Wu, M.L.; Varga, V.S.; Kamaras, V.; Ficsor, L.; Tagscherer, A.; Tulassay, Z.; Molnar, B. Three-dimensional virtual microscopy of colorectal biopsies. *Arch. Pathol. Lab. Med.* **2005**, *129*, 507–510. [CrossRef] [PubMed]

diagnostics

MDPI

Article

Automated Computer-Assisted Medical Decision-Making System Based on Morphological Shape and Skin Thickness Analysis for Asymmetry Detection in Mammographic Images

Rafael Bayareh-Mancilla [1,*], Luis Alberto Medina-Ramos [2], Alfonso Toriz-Vázquez [3], Yazmín Mariela Hernández-Rodríguez [1] and Oscar Eduardo Cigarroa-Mayorga [1,*]

1 Department Advanced Technologies, UPIITA-Instituto Politécnico Nacional, Av. IPN No. 2580, Mexico City C.P. 07340, Mexico; ymhernandez@ipn.mx
2 CICATA-National Polytechnic Institute, Legaria 694, Mexico City C.P. 11500, Mexico; lmedinar1503@alumno.ipn.mx
3 Academic Unit, Institute of Applied Mathematics and Systems Research of the State of Yucatan, National Autonomous University of Mexico, Merida C.P. 97302, Yucatan, Mexico; atoriz98@comunidad.unam.mx
* Correspondence: rafael.bayareh@gmail.com (R.B.-M.); ocigarroam@ipn.mx (O.E.C.-M.)

Abstract: Breast cancer is a significant health concern for women, emphasizing the need for early detection. This research focuses on developing a computer system for asymmetry detection in mammographic images, employing two critical approaches: Dynamic Time Warping (DTW) for shape analysis and the Growing Seed Region (GSR) method for breast skin segmentation. The methodology involves processing mammograms in DICOM format. In the morphological study, a centroid-based mask is computed using extracted images from DICOM files. Distances between the centroid and the breast perimeter are then calculated to assess similarity through Dynamic Time Warping analysis. For skin thickness asymmetry identification, a seed is initially set on skin pixels and expanded based on intensity and depth similarities. The DTW analysis achieves an accuracy of 83%, correctly identifying 23 possible asymmetry cases out of 20 ground truth cases. The GRS method is validated using Average Symmetric Surface Distance and Relative Volumetric metrics, yielding similarities of 90.47% and 66.66%, respectively, for asymmetry cases compared to 182 ground truth segmented images, successfully identifying 35 patients with potential skin asymmetry. Additionally, a Graphical User Interface is designed to facilitate the insertion of DICOM files and provide visual representations of asymmetrical findings for validation and accessibility by physicians.

Keywords: asymmetry analysis; breast cancer; computer-aided diagnosis; early detection; mammography

Citation: Bayareh-Mancilla, R.; Medina-Ramos, L.A.; Toriz-Vázquez, A.; Hernández-Rodríguez, Y.M.; Cigarroa-Mayorga, O.E. Automated Computer-Assisted Medical Decision-Making System Based on Morphological Shape and Skin Thickness Analysis for Asymmetry Detection in Mammographic Images. *Diagnostics* **2023**, *13*, 3440. https://doi.org/10.3390/diagnostics13223440

Academic Editors: Michał Strzelecki, Adam Piorkowski and Rafał Obuchowicz

Received: 5 July 2023
Revised: 28 July 2023
Accepted: 3 August 2023
Published: 14 November 2023

1. Introduction

Breast cancer is a multifaceted disease with diverse clinical manifestations and prognoses. Among the common presentations are Invasive Ductal Carcinoma, Invasive Lobular Carcinoma, and Inflammatory Breast Cancer. These specific types of cancer are often associated with asymmetry between the right and left breasts, serving as early indications of the disease. The early detection of breast cancer is crucial to ensure successful treatment and improved survival rates. Breast cancer is the most common cancer among women worldwide, accounting for over 2 million new cases and 600,000 deaths annually [1]. The incidence of breast cancer varies widely across regions, with higher rates observed in developed countries. In Mexico, breast cancer is the leading cause of cancer death in women, with an estimated 28,000 new cases and 6000 deaths each year [2]. In recent years, efforts have been made to improve breast cancer diagnosis by accessing the population screening services in Mexico. The establishment of national breast cancer screening programs and the introduction of new technologies, such as digital mammography and breast ultrasound, have contributed to improved detection and treatment outcomes [3]. Thus, the early detection of breast cancer is crucial to ensuring successful treatment and improved survival rates. The early detection of breast cancer is crucial to improving survival rates and reducing the burden of the disease. Breast cancer screening programs, including mammography and clinical breast examination, are effective in reducing breast cancer mortality. However, access to screening and treatment remains a challenge in many low- and middle-income countries, where disparities in access to healthcare and cancer services exist [2,4].

While histopathology images provide valuable insights into the cellular composition of breast tissue, obtaining such data can be challenging in certain regions, limiting the widespread application of AI-based techniques [5]. To address this limitation, our research focuses on developing a computer system solely based on computer vision and morphological features for asymmetry detection in mammographic images. By leveraging mammogram images, which are widely available and commonly used for breast cancer screening, our approach aims to support advances in AI for breast cancer detection. In this context, we explore mammogram images over histopathology images in terms of availability and practicality. While histopathology images offer detailed cellular information, they may not always be easily accessible, especially as per current medical guidelines, histopathology is not typically recommended for cases classified via the Breast Imaging Reporting and Data System (BIRADS category 1 or 2) [6]. Also, In BIRADS 1 and 2 cases, asymmetry is usually considered benign and is not a cause for immediate concern. However, it is still important for radiologists to identify and document any asymmetry observed during mammogram evaluation. Regular follow-up mammograms may be recommended to monitor the stability of these findings over time [7,8]. On the other hand, mammogram images provide a non-invasive and widely accessible imaging modality for breast cancer screening, making them a more practical choice for developing AI-based systems.

While discussing the use of mammogram images, we also acknowledge the significance of the AI-based histopathology techniques proposed in the literature. These techniques have demonstrated valuable insights into the cellular and molecular characteristics of breast cancer. However, our study focuses on harnessing the potential of computer vision and feature extraction to develop an innovative approach that complements existing methods and supports early breast cancer detection.

Currently, medicine has several auxiliary systems in terms of diagnosis and treatment. Computer-Aided Diagnosis (CAD) is an advanced tool in the field of medical diagnosis, with rapid acceptance among the medical community. The primary objective is to support evidence-based medical diagnoses by leveraging statistical analysis and data processing techniques [9]. Regarding early breast cancer detection, CAD systems have emerged as relevant systems in the detection and analysis of mammographic images. The development of a CAD system specifically designed for the identification of asymmetries in mammograms based on morphology and skin thickness is a significant contribution to this field. The integration of CAD technology in the early diagnosis of breast cancer has the potential to enhance detection rates and reduce false positives. It serves as a complementary tool that aids in the identification of suspicious areas, allowing for timely intervention and treatment planning [10]. Figure 1 depicts a schematic representation showcasing the integration of a CAD system with mammography equipment. This seamless connection enables the real-time diagnosis of potential breast cancer cases at an early stage.

Once the result is obtained, it is displayed in a PDF report so that only the physician will have that information. Asymmetry holds potential biological significance and can be influenced by various factors, including genetics, hormonal fluctuations, developmental abnormalities, and lifestyle choices, without necessarily indicating the presence of tumors. For instance, during puberty, hormonal changes can lead to the uneven growth of breast tissue, resulting in variations in breast size and shape. Moreover, genetic factors can also contribute to variations in breast size and shape, as some individuals may inherit larger or smaller breasts from their parents. Overall, breast asymmetry is a natural occurrence influenced by a range of biological and lifestyle factors. A study conducted by Kayar et al.

showed a correlation between morphological asymmetry and breast cancer. Statistically significant evidence was observed linking breast asymmetry and cancer when the breast asymmetry ratio exceeded 20%. This association held true in all age groups, specifically in the 40–69 age range, with a 99% confidence interval [11]. Although generally considered normal, significant asymmetry can cause physical and emotional discomfort, and more importantly, serve as an indicator of potential tumor development [11,12].

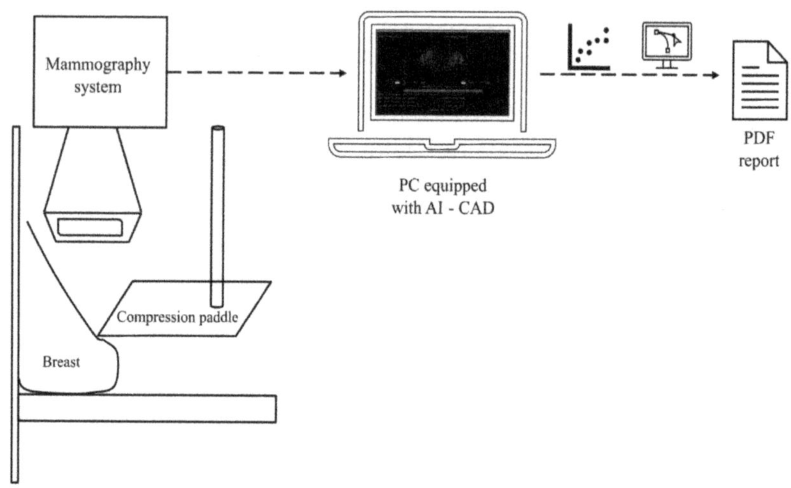

Figure 1. Schematic representation illustrating the integration of a CAD system with mammography equipment. The idea is that mammograms are extracted and submitted directly to the computer so that it can find patterns of asymmetry and interpret any abnormalities.

One of the key factors in the early detection of breast cancer is identifying asymmetry between the left and right breasts. Asymmetry can be a potential sign of breast cancer, as it may indicate changes in breast tissue composition and structure. Therefore, asymmetry analysis plays a critical role in breast cancer diagnosis and treatment [12,13]. Asymmetry is typically observed in one of the standard mammographic projections and is not visible in additional views. This finding commonly represents the overlap of fibroglandular tissue and summation artifacts, occurring in approximately 80% of cases. It often contains interspersed fat within the affected area. In recent years, several approaches have explored methods for asymmetry analysis, including shape analysis and skin thickness measurement. Breast asymmetry is an effective tool for breast cancer diagnosis and classification [14]. Centroid-based shape analysis involves the extraction of features related to breast shape from mammograms, which can then be used to train machine-learning models for breast cancer classification. In addition to shape analysis, skin thickness measurement has also been identified as an important factor in breast cancer asymmetry classification. Studies have shown that skin thickness measurements can provide valuable information about breast tissue composition and structure, which can help to identify potential areas of concern [15]. Table 1 presents a summary of state-of-the-art breast asymmetry studies.

Based on the information provided in Table 1, several approaches can be observed for assessing asymmetry based on certain contralateral features. However, the extraction of these features may rely on human criteria, which could lead to false positives due to subjectivity, overlapping features, human fatigue, or diagnostic challenges. On the other hand, Artificial Intelligence (IA) techniques such as Deep Learning may overcome these issues; however, the selection of feature extraction from digital mammograms plays a crucial role in training prediction models. Therefore, in this paper, we propose an approach for feature extraction related to breast shape based on Dynamic Time Warping (DTW) and skin thickness based on Growing Region Seed (GRS).

Table 1. The following table provides a summary of some of the most recent studies in which morphological asymmetries are proven to be signs of breast cancer.

Description	Characteristics	References
Results indicate that breast asymmetry is more likely to be a predictor of breast cancer rather than a result or consequence of the disease.	Breast volume and size.	Scutt D. Lancaster G. A., Manning J. T, 2005 [12,16]
The system employs medical imaging software to extract morphological features, involving the three-dimensional shape and size of both the breast and tumor.	Three-dimensional morphological features.	M. Singh, T. Singh, and S. Soni, 2021 [17]
Radiologists evaluated the images to assess the morphological features and enhancement characteristics.	Breast density.	Dawoud et al., 2022 [18]
Radiologists conducted a mammography descriptors evaluation such as shape and border characteristics, architectural distortion, and the presence of a hyperechoic rim and cystic complex structure.	Mass, chape, and microcalcifications.	Ghunaim et al., 2022 [19]
A deep learning model was proposed for bilateral asymmetrical detection based on Xception with two-dimensional tensors.	Bilateral asymmetrical detection.	Shimokawa et al., 2023 [20]

The computed features from our analysis focused on morphological analysis and breast skin segmentation, and serve as essential building blocks that could enhance current AI models. By leveraging advanced deep learning frameworks and multi-scale feature fusion approaches, we envision these features contributing to a more robust analysis of segmentation, tumor detection, and classification. Additionally, our findings hold promise for applications like AI-based classification using histopathological and textural images, as a perspective for this paper. These advancements demonstrate the potential of AI models to enhance accuracy and efficiency in diagnosing and classifying medical conditions, which is crucial for clinical decision-making and patient outcomes. These features can be integrated into a machine learning pipeline, such as in the approaches seen in the SegR-Net, RAAGR2-Net, and Deep Learning classification studies. By incorporating the identified features into machine learning models, we envision an improvement in the automated classification of mammograms, allowing for the accurate and efficient detection of potential breast asymmetry [21–24]. Additionally, our research lays the groundwork for exploring similar AI-based approaches in other medical-related research problems, such as retinal vessel segmentation and brain tumor classification. As AI continues to be a hot topic in medical research, our study contributes to the growing body of knowledge on the potential applications of AI in medical image analysis. The exploration of morphological shape and skin thickness analysis as valuable features for breast cancer detection opens avenues for further research and collaboration between medical professionals and technology experts. Our findings serve as a steppingstone toward developing more sophisticated AI models that can aid in early detection and improve treatment outcomes for breast cancer patients and potentially other medical conditions.

We aimed to research the potential use of morphological shape and skin thickness analysis in mammographic images for the early detection of breast cancer. Our hypothesis aims to explore the potential of a comparative analysis between the centroid–perimeter radius and breast skin segmentation as a reliable method for detecting potential breast asymmetry. We have utilized Dynamic Time Warping (DTW) for shape analysis and the Growing Seed Region (GRS) method for breast skin segmentation to assess their accuracy. Despite most mammograms not being classified with asymmetry by radiologists, the differences obtained with DTW and skin thickness analysis could indicate possible asymmetry patterns. It is essential to emphasize that our results are solely descriptive features that could be utilized in future works, particularly in the development of learning models for early breast cancer detection. The results of our investigation could have implications for the early diagnosis and monitoring of breast cancer, ultimately leading to improved patient outcomes. However, it is essential to consider the limitations of our study, such as the absence of a classification learning model and the reliance on radiologist-identified cases for validation. We discuss these limitations in more detail to provide a comprehensive understanding of the scope and potential future directions of our research.

The paper's structure is as follows: Section 2 describes the patient dataset, image preprocessing, breast structural and thickness analysis, and the Graphical User Interface (GUI) for the CAD system design. Section 3 of the paper presents the results and their corresponding analyses, while Sections 4 and 5 are dedicated to the discussion and conclusion, respectively.

2. Materials and Methods

This section provides a detailed explanation of the procedure employed to identify potential cases of asymmetry based on the analysis of breast morphology and skin thickness differences between contralateral breasts. The procedure involves several steps, including data collection, image preprocessing, feature extraction, and statistical analysis. The methodology ensures a systematic and rigorous approach to identifying and classifying asymmetry in breast images, contributing to the early detection of potential abnormalities. Figure 2 depicts a flowchart for this section.

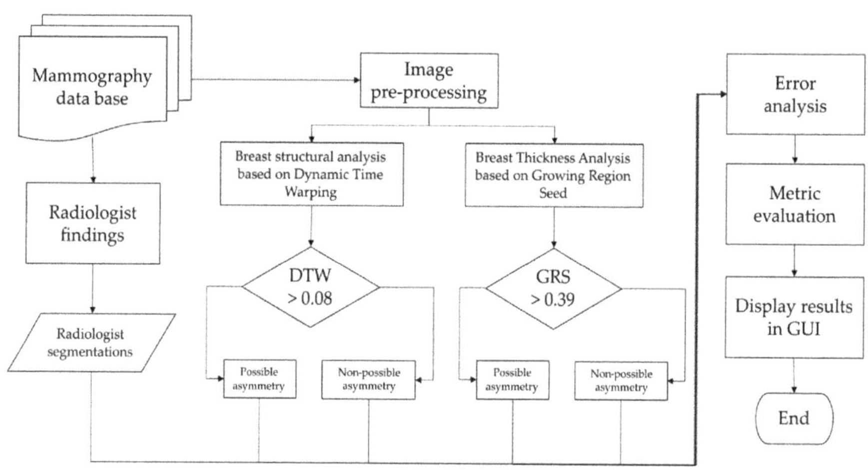

Figure 2. We propose hierarchical steps for breast image extraction from imaging to CAD system design. The process involves interpretation by specialists, image and data processing, evaluation of metrics, and presentation in a graphical interface.

2.1. Patient Data-Set

The clinical dataset includes the Medial Lateral Oblique (MLO) and Craniocaudal (CC) projections, which represent the breast shape along the sagittal and coronal planes, respectively. Additionally, the dataset includes views from the axial/transverse plane for both left and right breasts. The MLO projection provides an oblique view of the breast from the medial to lateral direction, while the CC projection offers a view from the cranial to the caudal direction. The dataset consists of 114 patients in DICOM format from 30 to 71 years old, diagnosed with BI-RADS from B1 to B6, with an average age of 55 years old. Table 2 presents the distribution of patients regarding the BI-RADS category. The data used in this study were sourced from the reports obtained from the University of Pittsburgh Text Information Extraction System (TIES) [25,26]. TIES is a database with anonymized information that includes around 24 million radiology reports generated between 2003 and 2015 at all 18 hospitals affiliated with the University of Pittsburgh Medical Center (UPMC).

Table 2. Patient distribution is based on BI-RADS labeling.

BI-RADS Category	No. of Patients	Age Means	Standard Deviation
B1	14	50	10
B2	5	57	11
B3	70	58	9
B4	19	56	9
B5	6	55	5

Spatial transformation through normalization aligns mammographic images to a standardized resolution, ensuring the consistent and accurate analysis of spatial descriptive features. This process enhances the reliability of comparisons between images, facilitating the detection of potential breast asymmetry patterns and contributing to early breast cancer detection.

2.2. Image Pre-Processing

Digital mammography is an imaging technique used for the early detection and diagnosis of breast cancer, replacing traditional film-based mammography by capturing digital images of the breast using specialized X-ray equipment. In digital mammography, the acquired images are immediately available for review on a computer screen, providing higher image quality, and enabling radiologists to visualize subtle abnormalities more clearly. However, digital images can be enhanced, manipulated, and magnified to assist in accurate interpretation and analysis. Commonly, digital mammography has a .dcm format extension, which stands for Digital Imaging and Communications in Medicine (DICOM), which is a standard format for medical images that enables the exchange of images and patient information.

DICOM images can be processed using image processing techniques for various applications such as diagnosis, treatment planning, and research. Nevertheless, due to variations in equipment standards, the spatial resolution of mammographic images may pose challenges in accurately describing statistical features related to spatial characteristics. To mitigate this issue, a normalization process was implemented to ensure consistency across all images obtained. The resolution of the images was standardized to 521 × 512 pixels, considering the variation in spatial resolution among the mammograms, as there was no uniform standard for acquisition. However, upon careful examination, it was deemed more appropriate to align the images to a resolution of 512 × 512 pixels to ensure consistent analysis of spatial descriptive features. Consequently, the normalization process was adjusted accordingly, aligning all mammograms to the common resolution of 512 × 512 pixels. This standardization facilitates reliable and meaningful comparisons by ensuring consistent spatial dimensions for the accurate analysis of spatial descriptive features. Spatial transformation through normalization aligns mammographic images to a standardized resolution, ensuring consistent and accurate analysis of spatial descriptive features. This process enhances the reliability of comparisons between images, facilitating the detection of potential breast asymmetry patterns and contributing to early breast cancer detection.

A second issue concerns annotations made by radiologists, physicians, or the imaging equipment itself, which include patient information, projections, and metadata. These interferences can disrupt feature extraction algorithms and bias prediction thresholds. To address this, a threshold value (T = 100) was calibrated on the grayscale level scale, ensuring the identification of all relevant regions above the threshold. Subsequently, the largest connected component, which consistently corresponds to the breast, was segmented. This approach aimed to mitigate the influence of annotation variability and establish a standardized basis for subsequent analyses. To identify the breast region, we utilized the concept of the largest connected component, which refers to the region with the highest number of connected pixels in the image. As annotations are typically small, isolated islands, the largest connected component consistently represented the breast region with the most extensive pixel count. By employing this approach, we ensured that the breast area was accurately segmented, mitigating the influence of annotation variability and establishing a standardized basis for subsequent analyses.

2.3. *Breast Structural Analysis Based on Dynamic Time Warping*

Breast asymmetry structure serves as a significant indicator that can potentially alert medical professionals to the presence of early-stage cancer. While it is natural for some degree of asymmetry to exist between the breasts of most individuals, a significant deviation from the expected level of asymmetry can raise concerns. In the context of cancer detection, an abnormal level of asymmetry can signify the presence of underlying malignancies [11].

The approach of DTW in our study is well-established and widely used in various fields, including speech recognition, image processing, and pattern recognition. DTW allows for finding the optimal alignment path between two sequences, considering temporal or spatial distortions between them. The algorithm follows a right, diagonal, and down approach related to three fundamental conditions: boundary, monotonicity, and continuity [27]. The boundary condition ensures that the warping path must begin with the starting points of both signals and end with their endpoints. The monotonicity condition preserves the relationship in time, meaning a point from signal A in position x_n must not be aligned with a point from signal B in position x_n − 1. This constraint is the main reason why the approach follows a right, diagonal, and down path. Lastly, the continuity condition limits the number of steps a point from signal A may take to align itself with a point in signal B. While other alignment methods like left, diagonal, and up approaches are theoretically possible, they are less commonly used in image analysis and sequence comparison. In our application of breast image asymmetry analysis, the right, diagonal, and down approach offers better alignment accuracy. We have conducted a thorough review of the related literature and found numerous studies that utilized DTW with the right, diagonal, and down approach for similar applications, such as feature extraction and sequence alignment in medical imaging and time-series analysis [27,28]. Introducing the left, diagonal, and up approach might lead to less accurate alignment and could compromise the reliability of our asymmetry analysis results. Therefore, in alignment with the existing literature and to ensure robust and accurate results, we maintain the use of the right, diagonal, and down approach in our study of breast image asymmetry analysis.

An approach to analyzing breast asymmetry is to compute a collection of radii formed between the centroid and the skin perimeter. Once the centroid is computed, the radius is retrieved based on the angle as the argument between $[0, \pi]$ radians in steps of one radian, creating a distance for each orientation between the centroid and perimeter of the skin. This supports uniformity and ensures that the image is created using the outermost breast pixels. To handle variations in breast size and shape among different DICOM images, the distances between the centroid and the skin perimeter were normalized to fall within the range of $[0, \pi]$. This normalization process ensured that the spatial features used for evaluation were consistent and comparable across all images, regardless of their individual radii. By generalizing this approach to the entire dataset, we were able to effectively evaluate the morphological asymmetry in breast skin thickness and its potential significance for early breast cancer detection.

Nevertheless, an important factor to consider is the orientation of the breast. Hence, both images (i.e., left, and right breast images) were adjusted to the same orientation, i.e., towards the right. The hypothesis proposes that the set of distances would differ between left and right views of the breast. To maintain consistency and ensure direct comparability, all images were oriented toward the right. Although orienting the image set towards the left is feasible, it would necessitate image reprocessing and algorithm adjustments to adjust the left-oriented images. Our specific focus on the right-oriented images allowed us to investigate the impact of asymmetry on the right breast and evaluate the effectiveness of our proposed approach. Figure 3 presents the flowchart used to obtain a collection of distances between the centroid and the edge of the breast, depicted in Figure 4. Appendix A provides the pseudocode for the algorithm.

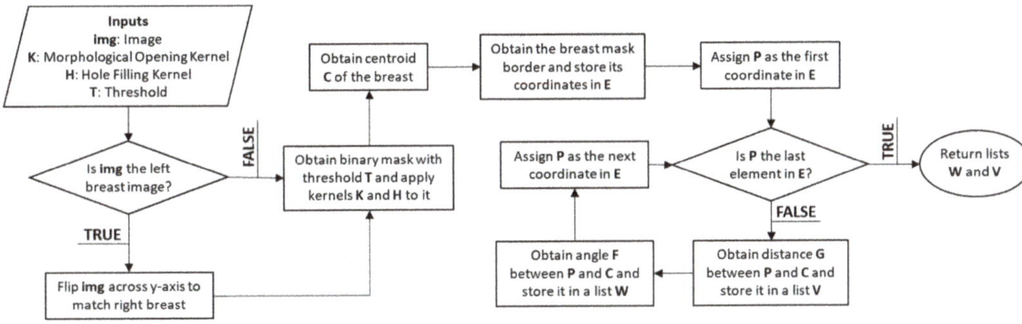

Figure 3. Flowchart illustrating the process of computing distances from the centroid to the breast's edge.

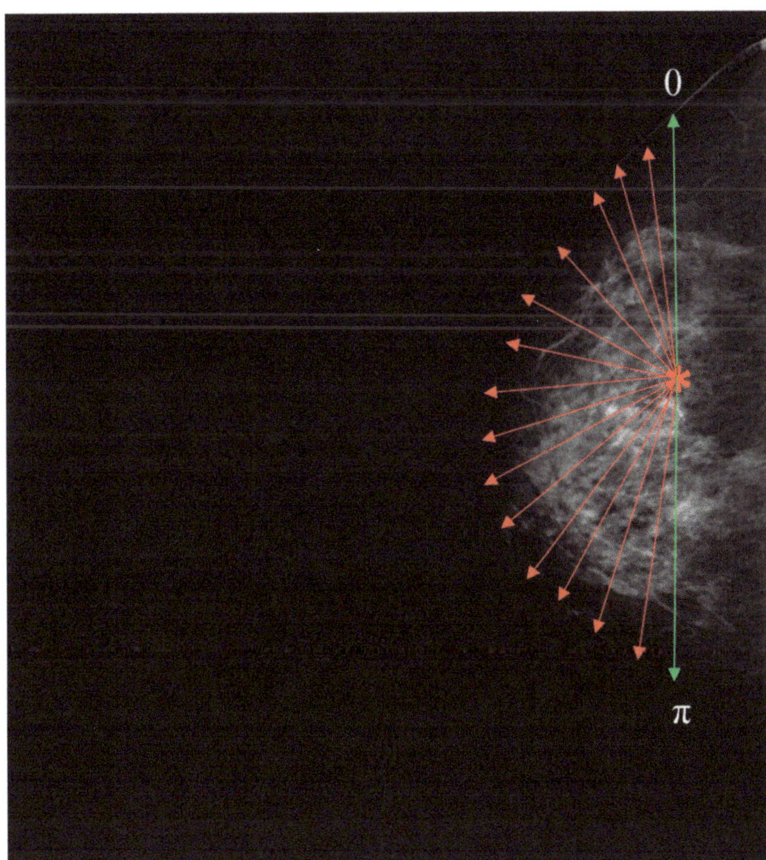

Figure 4. This diagram serves as an illustrative representation demonstrating the process of deriving a set of distances between the centroid and the periphery of the breast. The centroid is indicated by the asterisk symbol, while the red radius is depicted as arrows extending towards the edge of the breast and the green arrows indicate the beginning and ending [0–π].

After obtaining the set of distances between the centroid and the edge of the breast at the angles ranging from 0 to π radians, a vector comprising these distances is extracted. This vector serves the purpose of enabling a structural comparison of shape between the contralateral breasts. It is assumed that the presence of asymmetry is indicated when shape similarity between the two vectors is lacking. To assess the similarity between both vectors, Dynamic Time Warping (DTW) can be used to assess alikeness.

DTW is a technique used to measure the similarity between two time series with varying lengths. However, in the context of breast asymmetry detection, the vector of distances can be used to represent the shape characteristics of the breasts [29].

Breasts can exhibit variations in size while maintaining similar underlying structural characteristics. When analyzing the vector of distances, the objective is to identify corresponding points on the curves of the left and right breasts. These corresponding points represent areas where the radii's values align or coincide, indicating similarity in shape or structure. To achieve this, DTW allows a warping path between the curves. In the context of breast asymmetry detection, the angle values serve as the basis for constructing the time series. By employing DTW, the algorithm searches for the most similar alignment between the curves of two shapes, i.e., the left and right breasts. It identifies points along the curves that exhibit similar shapes and structures, regardless of their temporal or spatial position. These points of coincidence or matching correspondence indicate areas where the breasts demonstrate asymmetry. By analyzing the extent of similarity or dissimilarity in the correspondences, it becomes possible to quantify the degree of asymmetry between the breasts. The greater the deviation in the alignment or the lack of matching correspondences, the higher the likelihood of significant asymmetry. Therefore, through the application of DTW to the vector of distances, the algorithm aims to identify and characterize asymmetry by seeking points of coincidence that align the curves of the left and right breasts, even in the presence of size differences.

The DTW algorithm involves several steps. First, a cost matrix is computed to measure the dissimilarity between each pair of elements in the sequences. The cost between two elements is typically determined using a distance metric, such as the absolute difference or squared difference. Let $d = [d_1, d_2, \ldots, d_n]$ and $a = [a_1, a_2, \ldots, a_m]$ be two input sequences of distances and angles, respectively, of lengths n and m. The cost matrix C of size $(n + 1) \times (m + 1)$ is defined. Then, it is initialized to $C[0][0] = 0$, and the cost between x[i] and y[j] is computed for each C[i][j]. The cost between both elements x[i] and y[j] is typically defined as the absolute difference or squared difference.

Next, dynamic programming is applied to find the optimal alignment. The cumulative cost matrix is generated by considering three possible moves: moving horizontally, diagonally, or vertically. The cumulative cost at each position in the matrix is updated based on the cost of the current elements and the minimum of the adjacent cumulative costs. The cumulative cost starting from C[1][1] considers three possible moves: right, diagonal, and down. For each element C[i][j], the minimum cost is computed and C[i][j] is updated as the cost between x[i] and y[j] plus the minimum of the three adjacent elements.

Once the cumulative cost matrix is computed, the optimal path is determined by backtracking from the bottom-right corner to the top-left corner. This path represents the alignment that minimizes the total cost and provides information about the similarity or dissimilarity between both breasts. This alignment provides insights into the similarity or dissimilarity between two sequences and is useful for pattern recognition, which is the goal of this paper.

2.4. Breast Thickness Analysis Based on Growing Region Seed

Breast skin thickness has been found to correlate with breast asymmetry and cancer. Studies based on ultrasound imaging were found to be significantly different in patients with breast cancer compared to those without cancer and have shown increased thickness of the breast skin [16].

The process of skin segmentation involves setting a threshold for the breast since skin typically has a higher intensity than the internal tissues of the breast. While a few regions of the perimeter may stand out, they may not fully define the skin. However, these regions are used as "seeds" to expand the Region of Interest (RoI). To standardize the gray intensity of each mammogram, a normalization process was conducted, considering the number of bits present in the image, which was determined from its maximum intensity value.

The first step was to obtain a mask by the Otsu method to eliminate low-contrast tissue. Then, erosion was performed on the binary image using a kernel with size 8, which determines the border thickness. By comparing the mask concerning the border coordinates, we obtain the pixels that changed after erosion, thus retaining only the boundary of the breast. Then, a window with zero matrices with dimensions $(n + 2) \times (m + 2)$ was created, and a line of pixels at the top, bottom, left, and right, to evaluate the border pixels during region growth without requiring special functions. The output of these steps is an image with the same size as the original mammography but with these black border pixels, allowing post-processing instead of the original image without encountering border-related issues. To mitigate the issue of jagged edges observed in the mask when directly applying the Volume of Interest Look-Up Table (VOI LUT) technique, an alternative approach was adopted. However, the resulting mask was subsequently applied to the mammograms that had undergone normalization using the VOI LUT method. The VOI LUT is a technique employed in medical imaging to enhance the visualization of image data by mapping pixel values to a desired display intensity range. It is an important component in the post-processing pipeline for adjusting the windowing and leveling of medical images. The VOI LUT operates by applying a linear transformation to the pixel values, thereby adjusting their brightness and contrast characteristics. This allows medical professionals to better perceive subtle details and abnormalities in the image. The application of the VOI LUT is typically performed by specific standards and algorithms, ensuring consistent and reproducible results across different imaging systems. By effectively optimizing the display of medical images, the VOI LUT contributes to the accurate interpretation and diagnosis of various medical conditions [30].

Afterward, the coordinates where the seeds are located are obtained to proceed with the expansion function. At this point, we evaluate whether the pixel should belong to the growing region, or whether this pixel was already taken as part of the mask. Then, this pixel is stored in the mask sub-image, and a counter starts to mark a value of 1 for each pixel evaluated, preventing redundant evaluations, and significantly speeding up processing. At this point, a list of the seed coordinates and their neighboring pixels is created. Four conditions are evaluated for the neighboring pixels to determine if they should be part of the RoI:

1. The pixel must not be black.
2. Its difference from the seed pixel must be below 200, set as the high threshold.
3. Its difference from the seed pixel must be above 1, set as the low threshold.
4. Its Euclidean distance from the first seed in the cycle must not exceed 4 pixels deep.

If all these conditions are met, and the pixel has not been discarded before, it is attempted to add the pixel to the mask. Then, the current seed is removed, and the list is evaluated to find new seeds that may have appeared until no further derivations remain with all other seeds along the border. At the end of this loop, the border mask is retrieved. Figure 5 depicts the pseudocode for region growth by seed expansion. Appendix B provides the pseudocode for the algorithm.

In summary, the idea behind the algorithm is to use skin thickness as a preliminary indicator for detecting edges in an image. Skin thickness can be defined as the degree of brightness or darkness of the pixels in the skin region. By identifying the skin pixels in an image, the algorithm can then expand the mask to include neighboring pixels with similar intensity values. This assumes that skin pixels should have similar intensities due to the uniformity of skin tissue density. Several methods have been proposed for skin breast segmentation; however, unsupervised, or semi-supervised, segmentation processes must be validated against a set of ground truth images. This process was validated with the Average Symmetric Surface Distance (ASSD) method and a second validation was performed with the Relative Volumetric Distance (RVD) method. To assess the performance of the GRS algorithm for segmentation, ASSD was chosen as the first validation metric due to its ability to measure the average distance between corresponding points on segmented surfaces. By comparing the segmented results to ground truth data, ASSD provides valuable insights into the accuracy of surface delineation. However, since ASSD focuses solely on surface distances, it may not fully capture volumetric differences. Therefore, as a complementary measure, we employed RVD as the second validation metric. RVD evaluates the volumetric differences between segmented regions, considering both surface and internal variations. By using both ASSD and RVD, we gain a comprehensive understanding of how well our algorithm performs in terms of both surface delineation and volumetric consistency. This two-fold validation approach ensures a more robust evaluation of the growing region seed algorithm's effectiveness in accurately segmenting regions of interest.

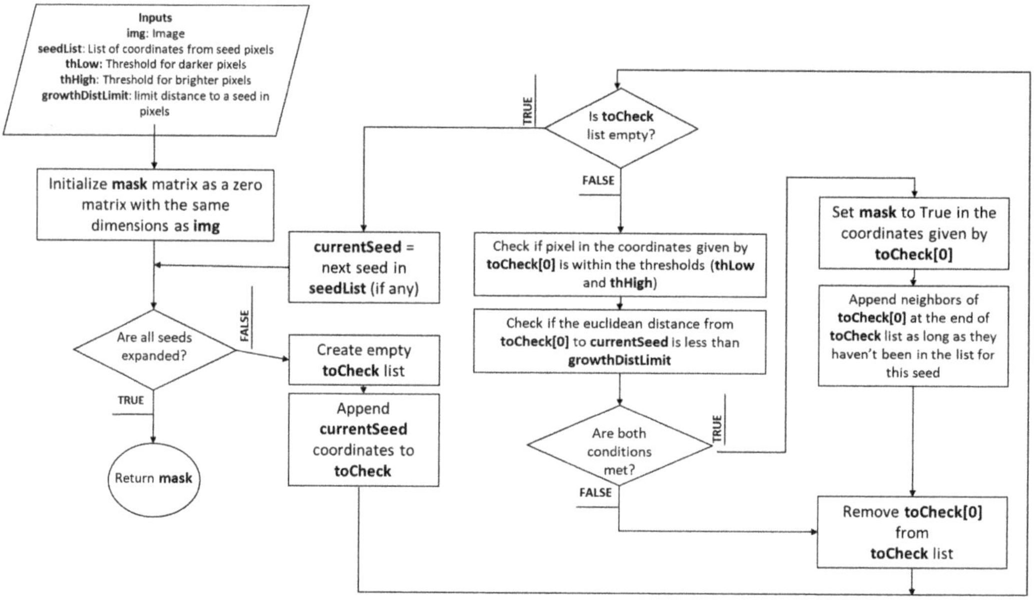

Figure 5. Flowchart representing the pseudocode for region growth through seed expansion.

ASSD is a metric used to evaluate the similarity between two shapes, typically in the field of medical imaging or computer vision [31]. It quantifies the average distance between corresponding points on segmented skin. Let S1 be the skin image segmented by GRS and $S2$ represent the radiologist segmentation, these being the two shapes that need to be compared. Then, the corresponding point should be identified. Let $P(S1)$ denote a set of surface points of $S1$, so $P1 = \{p_1, p_2, \ldots, p_n\}$ is the set of corresponding points on $S1$, and $P2 = \{q_1, q_2, \ldots, q_n\}$ is the set of corresponding points on $S2$. The Euclidian distance $d(p_i, q_i)$ between each pair of corresponding points is computed for calculating the sum of distances. The shortest distance of an arbitrary pixel p to $P(S1)$ is defined as Equation (1).

$$d(p,\ P(S1)) = min_{P1\in P(S1)}\ p - P1 \quad (1)$$

The ASSD is computed by dividing the sum of distances by the total number of corresponding points. The resulting value of ASSD represents the average distance between corresponding points on the two surfaces, indicating the overall similarity or dissimilarity between them, presented in Equation (2). A smaller ASSD value indicates a closer match and higher similarity, while a larger value suggests greater dissimilarity between the surfaces.

$$ASD(S1, S2) = \frac{1}{|P(S1)| + |P(S2)|} \left(\sum_{P1 \in P(S1)} d(P1, P2) + \sum_{P1 \in P(S1)} d(P2, P1) \right) \quad (2)$$

On the other hand, (RVD) is a metric used to assess the similarity or dissimilarity between two segmented RoIs in medical imaging [32]. It provides a quantitative measure of the volumetric differences between the two regions. Let $V1$ be the segmented skin image and $V2$ represent the radiologist segmentation. First, the volume difference (VD) between $V1$ and $V2$ is calculated by taking the absolute difference: $VD = |V1 - V2|$. This represents the absolute volume difference between the two regions. Next, the average volume (AV) of the two regions is computed as $AV = (V1 + V2)/2$. The AV serves as a reference value for normalization. Finally, the RVD is obtained by dividing the volume difference (VD) by the average volume (AV) in Equation (3). Also, let A and B be two classes.

$$RVD = \frac{|B| - |A|}{|A|} = \frac{VD}{AV} \quad (3)$$

The RVD value represents the proportion of the volume difference relative to the average volume. A smaller RVD value indicates a higher similarity or overlap between the volumes, while a larger value suggests greater dissimilarity. The RVD method provides researchers with a quantitative measure to assess the volumetric differences between segmented regions, enabling the objective evaluation of segmentation algorithms and the comparison of segmentation results.

2.5. Graphical User Interface for CAD System in Medical Diagnosis

In this section, we describe the Graphical User Interface (GUI) design for a specialized Computer-Aided Diagnosis (CAD) system in the context of medical imaging. Specifically, the GUI is designed to facilitate the diagnosis of breast asymmetry using mammography images. This section focuses on the key features and functionalities of the GUI, emphasizing its role in enabling seamless interaction, data input, image loading, and report generation.

The GUI encompasses various essential components tailored to the specific requirements of medical diagnosis. It offers an intuitive and user-friendly interface through which medical professionals can input vital patient information, including name, surname, age, weight, and height. These data play a crucial role in maintaining accurate patient records and facilitating personalized analysis. A fundamental aspect of the GUI is its ability to load mammography images for both the left and right breast. This feature streamlines the process of importing relevant images into the CAD system, ensuring efficient analysis and evaluation. Through a well-designed file selection mechanism, medical professionals can easily access and integrate the necessary image data into the CAD system for further examination.

Once the mammography images are loaded, the CAD system performs a comprehensive analysis using advanced image processing techniques, feature extraction, and pattern recognition algorithms. The GUI presents the diagnostic results in a concise and visually informative manner, allowing medical professionals to interpret and evaluate the findings effectively. The aim is to provide accurate and reliable assessments of breast asymmetry, aiding in the subsequent diagnosis and treatment planning process. In addition to real-time analysis, the GUI enables the generation of detailed PDF reports summarizing the patient's information, image analysis results, and diagnostic outcomes related to breast asymmetry. These reports serve as valuable resources for medical practitioners, providing a comprehensive overview of the patient's condition and facilitating effective communication and collaboration among healthcare professionals. Figure 6 depicts the steps of the GUI system.

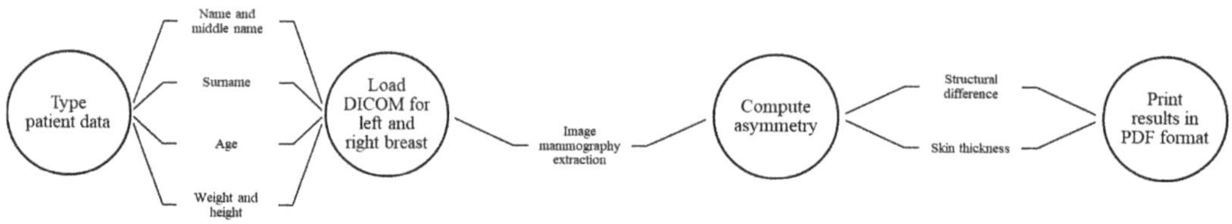

Figure 6. Flowchart depicting the GUI design for a CAD system used in medical diagnosis. The flowchart illustrates the sequential steps involved in the graphical user interface, including input fields for patient name, surname, age, weight, height, and the option to load mammography images for the left and right breast. The system generates a PDF report that includes the patient data, the loaded images, and the diagnosis result for possible breast asymmetry.

In conclusion, the GUI developed for the CAD system in this study offers an intuitive and user-friendly interface specifically designed for medical professionals involved in breast asymmetry diagnosis using mammography images. By streamlining data input, image loading, analysis, and report generation, the GUI enhances the overall functionality and usability of the CAD system. The integration of advanced image processing algorithms within the CAD system ensures accurate and reliable diagnostic outcomes. Future research in this domain should focus on expanding the system's capabilities, exploring additional diagnostic parameters, and leveraging emerging technologies to further improve the accuracy and efficiency of breast asymmetry detection.

3. Results

3.1. Breast Shape Asimetry Analysis

The results point to several cases with asymmetry as inferred from the centroid–perimeter radius analysis, pointing to possible asymmetry for a DTW > 0.08 in an index from [0, 1] with a standard deviation of 0.19. From a total of 114 patients, 23 cases were found that could indicate asymmetry, and these are presented in Table 3.

Table 3. Cases found for asymmetry with respect to shape analysis with DTW.

BI-RADS	Number of Cases	Number of Asymmetries
1	24	6
2	10	2
3	34	7
4	34	6
5	11	1
6	1	1

Figure 7 shows mammography images of a patient with BI-RADS 4 exhibiting contralateral asymmetry. In this particular case, the result reveals the presence of noticeable asymmetry between both breast pairs, primarily within the first quadrant of the tomography. This observation implies that a substantial disparity should exist when comparing the radius projected from the centroid to the perimeter. To facilitate this analysis, it becomes essential to define a specific region where the centroid of a closed and solid geometry can be accurately calculated. In this regard, the previously established segmentation mask proves useful as it encompasses the relevant coordinates necessary for this calculation.

To test the hypothesis regarding the disparities in centroid–perimeter distances, a comparison was conducted between the two geometries, as depicted in Figure 8. In Figure 8a, the centroid was determined to have coordinates (433,202), while in Figure 8b, the centroid coordinates were found to be (95,288). These coordinate values provide crucial information for further analysis and assessment of the observed differences in the context of the hypothesis.

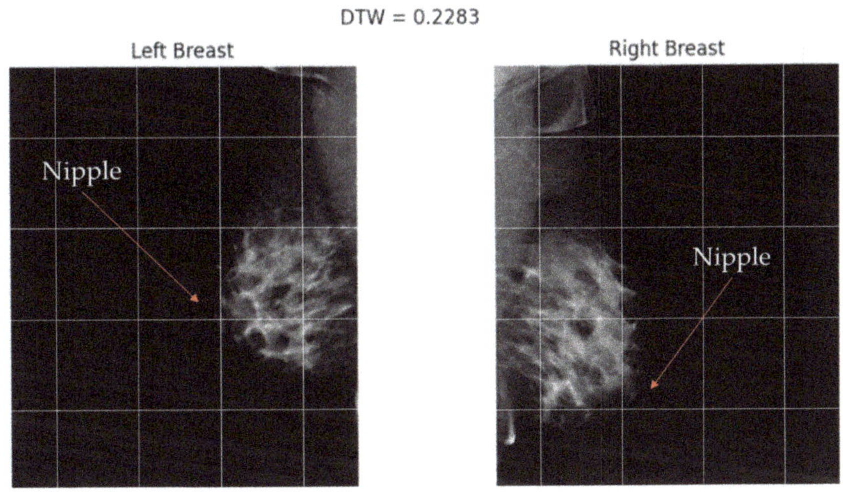

(**a**) (**b**)

Figure 7. Shape asymmetry can be visible in the sample. Also, the disparity of the nipple is observable; (**a**) left CC projection and (**b**) right CC projection. The red arrow highlights the nipple region. By comparing the different regions relative to the grid, noticeable differences in shape and position can be observed. Additionally, using the pectoral muscles as a reference, it becomes evident that the right breast is positioned lower than the left one, further accentuating the asymmetry between the two breasts.

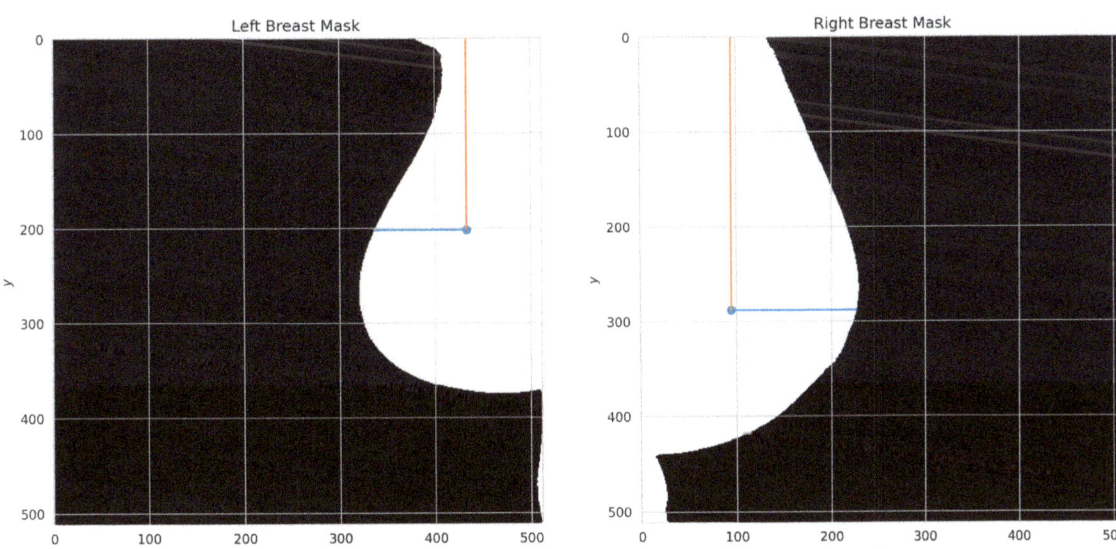

(**a**) (**b**)

Figure 8. A plot of centroid–perimeter distances between the left and right breasts, in which it is possible to retrieve metrics such as Euclidian distance to compare as a first approach. If both breasts were similar, the distances should be similar; (**a**) centroid in left breast masking, (**b**) centroid in right breast masking.

Figure 9 shows the same BIRADS 4 representative cases of the presence of symmetry with a 0.24×10^{-4} DTW index; on the other hand, we can observe in Figure 10 the inequality between curves that reflect the geometry of the breast for the patient with BIRADS 2—even the retraction of the nipple is appreciable in the right breast, with a 0.22 DTW index. By comparing these parameters between both breasts, healthcare can identify differences in breast shape and size, and determine a degree of asymmetry.

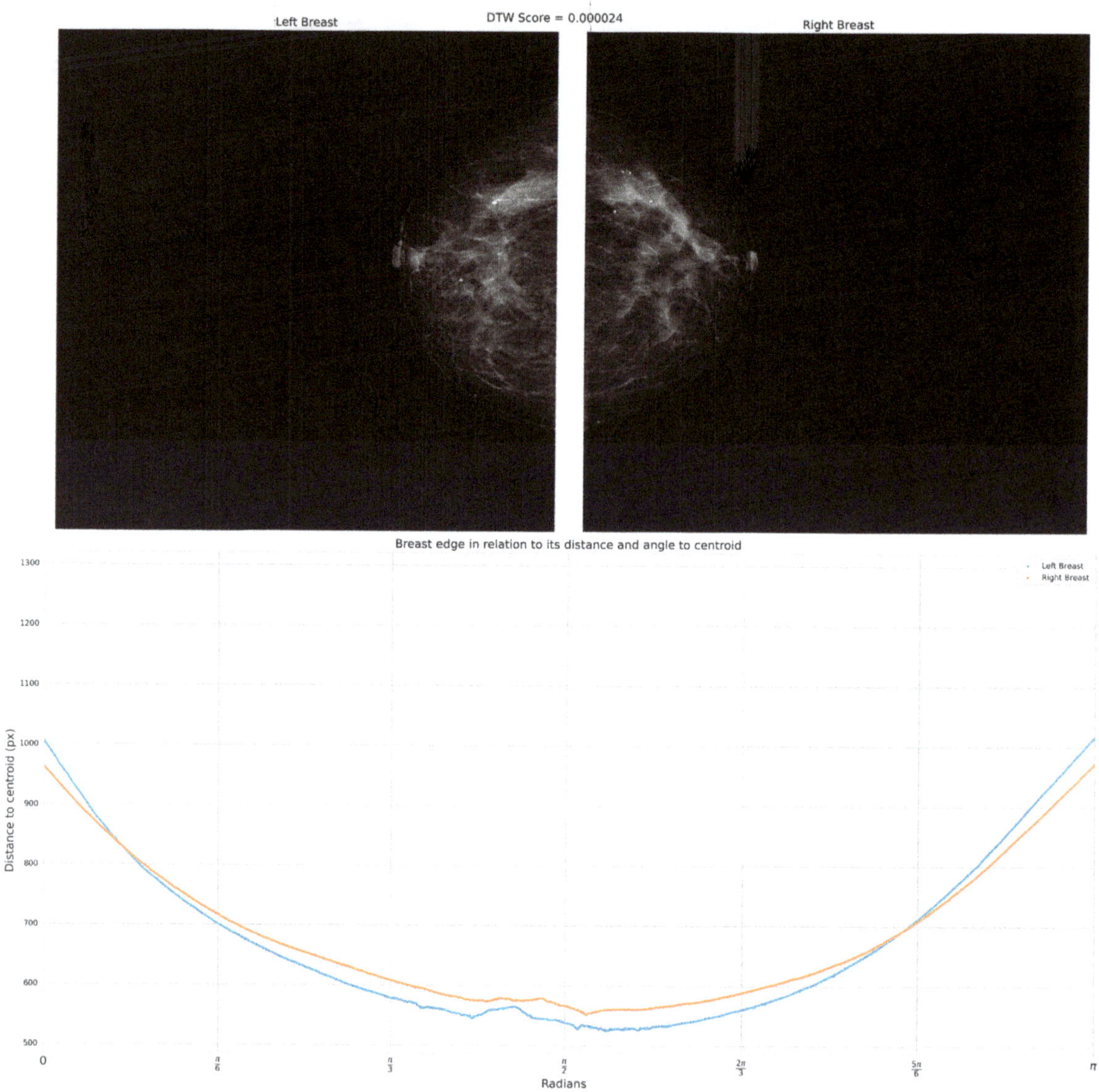

Figure 9. The results demonstrate a high degree of structural similarity between the breasts, indicating a lack of significant asymmetry. The curves generated from the analysis visually depict the close resemblance and alignment of the structural features. This finding supports the notion that no pronounced differences or variations exist between the corresponding regions of the breasts.

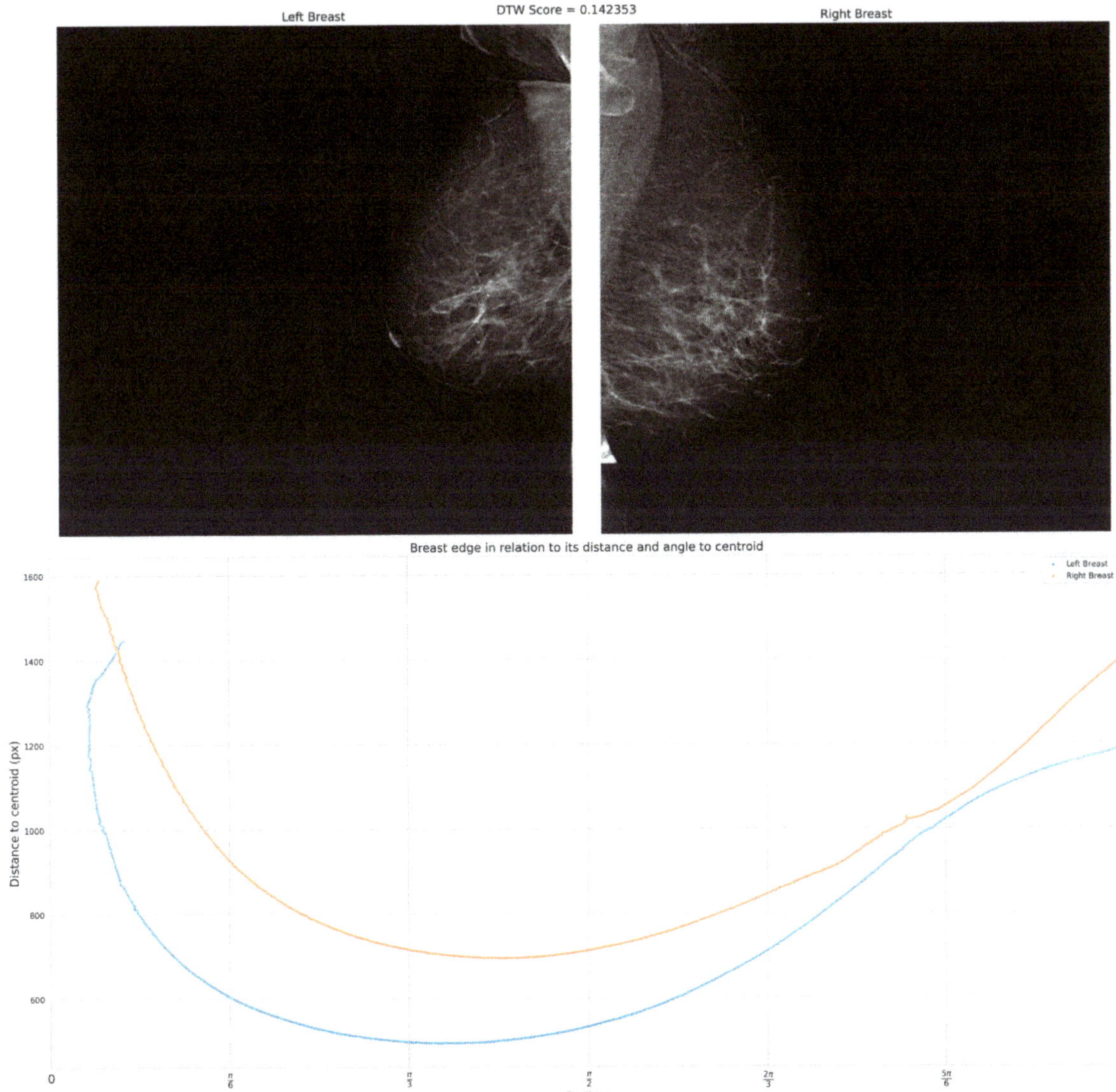

Figure 10. The analysis of this case reveals the presence of significant asymmetry, as indicated by a DTW index of 0.22. The generated result demonstrates notable differences and variations between the corresponding regions of the breasts. The observed asymmetry is visually apparent in the distinct patterns and shapes depicted in the curves.

According to the results of the database analysis, a DTW index less than 0.08 could indicate asymmetry due to shape or even factors such as nipple retraction. Figure 11 displays the results of the DTW analysis conducted on each patient, while Figure 7 shows validation results according to the Confusion matrix. The accuracy of the algorithm was 82%.

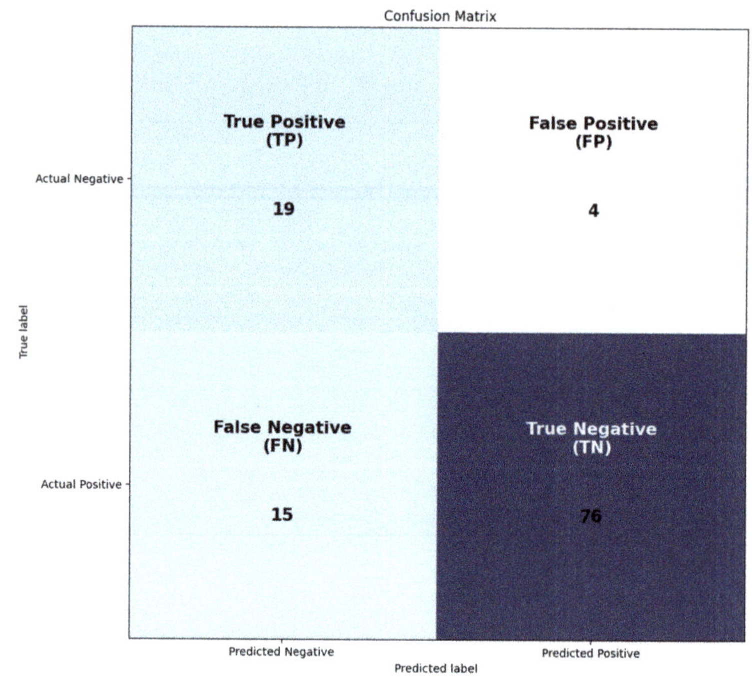

Figure 11. Confusion matrix for the DTW-based breast structure analysis algorithm.

Figure 12 presents the results according to the DTW index for each of the 114 patients.

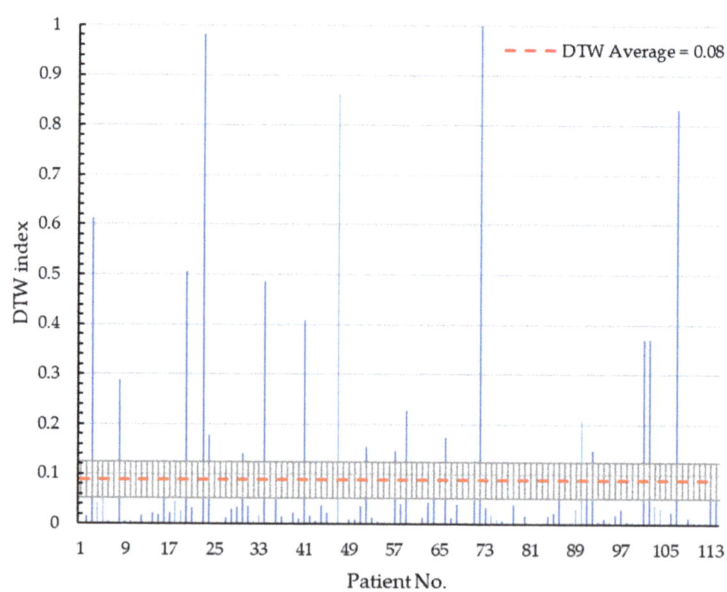

Figure 12. The DTW index results for each patient in the database are displayed in the graph. The graph includes a red segmented line representing the average value of $\mu = 0.08$, with a standard deviation of std = 0.19. The DTW index values for each patient can be observed concerning the average line, providing insights into the level of dissimilarity or asymmetry present in the data.

3.2. Breast Skin Segmentation Analysis

Breast skin segmentation enables a comprehensive evaluation of breast morphology, for the detection and characterization of asymmetrical regions. This information is crucial in the early detection and diagnosis of breast cancer, as asymmetry can be a significant indicator of the presence of tumors or other abnormalities [14]. Figure 13 presents the results of the GSR method.

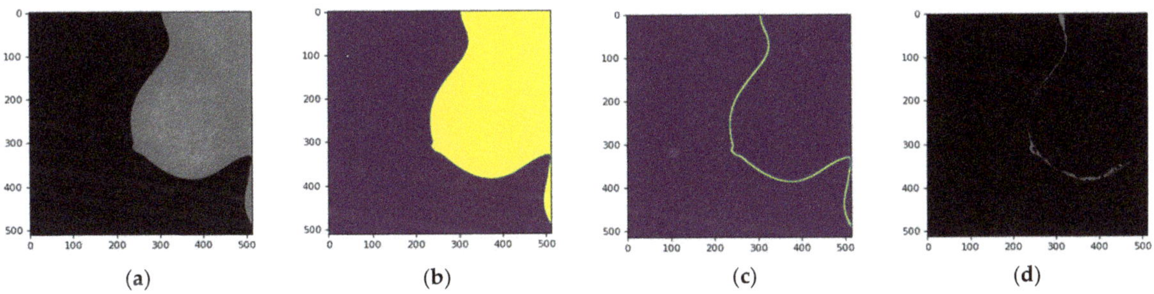

Figure 13. Skin segmentation results using GRS technique, (**a**) original projective CC-R image of a patient diagnosed with BI-RADS 1, (**b**) RoI detection mask, (**c**) seed placement based on threshold change detection, (**d**) seed growth based on average intensities of neighboring locations.

To identify the optimal combination of upper and lower thresholds, erosion kernel size for seed generation, and growth distance, a parametric study was conducted using the data presented in Table 4. This study aimed to determine the parameter values that yield the most accurate and robust results in the segmentation process.

Table 4. Parameters used to define the most optimal option.

Variable	Parameters
Upper intensity threshold	50, 100, 150, 255
Lower intensity threshold	1, 4, 7
Size n of the erosion kernel $n \times n$ for obtaining seeds	3, 6, 9, 12
Growing distance limit (maximum seed separation of seed from the edge)	3, 5, 7, 9

Table 5 presents the validation results of GRS-segmented images compared to those segmented by specialists. The validation metrics of ASSD and RVD were used for comparison to identify the most favorable combination that closely resembles the specialist-segmented images.

Once the optimal parameter combination was determined, the study focused on identifying cases of asymmetry based on the difference in skin thickness between contralateral breasts. Out of the 114 cases analyzed, the GRS algorithm detected 35 cases of asymmetry, among which 20 patients were previously labeled according to the radiologist's criteria. At this juncture, it is crucial to highlight that certain cases of asymmetry attributed to variations in skin thickness may evade detection through mere visual examination by specialists. This underscores the significance of employing the algorithm, as it holds the potential to provide valuable alerts about the early diagnosis of breast cancer. The detection of these asymmetrical cases further validates the effectiveness of the GRS algorithm in assisting radiologists and facilitating the early detection of potential breast abnormalities.

Table 5. Results of the parametric study in which the kernel size was varied for the erection and maximum growth distance. The best result is presented in bold—a 9 × 9 kernel 3 pixels deep.

Kernel for Erosion	Maximum Distance to Seed	RVD	ASSD
3.00	3.00	−0.8763	9.2908
3.00	5.00	−0.8441	9.1213
3.00	7.00	−0.8299	9.0694
3.00	9.00	−0.8236	9.0689
6.00	3.00	−0.3492	2.0468
6.00	5.00	−0.3014	2.0238
6.00	7.00	−0.2771	2.0334
6.00	9.00	−0.2651	2.0492
9.00	**3.00**	**0.1589**	**1.9182**
9.00	5.00	0.2380	1.9568
9.00	7.00	0.2848	2.0087
9.00	9.00	0.3096	2.0514
12.00	3.00	0.6451	2.0817
12.00	5.00	0.7327	2.1486
12.00	7.00	0.7857	2.2193
12.00	9.00	0.8129	2.2712

Figure 14a presents the case with the lowest asymmetry in skin thickness difference, with an average difference of 0.04×10^{-2}, while Figure 14b shows the case with the highest asymmetry, with an average of 1.119.

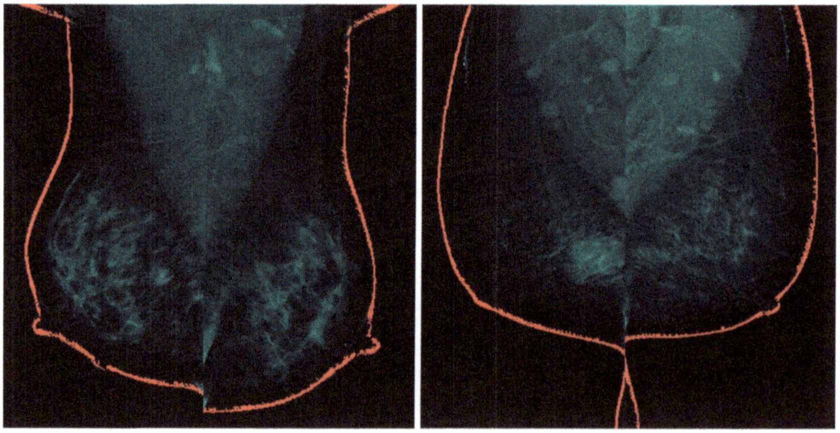

(a) (b)

Figure 14. The results of the skin segmentation are depicted by the red-colored region, where (**a**) shows a patient with BIRADS 3, exhibiting an average difference of 0.004, indicating the best level of symmetry. On the other hand, (**b**) presents a patient with BIRADS 5, displaying an average difference of 1.199, suggesting the presence of asymmetry due to skin thickness. These findings highlight the ability of the segmentation algorithm to accurately identify and quantify variations in skin thickness, enabling the characterization of asymmetry in breast tissue.

Table 6 presents the performance of the GRS algorithm compared to reference images (182 images). The validation of the algorithm was conducted using the ASSD and RVD metrics. This analysis allows for a comprehensive assessment of the algorithm's accuracy and effectiveness in quantifying the similarity between segmented images and reference data.

Table 6. The accuracy results are presented in terms of the average similarity to manually segmented images, along with the corresponding standard deviation.

Metric	Similarity Mean	Std. Deviation	Acceptable Similarity Cases (within Std. Deviation)	Non-Acceptable Similarity Cases (beyond Std. Deviation)
ASSD	1.887	2.313	171 (90.47%)	18 (9.52%)
RVD	0.158	0.344	126 (66.66%)	63 (33.33%)

Figure 15 illustrates the level of asymmetry based on the average difference in thickness for each patient, where the line indicates the average difference that serves as the decision threshold.

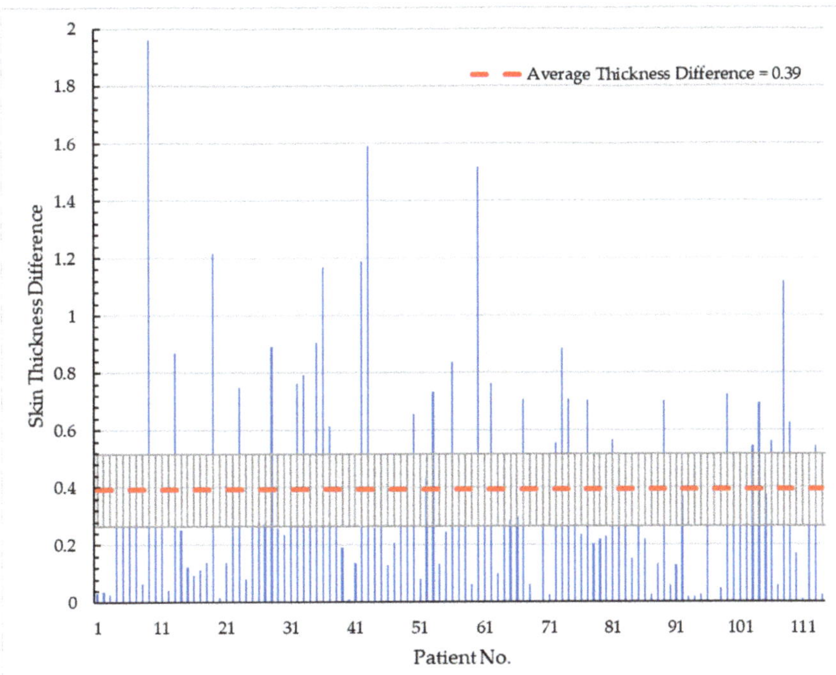

Figure 15. The results of the level of difference in skin thickness between contralateral sides for each patient are presented. An average threshold of $\mu = 0.39$ was identified, with a standard deviation of std = 0.35. These findings indicate the extent of variation in skin thickness between the corresponding sides of the breasts. The threshold value provides a reference point for determining the presence of significant differences in skin thickness, aiding in the detection of potential asymmetries or abnormalities.

This analysis offers valuable insights into the magnitude of asymmetry observed in the measurements of breast skin thickness. By setting a threshold based on the average skin thickness difference, it becomes feasible to classify patients based on their asymmetrical or symmetrical characteristics, thereby assisting in the detection of potential abnormalities. These findings may contribute to a better understanding of the variations in breast skin thickness and provide a useful tool for assessing breast health and identifying potential pathological conditions.

3.3. Graphical User Interface Design for Computer-Aided Diagnosis

This section presents key features of the GUI, including successfully capturing and organizing essential patient information, such as name, surname, age, weight, and height, ensuring accurate association with the corresponding medical images. Users can conveniently load mammography images for both the left and right breasts, enabling detailed examination and analysis. Advanced image processing techniques were incorporated into the GUI to preprocess the loaded images, effectively removing unwanted annotations, and enhancing the clarity of the breast structure (Figure 16).

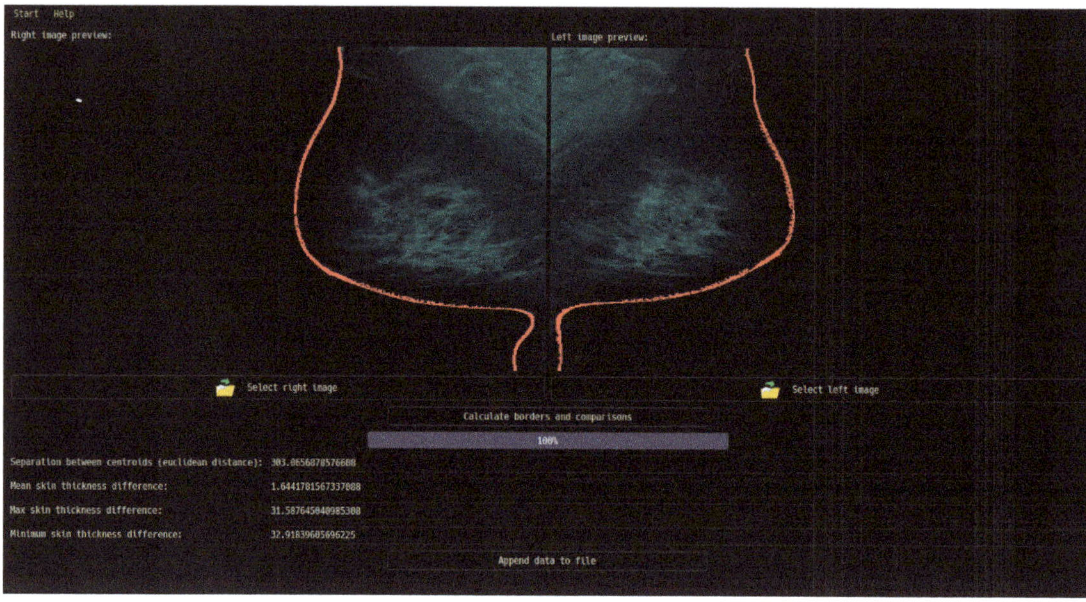

Figure 16. The GUI of the system enables the visualization of potential asymmetry cases. Users can upload mammograms in DICOM format, and the GUI seamlessly integrates with preprocessing algorithms, structural asymmetry analysis, and skin thickness difference computation. The specific case being referred to in this context involves a patient labeled as BIRADS 2, indicating a low suspicion of malignancy, yet exhibiting signs of possible asymmetry.

A significant feature of the GUI was the generation of PDF reports, which encompassed the loaded images, patient data, and diagnostic results. This streamlined the reporting process, allowing medical professionals to access a comprehensive summary of the analysis performed by the CAD system. The reports provide a clear representation of the identified asymmetry, or lack thereof, aiding in the detection of potential breast abnormalities.

Overall, the evaluation of using the GUI in the CAD system in medical diagnosis shows its efficacy in enhancing the workflow of medical professionals. The intuitive design and integration of essential features, such as patient data input, image preprocessing, and report generation, contributed to an efficient and accurate diagnosis. The GUI's successful implementation paves the way for improved medical decision-making and enhanced patient care in the field of breast diagnostics.

4. Discussion

To improve the comparison analysis for the centroid–perimeter radius and breast skin segmentation with the GRS method, several approaches can be employed. One possible method is the use of machine learning algorithms, which can learn from a large dataset of mammograms to identify potentially cancerous areas accurately. Another approach is to incorporate texture analysis techniques, such as the co-occurrence matrix, into the segmentation process to improve the accuracy of the segmentation. Moreover, using a combination of different segmentation techniques can also enhance the accuracy of the analysis.

To validate the results, ground-truth annotations are necessary, which can be obtained through biopsy or histopathology. Furthermore, the use of multiple radiologists to assess mammograms can increase confidence in the results. The integration of CAD systems can also improve the validation process by providing a second opinion on potential cancerous areas. In terms of medical correlation, the early diagnosis of breast cancer can significantly increase the chances of successful treatment. The accuracy of the comparative analysis of the centroid–perimeter radius and breast skin segmentation can aid in the detection of potential cancerous cases, which can then be confirmed through biopsy or histopathology.

Our study primarily focused on feature extraction and validation, rather than the implementation of a classification learning model. As a result, we did not have conventional control data. Instead, we applied the cases identified by radiologists as potential asymmetry, using them as reference points for our validation process. This approach allowed us to assess the accuracy and performance of the feature extraction methods utilized in our system. By employing the identified cases as validation data, we could evaluate the effectiveness of our proposed techniques in detecting possible breast asymmetry. It is crucial to underscore that the main goal of our study was to demonstrate the feasibility and reliability of the feature-based preliminary asymmetry detection system, paving the way for future advancements in developing more advanced classification models.

From a medical standpoint, the early detection of breast cancer is crucial for successful treatment outcomes. The accurate comparison analysis may serve as an auxiliary tool in identifying suspicious cancer cases that specialists may not assess due to human factors. However, it is important to note that the analysis presented in this study should be complemented by further clinical investigations and follow-up studies. Biopsy or histopathology remains the gold standard for definitive diagnosis, and the integration of our CAD system with these procedures can provide a comprehensive diagnostic approach. Overall, this research highlights the potential of both characteristics for early breast cancer detection and underscores the relevance of their integration into a Graphical User Interface designed for effective utilization in the healthcare field. The integration of machine learning algorithms, texture analysis techniques, and multiple segmentation methods may enhance analysis accuracy. Accurate and reliable analysis methods may enable an early diagnosis to have the potential to significantly improve treatment outcomes for breast cancer patients.

While our research has demonstrated the effectiveness of feature extraction in detecting possible breast asymmetry, there are certain limitations to consider. One of the key limitations is the lack of implementation of a classification learning model in our study. While we focused on feature extraction and validation, the absence of a classification model may limit the direct application of our findings for real-time diagnosis. Future work should aim to integrate machine learning algorithms into our system to enable the automated classification and accurate identification of potentially cancerous areas. Another limitation is the reliance on radiologist-identified cases as reference points for validation. While this approach served our purpose, it introduces a level of subjectivity that may affect the generalizability of our results. To address this limitation, future studies could incorporate ground-truth annotations obtained through biopsy or histopathology, providing a more objective basis for evaluation. Furthermore, the current study employed a specific set of segmentation techniques for breast skin and shape analysis. Alternative approaches, such as texture analysis techniques using co-occurrence matrices, could be explored to enhance the accuracy of segmentation and feature extraction. Lastly, our research employed a dataset with a limited number of patients, and may not fully represent the diversity and complexity of breast cancer cases. A larger and more diverse dataset would provide a more comprehensive evaluation of the proposed system's performance and generalizability.

5. Conclusions

In conclusion, our research contributes to the advancement of early breast cancer detection through feature extraction and validation. The study underscores the significance of comparing contra-lateral breasts for asymmetrical analysis, focusing on structural shape and skin thickness. The findings emphasize the potential clinical applications of our proposed system and the need for further exploration in this field. Advancements in technology and machine learning offer the opportunity to enhance structural analysis for early detection and monitoring. While our study primarily focused on feature extraction and validation, future research should explore the integration of machine learning algorithms for automated diagnosis and improved accuracy.

Our study focused on developing a computer system for asymmetry detection in mammographic images, using Dynamic Time Warping (DTW) for shape analysis and the Growing Seed Region (GRS) method for breast skin segmentation. We observed a notable presence of asymmetry in 23 cases subjected to structural analysis, with a DTW index > 0.08 and an accuracy of 83%. The GRS-based skin segmentation achieved an average accuracy of 90.47% (ASSD) and 66.66% (RVD), successfully identifying 35 patients with potential skin asymmetry. However, we acknowledge certain limitations in our research. Notably, the absence of a classification learning model may limit the real-time application of our findings. Future work should aim to incorporate machine learning algorithms to enable the automated classification and precise identification of potentially cancerous areas. Additionally, the reliance on radiologist-identified cases introduces subjectivity, which could affect the generalizability of our results. Including ground-truth annotations obtained through biopsy or histopathology in future studies would provide a more objective basis for evaluation. Furthermore, exploring alternative segmentation techniques, such as texture analysis using co-occurrence matrices, may enhance segmentation accuracy. Lastly, our study employed a dataset with a limited number of patients, potentially affecting the representation of diverse breast cancer cases. A larger and more diverse dataset would provide a comprehensive evaluation of the proposed system's performance and generalizability. Despite these limitations, our research demonstrates the potential of feature-based asymmetry analysis and its relevance in contributing to early breast cancer detection. Continued research and collaboration between medical professionals and technological experts holds promise for advancing diagnostic capabilities, ultimately improving patient outcomes in the fight against breast cancer.

In conclusion, our research contributes to the advancement of early breast cancer detection through feature extraction and validation. The study underscores the significance of comparing contra-lateral breasts for asymmetry analysis, focusing on structural shape and skin thickness. The findings emphasize the potential clinical applications of our proposed system and the need for further exploration in this field. Advancements in technology and machine learning offer the opportunity to enhance structural analysis for early detection and monitoring. While our study primarily focused on feature extraction and validation, future research should explore the integration of machine learning algorithms for automated diagnosis and improved accuracy.

Author Contributions: R.B.-M., L.A.M.-R., A.T.-V., Y.M.H.-R. and O.E.C.-M.; conceptualization, R.B.-M., L.A.M.-R., A.T.-V., Y.M.H.-R. and O.E.C.-M.; methodology, R.B.-M., L.A.M.-R. and A.T.-V.; software, R.B.-M., L.A.M.-R. and A.T.-V.; validation, R.B.-M. and A.T.-V.; formal analysis, R.B.-M., L.A.M.-R. and A.T.-V.; investigation, R.B.-M., L.A.M.-R., A.T.-V., Y.M.H.-R. and O.E.C.-M.; resources, Y.M.H.-R. and O.E.C.-M.; data curation, Y.M.H.-R. and O.E.C.-M.; writing—original draft preparation, R.B.-M.; writing—review and editing, Y.M.H.-R. and O.E.C.-M.; visualization, Y.M.H.-R. and O.E.C.-M.; supervision, Y.M.H.-R. and O.E.C.-M.; project administration, Y.M.H.-R. and O.E.C.-M.; funding acquisition, Y.M.H.-R. and O.E.C.-M. All authors have read and agreed to the published version of the manuscript.

Funding: This research was funded by the Mexican Consejo Nacional de Humanidades, Ciencia y Tecnilogía (CONAHCYT), grant number FORDECYT-PRONACES/6005/20, and the Secretaría de Investigación y Posgrado of the Instituto Politécnico Nacional.

Institutional Review Board Statement: Not applicable.

Informed Consent Statement: Not applicable.

Data Availability Statement: All data used in this research are available upon request from the corresponding author.

Acknowledgments: The authors acknowledge the Mexican *Consejo Nacional de Humanidades, Ciencia y Tecnología (CONAHCYT)* for the economic support given to this research through the project FORDECYT-PRONACES/6005/20. The authors also thank the *Secretaría de Investigación y Posgrado* of the *Instituto Politécnico Nacional (IPN)* for the partial economic support given to this research. The research was carried out in the *Laboratorio de Sistemas para Diagnóstico y Tratamiento de Cáncer, UPIITA-IPN*.

Conflicts of Interest: The authors declare no conflict of interest.

Appendix A. Algorithm to Compute the Distance between the Centroid and the Breast Perimeter

Algorithm A1 Estimation of the radius formed between the centroid and the edge of the breast

Inputs:

- Breast Image *Img*
- Morphological Opening Kernel K = ones(7,7)
- Hole-filling kernel H = ones(2,2)
- Threshold $T = 1$

Outputs:

- Vector V which contains a collection of distances correlated to the orientations W.

if left breast image **do**
    flip *Img* across y axis to match the right position;
Create a binary matrix image A based on the threshold T;
Apply a binary opening (erosion + dilation) using kernel K to update A;
Fill any holes inside matrix A using kernel H;
Compute the area of all regions found in binary image A;
Compute the centroid coordinates $C = (C_x, C_y)$ for the region with the biggest area;
Find border of biggest area and store it in E as pixels coordinates (P_x, P_y);
Initialize empty distance list V;
Initialize empty angle list W;
for each coordinate (P_x, P_y) in E **do**
    Define distance $G = sqrt((C_x - P_x)\hat{}2 + (C_y - P_y)\hat{}2)$;
    Append G to V;
    Define angle $F = atan2((C_x - P_x)/(C_y - P_y))$;
    Append F to W;
Return V and W.

Appendix B. Algorithm for Growing Region Seed

Algorithm A2 Growing Region Seed

Inputs:

- *img*: An image represented as a matrix
- *seedsList*: A list of seed coordinates
- *thLow*: Lower threshold = 1
- *thHigh*: Higher threshold value = 255
- *growthDistLimit*: maximum distance to seed = 3

Outputs:

- mask: A binary mask indicating the region of seed growth

```
Initialize an empty binary mask called 'mask' with the same shape as img;
for each currentSeed in seedsList:
    Initialize an empty binary matrix called 'revised' with the same shape as img;
    Create an empty list called 'check' and add currentSeed to it;
    Create an empty list called 'comparators' and append img[currentSeed[0],
    currentSeed[1]] to it;
    Set originCoords to [currentSeed[0], currentSeed[1]];
    While 'check' list is not empty:
        Set currentValue to img[revisar[0][0], revisar[0][1]];
        Set valComp to comparators[0];
        Set condLow to (currentValue − valComp) < thigh;
        Set condHigh to (currentValue − valComp) > -thLow;
        Set condDist to sqrt((check[0][0] − originCoords[0])^2 + (check[0][1] −
        originCoords[1])^2) < growthDistLimit;
        if img[check[0][0], check [0][1]] > 0 and condLow and condHigh and condDist:
            Set mask[check [0][0], check [0][1]] to True;
            for i from check [0][0] − 1 to check [0][0] + 1:
                for j from check[0][1] − 1 to check[0][1] + 1:
                    if revised[i, j] is False and [i, j] is not in 'check':
                        Add [i, j] to 'check';
                        Add currentValue to comparators;
        Set revised[check[0][0], check[0][1]] to True;
        Remove the first element from 'check';
        Remove the first element from comparators;
Return mask
```

References

1. Bray, F.; Ferlay, J.; Soerjomataram, I.; Siegel, R.L.; Torre, L.A.; Jemal, A. Global Cancer Statistics 2018: GLOBOCAN Estimates of Incidence and Mortality Worldwide for 36 Cancers in 185 Countries. *CA Cancer J. Clin.* **2018**, *68*, 394–424. [CrossRef] [PubMed]
2. Chávarri-Guerra, Y.; Villarreal-Garza, C.; Liedke, P.E.R.; Knaul, F.; Mohar, A.; Finkelstein, D.M.; Goss, P.E. Breast Cancer in Mexico: A Growing Challenge to Health and the Health System. *Lancet Oncol.* **2012**, *13*, e335–e343. [CrossRef] [PubMed]
3. Kim, M.Y.; Choi, N. Mammographic and Ultrasonographic Features of Triple-Negative Breast Cancer: A Comparison with Other Breast Cancer Subtypes. *Acta Radiol.* **2013**, *54*, 889–894. [CrossRef]
4. Marmot, M.G.; Altman, D.G.; Cameron, D.A.; Dewar, J.A.; Thompson, S.G.; Wilcox, M. The Benefits and Harms of Breast Cancer Screening: An Independent Review: A Report Jointly Commissioned by Cancer Research UK and the Department of Health (England) October 2012. *Br. J. Cancer* **2013**, *108*, 2205. [CrossRef] [PubMed]
5. Xu, F.; Li, M.; Li, J.; Jiang, H. Diagnostic Accuracy and Prognostic Value of Three-Dimensional Electrical Impedance Tomography Imaging in Patients with Breast Cancer. *Gland Surg.* **2021**, *10*, 2673–2685. [CrossRef] [PubMed]
6. Aziz, S.; Mohamad, M.A.; Zin, R.R.M. Histopathological Correlation of Breast Carcinoma with Breast Imaging-Reporting and Data System. *Malays. J. Med. Sci.* **2022**, *29*, 65. [CrossRef]
7. Kwan, M.L.; Kushi, L.H.; Weltzien, E.; Maring, B.; Kutner, S.E.; Fulton, R.S.; Lee, M.M.; Ambrosone, C.B.; Caan, B.J. Epidemiology of Breast Cancer Subtypes in Two Prospective Cohort Studies of Breast Cancer Survivors. *Breast Cancer Res.* **2009**, *11*, R31. [CrossRef]
8. Paolini, B.; Leddy, R.; Irshad, A. Disappearing Grouped Breast Calcifications: An Ominous Sign. *Radiol. Case Rep.* **2020**, *15*, 2453–2458. [CrossRef]

9. Vállez, N.; Bueno, G.; Déniz, Ó.; Fernández, M.D.M.; Pastor, C.; Rienda, M.Á.; Esteve, P.; Arias, M. CADe System Integrated within the Electronic Health Record. *Biomed Res. Int.* **2013**, *2013*, 219407. [CrossRef]
10. Hassan, N.M.; Hamad, S.; Mahar, K. Mammogram Breast Cancer CAD Systems for Mass Detection and Classification: A Review. *Multimed. Tools Appl.* **2022**, *81*, 20043–20075. [CrossRef]
11. Kayar, R.; Çilengiroğlu, Ö.V. Breast Volume Asymmetry Value, Ratio, and Cancer Risk. *Breast Cancer* **2015**, *9*, 87–92. [CrossRef]
12. Scutt, D.; Lancaster, G.A.; Manning, J.T. Breast Asymmetry and Predisposition to Breast Cancer. *Breast Cancer Res.* **2006**, *8*, R14. [CrossRef]
13. Skaane, P. Malignant Mammographic Asymmetric Densities without US Correlate Remain Challenging. *Radiology* **2022**, *302*, 533–534. [CrossRef] [PubMed]
14. Tariq, M.; Iqbal, S.; Ayesha, H.; Abbas, I.; Ahmad, K.T.; Niazi, M.F.K. Medical Image Based Breast Cancer Diagnosis: State of the Art and Future Directions. *Expert Syst. Appl.* **2021**, *167*, 114095. [CrossRef]
15. Yoon, J.H.; Kim, E.K. Deep Learning-Based Artificial Intelligence for Mammography. *Korean J. Radiol.* **2021**, *22*, 1225–1239. [CrossRef]
16. Scutt, D.; Manning, J.T.; Whitehouse, G.H.; Leinster, S.J.; Massey, C.P. The Relationship between Breast Asymmetry, Breast Size and the Occurrence of Breast Cancer. *Br. J. Radiol.* **2014**, *70*, 1017–1021. [CrossRef]
17. Singh, M.; Singh, T.; Soni, S. Pre-Operative Assessment of Ablation Margins for Variable Blood Perfusion Metrics in a Magnetic Resonance Imaging Based Complex Breast Tumour Anatomy: Simulation Paradigms in Thermal Therapies. *Comput. Methods Programs Biomed.* **2021**, *198*, 105781. [CrossRef]
18. Dawoud, B.M.; Darweesh, A.N.; Hefeda, M.M.; Kamal, R.M.; Younis, R.L. Diagnostic Value of Contrast-Enhanced Mammography in the Characterization of Breast Asymmetry. *Egypt. J. Radiol. Nucl. Med.* **2022**, *53*, 259. [CrossRef]
19. Ghunaim, H.; Laenen, A.; De Keyzer, F.; Soens, J.; Keupers, M.; Postema, S.; Neven, P.; Van Ongeval, C. Comparing Breast Cancer Imaging Characteristics of CHEK2 with BRCA1 and BRCA2 Gene Mutation Carriers. *Eur. J. Radiol.* **2022**, *146*, 110074. [CrossRef] [PubMed]
20. Shimokawa, D.; Takahashi, K.; Kurosawa, D.; Takaya, E.; Oba, K.; Yagishita, K.; Fukuda, T.; Tsunoda, H.; Ueda, T. Deep Learning Model for Breast Cancer Diagnosis Based on Bilateral Asymmetrical Detection (BilAD) in Digital Breast Tomosynthesis Images. *Radiol. Phys. Technol.* **2023**, *16*, 20–27. [CrossRef] [PubMed]
21. Mangj, S.M.; Hussan, P.H.; Shakir, W.M.R. Efficient Deep Learning Approach for Detection of Brain Tumor Disease. *Int. J. Online Biomed. Eng.* **2023**, *19*, 66–80. [CrossRef]
22. Ryu, J.; Rehman, M.U.; Nizami, I.F.; Chong, K.T. SegR-Net: A Deep Learning Framework with Multi-Scale Feature Fusion for Robust Retinal Vessel Segmentation. *Comput. Biol. Med.* **2023**, *163*, 107132. [CrossRef] [PubMed]
23. Rehman, M.U.; Ryu, J.; Nizami, I.F.; Chong, K.T. RAAGR2-Net: A Brain Tumor Segmentation Network Using Parallel Processing of Multiple Spatial Frames. *Comput. Biol. Med.* **2023**, *152*, 106426. [CrossRef] [PubMed]
24. Attallah, O.; Zaghlool, S. AI-Based Pipeline for Classifying Pediatric Medulloblastoma Using Histopathological and Textural Images. *Life* **2022**, *12*, 232. [CrossRef]
25. Castro, S.M.; Tseytlin, E.; Medvedeva, O.; Mitchell, K.; Visweswaran, S.; Bekhuis, T.; Jacobson, R.S. Automated Annotation and Classification of BI-RADS Assessment from Radiology Reports. *J. Biomed. Inform.* **2017**, *69*, 177–187. [CrossRef]
26. The TIES Cancer Research Network (TCRN) Homepage. Available online: http://ties.dbmi.pitt.edu/tcrn (accessed on 8 February 2023).
27. Müller, M. Dynamic Time Warping. In *Information Retrieval for Music and Motion*; Springer: Berlin/Heidelberg, Germany, 2007; pp. 69–84. [CrossRef]
28. Senin, P. *Dynamic Time Warping Algorithm Review*; Information and Computer Science Department University of Hawaii at Manoa Honolulu: Manoa, HI, USA, 2008.
29. Gold, O.; Sharir, M. Dynamic Time Warping and Geometric Edit Distance. *ACM Trans. Algorithms* **2018**, *14*, 1–17. [CrossRef]
30. DicomSDL & VOI-LUT | Kaggle. Available online: https://www.kaggle.com/code/crischir/dicomsdl-voi-lut (accessed on 24 June 2023).
31. Reinke, A.; Tizabi, M.D.; Sudre, C.H.; Eisenmann, M.; Rädsch, T.; Baumgartner, M.; Acion, L.; Antonelli, M.; Arbel, T.; Bakas, S.; et al. Common Limitations of Image Processing Metrics: A Picture Story. *arXiv* **2021**, arXiv:2104.05642.
32. Yeghiazaryan, V.; Voiculescu, I. Family of Boundary Overlap Metrics for the Evaluation of Medical Image Segmentation. *J. Med. Imaging* **2018**, *5*, 1. [CrossRef]

cancers

MDPI

Article

Dynamic Chest Radiograph Simulation Technique with Deep Convolutional Neural Networks: A Proof-of-Concept Study

Dongrong Yang [1,2], Yuhua Huang [1], Bing Li [1,3], Jing Cai [1,*] and Ge Ren [1,4,*]

[1] Department of Health Technology and Informatics, The Hong Kong Polytechnic University, Kowloon, Hong Kong; dongrongyang@outlook.com (D.Y.); yu-hua.huang@connect.polyu.hk (Y.H.); zlyylibing4127@zzu.edu.cn (B.L.)

[2] Department of Radiation Oncology, Duke University Medical Center, Durham, NC 27708, USA

[3] The Affiliated Cancer Hospital of Zhengzhou University & Henan Cancer Hospital, Zhengzhou 450008, China

[4] The Hong Kong Polytechnic University Shenzhen Research Institute, Shenzhen 518000, China

[*] Correspondence: jing.cai@polyu.edu.hk (J.C.); gary-ge.ren@polyu.edu.hk (G.R.)

Simple Summary: Dynamic chest radiographs offer a distinct advantage over traditional chest radiographs by integrating motion and functional data, elevating their significance in clinical diagnostics. This study introduces a pioneering technique employing deep neural networks to simulate respiratory lung motion and extract local functional details from single-phase chest X-rays, thereby enhancing lung cancer clinical diagnostic capabilities. Our research establishes the viability of generating patient-specific respiratory motion profiles from single-phase chest radiographs. The evaluation of results from the network developed here underscores its substantial accuracy and fidelity, affirming its robustness in providing valuable supplementary insights into pulmonary function.

Abstract: In this study, we present an innovative approach that harnesses deep neural networks to simulate respiratory lung motion and extract local functional information from single-phase chest X-rays, thus providing valuable auxiliary data for early diagnosis of lung cancer. A novel radiograph motion simulation (RMS) network was developed by combining a U-Net and a long short-term memory (LSTM) network for image generation and sequential prediction. By utilizing a spatial transformer network to deform input images, our proposed network ensures accurate image generation. We conducted both qualitative and quantitative assessments to evaluate the effectiveness and accuracy of our proposed network. The simulated respiratory motion closely aligns with pulmonary biomechanics and reveals enhanced details of pulmonary diseases. The proposed network demonstrates precise prediction of respiratory motion in the test cases, achieving remarkable average Dice scores exceeding 0.96 across all phases. The maximum variation in lung length prediction was observed during the end-exhale phase, with average deviation of 4.76 mm ($\pm$6.64) for the left lung and 4.77 mm ($\pm$7.00) for the right lung. This research validates the feasibility of generating patient-specific respiratory motion profiles from single-phase chest radiographs.

Keywords: chest radiograph; deep learning; motion simulation; lung nodule

Citation: Yang, D.; Huang, Y.; Li, B.; Cai, J.; Ren, G. Dynamic Chest Radiograph Simulation Technique with Deep Convolutional Neural Networks: A Proof-of-Concept Study. *Cancers* **2023**, *15*, 5768. https://doi.org/10.3390/cancers15245768

Academic Editors: Michał Strzelecki, Adam Piórkowski and Rafał Obuchowicz

Received: 9 November 2023
Revised: 30 November 2023
Accepted: 4 December 2023
Published: 8 December 2023

1. Introduction

Chest radiography is one of the most frequently used modalities for routine pulmonary disease diagnosis and is commonly utilized as the first exploratory study, as it has the advantages of high availability, low radiation dose, and an efficient workflow. It can suffice as a preliminary examination for conditions including pneumonia, lung cancer, emphysema, and pulmonary fibrosis [1]. In the clinical process, radiologists manually locate anomalies on the radiograph and diagnose the pathology after the picture has been acquired. Then, depending on the circumstances, further diagnostic imaging procedures may be recommended to obtain specialized data for a more precise diagnosis.

Dynamic chest X-ray (DXR) is a functional imaging technique that uses a flat-panel detector (FPD) to generate sequential images. The large field of view and real-time observation can provide information about diaphragm kinetics, pulmonary ventilation, and circulation [2]. It can be deployed as a simple and rapid means of functional imaging, but currently DXR machines have not been widely deployed in clinical practice. Furthermore, the total patient dose of a DXR image is 0.23 mSv, about twice that of a conventional chest X-ray (CXR) [3]. Although dynamic chest X-rays may provide extra functional information, the high dose and scarcity of clinical deployment make it difficult to apply them in routine medical practice. In contrast, conventional chest X-ray machines are widely deployed and expose patients to a lower radiation dose [4].

In recent years, advances in engineering and computer science have enabled researchers to reveal a wealth of hidden information encoded in chest radiographs [5], and thereby improve clinical diagnosis. A decrease in X-ray pixel value suggests the presence of a localized air space, which might indicate abnormalities such as a lung cyst, emphysema, bulla, or pneumothorax. Conversely, an increase in the pixel value may indicate a reduction in pulmonary air or intensified tissues, pointing to conditions such as pulmonary inflammation, fibrosis, edema, or sclerosis [6]. By leveraging the capabilities of artificial intelligence (AI), researchers are able to extract a significantly greater amount of information compared to relying solely on observations with the human eye [7]. Approaches like image enhancement [8–11], organ segmentation [12–14], and feature extraction [15–17] have been investigated and have achieved promising results. Previous studies have considered whether chest X-rays contain rich three-dimensional and functional information that could be extracted by AI to provide extra information that would improve clinical diagnosis [18]. One previous study tried to predict patients' lung function to evaluate the need for mechanical ventilation in hospitalized patients based on chest X-ray images and achieved 90.06% accuracy [19]. Liyue et al. explored a deep learning approach for tomographic X-ray imaging with a single-view projection data input, introducing the concept of dimension transformation in the image feature domain to enable volumetric imaging using either single or multiple 2D projections [18]. The feasibility of transforming inter-modality medical imaging with deep learning has also been studied by earlier investigators [20,21].

Motivated by such earlier research, in this study, we utilize the power of deep learning and convolutional neural networks to simulate respiratory lung motion and relative functional information from single-phase digital reconstructed radiographs (DRRs). This technique could be used on conventional chest X-rays to provide auxiliary information for clinical lung cancer diagnosis like dynamic chest X-rays while maintaining the convolutional chest X-ray's low dosage and high availability.

2. Methods

2.1. Datasets

In this study, 60 patients' 4D CT sequences were obtained from several online and local datasets [22–24]. The distribution of cases from each dataset is summarized in Table 1. The present study was conducted with the approval of the Institutional Review Board of the Affiliated Cancer Hospital of Zhengzhou University and Henan Cancer Hospital. Due to the scarcity of DXR FPD and DXR data, we generated sequential DRRs from 4D CT to simulate the DXR. During the preprocessing stage, thresholding was first applied to segment the lung volume from the patients' bodies. With less interference from the contiguous bone components, abnormalities could be readily visualized, allowing the information to be interpreted more precisely. After that, beam eye view (BEV) ray-casting was utilized to generate the DRR. Then, minimum bounding boxes were generated by detecting the boundary of the projected lung volume, and these served as the basis for DRR segmentation. To simulate the screening procedure of chest X-rays, the end-of-inhale (EOI) phase served as the input to the network and the subsequent phases were set as targets. We randomly selected 40 out of 60 patients as the training set, while the remaining 20 patients served as the test set to evaluate the effectiveness of the model.

Table 1. Dataset information.

Dataset Resource	Number of Patients
DIR Lab 4D-CT dataset [22]	10
POPI dataset [24]	5
VAMPIRE challenge 4D CT dataset [23]	12
Henan Cancer Hospital	33

2.2. Network Design

In this study, a novel deep neural network was developed to predict the entire sequence of phases of a complete respiration cycle and the corresponding motion pattern from a single-phase chest X-ray. This technique could be applied to generate auxiliary diagnostic information from a chest X-ray without unnecessary radiation doses or screening time.

The general workflow is shown in Figure 1. The development of the proposed radiograph motion simulation (RMS) network mainly comprised two steps: image generation and sequential prediction. To predict the succeeding phases of the respiration cycle $(p_1, p_2, \ldots p_9)$ from the EOI phase p_0, we adopted a recurrent neural network (RNN) embedded U-Net structure [25]. First, the input image p_0 was down-sampled by the encoder block of the U-Net to a one-dimensional feature vector x_0, which was fed into the RNN for sequential prediction. The decoder block then up-sampled the predicted vector to generate the subsequent phases. Instead of explicitly predicting the images, we used the decoder to generate deformation vector fields (DVFs) and a spatial transformer network to deform the input image to generate accurate medical imaging without sacrificing anatomical detail.

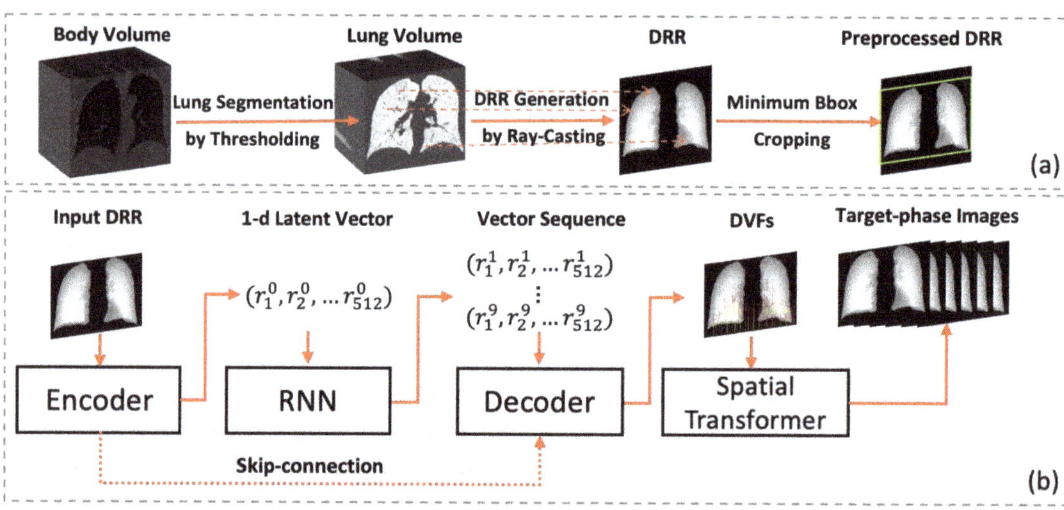

Figure 1. Illustration of the general workflow. (**a**) Preprocessing steps: The lung volumes were initially segmented from the 4D CT body volumes using thresholding. Subsequently, ray-casting was employed to generate the sequential DRR images, which were then cropped to fit the minimum bounding box. (**b**) Sequential prediction steps: The preprocessed DRR images were down-sampled by the encoder to 1D latent vectors. The vectors were then fed into the RNN for sequential prediction. The predicted latent vectors for nine phases were decoded by the decoder to predict the corresponding DVFs. These DVFs were registered to the initial DRR to generate target phase images using a spatial transformer. Abbreviations: DRR—digital reconstructed radiograph; Bbox—bounding box; RNN—recurrent neural network; DVF—deformation vector field.

2.2.1. Medical Imaging Generation Model

The schematic of the proposed RMS network is shown in Figure 2. The U-Net contracting path consists of four down-sampling blocks. Each block comprises the repeated application of two 3 × 3 convolutions (unpadded convolutions) followed by a rectified linear unit (ReLU) and a 2 × 2 max-pooling operation with stride 2 for down-sampling. Each convolutional operation involves a doubling of the feature channels. After the convolution and max-pooling operation in the contracting path, the down-sampled feature maps are flattened and sent to a fully connected layer to generate a 1 × 512 latent representation. A one-to-many long short-term memory (LSTM) [26] network is then utilized to predict the sequential latent representations. As a kind of recurrent neural network that uses a single input to generate a sequence of outputs, it utilizes the LSTM cells' ability to maintain an internal state across time steps to selectively retain or discard information. The two-layer network takes as input the previous phase's latent vector x_{t-1}, and outputs the latent vector of the next time step x_t, which is then fed into the network as the input for the next time step until the maximum time step $t = 9$ is reached. From the LSTM, a sequence of latent representations with 9 phases was generated.

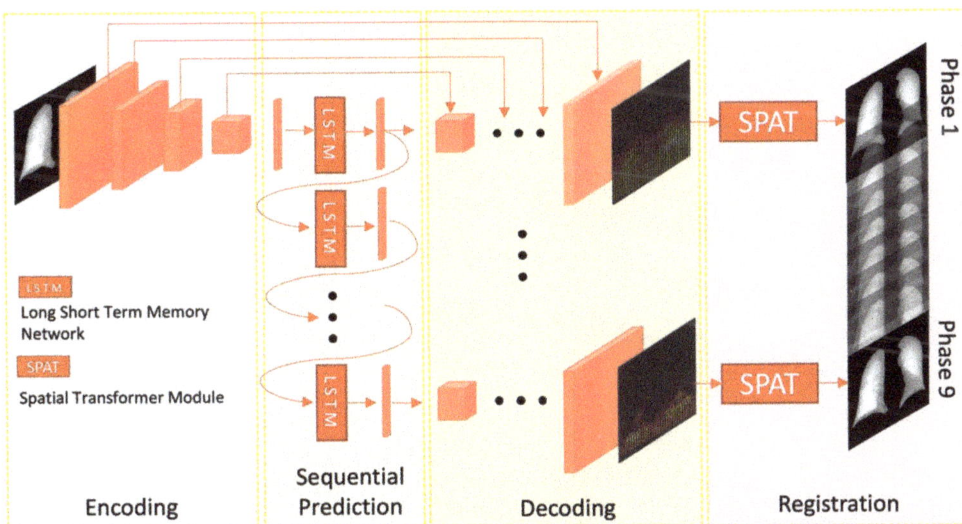

Figure 2. Schematic representation of the advanced RMS framework. The encoding path comprises four successive down-sampling blocks utilizing convolution and max-pooling operations. Sequential prediction is accomplished through a two-layer one-to-many LSTM network. The decoding path is constructed with four symmetrical up-sampling blocks employing de-convolution operations. Additionally, motion registration is achieved simultaneously through a differentiable spatial transformer.

After sequential prediction by the LSTM, the latent vectors were recovered to the same size as the input image through a symmetrically expanding path. Skip-connections between the encoders and decoders ensured information fusion and precise prediction. The expanding path of the U-Net output 9 DVFs ($D_0, D_1, D_2 \ldots D_9$) for the subsequent phases of the respiration cycle. A differentiable spatial transformer module [27] was concatenated after the U-Net to deform the initial phase image p_0 to p_{1-9} according to the predicted DVFs. For each pixel V_i^t in the target phase image p_t,

$$V_i^t = \sum_n^H \sum_m^W p_0(n,m)k(h_i^t - m; \Phi_x)k(w_i^t - n; \Phi_y) \;\; \forall\, i \in [1 \ldots HW]$$

where H and W are the height and width of the initial phase image, and $p_0(n,m)$ is the value at location (n,m) in the input image. Each (h_i^t, w_i^t) coordinate in the corresponding phase's DVF defines the spatial location in the input image at which a sampling kernel is applied, to obtain the value at a specific output pixel. The Φ_x and Φ_y are parameters of the sampling kernel; we use bilinear interpolation in this study.

After spatial transformation, the discrepancy between predicted phases and the ground truth images was computed and backpropagated to optimize the network parameter.

2.2.2. Loss Function Design

During the optimization, three loss functions were combined to promote precise motion simulation. The first loss function L_{smooth} penalizes the gradient in the predicted DVFs, which aims to foster smoother spatial variations, aligning with the principles of physical feasibility and promoting a more realistic representation of the lung motion.

$$L_{smooth} = \frac{1}{2N}\sum_{t=1}^{9}\sum_{i=0}^{N-1}\left(\left(\frac{\partial D_t^i}{\partial h}\right)^2 + \left(\frac{\partial D_t^i}{\partial w}\right)^2\right)$$

This computes the L2 loss of the DVF's gradient in both directions. N denotes all the pixels in the DVFs and D_t^i is the ith pixel value in the phase t DVF.

The second loss function is the mean squared error (MSE) loss, measuring the discrepancy between the predicted phase image and the ground truth phase image. It can be expressed as

$$L_{MSE} = \sum_{t=1}^{9}\left(p_t^{gt} - p_t^{pre}\right)^2$$

where p_t^{gt} is the ground truth image for phase t while p_t^{pre} represents the predicted image for phase t. Beyond the MSE loss, we also utilized the local cross-correlation loss to improve model robustness, which is computed as

$$L_{NCC} = \frac{\left(\sum_i\left(p_{t,i}^{gt} - \hat{p}_{t,i}^{gt}\right)\left(p_{t,i}^{pre} - \hat{p}_{t,i}^{pre}\right)\right)^2}{\left(\sum_i\left(p_{t,i}^{gt} - \hat{p}_{t,i}^{gt}\right)^2\right)\left(\sum_i\left(p_{t,i}^{pre} - \hat{p}_{t,i}^{pre}\right)^2\right)}$$

where $p_{t,i}^{gt}$, $p_{t,i}^{pre}$ denote the ground truth and predicted value of pixel i in phase t, and $\hat{p}_{t,i}^{gt}$, $\hat{p}_{t,i}^{pre}$ are the corresponding pixel values calculated using the sliding window technique. The final loss was computed as the sum of these three loss functions:

$$L = L_{smooth} + L_{MSE} + L_{NCC}$$

2.2.3. Evaluation Metrics

The accuracy and authenticity of the simulated motion were evaluated using both qualitative and quantitative methods. For qualitative evaluation, the Jacobian determinant of the DVFs was computed. The Jacobian determinant for each pixel shows how much the pixel was stretched or compressed during the deformation, which provides regional information regarding the expansion and contraction of the lung tissue during the respiration cycle to help further identify regions with inadequate or excessive ventilation and assess overall lung function.

$$v(x,y) = J(x,y) - 1 = \begin{bmatrix} 1 + \frac{\partial u_x(x,y)}{\partial x} & \frac{\partial u_x(x,y)}{\partial y} \\ \frac{\partial u_y(x,y)}{\partial x} & 1 + \frac{\partial u_y(x,y)}{\partial y} \end{bmatrix}$$

where $v(x, y)$ is the ventilation of a volume at the point (x, y); $J(x, y)$ is the Jacobian of the volume at the point (x, y); and u_x, u_y correspond to the x, y components of the DVF, respectively.

For quantitative evaluation, the Dice similarity coefficient (DSC) was computed between the predicted images and the ground truth images for each phase.

$$DSC = \frac{2\times\left|p_t^{gt} \cap p_t^{pre}\right|}{\left|p_t^{gt}\right| + \left|p_t^{pre}\right|}$$

Diaphragmatic positions (i.e., the difference between the lung apex and diaphragm dome) were also measured to further analyze the accuracy of the predicted respiratory movements.

2.2.4. Experiment Setup

In this study, we implemented the proposed RMS framework using Python 3.8 and Pytorch 1.12.1 with CUDA 11.6 on a workstation equipped with an NVIDIA GeForce RTX 2080 Ti GPU. The training took approximately two hours. After training, it took less than three seconds to generate the simulated motion per patient.

3. Results

For qualitative evaluation, Figure 3 displays the predicted respiratory phases' images and corresponding Jacobian determinant distribution for one test patient. The movement of the diaphragm can be interpreted by observing the distance between the diaphragm dome and the reference line. During the EOI phase (p_0) and the end of exhale (EOX) phase (p_4), as the lung muscles contract, the diaphragm moves upwards, resulting in a significant increase in the distance between the diaphragm dome and the reference. Subsequently, as air is inhaled and the lungs expand, the diaphragm gradually moves downwards, and the distance progressively decreases.

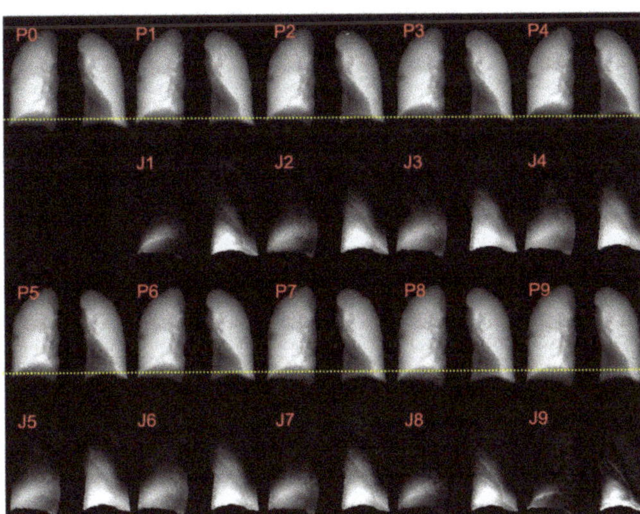

Figure 3. Example test case of respiratory motion simulation. The first and third rows represent the phase images, while the second and final rows display corresponding Jacobian determinant visualizations. Reference lines are shown in the phase images to facilitate a comparison of diaphragm movement between different phases. Abbreviations: P phase image; J Jacobian determinant distribution.

When analyzing the Jacobian determinant distribution, it becomes evident that all phases exhibit a higher contraction factor in the lower lung region compared to the upper lung region. Moreover, phases 4 and 5 demonstrate an overall higher contraction factor compared to the other phases. This observation highlights the regional variations in lung expansion and contraction throughout the respiratory cycle, providing valuable insights into the dynamics of lung function.

The phase-based Dice coefficient distribution of the test set is shown in Figure 4. For each phase, a normal distribution curve of best fit, with the corresponding mean and standard deviation (std), is included. As shown in the figure, the Dice scores of the phases close to EOI are generally better than those of the EOX phases. Meanwhile, the minimum Dice score of all phases is above 0.93, which indicates that the RMS framework is able to accurately predict the respiratory motion based on the input EOI image.

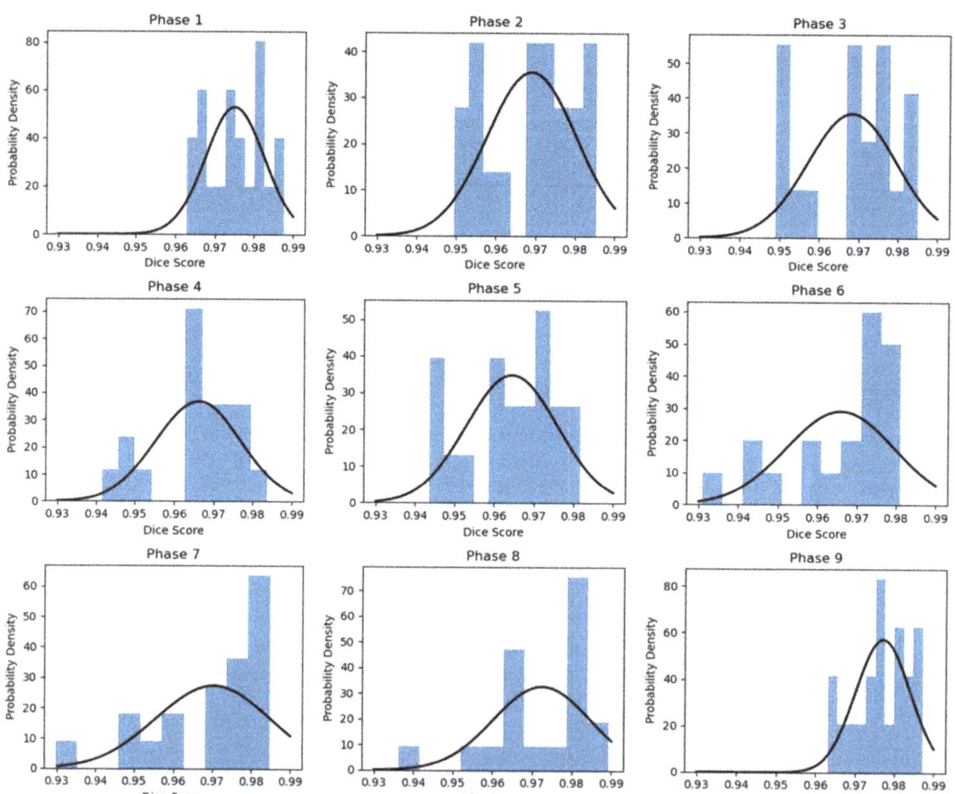

Figure 4. Histogram of phase-based Dice coefficient distribution. For each subfigure, the x-axis represents the Dice score of the predicted image, and the y-axis represents the probability density.

The variation in predicted and ground truth diaphragmatic positions is presented in Figure 5. It can be seen that for all phases, the majority of the cases' variation is less than 10 mm for both lungs. The trend of inter-phase variation is also consistent with the Dice score; the prediction of the end of the sequence is more accurate and precise than the middle of the sequence.

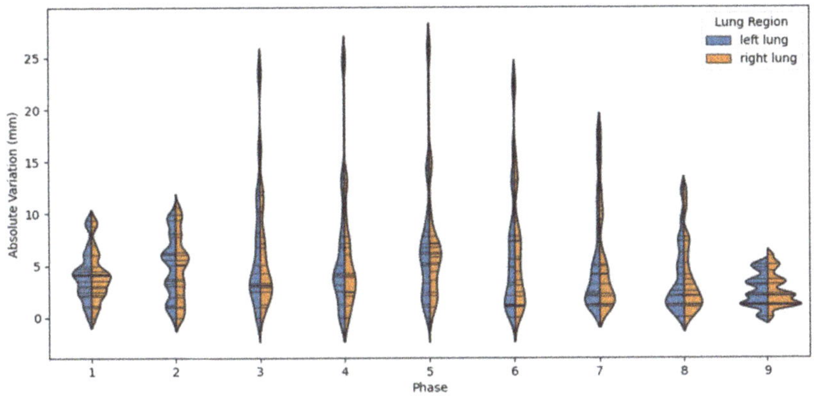

Figure 5. Violin plot of phase-based diaphragmatic position variation. Blue denotes left lung and orange denotes right lung.

Table 2 presents the mean and std of the Dice score and diaphragmatic position variation for all phases.

Table 2. Phase-based prediction statistics.

Phase #	1	2	3	4	5	6	7	8	9
Dice: *mean* (±*std*)	0.975 (±0.0075)	0.969 (±0.0112)	0.968 (±0.0112)	0.966 (±0.0108)	0.964 (±0.0114)	0.966 (±0.0137)	0.970 (±0.0146)	0.972 (±0.0122)	0.977 (±0.0070)
LLE (mm): *mean* (±*std*)	4.00 (±2.60)	4.73 (±3.29)	5.80 (±6.42)	4.77 (±6.64)	4.36 (±7.72)	3.09 (±7.98)	1.30 (±7.00)	0.28 (±4.93)	0.36 (±2.89)
RLE (mm): *mean* (±*std*)	3.93 (±2.59)	4.62 (±3.47)	5.75 (±6.42)	4.76 (±7.00)	4.29 (±7.92)	3.03 (±7.95)	1.35 (±7.07)	0.18 (±5.00)	0.31 (±2.98)

Abbreviations: LLE—left lung diaphragmatic position variation; RLE—right lung diaphragmatic position variation.

Given the absence of dynamic chest X-ray data, we conducted our network training using digital reconstructed radiograph (DRR) images to establish proof of concept. Subsequently, we sourced chest X-rays from the JSRT dataset [28] for external validation. These chest X-ray images were processed and integrated into the RMS framework for motion synthesis without further training. A representative example of the motion simulation results from one chest X-ray is illustrated in Figure 6. Notably, the relative motion of a nodule in the right upper lobe becomes readily apparent, alerting radiologists to the presence of abnormal lung function in this specific region.

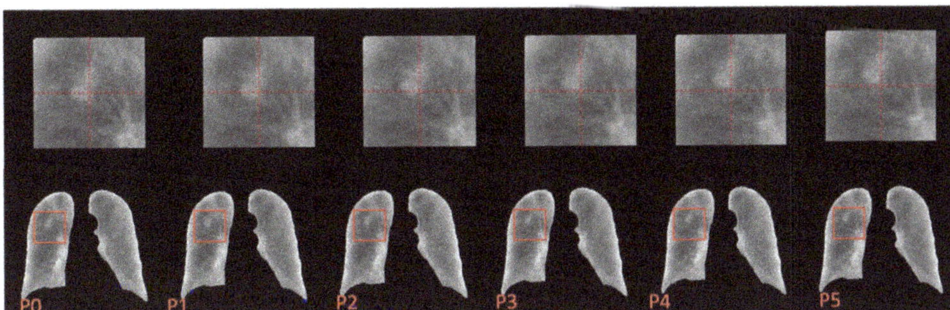

Figure 6. Simulated phase images of a JSRT chest X-ray sample. The region around the nodule was enlarged to improve the visibility of the relative motion.

4. Discussions

The chest radiograph continues to be one of the most widely used diagnostic tools for routine pulmonary diseases, for which there has traditionally been a reliance on manual interpretation by radiologists. With the advent of artificial intelligence and computer-assisted diagnosis (CAD) systems, there is persuasive evidence that a wealth of valuable information may be retrieved to improve diagnosis and clinical efficiency [29,30]. While numerous investigations have been made into developing automated classification models based on extracted deep features, the application of AI for providing lung functional information to clinicians remains relatively unexplored. In contrast to directly generating diagnostic decisions, generative models offer additional information that can be harnessed by human radiologists, fostering a more human-centered clinical decision-making process. In this study, we aimed to develop an innovative generative Respiratory Motion Synthesis (RMS) system to generate patient-specific respiratory motion patterns and corresponding deformation vector fields using chest radiographs. The primary objective of this system is to provide clinicians with supplementary functional information, enhancing their understanding of patients' respiratory dynamics.

The backbone of the RMS framework is a four-layer U-Net. Initially, the input chest radiographs undergo sequential convolution and max-pooling operations in the encoder blocks, resulting in down-sampled one-dimensional vectors with a length of 512. Subsequently, these vectors are fed into a one-to-many LSTM network, predicting the latent representation of subsequent phases. Then, they are up-sampled by the decoder blocks to generate DVFs with the same size as the input radiographs. Throughout the up-sampling process, skip connections are incorporated to pass features from the encoder path to the decoder path, enabling the recovery of lost spatial information and enhancing model performance. A subsequent spatial transformer network is utilized to deform the input radiograph to the target phases with the predicted DVFs. Compared with directly reconstructing the images from the decoder block, this prevents the loss of anatomical details in the images.

The prediction results of one patient randomly selected from the test set are visualized in Figure 3. It can be quantitively observed that the anatomical details of the lung are well-preserved in the subsequent phases. And the diaphragm motion is shown by the distance between the diaphragm dome and the reference line. The lung tissue's two-dimensional contraction is visualized by calculating the Jacobian determinant. In the process of exhalation, as respiratory gases are gradually expelled from the lungs, a progressive diminution of the lung volume is observed. This progress of contraction was visualized by sequentially plotting the increasing value of the Jacobian determinant. It can also be observed that the Jacobian determinant of the lower lung region is larger than the upper lung region, representing a higher expansion or contraction factor during respiration, which is consistent with pulmonary biomechanics. The distribution appears smooth without significant fluctuations. The simulation results presented in Figure 6 further substantiate the practicality of integrating this technique into clinical workflows, offering radiologists valuable supplementary information to enhance diagnosis.

For quantitative analysis, the statistics of Dice coefficients between the ground truth and predicted phase images are presented in Figure 4 and Table 2. In the test set, our model consistently attained an average Dice score exceeding 0.96 across all phases, demonstrating its capability to accurately predict overall pulmonary deformation during the respiratory cycle. Notably, when comparing different phases, the inhale phases exhibited generally higher Dice coefficients than the exhale phases. Figure 5 further illustrates a similar pattern in the measurements of diaphragmatic positions. The observed trend could be attributed to increased variation in pulmonary motion and inconsistencies in phase gating. This issue may potentially be mitigated with sequences having a higher imaging frequency and a larger dataset. It also reveals that while most of the test data exhibited errors conforming to a normal distribution with modest mean values, a few data points displayed significant variation across all phases. This discrepancy is likely due to the presence of severe pulmonary diseases, where patients demonstrate relatively extreme motion dynamics that differ from the training dataset, and consequently elude accurate capture by the model. In conclusion, the developed model demonstrates the ability to effectively generate respiratory motion based on single-phase DRR, thereby providing valuable supplementary functional information for diagnosis. This capability holds the potential to enhance clinical efficiency, reduce costs, and facilitate improved patient outcomes.

However, it is important to acknowledge several limitations of this proof-of-concept study. Firstly, the model was trained using digitally reconstructed radiographs (DRRs) generated from 4D CT sequences, rather than sequential chest radiograph datasets. This disparity between the chest radiograph and DRRs may have resulted in a potential decrease in model performance. Secondly, the current dataset size is limited, which could potentially hinder the model's generalizability across a wide range of clinical conditions. To address these limitations, future efforts will involve acquiring sequential chest radiograph data to fine-tune the model and comprehensively evaluate its performance. By incorporating sequential chest radiographs and expanding the dataset size, we aim to improve the model's accuracy, robustness, and applicability in real-world clinical scenarios. Another potential area for enhancement is to include a nodule segmentation module in the proposed framework. Currently, in the absence of nodule-specific ground truth annotations within our dataset, our assessment relies on the aggregate motion of the entire lung. Consequently, this approach does not capture the distinct motion characteristics of the nodules. To address this, we intend to collect a dataset with nodule annotations and integrate a dedicated nodule segmentation module to refine the simulation's precision and improve the accuracy and robustness of the simulation framework.

5. Conclusions

This study validated the feasibility of synthesizing patient-specific respiratory motion from a single-phase chest radiograph. Specifically, a novel deep convolutional neural network (RMS) was developed to predict the deformation vector fields of the subsequent phases and deform the initial radiograph with a spatial transformer. The test results of the proposed network have shown promising accuracy and authenticity, confirming its reliability as a tool for extracting additional pulmonary function information. Efficiently providing this information in a human-centered manner, without incurring any additional costs, gives this network the potential to significantly augment the clinical decision-making process.

Author Contributions: Conceptualization, D.Y., J.C. and G.R.; Methodology, D.Y., B.L. and Y.H.; Software, D.Y. and Y.H.; Validation, Y.H.; Resources, B.L.; Data curation, B.L.; Writing—original draft, D.Y.; Writing—review & editing, J.C. and G.R.; Supervision, J.C. and G.R.; Funding acquisition, B.L., J.C. and G.R. All authors have read and agreed to the published version of the manuscript.

Funding: This research was partly supported by the General Research Fund (15103520) of the University Research Committee, by the Health and Medical Research Fund (07183266, 09200576) of the Health Bureau, PolyU (UGC), by the RI-IWEAR Seed Project (P0044802), by the Shenzhen Science and Technology Program (JCYJ 20230807140403007),by the Henan Provincial Medical Science and Technology Research Project (SBGJ202103038), and by the Natural Science Foundation of Henan Province of China (222300420575 and 232300420231).

Institutional Review Board Statement: The present study was conducted with the approval of the Institutional Review Board of the Affiliated Cancer Hospital of Zhengzhou University and Henan Cancer Hospital.

Informed Consent Statement: Informed consent was obtained from all subjects involved in the study.

Data Availability Statement: The datasets generated or analyzed during this study are not publicly available due to restrictions related to patient privacy but are available from the corresponding author on reasonable request.

Conflicts of Interest: The authors declare no conflict of interest.

References

1. Wielputz, M.O.; Heussel, C.P.; Herth, F.J.; Kauczor, H.U. Radiological diagnosis in lung disease: Factoring treatment options into the choice of diagnostic modality. *Dtsch. Arztebl. Int.* **2014**, *111*, 181–187. [CrossRef] [PubMed]
2. Tanaka, R.; Sanada, S.; Okazaki, N.; Kobayashi, T.; Suzuki, M.; Matsui, T.; Matsui, O. Detectability of regional lung ventilation with flat-panel detector-based dynamic radiography. *J. Digit. Imaging* **2008**, *21*, 109–120. [CrossRef] [PubMed]
3. Tamura, M.; Matsumoto, I.; Saito, D.; Yoshida, S.; Takata, M.; Tanaka, R.; Takemura, H. Dynamic chest radiography: Novel and less-invasive imaging approach for preoperative assessments of pleural invasion and adhesion. *Radiol. Case Rep.* **2020**, *15*, 702–704. [CrossRef]
4. Johnston, D.A.; Brennan, P.C. Reference dose levels for patients undergoing common diagnostic X-ray examinations in Irish hospitals. *Br. J. Radiol.* **2000**, *73*, 396–402. [CrossRef] [PubMed]
5. van Ginneken, B.; Hogeweg, L.; Prokop, M. Computer-aided diagnosis in chest radiography: Beyond nodules. *Eur. J. Radiol.* **2009**, *72*, 226–230. [CrossRef]
6. Fraser, R.S.; Müller, N.L.; Colman, N.; Pare, P. *Fraser and Paré's Diagnosis of Diseases of the Chest*; Wb Saunders: Philadelphia, PA, USA, 1999; Volumes 1–4.
7. Qin, C.; Yao, D.; Shi, Y.; Song, Z. Computer-aided detection in chest radiography based on artificial intelligence: A survey. *Biomed. Eng. Online* **2018**, *17*, 113. [CrossRef] [PubMed]
8. Anand, S.; Roshan, R.K.; Deiveega, S.M. Chest X ray image enhancement using deep contrast diffusion learning. *Optik* **2023**, *279*, 170751. [CrossRef]
9. Lin, C.-H.; Wu, J.-X.; Li, C.-M.; Chen, P.-Y.; Pai, N.-S.; Kuo, Y.-C. Enhancement of Chest X-ray Images to Improve Screening Accuracy Rate Using Iterated Function System and Multilayer Fractional-Order Machine Learning Classifier. *IEEE Photonics J.* **2020**, *12*, 1–18. [CrossRef]
10. Xu, L.; Zeng, X.; Huang, Z.; Li, W.; Zhang, H. Low-dose chest X-ray image super-resolution using generative adversarial nets with spectral normalization. *Biomed. Signal Process. Control* **2020**, *55*, 101600. [CrossRef]
11. Ghosh, S.K.; Ghosh, A. ENResNet: A novel residual neural network for chest X-ray enhancement based COVID-19 detection. *Biomed. Signal Process. Control* **2022**, *72*, 103286. [CrossRef]
12. Rahman, T.; Khandakar, A.; Kadir, M.A.; Islam, K.R.; Islam, K.F.; Mazhar, R.; Hamid, T.; Islam, M.T.; Kashem, S.; Mahbub, Z.B.; et al. Reliable Tuberculosis Detection Using Chest X-ray With Deep Learning, Segmentation and Visualization. *IEEE Access* **2020**, *8*, 191586–191601. [CrossRef]
13. Souza, J.C.; Diniz, J.O.B.; Ferreira, J.L.; Da Silva, G.L.F.; Silva, A.C.; de Paiva, A.C. An automatic method for lung segmentation and reconstruction in chest X-ray using deep neural networks. *Comput. Methods Programs Biomed.* **2019**, *177*, 285–296.
14. Stirenko, S.; Kochura, Y.; Alienin, O.; Rokovyi, O.; Gordienko, Y.; Gang, P.; Zeng, W. Chest X-ray Analysis of Tuberculosis by Deep Learning with Segmentation and Augmentation. In Proceedings of the 2018 IEEE 38th International Conference on Electronics and Nanotechnology (ELNANO), Kyiv, Ukraine, 24–26 April 2018; pp. 422–428.
15. Yang, D.; Ren, G.; Ni, R.; Huang, Y.H.; Lam, N.F.D.; Sun, H.; Wan, S.B.N.; Wong, M.F.E.; Chan, K.K.; Tsang, H.C.H.; et al. Deep learning attention-guided radiomics for COVID-19 chest radiograph classification. *Quant. Imaging Med. Surg.* **2023**, *13*, 572–584. [CrossRef] [PubMed]
16. Shankar, K.; Perumal, E.; Tiwari, P.; Shorfuzzaman, M.; Gupta, D. Deep learning and evolutionary intelligence with fusion-based feature extraction for detection of COVID-19 from chest X-ray images. *Multimed. Syst.* **2022**, *28*, 1175–1187. [CrossRef] [PubMed]
17. Sharma, H.; Jain, J.S.; Bansal, P.; Gupta, S. Feature Extraction and Classification of Chest X-ray Images Using CNN to Detect Pneumonia. In Proceedings of the 2020 10th International Conference on Cloud Computing, Data Science & Engineering (Confluence), Noida, India, 29–31 January 2020; pp. 227–231.
18. Shen, L.; Zhao, W.; Xing, L. Patient-specific reconstruction of volumetric computed tomography images from a single projection view via deep learning. *Nat. Biomed. Eng.* **2019**, *3*, 880–888. [CrossRef] [PubMed]
19. Kulkarni, A.R.; Athavale, A.M.; Sahni, A.; Sukhal, S.; Saini, A.; Itteera, M.; Zhukovsky, S.; Vernik, J.; Abraham, M.; Joshi, A.; et al. Deep learning model to predict the need for mechanical ventilation using chest X-ray images in hospitalised patients with COVID-19. *BMJ Innov.* **2021**, *7*, 261–270. [CrossRef]
20. Li, W.; Li, Y.; Qin, W.; Liang, X.; Xu, J.; Xiong, J.; Xie, Y. Magnetic resonance image (MRI) synthesis from brain computed tomography (CT) images based on deep learning methods for magnetic resonance (MR)-guided radiotherapy. *Quant. Imaging Med. Surg.* **2020**, *10*, 1223–1236. [CrossRef] [PubMed]
21. Boulanger, M.; Nunes, J.C.; Chourak, H.; Largent, A.; Tahri, S.; Acosta, O.; De Crevoisier, R.; Lafond, C.; Barateau, A. Deep learning methods to generate synthetic CT from MRI in radiotherapy: A literature review. *Phys. Med.* **2021**, *89*, 265–281. [CrossRef]
22. Castillo, R.; Castillo, E.; Guerra, R.; Johnson, V.E.; McPhail, T.; Garg, A.K.; Guerrero, T. A framework for evaluation of deformable image registration spatial accuracy using large landmark point sets. *Phys. Med. Biol.* **2009**, *54*, 1849–1870. [CrossRef] [PubMed]
23. Kipritidis, J.; Tahir, B.A.; Cazoulat, G.; Hofman, M.S.; Siva, S.; Callahan, J.; Hardcastle, N.; Yamamoto, T.; Christensen, G.E.; Reinhardt, J.M.; et al. The VAMPIRE challenge: A multi-institutional validation study of CT ventilation imaging. *Med. Phys.* **2019**, *46*, 1198–1217. [CrossRef]
24. Vandemeulebroucke, J.; Rit, S.; Kybic, J.; Clarysse, P.; Sarrut, D. Spatiotemporal motion estimation for respiratory-correlated imaging of the lungs. *Med. Phys.* **2011**, *38*, 166–178. [CrossRef] [PubMed]

25. Ronneberger, O.; Fischer, P.; Brox, T. U-Net: Convolutional Networks for Biomedical Image Segmentation. In Proceedings of the Medical Image Computing and Computer-Assisted Intervention–MICCAI 2015, Munich, Germany, 5–9 October 2015; Springer International Publishing: Cham, Switzerland, 2015; pp. 234–241.
26. Hochreiter, S.; Schmidhuber, J. Long Short-Term Memory. *Neural Comput.* **1997**, *9*, 1735–1780. [CrossRef] [PubMed]
27. Jaderberg, M.; Simonyan, K.; Zisserman, A.; Kavukcuoglu, K. Spatial transformer networks. *Adv. Neural Inf. Process. Syst.* **2015**, *2015*, 2017–2025.
28. Shiraishi, J.; Katsuragawa, S.; Ikezoe, J.; Matsumoto, T.; Kobayashi, T.; Komatsu, K.-I.; Matsui, M.; Fujita, H.; Kodera, Y.; Doi, K. Development of a Digital Image Database for Chest Radiographs With and Without a Lung Nodule. *Am. J. Roentgenol.* **2000**, *174*, 71–74. [CrossRef] [PubMed]
29. Singh, R.; Kalra, M.K.; Nitiwarangkul, C.; Patti, J.A.; Homayounieh, F.; Padole, A.; Rao, P.; Putha, P.; Muse, V.V.; Sharma, A.; et al. Deep learning in chest radiography: Detection of findings and presence of change. *PLoS ONE* **2018**, *13*, e0204155. [CrossRef] [PubMed]
30. Ahn, J.S.; Ebrahimian, S.; McDermott, S.; Lee, S.; Naccarato, L.; Di Capua, J.F.; Wu, M.Y.; Zhang, E.W.; Muse, V.; Miller, B.; et al. Association of Artificial Intelligence–Aided Chest Radiograph Interpretation with Reader Performance and Efficiency. *JAMA Netw. Open* **2022**, *5*, e2229289. [CrossRef]

diagnostics

MDPI

Article

Combining State-of-the-Art Pre-Trained Deep Learning Models: A Noble Approach for Skin Cancer Detection Using Max Voting Ensemble

Md. Mamun Hossain [1], Md. Moazzem Hossain [1], Most. Binoee Arefin [1], Fahima Akhtar [1] and John Blake [2,*]

[1] Department of Computer Science and Engineering, Bangladesh Army University of Science and Technology, Saidpur 5310, Bangladesh
[2] School of Computer Science and Engineering, University of Aizu, Aizuwakamatsu 965-8580, Japan
* Correspondence: jblake@u-aizu.ac.jp

Abstract: Skin cancer poses a significant healthcare challenge, requiring precise and prompt diagnosis for effective treatment. While recent advances in deep learning have dramatically improved medical image analysis, including skin cancer classification, ensemble methods offer a pathway for further enhancing diagnostic accuracy. This study introduces a cutting-edge approach employing the Max Voting Ensemble Technique for robust skin cancer classification on ISIC 2018: Task 1-2 dataset. We incorporate a range of cutting-edge, pre-trained deep neural networks, including MobileNetV2, AlexNet, VGG16, ResNet50, DenseNet201, DenseNet121, InceptionV3, ResNet50V2, InceptionResNetV2, and Xception. These models have been extensively trained on skin cancer datasets, achieving individual accuracies ranging from 77.20% to 91.90%. Our method leverages the synergistic capabilities of these models by combining their complementary features to elevate classification performance further. In our approach, input images undergo preprocessing for model compatibility. The ensemble integrates the pre-trained models with their architectures and weights preserved. For each skin lesion image under examination, every model produces a prediction. These are subsequently aggregated using the max voting ensemble technique to yield the final classification, with the majority-voted class serving as the conclusive prediction. Through comprehensive testing on a diverse dataset, our ensemble outperformed individual models, attaining an accuracy of 93.18% and an AUC score of 0.9320, thus demonstrating superior diagnostic reliability and accuracy. We evaluated the effectiveness of our proposed method on the HAM10000 dataset to ensure its generalizability. Our ensemble method delivers a robust, reliable, and effective tool for the classification of skin cancer. By utilizing the power of advanced deep neural networks, we aim to assist healthcare professionals in achieving timely and accurate diagnoses, ultimately reducing mortality rates and enhancing patient outcomes.

Keywords: skin cancer; medical imaging; classification; deep learning; transfer learning; max voting

Citation: Hossain, M.M.; Hossain, M.M.; Arefin, M.B.; Akhtar, F.; Blake, J. Combining State-of-the-Art Pre-Trained Deep Learning Models: A Noble Approach for Skin Cancer Detection Using Max Voting Ensemble. *Diagnostics* **2024**, *14*, 89. https://doi.org/10.3390/diagnostics14010089

Academic Editors: Michał Strzelecki, Adam Piórkowski and Rafał Obuchowicz

Received: 3 October 2023
Revised: 21 December 2023
Accepted: 22 December 2023
Published: 30 December 2023

1. Introduction

Skin cancer, a prevalent yet often misdiagnosed disease poses a significant challenge in medical diagnostics. Skin cancer is characterized by the uncontrolled growth of abnormal cells in the skin. The three most common types of skin cancer are basal cell carcinoma, squamous cell carcinoma, Merkel cell carcinoma, and melanoma [1]. Non-melanoma skin cancer is primarily caused by DNA damage from ultraviolet radiation exposure [2,3]. Skin cancer ranks among the most prevalent forms of cancer globally, making up around a third of all reported cancer diagnoses, with its frequency steadily rising each year [4]. Just in the United States, it is estimated that more than 9500 individuals receive a skin cancer diagnosis daily [5,6].

While skin cancer is often treatable, early detection and precise diagnosis play a pivotal role in achieving favourable treatment results and enhancing patient survival rates [7–9]. Skin cancer detection has traditionally relied on a mix of visual examination and histopathological analysis, methods which are fraught with limitations in accuracy and scalability. Unfortunately, traditional methods for diagnosing skin cancer, such as visual inspection and histopathological examination, can be time-consuming and subjective, and have high inter-observer variability.

Numerous non-invasive imaging technologies for skin cancer detection and monitoring have been developed [10] in recent years. One notable example is the use of multi-spectral sensors to detect differences in the refraction index using millimeter-wave to terahertzphotonic near-field imaging [11]. Recent advancements in different areas of artificial intelligence (AI) have significantly impacted many fields, suggesting that AI could greatly improve cancer diagnosis. The field of medical diagnostics is now benefiting from the rapid evolution of deep learning technologies, employing advanced models like Xception, InceptionResNetV2, and ResNet50V2 [12–14].

Despite these technological advancements, the difficulty in differentiating malignant from benign cases continues to hinder diagnostic accuracy. In recent years, deep neural networks, especially convolutional neural networks (CNNs), [15] have demonstrated significant potential in precisely detecting and categorizing skin cancer from medical images [7,16,17], such as those shown in Figure 1. CNNs are a specialized class of neural networks ideally suited for tasks involving image classification, as they can autonomously acquire hierarchical representations of image features directly from raw pixel values [18,19].

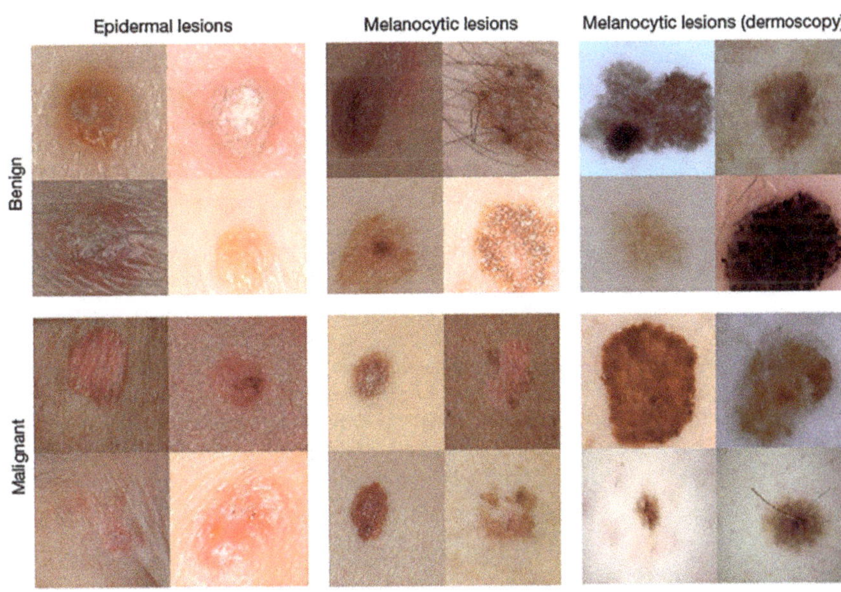

Figure 1. Skin Cancer—Malignant and benign sample images [3].

The challenge in skin cancer detection lies in the high variability of lesions and the limitations of current diagnostic methods, which often lead to inaccuracies and inconsistent diagnoses. Transfer learning, which entails the utilization of pre-trained CNN models for feature extraction from medical images, has additionally demonstrated enhancements in classification accuracy [20–22]. The application of deep learning methods in the domain of skin cancer detection and classification has emerged as a vibrant research area in recent years. Numerous investigations have substantiated that CNN models can achieve high accuracy in detecting and classifying skin cancer from medical images, and that transfer learning can further improve classification accuracy [23–25].

Although a substantial research gap still exists in the investigation of a holistic and cohesive methodology, the extant body of literature on skin cancer diagnosis utilizing deep learning models is rapidly growing with many researchers vying to create increasingly more accurate detection methods. There are, however, few studies that harness the possible synergies that may be achieved using a Max Voting Ensemble approach, which has not—as far as we know—been thoroughly investigated on a diverse dataset in the context of skin cancer diagnosis.

While deep learning for medical image analysis has made significant strides, a particular area of study interest is the application of cutting-edge pre-trained deep learning models to skin cancer detection. The primary aim of this work is to improve the precision and dependability of skin cancer diagnostic systems by examining the special use of the Max Voting Ensemble approach. This study presents an innovative approach for skin cancer detection and classification, utilizing the max voting ensemble technique with cutting-edge pre-trained deep learning models. The method proposed here combines the strengths of multiple pre-trained deep learning models, including Xception, InceptionResNetV2, ResNet50V2, InceptionV3, DenseNet121, DenseNet201, ResNet50, VGG16, AlexNet, and MobileNetV2, to create an ensemble with enhanced accuracy and robustness. The cutting-edge pre-trained models employed in this investigation have been extensively trained on large-scale skin cancer datasets, enabling them to capture complex features and patterns specific to skin lesions. By combining these models through the max voting ensemble technique, we aim to harness the diversity and complementarity of their predictive abilities, leading to improved generalization and discrimination power.

Our approach integrates multiple deep learning models, each selected for its proven strengths in image classification and feature extraction. Models such as Xception, ResNet50V2, and InceptionResNetV2 are renowned for their robustness in handling complex image data. This diverse selection ensures a comprehensive analysis of skin lesions, capturing a wide range of features from color and texture to shape and size variations. The synergy of these architectures enables a detailed analysis of dermatological images. The models underwent extensive training with augmented datasets, fine-tuning hyperparameters to balance bias and variance effectively, ensuring that the models are adept at identifying the subtleties of skin cancer lesions.

The max voting ensemble technique combines predictions from multiple deep learning models to derive a diagnosis. This technique harnesses the collective intelligence of the models. By aggregating the outputs, the ensemble approach effectively mitigates individual model biases and errors, leading to a more robust and reliable diagnostic tool. The Max Voting Ensemble is particularly adept at handling diverse and challenging skin lesion cases, offering adaptability and accuracy in various clinical scenarios. This adaptability is crucial in addressing the wide variability seen in skin cancer presentations. Empirical validation of our ensemble model underscores its superiority over individual models. The ensemble consistently demonstrated higher accuracy and diagnostic precision in our tests, which included a comprehensive comparison with the performance of standalone deep learning models.

Our research contributes to the ongoing evolution of artificial intelligence in medical diagnostics by addressing the demand for enhanced detection methods in dermatological health by examining the unexplored terrain of merging these advanced models. Our proposed noble approach holds the potential to transform the field of skin cancer identification and categorization, contributing to early diagnosis and more effective treatment strategies. With the integration of cutting-edge pre-trained deep learning models and the power of ensemble techniques, we aim to make a meaningful impact on healthcare and lead the way towards improved patient care and outcomes.

This research makes four significant contributions to the field of skin cancer detection and classification.

- Firstly, it successfully utilizes the max voting ensemble technique to effectively distinguish between benign and malignant skin cancer cases. This approach is notable for its ability to amalgamate the predictions of multiple models, thereby enhancing accuracy and demonstrating superior performance compared to individual models.

- Secondly, the study offers a comprehensive analysis of various deep learning models used for skin cancer classification. This analysis is rigorously assessed using the area under the receiver operating characteristic (ROC) curve (AUC), providing valuable insights into the effectiveness of these models in accurate disease identification.
- Furthermore, the ensemble method developed in this study not only surpasses the accuracy of single models but also showcases remarkable robustness and broad applicability across different datasets. This is evidenced by its ability to detect malignant cases with high precision in varied scenarios.
- In addition, the study introduces an innovative algorithm based on the max voting ensemble technique, integrating ten pre-trained individual models. These models collectively exhibit robustness and reliability, significantly outperforming most comparable alternatives. The efficacy of this ensemble approach is exemplified by achieving an AUC score of 0.932, underscoring its potential in medical diagnostics.

The results of this study can help guide the development of more accurate and efficient skin cancer detection and classification algorithms, which can ultimately lead to improved patient outcomes and survival rates. Beyond its immediate applications in skin cancer detection, our work carries broader implications for the field of medical image analysis. The successful integration of ensemble techniques could be extended to various medical imaging modalities and disease classifications. By leveraging the collective intelligence of multiple models, we can enhance the accuracy and reliability of diagnostic tools across diverse healthcare domains.

The organization of the remainder of this paper is as follows: the subsequent section presents an extensive review of the literature. Then, we introduce the materials and methods, describing the dataset, model architectures, and experimental setup. Subsequently, we present the results and analysis, showcasing the AUC scores of individual models and the Max Voting Ensemble. Finally, we discuss the ramifications of our findings, highlighting the significance of ensemble techniques in medical image analysis, particularly for skin cancer detection, and conclude with future research directions in this promising field.

2. Literature Review

Skin cancer ranks as the most prevalent form of cancer in the United States, accounting for approximately 5.4 million diagnosed cases each year. Skin cancer is often curable, but effective treatment is contingent on early detection. Delayed diagnosis and treatment can result in adverse outcomes, such as increased rates of metastasis, as well as heightened morbidity and mortality. Conventional approaches to skin cancer diagnosis, including visual inspection and biopsy, are characterized by subjectivity, time-intensive procedures, and the potential for significant inter-observer variation. Consequently, there has been a rising interest in leveraging machine learning methods, notably deep learning techniques, to enhance the precision and efficiency of skin cancer diagnosis. Over the past few years, deep learning methods [26], particularly convolutional neural networks (CNNs) [17,23,26–28], have demonstrated significant potential for accurately identifying and categorizing skin cancer from medical images of skin lesions [7,16,24,25]. In this review of relevant literature, we explore recent studies focusing on the detection and classification of skin cancer through the application of deep learning and transfer learning methodologies.

The advancements in deep learning techniques have spurred considerable progress in skin cancer detection. Recent review articles [29–33] explore key contributions in this domain, with a focus on the application of deep learning methods. For example, the comprehensive review by Dildar et al. (2021) [33] provides a contemporary overview of skin cancer detection leveraging deep learning techniques. The authors systematically analyze the latest advancements and methodologies in the field, offering valuable insights into the evolving landscape of skin cancer diagnostics. Brinker et al.'s [29] systematic review focuses on the application of convolutional neural networks (CNNs) for skin cancer classification. The paper critically assesses the state of the art in CNN-based skin cancer diagnosis, providing a comprehensive synthesis of existing literature and highlighting key trends in this rapidly evolving field. Adegun and Viriri (2021) conduct an in-depth exploration of deep learning techniques applied to skin lesion analysis and melanoma cancer detection [30]. Their survey not only summarizes existing methodologies but also critically examines the state of the art, offering insights into challenges and opportunities for further research in this domain. Munir et al.'s bibliographic review [31], featured in *Cancers*, presents a holistic examination of cancer diagnosis using deep learning, encompassing various cancer types, including skin cancer. The review consolidates knowledge from diverse studies, shedding light on the broad applications of deep learning in cancer diagnosis and underscoring its potential impact on improving diagnostic accuracy. Li et al. [32] conscientiously discuss the challenges faced in the application of deep learning to skin disease diagnosis. From data limitations to interpretability issues, the authors provide a balanced perspective on the obstacles that researchers and practitioners must navigate. Importantly, the review concludes with insights into potential future research directions, guiding the trajectory of advancements in this evolving field.

A number of recent studies of ensemble deep learning for biomedical imaging have advanced the field [34,35]. A case in point is Shokouhifar et al. [34], which used swarm intelligence empowered three-stage ensemble deep learning to measure arm volume as a difference in arm volume is an indicator of the presence of and change in the status of lymphedema. Another prime example is Bao et al. [35], who utilized integrated stack-ensemble deep learning to enhance the preoperative prediction of prostate cancer Gleason grade.

Numerous investigations have delved into the utilization of CNNs for the detection and categorization of skin cancer over the past two decades [7,32,35,36]. As an illustration, Esteva et al. (2017) trained a CNN model on a dataset comprising more than 129,000 clinical images to identify skin cancer. This model achieved a sensitivity of 95%, a specificity of 85%, and classification accuracy of 91% for melanoma detection, a performance level on par with that of dermatologists [23]. Likewise, Tschandl et al. (2018) employed a CNN model for the classification of skin lesions into benign or malignant categories, achieving an Area Under the Curve (AUC) score of 0.94 on an independent test dataset [37]. Brinker et al. (2019) conducted an assessment of various CNN models to detect melanoma, the most lethal type of skin cancer. Their findings highlighted a substantial enhancement in classification accuracy through the application of transfer learning with pre-trained CNN models [29].

Transfer learning, which encompasses the utilization of pre-trained CNN models to extract features from medical images, has also demonstrated an enhancement in classification accuracy when applied to the detection of skin cancer [23–25]. Han et al. (2018) developed a transfer learning-based CNN model that achieved a classification accuracy of 89.1% for melanoma detection using dermoscopic images [38]. Haenssle et al. (2018) used transfer learning to fine-tune a pre-trained VGG-19 model on a dataset of dermoscopic images to detect melanoma, achieving an AUC of 0.86 on an independent test set [38]. Codella et al. (2018) similarly used transfer learning to fine-tune a pre-trained Inception-v3 model on a dataset of dermoscopic images, achieving an AUC of 0.93 on an independent test set [39].

While numerous studies have showcased the potential of deep learning techniques for skin cancer detection and classification [29–33], it is necessary to compare the performance of various CNN models dedicated to this task. Comparative studies have focused specifically on comparing the performance of different CNN models for skin cancer detection and classification [7,9,14,40,41]. For example, Codella et al. (2018) compared the performance of three different CNN models (Inception-V3, ResNet50, and DenseNet-121) for classifying skin lesions as either benign or malignant. The authors found that DenseNet-121 achieved the highest classification accuracy, with an area under the receiver operating characteristic curve (AUC-ROC) of 0.91 [39,42].

Kawahara et al. (2018) used a pre-trained CNN model to extract features from skin lesion images and then trained a Support Vector Machine (SVM) classifier to distinguish between benign and malignant lesions. The authors found that their SVM classifier achieved an accuracy of 83.6%, which outperformed several other classification methods [43]. Brinker et al. (2019) compared the performance of five different CNN models on a dataset of dermoscopic images, finding that the Inception-v3 model achieved the highest AUC of 0.90 [29]. Patel et al.'s (2021) study suggests that transfer learning can improve the performance of CNN models for skin cancer detection and classification. Their findings also indicate that the InceptionV3 model may be particularly effective for this task [44]. Zhang et al. (2020) similarly compared the performance of six different CNN models on a dataset of dermoscopic images, finding that the DenseNet-121 model achieved the highest AUC of 0.95 [45]. Zaidan et al. (2021) reviewed the use of deep learning techniques for skin cancer detection and classification in their article published in the Journal of Healthcare Engineering. They discussed the effectiveness of different CNN models and transfer learning methods and the need for further research on larger and more diverse datasets [33].

Recent research endeavors have additionally investigated the application of alternative deep learning methodologies, including Generative Adversarial Networks (GANs), in the context of skin cancer detection and classification [46–48]. An illustrative example is Bi et al. (2020), who employed a GAN to generate synthetic skin lesion images and subsequently harnessed a CNN for their classification into benign or malignant categories. The researchers reported that their GAN-CNN model achieved a classification accuracy of 83% [49].

More recently, Guergueb and Akhloufi (2022) [50] used ensemble learning to achieve a predictive accuracy of just under 98% for melanoma disease, the most dangerous form of skin cancer. The authors used only one image dataset for training and testing and so the generalizability of their model to other datasets and other forms of skin cancer is untested. Avanija et al. (2023) [51] recorded an accuracy rate of 86% on the ISIC Skin Cancer Dataset using an ensemble learning approach harnessing three deep learning algorithms, namely, VGG16, CapsNet, and ResUNet. Even more recently, Sethanan et al. (2023) [8] report a cancer detection rate of 99.7% and cancer classification rates of approximately 96%. Their ensemble model harnesses modified CNN architectures, refined image segmentation techniques, and an artificial multiple intelligence system algorithm for optimized decision fusion. The authors, however, note that further research is needed to establish the generalizability, robustness, and clinical applicability of their model.

The max voting ensemble technique is a powerful approach within ensemble learning [8,34,35]. It aggregates predictions from multiple models and selects the class with the highest frequency of votes as the final prediction. This technique leverages the wisdom of crowds to arrive at a more accurate and stable prediction [52,53]. In the context of skin cancer detection, the max voting ensemble technique presents an opportunity to harness the collective intelligence of deep learning models and enhance classification outcomes [54]. For example, the study by Kausar et al. (2021) introduces deep learning-based ensemble models, achieving accuracies up to 91.8% using an individual-based model. However, using ensemble techniques boosted accuracy to 98% and 98.6%. The proposed models outperformed recent approaches, offering significant potential for enhanced multiclass skin cancer classification [55].

In summary, deep learning techniques, particularly CNNs and transfer learning, have shown great promise in accurately detecting and classifying skin cancer from medical images. Numerous studies have showcased the efficacy of CNN models for skin cancer detection and classification, and comparative studies have shown that certain CNN models, such as DenseNet-121 and Xception, achieve higher accuracy than others [56]. Transfer learning has also been shown to improve classification accuracy for skin cancer detection. However, further research is needed to evaluate the performance of different CNN models and transfer learning methods for skin cancer detection and classification on larger and more diverse datasets.

Table 1 below provides an overview of recent research conducted on diverse skin cancer datasets.

Table 1. Recently conducted research on various skin cancer datasets.

Authors and Paper	Dataset	Model	Published Year	Performance
Gajera et al. [57]	ISIC 2016, 2017, PH2, HAM10000	AlexNet, VGG16, VGG19	2023	Accuracy = 98.33%, F1 score = 96%
Alenezi et al. [58]	ISIC 2017, HAM10000	deep residual network	2023	Accuracy = 96.97%
Inthiyaz et al. [59]	Xiangya-Derm	CNN	2023	AUC = 0.87
Alwakid et al. [60]	HAM10000	CNN, ResNet50	2023	F1-score = 0.859 (CNN), 0.852 (ResNet50)
Alenezi et al. [61]	ISIC 2019, 2020	ResNet-101 and SVM	2023	Accuracy = 96.15% (ISIC19), 97.15% (ISIC20)%
Abbas and Gul [62]	ISIC 2020	NASNet	2022	Accuracy = 97.7%, F1-score = 0.97%
Abdar et al. [12]	ISIC 2019	ResNet15V2, MobileNetV2	2021	Accuracy = 89%, F1-score = 0.91
Jain et al. [13]	HAM10000	Xception, InceptionV3, VGG19, ResNet50, and MobileNet	2021	Accuracy = 90.48% (Xception)
Aljohani and Turki [14]	ISIC 2019	Xception, DenseNet201, ResNet50V2, MobileNetV2, VGG16, VGG19, and GoogleNet	2022	Accuracy = 76.09%
Bechelli and Delhommelle [63]	HAM10000	CNN, VGG16, Xception, ResNet50	2022	Accuracy = 88% (VGG16)
Demir et al. [64]	ISIC archive	ResNet101 and InceptionV3	2019	F1-score = 84.09% (ResNet101) and 87.42% (InceptionV3)
Rashid et al. [65]	ISIC 2020	MobileNetV2	2022	accuracy = 98.20%
Reis et al. [66]	HAM10000, ISIC 2019, 2020	InSiNet, U-Net	2022	Accuracy= 94.59% (HAM10000), 91.89% (ISIC2019), and 90.54% (ISIC2020)
Khan et al. [67]	ISBI 16, 17, 18, PH2, HAM10000	ResNet101, DenseNet201	2021	Accuracy = 98.70% (PH2), Accuracy = 98.70% (HAM10000)
Khan et al. [68]	ISBI 2018,	A hybrid model	2021	Accuracy = 92.70%
Kaggle Compt. [69]	ISIC 2018	Top 10 model Average	2020	Accuracy =86.7%
Gouda et al. [70]	ISIC 2018	CNN	2022	Accuracy = 83.2%

In Table 1, it is evident that researchers have conducted experiments utilizing a diverse assortment of pre-trained deep learning models across multiple skin cancer datasets, with a particular emphasis on ISIC 16, 17, 18, 19, 20, PH2, and HAM10000. These investigations have yielded a spectrum of accuracy scores. Notably, all datasets except ISIC 2018: Task 1-2 have demonstrated accuracy ranging from 95% to 99%. Conversely, ISIC 2018 falls short of this range with an accuracy below 92%. Consequently, our research focuses on the ISIC 2018: Task 1-2 dataset, aiming to enhance its accuracy to a level that meets acceptable standards.

3. Proposed Methodology

Skin cancer detection is a crucial task in medical diagnostics. This approach delineates an innovative method that harnesses cutting-edge pre-trained deep learning models within an ensemble framework, aimed at achieving precise skin cancer detection. The methodology entails the amalgamation of multiple pre-trained convolutional neural network (CNN) models through a maximum voting ensemble technique, thereby augmenting the overall classification performance.

3.1. Abstract View of Proposed Model

The primary objective of this research is to investigate whether the proposed Max Voting-based ensemble model can more accurately distinguish between benign and malignant skin lesions than the individual models. Figure 2 shows the abstract view of the proposed Max Voting-based methodology using pre-trained models.

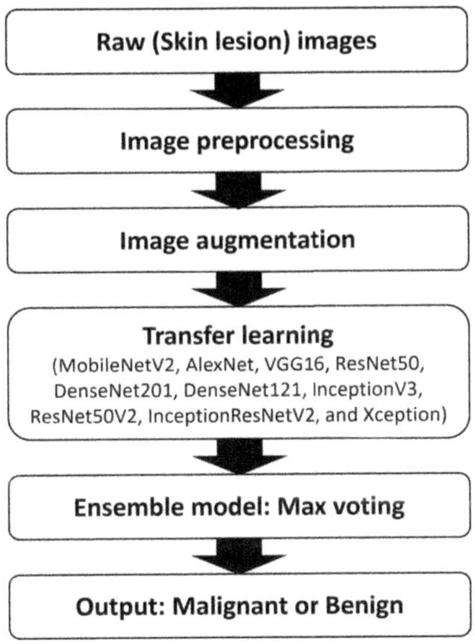

Figure 2. Abstract view of proposed Max Voting-based skin cancer classification.

The essential stages of the method we propose are outlined below:

1. Dataset: Select a comprehensive skin cancer dataset containing a variety of benign and malignant skin lesion images. Common datasets including ISIC (International Skin Imaging Collaboration) are used to conduct the experiment.
2. Preprocessing: Ensure proper data preprocessing, including resizing, normalization, and data augmentation.
3. Data augmentation: During training, data augmentation techniques are employed to enhance learning outcomes and prevent overfitting.
4. Pre-trained Models: Choose a set of diverse pre-trained deep learning models for feature extraction. Models like MobileNetV2, ResNet50, InceptionV3, DenseNet201, and Xception are ideal due to their strong performance on image classification tasks. For each input image, extract features from intermediate layers of the selected pre-trained models. These features capture high-level representations of the image content.
5. Ensemble Construction: Implement a maximum voting ensemble technique to combine the predictions of individual models. For each test image, the ensemble generates predictions based on the majority vote from predictions made by the individual models.
6. Assess the ensemble's performance on the test set using metrics including accuracy, precision, recall, F1-score, and ROC-AUC.

3.2. Detailed View of Proposed Model

The process of detection of benign and malignant lesions starts with image data collection, followed by steps such as image preprocessing, data augmentation, feature extraction through pre-trained models, and finally, classification by combining the outputs of individual pre-trained models through the proposed max voting-based ensemble model. Transfer learning was used in our models to reduce the computation time and encourage better results and performance.

This section outlines the detailed methodology we used to train and test our CNN models and predict the final outcome using the proposed max voting-based model as shown in Figure 3.

Figure 3 depicts the overall process of the proposed model for a skin lesion detection system. This process employs a max voting ensemble technique for classifying skin cancer using pre-trained deep-learning models. It begins by selecting a range of models, preprocessing input images, loading the chosen models, generating predictions for each model, aggregating predictions via majority voting, evaluating ensemble accuracy, and finally displaying the results. Algorithm 1 illustrates the data flow diagram of the model we propose, shown in Figure 3.

Algorithm 1 Max Voting Ensemble Technique for Skin Cancer Classification

Input:- List of pre-trained deep learning models (SelectedModels)
List of skin lesion images (InputImages)
Corresponding ground truth labels (groundTruthLabels)
Output: Ensemble predictions for each input image (ensemble predictions)

Step 1: Model Selection
SelectedModels $\leftarrow [MobileNetV2, AlexNet, VGG16, ResNet50, DenseNet201, DenseNet121, InceptionV3, ResNet50V2, InceptionResNetV2, Xception]$

Step 2: Preprocessing
while $InputImages \neq eachImage$ **do**
  $PreprocessedImage \leftarrow preprocessImage(image)$
  $PreprocessedImages.append(PreprocessedImage)$

Step 3: Load Pre-trained Models
$LoadedModels \leftarrow loadModels(SelectedModels)$

Step 4: Generate Predictions
while $eachpreprocessed \neq PreprocessedImages$ **do**
  $predictions \leftarrow []$
  while $eachloadedmodel \neq LoadedModels$ **do**
    $ModelPrediction \leftarrow PredictImage(LoadedModel, PreprocessedImage)$
    $predictions.append(ModelPrediction)$
  $individualPredictions.append(predictions)$

Step 5: Max Voting Ensemble Technique
while $eachIndividualPredictions \neq individualPredictions$ **do**
  $aggregatedPredictions \leftarrow []$
  while $eachImagePrediction \neq setofpredictions$ **do**
    $majorityVote \leftarrow calculateMajorityVote(imagePrediction)$
    $aggregatedPredictions.append(majorityVote)$
  $ensemblePredictions.append(aggregatedPredictions)$

Step 6: Evaluate Ensemble Performance
$accuracy \leftarrow evaluateAccuracy(ensemblePredictions, groundTruthLabels)$
Step 7: Display Results

$print("EnsembleAccuracy: ", accuracy)$

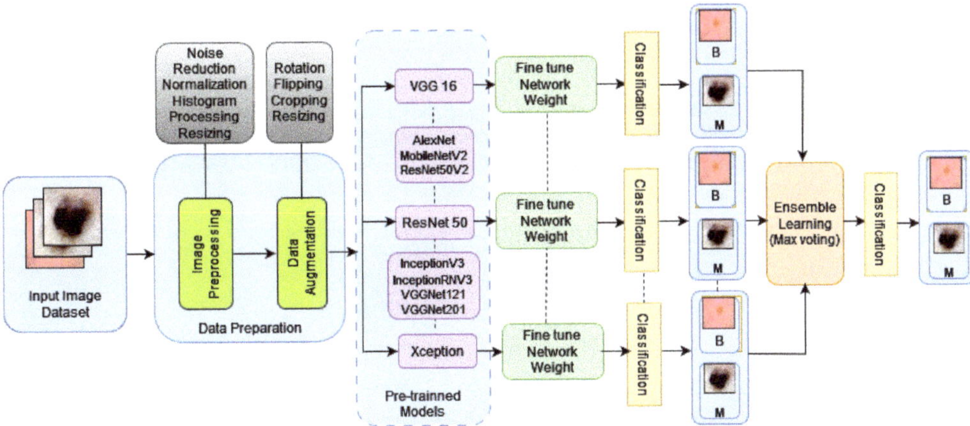

Figure 3. Overall workflow architecture of the proposed max voting-based ensemble technique.

The max voting ensemble technique is a powerful and reliable method for skin cancer classification. The integration of pre-trained deep learning models aids in accurate and early diagnosis, contributes to improved patient outcomes and reduced mortality rates, and holds promising implications for medical image analysis and other diagnostic tasks.

This algorithm provides a conceptual overview of the approach outlined in the abstract. The implementation details, such as the specific code for loading models and conducting experiments, would require further elaboration based on the deep learning framework and tools being used.

3.3. *ISIC 2018 Dataset: Task 1-2*

The ISIC 2018: Task 1-2 Skin Cancer Challenge Dataset publicly available at the ISIC Archive https://challenge.isic-archive.com/data/#2018 (accessed on 2 October 2023) [71] comprises an extensive compilation of dermoscopic images depicting skin lesions. These images are obtained from various sources and cover a range of skin conditions, including both benign and malignant cases. The dataset includes images with different lighting conditions, scales, and levels of image quality. The ISIC 2018 challenge is divided into three distinct tasks: Task 1 focuses on Lesion Segmentation, Task 2 involves Lesion Attribute Detection, and Task 3 addresses Disease Classification. The ISIC Challenge 2018: Task 1-2 datasets are systematically categorized into three segments, encompassing both benign and malignant cases. The Training dataset comprises 2597 images, with 2068 benign images, 519 malignant images, and 7 undetermined images. The Testing dataset includes 1000 images, featuring 794 benign images and 204 malignant images. The Validation dataset consists of 100 images. Figure 4 visually presents sample images depicting both malignant and benign cases from the ISIC 2018 Dataset of Task 1-2.

Here is a concise overview of the primary features of the ISIC 2018: Task 1-2 dataset:

- Annotations: Each image in the dataset is annotated with ground truth labels that indicate the type of skin lesion it depicts. These labels are typically categorized into classes such as melanoma, nevus, basal cell carcinoma, and others.
- Purpose: The dataset was released in conjunction with the ISIC Skin Cancer Detection Challenge in 2018. It was intended to serve as a benchmark for evaluating and comparing machine learning algorithms for skin cancer diagnosis.
- Usage: Researchers and participants in the challenge use this dataset to develop and train machine learning models, particularly deep learning models like convolutional neural networks (CNNs), for the accurate classification of skin lesions into different categories.
- Challenges: The ISIC community often organizes challenges based on these datasets to foster innovation in the field of skin cancer detection. Participants submit their

algorithms for evaluation, and the results help advance the state of the art in automated skin lesion diagnosis.

- Availability: The dataset is made available to participants of the challenge and researchers interested in the field. It is often accessible through the ISIC website https://challenge.isic-archive.com/data/#2018 or challenge-specific platforms https://challenge.isic-archive.com/data/#2018 (accessed on 2 October 2023).

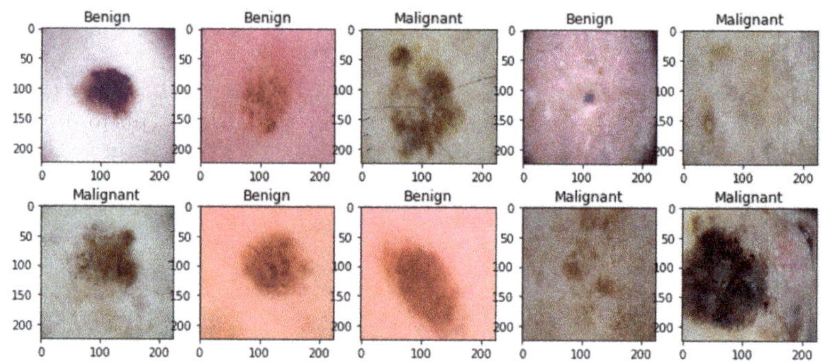

Figure 4. Skin Cancer—Malignant and benign sample images.

3.4. *ISIC 2018 Dataset: Task 3 (HAM10000)*

Discover the HAM10000 Skin Cancer Dataset [37]—a comprehensive collection of high-resolution images meticulously curated to aid in the early detection and diagnosis of various skin conditions. The HAM10000 dataset contains exactly 10,015 labelled images of seven skin disease classes: melanoma, melanocytic nevus, basal cell carcinoma, actinic keratosis, benign keratosis, dermatofibroma, and vascular lesion. It is possible to unleash the power of AI and dermatology with this invaluable resource, advancing skin cancer research and improving patient care worldwide. In this study, we performed experiments on the HAM10000 dataset to validate and extend the applicability of our proposed max voting-based ensemble technique to the ISIC 2018 dataset [71].

3.5. *Image Preprocessing*

Image preprocessing is the initial step where raw images are transformed to improve their quality and suitability for analysis. Preprocessing techniques for skin cancer classification include:

- Resizing and Cropping: Images are resized to a consistent resolution to ensure uniformity during model training.
- Rescale: Rescaling an image by dividing all pixel values by 255 is a common preprocessing step in image analysis and deep learning. The rescaling operation is particularly important when working with images represented in the RGB color space, where pixel values can range from 0 to 255 for each color channel.
- Normalization: Pixel values are normalized to a common range, such as [0, 1] through rescaling, to improve convergence during model training.
- Color Correction: Adjusting color balance and correcting lighting variations to reduce the impact of different acquisition conditions.
- Noise Reduction: Applying filters to reduce noise and enhance image clarity.

Table 2 summarizes the preprocessing required to conduct this research.

Table 2. Preprocessing operations for test dataset.

Processing	Mode (Range)
Resizing	224 × 224
Rescale	1.0/255
Normalize	[0, 1]

3.6. Data Augmentation

To enhance the performance and generalization capabilities of our image categorization model, we employed data augmentation techniques to artificially increase the diversity of training data by applying various transformations to the existing data. The goal is to improve the generalization and robustness of a machine learning model. This involved the strategic alteration of existing images in the training dataset. The augmentation process was carried out with careful consideration of several transformations, inspired by foundational works in the field [72–75]. The augmentation operations applied to the training dataset are summarized in Table 3. These operations were chosen based on their effectiveness in diversifying the dataset and improving the robustness of the model.

Table 3. Augmentation operations for training set.

Processing	Mode (Range)
Rescale	1.0/255
Rotation range	0.2
Width shift range	0.2
Shear range	0.3
Zoom range	0.3
Fill mode	Nearest
Horizontal flip	True

Our data augmentation strategy aligns with the findings and recommendations of key works in the literature. Mikołajczyk and Grochowski (2018) explored the benefits of data augmentation for improving deep learning in image classification problems, emphasizing the importance of exposing models to diverse data variations [72]. Perez and Wang (2017) further validated the effectiveness of data augmentation in image classification using deep learning through their comprehensive study [73]. In addition to these studies, the research report by Luis and Jason from Stanford University (2017) provided valuable insights into the practical effectiveness of data augmentation techniques in the context of image classification [74]. The survey by Shorten and Khoshgoftaar (2019) offered a broader perspective, summarizing various image data augmentation approaches and their impact on deep learning [75].

Our decision to employ data augmentation stems from its three-fold advantage. Firstly, it significantly expands the effective size of the training dataset, a particularly valuable asset when working with constrained data resources. Secondly, the exposure of the model to diverse variations in the training data enhances its ability to generalize to unseen scenarios. Thirdly, data augmentation acts as a regularization technique, mitigating overfitting risks by introducing controlled noise and diversity into the training data.

In conclusion, the augmentation strategy, informed by seminal works and a comprehensive survey, significantly contributes to the model's performance, reliability, and generalization capabilities.

3.7. Usage of Pre-Trained CNN Models

A convolutional neural network (CNN) is a dedicated neural network architecture that integrates both convolutional layers and conventional neural network layers. Convolution, an operation that involves filtering images to extract features, is employed to enhance the model's performance. Subsequently, a 3 × 3 MaxPooling filter is applied to obtain the most critical features. To enhance the model's effectiveness, batch normalization and dropout layers are implemented to normalize the values and prevent overfitting.

The output from these layers is presented in a multi-dimensional matrix, which requires conversion to a one-dimensional form for input into the dense layer. This is accomplished using a flattened layer. The model was trained using the Rectified Linear Unit (ReLU) activation function within the initial dense layer. This choice was made to facilitate computation, as ReLU allows for the activation of only a few neurons at a time. In the second dense layer, the sigmoid function was employed, as it outputs a value between 0 and 1. This choice of activation function was made based on the specific requirements of the model and its intended use case. If the result exceeds 0.5, the classification is deemed malignant; otherwise, it is categorized as benign.

ImageNet was used, which we could use to load a pre-trained version of the network trained on more than a million images from the ImageNet database. In our preprocessing step, the batch size for our dataset was set to 32 and the image dimension to 224 × 224. A categorical class mode for our training data, which consisted of 2637 images, was used, while our testing dataset had 660 images. We then plotted our preprocessed images. Next, a pre-trained Model function from the Keras Applications class was imported and assigned to the variable 'dense'. We specified "ImageNet" as the weights to be used and set the input shape to (224, 224, 3).

Figure 5 shows the flowchart depicting the steps involved in individual training and testing the model.

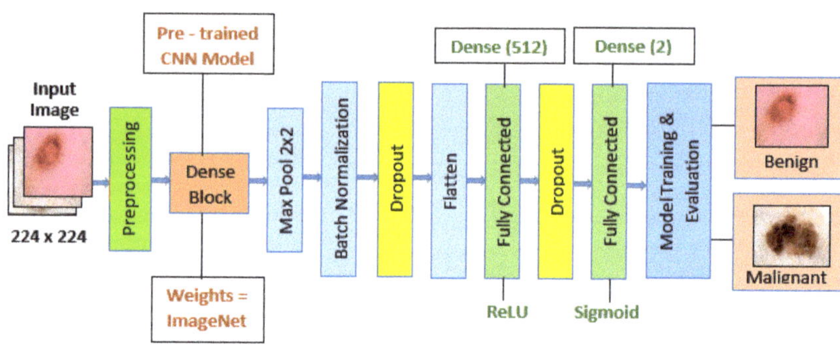

Figure 5. Flowchart depicting the steps involved in individual training and testing the models (Models: MobileNetV2, AlexNet, VGG16, ResNet50, DenseNet201, DenseNet121, InceptionV3, ResNet50V2, InceptionResNetV2, and Xception).

We formulated our model through a systematic layering approach within a sequential architecture, emphasizing the methodology without delving into the specifics of code functions. Firstly, we incorporated a pre-trained convolutional neural network (CNN) model, harnessing its feature extraction capabilities. This pre-trained model served as the foundation for our architecture. Subsequently, we applied max pooling with a 2 × 2 filter to facilitate spatial downsampling. This pooling operation allowed us to retain essential features while reducing computational complexity. By adhering to this structured methodology, we aimed to provide a comprehensive understanding of our model construction process, treating this document as an academic paper rather than a manual for code tools.

The subsequent subsection briefly outlines the pre-trained models employed in this study:

3.7.1. AlexNet

AlexNet is a trailblazing deep convolutional neural network architecture renowned for its revolutionary advancements in image classification endeavors. It comprises a series of convolutional and pooling layers, culminating in fully connected layers. AlexNet distinguishes itself through the pioneering incorporation of ReLU activation functions and dropout regularization, contributing significantly to the prevention of overfitting. Although originally designed for general image classification, its principles have been adapted for skin cancer classification. By learning hierarchical features from skin images, AlexNet demonstrates effectiveness in distinguishing between different types of skin lesions, aiding in accurate diagnosis and classification [76].

3.7.2. VGG16

VGG16 is a popular choice for tasks related to image recognition and classification. Its architecture comprises a total of 16 layers, encompassing 13 convolutional layers and 3 fully connected layers. Notably, all the convolutional layers employ a compact filter size of 3×3 with a stride of 1, facilitating precise feature localization within the input image. Additionally, the network incorporates max pooling layers with a 2×2 filter size to downsize the spatial dimensions of the output, effectively mitigating the risk of overfitting [77].

3.7.3. InceptionV3

InceptionV3 is a convolutional neural network architecture renowned for its prowess in image classification tasks. In this context, InceptionV3 serves as a feature extractor, with the last layer substituted by a fully connected layer coupled with a sigmoid activation function, enabling the prediction of skin cancer probabilities. Training involves the utilization of a binary cross-entropy loss function, with optimization achieved through stochastic gradient descent [78].

3.7.4. ResNet50

ResNet50 consists of 50 layers and is a variant of the ResNet architecture. It has a similar structure to other deep learning models. However, the unique feature of ResNet50 is its residual connection structure. The residual connections in ResNet50 help mitigate the problem of vanishing gradients by allowing the gradients to flow directly through the network. This enables the network to be trained deeper and more effectively, which has been shown to improve accuracy on image classification tasks [79].

3.7.5. ResNet50V2

ResNet50V2, short for Residual Network 50 version 2, is a specific variant of the ResNet architecture that was introduced as an improvement over the original ResNet-50 model. The ResNet-50v2 architecture follows the same fundamental principles as the original ResNet, including the use of residual blocks and skip connections. However, it incorporates some modifications to the architecture and training process.ResNet-50v2 incorporates several enhancements aimed at improving training efficiency, convergence speed, and overall performance. Due to these enhancements, ResNet-50v2 generally achieves better accuracy and convergence speed compared to the original ResNet-50, especially on tasks like image classification [80].

3.7.6. DenseNet121

DenseNet121 consists of 121 layers and is a variant of the DenseNet architecture. It has a similar structure to other deep learning models, with alternating convolutional and pooling layers, followed by a global average pooling layer and a fully connected layer for classification. However, the unique feature of DenseNet121 is its dense connectivity structure. The dense connectivity structure of DenseNet121 also enables it to efficiently learn complex features with fewer parameters compared to other deep learning models [81].

3.7.7. DenseNet201

DenseNet201 is a powerful neural network architecture commonly used for skin cancer classification tasks. It is an extension of the DenseNet family, designed to enhance feature reuse and information flow across layers. With its densely connected blocks and efficient gradient propagation, DenseNet201 excels at capturing intricate patterns in skin images, aiding the accurate classification of various skin cancer types. Its deep structure and skip connections make it well suited for complex medical image analysis tasks like skin cancer detection [81].

3.7.8. Xception

Xception shares similarities with InceptionV3 and has shown promising results for image classification tasks. One advantage of Xception is its ability to learn highly discriminative features with fewer parameters than other deep neural networks, which could lead to improved efficiency and reduced computational resources [82].

3.7.9. MobileNetV2

MobileNetV2 stands out as a neural network architecture meticulously crafted for swift and efficient image classification, particularly on mobile and resource-constrained devices. It capitalizes on depthwise separable convolutions and linear bottlenecks to curtail computational complexity without compromising performance. MobileNetV2 strikes an optimal equilibrium between model size and accuracy, rendering it apt for real-time diagnosis and classification, even on devices with restricted resources. Its architectural design aligns with applications like skin cancer classification [83].

3.7.10. InceptionResNetV2

InceptionResNetV2 is a sophisticated neural network architecture that combines elements from both Inception and ResNet designs. It leverages the advantages of both architectures, incorporating multi-scale feature extraction and residual connections. In skin cancer classification, InceptionResNetV2 excels at capturing intricate patterns within skin images and provides strong predictive capabilities. Its deep structure and complex modules make it well suited for detecting subtle characteristics indicative of various skin cancer types, enhancing classification accuracy and aiding in medical diagnoses [84,85]

The methodology employed in this study incorporates a diverse set of pre-trained deep learning models for skin cancer classification. Each model is selected based on its proven efficacy in image classification tasks and is fine-tuned to enhance its performance specifically for skin cancer diagnosis. The following fine-tuning and modifications are incorporated for the Skin Cancer Classification task:

1. Transfer Learning: The individual base model (e.g., Xception) is pre-trained on a large dataset for general image classification tasks, such as ImageNet. The knowledge gained from this pre-training is transferred to the skin cancer classification task.
2. Feature Extraction: Individual model (e.g., Xception) layer serves as a feature extractor, capturing hierarchical features from the input skin lesion images.
3. Additional Layers: Following the particular individual model (e.g., Xception) layer, additional layers such as convolutional, pooling, normalization, dropout, and dense layers are added. These layers contribute to further feature extraction, fine-tuning, and classification.
4. Dropout and Batch Normalization: Dropout layers are added to mitigate overfitting during training. Batch normalization is utilized for stable and accelerated training.
5. Output Layer: The final dense layer with two neurons (binary classification) is added for the skin cancer classification.

Table 4 shows the architectural summary of Xception (Functional) model along with parameters details as an example. The used Hyperparameters are—Learning Rate: 0.0001, Image Preprocessing Batch Size: 32, Varying Training Batch Sizes, Dropout Rates: 0.3 and 0.5, Optimizer: Adam, Loss Function: Categorical Cross-Entropy, callback Function: ReduceLROnPlateau with patience 1 and factor 0.5.

Table 4. Architecture of Xception (Functional) model.

Layer (Type)	Output Shape	Number of Parameter(s)
xception (Functional)	(None, 7, 7, 2048)	20,861,480
conv2d_14 (Conv2D)	(None, 5, 5, 32)	589,856
max_pooling2d_2 (MaxPooling 2D)	(None, 2, 2, 32)	0
batch_normalization_14 (BatchNormalization)	(None, 2, 2, 32)	128
dropout_6 (Dropout)	(None, 2, 2, 32)	0
flatten_2 (Flatten)	(None, 128)	0
dense_6 (Dense)	(None, 512)	66,048
dropout_7 (Dropout)	(None, 512)	0
dense_7 (Dense)	(None, 256)	131,328
dropout_8 (Dropout)	(None, 256)	0
dense_8 (Dense)	(None, 2)	514
Total Parameter:		21,649,354
Trainable Parameter:		21,594,762
Non-trainable Parameter:		54,592

In summary, for each of these models, the process of adaptation generally involves:

- Pre-training: The models are pre-trained on a large dataset (e.g., ImageNet) to learn generic features.
- Fine-tuning: The pre-trained models are fine-tuned on a skin cancer dataset to adapt them to the specific task.

3.8. Max Voting Mechanism

Majority-based voting, often referred to as plurality voting, constitutes a prevalent approach within ensemble classification [86]. In this context, the EnsembleVoteClassifier technique stands out for its capacity to enhance overall performance and bolster prediction robustness through the amalgamation of multiple machine learning models—specifically, neural network classifiers. The mechanism hinges on a majority voting strategy, where each individual model in the ensemble contributes its prediction, ultimately culminating in a final prediction decided by the collective majority vote among these models.

The proposed approach herein revolves around the application of ten aforementioned pre-trained classification models, namely, MobileNetV2, AlexNet, seven instances of VGG16, ResNet50, DenseNet201, DenseNet121, InceptionV3, ResNet50V2, Inception-ResNetV2, and Xception. To augment classification outcomes, this approach employs a majority-based voting mechanism. For each test instance, the classification results are independently computed by each of the specified models, and the ultimate output is predicted based on the outcomes that achieve majority representation [87–90]. In the context of majority voting, the class label y is forecasted by virtue of a majority (plurality) consensus reached by the individual classifiers, denoted as C.

$$\hat{\mathbf{y}} = \mathrm{mode}\{C_1(x), C_2(x), \ldots, C_m(x)\}$$

By following this comprehensive methodology, the proposed approach aims to provide a robust and accurate method for skin cancer detection using an ensemble of pre-trained deep learning models with a maximum voting scheme.

In summary, the max voting ensemble technique is chosen to address the limitations of individual models by combining their strengths, mitigating weaknesses, and providing a more robust and accurate solution for skin cancer detection. The ensemble approach capitalizes on the collective intelligence of multiple models to enhance diagnostic reliability and performance.

4. Experimental Evaluation

The subsequent subsection furnishes a comprehensive account of the experimental evaluation, elucidating the experimental setup, encompassing details such as hardware requirements, software environment, hyperparameters, and other relevant aspects.

4.1. Experimental Setup

The hardware environment utilized comprised Kaggle GPU resources, specifically T4x2, P100, and TPU: VM v3-8, each with a session duration of 12 hours, a disk capacity of 73 GB, RAM of 29 GB, and GPU memory of 16 GB. The software environments employed were Kaggle and Colab. The dataset utilized in this study was sourced from ISIC Challenge Datasets 2018: Task 1-2 [39,71], as obtaining skin lesion images from hospitals posed privacy and confidentiality concerns. For model training, 1440 benign and 1197 malignant skin lesion images from the ISIC Archive were employed. Model testing utilized 660 images, with 360 benign and 300 malignant images. Given the initial imbalance in the original image from the ISIC Archive https://challenge.isic-archive.com/data/#2018 (accessed on 2 October 2023), necessary adjustments were made to create a balanced set for both the training and testing sets. The original images were resized to 224 by 224, a measure taken to enhance model speed and performance. For ISIC 2018: Task 3 (HAM10000) Datasets, which are extremely imbalanced, we perform an oversampling operation to overcome the class imbalance problem and to ensure all the classes contain equal amount of images. Then, we employ Hold-Out Cross-Validation. The dataset is divided into training, validation, and test sets using an 80-10-10 split, with 80% of the data allocated to the training set, 10% to the test set, and 10% to the validation set.

Leveraging metrics such as training accuracy, training loss, validation accuracy, validation loss, true positive, true negative, false positive, false negative, precision, recall, and f1-score, we assessed the outcomes of our training process for the compiled model. The optimization was performed using the Adam optimizer with a learning rate set to 0.0001, and the loss function employed was categorical cross-entropy. We compiled the model with 11.6 million trainable and 85,000 non-trainable parameters for the DenseNet-121 model.

In our callback function, patience one and factor 0.5 for the ReduceLROnPlateau function were used. The model was then trained using both the train and validation datasets. As cross-validation was utilized, the test dataset was used as our validation dataset. For the DenseNet-121 and Xception models, we increased the training batch size to 64 and set the number of epochs to 20, which yielded improved accuracies. However, we did not specify any training batch size for other models and ran for 20 epochs.

Following the completion of training, the results of training and validation were gathered as seen in Tables 5 and 6, respectively.Our model underwent evaluation using the test dataset and achieved accuracy consistent with the validation set. The results encompassed accuracy, precision, recall, F1-score, and loss metrics. Additionally, Mathews Correlation Coefficient, true positive, true negative, false positive, and false negative values were computed to assess the model's capability. Accuracy, precision, recall, and F1-score are typically presented as percentages, providing a comprehensive evaluation of the model's performance. On the other hand, the loss value is expressed as a scalar, representing the extent of the model's error in predicting target values. Unlike other metrics, it is not constrained to a specific percentage range.

To evaluate the model's predictions on the test dataset, we generated a list of predicted outputs. Subsequently, we visualized the accuracy and loss of our model through graphs. Additionally, a confusion matrix was generated to offer further insights into the model's performance.

The hyperparameters used for the algorithm were a learning rate of 0.0001, an image preprocessing batch size of 32, and varying training batch sizes, with dropout rates of 0.3 and 0.5. The results obtained for all models are noted and compared in Tables 5 and 6, respectively, and it was found that the highest accuracy was achieved using the Xception model. To further investigate the impact of different variables on the accuracy of the

Xception model, several operations were performed and the resulting accuracy changes were recorded.

Table 5. Performance of the training dataset.

Metrics	MobileNet V2	AlexNet	vgg16	ResNet50	DenseNet 121	DenseNet 201	Inception V3	ResNet 50V2	Inception ResNetV2	Xception
Accuracy	88.14%	89.40%	93.70%	94.50%	95.10%	95.80%	95.37%	96.30%	97.20%	99.20%
Precision	87.74%	89.10%	93.20%	93.65%	94.55%	95.20%	94.44%	96.00%	96.44%	99.20%
Recall	88.44%	90.14%	94.17%	95.30%	95.90%	96.00%	96.02%	96.95%	98.10%	99.40%
F1	88.94%	89.20%	93.60%	94.30%	95.05%	95.70%	95.34%	96.21%	96.70%	99.20%
Loss	0.1715	0.1646	0.1555	0.1423	0.1306	0.1237	0.1142	0.0820	0.0703	0.0037

Table 6. Performance of the testing dataset.

Metrics	MobileNet V2	AlexNet	vgg16	ResNet50	DenseNet 121	DenseNet 201	Inception V3	ResNet 50V2	Inception ResNetV2	Xception
Accuracy	77.20%	80.10%	83.80%	86.05%	88.30%	88.80%	89.50%	89.30%	90.20%	91.90%
Precision	77.00%	79.80%	83.60%	85.65%	88.10%	88.20%	88.94%	89.10%	89.80%	91.20%
Recall	78.10%	81.00%	84.20%	86.95%	88.80%	89.20%	90.15%	90.10%	91.10%	92.10%
F1	77.10%	80.00%	83.70%	86.01%	88.10%	88.50%	89.47%	89.20%	90.10%	91.80%
Loss	0.3310	0.3156	0.3045	0.3023	0.2706	0.2657	0.2481	0.2427	0.1213	0.0943

As part of our experimentation, we sought to determine how changes to the Training batch size affected the accuracy of our model. By systematically varying the batch size and recording the corresponding accuracy, it was found that increasing the batch size to 64 resulted in the highest accuracy, with a recorded value of 86%, as seen in Figure 6a.

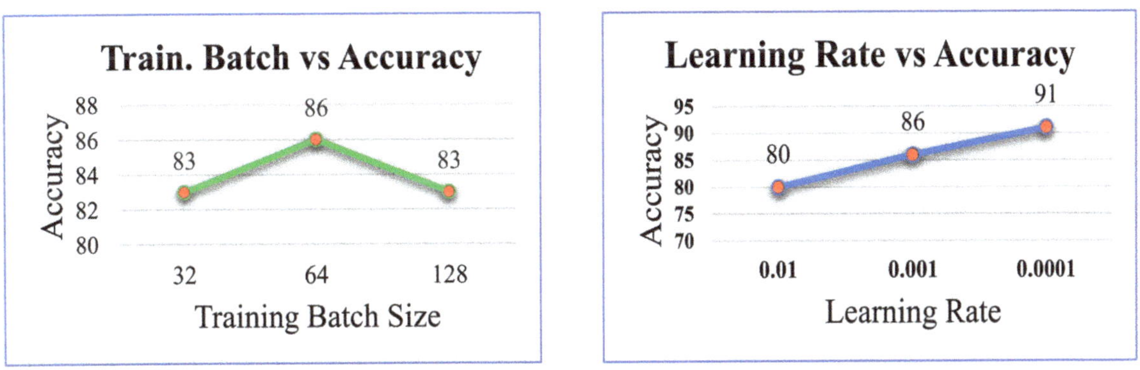

(**a**) Training Batch size vs. Accuracy (**b**) Learning Rate vs. Accuracy

Figure 6. Training Batch size and Learning Rate versus Accuracy.

For the next operation, the training batch size was fixed at 64 and the learning rate was varied to observe its effect on the model's accuracy. Through this operation, we found that the highest accuracy of 91% was obtained when using a learning rate of 0.0001, as seen in Figure 6b.

Subsequently, experiments were conducted to evaluate the impact of varying the image size on the model's performance. Keeping the learning rate and training batch size fixed at 0.0001 and 64, respectively, the model was evaluated using different image sizes. It was found that for both 244 and 300 sizes, the model achieved similar accuracy results of 91% as seen in Figure 7a. However, as a smaller image size takes less time to process, we decided to keep the initial image size of 244 for our further experiments.

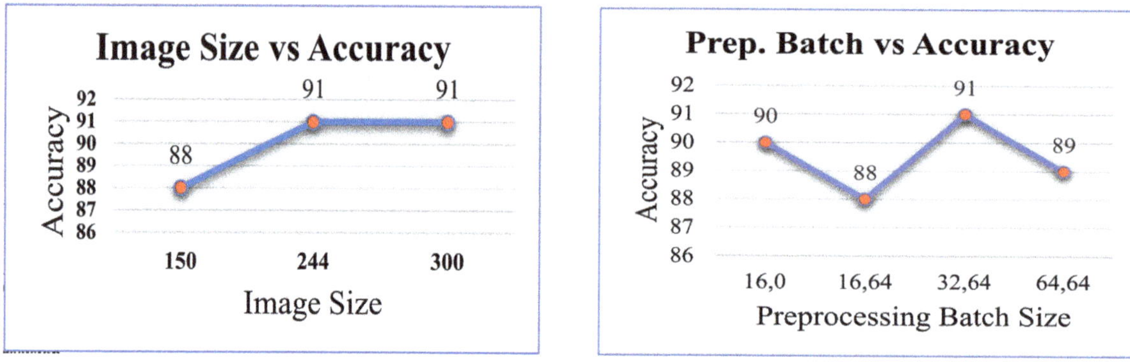

(**a**) Image size vs. accuracy (**b**) Preprocessing batch size vs. accuracy

Figure 7. Image size vs. accuracy and preprocessing batch size vs. accuracy.

In order to optimize our model's performance, we explored different combinations of inner and outer batch sizes while preprocessing the images. After testing various configurations, it was found that the initial combination of 32 image preprocessing and 64 training batch size resulted in the highest accuracy of 91% as seen in Figure 7b.

To further improve the model's performance and address the issue of overfitting, we experimented with different dropout rates as seen in Figure 8a. It was observed that using a lower rate for the first dropout and a higher rate for the second dropout helped overcome the overfitting problem. However, it was also found that increasing the dropout rate beyond a certain point led to a decrease in accuracy to some extent. These findings suggest that careful selection of dropout rates is crucial in achieving optimal model performance.

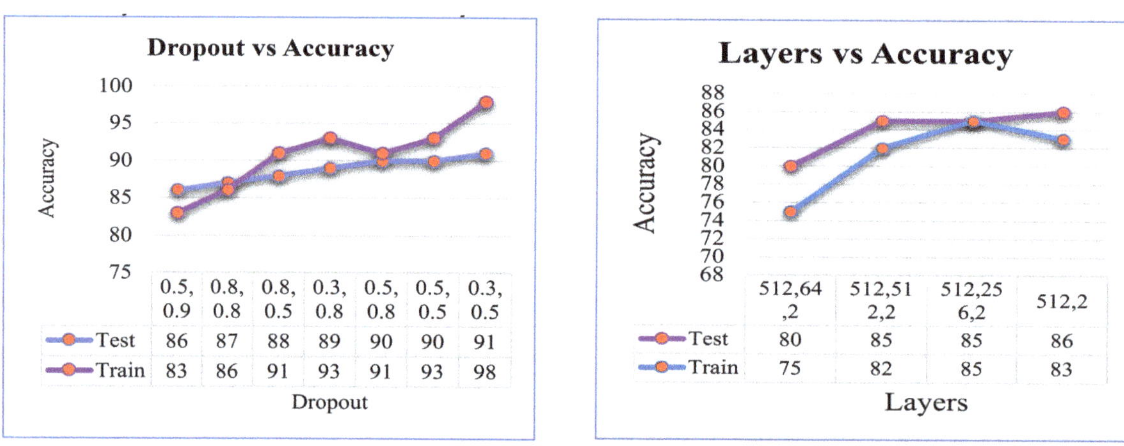

(**a**) Drop out vs. accuracy (**b**) Layers vs. Accuracy

Figure 8. Dropout vs. accuracy and layers vs. accuracy.

To improve the performance of our model, we added some more layers with dropout rates of 0.5 and 0.9, respectively, as seen in Figure 8b. Despite no overfitting issues, we did not observe any significant improvement in accuracy. In fact, increasing the number of layers resulted in a decrease in accuracy. There are several possible explanations for this, including diminishing returns, vanishing gradients, suboptimal hyperparameters, or limitations in the size or diversity of our dataset.

The aforementioned images correspond to the results obtained from running the Xception model for 30 epochs without specifying any training batch size. The identical model underwent several iterations with an unaltered dataset and a constant training batch size of 32 for 50, 100, 150, and 200 epochs, resulting nearly identical outputs throughout these runs.

4.2. Performance with ISIC 2018 Dataset: Task 1-2

In this study, following the application of various hyperparameter combinations, we evaluate the performance of ten distinct pre-trained deep learning models (MobileNetV2, AlexNet, VGG16, ResNet50, DenseNet201, DenseNet121, InceptionV3, ResNet50V2, InceptionResNetV2, and Xception). After fine-tuning all the hyperparameters, we assess the performance of the skin cancer classification method based on max voting through multiple evaluation metrics detailed in the subsequent subsections. We calculated the performance for both the training set and test set, which is noted in Table 7, and compared it with the proposed max voting ensemble technique.

The data aim to present the initial accuracies of various pre-trained CNN models on both the Training and Test sets. Each row in the dataset corresponds to a different model, and each column represents a specific dataset. The values in each cell of the table indicate the accuracy of the respective model on the corresponding dataset. For example, the accuracy of MobileNet V2 on the Train Set is 88.14%, and on the Test Set it is 77.20%. The accuracy of AlexNet on the Train Set is 89.40%, and on the Test Set it is 80.10%. Similarly, the percentages show the accuracy of each model on both the Train and Test sets.

Table 7. Initial accuracies of Train and Test set.

Set	Mobile-NetV2	AlexNet	vgg16	ResNet50	DenseNet 201	DenseNet 201	Inception V3	ResNet 50V2	Inception ResNetV2	Xception	Max Voting
Train	88.14%	89.40%	93.70%	94.50%	95.10%	95.80%	95.70%	96.30%	97.20%	99.90%	-
Test	77.20%	80.10%	83.80%	86.50%	88.30%	88.80%	88.70%	89.30%	90.20%	91.90%	93.18%

The table is meant to provide a quick overview of how well each model performs on the given datasets, allowing us to compare their performance easily. This information could be used to evaluate and choose the most suitable model for the given task based on its performance on both the Train and Test sets. The data in the table highlight that the ensemble technique based on max voting yields superior accuracy compared to the individual models.

4.2.1. Training and Validation Accuracy and Loss

Figure 9 illustrates the progression of training and validation accuracy across epochs for the Xception model, which stands as the highest-performing individual model within the group of ten. From the outset, the training accuracy surpasses the validation accuracy, and this trend remains consistent up to epoch 30. Between epochs 1 and 20, both training and validation accuracy gradually ascend, with training accuracy reaching 100% and validation accuracy reaching 91%. Subsequently, the training accuracy experiences a sharp ascent from epoch 1 to 3, undergoes a substantial increase from epoch 3 to 20, and then stabilizes at 100% accuracy beyond the 20th epoch, depicted by the green line in Figure 10a. Similarly, the validation accuracy exhibits a sharp increase from epoch 1 to 3 but undergoes substantial fluctuation from epoch 3 to 20, and ultimately stabilizes beyond the 20th epoch, achieving a 91% accuracy level, represented by the red line in the same Figure. Between epochs 20 and 30, both training and validation accuracy maintain stability. Ultimately, the classifier model attains its highest accuracy at epoch 30.

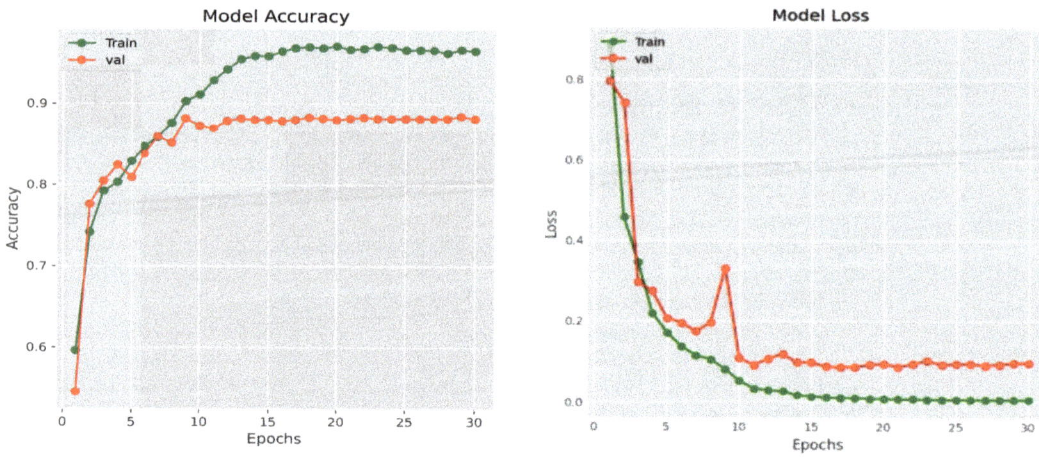

(**a**) Model accuracy (**b**) Model loss

Figure 9. Accuracy and Loss Curve of Xception Model for 30 Epochs.

In a similar vein, right from the start, the training loss consistently outperformed the validation loss, maintaining this pattern throughout all epochs. The training loss showed a swift decline from the initial epoch to the fourth, reaching a plateau at approximately 10 after the tenth epoch. In comparison, the validation loss experienced a rapid decrease from the first to the third epoch, followed by fluctuations between the fifth and tenth epochs, eventually settling at around 0.10, as depicted by the green and red lines in Figure 10b. Beyond the 30-epoch mark, the training loss converged to around 0.0037 while the validation loss stabilized at roughly 0.0943., with both values remaining constant.

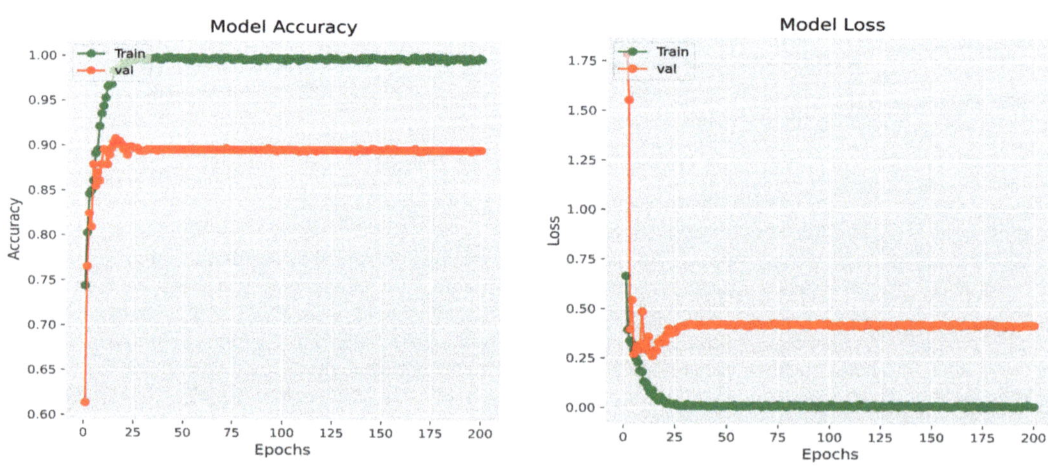

(**a**) Model accuracy for 200Epochs (**b**) Model loss for 200Epochs

Figure 10. Accuracy and Loss Curve of Xception Model for 200 Epochs.

The proposed system exhibited complete convergence by the 10th epoch, with the training loss hovering around 0 and the validation loss approximately at 0.10. These loss values remained nearly constant and exhibited a linear trend between epochs 10 and 30. The identical model underwent multiple executions utilizing the same dataset and a constant training batch size for 50, 100, 150, and 200 epochs, producing nearly identical outputs throughout these iterations. Figure 10 shows the accuracy and loss curve of Xception Model for 200 Epochs.

4.2.2. Receiver Operator Characteristic (ROC)

The Receiver Operating Characteristic (ROC) is a graphical representation used to assess the performance of a classification model. In Figure 11, we depict the ROC curve for each category in our skin cancer dataset, along with the corresponding Area Under the Curve (AUC) values. On the graph, the x-axis represents the false positive rate, while the y-axis represents the true positive rate. The AUC values associated with each category provide insight into the probability values for those respective categories. The diagonal curve signifies random class probability selection. Among the individual models, the highest AUC value achieved is 91.90%, while the lowest is 77.00%, recorded by the Xception and MobileNetV2 models, respectively, as shown in Figure 11. These values are particularly relevant for two categories, namely malignant, and benign.

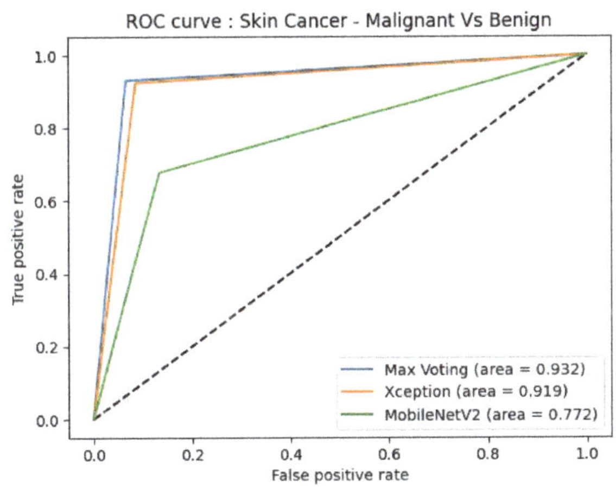

Figure 11. ROC Curve and AUC Scores for the MobileNetV2, Xception, and Max Voting.

Figure 11 unmistakably demonstrates that the proposed ensemble model, based on majority voting, surpasses the performance of the individual models by achieving an AUC score of 93.20%.

4.2.3. Error Analysis and Confusion Matrix (CM)

The confusion matrix (CM) serves as a tabular representation for summarizing a classifier's performance [91]. In Figure 12, we present the confusion matrices for three distinct models: MobileNetV2 (Figure 12a), Xception (Figure 12b), and our proposed max voting-based ensemble model (Figure 12c). These matrices offer a consolidated view of prediction outcomes, distinguishing correct from incorrect classifications for two distinct classes, benign and malignant.

Within the matrix, counts are used to depict the number of accurate and erroneous predictions, categorized by class. The primary purpose of this matrix is to visualize classification errors and confusion that may arise during predictions. Specifically, the diagonal cells signify the count of accurate predictions aligning with the actual class, while the off-diagonal cells denote misclassifications into other classes. For instance, in Figure 12c, we assess the Benign class across a total of 360 unlabeled images. Among these, 336 images are correctly predicted as belonging to the Benign class, as evidenced by the first diagonal cell in Figure 12c. Conversely, 24 images are misclassified as malignant class. Similarly, we evaluate the Malignant class across 300 unlabeled images. Here, 279 images are correctly predicted as members of the Malignant class, shown in the fourth diagonal cell of Figure 12c, while 21 images are erroneously classified as benign classes.

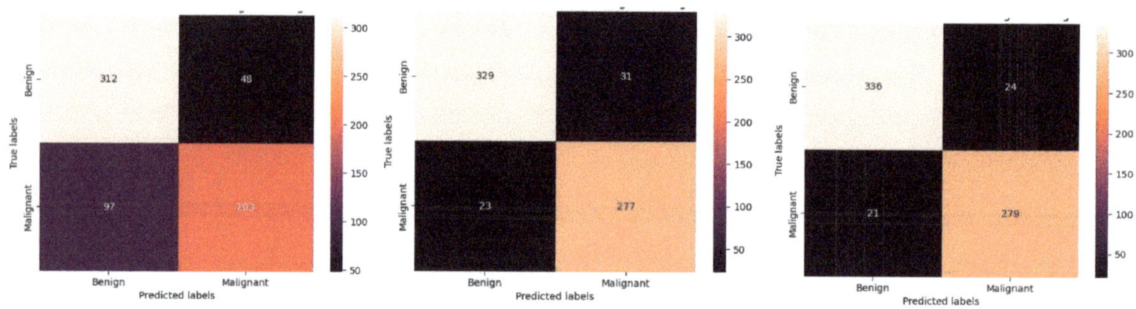

(**a**) Confusion Matrix: MobileNetV2 (**b**) Confusion Matrix: Xception (**c**) Confusion Matrix: Max Voting

Figure 12. Confusion Matrix: Individual Models (MobileNetV2, Xception) Versus Proposed (Max Voting-based Ensemble) Model.

4.2.4. Comparative Analysis of Individual Models versus Proposed (Max Voting) Model

The results of the skin cancer classification performance, measured by the area under the receiver operating characteristic (ROC) curve (AUC), for various deep learning models are depicted in Figure 13.

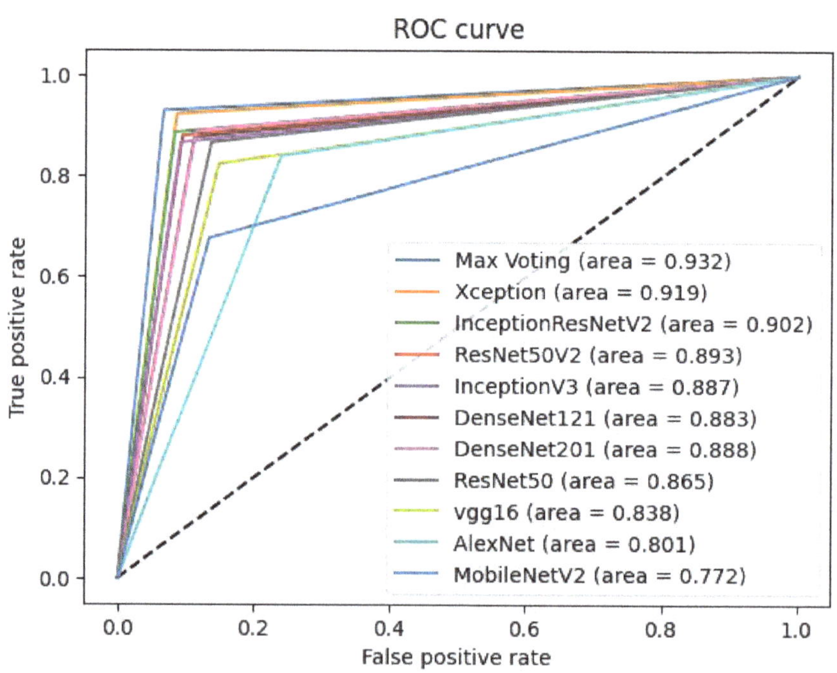

Figure 13. ROC Curve and AUC Scores for the models (Models: MobileNetV2, AlexNet, VGG16, ResNet50, DenseNet201, DenseNet201, InceptionV3, ResNet50V2, InceptionResNetV2, Xception and Max Voting).

The performance of various deep learning models for skin cancer classification is quantified using the area under the receiver operating characteristic (ROC) curve (AUC) in which a higher AUC score indicates better classification performance. Table 8 provides a summary of the AUC scores for each model.

Table 8. Summary of AUC Scores and Key Features of Deep Learning Models for ISIC 2018 dataset.

Model	AUC Score	Key Features
Max Voting	0.932	Ensemble Technique
Xception	0.919	Modified Inception Architecture
InceptionResNetV2	0.902	Inception + ResNet
ResNet50V2	0.893	Improved ResNet with Skip Connections
InceptionV3	0.887	Multi-level Feature Extraction
DenseNet121	0.883	Dense Connectivity
DenseNet201	0.888	Dense Connectivity
ResNet50	0.865	Skip Connections
VGG16	0.838	Classic Deep Learning Model
AlexNet	0.801	Pioneering Deep Learning Model
MobileNetV2	0.772	Lightweight Architecture

Three main similarities can be identified among these models, namely:

1. All models have AUC scores above 0.75, suggesting reasonably good classification abilities.
2. Architectural innovations, such as skip connections and dense connectivity, are common across most models.
3. Both ensemble and standalone models demonstrate strong performance.

Four differences may also be discerned. These are:

1. AUC scores range from 0.932 (Max Voting) to 0.772 (MobileNetV2), indicating different levels of effectiveness.
2. Computational complexity varies, affecting AUC scores.
3. Max voting, an ensemble technique, outperforms standalone models.
4. Newer architectures generally surpass older ones like AlexNet.

Our analysis reveals that while there are commonalities in high AUC scores and architectural innovations, the models also differ in computational complexity, intended use cases, and overall performance. Understanding these nuances can guide the choice of model for specific classification tasks in skin cancer detection.

The AUC metric is commonly used to evaluate the performance of binary classification models like skin cancer detection. A higher AUC indicates better discriminative power and better performance of the model in distinguishing between positive and negative samples. The provided AUC scores reveal the strengths and relative performance of different deep-learning models for skin cancer classification. Based on the AUC scores, the max voting ensemble achieves the highest performance with an AUC of 0.932, outperforming all individual models. The Xception model comes second with an AUC of 0.919, followed by InceptionResNetV2 with an AUC of 0.902.

These results indicate that the max voting ensemble technique, which combines predictions from multiple models, demonstrates superior performance in skin cancer classification compared to using individual models alone. The ensemble approach effectively leverages the strengths of different models, resulting in improved accuracy and generalization capabilities for skin cancer detection tasks.

Overall, the findings suggest that the max voting ensemble technique offers a promising approach to enhance the performance of skin cancer classification models using pre-trained deep learning models, potentially leading to better diagnostic capabilities in real-world applications. Researchers and practitioners can use these findings to select appropriate models for skin cancer classification tasks based on their specific requirements and resource constraints.

4.3. Classification Report Showing Accuracy, Precision, and Recall Score for the ISIC 2018 Datasets

The classification report, as depicted in Table 9, provides an overview of the performance of the proposed max voting ensemble model on the ISIC 2018 test datasets, encompassing a total of 660 images, including 360 benign and 300 malignant images, specifically highlighting the model's performance in distinguishing between benign and malignant skin lesions. The report provides essential metrics such as precision, recall, and F1-score for each class, as well as overall accuracy, macro average, and weighted average values.

Table 9. Classification Report for the ISIC 2018 datasets.

	Precision	Recall	F1-Score	Support
Benign	94.12	93.33	93.72	360
Malignant	92.08	93.00	92.54	300
Accuracy			93.18	660
Macro avg	93.10	93.17	93.13	660
Weighted avg	93.19	93.18	93.18	660

In the benign class (0), the model achieved a precision of 94.12%, indicating a high accuracy of positive predictions. The recall for benign instances stands at 93.33% and corresponding F1-score is reported at 93.72%. For the malignant class (1), the model achieved a precision of 92.08%, ensuring reliability in identifying malignant cases. The recall for malignant instances is 93.00% and the F1-score for the malignant class is 92.54%.

The overall accuracy of the model is reported at 93.18%, reflecting its correctness in classifying skin lesions. Macro average values of 93.10%, 93.17%, and 93.13% for precision, recall, and F1-score, respectively, provide an unbiased evaluation across both classes. The weighted averages, accounting for class imbalances are 93.19%, 93.18%, and 93.18% for precision, recall, and F1-score respectively.

In summary, the classification report offers a comprehensive assessment of the model's effectiveness in classifying skin lesions in the ISIC 2018 datasets. The high precision, recall, and F1-score values for both benign and malignant classes, coupled with a notable accuracy of 93.18%, demonstrate the model's robust performance in dermatological image analysis. The macro and weighted averages provide detailed insights on the distribution of instances across classes and offer a holistic view of the model's capabilities in skin cancer classification. The reported accuracy of 93.18% indicates the overall correctness of the model's predictions across classes.

4.4. State-of-the-Art Comparison for the ISIC 2018: Task 1-2 Datasets

The performance and accuracy of machine learning and deep learning models for medical image analysis, specifically in the context of detecting and classifying skin lesions, can vary depending on several factors, including the dataset used, the choice of model architecture, and the preprocessing techniques applied. The following subsection describes the performance of the proposed model with existing models. In recent years, few research activities have been carried out on skin cancer classification on the ISIC 2018: Task 1-2 dataset. The performance of the proposed system is compared with the previous approaches in terms of applied models and accuracy. Table 10 presents the testing accuracy values in existing approaches with our proposed method.

Table 10. Comparative analysis of ISIC 2018: Task 1-2 dataset used for skin cancer classification.

Authors and Paper	Year	Model	Accuracy
Gouda et al. [70]	2021	CNN	83.2%
Kaggle Compt. (3rd Place) [69]	2020	Individual Single Model	84.5%
Kaggle Compt. (2nd Place) [69]	2020	Meta Ensemble	86.3%
Kaggle Compt. (1st Place) [69]	2020	Top 10 model Average	86.7%
Khan et al. [68]	2022	A hybrid CNN Model	92.70%
Proposed method	2023	Max Voting ensemble model	93.18%

Table 10 indicates that the proposed approach achieves higher training and testing accuracy than the existing approaches of skin cancer classification using the ISIC 2018 dataset.

4.5. Performance with ISIC 2018: Task 3 (HAM10000) Dataset

The max voting ensemble technique introduced in this study for skin cancer classification exhibits significant potential for generalization to other related tasks within medical image analysis. Leveraging a comprehensive ensemble of pre-trained deep neural networks, including MobileNetV2, AlexNet, VGG16, ResNet50, DenseNet201, DenseNet121, InceptionV3, ResNet50V2, InceptionResNetV2, and Xception, this method showcases the adaptability of ensemble techniques for combining diverse model architectures. The success of the ensemble in elevating classification performance by aggregating predictions from individual models also suggests its applicability beyond skin cancer classification. With proper adaptation and training on relevant datasets, the max voting ensemble technique could be seamlessly extended to address diagnostic challenges in other medical domains, providing a versatile and effective tool for healthcare professionals in diverse diagnostic scenarios.

Adapting the max voting ensemble technique for the HAM1000 dataset, which is a widely recognized dataset for skin image analysis, holds promise for enhancing the diagnostic capabilities of the ensemble method as shown in Figure 14. The ensemble's incorporation of pre-trained deep neural networks, initially developed for skin cancer classification, aligns well with the diversity and complexity of the HAM1000 dataset, which comprises a large variety of skin lesions. Similar to the ISIC 2018: Task 1-2 dataset, the HAM10000 dataset attains improved accuracy through the amalgamation of individual models using the proposed technique.

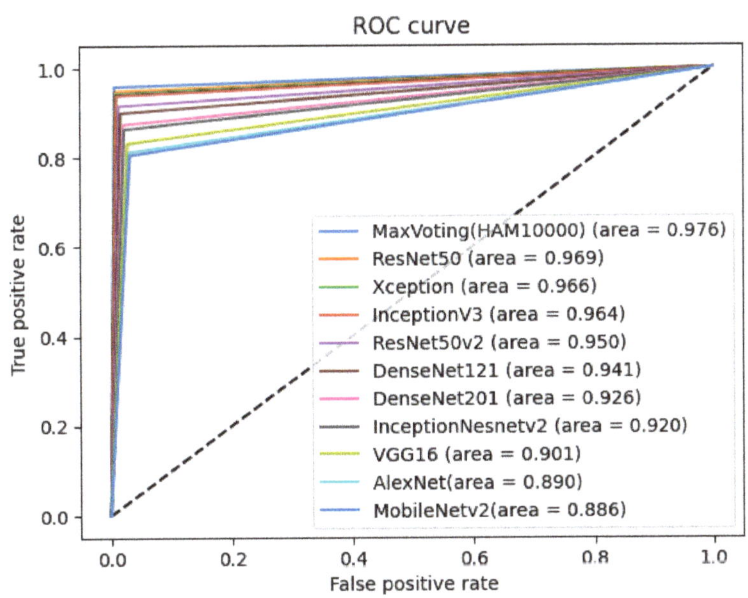

Figure 14. ROC Curve and AUC Scores for HAM10000 Dataset using proposed Max Voting Ensemble Technique.

Figure 15 shows the ROC curve with individual classes and confusion matrix for the max voting ensemble technique on the HAM10000 dataset. The proposed model is able to classify all seven skin decease classes, namely, melanoma, melanocytic nevus, basal cell carcinoma, actinic keratosis, benign keratosis, dermatofibroma, and vascular lesion with high accuracy, as shown in Figure 15a. The diagonal of the confusion matrix shown in Figure 15b indicates the high accuracy rate of the model as well.

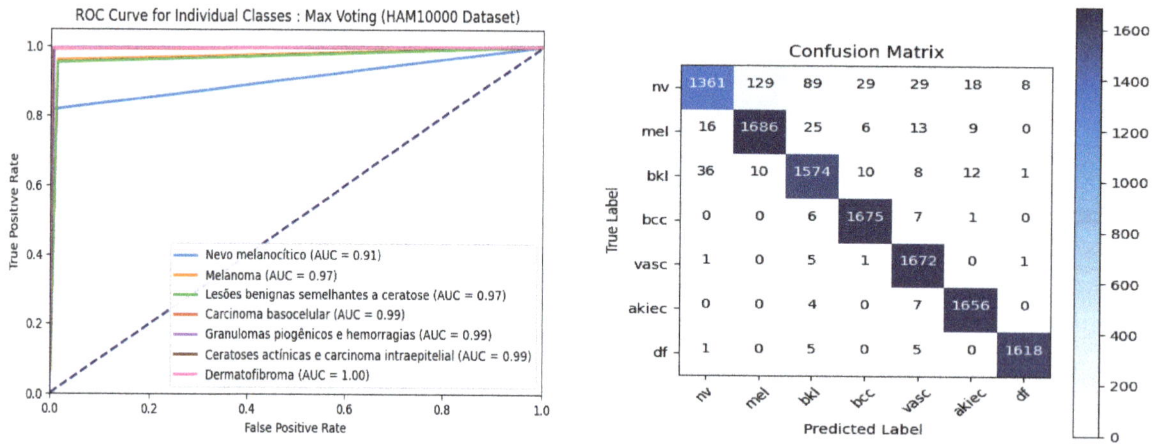

(a) ROC Curve for Individual classes (b) Confusion Matrix

Figure 15. ROC Curve and Confusion Matrix using Max Voting Technique for HAM10000 Dataset.

4.6. Classification Report for the HAM10000 Datasets

Table 11 presents a detailed Classification Report for the HAM10000 datasets, summarizing the performance metrics for seven different classes (akiec, bcc, bkl, df, nv, vasc, and mel). For each class, the table includes precision, recall, F1-score, and support values, providing a comprehensive assessment of the model's ability to classify skin lesions within each category. Additionally, the table reports the overall accuracy of the model, which stands at 95%, indicating the percentage of correctly predicted instances out of the total. Both macro and weighted averages are presented, offering unweighted and weighted evaluations across all classes, with each average showing consistent values of 0.95 for precision, recall, and F1-score. The support column reveals the distribution of instances among the different classes, totaling 11,734 instances in the dataset. This classification report offers valuable insights into the model's performance on the HAM10000 datasets, crucial for understanding its effectiveness in dermatological image classification.

Table 11. Classification Report for the HAM10000 dataset.

	Precision	Recall	F1-Score	Support
class 0: akiec	0.98	0.98	0.98	1667
class 1: bcc	0.96	0.98	0.97	1689
class 2: bkl	0.94	0.94	0.94	1651
class 3: df	0.99	0.98	0.99	1629
class 4: nv	0.89	0.82	0.85	1663
class 5: vasc	0.94	0.98	0.96	1680
class 6: mel	0.92	0.95	0.94	1755
Accuracy			0.95	11,734
Macro Avg.	0.95	0.95	0.95	11,734
Weighted Avg.	0.95	0.95	0.95	11,734

This approach demonstrates the potential for the broader application of ensemble methods in medical image analysis, contributing to the development of robust and reliable diagnostic tools across various healthcare domains.

5. Conclusions

In this study, we proposed a noble approach of employing the max voting ensemble technique with cutting-edge pre-trained deep learning models for skin cancer detection and classification. This study has made significant contributions to the field of skin cancer detection and medical image analysis. Our primary focus has been on enhancing both the accuracy and reliability of skin cancer diagnosis and we have achieved this through the innovative application of the max voting ensemble technique in conjunction with state-of-the-art pre-trained deep learning models. The main contribution of our research lies in the successful utilization of the max voting ensemble technique. By combining the predictions of multiple models, we have demonstrated its potential to significantly elevate the diagnostic performance in the context of skin cancer detection. Through extensive experimentation and rigorous analysis, we have shown that our ensemble method outperforms individual models. With an AUC of 0.932, it has established itself as a powerful tool for improving the accuracy and reliability of skin cancer classification. Our study offers insights into the selection of cutting-edge pre-trained deep learning models for single lesions. Models such as Xception, InceptionResNetV2, and ResNet50V2 have proven to be highly effective in capturing intricate features and patterns relevant to skin cancer detection. Beyond its immediate applications in skin cancer detection, our work carries broader implications for the field of medical image analysis. The successful integration of ensemble techniques can be extended to various medical imaging modalities and disease classifications. By leveraging the collective intelligence of multiple models, the accuracy and reliability of diagnostic tools across diverse healthcare domains can be enhanced. Moreover, the max voting ensemble technique serves as a testament to the potential of ensemble learning in medical image analysis. As we continue to advance in this field, ensemble techniques are likely to play an increasingly vital role in improving the quality of healthcare, aiding healthcare professionals in making more accurate and timely diagnoses, and ultimately contributing to better patient outcomes. Our research not only addresses the critical challenge of skin cancer detection but also provides a framework for the integration of ensemble techniques in medical image analysis.

The study primarily focuses on skin cancer detection using the ISIC 2018: Task 1-2 dataset. While this dataset is widely used and provides valuable insights, the generalizability of the proposed max voting ensemble technique may be influenced by several factors. These include the characteristics of the dataset such as image quality, class imbalance, and algorithmic diversity, as well as evaluation metrics and computational resources. Additionally, potential biases in the dataset, particularly the underrepresentation of certain skin types, ages, or ethnic groups, could limit the accuracy of the model across diverse populations. Although we assess the validity and generalizability of the proposed model with the HAM10000 dataset, further research could explore the effectiveness of the technique on diverse datasets representing different demographics and skin conditions. Furthermore, the challenges in ensuring algorithmic transparency and interpretability are crucial, especially in healthcare applications, for gaining trust from professionals and patients. The study mainly evaluates the proposed approach using quantitative metrics. To establish its real-world clinical utility, further validation through collaboration with healthcare professionals and clinical trials is necessary. This should include continuous evaluation of the performance of the model over time and its adaptability to evolving medical guidelines and emerging skin conditions. Moreover, regulatory challenges and ethical considerations, such as obtaining approval and addressing patient consent and data privacy, are essential factors to consider. The resource requirements for implementing and maintaining such a system in clinical settings, especially in resource-limited environments, also present significant limitations. The integration of AI tools into clinical workflows might face challenges, including the reception by healthcare professionals and the need for training and adaptation to new technologies. Lastly, a comparison with existing diagnostic methods in terms of accuracy, efficiency, and cost is imperative to understand the potential limitations in outperforming or integrating with current diagnostic processes.

Our research focused on the categorization of images of single lesions as either benign or cancerous. However, it should be noted that other approaches place the focus on wide-field or total-body imaging. Soenksen et al. (2021) cogently argue that a whole body approach may better replicate the practise of medical practitioners, who take into account a multitude of factors, such as the context, number and type of lesions when identifying lesions as suspicious [92]. Several studies using total-body imaging have shown fruitful results. Betz et al. (2022) on a study of 10 patients and analysis of slightly under 5000 lesions achieved 70% agreement with the gold standard. Their focus, however, was on counting the number of naevi [93] among the lesions. Strzelecki et al. (2021) studied the effectiveness of three algorithms developed to detect and segment lesions using whole-body imaging, and report promising results particularly for lesions greater than 3 mm [94]. Birkenfeld et al. (2020) used whole-body visual examinations to classify pigmented lesions as suspicious or non-suspicious, and propose that this approach may be adopted as a preliminary method to screen a population for further testing [95]. A direct comparison between total-body-imaging approaches and single-lesion approaches is difficult, given the difference in tasks. Single-lesion approaches tend to adopt dichotomous classification tasks of either benign or cancerous, while total-body imaging studies do not tend to assign such values to individual lesions, but adopt different tasks, such as calculating the number of naevi [93], and comparing the performance of different algorithms [94]. A single-lesion approach is highly effective in offering a detailed and focused analysis of individual lesions, which is crucial for precise diagnosis of specific skin conditions. However, its limitation lies in the potential oversight of the broader context of skin health, as it fails to consider the overall distribution, number, or variety of lesions across the body, which can be critical in identifying patterns indicative of systemic health issues. On the other hand, the total-body imaging approach stands out for its ability to provide a comprehensive overview of the skin, facilitating the detection of patterns and multiple lesions that are essential in diagnosing conditions spread over larger skin areas. This method, however, may compromise on the level of detail for individual lesions, potentially affecting the accuracy in diagnosing certain types of skin anomalies that require further examination.

In future work, we aim to seek out ways to address these limitations and to explore more sophisticated algorithms. Our next step will be to test the efficacy of including the integration of a weighted averaging technique, and assess its impact on classification performance with respect to the current state-of-the-art models. Another key focus will be on diversifying the datasets used in our models to include a broader range of skin types, ages, and ethnic backgrounds, thereby reducing bias and improving the generalizability of our findings. We plan to continue advancing the field of medical image analysis and enhancing healthcare outcomes. In addition to addressing the pressing challenge of skin cancer detection, our work serves as a catalyst for future innovation, ultimately contributing to the improvement of healthcare delivery and the preservation of lives through enhanced diagnostic accuracy, and achieving the ultimate goal of enhancing healthcare delivery and saving lives through improved diagnostic accuracy.

Author Contributions: Conceptualization, M.M.H. (Md. Mamun Hossain) and M.M.H. (Md. Moazzem Hossain); methodology, M.M.H. (Md. Mamun Hossain), M.M.H. (Md. Moazzem Hossain), M.B.A. and F.A.; software, M.M.H. (Md. Mamun Hossain), M.B.A. and F.A.; validation, M.M.H. (Md. Mamun Hossain), M.B.A. and F.A.; formal analysis, M.M.H. (Md. Mamun Hossain), M.B.A. and M.M.H. (Md. Moazzem Hossain); investigation, M.M.H. (Md. Mamun Hossain), M.B.A. and M.M.H. (Md. Moazzem Hossain); resources, M.M.H. (Md. Mamun Hossain), M.M.H. (Md. Moazzem Hossain), F.A. and M.B.A.; data curation, M.M.H. (Md. Mamun Hossain), F.A. and M.B.A.; writing—original draft preparation, M.M.H. (Md. Mamun Hossain), F.A. and M.B.A.; writing—review and editing, J.B.; visualization, M.M.H. (Md. Mamun Hossain) and J.B.; supervision, J.B.; project administration, M.M.H. (Md. Mamun Hossain) and J.B.; funding acquisition, J.B. All authors have read and agreed to the published version of the manuscript.

Funding: This research received no external funding.

Data Availability Statement: The data presented in this study are openly available at the ISIC Challenge Datasets website: https://challenge.isic-archive.com/data/#2018 (accessed on 2 October 2023) and, the ISIC Archive Datasets websites: https://api.isic-archive.com/collections/63/ and https://api.isic-archive.com/collections/212/ (accessed on 2 October 2023).

Acknowledgments: We wish to express our heartfelt gratitude to Abu Sahel Musa Miah for his invaluable support throughout the research process. His contributions have unquestionably played a pivotal role in shaping the outcome of this work. His dedication and assistance have been indispensable, and we extend our sincere appreciation for his unwavering commitment to the success of this research endeavor.

Conflicts of Interest: The authors declare no conflict of interest.

Abbreviations

The following abbreviations are used in this manuscript:

AlexNet	Alex Krizhevsky's Network
AUC	Area Under the Curve
CNN	Convolutional Neural Networks
DenseNet121	Densely Connected Network 121
DenseNet201	Densely Connected Network 201
GAN	Generative Adversarial Network
InceptionResNetV2	Inception + ResNet Variant 2
InceptionV3	Inception Variant 3
MobileNetV2	Mobile Network Variant 2
ReLU	Rectified Linear Unit
ResNet50	Residual Network 50
ResNet50V2	Residual Network 50 Variant 2
ROC	Receiver Operating Characteristic
SVM	Support Vector Machine
VGG16	Visual Geometry Group 16
Xception	Extreme Inception

References

1. Zelin, E.; Zalaudek, I.; Agozzino, M.; Dianzani, C.; Dri, A.; Di Meo, N.; Giuffrida, R.; Marangi, G.F.; Neagu, N.; Persichetti, P.; et al. Neoadjuvant therapy for non-melanoma skin cancer: Updated therapeutic approaches for basal, squamous, and merkel cell carcinoma. *Curr. Treat. Options Oncol.* **2021**, *22*, 1–24. [CrossRef] [PubMed]
2. Ichihashi, M.; Ueda, M.; Budiyanto, A.; Bito, T.; Oka, M.; Fukunaga, M.; Tsuru, K.; Horikawa, T. UV-induced skin damage. *Toxicology* **2003**, *189*, 21–39. [CrossRef] [PubMed]
3. Kim, I.; He, Y.Y. Ultraviolet radiation-induced non-melanoma skin cancer: Regulation of DNA damage repair and inflammation. *Genes Dis.* **2014**, *1*, 188–198. [CrossRef] [PubMed]
4. US Department of Health and Human Services. Skin Cancer as a Major Public Health Problem. In *The Surgeon General's Call to Action to Prevent Skin Cancer*; Office of the Surgeon General (US): Washington, DC, USA, 2014. Available online: https://www.ncbi.nlm.nih.gov/books/NBK247164/ (accessed on 22 August 2023).
5. Zhang, B.; Zhou, X.; Luo, Y.; Zhang, H.; Yang, H.; Ma, J.; Ma, L. Opportunities and challenges: Classification of skin disease based on deep learning. *Chin. J. Mech. Eng.* **2021**, *34*, 112. [CrossRef]
6. de Zarzà, I.; de Curtò, J.; Hernández-Orallo, E.; Calafate, C.T. Cascading and Ensemble Techniques in Deep Learning. *Electronics* **2023**, *12*, 3354. [CrossRef]
7. Naqvi, M.; Gilani, S.Q.; Syed, T.; Marques, O.; Kim, H.C. Skin Cancer Detection Using Deep Learning—A Review. *Diagnostics* **2023**, *13*, 1911. [CrossRef]
8. Sethanan, K.; Pitakaso, R.; Srichok, T.; Khonjun, S.; Thannipat, P.; Wanram, S.; Boonmee, C.; Gonwirat, S.; Enkvetchakul, P.; Kaewta, C.; et al. Double AMIS-ensemble deep learning for skin cancer classification. *Expert Syst. Appl.* **2023**, *234*, 121047. [CrossRef]
9. Alzubaidi, L.; Zhang, J.; Humaidi, A.J.; Al-Dujaili, A.; Duan, Y.; Al-Shamma, O.; Santamaría, J.; Fadhel, M.A.; Al-Amidie, M.; Farhan, L. Review of deep learning: Concepts, CNN architectures, challenges, applications, future directions. *J. Big Data* **2021**, *8*, 53. [CrossRef]
10. Giuffrida, R.; Conforti, C.; Di Meo, N.; Deinlein, T.; Guida, S.; Zalaudek, I. Use of noninvasive imaging in the management of skin cancer. *Curr. Opin. Oncol.* **2020**, *32*, 98–105. [CrossRef]
11. Mohammad, I.; Rymanov, V.; Makhlouf, S.; Stoffels, I.; Klode, J.; Tang, X.; Ali, M.; Stöhr, A. Photonic non-invasive mmw and THz imaging for non-melanoma skin cancer detection. In Proceedings of the 2022 Fifth International Workshop on Mobile Terahertz Systems (IWMTS), Duisburg, Germany, 4–6 July 2022; pp. 1–5.

12. Abdar, M.; Samami, M.; Mahmoodabad, S.S.S.; Doan, T.; Mazoure, B.; Hashemifesharaki, R.; Liu, L.; Khosravi, A.; Acharya, U.R.; Makarenkov, V.; et al. Uncertainty Quantification in Skin Cancer Classification Using Three-way Decision-based Bayesian Deep Learning. *Comput. Biol. Med.* **2021**, *135*, 104418. [CrossRef]
13. Jain, S.; Singhania, U.; Tripathy, B.; Nasr, E.A.; Aboudaif, M.K.; Kamrani, A.K. Deep Learning-based Transfer Learning for Classification of Skin Cancer. *Sensors* **2021**, *21*, 8142. [CrossRef] [PubMed]
14. Aljohani, K.; Turki, T. Automatic Classification of Melanoma Skin Cancer with Deep Convolutional Neural Networks. *AI* **2022**, *3*, 512–525. [CrossRef]
15. Miah, A.S.M.; Shin, J.; Hasan, M.A.M.; Rahim, M.A. Bensignnet: Bengali sign language alphabet recognition using concatenated segmentation and convolutional neural network. *Appl. Sci.* **2022**, *12*, 3933. [CrossRef]
16. Raval, D.; Undavia, J.N. A Comprehensive assessment of Convolutional Neural Networks for skin and oral cancer detection using medical images. *Healthc. Anal.* **2023**, *3*, 100199. [CrossRef]
17. Iqbal, S.; NQureshi, A.; Li, J.; Mahmood, T. On the Analyses of Medical Images Using Traditional Machine Learning Techniques and Convolutional Neural Networks. *Arch. Comput. Methods Eng.* **2023**, *30*, 3173–3233. [CrossRef] [PubMed]
18. Yamashita, R.; Nishio, M.; Do, R.K.; Togashi, K. Convolutional Neural Networks: An Overview and Application in Radiology. *Insights Imaging* **2018**, *9*, 611–629. [CrossRef] [PubMed]
19. Taye, M.M. Theoretical Understanding of Convolutional Neural Network: Concepts, Architectures, Applications, Future Directions. *Computation* **2023**, *11*, 52. [CrossRef]
20. Krishnapriya, S.; Karuna, Y. Pre-trained deep learning models for brain MRI image classification. *Front. Hum. Neurosci.* **2023**, *17*, 1150120. [CrossRef]
21. Morid, M.A.; Borjali, A.; Del Fiol, G. A scoping review of transfer learning research on medical image analysis using ImageNet. *Comput. Biol. Med.* **2021**, *128*, 104115. [CrossRef]
22. Salehi, A.W.; Khan, S.; Gupta, G.; Alabduallah, B.I.; Almjally, A.; Alsolai, H.; Siddiqui, T.; Mellit, A. A Study of CNN and Transfer Learning in Medical Imaging: Advantages, Challenges, Future Scope. *Sustainability* **2023**, *15*, 5930. [CrossRef]
23. Esteva, A.; Kuprel, B.; Novoa, R.A.; Ko, J.; Swetter, S.M.; Blau, H.M.; Thrun, S. Dermatologist-level classification of skin cancer with deep neural networks. *Nature* **2017**, *542*, 115–118. [CrossRef] [PubMed]
24. Chaturvedi, S.S.; Tembhurne, J.V.; Diwan, T. A multi-class skin Cancer classification using deep convolutional neural networks. *Multimed. Tools Appl.* **2020**, *79*, 28477–28498. [CrossRef]
25. Asadi, O.; Yekkalam, A.; Manthouri, M. MDDC: Melanoma detection using discrete wavelet transform and convolutional neural network. *J. Ambient. Intell. Humaniz. Comput.* **2023**, *14*, 12959–12966. [CrossRef]
26. Miah, A.S.M.; Hasan, M.A.M.; Shin, J.; Okuyama, Y.; Tomioka, Y. Multistage spatial attention-based neural network for hand gesture recognition. *Computers* **2023**, *12*, 13. [CrossRef]
27. Shin, J.; Musa Miah, A.S.; Hasan, M.A.M.; Hirooka, K.; Suzuki, K.; Lee, H.S.; Jang, S.W. Korean Sign Language Recognition Using Transformer-Based Deep Neural Network. *Appl. Sci.* **2023**, *13*, 3029. [CrossRef]
28. Miah, A.S.M.; Hasan, M.A.M.; Shin, J. Dynamic Hand Gesture Recognition using Multi-Branch Attention Based Graph and General Deep Learning Model. *IEEE Access* **2023**, *11*, 4703–4716. [CrossRef]
29. Brinker, T.J.; Hekler, A.; Utikal, J.S.; Grabe, N.; Schadendorf, D.; Klode, J.; Berking, C.; Steeb, T.; Enk, A.H.; von Kalle, C. Skin Cancer Classification Using Convolutional Neural Networks: Systematic Review. *J. Med. Internet Res.* **2018**, *20*, e11936. [CrossRef]
30. Adegun, A.; Viriri, S. Deep learning techniques for skin lesion analysis and melanoma cancer detection: A survey of state-of-the-art. *Artif. Intell. Rev.* **2021**, *54*, 811–841. [CrossRef]
31. Munir, K.; Elahi, H.; Ayub, A.; Frezza, F.; Rizzi, A. Cancer Diagnosis Using Deep Learning: A Bibliographic Review. *Cancers* **2019**, *11*, 1235. [CrossRef]
32. Li, H.; Pan, Y.; Zhao, J.; Zhang, L. Skin disease diagnosis with deep learning: A review. *Neurocomputing* **2021**, *464*, 364–393. [CrossRef]
33. Dildar, M.; Akram, S.; Irfan, M.; Khan, H.U.; Ramzan, M.; Mahmood, A.R.; Alsaiari, S.A.; Saeed, A.H.M.; Alraddadi, M.O.; Mahnashi, M.H. Skin cancer detection: A review using deep learning techniques. *Int. J. Environ. Res. Public Health* **2021**, *18*, 5479. [CrossRef] [PubMed]
34. Shokouhifar, A.; Shokouhifar, M.; Sabbaghian, M.; Soltanian-Zadeh, H. Swarm intelligence empowered three-stage ensemble deep learning for arm volume measurement in patients with lymphedema. *Biomed. Signal Process. Control* **2023**, *85*, 105027. [CrossRef]
35. Bao, J.; Hou, Y.; Qin, L.; Zhi, R.; Wang, X.M.; Shi, H.B.; Sun, H.Z.; Hu, C.H.; Zhang, Y.D. High-throughput precision MRI assessment with integrated stack-ensemble deep learning can enhance the preoperative prediction of prostate cancer Gleason grade. *Br. J. Cancer* **2023**, *128*, 1267–1277. [CrossRef] [PubMed]
36. Miah, A.S.M.; Shin, J.; Al Mehedi Hasan, M.; Rahim, M.A.; Okuyama, Y. Rotation, Translation and Scale Invariant Sign Word Recognition Using Deep Learning. *Comput. Syst. Sci. Eng.* **2023**, *44*, 2522–2536.
37. Tschandl, P.; Rosendahl, C.; Kittler, H. The HAM10000 dataset, a large collection of multi-source dermatoscopic images of common pigmented skin lesions. *Sci. Data* **2018**, *5*, 180161. [CrossRef] [PubMed]

38. Han, S.S.; Kim, M.S.; Lim, W.; Park, G.H.; Park, I.; Chang, S.E. Classification of the clinical images for benign and malignant cutaneous tumors using a deep learning algorithm. *J. Investig. Dermatol.* **2018**, *138*, 1529–1538. [CrossRef] [PubMed]
39. Codella, N.C.; Gutman, D.; Celebi, M.E.; Helba, B.; Marchetti, M.A.; Dusza, S.W.; Kalloo, A.; Liopyris, K.; Mishra, N.; Kittler, H.; et al. Skin lesion analysis toward melanoma detection: A challenge at the 2017 international symposium on biomedical imaging (isbi), hosted by the international skin imaging collaboration (isic). In Proceedings of the 2018 IEEE 15th International Symposium on Biomedical Imaging (ISBI 2018), Washington, DC, USA, 4–7 April 2018; pp. 168–172.
40. Hasan, M.R.; Fatemi, M.I.; Khan, M.M.; Kaur, M.; Zaguia, A. Comparative Analysis of Skin Cancer (Benign vs. Malignant) Detection Using Convolutional Neural Networks. *J. Healthc. Eng.* **2021**, *2021*, 5895156. [CrossRef]
41. Balaha, H.; Hassan, A. Skin cancer diagnosis based on deep transfer learning and sparrow search algorithm. *Neural Comput. Appl.* **2023**, *35*, 815–853. [CrossRef]
42. Gutman, D.; Codella, N.C.; Celebi, E.; Helba, B.; Marchetti, M.; Mishra, N.; Halpern, A. Skin lesion analysis toward melanoma detection: A challenge at the international symposium on biomedical imaging (ISBI) 2016, hosted by the international skin imaging collaboration (ISIC). *arXiv* **2016**, arXiv:1605.01397.
43. Barata, C.; Celebi, M.E.; Marques, J.S. Improving dermoscopy image classification using color constancy. *IEEE J. Biomed. Health Inform.* **2014**, *19*, 1146–1152.
44. Sahinbas, K.; Catak, F.O. Transfer learning-based convolutional neural network for COVID-19 detection with X-ray images. In *Data science for COVID-19*; Elsevier: Amsterdam, The Netherlands, 2021; pp. 451–466.
45. Goyal, M.; Knackstedt, T.; Yan, S.; Hassanpour, S. Artificial intelligence-based image classification methods for diagnosis of skin cancer: Challenges and opportunities. *Comput. Biol. Med.* **2020**, *127*, 104065. [CrossRef] [PubMed]
46. Ahmad, B.; Jun, S.; Palade, V.; You, Q.; Mao, L.; Zhongjie, M. Improving Skin Cancer Classification Using Heavy-Tailed Student T-Distribution in Generative Adversarial Networks (TED-GAN). *Diagnostics* **2021**, *11*, 2147. [CrossRef] [PubMed]
47. Rashid, H.; Tanveer, M.A.; Aqeel Khan, H. Skin Lesion Classification Using GAN based Data Augmentation. In Proceedings of the 2019 41st Annual International Conference of the IEEE Engineering in Medicine and Biology Society (EMBC), Berlin, Germany, 23–27 July 2019; pp. 916–919. [CrossRef]
48. Al-Rasheed, A.; Ksibi, A.; Ayadi, M.; Alzahrani, A.I.A.; Elahi, M.M. An Ensemble of Transfer Learning Models for the Prediction of Skin Lesions with Conditional Generative Adversarial Networks. *Contrast Media Mol. Imaging* **2023**, *2023*, 5869513. [CrossRef]
49. Kebaili, A.; Lapuyade-Lahorgue, J.; Ruan, S. Deep Learning Approaches for Data Augmentation in Medical Imaging: A Review. *J. Imaging* **2023**, *9*, 81. [CrossRef] [PubMed]
50. Guergueb, T.; Akhloufi, M.A. Skin Cancer Detection using Ensemble Learning and Grouping of Deep Models. In Proceedings of the 19th International Conference on Content-Based Multimedia Indexing, Graz, Austria, 14–16 September 2022; pp. 121–125.
51. Avanija, J.; Reddy, C.C.M.; Reddy, C.S.C.; Reddy, D.H.; Narasimhulu, T.; Hardhik, N.V. Skin Cancer Detection using Ensemble Learning. In Proceedings of the 2023 International Conference on Sustainable Computing and Smart Systems (ICSCSS), Coimbatore, India, 14–16 June 2023; pp. 184–189.
52. Naji, M.A.; Filali, S.E.; Bouhlal, M.; Benlahmar, E.H.; Abdelouhahid, R.A.; Debauche, O. Breast Cancer Prediction and Diagnosis through a New Approach based on Majority Voting Ensemble Classifier. *Procedia Comput. Sci.* **2021**, *191*, 481–486. [CrossRef]
53. Abdar, M.; Zomorodi-Moghadam, M.; Zhou, X.; Gururajan, R.; Tao, X.; Barua, P.D.; Gururajan, R. A new nested ensemble technique for automated diagnosis of breast cancer. *Pattern Recognit. Lett.* **2020**, *132*, 123–131. [CrossRef]
54. Zolfaghari, B.; Mirsadeghi, L.; Bibak, K.; Kavousi, K. Cancer Prognosis and Diagnosis Methods Based on Ensemble Learning. *ACM Comput. Surv.* **2023**, *55*, 262. [CrossRef]
55. Kausar, N.; Hameed, A.; Sattar, M.; Ashraf, R.; Imran, A.S.; Abidin, M.Z.u.; Ali, A. Multiclass Skin Cancer Classification Using Ensemble of Fine-Tuned Deep Learning Models. *Appl. Sci.* **2021**, *11*, 10593. [CrossRef]
56. Ali, A.R.A.; Deserno, T.M. A systematic review of automated melanoma detection in dermatoscopic images and its ground truth data. In *Medical Imaging 2012: Image Perception, Observer Performance, and Technology Assessment*; SPIE: Bellingham, WA, USA, 2012; Volume 8318, pp. 421–431.
57. Gajera, H.K.; Nayak, D.R.; Zaveri, M.A. A Comprehensive Analysis of Dermoscopy Images for Melanoma Detection via Deep CNN Features. *Biomed. Signal Process. Control* **2023**, *79*, 104186. [CrossRef]
58. Alenezi, F.; Armghan, A.; Polat, K. Wavelet Transform-Based Deep Residual Neural Network and ReLU-Based Extreme Learning Machine for Skin Lesion Classification. *Expert Syst. Appl.* **2023**, *213*, 119064. [CrossRef]
59. Inthiyaz, S.; Altahan, B.R.; Ahammad, S.H.; Rajesh, V.; Kalangi, R.R.; Smirani, L.K.; Hossain, M.A.; Rashed, A.N.Z. Skin Disease Detection Using Deep Learning. *Adv. Eng. Softw.* **2023**, *175*, 103361. [CrossRef]
60. Alwakid, G.; Gouda, W.; Humayun, M.; Sama, N.U. Melanoma Detection Using Deep Learning-Based Classifications. *Healthcare* **2022**, *10*, 2481. [CrossRef] [PubMed]
61. Alenezi, F.; Armghan, A.; Polat, K. A Multi-Stage Melanoma Recognition Framework with Deep Residual Neural Network and Hyperparameter Optimization-Based Decision Support in Dermoscopy Images. *Expert Syst. Appl.* **2023**, *215*, 119352. [CrossRef]
62. Abbas, Q.; Gul, A. Detection and Classification of Malignant Melanoma Using Deep Features of NASNet. *SN Comput. Sci.* **2022**, *4*, 21. [CrossRef]
63. Bechelli, S.; Delhommelle, J. Machine Learning and Deep Learning Algorithms for Skin Cancer Classification from Dermoscopic Images. *Bioengineering* **2022**, *9*, 97. [CrossRef] [PubMed]

64. Demir, A.; Yilmaz, F.; Kose, O. Early Detection of Skin Cancer Using Deep Learning Architectures: ResNet-101 and Inception-v3. In Proceedings of the 2019 Medical Technologies Congress (TIPTEKNO), Izmir, Turkey, 3–5 October 2019; pp. 1–4. [CrossRef]
65. Rashid, J.; Ishfaq, M.; Ali, G.; Saeed, M.R.; Hussain, M.; Alkhalifah, T.; Alturise, F.; Samand, N. Skin Cancer Disease Detection Using Transfer Learning Technique. *Appl. Sci.* **2022**, *12*, 5714. [CrossRef]
66. Reis, H.; Turk, V.; Khoshelham, K.; Kaya, S. InSiNet: A Deep Convolutional Approach to Skin Cancer Detection and Segmentation. *Med. Biol. Eng. Comput.* **2022**, *60*, 643–662. [CrossRef]
67. Khan, M.A.; Sharif, M.; Akram, T.; Damaševičius, R.; Maskeliunas, R. Skin Lesion Segmentation and Multiclass Classification Using Deep Learning Features and Improved Moth Flame Optimization. *Diagnostics* **2021**, *11*, 811. [CrossRef]
68. Khan, M.A.; Muhammad, K.; Sharif, M.; Akram, T.; Kadry, S. Intelligent Fusion-Assisted Skin Lesion Localization and Classification for Smart Healthcare. *Neural Comput. Appl.* **2021**, *11*, 811. [CrossRef]
69. kaggle competitions. SIIM-ISIC Melanoma Classification. SIC 2018–Winners Final 3 Submissions, 2020. SIIM-ISIC Melanoma Classification-Identify Melanoma in Lesion Images. Available online: https://www.kaggle.com/competitions/siim-isic-melanoma-classification/discussion/173086w (accessed on 22 August 2023).
70. Gouda, W.; Sama, N.U.; Al-Waakid, G.; Humayun, M.; Jhanjhi, N.Z. Detection of Skin Cancer Based on Skin Lesion Images Using Deep Learning. *Healthcare* **2022**, *10*, 1183. [CrossRef]
71. Codella, N.; Rotemberg, V.; Tschandl, P.; Celebi, M.E.; Dusza, S.; Gutman, D.; Helba, B.; Kalloo, A.; Liopyris, K.; Marchetti, M.; et al. Skin Lesion Analysis Toward Melanoma Detection 2018: A Challenge Hosted by the International Skin Imaging Collaboration (ISIC). *arXiv* **2019**, arXiv:1902.03368.
72. Mikołajczyk, A.; Grochowski, M. Data augmentation for improving deep learning in image classification problem. In Proceedings of the 2018 International Interdisciplinary PhD Workshop (IIPhDW), Swinoujscie, Poland, 9–12 May 2018; pp. 117–122.
73. Perez, L.; Wang, J. The effectiveness of data augmentation in image classification using deep learning. *arXiv* **2017**, arXiv:1712.04621.
74. Perez, L.; Wang, J. *The Effectiveness of Data Augmentation in Image Classification Using Deep Learning*; Research report; Stanford University: Stanford, CA, USA, 2017.
75. Shorten, C.; Khoshgoftaar, T.M. A survey on Image Data Augmentation for Deep Learning. *J. Big Data* **2019**, *6*, 60. [CrossRef]
76. Krizhevsky, A.; Sutskever, I.; Hinton, G.E. ImageNet Classification with Deep Convolutional Neural Networks. In Proceedings of the Advances in Neural Information Processing Systems (NIPS) 25, Lake Tahoe, NV, USA, 3–6 December 2012.
77. Simonyan, K.; Zisserman, A. Very Deep Convolutional Networks for Large-Scale Image Recognition. *arXiv* **2014**, arXiv:1409.1556.
78. Szegedy, C.; Vanhoucke, V.; Ioffe, S.; Shlens, J.; Wojna, Z. Rethinking the Inception Architecture for Computer Vision. *arXiv* **2015**, arXiv:1512.00567.
79. He, K.; Zhang, X.; Ren, S.; Sun, J. Identity Mappings in Deep Residual Networks. *arXiv* **2016**, arXiv:1603.05027.
80. He, K.; Zhang, X.; Ren, S.; Sun, J. Deep Residual Learning for Image Recognition. *arXiv* **2015**, arXiv:1512.03385.
81. Huang, G.; Liu, Z.; van der Maaten, L.; Weinberger, K.Q. Densely Connected Convolutional Networks. *arXiv* **2016**, arXiv:1608.06993.
82. Chollet, F. Xception: Deep Learning with Depthwise Separable Convolutions. *arXiv* **2016**, arXiv:1610.02357.
83. Sandler, M.; Howard, A.; Zhu, M.; Zhmoginov, A.; Chen, L.C. Inverted Residuals and Linear Bottlenecks: Mobile Networks for Classification, Detection and Segmentation. *arXiv* **2018**, arXiv:1801.04381.
84. Szegedy, C.; Ioffe, S.; Vanhoucke, V. Inception-v4, Inception-ResNet and the Impact of Residual Connections on Learning. *arXiv* **2016**, arXiv:1602.07261.
85. Szegedy, C.; Ioffe, S.; Vanhoucke, V.; Alemi, A. Inception-v4, Inception-ResNet and the Impact of Residual Connections on Learning. In Proceedings of the Thirty-First AAAI Conference on Artificial Intelligence, San Francisco, CA, USA, 4–9 February 2017.
86. Delgado, J.; Ishii, N. Memory-based weighted majority prediction. In Proceedings of the SIGIR Workshop Recommender Systems, Berkeley, CA, USA, 15–19 August 1999.
87. Dogan, A.; Birant, D. A Weighted Majority Voting Ensemble Approach for Classification. In Proceedings of the 2019 4th International Conference on Computer Science and Engineering (UBMK), Samsun, Turkey, 11–15 September 2019; pp. 1–6. [CrossRef]
88. mlxtend. EnsembleVoteClassifier: A Majority Voting Classifier. 2023. Available online: https://rasbt.github.io/mlxtend/user_guide/classifier/EnsembleVoteClassifier/ (accessed on 27 August 2023).
89. Yang, Y. Chapter 4—Ensemble Learning. In *Temporal Data Mining Via Unsupervised Ensemble Learning*; Yang, Y., Ed.; Elsevier: Amsterdam, The Netherlands, 2017; pp. 35–56. [CrossRef]
90. Akanbi, O.A.; Amiri, I.S.; Fazeldehkordi, E. Chapter 6— Conclusions. In *A Machine-Learning Approach to Phishing Detection and Defense*; Akanbi, O.A., Amiri, I.S., Fazeldehkordi, E., Eds.; Syngress: Boston, MA, USA, 2015; pp. 75–79. [CrossRef]
91. Stehman, S.V. Selecting and interpreting measures of thematic classification accuracy. *Remote Sens. Environ.* **1997**, *62*, 77–89. [CrossRef]
92. Soenksen, L.R.; Kassis, T.; Conover, S.T.; Marti-Fuster, B.; Birkenfeld, J.S.; Tucker-Schwartz, J.; Naseem, A.; Stavert, R.R.; Kim, C.C.; Senna, M.M.; et al. Using deep learning for dermatologist-level detection of suspicious pigmented skin lesions from wide-field images. *Sci. Transl. Med.* **2021**, *13*, eabb3652. [CrossRef] [PubMed]

93. Betz-Stablein, B.; D'Alessandro, B.; Koh, U.; Plasmeijer, E.; Janda, M.; Menzies, S.W.; Hofmann-Wellenhof, R.; Green, A.C.; Soyer, H.P. Reproducible naevus counts using 3D total body photography and convolutional neural networks. *Dermatology* **2022**, *238*, 4–11. [CrossRef]
94. Strzelecki, M.H.; Strąkowska, M.; Kozłowski, M.; Urbańczyk, T.; Wielowieyska-Szybińska, D.; Kociołek, M. Skin lesion detection algorithms in whole body images. *Sensors* **2021**, *21*, 6639. [CrossRef]
95. Birkenfeld, J.S.; Tucker-Schwartz, J.M.; Soenksen, L.R.; Avilés-Izquierdo, J.A.; Marti-Fuster, B. Computer-aided classification of suspicious pigmented lesions using wide-field images. *Comput. Methods Programs Biomed.* **2020**, *195*, 105631. [CrossRef]

MDPI
St. Alban-Anlage 66
4052 Basel
Switzerland
www.mdpi.com

MDPI Books Editorial Office
E-mail: books@mdpi.com
www.mdpi.com/books

www.ingramcontent.com/pod-product-compliance
Lightning Source LLC
Chambersburg PA
CBHW061450220926
51850CB00004B/77
9783725812592